T0253175

Plasma Science and Technology

Plasma Science and Technology

Lectures in Physics, Chemistry, Biology, and Engineering

Alexander Fridman

WILEY-VCH

Author

Professor Alexander Fridman
Nyheim Chair Professor, MEM, Drexel
University
Director of the C&J Nyheim Plasma
Institute
US National Academy of Inventors
Fellow

Cover Image: © Fortis Design/Shutterstock

Library of Congress Card No.: applied for

British Library Cataloguing-in-Publication Data
A catalogue record for this book is available from the British Library.

Bibliographic information published by the Deutsche Nationalbibliothek The Deutsche Nationalbibliothek lists this publication in the Deutsche Nationalbibliografie; detailed bibliographic data are available on the Internet at <http://dnb.d-nb.de>.

Print ISBN: 978-3-527-34954-8
ePDF ISBN: 978-3-527-83412-9
ePub ISBN: 978-3-527-83411-2

Typesetting   Straive, Chennai, India
Printing and Binding   CPI Group (UK) Ltd, Croydon, CR0 4YY

C9783527349548_061223

To my wife Irene

Contents

Part IV Organic and Polymer Plasma Chemistry, Plasma Medicine, and Agriculture *523*

Lecture 25 Organic Plasma Chemistry: Synthesis and Conversion of Organic Materials and Their Compounds, Synthesis of Diamonds and Diamond Films *525*

Preface

Plasma science and technology inspired more than 50 years of the author's life, from January 1972, when as a student at the Moscow Institute of Physics and Technology, he approached his first plasma physics problem on propagation of relativistic electron beams in plasma under the supervision of Professor Rudakov (Kurchatov Institute of Atomic Energy), to now, when he leads research at the Nyheim Plasma Institute of Drexel University. During this half-century, plasma has become a rapidly growing area of scientific endeavor that has spread from fusion, microelectronics, and thermal spraying to multiple novel applications in industrial and medical fields. Plasma has become a ubiquitous element that pervades many aspects of our lives. For example, the public is aware of plasma TV, fluorescent lamps, and plasma thrusters, as well as popular-culture concepts such as plasma guns and plasma shields from Star Trek. Not many are aware, however, that computers, cell phones, and other modern electronic devices are manufactured using mostly plasma-enabled processing equipment; that most of the synthetic fibers used in clothing and advanced packaging materials are plasma treated; that a significant amount of potable water in the world is purified using ozone-plasma technology; and that many different tools and special surfaces are plasma coated to protect and provide them with new extraordinary properties.

Today, plasma plays an important role in a wide variety of modern industrial processes including material processing, energy systems and environmental control, electronic chip manufacturing, chemical technologies, metallurgy, light sources and lasers, green energy, fuel conversion and hydrogen production, aerodynamics and flow control, catalysis, space propulsion, and even recently biotechnologies, agriculture, food processing, and medicine. Plasma is also central to understanding most of the universe outside of Earth.

As such, the focus of this book of lectures is to provide a thorough introduction to the subject and to serve different professionals, surely engineers and scientists already involved, but also to serve as a textbook for numerous students interested in plasma and attending plasma courses in universities all over the world. There are a number of books focused on different, specific plasma topics. The author himself wrote eight of them, focused separately on plasma chemistry, plasma physics and engineering, plasma liquids, and plasma medicine. Those are mostly reference books, not so easy to read for students, often used in an encyclopedic way to better learn specific plasma-related subjects. This book of lectures, in contrast, is written as a comprehensive plasma textbook covering on the same non-overcomplicated basis all major subjects of plasma science and technology: plasma physics, chemistry, engineering, as well as plasma biology and medicine. Although the author tried not to overcomplicate this book, the reader is expected to have the standard engineering background in thermodynamics, chemistry, physics, and fluid mechanics upon which to build an understanding of the plasma subject.

The author has been teaching plasma courses for more than 40 years. Based on the teaching experience, the book of lectures is divided into four parts. They correspond to the four plasma courses included in the plasma course sequence accepted in the Drexel University Curriculum and taught continuously from 2002 until the present. These four parts of the book cover: (i) plasma kinetics, thermodynamics, fluid mechanics, and electrodynamics; (ii) physics and engineering of electric discharges; (iii) inorganic plasma chemistry, energy systems, and environmental control; (iv) organic and polymeric plasma chemistry, plasma biology, and medicine. Each part is covered at Drexel University during one semester/term and includes eight lectures, bringing the total number of lectures in the book to 32.

The content included in each lecture is quite extensive to make it easier to read and comprehend, some portions of the text are marked with special icons:

This icon means ***Especially Important!,*** and is absolutely required to read, to pay attention, and to understand. Text relevant to this icon is usually marked in italics.

This icon means ***If you want to go deeper and learn more…*** – somewhat more complicated and challenging subjects, not required is usually presented with smaller letter sizes and shorter distance between lines.

This icon means: ***If you want to read and learn more…*** – materials and references recommended for further reading. Generally, most of the references in this book of lectures (especially in the first 26 lectures) are collected to indicate additional recommended educational materials. The last six lectures focused on plasma biology and medicine have more detailed data information and therefore include more traditional references not exclusively focused on educational purposes.

The illustrations in this book of lectures are almost always accompanied by either a "lecturing professor" pointing out that the major objective of this figure is the explanation of a concept or a "PowerPoint presentation" pointing out that the major objective of this figure is presentation of some data. Finally, each lecture is accompanied by a set of "problems and concept questions" to be used by instructors as well as individual readers. So with that, we can start the story in 32 lectures of plasma physics, plasma chemistry, plasma biomedicine, and engineering.

Part I

Plasma Fundamentals: Kinetics, Thermodynamics, Fluid Mechanics, and Electrodynamics

Lecture 1

The Major Component of the Universe, the Cornerstone of Microelectronics, The High-Tech Magic Wand of Technology

1.1 The Forth State of Matter: Plasma in Nature, in Lab, in Technology

Walking the streets of a beautiful city of Belgrade with an excellent plasma scientist and friend Zoran Petrovic, I have been shocked by and took pictures of the huge billboards claiming "Plasma, that's all you need," see Figure 1.1. When asking people on the streets of the Serbian capital: "Why all you need is plasma?", the best answer was "because… plasma is good!". Surely, they meant their popular brand of tasty soft biscuits, but still, I agree with them: plasma is good! Before asking ourselves what's so good about plasma, let's introduce the term plasma, which means for us neither the Serbian cookies nor component of blood, but the very special fourth state of matter.

Plasma is an ionized gas, a distinct fourth state of matter. "Ionized" means that at least one electron is not bound to an atom or molecule, converting the atoms or molecules into positively charged ions. As temperature increases, molecules become more energetic and transform matter in the sequence: solid, liquid, gas, and finally plasma, which justifies the title **"fourth state of matter."** The free electric charges, electrons, and ions, make plasma electrically conductive, sometimes more than gold and copper, internally interactive, and strongly responsive to electromagnetic fields. *The ionized gas is called plasma when it is electrically neutral (i.e. electron density is balanced by that of positive ions) and contains a sufficiently high number of the electrically charged particles to affect its electrical properties and behavior.* In addition to being important in many aspects of our daily lives, *plasmas are estimated to constitute more than 99% of the visible universe.*

The term plasma was first introduced by Irving Langmuir in 1928. The multicomponent, strongly interacting ionized gas reminded him of blood plasma (thus at least no direct connection with the tasty soft biscuits). Langmuir wrote: "Except near the electrodes, where there are sheaths containing very few electrons, the ionized gas contains ions and electrons in about equal numbers so that the resultant space charge is very small. We shall use the name plasma to describe this region containing balanced charges of ions and electrons." There is usually not much confusion between the fourth state of matter (plasma) and blood plasma; probably the only exception is the process of plasma-assisted blood coagulation, where the two concepts meet.

Plasmas occur naturally but also can be effectively man-made in laboratory and in industry, which provides opportunities for numerous applications, including thermonuclear synthesis, electronics, lasers, fluorescent lamps, and many others. To be more specific, most computer and cell phone hardware is made based on plasma technologies, not to forget about plasma TV. Plasma offers three major attractive features: (i) temperatures of at least some plasma components and energy density can significantly exceed those in conventional technologies, (ii) plasmas can produce very high concentrations of energetic and chemically active species (e.g. electrons, ions, atoms and radicals, excited states, and different wavelength photons), and (iii) plasma systems can essentially be far from thermodynamic equilibrium, providing extremely high density of active species keeping bulk temperature as low as room temperature.

These plasma features permit intensification of conventional chemical processes, increases their efficiency, and often even permit reactions impossible in conventional chemistry. Plasma today is a rapidly expanding area of science and engineering, with technological applications widely spread from micro-fabrication in electronics to making protective coatings for aircrafts, from treatment of polymer fibers and films before painting to medical cauterization for stopping blood and wound healing, from food safety to treatment of cancer, and from production of ozone to the plasma TVs. Summarizing, "plasma is good."

Plasma Science and Technology: Lectures in Physics, Chemistry, Biology, and Engineering, First Edition. Alexander Fridman.
© 2024 WILEY-VCH GmbH. Published 2024 by WILEY-VCH GmbH.

Figure 1.1 "Plasma, that's all you need," how about that?!

Plasma comprises most of the universe: the solar corona, solar wind, nebula, and Earth's ionosphere are all plasmas. Lightning is the plasma phenomenon in Earth's atmosphere, well observed and used by humans from their early days. The breakthrough experiments with this natural form of plasma were performed long ago by Benjamin Franklin, which probably explains the special interest to plasma research in Philadelphia, where the author of this book works at the Nyheim Plasma Institute of Drexel University. At altitudes of approximately 100 km, the atmosphere no longer remains nonconducting due to ionization and formation of plasma by solar radiation. As one progresses further into near-space altitudes, the Earth's magnetic field interacts with charged particles streaming from the sun. These particles are diverted and often become trapped by the Earth's magnetic field. The trapped particles are most dense near the poles and account for the aurora borealis. **Lightning and the aurora borealis** are the most common natural plasmas observed on Earth.

Natural and man-made plasmas (generated in gas discharges) occur over a wide range of pressures, electron temperatures, and electron densities, see Figure 1.2. The temperatures of man-made plasmas range from slightly above room temperature to temperatures comparable to the interior of stars, and electron densities span over 15 orders of magnitude. Most plasmas of practical significance, however, have electron temperatures of 1–20 eV, with electron densities in the range 10^6–10^{18} cm^{-3}. *The high temperatures are conventionally expressed in electron volts; 1 eV approximately equals 11 600 K.* Not all particles need to be ionized in plasma, a common condition is for the gases to be only partially ionized. The ionization degree (i.e. ratio of density of major charged species to that of neutral gas) in the conventional plasma engineering systems is in the range 10^{-7}–10^{-4}. When the ionization degree is close to unity, such plasma is called **completely ionized plasma**. Completely ionized plasmas are conventional for thermonuclear plasma systems: tokamaks, stellarators, plasma pinches, focuses, and so on. When the ionization degree is low, the plasma is called **weakly ionized plasma**, which is the focus of most of this book. Both natural and

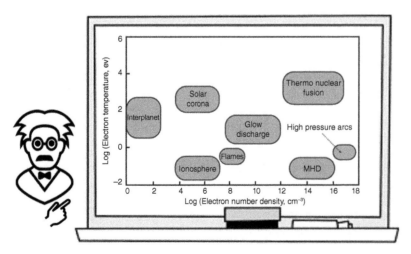

Figure 1.2 Electron temperatures and number densities in different plasmas.

man-made laboratory plasmas are quasi-neutral, which means that concentrations of positively charged particles (positive ions) and negatively charged particles (electrons and negative ions) are well balanced. Langmuir was one of the pioneers who studied gas discharges and defined plasma to be a region not influenced by its boundaries. The transition zone between the plasma and its boundaries was termed the **plasma sheath**. The properties of the sheath differ from those of the plasma, and these boundaries influence the motion of the charged particles in this sheath. The particles form an electrical screen for the plasma from influences of the boundary.

1.2 Multiple Plasma Temperatures, Plasma Nonequilibrium, Thermal and Nonthermal Plasmas

Temperature in plasma is determined by the average energies of the plasma particles (neutral and charged) and their relevant degrees of freedom (translational, rotational, vibrational, and those related to electronic excitation). Thus, plasmas, as multi-component systems, can exhibit **multiple plasma temperatures**. In the electric discharges common for plasma generation in the laboratory, energy from the electric field is first accumulated by the electrons between collisions and, subsequently, is transferred from the electrons to the heavy particles. Electrons receive energy from the electric field during their mean free path and, during the following collision with a heavy particle,

 lose only a small portion of that energy (because electrons are much lighter than the heavy particles). That is why *the electron energy and temperature in plasma can be higher than those of heavy particles. Subsequently, collisions of electrons with heavy particles (Joule heating) can equilibrate their temperatures. This "equilibration" doesn't happen if time or energy are not sufficient (which occurs, for example, in coronas and pulsed discharges) or if there is an intensive cooling preventing heating of the entire gas (which occurs, for example, in wall-cooled low-pressure discharges). This* determines the **basic plasma nonequilibrium** $(T_e \gg T_0)$. The temperature difference between electrons and heavy neutral particles due to Joule heating in the collisional weakly ionized plasma is usually proportional to the square of the ratio of electric field (E) to pressure (p). Only in the case of low E/p, the temperatures of electrons and heavy particles approach each other. This is a basic requirement for **local thermodynamic equilibrium (LTE)**. The LTE conditions also require chemical equilibrium and restrictions on gradients. The *LTE plasma follows the major laws of equilibrium thermodynamics and can be characterized by a single temperature at each point of space. Ionization and chemical processes in such plasmas are determined by a single temperature (and only indirectly by E/p through Joule heating).* The quasi-equilibrium plasmas of this kind are usually called **thermal plasmas**. Thermal plasmas in nature are represented by solar plasma and lightning.

Numerous plasmas exist very far from the thermodynamic equilibrium and are characterized by multiple different temperatures related to different plasma particles and different degrees of freedom. It is the electron temperature that often significantly exceeds that of heavy particles $(T_e \gg T_0)$. Ionization and chemical processes in such **nonequilibrium plasmas** are directly determined by electron temperature and, therefore, are not so sensitive to thermal processes and temperature of the gas. The nonequilibrium plasma of this kind is called **nonthermal plasma**. An example of nonthermal plasmas in nature is the aurora borealis. Although the relationship between different plasma temperatures in nonthermal plasmas can be quite sophisticated, it can be conventionally presented in the collisional weakly ionized plasmas as $T_e > T_v > T_r \approx T_i \approx T_0$. Electron temperature ($T_e$) is the highest in the system, followed by the temperature of vibrational excitation of molecules (T_v). The lowest temperature is usually shared by heavy neutrals (T_0), temperature of translational degrees of freedom or simply gas temperature), ions (T_i), as well as rotational degrees of freedom of molecules (T_r).

In many **nonthermal plasmas**, electron temperature is about 1 eV (about 10 000 K), whereas the gas temperature is close to room temperature. Nonthermal plasmas are usually generated either at low pressures or at lower power levels, or in different kinds of pulsed discharge systems. The engineering aspects and application areas are quite different for thermal and nonthermal plasmas. For example, thermal plasmas are usually more powerful, whereas nonthermal plasmas are more chemically selective.

 It is interesting to note *that both thermal and nonthermal plasmas usually have the highest temperature (T_e in one case, and T_0 in the other) on the order of 1 eV, which is about 10% of the total energy required for ionization (ionization potential, about 10 eV). It reflects the general rule found by Zeldovich and Frank-Kamenetsky for atoms and small molecules in chemical kinetics: the temperature required for a chemical process (including ionization) is typically about 10% of the total required energy, which is the Arrhenius activation energy*

(or ionization potential in plasma). Funny, a similar **rule of 10%** can usually be applied to determine a down payment to buy a house or a new car. Thus, the plasma temperatures can be somewhat identified as the "down payment for the ionization process." Thus, in the thermal quasi-equilibrium plasmas, the single temperature is about 1 eV. In the strongly nonequilibrium nonthermal plasmas, the electron temperature is on the order of 1 eV, while gas temperature can be from the room temperature and above simply following the heating/cooling balance.

1.3 Plasma Sources: Nonthermal, Thermal, and Transitional "Warm" Discharges, Discharges in Gases and Liquids

Plasma sources, which in most of practical cases are the electric discharges, represent the engineering basis of the plasma science and technology. For simplicity, an electric discharge can be first viewed as two electrodes inserted into a tube and connected to a power supply. It was Michel Faraday, who was observing and investigating such plasma sources as early as in 1830s. The tube can be filled with various gases or evacuated. As the voltage applied across the two electrodes increases, the current suddenly increases sharply at a certain voltage required for sufficiently intensive electron avalanches. If the pressure is low, on the order of a few Torrs, and the external circuit has a large resistance to prohibit a large current, a **glow discharge** develops. This is a low-current, high voltage discharge widely used to generate nonthermal plasma with T_e about 1 eV and above, and T_0 about room temperature. A similar discharge is known to everyone as the plasma source in fluorescent lamps (you can touch it to double-check that T_0 is about room temperature). *For educational purposes, the glow discharge can be considered as a major example of low-pressure nonthermal plasma sources*, see Figure 10.1 in Lecture 10.

A nonthermal **corona discharge** occurs at high pressures (including atmospheric pressure) only in regions of sharply nonuniform electric fields. Glowing powerlines and spikes on top of high buildings are examples of the coronas, which originated the name of the discharges meaning "crowns" in several languages. The electric field near one or both electrodes of the corona discharges must be stronger than in the rest of the gas. This occurs near sharp points, edges, or small-diameter wires. These discharges tend to be low-power plasma sources limited by the onset of electrical breakdown of the gas. It is possible, however, to circumvent this restriction using short-pulse power supplies and organizing large-scale arrays of the more powerful **pulsed coronas**. Figure 1.3 shows example of the 10-kW pulsed corona. The picture is made through a window of the Mobile Environmental Laboratory, see Figure 24.5 in the Lecture 24. Electron temperature in the corona discharges exceeds 1 eV, whereas the gas remains at room temperature. The corona discharges are widely applied in the treatment of polymer materials: most synthetic fabrics applied to make clothing have been treated before dying in corona-like discharges to provide sufficient adhesion. *For educational purposes, the corona discharges can be considered as major example of atmospheric pressure nonthermal plasma sources*, see Figures 13.2 and 13.3 in Lecture 13.

If the pressure is high, on the order of an atmosphere, and the external circuit resistance is low, an **electric arc discharge** can be organized between two electrodes. This discharge is also one of the longtimers, it has been invented

Figure 1.3 Photo of the 10-kW pulsed corona discharge through a window of the Mobile Environmental Laboratory.

by Sir Humphry Davy in 1800. Thermal arcs usually carry large currents, greater than 1 A at voltages of the order of tens of volts. Furthermore, they release large amounts of thermal energy at very high temperatures often exceeding 10 000 K. The arcs are often coupled with a gas flow to form high-temperature plasma jets. The arc discharges are well known not only to scientists and engineers but also to the public because of their wide applications in welding devices. *For educational purposes, the arc discharge can be considered a major example of thermal plasma sources*, see Figure 11.2 in Lecture 11.

Between other electric discharges widely applied in plasma science and technology, we can point out the **nonequilibrium, low-pressure radiofrequency discharges** playing the key roles in sophisticated etching and deposition processes of modern micro-electronics, micro-electro-mechanics, as well as in treatment of polymer materials, see Figure 12.7 in Lecture 12. It is these discharges manufacture for us most our cellphones or computers, which we are enjoying so much. Between less traditional but very practically interesting discharges, we can point out the nonthermal, high-voltage, atmospheric-pressure, **floating-electrode dielectric barrier discharge (FE-DBD)**, which can use the human body as a second electrode without damaging the living tissue. Such discharge obviously provides very interesting opportunities for direct plasma applications in biology and medicine, see Figure 1.4. FE-DBDs are younger members of the big family of the dielectric barrier discharges (DBD), introduced as early as 1857 by Werner von Siemens, and playing an important role in generation of ozone, plasma TVs, plasma aerodynamics, plasma medicine, and agriculture, see Figures 13.5 and 13.6.

 As it was explained above in the Section 1.2, *electron temperature in most of plasma systems stay on the level of 10 000 K (about 1 eV, which is 10% of the ionization potential). Gas temperature controlled by the gas heat balance can be close to room temperature (nonthermal plasma) or vice versa close to electron temperature (thermal plasma). Although plasma is often goes to these extremes, it can be also organized in the intermediate regime of the so-called* ***transitional or warm plasmas****, where T_e is still around 1 eV, but gas temperature can be "warm", around 1000 K and below*. Members of this group include **microwave discharges, sparks** (see the transitional PHD spark discharge in Figure 1.5), as well as different types of the **gliding arc discharges**. The most known recently are the gliding arc discharges stabilized in reverse vortex "tornado" flow generating high power nonequilibrium atmospheric pressure plasma, see Figure 1.6. These discharges provide a unique opportunity of combining the high power and atmospheric pressure typical for arc discharges with the relatively high level of nonequilibrium and therefore selectivity typical for nonthermal discharges.

While majority of electric discharges generate plasma in gas phase, some discharges are operating effectively to generate **plasma in liquid phase**. Most of them like gliding arcs and sparks generate plasma in bubbles or voids in the liquid, see Figure 1.7. These types of "liquid discharges" are applied for chemical and biological cleaning of water, as well as for medical, agricultural, and food processing purposes. Some plasma sources, usually based on the nano-second and sub-nanosecond-based high-voltage pulsing with extremely short voltage rise time, are able, however, directly ionize liquids without bubbles and voids. *So, plasma is not necessarily only "ionized gas," but it can be also "directly ionized liquid."* All these plasma sources, and many others are going to be considered in the Part 2 of this book, Lectures 9–16.

Figure 1.4   Floating-electrode dielectric barrier discharge, FE-DBD.

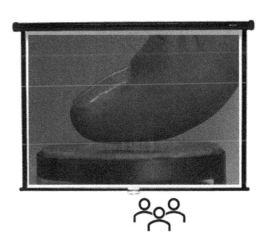

Figure 1.5 Pin-to-hole (PHD) spark discharge.

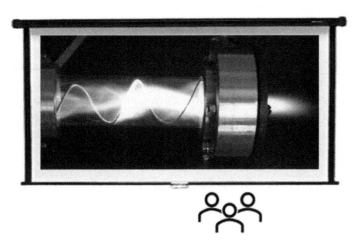

Figure 1.6 Gliding arc discharge stabilized in reverse-vortex "tornado" flow.

Figure 1.7 Spark discharge in water.

1.4 Plasma Processes: Major Plasma Components, High Selectivity and Controllability of Nonequilibrium Reactions, "Multidisciplinarity Without Borders"

Chemically and biologically active plasmas are multi-component systems highly reactive due to large concentrations of charged particles (electrons, negative and positive ions), excited atoms and molecules (electronic and vibrational excitation make a major contribution), active atoms and radicals, and ultraviolet (UV) photons. In thermal plasmas, the main global effect is mostly due to temperature and relevant quasi-equilibrium plasma components. In nonthermal nonequilibrium plasma, the processes are way more sophisticated. Each component of the chemically active plasma plays its own specific role in plasma-chemical kinetics. The **plasma electrons**, for example, are usually first to receive the energy from an electric field and then distribute it between other plasma components and specific degrees of freedom of the system. *Changing parameters of the electron gas (density, temperature, electron energy distribution function) often permit control and optimization of plasma-chemical processes.*

The **plasma ions** are charged heavy particles, that can make a significant contribution to plasma-chemical kinetics either due to their high energy (as in the case of sputtering and reactive ion etching) or due to their ability to suppress activation barriers of chemical reactions. This second feature of the plasma ions results in the so-called **ion or plasma catalysis**, which is particularly important in the plasma-assisted ignition and flame stabilization, fuel conversion, hydrogen production, exhaust gas cleaning, and even in plasma medicine and agriculture due to the direct plasma-catalytic treatment of living tissue. The energetic plasma ions also significantly contribute to the **reactive ion etching processes** crucial in the microelectronics for the deep ditch anisotropic etching and fabrication of the sophisticated relevant features in the integrated circuits.

The **vibrational excitation of molecules in plasma** often makes a major contribution to the plasma-chemical kinetics and energy balance because the plasma electrons with energies around 1 eV primarily transfer most of their energy in such gases as N_2, CO, CO_2, H_2, CH_4 and so forth into vibrational excitation. The vibrational excitation of molecules is also the most effective in overcoming the activation barriers of the endothermic chemical reactions. Stimulation of plasma-chemical processes through vibrational excitation permits the highest values of energy efficiency to be reached, which occurs for example in the plasma-chemical dissociation of CO_2, and plasma synthesis of NO in air.

Electronic excitation of atoms and molecules in plasma can also play a significant role in plasma kinetics, especially when the lifetime of these excited particles is quite long (as in the case of metastable electronically excited atoms and molecules). As an example, we can mention plasma-generated **metastable low-energy electronically excited oxygen** molecules $O_2(^1\Delta_g)$, singlet oxygen, which effectively participate in the plasma-stimulated oxidation process in polymer processing, and biological and medical applications of plasma. The singlet oxygen $O_2(^1\Delta_g)$ metastable molecules have very long radiative lifetime, about an hour. As an example of the **higher energy metastable electronically excited states of molecules** generated in plasma, we can mention those of nitrogen $N_2\left(A^3\Sigma_u^+\right)$ characterized with radiative lifetime of 13 s and playing significant role, for example, in ozone synthesis in air and plasma-assisted combustion and flame stabilization. In both applications, these quite energetic metastable nitrogen molecules (energy exceeding 6 eV) are able directly dissociate O_2 molecules producing oxygen atoms. Also, they can dissociate the passive radicals HO_2 to suppress the chain termination of oxidation of H_2 and hydrocarbons crucial for plasma-stimulated ignition, combustion, and flame stabilization.

The contribution of the **plasma-generated active atoms and radicals** is obviously also very significant. As an example, we can point out that O atoms and OH radicals effectively generated in atmospheric air discharges which play a key role in numerous plasma-stimulated oxidation processes. They are part of the family of the so-called **reactive oxygen species (ROS)**, which together with **reactive nitrogen species RNS** (like NO, peroxynitrite, nitrosylated organics, etc.) are the key players in the plasma-stimulated biochemistry especially in the liquid phase. F-atoms and the fluorocarbon radicals generated in low-pressure plasmas in C_2F_6, C_3F_8, NF_3, etc. play an important role in etching and other material processing reactions in microelectronics. We should mention that relatively stable long-living active chemicals like ozone, H_2O_2, NO_x compounds, and acids are also crucial in several plasma-initiated chemical and biochemical processes. **Plasma-generated photons** play a key role in a wide range of applications, from plasma light sources to UV sterilization of water. Sometimes strong plasma effects are due to **plasma-related**

electric fields like in the case of plasma-stimulated electroporation of cells so much important in plasma medicine and agriculture.

 Plasma is not only a multi-component system, but often a **strongly nonequilibrium system**, like it was already discussed above in the Section 1.3. Concentrations of the active species described earlier can exceed those of quasi-equilibrium systems by many orders of magnitude at the same gas temperature. Also, these *nonequilibrium concentrations of active species are very sensitive to electric discharge and plasma parameters, like electric fields, currents, energy input, composition, etc. It opens possibility of very flexible control of the plasma processes from plasma microelectronics to plasma treatment of cancer. The high level of* **controllability of the nonequilibrium reactions** in plasma permits achievement of very **high selectivity of the plasma processes**.

The successful control of plasma permits chemical and biochemical processes to be directed in a desired direction, selectively, and through optimal desired mechanism. Thus, plasma at different regimes can effectively produce NO in air for production of fertilizers and explosives and effectively destroy it in air for the environmental control purposes. Plasmas at different regimes can heal human tissue for treatment of chronic wounds and can selectively destroy the tissue to cancer treatment and for tissue ablation purposes. Surely, plasma is simply a tool but with very high level of controllability and selectivity. A hammer and a computer are also tools: hammer is an excellent tool but focused on one specific application to hit something, while computers are way more controllable and can be used for very many purposes from checking e-mails, participating in ZOOM meetings, and watching movies to writing books, and even hitting something if necessary. Thus, plasma as a tool due to its high controllability and selectivity is way closer to computers from this perspective. Surely effective control of the plasma systems requires detailed understanding of physics, chemistry, if necessary, biology, and surely engineering of the plasma processes. It makes plasma "multidisciplinary without borders" and creates challenge for scientists and engineers working with the "fourth state of matter". Meeting this challenge of the **"multidisciplinarity without borders"** is probably the major objective of this book of lectures.

1.5 Plasma Technologies: The Cornerstone of Microelectronics, the Major Successes Stories

The plasma technologies today are numerous and involve many industries. Discussion of all major plasma applications covers the whole second half of this book (Lectures 17–32). Between those, we can clearly point out the **plasma application to processing of electronic materials**, which can be proudly called the cornerstone of modern microelectronics. Plasma etching, especially deep ditch etching, sputtering, plasma-enhanced chemical vapor deposition (PE-CVD), ion implantation processes, etc. (see Lecture 21), today these plasma technologies determine the success of modern microelectronics, the so wide use of computers, cell phones, and entire almost infinite family of electronic devices, which represent our today's civilization. Just this one plasma application would be sufficient to justify importance of plasma science and technology to modern mankind. There are, however, several other significant plasma success stories.

In this regard, we can point out the **plasma technologies of production and spraying of powders**, and deposition of special coatings. These thermal plasma technologies are usually focused on the protective and specially functionalized coatings. Majority of the parts constituting modern aerospace and automotive engines, construction parts, and other elements undergoing today's thermal spraying for special coatings, significant percentage of those are plasma-related (see Lecture 20). This is today, probably, the number one industrial application of the thermal plasma systems. The thermal plasma spraying, and coating systems can be quite big, like the Drexel vacuum arc coating chamber built by Prof. R. Knight and his team shown in Figure 1.8. Between other successful **large-scale applications of thermal plasma**, we can mention conversion of natural gas to acetylene and ethylene, different ignition schemes, commutation devices, UV sources, plasma metallurgy, and plasma cutting, as well as plasma-stimulated treatment of waste, especially municipal waste, and radioactive waste.

The most successful large-scale applications of nonthermal plasma, outside of electronics, is **plasma treatment of synthetic fibers, fabrics, films**, etc., see Lecture 26. Most of these synthetic materials are plasma treated today to increase adhesion before printing, dying, etc. The very large-scale nonthermal plasma technology is **plasma generation of ozone**, see Lecture 18. The old but impressive photo of the large-scale ozone generator at the Los

Figure 1.8 Drexel vacuum arc plasma coating chamber.

Figure 1.9 Large scale industrial ozone generator.

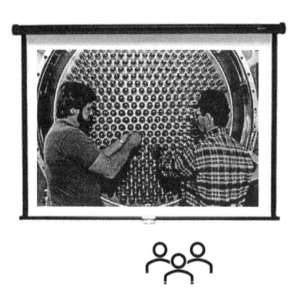

Angeles Aqueduct Filtration Plant is shown in Figure 1.9. Plasma-generated ozone is widely used in the world for water cleaning. Another plasma-based environmental technology is **plasma cleaning of air and exhaust gases**, industrially applied now in quite large scale in power plants (abatement of NO_x) and automotive tunnels (abatement of automotive exhaust), small units are used to suppress the automotive exhaust inside of cars and trucks, see Lecture 24. First impressive steps are made in **plasma cleaning and disinfection of water**. Not to forget is the large-scale commercial application of nonthermal plasma in different kind of **light sources** from common fluorescent lamps to plasma TVs and plasma-based lasers, see Lecture 23. As a reminder, less than 20 years ago, the incandescent light bulbs dominated the lighting sections of our supermarkets, and now it is very difficult even to find them in the store. The mass-market lighting is now almost completely based on plasma and light-emitting diodes (LED).

Exciting novel application of plasma is **plasma medicine**, that is direct application of plasma to human body to treat diseases, see Lectures 30–32. Largely started only in 2003, it came now to hospitals to treat diseases not effectively treated before. The best results are demonstrated so far in treatment of chronic wounds, especially ulcers, as well as in dermatology. First impressive results are demonstrated in oncological hospitals, in treatment of cancer. Promising results are demonstrated in clinical dermatology, first interesting research data collected in dentistry, gastroenterology, and ophthalmology, see illustration of the animal studies in ophthalmology in Figure 1.10. We should mention, that although the nonthermal plasma medicine itself is a relatively newcomer to the hospitals, the thermal

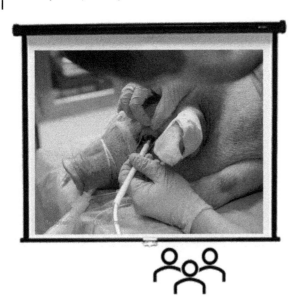

Figure 1.10 Plasma ophthalmology, animal studies.

plasma-based **blood cauterization technology** is widely used in hospitals already for many decades. Also, plasma technologies have relatively long success story in the thermal **plasma-induced tissue ablation**, as well as in non-thermal plasma-induced **tissue engineering** and sterilization of **medical devices**.

Closely related to the plasma medicine are first impressive technological results in **plasma agriculture and food processing** (Lecture 29), see illustration in Figure 1.11, Plasma has been demonstrated as a reliable tool to treat fresh produce and other foods increasing their shelf life and suppressing dangerous pathogens. Plasma, especially the dielectric barrier discharges, DBDs, has also proven to be effective in disinfection of already packaged food. No miracles, these discharges operate always through dielectric barriers, and the packaging material is just an additional barrier (if it is surely not conductive). The plasma agriculture and food processing technologies are organized not only directly but also through plasma activation of water to stimulate plant growth (especially in hydroponics) and to wash the fresh produce. Talking about plasma technologies, we should at least mention the nuclear fusion. This is a "big one" requiring special consideration of the relevant science and technology but stays outside of the scope of this book.

Figure 1.11 Plasma treatment of fresh produce.

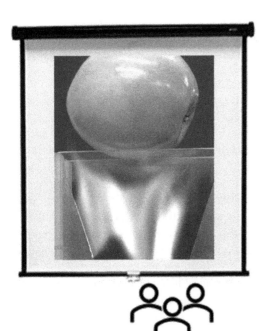

Thus summarizing, the plasma technology has a lot of the success stories to present today and hopefully way more tomorrow. Most of these technologies effectively use the two key advantages of plasma, explained in the previous Section 1.4: high plasma selectivity and controllability. Obviously, not everything is so smooth with plasma technologies, surely, they have challenges sometimes very serious. The most general challenges and pathways to meet them are going to be shortly discussed below.

1.6 Electric Energy Consumption as a Challenge of Plasma Technologies, Plasma is the Future Because the Future is Electric

The success stories of plasma technology described above are mostly related to the major advantages of plasma processes, namely high selectivity, and high controllability. There is another great advantage of the plasma technologies, very high specific productivity (productivity per unit volume of reactor). As an example, for the CO_2 dissociation in nonequilibrium plasma under supersonic flow conditions, it is possible to selectively introduce up to 90% of the total discharge power into CO production when the vibrational temperature is about 4000 K and the translational temperature is only about 100 K. The specific productivity of such a supersonic reactor achieves $1\,000\,000\,l\,h^{-1}$, with power levels up to 1 MW. To compare, this specific productivity exceeds that of the relevant electrolytic and thermos-catalytic about 1000 times. This plasma process has been examined for the fuel production on Mars, where the atmosphere mostly consists of CO_2. On Earth, it was applied as a plasma stage in a two-step process for hydrogen production from water, as well as simply for elimination and sequestration of CO_2 from exhaust gases.

 This very important feature of the **extremely high specific productivity (equipment compactness) of plasma technologies**, *three orders of magnitude above that of the conventional chemical approaches, attracts significant interest to large-scale plasma applications in chemical and environmental technologies, metallurgy, energy systems, fuel reforming and hydrogen production processes. In addition to plasma dissociation of CO_2, it includes, for example, fixation of nitrogen (NO) from air, liquefaction, and direct production of valuable organics from natural gas, plasma metallurgy, plasma stimulated waste treatment, plasma cleaning and disinfection of water, plasma activation for agriculture (stimulation of plant growth), for fresh produce washing, food processing, etc.* All these technologies are very much interested in application of plasma because of significant intensification of the processes, making equipment way more compact, as well as possibility of significant simplification of their maintenance. The wide commercialization of these plasma technologies is limited, however, today by a **major key challenge of the large-scale plasma processes, their energy cost**. While in plasma microelectronics to fabricate integrated circuits, or in plasma medicine to treat ulcers or cancers, energy cost of technology is not a crucial issue, in the large-scale chemical, environmental, and energy systems it is a crucial issue. As an example, the large-scale plasma nitrogen fixation for production of fertilizers has been successful in early 1900 but gave up to thermo-catalytic Haber–Bosch process exclusively due to energy cost competition.

Other general challenges of the plasma-based large-scale chemical, environmental, and energy systems are **scaling up, and by-products of the processes**. The scaling up challenge can be addressed by choosing discharges the most relevant for the scaling up, for example, the nonequilibrium "warm" discharges (microwave, spark, gliding arc discharges, etc.) as well as electron beams in some cases. The challenge of by-products can be addressed by choosing optimal regimes of the discharges, as well as by combination of the plasma technologies with conventional ones (like scrubbing, absorption, product separation technics, etc.). Thus, the challenges of scaling up and of by-products can be solved by the advancement in engineering, while the challenge of the electric energy cost stays as the most critical requirement for the plasma-based large-scale chemical, environmental, and energy system technologies. The large-scale plasma technologies are often still considered now as energy expensive.

 Two important problems should be solved to meet the electric energy challenge of the large-scale plasma technologies. First, is absolute **minimization of the electric energy cost of the plasma processes**. Significant progress here has been achieved here recently in the large-scale plasma cleaning of exhaust gases and in the plasma metallurgy due to engineering optimization of these technologies. Second, is **decrease the cost of electricity and development of safe more environmentally friendly sources of electric energy**. It requires further development of safe nuclear and thermonuclear reactors as well as progress in large-scale development of the renewable energy sources, like solar energy, wind energy, geothermal energy, hydropower, ocean energy, bioenergy, etc. This pathway is not fast and easy to accomplish, but the end point of the path is clear. **The future is electric**, there is no alternative to that for our civilization. We move already in this direction optimizing the worldwide energy distribution. Even automotive industry is getting converted now to electric and hybrid cars and trucks. Thus, future is going to be electric, and if so, plasma and electrochemical technologies would take initiative to convert the electricity into all other human needs (now this crucial niche is kept by crude oil and oil accompanied gases). *Keeping in mind that electrochemistry today is three orders of magnitude less energy intensive and compact than plasma, we have good chances to sustain this leadership. Thus, plasma is the future because the future is electric. Good prognosis and hopes for the future, but what about today?*

1.7 Plasma Today is a High-Tech Magic Wand of Modern Technology

In many of today's practical applications, plasma technology competes with other engineering approaches and sometimes successfully finds its specific niche in the modern industrial environment. Such situation takes place, for example, in thermal plasma spraying and deposition of protective coatings, in plasma stabilization of flames, in plasma conversion of fuels, in plasma light sources, lasers, in plasma cleaning of exhaust gases, in plasma sterilization of water, in plasma activation of wash water, in plasma hydroponics, and so on. All these plasma technologies are practically interesting, commercially viable, and generally make an important contribution to the successful development of our society.

 The most exciting applications of plasma, however, are related not to the aforementioned competing technologies but to those technologies which have no analogies and no (or almost no) competitors in modern industrial environment. A good relevant example is plasma applications in micro-electronics, especially for the case of etching deep trenches (at maximum 0.2 µm wide and at minimum 4 µm deep) in single crystal silicon, which is so much important in the fabrication of integrated circuits. Capabilities of plasma processing in micro-electronics are extraordinary and unique. We probably would not have computers and cell phones as we have now without plasma processing. When all alternatives fail, plasma can still be utilized; plasma chemistry in this case plays the role of **the high-tech magic wand of modern technology.**

Among other **examples, when plasma abilities are extraordinary and unique,** we can point out (i) plasma production of ozone where no other technologies can challenge plasma for more than 100 years; (ii) thermonuclear plasma reactors as a major future source of energy; (iii) low-temperature fossil fuel conversion where hydrogen is produced without CO_2 exhaust (see Section 22.10); (iv) direct liquefaction of natural gas by its incorporation into low quality usually nonsaturated hydrocarbon liquid fuels (see Section 22.12); (v) nonoxidative disinfection of fresh

Figure 1.12   FE-DBD plasma device for direct treatment of wounds, skin sterilization, and treatment of skin diseases.

produce by plasma-activated water (see Lecture 29); (vi) synthesis of polymetric nitrogen in the cryogenic plasma of liquid nitrogen (see Section 16.7); and finally sure (vii) plasma medicine with its healing of cancers, complicated ulcers, and other diseases not effectively treated before.

In Figure 1.12, Dr. Gregory Fridman, at that time still a student of the Nyheim Plasma Institute of Drexel University, holds in his hands the pencil-like active 35-kV FE-DBD electrode, which was safely and directly applied to the human body (see Lectures 30–32), and opened possibilities to cure diseases that were previously incurable. This plasma medical device, which is in use till now in dermatological practice, even looks like a magic wand. Each type of magic, however, requires a well-prepared magician. With these words, we now can make a step from the first introductive lecture to the following ones focused on the entire scope of plasma science and technology, including most of aspects of plasma physics, plasma chemistry, plasma biomedicine, and plasma engineering.

Lecture 2

Elementary Processes of Charged Particles in Plasma

Plasma processes are usually quite complex and determined by synergistic contribution of multiple elementary reactions taking place simultaneously. Each elementary reaction (or elementary process) involves interaction of individual species at certain conditions, such as energy, and level of excitation. Elementary reaction rates in plasma are determined by microkinetic characteristics or probabilities of these interactions as well as by relevant kinetic distribution or population functions representing probability of certain energy or excitation states of the interacting plasma species. The sequence of transformations of individual plasma species, involved substances, and types of energy are usually referred to as a *mechanism of the plasma process*.

Plasma is an ionized medium; therefore, the key plasma processes are related to ionization, conversion of charged particles, and their recombination. For this reason, we will start consideration of mechanisms and kinetics of fundamental plasma processes with analysis of elementary processes of charged species.

2.1 Elementary Charged Plasma Species and Their Transformation Pathways

Elementary charged plasma particles and major fundamental processes of their formation are illustrated in Figure 2.1. Ionization is the key process of plasma generation, which is actually conversion of a neutral atom or a molecule into an electron and positive ion. This process usually stimulated by electron impact is shown in Figure 2.1a. The two participants of this process, ***electrons and positive ions***, are the most important charged particles in plasma. In hot thermonuclear plasmas, the positive ions are multi-charged, but in relatively cold plasmas of technological interest, their charge is usually equal to +1e.

Normally, number densities of the electrons and the positive ions are equal or near equal in quasi-neutral plasmas, but in "electronegative" gases it can be different. Electronegative gases including air, O_2, and gaseous halogen compounds consist of atoms and molecules with a high electron affinity, which strongly attract electrons, resulting in formation of ***negative ions***, the third important group of charged particles in plasmas. Number densities of negative ions can exceed those of electrons in electronegative gases. Illustration of the electron attachment process and negative ion formation is given in Figure 2.1b. Electron attachment leads to the formation of negative ions with charge −1e. Attachment of another electron and formation of multi-charged negative ions is actually impossible in the gas phase because of electric repulsion (in contrast to electrochemical processes in liquids).

In high-pressure cold plasmas, charged particles can also appear in more complicated forms. Positive and negative ions attach neutral atoms or molecules to form quite large ***complex ions*** or ion clusters, e.g. $N_2^+N_2$ (N_4^+), O^-CO_2 (CO_3^-), H^+H_2O (H_3O^+), which is illustrated in Figure 2.1c. Ion-molecular bonds in such complexes are usually less strong than conventional chemical bonds, but stronger than inter-molecular ones in neutral clusters.

Electrons are usually first in getting energy from electric fields, because of their low mass and high mobility. Electrons then transmit their energy to all other plasma components, providing energy in particular for ionization, excitation, and dissociation of atoms and molecules. The kinetic rates of such processes depend on how many electrons have enough energy to do the job. It can be described by means of the ***electron energy distribution function (EEDF)*** $f(\varepsilon)$, which is the probability density for an electron to have energy ε. The EEDF strongly depends on electric field and gas composition in plasma and often can be very far

Plasma Science and Technology: Lectures in Physics, Chemistry, Biology, and Engineering, First Edition. Alexander Fridman.
© 2024 WILEY-VCH GmbH. Published 2024 by WILEY-VCH GmbH.

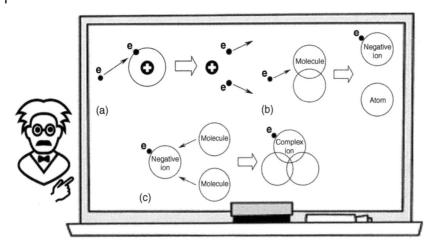

Figure 2.1 Charged plasma particles: electrons, negative, positive and complex ions. Illustration of elementary plasma processes: (a) ionization, (b) dissociative electron attachment, (c) complex ion formation.

from the equilibrium. Sometimes, however, (even in nonequilibrium plasmas), the EEDF depends mostly on electron temperature T_e, and can then be defined by the quasi-equilibrium **Maxwell–Boltzmann distribution function**:

$$f(\varepsilon) = 2\sqrt{\varepsilon/(\pi^3 k T_e)^3}\, \exp\left(-\frac{\varepsilon}{kT_e}\right), \tag{2.1}$$

where k is a Boltzmann constant (when temperature is given in energy units, then $k = 1$ and can be omitted). The **mean electron energy**, which is the first moment of the EEDF, is proportional to temperature in the conventional way:

$$<\varepsilon> = \int_0^\infty \varepsilon f(\varepsilon) dE = \frac{3}{2} T_e. \tag{2.2}$$

In most of the plasmas under consideration, the mean electron energy is 1–5 eV. Positive and negative ions are heavy particles like neutrals, so usually they cannot receive high energy directly from electric fields because of intensive collisional energy exchange with other plasma components (unless gas pressure is extremely low). The collisional nature of the energy transfer results in this case in the ion energy distribution function being not far from the Maxwellian with ion temperature T_i close to neutral gas temperature T_0.

Ionization of atoms and molecules, electron attachment to atoms and molecules, and ion-molecular reactions (see Figure 2.1) are examples of elementary plasma processes of charged species. These elementary reactive collisions, as well as others, e.g. electron–ion and ion–ion recombination, electron detachment and destruction of negative ions, surface- and photo-chemical processes – altogether determine balance of charged plasma particles. Before going to discussion of the specific elementary plasma processes of charged species, let us recall major fundamental parameters required for their description.

2.2 Fundamental Characteristics and Parameters of Elementary Processes

First of all, elementary reactions in plasma can be divided according to their energy balance into two classes – elastic and nonelastic processes. The **elastic collisions** are those in which the internal energies of colliding particles do not change, therefore the total kinetic energy is conserved as well. The elastic collisions actually result only in scattering. **Inelastic collisions** result in the transfer of energy from the kinetic energy of colliding partners into internal energy. For example, processes of excitation, dissociation, and ionization of molecules by electron impact are inelastic, providing transfer of high kinetic energy of plasma electrons into the internal degrees of freedom of the molecules. In some instances, the internal energy of excited atoms or molecules can be transferred back into kinetic energy (in particular, into kinetic energy of plasma electrons). Such elementary processes are usually referred to as **superelastic collisions**.

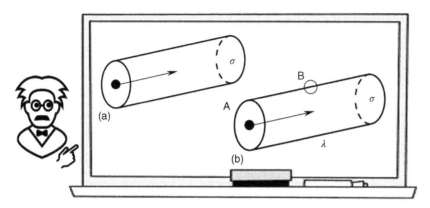

Figure 2.2 (a) Cross section of an elementary process A + B; (b) mean free path of particle A with respect to the process.

According to kinetic theory, the elementary processes including those in plasma can be described in terms of six main collision parameters: (i) cross section, (ii) probability, (iii) mean free path, (iv) interaction frequency, (v) reaction rate, and (vi) reaction rate coefficient. The most fundamental characteristics of elementary processes are their cross sections. The ***cross section of an elementary process*** between two particles can be interpreted as an imaginary circle with area σ moving together with one of the collision partners (see Figure 2.2a). If the center of the other collision partner crosses the circle, elementary reaction takes place. If two colliding particles can be considered as hard elastic spheres of radii r_1 and r_2, their collisional cross section is equal to $\pi (r_1 + r_2)^2$. The interaction radius and cross section can exceed the corresponding geometrical sizes when forces between particles have a long-distance nature (like in the case of electric interactions). Alternately, if only a few out of many collisions result in a chemical reaction, the cross section is considered to be less than the geometrical one.

Typical sizes of atoms and molecules are of the order of 1–3 Å, therefore cross section of simple elastic collisions between plasma electrons (energy 1–3 eV) and neutral particles is usually about 10^{-16}–10^{-15} cm^2. Cross sections of inelastic, endothermic, and electron-neutral collisions are normally lower. When the electron energy is very large, the cross section can also decrease because of reduction of the interaction time. The ratio of the inelastic collision cross section to corresponding cross section of elastic collision under the same condition is often called the dimensionless ***probability of the elementary process***.

The ***mean free path*** λ of a collision partner A with respect to the elementary process A + B represents the distance A travels before the reactive collision with B, which can be presented as:

$$\lambda = \frac{1}{(n_B \sigma)},$$ (2.3)

where n_B is number density of the particles B. This concept is simply illustrated in Figure 2.2b. During the mean free path λ, the particle A traverses the cylindrical volume $\lambda \sigma$. An elementary reaction occurs if this cylindrical volume contains at least one B-particle meaning $\lambda \sigma n_B = 1$, which leads to Eq. (2.3). In the above example of electron-neutral collisions, an electron mean free path before elastic scattering on neutrals at atmospheric pressure and $n_B = 3 * 10^{19}$ cm^{-3} is about $\lambda = 1 \mu$.

The ***interaction frequency*** v of a collision partner A (for example electrons moving with velocity v) with the other partner B (for example a heavy neutral) can be defined as v/λ or taking into account Eq. (2.3) as:

$$v_A = n_B \sigma v.$$ (2.4)

Actually, this relation should be averaged taking into account the velocity distribution function $f(v)$ and the dependence of the cross section σ on the collision partners' velocity. When the collision partners' velocity can be attributed mostly to one light particle (e.g. electron) Eq. (2.4) can be rewritten as:

$$v_A = n_B \int f(v)\sigma(v)v dv = <\sigma v> n_B.$$ (2.5)

Numerically, for the above example of atmospheric pressure electron-neutral collisions, the interaction frequency is about $v = 10^{12}$ 1 s^{-1}.

The number of elementary processes, which take place in unit volume per unit time is called the **elementary reaction rate w**. For binary (for example, bimolecular) processes A + B, the reaction rate can be calculated by multiplying the interaction frequency of partner A with partner B–"v_A" and number of particles A in the unit volume (which is their number density- n_A). Thus: $w = v_A n_A$. Taking into account Eq. (2.5) this results in:

$$w_{A+B} = <\sigma v> n_A n_B. \tag{2.6}$$

The factor $<\sigma v>$ is termed the **reaction rate coefficient k**, one of the most useful concepts of plasma-chemical kinetics. For binary reactions, it can be expressed following Eq. (2.5) as:

$$k_{A+B} = \int \sigma(v)vf(v)dv = <\sigma v>. \tag{2.7}$$

In contrast to the reaction cross section σ, which is a function of the partners' energy, the reaction rate coefficient k is an integral factor, which includes information on energy distribution functions and depends on temperatures or mean energies of the collision partners. Actually, Eq. (2.7) establishes the relation between micro-kinetics (which is concerned with elementary processes) and macro-kinetics (which takes into account real energy distribution functions). Numerically, for the above-considered example of electron-neutral elastic binary collisions, the reaction rate coefficient is approximately $k = 3*10^{-8}$ cm^3 s^{-1}.

The concept of reaction rate coefficients can be applied not only for binary processes but also for mono-molecular (A→ products) and three-body processes (A + B + C → products). Then the reaction rate is usually proportional to a product of concentrations of participating particles, and the reaction rate coefficient is just a coefficient of the proportionality:

$$w_A = k_A n_A \tag{2.8a}$$

$$w_{A+B+C} = k_{A+B+C} n_A n_B n_C. \tag{2.8b}$$

Obviously, the reaction rate coefficients cannot be calculated in this case using Eq. (2.7), they even have different dimensions. For example, for mono-molecular processes Eq. (2.8a), the reaction rate coefficient can be interpreted as the reaction frequency.

2.3 Classification of Ionization Processes, Elastic Scattering and Energy Transfer in Coulomb Collisions

The most typical ionization process provided by electron impact was illustrated in Figure 2.1a. Other ionization processes can dominate in specific conditions and plasma sources. While mechanisms of ionization can be very different, they can be generally subdivided into the following five groups.

1. **Direct ionization by electron impact** is ionization of preliminarily not excited neutrals by an electron, whose energy is high enough to provide the ionization act in one collision. These processes are the most important in cold discharges, where electric fields and therefore electron energies are quite high, but excitation level of neutral species is relatively moderate.
2. **Stepwise ionization by electron impact** is ionization of preliminary excited neutral species. These processes are important mostly in thermal or energy-intense discharges, when ionization degrees n_e/n_0 as well as concentration of excited neutral species are high.
3. **Ionization by collision of heavy particles** takes place during ion-neutral collisions, as well as in collisions of electronically or vibrationally excited species, when the total energy of the collision partners sufficiently exceeds the ionization potential. The chemical energy of colliding species can also contribute to such ionization in the so-called associative ionization processes.
4. **Photoionization** takes place in collisions of neutrals with photons, which result in formation of an electron–ion pair. It can be important in thermal plasmas, as well as in propagation of nonthermal discharges.
5. **Electron emission (surface ionization)** is provided by electron, ion, and photon collisions with different surfaces, or simply by surface heating.

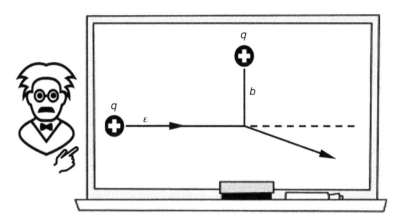

Figure 2.3 Illustration of Coulomb collisions of charged particles, and their cross section dependence on energy.

To analyze kinetics of the above ionization processes, it is useful to first introduce a concept of elastic collisions in general and those of charged particles in particular. **Elementary elastic collisions** (or elastic scattering processes) do not result in a change of chemical composition or excitation level of the colliding partners and for this reason, they are not able to influence directly plasma-chemical processes. From another hand, due to their high cross sections, elastic collisions are responsible for kinetic energy and momentum transfer between colliding partners. The typical value for cross sections of the elastic, electron-neutral collisions for electron energy of 1–3 eV is about $\sigma = 10^{-16}$–10^{-15} cm^2; the relevant rate coefficients are about $k = 3*10^{-8}$ cm^3 s^{-1}. For the elastic, ion-neutral collisions at room temperature, the typical value of cross section is about $\sigma = 10^{-14}$ cm^2; and relevant rate coefficient are about $k = 10^{-9}$ cm^3 s^{-1}.

Electron–electron, electron–ion, and ion–ion elastic scattering processes are usually called **coulomb collisions.** While their cross sections are quite large with respect to those of collisions with neutral partners, these are relatively infrequent processes in plasma with low degree of ionization. An important feature of the coulomb collisions is a strong dependence of their cross sections on the kinetic energy of colliding particles. This effect is illustrated in Figure 2.3 by interaction of two same-charge particles assuming for simplicity one collision partner at rest. Scattering takes place if the coulomb interaction energy (order of $U \sim q^2/b$, where b is the impact parameter) is approximately equal to the kinetic energy ε of a collision partner. Then the impact parameter $b \sim q^2/\varepsilon$ and the reaction cross section σ can be estimated as πb^2:

$$\sigma(\varepsilon) \sim \pi \, q^4/\varepsilon^2 \tag{2.9}$$

The electron–electron scattering cross sections at room temperature are 1000 times greater than those at electron temperature of 1 eV. Similar consideration of the charged particle scattering on neutral molecules with a permanent dipole moment (interaction energy $U \sim 1/r^2$) and an induced dipole moment ($U \sim 1/r^4$) results respectively in cross sections $\sigma(\varepsilon) \sim 1/\varepsilon$ and $\sigma(\varepsilon) \sim 1/\varepsilon^{1/2}$.

Energy transfer during elastic collisions is possible only as a transfer of kinetic energy. The average fraction γ of kinetic energy, transferred from one particle of mass m to another one of mass M is:

$$\gamma = 2mM/(m + M)^2 \tag{2.10}$$

In elastic collision of electrons with heavy neutrals or ions $m \ll M$, and hence $\gamma = 2m/M$, which means the fraction of transferred energy is negligible ($\gamma \sim 10^{-4}$). It explains, in particular, why the direct electron impact ionization is mostly determined by incident electron interaction not with a whole atom but specifically with a valence electron of the atom (see Figure 2.1a).

2.4 Direct Ionization of Atoms and Molecules by Electron Impact: Thomson Formula, Franck–Condon Principle

Direct ionization is a result of interaction of an incident electron, having high energy ε, with a valence electron of a neutral atom or molecule. The act of ionization occurs when energy $\Delta\varepsilon$ transferred to the valence electron exceeds the ionization potential I (see Figure 2.1a). Clear physical and even quantitative description of the process was suggested by Thomson in 1912. The Thomson model supposes that the valence electron is at rest, and also neglects the interaction of the two colliding electrons with the rest part of the initially neutral particle. The differential cross section of the incident electron scattering with energy transfer $\Delta\varepsilon$ to the valence electron can be defined then by the Rutherford formula (compare with Eq. (2.9)):

$$d\sigma_i = \frac{1}{(4\pi\varepsilon_0)^2} \frac{\pi e^4}{\varepsilon(\Delta\varepsilon)^2} d(\Delta\varepsilon). \tag{2.11}$$

Integration over $\Delta\varepsilon$, taking into account that for ionization acts, the transferred energy should exceed the ionization potential $\Delta\varepsilon \geq I$, gives the ***Thomson formula***:

$$\sigma_i = \frac{1}{(4\pi\varepsilon_0)^2} \frac{\pi e^4}{\varepsilon} \left(\frac{1}{I} - \frac{1}{\varepsilon}\right). \tag{2.12}$$

Obviously, it should be multiplied, in general, by number of valence electrons Z_v. According to the Thomson formula, the direct ionization cross section is growing linearly near the threshold of the elementary process $\varepsilon = I$. In the case of high electron energies $\varepsilon \gg I$, the Thomson cross section (2.12) decreases as $\sigma_i \sim 1/\varepsilon$. Quantum mechanics gives a more accurate asymptotic approximation for the high-energy electrons $\sigma_i \sim \ln\varepsilon/\varepsilon$. When $\varepsilon = 2I$, the Thomson cross section reaches the maximum value:

$$\sigma_i^{max} = \frac{1}{(4\pi\varepsilon_0)^2} \frac{\pi e^4}{4I^2} \tag{2.13}$$

The Thomson formula can be rewritten more accurately, taking into account kinetic energy ε_v of the valence electron (Smirnov 2001):

$$\sigma_i = \frac{1}{(4\pi\varepsilon_0)^2} \frac{\pi e^4}{\varepsilon} \left(\frac{1}{\varepsilon} - \frac{1}{I} + \frac{2\varepsilon_v}{3}\left(\frac{1}{I^2} - \frac{1}{\varepsilon^2}\right)\right). \tag{2.14}$$

The Thomson formula (2.12) can be derived from Eq. (2.14) by assuming that the valence electron is at rest and $\varepsilon_v = 0$. Useful variation of the Thomson formula can be obtained, assuming the valence electron interaction with the rest of atom as a Coulomb interaction. Then $\varepsilon_v = I$ and Eq. (2.14) is modified to:

$$\sigma_i = \frac{1}{(4\pi\varepsilon_0)^2} \frac{\pi e^4}{\varepsilon} \left(\frac{5}{3I} - \frac{1}{\varepsilon} - \frac{2I}{3\varepsilon^2}\right) \tag{2.15}$$

The discussed modifications of Thomson formula can be generalized as:

$$\sigma_i = \frac{1}{(4\pi\varepsilon_0)^2} \frac{\pi e^2}{I^2} Z_v f\left(\frac{\varepsilon}{I}\right). \tag{2.16}$$

where Z_v is a number of valence electrons, and $f(\varepsilon/I) = f(x)$ is a general function, which is common for all atoms. Thus, for the Thomson formula (2.12):

$$f(x) = \frac{1}{x} - \frac{1}{x^2}. \tag{2.17}$$

Equation (2.16) is in a good agreement with experimental data for different atoms and molecules (Smirnov 2001) if:

$$\frac{10(x-1)}{\pi(x+0.5)(x+8)} < f(x) < \frac{10(x-1)}{\pi x(x+8)}. \tag{2.18}$$

Electron impact ionization of molecules ($e + AB \rightarrow AB^+ + e + e$) has some specific features illustrated in Figure 2.4a by potential energy curves for a molecule AB and ion AB^+. The fastest internal motion of atoms inside of molecules is molecular vibration. However, even these vibrations have typical times of 10^{-14}–10^{-13} s^{-1}, which is much longer, than the interaction time between plasma electrons and the molecules $a_0/v_e \sim 10^{-16}$–10^{-15} s^{-1} (a_0 is the length, v_e is the mean electron velocity). Therefore, all electronic excitation processes induced by electron impact (including

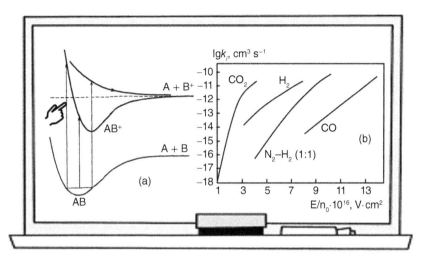

Figure 2.4 Dissociative and non-dissociative ionization of molecules by direct electron impact: (a) molecular and ionic potential energy curves (finger points to Franck–Condon transition), (b) total ionization rate coefficients in molecular gases as a function of reduced electric field.

molecular ionization, see Figure 2.4a) are much faster, than all kind of atomic motions inside of the molecules. As a result, all the atoms inside of molecules can be considered as frozen during the process of electronic transition, stimulated by electron impact. This fact is known as the **_Franck–Condon Principle_**. In accordance with this principle, molecular ionization is presented in Figure 2.4a by vertical lines. As one can see, the non-dissociative ionization process ($e + AB \rightarrow AB^+ + e + e$) results in formation of a vibrationally excited ion $(AB^+)^*$ and requires more energy, than corresponding atomic ionization. When the electron energy significantly exceeds the ionization potential, the dissociative ionization process can take place $e + AB \rightarrow A + B^+ + e + e$. This process corresponds to electronic excitation of AB into a repulsive state of ion $(AB^+)^*$, followed by a decay of this molecular ion.

The **_ionization rate coefficient k_i (T_e)_** can be calculated by integration of the cross section $\sigma_i(\varepsilon)$ over the EEDF, see Eq. (2.7). Assuming Maxwellian EEDF, the direct electron impact ionization rate coefficient can be presented as:

$$k_i(T_e) = \sqrt{8T_e/\pi m}\,\sigma_0 \exp\left(-\frac{I}{T_e}\right).$$ (2.19)

The cross section $\sigma_0 = Z_v \pi e^4 / I^2 (4\pi\varepsilon_0)^2$ corresponds to the characteristic atomic or molecular sizes, that is about $(1\text{–}3)10^{-16}$ cm^2. Kinetic data on the electron impact direct ionization for some molecular gases are presented in Figure 2.4b as a function of reduced electric field E/n_0, which is the ratio of electric field over neutral gas density.

2.5 Stepwise Ionization by Electron Impact

Ionization in energy-dense especially thermal plasmas can be dominated by a stepwise mechanism with preliminary electronic excitation of neutrals followed by act of their ionization with low-energy electrons (usually energies about T_e). Stepwise ionization (sometimes called thermal ionization) can be way faster than direct one in thermal plasmas because the statistical weight of the electronically excited neutrals (making here major contribution to always the same ionization energy I) is significantly greater than that of free plasma electrons. In other words, in thermal ionization conditions with $T_e \ll I$, the probability of obtaining the energy I and ionize is much lower for free plasma electrons (direct mechanism), than for excited atoms and molecules (stepwise mechanism).

 In thermodynamic equilibrium, the ionization processes $e + A \rightarrow A^+ + e + e$ are reverse to three-body recombination $A^+ + e + e \rightarrow A^* + e \rightarrow A + e$, which proceeds through a set of excited states. According to the principle of detailed equilibrium, it means, that the ionization $e + A \rightarrow A^+ + e + e$ should go through the set of electronically excited states as well, which means the ionization should be a stepwise process in the quasi-equilibrium (thermal plasmas). The stepwise ionization rate coefficient k_i^s can be found in this case by summation of the partial rate coefficients $k_i^{s,n}$, corresponding to the nth electronically excited state, over all states of excitation, taking into account their concentrations:

$$k_i^s = \sum_n k_i^{s,n} N_n(\varepsilon_n)/N_0 \tag{2.20}$$

To calculate the maximum stepwise ionization rate, Smirnov (2001), assumed that the electronically excited neutrals are in quasi-equilibrium with plasma electrons and have an energy distribution function, corresponding to the Boltzmann law with the electron temperature T_e:

$$N_n = \left(\frac{g_n}{g_0}\right) N_0 \exp\left(-\frac{\varepsilon_n}{T_e}\right). \tag{2.21}$$

Here N_n, g_n, and ε_n are number density, statistical weight, and energy of the electronically excited neutrals; the index "n" is the principal quantum number. From statistical thermodynamics, the statistical weight of an excited particle is $g_n = 2g_i n^2$, g_i is statistical weight of ion. N_0 and g_0 – are the concentration and statistical weight of ground state particles. Typical energy transfer from a plasma electron to an electron, sitting on an excited atomic level is about T_e. This means that excited particles with energy $\varepsilon_n = I - T_e$ make the most important contribution into the sum (2.20). Taking into account that $I_n \sim 1/n^2$, the number of states with energy about $\varepsilon_n = I - T_e$ and ionization potential about $I_n = T_e$ has an order of n. Thus, from Eqs. (2.20) and (2.21):

$$k_i^s \approx \frac{g_i}{g_0} n^3 <\sigma v> \exp\left(-\frac{I}{T_e}\right) \tag{2.22}$$

The cross section σ corresponds here to energy transfer (about T_e) between electrons and can be estimated as $e^4/T_e^2(4\pi\varepsilon_0)^2$, velocity $v \sim \sqrt{T_e/m}$ and the quantum number can be taken from:

$$I_n \approx \frac{1}{(4\pi\varepsilon_0)^2} me^4/\hbar n^2 \approx T_e \tag{2.23}$$

As a result, the stepwise ionization rate coefficient can be finally expressed as:

$$k_i^s \approx \frac{g_i}{g_0} \frac{1}{(4\pi\varepsilon_0)^5} \left(me^{10}/\hbar^3 T_e^3\right) \exp\left(-\frac{I}{T_e}\right) \tag{2.24}$$

The most interesting and useful in this perspective is comparison of the stepwise (2.24) and direct (2.19) ionization rate coefficients. Equations (2.19) and (2.24) can be presented together as:

$$\frac{k_i^s(T_e)}{k_i(T_e)} \approx \frac{g_i a_0^2}{g_0 \sigma_0} \left(\frac{1}{(4\pi\varepsilon_0)^2} me^4/\hbar T_e\right)^{7/2} \approx \left(\frac{I}{T_e}\right)^{7/2} \tag{2.25}$$

where a_0 is the characteristic atomic length, $\sigma_0 \sim a_0^2$, ionization potential $I \approx 1/(4\pi\varepsilon_0)^2 me^4/\hbar^2$. Eq. (2.25) shows that for typical quasi-equilibrium discharges with $I/T_e \sim 10$, the stepwise ionization can be 10^3–10^4 times faster than the direct one.

 Equation (2.25) also permits to calculate contribution of stepwise ionization in nonthermal discharges where concentrations of electronically excited neutrals are relatively high but way lower than in equilibrium systems. Kinetic analysis of these processes, especially important in the so-called "warm" or transitional discharges like gliding arcs and moderate pressure microwave discharges, can be found in Fridman and Kennedy (2021).

2.6 Ionization by High-energy Electron Beams; Photoionization

Ionization processes induced by high-energy particles, including photons, have their own peculiar features. In the case of electron beams, the peculiarities are due to high energy of electrons, which usually varies from 50 keV to 1–2 MeV. Typical energy losses of the beams in atmospheric pressure air are about 1 MeV per 1 m. The beams with electron energies exceeding 500 keV (which is $E = mc^2$ for electrons at rest) are referred to as **relativistic electron beams**. Typical energy losses of such beams in atmospheric pressure air are about 1 MeV per 1 m. In the frameworks of nonrelativistic **Born approximation**, electron energy losses per unit length can be evaluated by **Bethe-Bloch formula**:

$$-\frac{dE}{dx} = \frac{2\pi Ze^4}{(4\pi\varepsilon_0)^2 mv^2} n_0 \ln\frac{2mEv^2}{I^2}, \tag{2.26}$$

Table 2.1 Photoionization cross sections.

Neutrals	λ, Å	σ, cm^2	Neutrals	λ, Å	σ, cm^2	Neutrals	λ, Å	σ, cm^2
Ar	787	$3.5*10^{-17}$	H$_2$	805	$0.7*10^{-17}$	O$_2$	1020	$0.1*10^{-17}$
N$_2$	798	$2.6*10^{-17}$	H	912	$0.6*10^{-17}$	Cs	3185	$2.2*10^{-19}$
N	482	$0.9*10^{-17}$	Ne	575	$0.4*10^{-17}$	Na	2412	$1.2*10^{-19}$
He	504	$0.7*10^{-17}$	O	910	$0.3*10^{-17}$	K	2860	$1.2*10^{-20}$

where Z is an atomic number of neutral particles, providing the beam stopping; n_0 – their number density; v- is the stopping electron velocity. In the case of relativistic electron beams with the electron energies 0.5–1 MeV, the energy losses can be numerically calculated as:

$$-\frac{dE}{dx} = 2 \times 10^{-22} n_0 Z \ln \frac{183}{Z^{1/3}}, \tag{2.27}$$

where energy losses dE/dx are expressed in MeV cm^{-1}, and density of neutrals n_0 is in cm^{-3}. Equation (2.27) can be rewritten in terms of effective ionization rate coefficient k_i^{eff} for relativistic electrons:

$$k_i^{eff} \approx 3 \times 10^{-10} \text{cm}^3 \text{ s}^{-1} \, Z \ln \frac{183}{Z^{1/3}}. \tag{2.28}$$

Numerically, this ionization rate coefficient is about 10^{-8}–10^{-7} cm^3 s^{-1}. The rate of ionization by relativistic electron beams can be expressed in this case as a function of the electron beam concentration n_b or the electron beam current density j_b (c is the speed of light):

$$q_e = k_i^{eff} n_b n_0 \approx k_i^{eff} \frac{1}{ec} n_0 j_b. \tag{2.29}$$

Contribution of photoionization is usually less than that of electrons (because of deficiency of high-energy photons), but still can make a significant contribution especially in dense thermal plasmas and discharge propagation processes. The **photoionization** of a particle A with ionization potential I (in eV) by a photon $\hbar\omega$ with wavelength λ can be illustrated as:

$$\hbar\omega + \text{A} \rightarrow \text{A}^+ + e, \lambda < \frac{12\ 400}{I(\text{eV})} \text{Å} \tag{2.30}$$

The photoionization requires usually energetic UV photon with wavelength below 100 nm (1000 Å). The process cross section increases sharply from zero at the threshold wavelength to the quite high values shown in Table 2.1 for some specific atoms and molecules.

2.7 Ionization in Collisions of Heavy Particles: Adiabatic Principle, Massey Parameter, Penning Ionization

A free electron with kinetic energy even slightly above the ionization potential is able to ionize (see Figure 2.1a). It is not true for ionization in collisions of heavy particles – ions and neutrals. Even when the heavy particles have enough kinetic energy, the ionization probability is small because their velocities are much lower than those of atomic electrons, and energy transfer is far from resonance. This effect is a reflection of a general **adiabatic principle**: slow motion is "adiabatic," – reluctant to transfer energy to a fast motion. In the case of ionization (as well as electronic excitation) in collision of heavy particles, it is due to slowness of the heavy particles collisions (characteristic frequency $\omega_{int} = \alpha v$, which is reverse time of interaction) and high frequency of electronic transfer processes inside of an atom $\omega_{tr} = \Delta E/\hbar$. Here $1/\alpha$-is a characteristic size of the interacting heavy particles, v is their velocity, ΔE is a change of electronic energy in atom, $\hbar$ is the Planck's constant. Only fast Fourier components of the slow interaction potential between heavy particles (with frequencies about $\omega_{tr} = \Delta E/\hbar$) contribute to the intra-atomic electron transfer leading to ionization (or electronic excitation). The relative fraction of these fast Fourier components is about $\exp(-\omega_{tr}/\omega_{int})$, which is very low if $\omega_{tr} \gg \omega_{int}$. Therefore, probability P_{EnTr} and cross sections of such energy transfer processes (including ionization and electronic excitation in collisions of heavy particles) can be expressed as:

$$P_{EnTr} \propto \exp\left(-\frac{\omega_{tr}}{\omega_{int}}\right) \propto \exp\left(-\frac{\Delta E}{\hbar \alpha v}\right) = \exp(-P_{Ma}). \tag{2.31}$$

Here $P_{Ma} = \Delta E / \hbar \alpha v$ is the **adiabatic Massey parameter**. If $P_{Ma} \gg 1$ the process of energy transfer is adiabatic, and its probability is exponentially low. To get the Massey parameter close to one and eliminate the adiabatic suppression of ionization, the kinetic energy of the colliding heavy particle has to be about 10–100 keV, which is about three orders of magnitude greater than ionization potential.

While kinetic energy of heavy particles in ground state is ineffective for ionization, the situation is different if they are electronically excited. If the total excitation energy is close to the ionization potential, it can result in resonant energy transfer and effective ionization. As an example, when electronic excitation energy of a metastable atom A^* exceeds the ionization potential of another atom B, their collision can lead to the so-called **Penning ionization**. It usually occurs through intermediate formation of an excited molecule (in the state of auto-ionization). The cross sections for the Penning ionization of N_2, CO_2, Xe, and Ar by metastable helium atoms He (2^3S) with excitation energy 19.8 eV reach gas-kinetic values of 10^{-15} cm^2. Similar cross sections can be attained in collisions of metastable neon atoms (excitation energy 16.6 eV) with argon atoms (ionization potential 15.8 eV). Exceptionally high cross section $1.4*10^{-14}$ cm^2 can be achieved in Penning ionization of mercury atoms (ionization potential 10.4 eV) by collisions with the metastable helium atoms He (2^3S, 19.8 eV). If the total electronic excitation energy of colliding particles is not sufficient, ionization is still possible when heavy neutrals recombine forming a molecular ion and electron, which is called **associative ionization**. This process (reverse to dissociative recombination) differs from the Penning ionization by the stability of the molecular ion, and therefore contribution of chemical bonding energy to ionization. For example, the associative ionization in collision of two metastable mercury atoms:

$$\text{Hg}\,(6^3P_1, E = 4.9\,\text{eV}) + \text{Hg}\,(6^3P_0, E = 4.7\,\text{eV})\,\text{Hg}_2^+ + e \rightarrow \text{Hg}_2^+ + e \tag{2.32}$$

The total excitation energy here, 9.6 eV, is less than ionization potential of mercury atom (10.4 eV), but sufficient to form Hg_2^+ molecular ion. Cross sections of the associative ionization (similar to the Penning ionization) are high and close to the gas-kinetic one (10^{-15} cm^2).

 Ionization, and in particular the associative ionization, can also occur in collision of vibrationally excited molecules like: $N_2^* \left(^1\Sigma_g^+, v_1 \approx 32\right) + N_2^* \left(^1\Sigma_g^+, v_2 \approx 32\right) \rightarrow N_4^+ + e$. Reaction rate coefficients are low in this case [about $10^{-15}\exp\left(-2000\,K/T\right)$, cm^3 s^{-1}] because of small Franck–Condon factors. Details regarding kinetics of associative ionization, including ionizing collisions of vibrationally excited molecules, can be found in particular in Rusanov and Fridman (1984), Smirnov (2001), Adamovich et al. (1993), and Fridman and Kennedy (2021).

2.8 Losses of Charged Particles: Mechanisms of Electron–Ion Recombination

Losses of charged particle in plasma can be subdivided into three major groups. The first group includes different types of **electron–ion recombination processes** (to be discussed in this section). Electron losses related to their "sticking" to neutrals and formation of negative ions (see Figure 2.1b) form the second group of volumetric losses, the so-called **electron attachment processes** (to be discussed later in this lecture). Note, that the actual losses of charged particles take place here in the following fast processes of **ion–ion recombination** (neutralization during collision of negative and positive ions). Finally, the third group of charged particle losses is not volumetric but is due to **surface recombination** (to be discussed in the following lectures).

The electron–ion recombination is a highly exothermic process requiring a channel of accumulation of the energy released during the neutralization of a positive ion and an electron. Most of these channels in plasma volume are related to either dissociation of molecules, or to radiation, or to three-body collisions.

A. ***Dissociative electron–ion recombination*** is the fastest electron–ion neutralization mechanism in molecular gases (or just in the presence of molecular ions):

$$e + AB^+ \rightarrow (AB)^* \rightarrow A + B^* \tag{2.33}$$

In these processes, the recombination energy goes to dissociation of the intermediate molecules (AB)* and to electronic excitation of products. Reaction rate coefficients for most of diatomic and three-atomic ions are on the level of 10^{-7} cm^3 s^{-1} (see Table 2.2).

In a group of similar ions, such as molecular ions of noble gases, the recombination reaction rate increases with a number of internal electrons: recombination of Kr_2^+ and Xe_2^+ is about 100 times faster than that of helium. The rate coefficient of dissociative electron–ion recombination has no activation energy and only slightly decreases with both electron T_e and gas T_0 temperatures:

$$k_r^{ei}(T_e, T_0) \propto \frac{1}{T_0 \sqrt{T_e}}. \tag{2.34}$$

While dissociative electron–ion recombination is common in molecular gases, it also occurs in atomic gases especially at high pressures because of preliminary formation of molecular ions in the ***ion conversion processes***: $A^+ + A + A \rightarrow A_2^+ + A$. Then the molecular ion can be fast neutralized in dissociative recombination [Table 2.2, see examples of noble gases, and complex ions like $N_4^+, O_4^+, (NO)_2^+$]. The ion conversion is usually a rapid reaction (see Table 2.3): when pressure exceeds 10 Torr, it is usually faster than the following process of dissociative recombination, which becomes a limiting stage in the overall kinetics. The ion-conversion three-body reaction rate coefficient can be analytically estimated as (Smirnov 2001):

$$k_{ic} \propto \left(\frac{\beta e^2}{4\pi \varepsilon_0} \right)^{5/4} \frac{1}{M^{1/2} T_0^{3/4}}, \tag{2.35}$$

Table 2.2 The dissociative electron–ion recombination rate coefficients at room gas temperature $T_0 = 300$ K, and electron temperatures $T_e = 300$ K and $T_e = 1$ eV.

Process	k_r^{ei}, cm^3 s^{-1}, ($T_e = 300$ K)	k_r^{ei}, cm^3 s^{-1}, ($T_e = 1$ eV)	Process	k_r^{ei}, cm^3 s^{-1}, ($T_e = 300$ K)	k_r^{ei}, cm^3 s^{-1}, ($T_e = 1$ eV)
$e + N_2^+ \rightarrow N + N$	$2*10^{-7}$	$3*10^{-8}$	$e + NO^+ \rightarrow N + O$	$4*10^{-7}$	$6*10^{-8}$
$e + O_2^+ \rightarrow O + O$	$2*10^{-7}$	$3*10^{-8}$	$e + H_2^+ \rightarrow H + H$	$3*10^{-8}$	$5*10^{-9}$
$e + H_3^+ \rightarrow H_2 + H$	$2*10^{-7}$	$3*10^{-8}$	$e + CO^+ \rightarrow C + O$	$5*10^{-7}$	$8*10^{-8}$
$e + CO_2^+ \rightarrow CO + O$	$4*10^{-7}$	$6*10^{-8}$	$e + He_2^+ \rightarrow He + He$	10^{-8}	$2*10^{-9}$
$e + Ne_2^+ \rightarrow Ne + Ne$	$2*10^{-7}$	$3*10^{-8}$	$e + Ar_2^+ \rightarrow Ar + Ar$	$7*10^{-7}$	10^{-7}
$e + Kr_2^+ \rightarrow Kr + Kr$	10^{-6}	$2*10^{-7}$	$e + Xe_2^+ \rightarrow Xe + Xe$	10^{-6}	$2*10^{-7}$
$e + N_4^+ \rightarrow N_2 + N_2$	$2*10^{-6}$	$3*10^{-7}$	$e + O_4^+ \rightarrow O_2 + O_2$	$2*10^{-6}$	$3*10^{-7}$
$e + H_3O^+ \rightarrow H_2 + OH$	10^{-6}	$2*10^{-7}$	$e + (NO)_2^+ \rightarrow 2NO$	$2*10^{-6}$	$3*10^{-7}$

Table 2.3 Ion conversion reaction rate coefficients k_{ic} at room temperature, $T_0 = 300$ K.

Ion conversion process	k_{ic}, cm^6 s^{-1}	Ion conversion process	k_{ic}, cm^6 s^{-1}
$N_2^+ + N_2 + N_2 \rightarrow N_4^+ + N_2^+$	$8*10^{-29}$	$O_2^+ + O_2 + O_2 \rightarrow O_4^+ + O_2$	$3*10^{-30}$
$H^+ + H_2 + H_2 \rightarrow H_3^+ + H_2$	$4*10^{-29}$	$Cs^+ + Cs + Cs \rightarrow Cs_2^+ + Cs$	$1.5*10^{-29}$
$He^+ + He + He \rightarrow He_2^+ + H$	$9*10^{-32}$	$Ne^+ + Ne + Ne \rightarrow Ne_2^+ + Ne$	$6*10^{-32}$
$Ar^+ + Ar + Ar \rightarrow Ar_2^+ + Ar$	$3*10^{-31}$	$Kr^+ + Kr + Kr \rightarrow Kr_2^+ + Kr$	$2*10^{-31}$
$Xe^+ + Xe + Xe \rightarrow Xe_2^+ + Xe$	$4*10^{-31}$	$O_2^+ + O_2 + N_2 \rightarrow O_4^+ + N_2$	$3*10^{-30}$

where M and β are mass and polarization coefficient of colliding atoms, T_0 is the gas temperature. Polyatomic ions (see Table 2.2) have very high recombination rates, often exceeding 10^{-6} cm^3 s^{-1} at room temperature. As a result, recombination of even molecular ions like N_2^+, and O_2^+ at elevated pressures sometimes goes through intermediate formation of dimers like N_4^+, and O_4^+.

B. **Three-body electron–ion recombination** is especially important in the absence of molecular ions, and usually involves two electrons:

$$e + e + A^+ \rightarrow A^* + e. \tag{2.36}$$

The released recombination energy is going to a free electron, "a third-body partner". Heavy particles (ions and neutrals) are unable to accumulate the high recombination energy fast enough in their kinetic energy and therefore are ineffective as the third-body partners [rate coefficients of the recombination with atoms as third-body partners are 10^8 times lower than those of (2.36)]. The recombination (2.36) is especially important in dense thermal plasmas, where molecular ion concentration is low because of high temperature. The process (2.36) starts with a capture of an electron by a positive ion and formation of a very highly excited atom with binding energy about T_e. This highly excited atom then gradually loses its energy in stepwise deactivation by electron impacts. The three-body electron–ion recombination is a reverse process with respect to the stepwise ionization (see Section 2.5), and therefore its rate coefficient can be derived from Eq. (2.24) and the Saha formula for ionization-recombination balance as:

$$k_r^{eei} = k_i^s \frac{n_0}{n_e n_i} = k_i^s \frac{g_0}{g_e g_i} \left(\frac{2\pi\hbar}{mT_e} \right)^{3/2} \exp\left(\frac{I}{T_e} \right) \approx \frac{e^{10}}{(4\pi\varepsilon_0)^5 \sqrt{mT_e^9}}, \tag{2.37}$$

where n_e, n_i, n_0 are number densities of electrons, ions, and neutrals, g_e, g_i, g_0 are their statistical weights, e, m are electron charge and mass, electron temperature T_e is taken as usually in energy units, I is an ionization potential. For numerical calculations, Eq. (2.37) to presented as:

$$k_r^{eei}, \text{cm}^6 \text{ s}^{-1} = \frac{\sigma_0}{I} 10^{-14} \left(\frac{I}{T_e} \right)^{4.5}, \tag{2.38}$$

where σ_0, cm^2, is the gas-kinetic cross section (see Eq. (2.19)), I and T_e are the ionization potential and electron temperature in eV. Typically, k_r^{eei} at room temperature is about 10^{-20} cm^6 s^{-1}, at $T_e = 1$ eV about 10^{-27} cm^6 s^{-1}. At room temperature, the three-body recombination competes with dissociative recombination when electron concentration is quite high and exceeds 10^{13} cm^{-3}. If the electron temperature is about 1 eV, three-body recombination can compete with the dissociative recombination only at extremely high electron densities above 10^{20} cm^{-3}.

C. **Radiative electron–ion recombination** converts recombination energy into radiation:

$$e + A^+ \rightarrow A^* \rightarrow A + \hbar\omega, \tag{2.39}$$

and it is a relatively slow process, because of requirement of a photon emission during a short interval of the electron–ion interaction. Typical values of cross sections are about only 10^{-21} cm^2. This type of recombination is significant only when plasma density is low suppressing the three-body mechanisms: $n_e < 3 \times 10^{13}(T_e, \text{eV})^{3.75}$, cm^3 s^{-1}, and molecular ions are absent (no dissociative recombination). Rate coefficients of (2.39) can be estimated as (Fridman and Kennedy 2021):

$$k_{rad.rec.}^{ei} \approx 3 \times 10^{-13}(T_e, \text{eV})^{-3/4}, \text{cm}^3 \text{ s}^{-1}. \tag{2.40}$$

2.9 Electron Losses in Electronegative Gases, Electron Attachment Processes

A simple illustration of the **electron attachment and negative ion formation** was presented in Figure 2.1b). The electron balance in electro-negative gases with high electron affinity (oxygen, different halogens and their compounds, etc.) can be strongly affected by these processes. In some other gases, such as CO_2 and H_2O, the electron attachment can be equally important, because of high electron affinity of decomposition products and formation of negative ions like O^- and H^- during dissociation. The negative ions can be then neutralized by positive ions in the following very fast processes of **ion–ion recombination** leading to significant losses of charged particles in general. The electron attachment, however, can be effectively suppressed by the reverse processes of negative ion decay through the so-called ***electron detachment*** (to be considered in the next section).

A. ***Dissociative electron attachment*** is a major mechanism of negative ion formation in electronegative molecular gases like CO_2, H_2O, SF_6, and CF_4 when dissociation products have positive electron affinities:

$$e + AB \rightarrow (AB^-)^* \rightarrow A + B^- \tag{2.41}$$

Mechanism of this process is somewhat similar to the dissociative recombination (Eq. (2.33)) and proceeds by intermediate formation of an auto-ionization state $(AB^-)^*$. This excited state is unstable, and its decay leads either to the reverse process of auto-detachment $(AB + e)$, or to dissociation $(A + B^-)$. An electron here is captured without balancing energy of the elementary process. For this reason, the dissociative attachment is a resonant process, requiring quite definite values of the electron impact energy.

The most typical potential energy curves, illustrating the dissociative attachment, are presented in Figure 2.5a. The electron attachment process starts in this case with a vertical transition from AB molecular ground state electronic term to a repulsive state of AB^- (following the Franck–Condon Principle, see Section 2.4). During the repulsion, before the $(AB^-)^*$ reaches an intersection point of AB and AB^- electronic terms, the reverse auto-detachment reaction $(AB + e)$ is very possible. But after passing the intersection, the AB potential energy exceeds that of AB^- and further repulsion results in dissociation $(A + B^-)$. To estimate the cross section of the dissociative attachment we should take into account, that the repulsion time with possible auto-detachment is proportional to square root of reduced mass of AB molecule $\sqrt{M_A M_B/(M_A + M_B)}$. The characteristic electron transition time is much shorter and proportional to square root of its mass m. For this reason, the maximum cross section of dissociative attachment with the described configuration of electronic terms can be estimated as:

$$\sigma_{d.a.}^{max} \approx \sigma_0 \sqrt{\frac{m(M_A + M_B)}{M_A M_B}} \tag{2.42}$$

The maximum cross section is two orders of value less than the gas-kinetic cross section σ_0 and numerically is about 10^{-18} cm^2. Some halogens and their compounds have electron affinity of a product exceeding the dissociation energy, see potential energy curves in Figure 2.5b. In this case, even low-energy electrons can effectively provide the dissociative attachment. Also, if the electron affinity of a product exceeds the dissociation energy, the intersection point of AB and AB^- electronic terms is actually located inside of the so-called geometrical sizes of the dissociating molecules. As a result, during the repulsion of $(AB^-)^*$, probability of the reverse auto-detachment reaction $(AB + e)$ is low, and cross section of the dissociative attachment can reach the gas-kinetic cross section σ_0 about 10^{-16} cm^2.

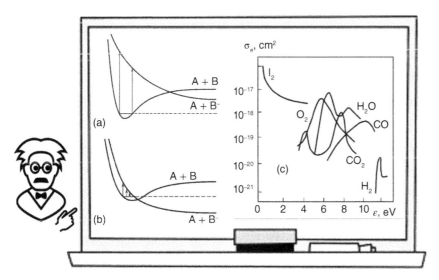

Figure 2.5 Dissociative attachment of electrons to molecules: (a) potential energy curves at low electron affinity, (b) potential energy curves at high electron affinity, (c) cross sections as a function of electron energy.

Table 2.4 Dissociative attachment of electrons: resonance parameters of the process.

Dissociative attachment process	ε_{max}, eV	$\sigma_{d.a}^{max}$, cm^2	$\Delta\varepsilon$, eV
$e + O_2 \rightarrow O^- + O$	6.7	10^{-18}	1
$e + H_2 \rightarrow H^- + H$	3.8	10^{-21}	3.6
$e + NO \rightarrow O^- + N$	8.6	10^{-18}	2.3
$e + CO \rightarrow O^- + C$	10.3	$2*10^{-19}$	1.4
$e + HCl \rightarrow Cl^- + H$	0.8	$7*10^{-18}$	0.3
$e + H_2O \rightarrow H^- + OH$	6.5	$7*10^{-18}$	1
$e + H_2O \rightarrow O^- + H_2$	8.6	10^{-18}	2.1
$e + H_2O \rightarrow H + OH^-$	5	10^{-19}	2
$e + D_2O \rightarrow D^- + OD$	6.5	$5*10^{-18}$	0.8
$e + CO_2 \rightarrow O^- + CO$	4.35	$2*10^{-19}$	0.8

Cross sections of the dissociative attachment for several molecular gases are presented in Figure 2.4c as a function of electron energy, reflecting the resonance nature of the process. The resonant structure of $\sigma_a(\varepsilon)$ permits estimating the dissociative attachment rate coefficient k_a as a following function of electron temperature T_e:

$$k_a(T_e) \approx \sigma_{d.a.}^{max}(\varepsilon_{max})\sqrt{\frac{2\varepsilon_{max}}{m}}\frac{\Delta\varepsilon}{T_e}\exp\left(-\frac{\varepsilon_{max}}{T_e}\right), \tag{2.43}$$

where a single resonance of the $\sigma_a(\varepsilon)$ is taken into account; ε_{max} and $\sigma_{d.a}^{max}$ are the electron energy and maximum cross section, corresponding to this resonance, $\Delta\varepsilon$ is its energy width. Equation (2.43) is quite convenient for numerical calculations; the necessary parameters: ε_{max}, $\sigma_{d.a}^{max}$, and $\Delta\varepsilon$ are provided in Table 2.4.

B. ***Three-body electron attachment to molecules*** occurs in collision of an electron with two heavy particles, when at least one of them has positive electron affinity:

$$e + A + B \rightarrow A^- + B. \tag{2.44}$$

The three-body electron attachment can be a principal channel of electron losses through formation of negative ions, when electron energies are not sufficient for dissociative attachment, and pressure is elevated (usually above 0.1 atm) and the third kinetic order processes are preferable. The three-body process is exothermic; therefore, its rate coefficient does not depend strongly on electron temperature (at least, within the temperature range about 1 eV). In this case, heavy particles are responsible for consumption of energy released during the attachment. It is important, that electrons are usually kinetically not effective as third-body B partners, because of low degree of ionization and low energy release during the attachment (in contrast to the three-body electron–ion recombination Eq. (2.36)). As an example, the three-body attachment plays a key role in balance of charged particles in atmospheric pressure nonthermal discharges in air:

$$e + O_2 + M \rightarrow O_2^- + M. \tag{2.45}$$

The three-body electron attachment proceeds by the two-stage ***Bloch-Bradbury mechanism***. The first stage is an attachment with formation of a negative ion in an unstable auto-ionization state:

$$e + O_2 \xleftrightarrow{k_{att},\tau} \left(O_2^-\right)^*, \tag{2.46}$$

characterized by k_{att} – rate coefficient of the intermediate electron trapping, and τ – lifetime of the excited unstable ion. The second stage of the Bloch–Bradbury mechanism includes collision with the third-body partner M with either stabilization of O_2^- (rate coefficient k_{st}) or collisional decay of the unstable ion into initial state (rate coefficient k_{dec}):

$$\left(O_2^-\right)^* + M \xrightarrow{k_{st}} O_2^- + M. \tag{2.47}$$

$$\left(O_2^-\right)^* + M \xrightarrow{k_{dec}} O_2 + e + M. \tag{2.48}$$

Table 2.5 Electron attachment to O_2 at room with different third-body partners, $T_0 = 300$ K.

Three-body attachment	Rate coefficient	Three-body attachment	Rate coefficient
$e + O_2 + Ar \rightarrow O_2^- + Ar$	$3*10^{-32}$ cm^6 s^{-1}	$e + O_2 + Ne \rightarrow O_2^- + Ne$	$3*10^{-32}$ cm^6 s^{-1}
$e + O_2 + N_2 \rightarrow O_2^- + N_2$	$1.6*10^{-31}$ cm^6 s^{-1}	$e + O_2 + H_2 \rightarrow O_2^- + H_2$	$2*10^{-31}$ cm^6 s^{-1}
$e + O_2 + O_2 \rightarrow O_2^- + O_2$	$2.5*10^{-30}$ cm^6 s^{-1}	$e + O_2 + CO_2 \rightarrow O_2^- + CO_2$	$3*10^{-30}$ cm^6 s^{-1}
$e + O_2 + H_2O \rightarrow O_2^- + H_2O$	$1.4*10^{-29}$ cm^6 s^{-1}	$e + O_2 + H_2S \rightarrow O_2^- + H_2S$	10^{-29} cm^6 s^{-1}
$e + O_2 + NH_3 \rightarrow O_2^- + NH_3$	10^{-29} cm^6 s^{-1}	$e + O_2 + CH_4 \rightarrow O_2^- + CH_4$	$>10^{-29}$ cm^6 s^{-1}

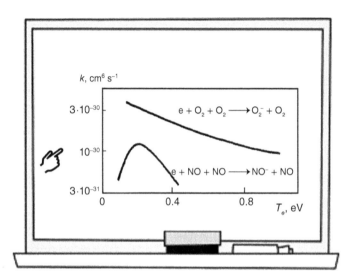

Figure 2.6 Three-body electron attachment as a function of electron temperature, $T_0 = 300$ K.

Taking into account the steady state conditions for number density of the intermediate excited ions $(O_2^-)^*$, the rate coefficient for the total three-body electron attachment (2.45) is:

$$k_{3M} = \frac{k_{att}k_{st}}{\frac{1}{\tau} + (k_{st} + k_{dec})n_0} \qquad (2.49)$$

where n_0 is concentration of the third-body heavy particles M. When the pressure is not too high, $(k_{st} + k_{dec})n_0 \ll \tau^{-1}$, the Eq. (2.49) becomes $k_{3M} \approx k_{att}k_{st}\tau$ determining the third kinetic order of the attachment. The rate coefficient (2.49) equally depends on formation and stabilization of negative ions on the third particle. The more complicated molecule is the third body (M), the easier it stabilizes the $(O_2^-)^*$ and the higher is the total reaction rate coefficient $k_{3M,}$ which is illustrated in Table 2.5.

The three-body attachment rate coefficients are shown in Figure 2.6 as a function of electron temperature ($T_0 = 300$ K). For simple estimations, when $T_e = 1$ eV, $T_0 = 300$ K: $k_{3M} \approx 10^{-30}$ cm^6 s^{-1}. The rate of three-body process is greater than dissociative attachment (k_a) when the gas number density exceeds a critical value $n_0 > k_a(T_e)/k_{3M}$. Numerically in oxygen with $T_e = 1$ eV, $T_0 = 300$ K, this means $n_0 > 10^{18}$ cm^{-3}, or in pressure units at room temperature $p > 30$ Torr.

C. **Other mechanisms of formation of negative ions**, which are less significant, include **the polar dissociation**:

$$e + AB \rightarrow A^+ + B^- + e. \qquad (2.50)$$

This process combines both ionization and dissociation, therefore the threshold energy is quite high. The electron is not captured here, and the process is not resonant. For example, in molecular oxygen, the maximum value of cross section of the polar dissociation is about $3*10^{-19}$ cm^2, and corresponds to electron energy 35 eV. Another process is the radiative attachment:

$$e + M \rightarrow (M^-)^* \rightarrow M^- + \hbar\omega. \tag{2.51}$$

Such electron capture occurs at low electron energies but with low probability about 10^{-5}–10^{-7} (cross sections are about 10^{-21}–10^{-23} cm^2). Finally, some electronegative polyatomic molecules like SF$_6$ have a negative ion state very close to a ground state (only 0.1 eV for SF$_6^-$). As a result, lifetime of metastable negative ions is long, and electron attachment in this case can be ***direct in a simple binary collision***. This attachment process is resonant and for very low electron energies has high maximum cross sections of about 10^{-15} cm^2.

2.10 Electron Detachment from Negative Ions

Negative ions can be neutralized in the fast ion–ion recombination (to be discussed in next section) or can release an electron in different detachment processes. In nonthermal discharges, especially important negative ions destruction mechanism is the ***associative detachment***:

$$A^- + B \rightarrow (AB^-)^* \rightarrow AB + e. \tag{2.52}$$

This process is a reverse with respect to dissociative attachment (2.41) and can also be illustrated by Figure 2.5. The associative detachment is a nonadiabatic process, which occurs by intersection of electronic terms of a complex negative ion A$^-$–B and corresponding molecule AB, which results in high values of rate coefficients $k_d = 10^{-10}$–10^{-9} cm^3 s^{-1}, see Table 2.6.

When ionization degree and electron density are high, destruction of negative ions can be dominated by the ***electron impact detachment***:

$$e + A^- \rightarrow A + e + e \tag{2.53}$$

Table 2.6 Associative electron detachment processes at room temperature.

Associative attachment process	Reaction enthalpy	Rate coefficient
$H^- + H \rightarrow H_2 + e$	-3.8 eV (exothermic)	$1.3*10^{-9}$ cm^3 s^{-1}
$H^- + O_2 \rightarrow HO_2 + e$	-1.25 eV (exothermic)	$1.2*10^{-9}$ cm^3 s^{-1}
$O^- + O \rightarrow O_2 + e$	-3.8 eV (exothermic)	$1.3*10^{-9}$ cm^3 s^{-1}
$O^- + N \rightarrow NO + e$	-5.1 eV (exothermic)	$2*10^{-10}$ cm^3 s^{-1}
$O^- + O_2 \rightarrow O_3 + e$	0.4 eV (endothermic)	10^{-12} cm^3 s^{-1}
$O^- + O_2(^1\Delta_g) \rightarrow O_3 + e$	-0.6 eV (exothermic)	$3*10^{-10}$ cm^3 s^{-1}
$O^- + N_2 \rightarrow N_2O + e$	-0.15 eV (exothermic)	10^{-11} cm^3 s^{-1}
$O^- + NO \rightarrow NO_2 + e$	-1.6 eV (exothermic)	$5*10^{-10}$ cm^3 s^{-1}
$O^- + CO \rightarrow CO_2 + e$	-4 eV (exothermic)	$5*10^{-10}$ cm^3 s^{-1}
$O^- + H_2 \rightarrow H_2O + e$	-3.5 eV (exothermic)	10^{-9} cm^3 s^{-1}
$O^- + CO_2 \rightarrow CO_3 + e$	Endothermic	10^{-13} cm^3 s^{-1}
$C^- + CO_2 \rightarrow CO + CO + e$	-4.3 (exothermic)	$5*10^{-11}$ cm^3 s^{-1}
$C^- + CO \rightarrow C_2O + e$	-1.1 (exothermic)	$4*10^{-10}$ cm^3 s^{-1}
$Cl^- + O \rightarrow ClO + e$	0.9 (endothermic)	10^{-11} cm^3 s^{-1}
$O_2^- + O \rightarrow O_3 + e$	-0.6 eV (exothermic)	$3*10^{-10}$ cm^3 s^{-1}
$O_2^- + N \rightarrow NO_2 + e$	-4.1 eV (exothermic)	$5*10^{-10}$ cm^3 s^{-1}
$OH^- + O \rightarrow HO_2 + e$	-1 eV (exothermic)	$2*10^{-10}$ cm^3 s^{-1}
$OH^- + N \rightarrow HNO + e$	-2.4 eV (exothermic)	10^{-11} cm^3 s^{-1}
$OH^- + H \rightarrow H_2O + e$	-3.2 eV (exothermic)	10^{-9} cm^3 s^{-1}

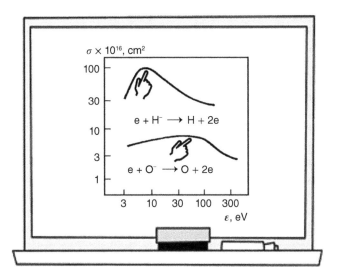

Figure 2.7 Electron detachment from H^-, O^- ions as a function of electron impact energy.

This process is somewhat similar to the direct ionization by an electron impact (Thomson mechanism, see Section 2.4). The main difference is repulsive Coulomb force acting between the incident electron and the negative ion. For incident electron energies about 10 eV, the cross section of the detachment process can be high, about 10^{-14} cm². Typical detachment cross section dependence (see Fridman and Kennedy 2021) on the incident electron velocity v_e can be illustrated by that for the detachment from negative hydrogen ion ($e + H^- \rightarrow H + 2e$):

$$\sigma(v_e) \approx \frac{\sigma_0 e^4}{(4\pi\varepsilon_0)^2 \hbar^2 v_e^2} \left(-7.5 \ln \frac{e^2}{4\pi\varepsilon_0 \hbar v_e} + 25 \right), \tag{2.54}$$

where σ_0 is the geometric atomic cross section. The detachment cross section as a function of electron energy is presented in Figure 2.7 for hydrogen and oxygen ions. The maximum cross sections of about 10^{-15}–10^{-14} cm² correspond to electron energies 10–50 eV, exceeding electron affinities more than 10 times (in contrast to the Thomson mechanism, where the ionization cross section maximum occurs at energies exceeding ionization potential about twice, which can be explained by the electron–ion Coulomb repulsion).

Finally, the ***electron detachment in collisions with excited particles*** occurs when electronic excitation energy of a collision partner B exceeds the electron affinity of particle A:

$$A^- + B^* \rightarrow A + B + e \tag{2.55}$$

This process is somewhat similar to the Penning ionization (see Section 2.7), it has the electronically nonadiabatic nature and therefore is quite fast. For example, the destruction of a negative oxygen ion by a metastable electronically excited oxygen with excitation energy 0.98 eV:

$$O_2^- + O_2(^1\Delta_g) \rightarrow O_2 + O_2 + e, \Delta H = -0.6 \text{ eV} \tag{2.56}$$

has high-rate coefficient of about $2*10^{-10}$ cm³ s⁻¹ at room temperature. Electron detachment can be also effective in ***collisions with vibrationally excited molecules***, for example $O_2^- + O_2^*(v > 3) \rightarrow O_2 + O_2 + e$, which can be described by Arrhenius formula: $k_d \propto \exp(-E_a/T_v)$ with activation energy $E_a \approx 0.44$ eV and gas-kinetic preexponential factor (in thermal plasmas, vibrational temperature T_v can be replaced here by gas temperature T_0).

2.11 Losses of Charged Particles: Mechanisms of Ion–Ion Recombination

Actual losses of charged particles in electro-negative gases at elevated pressures are mostly due to ion–ion recombination, that is mutual neutralization of positive and negative ions in binary or three-body collisions. At high pressures (usually more than 30 Torr), a three-body mechanism dominates the recombination with maximum rate coefficient about 1–$3*10^{-6}$ cm³ s⁻¹ at pressures near atmospheric and room temperature. Traditionally the rate coefficient of

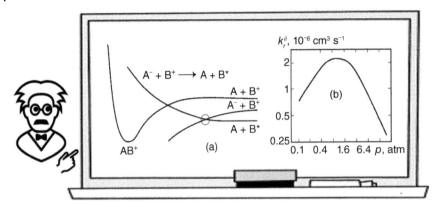

Figure 2.8 Ion–ion recombination: (a) correlation of energy terms for binary process, (b) pressure dependence of the total rate coefficient in air at room temperature.

ion–ion recombination is recalculated with respect to concentrations of positive and negative ions, that is to the second kinetic order. Due to the three-molecular nature of the recombination at high pressures, the recalculated (second kinetic order) recombination rate coefficient depends on pressure. Both an increase and a decrease of pressure from optimal (near 1 atm) results in proportional reduction of the three-body ion–ion recombination rate coefficient. At low pressures, the three-body mechanism becomes relatively slow, and the binary collisions with transfer of energy into electronic excitation make a major contribution to the ion–ion recombination.

A. **Ion–ion recombination in binary collisions** is a neutralization reaction of a negative and a positive ion with the released energy going to electronic excitation of a product:

$$A^- + B^+ \rightarrow A + B^* \tag{2.57}$$

This reaction has a second kinetic order and dominates the ion–ion recombination in low-pressure discharges ($p < 10–30\,\text{Torr}$).

Electronic terms, illustrating the recombination, are presented in Figure 2.8a. The ions A^- and B^+ are approaching each other following the attractive A^-–B^+ Coulomb potential curve. When distance between the heavy particles is large, this potential curve lies below the term of A^-–B^+ and above the final A–B^* electronic term on the energy interval:

$$\Delta E \approx \frac{I_B}{n^2} - EA_A, \tag{2.58}$$

where I_B is the ionization potential of particle B. If the principal quantum number of particle B after recombination is high $n = 3–4$, then ΔE is low. This means that the electronic terms of A^-–B^+ and A–B^* are relatively close when the ions A^- and B^+ are approaching each other. When the principal quantum number n is not specifically defined, then the affinity EA_A can be taken as a reasonable estimation for the energy interval, that is $\Delta E \approx EA_A$. As can be seen from Figure 2.5, the low value of ΔE results in the possibility of effective transition between the electronic terms (from A^-–B^+ to A–B^*), when distance R_{ii} between ions is still large. Even for this long-distance R_{ii} between the ions, the Coulomb attraction energy is already sufficient to compensate the initial energy gap ΔE between the terms:

$$R_{ii} \approx \frac{e^2}{4\pi\varepsilon_0} \left[\frac{I_B}{n^2} - EA_A \right]^{-1} \approx \frac{e^2}{4\pi\varepsilon_0 EA_A} \tag{2.59}$$

The high value of R_{ii} characterizing interaction radius results in large cross sections of the ion–ion recombination (2.57), which can be estimated from conservation of angular momentum during the Coulomb collision and assuming maximum kinetic energy equal to the potential energy. The impact parameter b can be then found as a function of the ion kinetic energy ε:

$$b \approx R_{ii} \frac{\sqrt{EA_A}}{\sqrt{\varepsilon}} \tag{2.60}$$

Based on Eqs. (2.59) and (2.60), the ion–ion recombination cross section can be expressed as:

$$\sigma^{ii}_{rec} = \pi b^2 \approx \pi \frac{e^4}{(4\pi\varepsilon_0)^2} \frac{1}{\varepsilon\alpha_A} \frac{1}{\varepsilon} \tag{2.61}$$

Table 2.7 Rate coefficients of the ion–ion recombination in binary collisions, $T_0 = 300$ K.

Recombination process	Released energy, $I - EA$	Rate coefficient
$H^- + H^+ \rightarrow H + H$	12.8 eV	$3.9*10^{-7}$ cm^3 s^{-1}
$O^- + O^+ \rightarrow O + O$	12.1 eV	$2.7*10^{-7}$ cm^3 s^{-1}
$O^- + N^+ \rightarrow O + N$	13.1 eV	$2.6*10^{-7}$ cm^3 s^{-1}
$O^- + O_2^+ \rightarrow O + O_2$	11.6 eV	10^{-7} cm^3 s^{-1}
$O^- + NO^+ \rightarrow O + NO$	7.8 eV	$4.9*10^{-7}$ cm^3 s^{-1}
$SF_6 + SF_5^+ \rightarrow SF_6 + SF_5$	15.2 eV	$4*10^{-8}$ cm^3 s^{-1}
$NO_2^- + NO^+ \rightarrow NO_2 + NO$	5.7 eV	$3*10^{-7}$ cm^3 s^{-1}
$O_2^- + O^+ \rightarrow O_2 + O$	13.2 eV	$2*10^{-7}$ cm^3 s^{-1}
$O_2^- + O_2^+ \rightarrow O_2 + O_2$	11.6 eV	$4.2*10^{-7}$ cm^3 s^{-1}
$O_2^- + N_2^+ \rightarrow O_2 + N_2$	15.1 eV	$1.6*10^{-7}$ cm^3 s^{-1}

Quantum mechanical calculation of this cross section taking into account the effect of electron tunneling (Smirnov 2001) results in $\sigma_{rec}^{ii} \propto 1/\sqrt{\varepsilon}$.

Equation (2.61) can be presented as $\sigma_{rec}^{ii} \approx \sigma_0 \frac{I}{EA} \frac{I}{\varepsilon}$, where I is ionization energy, showing that the recombination cross section significantly exceeds the gas-kinetic one σ_0. The same is true for the rate coefficients, which are about 10^{-7} cm^3 s^{-1}. Some binary recombination rate coefficients are presented in Table 2.7 together with energy released in the process.

B. **Three-body Ion–ion recombination** is a neutralization process in a triple collision of a heavy neutral with a negative and a positive ion:

$$A^- + B^+ + M \rightarrow A + B + M, \tag{2.62}$$

and dominates the ion–ion recombination at moderate and high pressures ($p > 10$–30 Torr). This reaction has the third kinetic order only in the moderate pressure range, usually less than 1 atm (for high pressures the process is limited by ion mobility). It means that only at moderate pressures the reaction rate is proportional to the product of concentrations of all three collision partners.

 At moderate pressures (0.01–1 atm), the three-body ion–ion recombination can be described in framework of the **Thomson theory**. According to this theory, the recombination occurs if negative and positive ions approach each other closer than the critical distance $b \approx e^2/4\pi\varepsilon_0 T_0$. Then Coulomb interaction energy of ions reaches the level of thermal energy (T_0) and can be effectively absorbed by the third body (a heavy neutral particle, concentration n_0, velocity v_t), which enables recombination. The probability P_+ for a positive ion (concentration n_+) to be closer than $b \approx e^2/4\pi\varepsilon_0 T_0$ to a negative ion (n_-) is:

$$P_+ \approx n_- b^3 \approx n_- \frac{e^6}{(4\pi\varepsilon_0)^3 T_0^3}. \tag{2.63}$$

The frequency of the collisions of a positive ion with neutral particles, resulting in ion–ion recombination, can then be found as $\nu_{ii} = (\sigma v_t n_0) P_+$. The total recombination rate then is:

$$w_{ii} \approx (\sigma v_t) \frac{e^6}{(4\pi\varepsilon_0)^3 T_0^3} n_0 n_- n_+. \tag{2.64}$$

Summarizing, the process has third kinetic order and rate coefficient decreasing with T_0:

$$k_{r3}^{ii} \approx \left(\sigma \sqrt{\frac{1}{m}}\right) \frac{e^6}{(4\pi\varepsilon_0)^3} \frac{1}{T_0^{5/2}} \tag{2.65}$$

These third-kinetic-order coefficients are about 10^{-25} cm^6 s^{-1} at $T_0 = 300$ K, see Table 2.8. Following tradition, same rate coefficients are also shown in the table as second-order coefficients k_{r2}^{ii} ($k_{r2}^{ii} = k_{r3}^{ii} n_0$), assuming normal

Table 2.8 Rate coefficients of the three-body ion–ion recombination, $T_0 = 300$ K.

Three-body ion–ion recombination	Rate coefficients, third kinetic order	Rate coefficients, 1 atm, recalculated to second order
$O_2^- + O_4^+ + O \rightarrow O_2 + O_2 + O_2 + O_2 + O_2$	$1.55*10^{-25}$ cm^6 s^{-1}	$4.2*10^{-6}$ cm^3 s^{-1}
$O^- + O_2^+ + O_2 \rightarrow O_3 + O_2$	$3.7*10^{-25}$ cm^6 s^{-1}	10^{-5} cm^6 s^{-1}
$NO_2^- + NO^+ + O_2 \rightarrow NO_2 + NO + O_2$	$3.4*10^{-26}$ cm^6 s^{-1}	$0.9*10^{-6}$ cm^6 s^{-1}
$NO_2^- + NO^+ + N_2 \rightarrow NO_2 + NO + N_2$	10^{-25} cm^6 s^{-1}	$2.7*10^{-6}$ cm^6 s^{-1}

conditions $n_0 = 2.7*10^{19}$ cm^{-3}. Tables 2.7 and 2.8 show that the binary and triple collisions contribute equally to the recombination at pressures about 10–30 Torr.

The three-body ion–ion recombination rate coefficient recalculated to the second kinetic order $k_{r2}^{ii} = k_{r3}^{ii} n_0$ grows linearly with gas density. This growth is limited to moderate pressures (below 1 atm) by the frameworks of the Thomson's theory, which requires the capture distance b to be less than an ion mean free path ($1/n_0\sigma$):

$$n_0\sigma b \approx n_0 \frac{e^2}{(4\pi\varepsilon_0) T_0} \sigma < 1. \tag{2.66}$$

At high pressures, $n_0\sigma b > 1$ (atmospheric and above), the recombination is limited by a positive and negative ion electrical drift to each other overcoming multiple collisions with neutrals. The following **Langevin model** describes the ion–ion recombination in this high-pressure range.

According to Langevin, motion of a positive and negative ion to collide in recombination can be considered as their drift (with mobilities μ_+ and μ_-) in the Coulomb field $e/(4\pi\varepsilon_0)r^2$, where r is the distance between them. Then the ion drift velocity to meet each other is:

$$v_d = \frac{e}{(4\pi\varepsilon_0)^2}(\mu_+ + \mu_-). \tag{2.67}$$

Consider sphere with radius r, surrounding positive ion. Flux of negative ions with concentration n_- approaching the positive one and the recombination frequency v_{r+} for the positive ion are:

$$v_{r+} = 4\pi r^2 v_d n_- \tag{2.68}$$

The total ion–ion recombination rate then found is: $w = n_+ v_{r+} = 4\pi r^2 v_d n_- n_+$, showing the second kinetic order of the process. The final Langevin expression for the ion–ion recombination rate coefficient $k_r^{ii} = w/n_+ n_-$ in the limit of high pressures (more than 1 atm) gives:

$$k_r^{ii} = 4\pi e \left(\mu_+ + \mu_-\right), \tag{2.69}$$

showing the rate coefficient proportionality to the ion mobilities, and therefore reverse proportionally to the pressure.

Summarizing, the three-body recombination rate coefficients grow with pressure $k_{r2}^{ii} = k_{r3}^{ii} n_0$ at moderate pressures, and then at high pressures begin to decrease as $1/p$ together with ion mobility. The highest recombination rate coefficient occurs at pressures close to atmospheric (concentration of neutrals: $n_0 \approx \frac{4\pi\varepsilon_0 T_0}{\sigma e^2}$) and can be expressed as:

$$k_{r,max}^{ii} \approx \frac{e^4}{(4\pi\varepsilon_0)^2} \frac{v}{T_0^2}, \tag{2.70}$$

where v is ionic thermal velocity. Numerically, it is about $1-3*10^{-6}$ cm^3 s^{-1}. This coefficient decreases for both an increase and decrease of pressure from the atmospheric one. The typical total experimental dependence of the k_r^{ii} on pressure is presented in Figure 2.8b.

2.12   Ion-molecular Processes: Polarization Collisions, Langevin Capture

The ion-molecular processes start with scattering in the polarization potential, leading to the so-called Langevin capture of a charged particle and formation of an intermediate ion-molecular complex. If a neutral particle itself has no permanent dipole moment, the ion-neural charge-dipole interaction and scattering is due to the dipole moment p_M, induced in the neutral particle by the electric field E of an ion $p_m = \alpha \varepsilon_0 E = \alpha \frac{e}{4\pi r^2}$, where r is the distance between the interacting particles, α is the polarizability of a neutral atom or molecule (see Table 2.9).

Typical orbits of relative ion and neutral motion during polarization scattering are shown in Figure 2.9. When the impact parameter b is high, the orbit is hyperbolic, but when it is low, the scattering leads to the **Langevin polarization capture** and formation of an ion-molecular complex. The capture occurs when the interaction energy between charge and induced dipole $p_m E = \alpha \frac{e}{4\pi r^2} \frac{e}{4\pi \varepsilon_0 r^2}$ becomes comparable with kinetic energy of the colliding partners $\frac{1}{2} M v^2$, where M is their reduced mass, and v is their relative velocity. The **Langevin cross section** of the polarization capture can be found as $\sigma_L \approx \pi r^2$:

$$\sigma_L = \sqrt{\frac{\pi \alpha e^2}{\varepsilon_0 M v^2}}, \tag{2.71}$$

Table 2.9   Polarizability coefficients of some atoms and molecules.

Atom or molecule	α, 10^{-24} cm^3	Atom or molecule	α, 10^{-24} cm^3
Ar	1.64	H	0.67
C	1.78	N	1.11
O	0.8	CO	1.95
Cl$_2$	4.59	O$_2$	1.57
CCl$_4$	10.2	CF$_4$	1.33
H$_2$O	1.45	CO$_2$	2.59
SF$_6$	4.44	NH$_3$	2.19

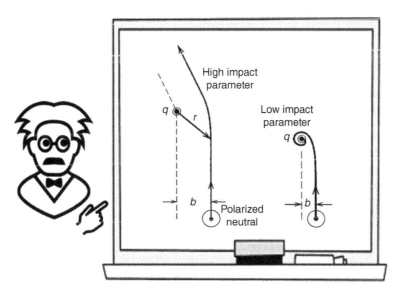

Figure 2.9   Ion-neutral polarization collisions at different impact parameters: from Langevin scattering to Langevin capture.

The **Langevin rate coefficient** $k_L \approx \sigma_L v$ does not depend on velocity v (and temperature):

$$k_L = \sqrt{\frac{\pi \alpha e^2}{\varepsilon_0 M}} 2.3 \times 10^{-9} \text{ cm}^3 \text{ s}^{-1} \times \sqrt{\alpha, \frac{10^{-24} \text{ cm}^3}{M}}, \text{amu}, \tag{2.72}$$

and is usually about 10^{-9} cm^3 s^{-1}, which 10 times exceeds the gas-kinetic neutral–neutral rate coefficient: $k_0 \approx 10^{-10}$ cm^3 s^{-1}. Here α/M reflects the "specific volume" of an atom or molecule, because the polarizability α actually corresponds to the particle volume. It means that the Langevin capture rate coefficient grows with a decrease of "density" of an atom or molecule. If ion interacts with a molecule having also a permanent dipole moment μ_D, the Langevin capture becomes about $\sqrt{I/T_0}$ times faster (see Fridman and Kennedy 2021):

$$k_L = \sqrt{\frac{\pi e^2}{\varepsilon_0 M}} \left(\sqrt{\alpha} + c\mu_D \sqrt{\frac{2}{\pi T_0}} \right). \tag{2.73}$$

Derivation of the Langevin formula (2.72) did not specify the mass of a charged particle, and therefore can be also applied for *Langevin electron-neutral collisions*:

$$\kappa_L^{electron/neutral} = 10^{-7} \text{cm}^3 \text{ s}^{-1} \times \sqrt{\alpha, 10^{-24} \text{cm}^3}, \tag{2.74}$$

which numerically is about 10^{-7} cm^3 s^{-1} and also does not depend on temperature.

2.13   Resonant and Nonresonant Ion-atomic Charge Transfer Processes

An electron can be transferred during a collision from a neutral particle to a positive ion, or from negative ion to a neutral particle, which is referred to as charge transfer or charge exchange. These processes occurring without significant defect of the electronic energy ΔE are called the **resonant charge transfer**. Otherwise, they are called **nonresonant**. The resonant charge transfer is a nonadiabatic process and has a large cross section (about 10^{-14} cm^2).

Consider a ***resonant charge exchange*** between a neutral particle B and a positive ion A$^+$ assuming the particle B/B$^+$ being at rest:

$$A^+ + B \rightarrow A + B^+ \tag{2.75}$$

The energy schematic of this reaction is illustrated in Figure 2.10a by potential energy of an electron in the Coulomb field of A$^+$ and B$^+$ (r_{AB} is a distance between centers of A and B):

$$U(z) = -\frac{e^2}{4\pi\varepsilon_0 z} - \frac{e^2}{4\pi\varepsilon_0 |r_{AB} - z|}, \tag{2.76}$$

The maximum potential energy corresponds to $z = r_{AB}/2$ and can be expressed as:

$$U_{max} = -\frac{e^2}{\pi\varepsilon_0 r_{AB}} \tag{2.77}$$

The classical charge transfer (2.75) is possible (see Figure 2.10a) if the maximum of potential energy U_{max} is lower than the initial energy E_B of an electron which is going to be transferred from level n of particle B:

$$E_B = -\frac{I_B}{n^2} - \frac{e^2}{4\pi\varepsilon_0 r_{AB}} \geq U_{max} \tag{2.78}$$

Here I_{AB} is the ionization potential of atom B. From (2.77) and (2.78), the maximum distance between the interacting heavy particles when the classical charge transfer is still permitted is:

$$r_{AB}^{max} = \frac{3e^2 n^2}{4\pi\varepsilon_0 I_B} \tag{2.79}$$

If the charge exchange is resonant and therefore not limited by the defect of energy, the classical reaction cross section can be expressed from (2.79) as πr_{AB}^2:

$$\sigma_{chtr}^{class} = \frac{9e^4 n^4}{16\pi\varepsilon_0^2 I_B^2}. \tag{2.80}$$

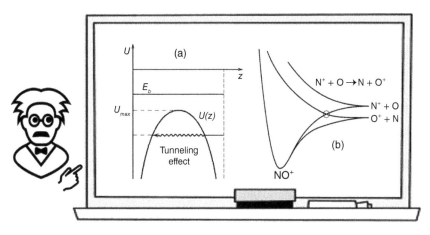

Figure 2.10 Energy term correlation for the charge transfer reactions: (a) resonant process, (b) nonresonant process.

The classical cross section of charge exchange does not depend on the kinetic energy of the interacting species and its numerical value for ground state transfer ($n = 1$) is approximately the gas-kinetic cross section for collisions of neutrals. The actual cross section of a resonant charge transfer can be much higher taking into account the quantum mechanical effect of electron tunneling from B to A^+ (see Figure 2.10a). This effect can be estimated by calculating the electron tunneling probability P_{tunn} across a potential barrier of height about I_B and width d:

$$P_{tunn} \approx \exp\left(-\frac{2d}{\hbar}\sqrt{2meI_B}\right). \tag{2.81}$$

The frequency of electron oscillation in the ground state atom B is about $I_B/\hbar$, so the frequency of the tunneling is $I_B P_{tunn}/\hbar$. Taking into account (2.81), the maximum barrier width d_{max}, when the tunneling frequency still exceeds the reverse ion-neutral collision time: $I_B P_{tunn}/\hbar > v/d$ and as a result, the tunneling can take place. Here v is the relative velocity of the colliding partners. Then the cross section πd_{max}^2 of the electron tunneling from B to A^+, leading to the resonance charge exchange is:

$$\sigma_{ch.tr}^{tunn} \approx \frac{1}{I_B}\left(\frac{\pi\hbar^2}{8me}\right)\left(\ln\frac{I_B d}{\hbar} - \ln v\right)^2, \tag{2.82}$$

which can be numerically expressed as:

$$\sqrt{\sigma_{ch.tr}^{tunn}\text{cm}^2} \approx \frac{1}{\sqrt{I_B\text{eV}}}(6.5 \times 10^{-7} - 3 \times 10^{-8}\ln v, \text{cm s}^{-1}) \tag{2.83}$$

More details on the subject, especially analyzing effects of collision partner velocities and Langevin capturing on quantum mechanical and classical resonant charge transfer processes can be found in particular in Fridman and Kennedy (2021).

As an example of ***nonresonant charge exchange processes***, consider an electron transfer between oxygen and nitrogen atoms:

$$N^+ + O \rightarrow N + O^+, \quad \Delta E = -0.9 \text{ eV}, \tag{2.84}$$

illustrated in Figure 2.10b. The ionization potential of oxygen ($I = 13.6$ eV) is lower than that of nitrogen ($I = 14.5$ eV). Therefore, the electron transfer from oxygen to nitrogen is an exothermic process and the separated $N + O^+$ energy level is located 0.9 eV lower than the separated $O + N^+$ energy level. The reaction (2.84) begins with N^+ approaching O by following the attractive NO^+ term. Then this term is crossing with the repulsive NO^+ term and the system experiences a nonadiabatic transfer, which results in formation of $O^+ + N$. The excess energy of approximately 1 eV goes into the kinetic energy of the products. The cross section of such exothermic charge exchange reactions at low thermal energies is of the order of the resonant cross sections of tunneling or Langevin capture (see above). The endothermic reactions of charge exchange, such as the reverse process $N + O^+ \rightarrow N^+ + O$, are usually very slow at low gas temperatures.

The importance of the nonresonant charge exchange processes can be illustrated by effect of **acidic behavior of nonthermal air plasma**. Ionization of air in cold discharges leads to large amount of N_2^+ ions because of the high molar fraction of nitrogen in air. Later, low ionization potential and high dipole moment of water molecules enable the nonresonant charge exchange:

$$N_2^+ + H_2O \rightarrow N_2 + H_2O^+, \quad k(300\,K) = 2.2 \times 10^{-9} cm^3 s^{-1}, \tag{2.85}$$

generating water ions, which further reacts with water-forming H_3O^+-ions and OH-radicals:

$$H_2O^+ + H_2O \rightarrow H_3O^+ + OH \ \Delta H = -12\,kcal\,mol^{-1}, k(300\,K) = 0.5 \times 10^{-9} cm^3\,s^{-1}. \tag{2.86}$$

The acidic behavior of air plasma contacting liquid water can be also significantly due to formation of NO_x compounds leading to formation of nitric and nitrous acids. If water-contacting plasma is generated not in air but in different gases (like, for example, noble gases), the above-described nonresonant charge exchange with H_2O becomes crucial in forming plasma acids.

2.14 Ion-molecular Reactions with Rearrangement of Chemical Bonds, Plasma Catalysis

Elementary reaction (2.85) responsible for the plasma acidic effect was our first example of the ion-molecular reactions with rearrangement of chemical bonds. This class of elementary plasma processes can be subdivided into the following groups:

- $(A)B^+ + C \rightarrow A + (C)B^+$, reactions with an ion transfer,
- $A(B^+) + C \rightarrow (B^+) + AC$, reactions with a neutral transfer,
- $A(B^+) + (C)D \rightarrow (A)D + (C)B^+$, double exchange reactions,
- $A(B^+) + (C)D \rightarrow AC^+ + BD$, reconstruction processes.

These groups of processes are shown above with positive ions, but similar reactions take place with negative ions.

The special feature of the ion molecular reactions first stated more than 70 years ago by Talrose and making them very distinctive from conventional atom-molecular processes is: ***most of the exothermic ion-molecular reactions have no activation energy***. The quantum mechanical repulsion between molecules, which provides the activation barrier even in the exothermic reactions of neutrals, is effectively suppressed by the charge-dipole attraction in ion-molecular reactions. This effect is especially important at low temperatures when even small activation barriers can dramatically slow down exothermic reactions. The rate coefficients of the exothermic ion-molecular reactions are very high and often correspond to Langevin capture (2.72) and (2.73). Details on the subject and an extensive listing of the relevant kinetic data can be found in particular in the monograph of Virin et al. (1978).

The absence of activation energies in exothermic ion-molecular reactions facilitates organization of chain reactions in ionized media. This phenomenon is known as **ionic catalysis** and play significant role in **plasma catalysis**, which is going to be a subject of Lecture 22. As an example, we can mention an important process of SO_2 oxidation to SO_3 (and sulfuric acid), which is highly exothermic, but cannot be arranged as a chain process without catalyst, which can be platinum providing charged surface centers operation similar to the plasma ions and removing the energy barriers of intermediate reactions between neutrals.

If noncatalytic surface (for example, quartz) is located nearby platinum during the process, the charged catalytic centers (ions) can be "spilled over" from platinum to quartz, making the quartz catalytically active for some time. This "mysterious conversion of quartz into platinum" in heterogeneous catalysis is referred to as **the spillover effect**. The long-length SO_2 chain oxidation occurs in the same way in the cold plasma systems through the negative-ionic mechanism during plasma-water or plasma-droplet interaction, which is to be discussed in Lecture 24.

2.15   Problems and Concept Questions

2.15.1   Maxwell–Boltzmann Distribution Function

Assuming EEDF as Maxwellian (2.1) with electron temperature $T_e = 1\,\text{eV}$, calculate (i) mean electron energy $m<v>^2/2$; and (ii) characteristic electron velocity $\sqrt{<v^2>}$. Discuss difference between the characteristic velocity $\sqrt{<v^2>}$ and mean electron velocity $<v>$ in terms of standard deviation.

2.15.2   Positive and Negative Ions

Why typical ionization potentials (related to formation of positive ions) are greater than typical electron affinities (related to formation of negative ions)? Why double- or multi-charged positive ions such as A^{++} or A^{+++} do exist in plasma, and double or multi-charged negative ions such as A^{--} or A^{---} do not?

2.15.3   Direct Ionization by Electron Impact

Using Eq. (2.18) for the general ionization function $f(x)$, find the electron energy corresponding to maximum value of direct ionization cross section. Compare this energy with the electron energy optimal for direct ionization according to the Thomson formula.

2.15.4   Stepwise Ionization

Estimate a stepwise ionization reaction rate coefficient in Ar at electron temperature 1 eV, assuming quasi-equilibrium between plasma electrons and electronic excitation of atoms and using Eqs. (2.19) and (2.25). Why the direct ionization dominates in nonthermal discharges while in thermal plasmas stepwise ionization is crucial?

2.15.5   Electron Beam Propagation in Gases

How does the propagation distance of a high-energy electron beam depend on pressure at fixed temperature? How does it depend on temperature at a fixed pressure?

2.15.6   Ionization in Ion-Neutral Collisions, Massey Parameter

Estimate the adiabatic Massey parameter for ionization of Ar-atom at rest by impact of Ar^+-ion, having kinetic energy exceeding the ionization potential: twice, 10 times. What are the relevant ionization probabilities?

2.15.7   Dissociative Electron–Ion Recombination

According to (2.34), rate coefficient of dissociative recombination is reversibly proportional to gas temperature. However, in reality, when the electric field is fixed, this rate coefficient is decreasing faster than the reverse proportionality. How can you explain that?

2.15.8   Dissociative Attachment

Using Eq. (2.43) and Table 2.4, calculate dissociative attachment rate coefficients for molecular hydrogen and molecular oxygen at electron temperatures 1 eV. Compare and comment on the results.

2.15.9 Detachment by Electron Impact

Detachment of an electron from negative ion by electron impact can be considered as "ionization" of the negative ion by electron impact. Why can't the Thomson formula be used in this case instead of Eq. (2.54)?

2.15.10 Langevin Capture Cross Section

Calculate the Langevin cross section for Ar^+-ion collision with an argon atom and with water molecule. Comment on your result. Compare the contribution of the polarization term and the charge–dipole interaction term in the case of Ar^+-ion collision with a water molecule.

2.15.11 Nonresonant Charge Exchange

Most negative ions in nonthermal atmospheric pressure air discharges are in form of O_2^-. What will happen to the negative ion distribution if a small amount of fluorine compounds is added to air?

Lecture 3

Elementary Processes of Excited Atoms and Molecules in Plasma

High chemical and biological reactivity of plasma are often due to very high and super-equilibrium concentration of active neutral species. The active neutral species generated in plasma include chemically aggressive atoms and radicals in ground and electronically excited states, as well as vibrationally and electronically excited molecules. Elementary processes and kinetics of atoms and radicals are traditionally considered in a framework of chemical kinetics.

We would recommend on this subject, very important for understanding of plasma processes, in particular books of Henriksen and Hansen (2019), and Kondratiev and Nikitin (1980). In this lecture, we'll focus on plasma-specific elementary processes of excitation, relaxation, and chemical reactions of vibrationally, rotationally, and electronically excited neutrals. Those interested in introductory details on physics of these excited neutral species can be in particular referred to Fridman and Kennedy (2021).

3.1 Vibrational Excitation of Molecules by Electron Impact

Vibrational excitation is one of the most important elementary processes in nonthermal molecular plasma. It is responsible for the major portion of energy exchange between electrons and molecules, often more than 95% of the discharge energy in molecular gases can be focused on vibrational excitations by electron impact. The elastic collisions of electrons and molecules are not effective in the process of vibrational excitation because of the significant difference in their masses ($m/M \ll 1$). Typical energy transfer from an electron with kinetic energy ε to a molecule in the elastic collision is about $\varepsilon(m/M)$; vibrational quantum is about $\hbar\omega \approx I\sqrt{\frac{m}{M}}$ and exceeds this energy. Therefore, classical cross section of the vibrational excitation in an elastic electron-molecular collision is low ($\sigma_0 \sim 10^{-16}$ cm^2 is the gas-kinetic cross section, I is ionization potential):

$$\sigma_{vib}^{elastic} \approx \sigma_0 \frac{\varepsilon}{I}\sqrt{\frac{m}{M}}, \tag{3.1}$$

numerically about 10^{-19} cm^2 for $\varepsilon = 1$ eV. Experiments, however, show that the vibrational excitation cross sections can be much larger, about the same as atomic ones (10^{-16} cm^2). Also, these cross sections are nonmonotonic functions of electron energy, and probability of multi-quantum excitation is not very low. It indicates that vibrational excitation of a molecule AB from vibrational level v_1 to v_2 is not a direct elastic process, but rather a resonance one, proceeding through formation of an intermediate nonstable negative ion:

$$AB(v_1) + e \xleftrightarrow{\Gamma_{1i},\Gamma_{i1}} AB^-(v_i) \xrightarrow{\Gamma_{i2}} AB(v_2) + e. \tag{3.2}$$

In this relation, v_i is the vibrational quantum number of the nonstable negative ion, and $\Gamma_{\alpha\beta}$ (in s^{-1}) is the probabilities of corresponding transitions between different vibrational states. Cross sections of the resonance vibrational excitation process (3.2) can be found in the quasi-steady-state approximation using the **Breit–Wigner formula**:

$$\sigma_{12}(v_i, \varepsilon) = \frac{\pi\hbar^2}{2m\varepsilon} \frac{g_{AB^-}}{g_{AB}g_e} \frac{\Gamma_{1i}\Gamma_{i2}}{\frac{1}{\hbar^2}(\varepsilon - \Delta E_{1i})^2 + \Gamma_i^2}. \tag{3.3}$$

Here ε is the electron energy, ΔE_{1i} is energy of transition to the intermediate state $AB(v_1) \rightarrow AB^-(v_i)$; g_{AB^-}, g_{AB} and g_e are corresponding statistical weights; Γ_i is the total probability of $AB^-(v_i)$ decay through all channels. Equation (3.3)

Plasma Science and Technology: Lectures in Physics, Chemistry, Biology, and Engineering, First Edition. Alexander Fridman.
© 2024 WILEY-VCH GmbH. Published 2024 by WILEY-VCH GmbH.

Table 3.1 Cross sections of vibrational excitation of molecules by electron impact.

Molecule	Most effective electron energy (eV)	Maximum cross section (cm^2)	Molecule	Most effective electron energy (eV)	Maximum cross section (cm^2)
N_2	1.7–3.5	3×10^{-16}	NO	0–1	10^{-17}
CO	1.2–3.0	$3.5 * 10^{-16}$	NO_2	0–1	—
CO_2	3–5	$2 * 10^{-16}$	SO_2	3–4	—
C_2H_4	1.5–2.3	$2 * 10^{-16}$	C_6H_6	1.0–1.6	—
H_2	~3	$4 * 10^{-17}$	CH_4	Thresh. 0.1	10^{-16}
N_2O	2–3 eV	10^{-17}	C_2H_6	Thresh. 0.1	$2 * 10^{-16}$
H_2O	5–10 eV	$6 * 10^{-17}$	C_3H_8	Thresh. 0.1	$3 * 10^{-16}$
H_2S	2–3	—	Cyclo-propane	Thresh. 0.1	$2 * 10^{-16}$
O_2	0.1–1.5	10^{-17}	HCl	2–4	10^{-15}

illustrates the resonance structure of vibrational excitation cross section dependence upon electron energy with energy width of the resonance spikes $\hbar\Gamma_i$, which is related to the lifetime of the nonstable, intermediate, negative ion $AB^-(v_i)$. The maximum value of the cross section (3.3) is about the atomic one (10^{-16} cm^2), see Table 3.1.

Energy dependence of the vibrational excitation cross section is determined by the lifetime of the intermediate ionic states, the so-called resonances. First, consider **short-lifetime resonances** (e.g. H_2, N_2O, H_2O), where the lifetime of the auto-ionization states $AB^-(v_i)$ is much shorter than the period of oscillation of a molecule's nuclei ($\tau \ll 10^{-14}$ s). The energy width of the auto-ionization level $\sim\hbar\Gamma_i$ is very large for the short-lifetime resonances in accordance with the uncertainty principle. This results in relatively wide maximum pikes (several eV) and absence of fine energy structure of $\sigma_{12}(\varepsilon)$, see Figure 3.1a. Because of the short lifetime of the auto-ionization state $AB^-(v_i)$, displacement of nuclei during the lifetime period is small. As a result, decay of the nonstable, negative ion leads to excitation mostly to low vibrational levels. If an intermediate ion lifetime is about molecular oscillation period ($\sim10^{-14}$ s), such a resonance is referred to as **"the boomerang resonance."** In particular, this boomerang type of vibrational excitation takes place for low-energy resonances in N_2, CO, and CO_2. It includes formation and decay of the negative ion during one oscillation as interference of the coming and reflected waves. The interference of the nuclear wave packages results in an oscillating dependence of excitation cross section on the electron energy with typical peak period about 0.3 eV (see Figure 3.1b). The boomerang resonances require usually larger electron energies for excitation of higher vibrational levels. For example, the excitation threshold of $N_2(v=1)$ is 1.9 eV, and that of $N_2(v=10)$ from ground state is about 3 eV. Excitation of CO($v=1$) requires a minimal electron energy of 1.6 eV, while the threshold for CO($v=10$) excitation from the ground state is about 2.5 eV. The maximum value of the vibrational excitation cross section for boomerang resonances usually decreases with growth in the vibrational quantum number. The **long-lifetime resonances** correspond to auto-ionization states $AB^-(v_i)$ with lifetime ($\tau = 10^{-14}$–10^{-10} s) longer than a period of oscillation of the nuclei in a molecule. In particular, this type of vibrational excitation takes place for low-energy resonances in such molecules as O_2, NO, and C_6H_6. The long-lifetime resonances result in quite narrow isolated peaks (about 0.1 eV) in the cross section dependence on electron energy (see Figure 3.1c). In contrast to the boomerang resonances, the maximum value of the vibrational excitation cross section remains about the same for different vibrational quantum numbers.

3.2 Rate Coefficients of Vibrational Excitation in Plasma, Fridman Approximation

Electron energies most effective in vibrational excitation are 1–3 eV (see Table 3.1), which usually corresponds to the maximum in the electron energy distribution function in nonthermal plasmas. The vibrational excitation rate coefficients, which are the results of integration of the cross sections over the excitation energy distribution functions (EEDF), are therefore very high and reach 10^{-7} cm^3 s^{-1}, see Table 3.2. For such molecules as N_2, CO, and CO_2, almost every electron-molecular collision leads to vibrational excitation, resulting in large fraction of electron energy (sometimes above 95%) going in nonthermal discharges to molecular vibration at $T_e = 1$–3 eV.

Vibrational excitation by electron impact is often a one-quantum process, but multi-quantum excitation can be also significant. The rate coefficients $k_{eV}(v_1, v_2)$ from an initial vibrational level v_1 to a final level v_2 can be calculated

Table 3.2 Rate coefficients of vibrational excitation by electron impact, $T_e = 0.5$–2 eV.

Molecule	$T_e = 0.5$ eV	$T_e = 1$ eV	$T_e = 2$ eV
H_2	$2.2 * 10^{-10}$ cm^3 s^{-1}	$2.5 * 10^{-10}$ cm^3 s^{-1}	$0.7 * 10^{-9}$ cm^3 s^{-1}
D_2	—	—	10^{-9} cm^3 s^{-1}
N_2	$2 * 10^{-11}$ cm^3 s^{-1}	$4 * 10^{-9}$ cm^3 s^{-1}	$3 * 10^{-8}$ cm^3 s^{-1}
O_2	—	—	10^{-10}–10^{-9} cm^3 s^{-1}
CO	—	—	10^{-7} cm^3 s^{-1}
NO	—	$3 * 10^{-10}$ cm^3 s^{-1}	—
CO_2	$3 * 10^{-9}$ cm^3 s^{-1}	10^{-8} cm^3 s^{-1}	$3 * 10^{-8}$ cm^3 s^{-1}
NO_2	—	—	10^{-10}–10^{-9} cm^3 s^{-1}
H_2O	—	—	10^{-10} cm^3 s^{-1}
C_2H_4	—	10^{-8} cm^3 s^{-1}	—

using the **Fridman approximation**:

$$k_{eV}(v_1, v_2) = k_{eV}(0,1) \frac{\exp[-\alpha(v_2 - v_1)]}{1 + \beta v_1} \tag{3.4}$$

Numerical values of the parameters α and β are shown in Table 3.3 for several molecular gases assuming Maxwellian EEDF, and $T_e = 1$–3 eV. Additional information regarding the Fridman approximation can be found in particular in Rusanov et al. (1981) and Rusanov and Fridman (1984).

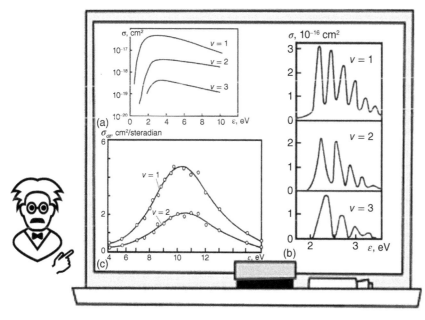

Figure 3.1 Cross sections of vibrational excitation by electron impact as a function of electron energy: (a) short-lifetime resonance (H_2, vibrational levels v = 1,2,3); (b) boomerang resonance (N_2, vibrational levels v = 1,2,3); (c) long-lifetime resonance (O_2, vibrational levels v = 1,2).

Table 3.3 Parameters for the Fridman approximation of the multi-quantum vibrational excitation of molecules by electron impact in nonthermal plasma, $T_e = 1-3$ eV.

Molecule	α	β	Molecule	α	β
N_2	0.7	0.05	H_2	3	—
CO	0.6	—	O_2	0.7	—
$CO_2(\nu_3)$	0.5	—	NO	0.7	—

3.3 Rotational Excitation of Molecules by Electron Impact

The rotational excitation can be significant in plasma energy balance at lower electron energies, when the resonant vibrational excitation is ineffective, but rotational transitions are still possible by long-distance electron-molecular interaction. When electron energies exceed 1 eV, the rotational excitation can proceed resonantly through auto-ionization states of a negative ion similarly to vibrational excitation. Contribution of this mechanism is usually nonsignificant because rotational quanta are much smaller than vibrational, which results in domination of nonresonant "quasi-elastic" rotational excitation. The **nonresonant rotational excitation of molecules** by electron impact can be illustrated using the classical approach. Typical energy transfer from an electron with kinetic energy ε to a molecule in an elastic collision, inducing rotational excitation, is about $\varepsilon(m/M)$. Typical distance between rotational levels, rotational quantum, is: $I(m/M)$; where I is the ionization potential. Thus, the classical cross section of the nonresonant rotational excitation can be expressed as: related to – in the following manner:

$$\sigma_{rotational}^{elastic} \approx \sigma_0 \frac{\varepsilon}{I}, \tag{3.5}$$

where $\sigma_0 \sim 10^{-16}$ cm^2 is the gas-kinetic collisional cross section. The cross section (3.5) of rotational excitation exceeds that of nonresonant vibrational excitation by a factor of 100, see (3.1), keeping in mind however that rotational quanta are quite small.

To evaluate the elastic or "quasi-elastic" energy transfer from electrons to neutral molecules, *the rotational excitation can be taken into account combined together with the elastic collisions*. The process is then characterized by the gas-kinetic rate coefficient $k_{e0} \approx \sigma_0 <v_e> \approx 3 \cdot 10^{-8}$ cm^3 s^{-1} ($<v_e>$ is the average thermal velocity of electrons) and each collision is considered as a loss of about $\varepsilon(m/M)$ of electron energy.

A quantum mechanical approach leads to similar results and conclusions regarding the rotational excitation in plasma. An electron collision with a dipole molecule induces the rotational transitions with a change of the rotational quantum number $\Delta J = 1$. Quantum mechanical cross sections of rotational excitation of a linear dipole molecule by a low-energy electron can be expressed as:

$$\sigma(J \to J+1, \varepsilon) = \frac{d^2}{3\varepsilon_0 a_0 \varepsilon} \frac{J+1}{2J+1} \frac{\sqrt{\varepsilon} + \sqrt{\varepsilon'}}{\sqrt{\varepsilon} - \sqrt{\varepsilon'}}, \tag{3.6}$$

where d is a dipole moment, a_0 is the Bohr radius, $\varepsilon' = \varepsilon - 2B(J+1)$ is the electron energy after collision, and B is the rotational constant. Numerically, this cross section is approximately $1-3 * 10^{-16}$ cm^2 when the electron energy is about 0.1 eV. Homonuclear molecules, such as N_2 or H_2, have no dipole moment, and any rotational excitation of such molecules is due to electron interaction with their quadrupole moment Q. In this case, the rotational transition takes place with the change of rotational quantum number $\Delta J = 2$. Cross section of the rotational excitation of homonuclear molecules by a low-energy electron is obviously lower:

$$\sigma(J \to J+2, \varepsilon) = \frac{8\pi Q^2}{15 e^2 a_0^2} \frac{(J+1)(J+2)}{(2J+1)(2J+3)} \ln \sqrt{\frac{\varepsilon}{\varepsilon'}}, \tag{3.7}$$

and numerically is about $1-3 * 10^{-17}$ cm^2 at electron energy 0.1 eV.

3.4 Electronic Excitation by Electron Impact: Metastable States, Dissociation of Molecules

Electronically excited neutrals make significant contribution to plasma reactivity if they have sufficient lifetime. The most important factor defining stability of the excited species is their radiation, which depends on whether the related transition is forbidden or not by the selection rules. Such selection rules in the case of electric dipole radiation of excited molecules require:

$$\Delta \Lambda = 0, \pm 1; \quad \Delta S = 0; \tag{3.8}$$

additionally, for transitions between Σ-states, and homonuclear molecules, selection rules require:

$$\Sigma^+ \to \Sigma^+ \ \text{ or } \ \Sigma^- \to \Sigma^-, \ \text{ and } \ g \to u \ \text{ or } \ u \to g. \tag{3.9}$$

If radiation is allowed, its frequency can be as high as $10^9 \, \text{s}^{-1}$, and effectiveness of the relevant excited states in chemical reactivity is limited. In contrast to the optically permitted "resonance" states, the **electronically excited metastable species** corresponding to optically forbidden transitions, have very long lifetimes: seconds, minutes, and sometimes even hours resulting in their significant contribution to plasma reactivity, see Table 3.4.

In contrast to the vibrational and rotational excitation discussed above, the electronic excitation by electron impact requires usually higher electron energies ($\varepsilon > 10 \, \text{eV}$), which justifies use of the **Born approximation** ($\varepsilon \gg \Delta E_{ik}$). For *excitation of optically permitted transitions* from an atomic state "i" to another state "k," the Born approximation gives the cross section:

$$\sigma_{ik}(\varepsilon) = 4\pi a_0^2 f_{ik} \left(\frac{Ry}{\Delta E_{ik}} \right)^2 \frac{\Delta E_{ik}}{\varepsilon} \ln \frac{\varepsilon}{\Delta E_{ik}}, \tag{3.10}$$

where Ry is the Rydberg constant, a_0 is the Bohr radius, f_{ik} is the force of oscillator for the transition $i \to k$, ΔE_{ik} is energy of the transition. The cross sections of electronic excitation in a wide range of electron energies can be calculated by semi-empirical formula Smirnov (2001):

$$\sigma_{ik}(\varepsilon) = 4\pi a_0^2 f_{ik} \left(\frac{Ry}{\Delta E_{ik}} \right)^2 \frac{\ln(0.1x + 0.9)}{x - 0.7}, \tag{3.11}$$

where $x = \varepsilon/\Delta E_{ik}$. Obviously, the semi-empirical formula corresponds to the Born approximation at high electron energies ($x \gg 1$). Numerically, the maximum cross sections of excitation of optically permitted transitions are about the gas-kinetic cross sections, $\sigma_0 \sim 10^{-16} \, \text{cm}^2$. To reach the maximum cross section of the electronic excitation, the electron energy should be 2–3 times greater than the transition energy ΔE_{ik}. The dependence $\sigma_{ik}(\varepsilon)$ is quite different for *excitation of optically forbidden transitions to "metastable" states*, where the exchange interaction and details of electron shell structures become important. As a result, the maximum cross section, which is also about the size of the atomic one, $\sigma_0 \sim 10^{-16} \, \text{cm}^2$, can be reached at much lower electron energies $\varepsilon/\Delta E_{ik} \approx 1.2 - 1.6$.

Table 3.4 Excitation energies of the metastable states of diatomic molecules (on the lowest vibrational level), and lifetimes of these metastable molecules.

Metastable molecule	Electronic state	Energy of the state (eV)	Radiative lifetime (s)
N_2	$A^3\Sigma_u^+$	6.2	13
N_2	$a'^1\Sigma_u^-$	8.4	0.7
N_2	$a^1\Pi_g$	8.55	2×10^{-4}
N_2	$E^3\Sigma_g^+$	11.9	300
O_2	$a^1\Delta_g$	0.98	$3 * 10^3$
O_2	$b^1\Sigma_g^+$	1.6	7
NO	$a^4\Pi$	4.7	0.2

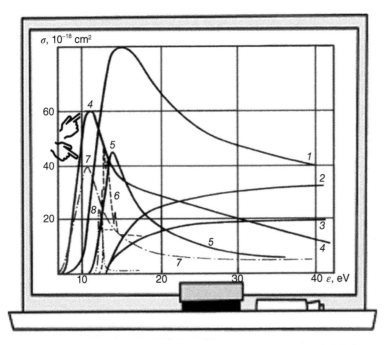

Figure 3.2 Cross-section of excitation of different electronic states of $N_2(X^1\Sigma_g^+, v = 0)$ by electron impact: $1 - a^1\Pi_g$; $2 - b^1\Pi_u$ $(v_k = 0-4)$; 3 – transitions 12,96 eV; $4 - B^3\Pi_g$; $5 - C^3\Pi_u$; $6 - a''^1\Sigma_g^+$; $7 - A^3\Sigma_u^+$; $8 - E^3\Sigma_g^+$; pointer indicates examples of effective excitation of the metastables.

 This leads to an **effect of predominant excitation of the optically forbidden, metastable states** by electron impact in nonthermal discharges, where electron temperature T_e is usually much less than the transition energy ΔE_{ik}. Obviously, this effect requires availability of the optically forbidden, metastable states at relatively low energies. Some cross sections of electronic excitation by electron impact are presented in Figure 3.2, pointing to the effect of predominant excitation of the metastable.

The rate coefficients of the electronic excitation are the result of the cross sections $\sigma_{ik}(\varepsilon)$ integration over EEDF, which in the case of Maxwellian distribution with $T_e \ll \Delta E_{ik}$ gives:

$$k_{el.excit.} \propto \exp\left(-\frac{\Delta E_{ik}}{T_e}\right),\tag{3.12}$$

and for numerical calculations can be presented in a semi-empirical form:

$$\lg k_{el.excit.} = -C_1 - \frac{C_2}{E/n_0},\tag{3.13}$$

where, $k_{el.excit.}$ is expressed in $cm^3\,s^{-1}$; E is the electric field in $V\,cm^{-1}$; n_0 is gas density in $1/cm^3$; parameters C_1 and C_2 are presented in the Table 3.5.

If vibrational excitation is significant, superelastic collisions provide higher electron energies and higher electronic excitation rate coefficients for the same value of the reduced electric field E/n_0. It can be taken into account by including in Eq. (3.13) two additional terms:

$$\lg k_{el.excit.} = -C_1 - \frac{C_2}{E/n_0} + \frac{40z + 13z^2}{[(E/n_0)*10^{16}]^2} - 0.02\left(\frac{T_v}{1000}\right)^{2/3},\tag{3.14}$$

where vibrational temperature T_v is in degrees Kelvin, and $z = \exp\left(-\frac{\hbar\omega}{T_v}\right)$.

Dissociation of molecules through their electronic excitation by electron impact can proceed "directly" in a single collision, but through different mechanisms (see Figure 3.3a):

- *Mechanism A* starts with electronic excitation of a molecule from ground state to a repulsive state followed by dissociation. The required electron energy can significantly exceed here the dissociation energy. This mechanism can produce high-energy neutral fragments.

Table 3.5 Semi-empirical approximation parameters for rate coefficients of electronic excitation and ionization of CO_2 and N_2 by electron impact.

Mol.	Excitation level or ionization	C_1	$C_2 * 10^{16}$, $V * cm^2$	Mol.	Excitation level or ionization	C_1	$C_2 * 10^{16}$, $V * cm^2$
N_2	$A^3\Sigma_u^+$	8.04	16.87	N_2	$c^1\Pi_u$	8.85	34.0
N_2	$B^3\Pi_g$	8.00	17.35	N_2	$a^1\Pi_u$	9.65	35.2
N_2	$W^3\Delta_u$	8.21	19.2	N_2	$b'^1\Sigma_u^+$	8.44	33.4
N_2	$B'^3\Sigma_u^-$	8.69	20.1	N_2	$c^3\Pi_u$	8.60	35.4
N_2	$a'^1\Sigma_u^-$	8.65	20.87	N_2	$F^3\Pi_u$	9.30	32.9
N_2	$a^1\Pi_g$	8.29	21.2	N_2	Ionization	8.12	40.6
N_2	$W^1\Delta_u$	8.67	20.85	CO_2	$^3\Sigma_u^+$	8.50	10.7
N_2	$C^3\Pi_u$	8.09	25.5	CO_2	$^1\Delta_u$	8.68	13.2
N_2	$E^3\Sigma_g^+$	9.65	23.53	CO_2	$^1\Pi_g$	8.84	14.8
N_2	$a''^1\Sigma_g^+$	8.88	26.5	CO_2	$^1\Sigma_g^+$	8.23	18.9
N_2	$b^1\Pi_u$	8.50	31.88	CO_2	Other levels	8.34	20.9
N_2	$c'^1\Sigma_u^+$	8.56	35.6	CO_2	Ionization	8.38	25.5

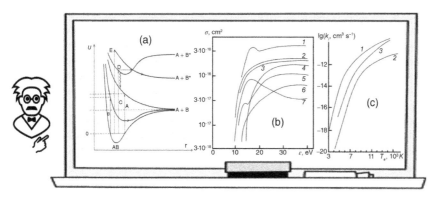

Figure 3.3 Dissociation of molecules through electronic excitation by electron impact: (a) electronic terms and transitions corresponding to different process mechanisms (A–E); (b) cross sections as a function of electron energy (1-CH_4, 2-O_2, 3-NO, 4-N_2, 5-CO_2, 6-CO, 7-H_2); (c) rate coefficients in N_2 plasma (1-total stepwise electronic excitation, 2-dissociation by direct electron impact, 3-dissociation in stepwise electronic excitation sequence).

- *Mechanism B* includes the excitation from ground state to an attractive state with energy exceeding the dissociation threshold. Energy of the dissociation fragments is lower here.
- *Mechanism C* consists in the direct electronic excitation of a molecule from ground state to an attractive state corresponding to electronically excited dissociation products. The excitation of the state can lead to radiative transition to a low-energy repulsive state followed by dissociation. Energy of the dissociation fragments in this case is similar to those of mechanism A.
- *Mechanism D* starts with electronic excitation of a molecule from ground state to an attractive state corresponding to electronically excited dissociation products. In contrast to mechanism C, excitation leads here to nonradiative transfer to a highly excited repulsive state followed by dissociation. This mechanism is usually referred to as **predissociation**.
- *Mechanism E* is similar to the mechanism A but with dissociation into electronically excited fragments. This mechanism requires the largest values of electron energies, and therefore the corresponding rate coefficients are relatively low.

Cross sections and rate coefficients of the molecular dissociation by direct electron impact are presented in Figure 3.3b,c as functions of electron energy and electron temperature, respectively.

3.5 Distribution of Electrons Energy in Nonthermal Discharges Between Different Channels of Excitation and Ionization

Electron energy received from the electric field in nonthermal plasma is distributed between elastic energy losses and different channels of excitation and ionization. Specifics of this distribution are determined by reduced electric field E/n_0 as well as plasma gas composition and are characterized by some important general features to be summarized below.

3.5.1 Elastic vs Inelastic Collisions

Contribution of elastic energy losses (energy transfer from electrons to translational degrees of freedom of neutrals) is significant only at low reduced electric fields E/n_0, and hence at low electron temperatures. This is natural since these processes are nonresonant and take place at low electron energies ($\ll 1$ eV). As it was discussed in Section 3.3, the rotational excitation can be combined with translational energy transfer in the generalized group elastic collisions. This effect is illustrated in Figure 3.4, also reflecting contribution of the superelastic collision transferring energy back from excited neutrals to plasma electrons.

3.5.2 Exclusive Contribution of Discharge Energy to Vibrational Excitation

 At electron temperatures about 1 eV, conventional for nonthermal plasmas, very significant portion of electron energy and hence most of discharge power can be localized on vibrational excitation of molecules (see Figure 3.5), which makes the vibrational excitation exceptionally important in the nonequilibrium plasma chemistry of molecular gases.

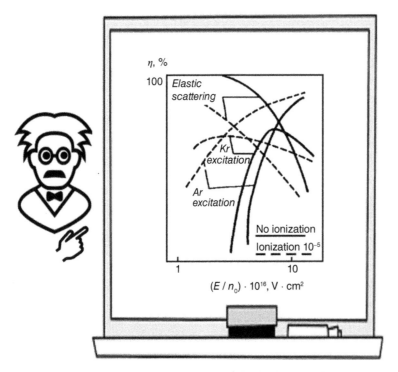

Figure 3.4 Electron energy distribution between electronic excitation channels in noble gases (5% Kr, 95% Ar), presence of plasma electrons reflects contribution of superelastic collisions.

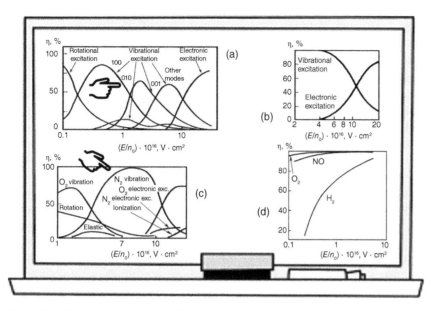

Figure 3.5 Electron energy distribution between excitation channels in molecular gasses, significant contribution of discharge energy to vibrational excitation: (a) CO_2; (b) CO; (c) air; (d) fraction of electron energy going to vibrational excitation in O_2, NO, H_2. Pointer indicates the most significant channels of vibrational excitation.

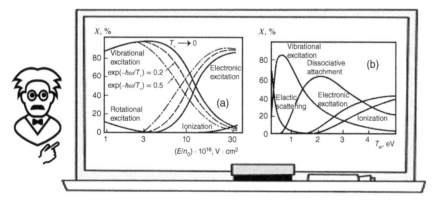

Figure 3.6 Electron energy distribution between excitation channels in molecular gasses: (a) effect of superelastic collisions in N_2 plasma at different vibrational temperatures; (b) illustration of contribution of dissociative attachment in H_2O vapor plasma.

3.5.3 Effect of Superelastic Collisions on Contribution of Discharge Energy to Vibrational Excitation

When level of vibrational excitation and therefore vibrational temperature T_v are already high, efficiency of the further vibrational excitation becomes lower. It can be interpreted as a contribution of superelastic collisions at higher T_v transferring energy back to electrons, stimulating the high-energy tail of EEDF, and intensifying electronic excitation and ionization processes, see Figure 3.6a.

3.5.4 Contribution of Electron Attachment, Electronic Excitation, and Ionization Processes

Contribution of the electron attachment processes, especially dissociative attachment, is able to compete with vibrational excitation at similar electron temperatures, but only in electro-negative gases, see Figure 3.6b. Finally, the contribution of electronic excitation and ionization becomes significant at higher values of E/n_0 and higher electron temperatures, because of high-energy thresholds of these processes.

3.6 Vibrational-to-translational Energy Transfer Processes, VT-relaxation

Exceptional role of vibrational excitation in nonthermal molecular plasma is not only due to the "exclusive contribution of discharge energy to vibrational excitation" (see Section 3.5.2), but also due to ability of molecules to keep their vibrational excitation for sufficiently long time without losing this energy through collisions to other degrees of freedom and especially to translational ones (VT-relaxation). This, the so-called **Landau–Teller effect**, is a manifestation of the adiabatic nature of the VT-relaxation (see Section 2.7), high value of the Massey parameter (see Eq. 2.37), and therefore often very low probability of the process.

3.6.1 Slow Adiabatic VT-relaxation of Harmonic Oscillators

Main features of the VT-relaxation can be demonstrated by classical analysis of collision of vibrationally excited diatomic molecule (represented as a harmonic oscillator) with an atom or molecule (represented as an external force $F(t)$, see Figure 3.7). The one-dimensional motion of the harmonic oscillator can then be described in the center of mass system by the Newton equation:

$$\frac{d^2y}{dt^2} + \omega^2 y = \frac{1}{\mu_0} F(t),$$

(3.15)

where, y is the vibrational coordinate, ω is the oscillator frequency, and μ_0 is its reduced mass. The oscillator is initially $(t \to -\infty)$ not excited: $y(t \to -\infty) = 0$, $\frac{dy}{dt}(t \to -\infty) = 0$. Then the vibrational energy transferred to the oscillator during the collision can be expressed as:

$$\Delta E_v = \frac{\mu_0}{2}\left[\left(\frac{dy}{dt}\right)^2 + \omega^2 y^2\right]_{t=\infty}.$$

(3.16)

To calculate the energy transfer ΔE_v, a complex variable $\xi(t) = \frac{dy}{dt} + i\omega y$ can be introduced. The oscillator energy transfer can be found as square of module of this variable $\Delta E_v = \frac{\mu_0}{2}|\xi(t)|^2_{t=\infty}$. The complex function $\xi(t)$ follows the first order differential equation:

$$\frac{d}{dt}\left(\frac{dy}{dt} + i\omega y\right) - i\omega\left(\frac{dy}{dt} + i\omega y\right) = \frac{1}{\mu_0}F(t); \frac{d\xi}{dt} - i\omega\xi = \frac{1}{\mu_0}F(t).$$

(3.17)

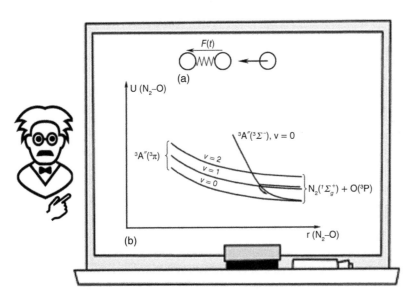

Figure 3.7 VT-relaxation processes: (a) illustration of the Landau–Teller effect for adiabatic VT relaxation; (b) nonadiabatic VT relaxation of N_2 molecules in collisions with O-atoms.

Solution of the linear nonuniform differential equation with initial conditions given above $y(t \to -\infty) = 0$, $\frac{dy}{dt}$ $(t \to -\infty) = 0$ can easily be presented as:

$$\xi(t) = \exp(i\omega t) \int_{-\infty}^{t} \frac{1}{\mu_0} F(t') \exp(-i\omega t') dt', \tag{3.18}$$

which leads to the following expression for the energy transfer during the act of VT-relaxation:

$$\Delta E_v = \frac{1}{2\mu_0} \left| \int_{-\infty}^{+\infty} F(t) \exp(-i\omega t) dt \right|^2. \tag{3.19}$$

Only small Fourier component of the perturbation force on the oscillator frequency is effective in collisional excitation (or deactivation) of vibrational energy of molecules. Considering time "t" as a complex variable and supposing that $F(t) \to \infty$ in a singularity point $t = \tau + i\tau_{col}$ ($\tau_{col} = 1/\alpha v$ – is time of the collision, α is the reverse radius of interaction between molecules, v is a relative velocity of the colliding particles), integration (3.19) gives:

$$\Delta E_v \propto \exp(-2\omega\tau_{col}) \tag{3.20}$$

This expression of the **Landau–Teller effect** shows the adiabatic behavior of the vibrational relaxation. The Massey parameter $\omega\tau_{col}$ (see Section 2.7) is very high $\omega\tau_{col} \gg 1$ at moderate gas temperatures, which explains the slowness of the adiabatic VT-relaxation. Molecular vibrations for some gases such as N_2, CO_2, H_2, and CO are able "to trap" energy of nonthermal discharges; it is easy to activate them in cold plasma and difficult to deactivate. Quantitatively, the classical VT energy transfer between molecule BC and atom A is:

$$\Delta E_v = \frac{8\pi^2\omega^2\mu^2\lambda^2}{\alpha^2\mu_0^2} \exp\left(-\frac{2\pi\omega}{\alpha v}\right). \tag{3.21}$$

In this relation μ is a reduced mass of A and BC, μ_0 is a reduced mass of the molecule BC, and $\lambda = m_C/(m_B + m_C)$. Probability of transfer of one vibrational quantum $\hbar\omega$ can be then presented as:

$$P_{01}^{VT}(v) = \frac{\Delta E_v}{\hbar\omega} = \frac{8\pi^2\omega\mu^2\lambda^2}{\hbar\alpha^2\mu_0^2} \exp\left(-\frac{2\pi\omega}{\alpha v}\right), \tag{3.22}$$

which can describe both vibrational activation and deactivation. Quantum mechanics generalizes this equation to describe probability $P_{mn}^{VT}(v)$ of an oscillator transition from an initial state with vibrational quantum number "m" to a final one "n":

$$P_{mn}^{VT}(v) = \frac{16\pi^2\mu^2\omega_{mn}^2\lambda^2}{\alpha^2\hbar^2} \langle m|y|n \rangle^2 \exp\left(-\frac{2\pi|\omega_{mn}|}{\alpha v}\right), \tag{3.23}$$

where $\hbar\omega_{mn} = E_m - E_n$ is the transition energy, $\langle m|y|n \rangle$ is a matrix element corresponding to the eigenfunctions (m and n) of nonperturbed Hamiltonian of the oscillator, which is not equal to zero for harmonic oscillators only for one-quantum transitions $n = m \pm 1$:

$$\langle m|y|n \rangle = \sqrt{\frac{\hbar}{2\mu_0\omega}} (\sqrt{m}\delta_{n,m-1} + \sqrt{m+1}\delta_{n,m+1}). \tag{3.24}$$

Here the symbol $\delta_{ij} = 1$ if $i = j$, and $\delta_{ij} = 0$ if $i \neq j$. Therefore, harmonic oscillators are permitted only for one-quantum VT-relaxation. Multi-quantum VT relaxation is possible only due to anharmonicity of oscillations and has much lower probability.

3.6.2 VT-relaxation Rate Coefficients for Harmonic Oscillators, Landau–Teller Formula

To calculate the rate coefficient of VT-relaxation from vibrational level $n + 1$ to n, the relaxation probability ((3.22) and (3.23)) as a function of the relative velocity v can be presented as:

$$P_{n+1,n}^{VT}(v) \propto (n+1) \exp\left(-\frac{2\pi\omega}{\alpha v}\right), \tag{3.25}$$

and then integrated over the Maxwellian distribution $f(v)$ of molecules with temperature T_0:

$$P_{n+1,n}^{VT}(T_0) = (n+1)P_{1,0}^{VT}(T_0), \quad P_{10}^{VT} \propto \exp\left[-3\left(\frac{\hbar^2\mu\omega^2}{2\alpha^2 T_0}\right)^{1/3}\right]. \tag{3.26}$$

Correspondent expression for the rate coefficient of a single-quantum VT-relaxation exponentially growing up with temperature T_0 is referred to as the **Landau–Teller formula**:

$$k_{VT}^{10} \propto \exp\left(-\frac{B}{T_0^{1/3}}\right), B = \sqrt[3]{\frac{27\hbar^2\mu\omega^2}{2\alpha^2 T_0}}. \tag{3.27}$$

For numerical calculations, the following semi-empiric relations can be used for k_{VT}^{10}:

$$k_{VT}^{10} = 3.03*10^6 (\hbar\omega)^{2.66} \mu^{2.06} \exp\left[-0.492(\hbar\omega)^{0.681}\mu^{0.302}T_0^{-1/3}\right], \tag{3.28}$$

and for the VT-relaxation time τ_{VT} (see details for example in Fridman and Kennedy 2021):

$$\ln(p\tau_{VT}) = 1.16*10^{-3}\mu^{1/2}(\hbar\omega)^{4/3}\left(T_0^{-1/3} - 0.015\mu^{1/4}\right) - 18.42, \tag{3.29}$$

where k_{VT}^{10} is in cm^3 mol^{-1} * s, T_0 and $\hbar\omega$ is in degrees Kelvin, μ is the reduced mass of colliding particles in atomic mass units, pressure p is in atm, τ_{VT} in s. Numerical values of the Landau–Teller vibrational relaxation rate coefficients at room temperature and as a function of temperature are presented in Table 3.6.

3.6.3 VT-relaxation of Anharmonic Oscillators

 The key peculiarity of relaxation of anharmonic oscillators is reduction of the transition energy with an increase of vibrational quantum number n:

$$\omega_{n,n-1} = \omega(1 - 2x_e n), \tag{3.30}$$

where x_e is the coefficient of anharmonicity. As seen from Eq. (3.25), the probability and rate coefficient of vibrational VT-relaxation increases with n even in the case of harmonic oscillators. Furthermore, in the case of anharmonic ones, the increase of vibrational quantum number "n" leads also to a reduction of the Massey parameter, making relaxation less adiabatic and exponentially faster:

$$P_{n+1,n}^{VT}(T_0) = (n+1)P_{1,0}^{VT}(T_0)\exp(\delta_{VT}n). \tag{3.31}$$

The temperature dependence of the probability of VT-relaxation of anharmonic oscillators is similar to that of harmonic oscillators and corresponds to the Landau–Teller formula. The exponential parameter δ_{VT} can be determined in the case of anharmonic oscillators as:

$$\delta_{VT} = 4\gamma_n^{2/3}x_e, \quad \text{if } \gamma_n \geq 27 \tag{3.32}$$

$$\delta_{VT} = \frac{4}{3}\gamma_n x_e, \quad \text{if } \gamma_n < 27. \tag{3.33}$$

The adiabatic factor γ_n is the Massey parameter for the vibrational relaxation transition $n+1 \to n$ with the transition energy $E_{n+1} - E_n$. This adiabatic factor can be expressed as:

$$\gamma_n(n+1 \to n) = \frac{\pi(E_{n+1} - E_n)}{\hbar\alpha}\sqrt{\frac{\mu}{2T_0}}, \tag{3.34}$$

and numerically calculated as (the reduced mass μ is in atomic units, reverse radius of interaction between colliding particles α is in A^{-1}, vibrational quantum $\hbar\omega$ and gas temperature T_0 are in degrees Kelvin):

$$\gamma_n = \frac{0.32}{\alpha}\sqrt{\frac{\mu}{T_0}}\hbar\omega(1 - 2x_e(n-1)). \tag{3.35}$$

Table 3.6 Adiabatic VT-relaxation rate coefficients $k_{VT}^{10}(T_0)$ for one-component gases at room temperature and as a function of temperature.

Molecule	$k_{VT}^{10}(T_0 = 300\ \text{K}),\ \text{cm}^3\ \text{s}^{-1}$	Molecule	$k_{VT}^{10}(T_0 = 300\ \text{K}),\ \text{cm}^3\ \text{s}^{-1}$
O_2	$5 * 10^{-18}$	F_2	$2 * 10^{-15}$
Cl_2	$3 * 10^{-15}$	D_2	$3 * 10^{-17}$
Br_2	10^{-14}	$CO_2(01^10)$	$5 * 10^{-15}$
J_2	$3 * 10^{-14}$	$H_2O(010)$	$3 * 10^{-12}$
N_2	$3 * 10^{-19}$	N_2O	10^{-14}
CO	10^{-18}	COS	$3 * 10^{-14}$
H_2	10^{-16}	CS_2	$5 * 10^{-14}$
HF	$2 * 10^{-12}$	SO_2	$5 * 10^{-14}$
DF	$5 * 10^{-13}$	C_2H_2	10^{-12}
HCl	10^{-14}	CH_2Cl_2	10^{-12}
DCl	$5 * 10^{-15}$	CH_4	10^{-14}
HBr	$2 * 10^{-14}$	CH_3Cl	10^{-13}
DBr	$5 * 10^{-15}$	$CHCl_3$	$5 * 10^{-13}$
HJ	10^{-13}	CCl_4	$5 * 10^{-13}$
HD	10^{-16}	NO	10^{-13}
Molecule	Temperature dependence $k_{VT}^{10}(T_0),\ \text{cm}^3\ \text{s}^{-1}$; temperature T_0 in Kelvin		
O_2	$10^{-10} \exp\left(-129^* T_0^{-1/3}\right)$		
Cl_2	$2^* 10^{-11} \exp\left(-58^* T_0^{-1/3}\right)$		
Br_2	$2^* 10^{-11} \exp\left(-48^* T_0^{-1/3}\right)$		
J_2	$5^* 10^{-12} \exp\left(-29^* T_0^{-1/3}\right)$		
CO	$10^{-12} T_0 \exp\left(-190^* T_0^{-1/3} + 1410^* T_0^{-1}\right)$		
NO	$10^{-12} \exp\left(-14^* T_0^{-1/3}\right)$		
HF	$5^* 10^{-10} T_0^{-1} + 6^* 10^{-20} T_0^{2.26}$		
DF	$1.6^* 10^{-5} T_0^{-3} + 3.3^* 10^{-16} T_0$		
HCl	$2.6^* 10^{-7} T_0^{-3} + 1.4^* 10^{-19} T_0^2$		
F_2	$2^* 10^{-11} \exp\left(-65^* T_0^{-1/3}\right)$		
D_2	$10^{-12} \exp\left(-67^* T_0^{-1/3}\right)$		
$CO_2\ (01^10)$	$10^{-11} \exp\left(-72^* T_0^{-1/3}\right)$		

3.6.4 Fast Nonadiabatic Mechanisms of VT-relaxation

The vibrational relaxation is quite slow in adiabatic collisions when there is no chemical interaction between colliding partners. The probability of a vibrationally excited N_2 deactivation in collision with another N_2 molecule at room temperature can be as low as 10^{-9}. If the colliding partners interact chemically, VT-relaxation can be nonadiabatic and therefore proceed much faster, see the mechanisms below.

(a) ***VT-relaxation in molecular collisions with atoms and radicals*** can be illustrated by the relaxation of vibrationally excited N_2-molecules on atomic oxygen, see Figure 3.7b. The interval between degenerated electronic terms grows as a molecule and an atom approach each other. Finally, when this energy of electronic transitions becomes equal to a vibrational quantum, the nonadiabatic relaxation (the so-called **vibronic transition**) can take place. Temperature dependence of the process is not significant, and typical rate coefficients are high,

Table 3.7 Accommodation coefficients for the heterogeneous VT-relaxation.

Mol.	Mode	$\hbar\omega$, cm^{-1}	T_0, K	Surface	Acc. coefficient
CO_2	$\nu_2 = 1$	667	277–373	Platinum, NaCl	0.3–0.4
CO_2	$\nu_3 = 1$	2349	300–350	Pyrex, brass, teflon, mylar, quartz	0.2–0.4
CO_2	$\nu_3 = 1$	2349	300–560	Molybdenum glass	0.3–0.4
CH_4	$\nu_4 = 1$	1306	273–373	Platinum	0.5–0.9
H_2	$\nu = 1$	4160	300	Molybdenum glass, pyrex, quartz	$1.3 * 10^{-4}$
D_2	$\nu = 1$	2990	300	Quartz, molybdenum glass	10^{-4}
N_2	$\nu = 1$	2331	300	Steel, Al, Cu, molybdenum glass	$3 * 10^{-3}$
N_2	$\nu = 1$	2331	300	Teflon, alumina, pyrex	10^{-3}
N_2	$\nu = 1$	2331	295	Silver	$1.4 * 10^{-2}$
CO	$\nu = 1$	2143	300	Pyrex	$1.9 * 10^{-2}$
HF	$\nu = 1$	3962	300	Molybdenum glass	10^{-2}
HCl	$\nu = 1$	2886	300	Pyrex	0.45
OH	$\nu = 9$	—	300	Boron acid	1
N_2O	$\nu_2 = 1$	589	273–373	Platinum, NaCl	0.3
N_2O	$\nu_3 = 1$	2224	300–350	Pyrex, quartz	0.2
N_2O	$\nu_3 = 1$	2224	300–560	Molybdenum glass	0.01–0.03
CF_3Cl	$\nu_3 = 1$	732	273–373	Platinum	0.5–0.6

usually about 10^{-13}–10^{-12} cm^3 s^{-1}. Sometimes, as in the case of relaxation of alkaline atoms, the nonadiabatic VT-relaxation rate coefficients reach those for gas-kinetic collisions, about 10^{-10} cm^3 s^{-1}.

(b) **VT-relaxation through intermediate formation of long-living complexes** occurs, in particular, in collisions such as H_2O^*–H_2O, CO_2^*–H_2O, CH_4^*–CO_2, CH_4^*–H_2O, $C_2H_6^*$–O_2, NO^*–Cl_2. Relaxation rate coefficients also can reach here the gas-kinetic values about 10^{-10} cm^3 s^{-1}. These relaxation rate coefficients usually decrease with temperature growth, and probabilities of one-quantum and multi-quantum transfers are about the same.

(c) **VT-relaxation in symmetrical exchange reactions** such as:

$$A' + (BA'') * (n = 1) \rightarrow A'' + BA'(n = 0). \tag{3.36}$$

(d) are very fast, when activation energies of corresponding chemical processes are low. An important example is the barrierless oxygen exchange reaction: $O_2^* + O \rightarrow O_3^* \rightarrow O + O_2$ with the rate coefficient about 10^{-11} cm^3 s^{-1} and no temperature dependence. Similar fast nonadiabatic mechanisms determine VT-relaxation of H_2-molecules on H-atoms, halogen molecules on halogen atoms, and hydrogen halides on the hydrogen atoms.

(e) **Fast heterogeneous VT-relaxation and losses of vibrational energy in surface collisions** occur during the adsorption of vibrationally excited molecules. Probability of the process is usually referred to as the **accommodation coefficient**, see Table 3.7.

3.7 Vibrational Energy Exchange Between Molecules, VV-relaxation

Generation of highly vibrationally excited and therefore reactive molecules is usually due to not direct electron excitation, but collisional vibrational energy exchange between the molecules referred to as VV-relaxation and subdivided into fast resonant and nonresonant processes.

3.7.1 VV-relaxation Close to Resonant

VV-relaxation close to resonant is quite fast, often implies vibrational energy exchange between molecules of the same kind, for example: $N_2^*(\nu = 1) + N_2(\nu = 0) \rightarrow N_2(\nu = 0) + N_2^*(\nu = 1)$, and is characterized by the probability $q_{mn}^{sl}(\nu)$

of a collisional transition, when one oscillator changes its vibrational quantum number from "s" to "l", and other from "m" to "n".

Quantum mechanics determines the probability $q_{mn}^{sl}(v)$ similarly to ((3.22) and (3.23)) but with the following defect of resonance replacing ω_{mn} (more detailed information in Nikitin 1974):

$$\omega_{ms,nl} = \frac{1}{\hbar}(E_m + E_s - E_l - E_n) = \frac{\Delta E}{\hbar}. \tag{3.37}$$

Also, instead of the matrix elements of transition $m \rightarrow n$ in ((3.22) and (3.23)), the probability of the VV-exchange is determined by a product of matrix elements of transitions $m \rightarrow n$ and $s \rightarrow l$ in both two interacting oscillators:

$$q_{mn}^{sl}(v) = \frac{\langle m|y_1|n\rangle^2 \langle s|y_2|l\rangle^2}{\hbar^2} \left| \int_{-\infty}^{+\infty} F(t)\exp(i\omega_{ms,nl}t)dt \right|^2. \tag{3.38}$$

The Fourier component of interaction force $F(t)$ on a transition frequency (3.37) characterizes here the level of resonance, and nonadiabatic behavior of the process in contrast to VT-relaxation. The matrix elements for harmonic oscillators $\langle m|y|n\rangle$ are nonzero only for single-quantum transitions with $n = m \pm 1$:

$$Q_{n+1,n}^{m,m+1}(T_0) = (m+1)(n+1)Q_{10}^{01}(T_0). \tag{3.39}$$

The higher powers in expansion of intermolecular interaction potential permit the multi-quantum VV-exchange even for harmonic oscillators, but obviously with lower probability:

$$Q_{0,k}^{m,m-k} \approx \frac{m!}{(m-k)!k!2^{k-1}}(\Omega\tau_{col})^{2k}, \tag{3.40}$$

where τ_{col} is a collision time, Ω is the VV-transition frequency (3.37).

The factor $\Omega\tau_{col} \approx 0.1$–$0.01$ is usually small due to the smallness of the vibrational amplitude with respect to intermolecular interaction radius, and probability of a single quantum transfer is about $Q_{10}^{01} \approx (\Omega\tau_{col})^2 \approx 10^{-2} - 10^{-4}$, see Table 3.8. The probability of multi-quantum VV-exchange in this case according to Eq. (3.40) is very low (it is about 10^{-9} even for resonant three-quantum exchange). For some molecules, however, such as CO_2 and N_2O, VV-relaxation is due to dipole interaction and long-distance forces, leading to $Q_{10}^{01} \approx (\Omega\tau_{col})^2 \approx 1$ (see Table 3.8). The cross sections of the resonance VV-exchange between highly vibrationally excited molecules can exceed in this case the gas-kinetic cross section.

Temperature dependence of the VV-relaxation probability $Q_{10}^{01} \approx (\Omega\tau_{col})^2$ is different for dipole and exchange interactions of colliding molecules. The transition frequency Ω is proportional to the average interaction energy, hence in the case of the short distance exchange interaction $\Omega \propto T_0$, $\tau_{col} \propto 1/v \propto T_0^{1/2}$, and therefore $Q_{10}^{01} \approx (\Omega\tau_{col})^2 \propto T_0$. In the case of long-distance dipole interactions the transition frequency Ω does not depend on temperature and therefore $Q_{10}^{01} \approx (\Omega\tau_{col})^2 \propto \tau_{col}^2 \propto 1/T_0$. In general, the resonant VV-exchange is much faster at low temperatures than VT-relaxation (compare Tables 3.7 and 3.8), leading to generation of highly vibrationally excited and therefore very reactive molecules in nonthermal discharges.

Table 3.8 Resonant VV-relaxation rate coefficients at room temperature and their ratio to relevant rate coefficients of gas-kinetic collisions (k_{VV}/k_0).

Molecule	k_{VV}, cm^3 s^{-1}	k_{VV}/k_0	Molecule	k_{VV}, cm^3 s^{-1}	k_{VV}/k_0
CO_2 (001)	$5*10^{-10}$	4	HF	$3*10^{-11}$	0.5
CO	$3*10^{-11}$	0.5	HCl	$2*10^{-11}$	0.2
N_2	10^{-13}–10^{-12}	10^{-3}–10^{-2}	HBr	10^{-11}	0.1
H_2	10^{-13}	10^{-3}	DF	$3*10^{-11}$	0.3
N_2O (001)	$3*10^{-10}$	3	HJ	$2*10^{-12}$	0.02

3.7.2 VV-relaxation of Anharmonic Oscillators

VV-relaxation of anharmonic oscillators is slightly nonresonant, slightly adiabatic, and therefore a little slower than the resonant one (more details in Nikitin 1974):

$$Q_{n+1,n}^{m,m+1} = (m+1)(n+1)Q_{10}^{01} \exp(-\delta_{VV}|n-m|) \; [2/3 - 1/2\exp(-\delta_{VV}|n-m|)]$$

$$\delta_{VV} = \frac{4}{3}x_e\gamma_0 = \frac{4}{3}\frac{\pi\omega x_e}{\alpha}\sqrt{\frac{\mu}{2T_0}}. \tag{3.41}$$

In one-component gases $\delta_{VV} = \delta_{VT}$, and numerically: $\delta_{VV} = \frac{0.427}{\alpha}\sqrt{\frac{\mu}{T_0}}\, x_e\hbar\omega$, where the reduced mass of colliding molecules μ is in atomic units, reverse intermolecular interaction radius α – in A^{-1}, translational gas temperature T_0 and vibrational quantum $\hbar\omega$ in degrees Kelvin.

It is interesting to compare the rate coefficients of VV-exchange and VT relaxation taking into account the anharmonicity. The effect of anharmonicity on the VV-exchange is negligible in resonant collisions, when $|m-n|\,\delta_{VV} \ll 1$, but become significant in collisions of highly excited molecules and molecules on low vibrational levels. The rate coefficient of VV-exchange process $k_{VV}(n) = k_0 Q_{n+1,n}^{0,1}$ decreases with growth of the vibrational quantum number "n"; in contrast, the VT-relaxation rate coefficient $k_{VT}(n) = k_0 P_{n+1,n}^{VT}$ increases with "n" ($k_0 \approx 10^{-10}\text{cm}^3\,\text{s}^{-1}$ is the rate coefficient of gas-kinetic collisions). VT-to-VV rate ratio:

$$\xi(n) = \frac{k_{VT}(n)}{k_{VV}(n)} \approx \frac{P_{10}^{VT}}{Q_{10}^{01}} \exp[(\delta_{VV} + \delta_{VT})n]. \tag{3.42}$$

is small ($\xi \ll 1$) at low excitation levels because $P_{10}^{VT}/Q_{10}^{01} \ll 1$. It means that the population of highly vibrationally excited states can increase in nonthermal plasma much faster than losses of vibrational energy from these levels. The ratio $\xi(n)$, however, is growing exponentially with "n" and the VT-relaxation catches up with the VV-exchange ($\xi = 1$) at some critical vibrational energy $E^*(T_0)$, reflecting a maximum level of effective vibrational excitation:

$$E^*(T_0) = \hbar\omega\left(\frac{1}{4x_e} - b\sqrt{T_0}\right). \tag{3.43}$$

The first term in this relation corresponds to the dissociation energy D of a diatomic molecule; parameter "b" depends on the gas: for CO $b = 0.90\,\text{K}^{-0.5}$, for N_2 $b = 0.90\,\text{K}^{-0.5}$, for HCl $b = 0.90\,\text{K}^{-0.5}$. More information on the subject can be found in Rusanov and Fridman (1984).

3.7.3 Intermolecular VV′-relaxation

Intermolecular VV′-relaxation can be considered first as a vibrational exchange in a mixture of diatomic molecules A and B with only slightly different quanta $\hbar\omega_A > \hbar\omega_B$. The adiabatic factors determine the small probability of the process when a molecule A transfers a quantum ($v_A + 1 \to v_A$) to a molecule B ($v_B + 1 \to v_B$):

$$Q_{v_A+1,v_A}^{v_B,v_B+1} = (v_A + 1)(v_B + 1)\, Q_{10}^{01}(AB)\; \exp(-|\delta_B v_B - \delta_A v_A + \delta_A p|)\; \exp(\delta_A p). \tag{3.44}$$

Here $Q_{10}^{01}(AB)$ is the probability of a quantum transfer from A to B for the lowest levels; parameters δ_A and δ_B can be found from (3.41), taking for each molecule a separate coefficient of anharmonicity: x_{eA} and x_{eB}. Parameter $p > 0$ is vibrational level of the oscillator A, corresponding to the resonant transition $A(p + 1 \to p) - B(0 \to 1)$ of a quantum from molecule A to B:

$$p = \frac{\hbar(\omega_A - \omega_B)}{2x_{eA}\hbar\omega_A}. \tag{3.45}$$

$Q_{10}^{01}(AB) \exp(\delta_A p)$ in (3.45) corresponds to the resonant exchange and does not include the adiabatic smallness, that is $Q_{10}^{01}(AB) \exp(\delta_A p) \approx (\Omega\tau_{col})^2$, where the factor $(\Omega\tau_{col})^2$ is about 10^{-2}–10^{-4}. Then:

$$Q_{v_A+1,v_A}^{v_B,v_B+1} = (v_A + 1)(v_B + 1)\,(\Omega\tau_{col})^2\; \exp(-|\delta_B v_B - \delta_A v_A + \delta_A p|), \tag{3.46}$$

For a nonresonant single-quantum transfer from a molecule A to B:

$$Q_{10}^{01}(AB) = (\Omega\tau_{col})^2 \exp(-\delta p), \tag{3.47}$$

Table 3.9 Rate coefficients of nonresonant VV'-relaxation, $T_0 = 300$ K.

VV'-exchange process	$k_{VV'}$, cm^3 s^{-1}
$O_2(v=0) + CO(v=1) \rightarrow O_2(v=1) + CO(v=0)$	$4 * 10^{-13}$
$N_2(v=0) + CO(v=1) \rightarrow N_2(v=1) + CO(v=0)$	10^{-15}
$NO(v=0) + CO(v=1) \rightarrow NO(v=1) + CO(v=0)$	$3 * 10^{-14}$
$CO(v=0) + H_2(v=1) \rightarrow CO(v=1) + H_2(v=0)$	10^{-16}
$NO(v=0) + CO(v=1) \rightarrow NO(v=1) + CO(v=0)$	10^{-14}
$H_2O(000) + N_2(v=1) \rightarrow H_2O(010) + N_2(v=0)$	10^{-15}
$CO(v=1) + CH_4 \rightarrow CO(v=1) + CH_4{}^*$	10^{-14}
$O_2(v=0) + CO_2(01^10) \rightarrow O_2(v=1) + CO_2(00^00)$	$3 * 10^{-15}$
$CO_2(00^00) + N_2(v=1) \rightarrow CO_2(00^01) + N_2(v=0)$	10^{-12}
$CO(v=1) + CF_4 \rightarrow CO(v=1) + CF_4{}^*$	$2 * 10^{-16}$
$CO(v=1) + SO_2 \rightarrow CO(v=1) + SO_2{}^*$	10^{-15}
$CO(v=1) + SF_6 \rightarrow CO(v=1) + SF_6{}^*$	10^{-15}
$CO(v=1) + CO_2 \rightarrow CO(v=1) + CO_2{}^*$	$3 * 10^{-13}$
$H_2O^* + CO_2 \rightarrow H_2O + CO_2{}^*$	10^{-12}
$CO(v=0) + CO_2(00^01) \rightarrow CO(v=1) + CO_2(00^00)$	$3 * 10^{-15}$
$O_2(v=0) + CO_2(10^00) \rightarrow O_2(v=1) + CO_2(00^00)$	10^{-13}

which can be numerically expressed as (see Fridman and Kennedy 2021):

$$Q_{10}^{01}(AB) = 3.7 {}^* 10^{-6} T_0 \ ch^{-2}(0.174 \Delta \omega \hbar / \sqrt{T_0}); \tag{3.48}$$

here the defect of resonance $\hbar\Delta\omega$ is expressed in cm^{-1}, and temperature T_0 in Kelvin. The probability of VV'-exchange decreases with defect of resonance $\hbar\Delta\omega$. However, when the defect of resonance is quite large $\hbar\Delta\omega \approx \hbar\omega$, different multi-quantum resonant VV'-exchange processes are become important (Nikitin 1974):

$$Q_{n,n-s}^{m,m+r} = \frac{1}{r!s!} \frac{n!(m+r)!}{(n-s)!m!} \ Q_{s0}^{0r}. \tag{3.49}$$

The probability of a multi-quantum resonant exchange $Q_{s0}^{0r} \propto (\Omega \tau_{col})^{r+s}$, and therefore:

$$Q_{n,n-s}^{m,m+r} = \frac{1}{r!s!} \frac{n!(m+r)!}{(n-s)!m!} \ (\Omega \tau_{col})^{r+s}. \tag{3.50}$$

Rate coefficients of some nonresonant VV'-exchange processes are presented in Table 3.9.

3.8 VT- and VV-relaxation of Highly Vibrationally Excited Polyatomic Molecules

Vibrational energy transfer in nonthermal CO$_2$ plasma can be a good illustration of such processes. Plasma electrons at $T_e = 1$–2 eV excite lower vibrational levels of CO$_2$, and predominantly the asymmetric mode of the vibrations. The VV-exchange leading to population of higher-excited states occurs at low vibrational levels of polyatomic molecules independently along the different modes (see previous Sections 3.7 of this chapter). As the level of excitation of polyatomic molecules increases, the vibrations of different types are able to mix collisionlessly due to the inter-mode anharmonicity and the Coriolis interaction. This intramolecular VV' exchange results in the **vibrational quasi-continuum** of the highly excited vibrational states of polyatomic molecules, which significantly influences the vibrational energy transfer processes in plasma.

3.8.1 Shchuryak Model of Transition to the Vibrational Quasi-continuum

Shchuryak model of transition to the vibrational quasi-continuum is based on the effect of beating. In the case of CO_2, energy on of the asymmetric mode (frequency ω_3, internal anharmonicity x_{33}) changes during the beating in the interaction with symmetric modes ($\nu_3 \rightarrow \nu_1 + \nu_2$), and as a result the effective frequency of this oscillation mode changes as well:

$$\Delta\omega_3 \approx (x_{33}\omega_3)^{1/3}(A_0\omega_3 n_{sym})^{2/3}; \tag{3.51}$$

here $A_0 \approx 0.03$ is a dimensionless characteristic of the interaction between modes ($\nu_3 \rightarrow \nu_1 + \nu_2$), and n_{sym} is the total number of quanta on the symmetric modes (ν_1 and ν_2), assuming quasi-equilibrium between modes inside of a CO_2 molecule. As the level of excitation n_{sym} increases, the value of $\Delta\omega_3$ grows and at a certain critical number of quanta n_{sym}^{cr}, it covers the defect of resonance $\Delta\omega$ of the asymmetric-to-symmetric quantum transition:

$$n_{sym}^{cr} \approx \frac{1}{\sqrt{x_{33}A_0^2}}\left(\frac{\Delta\omega}{\omega_3}\right)^{3/2}. \tag{3.52}$$

When the number of quanta exceeds the critical value (3.52), which corresponds to the **Chirikov stochasticity criterion**, the motion becomes quasi-random and the modes become mixed in the vibrational quasi-continuum. Generally for polyatomic molecules, the critical excitation level n^{cr} decreases with the number N of vibrational modes: $n^{cr} \approx 100/N^3$. This means, that for molecules with four and more atoms, the transition to the vibrational quasi-continuum takes place at a quite low level of excitation.

The polyatomic molecules in the state of vibrational quasi-continuum can be considered as an infinitely high number of small oscillators. These oscillators can be characterized by the distribution function of the squares of their amplitudes $I(\omega)$, which is actually the vibrational Fourier spectrum of the system. $I(\omega)\hbar d\omega = \frac{dE}{\omega}$ is an adiabatic invariant (shortened action) for oscillators with energy dE in the frequency range from ω to $\omega + \Delta\omega$ and, according to the correspondence principle, is an analogue of the number of quanta for these oscillators. If a polyatomic molecule is considered as a set of harmonic oscillators with vibrational frequencies ω_{0i} and quantum numbers n_i, then the vibrational Fourier spectrum $I(\omega)$ can be presented as a sum of δ-functions:

$$I(\omega) = \sum_i n_i\delta(\omega - \omega_{0i}). \tag{3.53}$$

Anharmonicity and interaction between modes cause the vibrational spectrum $I(\omega)$ to differ from (3.53) in two ways. First, the anharmonicity can be considered as a shift of the fundamental vibration frequencies:

$$\omega_i(n_i) = \omega_{0i} - 2\omega_{0i}x_{0i}n_i \tag{3.54}$$

$$x_{0i} = \frac{1}{2}\sum_{j=1}^{j=N}x_{ij}(1 + \delta_{ij})\frac{n_j}{n_i}, \tag{3.55}$$

where x_{ij} is the anharmonicity coefficients, N is the number of vibrational modes, and δ_{ij} is the Kronecker δ-symbol. Second, the energy exchange between modes with characteristic frequency δ_i leads to a broadening of the vibrational spectrum lines, which can be described by the Lorentz profile:

$$I(\omega) = \frac{1}{\pi}\sum_i \frac{n_i\delta_i}{[\omega - \omega_i(n_i)]^2 + \delta_i^2}. \tag{3.56}$$

When the third-order resonances prevail in the inter-mode exchange of vibrational energy, the frequency δ_i is growing with vibrational energy as $\delta_i \sim E^{3/2}$. The frequency δ_i can be estimated for CO_2 as the inter-mode anharmonicity $\delta \approx x_{23}n_3 n_{sym}/\hbar$. With the growing number of atoms in a molecule, the vibrational energy of transition to the quasi-continuum decreases, and the line width at a fixed energy increases. Also, energy exchange frequency δ (at least for molecules like CO_2 and not the highest excitation levels) is less than difference in fundamental frequencies. It means, that in spite of rapid mixing, the modes in vibrational quasi-continuum still keep their individuality.

3.8.2 VT-relaxation of Polyatomic Molecules in Vibrational Quasi-continuum

VT-relaxation of polyatomic molecules in vibrational quasi-continuum can be considered in terms of mean square of the vibrational energy transferred to translational degrees of freedom $< \Delta E_{VT}^2 >$ and presented by averaging of the exponential Landau–Teller factors over the vibrational Fourier spectrum of the system $I(\omega)$:

$$< \Delta E_{VT}^2 > = \int_0^\infty (\Omega\tau_{col})_{VT}^2(\hbar\omega)^2 I(\omega)\exp\left(-\frac{\omega}{\alpha v}\right)d\omega \tag{3.57}$$

The factor $(\Omega\tau_{col})^2_{VT} \approx 0.01$ characterizes the smallness of the probability of transition due to the smallness of the amplitude of vibrations relative to the interaction radius. Assuming the Lorentz profile (3.56) for the vibrational Fourier spectrum $I(\omega)$, integration (3.57) gives:

$$< E^2_{VT} >= (\Omega\tau_{col})^2_{VT} n \left[\frac{\alpha v}{\alpha v + \delta} \exp\left(-\frac{\omega_n - \delta}{\alpha v} \right) + \frac{\delta}{3\pi\omega_n} \left(\frac{\alpha v}{\omega_n} \right)^3 \right] (\hbar\omega_n)^2, \tag{3.58}$$

where δ is the inter-mode vibrational energy exchange frequency, factor "n" is the number of quanta on the mode. The VT-relaxation of polyatomic molecules in quasi-continuum is determined by two effects: an adiabatic effect and a quasi-resonant effect First term in (3.58) is the adiabatic effect and is somewhat similar to the case of diatomic molecules. It is, however, growing faster with "n" because of the effective reduction of the vibrational frequency and the Massey parameter $\frac{\omega_n - \delta(n)}{\alpha v}$ due to broadening of the given mode line in the vibrational spectrum $I(\omega)$. Second term in (3.58) corresponds to quasi-resonance (nonadiabatic) relaxation of polyatomic molecules at low frequencies $\omega \approx \alpha v$. Comparison of these two relaxation effects shows, that the quasi-resonant VT-relaxation has no exponentially small factor and can exceed the adiabatic relaxation.

 Thus VT-relaxation of polyatomic molecules becomes nonadiabatic and very fast when high levels of their vibrational excitation lead to transition to quasi-continuum of vibrational spectrum. Also, polyatomic molecules have specific effect of rotations on vibrational relaxation, the so-called VRT-relaxation process. Degenerated vibrational modes can have "circular" polarization and angular momentum of the quasi-rotations, which leads to fast nonadiabatic VT relaxation through intermediate rotations of polyatomic molecules.

3.8.3 VV-exchange of Polyatomic Molecules in Vibrational Quasi-continuum

 VV-Exchange of Polyatomic Molecules in vibrational quasi-continuum can be presented similarly to (3.57) in terms of mean square of the transferred vibrational energy:

$$< \Delta E^2_{VV} >_{12} = \iint\limits_{0,\infty} (\Omega\tau_{col})^2_{VV} I_1(\omega_1) I_2(\omega_2) \exp\left(-\frac{|\omega_1 - \omega_2|}{\alpha v} \right) (\hbar\omega_1)^2 d\omega_1 \, d\omega_2, \tag{3.59}$$

where indices 1 and 2 correspond to quantum transferring and accepting molecules. For VV-exchange between a molecule from the discrete spectrum (in the first excited state, frequency ω_0) and a molecule in the quasi-continuum:

$$< \Delta E^2_{VV} >= (\Omega\tau_{col})^2_{VV} (\hbar\omega_0)^2 \int_0^\infty \frac{1}{\pi} \frac{n\delta}{(\omega - \omega_n)^2 + \delta^2} \exp\left(-\frac{|\omega_0 - \omega|}{\alpha v} \right) d\omega, \tag{3.60}$$

where "n" and ω_n are the number of quanta and corresponding frequency. When one of the molecules is in quasi-continuum, the line width δ exceeds usually the anharmonic shift $|\omega_0 - \omega|$, and the VV-exchange is quasi-resonant:

$$< \Delta E^2_{VV} >= (\Omega\tau_{col})^2_{VV} n \frac{\alpha v}{\delta + \alpha v} (\hbar\omega_0)^2, \tag{3.61}$$

and the corresponding rate coefficients do not decrease with the excitation level. Similarly to (3.42), we can compare the rate coefficients of the VV-exchange (with a molecule in the first excited state) and VT relaxation:

$$\xi(n) = \frac{< \Delta E^2_{VT} > (n)}{< \Delta E^2_{VV} > (n)} \approx \frac{(\Omega\tau_{col})^2_{VT}}{(\Omega\tau_{col})^2_{VV}} \exp\left(-\frac{\omega_n - \delta(n)}{\alpha v} \right). \tag{3.62}$$

 The VV-exchange proceeds faster than the VT-relaxation $\xi(n) \ll 1$ at low levels of excitation (similarly to diatomic molecules). Inter-mode exchange frequency $\delta(n)$ is growing with "n" faster than $\Delta\omega = \omega_0 - \omega_n$. For this reason, $\xi(n)$ reaches unity and the VT-relaxation catches up with the VV-exchange at lower levels of vibrational excitation than in the case of diatomic molecules. Summarizing, it is much more difficult to sustain the high population of upper vibrational levels of polyatomic molecules in quasi-continuum, than those of diatomic molecules. More information on vibrational kinetics of polyatomic molecules in quasi-continuum can be found in Capitelli (1986) and Fridman and Kennedy (2021).

Table 3.10 Number of collisions Z_{rot} for RT-relaxation in some one-component gases.

Molecule	$Z_{rot}^\infty(T_0 \to \infty)$	$Z_{rot}(T_0 = 300\,K)$	Molecule	$Z_{rot}^\infty(T_0 \to \infty)$	$Z_{rot}(T_0 = 300\,K)$
Cl_2	47.1	4.9	N_2	15.7	4.0
O_2	14.4	3.45	H_2	—	~500
D_2	—	~200	CH_4	—	15
CD_4	—	12	$C(CH_3)_4$	—	7
SF_6	—	7	CCl_4	—	6
CF_4	—	6	SiH_4	—	28

3.9 Rotational Relaxation Processes, Parker Formula

Both rotational–rotational (RR) and rotational–translational (RT) energy transfer (relaxation) processes are usually nonadiabatic and fast, because rotational quanta and therefore Massey parameters are small. As a result, collision of a rotator with an atom or another rotator can be considered as a free, classical collision with high probability of RT and RR relaxation. Number of collisions required for RT relaxation follows the so-called **Parker formula**:

$$Z_{rot} = \frac{Z_{rot}^\infty}{1 + (\pi/2)^{3/2}\sqrt{\frac{T^*}{T_0}} + (2 + \pi^2/4)\,\frac{T^*}{T_0}}, \tag{3.63}$$

where T^* is energy of inter-molecular attraction; T_0 is gas temperature, $Z_{rot}^\infty = Z_{rot}(T_0 \to \infty)$. The increase of Z_{rot} with T_0 is due to inter-molecular attraction accelerating the RT energy exchange. The attraction effect becomes less significant at higher temperatures. Numerical values of Z_{rot} are presented in Table 3.10 The factors Z_{rot} are usually small (3.10) except hydrogen and deuterium, where Z_{rot} is approximately 200–500 due to higher values of rotational quanta and Massey parameters. The rotational quanta $2B(J+1)$ grow with the level of rotational excitation, therefore RT-relaxation of the highly rotationally excited molecules becomes more adiabatic and slower. Generally, RT relaxation (and furthermore RR relaxation) are fast processes with rates comparable to thermalization. Thermalization (TT-relaxation) sustains equilibrium inside of the translational degrees of freedom. Therefore, rotation can be mostly considered in quasi-equilibrium with the translational degrees of freedom and characterized by the same temperature T_0.

3.10 Relaxation of Electronically Excited Atoms and Molecules

3.10.1 Relaxation of Electronic Excitation (ET Process)

Relaxation of electronic excitation (ET process) is usually related to transfer of significant energy (several eV) into translational degrees of freedom, and therefore is a strongly adiabatic process with very high Massey parameters ($\omega\tau_{col} \sim 100$–1000). In this case, the relaxation is very slow. For example, the relaxation $Na(3^2P) + Ar \to Na + Ar$ has probability 10^{-9} or less. Sometimes, however, the ET relaxation can be quite fast as in the case of relaxation of electronically excited oxygen $O(^1D)$ on heavy atoms of noble gases requiring only several collisions because of formation of an intermediate quasi-molecule. Electronically excited atoms and molecules can also transfer energy into vibrational and rotational degrees of freedom. These processes can proceed with higher probability because of lower energy losses from the internal degrees of freedom and therefore smaller Massey parameters. Even faster relaxation occurs due to formation of intermediate ionic complexes. For example, the relaxation probability is close to unity for electronic energy transfer from excited metal atoms Me* to vibrational excitation of nitrogen molecules $Me^* + N_2(v=0) \to Me^+N_2^- \to Me + N_2(v>0)$. Maximum cross sections of such fast processes of relaxation of excited sodium atoms are presented in Table 3.11.

Table 3.11 Cross sections of nonadiabatic relaxation of electronically excited sodium.

Relaxation processes	Cross sections
$Na^* + N_2(v = 0) \rightarrow Na + N_2^*(v > 0)$	$2^* 10^{-14}$ cm^2
$Na^* + CO_2(v = 0) \rightarrow Na + CO_2^*(v > 0)$	10^{-14} cm^2
$Na^* + Br_2(v = 0) \rightarrow Na + Br_2^*(v > 0)$	10^{-13} cm^2

3.10.2 The Electronic Excitation Energy Transfer Processes

The Electronic Excitation Energy Transfer Processes can be effective only very close to resonance, within about 0.1 eV or less. The He–Ne gas laser provides practically important examples of such kind of highly resonant processes:

$$He(^1S) + Ne \rightarrow He + Ne(5s), \tag{3.64}$$

$$He(^3S) + Ne \rightarrow He + Ne(4s). \tag{3.65}$$

Interaction radius here is large (up to 1 nm), and cross sections reach 10^{-14} cm^2. When the degree of ionization in plasma exceeds 10^{-6}, fast electronic excitation transfer can occur by interaction with the electron gas. Such excitation energy transfer, conventional for thermal plasmas, proceeds through deactivation of the excited neutrals in super-elastic collisions with electrons followed by excitation of other neutrals by electron impact.

3.11 Elementary Chemical Reactions of Excited Molecules

3.11.1 Arrhenius Formula for Excited Molecules

Arrhenius formula for excited molecules with vibrational energy E_v express acceleration of the reactions in terms of decrease of activation energy E_a (can be also used for other excitations, see Levine–Bernstein theoretical-informational approach):

$$k_R(E_v, T_0) = k_{R0} \exp\left(-\frac{E_a - \alpha E_v}{T_0}\right) \theta(E_a - \alpha E_v), \tag{3.66}$$

where coefficient "α" is the *efficiency of excitation energy in overcoming of the activation energy barrier*; $\theta(x - x_0)$ is the Heaviside function ($\theta(x - x_0) = 1$, when $x > 0$; and $\theta(x - x_0) = 0$, when $x < 0$). If the vibrational temperature exceeds the translational one $T_v \gg T_0$ and the chemical reaction is mostly determined by the excited molecules, then (3.66) can be simplified to $k_R(E_v) = k_{R0} \theta(\alpha v - E_a)$. The effective activation energy, in this case, is E_a/α. The pre-exponential factor can be calculated using the **transition state theory**, or simply taken as the gas-kinetic rate coefficient $k_{R0} \approx k_0$.

Equation (3.66) describes the rate coefficients of reactions of vibrationally excited molecules after averaging over the Maxwellian distribution function for translational degrees of freedom. Rate of these reactions without the averaging can be found using the **Le Roy formula**, giving the probability $P_v(E_v, E_t)$ as a function of vibrational E_v and translational E_t energies:

$$P_v(E_v, E_t) = 0, \quad \text{if} \quad E_t < E_a - \alpha E_v, \ E_a \geq \alpha E_v, \tag{3.67a}$$

$$P_v(E_v, E_t) = 1 - \frac{E_a - \alpha E_v}{E_t}, \quad \text{if} \quad E_t > E_a - \alpha E_v, \ E_a \geq \alpha E_v, \tag{3.67b}$$

$$P_v(E_v, E_t) = 1, \quad \text{if} \quad E_a < \alpha E_v, \quad \text{any} \ E_t. \tag{3.67c}$$

Averaging of $P_v(E_v, E_t)$ over the Maxwellian distribution over E_t obviously gives (3.66).

3.11.2 Activation Energy

Activation energy is still the major exponential factor determining the plasma-chemical kinetics of excited molecules. If experimental and quantum-chemical data are not sufficient, the following semi-empirical methods can be applied to find the required activation energies.

Polanyi–Semenov rule predicts activation energies for exothermic reactions as found as:

$$E_a = \beta + \alpha\,\Delta H, \tag{3.68}$$

where ΔH is negative enthalpy, $\alpha = 0.25$–0.27 and $\beta = 11.5\,\text{kcal mol}^{-1}$ for the following groups of exchange reactions: $H + RH \rightarrow H_2 + R$, $D + RH \rightarrow DH + R$, $H + RCHO \rightarrow H_2 + RCO$, $H + RCl \rightarrow HCl + R$, $H + RBr \rightarrow HBr + R$; $Na + RCl \rightarrow NaCl + R$, $Na + RBr \rightarrow NaBr + R$; $CH_3 + RH \rightarrow CH_4 + R$, $CH_3 + RCl \rightarrow CH_3Cl + R$, $CH_3 + RBr \rightarrow CH_3Br + R$; $OH + RH \rightarrow H_2O + R$. Relevant endothermic reactions with positive ΔH, can be considered as a reverse processes with the same coefficients α and β: $E_a = \Delta H + (\beta + \alpha(-\Delta H)) = \beta + (1 - \alpha)\Delta H$.

Kagija method can be applied for exchange reactions $A + BC \rightarrow AB + C$, based on the dissociation energy D of the molecule BC, the reaction enthalpy ΔH, and a factor "γ" (energies are in kcal mol^{-1}):

$$E_a = \frac{D}{(2\gamma D - 1)^2} + \frac{2\gamma D(2\gamma^2 D^2 - 3\gamma D + 2)}{(2\gamma D - 1)^3}\Delta H. \tag{3.69}$$

The Kagija parameter γ is fixed for the following groups of reactions: (i) for detachment of hydrogen atoms by alkyl radicals $\gamma = 0.019$; (ii) for detachment of hydrogen atoms by halogen atoms $\gamma = 0.025$; (iii) for detachment of halogen atoms from alkyl-halides the parameter γ grows from $\gamma = 0.019$ for chlorides to $\gamma = 0.03$ for iodides.

Sabo method determines the activation energy based upon the sum of bond energies of initial molecules $\sum D_i$ and the that of final molecules $\sum D_f$:

$$E_a = \sum D_i - a \sum D_f. \tag{3.70}$$

For exothermic substitution exchange reactions, the Sabo parameter $a = 0.83$; for endothermic substitution exchange reactions $a = 0.96$; for disproportioning $a = 0.60$ and for exchange reactions of inversion $a = 0.84$.

Alfassi–Benson method predicts activation energies of the exchange reactions $R + AR' \rightarrow AR + R'$ as a function of the sum A_e of electron affinities to the particles R and R' and the reaction enthalpy ΔH (all in kcal mol^{-1}):

$$E_a = \frac{14.8 - 3.64^* A_e}{1 - 0.025^* \Delta H}. \tag{3.71}$$

3.12 Efficiency of Vibrational Energy in Overcoming Activation Energy of Chemical Reactions, Fridman–Macheret α-formula

3.12.1 Efficiency α of Molecular Excitation Energy

Efficiency α of molecular excitation energy in overcoming activation barriers is the key parameter (see 3.65) determining high chemical reaction rates of excited molecules in plasma. Some experimental values of the α-coefficients are presented in Table 3.12. Extensive database on this subject can be found in Rusanov and Fridman (1984) and Fridman and Kennedy (2021), and permits classification of chemical reactions with regard to efficiency α, which is also presented at the end of Table 3.12.

3.12.2 Fridman–Macheret α-formula

Fridman–Macheret α-Formula permits to calculate the efficiency α of vibrational energy in overcoming activation barriers of chemical processes based on activation energies of the corresponding direct and reverse reactions (Macheret et al. 1984).

Fridman–Macheret model for the exchange reaction $A + BC \rightarrow AB + C$ is illustrated in Figure 3.8. Vibration of a molecule BC is taken into account here using the so-called vibronic terms shown by dash-dotted lines. The energy profile, corresponding to reaction of vibrationally excited molecule with energy E_v (the vibronic term), can be obtained by a parallel shift of the initial profile $A + BC$ up on the value of E_v. Part of the reaction profile corresponding to products $AB + C$ remains the same. Simple geometry shows here the effective decrease of activation energy due to vibrational excitation E_v as:

Table 3.12 Efficiency of vibrational energy in overcoming activation energy: α_{exp} – experimental, α_{MF} – from the Fridman–Macheret α-formula, α-classification in the end.

Reaction	α_{exp}	α_{MF}	Reaction	α_{exp}	α_{MF}
$F + HF^* \rightarrow F_2 + H$	0.98	0.98	$NO + HCl^* \rightarrow NOCl + H$	1.0	0.98
$O_2 + OH^* \rightarrow O_3 + H$	1.0	1.0	$NO + OH^* \rightarrow NO_2 + H$	1.0	1.0
$H + HF^* \rightarrow H_2 + F$	1.0	0.95	$HCO + HF^* \rightarrow H_2CO + F$	1.0	0.99
$OH + HF^* \rightarrow H_2O + F$	1.0	0.95	$HS + HF^* \rightarrow H_2S + F$	1.0	0.98
$HO_2 + HF^* \rightarrow H_2O_2 + F$	1.0	0.98	$NH_2 + HF^* \rightarrow NH_3 + F$	1.0	0.97
$S + CO^* \rightarrow CS + O$	1.0	0.99	$F + CO^* \rightarrow CF + O$	1.0	1.0
$SO + CO^* \rightarrow COS + O$	0.96	0.92	$F_2 + CO^* \rightarrow CF_2 + O$	1.0	1.0
$C_2H_4 + CO^* \rightarrow (CH_2)_2 C + O$	0.94	0.98	$CH_2 + CO^* \rightarrow C_2H_2 + O$	0.90	0.94
$CH_3 + OH^* \rightarrow CH_4 + O(^1D_2)$	1.0	1.0	$OH + OH^* \rightarrow H_2O + O(^1D_2)$	1.0	0.97
$O + NO^* \rightarrow O_2 + N$	0.94	0.86	$CO^* + HF \rightarrow CHF + O$	1.0	1.0
$H_2CO + HF^* \rightarrow CH_3F + O(^1D_2)$	1.0	1.0	$O + N_2 \rightarrow NO + N$ ad(non-ad)	0.6(1)	1(1)
$H + HCl^* \rightarrow H_2 + Cl$	0.3	0.4	$NO + O_2^* \rightarrow NO_2 + O$	0.9	1.0
$O + H_2^* \rightarrow OH + H$	0.3	0.5	$O + HCl^* \rightarrow OH + Cl$	0.60	0.54
$H + HCl^* \rightarrow HCl + H$	0.3	0.5	$H + H_2^* \rightarrow H_2 + H$	0.4	0.5
$NO + O_3^* \rightarrow NO_2(^3B_2) + O_2$	0.5	0.3	$SO + O_3^* \rightarrow SO_2(^1B_1) + O_2$	0.25	0.15
$OH + H_2^* \rightarrow H_2O + H$	0.24	0.22	$N + O_2^* \rightarrow NO + O$	0.24	0.19
$H + N_2O^* \rightarrow OH + N_2$	0.4	0.2	$O_3 + OH^* \rightarrow O_2 + O + OH$	0.02	0
$H_2 + OH^* \rightarrow H_2O + H$	0.03	0	$Cl_2 + NO^* \rightarrow ClNO + Cl$	0	0
$O_3 + NO^* \rightarrow NO_2(^2A_1) + O_2$	0	0	$O_3 + OH^* \rightarrow HO_2 + O_2$	0.02	0
Reaction type	Simple exchange Bond break in excited molecule Through complex	Simple exchange Bond break in not-excited molecule Direct	Simple exchange Bond break in not-excited molecule Direct	Double exch.	
Endothermic	0.9–1.0	0.8	<0.04	0.5–0.9	
Exothermic	0.2–0.4	0.2	0	0.1–0.3	
Thermoneutral	0.3–0.6	0.3	0	0.3–0.5	

$$\Delta E_a = E_v \frac{F_{A+BC}}{F_{A+BC} + F_{AB+C}}, \tag{3.72}$$

where F_{A+BC} and F_{AB+C} are slopes of the terms $A + BC$ and $AB + C$. If the energy terms exponentially depend on reaction coordinates with the decreasing parameters γ_1 and γ_2 (reverse radii of the corresponding exchange forces), then:

$$\frac{F_{A+BC}}{F_{AB+C}} = \frac{\gamma_1 E_a^{(1)}}{\gamma_2 E_a^{(2)}}, \tag{3.73}$$

The decrease of activation energy (3.72) together with (3.73) not only explains the main kinetic relation for reactions of excited molecules (3.66) but also determines the value of the coefficient α (which is actually $\alpha = \Delta E_a/E_v$):

$$\alpha = \frac{\gamma_1 E_a^{(1)}}{\gamma_1 E_a^{(1)} + \gamma_2 E_a^{(2)}} = \frac{E_a^{(1)}}{E_a^{(1)} + \gamma_1/\gamma_2 \; E_a^{(2)}}. \tag{3.74}$$

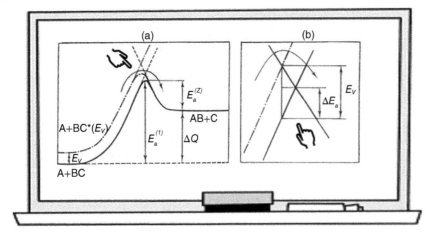

Figure 3.8 Fridman-Macheret model, efficiency of vibrational energy in exchange reaction $A + BC \rightarrow AB + C$: (a) solid curve – reaction profile; dashed line – a vibronic term, corresponding to A interaction with a vibration excited molecule $BC^*(E_v)$; (b) part of the reaction profile near the barrier summit (shown with a pointer).

The above geometric derivation is similar to that of Polanyi–Semenov rule (see Section 3.11.2). The exchange force parameters for direct and reverse reactions γ_1 and γ_2 are usually close ($\gamma_1/\gamma_2 \approx 1$). It results in a very convenient **Fridman–Macheret α-formula**:

$$\alpha \approx \frac{E_a^{(1)}}{E_a^{(1)} + E_a^{(2)}}, \tag{3.75}$$

which is in good numerical agreement with the experimental data (see Table 3.12) and reflects the three most important tendencies of the α-coefficient:

- efficiency α of the vibrational energy is the highest, close to 100%, for strongly endothermic reactions with activation energies close to the reaction enthalpy;
- efficiency α of the vibrational energy is the lowest, close to zero, for exothermic reactions without activation energies;
- sum of the α-coefficients for direct and reverse reactions is equal to unity $\alpha^{(1)} + \alpha^{(2)} = 1$.

The Fridman-Macheret formula does not include any detailed information on the dynamics of elementary reaction and type of excitation. It can be applied, therefore to wide range of chemical reactions of excited species in plasma.

3.12.3 Reactions Proceeding Through Intermediate Complexes

Reactions proceeding through intermediate complexes include two stages. The first one is formation of the complexes, which is exothermic and not so much sensitive to molecular excitation in plasma. The second stage is the complex dissociation into products with probability:

$$P_{st} \propto \left(\frac{E - E_a}{E}\right)^{s-1}, \tag{3.76}$$

following the **statistical theory of chemical reactions**. This statistical probability is determined by number "s" of degrees of freedom and total energy E of the intermediate complexes effective in overcoming activation energy E_a, and therefore is very much sensitive to the molecular excitation in plasma. Extensive information with specific examples on the subject can be found, in particular, in Nikitin (1974), Rusanov and Fridman (1984), and Fridman and Kennedy (2021). Here, we only point out information on contribution of different degrees of freedom to the total effective energy E of the complex:

$$E = \alpha_v E_v + \alpha_r E_r + \alpha_t E_t. \tag{3.77}$$

Efficiency α_v of vibrational energy E_v in dissociation is close to 100% in accordance with the statistical theory of monomolecular reactions. However, the efficiencies α_r and α_t of the rotational E_r and translational E_t energies in overcoming the activation energy barrier are less (which exponentially decreases their contribution to the reaction rate):

$$\alpha_r \approx \alpha_t \approx 1 - \frac{r_e^2}{R_a^2}, \tag{3.78}$$

mostly due to conservation of angular momentum during the reaction. Portion of the rotational and translational energies remains in rotation at the moment of the complex decay when distance between products is R_a (r_e is characteristic size of the initial molecule).

3.12.4 Chemical Reactions of Two Vibrationally Excited Molecules

Chemical Reactions of Two Vibrationally Excited Molecules can be considered using as an example the disproportioning of CO (enthalpy 5.5 eV, activation energy 6 eV):

$$CO^*(v_1) + CO^*(v_2) \rightarrow CO_2 + C \tag{3.79}$$

General Eq. (3.66) can be applied to the reaction of two excited molecules. In this case, however, both molecules contribute their energy to the decrease the activation energy:

$$\Delta E_a = \alpha_1 E_{v1} + \alpha_2 E_{v2}. \tag{3.80}$$

Here, subscript 1 corresponds to the donating molecule, and subscript 2 is related to another one accepting the atom. It was shown using the vibronic term approach that $\alpha_1 = 1$ and $\alpha_2 \approx 0.2$, meaning that vibrational energy of the donating molecule is more efficient in the stimulation of endothermic exchange reactions (Nester et al. 1983).

3.13 Nonequilibrium Dissociation of Molecules with Essential Contribution of Vibrational and Translational Energy

The nonequilibrium factor Z describes dissociation kinetics, when both vibrational T_v and translational T_0 temperatures make significant contribution into stimulation of dissociation of molecules in plasma or after shock waves (Losev et al. 1996):

$$Z(T_0, T_V) = \frac{k_R(T_0, T_v)}{k^0(T_0, T_0 = T_v)} = Z_h(T_0, T_v) + Z_l(T_0, T_v). \tag{3.81}$$

In this formula $k_R(T_0, T_v)$ is the dissociation rate coefficient; $k_R^0(T_0, T_v = T_0)$ is the corresponding equilibrium rate coefficient for temperature T_0; the nonequilibrium factors Z_h and Z_l are related to dissociation from high "h" and low "l" vibrational levels. They can be expressed using the **Macheret–Fridman model** (Macheret et al. 1994), based on the assumption of classical impulsive collisions. Dissociation occurs mainly through an optimum collisional configuration minimizing the energy barrier. If D is the dissociation energy and L is the pre-exponential factor related to configuration of collisions, the Macheret–Fridman model gives:

$$Z_h(T_0, T_v) = \left\{ \frac{1 - \exp\left(-\frac{\hbar\omega}{T_v}\right)}{1 - \exp\left(-\frac{\hbar\omega}{T_0}\right)} (1 - L) \right\} \exp\left[-D\left(\frac{1}{T_v} - \frac{1}{T_0}\right)\right], \tag{3.82}$$

$$Z_l(T_0, T_v) = L \exp\left[-D\left(\frac{1}{T_{eff}} - \frac{1}{T_0}\right)\right]. \tag{3.83}$$

The Macheret–Fridman model introduces the effective temperature combining contribution of vibrational and translational energies into collisional dissociation of molecules:

$$T_{eff} = \alpha T_v + (1 - \alpha) T_0, \quad \alpha = \left(\frac{m}{M + m}\right)^2, \tag{3.84}$$

where m is mass of an atom in dissociating molecule, and M is mass of an atom in another colliding partner. Experimental and theoretical values of the nonequilibrium factor Z are in a good agreement and discussed in Losev et al. (1996), see also Fridman and Kennedy (2021). In equilibrium $T_v = T_0$, and the effective temperature surely is $T_{eff} = T_v = T_0$. Several semi-empirical models well correlated with the Macheret–Fridman model were proposed to describe the nonequilibrium dissociation, especially for the shock wave conditions ($T_0 > T_v$):

Park Model (suggested by Park in 1987) applies the Arrhenius formula for reaction rates simply replacing the temperature by a new effective value:

$$T_{eff} = T_0^s T_v^{1-s}, \tag{3.85}$$

here the power "s" is a fitting parameter recommended to be taken as $s = 0.7$ The Macheret–Fridman model permits to find theoretically the Park fitting parameter as:

$$s = 1 - \alpha = 1 - \left(\frac{m}{m+M}\right)^2. \tag{3.86}$$

If two atomic masses are not very different $m \sim M$, then the parameter $s \approx 0.75$ according to (3.86), which is pretty close to the value $s = 0.7$, recommended by Park.

Losev Model (suggested by Losev in 1961) applies the Arrhenius formula with vibrational temperature T_v and effective value D_{eff} of dissociation energy:

$$D_{eff} = D - \beta T_0. \tag{3.87}$$

The coefficient $\beta \approx 1.0$–1.5 is a fitting parameter for a narrow range of temperatures. The Macheret–Fridman model predicts the Losev parameter as:

$$\beta \approx \frac{D(T_0 - T_v)}{T_0^2}, \quad T_0 > T_v. \tag{3.88}$$

Marrone–Treanor Model (suggested in 1963) assumes exponential distribution (with the parameter U) of the probabilities of dissociation from different vibrational levels (If $U \to \infty$, these probabilities are equal. The nonequilibrium factor Z can be then expressed as:

$$Z(T_0, T_v) = \frac{Q(T_0)Q(T_f)}{Q(T_v)Q(-U)}, \quad T_0 > T_v, \tag{3.89}$$

where all the statistical factors are $Q(T_j) = \dfrac{1 - \exp\left(-\frac{D}{T_j}\right)}{1 - \exp\left(-\frac{\hbar\omega}{T_j}\right)}$, $T_f = \dfrac{1}{\frac{1}{T_v} - \frac{1}{T_0} - \frac{1}{-U}}$, and the negative temperature $-U$ is a fitting parameter of the model with recommended $U = 0.6\,T_0 - 3T_0$. The Macheret–Fridman model predicts this parameter as

$$U = \frac{T_0[\alpha T_0 + (1-\alpha)T_v]}{(T_0 - T_v)(1-\alpha)}. \tag{3.90}$$

3.14 Problems and Concept Questions

3.14.1 Multi-quantum Vibrational Excitation by Electron Impact

Based on the Fridman approximation (3.4), derive formula for the total rate coefficient of the multi-quantum vibrational excitation from an initial state v_1 to all other vibrational levels $v_2 > v_1$.

3.14.2 Influence of Vibrational Temperature on Electronic Excitation Rate Coefficients

Calculate the relative increase of electronic excitation rate coefficient for molecular nitrogen in a nonthermal discharge with reduced electric field $E/n_0 = 3*10^{-16}\,\mathrm{V\,cm^2}$, when the vibrational temperature increases from room temperature to $T_v = 3000$ K, use relation (3.14).

3.14.3 Dissociation of Molecules Through Electronic Excitation by Direct Electron Impact

Why electron energy threshold of the dissociation through electronic excitation often exceeds the dissociation energy in contrast to dissociation through vibrational excitation, where the threshold is usually equal to the dissociation energy? Use diatomic molecules as an example.

3.14.4 Distribution of Electron Energy Between Different Channels of Excitation and Ionization

Analyzing Figures 3.4–3.6, identify the molecular gases with the highest possibility to localize selectively discharge energy on vibrational excitation. What is the highest percentage of electron energy, which can be selectively focused on vibrational excitation? Which reduced electric field or electron temperatures are optimal to reach this selective regime?

3.14.5 VT-relaxation Rate Coefficient as a Function of Vibrational Quantum Number

VT-relaxation rate increases with number of quanta for two reasons: (i) the matrix element increases as $(n+1)$, and (ii) vibrational frequency decreases because of anharmonicity. Which effect dominates the acceleration of VT-relaxation with level of vibrational excitation?

3.14.6 The Resonant Multi-quantum VV-exchange

Estimate the smallness of the double-quantum resonance VV-exchange $A_2^*(v=2) + A_2(v=0) \rightarrow A_2(v=0) + A_2^*(v=2)$ and triple-quantum resonance VV-exchange $A_2^*(v=3) + A_2(v=0) \rightarrow A_2(v=0) + A_2^*(v=3)$ in diatomic molecules, taking the matrix element factor $(\Omega\tau_{col})^2 = 10^{-3}$, see Eq. (3.40).

3.14.7 VV-relaxation of Polyatomic Molecules

Why excitation of high vibrational levels of polyatomic molecules in nonthermal discharges is less effective than in diatomic molecules?

3.14.8 Transition to the Vibrational Quasi-continuum

Calculate based on Eq. (3.52), at which level of vibrational energy CO_2 modes cannot be considered any more individually?

3.14.9 Rotational RT-relaxation

Using the Parker formula and data from Table 3.10, calculate the energy depth of the intermolecular attraction potential T^* and probability of the RT-relaxation in pure molecular nitrogen at translational gas temperature 400 K.

3.14.10 LeRoy Formula and α-model

Derive the general Eq. (3.66), describing stimulation of chemical reactions by vibrational excitation, averaging the LeRoy formula over Maxwellian distribution for translational energies E_t.

3.14.11 Contribution of Translational Energy in Dissociation of Molecules Under Nonequilibrium Conditions

Using expression for the nonequilibrium factor Z of the Fridman-Macheret model, derive the formula for dissociation of molecules at translational temperature much exceeding the vibrational one.

Lecture 4

Physical Kinetics and Transfer Processes of Charged Particles in Plasma

4.1 Boltzmann Kinetic Equation for Distribution Functions of Charged Particles: Vlasov Equation, Collisional Integral, Quasi-equilibrium Maxwellian EEDF

 The exceptional significance of kinetic distribution functions in plasma, and especially *electron energy distribution functions (EEDF),* has been discussed in Section 2.1. These distributions determine fraction of the highly energetic charged species, which often make major contribution in kinetics of plasma processes. Because of the exponential nature of the kinetic distribution functions (see Sections 2.1 and 2.2), their contribution to kinetics of plasma processes is often even more important than characteristics of the individual elementary reactions.

Evolution of a distribution function $f(\vec{r}, \vec{v}, t)$ in the six-dimensional **phase space** $(\vec{r}, \vec{v})$ of particle positions and velocities is described by the Boltzmann equation. Quantum numbers related to internal degrees of freedom are not included here because of primary interest in the plasma electron distributions. In the absence of collisions between particles:

$$\frac{df}{dt} = \frac{f(\vec{r} + \vec{dr}, \vec{v} + \vec{dv}, t + dt) - f(\vec{r}, \vec{v}, t)}{dt}. \tag{4.1}$$

Number of particles in a given state is fixed: $\frac{df}{dt} = 0$, $\frac{\vec{dv}}{dt} = \frac{\vec{F}}{m}$ and $\frac{\vec{dr}}{dt} = \vec{v}$ in the collisionless case; here m is the particle mass and $\vec{F}$ is an external force; then:

$$\frac{df}{dt} = \frac{\partial f}{\partial t} + \vec{v} \frac{\partial f}{\partial \vec{r}} + \frac{\vec{F}}{m} \frac{\partial f}{\partial \vec{v}} = 0. \tag{4.2}$$

The external electric E and magnetic B forces for electrons can be introduced in this relation as: $\vec{F} = -e(\vec{E} + \vec{v} \times \vec{B})$. It brings us to the **collisionless Vlasov equation** for electrons:

$$\frac{\partial f}{\partial t} + \vec{v} * \nabla_r f - \frac{e}{m}(\vec{E} + \vec{v} \times \vec{B}) * \nabla_v f = 0, \tag{4.3}$$

where operators ∇_r and ∇_v denote the electron distribution function gradients related to space and velocity coordinates. Binary collisions over very short time, velocities can be changed practically "instantaneously," and $\frac{df}{dt}$ is no longer zero. The corresponding evolution of the distribution function can be described by adding the special term $I_{col}(f)$, the so-called **collisional integral**:

$$\frac{df}{dt} = \frac{\partial f}{\partial t} + \vec{v} * \nabla_r f + \frac{\vec{F}}{m} * \nabla_v f = I_{col}(f). \tag{4.4}$$

This is the well-known **Boltzmann kinetic equation**, which is used the most to describe evolution of distribution functions.

Plasma Science and Technology: Lectures in Physics, Chemistry, Biology, and Engineering, First Edition. Alexander Fridman.
© 2024 WILEY-VCH GmbH. Published 2024 by WILEY-VCH GmbH.

The **collisional integral** of the Boltzmann kinetic equation is a nonlinear function of $f(\vec{v})$:

$$I_{col}(f) = \int \left[f(v')f(v_1') - f(v)f(v_1) \right] |v - v_1| \frac{1}{\varepsilon} \, d\omega \, dv_1, \tag{4.5}$$

where v_1 and v are velocities before collision, v_1' and v' those after collision, ε is the mean free path ratio to the characteristic system size, and $d\omega$ is area element in the plane perpendicular to the $(v_1 - v)$ vector. In equilibrium, the collisional integral (4.5) should be equal to zero:

$$f(v')f(v_1') = f(v)f(v_1) \quad \text{or} \quad \ln f(v') + \ln f(v_1') = \ln f(v) + \ln f(v_1), \tag{4.6}$$

which corresponds to the conservation of kinetic energy during the elastic collision:

$$\frac{m(v')^2}{2} + \frac{m(v_1')^2}{2} = \frac{mv^2}{2} + \frac{mv_1^2}{2}, \tag{4.7}$$

and explains the proportionality $\ln f \propto \frac{mv^2}{2}$. It results in **the quasi-equilibrium Maxwell distribution**:

$$f_0(\vec{v}) = B \exp\left(-\frac{mv^2}{2T} \right), \quad f_0(\vec{v}) = \left(\frac{m}{2\pi T} \right)^{3/2} \exp\left(-\frac{mv^2}{2T} \right), \tag{4.8}$$

surely corresponding to the Maxwellian EEDF for electrons discussed in Section 2.1.

The nonequilibrium ideal gas and therefore **plasma electrons entropy** is $S = \int f \ln \frac{e}{f} \vec{dv}$. Taking into account the Boltzmann equation, kinetic time evolution of the entropy can be expressed as:

$$\frac{dS}{dt} = -\int \ln f * I_{col}(f) \, \vec{dv}. \tag{4.9}$$

In equilibrium, when the collisional integral $I_{col}(f)$ (4.5) equals to zero, the entropy reaches its maximum value, which is known as the **Boltzmann H-Theorem.**

The crucial part of the Boltzmann equation is the collisional integral (4.5), which is quite complicated. Convenient simplification of the collisional integral is the so-called τ-**approximation**:

$$I_{col}(f) = -\frac{f - f_0}{\tau}, \tag{4.10}$$

based on the fact, that the collisional integral should be equal to zero, when the distribution function is the Maxwellian f_0. The characteristic time of evolution of the equilibrium distribution is the collisional time τ, which is the reverse collisional frequency (2.4). The Boltzmann equation in the frameworks of the τ-approximation becomes simple:

$$\frac{\partial f}{\partial t} = -\frac{f - f_0}{\tau}, \tag{4.11}$$

and illustrates how the initial distribution $f(\vec{v}, 0)$ exponentially approaches the Maxwellian one:

$$f(\vec{v}, t) = f_0 + [f(\vec{v}, 0) - f_0] \exp\left(-\frac{t}{\tau} \right); \tag{4.12}$$

Details on the important subject of the Boltzmann kinetic equation, approximations of the collisional integral, and solutions of the Boltzmann equation can be found in particular in a book of Landau and Lifshitz (2002).

4.2 Microscopic Consequences of the Boltzmann Kinetic Equation: Continuity, Momentum, and Energy Conservation Equations

4.2.1 The Continuity Equation

The continuity equation is the first microscopic consequence of the Boltzmann kinetic equation. Integrating the Boltzmann equation (4.4) over the particle velocity, the right-hand side represents the rate of changing of particle density, "$G - L$" (generation rate minus rate of losses):

$$\int \vec{dv} \frac{\partial f}{\partial t} + \int \vec{dv} * \left(\vec{v} \frac{\partial f}{\partial \vec{r}} \right) + \frac{\vec{F}}{m} \int \frac{\partial f}{\partial \vec{v}} \vec{dv} = G - L. \tag{4.13}$$

The distribution function f is normalized here on the number of particles $\int f \, d\vec{v} = n$, therefore the first term in (4.13) corresponds to $\frac{\partial n}{\partial t}$. The particles flux is determined by $\int \vec{v} f \, d\vec{v} = n\vec{u}$, where $\vec{u}$ is the particles mean velocity normally referred to as the **drift velocity.** Then the second term in (4.13) can be presented as $\frac{\partial}{\partial \vec{r}}(n\vec{u})$. The third term in (4.13) equals to zero, because $f = 0$ at infinitely high velocities. Summarizing, (4.13) can be expressed as:

$$\frac{\partial n}{\partial t} + \nabla_r(n\vec{u}) = G - L, \tag{4.14}$$

which reflects the particles conservation and is known as **the continuity equation.** The right-hand side of the equation is equal to zero if the particles generation and loss are absent or balanced.

4.2.2 The Momentum Conservation Equation

The momentum conservation equation can be derived from the Boltzmann equation, multiplying (4.4) by $\vec{v}$ and then integrating over velocities. Then for charged particles in plasma:

$$mn\left[\frac{\partial \vec{u}}{\partial t} + (\vec{u} * \nabla)\vec{u}\right] = qn(\vec{E} + \vec{u} \times \vec{B}) - \nabla * \Pi + F_{col}. \tag{4.15}$$

The second term from the right-hand side is the divergence of pressure tensor Π_{ij}, which can be replaced in plasma by pressure gradient ∇p; q is particle charge, E and B are the electric and magnetic fields. The last term F_{col} represents the collisional force, which is the rate of momentum transfer due to collisions with other species (i). The collisional force for plasma electrons and ions is usually mostly due to collisions with neutrals. The **Krook collision operator** approximates the collisional force by summation over the "other species i":

$$\vec{F_{col}} = -\sum_i mn\nu_{mi}(\vec{u} - \vec{u_i}) - m\vec{u}(G - L); \tag{4.16}$$

here ν_{mi} is the momentum transfer frequency for collisions with the species "i," $\vec{u}$ and $\vec{u_i}$ are the mean or drift velocities. The first term in the Krook collision operator can be interpreted as "friction." The second term corresponds to the momentum transfer due to creation or destruction of particles and is usually small. Equation (4.16) for plasma in electric field can be simplified considering in the Krook operator only for neutral species, assuming they are at rest ($u_i = 0$); and also neglecting the inertial force (($\vec{u} * \nabla)\vec{u} = 0$). This results in the most common form of the **momentum conservation equation** for plasma in electric field:

$$m\frac{\partial \vec{u}}{\partial t} = q\vec{E} - \frac{1}{n}\nabla p - m\nu_m\vec{u}. \tag{4.17}$$

4.2.3 The Energy Conservation Equation

The energy conservation equation for the electron and ion fluids in plasma can be derived by multiplying the Boltzmann equation (4.4) by $\frac{1}{2}mv^2$ and integrating over velocities:

$$\frac{\partial}{\partial t}\left(\frac{3}{2}p\right) + \nabla * \frac{3}{2}(p\vec{u}) + p\nabla * \vec{u} + \nabla * \vec{q} = \frac{\partial}{\partial t}\left(\frac{3}{2}p\right)_{col}. \tag{4.18}$$

In this relation $\vec{q}$ is the heat flow vector, $\frac{3}{2}p$ corresponds to the energy density, and the term $\frac{\partial}{\partial t}\left(\frac{3}{2}p\right)_{col}$ represents changes of the energy density due to collisional processes, including Joule heating, excitation, and ionization. The energy balance of the steady-state discharges can be usually described by the simplified form of the equation (Eq. 4.18):

$$\nabla * \frac{3}{2}(p\vec{u}) = \frac{\partial}{\partial t}\left(\frac{3}{2}p\right)_{col}, \tag{4.19}$$

which balances macroscopic energy flux with rate of energy change due to collisional processes.

4.3 The Fokker–Planck Kinetic Equation for the Electron Energy Distribution Functions (EEDF)

Nonthermal plasma EEDF can be very sophisticated, quite different from the quasi-equilibrium Maxwellian distribution (see Sections 2.1 and 4.1); and can be surely determined using the Boltzmann kinetic equation (see Lieberman and Lichtenberg 2005). Better physical interpretation of the nonequilibrium EEDF can be achieved, however, using the Fokker–Planck approach considering the *EEDF evolution as an electron diffusion and drift in the space of electron energy* (see, Raizer 2011; Rusanov and Fridman 1984). To derive the Fokker–Planck kinetic equation for EEDF $f(\varepsilon)$, let us consider the dynamics of energy transfer from the electric field to electrons. An electron has velocity $\vec{v}$ after collision with a neutral particle, and then receives an additional velocity during the free motion between collisions (see Figure 4.1):

$$\vec{u} = -\frac{e\vec{E}}{mv_{en}},$$ (4.20)

which corresponds to its drift in the electric field $\vec{E}$. Here v_{en} is frequency of the electron–neutral collisions, e and m are an electron charge and mass. The corresponding change of electron kinetic energy between two collisions is equal to:

$$\Delta\varepsilon = \frac{1}{2}m(\vec{v}+\vec{u})^2 - m\vec{v}^2 = m\vec{v}\vec{u} + \frac{1}{2}m\vec{u}^2.$$ (4.21)

Usually, the drift velocity is much lower than the thermal velocity $u \ll v$, and the absolute value of the first term $m\vec{v}\vec{u}$ in (4.21) much significantly exceeds the second one. However, the thermal velocity vector $\vec{v}$ is isotropic, and the value of $m\vec{v}\vec{u}$ can be positive and negative with the same probability. Therefore, average contribution of the first term into $\Delta\varepsilon$ is zero, and the average electron energy increase between two collisions is only due to the square of drift velocity:

$$<\Delta\varepsilon> = \frac{1}{2}m\vec{u}^2.$$ (4.22)

Summarizing, *the electron motion along the energy spectrum (or in energy space) can be considered as a diffusion process.* An electron with energy $\varepsilon = \frac{1}{2}mv^2$ receives or loses per one collision an energy portion about "*mvu*," depending on the direction of its motion – along or opposite to the electric field (see Figure 4.1). The energy portion "*mvu*" can be considered in this case as the electron "mean free path along the energy spectrum" (or in the energy space). Taking into account also the possibility of an electron motion across the electric field, the corresponding coefficient of electron diffusion along the energy spectrum can be introduced as:

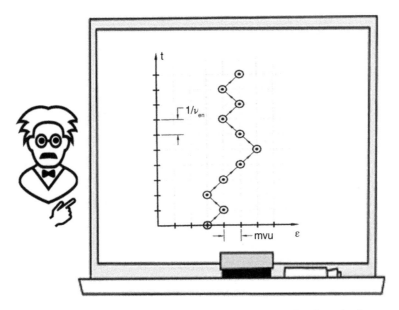

Figure 4.1 Diffusion and drift of electrons in energy space (or along the "energy spectrum").

$$D_\varepsilon = \frac{1}{3}(mvu)^2 v_{en} = \frac{2}{3}mu^2\,\varepsilon\,v_n \tag{4.23}$$

Besides the diffusion along the energy spectrum, there is also *drift in the energy space*, related to the permanent average energy gain and losses. Such energy consumption from the electric field is described by Eq. (4.22). The average energy losses per one collision are mostly due to the elastic scattering $\frac{2m}{M}\varepsilon$ and the vibrational excitation $P_{eV}(\varepsilon)\,\hbar\omega$ of molecular gases. Here M and m are the neutral particle and electron mass, $P_{eV}(\varepsilon)$ is the probability of vibrational excitation by electron impact. The mentioned three effects determine the *electron drift velocity in the energy space* neglecting the super-elastic collisions as:

$$u_\varepsilon = \left[\frac{mu^2}{2} - \frac{2m}{M}\varepsilon - P_{eV}(\varepsilon)\,\hbar\omega\right]v_{en}. \tag{4.24}$$

The EEDF $f(\varepsilon)$ can be considered then as number density of electrons in the energy space and can be found from the continuity equation (4.14) in energy space. Based upon the expressions for diffusion coefficient (4.23) and drift velocity (4.24), the *continuity equation for an electron motion along the energy spectrum* can be presented as the **Fokker–Planck kinetic equation:**

$$\frac{\partial f(\varepsilon)}{\partial t} = \frac{\partial}{\partial \varepsilon}\left[D_\varepsilon \frac{\partial f(\varepsilon)}{\partial \varepsilon} - f(\varepsilon)\,u_\varepsilon\right]. \tag{4.25}$$

The Fokker–Plank kinetic equation for electrons is much easier to interpret and to solve for different nonequilibrium plasma conditions than the Boltzmann kinetic equation.

4.4 Maxwellian, Druyvesteyn, and Margenau Electron Energy Distribution Functions

The Fokker–Plank kinetic equation permits to introduce the major types of EEDF in nonthermal plasma. Keeping in mind boundary conditions $f(\varepsilon \to \infty) = 0$, $df/d\varepsilon(\varepsilon \to \infty) = 0$, the steady-state solution of Eq. (4.25) can be presented as:

$$f(\varepsilon) = B\exp\left\{\int_0^\varepsilon \frac{u_\varepsilon}{D_\varepsilon}\,d\varepsilon'\right\}, \tag{4.26}$$

where B is the preexponential normalizing factor. Using expressions ((4.23) and (4.24)) for the diffusion coefficient and drift velocity in the energy space, the EEDF appears in the following integral form:

$$f(\varepsilon) = B\exp\left[-\int_0^\varepsilon \frac{3m^2}{Me^2E^2}v_{en}^2\left(1 + \frac{M}{2m}\frac{\hbar\omega}{\varepsilon'}P_{eV}(\varepsilon)\right)d\varepsilon'\right]. \tag{4.27}$$

This EEDF assumed constant electric field but can be generalized to alternating one with average strength E and frequency ω by replacing E in the Eq. (4.27) with the effective strength:

$$E_{eff}^2 = E^2\frac{v_{en}^2}{\omega^2 + v_{en}^2}. \tag{4.28}$$

Therefore, the electric field determining EEDF (4.27) can be considered quasi-constant at room temperature and pressure 1 Torr when the field frequencies ω are below 2 GHz. Let us consider now three specific types of EEDF, which can be derived from integration in Eq. (4.27):

4.4.1 Maxwellian Distribution

If elastic collisions dominate electron energy losses, $\left(P_{eV} \ll \frac{2m}{M}\frac{\varepsilon}{\hbar\omega}\right)$; and electron–neutral collision frequency can be approximated as constant $v_{en}(\varepsilon) = const$, then integration in Eq. (4.27) gives the Maxwellian distribution (2.1), (4.8) with:

$$T_e = \frac{e^2E^2M}{3m^2v_{en}^2}. \tag{4.29}$$

It is interesting that the Maxwellian distribution, in this case, is not due to the quasi-equilibrium of electron gas, but the result of $v_{en}(\varepsilon) = const$ and therefore proportionality of drift and diffusion (4.26) in energy space permitting introduction of temperature T_e, see below the general Einstein relation between diffusion coefficient, mobility, and mean energy.

4.4.2 Druyvesteyn Distribution

If elastic collisions dominate energy losses; and electrons mean free path λ is assumed constant $v_{en} = {}^v/_\lambda$ (which is more realistic than $v_{en}(\varepsilon) = const$), then integration in Eq. (4.27) gives the exponential-parabolic Druyvesteyn EEDF, derived in 1930:

$$f(\varepsilon) = B \exp\left[-\frac{3m}{M} \frac{\varepsilon^2}{(eE\lambda)^2}\right]. \tag{4.30}$$

The Druyvesteyn EEDF decreases with energy much faster than the Maxwellian one for the same mean energy, see Figure 4.2. The factor $eE\lambda$ in the Druyvesteyn exponent corresponds to the energy, which an electron receives during mean free path λ along the electric field E. An electron requires about $\sqrt{M/m} \approx 100$ of the mean free paths λ to get the average energy.

4.4.3 Margenau Distribution

This EEDF introduced in 1946 occurs in conditions similar to those of the Druyvesteyn distribution, but for alternating electric field with frequency ω:

$$f(\varepsilon) = B \exp\left[-\frac{3m}{M} \frac{1}{(eE\lambda)^2}(\varepsilon^2 + \varepsilon \, m\omega^2\lambda^2)\right]. \tag{4.31}$$

This EEDF obviously corresponds to the Druyvesteyn distribution when $\omega = 0$.

4.5 Effect of Electron–molecular and Electron–electron Collisions on EEDF

 Actual EEDF in specific plasma conditions are considered in particular by Raizer (2011), and are surely more sophisticated than those of Maxwellian, Druyvesteyn, and Margenau, see Figure 4.3. *The EEDF in nonequilibrium discharges in noble gases are usually close to the Druyvesteyn distribution.* At the same mean electron energies, $f(\varepsilon)$ in molecular gases can be closer to the Maxwellian function at low energies,

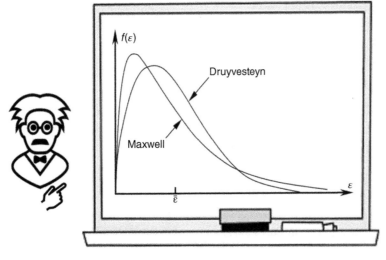

Figure 4.2 Maxwell and Druyvesteyn EEDF at the same of mean electron energy (statistical weight related to the pre-exponential factor B and resulting in $f(0) = 0$ is taken into account).

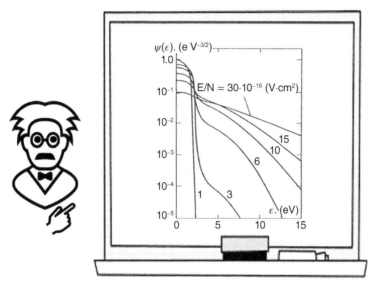

Figure 4.3 EEDF in nitrogen at different values of the reduced electric field E/N, presented as $\psi(\varepsilon) = n(\varepsilon)/(n_e \sqrt{\varepsilon})$.

but with significant deviation at higher electron energies mostly due to the effect of vibrational excitation $(P_{eV} \gg \frac{2m}{M} \frac{\varepsilon}{\hbar\omega}$, see Eq. (4.27)). Assuming $P_{eV}(\varepsilon) = const$ at $\varepsilon_1 < \varepsilon < \varepsilon_2$, $\lambda = const$ $(v_{en} = v/\lambda)$, the EEDF in the energy interval $\varepsilon_1 < \varepsilon < \varepsilon_2$ can be presented based on (4.27) as:

$$f(\varepsilon) \propto \exp\left[-\frac{6P_{eV}\hbar\omega}{(eE\lambda)^2}\varepsilon\right].\tag{4.32}$$

Vibrational excitation takes much more electron energy than elastic collisions $P_{eV} \gg \frac{2m}{M} \frac{\varepsilon}{\hbar\omega}$, and therefore results in *sharp and significant EEDF decrease* in the electron energy interval $\varepsilon_1 < \varepsilon < \varepsilon_2$ corresponding to significant vibrational excitation (usually between 1 and 3 eV):

$$\alpha_v = \frac{f(\varepsilon_2)}{f(\varepsilon_1)} \approx \exp\left(-\frac{2M}{m}P_{eV}\hbar\omega\frac{\varepsilon_2 - \varepsilon_1}{<\varepsilon>^2}\right),\tag{4.33}$$

where $<\varepsilon>$ is average electron energy. The EEDF parameter α_v actually describes the probability for an electron to pass through the energy interval of intensive vibrational excitation. Therefore, the rate coefficients of high-energy electron impact processes, including electronic excitation, ionization, and dissociation of molecules are proportional to the parameter α_v.

If vibrational temperature T_v is high, the superelastic collisions (energy transfer from molecular vibration back to electrons) are able to balance the EEDF drop (4.33) related to vibrational excitation. The probability P_{eV} due to the superelastic collisions is proportional to the number density of vibrationally nonexcited molecules $N(v = 0) \propto 1 - \exp\left(-\frac{\hbar\omega}{T_v}\right)$. As a result, the vibrational temperature T_v effect on ionization, electronic excitation, and dissociation rate coefficients can be expressed as the double exponential function:

$$\alpha_v \approx \exp\left[-\frac{2M}{m}P_{eV}^0\hbar\omega\frac{\varepsilon_2 - \varepsilon_1}{<\varepsilon>^2}\left(1 - \exp\left(-\frac{\hbar\omega}{T_v}\right)\right)\right];\tag{4.34}$$

here P_{eV}^0 is the vibrational excitation probability of non-excited molecules ($T_v = 0$). Relation (4.34) illustrates *effect of the strong increase of EEDF's high-energy tail with vibrational temperature.*

Thus, the EEDF high-energy tail is significantly suppressed in molecular gases with respect to noble gases. *Even small admixture of a molecular gas (about 1%) into a noble gas dramatically changes the EEDF, strongly decreasing the fraction of high-energy electrons.* The influence of molecular gas admixture is the strongest at relatively low reduced electric fields E/n_0 (n_0 is gas number density), when the Ramseur effect is essential (see below in this lecture). In this case, for example, addition of only 10% of air into argon makes the EEDF look like that of molecular gas.

The electron–electron collisions can influence EEDF and make it Maxwellian, when the ionization degree n_e/n_0 is high, which can be characterized by the following numerical factor:

$$a = \frac{v_{ee}}{\delta v_{en}} \approx 10^8 \frac{n_e}{n_0} \times \frac{1}{<\varepsilon, eV>^2} \times \frac{10^{-16} cm^2}{\sigma_{en}, cm^2} \times \frac{10^{-4}}{\delta}; \tag{4.35}$$

here v_{ee} and v_{en} are frequencies of electron–electron and electron–neutral collisions, corresponding to the average electron energy $<\varepsilon>$; σ_{en} is the cross-section of the electron–neutral collisions at the same electron energy; δ is the average fraction of electron energy transferred to a neutral particle during their collision. The electron–electron collisions make the EEDF Maxwellian when $a \gg 1$. Maxwellization of electrons in plasma of noble gases (smaller δ) requires lower degrees of ionization than in case of molecular gases (higher δ). For example, the effective Maxwellization in argon with $E/n_0 = 10^{-16} V * cm^2$ begins at the ionization degree $n_e/n_0 = 10^{-7} \div 10^{-6}$. For molecular nitrogen at the same $E/n_0 = 10^{-16} V * cm^2$ and ionization degrees up to 10^{-4}, the EEDF deviations due to the electron–electron collisions are still not significant.

4.6 Relations Between Electron Temperature *Te* and the Reduced Electric Field *E/n₀*

Average electron energy and the effective electron temperature $<\varepsilon> = \frac{3}{2}T_e$ can be related using Eqs. (4.20, 4.22 and 4.24) to the reduced electric field E/n_0 even for non-Maxwellian EEDF as:

$$\frac{e^2 E^2}{m v_{en}^2} = \delta * \frac{3}{2}T_e. \tag{4.36}$$

Factor δ characterizes the fraction of electron energy lost in a collision with neutral particle:

$$\delta \approx \frac{2m}{M} + <P_{eV}> \frac{\hbar\omega}{<\varepsilon>}. \tag{4.37}$$

In atomic gases $\delta = 2m/M$. In molecular gases, however, this factor is usually considered as a semi-empirical parameter. For example, $\delta \approx 3 * 10^{-3}$ for typical nonequilibrium discharges in nitrogen. Assuming $v_{en} = n_0 <\sigma_{en}v>$, $\lambda = \frac{1}{n_0\sigma_{en}}, <v> = \sqrt{\frac{8T_e}{\pi m}}$, the relation between electron temperature, electric field, and the mean free path of electrons can be presented as:

$$T_e = \frac{eE\lambda}{\sqrt{\delta}}\sqrt{\pi/12}. \tag{4.38}$$

This relation is in agreement with both Maxwell and Druyvesteyn EEDF, and can be rewritten as a relation between electron temperature T_e and the reduced electric field E/n_0:

$$T_e = \left(\frac{E}{n_0}\right)\frac{e}{<\sigma_{en}>}\sqrt{\frac{\pi}{12\delta}}. \tag{4.39}$$

This linear relation between electron temperature and the reduced electric field is obviously only a qualitative one. Figure 4.4 illustrates some actual functions $T_e\left(\frac{E}{n_0}\right)$ for molecular gases.

4.7 Plasma Electron Conductivity: Isotropic and Anisotropic Parts of EEDF

The electron velocity distribution $f(\vec{v})$ is anisotropic in electric field, due to preferential electron motion in the field direction. This anisotropy determines electric current, plasma conductivity, and related effects. To describe the anisotropy, we should focus on the velocity $\vec{u}$ received by an electron during free motion between collisions, see Eq. (4.20) and Figure 4.5.

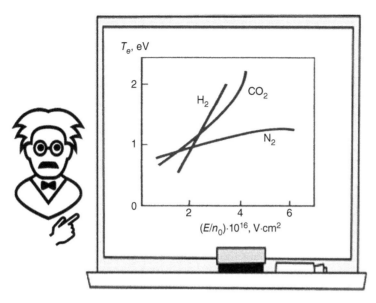

Figure 4.4 Electron temperature T_e as a function of reduced electric field E/n_0 in molecular gases.

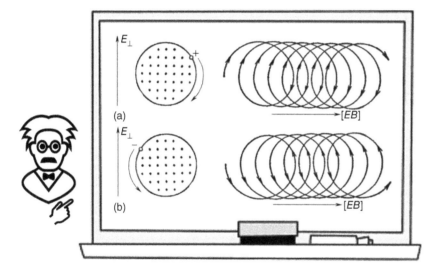

Figure 4.5 Illustration of drift of positively (a) and negatively (b) charged particles in crossed electric and magnetic fields.

If the anisotropy is not strong $u \ll v$, we can assume that fraction of electrons in the point $\vec{v}$ of the real anisotropic distribution f is directly related to the fraction of electrons in the point $\vec{v} - \vec{u}$ of the correspondent isotropic distribution $f^{(0)}$. Directions of $\vec{E}$ and $\vec{u}$ are opposite for electrons, and the electron velocity distribution function can be expressed as:

$$f(\vec{v}) = f^{(0)}(\vec{v} - \vec{u}) \approx f^{(0)}(\vec{v}) - \vec{u}\,\frac{\partial}{\partial \vec{v}} f^{(0)}(\vec{v}) = f^{(0)}(\vec{v}) + u\cos\theta\,\frac{\partial f^{(0)}(v)}{\partial v}; \tag{4.40}$$

where θ is the angle between directions of the velocity $\vec{v}$ and the electric field $\vec{E}$. Taking into account the azimuthal symmetry of $f(\vec{v})$, and omitting the dependence on θ for the isotropic function $f^{(0)}$, the Eq. (4.40) can be rewritten in the following scalar form:

$$f(v, \theta) = f^{(0)}(v) + \cos\theta * f^{(1)}(v). \tag{4.41}$$

The function $f^{(1)}(v)$ determines the anisotropy of the electron velocity distribution $f(\vec{v})$:

$$f^{(1)}(v) = u\frac{\partial f^{(0)}(v)}{\partial v} = \frac{eE}{mv_{en}}\frac{\partial f^{(0)}(v)}{\partial v}. \tag{4.42}$$

Relation (4.41) can be interpreted as first two terms of the series expansion of $f(\vec{v})$ using the ortho-normalized system of Legendre polynomials. The first term of the expansion is the isotropic part of the distribution, corresponding to EEDF as:

$$f(\varepsilon)\, d\varepsilon = f^{(0)}(v) * 4\pi v^2\, dv. \tag{4.43}$$

The second term in (4.41) includes electric field and is related to electron current density:

$$j = -e \int \vec{v} f(\vec{v})\, \overrightarrow{dv} = -\frac{4\pi}{3} e \int_0^\infty v^3 f^{(1)}(v)\, dv. \tag{4.44}$$

The anisotropic effects in electron distribution function $f(\vec{v})$ could be generalized for the case of high frequency fields by replacing electric field with the effective one (4.28). The Eq. (4.63) permits to compare the anisotropic and isotropic parts of the electron velocity distribution function:

$$f^{(1)} = u \frac{\partial f^{(0)}}{\partial v} \sim \frac{u}{<v>} * f^{(0)}, \tag{4.45}$$

illustrating that the smallness of the anisotropy of the distribution function is directly related to the smallness of the electron drift velocity u with respect to their thermal velocity v.

Based on the anisotropic part of distribution function $f^{(1)}(v)$ (Eqs. 4.42 and 4.44) and the relation between current density and electric field $j = \sigma E$, we can express the **electron conductivity** in plasma in terms of an EEDF as:

$$\sigma = \frac{4\pi e^2}{3m} \int_0^\infty \frac{v^3}{v_{en}(v)} \left[-\frac{\partial f^{(0)}(v)}{\partial v} \right] dv. \tag{4.46}$$

Assuming $v_{en}(v) = v_{en} = const$ and the integrating (4.46) by parts, we come to the conventional formula for electron conductivity:

$$\sigma = \frac{n_e e^2}{m v_{en}}, \tag{4.47}$$

which can be presented in a convenient numerical form:

$$\sigma = 2.82 * 10^{-4} \frac{n_e(\text{cm}^{-3})}{v_{en}(\text{s}^{-1})}, \quad \text{Ohm}^{-1}\,\text{cm}^{-1}. \tag{4.48}$$

4.8 Joule Heating and Electron Mobility, Similarity Parameters in Nonthermal Discharges

Power transferred from the electric field to plasma electrons, the so-called **Joule heating**, can be expressed based on the Eq. (4.47) as:

$$P = \sigma E^2 = \frac{n_e e^2 E^2}{m v_{en}}. \tag{4.49}$$

Another important plasma parameter, the coefficient of proportionality between electron drift velocity v_d and electric field E, the so-called **electron mobility** μ_e, can be presented as:

$$v_d = \mu_e E, \quad \mu_e = \frac{\sigma_e}{e n_e} = \frac{e}{m v_{en}}. \tag{4.50}$$

For numerical calculations of the electron mobility, one can use the relation:

$$\mu_e = \frac{1.76 * 10^{15}}{v_{en}(\text{s}^{-1})}, \text{cm}^2\,\text{V}^{-1}\,\text{s}^{-1}. \tag{4.51}$$

The relation (4.50) reflects the proportionality of drift velocity and reduced electric field E/n_0, which assumes that the electron–neutral collision cross-section is not changing significantly. It is not always true. This cross-section, for example, decreases significantly in noble gases at low E/n_0, which is known as the **Ramsauer effect**. As a result, the effective electron mobility in noble gases can be relatively high at low reduced electric fields, in argon it takes

Table 4.1 Nonthermal plasma similarity parameters: electron–neutral collision frequency, electron mean free path, electron mobility, and conductivity at $\frac{E}{p} = 1 \div 30\ \text{V cm}^{-1}\ \text{Torr}^{-1}$.

Gas	$\lambda * p,\ 10^{-2}\text{cm}$ Torr^{-1}	$\frac{v_{en}}{p},\ 10^{9}\text{s}^{-1}$ Torr^{-1}	$\mu_e * p,\ 10^{6}\ \text{cm}^2$ $\text{Torr V}^{-1}\ \text{s}^{-1}$	$\frac{\sigma * p}{n_e},\ 10^{-13}$ $\text{Torr cm}^2\ \text{Ohm}^{-1}$
Air	3	4	0.45	0.7
N_2	3	4	0.4	0.7
H_2	2	5	0.4	0.6
CO_2	3	2	1	2
CO	2	6	0.3	0.5
Ar	3	5	0.3	0.5
Ne	12	1	1.5	2.4
He	6	2	0.9	1.4

place at E/p about $10^{-3}\ \text{V cm}^{-1}\ \text{Torr}^{-1}$. We should also note that electron mobility and plasma conductivity can be significantly affected by electron collisions with charged particles at ionization degrees exceeding 10^{-3}, see Fridman and Kennedy (2021).

Taking into account the Eqs. (4.48 and 4.50), it is convenient to construct the so-called **similarity parameters** $\frac{v_{en}}{p}$, λp, $\mu_e p$, $\frac{\sigma p}{n_e}$. These similarity parameters are approximately constant for each gas, at room temperature and $\frac{E}{p} = 1 \div 30\ \text{V cm}^{-1}\ \text{Torr}^{-1}$. Therefore, they can be applied for determination of the electron–neutral collision frequency v_{en}, electron mean free path λ, electron mobility μ_e, and conductivity σ. Numerical values of these similarity parameters for several gases are collected in Table 4.1.

4.9 Plasma Conductivity in Crossed Electric and Magnetic Fields

Plasma current in the presence of magnetic field becomes not collinear with the electric field and conductivity should be considered as the tensor σ_{ij}:

$$j_i = \sigma_{ij}E_j. \tag{4.52}$$

This conductivity is sophisticated and can be first illustrated (see Figure 4.5) by collisionless motion of a particle with charge q in the uniform perpendicular electric E and magnetic B fields:

$$m\frac{\overrightarrow{dv}}{dt} = q\,(\overrightarrow{E} + \overrightarrow{v} \times \overrightarrow{B}). \tag{4.53}$$

Consider this motion in a reference frame moving with velocity $\overrightarrow{v_{EB}}$. The particle velocity in this fame is: $\overrightarrow{v'} = \overrightarrow{v} + \overrightarrow{v_{EB}}$. Then Eq. (4.53) in the moving reference frame can be rewritten as:

$$m\frac{\overrightarrow{dv'}}{dt} = q\,(\overrightarrow{E} + \overrightarrow{v_{EB}} \times \overrightarrow{B}) + q\,\overrightarrow{v'} \times \overrightarrow{B}. \tag{4.54}$$

We can find the velocity $\overrightarrow{v_{EB}}$, which makes the first right-hand-side term in Eq. (4.54) equal to zero: $\overrightarrow{E} + \overrightarrow{v_{EB}} \times \overrightarrow{B} = 0$ and therefore reduces the particle motion to $m\frac{\overrightarrow{dv'}}{dt} = q\,\overrightarrow{v'} \times \overrightarrow{B}$, that is to a circular motion, rotation around the magnetic field lines (see Figure 4.5). The frequency of this rotation is the so-called **cyclotron frequency**:

$$\omega_B = \frac{eB}{m}. \tag{4.55}$$

Multiplying the requirement $\vec{E} + \vec{v_{EB}} \times \vec{B} = 0$ by $\vec{B}$, a formula for the **drift velocity** $\vec{v_{EB}}$ can be derived. *The "spiral" moves (see Figure 4.5) with the drift velocity in the direction perpendicular to both electric and magnetic fields:*

$$\vec{v_{EB}} = \frac{\vec{E} \times \vec{B}}{B^2}. \tag{4.56}$$

This motion of a charged particle referred to as the **drift in crossed electric and magnetic fields**. Direction of the drift velocity does not depend on a charge sign; it is the same for electrons and ions. Collisions with neutral particles make the charged particles to move additionally along the electric field. This combined motion can be described by the conductivity tensor (4.52), which includes two components: the first one represents conductivity along the electric field $\sigma_\parallel$, and the second one corresponds to the current perpendicular to both electric and magnetic fields $\sigma_\perp$:

$$\sigma_\parallel = \frac{\sigma_0}{1 + \left(\frac{\omega_B}{v_{en}}\right)^2}, \quad \sigma_\perp = \sigma_0 \frac{\left(\frac{\omega_B}{v_{en}}\right)}{1 + \left(\frac{\omega_B}{v_{en}}\right)^2}. \tag{4.57}$$

In these relations σ_0 is the conventional conductivity (4.47) in the absence of magnetic field. When the magnetic field is low or pressure high: $\omega_B \ll v_{en}$, the transverse conductivity $\sigma_\perp$ can be neglected, and the longitudinal conductivity $\sigma_\parallel$ coincides with the conventional one σ_0. In the case of high magnetic fields and low pressures ($\omega_B \gg v_{en}$), the electrons become "trapped" in magnetic field and start drifting across the electric and magnetic fields. The longitudinal conductivity then can be neglected, and the transverse conductivity becomes independent of pressure and mass of a charged particle:

$$\sigma_\perp \approx \frac{n_e e^2}{m \omega_B} = \frac{n_e e}{B}. \tag{4.58}$$

If magnetic field is much larger than it is required for $\omega_B \gg v_{en}$, the ion motion can also become quasi-collisionless, and they move together with electrons (4.56) without any net current.

4.10 Electric Conductivity of the Strongly Ionized Plasma

The electric conductivity in weakly ionized plasma (4.47) is related to electron–neutral collisions. In strongly ionized plasma with high degrees of ionization, electron–ion scattering also contributes to electric conductivity. In this case, the electron collision frequency v_{en} in (4.47) has to be replaced by the total frequency, including the electron–ion collisions:

$$v_\Sigma = v_{en} + n_e <v> \sigma_{Coul}. \tag{4.59}$$

It is assumed, that the ion and electron densities are equal (n_e); $<v>$ is the average electron velocity; and σ_{Coul} is the averaged **Coulomb cross-section** of the electron–ion collisions:

$$\sigma_{Coul} = \frac{4\pi}{9} \frac{e^4 \ln\Lambda}{(4\pi\varepsilon_0 T_e)^2} = \frac{2.87 * 10^{-14} \ln\Lambda}{(T_e, eV)}, cm^2. \tag{4.60}$$

Though trajectory deflection in the electron–ion interaction is low when they are relatively far, the large distance collisions make significant contribution to the momentum transfer cross-section because of long-range nature of Coulomb forces. This effect is taken into account in (4.60) by multiplication of the normal cross-section for interaction of charged particles $\frac{e^4}{(4\pi\varepsilon_0 T_e)^2}$ by the so-called **Coulomb logarithm:**

$$\ln\Lambda = \ln\left[\frac{3}{2\sqrt{\pi}} \frac{(4\pi\varepsilon_0 T_e)^{3/2}}{e^3 n_e^{1/3}}\right] = 13.57 + 1.5 \log(T_e, eV) - 0.5 \log n_e. \tag{4.61}$$

The electron–ion collisions become significant in the total electron–collision frequency and, therefore, in the electric conductivity when the degree of ionization exceeds critical value about: $\frac{n_e}{n_0} \geq 10^{-3}$. When the electron–ion collisions are dominant, the electric conductivity becomes independent on the electron density and reaches its maximum value:

$$\sigma = \frac{9\varepsilon_0 T_e^2}{m <v> e^2 \ln\Lambda} = 1.9 * 10^2 \frac{(T_e, \text{eV})^{3/2}}{\ln\Lambda}, \text{Ohm}^{-1}\text{cm}^{-1}. \tag{4.62}$$

4.11 Ion Energy and Ion Drift in Electric Field

The relationship between the ions' average energy $<\varepsilon_i>$, gas temperature T_0, and electric field E is determined by balance of ion-molecular collisions:

$$<\varepsilon_i> = \frac{3}{2} T_0 + \frac{M}{2M_i} \left(1 + \frac{M_i}{M}\right)^3 \frac{e^2 E^2}{M_i v_{in}^2}; \tag{4.63}$$

here M_i and M are masses of an ion and a neutral particle; v_{in} is frequency of ion-neutral collisions. If the electric field is not too high (usually $E/p < 10 \, \text{V cm}^{-1} \text{Torr}^{-1}$), the ion energy only slightly exceeds that of neutrals. For electrons, because of their low mass, such a situation takes place only at very low electric fields $E/p < 0.01 \, \text{V cm}^{-1} \text{Torr}^{-1}$. When the reduced electric fields are very high ($E/p \gg 10 \, \text{V cm}^{-1} \text{Torr}^{-1}$), the ion velocities, collision frequencies v_{in}, and the ion energies increase with the electric field (λ is the ion mean free path), for example:

$$<\varepsilon_i> \approx \frac{1}{2}\sqrt{\frac{M}{M_i}} \left(1 + \frac{M_i}{M}\right)^{3/2} eE\lambda; \tag{4.64}$$

In the case of weak and moderate electric fields ($E/p < 10 \, \text{V cm}^{-1} \text{Torr}^{-1}$), the ion drift velocity is proportional to the electric field and the ion mobility is constant:

$$\overrightarrow{v_{id}} = \frac{e\overrightarrow{E}}{v_{in} * MM_1/(M + M_1)}, \quad \mu_i = \frac{e}{v_{in} * MM_1/(M + M_1)}. \tag{4.65}$$

The convenient numerical relation can be used for calculations of the ion mobility:

$$\mu_i = \frac{2.7 * 10^4 \sqrt{1 + M/M_i}}{p(\text{Torr})\sqrt{A * (\alpha/a_0^3)}}; \tag{4.66}$$

here α is polarizability of a neutral particle; "A" is its molecular mass; a_0 is the Bohr radius. When $E/p \gg 10 \, \text{V cm}^{-1} \text{Torr}^{-1}$, the ion drift velocities are proportional to square root of the electric field:

$$v_{id} \approx \sqrt[4]{\frac{M_i}{M}\left(1 + \frac{M_i}{M}\right)}\sqrt{\frac{eE\lambda}{M_i}}. \tag{4.67}$$

4.12 Free Diffusion of Electrons and Ions and Continuity Equation for Charged Particles; Fick's Law and Einstein Relation Between Diffusion, Mobility, and Mean Energy

If electron (e) and ion (i) concentrations are low, their diffusion in plasma can be considered "free" and independent. The total flux of charged particles includes, in this case, their drift in the electric field (see above) and diffusion, following the **Fick's law**:

$$\overrightarrow{\Phi_{e,i}} = \pm n_{e,i} \mu_{e,i} \overrightarrow{E} - D_{e,i} \frac{\partial n_{e,i}}{\partial \overrightarrow{r}}. \tag{4.68}$$

The continuity equation (4.14) then can be rewritten for charged particles as:

$$\frac{\partial n_{e,i}}{\partial t} + \frac{\partial}{\partial \overrightarrow{r}} \left(\pm n_{e,i} \mu_{e,i} \overrightarrow{E} - D_{e,i} \frac{\partial n_{e,i}}{\partial \overrightarrow{r}}\right) = G - L. \tag{4.69}$$

The characteristic diffusion time corresponding to the diffusion distance R can be estimated based on this relation as $\tau_D \approx R^2/D_{e,i}$. The source term $G - L$ stands here for particles "generation minus losses." The diffusion coefficients

Table 4.2 Free diffusion coefficients D of electrons and ions in different gases, $T_0 = 300$ K.

Diffusion	$D*p$, cm^2 s^{-1} Torr^{-1}	Diffusion	$D*p$, cm^2 s^{-1} Torr^{-1}	Diffusion	$D*p$, cm^2 s^{-1} Torr^{-1}
"e" in He	$2.1 * 10^5$	"e" in Ne	$2.1 * 10^6$	"e" in Ar	$6.3 * 10^5$
"e" in Kr	$4.4 * 10^4$	"e" in Xe	$1.2 * 10^4$	"e" in H$_2$	$1.3 * 10^5$
"e" in N$_2$	$2.9 * 10^5$	"e" in O$_2$	$1.2 * 10^6$	N$_2^+$ in N$_2$	40

can be estimated as a function of thermal velocity, mean free path and collisional frequency with neutrals ($<v_{e,i}>$, $\lambda_{e,i}$, $v_{en,in}$):

$$D_{e,i} = \frac{<v_{e,i}^2>}{3v_{en,in}} = \frac{\lambda_{e,i} <v_{e,i}>}{3}. \tag{4.70}$$

The diffusion coefficients are inversely proportional to pressure at room temperature. Therefore, the similarity parameter $D_{e,i} * p$ (Section 4.8) can be used for calculation of free electrons and ions diffusion coefficients in this case, see Table 4.2.

The coefficient of free electron diffusion (4.70) grows with temperature at constant pressure as $D_e \propto T_e^{1/2} T_0$. Assuming, ions and neutrals have the same temperature, the coefficient of free ions diffusion (4.70) grows with temperature at constant pressure as $D_i \propto T_0^{3/2}$.

It is interesting that ratio of the electron free diffusion coefficient (4.70) and electron mobility (4.50) corresponds to the average electron energy:

$$\frac{D_e}{\mu_e} = \frac{2/3 <\varepsilon_e>}{e}. \tag{4.71}$$

A similar relation is valid for ions as well. In the case of Maxwellian distribution function, the average energy is equal to $3/2 T$ (assuming like always the Boltzmann coefficient $k = 1$). The formula (4.71) can be therefore expressed in a more general form valid for different particles and known as **the Einstein relation** between diffusion and mobility:

$$\frac{D}{\mu} = \frac{T}{e}. \tag{4.72}$$

4.13 Ambipolar Diffusion, Debye Radius, and Definition of Plasma

When the ionization degree is high, electron and ion diffusion cannot be considered "free" and independent like it was assumed above. The electrons are moving faster than ions and form charge separation zone with a strong polarization field, accelerating ions, slowing down electrons, and equalizing their fluxes. This phenomenon known as the **ambipolar diffusion** is illustrated in Figure 4.6. The equality of the fluxes means:

$$\vec{\Phi_e} = -\mu_e \vec{E} n_e - D_e \frac{\partial n_e}{\partial \vec{r}}, \quad \vec{\Phi_i} = \mu_i \vec{E} n_i - D_i \frac{\partial n_i}{\partial \vec{r}}, \quad \vec{\Phi_e} = \vec{\Phi_i}. \tag{4.73}$$

Dividing the first relation by μ_e, second one by μ_i, adding up the results and assuming plasma electroneutrality gives the general relation for both electron and ion fluxes:

$$\vec{\Phi_{e,i}} = -\frac{D_i \mu_e + D_e \mu_i}{\mu_e + \mu_i} \frac{\partial n_{e,i}}{\partial \vec{r}}. \tag{4.74}$$

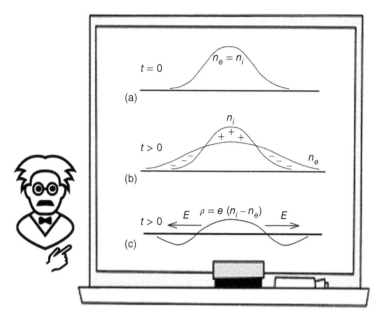

Figure 4.6 Illustration of the ambipolar diffusion effect due to initial plasma density gradient (a), followed by local separation of electrons and ions (b, c).

It permits introduction of the coefficient of the ambipolar diffusion $\overrightarrow{\Phi_{e,i}} = -D_a \frac{\partial n_{e,i}}{\partial \vec{r}}$, which describes the collective diffusive motion of electrons and ions:

$$D_a = \frac{D_i \mu_e + D_e \mu_i}{\mu_e + \mu_i}.$$ (4.75)

Taking into account the Einstein relation (4.72) and noting that $\mu_e \gg \mu_i$, $D_e \gg D_i$:

$$D_a = D_i + D_e \mu_i / \mu_e, \quad D_a \approx D_i + \frac{\mu_i}{e} T_e.$$ (4.76)

The ambipolar diffusion is faster than that of free ions and slower than that of free electrons. In quasi-equilibrium plasmas with equal electron and ion temperatures $D_a = 2D_i$. In nonthermal plasmas, $T_e \gg T_i$, and the ambipolar diffusion coefficient $D_a = \frac{\mu_i}{e} T_e$ corresponds to the temperature of the fast electrons and mobility of the slow ions.

To determine the conditions of the ambipolar diffusion with respect to the free diffusion of electrons and ions, we estimate first the ambipolar polarization field from (4.73) as:

$$E \approx \frac{D_e}{\mu_e} \frac{1}{n_e} \frac{\partial n_e}{\partial r} = \frac{T_e}{e} \frac{\partial \ln n_e}{\partial r} \propto \frac{T_e}{eR};$$ (4.77)

where R is the characteristic length of change of the electron density. The difference between the ion and electron densities $\Delta n = n_i - n_e$ characterizing the space charge is related to the electric field by the Maxwell equation $\frac{\partial}{\partial \vec{r}} \vec{E} = \frac{e \Delta n}{\varepsilon_0}$, which estimations means $\frac{E}{R} \propto \frac{e \Delta n}{\varepsilon_0}$. Combining this relation with (4.77), the deviation from plasma quasi-neutrality as a function of scale R is:

$$\frac{\Delta n}{n_e} \approx \frac{T_e}{e^2 n_e} \frac{1}{R^2} = \left(\frac{r_D}{R}\right)^2, \quad r_D = \sqrt{\frac{T_e \varepsilon_0}{e^2 n_e}}.$$ (4.78)

The **Debye radius** r_D is an important parameter characterizing the plasma quasi-neutrality and representing the characteristic size of charge separation. If the electron density is high and the Debye radius is small with respect to size of the system ($r_D \ll R$), then deviation from quasi-neutrality is small, the electrons and ions move "as a group" and the diffusion is ambipolar. If, vice versa, the electron density is low and the Debye radius is large $r_D \geq R$, then the ionized medium is not quasi-neutral, the electrons and ions move separately, and their diffusion is independent. Such ionized medium cannot be officially called plasma.

Plasma is "officially defined" as a quasi-neutral ionized medium with size exceeding the Debye radius $r_D \ll R$. For calculations of the Debye radius, one can use the numerical formula:

$$r_D = 742 \sqrt{\frac{T_e, \text{eV}}{n_e, \text{cm}^{-3}}}, \text{cm}. \tag{4.79}$$

For example, if T_e is 1 eV and electron density exceeds 10^8cm^{-3}, the Debye radius is less than 0.7 mm and ionized medium above 1–3 mm can be considered as quasi-neutral, that is plasma.

4.14 Problems and Concept Questions

4.14.1 The Fokker–Planck Kinetic Equation

Using the Fokker–Planck kinetic equation (4.25), estimate time required to reestablish the steady-state EEDF after electric field fluctuation.

4.14.2 The Druyvesteyn Electron Energy Distribution Function

Calculate the average electron energy for the Druyvesteyn distribution. Define the effective electron temperature of the distribution and compare it with that of the Maxwellian distribution function.

4.14.3 The Margenau EEDF

Estimate the average electron energy and effective electron temperature for the Margenau EEDF. Consider case of very high frequencies, when the Margenau distribution becomes exponentially linear with respect to electron energy.

4.14.4 Effect of Vibrational Temperature on EEDF

Simplify Eq. (4.34) for the case of high vibrational temperatures $T_v \gg \hbar\omega$. Using this relation, estimate acceleration of ionization rate coefficients corresponding to 10% increase of vibrational temperature.

4.14.5 Electron–Electron Collisions and EEDF Maxwellization

Calculate the minimum ionization degree $\frac{n_e}{n_0}$, when EEDF Maxwellization by electron–electron collisions (4.35) becomes essential in (i) argon-plasma, $T_e = 1$ eV; and (ii) nitrogen-plasma $T_e = 1$ eV.

4.14.6 Similarity Parameters

Based on the similarity parameters, presented in Table 4.1, find the electron mean-free-pass, electron–neutral collision frequency, and electron mobility in atmospheric pressure air at room temperature.

4.14.7 Electron Drift in the Crossed Electric and Magnetic Fields

Explain why the electron drift in crossed electric and magnetic fields v_{EB} (4.56) is impossible at low magnetic fields and relatively high electric fields. Estimate the criteria for the critically high electric fields. Illustrate the electron motion if the electric field exceeding the critical one.

4.14.8 Plasma Rotation in the Crossed Electric and Magnetic Fields, Plasma Centrifuge

Describe plasma motion in the electric field $\vec{E}(r)$ created by a long-charged cylinder and uniform magnetic field $\vec{B}$ parallel to the cylinder. Find out maximum operational pressure of such plasma centrifuge. Is it necessary to trap ions in the magnetic field in such centrifuge?

4.14.9 Ambipolar Diffusion

Estimate the ambipolar diffusion coefficient in room temperature atmospheric pressure nitrogen at electron temperature about 1 eV. Compare this diffusion coefficient with that for free diffusion of electrons and free diffusion of nitrogen molecular ions (take the ion temperature as room temperature).

4.14.10 Debye Radius and Ambipolar Diffusion

Estimate the Debye radius and, therefore, the system size when diffusion of charged particles can be considered as not ambipolar but independent. Take as an example the electron temperature 1 eV and electron density 10^{12}cm^{-3}. Which plasma systems can involve such size scales?

Lecture 5

Physical and Chemical Kinetics of Excited Atoms and Molecules in Plasma

5.1 Excitation Energy Distribution in Nonequilibrium Plasma: Vibrational Kinetics, Fokker–Planck Kinetic Equation for Vibrational Distribution Functions

Excited atoms and molecules make significant contribution to variety of plasma processes (see Lecture 3). Relevant elementary reactions surely depend on excitation energies of atoms and molecules (see e.g., Section 3.11). Therefore, physical and chemical kinetics of the plasma processes are very much determined by the **excitation energy distribution functions (EEDF)** showing fraction of atoms and molecules with certain excitation energies, which in nonequilibrium plasma conditions can be very far from the Boltzmann distribution (similarly to the case of EEDF, see Section 4.3).

 Let us start with analysis of **vibrational energy distribution functions**, the so-called **vibrational kinetics**, playing very significant role in plasma processes. Electrons in nonthermal discharges mostly provide excitation of low vibrational levels, which determines vibrational temperature T_v. Formation of highly excited molecules depends on variety of processes, but mostly on competition of multi-step VV-exchange processes and vibrational quanta losses in vibrational-translational (VT)-relaxation (see Sections 3.6 and 3.7). Simultaneous consideration of these processes is convenient in the frameworks of the Fokker–Planck approach similar to the case of EEDF (see Section 4.3). In the Fokker–Planck approach, *evolution of the vibrational distribution function is considered as diffusion and drift of molecules in the space of vibrational energies*. The continuous distribution $f(E)$ of excited molecules over vibrational energies E (the vibrational distribution, where for harmonic oscillators $E = \hbar\omega v$) is presented in this case as density in energy space and determined from the continuity equation for the "motion along the energy spectrum" (compare to 4.25):

$$\frac{\partial f(E)}{\partial t} + \frac{\partial}{\partial E} J(E) = 0, \tag{5.1}$$

In the Fokker–Planck equation (5.1), $J(E)$ is the flux of molecules in the energy space, which includes all relaxation and energy exchange processes, as well as chemical reactions from different vibrationally excited states. In contrast to EEDF, the vibrational energy distribution functions are actually discrete because of quantum nature of molecular excitation. However, the continuous distribution approximation is quite correct when $f(E)$ changes not significantly on the scale of one quantum $\hbar\omega$: $\left| \hbar\omega \frac{\partial \ln f(E)}{\partial E} \right| \ll 1$.

We are going next to consider the Fokker–Planck evolution (5.1) of $f(E)$ starting with the case of one-component diatomic molecular gases, reflecting the major features of the vibrational kinetics in nonequilibrium plasma.

5.2 VT- and VV-fluxes of Excited Molecules in Energy Space

5.2.1 Energy-space-diffusion Related to VT-relaxation

Energy-space-diffusion related to VT-relaxation is illustrated in Figure 5.1 for the case of excited diatomic molecules with vibrational energy E and quantum $\hbar\omega$. The VT-component of total flux $J(E)$ in the Fokker–Planck equation (5.1) can be expressed as:

$$j_{VT} = f(E) k_{VT}(E, E + \hbar\omega) n_0 \hbar\omega - f(E + \hbar\omega) k_{VT}(E + \hbar\omega, E) n_0 \hbar\omega, \tag{5.2}$$

Plasma Science and Technology: Lectures in Physics, Chemistry, Biology, and Engineering, First Edition. Alexander Fridman.
© 2024 WILEY-VCH GmbH. Published 2024 by WILEY-VCH GmbH.

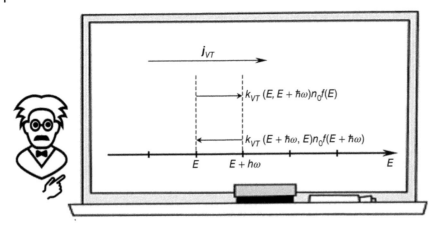

Figure 5.1 Illustration of *VT*-flux in of excited molecules in vibrational energy space.

where $k_{VT}(E + \hbar\omega, E)n_0$ and $k_{VT}(E, E + \hbar\omega)n_0$ are the frequencies of the direct and reverse processes of *VT*, whose ratio is $\exp\frac{\hbar\omega}{T_0}$; n_0 is gas density. Expanding the vibrational distribution in series $f(E + \hbar\omega) = f(E) + \hbar\omega\frac{\partial f(E)}{\partial E}$, and denoting the rate coefficient $k_{VT}(E + \hbar\omega, E) \equiv k_{VT}(E)$, the *VT*-flux in energy space comes to:

$$j_{VT} = -D_{VT}(E)\left[\frac{\partial f(E)}{\partial E} + \tilde{\beta}_0 f(E)\right]. \tag{5.3}$$

VT-diffusion coefficient D_{VT} of molecules in energy space is introduced here:

$$D_{VT}(E) = k_{VT}(E)n_0 \cdot (\hbar\omega)^2; \tag{5.4}$$

and corresponds to the conventional diffusion coefficient with mean free path λ replaced by vibrational quantum $\hbar\omega$ (mean free path in energy space), and frequency of collisions ν in the coordinate space replaced by the quantum transfer frequency $k_{VT}(E)n_0$. Parameter $\tilde{\beta}_0$ in (5.3) is $\tilde{\beta}_0 = \dfrac{1 - \exp\left(-\dfrac{\hbar\omega}{T_0}\right)}{\hbar\omega}$. When $T_0 \gg \hbar\omega$, $\tilde{\beta}_0 = \beta_0 = 1/T_0$. The first term in the flux (5.3) can be interpreted as diffusion and the second term as the drift in energy space; therefore $\tilde{\beta}_0 = \beta_0 = 1/T_0$ is the ratio of diffusion coefficient over mobility completely in accordance with the Einstein relation but this time in energy space (see Section 4.14).

When *VT*-relaxation is the dominating vibrational kinetic process, the Fokker–Planck kinetic equation (5.1) for vibrational distribution can be presented as:

$$\frac{\partial f(E)}{\partial t} = \frac{\partial}{\partial E}\left\{D_{VT}(E)\left[\frac{\partial f(E)}{\partial E} + \frac{1}{T_0}f(E)\right]\right\}. \tag{5.5}$$

At steady-state ($\partial/\partial t = 0$), integration (5.5) gives $\frac{\partial f(E)}{\partial E} + \frac{1}{T_0}f(E) = const(E)$. Taking into account the boundary conditions at $E \to \infty$: $\frac{\partial f(E)}{\partial E} = 0, f(E) = 0$, results in $const(E) = 0$. Therefore, the domination of the *VT*-relaxation, leads to the quasi-equilibrium Boltzmann distribution with temperature T_0: $f(E) \propto \exp(-E/T_0)$.

5.2.2 Energy-space-diffusion Related to VV-exchange

Energy-space-diffusion related to VV-exchange in contrast to *VT*-relaxation involves two vibrationally excited molecules and, therefore, relevant *VV*-flux in energy space is nonlinear with respect to vibrational distribution function:

$$j_{VV} = k_0 n_0 \hbar\omega \int_0^\infty \left[Q_{E+\hbar\omega,E}^{E',E'+\hbar\omega} f(E + \hbar\omega)f(E') - Q_{E,E+\hbar\omega}^{E'+\hbar\omega,E'} f(E)f(E' + \hbar\omega)\right] dE'. \tag{5.6}$$

The *VV*-exchange probabilities Q correspond to those presented in Section 3.7, k_0 is rate coefficient of neutral–neutral gas-kinetic collisions. The defect of vibrational energy in the *VV*-exchange process is $2x_e(E' - E)$, where x_e

is the coefficient of anharmonicity. Therefore, probabilities Q are related to each other by the detailed equilibrium relation:

$$Q_{E,E+\hbar\omega}^{E'+\hbar\omega,E'} = Q_{E+\hbar\omega,E}^{E',E'+\hbar\omega} \exp\left[-\frac{2x_e(E'-E)}{T_0}\right].$$

(5.7)

To analyze the total VV-flux along the vibrational energy spectrum (5.6 and 5.7), it is convenient to divide it into linear $j^{(0)}$ and nonlinear $j^{(1)}$ flux components:

$$j_{VV}(E) = j_{VV}^{(0)}(E) + j_{VV}^{(1)}(E).$$

(5.8)

5.2.3 The Linear VV-flux Component

The linear VV-flux component $j_{VV}^{(0)}(E)$ corresponds to the nonresonant VV-exchange of a highly vibrationally excited molecule of energy E with the bulk of low vibrational energy molecules, corresponding in (5.6) to integration domain $0 < E' < T_v$. The distribution function for the low vibrational energy domain is $f(E') \approx \frac{1}{T_v}\exp\left(-\frac{E'}{T_v}\right)$. Also $f(E+\hbar\omega) = f(E) + \hbar\omega\frac{\partial f(E)}{\partial E}$, $\exp\left(-\frac{\hbar\omega}{T_v} + \frac{2x_e E}{T_0}\right) \approx 1 - \frac{\hbar\omega}{T_v} + \frac{2x_e E}{T_0}$. Then relations ((5.6) and (5.7)) give the linear component of the VV-flux in the vibrational energy space as:

$$j_{VV}^{(0)} = -D_{VV}(E)\left[\frac{\partial f(E)}{\partial E} + \left(\frac{1}{T_v} - \frac{2x_e E}{T_0 \hbar\omega}\right)f(E)\right].$$

(5.9)

The diffusion coefficient D_{VV} in the energy space, related to the nonresonant VV-exchange of a molecule of vibrational energy E with the bulk of low energy molecules is:

$$D_{VV}(E) = k_{VV}(E)n_0(\hbar\omega)^2.$$

(5.10)

The relevant VV-exchange rate coefficient can be expressed (see Section 3.7) as: $k_{VV}(E) \approx \frac{ET_v}{(\hbar\omega)^2}Q_{01}^{10}k_0\exp(-\delta_{VV}E)$, where δ_{VV} is relevant adiabatic parameter. The linear VV-diffusion coefficient along the energy spectrum (5.10) surely corresponds to the diffusion coefficient in the coordinate space but with mean free path λ replaced by vibrational quantum $\hbar\omega$, and collisional frequency v replaced by the nonresonant quantum exchange frequency $k_{VV}(E)n_0$.

5.2.4 The Nonlinear Flux Component

The nonlinear flux component $j_{VV}^{(1)}(E)$ in (5.8) corresponds to the resonant VV-exchange of two highly vibrationally excited molecules. Vibrational energies of the two molecules are mostly in the interval confined by the adiabatic parameter δ_{VV} (see Section 3.7): $|E - E'| \leq \delta_{VV}^{-1}$, then ((5.6) and (5.7)) can be presented as:

$$j_{VV}^{(1)} = k_0 n_0 Q_{10}^{01}\frac{E+\hbar\omega}{\hbar\omega}\int_{T_v}^{\infty}\frac{E'+\hbar\omega}{\hbar\omega}\exp(-\delta_{VV}|E'-E|)$$
$$\times \left\{\hbar\omega\left[f(E')\frac{\partial f(E)}{\partial E} - f(E)\frac{\partial f(E')}{\partial E'}\right] + \frac{2x_e(E'-E)}{T_0}f(E)f(E')\right\}\hbar\omega\, dE'.$$

(5.11)

The under-integral function has a sharp maximum at $E = E'$. Therefore, assuming $f(E) \approx f(E')$ and $\frac{\partial \ln f(E')}{\partial E'} - \frac{\partial \ln f(E)}{\partial E} \approx \frac{\partial^2 \ln f(E)}{\partial E^2}(E'-E)$, the integration of (5.11) over $(E - \delta_{VV}^{-1}, E + \delta_{VV}^{-1})$ gives the final differential expression of the nonlinear VV-flux along the vibrational energy spectrum:

$$j_{VV}^{(1)} = -D_{VV}^{(1)}\frac{\partial}{\partial E}\left[f^2(E)E^2\left(\frac{2x_e}{T_0} - \hbar\omega\frac{\partial^2 \ln f(E)}{\partial E^2}\right)\right].$$

(5.12)

The energy-space diffusion coefficient $D_{VV}^{(1)}$ describing the resonance VV-exchange is:

$$D_{VV}^{(1)} = 3k_0 n_0 Q_{10}^{01}(\delta_{VV}\hbar\omega)^{-3}\hbar\omega.$$

5.3 Nonequilibrium Vibrational Distribution Functions Dominated by VV-exchange, the Treanor Distribution

When *VT*-relaxation dominates vibrational kinetics, the Fokker–Planck kinetic equation (5.5) is determined by a single temperature T_0 resulting in the quasi-equilibrium Boltzmann distribution in steady state conditions. In contrast to that, if the vibrational kinetics is mostly controlled by *VV*-exchange (which is the case for many molecular gases, including N_2, H_2, CO, CO_2, CH_4, etc.), the relevant linear and nonlinear fluxes ((5.9) and (5.12)) are determined by both vibrational T_v and translational T_0 temperatures, resulting in steady state distributions very far from the Boltzmann.

Specific physics of these non-Boltzmann vibrational distributions can be best interpreted for the most typical case of not very high vibrational temperatures (the so-called case of weak excitation $\delta_{VV} T_v < 1$, $\frac{x_e T_v^2}{T_0 \hbar \omega} < 1$, see Section 5.5), when vibrational kinetics is dominated by the nonlinear *VV*-flux (5.9). The Fokker–Planck equation (5.1) at steady-state ($\partial/\partial t = 0$) dominated by linear nonresonant *VV*-exchange (keeping in mind the boundary conditions at $E \to \infty$: $\frac{\partial f(E)}{\partial E} = 0$, $f(E) = 0$) results in the linear kinetic equation $j_{VV}^{(0)}(E) = 0$ with the *VV*-flux (5.9):

$$j_{VV}^{(0)} = -D_{VV}(E) \left[\frac{\partial f(E)}{\partial E} + \left(\frac{1}{T_v} - \frac{2x_e E}{T_0 \hbar \omega} \right) f(E) \right] = 0. \tag{5.13}$$

 Solution of this equation is the **Treanor vibrational distribution function**:

$$f(E) = B \exp \left(-\frac{E}{T_v} + \frac{x_e E^2}{T_0 \hbar \omega} \right), \quad \frac{1}{T_v} = \left. \frac{\partial \ln f(E)}{\partial E} \right|_{E \to 0}, \tag{5.14}$$

the exponentially parabolic distribution describing significant overpopulation of highly vibrationally excited states and illustrated in Figure 5.2a in comparison with the Boltzmann distribution. Replacing simplified vibrational energies $E = \hbar \omega v$ by vibrational quantum numbers v, the Treanor distribution (5.14) can be presented in more correct discrete form:

$$f(v, T_v, T_0) = B \exp \left(-\frac{\hbar \omega v}{T_v} + \frac{x_e \hbar \omega v^2}{T_0} \right), \tag{5.15}$$

where B is the normalizing factor. The Treanor distribution (5.15) corresponds to the Boltzmann distribution in the following two cases. First, if molecules can be considered as harmonic oscillators and the anharmonicity coefficient is zero $x_e = 0$, then (5.15) gives the Boltzmann distribution $f(v) = B \exp \left(-\frac{\hbar \omega v}{T_v} \right) = B \exp \left(-\frac{E_v}{T_v} \right)$ even if $T_v > T_0$. Second, if the translational and vibrational temperatures are equal $T_v = T_0 = T$, we come back from Treanor to Boltzmann distribution even for anharmonic oscillators:

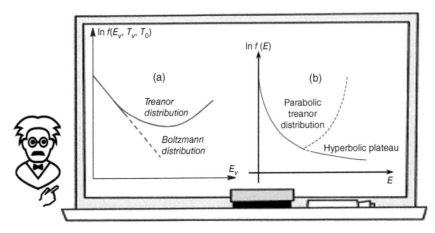

Figure 5.2 Comparison of the exponential parabolic Treanor vibrational energy distribution function at $T_v > T_0$ with the Boltzmann distribution (a), and hyperbolic plateau distribution (b).

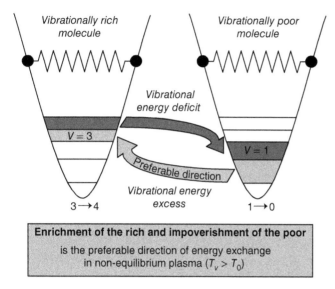

Figure 5.3 Illustration of the Treanor effect: overpopulation of the highly vibrationally excited states using energy of the lower excited states in nonequilibrium plasma conditions ($T_v > T_0$).

$$f(v, T) = B \exp\left(-\frac{\hbar\omega v}{T} + \frac{x_e\hbar\omega v^2}{T}\right) = B \exp\left(-\frac{\hbar\omega v - x_e\hbar\omega v^2}{T}\right) \propto \exp\left(-\frac{E_v}{T}\right). \tag{5.16}$$

Physics of the **Treanor effect** (5.15) can be interpreted by analyzing a collision of two oscillators, illustrated in Figure 5.3. VV-exchange of harmonic oscillators is resonant, probabilities of direct and reverse processes of vibrational energy transfer to highly excited molecules are equal, which results in the Boltzmann distribution even when the vibrational temperature exceeds the translational one ($T_v > T_0$). In contrast to that, VV-exchange of anharmonic oscillators is not completely resonant (see Figure 5.3) and gives priority to a quantum transfer from lower excited molecules (having higher vibrational quantum) to higher excited molecules (having smaller vibrational quantum) because of deficiency of translational energy ($T_v > T_0$) required to cover the defect of vibrational energy.

 As a result of such VV-exchange, the higher excited molecules keep receiving more energy than those with small number of quanta. In other words, the rich molecules become even richer, and the poor molecules become even poorer. That is why the Treanor effect of over-population of highly excited vibrational states in nonequilibrium conditions is sometimes referred to as "*the capitalism in molecular life.*"

The Treanor distribution (5.16) has a minimum at a specific number of quanta:

$$v_{\min}^{Tr} = \frac{1}{2x_e}\frac{T_0}{T_v}. \tag{5.17}$$

Vibrational population (5.16) corresponding to the Treanor minimum (5.17) is:

$$f_{\min}^{Tr}(T_v, T_0) = B \exp\left(-\frac{\hbar\omega}{4x_e}\frac{T_0}{T_v^2}\right). \tag{5.18}$$

At levels above the Treanor minimum (5.17), the population (5.16) of vibrational states becomes "inverse" (see Figure 5.2a), which is usually limited in reality by contribution of VT-relaxation prevailing the VV-exchange at very high excitation levels (see below).

The Treanor distribution function ((5.14) and (5.16)) has been derived and interpreted above for the case of domination of the linear nonresonant VV-exchange $j_{VV}^{(0)}(E) = 0$. However, domination of the resonant nonlinear VV-exchange $j_{VV}^{(1)}(E) = 0$ also can result in the Treanor distribution, which satisfies the equality: $\frac{2x_e}{T_0} - \hbar\omega\frac{\partial^2 \ln f(E)}{\partial E^2} = 0$ and makes the flux (5.12) equal to zero. Even the most general VV-flux ((5.6) and (5.7)) including both linear and nonlinear components results in the Treanor distribution as one of the possible solutions of the Fokker–Planck kinetic equation (see Fridman and Kennedy 2021). However, in the case of domination of the nonlinear resonant VV-exchange, the Treanor distribution function is not the only solution of the nonlinear kinetic equation $j_{VV}^{(1)}(E) = 0$; and the other plateau-like distribution can be more expected, see next Section 5.4.

5.4 Vibrational Distributions at the Strong Excitation Regime Dominated by Nonlinear Resonant VV-exchange, the Hyperbolic Plateau Distribution

The regime of strong excitation takes place at high vibrational temperatures $\delta_{VV} T_v \geq 1$, when exponential $f(E)$ decrease is less significant than slowing down of VV-relaxation due to deviation from resonance in VV-exchange between highly and lower excited molecules. It results in domination of the nonlinear resonant VV-exchange between highly excited molecules on evolution of vibrational distribution functions $f(E)$. The adiabatic parameter δ_{VV} at the room temperature is about $\delta_{VV} \approx (0.2 \div 0.5)/\hbar\omega$ (see Section 3.7), which means that the strong excitation regime requires vibrational temperatures exceeding 5000–10 000 K. Using resonant VV-flux (5.12) and neglecting VT-relaxation, the Fokker–Planck equation (5.1) can be then expressed at steady state as:

$$E^2 f^2(E) \left(\frac{2x_e}{T_0} - \hbar\omega \frac{\partial^2 \ln f(E)}{\partial E^2} \right) = F. \tag{5.19}$$

Here F is a constant quantum flux from low vibrational levels (where the quanta appear by electron impact) to high levels (where they disappear in VT-relaxation and chemical reactions). At low vibrational energies, $Ef(E)$ is large: $\frac{2x_e}{T_0} - \hbar\omega \frac{\partial^2 \ln f(E)}{\partial E^2} \approx 0$ and the vibrational distribution is close to the Treanor function.

 Distribution $f(E)$ decreases at higher energies, $Ef(E)$ becomes small, and $\frac{2x_e}{T_0} - \hbar\omega \frac{\partial^2 \ln f(E)}{\partial E^2} = \frac{F}{E^2 f^2(E)}$ cannot be taken as zero. Vibrational distribution deviates from the Treanor function, becomes flatter, and $\frac{\partial^2 \ln f(E)}{\partial E^2} \approx 0$. It leads based on (5.19) to the **hyperbolic plateau distribution** (see Figure 5.2b):

$$\frac{2x_e}{T_0} = \frac{F}{E^2 f^2(E)}, \quad f(E) = \frac{C}{E}, \tag{5.20}$$

where "plateau level" C is determined by power P_{eV} of vibrational excitation per molecule:

$$C(P_{eV}) = \frac{1}{\hbar\omega} \sqrt{\frac{P_{eV} T_0 (\delta_{VV} \hbar\omega)^3}{4x_e k_0 n_0 Q_{10}^{01}}}. \tag{5.21}$$

The vibrational distribution in the strong excitation regime first follows the Treanor function at relatively low energies (see Section 4, details in Fridman and Kennedy 2021):

$$E < E_{Tr} - \hbar\omega \sqrt{\frac{T_0}{2x_e \hbar\omega}}. \tag{5.22}$$

Here $E_{Tr} = \hbar\omega \cdot v_{\min}^{Tr}$ is the Treanor minimum. At vibrational energies exceeding the Treanor minimum ($E > E_{Tr}$), the distribution converts to the hyperbolic (see Figure 5.2b):

$$f(E) = B \frac{E_{Tr}}{E} \exp\left(-\frac{T_0 \hbar\omega}{4x_e T_v^2} - \frac{1}{2} \right), \tag{5.23}$$

where B is the Treanor normalization factor. At even higher vibrational energies:

$$E > E(\text{plateau-VT}) = \frac{1}{\delta_{VT}} \left[\ln \frac{k_0 Q_{10}^{01}}{k_{VT}(E=0)} \frac{E_{Tr}}{\hbar\omega} - \frac{T_0 \hbar\omega}{4x_e T_v^2} - \frac{1}{2} \right], \tag{5.24}$$

VT-relaxation dominates over the resonant VV-exchange, resulting in the hyperbolic plateau (5.23) transition to sharp decreasing Boltzmann distribution with temperature T_0.

We should mention that vibrational distribution like the one described above (see Figure 5.2b) also takes place in the *regime of intermediate excitation*, when the vibrational temperature is less than the so-called VV-exchange radius $T_v < \delta_{VV}^{-1}$, which is about 5000–10 000 K, but sufficiently high to provide the strong Treanor effect:

$$\frac{x_e T_v^2}{T_0 \hbar\omega} \geq 1 \tag{5.25}$$

The vibrational population (5.18) at Treanor minimum $E = E_{Tr}$ is high in this case, resonant VV-exchange dominates, and also provides the hyperbolic plateau at $E > E_{Tr}$.

5.5 Steady-state Vibrational Distributions Controlled by Linear VV- and VT-relaxation Processes in Weak Excitation Regime, the Gordiets Distribution

The vibrational distributions in nonequilibrium plasma are usually controlled by VV-exchange at low excitation levels and by VT-relaxation at high excitation levels. The vibrational excitation of molecules by electron impact, chemical reactions, radiation, etc. is mainly related to the balance of average vibrational energy and temperatures. At steady-state conditions, the Fokker–Planck kinetic equation (5.1) gives $J(E) = const$. Considering that at $E \to \infty$: $\frac{\partial f(E)}{\partial E} = 0, f(E) = 0$, yields $const(E) = 0$. As a result, the steady-state vibrational distributions, provided by VT- and VV-relaxation processes is controlled by linear and nonlinear VV-fluxes ((5.9) and (5.12)), as well as by VT-relaxation flux (5.3):

$$j_{VV}^{(0)}(E) + j_{VV}^{(1)}(E) + j_{VT}(E) = 0. \tag{5.26}$$

In the previous section (Section 5.4), we considered the strong (and intermediate) excitation regimes with very high vibrational temperatures ($\delta_{VV}T_v \geq 1$, $\frac{x_e T_v^2}{T_0 \hbar \omega} \geq 1$), corresponding to domination of the nonlinear VV-flux and the hyperbolic plateau distribution. Let us consider now, a more common case of the so-called weak vibrational excitation:

$$\delta_{VV}T_v < 1, \quad \frac{x_e T_v^2}{T_0 \hbar \omega} < 1, \tag{5.27}$$

usually corresponding to vibrational temperatures in plasma $T_v = 1000 - 5000$ K, when the nonlinear term in (5.26) can be neglected and the Fokker–Planck equation simplified to:

$$\frac{\partial f(E)}{\partial E}(1 + \xi(E)) + f(E) \cdot \left(\frac{1}{T_v} - \frac{2x_e E}{T_0 \hbar \omega} + \xi(E)\tilde{\beta}_0 \right) = 0. \tag{5.28}$$

Here $\xi(E)$ is an exponentially growing ratio of the rate coefficients of VT-relaxation and nonresonant VV-exchange, see ((3.41) and (3.42)), see also (5.3) regarding the $\tilde{\beta}_0$ temperature parameter ($\tilde{\beta}_0 = \dfrac{1 - \exp\left(-\dfrac{\hbar \omega}{T_0}\right)}{\hbar \omega}$; when $T_0 \gg \hbar \omega$, $\tilde{\beta}_0 = \beta_0 = 1/T_0$). Solution of (5.28) gives the vibrational distribution for the weak excitation regime:

$$f(E) = B \exp \left[-\frac{E}{T_v} + \frac{x_e E^2}{T_0 \hbar \omega} - \frac{\tilde{\beta}_0 - \dfrac{1}{T_v}}{2\delta_{VV}} \ln(1 + \xi(E)) \right], \tag{5.29}$$

and corresponds to the discrete **Gordiets distribution** over vibrational quantum numbers:

$$f(v) = f_{Tr}(v) \prod_{i=0}^{v} \frac{1 + \xi_i \exp \dfrac{\hbar \omega}{T_0}}{1 + \xi_i}. \tag{5.30}$$

Here, $f_{Tr}(v)$ is the discrete form of the Treanor distribution (5.15). At low energies ($E < E^*(T_0)$, ((3.41) and (3.42)): $\xi(E) \ll 1$, and the vibrational distribution is close to the Treanor function. At higher energies: $\xi(E) \gg 1$, and the vibrational distribution is exponentially decreasing according to the Boltzmann law with temperature T_0 (see examples in Figure 5.4a). Figure 5.4b illustrates comparison of the actual vibrational distribution ((5.29), and Gordiets distribution) with simplified models: the Bray distribution considering VT-relaxation only when $\xi(E) \geq 1$ and overestimates the actual $f(E)$; and completely classical ($\tilde{\beta}_0 = \beta_0$) Brau distribution underestimating the actual $f(E)$.

Interesting that vibrational distribution function $f(E)$, controlled by the VV- and VT-relaxation processes in the regime of weak excitation always has **an inflection point** ($\frac{\partial^2 \ln f(E)}{\partial E^2} = 0$), see Figure 5.4, corresponding to the vibrational energy:

$$E_{\inf l} = E^*(T_0) - \frac{1}{2\delta_{VV}} \ln \frac{\delta_{VV} T_0}{x_e} < E^*(T_0); \tag{5.31}$$

and even can include the **domain of inverse population** ($\frac{\partial \ln f(E)}{\partial E} > 0$), if:

$$2\delta_{VV}(E^*(T_0) - E_{Tr}) > \ln \frac{x_e}{\delta_{VV} T_0}. \tag{5.32}$$

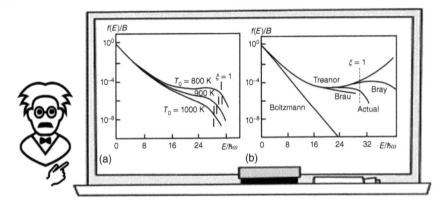

Figure 5.4 Nonequilibrium vibrational distribution functions in nitrogen at $T_v = 3000\,\text{K}$, $\xi = 1$ indicates equality of VV- and VT-relaxation rates: (a) actual distribution at different gas temperatures $T_0 = 800, 900, 1000\text{K}$; (b) comparison of different vibrational kinetic models, $T_0 = 800\,\text{K}$.

5.6 Direct Effect of Vibrational Excitation by Electron Impact on the Nonequilibrium Vibrational Distribution Functions

At high degrees of ionization, the frequency of vibrational excitation by electron impact (eV-processes) becomes comparable with that of VV-exchange, and eV processes are able not only determine energy balance and vibrational temperature of the system but directly influence evolution of the vibrational distribution functions. In contrast to VV- and VT-exchange, the eV-relaxation can be effective not only as a one-quantum, but also as a multi-quantum process, see the Fridman's approximation (3.4). While relevant eV-flux along the vibrational energy spectrum is quite sophisticated, the simplest linear eV-flux $j_{eV}^{(1)}$ describing transfer of one or a few quanta can be expressed similarly to the linear VV- and VT-fluxes in the Fokker–Planck form:

$$j_{eV}^{(1)}(E) = -D_{eV}\left(\frac{\partial f(E)}{\partial E} + \frac{1}{T_e}f(E)\right), \tag{5.33}$$

where D_{eV} is the one-quantum, vibrational excitation diffusion coefficient in energy space:

$$D_{eV} = \lambda k_{eV}^0 n_e(\hbar\omega)^2, \quad \lambda \approx \frac{2}{\alpha^3}. \tag{5.34}$$

The one-quantum excitation rate coefficient $k_{eV}^0 = k_{eV}(0,1)$ and parameter α correspond to the Fridman's approximation (3.4), numerically $\alpha \approx 0.5 \div 0.7$. The factor λ accounts for the transfer of a few quanta and numerically in nitrogen $\lambda \approx 10$. High ionization degrees can lead to domination of eV processes, which according to the Fokker–Planck equation (5.1) with the eV-flux (5.33) will lead to the Boltzmann vibrational distribution with temperature T_e: $f(E) \propto \exp(-E/T_e)$. Keeping in mind that electron temperature can significantly exceed vibrational one $T_e \gg T_v$, this Boltzmann distribution looks like hyperbolic plateau (see Section 5.4) but is determined by very different physics.

At intermediate excitation regimes, the eV-processes can build such "Boltzmann plateau" only at intermediate vibrational levels, while the low excitation levels follow the Treanor distribution and high excitation levels follow the Boltzmann distribution with gas temperature, see Figure 5.5.

Detailed information and references on this sophisticated subject can be found in Fridman and Kennedy (2021). Also, those who are interested can find in this book the kinetic specifics of relevant *non-steady-state vibrational distributions* in strongly nonequilibrium discharges.

5.7 Nonequilibrium Vibrational Distributions of Polyatomic Molecules

Vibrational kinetics of polyatomic molecules (see Section 3.8) at low excitation levels is discrete and similar to that of diatomic molecules. Interesting example of such discrete vibrational distributions for polyatomic molecules is presented in Figure 5.6, where even the Treanor effect can be seen for CO_2 molecules controlled by high vibrational

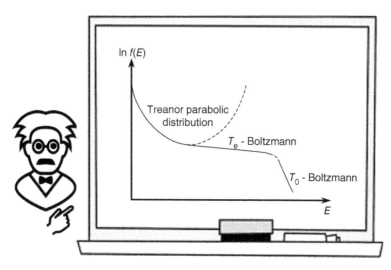

Figure 5.5 Illustration of direct effect of eV-processes on vibrational distribution functions at intermediate excitation regimes; eV-related "Boltzmann plateau."

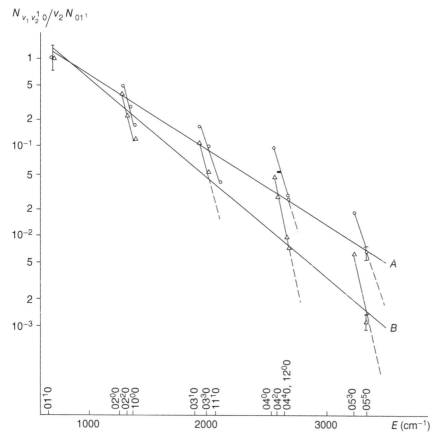

Figure 5.6 Vibrational distribution function of CO_2 at low levels (symmetric valence and deformation modes): (a) $T_{v1} = 780 \pm 40$ K, $T_{r1} = 150 \pm 15$ K; (b) $T_{v2} = 550 \pm 40$ K, $T_{r2} = 110 \pm 10$ K.

temperature. Also, these experiments demonstrate the Boltzmann distribution with low gas temperature (rotational T_r and translational T_0 are equal) for almost resonant symmetric valence and deformational modes (Baronov et al. 1989). Specific features of polyatomic molecules, however, manifest themselves at higher excitation levels, when interaction between vibrational modes is strong and the molecules are in the state of vibrational quasi-continuum (see Section 3.8).

Distribution $f(E)$ over the total vibrational energy E in the quasi-continuum can be then found from the Fokker–Planck kinetic equation:

$$\frac{\partial f(E)}{\partial E} + \frac{\partial}{\partial E}\left(j_{VV}^{poly} + j_{VT}^{poly}\right) = 0, \tag{5.35}$$

where VT-relaxation flux can be expressed for polyatomic molecules as:

$$j_{VT}^{poly} = -\sum_{i=1}^{N} D_{VT}^i(E)\rho(E)\left[\frac{\partial}{\partial E}\left(\frac{f(E)}{\rho(E)}\right) + \frac{1}{T_0}\frac{f(E)}{\rho(E)}\right]. \tag{5.36}$$

The main peculiarity of this flux with respect to diatomic molecules (5.3) is in the statistical factor $\rho(E) \propto E^{s-1}$, showing density of vibrational states and considering the effective number "s" of vibrational degrees of freedom. The summation is over all N vibrational modes "i" with relevant diffusion coefficients:

$$D_{VT}^i(E) = <\left(E_{VT}^i\right)^2 > k_0 n_0, \tag{5.37}$$

where $<\left(E_{VT}^i\right)^2 >$ is determined by (3.57). The VT-flux (5.35) equals to zero for the Boltzmann distribution function with the statistical weight factor: $f(E) \propto \rho(E)\exp(-E/T_0)$.

The VV-flux j_{VV}^{poly} in quasi-continuum (5.35), has only linear component $j_{VV}^{(0)}$ in contrast to diatomic molecules (5.8) and corresponds to VV-exchange between highly excited molecules with low excited molecules of the "thermal reservoir" (nonlinear VV-flux $j_{VV}^{(1)}$ can be neglected because of the resonant nature of VV-exchange in quasi-continuum):

$$j_{VV}^{poly}(E) = -\sum_{i=1}^{N} D_{VV}^i(E)\rho(E)\left[\frac{\partial}{\partial E}\left(\frac{f(E)}{\rho(E)}\right) + \frac{1}{T_v^i}\left(\frac{f(E)}{\rho(E)}\right)\right.$$
$$\left. -\left(\frac{f(E)}{\rho(E)}\right)\frac{E}{sT_0\hbar\omega_{0i}}\sum_{i=1}^{N}x_{ij}(1+\delta_{ij})g_i\frac{\omega_{0i}}{\omega_{0j}}\right]. \tag{5.38}$$

Here, T_{vi} and $\hbar\omega_{0i}$ are vibrational temperatures and quantum for different modes; s is number of effective vibrational degrees of freedom; δ_{ij} is the Kronecker delta-symbol; x_{ij} are the anharmonicity coefficients; g_i is the degree of degeneracy of a vibrational mode. The diffusion coefficient in energy space D_{VV}^i is related to the VV-exchange between excited molecules with molecules of the thermal reservoir with temperature T_v:

$$D_{VV}^i(E) = <\left(E_{VV}^i\right)^2 > k_0 n_0; \tag{5.39}$$

$<\left(E_{VV}^i\right)^2 >$ is determined by ((3.59) and (3.60)).

The Fokker–Planck equation (5.35) in steady state conditions is $j_{VV}^{poly} + j_{VT}^{poly} = 0$. Thus, in regimes controlled only by VV-exchange ($j_{VV}^{poly} = 0$), vibrational distribution is:

$$\frac{f(E)}{\rho(E)} = B\exp\left[-\frac{E}{T_{Va}} + \frac{E^2}{2sT_0}\sum_{j=1}^{N}\frac{x_{aj}}{\hbar\omega_{0j}}(1+\delta_{aj})g_j\right]. \tag{5.40}$$

This is the *Treanor distribution for polyatomic molecules in quasi-continuum*. Statistical weight factor $\rho(E) \propto E^{s-1}$ characterizes density of the vibrational states; B is normalization factor. Effective coefficient of anharmonicity for polyatomic molecules

$$x_m = \frac{1}{2s}\sum_{j=1}^{N}\frac{x_{aj}}{\hbar\omega_{0j}}(1+\delta_{aj})g_j \tag{5.41}$$

is less than that of diatomic molecules, which suppresses the Treanor effect, especially at high numbers of vibrational degrees of freedom.

The Treanor effect is still valid for polyatomic molecules even when vibrations are in quasi-continuum, which is due to the third term in (5.38) reflecting balance of direct and reverse VV-exchange processes (see Section 3.8). Each act of VV-exchange between weakly excited molecule and a highly excited one in quasi-continuum is actually nonresonant and leads to the transfer of the anharmonic defect of energy into translational motion. This effect can be interpreted in the quasi-classical approximation as conservation in the collision process of the adiabatic invariant-shortened action, which is the ratio of vibrational energy of each oscillator to its frequency:

$$\frac{E_{v1}}{\omega_1} + \frac{E_{v2}}{\omega_2} = const. \tag{5.42}$$

This conservation of the adiabatic invariant corresponds in quantum mechanics to conservation of the number of quanta during a collision. Therefore, transfer of vibrational energy E_v from a higher frequency oscillator to a lower frequency one results in a decrease of the total vibrational energy, losses of energy into translational motion even in the conditions of quasi-continuum, resulting in the Treanor effect, see more on this subject in Capitelli (1986), and Landau and Lifshitz (2002).

VT-relaxation (5.36) perturbs the vibrational distribution at higher excitation levels. The Fokker–Planck equation $j_{VV}^{poly} + j_{VT}^{poly} = 0$ results then in the integral:

$$f(E) = B\rho(E) \exp\left(-\int_{E_c}^{E} \frac{\frac{1}{T_{va}} - \frac{2x_m E'}{T_0 \hbar\omega_{0a}} + \frac{\xi(E')}{T_0}}{1 + \xi(E')} dE'\right). \tag{5.43}$$

Here E_c designates energy at where polyatomic molecule enters the quasi-continuum; factor $\xi(E) = \sum_i D_{VT}^i / D_{VV}^a$ characterizes ratio of VT- and VV-relaxation rates in quasi-continuum, see Section 3.8. This factor is much larger for polyatomic molecules than for diatomic ones. Therefore, transition to the Boltzmann distribution with translational temperature T_0 occurs in quasi-continuum at lower levels of vibrational excitation.

5.8 Macro-kinetics of Chemical Reactions of Vibrationally Excited Molecules

The macro-kinetic chemical reaction rates are self-consistent with the influence of these reactions on the vibrational distributions $f(E)$, which can be considered by introducing into the Fokker–Planck equation of an additional reaction-related flux:

$$j_R(E) = -\int_E^\infty k_R(E')n_0 f(E')dE' = -J_0 + n_0 \int_0^E k_R(E')f(E')dE'. \tag{5.44}$$

Here $J_0 = -j_R(E=0)$ is the total flux of the molecules in the chemical reaction (total reaction rate $w_R = n_0 J_0$); $k_R(E)$ is the microscopic reaction rate coefficient (3.65).

In the weak excitation regime controlled by nonresonant VV- and VT-relaxation processes: $j_{VV}^{(0)} + j_{VT} + j_R = 0$. In this case typical for plasma chemistry, the Fokker–Planck equation can be expressed as:

$$\frac{\partial f(E)}{\partial E}(1 + \xi(E)) + f(E) \cdot \left(\frac{1}{T_v} - \frac{2x_e E}{T_0 \hbar\omega} + \tilde{\beta}_0 \xi(E)\right) = \frac{1}{D_{VV}(E)} j_R(E). \tag{5.45}$$

The solution $f(E)$ of this nonuniform linear equation can be presented with respect to solution $f^{(0)}(E)$ of the corresponding uniform equation, see (5.29), as:

$$f(E) = f^{(0)}(E)\left[1 - \int_0^E \frac{-j_R(E')dE'}{D_{VV}(E')f(E')(1 + \xi(E'))}\right]. \tag{5.46}$$

The function $-j_R(E)$ determines the flux of molecules along the energy spectrum, which are going to participate in chemical reaction at $E \geq E_a$; E_a is activation energy. At low energies $E < E_a$, reaction can be neglected: $-j_R(E) = \int_{E_a}^\infty k_R(E')n_0 f(E')dE' = J_0 = const$. At these energies ($E < E_a$), the perturbation of the vibrational distribution $f^{(0)}(E)$ by reaction is:

$$f(E) = f^{(0)}(E)\left[1 - J_0 \int_0^E \frac{dE'}{D_{VV}(E')f^{(0)}(E')(1 + \xi(E'))}\right]. \tag{5.47}$$

At $E \geq E_a$: $j_R(E) \approx -k_R(E)n_0 f(E)\hbar\omega$. Then the integral equation (5.46) can be solved as:

$$f(E) \propto f^{(0)}(E) \exp\left[-\int_{E_a}^{E} \frac{k_R(E')\, \hbar\omega\, dE'}{D_{VV}(E')(1 + \xi(E'))}\right], \tag{5.48}$$

which determines the decrease of the vibrational distribution function at $E \geq E_a$. Relevant consideration for the strong excitation regime can be found in Fridman and Kennedy (2021).

The vibrational distribution functions considering chemical reactions are illustrated in Figure 5.7 for both weak and strong excitation regimes. The vibrational reaction kinetics can be discussed in two extreme limits of slow and fast chemical reactions.

(1) **The fast reaction limit** implies, that the chemical reaction is fast for $E \geq E_a$:

$$D_{VV}(E = E_a) \ll n_0 \cdot k_R(E + \hbar\omega) \cdot (\hbar\omega)^2, \tag{5.49}$$

and the chemical process is limited by the VV-diffusion along the vibrational spectrum to the threshold $E = E_a$. In this case, the distribution function $f(E)$ falls very fast at $E > E_a$, and one can assume in (5.46) that $f(E = E_a) = 0$, see Figure 5.7:

$$1 = \int_0^{E_a} \frac{-j_R(E')dE'}{D_{VV}(E')f^{(0)}(E')(1 + \xi(E'))}. \tag{5.50}$$

Considering $-j_R(E) = J_0 = const$ at $E < E_a$, the total chemical process rate is:

$$w_R = n_0 J_0 = n_0 \left\{\int_0^{E_a} \frac{dE'}{D_{VV}(E')f^{(0)}(E')(1 + \xi(E'))}\right\}^{-1}. \tag{5.51}$$

The chemical reaction rate in this case is determined by the frequency of the VV-relaxation and by the nonperturbed vibrational distribution function $f^{(0)}(E)$. It is not sensitive to the detailed characteristics of the elementary chemical reaction once it is sufficiently large. It is true, in particular, for *plasma-chemical monomolecular dissociation of polyatomic molecules* such as CO_2 and H_2O, where rate coefficient can be expressed as:

$$k_R^{macro} = \frac{k_{VV}^0}{\Gamma(s)} \frac{\hbar\omega}{T_v} \left(\frac{E_a}{T_v}\right)^s \exp\left(-\frac{E_a}{T_v}\right) \sum_{r=0}^{\infty} \frac{(s+r-1)!}{(s-1)!r!} \frac{\gamma(r+1, E_a/T_v)}{(E_a/T_v)^r}. \tag{5.52}$$

In this relation, $\Gamma(s)$ is the gamma-function; $\gamma(r+1, E_a/T_v)$ is the incomplete gamma-function; s is the number of vibrational degrees of freedom; k_{VV}^0 and $\hbar\omega$ are the lowest vibrational quantum and the corresponding low energy VV-exchange rate coefficient. The sum in (5.52) is not a strong function of E_a/T_v, and for numerical calculations can be taken as 1.1–1.3 if $E_a = 3 \div 5$ eV and $T_v = 1000 \div 4000$ K.

(2) **The slow reaction limit** is opposite to (5.49). VV-exchange population of the reactive states $E > E_a$ is faster in this case than the elementary chemical reaction itself. Vibrational distribution then is not perturbed by the chemical reaction $f(E) \approx f^{(0)}(E)$, see Figure 5.7, and the total macroscopic reaction rate coefficient can be expressed as:

$$k_R^{macro} = \int_0^{\infty} k_R(E')f(E')\, dE'. \tag{5.53}$$

The elementary reaction rates $k_R(E)$, see Sections 3.11 and 3.12, are crucial in this case.

5.9 Macro-kinetics of Vibrational Energy Losses Due to VT- and VV-relaxation

The macroscopic losses of average vibrational energy $\varepsilon_v = \int_0^{\infty} E f(E)\, dE$ due to VT- and VV-relaxation can be analyzed by multiplying the Fokker–Planck equation (5.1) by E with following integration leading to the energy balance equation:

$$\frac{d\varepsilon_v}{dt} = \int_0^{\infty} j_{VV}(E)dE + \int_0^{\infty} j_{VT}(E)dE. \tag{5.54}$$

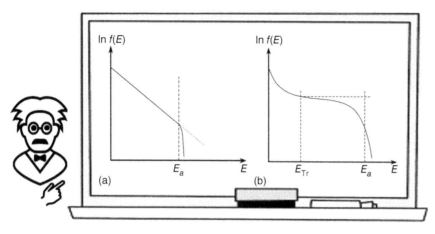

Figure 5.7 Illustration of influence of chemical reactions on the vibrational distribution functions: (a) week excitation regime, (b) strong excitation regime.

(1) Consider first **total energy losses related to VT-relaxation**, assuming the VT-diffusion coefficient in energy space as:

$$D_{VT}(E) = k_{VT}(E)\, n_0(\hbar\omega)^2 = k_{VT}^0 \left(\frac{E}{\hbar\omega} + 1\right) \exp(\delta_{VT}E)\, n_0(\hbar\omega)^2. \tag{5.55}$$

Then the energy balance equation (5.54) gives total macroscopic VT-related losses:

$$\left(\frac{d\varepsilon_v}{dt}\right)^{VT} = D_{VT}(0)f(0) - \int_0^\infty D_{VT}(0)\left[\frac{(\tilde{\beta}_0 - \delta_{VT})E - 1}{\hbar\omega} - (\tilde{\beta}_0 - \delta_{VT})\right] \exp(\delta_{VT}E)f dE. \tag{5.56}$$

These VT-vibrational energy losses can be subdivided into two classes – first one, related to the low vibrational levels, and prevailing in conditions of weak excitation; and second one, related to the high vibrational levels, and dominating when excitation is strong.

(2) **VT-relaxation losses from low vibrational levels** calculated from (5.56) are usually referred to as **Losev formula** (here $\varepsilon_{v0} = \varepsilon_v(T_v = T_0)$):

$$\left(\frac{d\varepsilon_v}{dt}\right)_L = -k_{VT}(0)n_0\left[1 - \exp\left(-\frac{\hbar\omega}{T_0}\right)\right] \cdot \left[\frac{1 - \exp\left(-\dfrac{\hbar\omega}{T_v}\right)}{1 - \exp\left(-\dfrac{\hbar\omega}{T_v} + \delta_{VT}\hbar\omega\right)}\right]^2 (\varepsilon_v - \varepsilon_{v0}). \tag{5.57}$$

Neglecting anharmonicity ($\delta_{VT} = 0$), it becomes the **Landau–Teller relation**:

$$\left(\frac{d\varepsilon_v}{dt}\right)_L = -k_{VT}(0)n_0\left[1 - \exp\left(-\frac{\hbar\omega}{T_0}\right)\right](\varepsilon_v - \varepsilon_{v0}). \tag{5.58}$$

In equilibrium between vibrational and translational degrees of freedom $T_v = T_0$, hence $\varepsilon_v = \varepsilon_0$, and therefore losses of vibrational energy ((5.57) and (5.58)) become zero.

(3) **VT-relaxation losses from high vibrational levels** are mostly related to the highest levels before the fast fall of the distribution function due to VT-relaxation or chemical reaction. In the fast reaction limit assuming low gas temperature T_0, (5.56) gives:

$$\left(\frac{d\varepsilon_v}{dt}\right)_H^{VT} \approx -D_{VT}(0)f(0) - \left[\frac{(\tilde{\beta}_0 - \delta_{VT})E_a - 1}{\hbar\omega} - (\tilde{\beta}_0 - \delta_{VT})\right] D_{VT}(0)\exp(\delta_{VT}E_a)f^{(0)}(E_a)\Delta$$

$$\approx -k_{VT}(E_a)\, n_0 \hbar\omega\, f^{(0)}(E_a)\,\Delta. \tag{5.59}$$

where $f^{(0)}(E)$-is the vibrational distribution not perturbed by chemical reaction, and the effective integration domain: $\Delta = \left|\frac{1}{T_v} - \delta_{VT} - \frac{2x_e E_a}{T_0\hbar\omega}\right|^{-1}$. Assuming $\xi(E_a) \ll 1$, these VT-losses are convenient to express per one act of the fast chemical reaction:

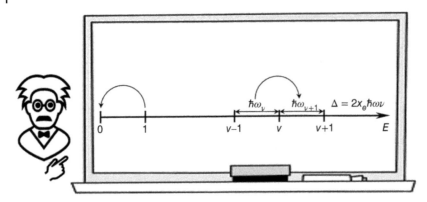

Figure 5.8 Vibrational energy losses due to nonresonant nature of VV-exchange processes.

$$\Delta\varepsilon_{VT} \approx \frac{k_{VT}(E_a)}{k_{VV}(E_a)}\Delta. \tag{5.60}$$

This relation reflects that the fast chemical reactions and VT-relaxation from high levels are related to the same excited molecules with energies slightly exceeding E_a; frequencies are, however, different – and proportional respectively to k_{VV} and k_{VT}.

(4) **Vibrational energy losses due to the non-resonance nature of VV-exchange.** The VV-losses per one act of chemical reaction can be calculated in frameworks of the model, illustrated in Figure 5.8. A diatomic molecule is excited by electron impact from the zero-level to the first vibrational level; so, a quantum comes to the system as $\hbar\omega$. Further population of higher excited levels is due to VV-exchange, with quanta becoming smaller and smaller due to anharmonicity. Each step up on the "vibrational ladder" requires energy transfer $2x_e\hbar\omega v$ from vibrational to translational degrees of freedom. The total losses corresponding to excitation of a molecule to the nth vibrational level is:

$$\Delta\varepsilon^{VV}(n) = \sum_{v=0}^{v=n-1} 2x_e\hbar\omega\,v = x_e\hbar\omega\,(n-1)\,n. \tag{5.61}$$

In particular, for dissociation of diatomic molecules $n = n_{max} \approx 1/2x_e$, the losses of vibrational energy per one act of dissociation, associated with anharmonicity and non-resonance nature of VV-relaxation, are equal to the dissociation energy:

$$\Delta\varepsilon_D^{VV}(n = n_{max}) \approx \frac{\hbar\omega}{4x_e} = D_0. \tag{5.62}$$

Therefore, the total vibrational energy necessary for dissociation provided by the VV-exchange is equal to not D_0, but $2D_0$. If a reaction is stimulated by vibrational excitation and has activation energy $E_a \ll D_0$, the VV-losses are:

$$\Delta\varepsilon_R^{VV}(E_a) \approx \frac{1}{4}D_0\left(\frac{E_a}{D_0}\right)^2 = x_e\frac{E_a^2}{\hbar\omega}, \tag{5.63}$$

and can be more significant than relevant losses due to VT-relaxation (5.60).

5.10 Vibrational Kinetics in Gas Mixtures, Treanor Isotopic Effect

Vibrationally excitation in molecular gas mixtures is controlled not only by VV-, VT-, and eV-processes but also by the nonresonant VV'-exchange between different molecular components. Even if the difference of their oscillation frequencies is small (like in case of isotopic mixtures), the VV'-exchange can result in significant differences in the level of their vibrational excitation. Usually, a molecular component with lower vibrational quanta becomes excited to the higher vibrational levels.

Consider the Fokker–Planck equation for vibrational distributions $f_i(E)$ of two-component mixture: subscripts $i = 1, 2$ correspond to the first (lower oscillation frequency) and second (higher frequency) molecular components:

$$\frac{\partial f_i(E)}{\partial t} + \frac{\partial}{\partial E}\left[j_{VV}^{(i)}(E) + j_{VT}^{(i)}(E) + j_{eV}^{(i)}(E) + j_{VV'}^{(i)}(E)\right]. \tag{5.64}$$

The VV-, VT- and eV-fluxes $j_{VV}^{(i)}(E), j_{VT}^{(i)}(E), j_{eV}^{(i)}(E)$ are like those for a one-component gas; the VV'-flux providing predominant population of a component with lower oscillation frequency ($i = 1$) can be expressed (Macheret et al. 1980(a,b)), as:

$$j_{VV'}^{(i)} = -D_{VV'}^{(i)}\left(\frac{\partial f_i}{\partial E} + \beta_i f_i - 2x_e^{(i)}\beta_0 \frac{E}{\hbar\omega_i}f_i\right). \tag{5.65}$$

Here, $x_e^{(i)}, \omega_i$ are the anharmonicity coefficient and the oscillation frequency for the molecular component "i"; $D_{VV'}^{(i)} = k_{VV'}^{(i)} n_{l\neq i} (\hbar\omega_i)^2$ is diffusion coefficient in energy space; $k_{VV'}^{(i)}$ is intercomponent VV' – exchange rate coefficient related to the resonant $k_{VV}^{(0)}$ at low vibrational levels of the component "i" as:

$$k_{VV'}^{(i)} = k_{VV}^{(0)}\exp\left[\frac{\delta_{VV'}\hbar(\omega_i - \omega_l)}{2x_e^{(i)}} - \delta_{VV'}E\right]. \tag{5.66}$$

The inverse temperature factor β_i in the flux (5.65) is $\beta_i = \frac{\omega_i}{\omega_l}\beta_{vl} + \frac{\omega_i - \omega_l}{\omega_i}\beta_0$; other inverse temperature parameters herein are conventional $\beta_0 = T_0^{-1}, \beta_{vi} = (T_{vi})^{-1}$. Isotope mixtures usually consist of a large fraction of a light gas component (higher frequency of molecular oscillations, concentration n_2), and only a small fraction of heavy component (lower oscillation frequency, concentration n_1). In this case, steady-state solution of the Fokker–Planck kinetic equation (5.64) gives the following vibrational distribution functions for two components of a gas mixture:

$$f_{1,2}(E) = B_{1,2}\exp\left[-\int_0^E \frac{\beta_{1,v2} - 2x_e^{(1,2)}\beta_0 \frac{E'}{\hbar\omega_{1,2}} + \tilde{\beta}_0^{(1,2)}\xi_{1,2} + \beta_e\eta_{1,2}}{1 + \xi_{1,2} + \eta_{1,2}}dE'\right]. \tag{5.67}$$

Here $B_{1,2}$ are the normalization factors; $\beta_e = T_e^{-1}$ is inverse electron temperature; factors ξ_i and η_i describe the relative contribution of VT- and eV-processes with respect to VV (VV') – exchange $\xi_i(E) = \xi_i(0)\exp(2\delta_{VV}E), \eta_i(E) = \eta_i(0)\exp(\delta_{VV}E)$.

VT- and eV-processes can be neglected ($\xi_i \ll 1, \eta_i \ll 1$) in (5.67) at low excitation levels. Then, both gas components ($i = 1, 2$) have Treanor distributions with the same translational temperature T_0, but with different vibrational temperatures T_{v1} and T_{v2}:

$$\frac{\omega_1}{T_{v1}} - \frac{\omega_2}{T_{v2}} = \frac{\omega_1 - \omega_2}{T_0}. \tag{5.68}$$

This relation is known as **the Treanor formula for isotopic mixture**. The Treanor formula shows, that under nonequilibrium conditions ($T_{v1,v2} > T_0$) of isotopic mixture, the component with the lower oscillation frequency (heavier isotope) has a higher vibrational temperature. This Treanor effect can be applied for isotope separation in plasma chemical reactions, stimulated by vibrational excitation. The ratio of rate coefficients of chemical reactions for two isotopes $\kappa = k_R^{(1)}/k_R^{(2)}$ is proportional to the ratio of population of their vibrational levels $E \geq E_a$. It is called the **coefficient of selectivity**, and based on (5.68) can be expressed as:

$$\kappa \approx \exp\left[\frac{\Delta\omega}{\omega}E_a\left(\frac{1}{T_0} - \frac{1}{T_{v2}}\right)\right], \tag{5.69}$$

where $\frac{\Delta\omega}{\omega} = \frac{\omega_2 - \omega_1}{\omega_2}$ is the relative defect of resonance. The coefficient of selectivity (5.69) only slightly depends on the vibrational temperature when $T_v \gg T_0$, and strongly depends on translational temperature T_0 and activation energy E_a.

Interesting that the selectivity "directions" of the nonequilibrium Treanor effect ($T_v \gg T_0$) and isotopic effect in convention quasi-equilibrium kinetics are opposite. The Treanor effect results in faster reactions of heavy isotopes (smaller vibrational quantum), while in quasi-equilibrium, the heavy isotopes with the lower value of $\frac{1}{2}\hbar\omega$ (zero-vibration level) have higher activation energies and therefore react slower. Also, the nonequilibrium ($T_v \gg T_0$) selectivity coefficients (5.69) in cold plasma are usually greater than those in quasi-equilibrium because of low values of gas temperature, see Figure 5.9.

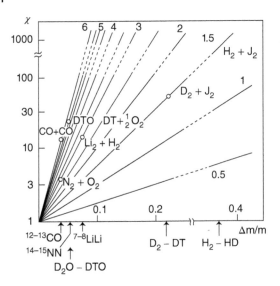

Figure 5.9 Coefficients of selectivity for the Treanor isotopic effect in cold plasma as a function of relative molecular mass difference of the isotopes. Number on curves (from 0.5 to 6) represents activation energies (in eV) of the specific plasma-chemical reactions indicated by circles.

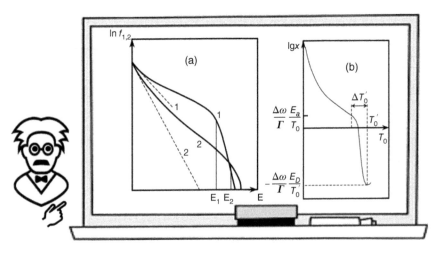

Figure 5.10 Vibrational distributions and isotopic effect in diatomic molecular plasma ($T_v \gg T_0$): (a). vibrational distributions of isotopes (1,2), dashed lines show Boltzmann functions, energies E1 and E2 indicate points of domination of VT-relaxation; (b). coefficient of selectivity as a function of T_0, temperature T_0' corresponds to the strongest reverse isotopic effect.

When contribution of VT-relaxation is essential, the vibrational distribution (5.67) for isotopic mixture after integration can be presented as:

$$f_{1,2}(E) = B_{1,2} \exp\left[-\beta_{1,v2}E + \frac{x_e^{(1,2)}\beta_0 E^2}{\hbar\omega_{1,2}} - \frac{\tilde{\beta}_0^{(1,2)}}{2\delta_{VV}}\ln(1+\xi_{1,2})\right], \tag{5.70}$$

which is illustrated in Figure 5.10a. The population $f_1(E)$ of low vibrational levels is higher for the heavier isotope ("1," usually small additive) corresponding to the Treanor effect. At higher levels of excitation, however, the vibrational population of light isotope exceeds that of the heavier one. This phenomenon determines the reverse isotopic effect in vibrational kinetics (Macheret et al. 1980). The function $f_1(E)$ for the heavier isotope (small additive) is determined by the VV'-exchange, which is slower than VV-exchange because of the defect of resonance. As a result, the VT-relaxation makes the vibrational distribution $f_1(E)$ start falling at lower energies $E_1(\xi_1 = 1)$ with respect to the distribution function $f_2(E)$ of the main isotope, which is determined by VV-exchange and starts falling at the higher vibrational energy $E_2(\xi_2) > E_1$:

$$E_1 = \frac{1}{2\delta_{VV}}\ln\frac{k_{VV}(0)}{k_{VT}(0)} - \frac{\hbar\Delta\omega}{4x_e}, \quad E_2 = \frac{1}{2\delta_{VV}}\ln\frac{k_{VV}(0)}{k_{VT}(0)}. \tag{5.71}$$

Thus, the reverse isotopic effect takes place if $E_1 < E_a < E_2$ (see Figure 5.10a) and the light isotope is excited stronger and reacts faster than the heavier one. The coefficient of selectivity for the reverse isotopic effect can be expressed as (Macheret et al. 1980):

$$\kappa \approx \exp\left(-\frac{\Delta\omega}{\omega}\frac{D_0}{T_0}\right).$$

(5.72)

where D_0 is the dissociation energy of the vibrationally excited diatomic molecules participating in the chemical reaction. The selectivity coefficient (5.72) for the reverse isotopic effect, does not depend directly on vibrational temperature. Also, the activation energy $E_a < D_0$ is not explicitly presented in (5.72). Considering that $E_a < D_0$, the reverse isotopic effect is much stronger than the direct one, but can be achieved only in a narrow range of translational temperatures:

$$\frac{\Delta T_0}{T_0} = 2\frac{\Delta\omega}{\omega}\left(1 - \frac{E_a}{D_0}\right)^{-1}.$$

(5.73)

The selectivity coefficient dependence on translational temperature is illustrated in Figure 5.10b. It shows that direct effect takes place at relatively low translational temperatures. By increasing the translational temperature one can find the narrow temperature range (5.73) of the reverse effect, where the isotopic effect changes "direction" and becomes much stronger. Additional information on this subject, as well as description of influence of eV processes on vibrational kinetics in mixtures and relevant integral effect of plasma-chemical isotope separation can be found in a relevant review in Fridman and Kennedy (2021).

5.11 Physical Kinetics of Population of Electronically Excited States in Plasma

The transfer of electronic excitation energy in collisions of heavy particles is effective, in contrast to VV-exchange, only for a limited number of specific electronically excited states close to resonance. The transitions between electronic states are mostly due to collisions with plasma electrons at degrees of ionization exceeding 10^{-6}. Population of the highly electronically excited states in plasma $n(E)$, due to energy exchange with electron gas, can be described in the framework of the Fokker–Planck equation:

$$\frac{\partial n(E)}{\partial t} = \frac{\partial}{\partial E}\left[D(E)\left(\frac{\partial n(E)}{\partial E} - \frac{\partial \ln n^0}{\partial E}n(E)\right)\right],$$

(5.74)

where $n^0(E)$ is the quasi-equilibrium population of the electronically excited states, corresponding to the electron temperature T_e:

$$n^0(E) \propto E^{-(5/2)}\exp\left(-\frac{E_1 - E}{T_e}\right);$$

(5.75)

E is the absolute value of the bonded electron energy; transition to continuum corresponds to the zero-electron energy $E = 0$; E_1 is the ground state energy ($E_1 \geq E$). For $E \gg T_e$, (5.74) can be simplified $\frac{\partial \ln n^0(E)}{\partial E} = \frac{1}{T_e}$. The diffusion coefficient $D(E)$ in energy space, related to the energy exchange between bonded electrons of highly electronically excited particles with plasma electrons, can be expressed as:

$$D(E) = \frac{4\sqrt{2\pi}\,e^4 n_e E}{3\sqrt{mT_e}(4\pi\varepsilon_0)^2}\Lambda,$$

(5.76)

where Λ is Coulomb logarithm for the excited state with ionization energy E.

Steady-state solution of the Fokker–Planck equation (5.74) is convenient to express using variable $y(E) = \frac{n(E)}{n^0(E)}$. Then boundary conditions are $y(E_1) = y_1$, $y(0) = y_e y_i$. Parameters: y_e, y_i are the electron and ion densities in plasma, divided by the corresponding equilibrium values. Then, solution of (5.74) can be expressed as:

$$y(E) = \frac{y_1\,\chi\left(\frac{E}{T_e}\right) + y_e y_i\left[\chi\left(\frac{E_1}{T_e}\right) - \chi\left(\frac{E}{T_e}\right)\right]}{\chi\left(\frac{E_1}{T_e}\right)},$$

(5.77)

where $\chi(x)$ is integral function: $\chi(x) = \frac{4}{3\sqrt{\pi}} \int_0^x t^{3/2} \exp(-t)\, dt$, having the following asymptotics:

$$\chi(x) \approx 1 - \frac{4}{3\sqrt{\pi}} e^{-x} x^{3/2}, \quad \text{if } x \gg 1, \tag{5.78}$$

$$\chi(x) \approx \frac{1}{2\sqrt{\pi}} x^{5/2}, \quad \text{if } x \ll 1. \tag{5.79}$$

For electronically excited levels $E \ll T_e \ll E_1$ close to continuum, the relative population is:

$$y(E) \approx y_e y_i \left[1 - \frac{1}{2\sqrt{\pi}} \left(\frac{E}{T_e} \right)^{5/2} \right] + y_1 \frac{1}{2\sqrt{\pi}} \left(\frac{E}{T_e} \right)^{5/2} \rightarrow y_e y_i. \tag{5.80}$$

This population of electronically excited states is decreasing exponentially with Boltzmann temperature T_e and the absolute value corresponding to equilibrium with continuum $y(E) \rightarrow y_e y_i$. For the opposite case $E \gg T_e$, population of the excited states is far from continuum and can be found as:

$$y(E) = y_1 + y_e y_i \frac{4}{3\sqrt{\pi}} \exp\left(-\frac{E}{T_e} \right) \cdot \left(\frac{E}{T_e} \right)^{3/2} \rightarrow y_1. \tag{5.81}$$

When electronic excitation energy is far from continuum, the population is also exponential with effective temperature T_e, but the absolute value corresponds to equilibrium with the ground state.

 In general, the Boltzmann distribution of electronically excited states with the temperature equal to the temperature of plasma electrons requires high degree of ionization $n_e/n_0 \geq 10^{-3}$, although domination of energy exchange with electron gas requires only $n_e/n_0 \geq 10^{-6}$. This is mostly due to influence of resonance transitions and the non-Maxwellian behavior of EEDF at the lower degrees of ionization. Additional information on the subject, including contribution of radiation, which is especially important at low pressures (below 1–10 Torr) can be found in the book of Biberman et al. (1987).

5.12 Physical Kinetics of the Rotationally Excited Molecules, Canonical Invariance

The rotational and translational degrees of freedom are usually close to equilibrium between themselves even in strongly nonequilibrium discharges and can be characterized by the same temperature T_0. Consider, as an example, the rotational and translational relaxation of small admixture of relatively heavy diatomic molecules (m_{BC}) in a light inert gas (m_A) in the frameworks of Fokker–Planck kinetic equation:

$$\frac{\partial f(E_t, E_r, t)}{\partial t} = \frac{\partial}{\partial E_t} \left[bE_t \left(\frac{\partial f}{\partial E_t} - f \frac{\partial \ln f^{(0)}}{\partial E_t} \right) \right] + \frac{\partial}{\partial E_r} \left[bE_r \left(\frac{\partial f}{\partial E_r} - f \frac{\partial \ln f^{(0)}}{\partial E_r} \right) \right]. \tag{5.82}$$

Here, $f(E_t, E_r, t)$ is the distribution function related to translational E_t and rotational E_r energies; $f^{(0)}(E_t, E_r)$ is the equilibrium distribution function corresponding to temperature T_0 of the monatomic gas; parameter $b = \frac{32}{3} \frac{m_A}{m_{BC}} N_A T_0 \Omega_{col}$ includes density N_A of the monatomic gas, and Ω_{col}-the dimensionless collisional integral. The averaged rotational distribution $f(E_r, t) = \int_0^\infty f(E_t, E_r, t) dE_t$ can be described by simplified Fokker–Planck equation including only rotational energy:

$$\frac{\partial f(E_r, t)}{\partial t} = \frac{\partial}{\partial E_r} \left[bE_r \left(\frac{\partial f(E_r, t)}{\partial E_r} - \frac{1}{T_0} f(E_r, t) \right) \right]. \tag{5.83}$$

Multiplication of (5.83) by E_r with following integrating from 0 to ∞ results in the macroscopic relation for the total rotational energy E_r^{total}:

$$\frac{dE_r^{total}}{dt} = \frac{E_r^{total} - E_{r,0}^{total}(T_0)}{\tau_{RT}}, \quad \tau_{RT} = \frac{T_0}{b}. \tag{5.84}$$

Here τ_{RT} is the RT-relaxation time, $E_r^{total}(T_0)$ is the equilibrium value of rotational energy at the inert gas temperature T_0. Solution of the Fokker–Planck equation (5.83) shows that the rotational distribution $f(E_r, t)$ maintains the same Boltzmann form during the relaxation to equilibrium, with the rotational temperature approaching translational one:

$$T_r(t) = T_0 + [T_r(t = 0) - T_0] \exp\left(-\frac{t}{\tau_{RT}}\right). \tag{5.85}$$

Maintaining the Boltzmann form of distribution during relaxation to equilibrium (with only temperature change) is usually referred to as the **canonical invariance**.

5.13 Nonequilibrium Translational Energy Distribution Functions, Effect of "Hot Atoms"

The relaxation of translational energy of neutral particles requires only a couple of collisions and usually determines the shortest timescale in plasma systems. Assumption of a local quasi-equilibrium is usually valid for the translational energy sub-system of neutral particles, even in strongly nonequilibrium plasmas. The translational energy distributions are then Maxwellian with one local temperature T_0 for all neutral components participating in plasma chemical processes. However, this general rule has some important exceptions, when highly energetic neutrals are formed in fast VT-relaxation or exothermic chemical reactions, resulting in perturbation of conventional Maxwellian distribution.

5.13.1 Effect of "Hot Atoms" in Fast VT-relaxation Processes

Different fast nonadiabatic VT-relaxation processes were discussed in the Section 3.6.4. The fastest is VT-relaxation of molecules, Mo, on alkaline atoms, Me, proceeding through intermediate formation of ionic complexes $[\text{Me}^+\text{Mo}^-]$. For example, VT-relaxation of nitrogen N_2^* on Li, K, and Na takes place at each collision, which makes these alkaline atoms "hot" when they are added to nonequilibrium nitrogen plasma ($T_v \gg T_0$).

Translational energy distribution function $f(E)$ of a small admixture of alkaline atoms into a nonequilibrium diatomic gas ($T_v \gg T_0$) is determined by kinetic competition of the fast VT-energy exchange between these atoms and the diatomic molecules, and the fast Maxwellization TT-processes. The $f(E)$ evolution can be expressed by Fokker–Planck kinetic equation describing diffusion of the atoms along the translational energy spectrum:

$$D_{VT}\left(\frac{\partial f}{\partial E} + \frac{f}{T_v}\right) + D_{TT}\left(\frac{\partial f}{\partial E} + \frac{f}{T_0}\right) = 0. \tag{5.86}$$

Here, D_{VT} and D_{TT} are the diffusion coefficients of the alkaline atoms, Me, in translational energy space related to VT-relaxation and Maxwellization respectively. The translational energy distribution $f(E)$ depends on ratio of the diffusion coefficients $\mu(E) = D_{TT}(E)/D_{VT}(E)$, which is proportional to ratio of the corresponding relaxation rate coefficients. Integration of the linear differential equation (5.86) gives:

$$f(E) = B\exp\left[-\int \frac{\frac{1}{T_v} + \frac{\mu(E)}{T_0}}{1 + \mu(E)}\,dE\right], \tag{5.87}$$

where B is the normalization factor. Typically, Maxwellization is faster than VT-relaxation ($\mu(E) \gg 1$), which results in Maxwellian distribution with temperature T_0. However, the situation can be different for the admixture of light alkaline atoms (like Li, atomic mass m) into a relatively heavy molecular gas (like N_2 or CO_2, molecular mass $M \gg m$). VT- and TT-relaxation frequencies are almost equal for this mixture, but energy transfer during Maxwellization can be lower than $\hbar\omega$ due to the significant difference in masses, which results in $\mu \ll 1$. Therefore, *translational temperature of the light alkaline atoms can be equal not to the translational T_0 but rather to the vibrational T_v temperature of the molecular gas*, see Figure 5.11a.

To get more details on the translational energy distribution function, the factor $\mu(E)$ when $m \ll M$ can be expressed as $\mu(E) \approx \frac{E+T_v}{\hbar\omega}\left(\frac{m}{M}\frac{E}{T_v}\right)^2$. Therefore, according to (5.87), the translation distribution function $f(E)$ can be Maxwellian with vibrational temperature T_v only at relatively low translational energies $E < E^*$. At higher energies $E > E^*$, $\mu \gg 1$, and the exponential decrease of $f(E)$ always corresponds to translational temperature T_0. The critical value of the translational energy $E^*(T_v)$, when the Maxwellian distribution changes characteristic temperature from T_v to T_0 is:

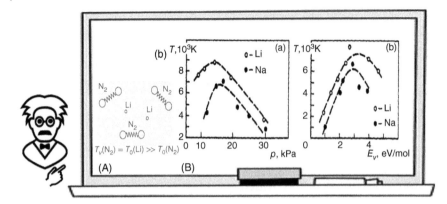

Figure 5.11 Effect of "hot atoms" in fast *VT*-relaxation: (a) quasi-equilibrium between vibrationally excited nitrogen molecules and translational motion of light Li atoms; (b) Li and Na atomic admixture temperatures in nonequilibrium microwave discharge in CO_2 as a function of (a) pressure at specific energy input $E_v = 3\,\mathrm{J\,cm^{-3}}$, and (b) specific energy input at pressure $p = 15.6\,\mathrm{kPa}$.

$$E^*(T_v) = \frac{M}{m}\sqrt{T_v\hbar\omega}, \quad \text{if } T_v > \hbar\omega\left(\frac{M}{m}\right)^2 \tag{5.88a}$$

$$E^*(T_v) = (\hbar\omega)^{1/3}\left(\frac{T_v M}{m}\right)^{2/3}, \quad \text{if } T_v < \hbar\omega\left(\frac{M}{m}\right)^2. \tag{5.88b}$$

5.13.2 Diagnostics of Nonequilibrium Molecular Gases Based on the Effect of "Hot Atoms" in Fast VT-relaxation

The described effect can be used to measure both vibrational and translational temperatures of molecular gases by using the Doppler broadening diagnostics of a small admixture of alkaline atoms. Such diagnostics are illustrated in Figure 5.11b by measurements of vibrational temperature of CO_2 in a nonequilibrium microwave discharge by adding a small amount of lithium and sodium atoms (Givotov et al. 1985). The Doppler broadening was observed by the Fourier analysis of the Li-spectrum line (transition $2^2s_{1/2}$–$2^2p_{1/2}$, $\lambda = 670.776\,\mathrm{nm}$) and the Na-spectrum line (transition $3^2s_{1/2}$–$3^2p_{1/2}$, $\lambda = 588.995\,\mathrm{nm}$). Separately measured gas temperature in this discharge was less than 1000 K, but the alkaline temperature (Figure 5.11b) was up to 10 times higher. The sodium temperature ($M/m = 1.9$) was always lower than temperature of lithium atoms ($M/m = 6.3$). More details on the effect of "hot atoms" in fast *VT*-relaxation, including discrete kinetic analyses of the processes and details on application of the effect to spectral characterization of molecular gases in nonequilibrium plasma can be found in the book of Givotov et al. (1985).

5.13.3 Generation of "Hot Atoms" in Fast Chemical Reactions

Energetic atoms can be generated in fast exothermic reactions and then react again before Maxwellization. They can significantly perturb translational distributions and significantly accelerate the exothermic chemical reactions. The fast laser-chemical chain reaction, stimulated in $H_2 - F_2$ mixture $H + F_2 \rightarrow HF^* + F$, $F + H_2 \rightarrow HF^* + H$ is a good example of generation of such hot atoms. In plasma systems such "hot atoms" can be generated in endothermic chemical reactions as well. For example, as vibrationally excited molecules participate in endothermic reaction with some excess of energy (see Figure 5.7), which depends on slope of the vibrational distribution $f(E)$ and on average equals to:

$$<\Delta E_v> = \left| \left[\frac{\partial \ln f}{\partial E}(E = E_a)\right]^{-1} \right| \tag{5.89}$$

This energy can be large and much exceeds the value of a vibrational quantum, mostly in slow reaction limit, see Section 5.8 (2).

5.14 Problems and Concept Questions

5.14.1 Diffusion of Molecules Along the Vibrational Energy Spectrum

Considering *VV*-exchange with frequency $k_{vv}n_0$ as diffusion of molecules along the vibrational energy spectrum, and using the relevant Fokker–Planck equation, estimate time for a molecule to get the vibrational energy, corresponding to the Treanor minimum E_{Tr}.

5.14.2 Flux of Molecules and Flux of Quanta Along the Vibrational Energy Spectrum

Show that the flux of molecules along the vibrational energy spectrum is equal to the derivative $\partial/\partial v$ of the flux of quanta in the energy space.

5.14.3 Hyperbolic Plateau Distribution Function

Find out a relation between the hyperbolic plateau coefficient "C" (5.20 and 5.21) and the degree of ionization n_e/n_0 in nonthermal plasma.

5.14.4 eV-flux Along the Vibrational Energy Spectrum

Prove, that eV-flux (5.33) in the energy space becomes equal to zero if the vibrational distribution is the Boltzmann function with temperature T_e: $f(E) \propto \exp(-E/T_e)$. Analyze, how this distribution is affected by non-Maxwellian behavior of EEDF.

5.14.5 Treanor Effect for Polyatomic Molecules

Analyze, how the Treanor effect, which is especially related to the discrete anharmonic structure of vibrational levels, can be achieved in polyatomic molecules in the quasi-continuum of vibrational states, where there is no discrete structure of the levels.

5.14.6 Treanor to Boltzmann Transition in Vibrational Distributions of Polyatomic Molecules

Explain why transition from the Treanor distribution with temperatures T_v and T_0 (at low energies) to the Boltzmann distribution with T_0 (at high energies) takes place in polyatomic molecules at lower vibrational levels than in the case of diatomic molecules.

5.14.7 VV- and VT-losses of Vibrational Energy of Highly Excited Molecules

Calculate the ratio of the vibrational energy losses related to *VT*-relaxation of highly excited diatomic molecules to the correspondent losses during series of the one-quantum *VV*-relaxation process, which provided the formation of the highly excited molecule.

5.14.8 Treanor Formula for Isotopic Mixtures

Using the Treanor formula (5.68) and assuming for an isotopic mixture $\frac{\Delta\omega}{\omega} = \frac{1}{2}\frac{\Delta m}{m}$, estimate the difference in vibrational temperatures of nitrogen molecules-isotopes $^{15}N^{14}N$ and $^{14}N^{14}N$ at room temperature and average vibrational temperature $T_v \approx 3000$ K.

5.14.9 Coefficient of Selectivity for Separation of Heavy Isotopes

Using relation (5.69) and Figure 5.9, estimate selectivity coefficient for separation of uranium isotopes $^{235}U/^{238}U$. Assume UF_6 chemical reaction with activation energy 5 eV at room temperature.

5.14.10 "Hot Atoms" Generated in Fast VT-relaxation

In nonthermal plasma of molecular gases $(T_v \gg T_0)$, admixture of light alkaline atoms can be in quasi-equilibrium with vibrational (not translational) degrees of freedom of molecules. This can be explained by high rate of *VT*-relaxation and the large difference in masses, which slows down the Maxwellization (*TT*-exchange) processes. Why the same effect cannot be achieved with addition of heavy alkaline atoms?

Lecture 6

Plasma Statistics and Thermodynamics, Heat and Radiation Transfer Processes

6.1 Complete (CTE) and Local (LTE) Thermodynamic Equilibrium in Plasma, Boltzmann Quasi-equilibrium Statistical Distribution

Evolution of plasma systems and rates of plasma processes, especially those in nonequilibrium nonthermal discharges, are generally determined by detailed physical and chemical kinetics, discussed in other lectures. However, application of different thermodynamic equilibrium approaches and quasi-equilibrium statistical distributions, wherever possible, is the easiest and clearest way to describe plasma systems and processes. The most consistent is concept of **complete thermodynamic equilibrium (CTE),** which is related to uniform homogeneous plasma, where plasma-chemical composition and properties are unambiguous functions of temperature. This temperature is supposed to be homogeneous and the same for all degrees of freedom, all the plasma system components, and all their possible reactions. The CTE plasma cannot be practically realized in laboratory. Nevertheless, thermal plasmas sometimes are modeled in this way for simplicity. To imagine a plasma in the CTE conditions, one should consider so large plasma volume that its central part is homogeneous and not sensitive to the boundaries. Electromagnetic radiation of the CTE plasma can be considered as black body radiation with a single temperature. Actual (even thermal) plasmas are quite far from these ideal conditions. Most plasmas are optically thin over a wide range of wavelengths, and radiation therefore is less intensive than that of a black body. Plasma nonuniformity leads to irreversible losses related to conduction, convection, and diffusion, which also disturb the CTE. A more realistic approximation is the so-called **local thermodynamic equilibrium (LTE).** Thermal LTE plasma is considered optically thin, and hence radiation is not required to be in equilibrium. However, the collisional (not radiative) processes are in local equilibrium like that of CTE but with temperature T differing from point to point in space and time, see book of Boulos et al. (1994).

The most general statistical distribution of particles over different states in quasi-equilibrium is the Boltzmann distribution. To describe it, consider an isolated system with total energy E, consisting of a big number N of particles in different states (i). Number "n_i" of particles in state "i", defined by a set of quantum numbers and energies E_i: $N = \sum_i n_i, E = \sum_i E_i n_i$. Differential form of these particles and energy conservation relations is:

$$\sum_i dn_i = 0, \quad \sum_i E_i dn_i = 0. \tag{6.1}$$

Objective of statistical approach is to find particles distribution over the different states "i" without details on transitions between the states. It is possible because probability to find n_i particles in state "i" is proportional to number of ways in which this distribution can be arranged. **Thermodynamic probability** $W(n_1, n_2, ..., n_i, ...)$ is the probability to have n_1 particles in the state "1", n_2 particles in the state "2" etc. N particles can be arranged in $N!$ different ways, but because n_i particles have the same energy, this number should be divided by the relevant factor to exclude repetitions. The thermodynamic probability can be then expressed as:

$$W(n_1, n_2, ..., n_i, ...) = A \frac{N!}{N_1! N_2! ... N_i! ...} = A \frac{N!}{\prod_i n_i!}, \tag{6.2}$$

where A is a normalizing factor. Let us find the most probable numbers of particles $\overline{n_i}$, when the thermodynamic probability (6.2) as well as its logarithm:

$$\ln W(n_1, n_2, ..., n_i, ...) = \ln(AN!) - \sum_i \ln n_i! \approx \ln(AN!) - \sum_i \int_0^{n_i} \ln x \, dx \tag{6.3}$$

Plasma Science and Technology: Lectures in Physics, Chemistry, Biology, and Engineering, First Edition. Alexander Fridman.
© 2024 WILEY-VCH GmbH. Published 2024 by WILEY-VCH GmbH.

have a maximum. The maximization of $\ln W$ requires:

$$0 = \sum_i \left(\frac{\partial \ln W}{\partial n_i} \right)_{n_i = \overline{n}_i} dn_i = \sum_i \ln \overline{n}_i \, dn_i. \tag{6.4}$$

Multiplying the two equations (6.1) respectively by parameters $-\ln C$ and $1/T$ and then adding (6.4), results in:

$$\sum_i \left(\ln \overline{n}_i - \ln C + \frac{E_i}{T} \right) dn_i = 0. \tag{6.5}$$

Sum (6.5) is supposed to be zero at any independent values of dn_i. This is possible only if the expression in parenthesis is zero, which determines the Boltzmann statistical distribution:

$$\overline{n}_i = C \exp\left(-\frac{E_i}{T} \right), \tag{6.6}$$

where C is the normalizing parameter related to total number of particles, and T is the temperature of the system, related to the average particle energy. As always, the same units are assumed for energy and temperature; therefore, Boltzmann coefficient is 1. If the state is degenerated, we should add the statistical weight "g", showing the number of states with the given quantum number:

$$\overline{n}_j = C \, g_j \exp\left(-\frac{E_j}{T} \right). \tag{6.7}$$

In this case, the subscript "j" corresponds to group of states with the statistical weight g_j and energy E_j. **Boltzmann statistical distribution** can be expressed then in terms of the number densities N_j and N_0 of particles in j-states and the ground state (0):

$$N_j = N_0 \frac{g_j}{g_0} \exp\left(-\frac{E_j}{T} \right). \tag{6.8}$$

In this relation, g_j and g_0 are the statistical weights of the "j" and ground states. The general Boltzmann statistical distribution (6.6) and (6.8) can be applied to derive many specific distribution functions, such as the Maxwell–Boltzmann (or just the Maxwellian) distribution (2.1), Boltzmann distributions discussed in Lectures 4 for charged particles and 5 for excited neutrals, as well as other statistical distributions important for plasma systems to be considered next.

6.2 Saha Ionization Equilibrium, Planck Formula, Stefan–Boltzmann Law, and Other Thermal-plasma-related Statistical Distributions

6.2.1 Saha Equation for Ionization Equilibrium in Thermal Plasma

The Boltzmann distribution (6.8) can be applied to determine the equilibrium ratio of number of electrons and ions $\overline{n}_e = \overline{n}_i$ to number of atoms $\overline{n}_a$ in a ground state, or in other words to describe the ionization equilibrium $A^+ + e \Leftrightarrow A$ in quasi-equilibrium thermal plasma:

$$\frac{\overline{n}_i}{\overline{n}_a} = \frac{g_e g_i}{g_a} \int \left[\frac{\overrightarrow{dp}\,\overrightarrow{dr}}{(2\pi\hbar)^3} \exp\left(-\frac{I + \frac{p^2}{2m}}{T} \right) \right]. \tag{6.9}$$

Here I is ionization potential; p is momentum of a free electron, then $I + \frac{p^2}{2m}$ is the energy necessary to produce the electron; g_a, g_i, and g_e are the statistical weights of atoms, ions, and electrons; $\overrightarrow{dp}\,\overrightarrow{dr}/(2\pi\hbar)^3$ is the statistical weight corresponding to continuous spectrum, which is the number of states in a given element of the phase volume $\overrightarrow{dp}\,\overrightarrow{dr} = dp_x dp_y dp_z \, dx\,dy\,dz$. Integration of (6.9) over the electron momentum gives:

$$\frac{\overline{n}_i}{\overline{n}_a} = \frac{g_e g_i}{g_a} \left(\frac{mT}{2\pi\hbar^2} \right)^{3/2} \exp\left(-\frac{I}{T} \right) \int \overrightarrow{dr}. \tag{6.10}$$

Here m is an electron mass, $\int \vec{dr} = V/\overline{n}_e$ is volume corresponding to one electron, and V is total system volume. Introducing the number densities of electrons $N_e = \overline{n}_e/V$, ions $N_i = \overline{n}_i/V$, and atoms $N_a = \overline{n}_a/V$, results in the **Saha equation**, describing the ionization equilibrium:

$$\frac{N_e N_i}{N_a} = \frac{g_e g_i}{g_a} \left(\frac{mT}{2\pi \hbar^2} \right)^{3/2} \exp\left(-\frac{I}{T} \right). \tag{6.11}$$

Statistical weight of the continuum spectrum is high. Therefore, the Saha equation predicts high degrees of ionization N_e/N_a, close to unity even at temperatures well below than ionization potential, $T \ll I$. Although the Saha formula refers to quasi-equilibrium thermal plasmas, it is sometimes used for estimations of nonthermal discharges assuming temperature T in (6.11) as electron temperature. It results in overestimation of N_e/N_a, because the Saha equilibrium assumes balance of electron impact ionization and three-body electron–ion recombination, while in nonthermal plasmas there are faster recombination mechanisms, see Section 2.8.

6.2.2 Statistical Relations for Radiation: Planck Formula, Stefan–Boltzmann Law

Probability to have "n" photons in a state with frequency ω based on (6.8) is $\exp(-\hbar\omega n)$. Average number of photons in the state with frequency ω can be then expressed by the **Plank distribution**:

$$\overline{n}_\omega = \frac{\sum\limits_n n \exp\left(-\frac{\hbar\omega n}{T} \right)}{\sum\limits_n \exp\left(-\frac{\hbar\omega n}{T} \right)} = \frac{1}{\exp\frac{\hbar\omega}{T} - 1}, \tag{6.12}$$

It can be rewritten in terms of the spectral density of radiation $U_\infty = \frac{1}{V} \frac{dE(V,\omega \div \omega+d\omega)}{d\omega}$, which is electromagnetic radiation energy per unit volume and unit interval of frequencies:

$$U_\omega = \frac{\hbar\omega^3}{\pi^2 c^3} \overline{n}_\omega = \frac{\hbar\omega^3}{\pi^2 c^3 \left(\exp\frac{\hbar\omega}{T} - 1 \right)}. \tag{6.13}$$

This is the **Planck formula** for spectral density of radiation. At high temperatures $\hbar\omega/T \ll 1$, it gives the classical **Rayleigh–Jeans formula** (without the Planck constant $\hbar$):

$$U_\omega = \frac{\omega^2 T}{\pi^2 c^3}, \tag{6.14}$$

and at low temperatures, $\hbar\omega/T \ll 1$ it gives the quantum-mechanical **Wien formula**:

$$U_\omega = \frac{\hbar\omega^3}{\pi^2 c^3} \exp\left(-\frac{\hbar\omega}{T} \right). \tag{6.15}$$

The Planck formula (6.13) permits calculation of the total radiation flux to the surface. In the case of a black body, this is equal to radiation flux from the surface:

$$J = \frac{c}{4} \int_0^\infty U_\omega d\omega = \frac{\pi^2 T^4}{60 c^2 \hbar^3} = \sigma T^4 \tag{6.16}$$

This is the **Stefan–Boltzmann law**, with the Stefan–Boltzmann coefficient σ:

$$\sigma = \frac{\pi^2}{60 c^2 \hbar^3} = 5.67 \times 10^{-12} \text{ W cm}^{-2} \text{ K}^{-4}. \tag{6.17}$$

6.2.3 Statistical Distribution of Diatomic Molecules Over Vibrational–Rotational States

Assume vibrational energy of diatomic molecules with respect to ground state is harmonic $E_v = \hbar\omega v$, then according to (6.8), number density of molecules with "v" vibrational quanta is:

$$N_v = N_0 \exp\left(-\frac{\hbar\omega v}{T} \right). \tag{6.18}$$

The total number density N of molecules is a sum of densities in different vibrational states:

$$N = \sum_{v=0}^{\infty} N_v = \frac{N_0}{1 - \exp\left(-\frac{\hbar\omega}{T}\right)}. \tag{6.19}$$

Therefore, the vibrational distribution (6.18) can be renormalized with respect to the total number density N of molecules:

$$N_v = N \left[1 - \exp\left(-\frac{\hbar\omega}{T}\right)\right] \exp\left(-\frac{\hbar\omega v}{T}\right). \tag{6.20}$$

In a similar way, taking rotational energy as $BJ(J+1)$ (B is the rotational constant, J is the rotational quantum number), and rotational statistical weight as $2J+1$, the equilibrium statistical distribution of diatomic molecules over vibrational–rotational states can be expressed as:

$$N_{vJ} = N\frac{B}{T}(2J+1) \left[1 - \exp\left(-\frac{\hbar\omega}{T}\right)\right] \exp\left(-\frac{\hbar\omega v + BJ(J+1)}{T}\right). \tag{6.21}$$

6.2.4 Dissociation Equilibrium in Molecular Gases

The Saha equation (6.11), derived above for the ionization equilibrium $A^+ + e \Leftrightarrow A$, can be generalized to describe the dissociation equilibrium $X + Y \Leftrightarrow XY$. Relation between densities N_X of atoms X, N_Y of atoms Y, and N_{XY} of the molecules XY in the ground vibrational–rotational state can be then expressed as:

$$\frac{N_X N_Y}{N_{XY}(v=0, J=0)} = \frac{g_X g_Y}{g_{XY}} \left(\frac{\mu T}{2\pi\hbar^2}\right)^{3/2} \exp\left(-\frac{D}{T}\right). \tag{6.22}$$

In this relation, g_X, g_Y, and g_{XY} are the relevant statistical weights; μ is the reduced mass of atoms X and Y; D is dissociation energy of the molecule XY. Most of the molecules are not in a ground state but rather in excited states at high plasma temperatures. Therefore, it worth to substitute the ground state concentration $N_{XY}(v=0, J=0)$ by the total N_{XY} one using (6.21):

$$N_{XY}(v=0, J=0) = \left[1 - \exp\left(-\frac{\hbar\omega}{T}\right)\right] \frac{B}{T} N_{XY}. \tag{6.23}$$

It results in the statistical relation for dissociation of molecules in thermal plasma:

$$\frac{N_X N_Y}{N_{XY}} = \frac{g_X g_Y}{g_{XY}} \left(\frac{\mu T}{2\pi\hbar^2}\right)^{3/2} \frac{B}{T} \left[1 - \exp\left(-\frac{\hbar\omega}{T}\right)\right] \exp\left(-\frac{D}{T}\right) \tag{6.24}$$

6.3 Thermodynamic Functions in Quasi-equilibrium Thermal Plasma: Partition Functions, Internal Energy, Helmholtz Free Energy, Gibbs Energy, Debye Corrections

Partition function Q of a particle system at temperature T can be expressed in general as a statistical sum over states "s" of the particles with energies E_s and statistical weights g_s:

$$Q = \sum_s g_s \exp\left(-\frac{E_s}{T}\right). \tag{6.25}$$

Translational and internal degrees of freedom of system components "i" can be considered independently. Therefore, their energy can be expressed as the sum $E_s = E^{tr} + E^{int}$, and the partition functions of a chemical species "i" and plasma volume V as the product:

$$Q_i = Q_i^{tr} Q_i^{int} = \left(\frac{m_i T}{\hbar^2}\right)^{3/2} V Q^{int}. \tag{6.26}$$

Plasma thermodynamic functions can be then calculated based on total partition function Q_{tot} of all the particles. For example, **Helmholtz free energy** F, related to reference free energy F_0 is (Landau and Lifshitz 2014):

$$F = F_0 - T \ln Q_{tot} \tag{6.27}$$

Assuming noninteracting particles, the total partition function can be expressed as product of partition functions Q_i of components "*i*": $Q_{tot} = \prod_i Q_i^{N_i} / \prod_i N_i!$, where N_i is total number of particles of the species "*i*". Using Stirling's formula for the above factorials ($\ln N! = N \ln N - N$), permits to express the Helmholtz free energy of the plasma system as:

$$F = F_0 - \sum_i N_i T \ln \frac{Q_i e}{N_i} \qquad (6.28)$$

Then, **internal energy** U and pressure in thermal plasma system can be found as:

$$U = U_0 + T^2 \left(\frac{\partial (F/T)}{\partial T} \right)_{V,N_i} = \sum_i N_i \left[\frac{3}{2} T + T^2 \left(\frac{\partial \ln Q_i^{int}}{\partial V} \right)_{T,N_i} \right], \qquad (6.29)$$

$$p = - \left(\frac{\partial (F/T)}{\partial V} \right)_{T,N} = \sum_i N_i T \left(\frac{\partial \ln Q_i}{\partial V} \right)_{T,N_i}. \qquad (6.30)$$

Plasma density significantly grows at higher temperatures and Coulomb interaction should be added to the thermodynamic functions. In the framework of Debye model, it corrects (6.28) to:

$$F = F_0 - \sum_i N_i T \ln \frac{Q_i e}{N_i} - \frac{TV}{12\pi} \left(\frac{e^2}{\varepsilon_0 TV} \sum_i Z_i^2 N_i \right)^{3/2}, \qquad (6.31)$$

where Z_i is the charges of a component "*i*" (for electrons $Z_i = -1$). The **Debye correction** of the Helmholtz free energy F obviously leads to corrections of the other thermodynamic functions. For example, the **Gibbs energy** G and pressure with the Debye correction become:

$$G = G_0 - \sum_i N_i T \ln \frac{Q_i}{N_i} - \frac{TV}{8\pi} \left(\frac{e^2}{\varepsilon_0 TV} \sum_i Z_i^2 N_i \right)^{3/2}, \qquad (6.32)$$

$$p = \frac{NT}{V} - \frac{T}{24\pi} \left(\frac{e^2}{\varepsilon_0 TV} \sum_i Z_i^2 N_i \right)^{3/2} \qquad (6.33)$$

Numerically, the Debye corrections become significant (around 3%) at thermal plasma temperatures above 14 000—15 000 K. The thermodynamic functions help to calculate transfer properties, as well as chemical and ionization composition of thermal plasmas. Important details on the subject including comprehensive numerical data can be found in the book of Boulos et al. (1994), as well as book of Nester et al. (1988).

6.4 Nonequilibrium Statistics and Thermodynamics of Thermal and Nonthermal Plasma Systems

Detailed description of nonequilibrium plasma systems and processes generally requires physical and chemical kinetics. Application of statistical and thermodynamic approaches far from equilibrium can lead to significant errors, see book of Slovetsky (1980). In some specific cases considered below, such approaches however cannot only be simple, but also quite successful.

6.4.1 Two-Temperature Statistics and Thermodynamics

Two-temperature statistics and thermodynamics can be useful in thermal discharges when deviation of electron temperature T_e from that of heavy particles T_0 is relatively small like in boundary layers separating quasi-equilibrium plasma from electrodes and walls. Relevant partition functions are then assumed to depend on both T_e (determining ionization processes) and T_0 (determining chemical and other plasma processes). These partition functions can be then used to calculate thermodynamic functions, plasma composition, and properties as a function of two temperatures T_e and T_0, see Figure 6.1 presenting such two-temperature consideration of Ar plasma. More details can be found in Boulos et al. (1994).

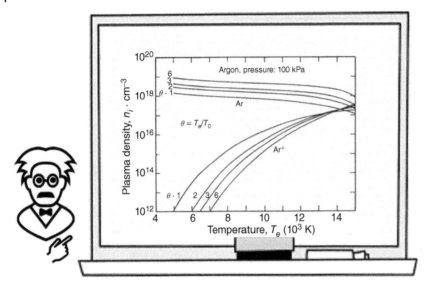

Figure 6.1 Number densities of Ar and Ar+ as function of electron temperature T_e in atmospheric pressure two-temperature argon plasma with $\theta = T_e/T_0 = 1,2,3,6$.

6.4.2 Strongly Nonequilibrium Thermodynamics of Plasma Gasification; Single Excited State Approach

Let us consider the nonthermal plasma gasification of solid surfaces:

$$A(solid) + bB^*(gas) \rightarrow cC(gas) \tag{6.34}$$

stimulated by particles B excited in strongly nonequilibrium discharges, predominantly in a single specific excited state (Veprek 1972; Legasov et al. 1978a, 1978b). Deviation from the conventional equilibrium constant $K = \frac{(Q_C)^c}{Q_A(Q_B)^b}$ is related then only to more general expression for the partition functions Q, accounting for differences in translational T_t, rotational T_r, and vibrational T_v temperatures, as well the nonequilibrium distribution function $f_e^X\left(\varepsilon_e^X\right)$ of electronically excited states for each reactants and products $X = A, B, C$:

$$Q_X = \sum_e g_e^X f_e^X\left(\varepsilon_e^X\right) \sum_{(t,r,v)} \prod_{k=t,r,v} g_{ke}^X \exp\left(-\frac{\varepsilon_{ke}^X}{T_k^X}\right), \tag{6.35}$$

where g and ε are statistical weights and energies of corresponding states.

Building the nonequilibrium thermodynamics by introduction of three independent temperatures (T_t, T_r, T_v) in (6.35) leads to internal inconsistencies due to breaks in detailed process equilibrium (Legasov et al. 1978a, 1978b), and can result in significant errors. The statistical thermodynamic approach can be consistently applied to the nonequilibrium reactions (6.34) only if it is limited to electronic excitation of a single state of a particle B with energy E_b. The population of the excited state can then be expressed by δ-function and the Boltzmann factor with an effective electronic temperature T^*:

$$f_e^B\left(\varepsilon_e^B\right) = \delta\left(\varepsilon_e^B - E_B\right) \exp\left(-\frac{E_B}{T^*}\right). \tag{6.36}$$

In this so-called **single excited state approach**, any other degrees of freedom of all the reaction participants should be considered in quasi-equilibrium with single gas temperature T_0:

$$f_e^{A,C}\left(\varepsilon_e^{A,C}\right) = 1, \quad T_{k=t,r,v}^X = T_0. \tag{6.37}$$

The *"quasi-equilibrium constant of nonequilibrium process"* can be then expressed as:

$$K \approx \exp\left\{-\frac{1}{T_0}[(\Delta H - \Delta F^A) - bE_B]\right\} = K_0 \exp\frac{bE_B}{T_0}, \tag{6.38}$$

where ΔH is the reaction enthalpy, ΔF_A is free energy change corresponding to heating to T_0, K_0 is equilibrium constant at temperature T_0. Thus, equilibrium constant can be significantly increased due to electronic excitation, and balance (6.34) significantly shifted toward the reaction products.

6.4.3 Nonequilibrium Two-Temperature Statistics of Vibrationally Excited Molecules, the Treanor Distribution

 The Treanor distribution reflecting "the capitalism in molecular life", when vibrational energy-rich molecules become even richer, and poor molecules further lose what they have, has been already introduced in Section 5.3. It was done through kinetic analysis of vibrational distribution functions dominated by the non-resonance VV-exchange of diatomic molecules in nonequilibrium conditions ($T_v > T_0$). The Treanor distribution describing population of N_v of a vibrational level "v", however, has more general, statistical nature, and even has been first derived this way by Treanor et al. (1968). The vibrational temperature in nonequilibrium statistics is defined by ratio of populations of ground state and first vibrationally excited level:

$$T_v = \frac{\hbar\omega}{\ln(N_0/N_1)}. \tag{6.39}$$

Considering vibrational quanta as quasi-particles, and using the Gibbs distribution with a variable number of quasi-particles "v", the relative population of the vibrational levels can be expressed (Kuznetzov 1971; Landau and Lifshitz 2014) as:

$$N_v = N_0 \exp\left(-\frac{\mu v - E_v}{T_0}\right), \tag{6.40}$$

where μ is chemical potential; E_v is energy of the vibrational level "v" with respect to zero-level. Comparing (6.40) and (6.39) for $v = 1$, the chemical potential of a quasi-particle μ is:

$$\mu = \hbar\omega\left(1 - \frac{T_0}{T_v}\right). \tag{6.41}$$

The Gibbs distribution (6.40) with chemical potential (6.41) gives the **Treanor distribution function** (x_e is the coefficient of anharmonicity, B is the normalizing factor):

$$f(v, T_v, T_0) = B \exp\left(-\frac{\hbar\omega v}{T_v} + \frac{x_e \hbar\omega v^2}{T_0}\right), \tag{6.42}$$

which is identical to one introduced in Section 5.3 and illustrated in Figure 5.2.

6.5 Thermal Conductivity in Quasi-equilibrium Plasma, Effect of Dissociation and Ionization on Plasma Heat Transfer

Thermal conductivity in moving one-component gas can be presented as:

$$\frac{\partial}{\partial t}(n_0 < \varepsilon >) + \nabla * \vec{q} = 0, \tag{6.43}$$

where n_0 and $<\varepsilon>$ are gas number density and average energy of a molecule; $\vec{q}$ is the heat flux. The average energy can be replaced here by gas temperature T_0 and specific heat c_v corresponding to one molecule. The heat flux includes two terms: one related to the gas motion with velocity $\vec{u}$ and another one related to the thermal conductivity with coefficient κ: $\vec{q} = < \varepsilon > n_0 \vec{u} - \kappa \nabla * T_0$, which results in **the heat conductivity equation for quasi-equilibrium thermal plasma**:

$$\frac{\partial}{\partial t}T_0 + \vec{u}\nabla * T_0 = \frac{\kappa}{c_v n}\nabla^2 T_0. \tag{6.44}$$

The thermal conductivity coefficient in a one-component gas without dissociation, ionization, and chemical reactions can be estimated as:

$$\kappa \approx \frac{1}{3}\lambda < v > n_0 c_v \propto \frac{c_v}{\sigma}\sqrt{\frac{T_0}{M}}, \tag{6.45}$$

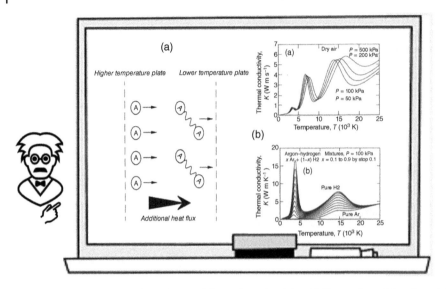

Figure 6.2 Temperature dependence of thermal conductivity: (a) Illustration of the dissociation-recombination effect; (b) "Roller coaster like" $\kappa(T_0)$ dependence in air and Ar-H$_2$ plasmas.

where σ is a typical cross-section of the molecular collisions, and M is the molecular mass. The thermal conductivity (6.45) does not depend on gas density and grows with temperature very slowly. Growth of thermal conductivity with T_0 in high temperature plasmas, however, can be much faster than (6.45), because of contribution of dissociation, ionization, and chemical reactions. Consider, for example, effect of dissociation and recombination $2A \Leftrightarrow A_2$ on accelerating of the temperature dependence of the thermal conductivity. Molecules are mostly dissociated at higher temperatures (see Figure 6.2a) and much less at lower temperature zones. Therefore, diffusion of the molecules (D_m) to the higher temperature zone leads to their intensive dissociation, consumption of dissociation energy E_D and to the related big heat flux:

$$\vec{q_D} = -E_D D_m \nabla n_m = -\left(E_D D_m \frac{\partial n_m}{\partial T_0} \right) \nabla T_0, \tag{6.46}$$

which can be interpreted as acceleration of thermal conductivity. When concentration of molecules is less than that of atoms $n_m \ll n_a$ and $T_0 \ll E_D$, the equilibrium relation (6.24) gives $\partial n_m / \partial T_0 = \left(E_D / T_0^2 \right) n_m$. Together with (6.46), it results in formula for coefficient of thermal conductivity in plasma related to the dissociation of molecules:

$$\kappa_D = D_m \left(\frac{E_D}{T_0} \right)^2 n_m. \tag{6.47}$$

Comparison of the coefficient of thermal conductivity coefficients with (6.47) and without (6.45) the dissociation-recombination effect gives:

$$\frac{\kappa_D}{\kappa} \approx \frac{1}{c_v} \left(\frac{E_D}{T_0} \right)^2 \frac{n_m}{n_a}. \tag{6.48}$$

The contribution of dissociation to temperature dependence of thermal conductivity related to dissociation can be very significant since $T_0 \ll E_D$. The strongest effect occurs in the narrow range of "dissociation" temperatures, when the concentrations of atoms and molecules are comparable. Contributions of ionization and chemical reactions to the thermal conductivity are similar to (6.48), and also significantly intensify the heat transfer at temperatures specific for these processes. It results in the *"roller coaster like"* $\kappa(T_0)$ *dependence in molecular gas plasmas*, illustrated in Figure 6.2b (see Boulos et al. 1994).

6.6 Nonequilibrium Effects in Thermal Conductivity

6.6.1 Fast Transfer of Vibrational Energy in Nonequilibrium Plasma ($T_v \gg T_0$): The Treanor Effect in Transfer Processes

Cold gas flowing around the high temperature plasma provides the vibrational-translational (VT) nonequilibrium in the area of their contact (Kurochkin et al. 1978). It is due, in particular, to a faster vibrational energy transfer from the quasi-equilibrium high temperature zone than that of translational energy, which is illustrated in Figure 6.3. Average vibrational quantum at higher T_v is lower because of anharmonicity. Therefore, fast VV-exchange during the quantum transfer occurs preferentially from higher T_v to lower T_v. This is **the Treanor effect in energy transfer** (see Section 6.4.3 for comparison), which results, in particular, in domination of the vibrational energy transfer over the transfer of translational energy. Relevant additional VV-flux of vibrational energy from higher T_v to lower T_v can be expressed as (Liventsov et al. 1984; Fridman and Kennedy 2021):

$$J_{VV} \approx -un_0\hbar\omega \iint Q_{v_2,v_2+1}^{v_1,v_1-1} \frac{2x_e\hbar\omega}{T_0}(v_2-v_1)f_1(v_1)f_2(v_2)dv_1dv_2, \tag{6.49}$$

where u is the average thermal velocity of molecules; $f(v)$ is the vibrational distribution function; n_0 is gas density; subscripts "1" and "2" are related to two planes (see Figure 6.3) separated by a mean free path. Integrating (6.49) using probabilities Q of VV-exchange (3.40) gives:

$$J_{VV} = -un_0\hbar\omega \frac{2x_e\hbar\omega}{T_0} Q_{10}^{01} \left(<v_1^2><v_2> - <v_1><v_2^2> \right), \tag{6.50}$$

where $<v>$ and $<v^2>$ are the first and second momentum of the vibrational distribution function. The additional VV-vibrational energy flux (6.50) can be recalculated assuming the Treanor distribution into relative increase of the coefficient of vibrational temperature-conductivity ΔD_v with respect to the conventional coefficient of temperature-conductivity D_0:

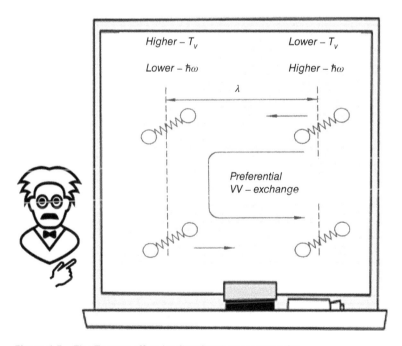

Figure 6.3 The Treanor effect in vibrational energy transfer.

$$\frac{\Delta D_v}{D_0} \approx 4Q_{01}^{10} q \frac{1 + 30q + 72q^2}{(1 + 2q)^2}, \quad q = \frac{x_e T_v^2}{T_0 \hbar\omega}. \tag{6.51}$$

The effect (5.61) is significant for molecules like CO_2, CO, and N_2O, where VV-exchange is provided by long-distance-forces and $Q_{01}^{10} \approx 1$. In room temperature CO_2 plasma, for example, with $T_v = 6000$ K, the transfer of vibrational energy (D_v) can be almost two orders of magnitude faster than that of translational energy (D_0).

6.6.2 Effect of Nonresonant VV-Exchange Close to the Vibrational-Translational (VT) Equilibrium

Vibrational quanta move in a system with a temperature gradient from higher T_v zone (where average quantum is lower because of anharmonicity) to the lower T_v zone (where average quantum is larger). The increase of the average quantum should be compensated from the translational energy reservoir due to the nonresonant VV-relaxation. Even in systems close to VT quasi-equilibrium, the vibrational energy transfer leads to some small but notable gas cooling ΔT and, hence, some VT nonequilibrium on the level of:

$$\frac{\Delta T_o}{T_0} \approx \frac{T_v - T_0}{T_v} \approx x_e \frac{T}{\hbar\omega}. \tag{6.52}$$

6.6.3 Nonequilibrium Effects Related to Recombination and Specific Heat

Diffusion of atoms and radicals from high to low temperature results in their recombination and over-equilibrium population of vibrationally excited states ($T_v > T_0$). Mixing of quasi-equilibrium hot and cold gases also can lead to $T_v > T_0$. Indeed, the mixing of hot and cold gases averages their energies and translational temperatures but not their vibrational temperatures (because of nonlinear dependence of vibrational energy on temperature, and growth of specific heat with temperature). The growth of vibrational specific heat with temperature results in VT nonequilibrium $T_v > T_0$ with temperature difference of about $1/2 \hbar\omega$.

6.7 Emission and Absorption of Continuous Spectrum Radiation in Plasma, Bremsstrahlung, and Radiative Electron–Ion Recombination Processes

Different types of plasma radiation can be classified according to the different nature of relevant electron transitions. Electron energy levels in the field of an ion as well as transitions between the levels are illustrated in Figure 6.4. The case when both initial and final electron states are in continuum is called the **free-free transition**. A free electron in this transition loses part of its kinetic energy in the Coulomb field of a positive ion or in interaction with neutrals. The emitted energy is a continuum usually infrared and called **bremsstrahlung emission** (stopping radiation). The reverse process is bremsstrahlung absorption. Electron transition between a free state in continuum and a bound state in atom is referred to as the **free-bound transition**. It corresponds to the process of continuous spectrum **radiative electron–ion recombination emission** and the reverse one of photoionization absorption. Such transitions also take place in electron-neutral collisions. In this case, they are related to photo-attachment and photo-detachment processes of formation and destruction of negative ions. Finally, **the bound-bound transitions** mean transition between discrete atomic levels and result in **emission and absorption of spectral lines**. Molecular spectra are obviously much more complex than those of single atoms because of possible transitions between different vibrational and rotational levels.

Intensity of continuous spectrum radiation is described by the emission spectral density J_ω meaning energy emitted in unit volume per unit time in the spectral interval $d\omega$. The *spectral density of the bremsstrahlung emission* due to electron collisions with plasma ions and neutrals can be expressed in quasi-equilibrium thermal plasmas as (Fridman and Kennedy 2021):

$$J_\omega^{brems} d\omega = \frac{16}{3} \left(\frac{2\pi}{3} \right)^{1/2} \frac{e^6 n_e n_i}{m^{3/2} c^3 (4\pi\varepsilon_0)^3 T^{1/2}} \exp\left(-\frac{\hbar\omega}{T} \right) d\omega = C \frac{n_i n_e}{T^{1/2}} \exp\left(-\frac{\hbar\omega}{T} \right) d\omega$$

$$= 1.08 \cdot 10^{-45} \text{W} \cdot \text{cm}^3 \cdot \text{K}^{1/2} \times \frac{n_i(1/\text{cm}^3) \cdot n_e(1/\text{cm}^3)}{(T, K)^{1/2}} \exp\left(-\frac{\hbar\omega}{T} \right) d\omega, \tag{6.53}$$

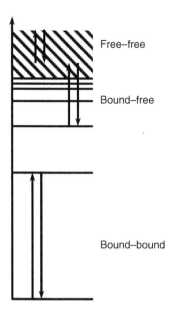

Figure 6.4 Schematic of energy levels and electron transitions induced by an ion field.

where n_e and n_i are electron and ion densities, $C = 1.08 \cdot 10^{-45} \text{W} \cdot \text{cm}^3 \cdot \text{K}^{1/2}$ is a constant parameter, and T is plasma temperature. The radiative recombination intensity, in turn, is determined by cross-section of the recombination emission in an electron–ion collision with formation of an atom in an excited state with the principal quantum number n:

$$\sigma_{RE} = \frac{4}{3\sqrt{3}} \frac{e^{10} Z^4}{(4\pi\varepsilon_0)^5 c^3 \hbar^4 m v_e^2 \omega} \frac{1}{n^3} = 2.1 \cdot 10^{-22} \text{cm}^2 \cdot \frac{(I_H Z^2)^2}{\varepsilon \, \hbar\omega \, n^3}. \tag{6.54}$$

Here ε, v_e is initial electron energy and velocity, $I_H = \frac{me^4}{2\hbar^2(4\pi\varepsilon_0)^2} \approx 13.6$ eV is the characteristic ionization potential, and Ze is the ion charge. The *spectral density of the recombination emission* $J_{\omega n} d\omega$, as a result of photorecombination of electrons with velocities in interval $v_e \div v_e + dv_e$ ($m v_e dv_e = \hbar d\omega$) and ions with $Z = 1$, leading to formation of an excited atom with the principal quantum number n is:

$$J_{\omega n} d\omega = \hbar\omega n_i n_e \sigma_{RE} f(v_e) v_e dv_e = C \frac{n_i n_e}{T^{1/2}} \frac{2 I_H}{T n^3} \exp\left(\frac{I_H}{T n^2} - \frac{\hbar\omega}{T} \right) d\omega. \tag{6.55}$$

The parameter $C = 1.08 \cdot 10^{-45} \text{W} \cdot \text{cm}^3 \cdot \text{K}^{1/2}$ is the same as in (6.53); $f(v_e)$ is the electron velocity distribution ($m v_e dv_e = \hbar d\omega$). When the radiation quanta and hence frequencies are not very large ($\hbar\omega < |E_g|$), the contributions of free–free transitions (bremsstrahlung) and free-bound transitions (recombination) in the total continuous emission are related to each other as:

$$J_\omega^{recomb} / J_\omega^{brems} = \exp\frac{\hbar\omega}{T} - 1. \tag{6.56}$$

Therefore, lower frequency emission $\hbar\omega < 0.7T$ is mostly due to the bremsstrahlung mechanism, while emission of larger quanta $\hbar\omega > 0.7T$ is mostly due to recombination. At thermal plasma temperatures about 10 000 K, this means only infrared radiation ($\lambda > 2\,\mu\text{m}$) is provided by bremsstrahlung, all other emission spectrum is due to the electron–ion recombination.

Total radiation losses over entire continuous emission spectrum can be found by integration of the spectral densities (6.53) and (6.55) and expressed by a convenient numerical formula:

$$J, \text{kW cm}^{-3} = 1.42 \cdot 10^{-37} \sqrt{T, K} \, n_e n_i (\text{cm}^{-3}) \cdot \left(1 + \frac{|E_g|}{T} \right), \tag{6.57}$$

where E_g is the lowest atomic exited state of with respect to transition to continuum. More details on the subject can be found in Fridman and Kennedy (2021).

6.8 Absorption of Continuous Spectrum Radiation in Plasma: the Kramers and Unsold-Kramers Formulas

Absorption of continuous spectrum radiation can be related to bremsstrahlung and photoionization processes, see Section 6.7. Relevant absorption coefficients can be found from corresponding emission characteristics using Einstein formulas, see Landau et al. (2013). First, **the coefficient of bremsstrahlung absorption** of a quantum $\hbar\omega$ per one electron with velocity v_e and one ion is (Fridman and Kennedy 2021):

$$a_\omega(v_e) = \frac{16\pi^3}{3\sqrt{3}} \frac{Z^2 e^6}{m^2 c (4\pi\varepsilon_0 \hbar\omega)^3 v_e}. \tag{6.58}$$

This is the **Kramers formula** of the bremsstrahlung absorption. Multiplying the Kramers formula by $n_e n_i$ and integrating over the Maxwellian distribution $f(v_e)$ gives the coefficient of bremsstrahlung absorption (the reverse length of absorption; frequency $v = \omega/2\pi$, $Z = 1$):

$$\kappa_\omega^{brems} = C_1 \frac{n_e n_i}{T^{1/2} v^3}, \quad C_1 = \frac{2}{3}\left(\frac{2}{3\pi}\right)^{1/2} \frac{e^6}{(4\pi\varepsilon_0)^3 m^{3/2} c\hbar}. \tag{6.59}$$

For numerical calculations of κ_ω^{brems} (cm^{-1}): $C_1 = 3.69 \cdot 10^8$ cm^{5}s$^{-3} \cdot$ K$^{1/2}$.

Another mechanism of plasma absorption of radiation in continuous spectrum is **photoionization**, the reverse process to recombination emission. Based on a detailed balance of the photoionization and recombination emission (6.54), the Saha equation (6.11) gives the photoionization cross-section for a photon $\hbar\omega$ and an atom with principal quantum number n:

$$\sigma_{\omega n} = \frac{8\pi}{3\sqrt{3}} \frac{e^{10} m Z^4}{(4\pi\varepsilon_0)^5 c\hbar^6 \omega^3 n^5} = 7.9 \cdot 10^{-18} \text{ cm}^2 \frac{n}{Z^2}\left(\frac{\omega_n}{\omega}\right)^3. \tag{6.60}$$

Here $\omega_n = |E_n|/\hbar$ is the minimum frequency sufficient for photoionization from the electronic energy level with energy E_n and principal quantum number n. The photoionization cross-section decreases with frequency as $1/\omega^3$ at $\omega > \omega_n$.

To determine total plasma absorption in continuum, the bremsstrahlung absorption (6.59) should be combined with the sum $\sum n_{0n}\sigma_{\omega n}$ related to the photoionization of atoms in different states of excitation (n) with concentrations n_{0n}. It results in the convenient formula for **the total plasma absorption coefficient in continuum** ($Z = 1$):

$$\kappa_\omega = C_1 \frac{n_e n_i}{T^{1/2} v^3} e^x \Psi(x) = 4.05 \cdot 10^{-23} \text{cm}^{-1} \frac{n_e n_i (\text{cm}^{-3})}{(T, K)^{7/2}} \frac{e^x \Psi(x)}{x^3}, \tag{6.61}$$

where factor C_1 is the same as in (6.59); parameter $x = \hbar\omega/T$; and the function $\Psi(x)$ is close to 1 at not very high frequencies. The product $n_e n_i$ in (6.61) can be replaced by the gas density n_0 using the Saha equation (6.11), leading to a simple expression for the total plasma absorption coefficient in continuum known as the **Unsold-Kramers formula**:

$$\kappa_\omega = \frac{16\pi}{3\sqrt{3}} \frac{e^6 T n_0}{\hbar^4 c\omega^3 (4\pi\varepsilon_0)^3} \frac{g_i}{g_a} \exp\left(\frac{\hbar\omega - I}{T}\right) = 1.95 \cdot 10^{-7} \text{cm}^{-1} \frac{n_0(\text{cm}^{-3})}{(T, K)^2} \frac{g_i}{g_a} \frac{e^{-(x_1 - x)}}{x^3}. \tag{6.62}$$

Here g_a, g_i are statistical weights of an atom and an ion, $x = \hbar\omega/T$, $x_1 = I/T$ and I the ionization potential. As an example, the absorption length κ_ω^{-1} of red light $\lambda = 0.65\,\mu$m in atmospheric pressure hydrogen plasma at 10 000 K is $\kappa_\omega^{-1} \approx 180$ m; so it is fairly transparent in continuum. It is important, that the Unsold-Kramers formula can be used only for quasi-equilibrium plasma conditions, while (6.61) can be applied for nonequilibrium plasmas as well.

6.9 Radiation Transfer in Plasma: Optically Thin and Optically Thick Systems, Plasma as a Gray Body

The intensity of radiation I_ω decreases along its path "s" due to absorption and increases because of spontaneous and stimulated emission. The **radiation transfer** in a quasi-equilibrium plasma can be then presented as:

$$\frac{dI_\omega}{ds} = \kappa_\omega'(I_{\omega e} - I_\omega), \quad \kappa_\omega' = \kappa_\omega\left[1 - \exp\left(-\frac{\hbar\omega}{T}\right)\right], \tag{6.63}$$

Here $I_{\omega e}$ is the radiation intensity corresponding to the Planck spectral density (6.12):

$$I_{\omega e} = \frac{\hbar\omega^3}{4\pi^3 c^2} \frac{1}{\exp(\hbar\omega/T) - 1}, \tag{6.64}$$

It is convenient to introduce the **optical coordinate** ξ, calculated from the plasma surface $x = 0$ with positive x directed into the plasma:

$$\xi = \int_0^x \kappa'_\omega(x)\, dx, \quad d\xi = \kappa'_\omega(x)\, dx. \tag{6.65}$$

The radiation transfer equation (6.63) can be then rewritten as:

$$\frac{dI_\omega(\xi)}{d\xi} - I_\omega(\xi) = -I_{\omega e}. \tag{6.66}$$

Let us assume plasma thickness is "d" in direction of a fixed ray; $x = 0$ corresponds to the plasma surface; and there is no source of radiation at $x > d$. Then the radiation intensity on the plasma surface $I_{\omega 0}$ can be expressed by solving of the radiation transfer equation as:

$$I_{\omega 0} = \int_0^{\tau_\omega} I_{\omega e}[T(\xi)] \exp(-\xi)\, d\xi, \quad \tau_\omega = \int_0^d \kappa'_\omega\, dx. \tag{6.67}$$

This relation introduces an important radiation transfer parameter τ_ω, usually referred to as **optical thickness of plasma.** If the optical thickness is small $\tau_\omega \ll 1$, plasma is **transparent or optically thin**, and the radiation intensity (6.67) on the plasma surface $I_{\omega 0}$ is:

$$I_{\omega 0} = \int_0^{\tau_\omega} I_{\omega e}[T(\xi)]\, d\xi = \int_0^{\tau_\omega} I_{\omega e}\kappa'_\omega\, dx = \int_0^d j_\omega\, dx. \tag{6.68}$$

Here the emissivity term $j_\omega = I_{\omega e}\kappa'_\omega$ corresponds to spontaneous emission. Radiation of optically thin plasma is the result of summation of independent emission from different intervals dx along the ray. In this case, all radiation generated in plasma volume is able to leave it. If plasma parameters are uniform, then the radiation intensity on the plasma surface can be expressed as:

$$I_{\omega 0} = j_\omega d = I_{\omega e}\kappa'_\omega d = I_{\omega e}\tau_\omega \ll I_{\omega e}. \tag{6.69}$$

The radiation intensity of the optically thin plasma is much less ($\tau_\omega \ll 1$) than equilibrium value (6.64) corresponding to the Planck formula.

 The opposite case takes place when the optical thickness is high $\tau_\omega \gg 1$, and system is **nontransparent or optically thick**. In this case, (6.67) gives: $I_{\omega 0} = I_{\omega e}(T)$. The emission density of the entire surface in all directions is the same and equal to the equilibrium Planck value (6.64). This is the case of the quasi-equilibrium **blackbody emission** corresponding to the Stefan–Boltzmann law, see Section 6.2.2. *It usually cannot be directly applied to continuous radiation of plasmas, which are mostly optically thin, and the Stephan–Boltzmann law requires special corrections.* On the other hand, plasma is not absolutely transparent and the total emission can be affected by **reabsorption of radiation**. This can be illustrated in (6.67) by introducing **the total emissivity coefficient** ε, which characterizes plasma as a **gray body** assuming fixed value of the absorption coefficient (6.61):

$$I_{\omega 0} = I_{\omega e}[1 - \exp(-\tau_\omega)] = I_{\omega e}\left[1 - \exp\left(-\kappa'_\omega d\right)\right] = I_{\omega e}\varepsilon. \tag{6.70}$$

In optically thin plasmas, the total emissivity coefficient coincides with the optical thickness: $\varepsilon = \tau_\omega$. If the total emissivity coefficient ε dependence on frequency can be neglected (assumption of the **ideal gray body**), the Stefan–Boltzmann formula (see Section 6.2.2) can be corrected and presented for plasma emission in continuous spectrum as:

$$J = \left[1 - \exp\left(-\kappa'_\omega d\right)\right] \sigma T^4 = \varepsilon\, \sigma T^4. \tag{6.71}$$

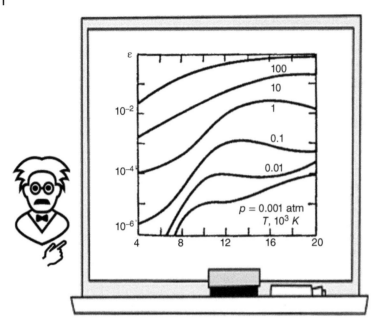

Figure 6.5 Total emissivity coefficient ε (grayness) of 1 cm layer of thermal plasma in air at different pressures and temperatures.

The total emissivity coefficients for the air plasma layer of 1 cm are illustrated in Figure 6.5 as a function of pressure and temperature. *The total emissivity coefficient approaches $\varepsilon = 1$ and therefore thermal plasma emission approaches that of the black body only at very high pressures (100 atm) and temperatures (20 000 K).* More details on the subject can be found in the book of Boulos et al. (1994).

6.10 Spectral Line Radiation in Plasma: Intensity, Natural Width, and Profile of Spectral Lines

Spontaneous discrete transition from an upper state E_n into a lower state E_k of a heavy plasma particle results in emission of a quantum (see Figure 6.4):

$$\hbar\omega_{nk} = \frac{2\pi\hbar c}{\lambda_{mn}} = E_n - E_k. \tag{6.72}$$

This emission determines a spectral line with frequency ω_{nk} and wavelength λ_{mn}. Intensity of the spectral line S can be characterized in quantum-mechanical terms by energy emitted by an exited atom or molecule per unit time in the line:

$$S = \hbar\omega_{nk}A_{nk} = \frac{1}{3\varepsilon_0}\frac{\omega_{nk}^4}{c^3}|\vec{d}_{nk}|^2 \tag{6.73}$$

where $\vec{d}_{nk}$ is the matrix element of dipole momentum of the atomic system, A_{nk} is the probability of the bound-bound transition in frequency units (A_{nk}^{-1} can be interpreted as the atomic lifetime with respect to the radiative transition $n \to k$). Typically, the transition probabilities for the intensive spectral lines are approximately $10^8\,\text{s}^{-1}$, numerical values of the coefficients A_{nk} for some strong atomic lines are given in Table 6.1.

We should note that classical electrodynamics gives an equivalent formula for the intensity of spectral line assuming that an electron is oscillating in an atom around equilibrium position ($\vec{r}$ is the electron radius-vector):

$$S = \frac{1}{6\pi\varepsilon_0 c^3}<\ddot{\vec{d}}^2> = \frac{1}{6\pi\varepsilon_0}\frac{\omega_0^4}{c^3}<\vec{d}^2> = \frac{e^2\omega_0^4}{6\pi\varepsilon_0 c^3}<\vec{r}^2> \tag{6.74}$$

where c is the speed of light, and ω_0 is frequency of the electron "oscillation" in an atom.

Table 6.1 Wavelengths, oscillator powers, and radiation frequencies (A_{nk}, s^{-1}) of some strong atomic lines in the visible spectrum.

Atomic line	Wavelength, λ, nm	Oscillator power, f	Coefficient A_{nk}, s^{-1}
H_α	656.3	0.641	4.4×10^7
H_β	486.1	0.119	8.4×10^6
He	587.6	0.62	7.1×10^7
Ar	696.5	—	6.8×10^6
Na	589.0	0.98	6.2×10^6
Hg	579.1	0.7	9.0×10^7

Excited states of an atom with energy E_n has finite lifetime with respect to radiation $\tau \approx A_{nk}^{-1}$. Therefore, according to the uncertainty principle, the energy level E_n actually is not absolutely thin but has a characteristic width of $\Delta E \propto \hbar/\tau \approx \hbar A_{nk}$. As a result, a spectral line related to transition from this energy level has a specific width $\Delta \omega \approx A_{nk}$, which is independent from external conditions and called the **natural spectral line width**. Typically, the natural width is very small with respect to the characteristic radiation frequency for electronic transitions $\Delta \omega \approx 10^8 s^{-1} \ll \omega_{nk} \approx 10^{15} s^{-1}$.

Profile of spectral lines can be determined as the photon distribution functions $F(\omega)$ over the radiation frequencies, which is usually normalized as: $\int_{-\infty}^{+\infty} F(\omega)\, d\omega = 1$. To describe the natural profile of the spectral line $F(\omega)$, we can express the corresponding oscillations of the electric and magnetic fields in an electromagnetic wave $f(t)$, taking into account the finite lifetime of both initial τ_n and final τ_k atomic states:

$$f(t) \propto \exp(i\omega_{nk}t - vt), \quad 2v = \frac{1}{\tau_n} + \frac{1}{\tau_k} = \frac{1}{\tau} \approx A_{nk}. \tag{6.75}$$

The frequency $v = \frac{1}{2} A_{nk}$ characterizes the photon emission probability (6.73) and describes attenuation of the electromagnetic wave because of the finite lifetime of atomic states. Frequency distribution for the electromagnetic wave can be found as a Fourier component f_ω of the function $f(t)$:

$$f_\omega = \frac{1}{2\pi} \int_{-\infty}^{+\infty} f(t) \exp(-i\omega t)dt \propto \frac{1}{v + i(\omega - \omega_{nk})}. \tag{6.76}$$

The radiation intensity of a spectral line at a fixed frequency ω and the photon distribution function $F(\omega)$ are both proportional to the square of the Fourier component $F(\omega) \propto |f_\omega|^2$. Then taking into account the normalization of the function $F(\omega)$, the following expression can be obtained for the photon distribution function and the natural profile of a spectral line related to finite lifetime of excited states:

$$F(\omega) = \frac{v}{\pi} \frac{1}{v^2 + (\omega - \omega_{nk})^2}, \quad v = \frac{1}{2}\left(\frac{1}{\tau_n} + \frac{1}{\tau_k}\right) = \frac{1}{2\tau} \approx \frac{1}{2} A_{nk}. \tag{6.77}$$

This shape of the photon distribution function and spectral line is usually referred to as the **Lorentz spectral line profile**. The Lorentz profile decreases slowly (hyperbolically) with deviation from the principal frequency $\omega = \omega_{nk}$, and has "wide hyperbolic wings", see Figure 6.6a. The parameter v in (6.77) is the so-called **spectral line half-width.** In this case, the width of a line at half the maximum intensity is presented as $2v$, which is a double half-width or the entire width of a spectral line (see Figure 6.6). It is interesting to remark that the natural width of a spectral line in the wavelength-scale is independent of frequency and can be estimated as:

$$\Delta \lambda = \frac{e^2}{3\varepsilon_0 mc^2} = 1.2 \cdot 10^{-5} \text{nm}. \tag{6.78}$$

6.11 The Doppler, Pressure, and Stark Broadening of Spectral Lines

Major mechanisms of spectral line broadening are due to thermal motion of the emitting particles (the Doppler effect), to collisions perturbing the emitting particles (pressure broadening), and to the influence of electric micro-fields (Stark effect).

6.11.1 Doppler Broadening

Electromagnetic wave emitted by a moving particle with frequency ω_{nk}, is observed by a detector (at rest) at frequency shifted by the Doppler effect:

$$\omega = \omega_{nk}\left(1 + \frac{v_x}{c}\right), \tag{6.79}$$

where v_x is the emitter velocity in the direction of wave propagation; c is the speed of light. The Doppler profile of a spectral line can be found assuming Maxwellian distribution for atomic velocities $f(v_x)$; $F(\omega)\,d\omega = f(v_x)\,dv_x$:

$$F(\omega) = \frac{1}{\omega_{mn}}\sqrt{\frac{Mc^2}{\pi T_0}}\exp\left[-\frac{Mc^2}{T_0}\frac{(\omega - \omega_{nk})^2}{\omega_{nk}^2}\right]; \tag{6.80}$$

where M is the mass of a heavy particle; T_0 is gas temperature. In contrast to the Lorentzian hyperbolic natural profile of spectral lines, the Doppler profile decreases much faster (exponentially) with frequency deviation from center of the line. Such shape of the curve $F(\omega)$ with relatively "narrow wings" is called the **spectral line Gaussian profile** (Figure 6.6b) The entire width (double half-width) of the Doppler profile of a spectral line can be expressed from (6.80) as:

$$\frac{\Delta\omega}{\omega_{nk}} = \frac{\Delta\lambda}{\lambda_{nk}} = 7.16 \cdot 10^{-7}\sqrt{T_0(K)/A}, \tag{6.81}$$

where A is the atomic mass of emitting particle. The Doppler broadening of spectral lines is most significant at high temperatures and for light atoms like hydrogen (for H_β line, $\lambda = 486.1$ nm, the Doppler width at a high gas temperature of 10 000 K is $\Delta\lambda = 0.035$ nm).

6.11.2 Pressure Broadening

The influence of atomic collisions on electromagnetic wave emission by excited atoms is illustrated in Figure 6.7. The phase of electromagnetic oscillations changes randomly during collision and therefore the oscillations can be

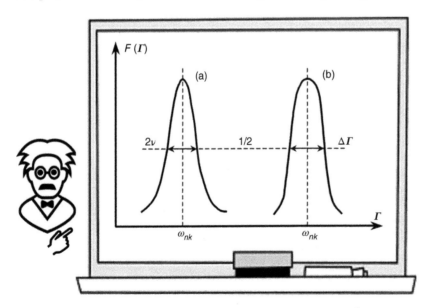

Figure 6.6 Profiles of spectral lines: (a) "wide wings" Lorentz profile; (b) "narrow wings" Gaussian profile.

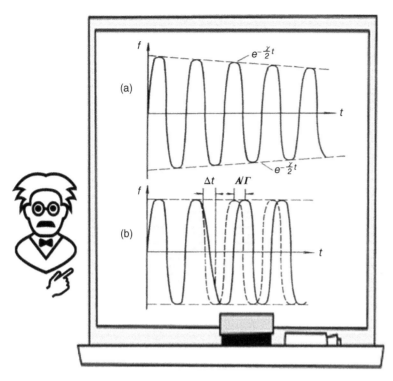

Figure 6.7 Time evolution of electromagnetic field amplitude: (a) finite state's lifetime; (b) collision of emitting and perturbing atoms; Δt – collision time; χ – collisional phase shift.

considered harmonic only between collisions. The effect of collisions is somewhat similar to the effect of spontaneous radiation (natural broadening), because it also restricts interval of harmonic oscillations. For this reason, collisions with all plasma components, including heavy neutrals, leads to this broadening effect. This broadening can be described by the same Lorentz profile (6.77) as in natural broadening, although in this case with frequency v related to frequency of the emitting atom collisions with other species (Smirnov 2001):

$$F(\omega) = \frac{n_0 <\sigma v>}{\pi} \frac{1}{(n_0 <\sigma v>)^2 + (\omega - \omega_{nk})^2}. \tag{6.82}$$

This effect is usually referred to as the **pressure broadening** of spectral lines. Here n_0 is neutral gas density; σ is the cross-section of perturbing collision (close to the gas-kinetic cross-section); and v is thermal velocity of collision partners. At high pressures, the collisional frequency obviously can exceed the frequency of spontaneous radiation; this can make the pressure broadening much more significant than the natural width of spectral lines.

6.11.3 Stark Broadening

When ionization degree exceeds about 10^{-2} and density of charged particles is high, their electric micro-fields perturb atomic energy levels providing the Stark broadening with the Lorentzian profile of spectral lines. It is the most significant for hydrogen atoms and some helium levels, where the effect is proportional to the electric field. In most other cases, the Stark effect is proportional to square of electric field and is not so strong. The Stark broadening induced by electric micro-fields of ions is qualitatively different from that of electrons. For slowly moving ions, Stark broadening can be described based on the statistical distribution of the electric micro-fields. In this **quasi-static approximation**, the electric micro-fields are provided by the closest charged particles and can be estimated as: $E \propto e^2/r^2 \propto n_e^{2/3}$, where $r \propto n_e^{-1/3}$ is the average distance between the charged particles. In the case of the strong linear Stark effect, the relevant width of a spectral line is also proportional to $n_e^{2/3}$. The Stark broadening induced by ions $\Delta \lambda_S$ in the specific case of the H_β line (486.1 nm, transition from $n = 4$ to $n = 2$) is:

$$\Delta \lambda_S = \Delta \omega_S \frac{\lambda_{nk}^2}{2\pi c} \approx \frac{3e^2 a_0 n(n-1)\lambda_{nk}^2}{8\pi^2 \varepsilon_0 \hbar c} n_e^{2/3} = 5 \cdot 10^{-12} \text{nm} \cdot (n_e, \text{cm}^{-3})^{2/3}, \tag{6.83}$$

where a_0 is the Bohr radius, λ_{nk} is the spectral line principal wavelength. The Stark broadening induced by fast-moving electrons can be described in framework of the **impact approximation,** when the emitting atom is unperturbed most of time. The entire Stark broadening, including the effects of ions and electrons, can be also approximated as proportional to $n_e^{2/3}$. For example, at electron temperatures $5000 \div 40\,000$ K and densities $10^{14} \div 10^{17}$ cm^{-3}, the total Stark broadening of the H_β line, 486.1 nm is:

$$\Delta\lambda_S \approx (1.8 \div 2.3)\cdot 10^{-11} \text{nm}\cdot(n_e, \text{cm}^{-3})^{2/3}. \tag{6.84}$$

6.11.4 Convolution of Lorentzian and Gaussian Profiles, the Voigt Profile of Spectral Lines

Three mechanisms of spectral line broadening in a plasma (Stark, pressure and natural) lead to the "wide wing" Lorentzian profile, and only the Doppler broadening results in the "narrow wing" Gaussian profile, see Figure 6.6. If the characteristic widths of the Lorentzian $\Delta\lambda_L$ and Gaussian $\Delta\lambda_G$ profiles are of the same order of value, the resulting total profile is not trivial: the central part of a spectral line can be for example related to Doppler broadening, while the wide wings can be provided by the Stark effect. The total shape of a spectral line is the result of convolution of the Gaussian and Lorentzian profiles. Assuming that the Doppler and Stark broadenings are independent, the result of the convolution is **the Voigt profile of spectral lines:**

$$F(\lambda) = \frac{1}{\Delta\lambda_G\sqrt{\pi}} \int_{-\infty}^{+\infty} \frac{a}{\pi} \frac{\exp(-y^2)}{(b-y)^2 + a^2} dy; \quad a = \frac{\Delta\lambda_L}{\Delta\lambda_G}(\ln 2)^{1/2}, \quad b = \frac{2\ln 2}{\Delta_G}(\lambda - \lambda_{nk} - \Delta). \tag{6.85}$$

The broadening of spectral lines plays significant role in plasma spectroscopy, and in plasma diagnostics, in general. This subject is covered in detail in several special books, including Griem (2005), and Hutchinson (2005).

6.12 Emission and Absorption of Radiation in Spectral Lines

6.12.1 Spectral Emissivity of a Line

Total energy emitted in a spectral line, corresponding to the transition $n \to k$, in unit volume per unit time, see (6.73), can be expressed as:

$$J_{nk} = \hbar\omega_{nk}A_{nk}n_n, \tag{6.86}$$

where n_n is the concentration of atoms in the state "n." Taking into account profile $F(\omega)$, the spectral emissivity (spectral density of J_{nk} taken per unit solid angle) can be presented as:

$$j_\omega d\omega = \frac{1}{4\pi}\hbar\omega_{nk}A_{nk}n_n F(\omega)\, d\omega. \tag{6.87}$$

The profile $F(\omega)$ is normalized: $\int_{-\infty}^{+\infty} F(\omega)\, d\omega = 1$. Therefore, *the spectral line broadening does not change the area of the spectral line.* Broadening makes a spectral line wider, but lower.

6.12.2 Selective Absorption of Radiation in Spectral Lines

Absorption of radiation by atoms and molecules are selective in spectral lines, but its efficiency depends on broadening. If the broadening is collisional, the absorption cross-section by one oscillator near the resonance is:

$$\sigma_\omega = \frac{e^2}{4mc\varepsilon_0} \frac{\tau^{-1} + n_0 <\sigma v>}{(\omega - \omega_{nk})^2 + \frac{1}{4}(\tau^{-1} + n_0 <\sigma v>)^2}. \tag{6.88}$$

This absorption–frequency curve has the Lorentz profile (see Figure 6.6) in accordance with **the Kirchhoff's law,** which requires correlation between emission and absorption processes; $n_0 <\sigma v>$ is collisional frequency, τ^{-1} is frequency of the spontaneous electronic transition $n \to k$, $\hbar\omega_{nk}$ is energy of this transition. The maximum resonance absorption cross-section (6.88) in the absence of collisions ($n_0 < \sigma v > \ll \tau^{-1}$, $\omega = \omega_{nk}$) is large for visible light:

$$\sigma_\omega^{max} = \frac{3\lambda_{nk}^2}{2\pi} \approx 10^{-9} \text{cm}^2. \tag{6.89}$$

If collisional broadening is significant, the absorption cross-section decreases with pressure: $\sigma_\omega \propto 1/n_0 < \sigma v > \propto 1/p$. The area of the absorption spectral line can be found by integrating the absorption cross-section σ_ω (6.88) over frequencies (traditionally, v):

$$\int \sigma_\omega dv = \frac{1}{2\pi} \int \sigma_\omega d\omega = \frac{e^2}{4mc\varepsilon_0} = 2.64 \cdot 10^{-2} \text{ cm}^2 \text{ s}^{-1}; \tag{6.90}$$

and describes the absorption in a spectral line of one classical oscillator.

6.12.3 The Oscillator Power

The total radiation energy absorbed in a spectral line per unit time and volume is independent of broadening:

$$\int 4\pi \kappa_\omega I_\omega d\omega = 4\pi I_{\omega nk} n_k \int \sigma_\omega d\omega; \tag{6.91}$$

here I_ω is the radiation intensity, κ_ω is the absorption coefficient (reverse length of absorption); n_k is density of the absorbing atoms. The total radiation energy absorbed in a spectral line (6.91) and (6.92) is determined by the area of the absorption line $\int \sigma_\omega d\omega$. Absorption factor $B_{kn} = \frac{1}{\omega_{nk}} \int \sigma_\omega dv$ as well as A_{nk} are usually called **the Einstein coefficients**. Based on (6.87) and Kirchhoff's law, the area of the absorption line is:

$$\sigma_\omega = \frac{g_n}{g_k} \frac{\pi^2 c^2}{\omega^2} A_{nk} F(\omega), \quad \int \sigma_\omega dv = \frac{g_n}{2g_k} \frac{\pi c^2}{\omega^2} A_{nk}. \tag{6.92}$$

Here, g_n, g_k are statistical weights of the upper and lower electronic states; A_{nk} is the Einstein coefficient for spontaneous emission in the spectral line (6.73). The actual area of the absorption line (6.91) and (6.92) differs from (6.89) corresponding to absorption of one classical oscillator. It is convenient to characterize the absorption in a spectral line by the ratio of the actual area of absorption line (6.91) and (6.92) over the area (6.89) corresponding to one classical oscillator:

$$f = \frac{g_n}{2g_k} \frac{4m\varepsilon_0 \pi c^3}{\omega^2 e^2} A_{nk}. \tag{6.93}$$

It is called **the oscillator power** and illustrates the number of classical oscillators providing the same absorption as the actual spectral line. Examples of the oscillator powers together with the Einstein coefficients A_{nk} for some spectral lines are presented in Table 6.1. The oscillator powers for the one-electron transitions are less than unity, and they approach the unity for the strongest spectral lines.

6.13 Radiation Transfer in Spectral Lines

The radiation transfer equation (6.63) for a spectral line in terms of radiative transitions $n \Leftrightarrow k$ in an atom or molecule can be presented as:

$$\frac{dI_\omega}{ds} = j_\omega + n_n \sigma_{b\omega} I_\omega - n_k \sigma_{a\omega} I_\omega, \tag{6.94}$$

where n_n, n_k are particle densities in upper "n" and lower "k" states; I_ω is radiation intensity; the coordinate "s" is along a ray propagation; j_ω is spontaneous emissivity at frequency ω. The cross sections of absorption $\sigma_{a\omega}$ and stimulated emission $\sigma_{b\omega}$ are related in accordance with the Kirchhoff and Boltzmann laws as (Landau et al. 2013):

$$\sigma_{b\omega} = \left(\frac{n_k}{n_n}\right)_{eq} \exp\left(-\frac{\hbar\omega}{T}\right) \sigma_{a\omega} = \frac{g_k}{g_n} \sigma_{a\omega}. \tag{6.95}$$

Here $(n_k/n_n)_{eq}$ is the equilibrium ratio of populations of lower and higher energy levels; g_k, g_n are the corresponding statistical weights. Then the radiation transfer equation (6.94) becomes:

$$\frac{dI_\omega}{ds} = j_\omega + (N_2 - N_1)g_k \sigma_{a\omega} I_\omega. \tag{6.96}$$

In this relation $N_2 = n_n/g_n$, $N_1 = n_k/g_k$ are the concentrations of atoms in the specific quantum states related to energy levels "n" and "k". These factors correspond in equilibrium to exponential factors of the Boltzmann distribution. In equilibrium, $N_2 = N_1 \exp(-\hbar\omega/T) < N_1$ and the radiation transfer equation (6.96) results in absorption corresponding to (6.63).

6.14 Inverse Population of Excited States in Nonequilibrium Plasmas and Principle of Laser Generation

 Thus, radiation transfer in spectral lines (6.96) results in the radiation absorption in quasi-equilibrium systems in general, and in thermal plasmas in particular. However, in nonequilibrium plasma conditions, the population of the higher energy states can exceed the population of those with lower energy $N_2 > N_1$. Such nonequilibrium conditions, quite typical for nonthermal plasmas, are usually referred to as the **inverse population of excited states**. *In the case of inverse population, according to (6.96), the nonequilibrium medium does not provide absorption but rather significant amplification of radiation in a spectral line.* This is the basis of the lasers. In homogeneous medium with $N_2 - N_1 = const > 0$, this exponential amplification of the radiation intensity along a ray can be expressed as:

$$I_\omega = I_{\omega 0} \exp[(N_2 - N_1)g_k\sigma_{a\omega}s]; \tag{6.97}$$

where the factor $(N_2 - N_1)g_k\sigma_{a\omega}$ in (6.97) is called the **laser amplification coefficient**. This provides physical basis for gas-discharge lasers.

The inverse population could be quite easily achieved in nonthermal plasmas due to intensive excitation of atoms and molecules by electron impact. It was discussed in detail in Lecture 5, especially regarding the VT nonequilibrium. When electron temperature significantly exceeds that of neutral species ($T_e \gg T_0$), the "excitation temperature" can be closer to T_e than to T_0. The nonequilibrium Treanor distribution, see Section 5.3, can be considered as one of the interesting examples of plasma-induced inverse population of excited states. The Treanor effect in plasma provides, in particular, the physical conditions for laser generation in the CO-lasers, see Bradley (1990). More information on the subject of laser principles can be found, in particular, in the book of Svelto (2009).

6.15 Problems and Concept Questions

6.15.1 Average Vibrational Energy

Calculate the average value of vibrational energy and related specific heat of a diatomic molecule using the equilibrium Boltzmann distribution function (6.20). Analyze the result for high ($T \gg \hbar\omega$) and low ($T \ll \hbar\omega$) temperatures.

6.15.2 Ionization Equilibrium, the Saha Equation

Estimate the degree of ionization in thermal 1 atm Ar-plasma at temperature $T = 20\,000$ K using the Saha equation. Explain, why the high ionization degree can be reached at temperatures much lower than ionization potential.

6.15.3 The Treanor Effect in Vibrational Energy Transfer

Using formula (6.51), estimate how big should be the parameter $q = \frac{x_e T_v^2}{T_0 \hbar\omega}$ in nitrogen plasma ($Q_{01}^{10} \approx 3 \times 10^{-3}$) to observe significant Treanor effect in vibrational energy transfer ($\Delta D_v \approx D_0$).

6.15.4 Vibrational-translational VT Nonequilibrium Caused by the Specific Heat Effect

Assuming no energy exchange between vibrational and other degrees of freedom, calculate temperatures T_v and T_0 after mixing 1 mol of air at room temperature with an equal amount of air at 1000 K. Explain and discuss difference in the temperatures.

6.15.5 Total Plasma Emission in Continuous Spectrum

Total plasma emission in continuous spectrum consists of bremsstrahlung and recombination components. The total plasma radiative energy losses per unit time and unit volume can be calculated using (6.57). Calculate the radiative power per unit volume for typical conditions of thermal plasma with $T = 10\,000$ K and compare with the typical value of specific thermal plasma power of about 1–10 kW cm^{-3}.

6.15.6 Natural Profile of Spectral Lines

The photon distribution and natural profile of spectral line, related to finite lifetime of excited states, can be described by the Lorentz profile (6.77). Show that the Lorentzian profile satisfies the normalization criterion: $\int_{-\infty}^{+\infty} F(\omega)\, d\omega = 1$.

6.15.7 Doppler Broadening of Spectral Lines

The Doppler broadening of spectral lines depends on gas temperature and atomic mass, while natural width does not. Estimate minimal temperature when Doppler broadening of argon line $\lambda = 696.5$ nm exceeds its natural width.

6.15.8 Pressure Broadening of Spectral Lines

Using formula (6.82) for the pressure broadening, compare the half-widths for this case with those typical for the Doppler broadening of spectral lines in Ar thermal plasma at 1 atm and $T = 5000$ K.

6.15.9 Absorption of Radiation in a Spectral Line by One Classical Oscillator

Integrate (6.90) to prove that the spectral line area $\int \sigma_\omega dv$ characterizing the absorption of one oscillator is constant and does not depend on frequencies $n_0 < \sigma v >$, τ^{-1} related to broadening.

6.15.10 Inverse Population of Excited States and the Laser Amplification Coefficient

Estimate the laser amplification coefficient (see Sections 6.13 and 6.14), assuming that the inverse population of excited states is due to the Treanor effect (6.42).

Lecture 7

Plasma Electrostatics and Electrodynamics, Waves in Plasma

7.1 Ideal and Nonideal Plasmas, Plasma Polarization and Debye Shielding of Electric Field

In most plasmas, somewhat similar to gases, charged particles move mostly in straight trajectories between collisions. It means that the inter-particle potential energy $U \propto e^2/4\pi\varepsilon_0 R$, corresponding to the average distance between electrons and ions $R \approx n_e^{-1/3}$, is much less than the electrons' kinetic energy (about T_e):

$$\frac{n_e e^6}{(4\pi\varepsilon_0)^3 T_e^3} \ll 1. \tag{7.1}$$

Plasma satisfying this condition is called **ideal plasma.** The nonideal plasmas, corresponding to the inverse inequality (7.1) and therefore very high density of charged particles, is not found in nature. Even creation of the nonideal plasmas in a laboratory is problematic with exception of dusty plasmas of large particulates, see Lecture 15.

The easiness of charged particle motion in the ideal plasmas explains its significant polarization in external electric field, which prevents penetration of the field inside of plasma. Such electrostatic "shielding or screening" of the electric field around a specified charged particle is illustrated in Figure 7.1. Space evolution of the potential φ follows the **Poisson's equation**:

$$div \vec{E} = -\Delta\varphi = \frac{e}{\varepsilon_0}(n_i - n_e), \tag{7.2}$$

where n_e and n_i are electron and positive ion densities; E is electric field. Assuming Boltzmann distribution for electrons (temperature T_e) and ions (temperature T_i), and quasi-neutral plasma concentration n_{e0}: $n_e = n_{e0} \exp\left(+\frac{e\varphi}{T_e}\right)$, $n_i = n_{e0}\exp\left(-\frac{e\varphi}{T_i}\right)$, the Poisson's equation becomes:

$$\Delta\varphi = \frac{\varphi}{r_D^2}, \quad r_D = \sqrt{\frac{\varepsilon_0}{n_{e0}e^2(1/T_e + 1/T_i)}}. \tag{7.3}$$

Here r_D is **the Debye radius**, an important electrostatic plasma parameter, which was already introduced in Section 4.13 regarding plasma electroneutrality, ambipolar diffusion, and definition of plasma. In 1D case, (7.3) gives: $d^2\varphi/d^2x = \varphi/r_D^2$, and describes the **Debye shielding** of external electric field E from E_0 at the plasma boundary ($x = 0$) along the axis "x" ($x > 0$):

$$\vec{E} = -\nabla\varphi = \vec{E_0}\exp\left(-\frac{x}{r_D}\right). \tag{7.4}$$

Similarly, the reduction of electric field of a specified charge "q" located in plasma (see Figure 7.1) can be also described by the Poisson's equation (7.3), but in spherical symmetry:

$$\Delta\varphi \equiv \frac{1}{r}\frac{d^2}{dr^2}(r\varphi) = \frac{1}{r_D^2}\varphi. \tag{7.5}$$

Plasma Science and Technology: Lectures in Physics, Chemistry, Biology, and Engineering, First Edition. Alexander Fridman.
© 2024 WILEY-VCH GmbH. Published 2024 by WILEY-VCH GmbH.

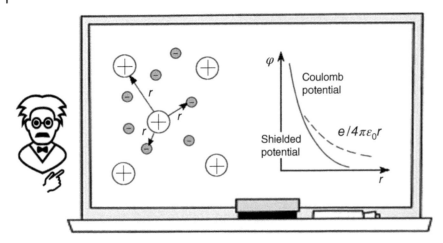

Figure 7.1 Illustration and electrostatic potential distribution for plasma polarization around a charged particle.

With boundary condition: $\varphi = q/4\pi\varepsilon_0 r$ at $r \to 0$. It again results in the exponential decrease of electric potential and field:

$$\varphi = \frac{q}{4\pi\varepsilon_0\, r}\, \exp\left(-\frac{r}{r_D}\right). \tag{7.6}$$

We should note some discrepancy between one-temperature T_e (4.77 and 4.78) and two-temperature T_e, T_i (7.3) relations for the Debye radius. According to (7.3), if $T_e \gg T_i$, the Debye radius depends mostly on the lower (ion) temperature; while according to (4.77 and 4.78) and to common sense it should depend on electron temperature. In reality, the heavy ions at low T_i are unable to establish the quasi-equilibrium Boltzmann distribution. It is more correct to assume at $T_e \gg T_i$, that the ions are at rest and $n_i = n_{e0} = const$. In this case, the term $1/T_i$ can be neglected and the Debye radius is determined by (4.77 and 4.78).

7.2 Quasi-neutral Plasma vs Sheath, Physics of DC Sheaths

Not all ionized gases are plasma, which is supposed to be quasi-neutral ($n_e \approx n_i$) and provide the above discussed Debye shielding. Thus, the typical size of plasma should be much larger than the Debye radius, which is illustrated in Table 7.1.

 Although plasma is quasi-neutral in general ($n_e \approx n_i$), it contacts with walls across nonquasineutral positively charged thin layers called **sheath**. The example of a sheath between plasma and zero-potential surfaces is illustrated in Figure 7.2. The sheath is formed because electron thermal velocity $\sqrt{T_e/m}$ exceeds that of ions about 1000 times. The fast electrons are able to stick to the walls leaving the area near the walls for positively charged ions alone. Corresponding potential profile is also shown in Figure 7.2. The bulk of plasma is quasi-neutral and therefore iso-potential ($\varphi = const$) according to the Poisson's equation (7.2). Near the discharge walls, the positive potential falls sharply, providing a high electric field, accelerating the ions and deceleration of electrons. Because of the ion acceleration in the sheath, the energy of ions bombarding the walls corresponds not to the ion temperature, but to the temperature of electrons.

 Let us first analyze a simple case of the direct-current **DC sheath**. If discharge operates at low gas temperatures ($T_e \gg T_0, T_i$) and pressures, conventional to micro-electronic applications (see Lecture 21), the sheath can be considered as collisionless. Then basic 1D equation governing the DC sheath potential φ in the direction perpendicular to the wall can be obtained from the Poisson's equation, energy conservation for ions, and the Boltzmann distribution for electrons:

$$\frac{d^2\varphi}{dx^2} = \frac{en_s}{\varepsilon_0}\left[\exp\frac{\varphi}{T_e} - \left(1 - \frac{e\varphi}{E_i}\right)^{-1/2}\right]. \tag{7.7}$$

Here n_s is plasma density at the sheath edge; $E_{is} = \frac{1}{2}Mu_{is}^2$ is the initial energy of an ion entering the sheath (u_{is} is the corresponding velocity); and the potential is assumed to be zero ($\varphi = 0$) at the sheath edge ($x = 0$). Multiplying

Table 7.1 Debye radius and typical size of different plasmas in lab and nature.

Type of plasma	Typical n_e (cm^{-3})	Typical T_e (eV)	Debye radius (cm)	Typical size (cm)
Earth ionosphere	10^5	0.03	0.3	10^6
Flames	10^8	0.2	0.03	10
He–Ne laser	10^{11}	3	0.003	3
Hg-lamp	10^{14}	4	$3 \cdot 10^{-5}$	0.3
Solar chromosphere	10^9	10	0.03	10^9
Lightning	10^{17}	3	$3 \cdot 10^{-6}$	100

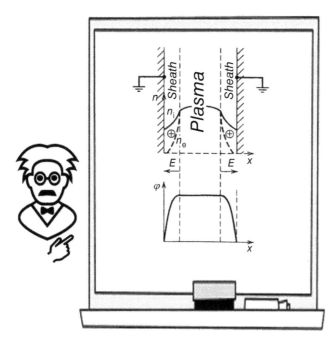

Figure 7.2 Plasma vs sheath illustration.

(7.7) by $d\varphi/dx$ and then integrating (assuming boundary conditions: $\varphi = 0$, $d\varphi/dx = 0$ at $x = 0$, see Figure 7.3) permits finding the electric field in the sheath (the potential gradient) as a function of potential:

$$\left(\frac{d\varphi}{dx}\right)^2 = \frac{2en_s}{\varepsilon_0}\left[T_e \exp\left(\frac{e\varphi}{T_e}\right) - T_e + 2E_{is}\left(1 - \frac{e\varphi}{E_{is}}\right)^{1/2} - 2E_{is}\right]. \tag{7.8}$$

The solution of (7.8) can exist only if its right-hand side is positive. Expanding (7.8) to the second order in a Taylor series, leads to the conclusion, that the sheath can exist only if the initial ion velocity exceeds the critical one; this is known as the **Bohm velocity** u_B:

$$u_{is} \geq u_B = \sqrt{T_e/M}. \tag{7.9}$$

The Bohm velocity equals to that of ions with energy corresponding to electron temperature. The requirement (7.9) of a sheath existence is referred to as the **Bohm sheath criterion**. To provide ions with the energy and directed velocity necessary to satisfy the Bohm criterion, there must be a quasi-neutral region (wider, than the sheath, e.g. several r_D) with some electric field. It is illustrated in Figure 7.3 and called **pre-sheath**, where minimum potential is:

$$\varphi_{presheath} \approx \frac{1}{2e}Mu_B^2 = \frac{T_e}{2e}. \tag{7.10}$$

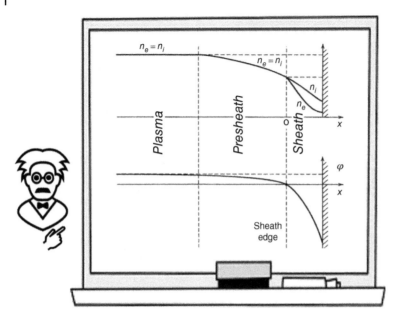

Figure 7.3 Illustration of sheath and presheath reflecting plasma contact with a wall.

Balancing ion and electron fluxes to the wall leads to change of potential across the sheath:

$$\Delta\varphi = \frac{1}{e}T_e \ln \sqrt{M/2\pi\, m}, \tag{7.11}$$

which is called the **floating potential**. Because the ion-to-electron mass ratio M/m is large, the floating potential) exceeds five to eight times the potential across the pre-sheath. Corresponding sheath width is about $s \approx 3r_D$.

7.3 Plasma Sheath Models: High Voltage Sheaths, Matrix, and Child Law Sheaths

Formula (7.11) corresponds to the floating potential which exceeds electron temperatures five to eight times. However, potential across a sheath (sheath voltage V_0) is often driven to be very large compared to electron temperature T_e/e. In this case, the electron concentration in the sheath can be neglected and only ions need to be taken into account. As an interesting consequence of the electron absence, the sheath region appears dark when visually observed. The simplest model of such a **high voltage sheath** assumes uniformity of the ion density in the sheath. This sheath is usually referred to as the **matrix sheath**. In the framework of the simple matrix sheath model, the sheath thickness can be expressed in terms of the Debye radius r_D corresponding to plasma concentration at the sheath edge:

$$s = r_D\sqrt{\frac{2V_0}{T_e}}. \tag{7.12}$$

When voltage V_0 is significant the matrix sheath can be large and exceed the Debye radius 10–50 times. More accurate approach takes into account the decrease of ion density as the ions accelerate across the sheath. This is done in the model of the so-called **Child law sheath**, where the ion current density $j_0 = n_s e u_B$ is taken equal to that of the **Child law of space-charge-limited current** in a plane diode:

$$j_0 = n_s e u_B = \frac{4\varepsilon_0}{9}\sqrt{\frac{2e}{M}}\frac{1}{s^2}\,V_0^{3/2}. \tag{7.13}$$

Solving (7.13) with respect to "s," gives the thickness of the Child law sheath:

$$s = \frac{\sqrt{2}}{3}r_D\left(\frac{2V_0}{T_e}\right)^{3/4}. \tag{7.14}$$

Numerically, the Child law sheath can be about 100 Debye lengths. More details regarding plasma sheaths, including collisional sheaths, sheaths in electronegative gases, radio-frequency plasma sheaths, and pulsed potential sheathes can be found in particular, in the book of Lieberman and Lichtenberg (2005).

7.4 Electrostatic Plasma Oscillations, Langmuir Frequency

While electrostatic plasma space-scale is characterized by Debye radius (see Section 7.1), the relevant time scale is related to electrostatic plasma oscillations with Langmuir frequency, which is illustrated in Figure 7.4. Let us assume 1D shift of all electrons at $x > 0$ to the right on the distance x_0 with respect to heavy ions at rest. This results in the occurrence of electric field pushing the electrons back. If $E = 0$ at $x < 0$, it follows the 1D Poisson's equation:

$$\frac{dE}{dx} = \frac{e}{\varepsilon_0}(n_i - n_e), \quad E = -\frac{e}{\varepsilon_0}n_{e0}x_0 \ \text{(at } x > x_0 \text{)}. \tag{7.15}$$

This electric field pushes the initially shifted electrons back to the left (Figure 7.4) together with their boundary (at $x = x_0$). The force is proportional to the shift, which according to the Newton's law, resulting in the harmonic electrostatic plasma oscillations:

$$\frac{d^2 x_0}{dt^2} = -\omega_p^2 x_0, \quad \omega_p = \sqrt{\frac{e^2 n_e}{\varepsilon_0 m}}. \tag{7.16}$$

Frequency of these electrostatic oscillations ω_p is called the **Langmuir frequency or plasma frequency**. Comparing the plasma frequency (7.16) and Debye radius (7.3), it is interesting to note that:

$$\omega_p \times r_D = \sqrt{2T_e/m} \tag{7.17}$$

Thus product of the Debye radius and the plasma frequency corresponds to thermal velocity of electrons. The time of a plasma reaction to an external perturbation ($1/\omega_p$) corresponds to time required for a thermal electron (with velocity about $\sqrt{2T_e/m}$) to travel the distance r_D necessary to provide screening of the electrostatic perturbation. The Langmuir frequency depends only on plasma density and can be calculated by a simple numerical formula:

$$\omega_p(\text{s}^{-1}) = 5.65 \cdot 10^4 \sqrt{n_e(\text{cm}^{-3})}. \tag{7.18}$$

While plasmas of different electrical discharges surely have different Langmuir frequencies, they are usually in the microwave frequency range of 1–100 GHz.

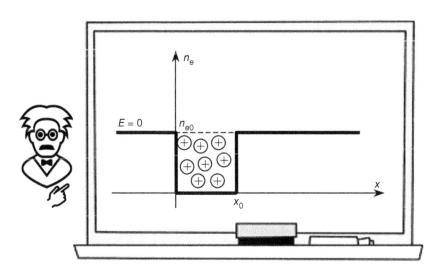

Figure 7.4 Illustration of electron density distribution in electrostatic plasma oscillations.

7.5 Plasma Skin Effect, Penetration of Slow-changing Fields into Plasma

Let us analyze the penetration of low-frequency electromagnetic field $\vec{E}$ ($\omega < \omega_p$) in plasma, applying the Ohm's law $\vec{j} = \sigma \vec{E}$ ($\vec{j}$ is current density, σ is plasma conductivity) and the relevant Maxwell equation for magnetic field:

$$curl\,\vec{H} = \vec{j} + \varepsilon_0 \frac{\partial \vec{E}}{\partial t}.$$
(7.19)

Assuming that the frequency of the field is low with respect to the plasma conductivity, the displacement current (second current term in (7.19)) can be neglected resulting in:

$$curl\,\vec{H} = \sigma \vec{E}.$$
(7.20)

Substituting the electric field (7.20) into Maxwell equation for electric field:

$$curl\,\vec{E} = -\mu_0 \frac{\partial \vec{H}}{\partial t}$$
(7.21)

leads to the differential equation for decrease of the electromagnetic field penetrating in plasma:

$$\frac{\partial \vec{H}}{\partial t} = -\frac{1}{\mu_0 \sigma} curl\,curl\,\vec{H} = -\frac{1}{\mu_0 \sigma} \nabla(div\,\vec{H}) + \frac{1}{\mu_0 \sigma} \Delta \vec{H} = \frac{1}{\mu_0 \sigma} \Delta \vec{H}.$$
(7.22)

 Electric field penetrating into a plasma follows a similar equation. In accordance with (7.22), amplitude of the low-frequency electric and magnetic fields decreases during their penetration in plasma with the characteristic space-scale, called the **skin layer**:

$$\delta = \sqrt{\frac{2}{\omega \mu_0 \sigma}}.$$
(7.23)

If the skin layer δ is small with respect to the plasma size, external fields, and currents are located only on the plasma surface layer (about δ), which is known as the **skin effect**. The depth of skin layer depends on frequency of the electromagnetic field ($f = \omega/2\pi$) and the plasma conductivity and can be calculated using simple numerical formula:

$$\delta\,(cm) = \frac{5.03}{\sigma^{1/2}(1/Ohm \cdot cm) \cdot f^{1/2}(MHz)}.$$
(7.24)

7.6 Electrostatic Plasma Waves and Their Collisional Damping

Electrostatic plasma oscillations (see Section 7.4) can propagate as longitudinal waves with electric fields in the propagation direction of the wave propagation (wave vector $\vec{k}$). Amplitude A^1 of oscillations of any macroscopic parameter $A(x,t)$ can be considered small ($A^1 \ll A_0$), and the oscillations $A(x,t)$ can be expressed as: $A = A_0 + A^1 \exp[i(kx - \omega t)]$, where A_0 corresponds to no oscillations, k is the wave number, ω is the wave frequency. To get the dispersion relation $\omega(k)$, the continuity equation, momentum conservation for the electron without dissipation, the adiabatic relation for the gas and Poisson equation can be linearized: $-i\omega n_e^1 + ikn_e u^1 = 0$, $-i\omega u^1 + ik\frac{p^1}{mn_e} + \frac{eE^1}{m} = 0$, $\frac{p^1}{p_0} = \gamma \frac{n_e^1}{n_e}$, $ikE^1 = -\frac{1}{\varepsilon_0}en_e^1$. Here: n_e^1, u^1, p^1, E^1 are the amplitudes of oscillations of the electron concentration, velocity, pressure, and electric field, n_e is the unperturbed plasma density; e, m, γ are charge, mass, and specific heat ratio for an electron gas; $p_0 = n_e m < v_x^2 > = n_e T_e$ is the electron gas pressure in the absence of oscillations; $< v_x^2 >$ averaged square of electron velocity in direction of oscillations; T_e is electron temperature. The **dispersion relation** $\omega(k)$ **for the electrostatic plasma waves** can be then expressed as:

$$\omega^2 = \omega_p^2 + \frac{\gamma T_e}{m} k^2, \quad \omega_p = \sqrt{\frac{n_e e^2}{\varepsilon_0 m}};$$
(7.25)

where ω_p is the plasma frequency. The electrostatic wave frequency is close to the plasma frequency if wavelength $2\pi/k$ exceeds the Debye radius. If the wavelengths are short, phase velocity of the electrostatic waves corresponds to the thermal speed of electrons.

Electron-neutral collisions (frequency ν_{en}) significantly influence the plasma oscillations, which even do not exist when $\omega < \nu_{en}$. At higher frequencies $\omega > \nu_{en}$, these collisions lead to damping of the electrostatic plasma oscillations, following the corrected dispersion relation (7.25):

$$\omega = \sqrt{\omega_p^2 + \frac{\gamma T_e}{m} k^2} - i\nu_{en}. \tag{7.26}$$

The amplitude of plasma oscillations decays exponentially $\propto \exp(-\nu_{en}t)$, which is called **collisional damping of the electrostatic plasma waves**. The wave frequencies are usually near the Langmuir frequency, and the numerical criterion of the existence of electrostatic plasma waves with respect to the collisional damping can be expressed based on (7.26) as:

$$\frac{\sqrt{n_e, \text{cm}^{-3}}}{n_0, \text{cm}^{-3}} \gg 10^{-12} \text{cm}^{3/2}; \tag{7.27}$$

here n_0 is gas density. At atmospheric pressure and room temperature $n_0 = 3 \cdot 10^{19}\text{cm}^{-3}$ and the criterion (7.27) requires relatively high electron densities $n_e \gg 10^{15}\text{cm}^{-3}$.

7.7 Ionic Sound in Plasma

The propagating electrostatic plasma oscillation related to motion of ions is called the ionic sound. These waves are longitudinal, and direction of electric field coincides with the direction of the wave vector $\vec{k}$. Dispersion relation $\omega(k)$ for of the ionic sound is determined by the Poisson equation for the potential φ of plasma oscillations:

$$\frac{\partial^2 \varphi}{\partial x^2} = \frac{e}{\varepsilon_0}(n_e - n_i). \tag{7.28}$$

Electrons quickly correlate their local instantaneous concentration in the wave $n_e(x, t)$ with the potential $\varphi(x,t)$ of the plasma oscillations in accordance with the Boltzmann distribution:

$$n_e = n_p \exp\left(+\frac{e\varphi}{T_e}\right) \approx n_p \left(1 + \frac{e\varphi}{T_e}\right). \tag{7.29}$$

Here the unperturbed density of the homogeneous plasma is n_p. Linearization $A = A_0 + A^1 \exp[i(kx - \omega t)]$ of (7.28 and 7.29) leads amplitude of oscillations of ion density:

$$n_i^1 = n_p \frac{e\varphi}{T_e}\left(1 + k^2 \frac{\varepsilon_0 T_e}{n_p e^2}\right). \tag{7.30}$$

Combining (7.30) with the linearized motion equation for ions in electric field of the wave:

$$M\frac{d\vec{u}_i}{dt} = e\vec{E} = -e\nabla\varphi, \quad M\omega u_i^1 = ek\varphi, \tag{7.31}$$

and with the linearized continuity equation for ions:

$$\frac{\partial n_i}{\partial t} + \nabla(n_i \vec{u}_i) = 0, \quad \omega n_i^1 = k n_p u_i^1, \tag{7.32}$$

results the **dispersion relation** $\omega(k)$ for of the ionic sound:

$$\left(\frac{\omega}{k}\right)^2 = c_{si}^2 \frac{1}{1 + k^2 r_D^2}, \quad c_{si} = \sqrt{\frac{T_e}{M}}; \tag{7.33}$$

here c_{si} is the speed of ionic sound; M is mass of an ion; $\vec{u}_i$ and u_i^1 are the ionic velocity and amplitude of its oscillation; r_D is the Debye radius. The ionic sound waves $\omega/k = c_{si}$ propagate in plasma with wavelengths exceeding the Debye radius ($k r_D \ll 1$). For shorter wavelengths ($k r_D \gg 1$), the dispersion relation (7.33) describes plasma oscillations with the frequency:

$$\omega_{pi} = \frac{c_{si}}{r_D} = \sqrt{\frac{n_p e^2}{M\varepsilon_0}}, \tag{7.34}$$

known as the **plasma-ion frequency**.

7.8 Magnetohydrodynamic Waves: Alfven Velocity, Alfven Wave, Magnetic Sound

Special types of waves occur when magnetic field is frozen in plasma (see Lecture 8). Any displacement of the magnetic field $\vec{H}$ leads then to plasma displacement, plasma oscillations, and propagation of magnetohydrodynamic waves. The propagation velocity of the elastic oscillations can be determined using the conventional relation for speed of sound: $v_A = \sqrt{\frac{\partial p}{\partial \rho}}$, where p is pressure, and $\rho = Mn_p = Mn_e$ is the plasma's density. The total pressure of the relatively cold plasma is equal to its magnetic pressure $p = \frac{\mu_0 H^2}{2}$, therefore:

$$v_A = \sqrt{\frac{\partial(\mu_0 H^2/2)}{\partial(Mn_p)}} = \sqrt{\frac{\mu_0 H}{M}\frac{\partial H}{\partial n_p}}. \tag{7.35}$$

Magnetic field is frozen in the plasma, and $\frac{\partial H}{\partial n_p} = \frac{H}{n_p}$, therefore the propagation velocity of the magnetohydrodynamic waves in plasma is:

$$v_A = \sqrt{\frac{\mu_0 H^2}{\rho}} = \frac{B}{\sqrt{\mu_0 \rho}}, \tag{7.36}$$

 which is known as the **Alfven velocity.** There are two types of the propagating plasma oscillations in magnetic field, which have the same Alfven velocity, but different directions of propagation. Those propagating along the magnetic field like a wave propagating along elastic string are referred to as the **Alfven wave** or magnetohydrodynamic wave. Oscillation of a magnetic line also induces oscillation of nearby magnetic lines, leading to the wave propagation perpendicularly to the magnetic field with Alfven velocity and called the **magnetic sound.** More details regarding magnetohydrodynamic, electrostatic, and other plasma waves can be found, for example, in Stix (2012).

7.9 Collisionless Interaction of Electrostatic Plasma Waves with Electrons, the Landau Damping, the Beam, and Buneman Kinetic Instabilities

Electrons and plasma oscillations can exchange energy without collisions, which is illustrated in Figure 7.5a in a reference frame moving with a wave. An electron can be trapped in a potential well created by the wave, which leads to their effective interaction. If an electron moves in the reference frame of electrostatic wave with velocity "u" along the wave, and after reflection changes direction to an opposite and its velocity to "$-u$," then change of electron energy due to the interaction with electrostatic wave is:

$$\Delta\varepsilon = \frac{m(v_{ph} + u)^2}{2} - \frac{m(v_{ph} - u)^2}{2} = 2mv_{ph}u; \tag{7.37}$$

here $v_{ph} = \omega/k$ is the phase velocity of the wave. If φ is the wave amplitude, then the typical velocity of a trapped electron in the reference frame of plasma wave is $u \approx \sqrt{e\varphi/m}$ and typical energy exchange between electrostatic wave and a trapped electron can be estimated as:

$$\Delta\varepsilon \approx v_{ph}\sqrt{me\varphi} = e\varphi\sqrt{\frac{mv_{ph}^2}{e\varphi}}. \tag{7.38}$$

In contrast to (7.38), typical energy exchange between an electrostatic wave and a non-trapped electron is about $e\varphi$. It means that at low amplitudes of oscillations, the collisionless energy exchange is mostly due to the trapped electrons. This collisionless energy exchange trends to equalize the electron distribution function $f(v)$ at the level of electron velocities near to their resonance with the phase velocity of the wave. The equalization (trend to form plateau on $f(v)$) takes place in the range of electron velocities between $v_{ph} - u$ and $v_{ph} + u$. It has frequency, corresponding to that of a trapped electron oscillation in potential well of the wave:

$$v_{ew} \approx uk \approx k\sqrt{\frac{e\varphi}{m}} \approx \sqrt{\frac{eE^1 k}{m}} \approx \sqrt{\frac{e^2 n_e^1}{\varepsilon_0 m}} = \omega_p \sqrt{\frac{n_e^1}{n_e}}; \tag{7.39}$$

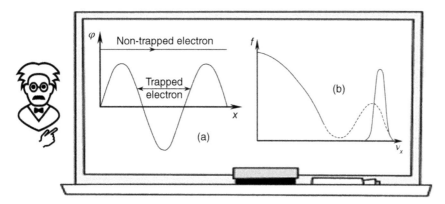

Figure 7.5 Collisional plasma wave's interaction with electrons: (a) electron interaction with plasma oscillations, (b) electron beam EEDF interaction with plasma electrons: solid curve – initial distribution, dashed curve – resulting distribution.

here k is the wave number; m is electron mass; E^1 is the amplitude of electric field oscillation; ω_p is the plasma frequency; n_e, n_e^1 are electron density and amplitude of its oscillation. While the electron's interaction with the Langmuir oscillations trends to form a plateau on $f(v)$ at electron velocities close to $v_p = \omega/k$, the electron–electron Maxwellization collisions trends to restore the electron distribution function $f(v)$. Electron energy distribution function (EEDF) changes during the collisionless interaction can be therefore neglected, only if relative perturbations of electron density are small: $\frac{n_e^1}{n_e} \ll \frac{n_e e^6}{(4\pi\varepsilon_0)^3 T_e^3}$.

Figure 7.5a and formula (7.37) illustrate that electrons with velocities $v_{ph} + u$ transfer energy to the plasma wave, while electrons with velocities $v_{ph} - u$ receive energy from the waves. Thus, the collisionless damping of electrostatic plasma oscillations takes place when:

$$\frac{\partial f}{\partial v_x}(v_x = v_{ph}) < 0; \tag{7.40}$$

where v_x is the electron velocity component in the direction of wave propagation; the derivative of distribution function is taken at electron velocity equal to phase velocity $v_{ph} = \omega/k$ of the wave. The electron distribution $f(v)$ decreases at high velocities corresponding to v_{ph}, the inequality (7.40) is satisfied and the collisionless damping of the electrostatic plasma waves takes place. The electrostatic plasma oscillations collisionlessly transfer energy in this case to electrons, which is known as the **Landau damping** $W = W_0 \exp(-2\gamma t)$:

$$\frac{\partial W}{\partial t} \approx [f(v+u) - f(v-u)]un_e \cdot v_{ew} \cdot \Delta\varepsilon \approx \frac{\partial f}{\partial v_x}\frac{n_e e^2}{m}\omega\varphi^2. \tag{7.41}$$

Considering $\varphi^2 \approx (E^1)^2/k^2 \approx W/\varepsilon_0 k^2$, the Landau damping increment is:

$$\frac{\partial W}{\partial t} = -2\gamma W, \quad \gamma \approx -\frac{\omega_p^2 \omega}{k^2}\frac{\partial f}{\partial v_x}\left(v_x = \frac{\omega}{k}\right) > 0. \tag{7.42}$$

The opposite situation of amplification of the electrostatic plasma waves due to the interaction with electrons is also possible. Injection of an electron beam in plasma creates distribution function, where the derivative (7.40) is positive (see Figure 7.5b); which corresponds to energy transfer from the electron beam and amplification of electrostatic plasma oscillations. The electron beam keeps transferring energy to plasma waves until the total electron distribution function becomes always decreasing (see Figure 7.5b). This so-called **beam instability** is used for stimulation of nonequilibrium plasma-beam discharges, see Lecture 15. The effect of beam instability is opposite with respect to Landau damping. The dispersion relation for plasma waves in the presence of electron beam can be expressed as:

$$1 = \frac{\omega_p^2}{\omega^2} + \frac{\omega_p^2}{(\omega - ku)^2}\frac{n_b}{n_p}; \tag{7.43}$$

here u is electron beam velocity. The strongest collisionless interaction of the electron beam with plasma occurs when phase velocity of plasma waves is close to the electron beam velocity $|\omega/k - u| \ll \omega/k$. The oscillation frequency ω is close to the plasma frequency ω_p, because the electron beam density is much less than the plasma density ($n_b \ll n_p$). Assuming $\omega = \omega_p + \delta$, $|\delta/\omega_p| \ll 1$, the dispersion relation (7.43) becomes:

$$\delta = \omega_p \left(\frac{n_b}{2n_p} \right)^{1/3} e^{2\pi im/3}, \tag{7.44}$$

where m is an integer. The amplification of the specific energy $W \propto (E^1)^2$ of plasma oscillations corresponds to the imaginary part of ω (and hence δ). This amplification of the Langmuir oscillation energy W again can be presented by negative increment γ:

$$\frac{\partial W}{\partial t} = -2\gamma W, \quad \gamma \approx -\frac{\sqrt{3}}{2} \left(\frac{n_b}{2n_p} \right)^{1/3} \omega_p = -0.69 \left(\frac{n_b}{n_p} \right)^{1/3} \omega_p < 0. \tag{7.45}$$

 The beam instability is an example of the **kinetic instabilities**, where amplification of plasma oscillations is due to difference in the motion of different groups of charged particles. Another example of kinetic instabilities is the **Buneman instability**, which occurs if the average electron velocity u exceeds that of ions. Similarly to the beam instability, electron energy dissipates here to generate and amplify electrostatic plasma oscillations following the dispersion relation:

$$1 = \frac{m}{M} \frac{\omega_p^2}{\omega^2} + \frac{\omega_p^2}{(\omega - ku)^2}. \tag{7.46}$$

Electrons are much lighter than ions $m/M \ll 1$, therefore, $\omega - ku$ is close to plasma frequency. Assuming: $\omega = \omega_p + ku + \delta$, $|\delta/\omega_p| \ll 1$, the dispersion relation gives:

$$\frac{2\delta}{\omega_p} = \frac{m}{M} \frac{\omega_p^2}{(\omega_p + ku + \delta)^2}. \tag{7.47}$$

The electrons interact efficiently with a wave having wave number: $k = -\omega_p/u$, then:

$$\delta = \left(\frac{m}{2M} \right)^{1/3} \omega_p e^{2\pi im/3}, \tag{7.48}$$

where m is an integer (the strongest plasma wave amplification corresponds to $m = 1$). The coefficient of amplification γ of the electrostatic plasma oscillations in this case is:

$$-\gamma = \operatorname{Im} \delta = \frac{\sqrt{3}}{2} \left(\frac{m}{2M} \right)^{1/3} \omega_p = 0.69 \left(\frac{m}{M} \right)^{1/3} \omega_p. \tag{7.49}$$

Because $k = -\omega_p/u$, the frequency of the amplified plasma oscillations for the Buneman instability is close to the coefficient of wave amplification: $\omega \approx |\gamma|$.

7.10 Dielectric Permittivity and Conductivity of Plasma in High-frequency Electric Fields

Electron motion in electric field $E = E_0 \cos \omega t = \operatorname{Re}(E_0 e^{i\omega t})$ can be described as:

$$m \frac{du}{dt} = -eE - mu\nu_{en}, \tag{7.50}$$

where ν_{en} is electron-neutral collision frequency, $u = \operatorname{Re}(u_0 e^{i\omega t})$ is the electron velocity, E_0 and u_0 are amplitudes of the corresponding oscillations. Relation between the amplitudes of electron velocity and electric field is complex, and based on (7.50) can be expressed as:

$$u_0 = -\frac{e}{m} \frac{1}{\nu_{en} + i\omega} E_0. \tag{7.51}$$

The imaginary part of the coefficient between u_0 and E_0 (complex electron mobility) reflects a phase shift between them. The Maxwell equation: $curl \vec{H} = \varepsilon_0 \frac{\partial \vec{E}}{\partial t} + \vec{j}$ allows the total current density to be presented as $\vec{j}_t = \varepsilon_0 \frac{\partial \vec{E}}{\partial t} + \vec{j}$. The first component here is related to displacement and its amplitude in complex form is $\varepsilon_0 i\omega E_0$. The second component corresponds to conduction current and has amplitude $-en_e u_0$. Thus, amplitude of the total current is:

$$j_{t0} = i\omega \varepsilon_0 E_0 - en_e u_0. \tag{7.52}$$

Considering the complex electron mobility (7.51) and the above Maxwell equation gives:

$$j_{t0} = i\omega\varepsilon_0 \left[1 - \frac{\omega_p^2}{\omega(\omega - iv_{en})} \right] \theta E_0, \quad curl\overrightarrow{H}_0 = i\omega\varepsilon_0 \left[1 - \frac{\omega_p^2}{\omega(\omega - iv_{en})} \right] \overrightarrow{E}_0, \tag{7.53}$$

where ω_p is plasma frequency. Assuming the Maxwell equation in the form $curl\overrightarrow{H}_0 = i\omega\varepsilon_0\varepsilon E_0$, we introduce the **complex dielectric permittivity of plasma in high frequency electric fields**:

$$\varepsilon = 1 - \frac{\omega_p^2}{\omega(\omega - iv_{en})}, \tag{7.54}$$

which is convenient to describe both plasma dielectric permittivity and conductivity. The complex dielectric permittivity can be rewritten by separating a real and imaginary parts:

$$\varepsilon = \varepsilon_\omega + i\frac{\sigma_\omega}{\varepsilon_0\omega}. \tag{7.55}$$

The real component ε_ω corresponds to high-frequency **dielectric permittivity of plasma**:

$$\varepsilon_\omega = 1 - \frac{\omega_p^2}{\omega^2 + v_{en}^2}. \tag{7.56}$$

The imaginary component corresponds to the high-frequency **plasma conductivity**:

$$\sigma_\omega = \frac{n_e e^2 v_{en}}{m \left(\omega^2 + v_{en}^2 \right)}. \tag{7.57}$$

Expressions for the high frequency, dielectric permittivity, and conductivity can be simplified in two cases: **collisionless plasma** $\omega \gg v_{en}$ and a static limit. For example, microwave plasma can be considered collisionless at low pressures 3 Torr and less. For collisionless plasmas:

$$\sigma_\omega = \frac{n_e e^2 v_{en}}{m\omega^2}, \quad \varepsilon_\omega = 1 - \frac{\omega_p^2}{\omega^2}. \tag{7.58}$$

The ratio of conduction to polarization (displacement) current can be estimated as:

$$\frac{j_{conduction}}{j_{polarization}} = \frac{\sigma_\omega}{\varepsilon_0\omega|\varepsilon_\omega - 1|} = \frac{v_{en}}{\omega}. \tag{7.59}$$

The polarization current in collisionless plasma ($v_{en} \ll \omega$) exceeds conductivity current. In opposite case of the **static limit** $v_{en} \gg \omega$, the conductivity and dielectric permittivity are:

$$\sigma_\omega = \frac{n_e e^2}{m v_{en}}, \quad \varepsilon = 1 - \frac{\omega_p^2}{v_{en}^2}. \tag{7.60}$$

In the static limit, the conductivity coincides with the conventional conductivity in DC conditions, and dielectric permittivity does not depend on frequency.

7.11 Propagation, Absorption, and Total Reflection of Electromagnetic Waves in Plasma: Bouguer Law and Critical Electron Density

Electromagnetic wave propagation in plasma follows the conventional wave equations: $\Delta\overrightarrow{E} - \frac{\varepsilon}{c^2}\frac{\partial^2\overrightarrow{E}}{\partial t^2} = 0$, $\Delta\overrightarrow{H} - \frac{\varepsilon}{c^2}\frac{\partial^2\overrightarrow{H}}{\partial t^2} = 0$, and corresponding dispersion relation:

$$\frac{kc}{\omega} = \sqrt{\varepsilon}. \tag{7.61}$$

Plasma peculiarity here is only due to complexity of dielectric permittivity ε (7.54 and 7.55). Assuming electric and magnetic fields as $\overrightarrow{E}, \overrightarrow{H} \propto \exp(-i\omega t + i\overrightarrow{k}\overrightarrow{r})$ with real wave frequency ω, the wave number k should be complex, because ε is complex in (7.61):

$$k = \frac{\omega}{c}\sqrt{\varepsilon} = \frac{\omega}{c}(n + i\kappa); \tag{7.62}$$

here n is **the refractive index**. The phase velocity of the wave is $v = \frac{\omega}{k} = \frac{c}{n}$, the wavelength in plasma is $\lambda = \lambda_0/n$ (λ_0 is corresponding wavelength in vacuum). The real wave number κ characterizes the **attenuation of electromagnetic wave in plasma**; the wave amplitude decreases e^κ times on the length $\lambda_0/2\pi$. Considering (7.55) for ε, relation between the refractive index n and attenuation of κ can be expressed through dielectric permittivity ε_ω and conductivity σ_ω:

$$n^2 - \kappa^2 = \varepsilon_\omega, \quad 2n\kappa = \frac{\sigma_\omega}{\varepsilon_0 \omega}. \tag{7.63}$$

These equations result in the explicit expression for the wave attenuation coefficient:

$$\kappa = \sqrt{\frac{1}{2}\left(-\varepsilon_\omega + \sqrt{\varepsilon_\omega^2 + \frac{\sigma_\omega^2}{\varepsilon_0^2 \omega^2}}\right)}. \tag{7.64}$$

Explicit expression for the refractive index can be presented from (7.63) as:

$$n = \sqrt{\frac{1}{2}\left(\varepsilon_\omega + \sqrt{\varepsilon_\omega^2 + \frac{\sigma_\omega^2}{\varepsilon_0^2 \omega^2}}\right)}. \tag{7.65}$$

 If the conductivity is small, the refractive index $n \approx \sqrt{\varepsilon_\omega}$, plasma polarization are negative, $\varepsilon_\omega < 1$, and $n < 1$. As a result, the phase velocity of electromagnetic waves exceeds the speed of light $v_{ph} = c/n > c$. The group velocity is less than the speed of light. The above expression for the refractive index $n \approx \sqrt{\varepsilon_\omega}$ in the low conductivity limit leads together with (7.60) to the **dispersion equation for electromagnetic waves in collisionless plasma**:

$$\frac{k^2 c^2}{\omega^2} = 1 - \frac{\omega_p^2}{\omega^2}, \quad \omega^2 = \omega_p^2 + k^2 c^2. \tag{7.66}$$

Corresponding dispersion curve for electromagnetic waves in plasma is compared in Figure 7.6a with that of electrostatic plasma waves (see Section 7.6). Differentiation of the dispersion equation (7.6) provides relation between the phase and group velocities of electromagnetic waves:

$$\frac{\omega}{k} \times \frac{d\omega}{dk} = v_{ph} v_{gr} = c^2. \tag{7.67}$$

The energy flux of electromagnetic waves can be described by the Poynting vector:

$$\vec{S} = \varepsilon_0 c^2 [\vec{E} \times \vec{B}], \tag{7.68}$$

where the electric and magnetic fields are averaged over the oscillation period. The electric and magnetic fields are related according to the Maxwell equations (when $\mu = 1$) as: $\varepsilon \varepsilon_0 E^2 = \mu_0 H^2$. As a result, attenuation of the energy flux (7.68) in plasma follows the **Bouguer law**:

$$\frac{dS}{dx} = -\mu_\omega S, \quad \mu_\omega = \frac{2\kappa\omega}{c} = \frac{\sigma_\omega}{\varepsilon_0 nc}, \tag{7.69}$$

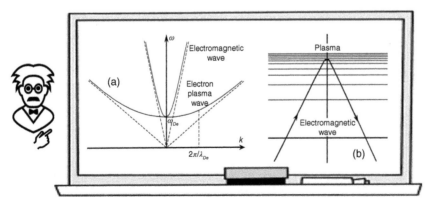

Figure 7.6 Propagation and total reflection of electromagnetic waves in plasma: (a) electromagnetic wave dispersion in comparison with electrostatic electron waves; (b) total reflection from plasma with density growing in vertical direction.

where μ_ω is the absorption coefficient. The product $\mu_\omega S$ presents the electromagnetic energy dissipated per unit volume of plasma, which corresponds to the Joule heating:

$$\mu_\omega S = \varepsilon_0 c^2 <EB> = \sigma <E^2>. \tag{7.70}$$

If the plasma degree of ionization and absorption are relatively low: $n \approx \sqrt{\varepsilon} \approx 1$, then expression for the absorption coefficient can be simplified:

$$\mu_\omega = \frac{n_e e^2 v_{en}}{\varepsilon_0 mc \left(\omega^2 + v_{en}^2\right)}. \tag{7.71}$$

For practical calculations (7.71) it is convenient to use the following numerical formula:

$$\mu_\omega, \mathrm{cm}^{-1} = 0.106 n_e (\mathrm{cm}^{-3}) \frac{v_{en}(\mathrm{s}^{-1})}{\omega^2(\mathrm{s}^{-1}) + v_{en}^2}. \tag{7.72}$$

At high frequencies $\omega \gg v_{en}$, the absorption coefficient is proportional to the square of wavelength $\mu_\omega \propto \omega^{-2} \propto \lambda^2$, therefore short waves easier propagate through plasma. If plasma conductivity is low $\sigma_\omega \ll \omega \varepsilon_0 |\varepsilon|$, electromagnetic wave easily propagates in plasma.

When frequency decreases, the dielectric permittivity $\varepsilon_\omega = 1 - \frac{\omega_p^2}{\omega^2}$ becomes negative and wave is unable to propagate. The negative dielectric permittivity makes the refractive index equal to zero ($n = 0$) and attenuation coefficient $\kappa \approx \sqrt{|\varepsilon|}$. The phase velocity tends to infinity and the group velocity becomes zero. Then depth of wave penetration in plasma is:

$$l = \frac{\lambda_0}{2\pi \sqrt{|\varepsilon_\omega|}} = \frac{\lambda_0}{2\pi} \left| 1 - \frac{\omega_p^2}{\omega^2} \right|^{-1/2}, \tag{7.73}$$

does not depend on conductivity and is not related to energy dissipation. Such not-dissipative stopping corresponds **to the total wave reflection from plasma**.

To illustrate the total wave reflection, let us consider a wave propagating in nonuniform plasma without significant dissipative absorption (see Figure 7.6b). The electromagnetic wave propagates from areas with low electron density to areas where the density increases. The wave frequency is fixed, but the plasma frequency increases together with the electron density leading to decrease of ε_ω. At the point when dielectric permittivity $\varepsilon_\omega = 1 - \frac{\omega_p^2}{\omega^2}$ becomes equal to zero, the total reflection takes place. This total reflection of electromagnetic waves takes place when the electron density reaches the critical value, which can be found from $\omega = \omega_p$ as:

$$n_e^{crit} = \frac{\varepsilon_0 m\omega^2}{e^2}, \quad n_e(\mathrm{cm}^{-3}) = 1.24 \cdot 10^4 \cdot [f(\mathrm{MHz})]^2. \tag{7.74}$$

Electromagnetic waves with frequency 3 GHz are reflected from plasma with density above $10^{11}\mathrm{cm}^{-3}$. Atmospheric air is slightly ionized by solar radiation at heights exceeding 100 km; electron density in the ionosphere is about $10^4 \div 10^5 \mathrm{cm}^{-3}$. This plasma reflects radio waves with frequencies about 1 MHz, providing their long-distance transmission.

7.12 Nonlinear Waves in Plasma: Modulation Instability, Lighthill Criterion, and Korteweg–de Vries Equation

Plasma is strongly nonlinear medium, and nonlinear phenomena play there an important role especially regarding plasma self-organization, self-adaptation, and structuring. Let us consider a perturbation of some plasma parameter as a "wave package" $a(x,t) = \sum_k a(k) \exp(ikx - i\omega t)$, which is a group of waves with different but close wave vectors ($\Delta k \ll k$) and relevant amplitudes $a(k)$. The dispersion relation $\omega(k)$ for the wave package can be expressed as:

$$\omega(k) = \omega(k_0) + \frac{\partial \omega}{\partial k}(k = k_0) \cdot (k - k_0) + \frac{1}{2} \frac{\partial^2 \omega}{\partial k^2}(k = k_0) \cdot (k - k_0)^2$$

$$= \omega_0 + v_{gr}(k - k_0) + \frac{1}{2} \frac{\partial v_{gr}}{\partial k}(k - k_0)^2, \tag{7.75}$$

where k_0 is the average value of the wave vector for the group of waves; ω_0 is frequency corresponding to k_0; v_{gr} is group velocity. Because of different group velocities for the waves with different values of wave vectors ($\partial v_{gr}/\partial k \neq 0$), the group of waves grows. If the waves are not interacting with each other (linear waves), the initial perturbation grows to the size $\Delta x \approx 1/\Delta k$ during the period about: $\tau \approx \left(\Delta k^2 \frac{\partial v_{gr}}{\partial k} \right)^{-1}$. Nonlinearity can be introduced by frequency dependence on amplitude $E(x)$: $\omega = \omega_0 - \alpha E^2$, where ω_0 is the wave frequency in the low amplitude limit. Then we can rewrite the plasma perturbation as:

$$a(x,t) = \sum_k a(k) \exp \left[i(k - k_0)(x - x_0) - ik_0 x_0 - i(k - k_0)^2 \frac{\partial v_{gr}}{\partial k} t - i\alpha E^2(x)t \right], \tag{7.76}$$

which shows modulation of the group of waves called the **modulation instability**. At some modes of the modulation instability, the "wave package" decays into smaller groups of waves or can be "compressed" and converted into specific single wave called the **soliton**. Competition between two last terms in the exponent (7.76) determines type of evolution of the wave package. The term $i(k - k_0)^2 \frac{\partial v_{gr}}{\partial k} t$ leads to expansion of the wave package, while the nonlinear term $i\alpha E^2(x)t$ can compensate the previous one. To compensate expansion of a perturbation and provide an opportunity of the wave package decay into smaller groups of waves, compression, or formation of solitons, the two terms should at least have different signs. This leads to the requirement:

$$\alpha \cdot \frac{\partial v_{gr}}{\partial k} < 0, \tag{7.77}$$

which is called the **Lighthill criterion** of modulation instability.

Influence of a weak nonlinearity on the wave package expansion due to dispersion $\omega(k)$ can be described using the **Korteweg–de Vries equation.** Let us consider propagation of the longitudinal long-wavelength oscillations (for example, ionic sound, see Section 7.7). These oscillations at relatively long wavelength ($r_0 k \ll 1$) can be described by the dispersion relation:

$$\omega = v_{gr} k \left(1 - r_0^2 k^2 \right). \tag{7.78}$$

In the case of ionic wave, the size-parameter r_0 corresponds to the Debye radius. The Euler equation for particle velocities in the longitudinal wave under consideration is:

$$\frac{\partial v}{\partial t} + v \frac{\partial v}{\partial x} - \frac{F}{M} = 0, \tag{7.79}$$

where $v(x, t)$ is particle velocity in the longitudinal wave along the axis "x"; F is force acting on the plasma particle with mass M. For linear approximation, $v = v_{gr} + v'$, where $v' \ll v_{gr}$ is the particle velocity with respect to the wave. The Euler equation can be then presented linearly:

$$\frac{\partial v'}{\partial t} + v_{gr} \frac{\partial v'}{\partial x} - \frac{F}{M} = 0, \tag{7.80}$$

with the function F/m considered as a linear operator of v'. In the harmonic approximation for the plasma particle velocities: $v' \propto \exp(-i\omega t + ikx)$, choose the linear operator $\frac{F}{m}(v')$ in a way to get from the linear equation (7.80) the dispersion relation (7.78). This leads to the linear operator $\frac{F}{m}(v') = -r_0^2 v_{gr} \frac{\partial^3 v'}{\partial x^3}$ and to the linear equations of motion in the form:

$$\frac{\partial v'}{\partial t} + v_{gr} \left(\frac{\partial v'}{\partial x} + r_0^2 \frac{\partial^3 v'}{\partial x^3} \right) = 0. \tag{7.81}$$

The last term in this linear equation describes dispersion of the long-wavelength oscillations. To consider the nonlinearity, we replace v' by the plasma particle velocity v. It results in the nonlinear motion equation for plasma particles:

$$\frac{\partial v}{\partial t} + v \frac{\partial v}{\partial x} + v_{gr} r_0^2 \frac{\partial^3 v}{\partial x^3} = 0. \tag{7.82}$$

It is known as the **Korteweg–de Vries equation** and is especially important for description of nonlinear dissipative processes because it considers both nonlinearity and dispersion of waves.

7.13 Langmuir Solitons in Plasma

The solitons are solitary nonharmonic waves with a particular ability to maintain their shape during propagation in a medium with dispersion. Although shorter harmonics are moving slower, nonlinearity compensates this effect, and wave package does not expand but rather keeps its shape. The solitons are determined by the following Korteweg–de Vries equation considering plasma particle velocities as $v = f(x - ut)$ and $\frac{\partial v}{\partial t} = -u\frac{\partial v}{\partial x}$ (u is the wave propagation velocity):

$$(v - u)\frac{dv}{dx} + v_{gr}r_0^2\frac{d^3v}{dx^3} = 0. \tag{7.83}$$

Integration of the Korteweg–de Vries equation considering absence of the plasma particles velocity at infinity ($v = 0$, $dv^2/dx^2 = 0$ at $x \to \infty$), leads to the second-order equation:

$$v_{gr}r_0^2\frac{d^2v}{dx^2} = uv - \frac{v^2}{2}. \tag{7.84}$$

The soliton, the solitary wave, is one of the solutions of this nonlinear equation:

$$v = \frac{3u}{\cosh^2\frac{x}{2r_0}\sqrt{\frac{u}{v_{gr}}}}; \tag{7.85}$$

here $\cosh \alpha$ is the hyperbolic cosine. Profiles of solitons with the same fixed parameters u, v_{gr}, and different amplitudes are illustrated in Figure 7.7. The solitary wave becomes narrower with increases of its amplitude. The product of the soliton's amplitude and square of the soliton's width remains constant during its evolution. If initial non-soliton perturbation has low amplitude, then the perturbation expands with time because of dispersion until the product of its amplitude and width square corresponds to that one for solitons (7.85). Then, the perturbation during its evolution converts into a soliton. Conversely, if the amplitude of an initial perturbation is relatively high, then during evolution of the perturbation it decays to form several solitons.

Formation of the solitons in plasma is due to electric fields induced by a wave, which confines a plasma perturbation in a local domain. The higher is the wave amplitude, the stronger electric field can be induced by this wave, resulting in stronger effect of perturbation compression. It can be illustrated by plasma oscillation leading to the formation of **Langmuir solitons**. The energy density of Langmuir oscillations at a point "x" is a function of time-averaged electric field:

$$W(x) = \frac{\varepsilon_0 < E^2(x, t) >}{2}. \tag{7.86}$$

Assuming plasma is completely ionized and quasi-neutral with equal electron and ion temperatures (T), the total effective pressure is also constant in space for the long-wavelength plasma oscillations with propagation velocities

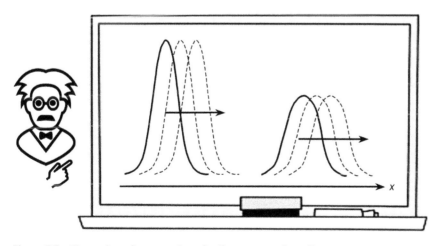

Figure 7.7 Illustration of propagation of solitary waves, the solitons.

much less than speed of sound:

$$2n_e(x)T + W(x) = 2n_{e0}T; \tag{7.87}$$

here n_{e0} is electron density at infinity, where there are no plasma oscillations. The dispersion equation (7.25) can be then rewritten in a nonlinear way with the effect of electric field amplitude:

$$\omega^2(x) = \omega_{p0}^2 \left(1 - \frac{\varepsilon_0 <E^2>}{4n_{e0}T}\right) + \frac{\gamma T}{m}k^2; \tag{7.88}$$

here ω_{p0} is plasma frequency far from the perturbation, T is plasma temperature, and γ is the specific heat ratio. Then considering $E = E_0 \cos \omega t, <E^2> = \frac{1}{2}E_0^2$, (7.78) gives:

$$\omega(x) = \omega_{p0} \left(1 - \frac{\varepsilon_0 E_0^2}{16n_{e0}T} + \gamma r_D^2 k^2\right). \tag{7.89}$$

The second and third terms are much less than unity; and (7.89) satisfies the Lighthill criterion (7.77). Differentiating (7.85) gives:

$$\alpha\frac{\partial v_{gr}}{\partial k} = \alpha\frac{\partial^2 \omega}{\partial k^2} = -\frac{\gamma \varepsilon_0 \omega_{p0}}{8mn_{e0}} < 0, \tag{7.90}$$

which explains formation of the solitary waves in plasma, called the Langmuir solitons.

7.14   Nonlinear Ionic Sound, Evolution of Strongly Nonlinear Oscillations

To describe the strongly nonlinear ionic sound, let us use the Euler equation of ionic motion, the continuity equation for ions, and the Poisson equation:

$$\frac{\partial v_i}{\partial t} + v_i\frac{\partial v_i}{\partial x} + \frac{e}{M}\frac{\partial \varphi}{\partial x} = 0, \tag{7.91}$$

$$\frac{\partial n_i}{\partial t} + \frac{\partial}{\partial x}(n_i v_i) = 0, \tag{7.92}$$

$$\frac{\partial^2 \varphi}{\partial x^2} = \frac{e}{\varepsilon_0}(n_e - n_i); \tag{7.93}$$

here v_i is the ion velocity in wave; φ is the electric field potential; n_e, n_i-are the electron and ion densities; e is an electron charge; M is an ion mass. Consider ionic motion as a wave propagating with velocity "u"; then the plasma parameters (v_i, n_i, φ) depend on coordinate x and time t as $f(x - ut)$. Consider electron mobility is high, which results in their Boltzmann quasi-equilibrium with electric field: $n_e = n_{e0} \exp\left(\frac{e\varphi}{T_e}\right)$, where T_e is the electron temperature and n_{e0} is the average density of charged particles. Then rewrite the above system of equations become:

$$-u\frac{dv_i}{dx} + v_i\frac{dv_i}{dx} + \frac{e}{M}\frac{d\varphi}{dx} = 0, \tag{7.94}$$

$$\frac{d}{dx}[n_i(v_i - u)] = 0, \tag{7.95}$$

$$\frac{\partial^2 \varphi}{\partial x^2} = \frac{e}{\varepsilon_0}\left[n_{e0}\exp\left(\frac{e\varphi}{T_e}\right) - n_i\right]. \tag{7.96}$$

Integrating (7.94 and 7.95) and assuming $n_i = n_{e0}$, $v_i = 0$ at $x \to \infty$ leads to the relations:

$$\frac{v_i^2}{2} - uv_i + \frac{e\varphi}{M} = 0, \tag{7.97}$$

$$n_i = n_{e0}\frac{u}{u - v_i} \tag{7.98}$$

The above Poisson equation can be then expressed as a nonlinear second order differential relation:

$$\frac{d^2\varphi}{dx^2} = \frac{en_{e0}}{\varepsilon_0}\left[\exp\left(\frac{e\varphi}{T_e}\right) - \frac{u}{\sqrt{u^2 - \frac{2e\varphi}{M}}}\right]; \tag{7.99}$$

and converted to first order equation multiplying (7.99) by $\frac{d\varphi}{dx}$, integrating, and assuming $\varphi = 0$, $d\varphi/dx = 0$:

$$\frac{1}{2}\left(\frac{d\varphi}{dx}\right)^2 + \frac{n_{e0}T_e}{\varepsilon_0}\left[1 - \exp\left(\frac{e\varphi}{T_e}\right)\right] + \frac{n_{e0}Mu^2}{\varepsilon_0}\left(1 - \sqrt{1 - \frac{2e\varphi}{Mu^2}}\right) = 0. \tag{7.100}$$

This nonlinear equation describes the solitary ionic sound waves, the **ionic sound solitons**, without limitations on their amplitudes. It illustrates the **effects of strong nonlinearity**, for example, relation between the maximum potential in the solitary wave $\varphi = \varphi_{max}$ and its propagation velocity u. Assuming $\varphi = \varphi_{max}$, $d\varphi/dx = 0$, and introducing dimensionless variables for the maximum potential $\xi = e\varphi_{max}/T_e$ and the wave velocity $\eta = Mu^2/2T_e$, (7.100) gives:

$$1 - \exp(\xi) + 2\eta(1 - \sqrt{1 - \xi/\eta}). \tag{7.101}$$

If amplitude of the ionic sound is small ($\xi \to 0$), (7.101) gives $\eta = 1/2$ and $u = \sqrt{T_e/M}$, which corresponds to the conventional ionic sound (see Section 7.7). In opposite limit of the maximum amplitude, ion energy in the soliton corresponds to potential energy in the wave $\xi = \eta$. In this case, the critical value of ξ according to (7.101) is determined by equation:

$$1 - \exp(\xi) + 2\xi = 0, \tag{7.102}$$

which gives: $\xi = 1.26$, $e\varphi_{max} = 1.26T_e$, $u = 1.58\sqrt{T_e/M}$. At amplitudes above the critical one, ions "are reflected" by the wave. It results in decay of the strongly nonlinear wave into smaller separate wave packages. Thus, solitons only exist at some limited levels of the wave amplitude. Large amplitudes and strong nonlinearity lead to decay of the solitary waves.

7.15 Problems and Concept Questions

7.15.1 Ideal and Nonideal Plasmas

What is the minimum electron density for plasma to be nonidea based on (7.1): (i) at electron temperature 1 eV, (ii) for room temperature electrons.

7.15.2 Charged Particles Inside Debye Sphere

Number of the particles necessary for "screening" of the electric field of a charged particle corresponds to amount of these particles in a Debye sphere. Show that this number is $\sqrt{T_e^3(4\pi\varepsilon_0)^3/e^6 n_e}$, and is large in ideal plasma.

7.15.3 Floating Potential

Micro-particles or aerosols are usually negatively charged in plasma and have negative floating potential (7.11) with respect to plasma. Estimate the negative charge of such spherical particles at $T_e = 1$ eV as a function of their radius.

7.15.4 Matrix and Child Law Sheaths

Calculate the matrix and Child law sheaths for nonthermal plasma with $T_e = 1$ eV, electron density 10^{12}cm^{-3}, and sheath voltage 300 V.

7.15.5 Electrostatic Plasma Waves

Using the dispersion equation (7.25) show that product of the phase and group velocities of the electrostatic plasma waves corresponds to the square of thermal electron velocities. Which of these two characteristic wave velocities is larger?

7.15.6 Ionic Sound

Based on the dispersion equation (7.33) derive relation for the group velocity of ionic sound. Compare the group and phase velocities of the ionic sound.

7.15.7 Landau Damping

Estimate the coefficient γ for Landau damping for electrostatic plasma oscillations (in ω_p units) in typical range of microwave frequencies. Assume that the plasma oscillation frequency is close to the Langmuir frequency, $k\, r_D \approx 0.1$.

7.15.8 High-frequency Dielectric Permittivity of Plasma

Explain why the high-frequency dielectric permittivity of plasma is less than one, while for conventional dielectric materials, the dielectric constant is greater than one?

7.15.9 Solitons as Solutions of the Korteweg–de Vries Equation

Prove that solitons (7.85) are solutions of the Korteweg–de Vries equation. Show that product of the soliton's amplitude and square of the soliton's width are constant value during its evolution.

7.15.10 Nonlinear Ionic Sound

Analyzing (7.94–7.98) for the nonlinear ionic sound show, that the ionic velocity here is always less than velocity of wave propagation ($v_i < u$).

7.15.11 Velocity of the Nonlinear Ionic-sound Waves

Analyzing (7.101) prove, that the velocity of ionic sound at low amplitudes is equal to $c_{si} = \sqrt{T_e/M}$.

7.15.12 The Ionic Sound Solitons

Using relation (7.101) determine the maximum amplitude of the potential in the solitary ionic sound wave. How does the nonlinear ionic-sound wave velocity depend on amplitude of the wave?

Lecture 8

Plasma Magneto-hydrodynamics, Fluid Mechanics and Acoustics

8.1 Plasma Magneto-hydrodynamics (MHD): Magnetic Field "Diffusion" in Plasma, Frozenness of Magnetic Field in Plasma

High-density plasma strongly interacts with magnetic fields: plasma motion induces electric currents, which together with the magnetic field influence the motion. The principal magnetohydrodynamic (MHD) equations describing such phenomena are:

(a) **Navier–Stokes equation** neglecting viscosity, but considering the magnetic force on the plasma current with density $\vec{j}$, B is magnetic induction, M is the mass of ions, $n_e = n_i$, $Mn_e = \rho$:

$$Mn_e \left[\frac{\partial \vec{v}}{\partial t} + (\vec{v} \nabla)\vec{v} \right] + \nabla p = [\vec{j} \, \vec{B}]. \tag{8.1}$$

(b) **Continuity equation** for electrons and ions, moving together with velocity $\vec{v}$:

$$\frac{\partial n_e}{\partial t} + div(n_e \vec{v}) = 0. \tag{8.2}$$

(c) **Maxwell equations** for magnetic field, neglecting the displacement current:

$$curl \, \vec{H} = \vec{j}, \quad div \, \vec{B} = 0, \tag{8.3}$$

(d) **Maxwell equation for electric field** $curl \, E = -\frac{\partial \vec{B}}{\partial t}$ together with (8.3) and Ohm's law ($\vec{j} = \sigma(\vec{E} + [\vec{v} \, \vec{B}])$) for plasma with conductivity σ give an additional equation:

$$\frac{\partial \vec{B}}{\partial t} = curl[\vec{v} \, \vec{B}] + \frac{1}{\sigma \mu_0} \Delta \vec{B}. \tag{8.4}$$

If plasma is at rest ($\vec{v} = 0$), the Eq. (8.4) can be reduced to that of diffusion:

$$\frac{\partial \vec{B}}{\partial t} = D_m \Delta \vec{B}, \quad D_m = \frac{1}{\sigma \mu_0}. \tag{8.5}$$

The factor D_m can be interpreted as a coefficient of "diffusion of magnetic field into plasma," and sometimes is called **magnetic viscosity**. If characteristic time of magnetic field change is $\tau = 1/\omega$, then length of the magnetic field diffusion is $\delta \approx \sqrt{2D_m \tau} = \sqrt{2/\sigma\omega\mu_0}$, which corresponds to the skin-layer depth (7.23). Damping time τ_m of currents and magnetic fields in a conductor with characteristic size L according to (8.5) is:

$$\tau_m = \frac{L^2}{D_m} = \mu_0 \sigma L^2; \tag{8.6}$$

it is infinitely long for superconductors and large objects (for solar spots more than 300 years).

Plasma Science and Technology: Lectures in Physics, Chemistry, Biology, and Engineering, First Edition. Alexander Fridman.
© 2024 WILEY-VCH GmbH. Published 2024 by WILEY-VCH GmbH.

If the plasma conductivity is high ($\sigma \to \infty$), the diffusion coefficient of magnetic field is small ($D_m \to 0$) and magnetic field is unable to "move" with respect to the plasma. In other words, **magnetic field is frozen in high conductivity plasma**.

To further explain the frozenness of magnetic field in plasma, let us consider displacement (correspondent to time interval dt) of the surface element ΔS, moving with velocity $\vec{v}$ together with plasma (see Figure 8.1a). Plasma conductivity is high ($\sigma \to \infty$) and (8.5) can be expressed as:

$$\frac{\partial \vec{B}}{\partial t} = curl\,[\vec{v}\,\vec{B}]. \tag{8.7}$$

Because $div\,\vec{B} = 0$, difference in fluxes $\vec{B}\,\frac{d\vec{S}}{dt}$ through surface elements ΔS_0 and ΔS_1 is equal to the flux $\oint \vec{B}[\vec{v}\,d\vec{l}\,] = \oint [\vec{B}\vec{v}]\,d\vec{l}$ through the side surface of the fluid element (Figure 8.1a). Then the time derivative of the magnetic field flux $\int \vec{B}\,d\vec{S}$ through the moving surface element ΔS can be expressed as:

$$\frac{d}{dt}\int \vec{B}\,d\vec{S} = \int \frac{\partial \vec{B}}{\partial t}\,d\vec{S} + \int \vec{B}\frac{d\vec{S}}{dt} = \int \frac{\partial \vec{B}}{\partial t}\,d\vec{S} + \oint [\vec{B}\vec{v}]\,d\vec{l} = \int curl\,[\vec{v}\,\vec{B}]\,d\vec{S} + \oint [\vec{B}\vec{v}]\,d\vec{l}. \tag{8.8}$$

According to the Stokes theorem, the last sum of (8.8) equals to zero, and therefore:

$$\frac{d\Phi}{dt} = \frac{d}{dt}\int \vec{B}\,d\vec{S} = 0, \quad \Phi = const. \tag{8.9}$$

It demonstrates that magnetic flux Φ through any surface element moving with plasma is constant, which proves again that *magnetic field is "frozen" in moving plasma with high conductivity*.

8.2 Plasma Equilibrium in Magnetic Field: Magnetic Pressure and Pinch Effect

The Navier–Stokes equation (8.11) can be simplified in steady-state ($d\,\vec{v}/dt = 0$) to:

$$grad\,p = [\,\vec{j}\,\vec{B}]. \tag{8.10}$$

It can be interpreted as a balance of hydrostatic pressure p and the Ampere force. Considering the first of Maxwell equations (8.3), we eliminate current from this balance of forces:

$$\nabla p = [\,\vec{j}\,\vec{B}] = \mu_0[curl\,\vec{H} \times \vec{H}] = -\frac{\mu_0}{2}\nabla H^2 + \mu_0(\vec{H}\nabla)\vec{H}. \tag{8.11}$$

Combining gradients in (8.11) leads to formula for **plasma equilibrium in magnetic field**:

$$\nabla\left(p + \frac{\mu_0 H^2}{2}\right) = \mu_0(\vec{H}\nabla)\vec{H} = \frac{\mu_0 H^2}{R}\vec{n}; \tag{8.12}$$

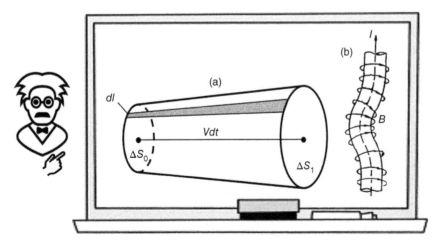

Figure 8.1 Illustrations of magneto-hydrodynamic effects in plasma: (a) magnetic field frozen in plasma; (b) pinch effect.

where R is the radius of curvature of magnetic field line; $\vec{n}$ is the normal vector to the line. The force $\frac{\mu_0 H^2}{R}\vec{n}$ is related to bending of the magnetic field lines and represents **tension of magnetic lines**. This tension tends to make the magnetic field lines straight and equal to zero when they are straight. The pressure term $\frac{\mu_0 H^2}{2}$ is called **magnetic pressure.** Sum of the hydrostatic and magnetic pressures $p + \frac{\mu_0 H^2}{2}$ is total pressure.

Plasma equilibrium in magnetic field can be interpreted as the dynamic balance of the gradient of total pressure and the tension of magnetic lines. If the magnetic field lines are straight and parallel, then $R \to \infty$ and the "tension" of magnetic lines is zero, leading to the equilibrium criterion:

$$p + \frac{\mu_0 H^2}{2} = const. \tag{8.13}$$

Generally, plasma equilibrium in a magnetic field means a balance of pressure in plasma with the outside magnetic pressure. Plasma equilibrium is also possible for special configurations of magnetic fields, when $p \ll \frac{\mu_0 H^2}{2}$. In this case, the outside magnetic pressure is compensated by the "tension" of magnetic lines. Such equilibrium configurations are called "forceless."

An interesting MHD phenomenon of plasma self-compression in its own magnetic field is called pinch effect. Let us consider this effect in a long cylindrical discharge with the electric current along the axis of the cylinder (Z-pinch, see Figure 8.1b). Equilibrium of the completely ionized Z-pinch plasma can be expressed by the **Bennet relation**:

$$N_L T = \frac{\mu_0}{8\pi} I^2; \tag{8.14}$$

where N_L is the plasma density per unit length of the cylinder; T is the plasma temperature. The Bennett relation shows that plasma temperature should grow proportionally to the square of the current to provide a balance of plasma pressure and magnetic pressure of the current. For example, to reach thermonuclear temperatures of about 100 keV in a plasma with density $10^{15}\,\mathrm{cm}^{-3}$ and cross-section $1\,\mathrm{cm}^2$, the necessary current according to the Bennet relation is about 100 kA. Such current can be achieved in Z-pinch discharges, which stimulated enthusiasm in the 1950s to achieve controlled thermonuclear fusion. However, the hopes for easily controlled fusion in Z-pinch discharges were shattered due to fast MHD instabilities of Z-pinch, related to plasma bent and nonuniform plasma compression, illustrated in Figure 8.1b. If the discharge column is bent, the magnetic field and magnetic pressure become larger on the concave side of the plasma, which leads to a break of the channel. This is usually referred to as the "wriggle" instability. Similarly, if the discharge channel becomes locally thinner, the magnetic field ($B \propto 1/r$) and magnetic pressure at this point grow leading to further compression and to a subsequent break of the channel.

8.3 Two-fluid Plasma MHD and the Generalized Ohm's Law

The plasma MHD approach in Section 8.1 assumed equal electron $\vec{v}_e$ and ion $\vec{v}_i$ velocities, which contradicts presence of the electric current $\vec{j} = en_e(\vec{v}_i - \vec{v}_e)$ in the quasi-neutral plasma with density n_e. The separate motion of electrons and ions is considered in the two-fluid plasma MHD, where the Navier–Stokes equation (8.1) for electrons includes their mass m, pressure p_e, velocity, and additionally takes into account friction between electrons and ions (which corresponds to the last term in the equation, where v_e is the frequency of electron collisions):

$$m n_e \frac{d\vec{v}_e}{dt} + \nabla p_e = -en_e \vec{E} - en_e[\vec{v}_e \vec{B}] - mn_e v_e(\vec{v}_e - \vec{v}_i). \tag{8.15}$$

Similar Navier–Stokes equation for ions includes the same friction term but with opposite sign. First term in (8.15) related to the electron inertia can be neglected because of their low mass. Denoting the ion's velocity as $\vec{v}$ and the plasma conductivity as $\sigma = n_e e^2 / m v_e$, we can rewrite (8.15) in the form known as the **generalized Ohm's law**:

$$\vec{j} = \sigma(\vec{E} + [\vec{v}\vec{B}]) + \frac{\sigma}{en_e}\nabla p_e - \frac{\sigma}{en_e}[\vec{j}\vec{B}]. \tag{8.16}$$

The generalized Ohm's law in contrast to the conventional one, considers electron pressure gradient and the $[\vec{j} \times \vec{B}]$ term related to the **Hall effect**. Both additional terms show that the current direction is not always simply correlated with the direction of the electric field. The Hall effect is related to the electron conductivity in the presence of magnetic field, which provides an electric current in the direction perpendicular to electric field.

Solution of (8.16) with respect to electric current is complicated because of the current presence in two terms of the generalized Ohm's law. This equation can be solved, however, if the plasma conductivity is sufficiently large ($\sigma \to \infty$), and the generalized Ohm's law is expressed as:

$$\vec{E} + [\vec{v}\vec{B}] + \frac{1}{en_e}\nabla p_e = \frac{1}{en_e}[\vec{j}\,\vec{B}]. \tag{8.17}$$

If electron temperature is uniform, the electron hydrodynamic can be used and:

$$\vec{E} = -[\vec{v}_e\vec{B}] - \frac{1}{e}\nabla(T_e \ln n_e). \tag{8.18}$$

Using (8.18) in the Maxwell equation $curl\, E = -\frac{\partial \vec{B}}{\partial t}$, and considering curl of grad as zero:

$$\frac{\partial \vec{B}}{\partial t} = curl\,[\vec{v}_e\vec{B}]. \tag{8.19}$$

This two-fluid plasma MHD equation is similar to (8.7).

8.4 The Generalized Ohm's Law and Plasma Diffusion Across Magnetic Field

The generalized Ohm's law (8.16) gives the coefficients of free electrons and ions diffusion perpendicular to magnetic field as (Kadomtsev and Shafranov 2000):

$$D_{\perp,e} = \frac{D_e}{1 + \left(\frac{\omega_{B,e}}{v_e}\right)^2}, \quad D_{\perp,i} = \frac{D_i}{1 + \left(\frac{\omega_{B,i}}{v_i}\right)^2}; \tag{8.20}$$

here: D_e and D_i are the coefficients of free diffusion of electrons and ions without magnetic field (see Section 4.12); v_e and v_i are the collisional frequencies of electrons and ions; $\omega_{B,e}$ and $\omega_{B,i}$ are the electron and ion cyclotron frequencies (see Section 4.9):

$$\omega_{B,e} = \frac{eB}{m}, \quad \omega_{B,i} = \frac{eB}{M}, \tag{8.21}$$

which show the frequencies of electron and ion collisionless rotation in magnetic field. Since ions are much heavier than electrons ($M \gg m$), the electron-cyclotron frequency is much greater than the ion-cyclotron frequency. If diffusion is ambipolar, which is the case in the highly ionized plasma, then formula (4.74) for D_a can still be applied in a magnetic field. Obviously, the free diffusion coefficients for electrons and ions should be replaced by those in the magnetic field. The conventional electron and ion mobilities μ_e, μ_i also should be replaced by those corresponding to the drift perpendicular to the magnetic field (4.56 and 4.57). It results in the following coefficient of ambipolar diffusion perpendicular to magnetic field:

$$D_\perp = \frac{D_a}{1 + \frac{\omega_{B,i}^2}{v_i^2} + \frac{\mu_i}{\mu_e}\left(1 + \frac{\omega_{B,e}^2}{v_e^2}\right)}; \tag{8.22}$$

here D_a is the conventional coefficient of ambipolar diffusion (4.74). The equation (8.22) can be simplified when electrons are magnetized $\omega_{B,e}/v_e \gg 1$ and trapped in the magnetic field, but heavy ions are not magnetized ($\omega_{B,i}/v_i \ll 1$):

$$D_\perp = \frac{D_a}{1 + \frac{\mu_i}{\mu_e}\frac{\omega_{B,e}^2}{v_e^2}}. \tag{8.23}$$

If magnetic field is very strong and $\frac{\mu_i}{\mu_e}\frac{\omega_{B,e}^2}{v_e^2} \gg 1$ the relation for ambipolar diffusion is:

$$D_\perp \approx D_a \frac{\mu_e}{\mu_i}\frac{v_e^2}{\omega_{B,e}^2} \approx D_e \frac{v_e^2}{(e/M)^2}\frac{1}{B^2}. \tag{8.24}$$

In strong magnetic fields, the coefficient $D_\perp$ significantly decreases: $D_\perp \propto 1/B^2$, the magnetic field is not "transparent for plasma" and can be used to prevent the plasma from decay. The slow ambipolar diffusion across the strong

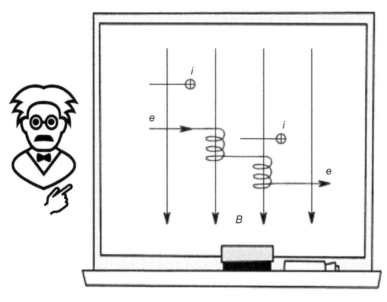

Figure 8.2 Ambipolar plasma diffusion across magnetic field.

magnetic field is illustrated in Figure 8.2. The magnetized electron is trapped by the magnetic field and rotates along the Larmor circles until a collision pushes the electron to another Larmor circle. The **electron Larmor radius**:

$$\rho_L = \frac{v_\perp}{\omega_{B,e}} = \frac{1}{eB}\sqrt{2T_e m}. \tag{8.25}$$

is the radius of circular motion of a magnetized electron (see Figure 8.2); here $v_\perp$ is component of electron velocity perpendicular to magnetic field. When magnetic field is strong, the Larmor radius plays the same role as mean free path in diffusion without magnetic field. The equation (8.24), therefore, can be rewritten in terms of the electron Larmor radius as:

$$D_\perp \approx D_e \frac{v_e^2}{\omega_{B,e}^2} \approx \rho_L^2 v_e. \tag{8.26}$$

When the magnetic field is high and electrons are magnetized ($\omega_{B,i}/v_i \ll 1$), then the electron Larmor radius is shorter than the electron mean free path ($\rho_L \ll \lambda_e$). Plasma diffusion across the magnetic field is slower and additionally decreases with the square of the field.

8.5 Magnetic Reynolds Number and Alfven Velocity: Conditions for Magneto-hydrodynamic (MHD) Behavior of Plasma

The MHD plasma behavior means that fluid dynamics is strongly coupled to magnetic field. It means that "diffusion" of magnetic field is less than the "convection," that is the curl in the (8.4) right-hand side exceeds the Laplacian in this equation:

$$\frac{vB}{L} \gg \frac{1}{\sigma\mu_0}\frac{B}{L^2}, \quad \text{or} \quad v \gg \frac{1}{\sigma\mu_0 L}; \tag{8.27}$$

here L is plasma size. Using the concept of magnetic viscosity D_m (8.5), we can rewrite (8.27) as:

$$Re_m = \frac{vL}{D_m} \gg 1. \tag{8.28}$$

Here Re_m is the **magnetic Reynolds number** corresponding to the conventional Reynolds number, but with kinematic viscosity replaced by the magnetic viscosity. The physical interpretation of the magnetic Reynolds can be

Table 8.1 The magnetic Reynolds numbers for some specific plasmas in nature and laboratory systems.

Type of plasma	B, Tesla	Space-scale (m)	ρ (kg m^{-3})	σ (1/Ohm·cm)	R_m
Ionosphere	10^{-5}	10^5	10^{-5}	0.1	10
Solar atmosphere	10^{-2}	10^7	10^{-6}	10	10^8
Solar corona	10^{-9}	10^9	10^{-17}	10^4	10^{11}
Hot interstellar gas	10^{-10}	10 light-years	10^{-21}	10	10^{15}
Arc discharge plasma	0.1	0.1	10^{-5}	10^3	10^3
Hot confined plasma, $n = 10^{15}\text{cm}^{-3}, T = 10^6\text{K}$	0.1	0.1	10^{-6}	10^3	10^4

further clarified considering that plasma velocity v in MHD systems usually corresponds to balance of dynamic and magnetic pressures:

$$\frac{\rho v^2}{2} \propto \frac{\mu_0 H^2}{2}, \quad \text{or} \quad v \propto v_A = \frac{B}{\sqrt{\rho \mu_0}}; \tag{8.29}$$

here $\rho = Mn_e$ is plasma density. The plasma velocity v_A when dynamic and magnetic pressures are equal is called the **Alfven velocity**. The criterion of the MHD plasma behavior then becomes:

$$Re_m = \frac{v_A L}{D_m} = BL\sigma \sqrt{\frac{\mu_0}{\rho}} \gg 1. \tag{8.30}$$

The magnetic Reynolds numbers for some specific plasmas are presented in Table 8.1.

8.6 Electromagnetic Wave Propagation in Magnetized Plasma

If electric field of a plane-polarized wave is codirected with external magnetic field, the magnetic field does not influence propagation of the wave. Interesting dispersion occurs when the electromagnetic wave propagates along the magnetic field. Let us analyze it.

Neglecting collisions ($v_{en} \ll \omega$), an electron motion equation for the transverse wave propagating along the magnetic field B_0 is:

$$\frac{d\vec{v}}{dt} = -\frac{e}{m}(\vec{E} + [\vec{v} \times \vec{B}_0]). \tag{8.31}$$

Neglecting ion motion and assuming $\vec{j} = -n_e e\vec{E}$, (8.31) can be expressed as:

$$\frac{d\vec{j}}{dt} = \frac{n_e e^2}{m}\vec{E} - \frac{e}{m}[\vec{j} \times \vec{B}_0]. \tag{8.32}$$

Together with the electromagnetic wave equation, relation (8.32) gives:

$$\Delta E - \frac{1}{c^2}\frac{\partial^2 E}{\partial t^2} - \mu_0\frac{\partial \vec{j}}{\partial t} = 0. \tag{8.33}$$

Let us look for a solution of (8.32 and 8.33) for a circularly polarized electromagnetic wave, when the electric vector rotates in a plane (x, y) perpendicular to direction z of the wave propagation: $E_x = E_0 \cos(\omega t - kz)$, $E_y = \pm E_0 \sin(\omega t - kz)$. Here E_0 is amplitude of electric field oscillations in the wave; signs (+) and (−) correspond to rotation of the electric field vector in opposite directions. Current density components can be also expressed in a similar way:

$$j_x = j_0 \cos(\omega t - kz), \quad j_y = \pm j_0 \sin(\omega t - kz), \tag{8.34}$$

where j_0 is amplitude of current density oscillations. Projection of the above equations on the axis "x" gives:

$$\frac{dj_x}{dt} = \varepsilon_0 \omega_p^2 E_x - \omega_B j_y, \tag{8.35}$$

$$\frac{\partial^2 E_x}{\partial z^2} - \frac{1}{c^2}\frac{\partial^2 E_x}{\partial t^2} - \mu_0 \frac{\partial j_x}{\partial t} = 0; \tag{8.36}$$

here ω_p is plasma frequency; $\omega_B = eB_0/m$ is electron-cyclotron frequency. Applying (8.34) for components of the electric field and current density, (8.35 and 8.36) result in the dispersion set of equations:

$$-\omega j_0 \sin(\omega t - kz) = \varepsilon_0 \omega_p^2 E_0 \cos(\omega t - kz) \pm \omega_B j_0 \sin(\omega t - kz), \tag{8.37}$$

$$-k^2 E_0 \cos(\omega t - kz) + \frac{\omega^2}{c^2} E_0 \cos(\omega t - kz) + \mu_0 \omega j_0 \sin(\omega t - kz) = 0. \tag{8.38}$$

This system of equations has a nontrivial solution only if the following relation between wavelength and frequency of the electromagnetic wave is valid:

$$\frac{k^2 c^2}{\omega^2} = 1 - \frac{\omega_p^2}{\omega^2}\frac{1}{\left(1 \pm \dfrac{\omega_B}{\omega}\right)}. \tag{8.39}$$

This is the **dispersion equation for the electromagnetic wave propagation in collisionless plasma along magnetic field**. In the absence of magnetic field, the electron-cyclotron frequency is zero $\omega_B = eB_0/m = 0$, and the dispersion Eq. (8.39) coincides with the conventional one (7.66) in non-magnetized plasma.

8.7 Ordinary and Extra-ordinary Polarized Electromagnetic Waves in Magnetized Plasma, Effect of Ionic Motion

Two signs (+) and (−) in the dispersion Eq. (8.39) correspond to two directions of rotation of $\vec{E}$. The (−) sign is related to the **right-hand-side circular polarization** of waves when the direction of $\vec{E}$ rotation coincides with the direction of an electron gyration in the magnetic field. In optics, such a wave is referred to as the **extra-ordinary wave**. The extra-ordinary wave has $\omega = \omega_B$ as the resonant frequency when the denominator in the dispersion equation tends to zero.

 This **electron-cyclotron resonance (ECR)** provides effective absorption of electromagnetic waves, which is used in the ECR-discharges. The (+) sign in (8.39) corresponds to the **left-hand-side circular polarization**. In optics, such a wave is referred to as the **ordinary wave** and it is not a resonant one. Phase velocity of the electromagnetic waves, propagating along the magnetic field, can be presented as:

$$v_{ph} = c\left[1 - \frac{\omega_p^2}{\omega^2}\frac{1}{\left(1 \pm \dfrac{\omega_B}{\omega}\right)}\right]^{-1/2}. \tag{8.40}$$

Propagation of electromagnetic waves in the absence of a magnetic field is possible according only with frequencies exceeding the plasma frequency $\omega > \omega_p$. It is different in magnetic field. The relevant dispersion curves (8.40) are shown in Figure 8.3, illustrating that propagation of both ordinary and extra-ordinary waves is possible at low frequencies ($\omega < \omega_p$). Also, in contrast to $B_0 = 0$, where phase velocities always exceed the speed of light, the electromagnetic waves can be "slower" than speed of light in the presence of magnetic field.

The extra-ordinary waves propagate in plasma at low frequencies (see Figure 8.3), when the ionic motion can be important. The ion-cyclotron frequency should be considered in this case:

$$\omega_{Bi} = \frac{e}{M}B, \tag{8.41}$$

corresponding to the ions (mass M) gyration in magnetic field B_0. The relevant dispersion equation for wave propagating along a uniform magnetic field B_0 can be then expressed as:

$$\frac{k^2 c^2}{\omega^2} = 1 - \frac{\omega_p^2}{\omega^2}\frac{1}{\left(1 \pm \dfrac{\omega_B}{\omega} - \dfrac{\omega_B \omega_{Bi}}{\omega^2}\right)}. \tag{8.42}$$

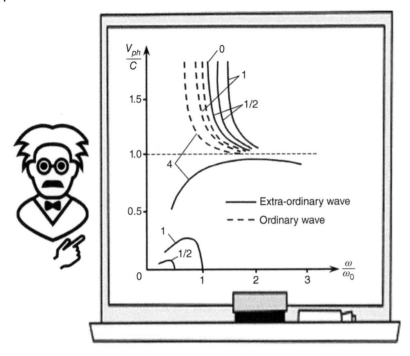

Figure 8.3 Electromagnetic wave propagation in magnetized plasma: dispersion of transverse waves, propagating along magnetic field; numbers correspond to values of ω_H/ω_p.

If the wave frequency exceeds the ion-cyclotron one ($\omega \gg \omega_B$), influence of the term $\omega_B\omega_{Bi}/\omega^2$ in the denominator is negligible and the dispersion (8.42) coincides with (8.39). If the wave frequency is low ($\omega \ll \omega_B$), then $\omega_B\omega_{Bi}/\omega^2$ dominates the dispersion and (8.42) becomes:

$$\frac{c^2}{(\omega/k)^2} = 1 + \frac{\omega_p^2}{\omega_B\omega_{Bi}} = 1 + \frac{\mu_0\rho}{B^2} \approx \frac{c^2}{v_A^2}, \tag{8.43}$$

introducing again the Alfven wave velocity v_A (8.29); here $\rho = n_e M$.

8.8 Dispersion and Amplification of Acoustic Waves in Nonequilibrium Plasma

Nonequilibrium molecular plasma can amplify acoustic waves due to different relaxation mechanisms. Let us analyze the dispersion $k(\omega)$ of acoustic waves considering heat released by vibrational relaxation and chemical reactions (see Sections 5.8 and 5.9). First, we should introduce relevant general dispersion equation (Kirillov et al. 1983; Fridman and Kennedy 2021).

Linearizing the continuity equation, momentum conservation, and balance of translational and vibrational energies, the dispersion equation for acoustic waves in the vibrationally nonequilibrium chemically active plasma can be expressed as:

$$\frac{k^2 c_s^2}{(\omega - \vec{k}\vec{v})^2} = 1 - \frac{i(\omega - \vec{k}\vec{v})[e_T(\gamma - 1) + e_n] - (e_v\bar{e}_T - e_T\bar{e}_v)\gamma(\gamma - 1) + (e_n\bar{e}_v - e_v\bar{e}_n)\gamma}{(\omega - \vec{k}\vec{v})^2 + i(\omega - \vec{k}\vec{v})(e_n - e_T - \gamma\bar{e}_v) + \gamma(e_v\bar{e}_T - e_T\bar{e}_v) + \gamma(e_n\bar{e}_v - e_v\bar{e}_n)}; \tag{8.44}$$

. here: $c_s = \sqrt{\gamma T_0/M}$ is the "frozen" speed of sound (vibrational degrees of freedom do not follow variations of T_0); M is mass of heavy particles; γ is the specific heat ratio; $\vec{v}$ is the gas velocity; $e_{n,T,v}$ and $\bar{e}_{n,T,v}$ are the of translational (e) and vibrational ($\bar{e}$) temperature changes related to perturbations of respectively gas density n_0, translational T_0, and vibrational T_v temperatures:

$e_n = 2(v_{VT} + \xi v_R)$, $e_T = \hat{k}_{VT}v_{VT} - k_{VT}n_0\frac{c_v^v(T_0)}{c_v}$, $e_v = k_{VT}n_0\frac{c_v^v(T_v)T_v}{c_vT_0} + \hat{k}_R\xi v_R$, $\bar{e}_n = v_{eV}\frac{c_vT_0}{c_v^v(T_v)T_v}\left(1 + \frac{\partial\ln n_e}{\partial\ln n_0}\right) - 2v_{VT}\frac{c_vT_0}{c_v^v(T_v)T_v} - 2\xi v_R\frac{c_vT_0}{c_v^v(T_v)T_v}$ $\bar{e}_T = -\hat{k}_{VT}v_{VT}\frac{c_vT_0}{c_v^v(T_v)T_v} + k_{VT}n_0\frac{T_0}{T_v}$, $\bar{e}_v = -k_{VT}n_0 - \hat{k}_Rv_R\frac{c_vT_0}{c_v^v(T_v)T_v}$. The characteristic frequencies of vibrational (VT) relaxation, chemical reaction, and vibrational excitation of molecules by electron impact can be expressed respectively by: $v_{VT} = \frac{\gamma-1}{\gamma} \cdot \frac{k_{VT} n_0^2 [\varepsilon_v(T_v) - \varepsilon_v(T_0)]}{p}$, $v_R = \frac{\gamma-1}{\gamma} \cdot \frac{k_R n_0^2 \Delta Q}{p}$, $v_{eV} = \frac{\gamma-1}{\gamma} \cdot \frac{k_{eV} n_e n_0 \hbar \omega}{p}$. Here $\varepsilon_v(T_v) = \frac{\hbar \omega}{\exp(\hbar\omega/T_v)-1}$ is the average vibrational energy of molecules; c_v, c_v^v are translational and vibrational heat capacities; $p = n_0 T_0$ is the gas pressure; k_{eV}, k_{VT}, and k_R are rate coefficients of vibrational excitation, vibrational relaxation, and chemical reaction; n_e and n_0 are electron and gas concentrations; ΔQ is the vibrational energy consumption per one act of chemical reaction; ξ is that part of the energy which is going into translational degrees of freedom (the chemical heat release); the dimensionless factors $\hat{k}_R = \frac{\partial \ln k_R(T_v)}{\partial T_v}$, $\hat{k}_{VT} = \frac{\partial \ln k_{VT}}{\partial \ln T_0}$ represents the logarithmic sensitivities of rate coefficients k_R and k_{VT} to vibrational and translational temperatures respectively. Without heat release, the dispersion Eq. (8.44) gives: $v_{ph} = v \pm c_s$ and describes sound waves propagating along gas flow.

Let us analyze the dispersion equation (8.44) in gas at rest ($\vec{k}\,\vec{v} = 0$), assuming the frequency ω as a real number, while the wave number $k = k_0 - i\delta$ as a complex one. Then k_0 characterizes the acoustic wavelength, and δ corresponds to the space amplification of sound.

8.8.1 Acoustic Waves in Molecular Gas at Equilibrium ($T_v = T_0$)

Assuming constant VT-relaxation time ($\tau_{VT} = const$), the general equation (8.44) then gives the well-known **dispersion equation of relaxation gas-dynamics**:

$$\frac{k^2 c_s^2}{\omega^2} = 1 + \frac{\frac{c_s^2}{c_e^2} - 1}{1 - i\omega\tau_{VT}}. \tag{8.45}$$

At high frequency limit, such acoustic wave propagates with the "frozen" sound velocity c_s (vibrational degrees of freedom are frozen, that is not participating in the sound propagation). The acoustic wave velocity at low frequencies ($\omega\tau_{VT} \ll 1$) is the conventional equilibrium speed of sound c_e. Maximum damping takes place at the wave resonance with VT-relaxation when $\omega\tau_{VT} \approx 1$. The phase velocity v_{ph} and damping δ/k_0 as functions of $\omega\tau_{VT}$ are shown in Figure 8.4.

8.8.2 Acoustic Waves in Nonequilibrium ($T_v > T_0$) Plasma, High-frequency Limit

When $T_v > T_0$, and sound frequency exceeds that of relaxation $\omega \gg \max\{\bar{e}_n, \bar{e}_T, \ldots, e_v\}$, the dispersion Eq. (8.44) can be simplified:

$$\frac{k^2 c_s^2}{\omega^2} = 1 - i\frac{(\gamma-1)\hat{k}_{VT}v_{VT} + 2(v_{VT} + \xi v_R)}{\omega}. \tag{8.46}$$

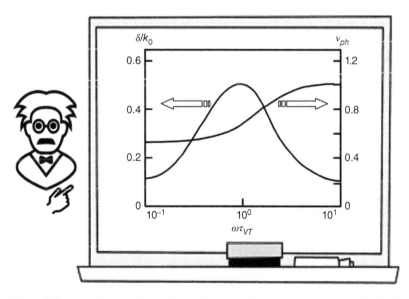

Figure 8.4 Acoustic wave in quasi-equilibrium molecular gas: phase velocity (in the units of speed of sound c_s) and attenuation coefficient as a function of relaxation parameter $\omega\tau_{VT}$.

The wave propagates in this limit at the "frozen" sound velocity with a slight dispersion:

$$v_{ph} = \frac{\omega}{k_0} = c_s \left[1 - \frac{((\gamma - 1)\hat{k}_{VT}v_{VT} + 2(v_{VT} + \xi v_R))^2}{8\omega^2} \right]. \tag{8.47}$$

The initial VT nonequilibrium results in amplification of acoustic waves in contrast to the relaxation gas dynamics, see (8.45). The amplification increment at the high frequency limit is:

$$\frac{\delta}{k_0} = \frac{(\gamma - 1)\hat{k}_{VT}v_{VT} + 2(v_{VT} + \xi v_R)}{2\omega}. \tag{8.48}$$

8.8.3 Acoustic Wave Dispersion in the Presence of Intensive Plasma-chemical Reaction

The reaction frequency in plasma-chemical systems can exceed that of vibrational relaxation, making heating mostly due to the chemical heat release ($\xi v_R \gg v_{VT}$). The dispersion (8.44) then becomes:

$$\frac{k^2 c_s^2}{\omega^2} = 1 - \frac{i\omega a v_R + b v_R^2}{\omega^2 + i\omega c v_R + b v_R^2}. \tag{8.49}$$

$$a = 2\xi, \quad b = -\xi\gamma \frac{c_v T_0}{c_v^v T_v} \frac{\partial \ln k_R}{\partial \ln T_v} \left(1 + \frac{\partial \ln n_e}{\partial \ln n_0}\right), \quad c = 2\xi + \gamma \frac{c_v T_0}{c_v^v T_v} \frac{\partial \ln k_R}{\partial \ln T_v}.$$

For example, in the nonequilibrium supersonic discharge in CO_2 (see Lecture 17): $T_v = 3500$ K, $T_0 = 100$ K, $E_a = 5.5$ eV, $\frac{\partial \ln k_R}{\partial \ln T_v} \approx \frac{E_a}{T_v} = 18$, $\xi = 10^{-2}$, $\frac{\partial \ln n_e}{\partial \ln n_0} \approx (-3) \div (-5)$; then $a \approx b \approx 0.2$, $c \approx 0.4$. Characteristic frequencies of chemical reaction v_R and VT-relaxation v_{VT} for this case are given in Table 8.2; relevant dispersion curves are illustrated in Figure 8.5.

An interesting physical effect can be seen in Figure 8.5c,d, illustrating the dependence $\delta/k_0(\omega)$. When frequency ω is decreasing, the amplification coefficient at first grows then passes maximum, and decreases reaching zero ($\delta = 0$) at frequency:

$$\omega = v_R \gamma \frac{c_v T_0}{c_v^v T_v} \sqrt{\xi \left| 1 + \frac{\partial \ln n_e}{\partial \ln n_0}\right|}. \tag{8.50}$$

It is due to an increase of the phase-shift between oscillations of heating rate and pressure when frequency is decreasing (at high frequency limit this phase-shift is about $2\xi v_R/\omega$).

Frequencies v_R, v_{VT} (Table 8.2) are in the ultra-sonic range of the acoustic waves ($\omega > 10^5$ Hz, $\lambda < 0.1$ cm), therefore the acoustic dispersion can be effectively applied in the ultra-sonic diagnostics of plasma-chemical systems (see Givotov et al. 1985; Fridman and Kennedy 2021).

Table 8.2 Characteristic frequencies of chemical reaction v_R, s^{-1} and VT-relaxation v_{VT}, s^{-1} at different T_0 and T_v; $n_0 = 3 \cdot 10^{18}$ cm^{-3}, $E_a = 5.5$ eV, $B = 72$ K$^{1/3}$ (B is the Landau–Teller coefficient for vibrational relaxation).

$T_0\downarrow$	$T_v = 2500$ K	$T_v = 3000$ K	$T_v = 3500$ K	$T_v = 4000$ K
100 K	$v_R = 7\cdot 10^3$	$v_R = 3\cdot 10^5$	$v_R = 4\cdot 10^6$	$v_R = 3\cdot 10^8$
	$v_{VT} = 3\cdot 10^2$	$v_{VT} = 4\cdot 10^2$	$v_{VT} = 5\cdot 10^2$	$v_{VT} = 7\cdot 10^2$
300 K	$v_R = 2\cdot 10^3$	$v_R = 10^5$	$v_R = 10^6$	$v_R = 10^8$
	$v_{VT} = 1.4\cdot 10^4$	$v_{VT} = 1.6\cdot 10^4$	$v_{VT} = 1.9\cdot 10^4$	$v_{VT} = 2.7\cdot 10^4$
700 K	$v_R = 10^3$	$v_R = 4\cdot 10^4$	$v_R = 5\cdot 10^5$	$v_R = 5\cdot 10^7$
	$v_{VT} = 10^5$	$v_{VT} = 1.2\cdot 10^5$	$v_{VT} = 1.3\cdot 10^5$	$v_{VT} = 1.9\cdot 10^5$

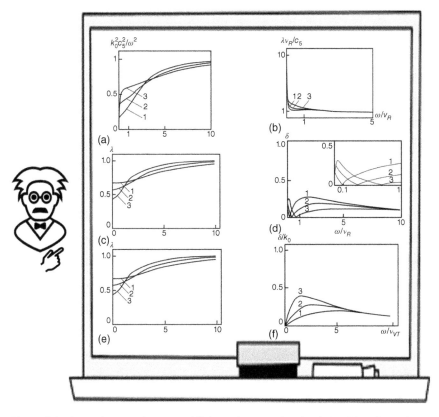

Figure 8.5 Acoustic waves in nonequilibrium plasma: refraction index (a) and wavelength (b) as functions of frequency, translational temperatures: 1 – 100 K, 2 – 300 K, 3 – 700 K; wave amplification coefficients (c, d) as functions of frequency, translational temperature: 1c, 1d – 100 K; 2c, 3d – 700 K; 2d – 300 K; wavelength λ (e, in units ω/c_s) and amplification coefficient δ/k_0 (f) as functions of frequency, translational temperatures: 1 – 100 K, 2 – 300 K, 3 – 700 K.

8.9 Evolution of Shock Waves in Plasma

Gas-dynamic perturbation with a sharp change of variable derivatives (while these variables themselves are changing continuously) is referred to as the **weak shock waves**. Their evolution and transition to the **strong shocks** in plasma can be controlled by VT relaxation and chemical heat release in reactions of vibrationally excited molecules (see Capitelli 2000).

The phenomena can be described by transport equation for the amplitude of weak shock wave $\alpha(x)$, derived based on the method of characteristics (Kirillov et al. 1983; Fridman and Kennedy 2021). The weak shock wave $\alpha(x)$ amplitude is introduced by the relative change of space derivatives of gas-dynamic functions (gas density n_0, velocity v_0, and pressure p) on the shock wave front: $n'^{(+)}_{0x} - n'^{(-)}_{0x} = \alpha(x)n_0$, $v'^{(+)}_{0x} - v'^{(-)}_{0x} = \alpha(x)c_s$, $p'^{(+)}_x - p'^{(-)}_x = \alpha(x)Mn_0c_s^2$, where c_s is the speed of sound; n_0 is gas density; and M is the mass of molecules. Solution of the transport equation results in:

$$\alpha(x) = \alpha_i \xi(x) \left[1 \pm \frac{\gamma+1}{2} \alpha_i \int_{x_0}^{x} \xi(x') \frac{dx'}{M \pm 1} \right]^{-1}. \tag{8.51}$$

Here α_i is initial amplitude of a weak shock wave ($x = x_0$); the sign "+" corresponds to a shock wave propagating downstream; the sign "−" corresponds to a shock wave propagating upstream; γ is the specific heat ratio; M is the Mach number. The special function $\xi(x)$ is defined as:

$$\xi(x) = \left(\frac{c_{si}}{c_s} \right)^{5/2} \left(\frac{M_i \pm 1}{M \pm 1} \right)^2 \exp \int_{x_0}^{x} \left\{ \delta_{lin}(x') + \frac{1}{c_s} \frac{\partial v}{\partial x'} \left(\frac{\pm \gamma M - 1}{M \pm 1} \right) \right\} \frac{dx'}{M \pm 1}, \tag{8.52}$$

where c_{si}, M_i are the initial values of the speed of sound and Mach number (at $x = 0$); c_s, M are the current values of speed of sound and Mach number at any arbitrary x; v is the gas velocity; $\delta_{lin}(x)$ is the increment of amplification of gas-dynamic perturbations in linear approximation. The equations (8.51 and 8.52) determine, in particular, the transition from a weak to a strong shock wave, which corresponds to $|\alpha(x)| \to \infty$. Strong shock wave appears at a critical point x_{cr}, where the denominator of (8.51) becomes equal to zero:

$$\frac{\gamma + 1}{2} \alpha_i \int_{x_0}^x \xi(x') \frac{dx'}{M \pm 1} = \mp 1. \tag{8.53}$$

The strong shock wave can be generated only from waves of compression (for compression waves moving downstream $\alpha_i < 0$, upstream $\alpha_i > 0$) with initial amplitude exceeding the critical one:

$$|\alpha_i| > \alpha_{cr} = \left| \frac{\gamma + 1}{2} \int_{x_0}^\infty \xi(x) \frac{dx}{M \pm 1} \right|^{-1}. \tag{8.54}$$

The threshold for generation of a strong shock wave (8.54) depends on the gradient of the background flow velocity dv/dx, where the perturbation propagates.

8.10 Nonthermal Plasma Fluid Mechanics in Fast Subsonic and Supersonic Flows

The optimal energy input in energy-effective plasma-chemical systems is usually about $1\,\text{eV}\,\text{mol}^{-1}$ at moderate and high pressures (see Fridman 2008, and Lecture 17). It determines proportionality between electric power and gas flow rate in relevant plasma-chemical systems: high discharge power requires high flow rates through the discharge. High continuous power (1 MW and above) of the nonequilibrium energy-effective discharges can be reached, therefore, in fast subsonic and supersonic gas flows (Legasov et al. 1983; Fridman and Kennedy 2021). Schematic and flow parameters of such microwave discharge (power 500 kW, Mach number $M = 3$) are shown, as an example, in Figure 8.6a. Ignition of the nonequilibrium discharge takes place in this case after a supersonic nozzle in a relatively low-pressure zone. In the after-discharge zone, pressure can be restored in a diffuser, so initial and final pressures can be above the atmospheric. Plasma and therefore heat release zone (Figure 8.6a) are located between nozzle and diffuser. The gas flow beyond the heating zone and shock in the diffuser are isentropic at high Reynolds numbers and smooth duct profile. Gas pressure p, density n_0, and temperature T are related in the isentropic zones to the Mach number M as:

$$p \left(1 + \frac{\gamma - 1}{2} M^2 \right)^{\frac{\gamma}{\gamma - 1}} = const, \tag{8.55}$$

$$n_0 \left(1 + \frac{\gamma - 1}{2} M^2 \right)^{\frac{1}{\gamma - 1}} = const, \tag{8.56}$$

$$T \left(1 + \frac{\gamma - 1}{2} M^2 \right) = const; \tag{8.57}$$

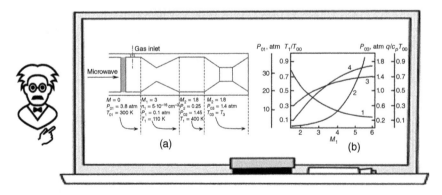

Figure 8.6 (a) Typical parameters of a nozzle system; subscripts 1,2 are related to inlet and exit of a discharge zone; subscript 3 is exit from the nozzle; subscript 0 is stagnation pressure and temperature. (b) Gas dynamic characteristics of a discharge in supersonic flow: discharge inlet temperature $T_1(1)$; initial tank pressure p_{01} (2); exit pressure p_{03} at the critical heat release (3); critical heat release q (4) as functions of Mach number in front of discharge. Initial gas tank temperature $T_{00} = 300\,\text{K}$; static pressure in front of discharge $p_1 = 0.1\,\text{atm}$.

here γ is the specific heat ratio. The Mach number M is determined by variation of the cross-section S of plasma-chemical system duct:

$$SM \left(1 + \frac{\gamma - 1}{2}M^2\right)^{\frac{\gamma + 1}{2(1 - \gamma)}} = const. \tag{8.58}$$

These equations relate the supersonic flow parameters at the beginning of the discharge with gas parameters in initial tank and Mach number after the nozzle, which is presented in Figure 8.6b for CO_2 plasma. For example, if initial tank pressure is about 5 atm at room temperature and $M = 0.05$, then in the supersonic flow after nozzle and before the discharge the gas pressure is 0.1 atm, Mach number $M \sim 3$ and gas temperature $T \sim 100$ K.

The supersonic gas motion strongly depends on the plasma heat release q, which leads to increase of stagnation temperature $\Delta T_0 = q/c_p$, and decrease of the flow velocity. If duct cross-section is constant: $\frac{\sqrt{T_0}}{f(M)} = const$, $f(M) = \frac{M\sqrt{1 + \frac{\gamma - 1}{2}M^2}}{1 + \gamma M^2}$, which determines the **critical heat release** for the supersonic reactor with constant cross-section corresponding to decrease of the initial Mach number $M > 1$ before the discharge to $M = 1$ afterward:

$$q_{cr} = c_p T_{00} \left[\frac{(1 + \gamma M^2)^2}{2(\gamma + 1)M^2 \left(1 + \frac{\gamma - 1}{2}M^2\right)} - 1\right]; \tag{8.59}$$

here T_{00} is the initial gas temperature. If initial Mach number is not close to one, the critical heat release can be estimated as $q_{cr} \approx c_p T_{00}$, see Figure 8.6b. Increase of the heat release above the critical value leads to flow perturbations like shock waves detrimental to the plasma systems. The **critical heat release effect** restricts the specific energy input and subsequently the degree of conversion of plasma-chemical processes. The maximum degree of conversion of CO_2 dissociation in a supersonic microwave discharge ($T_{00} = 300$ K and $M = 3$) does not exceed 15–20%, even at the extremely high energy efficiency of about 90% (see Lecture 17).

The effect of heat release on the supersonic plasma stability can be mitigated by increase of initial temperature T_{00}, by reagents dilution in noble gases, and by **profiling of the supersonic nozzle and discharge zone** (Potapkin et al. 1983). To sustain the constant Mach number at unrestricted heat release, the reactor cross-section S should increase with q as:

$$S = S_0 \left[1 + \frac{q}{T\left(c_p + \gamma M_0^2/2\right)}\right]^{\frac{\gamma M_0^2 + 1}{2}}; \tag{8.60}$$

here S_0, M_0, and T are the reactor cross-section, Mach number and temperature at the beginning of the discharge zone (energy, heat, and temperature are expressed here in the same units). At large Mach numbers, (8.60) can be simplified:

$$S \approx S_0 \left(1 + \frac{q}{c_p T_{00}}\right)^{\frac{\gamma M_0^2 + 1}{2}}. \tag{8.61}$$

Sustaining constant Mach number without restrictions of heat release looks very attractive but requires very significant increase of the discharge cross-section at high $q/c_p T_{00}$. More reasonable increases of the cross-section S/S_0 permit nonconstant Mach number but constant pressure during the heat release. The heat release, in this case, is not unlimited, but the critical value here is not as serious as that for the reactor with constant cross-section:

$$q_{cr}(p = const) = c_p T_{00} \frac{M_0^2 - 1}{1 + \frac{\gamma - 1}{2}M_0^2}. \tag{8.62}$$

At relatively high Mach numbers, this critical heat release at constant pressure is about $q_{cr}(p = const) \approx 5 c_p T_{00}$, that is about five times more than for the case of constant cross-section. Required increase of the reactor cross-section is more realistic here (about six times):

$$S = S_0 \left(1 + \frac{q}{c_p T}\right) = S_0 \left[1 + \frac{q}{c_p T_{00}}\left(1 + \frac{\gamma - 1}{2}M_0^2\right)\right]. \tag{8.63}$$

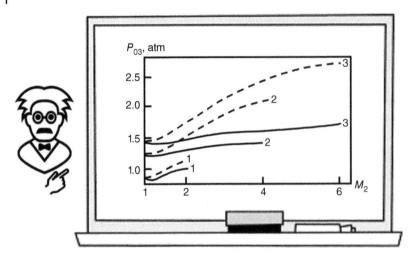

Figure 8.7 Pressure restoration in a diffuser (p_{03}) of supersonic plasma system: M_1, M_2 are Mach numbers in the discharge inlet and exit. $M_1 = 2$ (1), 4 (2), 6 (3). Dashed lines – ideal diffuser, solid lines – nonideal shock waves; static pressure at the discharge inlet 0.1 atm.

Supersonic plasma systems permit the discharge operation at moderate pressure, while keeping system inlet and exit at high pressures. This obviously requires an effective **supersonic diffuser** to restore the pressure after the discharge zone. Such pressure restoration after the supersonic discharge is illustrated in Figure 8.7, assuming before supersonic nozzle: pressure 3.8 atm and room temperature (heat release about half of the critical value).

8.11 Vibrational Relaxation in Fast Subsonic and Supersonic Flows of Nonthermal Reactive Plasmas

This phenomenon can be described by the following equations considering the fast plasma flows with gas compressibility and heat release due to plasma-chemical reactions:

(1) Continuity equation:

$$\rho u = const. \tag{8.64}$$

(2) Momentum conservation equation:

$$p + \rho u^2 = const. \tag{8.65}$$

(3) Translational energy balance:

$$\rho u \frac{\partial}{\partial x} \left(\frac{R}{\mu} \frac{\gamma}{\gamma - 1} T_0 + \frac{u^2}{2} \right) = \xi P_R(T_v) + P_{VT}. \tag{8.66}$$

(4) Vibrational energy balance:

$$\rho u \frac{\partial}{\partial x} [\varepsilon_v(T_v)] = P_{ex} - P_R(T_v) - P_{VT}. \tag{8.67}$$

These equations include specific powers of chemical reaction (P_R), VT-relaxation (P_{VT}), and vibrational excitation (P_{ex}), which can be expressed as: $P_R = k_R(T_v)\rho^2 \left(\frac{N_A}{\mu} \right)^2 \Delta Q$, $P_{VT} = k_{VT}(T_0)\rho^2 \left(\frac{N_A}{\mu} \right)^2 [\varepsilon_v(T_v) - \varepsilon_v(T_0)]$, $P_{ex} = k_{eV} n_e \rho \frac{N_A}{\mu} \hbar \omega$. Here: $k_R(T_v), k_{VT}(T_0), k_{eV}(T_e)$ are the rate coefficients of chemical reaction, VT-relaxation, and vibrational excitation by electron impact; n_e is electron density; $\gamma = c_p/c_v$ is the specific heat ratio including translational and rotational degrees of freedom; p, ρ, u, μ are pressure, density, velocity, and molecular mass of gas; N_A is the Avogadro number; ΔQ is vibrational energy spent per one act of chemical reaction; ξ is the fraction of this vibrational energy, which goes into translational degrees of freedom; T_v, T_0 are vibrational and translational temperatures. We can introduce a new dimensionless density variable $y = \rho_0/\rho(x)$, assume $T_v \approx \hbar\omega > T_0$ and use the ideal gas equation of state: $p = \frac{\rho}{\mu}RT_0$. Then the **relaxation dynamic equations** (8.64–8.67) can be expressed as:

$$u = u_0\, y, \tag{8.68}$$

$$T_0 = T_{00}\left[\left(1 + \gamma M_0^2\right)y - \gamma M_0^2 y^2\right], \tag{8.69}$$

$$y^2 \frac{\partial}{\partial x}\left[y\left(1 + \gamma M_0^2\right) - \frac{\gamma + 1}{2}M_0^2 y^2\right] = \xi Q_R + Q_{VT}, \tag{8.70}$$

$$y^2 \frac{\partial}{\partial x}\varepsilon_v(T_v) = \frac{k_{eV}n_e}{u_0}\hbar\omega y - c_p T_{00} Q_R - c_p T_{00} Q_{VT}. \tag{8.71}$$

In this system of equations: $Q_R = \frac{k_R(T_v)n_0}{u_0}\frac{\Delta Q}{c_p T_{00}}$, $Q_{VT} = \frac{k_{VT}(T_{00})n_0}{u_0}\frac{\hbar\omega}{c_p T_{00}}$, M is the Mach number; n_0 is gas concentration; the subscript "0" relates parameter to the inlet of plasma-chemical reaction zone.

Based on the above relaxation dynamic equations, let us first analyze *the vibrational relaxation in fast flows neglecting contribution of the chemical heat release* ($\xi \Delta Q \ll Q_{VT}$). In this case, the Eq. (8.70) can be solved with respect to the reduced gas density $y = \rho(x)/\rho_0$ and analyzed in the following integral form (Kirillov et al. 1984a, b):

$$\int_1^y \frac{(y')^2\left[1 + \gamma M_0^2 - (\gamma + 1)M_0^2 y'\right]\, dy'}{\exp\left[\frac{B}{T_{00}^{1/3}}\left(y'\left(1 + \gamma M_0^2\right) - (y')^2\gamma M_0^2\right)^{1/3}\right]} = \frac{k_0 n_0}{u_0}\frac{\hbar\omega}{c_p T_{00}}. \tag{8.72}$$

Here the Landau–Teller vibrational relaxation rate coefficient is $k_{VT}(T_0) = k_0\exp\left(-\frac{B}{T_0^{1/3}}\right)$. The density profiles $\rho(x)$, calculated from (8.72), are presented in Figure 8.8 for supersonic flows with different initial Mach numbers. Thus, density evolution during VT relaxation in supersonic flow is "explosive." Also "explosive" is growth of temperature and pressure as well as decrease of velocity during the relaxation in supersonic flow. Linearization of the relaxation dynamic equations allows finding the increment Ω describing the common time scale for the explosive temperature growth and corresponding explosive changes of density and velocity:

$$\Omega_{VT} = k_{VT}(T_{00})n_0\frac{\hbar\omega}{c_p T_{00}}\frac{2 + \hat{k}_{VT}\left(\gamma M_0^2 - 1\right)}{M_0^2 - 1}; \tag{8.73}$$

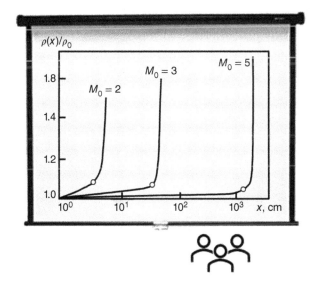

Figure 8.8 "Explosive" evolution of gas density in supersonic plasma flow with vibrational relaxation at different Mach numbers.

here T_{00}, n_0, and M_0 are initial translational temperature, gas density, and Mach number at the beginning of the relaxation process; $\hat{k}_{VT} = \frac{\partial \ln k_{VT}}{\partial \ln T_0}$ the logarithmic sensitivity of VT relaxation rate coefficient to translational temperature (normally $\hat{k}_{VT} \gg 2$). The dynamics of vibrational relaxation (8.73) are qualitatively different at different Mach numbers:

(1) *For subsonic flows with low Mach numbers*: $M_0 < \sqrt{\frac{1}{\gamma}\left(1 - \frac{2}{\hat{k}_{VT}}\right)}$, the increment (8.73) is positive ($\Omega_{VT} > 0$) and the explosive heating takes place. When $M \ll 1$:

$$\Omega_{VT} = k_{VT}(T_{00})n_0 \frac{\hbar\omega}{c_p T_{00}}(\hat{k}_{VT} - 2). \tag{8.74}$$

(2) *For transonic flows with Mach numbers* $\sqrt{\frac{1}{\gamma}\left(1 - \frac{2}{\hat{k}_{VT}}\right)} < M_0 < 1$, the increment (8.73) is negative ($\Omega_{VT} < 0$) and the effect of explosive heating does not take place. *The plasma system in this case is stable with respect to VT-relaxation overheating.*

(3) *For supersonic flows* ($M > 1$) the increment (8.73) again becomes positive ($\Omega_{VT} > 0$) and the effect of explosive overheating takes place. At high Mach numbers:

$$\Omega_{VT} \approx k_{VT}(T_{00})n_0 \frac{\hbar\omega}{c_p T_{00}} \gamma \hat{k}_{VT}, \tag{8.75}$$

which exceeds the increment (8.74) for subsonic flows. The overheating instability is faster in supersonic flows, because heating in this case not only increases temperature directly but also decelerates the flow, which leads to additional temperature growth.

The overall dependence of the overheating increment on Mach number $\Omega_{VT}(M_0)$ at fixed values of stagnation temperature and initial gas density is shown in Figure 8.9a. The same dependence $\Omega_{VT}(M_0)$ at fixed thermodynamic temperature and initial gas density is presented in Figure 8.9b. **The effect of plasma-chemical heat release on dynamics of vibrational relaxation in supersonic flows** results in modification of the expression for the overheating increment (8.73):

$$\Omega_{VT} = k_{VT}(T_{00})n_0 \frac{\hbar\omega}{c_p T_{00}} \frac{2 + \hat{k}_{VT}\left(\gamma M_0^2 - 1\right)}{M_0^2 - 1} + k_R(T_v)n_0 \frac{2\xi \Delta Q}{\left(M_0^2 - 1\right)c_p T_{00}}. \tag{8.76}$$

The chemical heat release stabilizes the overheating in subsonic flows where the second term in (8.76) is negative and accelerates the overheating in supersonic flows. The overheating increment Ω_{VT} can be recalculated into the vibrational relaxation length λ. Dependence of the reverse relaxation length λ^{-1} on the initial Mach number M_0 considering the chemical heat release in CO_2 plasma is presented in Figure 8.10 (Kirillov et al. 1984a, b).

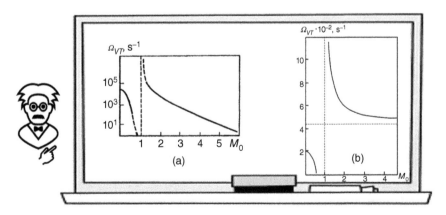

Figure 8.9 Relaxation frequency Ω_{VT} as a function of Mach number in a fast plasma flow: (a) at fixed stagnation temperature $T_{00} = 300\,K$, and initial gas density $n_0 = 3{\ast}10^{18}\,cm^{-3}$; (b) at fixed thermodynamic temperature $T_{00} = 300\,K$, and initial gas density $n_0 = 3{\ast}10^{18}\,cm^{-3}$FF.

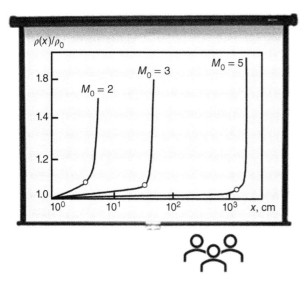

Figure 8.10 Inverse length of vibrational relaxation in CO_2 plasma as a function of Mach number at the discharge inlet; stagnation temperature $T_{00} = 300\,K$ is fixed. 1 – numerical calculation; 2,3 – different analytical models.

8.12 Spatial Nonuniformity and Space Structure of Unstable Vibrational Relaxation in Chemically Active Plasma

Nonthermal molecular plasma is very sensitive to spatial nonuniformity of relaxation processes especially at high and moderate pressures. The loss of uniformity significantly affects the plasma stability, structure, and efficiency (Kirillov et al. 1984a, b; Fridman and Kennedy 2021). Analysis of these phenomena requires understanding of evolution and amplification of plasma flow perturbations at different space scales.

Let us assume 1D distribution of vibrational $T_v(x)$ and translational $T_0(x)$ temperatures with fluctuations: $T_0(t = 0, x) = T_{00} + g(x)$, $T_v(t = 0, x) = T_{v0} + h(x)$. Because of the strong temperature dependence of VT-relaxation, heat transfer is unable to restore spatial uniformity of the temperatures when the density of vibrationally excited molecule is relatively high and heterogeneous vibrational relaxation can be neglected. At low Peclet numbers, evolution of the perturbations is controlled by balance of vibrational and translational energies:

$$n_0 c_v \frac{\partial T_0}{\partial t} = \lambda_T \frac{\partial T_0}{\partial x^2} + k_{VT} n_0^2 [\varepsilon_v(T_v) - \varepsilon_v(T_0)] + \xi k_R n_0^2 \Delta Q, \tag{8.77}$$

$$n_0 c_v^v \frac{\partial T_v}{\partial t} = \lambda_V \frac{\partial T_v}{\partial x^2} - k_{VT} n_0^2 [\varepsilon_v(T_v) - \varepsilon_v(T_0)] - k_R n_0^2 \Delta Q. \tag{8.78}$$

Here, $\lambda_V, \lambda_T, c_v^v, c_v$ are vibrational and translational coefficients of thermal conductivity and specific heat. To analyze time and spatial evolution of small perturbations $T_1(x, t)$, $T_{v1}(x, t)$ of the translational and vibrational temperatures, the system (8.77 and 8.78) can be linearized:

$$\frac{\partial T_1}{\partial t} = \omega_{TT} T_1 + \omega_{TV} T_{v1} + D_T \frac{\partial^2 T_1}{\partial x^2}, \tag{8.79}$$

$$\frac{\partial T_{v1}}{\partial t} = \omega_{VT} T_1 + \omega_{VV} T_{v1} + D_V \frac{\partial^2 T_{v1}}{\partial x^2}. \tag{8.80}$$

The following frequencies have been introduced here:

$$\omega_{TT} = k_{VT} n_0 \left[\frac{\varepsilon_v(T_v) - \varepsilon_v(T_0)}{c_v T_0} \hat{k}_{VT} - \frac{c_v^v(T_0)}{c_v} \right], \tag{8.81}$$

$$\omega_{VV} = -k_{VT} n_0 - k_R n_0 \frac{\Delta Q}{c_v^v T_v} \hat{k}_R. \tag{8.82}$$

They characterize changes of translational temperature due to perturbations of T_0 and changes of vibrational temperature due to perturbations of T_v. In a similar way, two other frequencies:

$$\omega_{TV} = k_{VT} n_0 \frac{c_v^v(T_v)}{c_v} + \xi k_R n_0 \frac{\Delta Q}{c_v T_v} \hat{k}_R, \tag{8.83}$$

$$\omega_{TT} = -k_{VT} n_0 \frac{c_v}{c_v^v(T_v)} \left[\frac{\varepsilon_v(T_v) - \varepsilon_v(T_0)}{c_v T_0} \hat{k}_{VT} - \frac{c_v^v(T_0)}{c_v} \right]. \tag{8.84}$$

describe changes of translational temperature due to perturbations of T_v and changes of vibrational temperature due to perturbations of T_0. Here, logarithmic sensitivity factors are $\hat{k}_R = \frac{\partial \ln k_R}{\partial \ln T_v}$ and $\hat{k}_{VT} = \frac{\partial \ln k_{VT}}{\partial \ln T_0}$; $D_T = \lambda_T / n_0 c_v$ and $D_V = \lambda_V / n_0 c_v^v$ are the reduced coefficients of translational and vibrational thermal conductivity. We can consider the temperature perturbations in the exponential form with amplitudes A, B: $T_1(x, t) = A \cos kx \cdot \exp(\Lambda t)$, $T_{v1}(x, t) = B \cos kx \cdot \exp(\Lambda t)$, which leads to the following dispersion equation relating increment Λ of amplification of the perturbations with their wave number k:

$$\left(\Lambda - \Omega_T^k \right) \left(\Lambda - \Omega_V^k \right) = \omega_{VT} \omega_{TV}. \tag{8.85}$$

In this dispersion equation: Ω_T^k and Ω_V^k corresponds to the so-called thermal and vibrational modes, which are determined by the wave number k as:

$$\Omega_T^k = \omega_{TT} - D_T k^2, \tag{8.86}$$

$$\Omega_V^k = \omega_{TT} - D_V k^2. \tag{8.87}$$

The factor $\omega_{VT} \omega_{TV}$ is a nonlinear coupling parameter between the vibrational and thermal modes. If this product is zero, the modes are independent. The solution of the dispersion equation (8.85) with respect to the increment $\Lambda(k)$ can be presented as:

$$\Lambda_\pm(k) = \frac{\Omega_T^k + \Omega_V^k}{2} \pm \sqrt{\frac{\left(\Omega_T^k - \Omega_V^k \right)^2}{4} + \omega_{VT} \omega_{TV}}. \tag{8.88}$$

This increment describes the linear phase of the spatial nonuniform vibrational relaxation in plasma. Typically, $\omega_{TT} > 0$, $\omega_{VV} < 0$, $\omega_{VT} \omega_{TV} < 0$, $\max \{ \omega_{VV}^2, \omega_{TT}^2 \} > |\omega_{VT} \omega_{TV}|$, and growth of the initial temperature perturbations is controlled by the thermal mode:

$$\Omega_T^k = \omega_{TT} - D_T k^2 = k_{VT} n_0 \left[\frac{\varepsilon_v(T_v) - \varepsilon_v(T_0)}{c_v T_0} \hat{k}_{VT} - \frac{c_v^v(T_0)}{c_v} \right] - D_T k^2 > 0. \tag{8.89}$$

The dispersion curves $\mathrm{Re}\,\Lambda(k^2)$ illustrating the evolution of temperature perturbations at $D_T > D_V$ are shown in Figure 8.11. $\mathrm{Re}\,\Lambda(k^2) > 0$ means amplification of initial perturbations, and $\mathrm{Re}\,\Lambda(k^2) < 0$, their stabilization. The unstable harmonics correspond to the thermal mode (8.89), which is related to Λ_+ line on the figure. During the linear phase of relaxation, short perturbations $k > k_{cr} = \sqrt{\omega_{TT} / D_T}$ decrease and disappear because of thermal conductivity, while perturbations with longer wavelength $\lambda > \lambda_{cr} = 2\pi / k_{cr}$ grow exponentially. When amplitudes of the temperature perturbations get large, the vibrational relaxation becomes nonlinear. This results in the formation of high-gradient structures consisting of periodical temperature zones with different relaxation times. Minimal distance between such zones is determined by the critical wavelength:

$$\lambda_{cr} = 2\pi \sqrt{\frac{D_T c_v T_0}{k_{VT} n_0 \hat{k}_{VT} (\varepsilon_v(T_v) - \varepsilon_v(T_0))}}. \tag{8.90}$$

Formation of the spatial structure of temperature provided by vibrational relaxation in CO_2 plasma ($n_0 = 3 \cdot 10^{18} \mathrm{cm}^{-3}$) on the nonlinear phase of evolution is shown in Figure 8.12 (Kirillov et al. 1984a, b). Initial and boundary ($x = 0, L$) conditions in this case are: $T_0(x, t = 0) = 100\mathrm{K}$, $T_v(x, t = 0) = 2500\mathrm{K}$, $\frac{\partial T_0}{\partial x}(x = 0, L; t) = \frac{\partial T_v}{\partial x}(x = 0, L; t) = 0$. The initial T_v perturbations of vibrational temperature were taken as "white noise" with the mean-square deviation $< \delta T_0^2 >= 0.8 \cdot 10^{-4} \mathrm{K}^2$. The evolution of perturbations in Figure 8.12a can be subdivided into two phases: linear ($t < 5 \cdot 10^{-4}\mathrm{s}$) and nonlinear ($t > 5 \cdot 10^{-4}\mathrm{s}$). The linear phase can be characterized by damping of the short-scale perturbations and, hence, by an approximate 100 times decrease of the mean-square level of the fluctuations $< \delta T_0^2 >$ (see Figure 8.12b). In the nonlinear phase, the perturbations grow significantly (see Figure 8.12b) forming the structures with characteristic sizes about 0.1–0.2 cm in a good agreement with (8.90).

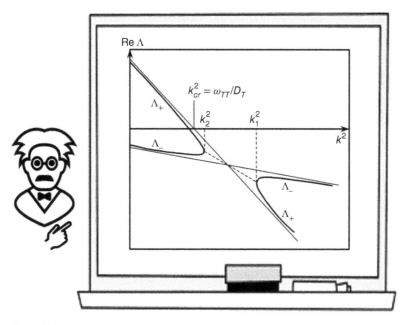

Figure 8.11 Linear evolution of the temperature perturbations in space structure of vibrational relaxation; Re Λ as a function of perturbation wave number.

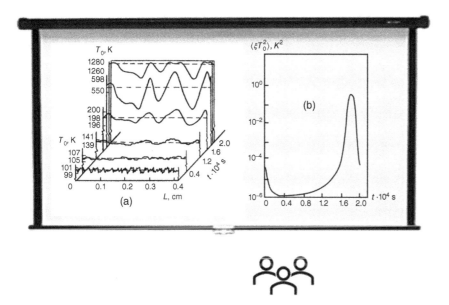

Figure 8.12 Space-non-uniform vibrational relaxation in plasma: (a) nonlinear evolution of the space nonuniformity (dashed line corresponds to the uniform evolution); (b) evolution of the translational temperature mean-square deviation.

8.13 Elements of Plasma Aerodynamics: Plasma Interaction with Fast Flows and Shocks, Ionic Wind

Plasma can strongly influence the high-speed gas flows and shock waves, stimulate the gas motion and mixing, as well as control the propagation in gas of different solid and liquid objects. These phenomena are in focus of plasma aerodynamics. For example, weak gas ionization can result in significant changes in the standoff distance ahead of a blunt body in ballistic tunnels, in reduced drag, and in modifications of traveling shocks. Energy addition to the flow leads to increase in the local sound speed, modifications of the flow, and changes to the pressure distribution around a vehicle due to the decrease in local Mach number.

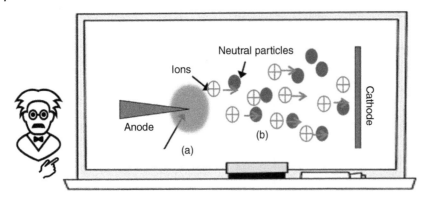

Figure 8.13 Illustration of the ionic wind effect: (a) ionization zone, (b) zone of the ion drift and the drag of neutral particles.

Although the heating is often global, plasmas can produce localized energy deposition effects more attractive for energy efficiency in flow control. Numerous schemes have been proposed recently for modifying and controlling the flow around a hypersonic vehicle. These schemes include approaches for plasma generation, MHD flow control and power generation, and other purely thermal approaches. Two major physical effects of heat addition to the flow field lead to drag reduction at supersonic speeds: (i) reduction of density in front of the body due to the temperature increase (that assumes either constant heating or a pulsed heating source having time to equilibrate in pressure before impinging on the surface); (ii) coupling of the low-density wake from the heated zone with the flow field around the body, which can lead to a dramatically different flow field (the effect is very large for blunt bodies, changing their flow field into something more akin to the conical flow). Efficiency of the effect grows with Mach number.

 Another strong aerodynamic effect is related to the surface dielectric barrier discharges (DBD) discharges (see Lecture 13) and permits rapid and selective heating of a boundary layer at (or slightly off) the surface, as well as creating plasma zones for effective MHD interactions. Such discharges are effective for adding thermal energy to a boundary layer. Also, the asymmetric DBD, with one electrode located inside (beneath) the barrier and another one mounted on top of the dielectric barrier and shifted aside, stimulate intensive "ion wind" along the barrier surface, influence boundary layers, and can be used as aerodynamic actuators for flow control, see Lecture 23. The ion wind, illustrated in Figure 8.13, pushes neutral gas to move with not very high velocities (typically, 5–10 m s^{-1} at atmospheric pressure). It is, however, sufficient to influence the flow near the surface, especially when the flow speed is not too high. The major contribution to the "ion wind" drag is mostly due to the ion motion between the streamers when ions move along the surface from one electrode to another. During the phase when the electrode mounted on the barrier is positive, the positive ions create the wind and drag the flow. During the following phase when the electrode mounted on the barrier is negative, the negative ions formed by electron attachment to oxygen (in air) make contribution to total ion wind and drag the flow in the same direction. Significant recent development in the plasma aerodynamics is reviewed in multiple publications. For those interested in the subject, we can recommend in particular the reviews and publications of Bletzinger et al. (2005), Macheret (2006), Samimi et al. (2007), Moreau (2007), Roth (1995, 2001), Roupassov et al. (2009), and Starikovskii et al. (2008, 2009).

8.14 Problems and Concept Questions

8.14.1 Magnetic Field Frozen in Plasma

Magnetic field is frozen in high conductivity plasmas. Using the Reynolds number criterion, estimate minimum plasma conductivity and corresponding level of plasma density required for the frozenness effect.

8.14.2 Magnetic Pressure and Plasma Equilibrium in Magnetic Field

Estimate magnetic field required for magnetic pressure to balance the hydrostatic pressure of hot confined plasma. Use typical parameters of the hot confined plasma from Table 8.1.

8.14.3 The Magnetic Reynolds Number

Compare magnetic and kinematic viscosity in plasma and analyze difference between magnetic and conventional Reynolds numbers. Compare the consequences of high magnetic and conventional Reynolds numbers.

8.14.4 Critical Heat Release in Supersonic Flows

Estimate maximum conversion degree for plasma-chemical process with $\Delta H \approx 1\,\mathrm{eV}$ in supersonic flow in a constant cross-section reactor with energy efficiency 30% and initial stagnation temperature 300 K. Assume the maximum heat release as a half of the critical one; assume specific heats as those for diatomic gases.

8.14.5 Electromagnetic Waves in Magnetized Plasma

Determine resonance frequencies for electromagnetic waves propagating along the magnetic field, also considering the ionic motion. Compare these frequencies with the electron-cyclotron frequency ($\omega = \omega_B$).

8.14.6 Profiling of Nonthermal Discharges in Supersonic Flow

Calculate the required increase of the reactor cross-section in the discharge zone to provide the conditions of the constant Mach number for $M = 3$ in CO_2 and initial temperature $T_{00} = 300\,\mathrm{K}$.

8.14.7 Dynamics of Vibrational Relaxation in Transonic Flows

Transonic flows can be stable with respect to thermal instability. In this case, however, the critical heat release is small because of nearness of speed of sound. Estimate the maximum possible heat release for the transonic flows with the Mach numbers sufficient for to the stabilization effect.

8.14.8 Space-nonuniform Vibrational Relaxation

Explain the qualitative physical difference between vibrational and translational modes of the nonuniform spatial vibrational relaxation. Why is the vibrational mode stable in most of practical discharge conditions?

8.14.9 Comparison of Linear and Nonlinear Approaches to Evolution of Perturbations

Analyzing (8.51 and 8.52) proves that this nonlinear approach becomes identical to the linear approach (8.48), if the flow can be considered as uniform $dv/dx = 0$ and the coordinate x is close to the initial one x_0 (which means small shock wave amplitude).

8.14.10 Generation of Strong Shock Waves and Detonation Waves in Plasma

Analyze the behavior of plasma with strong vibrational–translational nonequilibrium sustained in supersonic flow with heat release close to critical. Analyze (8.54) describing generation of strong shocks to determine possibility of propagation of a detonation wave in this case.

Part II

Plasma Physics and Engineering of Electric Discharges

Lecture 9

Electric Breakdown, Steady-state Discharge Regimes, and Instabilities

9.1 Electric Breakdown of Gases, the Townsend Mechanism

 Electric discharges start with breakdown, so the Part 2 of this book focused on plasma sources will start in the same way. The electric breakdown is threshold process, which occurs when applied electric field exceeds critical value. During the short breakdown period, usually $0.01 - 100\,\mu s$, the nonconducting gas becomes conductive resulting in generation of plasma. Breakdown usually starts with an electron avalanche, e.g. multiplication of some primary electrons in cascade ionization. Let us consider first the simplest breakdown mechanism in a plane gap with electrode separation d, connected to a DC power supply with voltage V, which provides electric field $E = V/d$, see Figure 9.1.

Occasional formation of primary electrons (initial density n_{e0}) provides initial low current i_0. Each primary electron drifts to the anode, ionizes the gas, and forms an avalanche (Figure 9.1). It is convenient to describe the avalanche by the **Townsend ionization coefficient α** that shows electron production per unit length along the electric field: $dn_e/dx = \alpha\, n_e$ or the same: $n_e(x) = n_{e0} \exp(\alpha x)$. The Townsend ionization coefficient is related to the ionization rate coefficient $k_i(E/n_0)$ and electron drift velocity v_d as:

$$\alpha = \frac{v_i}{v_d} = \frac{1}{v_d} k_i(E/n_0) n_0 = \frac{1}{\mu_e} \frac{k_i(E/n_0)}{E/n_0}, \tag{9.1}$$

where v_i is the ionization frequency with respect to one electron, μ_e is the electron mobility. The electron mobility at room temperature is inversely proportional to pressure, therefore it is convenient to express the Townsend coefficient α as *the similarity parameter* α/p depending on the reduced electric field E/p. The dependence $\alpha/p = f(E/p)$ is shown in Figure 9.2 for some inert and molecular gases (see Fridman and Kennedy 2021, for more examples).

According to definition of the Townsend coefficient α, each primary electron generated near cathode produces $\exp(\alpha d) - 1$ positive ions in the gap (Figure 9.1). All the $\exp(\alpha d) - 1$ positive ions produced per one electron are moving toward the cathode, and altogether provide extraction of $\gamma * [\exp(\alpha d) - 1]$ electrons from the cathode in the process of *secondary electron emission*. The Townsend coefficient γ, introduced here, is the **secondary electron emission coefficient**, characterizing the probability of an electron emission from cathode per single ion impact. The coefficient γ depends on cathode material, state of surface, type of gas, and reduced electric field E/p. While the secondary electron emission is attributed to the ion-cathode collisions, the electron extraction from cathode is mostly due to not low-energy-ion bombardment itself but to the electrostatic extraction, because ionization potential usually much exceeds the work function. Photons and meta-stables also contribute to the secondary electron emission, which is traditionally incorporated in the same "effective" γ coefficient. Typical values of γ are 0.001–0.01, see more data in Fridman and Kennedy (2021). Considering the current of primary electrons i_0 and electron current due to the secondary electron emission, the total electronic current from cathode i_{cath} is:

$$i_{cath} = i_0 + \gamma\, i_{cath}[\exp(\alpha d) - 1]. \tag{9.2}$$

The total current i in the external circuit equals to the electronic current at the anode (where there is no ionic current). Therefore $i = i_{cath} \exp(\alpha d)$, which leads to *the Townsend formula* (1902):

$$i = \frac{i_0 \exp(\alpha d)}{1 - \gamma[\exp(\alpha d) - 1]}. \tag{9.3}$$

Plasma Science and Technology: Lectures in Physics, Chemistry, Biology, and Engineering, First Edition. Alexander Fridman.
© 2024 WILEY-VCH GmbH. Published 2024 by WILEY-VCH GmbH.

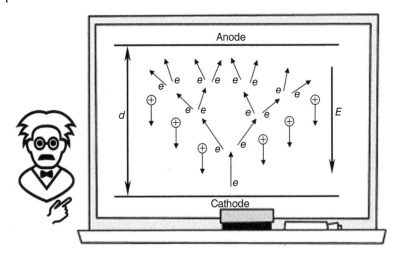

Figure 9.1 Illustration of the Townsend breakdown mechanism.

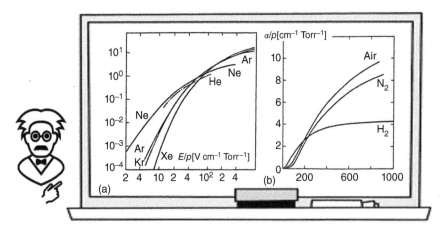

Figure 9.2 Ionization Townsend coefficient as a function of reduced electric field $\alpha/p = f(E/p)$ in different atomic (a) and molecular (b) gases. Source: Raizer (2011) and Fridman and Kennedy (2021).

 The current is non-self-sustained if denominator in (9.3) is positive. As soon as the electric field, and hence, the Townsend α coefficient become sufficiently high, the denominator in (9.3) goes to zero, resulting in transition to self-sustained current and the breakdown. Thus, the **Townsend uniform breakdown criterion** can be expressed as:

$$\gamma[\exp(\alpha d) - 1] = 1, \quad \alpha d = \ln\left(\frac{1}{\gamma} + 1\right). \tag{9.4}$$

This ignition mechanism of a self-sustained current in a gap, controlled by secondary electron emission from cathode, is referred to as **the Townsend breakdown mechanism**.

9.2 The Paschen Curves: Critical Breakdown Conditions, Breakdown of Larger Gaps and Effect of Electronegative Gases

To specify the Townsend breakdown voltages, it is convenient to rewrite the Townsend criterion (9.4) using semi-empirical relation between the similarity parameters α/p and E/p:

$$\frac{\alpha}{p} = A\exp\left(-\frac{B}{E/p}\right). \tag{9.5}$$

Table 9.1 Numerical parameters A and B for calculation of the Townsend coefficient α.

Gas	$A, \dfrac{1}{cm\,Torr}$	$B, \dfrac{V}{cm\,Torr}$	Gas	$A, \dfrac{1}{cm\,Torr}$	$B, \dfrac{V}{cm\,Torr}$
Air	15	365	N_2	10	310
CO_2	20	466	H_2O	13	290
H_2	5	130	He	3	34
Ne	4	100	Ar	12	180
Kr	17	240	Xe	26	350

The empirical parameters A and B at $E/p = 30 \div 500\ \mathrm{V\,cm^{-1}\,Torr^{-1}}$, are presented in Table 9.1. Based on ((9.4) and (9.5)), the breakdown voltage and reduced electric field can be expressed as functions of the similarity parameter pd:

$$V = \frac{B(pd)}{C + \ln(pd)}, \quad \frac{E}{p} = \frac{B}{C + \ln(pd)}, \tag{9.6}$$

where $C = \ln A - \ln \ln\left(\frac{1}{\gamma} + 1\right)$ and only very weakly affected by the secondary electron emission.

The breakdown voltage dependences on similarity parameter pd are referred to as the **Paschen curves**, see Figure 9.3, and have a minimum breakdown voltage point:

$$V_{min} = \frac{eB}{A} \ln\left(1 + \frac{1}{\gamma}\right), \quad \left(\frac{E}{p}\right)_{min} = B, \quad (pd)_{min} = \frac{e}{A} \ln\left(1 + \frac{1}{\gamma}\right), \tag{9.7}$$

where $e = 2.72$ is the base of natural logarithms. The reduced electric field at the Paschen minimum $(E/p)_{min} = B$ does not depend on γ and hence, on cathode material in contrast to the minimum voltage V_{min} and the corresponding similarity parameter $(pd)_{min}$. The typical minimum voltage is about 300 V, corresponding to E/p about 300 V cm * Torr and the parameter pd about 0.7 cm * Torr. The Paschen right-hand branch (pressure above 1 Torr for a gap about 1 cm) is related to the case when electron avalanches have enough distance and gas pressure to provide ionization even at moderate electric fields. In this case, the reduced electric field is almost constant. The left-hand branch of the Paschen curve is related to the case when ionization is limited by the avalanche size and gas pressure and requires high electric fields. Figure 9.3 illustrates that *breakdown of atomic gases usually requires lower electric fields than that of molecular gases*, which can be attributed to significant suppression of EEDF tails due to vibrational excitation, see Section 4.5.

The E/p at the Paschen curve minimum $(E/p)_{min} = B$ determines the lowest energy required to produce one electron–ion pair (minimum price of ionization), which is called the **Stoletov constant.** The energy price of ionization is $W = \frac{eE}{\alpha}$, and its minimum is $W_{min} = \frac{2.72*eB}{A}$. The Stoletov constant usually exceeds the ionization

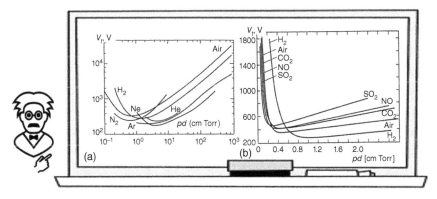

Figure 9.3 Paschen curves for atomic gases (shown on (a)), and molecular gases (b). Source: Raizer (2011) and Fridman and Kennedy (2021).

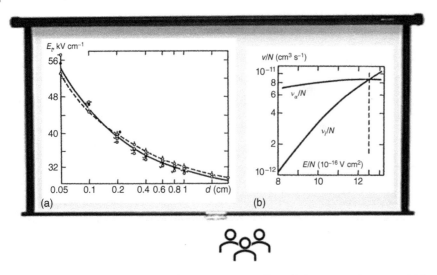

Figure 9.4 Townsend breakdown of larger gaps and effect of electronegative gases: (a) breakdown electric field for atmospheric air; (b) ionization and electron attachment frequencies in air. Source: Raizer (2011) and Fridman and Kennedy (2021).

Table 9.2 Electric fields required for the Townsend breakdown of larger centimeters-size gaps at atmospheric pressure.

Gas	E (kV cm^{-1})	Gas	E (kV cm^{-1})	Gas	E (kV cm^{-1})
Air	32	O_2	30	N_2	35
H_2	20	Cl_2	76	CCl_2F_2	76
CSF_8	150	CCl_4	180	SF_6	89
He	10	Ne	1.4	Ar	2.7

potentials several times because electrons dissipate energy to vibrational and electronic excitation per each act of ionization. The minimum ionization price in electric discharges with high electron temperatures is about 30 eV.

The E/p necessary for breakdown (9.6) is decreasing with pd in the framework of the Townsend breakdown mechanism ($pd < 4000\,\text{Torr cm}^{-1}$), see Figure 9.4a. For larger gaps and the larger avalanches, E/p is less sensitive to the secondary electron emission and cathode material. This reduction in electronegative gases is limited, however, by electron attachment processes which can be considered by introducing an additional Townsend coefficient β:

$$\beta = \frac{v_a}{v_d} = \frac{1}{v_d}k_a(E/n_0)\,n_0 = \frac{1}{\mu_e}\frac{k_a(E/n_0)}{E/n_0}, \tag{9.8}$$

where $k_a(E/n_0)$ and v_a are the attachment rate coefficient and frequency. Thus, altogether we have three coefficients α, β, and γ, describing the Townsend breakdown mechanism. The Townsend coefficient β shows electron losses due to attachment per unit length. Combining α and β gives:

$$\frac{dn_e}{dx} = (\alpha - \beta)\,n_e, \quad n_e(x) = n_{e0}\exp[(\alpha - \beta)\,x]. \tag{9.9}$$

The Townsend coefficient β is exponential function of reduced electric field, but not as sharp as that $\alpha(E/p)$, see Figure 9.4b. Therefore, ionization much exceeds attachment at high E/p and the β coefficient can be neglected with respect to α. When the gaps are relatively large (centimeters range at atmospheric pressure), the Townsend breakdown electric field in electronegative gases becomes constant and limited by attachment processes. Obviously, in this case, breakdown of electronegative gases requires higher E/p, see Table 9.2.

9.3 Spark Breakdown Mechanism, Physics of Avalanches and Streamers

The quasi-homogeneous Townsend breakdown occurs at relatively low pressures and distances ($pd < 4000\,\mathrm{Torr\,cm^{-1}}$, $d < 5\,\mathrm{cm}$ if $p = 1\,\mathrm{atm}$), when avalanches are independent. In larger gaps, the avalanches interact and disturb electric field leading to the so-called **spark breakdown** in a local narrow channel where current grow sometimes up to 10^4–10^5 A. The spark breakdown at high pd and overvoltage develops faster than ions crossing the gap to provide secondary emission and therefore does not depend much on the cathode material. The spark breakdown occurs due to the so-called **streamers**, which are the thin ionized channels growing fast along the positively charged trail between electrodes left by an intensive primary avalanche. This avalanche also generates photons, which in turn initiate numerous secondary avalanches in the vicinity of the primary one. Electrons of the secondary avalanches are pulled by the strong electric field into the positively charged trail of the primary avalanche, creating a streamer propagating between electrodes. If the distance between electrodes is multi-meter or even kilometers long like in lightning, the individual streamers are not sufficient to provide large-scale spark breakdown. In this case, a highly conductive **leader** containing multiple streamers propagates between electrodes. Thus, the spark breakdown mechanism is determined by intensive avalanches and their eventual transition to streamers, which is to be discussed below.

Evolution of an avalanche follows the Eq. (9.9) describing increase of the total number of electrons N_e, positive N_+, and negative N_- ions in an avalanche moving along the axis x:

$$\frac{dN_e}{dx} = (\alpha - \beta)N_e, \quad \frac{dN_+}{dx} = \alpha N_e, \quad \frac{dN_-}{dx} = \beta N_e. \tag{9.10}$$

If the avalanche starts from a single electron (at $x = 0$), the numbers electrons, positive and negative ions can be presented as:

$$N_e = \exp[(\alpha - \beta)x], \quad N_+ = \frac{\alpha}{\alpha - \beta}(N_e - 1), \quad N_- = \frac{\beta}{\alpha - \beta}(N_e - 1). \tag{9.11}$$

The avalanche electrons move altogether along the nondisturbed electric field E_0 (axis x) with the drift velocity $v_d = \mu_e E_0$. Concurrently, free diffusion (D_e) spread the group of electrons around the axis x in radial direction r. Taking into account both drift and diffusion, the electron density in avalanche $n_e(x, r, t)$ is:

$$n_e(x, r, t) = \frac{1}{(4\pi D_e t)^{3/2}} \exp\left[-\frac{(x - \mu_e E_0 t)^2 + r^2}{4D_e t} + (\alpha - \beta)\mu_e E_0 t\right]. \tag{9.12}$$

The electron density decreases radially following the Gaussian law. The avalanche's radius r_A (where the electron density is e times lower than on the axis x) grows with time and the distance x_0 of avalanche propagation:

$$r_A = \sqrt{4D_e t} = \sqrt{4D_e \frac{x_0}{\mu_e E_0}} = \sqrt{\frac{4T_e}{eE_0}x_0}. \tag{9.13}$$

The space distribution of positive and negative ion densities during the short interval of avalanche propagation, when the ions remain at rest, can be then presented as:

$$n_+(x, r, t) = \int_0^t \alpha \mu_0 E_0 n_e(x, r, t')dt', \quad n_-(x, r, t) = \int_0^t \beta \mu_0 E_0 n_e(x, r, t')dt'. \tag{9.14}$$

A simplified expression for the space distribution of positive ion density not too far from the axis x can be derived based on ((9.12) and (9.14)) in absence of attachment and in the limit $t \to \infty$ as:

$$n_+(x, r) = \frac{\alpha}{\pi r_A^2(x)} \exp\left[\alpha x - \frac{r^2}{r_A^2(x)}\right], \tag{9.15}$$

where $r_A(x)$ is the avalanche radius. This distribution reflects the fact that the ion concentration in the trail of the avalanche grows along the axis in accordance with the multiplication of electrons. Concurrently, the radial distribution is Gaussian with an effective avalanche radius $r_A(x)$ growing with x as (9.13).

The propagation of an avalanche is illustrated in Figure 9.5a. The qualitative change in the avalanche behavior occurs at large values of the charge amplification factor $\exp(\alpha x)$, when considerable space charge leads to significant electric field $\vec{E_a}$ perturbing the external field $\vec{E_0}$. The electrons are in the head of avalanche while the positive ions remain behind, creating a dipole with the characteristic length $\frac{1}{\alpha}$ and charge $N_e \approx \exp(\alpha x)$. For breakdown fields about $30\,\mathrm{kV\,cm^{-1}}$ in atmospheric pressure air, the α-coefficient is approximately $10\,\mathrm{cm^{-1}}$ and the ionization length

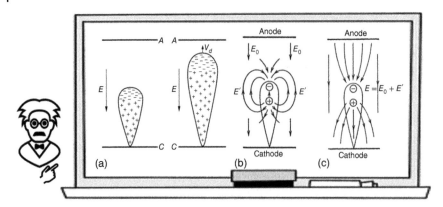

Figure 9.5 Propagation of an avalanche from cathode to anode (a); relevant electric field distribution: external and space charge fields separately (b) and combined (c).

can be estimated as $1/\alpha \approx 0.1$ cm. The external electric field distortion due to the space charge of the dipole is shown in Figure 9.5b. The external $\overrightarrow{E_0}$ and internal $\overrightarrow{E_a}$ electric fields complement each other in front of the avalanche head and behind the avalanche to make a total field stronger, which accelerates the ionization. Conversely, in between the separated charges or "inside of avalanche," the total electric field is lower than the external one, which slows down the ionization.

The space charge also creates a radial electric field (see Figure 9.5b). At a distance of the avalanche radius (9.13), the electric field of the charge $N_e \approx \exp(\alpha x)$ reaches the value of the external field $\overrightarrow{E_0}$ at some critical value of αx. During 1 cm-gap breakdown in air, the avalanche radius is about $r_A = 0.02$ cm and the critical value of αx when the avalanche electric field becomes comparable with E_0 is $\alpha x = 18$ (which corresponds to the Meek criterion, see next Section 9.4). When $\alpha x \geq 14$ the radial growth of an avalanche due to repulsive drift of electrons exceeds the diffusion effect and should be considered. In this case, the avalanche radius grows with x as:

$$r = \sqrt[3]{\frac{3e}{4\pi\varepsilon_0 \alpha E_0}} \exp\frac{\alpha x}{3} = \frac{3}{\alpha}\frac{E_a}{E_0}. \tag{9.16}$$

The growth of the transverse avalanche size restricts the electron density in avalanche to the maximum value:

$$n_e = \frac{\varepsilon_0 \alpha E_0}{e}. \tag{9.17}$$

When the transverse avalanche size reaches the characteristic ionization length $1/\alpha \approx 0.1$ cm, the broadening of the avalanche head slows down. Obviously, the avalanche electric field is about the external one in this case, see (9.16). Typical values of maximum electron density in an avalanche are about $10^{12} \div 10^{13}$ cm^{-3}.

As soon as the avalanche head reaches the anode, the electrons sink into the electrode and it is mostly the ionic trail that remains in the discharge gap. The electric field distortion due to the space charge in this case is shown in Figure 9.6. Because electrons are no longer present in the gap, the total electric field is due to the external field, the ionic trail, and the ionic charge "image" in the anode. The resulting electric field in the ionic trail near the anode is less than the external electric field, but further away from the electrode it exceeds E_0. The total electric field reaches the maximum value at the characteristic ionization distance $\frac{1}{\alpha} \approx 0.1$ cm from anode.

Thus, a strong primary avalanche can amplify the external electric field and form a thin, weakly ionized plasma channel, the **streamer**, which grows fast between electrodes. When the streamer channel connects the electrodes, current may be significantly increased to form the spark. The **avalanche-to-streamer transformation** occurs when ionization parameter αd is sufficiently large and the internal field of an avalanche becomes comparable with the external electric field. If the discharge gap is relatively small, the transformation occurs only when the avalanche reaches the anode. Such a streamer growth from anode to cathode is known as the **cathode-directed or positive streamer**. If the discharge gap and overvoltage are large, the avalanche-to-streamer transformation can occur far from anode. In this case the so-called **anode-directed or negative streamer** can propagate toward both electrodes.

The cathode-directed streamers are illustrated in Figure 9.7a,b. High-energy photons emitted from the primary avalanche provide photoionization in the vicinity, which initiates secondary avalanches. Electrons of the secondary

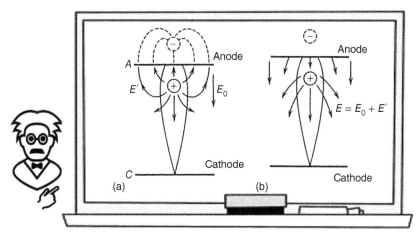

Figure 9.6 Distribution of electric fields when an avalanche approaches the anode: external and space charge fields separately (a), combined (b).

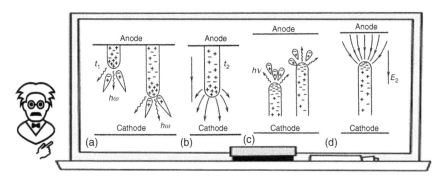

Figure 9.7 Propagation of the cathode-directed (a, b) and anode-directed (c, d) streamers: illustration of the streamer propagation (a, c), and electric fields near the streamer head (b, d).

avalanches are pulled into the ionic trail of the primary one (see the electric field distribution in Figure 9.6) and create a quasi-neutral plasma channel. Subsequently, the process repeats, providing growth of the streamer. The cathode-directed streamer begins near anode, where the positive charge and electric field of the primary avalanche are the highest. The streamer appears as a thin conductive needle growing from the anode. Electric field at the tip of the "anode needle" is very large, which results in fast streamer propagation to the cathode. Usually, the streamer propagation is limited by neutralization of ionic trail near tip of the needle. Here, the electric field is so high that electrons are drifting with velocities about 10^8 cm s^{-1}. This explains the high speed of the streamers, which is also about 10^8 cm s^{-1} and exceeds by an order of magnitude typical electron drift velocity 10^7 cm s^{-1} in the external field. Most streamer parameters are related to those of the primary avalanche. The streamer channel diameter is 0.01–0.1 cm and corresponds to maximum size of a primary avalanche head that is the ionization length $1/\alpha$. Plasma density in the streamer channel also corresponds to the maximum electron concentration in the head of primary avalanche, which is about $10^{12} \div 10^{13}$ cm^{-3}. Specific energy input (electron energy transferred to one molecule) in a streamer channel is small during the short period ($\sim$30 ns) of the streamer growth between electrodes. In molecular gases, it is about 10^{-3} eV mol^{-1}, which in temperature units corresponds to $\sim$10 K.

The anode-directed streamer occurs between electrodes if the primary avalanche becomes sufficiently strong even before reaching anode. Such streamer, which grows in two directions, is illustrated in Figure 9.7c,d. The mechanism of the streamer propagation in direction of the cathode is the same as that of cathode-directed streamers. The mechanism of the streamer growth in the direction of anode is also similar, but in this case, electrons from primary avalanche head neutralize the ionic trail of secondary avalanches. The secondary avalanches could be initiated here not only by photons but also by some electrons moving in front of the primary avalanche.

9.4 Meek Criterion of the Avalanche-to-streamer Transition and Spark Breakdown, Streamer Propagation Models, Concept of Leaders in Very Long Gaps

Avalanche-to-streamer transition requires the electric field of the avalanche space charge E_a to be about the external field E_0:

$$E_a = \frac{e}{4\pi\varepsilon_0 r_A^2} \exp\left[\alpha\left(\frac{E_0}{p}\right)*x\right] \approx E_0. \tag{9.18}$$

Assuming the avalanche head radius as: $r_a \approx 1/\alpha$, the criterion of streamer formation (9.18) requires the amplification parameter αd to exceed the critical value:

$$\alpha\left(\frac{E_0}{p}\right)*d = \ln\frac{4\pi\varepsilon_0 E_0}{e\alpha^2} \approx 20, \quad N_e = \exp(\alpha d) \approx 3\cdot 10^8, \tag{9.19}$$

which is known as the **Meek criterion** ($\alpha d \geq 20$); d is the distance between electrodes.

Electron attachment in electronegative gases slows down the electron multiplication in avalanches and increases the electric field required for a streamer formation. The ionization coefficient α in the Meek criterion should be then replaced by $\alpha-\beta$. When discharge gaps are not large (in air $d \leq 15$ cm), the Meek electric fields are high; then $\alpha \gg \beta$ and the attachment can be neglected. Increasing the distance d between electrodes in electronegative gases leads to some decrease of the breakdown electric field but is limited by the ionization-attachment balance $\alpha(E_0/p) = \beta(E_0/p)$. In atmospheric pressure air, this limit is about 26 kV cm^{-1}. In strongly electronegative gases such as SF$_6$ used for electric insulation, the balance $\alpha(E_0/p) = \beta(E_0/p)$ requires at atmospheric pressure very high electric field 117.5 kV cm^{-1}.

Breakdown voltage applied to the thin rod-electrode should provide the intensive ionization only near the electrode to initiate a streamer (due to nonuniformity of electric field). Once plasma channel is initiated, its further growth is controlled mostly by the high electric field of its own tip (propagation of the quasi-self-sustained streamer). This effect for very long (≥ 1 m) and nonuniform systems results in reduction of the average required breakdown electric field to as low as 2–5 kV cm^{-1} (see Fridman and Kennedy 2021). The breakdown threshold in the nonuniform electric field also depends on the polarity of the principal electrode, where the electric field is higher. The threshold voltage in a long gap between a negatively charged rod and a plane is about twice as large as in the case of a negatively charged rod. This polarity effect is due to the nonuniformity of the electric field near the electrodes. In the case of rod anode, the avalanches approach the anode where the electric field becomes stronger and stronger, which facilitates the avalanche-streamer transition. Also, the avalanche electrons easily sink into the anode in this case, leaving near the electrode the ionic trail enhancing the electric field. In the case of rod cathode, the avalanches move from the electrode into the low electric field zone, which requires higher voltage for the avalanche-streamer transition.

Physics of streamers can be interpreted by two models assuming different level of connection between the streamer head and electrode but leading however to similar predictions. The first model assumes propagation of the **quasi-self-sustained streamers** and is illustrated in Figure 9.8. It assumes very low conductivity of a streamer channel and makes the streamer propagation autonomous and independent from anode. Photons initiate avalanches at a distance x_1 from center of the positive charge zone of radius r_0. The avalanches are then growing in autonomous electric field of the positive space charge $E(x) = \frac{eN_+}{4\pi\varepsilon_0 x^2}$, the number of electrons increases by ionization: $N_e = \int_{x_1}^{x_2} \alpha(E)\,dx$, and the avalanche radius grows due to free diffusion:

$$\frac{dr^2}{dt} \approx 4D_e, \quad r(x_2) = \left[\int_{x_1}^{x_2}\frac{4D_e}{\mu_e E(x)}dx\right]^{1/2}. \tag{9.20}$$

To provide propagation of the self-sustained streamer, its positive space charge N_+ should be compensated by the negative charge of avalanche head $N_e = N_+$ at the meeting point of the avalanche and streamer: $x_2 = r_0 + r$. Also, radii of the avalanche and streamer should be correlated at this point $r = r_0$. It permits to describe the self-propagating streamer parameters, see Figure 9.9a.

In contrast to the first approach of the self-sustained propagation of streamers, the second model assumes the streamer channel as an ideal conductor connected to anode. In this model, the **ideally conducting streamer channel** is considered as an anode elongation in the direction of external electric field E_0 with the shape of an ellipsoid of revolution. The streamer propagation at each point of the ellipsoid is normal to its surface. The propagation velocity is equal to the electron drift velocity in the electric field E_m on the tip of the streamer with length l and radius r:

$$\frac{E_m}{E_0} = 3 + \left(\frac{l}{r}\right)^{0.92}, \quad 10 < \frac{l}{r} < 2000. \tag{9.21}$$

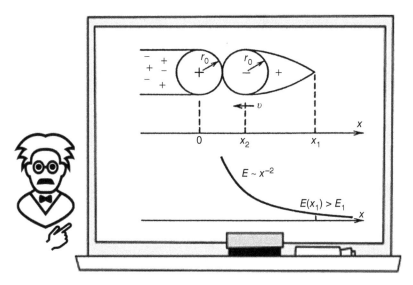

Figure 9.8 Propagation model of a self-sustaining streamer not connected to an electrode.

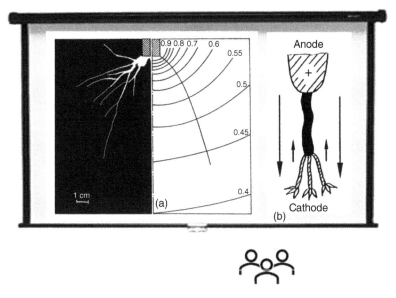

Figure 9.9 Streamers vs leaders: (a) streamer propagation from a positive 2 cm rod to a plane at distance of 150 cm, constant voltage 125 kV, equipotential surfaces are shown at right; (b) illustration of a leader combining multiple streamers.

Both models, although based on opposite assumptions regarding the streamer-electrode connection, are in good qualitative agreement with experimental observations when gaps are not too long, see Figure 9.9a. The streamer breakdown mechanisms, however, cannot be applied directly to very long gaps particularly in electronegative gases (including air). The spark cannot be formed in this case because the streamer channel conductivity is not sufficient to transfer the anode potential close to the cathode and there stimulate the return wave of intense ionization and spark. The streamers stop in the long nonuniform air gaps without reaching the opposite electrode. Usually, the streamer length does not exceed 0.1–1 m. Breakdown of longer gaps, including those with the multi-meter and kilometer long inter-electrode distances, are related to formation and propagation of the so-called **leaders**. The leader is highly ionized and highly conductive (with respect to streamer) plasma channel and grows from the active electrode along the path prepared by the preceding streamers. Because of the high conductivity, the leaders are more effective in transferring the anode potential close to the cathode and there stimulating the return wave of intense ionization and spark. The leaders were first investigated in connection with the natural phenomenon of lightning, and their propagation is illustrated in Figure 9.9b. Electric field on the leader head is high and able to create new streamers growing from the head and preparing the path for further propagation of the leader. The difference between leaders and streamers is more quantitative than qualitative: longer length, higher degree of ionization, higher conductivity, higher electric field; a streamer absorbs avalanches, a leader absorbs streamers (Figure 9.9). Heating effects of the relatively short centimeters long streamers are about 10 K. For meters long channels, the heating effect is 3000 K and more. The streamer-to-leader transition in air can be related to thermal detachment of electrons from the negative ions O_2^-, which are main products of electron attachment in the electronegative gas. The effective destruction of the ions O_2^- and as a result compensation of electron attachment becomes possible if temperature exceeds 1500 K in dry air and 2000 K in humid air (where the electron affinity of complexes $O_2^-(H_2O)_n$ is higher). Typical parameters of a leader growing from an anode are: leader current is about 100 A (for comparison streamer current is 0.1–10 mA), diameter of a plasma channel about 0.1 cm, quasi-equilibrium plasma temperature can be up to 20 000–40 000 K, electric conductivity is about 100 Ohm^{-1} cm^{-1}, propagation velocity is $2 * 10^6$ cm s^{-1}.

9.5 Steady-state Regimes of Nonequilibrium Discharges Controlled by Volume Reactions of Charged Particles and Surface Recombination Processes

The steady state regimes of nonequilibrium discharges occur due to balance of generation and losses of charged particles. The generation of electrons and positive ions is mostly due to volume ionization processes. To sustain the steady-state plasma, the *ionization should be quite intensive; this usually requires the electron temperature about 1/10 of ionization potential (~1 eV)*. The losses of charged particles can be due to volume processes such as recombination or attachment, as well as their diffusion to the walls with subsequent surface recombination. These two mechanisms of charge losses define two different regimes of sustaining the steady-state discharge: the first controlled by volume processes and the second controlled by diffusion to the walls. If the degree of ionization is relatively high and diffusion considered as ambipolar, the frequency of diffusion charge to the walls can be described as:

$$v_D = \frac{D_a}{\Lambda_D^2}. \tag{9.22}$$

where D_a is the coefficient of ambipolar diffusion; and Λ_D is the characteristic diffusion length, which can be calculated for different shapes of the discharge chambers:

– for a cylindrical discharge chamber of radius R and length L:

$$\frac{1}{\Lambda_D^2} = \left(\frac{2.4}{R}\right)^2 + \left(\frac{\pi}{L}\right)^2, \tag{9.23}$$

– for a parallelepiped with side lengths L_1, L_2, L_3:

$$\frac{1}{\Lambda_D^2} = \left(\frac{\pi}{L_1}\right)^2 + \left(\frac{\pi}{L_2}\right)^2 + \left(\frac{\pi}{L_3}\right)^2, \tag{9.24}$$

– for a spherical discharge chamber with radius R:

$$\frac{1}{\Lambda_D^2} = \left(\frac{\pi}{R}\right)^2. \tag{9.25}$$

Thus, criterion of predominantly volume-process-related charge losses and hence, the volume-process-related steady-state regime of the nonequilibrium plasma discharges is:

$$k_i(T_e)\, n_0 \gg \frac{D_a}{\Lambda_D^2}; \tag{9.26}$$

where $k_i(T_0)$ is the ionization rate coefficient, and n_0 is the neutral gas density. The criterion (9.26) restricts pressure, because $D_a \propto {}^1/_p$ and $n_0 \propto p$. When discharge pressure exceeds 10–30 Torr, the diffusion is relatively slow, and the balance of charged particles is due to volume processes. In this case, the kinetics of electrons as well as positive and negative ions can be characterized by the following set of balance equations:

$$\frac{dn_e}{dt} = k_i n_e n_0 - k_a n_e n_0 + k_d n_0 n_- - k_r^{ei} n_e n_+, \tag{9.27}$$

$$\frac{dn_+}{dt} = k_i n_e n_0 - k_r^{ei} n_e n_+ - k_r^{ii} n_+ n_-, \tag{9.28}$$

$$\frac{dn_-}{dt} = k_a n_e n_0 - k_d n_0 n_- - k_r^{ii} n_+ n_-. \tag{9.29}$$

Here: n_+, n_- are densities of positive and negative ions; n_e, n_0 are densities of electrons and neutral species; rate coefficients $k_i, k_a, k_d, k_r^{ei}, k_r^{ii}$ are related to the processes of ionization by electron impact, dissociative or other electron attachment processes, electron detachment from negative ions, electron–ion, and ion–ion recombination. The rate coefficients of processes involving neutral particles (k_i, k_a, k_d) are expressed with respect to the total gas density. The balance Eqs. (9.27–9.29) permit the following analysis of specific volume-process-related discharge regimes depending on the plasma gas electronegativity.

9.6 Steady-state Discharges Controlled by Volume Reactions of Charged Particles: Regimes Controlled by Electron–Ion Recombination, and by Electron Attachment

If moderate or high-pressure gas is not electronegative, then the volume balance ((9.27)–(9.29)) of electrons and positive ions in the nonequilibrium discharge can be reduced to the simple ionization-recombination balance. However, for electronegative gases, two qualitatively different self-sustained regimes can be achieved (at different effectiveness of electron detachment): one controlled by recombination, and the other controlled by electron attachment.

9.6.1 Regime Controlled by Electron–Ion Recombination

Even if negative ions are formed, their destruction, for example by electron detachment, can be faster than ion–ion recombination:

$$k_d n_0 \gg k_r^{ii} n_+. \tag{9.30}$$

In this case, the losses of charged particles are due to the electron–ion recombination similarly to nonelectronegative gases. For example, the associative electron detachment $O^- + CO \to CO_2 + e$, $O^- + NO \to NO_2 + e$, $O^- + H_2 \to H_2O + e$ during plasma dissociation of CO_2 and H_2O, and NO-synthesis, requires about 0.1 µs at CO, NO, and H_2 concentrations $\sim 10^{17}\,\mathrm{cm}^{-3}$. It is shorter than relevant time of ion–ion recombination, and discharge is controlled by electron–ion recombination. The electron attachment and detachment are in the dynamic quasi-equilibrium in the recombination regime when time sufficient for electron detachment ($t \gg 1/k_d n_0$). Then the concentration of negative ions is related to electron density as:

$$n_- = \frac{k_a}{k_d} n_e = n_e \varsigma, \tag{9.31}$$

where the parameter $\varsigma = {}^{k_a}/{}_{k_d}$ is quasi-constant. It permits to reduce ((9.27)–(9.29)) to a single kinetic equation for electron density in quasi-neutral plasma $n_+ = n_e + n_- = n_e(1 + \varsigma)$:

$$\frac{dn_e}{dt} = \frac{k_i}{1+\varsigma} n_e n_0 - \left(k_r^{ei} + \varsigma\, k_r^{ii} \right) n_e^2. \tag{9.32}$$

If $\varsigma \ll 1$, the balance (9.32) is equivalent to that for nonelectronegative gases. Eq. (9.32) includes the effective rate coefficient of ionization $k_i^{eff} = {}^{k_i}/{}_{1+\varsigma}$, which interprets the fraction of electrons lost in the attachment as not generated by ionization at all. In the same manner, the coefficient $k_r^{eff} = k_r^{ei} + \varsigma\, k_r^{ii}$ can be interpreted as effective coefficient of recombination describing both the direct electron losses through the electron–ion recombination, and the indirect electron losses through attachment and following ion–ion recombination. The kinetic Eq. (9.32) describes evolution of the ionization degree to the steady state, which together with k_i exponentially depends on electron temperature:

$$\frac{n_e}{n_0} = \frac{k_i^{eff}(T_e)}{k_r^{eff}} = \frac{k_i}{\left(k_r^{ei} + \varsigma k_r^{ii} \right)(1 + \varsigma)}. \tag{9.33}$$

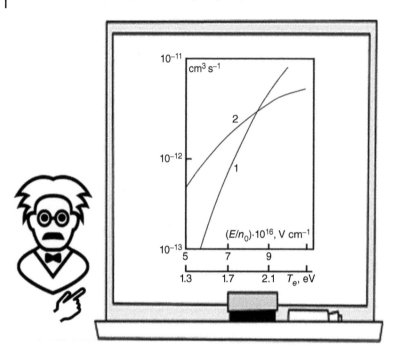

Figure 9.10 Nonequilibrium CO_2 discharge: rate coefficients of ionization by direct electron impact (1), and dissociative attachment (2).

We should note that the criterion of recombination-controlled regime (9.30) can be simplified considering (9.33) and the quasi-neutrality $n_+ = n_e + n_- = n_e(1 + \varsigma)$ to the form:

$$k_d \gg \frac{(k_i - k_a)\, k_r^{ii}}{k_r^{ei}}. \tag{9.34}$$

9.6.2 Discharge Regime Controlled by Electron Attachment

It occurs when the balance of charged particles is due to volume processes and the inequalities ((9.30) and (9.34)) are opposite. In this case, the negative ions produced by electron attachment go to ion–ion recombination, and electron losses are mostly due to the attachment. The steady-state solution of ((9.27)–(9.29)) then becomes:

$$k_i(T_e) = k_a(T_e) + k_r^{ei} \frac{n_+}{n_0}. \tag{9.35}$$

In the attachment-controlled regime, the electron attachment is usually faster than recombination and the Eq. (9.35) requires $k_i(T_e) \approx k_a(T_e)$. The exponential functions $k_i(T_e)$ and $k_a(T_e)$ usually have a single crossing T_{st}, (see Figure 9.10), which determines a single steady-state electron temperature. Ionization degree is usually not much controlled in this case by kinetic balance of charged particles.

9.7 Steady-state Discharges Controlled by Diffusion of Charged Particles to the Walls with Following Surface Recombination: the Engel–Steenbeck Relation

Balance of charged particles is governed by volume ionization and their diffusion to the walls when gas pressure is low and the inequality opposite to (9.26) is valid. The balance of direct ionization by electron impact and ambipolar diffusion to the walls of a long discharge chamber of radius R gives the following relation between electron temperature and pressure:

$$\left(\frac{T_e}{I}\right)^{1/2} \exp\left(\frac{I}{T_e}\right) = \frac{\sigma_0}{\mu_i p} \left(\frac{8I}{\pi m}\right)^{1/2} \left(\frac{n_0}{p}\right) (2.4)^2 (pR)^2, \tag{9.36}$$

Figure 9.11 Universal Engel–Steenbeck relation between electron temperature, pressure, and discharge tube radius.

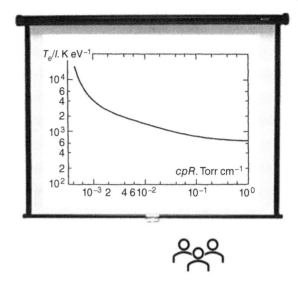

Table 9.3 The Engel–Steenbeck relation parameters.

Gas	C (Torr^{-2}cm^{-2})	c (Torr^{-1}cm^{-1})	Gas	C (Torr^{-2}cm^{-2})	c (Torr^{-1}cm^{-1})
N_2	$2 * 10^4$	$4 * 10^{-2}$	Ar	$2 * 10^4$	$4 * 10^{-2}$
He	$2 * 10^2$	$4 * 10^{-3}$	Ne	$4.5 * 10^2$	$6 * 10^{-3}$
H_2	$1.25 * 10^3$	10^{-2}			

which is called the **Engel–Steenbeck relation** for the diffusion-controlled regime of nonequilibrium discharges. Here I is the ionization potential, μ_i is the ion mobility, σ_0 is the electron-neutral gas-kinetic cross-section, and m is an electron mass. If the gas temperature is fixed, the parameters $\mu_i p$ and n_0/p are constant and the Engel–Steenbeck relation can be presented as:

$$\sqrt{\frac{T_e}{I}} \exp\left(\frac{I}{T_e}\right) = C(pR)^2. \tag{9.37}$$

where the constant C only depends on type of the gas and is provided for some gases in Table 9.3.

The universal relation between T_e/I and the similarity parameter cpR for the diffusion-controlled regime is presented in Figure 9.11. The gas type parameters c for this graph are also given in Table 9.3. According to the Engel–Steenbeck curve, the electron temperature in the diffusion-controlled regimes going down with increase of pressure and radius of the discharge tube. In contrast to the discharges sustained by volume processes, the diffusion-controlled regimes are sensitive to radial density distribution of charged particles. Such radial distribution for a long cylindrical discharge follows the Bessel function:

$$n_e(r) \propto J_0\left(\frac{r}{R}\right). \tag{9.38}$$

9.8 Propagation of Nonthermal Discharges, Ionization Waves

The breakdown electric fields are usually higher than those required to sustain the discharges. It is due to contribution of excited and chemically active species generated in plasma, which facilitate ionization and destruction of negative ions. In the case of thermal plasma, discharge propagation is due to the heat transfer providing high temperatures sufficient for thermal ionization of the incoming gas. Non-thermal plasma propagation at high electric fields can be provided by different mechanisms, including diffusion in front of the discharge of electrons, plasma excited species, or products of chemical reactions stimulating electron detachment.

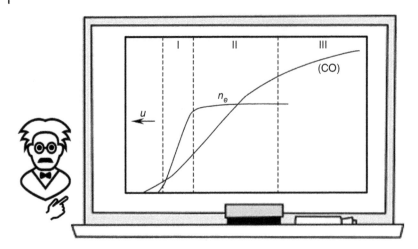

Figure 9.12 Electron and CO density distributions in front of the propagating discharge. I-low electron concentration zone; II-discharge zone where CO-diffusion provides effective detachment and sufficient electron density; III-effective CO_2 dissociation zone.

Consider as an example the nonequilibrium CO_2-discharge 1D propagation in uniform electric field, T_e about 1 eV (see Figure 9.12, Liventsov et al. 1981). CO_2 – breakdown is controlled by balance of ionization and dissociative attachment $e + CO_2 \rightarrow CO + O^-$, and therefore requires high electric fields and electron temperatures exceeding 2 eV (see Figure 9.10). CO_2 dissociation, however, produces enough CO for electron detachment (9.30) and the recombination-controlled regime corresponding to lower electric fields ($n_e = 10^{13} \text{cm}^{-3}$, $n_0 = 3 \cdot 10^{18} \text{cm}^{-3}$, $T_e = 1$ eV, $T_0 \approx 700$ K). The critical value of CO-concentration (9.30), separating the attachment and recombination-controlled regimes is:

$$[CO]_{cr} = \frac{k_a k_r^{ii}}{k_d k_r^{ei}} n_0 \approx (10^{-3} \div 10^{-4}) \cdot n_0. \tag{9.39}$$

If CO-density exceeds the critical value $[CO] > [CO]_{cr}$, the recombination-controlled regime occurs and electron density is high $n_e(T_e) = \frac{k_i(T_e)}{k_r^{ei}} n_0$. If the carbon monoxide concentration is below the critical limit, $[CO] < [CO]_{cr}$, then electron density is very low, controlled by the dissociative attachment, and is proportional to the CO concentration $n_e = \frac{k_i k_d}{k_a k_r^{ii}} [CO]$. Thus, propagation of the discharge is determined by to propagation of the CO density. Most of CO production occurs in the main plasma zone III (Figure 9.12). CO diffusion from the zone III into zone II provides the high CO concentration for sustaining the high electron density and therefore sufficient vibrational excitation for CO_2 dissociation in zone III. Further decrease of the CO concentration below the critical value in the zone I corresponds to dramatic fall of the electron density. Evolution of the electron density profile and velocity u of the ionization wave are then determined by differential equation with a single variable $\xi = x + ut$:

$$u \frac{\partial[CO]}{\partial \xi} = D \frac{\partial^2[CO]}{\partial \xi^2} + g(\xi, T_v, T_0, n_e, n_0). \tag{9.40}$$

Here T_v, T_0 are vibrational and translational gas temperatures; n_e, n_0 are electron and neutral gas densities; D is the diffusion coefficient of CO molecules, and g is a source of CO due to CO_2 dissociation:

$$g(\xi) = \frac{n_0}{\tau_{eV}} \exp\left\{ -\frac{[CO]_{cr}}{[CO](\xi - u\tau_{eV})} - \alpha \frac{[CO](\xi)}{[CO]_0} \right\}; \tag{9.41}$$

$\tau_{eV} = 1/k_{eV} n_e^{II}$ is vibrational excitation time in the zone II; $[CO]_0 = [CO](\xi_0)$ is the maximum CO density at the end of zone III, $\xi_0 = u\tau_{chem}$, τ_{chem} is chemical reaction time in zone *III*; parameter $\alpha \approx 3$ shows exponential smallness of the dissociation rate at the end of zone III $\xi \rightarrow \xi_0$, $t \rightarrow \tau_{chem}$. The boundary conditions for (9.41) are: $[CO](-\infty) = 0$, $[CO](\xi_0) = [CO]_0$. The source $g(\xi)$ is not significant when ξ is negative, and therefore perturbation theory can be applied. The non-perturbed equation ($g = 0$) gives the solution $[CO] = [CO]_0 \exp\left(\frac{\xi u}{D}\right)$. Contribution of the source $g(\xi)$ in the first order of the perturbation theory leads to the linear equation:

$$\frac{\partial}{\partial \xi} \left\{ u[CO](\xi) - D \frac{\partial}{\partial \xi}[CO](\xi) \right\} = \frac{n_0}{\tau_{eV}} \exp\left[-\aleph \exp\left(-\frac{\xi u}{D} \right) \right], \tag{9.42}$$

with the numerical parameter $\aleph$, which is equal to: $\aleph = \frac{[CO]_{cr}}{[CO]_0} \exp \frac{u^2 \tau_{eV}}{D}$. Considering the asymptotical decrease of the right-hand part of (9.42) as $\xi \to -\infty$, the solution of the equation can be presented as:

$$[CO](\xi) = [CO]_0 \exp\left(\frac{\xi u}{D}\right) \left\{ 1 - \frac{D}{u^2 \tau_{eV}} \frac{n_0}{[CO]_0} \exp\left(-\frac{\xi u}{D} - \aleph \exp\left(-\frac{\xi u}{D}\right)\right)\right\}. \tag{9.43}$$

First order perturbation theory in the interval $0 \le \xi \le \xi_0$ (at $\xi \to \xi_0$ the source $g(\xi)$ is very small) gives:

$$[CO](\xi) = [CO]_0 \left\{ 1 + \frac{n_0}{[CO]_0} \frac{\xi - \xi_0}{u \tau_{eV}} \exp(-\alpha)\right\}. \tag{9.44}$$

Distributions ((9.43) and (9.44)) are valid for both positive and negative values of the auto-model variable ξ, and should match at the ionization wave front $\xi = 0$. It is possible only when the velocity of the ionization wave is:

$$u^2 = \frac{D}{\tau_{eV}} \ln\left[\gamma \ln\left(\frac{De^\alpha}{u^2 \tau_{chem}}\right)\right], \tag{9.45}$$

where $\gamma = [CO]_0/[CO]_{cr} \approx 10^3 - 10^4$, and the parameter α is about 3.

Solution of the transcendent Eq. (9.46) gives the velocity of the ionization wave:

$$u^2 \approx \frac{D}{\tau_{eV}} \ln \gamma, \tag{9.46}$$

which can be interpreted as the velocity of diffusion of the detachment active heavy particles (CO) in front of the discharge to a distance required for vibrational excitation of CO_2 molecules with their further dissociation. For numerical calculations, it is convenient to rewrite (9.46) in terms of speed of sound c_s, Mach number M and the ionization degree in plasma n_e/n_0:

$$u \approx 30 \, c_s \sqrt{\frac{n_e}{n_0}}, \quad M = 30 \sqrt{\frac{n_e}{n_0}}. \tag{9.47}$$

Thus, the velocity of the considered ionization wave does not strongly depend on the details of propagation mechanism. It depends mostly on the degree of ionization in the main plasma zone and on the critical amount $(1/\gamma)$ of the ionization active species, which should be transported in front of the discharge to facilitate ionization. Such ionization waves usually propagate slower than speed of sound. Sometimes, however, discharge propagation can be sustained by much faster transfer processes like radiation transfer or electron drift in very high electric fields (as in nano-second pulsed discharges, see Lecture 13). It results in the discharge propagation significantly faster than the speed of sound, which is referred to as the **fast ionization waves, FIW**. Experimental, modeling, and application related data on this subject can be found, in particular, in Takashima et al. (2011); Starikovskiy (2011).

9.9 Nonequilibrium Behavior of Electron Gas: Electron-neutrals Temperature Difference, Deviations from the Saha Ionization Degree

Ionization in both nonthermal and thermal plasmas is mostly provided by electron impact, which requires the electron temperature T_e to be about 1/10 of the ionization potential (that is $\sim$1 eV). Control of gas temperature T_0 mostly determines the plasma's ability to be thermal or nonthermal. In thermal discharges T_0 is high and the system can be close to equilibrium, in nonthermal discharges T_0 is low (due to deficiency of energy or fast colling) and the degree of nonequilibrium T_e/T_0 can be high, sometimes up to 100. The nonequilibrium $T_e/T_0 \gg 1$ in low-pressure discharges, for example, can be due to fast heat losses to the walls. The difference between gas temperature in plasma T_0 and room temperature T_{00} can be then estimated as:

$$\frac{T_0 - T_{00}}{T_{00}} = \frac{P}{P_0}, \tag{9.48}$$

where P is the discharge power per unit volume; and P_0 is the critical value of the specific power, corresponding to the gas temperature increase over $T_{00} = 300$ K. The thermal conductivity and adiabatic discharge heating per one molecule are proportional to $1/p$, therefore the critical specific power does not depend on pressure p and numerically is $P_0 = 0.1 \div 0.3$ W cm^{-3}. In moderate and high-pressure nonthermal discharges (usually more than 20–30 Torr) heat losses to the wall are low. The neutral gas overheating can be prevented in this case either by high velocities and low residence times in the discharge or by shortness of the discharge pulses. In both cases, it restricts the specific energy input E_v, that is the discharge energy transferred to neutral gas per one molecule (it can be expressed in eV mol^{-1} or in J cm^{-3}). Similarly to (9.48), it can be expressed for moderate and high-pressure nonequilibrium discharges as:

$$\frac{T_0 - T_{00}}{T_{00}} = \frac{E_v(1 - \eta)}{c_p T_{00}}, \tag{9.49}$$

where c_p is the specific heat; and η is the energy efficiency of plasma chemical process. The energy efficiency η in moderate and high-pressure plasma chemical systems stimulated by vibrational excitation, can be large, up to 70–90%, which facilitates the conditions necessary for sustaining the nonequilibrium. In this case, (9.47) restricts the specific energy input by $E_v \approx 1.5$ eV mol^{-1}.

The quasi-equilibrium electron density is determined by a single temperature, based on the Saha formula (see Section 6.2). Although the ionization processes (both in thermal and nonthermal discharges) are mostly provided by electrons, for nonequilibrium discharges the Saha formula with electron temperature T_e gives the ionization degree n_e/n_0 several orders of value higher than the real one. Obviously, the Saha formula assuming the neutral gas temperature gives even lower electron density and much worse agrees with reality. This nonequilibrium effect is due to additional channels of charged particle losses in cold gas, which are faster than those reverse processes of recombination $e + A \Leftrightarrow A^+ + e + e$ assumed in the Saha-equilibrium. For example, the charged particle losses in moderate and high-pressure systems are mostly due to the dissociative recombination: $e + AB^+ \rightarrow A + B^*$, which is much faster than the three-body recombination. In nonequilibrium low-pressure discharges also, the ionization is usually not compensated by three-body recombination but rather by the fast electron diffusion to the walls.

9.10 Instabilities of Nonthermal Plasmas: Striations and Contractions, Ionization-overheating Thermal Instability in Monatomic Gases

The instabilities of nonthermal plasmas, where electron, vibrational, and translational temperatures differ: $T_e \gg T_v \gg T_0$, are related to the plasma tendency to restore the quasi-equilibrium between degrees of freedom. The homogeneous uniformity of the nonequilibrium discharges is possible only in a limited range of parameters, usually at low pressures, energy inputs, and powers per unit volume. Fluctuation of the plasma parameters beyond the stability range can grow exponentially, resulting in the discharge nonuniformity. The optimal specific energy input for several plasma-chemical processes is about $E_v \approx 1$ eV mol^{-1}, which exceeds at least 10 times the maximum one ($E_v \approx 0.1$ eV mol^{-1}) for gas lasers. The restriction of $E_v \approx 0.1$ eV mol^{-1} is due to gas heating instabilities, while the optimal energy input $E_v \approx 1$ eV mol^{-1} is related to requirements of effective vibrational excitation and chemistry, which can significantly affect the plasma stability, see in particular Fridman (2008). Let us first classify major types of the instabilities.

Striation is an instability related to formation of plasma structure, that looks like a series of alternating light and dark layers along the discharge current, see Figure 9.13. The striations can move fast with velocities up to 100 m s^{-1}, but also can be at rest, and do not significantly affect plasma parameters. They are related to the ionization instability (see below) and can be interpreted as ionization oscillations and waves.

 Contraction instability is the plasma "self-compression" into one or several bright current filaments. The contraction takes place when the pressure or specific energy input exceeds some critical values. In contrast to striation, contraction significantly changes plasma parameters. The plasma filaments formed because of contraction, are close to the quasi-equilibrium and seriously limit power and efficiency of gas lasers and nonequilibrium plasma-chemical systems. The contraction effect is usually related to the **ionization-overheating thermal instability**, which we are going now to analyze starting with the case of nonequilibrium discharges in monatomic gases. This so-called thermal instability is due to strong exponential dependence of the ionization rate hence the electron density on the reduced electric field E_0/n_0. The thermal instability can be illustrated by the following closed chain of causal links, which can start from fluctuation of the electron density:

$$\delta n_e \uparrow \rightarrow \delta T_0 \uparrow \rightarrow \delta n_0 \downarrow \rightarrow \delta \left(\frac{E}{n_0}\right) \uparrow \rightarrow \delta n_e \uparrow. \tag{9.50}$$

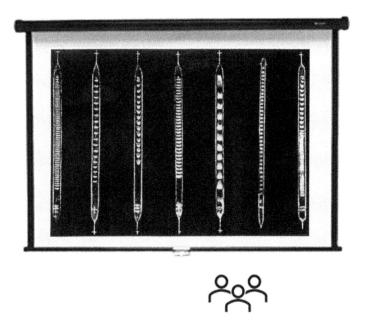

Figure 9.13 Illustration of the striations in a nonthermal plasma.

The local increase of electron density δn_e leads to intensification of gas heating by electron impact and, hence, to an increase of temperature δT_0. Taking into account that pressure $p = n_0 T_0$ is constant, the local increase of temperature δT_0 leads to decrease of gas density δn_0 and to an increase of the reduced electric field $\delta \left(\frac{E}{n_0} \right)$ (electric field $E = const$). Finally, the increase of the reduced electric field $\delta \left(\frac{E}{n_0} \right)$ results in the further increase of electron density δn_e, which makes the loop (9.50) closed and determines the positive feedback. The sequence (9.50) gives the physical interpretation of the thermal instability. Because of the strong dependence of the ionization rate on E/n_0, a small local initial overheating δT_{00} grows up exponentially resulting in contraction:

$$\delta T_0(t) = \delta T_{00} \cdot \exp \Omega\, t. \tag{9.51}$$

The parameter of exponential growth of initial perturbation is called **the instability increment**. If system is stable, the instability increment is negative ($\Omega > 0$), and the positive increment ($\Omega > 0$) means instability. The increment of thermal instability (9.51) can be found by linearization of the differential equations of heat and ionization balance (Fridman and Kennedy 2021). In the case of high or moderate pressures when the influence of walls can be neglected:

$$\Omega = \hat{k}_i \frac{\sigma E^2}{n_o c_p T_0} = \hat{k}_i \frac{\gamma - 1}{\gamma} \frac{\sigma E^2}{p}. \tag{9.52}$$

Here the dimensionless factor $\hat{k}_i = \frac{\partial \ln k_i}{\partial \ln T_e}$ is the sensitivity of the ionization rate to electron temperature (directly related to E/n_0); this factor is usually about 10; σ is plasma conductivity, p is pressure, c_p and γ are specific heat and specific heat ratio. $v_{Tp} = \frac{\gamma - 1}{\gamma} \frac{\sigma E^2}{p}$ is frequency of gas heating by electric current at constant pressure, therefore the instability increment (9.52) is determined by the heating frequency and exceeds v_{Tp} because of the strong sensitivity of ionization to electron temperature. The steady state discharges at high and moderate pressures are always unstable with respect to the thermal instability but can be suppressed for example by intensive cooling. Also, the thermal instability can be avoided if the specific energy input is simply low and not sufficient for the contraction (the gas residence time in discharge, or discharge duration, is small with respect to the thermal instability time $1/\Omega$).

9.11 Ionization-overheating Thermal Instability in Molecular Gases with Significant Vibrational Excitation, Effect of Plasma-chemical Reactions

The fluctuation of electron density δn_e in molecular gases leads to heating δT_0 not directly, but rather through intermediate vibrational excitation, which determines the key peculiarity of the thermal instability in this case.

Vibrational VT relaxation is relatively slow ($\tau_{VT} v_{Tp} \gg 1$, τ_{VT}-is the VT-relaxation time) in the highly effective laser and plasma-chemical systems. In this case, the thermal instability is more sensitive to VT-relaxation than to excitation itself as in monatomic gases (9.52). In the case of fast vibrational excitation and slow relaxation ($\tau_{VT} v_{Tp} \gg 1$), the increment of thermal instability (Fridman and Kennedy 2021) can be expressed as:

$$\Omega_T = \frac{b}{2} \pm \sqrt{\frac{b^2}{4} + c}, \tag{9.53}$$

where the parameters "*b*" and "*c*" are:

$$b = \frac{1}{\tau_{VT}} \left(1 - \frac{\partial \ln \tau_{VT}}{\partial \ln \varepsilon_v} \right) + v_{Tp}(2 + \hat{\tau}_{VT}), \tag{9.54}$$

$$c = \frac{v_{Tp}}{\tau_{VT}} \left(-\frac{\partial \ln n_e}{\partial \ln n_0} \right) \left(1 - \frac{\partial \ln \tau_{VT}}{\partial \ln \varepsilon_v} \right); \tag{9.55}$$

and the factor $\hat{\tau}_{VT} = \frac{\partial \ln \tau_{VT}}{\partial \ln T_0}$ is sensitivity of vibrational relaxation to translational temperature; numerically it is about $-3 \div -5$; ε_v is vibrational energy of a molecule. The thermal instability ((9.53)–(9.55)) follows two different pathways, the so-called thermal and vibrational modes:

(1) **The thermal mode** corresponds to $b < 0$. Considering the relatively high rate of vibrational excitation ($v_{Tp} \tau_{VT} \gg 1$), $b^2 \gg c$. Then ((9.53)–(9.55)) gives the increment:

$$\Omega_T = |b| \approx -v_{Tp}(2 + \hat{\tau}_{VT}) = k_{VT} n_0 \frac{\hbar \omega}{c_p T_0} (\hat{k}_{VT} - 2). \tag{9.56}$$

where $\hat{k}_{VT} = \frac{\partial \ln k_{VT}}{\partial \ln T_0}$ is sensitivity of the vibrational relaxation to translational temperature T_0, which is usually about $3 \div 5$. The increment of instability in the thermal mode corresponds to the frequency of heating due to vibrational VT-relaxation, multiplied by ($\hat{k}_{VT} - 2) \gg 1$. The "pure" thermal mode does not include ionization (no factors directly related to n_e). Hence, it is the "overheating" part of the entire "ionization-overheating" instability.

(2) **The vibrational mode** corresponds to $c > 0$ (at any sign of the parameter b). The instability increment at high rate of excitation ($v_{Tp} \tau_{VT} \gg 1$) can be expressed for this mode as:

$$\Omega_T = \frac{c}{|b|} \approx k_{VT} n_0 \frac{\hat{k}_i}{\hat{k}_{VT} - 2 - (v_{Tp} \tau_{Tp})^{-1}}, \tag{9.57}$$

and is mostly determined by frequency of vibrational relaxation. In contrast to thermal mode, the sensitivity term here includes the electron concentration and the overheating effect on ionization rate. This mode is like the instability in monatomic gases (9.52). Thus, it is the "ionization" part of the entire "ionization-overheating" instability. At slow excitation $v_{Tp} \tau_{VT} \ll 1$:

$$\Omega_T = \sqrt{\frac{v_{Tp}}{\tau_{VT}} \left(-\frac{\partial \ln n_e}{\partial \ln n_0} \right) \left(1 - \frac{\partial \ln \tau_{VT}}{\partial \ln \varepsilon_v} \right)}. \tag{9.58}$$

Both thermal and vibrational modes are illustrated in Figure 9.14a. Like in the monatomic gases (9.50), a local decrease of gas density in molecular gases leads to an increase of the reduced electric field and electron temperature. The T_e increase results in acceleration of ionization, growth of electron density, and vibrational temperature. The T_v growth makes VT-relaxation and heating more intensive, which leads to increase of the translational temperature and to a decrease of the gas density (because $p = n_0 T = const$). This is mechanism of the vibrational mode. Thermal mode, which is also shown in Figure 9.14a, is not directly related to electron density, temperature, and to ionization. This mode is due to the strong exponential dependence of VT-relaxation on translational temperature. Even small increase of T_0 leads to significant acceleration of the Landau–Teller VT-relaxation, intensification of heating and to further growth of the translational temperature. This phenomenon is called the *thermal explosion of vibrational reservoir*. In supersonic plasma flows, the gas heating leads not to a decrease but to an increase of gas density, and to a reduction of electron temperature, and reduction of further heating. Therefore, *the supersonic plasma is stable with respect to the vibrational mode, but the thermal mode of instability is still in place*. Plasma-chemical processes have both stabilizing

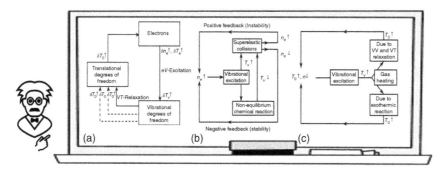

Figure 9.14 Illustration of ionization-overheating instabilities in molecular gas plasmas: (a) solid lines – vibrational mode, dashed lines – thermal mode; (b) effect of plasma-chemical reactions of vibrationally excited molecules; (c) effect of chemical heat release.

and destabilizing effect on the instabilities. Significant part of the vibrational energy in plasma-chemical systems can be consumed in endothermic reactions instead of heating, which stabilizes plasma. Destabilization in turn is due to fast heat release in exothermic reactions. Both effects are discussed below.

Both instability modes ((9.56) and (9.57)) leading to contraction, have typical time comparable with the time of VT relaxation ($\tau_{VT} = 1/k_{VT}n_0$). This makes the thermal instability less dangerous for effective plasma-chemical processes, where the reaction time is about time of vibrational excitation ($\tau_{eV} = 1/k_{eV}n_e$) and is shorter than time of VT relaxation. Effective plasma-chemical process can be completed before the development of the strong but slow thermal instability. One of "dangerous" **fast ionization instabilities** with a frequency approximately $\tau_{eV} = 1/k_{eV}n_e$ can occur due to direct increase of T_e and acceleration of ionization by an increase of T_v. The effect is provided by super-elastic collisions of electrons with vibrationally excited molecules. Illustration of the fast ionization instability is shown in Figure 9.12b. The initial small increase of electron density leads to intensification of vibrational excitation and growth of the vibrational temperature. Higher T_v results in acceleration of ionization because of super-elastic collisions and leads to further increase of the electron density. This instability includes neither VT relaxation nor any heating, and it is fast (controlled by vibrational excitation $\tau_{eV} = 1/k_{eV}n_e$). This instability can be stabilized by endothermic chemical reactions consuming vibrational energy and decreasing T_v, which is also illustrated in Figure 9.14b. Increment of this fast ionization instability can be calculated by linearization of the differential equations for the vibrational energy balance and balance of electrons, considering the endothermic reactions stimulated by vibrational excitation:

$$\Omega_v = k_{eV}n_e \frac{\hbar\omega}{c_v^v T_v}(\tilde{k}_i - \tilde{k}_r); \tag{9.59}$$

here $c_v^v = \frac{\partial \varepsilon_v}{\partial T_v}$ is the vibrational heat capacity; $\tilde{k}_i = \frac{\partial \ln k_i}{\partial \ln T_v}$ is sensitivity of the ionization to vibrational temperature; $\tilde{k}_r = \frac{\partial \ln k_r}{\partial \ln T_v}$ is sensitivity of the chemical reaction rate to vibrational temperature (in the case of weak excitation $\tilde{k}_r = \frac{E_a}{T_v}$, at strong excitation $\tilde{k}_r = \frac{T_0 \hbar\omega}{x_e T_v^2}$). The factor $\tilde{k}_i = \frac{\partial \ln k_i}{\partial \ln T_v}$ can be found considering the influence of vibrational temperature and hence, super-elastic processes on the ionization rate as: $\tilde{k}_i = \left(\frac{\hbar\omega}{T_e}\right)^2 \frac{\Delta \varepsilon}{T_v}$, $\Delta\varepsilon \approx 1 \div 3$ eV is the energy range of effective vibrational excitation and E_a is the activation energy. The fast ionization instability increment for the case of not very strong excitation can be expressed as:

$$\Omega_v = k_{eV}n_e \frac{\hbar\omega E_a}{c_v^v T_v^2}\left(\frac{\hbar^2\omega^2}{T_e^2}\frac{\Delta\varepsilon}{E_a} - 1\right). \tag{9.60}$$

Thus, reactions of vibrationally excited molecules can stabilize the perturbations of ionization (9.60). Unfortunately, they can also amplify the instabilities related to the plasma's direct overheating because each act of chemical reaction stimulated is accompanied by transfer of energy $\xi E_a = \Delta Q_\Sigma - \Delta H$ into heat. This so-called *chemical heat release* includes the effect of exothermic reactions, nonresonant VV-exchange, and VT-relaxation from high vibrational levels. This effect is illustrated in Figure 9.14c. The instability of chemical heat release is the thermal ionization-overheating instability. It is very fast because of the heating controlled by the chemical reactions themselves. The increment of the instability is (Fridman and Kennedy 2021):

$$\Omega_T = k_{eV}n_e \frac{\hbar\omega}{c_p T_0}\left\{\xi(\hat{k}_r - 2) - \tilde{k}_r \frac{c_p T_0}{c_v^v T_v}\right.$$

$$\left. + \sqrt{\left[\xi(\hat{k}_r - 2) + \tilde{k}_r \frac{c_p T_0}{c_v^v T_v}\right]^2 + 4\xi \tilde{k}_r \frac{c_p T_0}{c_v^v T_v}(\hat{k}_i - \hat{k}_r)}\right\}, \tag{9.61}$$

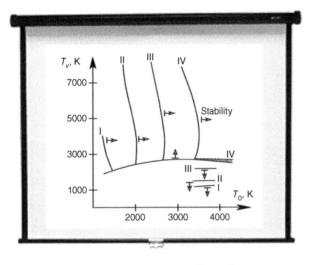

Figure 9.15 Plasma instability limits restricting specific energy input (I – 0.1 eV mol^{-1}, II – 0.3 eV mol^{-1}, III – 0.5 eV mol^{-1}, IV – 0.7 eV mol^{-1}) at different vibrational and translational temperatures. The range of stability parameters is crossed out.

where $\hat{k}_r = \frac{\partial \ln k_r}{\partial \ln T_0}$ is sensitivity of chemical reaction to translational temperature. The instability time is comparable with the time of vibrational excitation. This establishes very tough limit on the maximum specific energy input into continuous nonequilibrium discharge required for its uniform stability:

$$E_v^{\max} = k_{\mathrm{eV}} \frac{\hbar\omega}{\Omega_T} = c_p T_0 \left\{ \xi(\hat{k}_r - 2) - \tilde{k}_r \frac{c_p T_0}{c_v^v T_v} \right.$$

$$\left. + \sqrt{\left[\xi(\hat{k}_r - 2) + \tilde{k}_r \frac{c_p T_0}{c_v^v T_v} \right]^2 + 4\xi\, \tilde{k}_r \frac{c_p T_0}{c_v^v T_v}(\hat{k}_i - \hat{k}_r)} \right\}^{-1} . \tag{9.62}$$

This strong restriction of moderate or high-pressure discharges to provide their uniform stability is shown in Figure 9.15 for different vibrational and translational temperatures, and specific energy inputs.

9.12 Electron Attachment Instability and Other Ionization Instabilities of Nonthermal Plasma

9.12.1 Attachment Instability

Attachment instability occurs if electron detachment mostly compensates electron attachment. It can be observed in glow discharges when the perturbation of n_e does not affect current. The following sequence of perturbations can illustrate the instability:

$$\delta n_e \uparrow \to \delta T_e \downarrow \to v_a \downarrow \to \delta n_e \uparrow . \tag{9.63}$$

An increase of electron density leads to decrease of local electric field (current is not perturbed) and, hence to a decrease of electron temperature. It intensifies ionization, but the effect of weakening of attachment may have a stronger effect (because the ionization rate is much less than the rate of attachment in the presence of intensive detachment process). Before the perturbation, attachment and detachment processes were in balance. Therefore, weakening of the attachment rate at a constant level of detachment results in an increase of electron density and, finally, in the instability. Increment of the attachment instability is:

$$\Omega_a \approx k_a n_0 \left[\hat{k}_a \left(1 - \frac{k_i}{k_a} \frac{\hat{k}_i}{\hat{k}_a} \right) - \frac{n_+}{n_-} \right], \tag{9.64}$$

where k_a is the rate coefficient of electron attachment; $\hat{k}_a = \frac{\partial \ln k_a}{\partial \ln T_e}$ is sensitivity of this rate coefficient to T_e; n_+ and n_- are densities of positive and negative ions. Characteristic time of the attachment instability is, obviously, time of electron attachment. This instability leads to the formation of the **electric field domains**, which are a form of **striations**. An initial local fluctuation $\delta n_e > 0$, $\delta T_e < 0$ in the presence of high concentration of negative ions results in their decay, growth of electron density, and further decrease of a local electric field. This is called a *weak field domain*. An opposite local fluctuation $\delta n_e < 0$, $\delta T_e > 0$ leads to the formation of a *strong field domain*. The domains usually move toward anode slower than the electron drift velocity.

9.12.2 Ionization Instability Controlled by Dissociation of Molecules

The effective electron temperature is higher in monatomic gases than in corresponding molecular gases at the same value of reduced electric field. This is due to the excitation energy distribution functions (EEDF) reduction in an energy interval corresponding to intensive vibrational excitation. This effect explains the discharge instability:

$$\delta n_e \uparrow \to \delta(dissociation) \uparrow \to \delta T_e \uparrow \to \delta n_e \uparrow. \tag{9.65}$$

Increase of electron density (or temperature) leads to intensification of dissociation, conversion of molecules into atoms, and then results in further growth of the electron density and temperature. Increment of the instability (9.65) is determined by the dissociation time.

9.12.3 The Stepwise Ionization Instability

This instability is like the fast ionization instability related to vibrational excitation (9.59). The increase of electron density leads to overpopulation of electronically excited species, acceleration of ionization, and further increase of electron density. Increment of this instability corresponds to the frequency of electronic excitation. In contrast to the case of vibrational excitation (9.59), this one cannot be stabilized by chemical reactions.

9.12.4 Electron Maxwellization Instability

EEDF are restricted at high energies by a variety of channels of inelastic electron collisions. Maxwellization of electrons at higher electron densities provides larger amounts of high-energy electrons and therefore stimulate ionization instability:

$$\delta n_e \uparrow \to \delta(Maxwellization) \uparrow \to \delta f(E) \uparrow \to \delta n_e \uparrow. \tag{9.66}$$

9.12.5 Instability in Fast Oscillating Fields

The ionization instability of microwave plasma in low-pressure monatomic gases $v_{en} \ll \omega$ is like the modulation instability of hot plasma. An increase of electron density in a layer (perpendicular to electric field) provides growth of the plasma frequency, which approaches the microwave frequency ω and leads to an increase of the field. The growth of electric field results in intensification of ionization and further increase of electron density, which determines the instability. See Fridman and Kennedy (2021), for more details.

9.13 Problems and Concept Questions

9.13.1 Effect of Electron Attachment on Breakdown Conditions

Which attachment mechanism, the dissociative attachment or three-body attachment, prevents breakdown more?

9.13.2 Energy Input and Temperature in Streamers

Specific energy input in streamers is about 10^{-3} eV mol^{-1} in molecular gases, which corresponds to temperature increase 3–4 K. Estimate the vibrational temperature increase in the streamer channel assuming $\hbar\omega = 0.3\,eV$.

9.13.3 Streamer Propagation Velocity

In frameworks of the model of ideally conducting streamer, estimate the difference between streamer velocity and electron drift velocity.

9.13.4 Attachment-controlled Discharge Regime

In the attachment-controlled regime, the electron temperature and reduced electric field are fixed by the ionization-attachment balance $k_i(T_e) \approx k_a(T_e)$. What factors restrict electron density and degree of ionization in this case?

9.13.5 The Engel–Steenbeck Model

Calculate the electron temperature in nonthermal nitrogen discharge with chamber radius 1cm and pressure 1Torr.

9.13.6 Ionization Wave Propagation

What degree of ionization is required by (9.47) to provide the velocity of the ionization wave close to the speed of sound?

9.13.7 Thermal Instability in Monatomic Gases

The ionization-overheating instability follows the chain of causal links (9.50). Why is it not applicable to supersonic discharges?

9.13.8 Electron Attachment Instability

Why does the electron attachment instability affect plasma-chemical process much less than the ionization-overheating instability? Compare typical frequencies of the electron attachment instability with those of vibrational excitation by electron impact and VT-relaxation at room temperature.

Lecture 10

Nonthermal Plasma Sources: Glow Discharges

10.1 Major Types of Electric Discharges, Glow Discharge as a Conventional Nonthermal Plasma Source

 The term "gas discharge" initially reflected the process of "discharging" a capacitor into a circuit containing a gas gap between two electrodes. If voltage between the electrodes is sufficient for breakdown, a conductive plasma channel is generated, and the capacitor discharges. Now the term "gas discharge" is applied more generally to any system with ionization in a gap and plasma generation induced by electric field. Even electrodes are not necessary in general since discharges can occur by interaction of electromagnetic waves with gas. Electric discharges can be classified according to their physical features in many ways, for example, in the following five ways:

- **High-pressure discharges (usually atmospheric ones, like arcs or corona) and low-pressure discharges (10 Torr and less, like glow discharges)**. Differences between these plasma sources are related mostly to how discharge walls are involved in the kinetics of charged particles, energy, and mass balance. Low-pressure discharges are usually cold and not very powerful, while high-pressure ones can be both, very hot and powerful (arc) as well as cold and weak (corona).
- **Electrode discharges (like glow and arc) and electrodeless discharges (like inductively coupled radio-frequency RF and microwave)**. Differences between these are related mostly in the manner that electrodes contribute to sustaining electric current by surface ionization mechanisms and as a result closing the electric circuit.
- **Direct current (DC) discharges (like arc, glow, and pulsed corona) and non-DC discharges (like RF, microwave, and most of dielectric barrier discharges – DBD)**. The DC discharges can have either constant current (arc, glow) or can be sustained in pulse-periodic regime (pulsed corona). The pulsed periodic regime permits providing higher power in cold discharges at atmospheric pressure. Non-DC-discharges can be either low or high frequencies (including radiofrequency and microwave). Microwave discharges because of the skin effect can be sustained in a nonequilibrium way at moderate pressures at extremely high powers up to 1 MW.
- **Self-sustained and non-self-sustained discharges**. Non-self-sustained discharges can be externally supported by electron beams and ultraviolet radiation. Important aspect of these discharges is the independence of ionization and energy input to the plasma, which permits achieving large energy inputs at high pressures without instabilities.
- **Thermal (quasi-equilibrium) and nonthermal (nonequilibrium) discharges**. The thermal discharges are hot (5000–20 000 K), while the nonthermal discharges can operate close to room temperature. The difference between these two qualitatively different types of electric discharges is primarily related to different ionization mechanisms. Ionization in nonthermal discharges is mostly provided by direct electron impact (electron collisions with "cold" non-excited atoms and molecules), in contrast to thermal discharges where the stepwise ionization is due to electron collisions with preliminary excited hot atoms and molecules. Thermal discharges (the typical example is an electric arc) are usually powerful, easily sustained at high pressures, but operate close to thermodynamic equilibrium. The nonthermal discharges (the typical example is the glow discharge) can operate very far from thermodynamic equilibrium but usually with limited power.

Plasma Science and Technology: Lectures in Physics, Chemistry, Biology, and Engineering, First Edition. Alexander Fridman.
© 2024 WILEY-VCH GmbH. Published 2024 by WILEY-VCH GmbH.

Let us start consideration of plasma sources with the glow discharges, which were historically the most conventional sources of nonthermal plasma. The term "glow discharge" appeared to point out the discharge luminosity (in contrast to the low power dark discharges). According to physical definition: *glow discharge is the self-sustained continuous DC discharge with a cold cathode, which emits electrons due to secondary emission mostly induced by positive ions.* The **glow discharge structure** is illustrated in Figure 10.1. Distinctive feature of a glow discharge is **the cathode layer** with large positive space charge and strong electric field with a potential drop about 100–500 V. The thickness of cathode layer is inversely proportional to gas density and pressure. If the distance between electrodes is large, quasi-neutral plasma with a low electric field, the so-called positive column, is formed between the cathode layer and anode. The **positive column** is a traditional example of weakly ionized nonequilibrium low-pressure plasma. The positive column is separated from anode by an **anode layer** with negative space charge and elevated electric field. The conventional "tube" configuration of a glow discharge is shown in Figure 10.1; relevant parameters presented in Table 10.1. This glow configuration was widely applied for a century in fluorescent lamps as lighting devices.

Other **glow discharge configurations** are shown in Figure 10.2. For example, the coplanar magnetron glow discharge for sputtering and deposition including magnetic field for plasma confinement is presented in Figure 10.2a. The glow configuration, optimized as an electron bombardment plasma source is shown in Figure 10.2b; it is coaxial and includes the hollow cathode ionizer as well as a diverging magnetic field. The configurations applied for gas lasers are illustrated in Figure 10.2c,d. At high power levels (currents up to 20 A), it is usually a parallel plate discharge with a gas flow. The discharge can be transverse with the electric current perpendicular to the gas flow Figure 10.2c, or longitudinal if these are parallel to each other (Figure 10.2d).

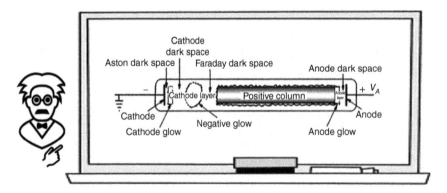

Figure 10.1 Illustration of the conventional glow discharge in tube and its structure.

Table 10.1 Typical parameters are the conventional glow discharge in tube.

Discharge tube radius	0.3–3 cm
Discharge tube length	10–100 cm
Plasma volume	About 100 cm^3
Gas pressure	0.03–30 Torr
Voltage between electrodes	100–1000 V
Electrode current	10^{-4}–0.5 A
Power level	Around 100 W
Electron temperature in positive column	1–3 eV
Electron density in positive column	10^9–10^{11} cm^{-3}

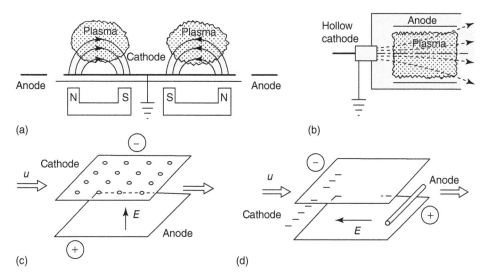

Figure 10.2 Different glow discharge configurations: (a) magnetron configuration; (b) hollow cathode configuration; (c) transverse configuration in gas flow; (d) longitudinal configuration in gas flow.

10.2 Plasma Parameters and Glow Pattern along the Glow Discharge

Glow discharge has a specific glow pattern, which is a sequence of dark and bright luminous layers distributed along the tube illustrated in Figure 10.3a. The glow size scale is determined by the electron mean free path $\lambda \propto 1/p$, and is inversely proportional to pressure. Therefore, it is easier to observe the glow pattern at low pressures: the layered pattern extends to centimeters at pressures about 0.1 Torr. Special individual names were given to each layer. Immediately adjacent to cathode is a dark layer known as the ***Aston dark space***. Then there is a relatively thin layer of the ***cathode glow***. This is followed by the ***dark cathode space***. The next zone is the so-called ***negative glow***, which is sharply separated from the dark cathode space. The negative glow gradually decreases in brightness toward the anode, becoming the ***Faraday dark space***. Only after that does the positive column begins. The ***positive column*** is bright (though not as bright as the negative glow), uniform, and can be long. In the anode layer, the positive column is transferred first into the ***anode dark space***, and finally into the narrow ***anode glow*** zone. The glow pattern can be interpreted based on the distribution of the discharge parameters shown in the Figure 10.4b–g. Electrons are ejected from cathode with low energies (about 1 eV) insufficient to excite atoms; this explains the Aston dark space. Then electrons receive from electric field the energy required for electronic excitation (and radiation) providing the cathode glow. Further acceleration of electrons in the cathode dark space leads mostly not to electronic excitation but to ionization which explains the low level of radiation and significant increase of electron density in the cathode dark space. Slowly moving ions have relatively high concentrations in the cathode layer and provide most of electric current. The high electron density at the end of the cathode dark space results in a decrease of the electric field, and hence a decrease of the electron energy and ionization rate but leads to significant intensification of radiation. It explains transition to the brightest layer of the negative glow. Moving further from cathode, electron energy decreases, which results in transition from the negative glow into the Faraday dark space. Plasma density decreases in the Faraday dark space and electric field again grows finally establishing the positive column. The electron energy in positive column is about 1–2 eV, which provides the light emission there. The cathode layer structure remains the same if electrodes are moved closer at fixed pressure, while the positive column shrinks. Anode repels ions and attracts electrons from the positive column, which creates the negative space charge and leads to some increase of electric field in the anode layer. Reduction of the electron density in this zone explains the anode dark space, while the electric field increase explains the anode glow.

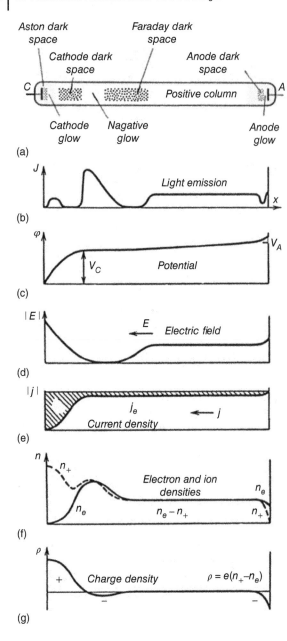

Figure 10.3 Distribution of glow pattern (a) and physical parameters (b–g) along a glow discharge tube.

10.3 Current–Voltage Characteristics of DC-discharges: Transition from Townsend Dark Discharge to Glow Discharge

When voltage between electrodes exceeds the breakdown threshold V_t, a self-sustained discharge is ignited. Current–voltage characteristics for wide range of currents are illustrated in Figure 10.4. Electric circuit of the discharge gap also includes an external ohmic resistance R. Then the Ohm's law for the circuit can be presented as **the load line**, also shown in Figure 10.4:

$$EMF = V + RI, \tag{10.1}$$

where EMF is the electromotive force, V is voltage on the discharge gap. Intersection of the current–voltage characteristic and the load line gives the discharge current and voltage. If the external ohmic resistance is large and the current is low (about 10^{-10}–10^{-5} A), then the electron and ion densities are also low and perturbations of the external electric field in plasma can be neglected. Such discharge is known as the **dark Townsend discharge**. The voltage

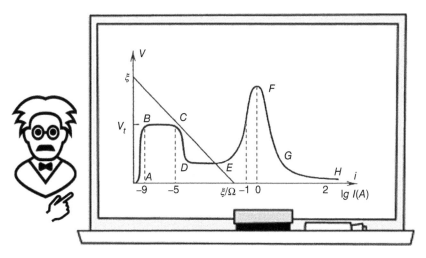

Figure 10.4 General current–voltage characteristics of DC discharges in wide range of currents.

necessary to sustain this discharge does not depend on current and coincides with the breakdown voltage. The dark Townsend discharge corresponds to the plateau BC. An increase of *EMF* or decrease of the external ohmic resistance *R* leads to growth of the discharge current and plasma density, which results in significant reconstruction of the electric field. This leads to reduction of voltage with current (interval CD in Figure 10.4) and to transition from a dark to a glow discharge. This low current regime of glow discharge is called the **sub-glow discharge**. Further *EMF* increase or *R* reduction leads to the lower voltage plateau DE corresponding to the **normal glow discharge** with currents 10^{-4}–0.1 A. The current density on the cathode is fixed in normal glow. Increase of the total current requires growth of the so-called **cathode spot** through which the current flows. When the current becomes so large that no additional free surface remains on the cathode, voltage increases to provide higher current densities. Such regime is called the **abnormal glow discharge** and corresponds to the interval *EF* on the current–voltage characteristics. Further increase of current accompanied by voltage growth in the abnormal glow regime leads to higher power levels and transition to arc discharge. The glow-to-arc transition usually occurs at currents about 1 A. To explain formation of the glow discharges, let us first consider physics of the dark discharges.

The distinctive feature of the dark discharges is smallness of its current and plasma density, which keeps the external electric field almost unperturbed. The steady state continuity equation for charged particles can be expressed considering their drift in electric field and ionization as:

$$\frac{dj_e}{dx} = \alpha j_e, \quad \frac{dj_+}{dx} = -\alpha j_e. \tag{10.2}$$

The direction from cathode to anode is chosen as the positive; j_e and j_+ are the electron and positive ion current densities; α is the Townsend coefficient. Adding equations (10.2) gives: $j_e + j_+ = j = const$, which reflects the constancy of the total current. The boundary conditions on the cathode ($x = 0$) relate the ion and electron currents on the surface due to the secondary electron emission with coefficient γ:

$$j_{eC}(x = 0) = \gamma j_{+C}(x = 0) = \frac{\gamma}{1+\gamma} j. \tag{10.3}$$

Anode boundary conditions ($x = d_0$, d_0 is the inter-electrode distance) reflect absence of ionic emission:

$$j_{+A}(x = d_0) = 0, \quad j_{eA}(x = d_0) = j. \tag{10.4}$$

Solution of (10.2) with the boundary condition on cathode can be expressed as:

$$j_e = \frac{\gamma}{1+\gamma} j \exp(\alpha x), \quad j_+ = j\left(1 - \frac{\gamma}{1+\gamma}\exp(\alpha x)\right). \tag{10.5}$$

To satisfy the anode boundary conditions, the electric field and Townsend coefficient α should be large:

$$\alpha(E)d_0 = \ln\frac{\gamma+1}{\gamma}. \tag{10.6}$$

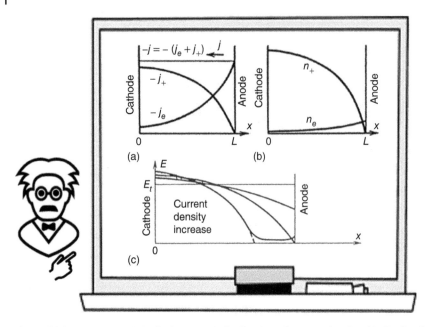

Figure 10.5 Townsend dark discharges: (a) distribution of current density; (b) distribution of electron and ion densities; (c) electric field evolution.

This formula describing the dark discharge self-sustainment, coincides with the breakdown condition in the gap, and gives the relation between currents as:

$$\frac{j_e}{j} = \exp[-\alpha(d_0 - x)], \quad \frac{j_+}{j} = 1 - \exp[-\alpha(d_0 - x)], \quad \frac{j_+}{j_e} = \exp[\alpha(d_0 - x)] - 1. \tag{10.7}$$

According to (10.6): $\alpha d_0 = \ln\frac{\gamma+1}{\gamma} \gg 1$, because $\gamma \ll 1(\alpha d_0 = 4.6$, if $\gamma = 0.01)$. From (10.7), the ion current exceeds electron current over a major part of discharge gap (Figure 10.5a). The electron and ion currents become equal only near the anode ($j_e = j_+$ at $x = 0.85d_0$). Difference in electrons and ions density is even stronger because of large differences in electron and ion mobilities (μ_e, μ_+). Electron and ion densities become equal at a point close to the anode (see Figure 10.5b), where:

$$1 = \frac{n_+}{n_e} = \frac{\mu_e}{\mu_+}\frac{j_+}{j_e} = \frac{\mu_e}{\mu_+}[\exp\alpha(d_0 - x) - 1]. \tag{10.8}$$

Assuming $\mu_e/\mu_+ \approx 100$, the electron and ion densities become equal at $x = 0.998$. Almost entire gap is charged positive in a dark discharge. However, the absolute value of the positive charge is not high because of low current and hence, low ion density in this discharge.

Transition from the dark to glow discharge is due to growth of the positive space charge and distortion of the external electric field, which results in formation of the cathode layer. To describe this transition, use the Maxwell equation:

$$\frac{dE}{dx} = \frac{1}{\varepsilon_0}e(n_+ - n_e). \tag{10.9}$$

Considering that $n_+ \approx j/e\mu_+E \gg n_e$, (10.9) gives the distribution of electric field:

$$E = E_c\sqrt{1 - \frac{x}{d}}, \quad d = \frac{\varepsilon_0\mu_+E_c^2}{2j}. \tag{10.10}$$

Here E_c is the electric field at the cathode. The electric field decreases near the anode with respect to the external field and grows in the vicinity of the cathode; see Figure 10.5c. The higher current densities lead to more distortion of the external electric field. The parameter "d" in (10.10) corresponds to a virtual point, where the electric field becomes equal to zero. This point is located far beyond the discharge gap ($d \gg d_0$) at low currents typical for dark discharges. When the current density becomes sufficiently high, this imaginary point of zero electric field can reach the anode ($d = d_0$). This critical current density is maximum for the dark discharge.

The maximum dark discharge current density corresponds to formation of the cathode layer and to transition from the dark to glow discharge:

$$j_{max} = \frac{\varepsilon_0\mu_+E_c^2}{2d_0}. \tag{10.11}$$

The electric field in the cathode's vicinity can be estimated as the breakdown electric field. Then, the maximum dark discharge current and therefore transition to glow discharge occurs at $j_{max} \approx 3 \cdot 10^{-5}$ A in the case of nitrogen at pressure 10 Torr, inter-electrode distance 10 cm, electrode area 100 cm^2, and secondary electron emission coefficient $\gamma = 10^{-2}$.

10.4 Cathode Layer of Glow Discharge, Engel–Steenbeck Model and Current–Voltage Characteristics

Organization of current in glow discharge requires high potential drop in the cathode layer, which is provided by positive space charge formed due to low ion mobility. Let us first consider the theory of a cathode layer developed in 1934 by von Engel and Steenbeck.

The electric field $E(x = d)$ on the "anode end" of a cathode layer ($x = d$, see Figure 10.5c) is much less than near the cathode $E(x = 0) = E_c$. The ion current into a cathode layer from a positive column can be neglected due to low ion mobility ($\mu_+/\mu_e \propto 10^{-2}$). For this reason, the Engel–Steenbeck model assumes zero electric field at the end of a cathode layer $E(x = d) = 0$ and considers the anode layer as an independent system of length d (10.10). Assuming constant electric field in the breakdown condition over the cathode layer $E(x) = E_c = const$, we can apply the relation (10.6) for the cathode layer. The inter-electrode distance d_0 can be replaced by the length d of the cathode layer; The cathode potential drop is $V_c = \int_0^d E(x)dx$, and can be simplified to $V_c = E_c d$. The Engel–Steenbeck model relates the electric field E_c, the cathode potential drop V_c, and the length of cathode layer pd, by formulas like those (9.6) describing breakdown of a gap:

$$V_c = \frac{B(pd)}{C + \ln(pd)}, \quad \frac{E_c}{p} = \frac{B}{C + \ln(pd)};$$ (10.12)

here $C = \ln A - \ln \ln \left(\frac{1}{\gamma} + 1 \right)$; A and B are the pre-exponential and exponential parameters of the function $\alpha(E)$ (see Table 9.1). The cathode potential drop V_c, electric field E_c, and the similarity parameter pd depend on the discharge current density j (which is close to the ion current density because $j_+ \gg j_e$ near the cathode, see Figure 10.4). To determine this dependence according to the Engel–Steenbeck model, we should first determine the positive ion density $n_+ \gg n_e$. It can be found based on the Maxwell equation (10.9), considering the linear decrease of electric field $E(x)$ along the cathode layer from $E(x = 0) = E_c$ to $E(x = d) = 0$:

$$n_+ \approx \frac{\varepsilon_0}{e} \left| \frac{dE(x)}{dx} \right| \approx \frac{\varepsilon_0 E_c}{ed}.$$ (10.13)

Total current density near cathode is close to the current density of positive ions:

$$j = en_+ \mu_+ E \approx \frac{\varepsilon_0 \mu_+ E_c^2}{d} \approx \frac{\varepsilon_0 \mu_+ V_c^2}{d^3}.$$ (10.14)

The cathode potential drop V_c as a function of the similarity parameter pd corresponds in frameworks of the Engel–Steenbeck approach to the Paschen curve (Figure 9.3) for breakdown. This function $V_c(pd)$ has a minimum V_n corresponding to the minimum breakdown voltage. The cathode potential drop V_c as a function of current density j also has the same minimum point V_n. The relations between V_c, E_c, pd, and j can be expressed using the dimensionless parameters:

$$\tilde{V} = \frac{V_c}{V_n}, \quad \tilde{E} = \frac{E_c/p}{E_n/p}, \quad \tilde{d} = \frac{pd}{(pd)_n}, \quad \tilde{j} = \frac{j}{j_n}.$$ (10.15)

Here electric field E_n/p and cathode layer length $(pd)_n$ correspond to the minimum point of the cathode voltage drop V_n. The subscript "n" stands here to denote the "normal" regime of a glow discharge. These three "normal" parameters: E_n/p, $(pd)_n$ as well as V_n can be found using the relation (9.7) for electric breakdown. Corresponding values of the normal current density can be expressed based on (10.14) using the similarity parameters:

$$\frac{j_n, \text{A cm}^{-2}}{(p, \text{Torr})^2} = \frac{1}{9 \cdot 10^{11}} \frac{(\mu_+ p), \text{cm}^2 \text{Torr}^{-1} \text{V}^{-1} \text{s}^{-1} \times (V_n, V)^2}{4\pi[(pd)_n, \text{cm Torr}]^3}.$$ (10.16)

Relations between V_c, E_c, j and the cathode layer length pd can be expressed as:

$$\tilde{V} = \frac{\tilde{d}}{1 + \ln \tilde{d}}, \quad \tilde{E} = \frac{1}{1 + \ln \tilde{d}}, \quad \tilde{j} = \frac{1}{\tilde{d}(1 + \ln \tilde{d})^2}$$ (10.17)

Dimensionless voltage $\tilde{V}$, electric field $\tilde{E}$, and cathode layer length $\tilde{d}$ are shown in Figure 10.6a as functions of current density, representing the cathode layer current–voltage characteristics.

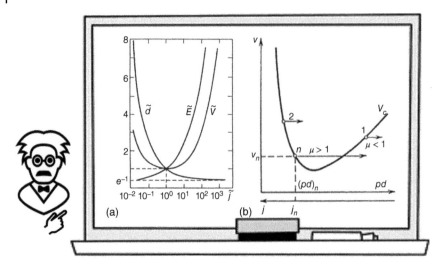

Figure 10.6 Current–voltage characteristics of glow discharges (a) specifying the normal regime; interpretation (b) of the normal current density effect.

10.5 The Normal Regime of Glow Discharges, Steenbeck Minimum Power Principle for Normal Cathode Current Density

According to the current–voltage characteristics (10.17) any current densities are possible in a glow discharge. In reality, however, *the discharge "prefers" to operate only at a single specific current density (10.16), the normal one j_n, which corresponds to the minimum of the cathode potential drop.* The total current I is controlled by the load line (10.1). The current-conducting channel occupies a cathode spot with area $A = I/j_n$, which provides the required normal current density. Other current densities are unstable. If due to perturbation $j > j_n$, the cathode spot grows until current density becomes normal; and vice versa. The glow discharge with normal current density on its cathode is referred to as the **normal glow discharge**. It has fixed current density j_n, fixed cathode layer thickness $((pd)_n)$, and voltage V_n, which only depending at room temperature on gas composition and cathode material. Typical values of these normal glow discharge parameters are presented in Table 10.2.

Typical normal cathode current density is about $100\,\mu A\ cm^{-2}$ at pressure about 1 Torr; typical thickness of the normal cathode layer at this pressure is about 0.5 cm, typical normal cathode potential drop is 200 V and does not depend on pressure and temperature.

The constancy of current density in the normal glow discharge is an impressive effect which was explained in different ways: analyzing current stability, modeling ionization kinetics and even applying the minimum power principle. First Engel and Steenbeck's explanation was related to instability of cathode layer with $j < j_n$, where the current–voltage characteristic is "falling" (see Figure 10.6b). The falling current–voltage characteristics are generally unstable in nonthermal discharges. For example, if a fluctuation results in a local increase of current density in some area of a cathode spot, the necessary voltage to sustain ionization in this area decreases. The actual voltage in this area then exceeds the required one, which leads to further increase of current until it becomes normal $j = j_n$. If a fluctuation results in local decrease of current density, the discharge extinguishes for the same reason. Such a mechanism explains the instability at $j < j_n$ and the growth of current density to reach the normal one.

 More detailed model, see Raizer (2011), describes establishing of the normal cathode current density $j = j_n$ starting from both lower $j < j_n$ and higher $j > j_n$ current densities in terms of the charge reproduction coefficient:

$$\mu = \gamma \left\{ \exp \int_0^{d(r)} \alpha[E(l)]\, dl - 1 \right\}. \tag{10.18}$$

Integration here is along the electric current line "l," which crosses the cathode in some point "r." The coefficient (10.18) shows multiplication of charged particles during single cathode layer ionization cycle. This cycle includes multiplication of the primary

Table 10.2 Normal current density j_n/p^2, $\mu A\,cm^{-2}\,Torr^{-2}$, normal thickness of cathode layer $(pd)_n$, cm Torr, and normal cathode potential drop V_n, V for different gases and cathode materials at room temperature.

Gas	Cathode material	Normal current density	Normal thickness of cathode layer	Normal cathode potential drop
Air	Al	330	0.25	229
Air	Cu	240	0.23	370
Air	Fe	—	0.52	269
Air	Au	570	—	285
Ar	Fe	160	0.33	165
Ar	Mg	20	—	119
Ar	Pt	150	—	131
Ar	Al	—	0.29	100
He	Fe	2.2	1.30	150
He	Mg	3	1.45	125
He	Pt	5	—	165
He	Al	—	1.32	140
Ne	Fe	6	0.72	150
Ne	Mg	5	—	94
Ne	Pt	18	—	152
Ne	Al	—	0.64	120
H_2	Al	90	0.72	170
H_2	Cu	64	0.80	214
H_2	Fe	72	0.90	250
H_2	Pt	90	1.00	276
H_2	C	—	0.90	240
H_2	Ni	—	0.90	211
H_2	Pb	—	0.84	223
H_2	Zn	—	0.80	184
Hg	Al	4	0.33	245
Hg	Cu	15	0.60	447
Hg	Fe	8	0.34	298
N_2	Pt	380	—	216
N_2	Fe	400	0.42	215
N_2	Mg	—	0.35	188
N_2	Al	—	0.31	180
O_2	Pt	550	—	364
O_2	Al	—	0.24	311
O_2	Fe	—	0.31	290
O_2	Mg	—	0.25	310

electrons formed on the cathode in an avalanche moving along a current line across the cathode layer and return of positive ions formed in the layer back to cathode to produce new electrons due to secondary electron emission with the coefficient γ. Sustaining the steady-state cathode layer requires $\mu = 1$. Excessive ionization then corresponds to $\mu > 1$, $\mu < 1$ means extinguishing the discharge. The curve $\mu = 1$ on the "voltage-cathode layer thickness" diagram represents the cathode layer potential drop V_c, with the near-minimum point $(pd)_n$ corresponding to the normal voltage V_n, see Figure 10.6b. All the area on the V-pd diagram above the cathode potential drop curve corresponds to $\mu > 1$; the area under this curve means $\mu < 1$. If current density is less than normal

$j < j_n (d > d_n$, for example, point "1" in Figure 10.6b, fluctuations can destroy the cathode layer by the instability. At the edges, however, the cathode layer decays in this case even without any fluctuations. Positive space charge is much less at the edges and the same potential corresponds to points located further from cathode, which can be illustrated as moving to the right from point "1" in the area where $\mu < 1$. As a result, current disappears from the edges and the discharge voltage increases in accordance with the load line. The increase of voltage over the line $\mu = 1$ in a central part of the cathode layer leads to growth of current density until it reaches the normal value j_n. Similarly, can be considered the cathode layer with supernormal current density $j > j_n$, which corresponds to point "2." In this case, the central part of the channel is stable with respect to fluctuations (see the Engel–Steenbeck stability analysis for $j < j_n$). At the edges of the channel, space charge is smaller and effective pd is greater for the fixed voltage, which brings us to the area of the V-pd diagram with the charge reproduction coefficient $\mu > 1$. The condition $\mu > 1$ leads to breakdown at the edges of the cathode spot, to an increase of total current, to a decrease of total voltage across the electrodes (in accordance with the load line), and finally to a decrease of current density in the major part of a cathode layer until it reaches the normal value j_n. Any deviations of current density from $j = j_n$ stimulate ionization at the edges of cathode spot, which bring current density to the normal value.

The effect of normal current density can be illustrated using **the minimum power principle**. The total power released in the cathode layer can be expressed as:

$$P_c(j) = A \int_0^d jE \, dx = AjV_C(j) = IV_c(j); \tag{10.19}$$

here A is area of a cathode spot, I is the total current. The current in glow is mostly determined by external resistance (see the load line) and can be considered as fixed. In this case, the cathode spot area can be varied together with the current density at fixed product $j \times A = const$. According to **the minimum power principle**, current density in glow discharge should minimize the power $P_c(j)$. As seen from (10.19), the minimization of the power $P(j)$ at constant current $I = const$ requires minimization of $V_c(j)$, which corresponds to the normal current density $j = j_n$. *The minimum power principle was proposed in 1932 by Steenbeck and is useful for illustrating discharge phenomena, including striation, thermal arcs, and the normal current in glow discharge. The minimum power principle cannot be derived from fundamental physical laws and hence, should be used for mostly for illustrations.*

10.6 Abnormal, Subnormal, and Obstructed Glow Discharge Regimes; the Hollow Cathode Discharge

Increase of current in a normal glow discharge is provided by growth of the cathode spot area at $j = j_n = const$. As soon as the entire cathode is covered, further current growth results in an increase of current density over the normal value. This discharge is called the **abnormal glow discharge**. The abnormal glow discharge corresponds to the right-hand-side branches ($j > j_n$) in Figure 10.6. The current–voltage characteristic of the abnormal discharge $\tilde{V}(\tilde{j})$ is growing. It corresponds to the interval EF in Figure 10.4. When the current density grows further ($j \to \infty$); the cathode layer thickness decreases asymptotically to a finite value $\tilde{d} = 1/e \approx 0.37$; while the cathode potential drop and electric field grow as:

$$\tilde{V} = \frac{1}{e^{3/2}} \sqrt{\tilde{j}}, \quad \tilde{E} \approx \frac{1}{e^{1/2}} \sqrt{\tilde{j}} \tag{10.20}$$

Growth of the current and cathode voltage is limited by cathode overheating. Significant cathode heating at voltages about $10 \, kV$ and current densities $10 - 100 \, A \, cm^{-2}$ results in transition of the abnormal glow discharge into an arc discharge. Normal glow discharge transition to a dark discharge takes place at low currents (about $10^{-5} A$) and starts with the so-called **subnormal discharge**. The subnormal discharge corresponds to the interval CD in Figure 10.5. The size of the cathode spot at low current becomes large and comparable with the total cathode layer thickness. This results in electron losses with respect to normal glow discharge and hence, requires higher voltages to sustain the discharge. Another glow discharge regime, different from the normal one, occurs at low pressures and narrow gaps between electrodes, when their product pd_0 is less than normal value $(pd)_n$ for a cathode layer. This discharge mode is called the **obstructed glow discharge**. Conditions in the obstructed discharge correspond to the left-hand branch of the Paschen curve, where voltage exceeds the minimum value V_n. Since short inter-electrode distance in the obstructed discharge is not sufficient for effective multiplication of electrons, the inter-electrode voltage is greater than the normal one.

Another interesting and practically important discharge is related to the negative glow region, which is a zone of intensive ionization and radiation, see Figure 10.3. Most electrons in the negative glow have moderate energies. However, quite a few electrons in this area are very energetic even though the electric field is relatively low. These energetic electrons are formed in the vicinity of the cathode and cross the cathode layer with only a few inelastic

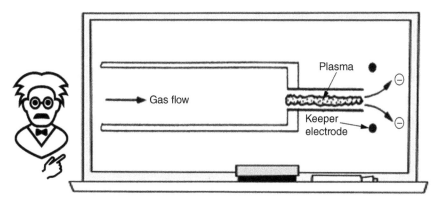

Figure 10.7 Illustration of the Lidsky capillary hollow cathode.

collisions. They provide a nonlocal ionization effect and lead to electron densities in a negative glow exceeding those in a positive column. The effect of intensive "nonlocal" ionization in a negative glow can be applied to form the so-called **hollow cathode discharge**. Let us imaging a glow discharge with a cathode arranged as two parallel plates with the anode on a side. If the distance between the cathodes gradually decreases, at some point the current can grow 100–1000 times without change of voltage. This effect takes place when two negative glow regions overlap, accumulating the energetic electrons from both cathodes. Effective accumulation of the high negative glow current can be reached if the cathode is arranged as a hollow cylinder and an anode lies further along the axis. Pressure is chosen in a way to have the cathode layer thickness comparable with internal diameter of the hollow cylinder. The most traditional configuration of this so-called **Lidsky hollow cathode** is shown in Figure 10.7. It is a narrow capillary-like nozzle, which operates with axially flowing gas. The hollow cathode is usually operated with the anode located about 1 cm downstream of the capillary nozzle. It can provide high electron currents with densities exceeding those corresponding to Child's law.

10.7 About Anode Layer of Glow Discharges

Positive ions are not emitted (but repelled) by the anode, therefore their concentration at the anode is close to zero. Thus, a negatively charged zone, called the anode layer, occurs between the anode and positive column (Figure 10.3). The ionic current density in the anode layer grows from zero at the anode to its value $j_{+c} = \frac{\mu_+}{\mu_e} j$ in positive column (here j is the total current density; μ_+, μ_e are mobilities of ions and electrons). In terms of the Townsend coefficient α:

$$\frac{dj_+}{dx} = \alpha j_e \approx \alpha j, \quad j_{+c} \approx j \times \int \alpha \, dx. \tag{10.21}$$

Based upon (10.21), it is sufficient for one electron to provide only a very small number of ionization acts to establish the necessary ionic current:

$$\int \alpha \, dx = \frac{\mu_+}{\mu_e} \ll 1. \tag{10.22}$$

Number of electrons produced in the anode layer is about three orders of magnitude smaller than the corresponding number of electrons produced in the cathode layer. For this reason, the anode layer's potential drop is less than the potential drop across the cathode layer. The anode potential drop is about the ionization potential of the gas in the discharge. The anode voltage slightly grows with pressure in the range of moderate pressures around 100 Torr. This dependence is stronger in electronegative gases, see Figure 10.8. Typical reduced electric fields E/p in the anode layer are about $E/p \approx 200 - 600$ V cm^{-1} Torr^{-1}. Thickness of the anode layer is of order of the electron mean free path and can be estimated as:

$$d_A(\text{cm}) \approx 0.05 \, \text{cm}/p(\text{Torr}). \tag{10.23}$$

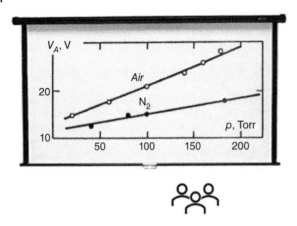

Figure 10.8 Anode voltage fall in nitrogen and airglow discharges at different pressures.

The current density j/p^2 at the anode is independent of the current in the same manner as for the normal cathode layer. Current densities in anode layer are about $100\,\mu A\,cm^{-2}$ at gas pressures about 1 Torr and coincide with those of the cathode layer, see Table 10.2.

10.8 Positive Column of Glow Discharges: Current–Voltage Characteristics and Heat Balance

Plasma parameters of long positive column are independent of the phenomena in cathode and anode layers. Plasma in the positive column is determined by local processes of charged particle formation and losses, and by the electric current controlled by external resistance and *EMF*. In conventional glow discharges, the balance of charged particles is due to diffusion to the walls, therefore the electric field E is determined by the Engel–Steenbeck relation (see Section 9.7) and does not depend on electron density and electric current. In this case, the current–voltage characteristic of positive column is almost horizontal (see Figure 10.4):

$$V(I) = V_n + E\,d_c \approx const;$$
(10.24)

here, d_c is the length of positive column; V_n is the normal potential drop in the cathode layer; E is the electric field in the column, which only depends on type of gas, pressure, and radius of discharge tube (see Section 9.7). When the current and hence electron density grows, the contribution of volumetric electron–ion recombination becomes significant, see Section 9.6. In this case, relation between current density and reduced electric field E/p becomes:

$$j = (\mu_e p)\frac{E}{p}e\frac{k_i(E/p)\ n_0}{\left(k_r^{ei} + \varsigma\,k_r^{ii}\right)(1+\varsigma)};$$
(10.25)

here $\mu_e p$ is the electron mobility presented as similarity parameter; $k_i(E/p)$ is ionization coefficient; n_0 is gas density; k_r^{ei}, k_r^{ii} are coefficients of electron–ion and ion–ion recombination; k_a, k_d are coefficients of electron attachment and detachment; $\varsigma = k_a/k_d$ is the factor characterizing the balance between electron attachment and detachment. Considering the sharp exponential behavior of the function $k_i(p)$, we can conclude that electric field E grows very slowly with current density in the recombination regime. Thus, the current–voltage characteristic of positive column is almost horizontal in this case as well. Dependences of the reduced electric field E/p on the similarity parameter pR (p is pressure, R is radius of a discharge tube) are presented in Figure 10.9. Some E/p decrease with current is due to increases in the gas temperature (ionization actually depends on E/n_0; increase of temperature leads to reduction of n_0 and growth of E/n_0 at constant pressure and electric field). Reduced electric fields E/p are about 10 times lower in inert gases than in molecular gases. It is due to effects of inelastic collisions, which significantly reduces the electron energy distribution function at the same values of reduced electric field. The electric fields sustaining glow discharges are much lower than those necessary for breakdown because electron losses to the walls during breakdown are provided by free diffusion, while the corresponding discharge is controlled by much slower ambipolar diffusion.

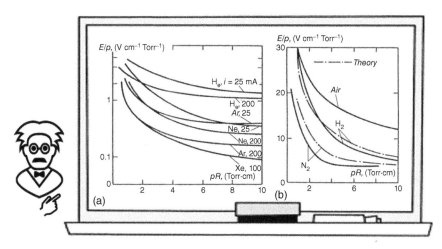

Figure 10.9 Reduced electric fields E/p in positive column of inert (a) and molecular (b) gases.

The power, which electrons receive from electric field and then transfer through collisions to atoms and molecules, can be expressed as Joule heating $jE = \sigma E^2$, which is usually balanced by conductive and convective energy transfer:

$$w = jE = n_0 c_p (T - T_0) \nu_T, \tag{10.26}$$

where w is the discharge power per unit volume; c_p is specific heat per one molecule; T is gas temperature in the discharge; T_0 is room temperature. ν_T is heat removal frequency, which for the cylindrical discharge tube of radius R and length d_0 can be determined as:

$$\nu_T = \frac{8}{R^2} \frac{\lambda}{n_0 c_p} + \frac{2u}{d_0}, \tag{10.27}$$

where λ is the coefficient of thermal conductivity, and u is the gas flow velocity. If heat removal is controlled by thermal conduction, the typical discharge power per unit volume, which doubles the gas temperature in the discharge $(T - T_0 = T_0)$, can be expressed as:

$$w = jE = \frac{8\lambda T_0}{R^2}. \tag{10.28}$$

The thermal conductivity coefficient λ does not depend on pressure and can be estimated as $\lambda \approx 3 \cdot 10^{-4}\,\text{W cm}^{-1}\,\text{K}^{-1}$. Therefore, specific discharge power also does not depend on pressure and for tubes with radius $R = 1\,\text{cm}$ can be estimated as $0.7\,\text{W cm}^{-3}$. Higher values of specific power result in higher gas temperatures and hence, in contraction of a glow discharge. Current density in the positive column with the heat removal controlled by conduction is inversely proportional to pressure and can be estimated as:

$$j = \frac{8\lambda T_0}{R^2} \frac{1}{(E/p)} \frac{1}{p}. \tag{10.29}$$

Assuming reduced electric field as $E/p = 3 - 10\,\text{V cm}^{-1}\,\text{Torr}^{-1}$: j, $\text{mA cm}^{-2} \approx 100/p$, Torr. Electron density in the positive column can be calculated considering Ohm's law $j = \sigma E$ and relevant expression for the electric conductivity σ:

$$n_e = \frac{w}{E^2} \frac{m\nu_{en}}{e^2} = \frac{w}{(E/p)^2} \frac{m\,k_{en}}{e^2 T_0} \frac{1}{p}; \tag{10.30}$$

here ν_{en}, k_{en} are the frequency and rate coefficient of electron-neutral collisions. The electron density in a positive column with conductive heat removal can be estimated as: n_e, $\text{cm}^{-3} = 3 \cdot 10^{11}/p$, Torr. The reduction of plasma degree of ionization with pressure is even more significant: $\frac{n_e}{n_0} \propto \frac{1}{p^2}$. *Hence, low pressures are more favorable for sustaining the steady state homogeneous nonthermal plasma.*

The convective heat removal represented by second term in (10.27) promotes pressure and power increases in the **fast flow glow discharges**, when gas velocities are $50 - 100$ m s^{-1}. Then current and electron densities corresponding to doubling of the gas temperature, do not depend on gas pressure:

$$j = \frac{2uc_p}{d_0(E/p)}, \qquad n_e = \frac{2uc_p m k_{en}}{e^2 d_0 T_0 (E/p)^2}. \tag{10.31}$$

Assuming $d_0 = 10\,\text{cm}$, $u = 50$ m s^{-1}, $E/p = 10\,\text{V cm}^{-1}\,\text{Torr}^{-1}$, the current density and electron concentration in in this case are: $j \approx 40\,\text{mA cm}^{-2}$, $n_e = 1.5 \cdot 10^{11}\,\text{cm}^{-3}$. The specific discharge power in the positive column with convective heat removal grows proportionally to pressure:

$$w = jE = c_p T_0 \frac{2u}{d_0} \cdot \frac{p}{T_o},$$

(10.32)

and is about w, W cm^{-3} = $0.4 \cdot p$, Torr. High gas flow velocity also results in voltage and reduced electric field growth (about $E/p = 10$–20 V^{-1} cm^{-1} Torr^{-1}) to intensify ionization and compensate charge losses. Significant increases of charge losses in fast gas flows can be related to turbulence, and can be estimated by replacing the ambipolar diffusion coefficient D_a by an effective one including a special turbulent term:

$$D_{eff} = D_a + 0.09Ru;$$

(10.33)

here u is the flow velocity, and R is the discharge tube radius (or half-distance between walls in plane geometry).

The current–voltage characteristic of glow discharges controlled by diffusion is slightly decreasing, which is due to Joule heating. The increase of current leads to some growth of gas temperature T_0, which at constant pressure results in decrease of gas density n_0. The ionization rate is a function of E/n, which is often only "expressed" as E/p assuming room temperature. For this reason, the decrease of gas density at a fixed ionization rate leads to a decrease of electric field and voltage, which finally explains the decreasing current–voltage characteristics. At fixed ionization rate: $E/n_0 \propto ET_0 \approx const$. In this case, the relation between current density and electric field can be expressed as:

$$\frac{j}{j_0} = \left(\frac{E_0}{E}\right)^{3/2} \left(\frac{E_0}{E} - 1\right),$$

(10.34)

where E_0 is electric field necessary to sustain low discharge current $j \to 0$, when gas heating is negligible; j_0 is the typical value of current density:

$$j_0 = n_0 c_p T_0 \frac{v_T}{E_0} = \frac{w_0}{E_0}.$$

(10.35)

10.9 Glow Discharge Instabilities: Contraction of the Positive Column

Contraction is a glow discharge instability related to instantaneous self-compression of a positive column into one or several bright current filaments when pressure or current are attempted to be increased. It can be illustrated by a glow discharge in room temperature neon at pressures 75–100 Torr sustained in a tube with 2.8 cm radius. The relevant current–voltage characteristic is shown in Figure 10.10a and demonstrates contraction when current exceeds about 100 mA. The current–voltage characteristic is decreasing at currents close to the critical one, which demonstrates the strong effect of Joule heating. The transition between the diffusive and contracted modes demonstrates hysteresis. At the critical current, the electric field in the positive column abruptly decreases with a related sharp transition of the discharge regime from the initial strongly nonequilibrium diffusive mode into the contracted mode. A brightly luminous filament appears along the axis of the discharge tube while the rest of the discharge becomes almost dark. Average current density at the transition point is 5.3 mA cm^{-2}, the corresponding current density on the axis of the discharge tube is 12 mA cm^{-2} with an electron density on the axis approximately 10^{11} cm^{-3}. Radial distributions of the relative electron density before and after contraction in the same glow discharge system at a fixed pressure $p = 113$ Torr are shown in Figure 10.10b. Corresponding changes of discharge parameters (electron and gas temperatures, electron density) during the contraction transition on axis of the discharge tube are presented in Table 10.3.

The diameter of a filament formed because of contraction is almost two orders of magnitude smaller than the diameter of a tube initially filled with the diffusive glow discharge, see Figure 10.10b. Electron concentration increases about 50 times. Gas temperature increases due to localized heat release, and the electron temperature decreases because of reduction of the electric field (Figure 10.10a). After contraction, a glow discharge is not a strongly nonequilibrium one (see Table 10.3). Therefore, the contraction phenomenon is sometimes referred to as **arcing**. Instability mechanisms leading to glow discharge contraction (see Section 9.10) include the thermal instability, the stepwise ionization instability, and the electron Maxwellization instability. These mechanisms provide nonlinear growth of ionization with electron density, which is the main cause of contraction. These instability mechanisms become significant at electron concentrations exceeding critical values of about 10^{11} cm^{-3}, and specific powers of about 1 W cm^{-3}.

Contraction of glow discharge is the **transverse instability;** plasma parameters are changed across the electric field. Taking into account that the tangential component of the electric field is always continuous, a sharp decrease of the electric field in the central filament occurs as a consequence of contraction and results in an overall voltage decrease (kind of a short circuit) and loss of nonequilibrium in the discharge as a whole (see Figure 10.10c). That is why the transverse instability, contraction, is so harmful for strongly nonequilibrium glow discharges.

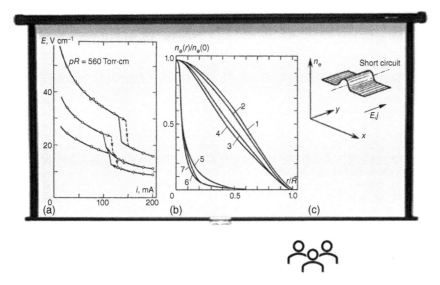

Figure 10.10 Illustration of glow discharge contraction: (a) $V-i$ characteristic for neon, tube radius $R = 2.8$ cm; (1) $pR = 210$ Torr cm^{-1}; (2) $pR = 316$ Torr cm^{-m}; (3) $pR = 560$ Torr cm^{-m} (the solid curve corresponds to decrease of current, and dashed curve to increase); (b) profiles of n_e (1) $i/R = 4.8$ mA cm^{-m}; (2) 15.4; (3) 26.8; (4) 37.5; (5) 42.9; (6) 57.2; (7) 71.5 mA cm^{-m} (the transition occurs between (4) and (5)); (c) electron density perturbation in transverse instabilities.

Table 10.3 Change of electron temperature, gas temperatures, and electron density during the glow discharge contraction.

Plasma parameters	Before contraction ($I = 96$ mA)	After contraction ($I = 120$ mA)
Electron temperature	3.7 eV	3.0 eV
Gas temperature	930 K	1200 K
Electron density	$1.2 \cdot 10^{11}$ cm^{-3}	$5.4 \cdot 10^{12}$ cm^{-3}

10.10 Glow Discharge Instabilities: The Striations

In contrast to contraction, **the striations** (see Section 9.10) are related to **longitudinal perturbations** of plasma parameters along a positive column. In this case, the electric current (and the current density in 1D approximation) remains fixed during a local perturbation of electron density δn_e and temperature δT_e. Local growth of electron density n_e (and conductivity σ) induces decrease of electric field E and vice versa; while the current density remains the same:

$$j = \sigma E \propto n_e E = const, \quad \frac{\delta n_e}{n_e} = -\frac{\delta E}{E}. \tag{10.36}$$

The electric field reduction in perturbations with elevated electron density and vice versa are illustrated in Figure 10.11a,b. Shift of electron density with respect to ion density is due to the electron drift in the electric field. The direction of polarization field δE is opposite to the direction of external electric field E if the fluctuation of electron concentration is positive. If the electron density is reduced, the polarization field is added to the external one. Such instability is unable to destroy the nonequilibrium discharge. The striations move fast up to 100 m s^{-1} from anode to cathode at pressures 0.1–10 Torr in inert gases, or they can remain at rest. Plasma parameters of a glow discharge with and without striations are nearly the same.

From a physical point of view, the striations can be interpreted as ionization oscillations and waves. They can be initiated by the stepwise ionization instability. In this case, an increase of electron density leads to the production of more excited species, which accelerates stepwise ionization and results in a further increase of the electron density. When the electron density becomes too large, super-elastic collisions deactivate the excited species. Further nonlinear growth of the ionization rate is usually due to the Maxwellization instability. Both of these ionization instability mechanisms are not directly related to gas overheating. Overheating requires high values of specific power and specific energy input. As a result, striations are observed at less intensive plasma parameters

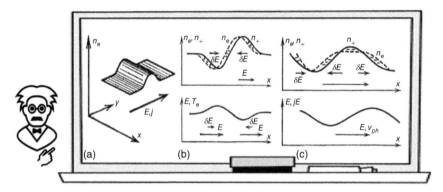

Figure 10.11 Physics of the glow discharge striations: (a, b) illustration of longitudinal instability; (c) illustration of the striation propagation mechanism.

(electron density, electric current, specific power) than those related to the thermal (ionization overheating) instability, which is responsible for contraction. Change of the electric field is a stabilizing factor for striations. Auto-acceleration of ionization and nonlinear growth of electron density in striations induces reduction of the electric field. It leads to reduction of the effective electron temperature after a short delay $\tau_f = 1/(v_{en}\delta)$ related to establishing the corresponding electron energy distribution function (EEDF) (here v_{en} is frequency of electron-neutral collisions, δ is the average fraction of electron energy transferred to a neutral particle during the collision). The reduction of electron temperature results in an exponential decrease of ionization and in suppression of the instability.

Striations are usually moving from anode to cathode, which is illustrated in Figure 10.11c. In a typical case of short wavelengths, the gradients of electron density in a perturbation δn_e are significant, and charge separation is mostly due to electron diffusion. The electric field of polarization δE, occurring because of this electron diffusion, determines the oscillation of the total electric field. Maximum of the electric field oscillations δE_{max} corresponds to the points on the wave where the electron density is not perturbed $\delta n_e = 0$. The maximum of the electric field oscillations δE_{max} is shifted with respect to the maximum of plasma density δn_e oscillations by one-quarter of a wavelength toward the cathode. The ionization rate is fastest at the point of maximum electric field (δE_{max}), resulting in moving the point of maximum plasma density δn_e toward the cathode. The striations propagate from anode to cathode as the ionization waves. To determine the striations velocity, we can assume perturbations of electric field, electron density, and temperature change as: $\delta E, \delta n_e, \delta T_e \propto \exp[i(\omega t - k_s x)]$. Then relation between perturbations of electric field and electron density is:

$$\delta E = ik_s \frac{T_e}{e} \frac{\delta n_e}{n_e}. \tag{10.37}$$

Electrons receive energy about T_e from the electric field during their drift over the length needed to establish the EEDF ($v_d \tau_f \approx \lambda/\sqrt{\delta}$), where λ is the electron mean free path, δ is the fraction of electron energy transferred during a collision). Thus, considering $eE(\lambda/\sqrt{\delta}) \approx T_e$, (10.37) can be rewritten as:

$$\frac{\delta E}{E} \approx ik_s \frac{\lambda}{\sqrt{\delta}} \frac{\delta n_e}{n_e}. \tag{10.38}$$

Relation between perturbations of electric field and electron temperature can be derived from balance of Joule heating and electron thermal conductivity (with coefficient λ_e) $j \cdot \delta E = k_s^2 \lambda_e \, \delta T_e$:

$$\frac{\delta T_e}{T_e} \approx \frac{1}{(k_s \lambda/\sqrt{\delta})^2} \frac{\delta E}{E}. \tag{10.39}$$

Here the Einstein relation, $\lambda_e/\mu_e = T_e/e$, the Ohm's law $j = en_e\mu_e E$, and the relation $eE(\lambda/\sqrt{\delta}) \approx T_e$ was taken into account. Acceleration of the ionization rate $\partial n_e/\partial t$ in striations related to the electron temperature increase δT_e can be expressed as:

$$\delta\left(\frac{\partial n_e}{\partial t}\right) \approx n_e n_0 \frac{\partial k_i}{\partial T_e}\delta T_e = k_i n_e n_0 \frac{\partial \ln k_i}{\partial \ln T_e} \frac{\delta T_e}{T_e}, \tag{10.40}$$

where n_e, n_0 are concentrations of electrons and neutral species; k_i is the ionization rate coefficient; and $\partial \ln k_i/\partial \ln T_e \approx I/T_e \gg 1$ is the logarithmic sensitivity of the ionization rate coefficient to the electron temperature. As illustrated in Figure 10.11c, the electron density grows with the amplitude perturbation δn_e during a quarter of a period (about $1/k_s v_{ph}$, where v_{ph} is the phase velocity of ionization wave, k_s is the wave number). This means: $\delta\left(\frac{\partial n_e}{\partial t}\right) \times \frac{1}{k_s v_{ph}} \approx \delta n_e$. Then, the phase velocity of the ionization wave is:

$$v_{ph} \approx \frac{k_i n_0}{k_s} \frac{\partial \ln k_i}{\partial \ln T_e} \left(\frac{\delta T_e}{T_e} \middle/ \frac{\delta n_e}{n_e}\right), \tag{10.41}$$

and can be simplified to:

$$v_{ph} = \frac{\omega_s}{k_s} = \frac{1}{k_s^2 \lambda/\sqrt{\delta}} k_i n_0 \frac{\partial \ln k_i}{\partial \ln T_e}. \tag{10.42}$$

The striations velocity is proportional to the square of the wavelength and is about $100\,\mathrm{m\,s^{-1}}$. The frequency of oscillations of electron density, temperature, and other plasma parameters in the striations can be expressed as:

$$\omega = \frac{1}{k_s \lambda/\sqrt{\delta}} k_i n_0 \frac{\partial \ln k_i}{\partial \ln T_e}. \tag{10.43}$$

The oscillation frequency in striations is proportional to wavelength ($2\pi/k_s$) and is about the ionization frequency $10^4 \div 10^5\,\mathrm{s^{-1}}$. According to the dispersion equation, the absolute value of group velocity of striations (v_{gr}) is equal to that of the phase velocity. Directions of these two velocities are opposite: $v_{ph} = \omega/k = -d\omega/dk = -v_{gr}$. For this reason, some special discharge marks (e.g. bright pulsed perturbations) move to the anode, in opposite direction with respect to striations themselves.

The striations can be illustrated using the **Steenbeck minimum power principle**. If the discharge current is fixed, the voltage drop related to a wavelength of striations is less than the corresponding voltage of a uniform discharge. This can be explained by the strong exponential dependence of the ionization rate on the electric field value. Because of this strong exponential dependence, an oscillating electric field provides a more intensive ionization rate than an electric field fixed at the average value. Hence, to provide the same ionization level in the discharge with striations requires less voltage, and consequently lower power at the same current.

10.11 About Energy Efficiency of Plasma-chemical Processes in Glow Discharges, Approaches to Glow Discharge Stabilization, Atmospheric Pressure Glow Discharges

Conventional glow discharges controlled by diffusion are of interest only for such chemical applications where energy cost-effectiveness is not an issue. Application of glow discharges for highly energy-effective plasma-chemical processes is limited by the following three major factors:

(1) Specific energy input in the glow discharges controlled by diffusion is about $100\,\mathrm{eV\,mol^{-1}}$ and exceeds the optimal value $E_v \approx 1\,\mathrm{eV\,mol^{-1}}$. Considering that the energy necessary for one act of chemical reaction is usually of about $3\,\mathrm{eV\,mol^{-1}}$, the maximum possible energy efficiency in these systems is about 3% even at complete 100%-conversion. These high specific energy inputs (about $100\,\mathrm{eV\,mol^{-1}}$) are due to low gas flows passing through the discharge.
(2) The most energy effective processes require high ionization degrees. Considering that in glow discharges $n_e/n_0 \propto 1/p^2$, this requirement leads to low gas pressures and hence, to further growth of the specific energy input and decrease of energy efficiency.
(3) Specific power of the conventional glow discharges controlled by diffusion does not depend on pressure and is quite low, about $0.3–0.7\,\mathrm{W\,cm^{-3}}$. For this reason, the specific productivity of such plasma-chemical systems is also relatively low.

Increase of the specific power, specific productivity, and energy efficiency of related plasma-chemical systems requires increasing gas pressure, suppression of the related glow discharge instabilities. At elevated currents, powers, pressures, and volumes, it requires application of **convective cooling**, fast flows, and other discharge stabilization approaches. The most applied approach is **segmentation of the cathode**. If a high conductivity plasma filament (contraction) occurs between two points on two large electrodes, current grows and the discharge voltage immediately drops. This can be suppressed by segmentation of an electrode, usually the cathode. Voltage is applied to each segment independently through an individual external resistance. If a filament occurs at one of the cathode segments, the discharge voltage related to other segments does not drop. Some practical ways of the cathode segmentation are illustrated in Figure 10.12. In the case of transverse discharges, the cathode segments are spread over a dielectric plate; in longitudinal discharges, the segments are arranged as a group of cathode rods at the gas inlet to the discharge chamber. Another approach to suppressing the contraction is related to the **gas flow in the discharge chamber**. Making the velocity field as uniform as possible prevents inception of instabilities. The high gas velocities also stabilize a discharge because of reduced residence time to values insufficient to contraction. Finally, discharge stabilization can be achieved by utilizing **intensive small-scale turbulence**, which provides damping of incipient perturbations.

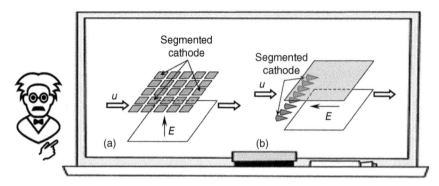

Figure 10.12 Cathode segmentation in transverse (a) and in longitudinal (b) glow discharge configurations.

Operating **atmospheric pressure glow discharges** is a challenging task, which has been accomplished, however, in transonic and supersonic flows. Application of special aerodynamic techniques permits sustaining the uniform steady-state glow discharges at atmospheric pressure and specific energy input up to $500\,\mathrm{J\,g^{-1}}$ due to suppression of the transverse diffusion influence on the temperature and current density distribution. Also, it becomes possible by using special gas mixtures and elaborating special types of electrodes. Such gas mixtures (usually of noble gases) are supposed to provide the necessary level of the ionization at low reduced electric field E/p. Mixtures of helium or neon with argon or mercury are very effective for ionization when the Penning effect takes place. In this case, hundreds of volts are sufficient to operate a glow discharge at atmospheric pressure. Use of helium is also helpful to increase heat exchange and cooling of the systems. The normal cathode current density is proportional to the square of pressure and becomes large at elevated pressures. Therefore, special types of electrodes like fine wires should be applied. Atmospheric pressure glow discharges are physically similar to such traditional nonthermal atmospheric pressure discharges as corona or DBD, however, their voltage can be much less.

More details on the subject can be found, in particular, in Kanazava et al. (1988), Yokayama et al. (1990), Okazaki and Kogoma (1994), and Babukutty et al. (1999).

10.12 Glow Discharges in Strong Magnetic Field: Penning Discharge, Plasma Centrifuge

The special feature of the **Penning discharge** (see Figure 10.13) is its strong magnetic field (up to 0.3 T), which permits magnetizing both electrons and ions. Low gas pressures in these discharges (10^{-6}–10^{-2} Torr) are required for the effective magnetization. The two cathodes in this scheme are grounded and the cylindrical anode has voltage about 0.5–5 kV. Even though the gas pressure in the Penning glow discharge is low, plasma densities in these systems can be relatively high, up to $6\cdot 10^{12}\mathrm{cm}^{-3}$. The plasma is so dense because radial electron losses are reduced by the strong magnetic field and axially the electrons are trapped in electrostatic potential well. Although the configuration of the Penning discharge is different from the traditional glow discharge in cylindrical tube, it is still a glow discharge because the electrode current is sustained by secondary electron emission from cathode provided by energetic ions. The ions in the Penning discharge are so energetic that they usually cause intensive sputtering from the cathode surface. Ion energy in these systems can reach several kilo-electron-volts and greatly exceeds the electron energy. Plasma between the two cathodes is almost equipotential so it plays the role of a second electrode inside of the cylindrical anode. As a result, the electric field inside of the cylindrical anode is close to radial. Both electrons and ions can be magnetized in the Penning discharge, which leads to azimuthal drift of charged particles in the crossed fields: radial electric E_r and axial magnetic B. The tangential velocity v_{EB} of the azimuthal drift is the same for electrons and ions, and the kinetic energy $E_K(e, i)$ of electrons and ions is proportional to their mass $M_{e,i}$:

$$E_K(e, i) = \frac{1}{2}M_{e,i}v_{EB}^2 = \frac{1}{2}M_{e,i}\frac{E_r^2}{B^2}. \tag{10.44}$$

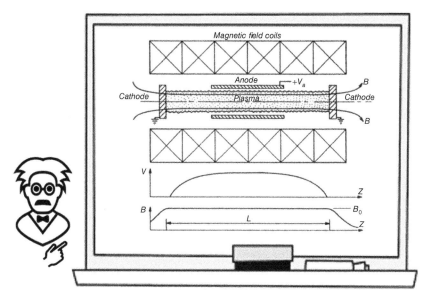

Figure 10.13 Illustration of a conventional Penning discharge with uniform magnetic field and electrostatic trapping of electrons.

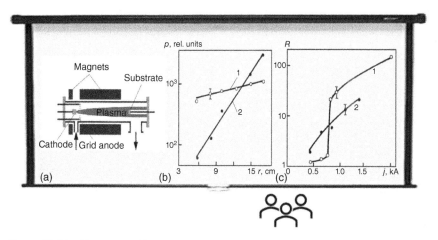

Figure 10.14 Plasma centrifuge: (a) illustration; (b) radial distribution of D_2 (1) and Ne (2) partial pressures; (c) separation coefficient for water decomposition products (1) and (2) for H_2–Ne mixture ($p = 0.3$ Torr, $H = 5$ kg, $\tau = 3$ ms)

This explains why the ion temperature in the Penning discharge much exceeds the electron temperature. The electron temperature in a Penning discharge is approximately 3–10 eV, while the ion temperature can be an order of magnitude higher (30–300 eV), see Roth (1995, 2001).

The **azimuthal drift in the crossed electric and magnetic fields** in the Penning discharge and the related fast plasma rotation allows creation of **plasma centrifuges**, see Figure 10.14a, where electrons and ions circulate fast (4.55) around the axial magnetic field. This "plasma wind" can drag neutral particles transferring to them the high energies of the charged particles. It is interesting to note that in the collisional regime, the gas rotation velocity is usually limited by the kinetic energy corresponding to ionization potential:

$$v_{rA} = \sqrt{2eI/M}, \tag{10.45}$$

where I and M are ionization potential and mass of heavy neutrals. The maximum rotation velocity v_{rA} in the plasma centrifuge is referred to as the **Alfven velocity for plasma centrifuge**.

There is no complete explanation of the critical Alfven velocity. It is clear, however, that acceleration of energetic ions becomes impossible in weakly ionized plasma when they have energy sufficient for ionization. Further energy transfer to ions does not go to their acceleration but rather to ionization. The problem is that ions in contrast to electrons are unable to ionize neutrals when their energy only slightly exceeds the ionization potential due to the adiabatic principle.

Gas rotation velocities in plasma centrifuges are very high and reach 2–$3 \cdot 10^6$ cm s^{-1} in the case of light atoms. For comparison, the maximum velocities in similar mechanical centrifuges are of about $5 \cdot 10^4$ cm s^{-1}. The fast gas rotation in a plasma centrifuge can be applied for isotope separation, which occurs because of diffusion (coefficient D) in a field of centrifugal forces. The separation time can be estimated as: $\tau_S \approx \frac{T_0 R_C^2}{D \Delta M v_\varphi^2}$, where T_0 is the gas temperature, v_φ is the maximum gas rotation velocity, R_C is the centrifuge radius, and $\Delta M = |M_1 - M_2|$ is the atomic mass difference of isotopes or components of gas in the mixture. Because of the high rotation velocities v_φ, the separation time is low, and the steady-state separation coefficient for binary mixture is significant even for isotopes with relatively small difference in their atomic masses $M_1 - M_2$:

$$R = \frac{(n_1/n_2)_{r=r_1}}{(n_1/n_2)_{r=r_2}} = \exp\left[(M_1 - M_2)\int_{r_1}^{r_2} \frac{v_\phi^2}{T_0}\frac{dr}{r}\right]. \tag{10.46}$$

Here $(n_1/n_2)_r$ is the density ratio of the binary mixture components at radius "r." For plasma centrifuges with $n_e/n_0 \approx 10^{-4}$–10^{-2}, the parameter $\frac{Mv_\perp^2}{2}/\frac{3}{2}T_0$ is about 3. The separation coefficient for mixture He–Xe in such centrifuges exceeds 300, for mixture ^{235}U–^{238}U it is about 1.1. Radial distribution of partial pressures of gas components for deuterium–neon separation in the plasma centrifuge is presented in Figure 10.14b. The plasma centrifuges can provide chemical process and product separation at once. Water can be dissociated this way $H_2O \rightarrow H_2 + \frac{1}{2}O_2$ with simultaneous separation of hydrogen and oxygen. It is illustrated in Figure 10.14c in comparison with H_2–Ne separation. For more details see Rusanov and Fridman (1984).

10.13 Magnetron Glow Discharges, Magnetic Mirror Effect

Schematic of the **magnetron discharge with parallel plate electrodes,** usually applied for sputtering, is shown in Figure 10.15a. For effective sputtering and film deposition, the mean free path of atoms must be large and gas pressure therefore should be low (10^{-3}–3 Torr). However in this system, because electrons are trapped in the magnetic field by the magnetic mirror, a plasma density on the level of 10^{10} cm^{-3} is achieved. Ions are not supposed to be magnetized here to provide sputtering. Typical voltage between electrodes is several hundred volts, magnetic induction approximately 5–50 mT. Negative glow electrons are trapped in the magnetron by the **magnetic mirror effect,** which causes "reflection" of electrons from areas with elevated magnetic field (see Figure 10.15c). The magnetic mirror effect is based on the fact, that if spatial gradients of the magnetic field are small, the magnetic moment of a charged particle gyrating around the magnetic lines is an approximate constant of the motion. The particle motion in such magnetic field is said to be the **adiabatic motion.** The electric field between the cathode and the negative glow zone is relatively strong, which provides ions with the energy necessary for effective sputtering of the cathode material. It is important that the drift in crossed electric and magnetic fields cause the plasma electrons in this system to drift around the closed plasma configuration. This makes the plasma of the magnetron discharge uniform, which is important for sputtering and film deposition. The magnetron discharges can be arranged in various configurations. For example, if the cathode location prevents effective deposition, it can be relocated. This leads to the so-called **co-planar configuration of the magnetron discharge,** shown in Figure 10.15b.

The magnetic mirror effect traps the negative glow electrons and providing sufficient plasma density for effective sputtering at relatively low pressures of the discharge. The magnetic moment μ of a charged particle (in this case it is an electron, mass m) gyrating in a magnetic field B is defined as the current I related to this circular motion multiplied by the enclosed area $\pi \rho_L^2$ of the orbit:

$$\mu = I \cdot \pi \rho_L^2 = \frac{mv_\perp^2}{2B} = const; \tag{10.47}$$

here ρ_L is the electron Larmor radius, $v_\perp$ is the component of electron thermal velocity perpendicular to magnetic field, see Figure 10.15c. When a gyrating electron moves adiabatically toward higher electric fields, its normal velocity component $v_\perp$ is growing proportional to the square root of magnetic field. The growth of $v_\perp$ is limited by energy conservation. For this reason, the electron drift into the zone with elevated magnetic field also should be limited leading to reflection of the electron back to the area with lower values of magnetic field. This generally explains the magnetic mirror effect. If the magnetic lines in the plasma zone are parallel to electrodes, energy transfer from electric field in this zone can be neglected. Then, the kinetic energy of the gyrating electron $mv^2/2$ and total electron velocity v can be considered as constant. Considering that $v_\perp = v\sin\theta$ (where θ is the angle between total electron velocity and magnetic field direction), the constant magnetic momentum (10.47) can be rewritten as:

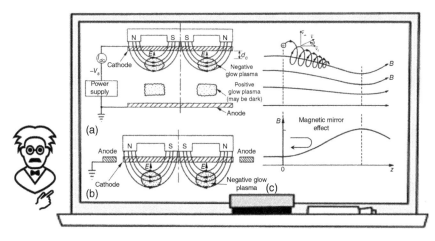

Figure 10.15 Magnetron glow discharge: (a) parallel plate configuration: the negative plasma glow is trapped in the magnetic mirror; (b) co-planar configuration: the negative plasma glow is trapped in the magnetic pole pieces; (c) magnetic mirror effect illustration.

$$\frac{\sin^2\theta}{B(z)} = \frac{2\mu}{mv^2} = const. \tag{10.48}$$

Thus, the angle θ between total electron velocity and magnetic field direction grows during the electron penetration into areas with higher values of magnetic field $B(z)$ until it reaches the "reflection" point $\theta = \pi/2$. From (10.48), electrons can be reflected if their initial angle θ_i is sufficiently large:

$$\sin\theta_i > \sqrt{B_{min}/B_{max}}, \tag{10.49}$$

where B_{min}, B_{max} are the maximum and minimum magnetic fields in the mirror, see Figure 10.15c. If the initial angle θ_i is not sufficiently large, electrons are not reflected by the mirror and are able to escape. Therefore, the minimum angle is referred to as the **escape cone angle**.

10.14 Problems and Concept Questions

10.14.1 Space Charges in Cathode and Anode Layers

Explain why the space charge of the glow discharge cathode layer is much greater than in the anode layer?

10.14.2 The Seeliger's Rule for Spectral Line Emission Sequence in Negative and Cathode Glows

The negative glow first reveals (closer to cathode) spectral lines emitted from higher excited atomic levels and then spectral lines related to lower excited atomic level. This sequence is reverse with respect to the cathode glow. Explain this so-called Seeliger's rule.

10.14.3 Glow Discharge in Tubes of Complicated Shapes

Glow discharges can be maintained in tubes of very complicated shapes. This effect is widely used in luminescent lamps. Explain the mechanism of sustaining the glow discharge uniformity in such a case.

10.14.4 Normal Cathode Potential Drop, Normal Current Density, and Normal Thickness of Cathode Layer

Prove that the normal cathode potential drop does not depend on gas pressure and temperature. Determine dependence of normal current density and normal thickness of cathode layer on temperature at constant pressure, and on pressure at constant temperature.

10.14.5 Glow Discharge with Hollow Cathode

For the Lidsky hollow cathode, explain how electric field can penetrate inside the metallic thin hollow cathode?

10.14.6 Contraction of Glow Discharge in Fast Gas Flow

Explain why transition to the contracted mode takes place in fast flow glow discharges when not specific power $(W\,cm^{-3})$, but rather the specific energy input $(eV\,mol^{-1}$ or $J\,cm^{-3})$ exceeds the critical value?

10.14.7 The Penning Discharge

Analyze the classical configuration of the Penning discharge and explain why the electric field in this system can be considered as radial. Compare the Penning discharge with the hollow cathode glow discharge, where the electric field configuration is also quite sensitive to the plasma presence.

10.14.8 The Alfven Velocity in Plasma Centrifuge

Estimate electric and magnetic field in plasma centrifuge when the ionic drift velocity in the crossed electric and magnetic fields reaches the critical Alfven velocity.

10.14.9 The Escape Cone Angle in Magnetic Mirror

If the initial angle θ_i between electron velocity and magnetic field direction is not sufficiently large, electrons are not reflected by the magnetic mirror and are able to escape. Derive the formula for the escape cone angle as a function of the maximum and minimum values of magnetic field in the magnetic mirror.

10.14.10 Atmospheric Pressure Glow Discharges

Explain challenges of glow discharge organization at atmospheric pressure. What is the difference between traditional nonthermal atmospheric pressure discharges (corona, DB) and the atmospheric pressure glow discharges?

Lecture 11

Thermal Plasma Sources: Arc Discharges

11.1 Arc Discharge as a Conventional Thermal Plasma Source: Types of Arcs, Plasma Parameters

 As glow discharges conventionally represent the nonthermal plasma sources, arcs are good example of thermal plasma sources. Arcs are self-sustaining direct current (DC) discharges, which in contrast to glow have low cathode fall voltage of about 10 eV, corresponding to ionization potential. Arc cathodes emit electrons by intensive **thermionic and field emission** providing high current without need in significant further multiplication of electrons in the cathode layer. Arc cathodes receive large amounts of Joule heating and can reach very high temperatures in contrast to the cold glow discharges. The main arc discharge zone located between electrode layers is called the **arc positive column**. The positive column can be either quasi-equilibrium or nonequilibrium depending on gas pressure. Nonequilibrium DC-plasma can be generated not only in glow discharges but also in arcs at low pressures, while quasi-equilibrium DC-plasma can be generated only in electric arcs. The principal cathode emission mechanism is thermionic in nonthermal regimes and mostly field emission in thermal arcs. The reduced electric field E/p is low in thermal arcs and relatively high in nonthermal arcs. The total voltage in any kind of arc is usually relatively low; it can be only couple of volts. Ranges of plasma parameters typical for the thermal and nonthermal arc discharges are outlined in Table 11.1.

Thermal arcs operating at high pressures are very energetic and called **high intensity arcs**. They have higher currents, current densities, and power per unit length. Generally, arcs can be classified into the following groups depending on peculiarities of the cathode processes, peculiarities of the positive column, and peculiarities of the working fluid.

11.1.1 Hot Thermionic Cathode Arcs

The cathode temperatures here are 3000 K and above providing high currents due to thermionic emission. The arcs are stationary to fixed and large cathode spot. Current is distributed over large cathode area and therefore its density is not high, about $10^2 - 10^4 \mathrm{A\,cm^{-2}}$. Only special refractory materials like carbon, tungsten, molybdenum, zirconium, tantalum, etc. can withstand such high temperatures. The hot thermionic cathode can be heated not only by the arc current but also from an external source.

11.1.2 Arcs with Hot Cathode Spots

If a cathode is made from low-melting-point metals like copper, iron, silver, or mercury, the high temperature necessary for emission cannot be sustained permanently. Electric current flows in this case through hot spots, which appear, move fast, and disappear on the cathode surface. Current density in the spots is high $10^4 - 10^7 \mathrm{A\,cm^{-2}}$. This leads to localized intensive heating and evaporation of the cathode material. The principal mechanism of electron emission from the spots is thermionic field emission to provide a high current density at temperatures limited by melting. The cathode spots appear not only on the low-melting-point cathodes but also on refractory metals at low currents and pressures.

Plasma Science and Technology: Lectures in Physics, Chemistry, Biology, and Engineering, First Edition. Alexander Fridman.
© 2024 WILEY-VCH GmbH. Published 2024 by WILEY-VCH GmbH.

Table 11.1 Typical arc plasma parameters in thermal and nonthermal regimes.

Plasma parameter	Thermal regime	Nonthermal regime
Gas pressure	$0.1 - 100$ atm	$10^{-3} - 100$ Torr
Arc current	30 A $- 30$ kA	$1 - 30$ A
Cathode current density	$10^4 - 10^7$ A cm^{-2}	$10^2 - 10^4$ A cm^{-2}
Voltage	$10 - 100$ V	$10 - 100$ V
Power per unit length	>1 kW cm^{-1}	<1 kW cm^{-1}
Electron density	$10^{15} - 10^{19}$ cm^{-3}	$10^{14} - 10^{15}$ cm^{-3}
Gas temperature	$1 - 10$ eV	$300 - 6000$ K
Electron temperature	$1 - 10$ eV	$0.2 - 2$ eV

11.1.3 Vacuum Arcs

These are low-pressure arcs, operating with the cathode spots, and working fluid provided by erosion and evaporation of the electrode material. The vacuum arc operates in a dense metal vapor, which is self-sustained in the discharge.

11.1.4 High Pressure Arc Discharges

These are quasi-equilibrium arcs operating at pressures exceeding 0.1–0.5 atm, see Table 11.1. At pressures exceeding 10 atm, the thermal plasma is so dense, that most of the discharge power, 80–90%, is converted into radiation.

11.1.5 Low Pressure Arc Discharges

At low pressures, $10^{-3} - 1$ Torr, the arc plasma is far from equilibrium, and somewhat like that of glow discharges. However, ionization degrees in such arcs are higher than in glow discharges because of higher currents, see Table 11.1.

The **current–voltage characteristics** of glow-to-arc transition and an arc itself are illustrated in Figure 11.1. They correspond to the classical **voltaic arc**, see Figure 11.2, which is a carbon arc in atmospheric air. Cathode and anode layer voltages in the arcs are about 10 V, the balance of voltage (see Figure 11.1b) corresponds to a positive column. Increase of the discharge length leads to linear growth of voltage, which means that the reduced electric field in the arc is constant at fixed current. When the discharge current grows, the electric field and voltage gradually decrease until a critical point of sharp explosive voltage reduction (see Figure 11.1b); this is followed by an almost horizontal current–voltage characteristic. The transition is accompanied by specific hissing noises associated with the formation of hot anode spots with intensive evaporation.

11.2 Electron Emission from Hot Cathode: Thermionic Emission, Sommerfeld formula, Schottky effect

Thermionic emission is extraction of energetic thermal electrons from high temperature metal surfaces like hot arc cathodes. This emission mechanism is the key for majority of arc discharges. Emitted electrons can remain in the surface vicinity creating a space charge resisting to further emission. The electric field is sufficient, however, to push the negative space charge out of the cathode and to reach the saturation current density, which is the main characteristic of the cathode thermionic emission.

To find the saturation current of thermionic emission, let's consider the electron distribution function in metals, which is the density of electrons in velocity interval $v_x \div v_x + dv_x$, etc. This distribution in metals is essentially quantum-mechanical, and can be described by the Fermi function:

$$f(v_x, v_y, v_z) = \frac{2m^3}{(2\pi\hbar)^3} \frac{1}{1 + \exp\frac{\varepsilon - \mu(T, n_e)}{T}};$$

(11.1)

where $\mu(T, n_e)$ is the chemical potential, which can be found from normalization of the Fermi distribution; m, ϵ are electron mass and total energy. The most energetic electrons can leave the metal if their kinetic energy $m v_x^2/2$ in the direction "x" perpendicular to the metal surface exceeds the absolute value of the potential energy $|\epsilon_p|$. The electric current density of these energetic electrons leaving the metal can be found by integrating (11.1):

$$j = e \int_{-\infty}^{+\infty} dv_y \int_{-\infty}^{+\infty} dv_z \int_{\sqrt{2|\epsilon_p|/m}}^{+\infty} v_x f(v_x, v_y, v_z) dv_x. \tag{11.2}$$

The energy distance from the highest electronic level in metal (the Fermi level) to the continuum is called **the work function** W; which corresponds to the minimum energy required to extract an electron. The integration (11.2) results in **the Sommerfeld formula** for the saturation current density for thermionic emission:

$$j = \frac{4\pi m e}{(2\pi \hbar)^3} T^2 (1 - R) \exp\left(-\frac{W}{T}\right); \tag{11.3}$$

here R is a quantum mechanical coefficient describing the reflection of electrons from the potential barrier related to the metal surface. It is convenient for practical calculations to use the numerical value of the Sommerfeld constant: $\frac{4\pi m e}{(2\pi \hbar)^3} = 120 \frac{A}{cm^2 K^2}$, and to consider typical values of the reflection coefficient $R = 0 \div 0.8$. The work function W for some cathode materials is given in Table 11.2.

The thermionic current grows with electric field until the negative space charge near cathode is eliminated and saturation is achieved. Further increase of the electric field gradually leads to an increase of the saturation current due to reduction of work function. The effect of electric field on the work function is known as the **Schottky effect.** The work function W is the binding energy of an electron to a metal surface. Neglecting the external electric field, this is work $W_0 = e^2/(4\pi\varepsilon_0)4a$ against the attractive image force $e^2/(4\pi\varepsilon_0)(2r)^2$; here a is the inter-atomic distance in metal. In the presence of an external electric field E, the total extracting force can be expressed as a function distance from the metal surface:

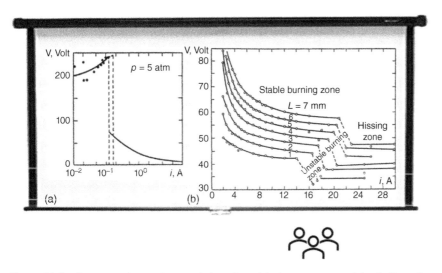

Figure 11.1 Current–voltage characteristics of arc: (a) glow-to-arc transition in Xe at 5 atm; (b) carbon arc in air, L is the distance between electrodes.

Table 11.2 Work functions of selected cathode materials.

Material	C	Cu	Al	Mo
Work function	4.7 eV	4.4 eV	4.25 eV	4.3 eV
Material	W	Pt	Ni	W/ThO$_2$
Work function	4.54 eV	5.32 eV	4.5 eV	2.5 eV

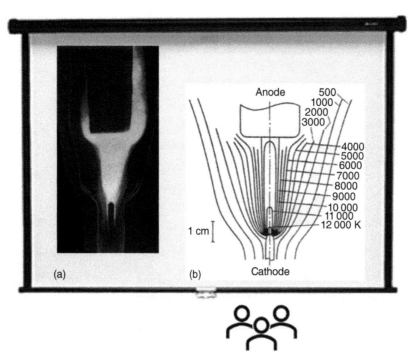

Figure 11.2 Illustration of the 200 A carbon arc in air (a) with gas temperature field (b).

$F(r) = \frac{e^2}{16\pi\varepsilon_0 r^2} - eE$. The extraction occurs if electron distance from the metal exceeds critical: $r_{cr} = \sqrt{e/16\pi\varepsilon_0 E}$, and attraction to the surface changes to repulsion. The work function can be then found by the **Schottky relation**:

$$W = \int_a^{r_{cr}} F(r)\, dr \approx \frac{e^2}{16\pi\varepsilon_0 a} - \frac{1}{\sqrt{4\pi\varepsilon_0}} e^{3/2}\sqrt{E}. \tag{11.4}$$

In the absence of electric field $E = 0$ and $W = W_0$. The Schottky decrease of the work function in an external electric field can be numerically presented as:

$$W, eV = W_0 - 3.8 \cdot 10^{-4} \cdot \sqrt{E, V\,cm^{-1}}. \tag{11.5}$$

The thermionic emission current dependence on electric field is presented in Table 11.3.

Table 11.3 Thermionic, field, and thermionic field emission currents as functions of electric field E (assuming $T = 3000$ K, $W = 4$ eV, $\varepsilon_F = 7$ eV, $A_0(1 - R) = 80$ A cm^{-2}K^2).

Electric field $(10^6$ V cm$^{-1})$	Schottky decrease of W (V)	Thermionic emission $(j,$ A cm$^{-2})$	Field emission $(j,$ A cm$^{-2})$	Thermionic field emission $(j,$ A cm$^{-2})$
0	0	$0.13 \cdot 10^3$	0	0
0.8	1.07	$8.2 \cdot 10^3$	$2 \cdot 10^{-20}$	$1.2 \cdot 10^4$
1.7	1.56	$5.2 \cdot 10^4$	$2.2 \cdot 10^{-4}$	$1.0 \cdot 10^5$
2.3	1.81	$1.4 \cdot 10^5$	1.3	$2.1 \cdot 10^5$
2.8	2.01	$3.0 \cdot 10^5$	130	$8 \cdot 10^5$
3.3	2.18	$6.0 \cdot 10^5$	$4.7 \cdot 10^3$	$2.1 \cdot 10^6$

11.3 Electron Emission from Cathode: Field, Thermionic Field, and Secondary Electron Emission Processes

If the external electric fields are very high (about $1 - 3 \cdot 10^6$ V cm^{-1}), they can not only decrease the work function but also directly extract electrons from cold metal due to the quantum-mechanical tunneling, which is called **field emission effect**. A simplified triangular potential barrier for electrons inside metal, considering external electric field E but neglecting the mirror forces is presented in Figure 11.3a. Electrons are able to escape from metal across the barrier due to the tunneling, leading to the field emission. The field electron emission current density can be calculated by the **Fowler–Nordheim formula**:

$$j = \frac{e^2}{4\pi^2\hbar}\frac{1}{(W_0 + \varepsilon_F)}\sqrt{\frac{\varepsilon_F}{W_0}}\exp\left[-\frac{4\sqrt{2m}\,W_0^{3/2}}{3e\hbar\,E}\right]; \tag{11.6}$$

here ε_F is the Fermi energy of a metal, W_0 is the work function not perturbed by external electric field. Electron tunneling across the potential barrier influenced by the Schottky effect is illustrated in Figure 11.3b. The Fowler–Nordheim formula can be then corrected by factor $\xi(\Delta W/W_0)$ depending on the relative Schottky decrease of work function:

$$j = 6.2 \cdot 10^{-6}\text{A cm}^{-2} \times \frac{1}{(W_0, eV + \varepsilon_F, eV)}\sqrt{\frac{\varepsilon_F}{W_0}}\exp\left[-\frac{6.85 \cdot 10^7 W_0^{3/2}(eV)\cdot\xi}{E, \text{V cm}^{-1}}\right]. \tag{11.7}$$

Some examples of the field emission current densities are shown in Table 11.3, relevant correction factors $\xi(\Delta W/W_0)$ are presented in Table 11.4.

The field emission current strongly depends on electric field: four-times change in electric field results in increase of the field emission by more than 23 orders of magnitude. According to the Fowler–Nordheim formula, the field emission becomes significant only when electric fields exceed 10^7 V cm^{-1}. However, the field emission starts making contribution at electric fields about $3 \cdot 10^6$ V cm^{-1} because of the field enhancement at the microscopic protrusions on metal surfaces.

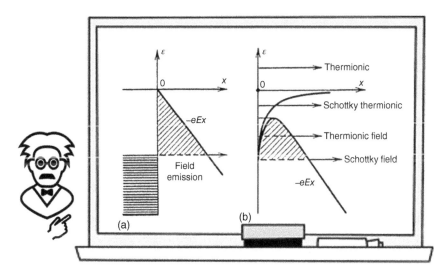

Figure 11.3 Electron potential energy curve on metal surface in the presence of electric field (a); illustration of thermionic and field emission mechanisms, and Schottky effect (b).

Table 11.4 Fowler–Nordheim correction factors $\xi(\Delta W/W_0)$ for the field emission.

Relative Schottky decrease of work function ($\Delta W/W_0$)	0	0.2	0.3	0.4	0.5
Fowler–Nordheim correction factor $\xi(\Delta W/W_0)$	1	0.95	0.90	0.85	0.78
Relative Schottky decrease of work function ($\Delta W/W_0$)	0.6	0.7	0.8	0.9	1
Fowler–Nordheim correction factor $\xi(\Delta W/W_0)$	0.70	0.60	0.50	0.34	0

Table 11.5 Coefficient γ for the potential electron emission induced by meta-stable atoms.

Meta-stable atom	Surface material	Secondary emission coefficient γ
He(2^3S)	Pt	0.24 electron atom^{-1}
He(2^1S)	Pt	0.4 electron atom^{-1}
Ar*	Cs	0.4 electron atom^{-1}

When cathode temperature and external electric field are high, both thermionic and field emission make significant contribution to the current of electrons escaping the metal. This emission mechanism is referred to as the **thermionic field emission** and plays an important role especially in the **cathode spots**. To compare the emission mechanisms, it is convenient to subdivide electrons escaping the metal surfaces into four groups, illustrated in Figure 11.3b. Electrons of the first group have energies below the Fermi level, so they can escape metal only through tunneling, or in other words by the field emission mechanism. Electrons of the fourth group leave metal by the thermionic emission mechanism without any support from the electric field. Electrons of the third group overcome the potential energy barrier because of its reduction in the external electric field. This Schottky effect of the electric field is classical. The second group of electrons can escape the metal only quantum-mechanically by tunneling similar to those from the first group. However, in this case, the potential barrier of tunneling is not so large because of the relatively high thermal energy. These electrons escape the cathode by the mechanism of the thermionic field emission. The thermionic field emission (see Table 11.3) dominates over other mechanisms at $T = 3000$ K and $E > 8 \cdot 10^6$ V cm^{-1}.

Electron emission from solids is possible even at low temperatures due to surface bombardment by different particles, which is referred to as the **secondary electron emissions** (see Sections 9.1 and 10.3). It is most important emission mechanism in glow discharges; however, it also contributes to arc and especially gliding arc discharges. **The secondary ion-electron emission** is induced by ion impact. The direct ionization in collisions of ions with neutral atoms is not effective because of the adiabatic principle (heavy ions are unable to transfer energy to light electrons). The secondary electron emission coefficient γ (electron yield per one ion) starts growing with ion energy only at energies exceeding 1 keV, when the Massey parameter becomes large. At lower ion energies, γ remains almost constant on the level of about 0.01. This can be explained by **the Penning mechanism of the secondary ion-electron emission**. According to the Penning mechanism, an ion approaching surface extracts an electron nonadiabatically because the ionization potential I exceeds the work function W. Empirically:

$$\gamma \approx 0.016(I - 2W).$$
(11.8)

Another secondary electron emission mechanism is related to the surface bombardment by excited meta-stable atoms with excitation energy exceeding the surface work function. This so-called **potential electron emission induced by meta-stable atoms** can have quite a high secondary emission coefficient γ, see Table 11.5.

Secondary electron emission also can be provided by a photo effect. This **photo-electron emission** is characterized by the **quantum yield** $\gamma_{\hbar\omega}$, which shows the number of emitted electrons per one quantum $\hbar\omega$ of radiation. Visible light and low energy UV-radiation give the quantum yield about $\gamma_{\hbar\omega} \approx 10^{-3}$, high energy UV-radiation results in the quantum yield 0.01–0.1. **Secondary electron–electron emission** usually does not significantly contribute to conventional DC-discharges but can be very important in heterogeneous discharges. It is characterized by the electron multiplication coefficient γ_e, which is usually about 2–3.

11.4 Cathode Layer of Arc Discharges: Physics and General Features

The cathode layer is supposed to provide the high current necessary for electric arc operation. Electron emission from the cathode in arcs is usually due to thermionic and field emission mechanisms. In the case of thermionic emission, ion bombardment provides cathode heating, which then leads to escape of electrons from the surface. The secondary emission usually gives about $\gamma \approx 0.01$ electrons per one ion, while thermionic emission can generate $\gamma_{eff} = 2 - 9$ electrons per one ion. The fraction of electron current near the cathode in glow discharge is small

$\frac{\gamma}{\gamma+1} \approx 0.01$. The fraction of electron current in the cathode layer of an arc is way higher:

$$S = \frac{\gamma_{eff}}{\gamma_{eff}+1} \approx 0.7 - 0.9. \tag{11.9}$$

The cathode's thermionic emission provides most of the arc current. From another hand, the current in positive column of both arc and glow discharges is almost completely due to electrons. In glow discharges, most of this current ($1 - \frac{\gamma}{\gamma+1} \approx 99\%$) is generated by electron-impact gas-phase ionization in the cathode layer. For this reason, the cathode layer voltage is high, on the level of hundreds of volts. In contrast, the electron-impact gas-phase ionization in the arc cathode layer should provide only a minor fraction of the total discharge current $1 - S \approx 10 - 30\%$; less than one generation of electrons should be born in the arc cathode layer. The necessary cathode voltage in this case is relatively low, about or even less than ionization potential. The arc cathode layer has several specific functions. First, sufficient number density of ions should be generated there to provide the necessary cathode heating for thermionic emission. Gas temperature near the cathode is the same as the cathode surface temperature and couple of times less than the temperature in positive column (see Figure 11.4). For this reason, thermal ionization alone is unable to provide the necessary degree of ionization. Therefore, additional nonthermal ionization mechanisms (direct electron impact, etc.) and hence, the elevated electric fields are required near the cathode (Figure 11.4). Also, the elevated electric field in the cathode vicinity stimulates electron emission by the Schottky effect and field emission. Intensive ionization in the cathode vicinity leads to high ion density, formation of a positive space charge, and elevates the electric field. The large positive space charge with high electric fields and most of cathode voltage drop is in the narrow layer near the cathode. This layer is even shorter than the ions and electrons mean free path, so it is usually referred to as **the collisionless zone of cathode layer.** Between the narrow collisionless layer and positive column, the longer **quasi-neutral zone of cathode layer** is located. While the electric field there is not so high, the ionization is quite intensive because electrons retain the high energy received in the collisionless layer. Most of ions carrying current and energy to the cathode are generated in this quasi-neutral zone of cathode layer. The electron and ion components of the total current are constant in the collisionless layer (Figure 11.5) where there are no sources of charge particles. In the following quasi-neutral zone of cathode layer, fraction of electron current grows from $S \approx 0.7 - 0.9$ to almost unity in the positive column (to ratio of mobilities $\mu_+/(\mu_e + \mu_+)$). Plasma density $n_e \approx n_+$ in the quasi-neutral zone of cathode layer steadily grows in the direction to the positive column because of intensive formation of electrons and ions.

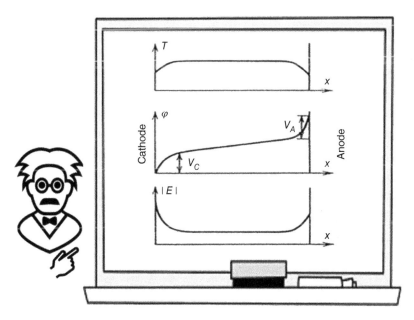

Figure 11.4 Arc temperature, potential, and electric field distribution from cathode to anode.

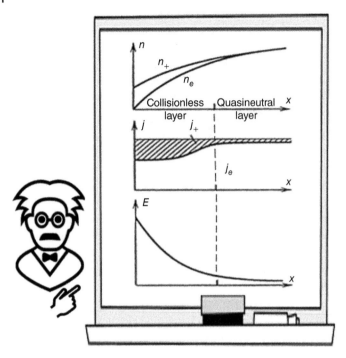

Figure 11.5 Arc cathode layer: typical distribution of charge density, current, and electric field.

To describe **electric field in the cathode vicinity**, let's consider the collisionless zone of the cathode layer. Electron j_e and ion j_+ components of the current density j are fixed in this zone:

$$j_e = S \cdot j = n_e e v_e, \quad j_+ = (1 - S)j = n_+ e v_+. \tag{11.10}$$

Electron and ion velocities v_e, v_+ are functions of voltage V:

$$v_e = \sqrt{2eV/m}, \quad v_+ = \sqrt{2e(V_C - V)/M}. \tag{11.11}$$

Voltage is zero $V = 0$ at the cathode and $V = V_C$ at the end of the collisionless layer (m and M are masses of electrons and positive ions). Then the Poisson's equation in the collisionless layer can be expressed as:

$$-\frac{d^2V}{dx^2} = \frac{e}{\varepsilon_0}(n_+ - n_e) = \frac{j}{\varepsilon_0\sqrt{2e}}\left[\frac{(1-S)\sqrt{M}}{\sqrt{V_C - V}} - \frac{S\sqrt{m}}{\sqrt{V}}\right]. \tag{11.12}$$

Considering $\frac{d^2V}{dx^2} = \frac{1}{2}\frac{dE^2}{dV}$, we can integrate the Poisson's equation assuming as a boundary condition that at the positive column side electric field is relatively low: $E \approx 0$ at $V = V_C$. This leads to the relation between the electric field near cathode, current density, and the cathode voltage drop (which can be represented by V_C):

$$E_C^2 = \frac{4j}{\varepsilon_0\sqrt{2e}}[(1 - S)\sqrt{M} - S\sqrt{m}]\sqrt{V_C}. \tag{11.13}$$

The first term in (11.13) is related to ion contribution to formation of positive space charge and enhancement of the electric field near the cathode. The second term is related to electron contribution to compensation of the ionic space charge. The fraction of electron current $S = 0.7 - 0.9$, therefore the second term can be neglected:

$$E_c, \text{V cm}^{-1} = 5 \cdot 10^3 \cdot A^{1/4}(1 - S)^{1/2}(V_C, V)^{1/4}(j, \text{A cm}^{-2})^{1/2}, \tag{11.14}$$

where A is the atomic mass of ions in a.m.u. For example, for an arc discharge in nitrogen ($A = 28$) at typical values of current density for hot cathodes $j = 3 \cdot 10^3 \text{A cm}^{-2}$, cathode voltage drop $V_C = 10 \text{ eV}$ and $S = 0.8$-gives the electric field near cathode: $E_c = 5.7 \cdot 10^5 \text{ V cm}^{-1}$. This electric field also provides reduction of the cathode work function about 0.27 eV, which permits the thermionic emission at 3000 K to triple. Integration of the Poisson's equation (11.13), neglecting the second term, permits finding length of the collisionless zone of the cathode layer as:

$$\Delta l = 4 V_C/3E_C. \tag{11.15}$$

Numerically, for this example it gives the length of the collisionless layer about $\Delta l \approx 2 \cdot 10^{-5}$cm.

11.5   Energy Balance of Cathode and Anode Layers: Electrode Erosion, Cathode Spots

The cathode layer characteristics are related to the S-factor depending on energy balance and showing the fraction (11.9) of electron current in the total one. We can assume that energy flux brought to the cathode by ions goes to electron emission. Each ion brings to the surface its kinetic energy (about the cathode voltage drop V_C) and also the energy released during neutralization (the difference $I - W$ between ionization potential and work function, necessary to provide an electron for the neutralization). Thus, simplified cathode energy balance is:

$$j_e \cdot W = j_+(V_C + I - W),\tag{11.16}$$

which leads to a relation for the S-factor, the fraction of electron current on cathode:

$$S = \frac{j_e}{j_e + j_+} = \frac{V_C + I - W}{V_C + I};\tag{11.17}$$

if $W = 4\,\text{eV}$, $I = 14\,\text{eV}$ and $V_C = 10\,\text{eV}$, the fraction of electron current is $S = 0.83$.

The high energy flux to the cathode results not only in thermionic emission but also in significant **erosion of the electrode material**. The erosion is very sensitive to presence of oxidizers: even 0.1% of oxygen or water vapor makes a significant effect. The process is usually characterized by the **specific erosion**, which shows the loss of electrode mass per unit charge passed through the arc. The specific erosion of tungsten rod cathodes at moderate and high pressures of inert gases and a current of about hundred amperes is about $10^{-7}\,\text{g}\,\text{C}^{-1}$. The most intensive erosion takes place in the **hot cathode spots**. At low pressures of about 1 Torr and less, the cathode spots are formed even on refractory materials.

 The cathode spots are localized current centers, which can appear on the cathode surface when current is very high, but the entire cathode cannot be heated to the required temperature. The most typical cause of the cathode spots is application of metals with relatively low melting point. Initially, the cathode spots are formed small ($10^{-4} - 10^{-2}\text{cm}$) and move very fast ($10^3 - 10^4\text{cm s}^{-1}$). These primary spots are nonthermal (relevant erosion is not significant) and related to micro-explosions due to localization of current on tiny protrusions on the cathode surface. After about 10^{-4}s, the small primary spots merge into larger spots ($10^{-3} - 10^{-2}\text{cm}$) with temperatures exceeding 3000 K and intensive thermal erosion. These spots also move slower $10 - 100\,\text{cm s}^{-1}$. Typical current through an individual spot is 1–300 A. Growth of current leads to splitting of the cathode spots and their multiplication. The minimum current through a single spot is about $I_{\min} \approx 0.1 - 1$ A, and can be estimated as:

$$I_{\min}, A \approx 2.5 \cdot 10^{-4} \cdot T_{boil}(K) \cdot \sqrt{\lambda, \text{W/cmK}};\tag{11.18}$$

here T_{boil}(in K) is electrode boiling temperature, and λ (in W/cmK) is heat conduction coefficient.

The cathode spots produce intensive jets of metal vapor. Emission of 10 electrons corresponds to an erosion of about one atom. The jet velocities can be very high: $10^5 - 10^6\text{cm s}^{-1}$, see Table 11.6. The current densities in cathode spots are also very high, 10^8A cm^{-2}, which can be explained by thermionic field emission and the **explosive electron emission**, related to localization of strong electric fields and following explosion of micro-protrusions on the cathode surface. An intriguing cathode spot paradox is related to the direction of its motion in external magnetic field. If the external magnetic field is applied along a cathode surface, the cathode spots move in the direction opposite to that corresponding to the magnetic force $\vec{I} \times \vec{H}$.

In the case of **externally heated cathodes**, heating by ion current is not necessary. The main function of a cathode layer then is acceleration of thermal electrons to energies sufficient for ionization and sustaining the necessary level of plasma density. Losses of charged particles in such thermal plasma systems cannot be significant. It results in low values of cathode voltage drop, which is often lower than ionization potential. For low pressure (about 1 Torr) inert gases, voltage is also low, about 1 V. The arc discharges with low total voltage of about 7–8 V are usually referred to as **the low-voltage arcs**. For example, such low values of voltage are sufficient for the non-self-sustained arc discharge in spherical chamber with a radius 5 cm in argon at 1–3 Torr pressure and currents 1–2 A. Such low-pressure gas discharges are used in diodes and thyratrons.

Arc can be connected to an anode also in two ways: by the diffuse connection or by the anode spots. The diffuse connection in **anode layer** occurs on large area anodes; with current densities about 100 A cm^{-2}. The **anode spots** appear on small and nonhomogeneous anodes, current density there is about $10^4 - 10^5$ A cm^{-2}. The number of the

Table 11.6 Conventional characteristics of the cathode spots.

Cathode material	Cu	Hg	Fe	W	Ag	Zn
Minimum current through a spot (A)	1.6	0.07	1.5	1.6	1.2	0.3
Average current through a spot (A)	100	1	80	200	80	10
Current density (A cm^{-2})	$10^4 \div 10^8$	$10^4 \div 10^6$	10^7	$10^4 \div 10^6$	—	$3 \cdot 10^4$
Cathode voltage drop (V)	18	9	18	20	14	10
Specific erosion at 100–200 A (g/C)	10^{-4}	—	—	10^{-4}	10^{-4}	—
Vapor jet velocity (10^5cm s^{-1})	1.5	1	0.9	3	0.9	0.4

spots grows with total current and pressure. Sometimes the anode spots are arranged in regular patterns and move. The **anode voltage drop** consists of two components. First one is related to negative space charge near the anode, which obviously repels ions. This small voltage drop (about ionization potential and less) stimulates some additional electron generation to compensate the absence of ion current in the region. The second component of the anode voltage drop is related to the arc discharge geometry. If the anode surface area is smaller than the positive column cross-section, electric current near the electrode should be provided only by electrons. This requires higher values of electric fields near the electrode and additional anode voltage drop, sometimes exceeding the space charge voltage by a factor of two. Each electron brings to the anode an energy of about 10 eV. The energy flux in an anode spot at current densities $10^4 - 10^5$ A cm^{-2} is about $10^5 - 10^6$ W cm^{-2}. Temperatures in anode spots of vacuum metal arcs are about 3000 K and in carbon arcs about 4000 K.

11.6 Positive Column of Arc Discharges: Elenbaas–Heller Equation, Steenbeck and Raizer "Channel" Models

The Joule heat released per unit length of positive column in the high-pressure arcs is quite significant, usually $0.2 - 0.5$ kW cm^{-1}. This heat release can be balanced in three different ways. If the Joule heat is balanced by heat transfer to cooled walls, it is referred to as **wall stabilized arc**. If the Joule heat is balanced by intensive (often rotating) gas flow, it is referred to as **flow stabilized arc**. Finally, if heat transfer to electrodes balances Joule heat in the short positive column, it is referred to as **electrode stabilized arc**. The arc discharge plasma of molecular gases at high pressures ($p \geq 1$ atm) is in quasi-equilibrium at any currents. In the case of inert gases, the electron-neutral energy exchange is less effective and requires high currents and electron densities to reach quasi-equilibrium at atmospheric pressure (see Figure 11.6a). The temperatures of electrons and neutrals can differ considerably in arc discharges at low pressures ($p \leq 0.1$ Torr) and currents ($I \approx 1$A). Typical **current–voltage characteristics** of arc discharges are illustrated in Figure 11.6b for different pressures. They are hyperbolic, which indicates that Joule heat per unit length $w = EI$ does not significantly change with current I. The Joule heating $w = EI$ grows with pressure due to intensification of heat transfer mostly by radiation of the high-density plasma. The contribution of radiation increases proportionally to the square of the plasma density, and therefore grows with pressure. The arc radiation losses in atmospheric air are only about 1% but become quite significant at pressures exceeding 10 atm and high powers. The highest level of radiation can be reached in Hg, Xe, and Kr, which is applied in mercury and xenon lamps. Empirical formulas for calculating plasma radiation are presented in Table 11.7.

The Elenbaas–Heller equation describes energy balance of an arc positive column considering a long cylindrical steady-state plasma column stabilized by walls in a tube of radius R. In the framework of the Elenbaas–Heller approach, – pressure and current are supposed to be not too high and plasma temperature does not exceed 1 eV. Radiation can be neglected in this case and heat transfer across the positive column can be reduced just to heat conduction with the coefficient $\lambda(T)$. According to Maxwell equation $curl\, E = 0$. For this reason, the electric field in a long homogeneous arc column is constant across its cross-section. Radial distributions of electric conductivity $\sigma(T)$, current density $j = \sigma(T)E$, and Joule heating density $w = jE = \sigma(T)E^2$ are determined only by the radial temperature distribution $T(r)$. Then plasma energy balance can be expressed by the heat conduction equation with the Joule heat source:

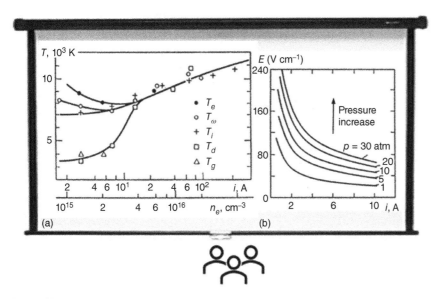

Figure 11.6 Arc positive column: (a) temperature separation in Ar at $p = 1$ atm as a function of current and electron density (T_e is electron temperature, T_w corresponds to the population of the upper electronic levels, the temperature T_i is related to n_e by the Saha formula, T_g is gas temperature, and T_d corresponds to the population of the lower electronic levels); (b) current–voltage characteristics in air at various pressures.

Table 11.7 Radiation power per unit length of positive column of arc discharges at different pressures and different values of Joule heating per unit length $w = EI$, W cm^{-1}.

Gas	Pressure (atm)	Radiation power per unit length (W cm^{-1})	$w = EI$, W cm^{-1}
Hg	≥ 1	$0.72 \cdot (w - 10)$	—
Xe	12	$0.88 \cdot (w - 24)$	>35
Kr	12	$0.72 \cdot (w - 42)$	>70
Ar	1	$0.52 \cdot (w - 95)$	>150

$$\frac{1}{r}\frac{d}{dr}\left[r\lambda(T)\frac{dT}{dr}\right] + \sigma(T)E^2 = 0. \tag{11.19}$$

This is known as the **Elenbaas–Heller equation**. Boundary conditions here are: $dT/dr = 0$ at $r = 0$, and $T = T_w$ at $r = R$ (the wall temperature T_w can be considered as zero). The electric field E is a parameter in the Elenbaas–Heller equation. Experimentally controlled parameter is however not electric field but current:

$$I = E\int_0^R \sigma[T(r)] \cdot 2\pi r\, dr. \tag{11.20}$$

The Elenbaas–Heller equation permits calculating $E(I)$, which is the current-voltage characteristic of plasma column. Electric and thermal conductivity $\sigma(T)$, $\lambda(T)$ are two material functions. To reduce number of material functions to one, we can introduce instead of temperature **the heat flux potential** $\Theta(T)$ as a parameter:

$$\Theta = \int_0^T \lambda(T)\, dT, \quad \lambda(T)\frac{dT}{dr} = \frac{d}{dr}\Theta. \tag{11.21}$$

Using the heat flux potential, the Elenbaas–Heller equation can be rewritten as:

$$\frac{1}{r}\frac{d}{dr}\left[r\frac{d\Theta}{dr}\right] + \sigma(\Theta)E^2 = 0. \tag{11.22}$$

This equation includes a single material function $\sigma(\Theta)$. The temperature dependence $\Theta(T)$ determined by (11.21) is smoother than $\lambda(T)$, which makes the material function $\mu(\Theta)$ also smooth. the Elenbaas–Heller equation cannot be solved analytically because of the complicated nonlinearity of the material function $\sigma(\Theta)$.

Qualitative and quantitative description of the positive column can be achieved by applying the Elenbaas–Heller equation in analytical **"channel" model** proposed by Steenbeck and illustrated in Figure 11.7. The model is based on strong exponential dependence of electric conductivity on the plasma temperature due to Saha equation. At relatively low temperatures (less than 3000 K), the quasi-equilibrium plasma conductivity is small. The electric conductivity grows significantly when temperature exceeds 4000–6000 K. The radial temperature decreases $T(r)$ is quite gradual (see Figure 11.7), while the electric conductivity change with radius $\sigma[T(r)]$ is very sharp. Therefore, arc current is located only in a "channel" of radius r_0, which is the principal physical basis of the Steenbeck model. Temperature and electric conductivity can be considered as constant inside of the arc channel and taken equal to their maximum value on the discharge axis: T_m and $\sigma(T_m)$. In this case, the total electric current of the arc can be expressed as:

$$I = E\sigma(T_m) \cdot \pi r_0^2. \tag{11.23}$$

Outside of the arc channel $r > r_0$ the electric conductivity can be neglected as well as the current and the Joule heat release. The Elenbaas–Heller equation can be integrated outside of the arc channel with boundary conditions: $T = T_m$ at $r = r_0$, and $T = 0$ by the walls at $r = R$. It leads to the relation between the heat flux potential $\Theta_m(T_m)$ in the arc channel and the discharge power (the Joule heating) per unit arc length $w = EI$:

$$\Theta_m(T_m) = \frac{w}{2\pi} \ln \frac{R}{r_0}. \tag{11.24}$$

The heat flux potential $\Theta_m(T_m)$ in the arc channel and power per unit arc length $w = EI$ are defined as:

$$w = \frac{I^2}{\pi r_0^2 \sigma_m(T_m)}, \quad \Theta_m(T_m) = \int_0^{T_m} \lambda(T)\, dT. \tag{11.25}$$

The channel model controls three discharge parameters: plasma temperature T_m, arc channel radius r_0, and electric field E. Electric current I and discharge tube radius R are experimentally controlled. To find the three unknown discharge parameters: T_m, r_0, and E, the Steenbeck channel model has only two equations ((11.23) and (11.24)). Therefore, Steenbeck suggested the **principle of minimum power** (later proved for arc discharges) to provide a lacking third equation to complete the system. According to the Steenbeck principle of minimum power, the plasma temperature T_m and the arc channel radius r_0 should minimize the specific discharge power w and electric field $E = w/I$ at fixed values of current I and discharge tube radius R. Application of the minimization requirement $\left(\frac{dw}{dr_0}\right)_{I=const} = 0$ gives the third equation of the **Steenbeck channel model**:

$$\left(\frac{d\sigma}{dT}\right)_{T=T_m} = \frac{4\pi \lambda_m(T_m)\, \sigma_m(T_m)}{w}. \tag{11.26}$$

The **Raizer channel model** does not require the minimum power principle to justify the system equations ((11.23), (11.24), and (11.26)). The "third" Eq. (11.26) is derived by analysis of the conduction heat flux J_0 from the channel: $w = J_0 \cdot 2\pi r_0$, which is provided by the temperature difference $\Delta T = T_m - T_0$ across the channel (see Figure 11.7):

$$J_0 \approx \lambda_m(T_m) \cdot \frac{\Delta T}{r_0} = \lambda_m(T_m) \cdot \frac{T_m - T_0}{r_0}. \tag{11.27}$$

This equation can be replaced by more accurate one, integrating the Elenbaas–Heller Eq. (8.23) inside of arc channel $0 < r < r_0$ and still assuming homogeneity of the Joule heating σE^2:

$$4\pi \Delta\Theta = w \approx 4\pi \lambda_m \Delta T, \quad \Delta\Theta = \Theta_m - \Theta_0. \tag{11.28}$$

The key of the Raizer modification of the channel model is definition of an arc channel as a region where electric conductivity decreases not more than "e" times with respect to the maximum value at the axis of the discharge. This definition permits specifying the arc channel radius r_0 and gives the "third" Eq. (11.26) of the channel model. The arc channel electric conductivity can be expressed from the Saha equation as the function of temperature:

$$\sigma(T) = C \exp\left(-\frac{I_i}{2T}\right); \tag{11.29}$$

here I_i ionization potential, and C is the constant conductivity parameter. The electric conductivity in air, nitrogen, and argon at atmospheric pressure and $T = 8000 \div 14\,000$ K can be expressed by the same numerical formula:

$$\sigma(T), \text{Ohm}^{-1}\text{cm}^{-1} = 83 \cdot \exp\left(-\frac{36\,000}{T, K}\right). \tag{11.30}$$

which corresponds to the effective ionization potential $I_{eff} \approx 6.2$ eV. Assuming $I/2T \gg 1$, the "e" times decrease of the conductivity corresponds to the following small temperature decrease:

$$\Delta T = T_m - T_0 = \frac{2T_m^2}{I_i}. \tag{11.31}$$

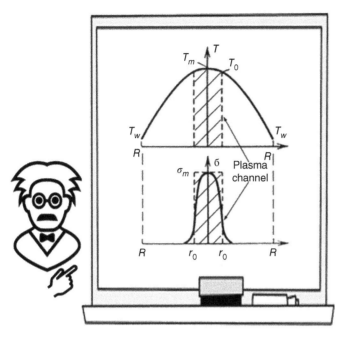

Figure 11.7 Steenbeck "channel" model of an arc positive column.

Combination of (11.28) and (11.32) gives the "third" equation of the arc channel model same as (11.26):

$$w = 8\pi\lambda_m(T_m)\frac{T_m^2}{I_i}.$$ (11.32)

11.7 Plasma Temperature, Specific Power, Electric Field, and Radius of the Arc Positive Column

Equation (11.32) of the Raizer channel model determines plasma temperature as a function of power w per unit length (I_i is ionization potential, λ_m is thermal conductivity coefficient):

$$T_m = \sqrt{w \cdot \frac{I_i}{8\pi\lambda_m}}.$$ (11.33)

Plasma temperature does not depend directly on discharge radius and mechanisms of the discharge cooling outside of the arc channel. It depends only on the specific power w, which in turn depends on intensity of the arc cooling. Dependence of plasma temperature T_m on the specific power is not very strong, less than $\sqrt{w}$ because the heat conductivity $\lambda(T)$ grows with T. The principal control parameter of an arc discharge is electric current. Assuming: $\lambda = const$, $\Theta = \lambda T$, the conductivity in the arc channel is almost proportional to electric current:

$$\sigma_m = I \cdot \sqrt{\frac{I_i C}{8\pi^2 R^2 \lambda_m T_m^2}}.$$ (11.34)

Plasma temperature in the arc grows with current I, but very slow, as logarithm:

$$T_m = \frac{I_i}{\ln\left(8\pi^2\lambda_m C T_m^2/I_i\right) - 2\ln(I/R)}.$$ (11.35)

The almost constancy of arc temperature reflects strong dependence of the degree of ionization on the temperature. *The weak plasma temperature growth with current leads to weak dependence on the electric current of the power w per unit length*:

$$w \approx \frac{const}{(const - \ln I)^2}.$$ (11.36)

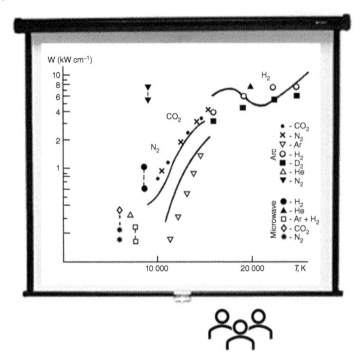

Figure 11.8 Discharge power per unit length as a function of maximum plasma temperature in thermal arcs and microwave discharges (solid lines represent calculations).

Figure 11.8 illustrates this relatively weak dependence of the arc discharge power w per unit length on arc conditions. The logarithmic constancy of the arc power $w = EI$ per unit length results in the hyperbolic decrease of electric field with current I:

$$E = \frac{8\pi \lambda_m T_m^2}{I_i} \cdot \frac{1}{I} \approx \frac{const}{I \cdot (const - \ln I)^2}. \tag{11.37}$$

The arc radius may be found in the framework of the channel model as:

$$r_0 = R\sqrt{\frac{\sigma_m}{C}} = R\sqrt{\frac{I}{R}} \sqrt[4]{\frac{I_i}{8\pi^2 \lambda_m T_m^2 C}}. \tag{11.38}$$

The arc radius grows as the square root of the discharge current I, and $I \propto r_0^2$, which means that the increase of current mostly leads to growth of the arc cross-section, while the current density is logarithmically fixed in the same manner as plasma temperature. All the relations include current in combination with the discharge tube radius R (the similarity parameter I/R). This is due to the assumption of "wall stabilization of arc" in the Elenbaas–Heller equation.

 When arc is stabilized in fast gas flow (velocity v), the cold outside gas low viscosity $v(T)$ does not penetrate much into the hot channel with high viscosity $v(T)$. *Because of very high viscosity, the arc channel behaves like "jam or jelly;" the heat transfer there is dominated by conduction.* This conductive heat transfer should be balanced by mostly convective cooling outside of the channel, which assumes the Nusselt number Nu to be close to one. Because $Nu \sim \sqrt{Re} \sim \sqrt{v r_0 / v}$, the intensive cooling in fast gas flows hyperbolically decreases the effective arc radius and increases plasma temperature in accordance with (11.35).

At low pressures and currents (higher electric fields E), the electron temperature T_e can exceed gas temperature T (see Figure 11.6a). It can be described by the electron energy balance:

$$\frac{3}{2}\frac{dT_e}{dt} = \left[\frac{e^2 E^2}{m v_{en}^2} - \delta(T_e - T)\right] \cdot v_{en}; \tag{11.39}$$

here v_{en} is frequency of electron-neutral collisions; factor δ characterizes the fraction of electron energy transferred to neutrals during collisions ($\delta = 2m/M$ in monatomic gases, and higher in molecular gases). The electrons receive energy from the electric field and transfer it to neutrals if $T_e > T$. If the difference $T_e - T$ is small, the excited neutrals transfer energy back to electrons in super-elastic collisions. In the steady state conditions:

$$\frac{T_e - T}{T} = \frac{2e^2 E^2}{3\delta T_e m v_{en}^2}. \tag{11.40}$$

Assuming monatomic gas ($\delta = 2m/M$), and replacing the electron-neutral frequency v_{en} by the electron mean free path $\lambda = v_e/v_{en}$ (v is the average electron velocity, A is atomic mass):

$$\frac{T_e - T}{T_e} \approx 200 \, A \cdot \left[\frac{E(\text{V cm}^{-1}) \cdot \lambda(\text{cm})}{T_e(\text{eV})} \right]^2. \tag{11.41}$$

11.8 Arc Dynamics: Bennet Pinch, Electrode Jets

High arc currents induce magnetic fields providing the arc compression. The body forces acting on axisymmetric arcs are illustrated in Figure 11.9. If the current density j in the arc channel is constant, the azimuthal magnetic field B_θ inside of arc is:

$$B_\theta(r) = \frac{1}{2}\mu_0 j r, \quad r \leq r_0. \tag{11.42}$$

It leads to radial body force (per unit volume) directed inward and tending to pinch the arc:

$$\vec{F} = \vec{j} \times \vec{B}, \quad F_r(r) = -\frac{1}{2}\mu_0 j^2 r. \tag{11.43}$$

A cylindrical arc channel, where plasma pressure p is balanced by the inward radial magnetic body force is called **the Bennet pinch**. This balance can be expressed as:

$$\nabla p = \vec{j} \times \vec{B}, \quad \frac{dp}{dr} = -jB_\theta. \tag{11.44}$$

Together with the formula (11.42) for azimuthal magnetic field B_θ, this leads to the relation for pressure distribution inside of the Bennet pinch of radius r_0:

$$p(r) = \frac{1}{4}\mu_0 j^2 \left(r_0^2 - r^2 \right). \tag{11.45}$$

The maximum pressure (on axis of the arc) depends on the total current $I = j \cdot \pi r_0^2$:

$$p_a = \frac{1}{4}\mu_0 j^2 r_0^2 = \frac{\mu_0 I^2}{4\pi^2 r_0^2}. \tag{11.46}$$

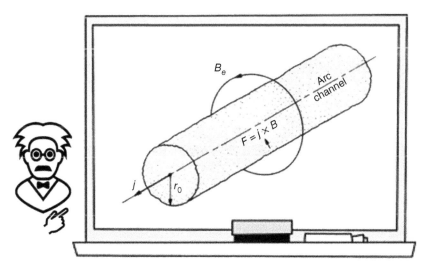

Figure 11.9 Radial body forces on cylindrical arc channel.

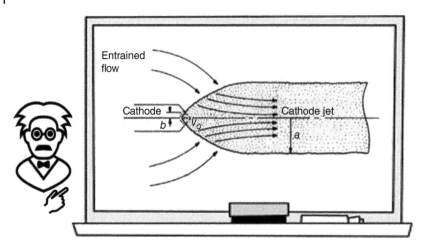

Figure 11.10 Illustration of the electrode jet formation.

The high arc current (about 10 000 A) is required, almost to provide an axial pressure of 1 atm in a channel with radius $r_0 = 0.5$ cm. Although the Bennet pinch pressure is usually small with respect to total pressure, it can initiate **electrode jets**, intensive gas streams flowing away from electrodes, see Figure 11.10. Additional gas pressure (11.46) is inversely proportional to the square of the arc radius. Also, the radius of arc channel attachment to electrode ($r_0 = b$) is less than that one corresponding to the positive column ($r_0 = a$). This results in development of an axial pressure gradient, which drives neutral gas along the arc axis away from electrodes:

$$\Delta p = p_b - p_a = \frac{\mu_0 I^2}{4\pi^2} \left(\frac{1}{b^2} - \frac{1}{a^2} \right). \tag{11.47}$$

Assuming $b \ll a$, the jet dynamic pressure and the jet velocity can be found from:

$$\Delta p \approx \frac{\mu_0 I^2}{4\pi^2 b^2} = \frac{1}{2} \rho v_{jet}^2; \tag{11.48}$$

here ρ is plasma density. The jet velocity is then a function of current and radius of arc attachment:

$$v_{jet} = \frac{I}{\pi b} \sqrt{\frac{\mu_0}{2\rho}}. \tag{11.49}$$

Typically, the electrode jet velocities are $v_{jet} = 3 - 300 \, \text{m s}^{-1}$.

11.9 Engineering Configurations of Arc Discharges

A. Free-Burning Linear Arcs are axisymmetric and can horizontal or vertical (see Figure 11.11 a,b). Buoyancy of hot gases in horizontal configuration leads to bowing up or "arcing" of the channel, which explains the name "arc." If the arc is vertical, cathode is placed at the top, and buoyancy provides better heating of cathode and more effective thermionic emission. The length of the free-burning linear arcs typically exceeds their diameter. Sometimes however, the arcs are shorter than their diameter (see Figure 11c) and referred to as the **obstructed arcs**. Distance between electrodes in such discharges is typically about 1 mm, voltage between electrodes nevertheless exceeds anode and cathode drops. The obstructed arcs are usually electrode-stabilized.

B. Wall-Stabilized Linear Arcs are widely used for gas heating. A simple configuration of the **wall-stabilized arc with a unitary anode** is illustrated in Figure 11.11d. The cathode here is axial and the unitary anode is hollow and coaxial. The arc channel is axisymmetric and stable with respect to asymmetric perturbations. If the channel approaches the coaxial anode, cooling intensifies, and temperature increase on the axis. It results in displacement of the channel back to the axis, which stabilizes the linear arc. The anode can be water-cooled. A better-defined discharge arrangement can be achieved in the **segmented wall-stabilized arc** configuration, see Figure 11.11e.

C. Linear Transferred Arcs with water-cooled cathodes are illustrated in Figure 11.11f. Electrons emission from inner walls of the hollow cathodes is due to field emission and permits megawatt power operation during thousands of hours. The circuit is completed by transferring the arc to an external anode, which is a conducting material where the arc is supposed to be applied.

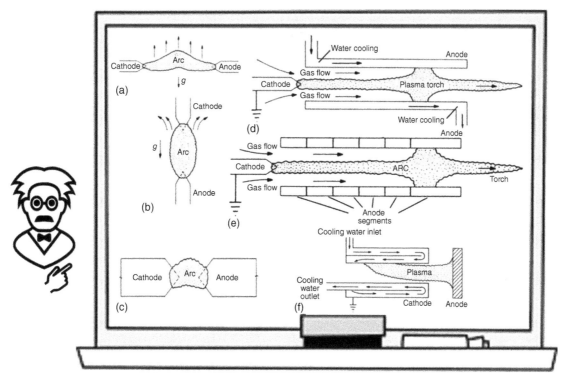

Figure 11.11 Configurations of linear arcs: (a) horizontal free-burning arc; (b) vertical free-burning arc; (c) obstructed electrode-stabilized arc; (d) wall-stabilized arc; (e) wall-stabilized arc with segmented anode; (f) nontransferred magnetic arc.

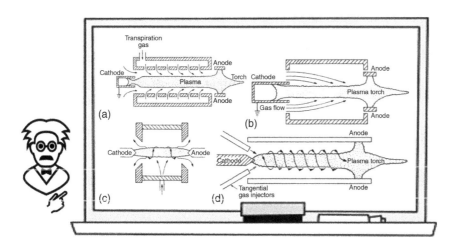

Figure 11.12 Flow stabilized arcs: (a) transpiration stabilized arc; (b) coaxial flow stabilized arc; (c, d) vortex flow stabilized arcs.

D. Flow-Stabilized Linear Arcs can be stabilized on the axis by radially inward injection of cooling water or gas, see Figure 11.12a, which is referred to as the **transpiration-stabilized arcs**. High power plasma can be also generated in **coaxial flow stabilized arc**, see Figure 11.12b. In this case, the anode is located far from the plasma channel and cannot provide the wall stabilization. Instead of walls, the arc channel is stabilized by a coaxial gas flow moving along the outer surface of the arc. Similarly, arc stabilization can be achieved by a flow fast rotating around the arc. Different configurations of **the vortex-stabilized arcs** are shown in Figure 11.12c,d. The vortex gas flow cools the edges of the arc and maintains the arc column confined to the axis of the discharge chamber. The vortex flow is very effective in promoting the heat flux from the thermal arc column to the walls of the discharge chamber.

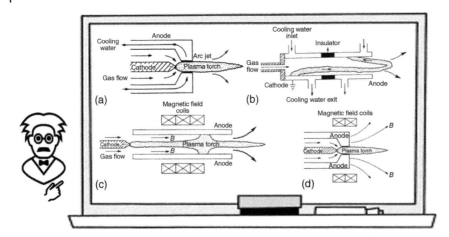

Figure 11.13 Plasma torch (a), magnetic nontransferred arc (b), magnetically stabilized arcs (c, d).

E. Plasma Torches or Thermal Plasma Jets are axisymmetric nontransferred arcs illustrated in Figure 11.13a. The arcs are generated in this case in a conical gap in the anode and is pushed out of this opening by the fast gas flow. The heated gas flow forms a very high temperature plasma jet sometimes at supersonic velocities.

F. Magnetically Stabilized Non-Transferred Arcs can be organized in different configurations. A **nonlinear non-transferred wall-stabilized arc** is shown in Figure 11.13b. This discharge supported by field emission consists of a cylindrical hollow cathode and coaxial hollow anode located in a water-cooled chamber and separated by an insulator. Gas flow blows the arc out of the anode opening. Magnetic $\vec{I} \times \vec{B}$ forces cause the arc roots to rotate around electrodes to protect them. **Magnetically stabilized rotating arcs** are illustrated in Figure 11.13c,d. The external axial magnetic field provides $\vec{I} \times \vec{B}$ forces and fast rotation of the arc protecting the anode from overheating. Figure 11.13c shows the case of additional magnetic stabilization of a wall-stabilized arc. Figure 11.13d presents a very practical configuration of the magnetically stabilized torches.

11.10 Gliding Arcs: Physics of the Flat Discharge Configuration

The flat gliding arcs are an auto-oscillating periodic discharges between two diverging electrodes in gas flow, see Figure 11.14a,b. Self-initiated in the upstream narrowest gap, the discharge forms a plasma column, which is dragged by gas flow towards the diverging downstream section. The discharge length grows with the increase of inter-electrode distance until it reaches a critical value, usually determined by the power supply limits. Then the discharge extinguishes but momentarily reignites at the minimum distance between the electrodes to start a new cycle, see series of snapshots in Figure 11.14c.

 The gliding arc has thermal or nonthermal properties depending on power input and flow rate. Along with completely thermal and completely nonthermal regimes, it is possible to organize *the transitional discharges starting with thermal plasma, but during the space and time evolution becoming the nonthermal one.* The powerful and energy-efficient transitional discharge combines the benefits of both equilibrium and nonequilibrium discharges providing plasma conditions typical for nonequilibrium cold plasmas but at elevated power levels. Gliding arcs can operate at atmospheric pressure or higher. In form of the **Jacob's ladder**, they are often used in science exhibits.

Let's consider a simple DC flat gliding arc in air with a conventional circuit illustrated in Figure 11.15a. One high-voltage generator (up to 5000 V) is used to ignite the discharge and the second one is a power supply (with the voltage up to 1 kV, and a total current I up to 60 A). A variable resistor $R = 0\text{--}25\,\Omega$ is in series with a self-inductance $L = 25\,\text{mH}$.

Initial breakdown of the gas begins the cycle at the shortest distance (1–2 mm) between electrodes (see Figure 11.15b). For atmospheric air and a distance between electrodes of about 1 mm, the breakdown voltage V_b is about 3 kV. The characteristic time of the arc formation τ_i can be found from the kinetic equation: $\frac{dn_e}{dt} = k_i n_e n_0 = \frac{n_e}{\tau_i}$, where k_i is the ionization rate coefficient; n_e and n_0 are electron and gas densities. At current $J = 1$, $\tau_i \approx 1\,\mu s$. Within the breakdown time about 1 µs, low resistance plasma is formed and the voltage between electrodes falls.

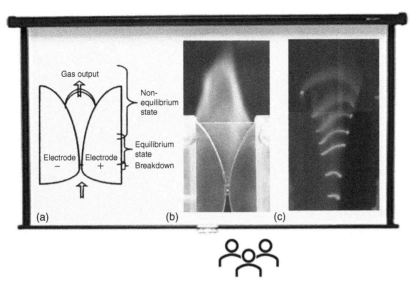

Figure 11.14 Flat gliding arc discharges: (a) schematic; (b) photo; (c) space-time evolution snapshots with clear recognizable transition from equilibrium to nonequilibrium phases.

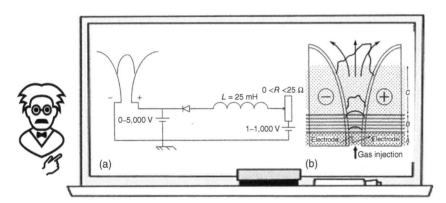

Figure 11.15 Flat gliding arc discharges: (a) typical DC electric circuit; (b) evolution phases: breakdown A, equilibrium B, and nonequilibrium C stages.

The equilibrium phase (see Figure 11.15b) starts after formation of a plasma channel. The gas flow pushes the plasma column with the velocity about $10\,\mathrm{m\,s^{-1}}$, and the arc length l increases together with the voltage. The current increases during formation of the quasi-equilibrium channel up to its maximum $I_m = V_0/R \approx 40\,\mathrm{A}$. The current-time dependence is sensitive to the inductance L: $I(t) = (V_0/R)\left(1 - e^{-t/\tau_L}\right)$; here $\tau_L = L/R \approx 1\,\mathrm{ms}$. The gas temperature T_0 does not change much $7000 \leq T_0 \leq 10\,000\,\mathrm{K}$ for our numerical example. At lower electric current and lower power, the temperature is lower: at 200 W and 0.1 A, it can be about 2500 K. The length of the arc l and the discharge power P increase up to the maximum permitted by power supply.

The nonequilibrium phase (see Figure 11.15b) begins when arc length exceeds its critical value l_{crit}. Heat losses from the thermal plasma column begin to exceed the energy supplied by the source, and plasma cools rapidly. The plasma conductivity is maintained by high electron temperature $T_e = 1\,\mathrm{eV}$ and stepwise ionization. After decay of the nonequilibrium discharge, a new breakdown takes place at the shortest distance between electrodes and the cycle repeats.

The **gliding arc's equilibrium phase evolution** can be described considering that the specific power w in the temperature range of 6000–12 000 K does not change significantly and remains about 50–70 kW m^{-1}. Neglecting the self-inductance L, the Ohm's law can be then expressed as:

$$V_0 = R\,I + w \cdot l/I;$$

(11.50)

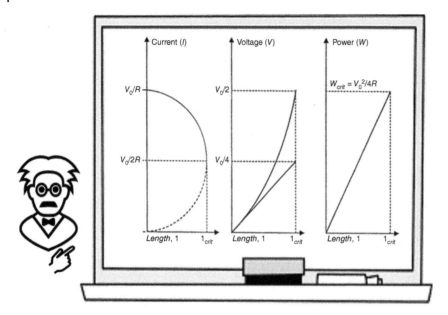

Figure 11.16 Evolution of current, voltage, and power in a flat gliding arc discharge.

here V_0, R, and I are the power supply voltage, the external resistance, and current. The arc current can be then presented as a function of the arc length l, which is growing during the arc evolution:

$$I = \left(V_0 \pm \left(V_0^2 - 4\,w\,l\,R \right)^{1/2} \right) /2R. \tag{11.51}$$

The solution with the sign "+" describes the steady state of the arc; and the solution with $I < V_0/2R$ corresponds to negative differential resistance ρ ($\rho = dV/dJ$) of and to an unstable arc regime:

$$\rho = R - wl/I^2 = 2R - V_0/I < 0. \tag{11.52}$$

The current slightly decreases during the quasi-equilibrium phase, while the arc voltage is growing as $W\,l\,/\,I$, and the total arc power $P = W\,l$ increases almost linearly with the length l. The absolute value of the differential resistance of the gliding arc grows. Arc discharges are generally stable and have descending current–voltage characteristics. It means that the differential resistance (dU/dI) of an arc as a part of the electric circuit is negative. To provide circuit stability, the total differential resistance should be positive. When the power dissipated in arc achieves its maximum, the differential resistance of arc becomes equal to the differential resistance of external part of the circuit resulting in instability. The power supply cannot provide the increasing arc power, and the system transfers to a nonequilibrium state. The current I, voltage V, and power P in the equilibrium phase are shown in Figure 11.16.

The quasi-equilibrium evolution is terminated at the **gliding arc critical parameters**. At the critical length:

$$l_{crit} = V_0^2/(4wR) \tag{11.53}$$

the square root in (11.51) and the differential resistance of the whole circuit becomes equal to zero. The current falls to its minimum $I_{crit} = V_0\,/2R$, which is a half of initial value. Plasma voltage, electric field E, and total power at the same critical point approach their maximum:

$$V_{crit} = V_0/2, E_{crit} = w/J_{crit}, W_{crit} = V_0^2/4R. \tag{11.54}$$

At the critical point, the growing plasma resistance becomes equal to the external one, and the maximum discharge power corresponds to half of the maximum power of generator. For the above numerical example, the critical parameters in air are: $I_{crit} = 20\,\text{A}$, $V_{crit} = V_0/2 = 400\,\text{V}$, $W_{crit} = 8\,\text{kW}$, $l_{crit} = 10\,\text{cm}$. The above results can be generalized for the AC and pulsed power supplies. In this case, the critical conditions are achieved when the growing plasma resistance starts matching impedance of the power supply. The maximum discharge power then can be much higher than half of the maximum generator power of generator, because of lower heat dissipation related to impedance.

11.11 Nonequilibrium Gliding Arcs, Fast Equilibrium-to-Nonequilibrium Transition

The "Fast Equilibrium-to-Nonequilibrium Transition" (the FENETRe phenomenon) in gliding arc is a kind of "discharge window" from the thermal plasma zone to a relatively cold one (Fridman et al., 1999). The arc instability is due to increase of the electric field $E = w/I$ and electron temperature T_e during the evolution:

$$T_e = T_0 \left(1 + E^2/E_i^2 \right); \tag{11.55}$$

here the electric field E_i corresponds to transition from thermal to direct electron impact ionization. The thermal arcs are usually stable. Indeed, a small temperature increase leads to growth of electron density and electric conductivity σ. Taking into account the fixed current density $j = \sigma E$, the conductivity increase results in a reduction of the electric field E and the heating power σE^2, and hence in stabilizing the initial temperature perturbations. In contrast to that, the direct electron impact ionization (typical for nonequilibrium plasma) is usually not stable. A small temperature increase leads, for a fixed pressure (Pascal law) and constant electric field, to a reduction of the gas density n_0 and to increase of the specific electric field E/n_0 and the ionization rate. This results in an increase of the electron concentration n_e, the conductivity σ, the Joule heating power σE^2 and hence in an additional temperature increase, and finally in discharge instability. When the conductivity σ depends mainly on the gas translational temperature, the discharge is stable. When the electric field is relatively high and the conductivity σ becomes depending mainly on the specific electric field, the ionization becomes unstable. The gliding arc passes such a critical point during its evolution and the related electric field grows. The electric field growth during the gliding arc evolution is:

$$E = 2wR/ \left(V_0 + \left(V_0^{\,2} - 4wlR \right)^{0,5} \right). \tag{11.56}$$

At high electric fields, the conductivity $\sigma(T_0, E)$ starts depending not only on gas temperature as in quasi-equilibrium but also on the field E. The corresponding logarithmic sensitivity corresponds to the Saha ionization:

$$\sigma_T = \partial \ln \sigma(T_0, E)/\partial \ln T_0 \approx I_i/2T_0; \tag{11.57}$$

here I_i is ionization potential. Considering (11.55), the logarithmic sensitivity of conductivity to the electric field is:

$$\sigma_E = \partial \ln \sigma(T_0, E)/\partial \ln E = I_i E^2/E_i^2 T_0. \tag{11.58}$$

To analyze the arc stability let's consider the thermal balance equation:

$$n_0 c_p \, dT_0/dt = \sigma(T_0, E) \, E^2 - 8\pi \lambda T_0^2/I_i S; \tag{11.59}$$

here S is the arc cross-section, n_0 and c_p – are the gas concentration and specific heat. The linearization procedure permits describing the time evolution of a temperature fluctuation ΔT in the exponential form:

$$\Delta T(t) = \Delta T_0 \exp(\Omega t); \tag{11.60}$$

here ΔT_0 is an initial temperature fluctuation; and exponential frequency parameter Ω is an instability decrement, that is a frequency of temperature fluctuation disappearance. In this case, the negative decrement $\Omega < 0$ corresponds to discharge stability, while the positive one $\Omega > 0$ corresponds to a case of instability. The linearization gives:

$$\Omega = -\omega \, \frac{\sigma_T}{1 + \sigma_E} \left(1 - \frac{E}{E_{crit}} \right). \tag{11.61}$$

In this relation, ω is the thermal instability frequency: $\omega = \frac{\sigma \, E^2}{n_0 c_p T_0}$.

The **gliding arc ionization instability**, illustrated in Figure 11.17a (see Fridman and Kennedy, 2021) results either in transition to nonequilibrium regime or simply in break of the arc channel. Initially, the gliding arc remains stable ($\Omega < 0$) at low electric field $E < E_{crit}$ ($l < l_{crit}$), but it becomes unstable ($\Omega > 0$) when the electric grows. The arc can lose stability even in the "theoretically stable" regime (see Figure 11.17a) when $E_i \, (T_0/I_i)^{0,5} < E < E_{crit}$. In this "quasi-instability" regime, Ω is still negative but decreases 10–30 times (compared to one in the initial regime of the gliding arc) due to the influence of the logarithmic sensitivity σ_E (factor $1/(1+\sigma_E)$). During the FENETRe-stage, the electric field and the electron temperature T_e in the gliding arc are increasing, and gas temperature decreases about three times. The electron density falls to about 10^{12} cm^{-3}. The decrease of the electron density and conductivity σ leads to the reduction of specific power σE^2. It cannot be compensated by the electric field growth as in the case of the stable quasi-equilibrium arc, because of the strong sensitivity of the conductivity to electric field. This is the main physical reason of the quasi-unstable discharge behavior in FENETRe conditions, see Figure 11.17a. During the transition, the specific heat losses w also decrease (to a smaller value w_{non-eq}) with the reduction of the translational temperature T_0. According to Ohm's law

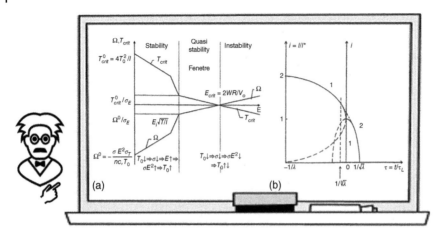

Figure 11.17 Gliding arc transition to the nonequilibrium phase: (a) instability decrement and critical thermal fluctuations; (b) effect of the self-inductance in the circuit.

($E < E_{crit}$), the total discharge power increases due to the growth of plasma resistance. This discrepancy between the total power and specific power (per unit of length or volume) results in a possible "explosive" increase in the arc length $l_{max}/l_{crit} = w/w_{non-eq} \approx 3$. After the fast FENETRe-transition, the **nonequilibrium gliding arc** can continue its evolution with $T_e \gg T_0$, and dissipation of up to 70–80% of the total arc power. Effective gliding arc evolution in the nonequilibrium phase is possible only if the electric field during the transition is sufficiently high, which is possible when the arc current is not too large. The critical gliding arc parameters before transition in atmospheric air are shown in Table 11.8 as a function of the initial electric current.

Even the relatively small electric field $\left(\frac{E}{n_0}\right)_{crit} \approx (0.5–1.0) \times 10^{-16}$ V × cm^2 is sufficient to sustain the nonequilibrium phase of a gliding arc, because of the influence of stepwise and possibly Penning ionization. Therefore:

(i) if $I_0 > 5$–10 A, then the arc discharge extinguishes after reaching the critical parameters;

(ii) for $J_0 < 5$–10 A, the electric field $\left(\frac{E}{n_0}\right)_{crit}$ is sufficient for nonequilibrium ionization regime.

It is possible to observe three different types of the gliding arc discharge. At relatively low currents and high gas flow rates, the gliding arc discharge is nonequilibrium throughout all stages of its development. At high currents and low gas flow rates, the discharge is thermal and just breaks at the critical point. Only at the intermediate currents and flow rates, the FENETRe-transition takes place, see Mutaf-Yardimci et al. (1999a,b).

The **effect of self-inductance L in gliding arc**, see Figure 11.15a (Fridman and Kennedy, 2021), is in slowing down current decrease and prolongation of the nonequilibrium arc evolution. The Ohm's law for power P is:

$$V_0 = RI + L \, {}^{dI}/_{dt} + P/I, P = P_{crit} + w_{crit} 2v\alpha \, t; \tag{11.62}$$

here P_{crit} and w_{crit} are total and critical power; v is arc velocity; 2α is angle between electrodes; $t = 0$ at transition point. With dimensionless variables: $i = I/I_{crit}, \tau = t/\tau_L, \lambda = 2v\tau_L\alpha/l_{max}$, the Ohm's law becomes:

$$(i - 1)^2 + idi/d\tau + \lambda t = 0; \tag{11.63}$$

here $\tau_L = L/R$ is the characteristic time of electric circuit. Typically, the λ is small ($\lambda = 0.003$). At the beginning of the process when $\tau < 0$ ($/\tau/ \gg 1$), neglecting the derivative $di/d\tau$: $i = 1 + (-\lambda\tau)^{1/2}$. When $\tau > -1/\lambda^{1/3}$:

$$i^2 + \lambda\tau^2 = 1. \tag{11.64}$$

The current corresponding to (11.64) is shown in Figure 11.17b. Thus, the self-inductance prolongs the transition phase by the time $\tau = 1/\lambda^{1/2}$ and increases the energy dissipated in this phase about 20–50%.

11.12 Gliding Arc Stabilized in Reverse Vortex (Tornado) Flow, and Other Special Gliding Arc Configurations

The gliding arcs can be organized in various ways. For example, **gliding arc in fluidized bed** is illustrated in Figure 11.18a,b, **gliding arc rotating in magnetic field** is shown in Figure 11.18c. An **expanding gliding arc configuration** is used in switchgears (Figure 11.19). The contacts start out closed; when they open, an arc is formed and glides along the opening until it extinguishes.

Table 11.8 Critical gliding arc parameters before transition as a function of initial current I_0.

I_{crit} (A)	50	40	30	20	10	5	1	0.5	0.1
I_0 (A)	100	80	60	40	20	10	2	1	0.2
T_0 (K)	10 800	10 300	9700	8900	7800	6900	5500	5000	4100
w_{crit} (W cm^{-1})	1600	1450	1300	1100	850	650	400	350	250
E_{crit} (V cm^{-1})	33	37	43	55	85	130	420	700	2400
$\left(\dfrac{E}{n_0}\right) 10^{-16}$ (V cm^2)	0.49	0.51	0.57	0.66	0.9	1.3	3.1	4.8	14

The most physically interesting is the **gliding arc stabilization in the reverse vortex (tornado) flow**, see Kalra et al. (2005). This approach is opposite to the conventional **forward-vortex stabilization**, where the swirl generator is placed upstream with respect to discharge and the rotating gas provides the walls protection from the heat flux. Some reverse axial pressure gradient and central reverse flow appear due to fast flow rotation and strong centrifugal effect near the gas inlet, which becomes a bit slower and weaker downstream. The hot reverse flow mixes with incoming cold gas and increases heat losses to the walls. More effective wall insulation and mixing are achieved by the **reverse-vortex stabilization**, physically identical to the tornado, see Figure 11.20a (Kalra et al., 2005). The plasma jet outlet is along the axis at the swirl generator side. Cold incoming gas moves at first by the walls providing their cooling and insulation, and only after that goes to the central plasma zone and becomes hot. In the case of reverse-vortex

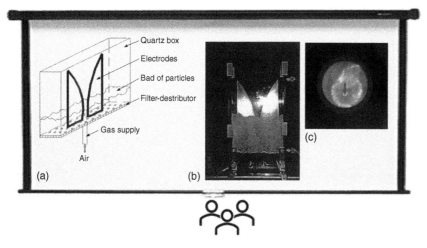

Figure 11.18 Gliding arc in fluidized bed (a, b), and gliding arc rotating in magnetic field.

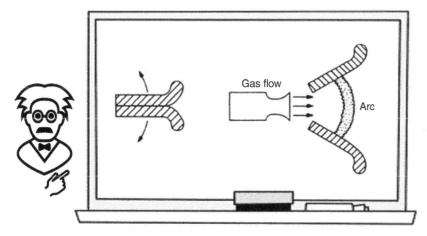

Figure 11.19 Illustration of a switch operating as gliding arc.

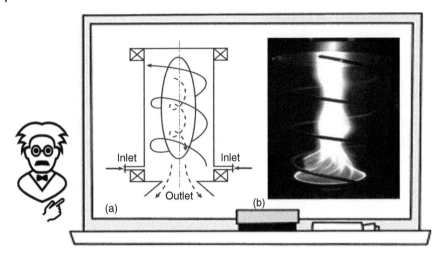

Figure 11.20 Illustration of the reverse vortex (tornado) flow (a) stabilizing the spiral electrode nonequilibrium gliding arc discharge (b).

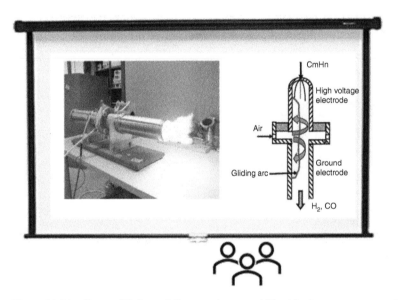

Figure 11.21 Nonequilibrium gliding arc plasma stabilized in the reverse vortex (tornado) flow.

stabilization, the incoming gas is entering discharge zone from all directions except the outlet side, which makes it effective for gas treatment, processing efficiency, and wall protection. The **gliding arc discharge with spiral electrode** stabilized in the reverse vortex (tornado) flow is shown in Figure 11.20b, see Kalra et al. (2005). The most robust configuration of the tornado-flow plasma stabilization is the so-called **gliding arc plasmatron** (Rabinovich et al., 2022), consisting of high-voltage cylindrical electrode, electrical insulator, and ground cylindrical electrode with tangential inlet holes, see Figure 11.21. Plasma gas enters the cylindrical reactor through tangential inlet holes. Gliding arc discharge starts in the gap between electrodes and stretches both ways (upward and downward) by incoming gas vortex.

The nonequilibrium gliding arc regime has been achieved with reverse-vortex stabilization at the current level about 1.5 A, voltage 1.5 kV, and power 2 kW. The FENETRe of the discharge has been observed at the arc channel length 15 cm, corresponding to the plasma power per unit length 0.15 kW cm^{-1}, average electric field 0.1 kV cm^{-1}, and reduced electric field 3 Td ($3*10{-}17$ V*cm^2). It permits to estimate expected physical limits for **scaling up of the nonequilibrium "cold" gliding arcs** from today's highest achieved power of 2 kW (the maximum thermal plasma power at the same configuration but higher currents is up to 15 kW). The reduced electric field before the FENETRe should remain during further scaling up about 3 Td ($3*10{-}17$ V*cm^2), and after the transition not less than 5–7 Td required to sustain the nonthermal ionization. Increase of current (from today's level of 1.5 A) and therefore power per unit length 0.15 kW cm^{-1}) is possible by cooling intensification to the maximum level of about 5 A and 0.5 kW cm^{-1}. It is 20 times lower than maximum power per unit length of the thermal arc discharges (achieving 10 kW cm^{-1}, corresponding to arc channel temperatures of 20 000 K. Maximum power of the nonequilibrium "cold" gliding arc is limited by maximum channel length (about 10 m). Therefore, the maximum one-unit "cold" gliding arc power is expected to be about 5 MW, which is still 20–40 times lower than the maximum one-unit power of thermal arcs, more details in Rabinovich et al. (2022).

11.13 Problems and Concept Questions

11.13.1 The Sommerfeld Formula for Thermionic Emission

Calculate the saturation current density of thermionic emission for tungsten cathode at 2500 K, compare with Table 11.3.

11.13.2 Secondary Electron Emission

Why the coefficient γ of the emission induced by the ion impact is almost constant at low energies when it is provided by the Penning mechanism.

11.13.3 Erosion of Hot Cathodes

Based on example from the Section 11.5, estimate the rate of cathode mass losses for the tungsten rod cathode in inert gases (1 atm) and current 300 A.

11.13.4 Radiation of the Arc Positive Column

Based on the formulas from Table 11.7, calculate the percentage of the arc power going to radiation for different gases at low and high values of the Joule heat per unit length $w = EI$, W cm^{-1}.

11.13.5 Arc Temperature in the Frameworks of the Channel Model

Why the arc temperature is close to a constant with weak logarithmic dependence on current and radius R.

11.13.6 Difference between Electron and Gas Temperatures in Arc Discharges

Estimate the critical electric current when the difference between electron and gas temperatures in atmospheric pressure argon arc becomes essential, compare your result with Figure 11.6b.

11.13.7 Electrode Jet Velocity

Estimate the electrode jet velocity corresponding to parameters of cathode spots given in Section 11.8 for atmospheric pressure arc discharges.

11.13.8 Stabilization of Linear Arcs Near Axis of the Discharge Tube

Why if arc is cooled on its edges, the thermal discharge temperature on axis rises?

11.13.9 Critical Length of Gliding Arc Discharge

Analyze the relation (11.51) for electric current in gliding arcs, and asymptotically simplify if in the vicinity of the critical point. Why this relation is unable to describe the current evolution for bigger lengths of the arc?

11.13.10 Quasi-Unstable Phase of Gliding Arc Discharge

Based on the stability diagram shown in Figure 11.17a, explain mechanism leading to instability of the formally stable regime of arc with length lower than the critical one.

Lecture 12

Radio-frequency, Microwave, and Optical Discharges

12.1 Thermal Plasma Generation in High-frequency Electromagnetic Fields

High frequency electromagnetic fields can generate thermal and nonthermal plasdensities exceeding the critical valuemas depending on gas pressure and discharge conditions. Low-pressure discharges produce mostly cold plasmas, while thermal plasma is mostly generated at near-atmospheric and higher pressures. Depending on the electromagnetic field frequency, we distinguish radio-frequency (RF, usually MHz and kHz range), microwave (GHz range), and optical (THz range) plasma sources. While the most traditional thermal plasma sources are arcs, the electrodeless RF and other high-frequency discharges are also often applied for thermal plasma generation at high pressures. Thermal plasma sustained by electromagnetic fields is characterized by quasi-equilibrium temperature T distributions determined by electromagnetic energy absorption and heat transfer:

$$\rho c_p \frac{dT}{dt} = -div\vec{J} + \sigma <\vec{E}^2> -\Phi, \quad \vec{J} = -\lambda\nabla T; \tag{12.1}$$

here $\vec{J}$ is the heat flux, c_p is the specific heat; λ is the thermal conductivity coefficient; σ is the high frequency conductivity; the square of electric field is averaged $<\vec{E}^2>$ over an oscillation period; the factor Φ describes radiation heat losses which can be neglected at atmospheric pressure and $T < 11\ 000 - 12\ 000\ K$; $\rho = Mn_0$ is the gas density, pressure $p = (n_0 + n_e)T$; M is the mass of heavy particles; n_e and n_0 are the number densities of electrons and heavy species. The material derivative dT/dt is related to fixed mass of gas (the Lagrangian flow description) and to the local Eulerian derivative $\partial T/\partial t$ as $dT/dt = \partial T/\partial t + (\vec{u}\cdot\vec{\nabla})T$, where $\vec{u}$ is velocity vector of the fixed mass of gas mentioned above. Neglecting effect of gas motion on plasma temperature in the energy release zone, the energy balance (12.1) in steady state becomes:

$$-div\vec{J} + \sigma <\vec{E}^2>= 0, \quad \vec{J} = -\lambda\nabla T. \tag{12.2}$$

The electric field E in (12.2) is determined from the electromagnetic field energy balance in terms of the Pointing vector, which is the flux density $\vec{S}$ of electromagnetic energy:

$$div<\vec{S}>= -\sigma <\vec{E}^2>, \quad \vec{S} = \varepsilon_0 c^2[\vec{E} \times \vec{B}]. \tag{12.3}$$

Combining (12.2) and (12.3) leads to the relation between energy fluxes:

$$div(\vec{J} + <\vec{S}>) = 0, \tag{12.4}$$

illustrating that total energy flux has no sources. It results in 1D in the **integral flux relation**:

$$\vec{J} + <\vec{S}>= \frac{const}{r^n}; \tag{12.5}$$

here "r" is the 1D-coordinate; and the power $n = 0$ for plane geometry, $n = 1$ for cylindrical geometry, and $n = 2$ for spherical one. The constant of integration in Eq. (10.1.5) can be determined from the boundary conditions. For example, radial thermal and electromagnetic fluxes are equal to zero at the axis of cylindrical plasma column; therefore, the constant is also zero in this case, and $J_r + S_r = 0$. This simple form of the integral flux is the third equation for the channel model discussed for arc discharges (see Section 11.6). It shows that thermal plasma columns sustained by DC and high frequency electromagnetic fields have any common features.

Plasma Science and Technology: Lectures in Physics, Chemistry, Biology, and Engineering, First Edition. Alexander Fridman.
© 2024 WILEY-VCH GmbH. Published 2024 by WILEY-VCH GmbH.

12.2 Thermal Plasma of Inductively Coupled RF Discharges, Metallic Cylinder Model

Plasma generation in the inductively coupled discharges is illustrated in Figure 12.1a. High frequency electric current passes through a solenoid coil inducing magnetic field along the axis and then the vortex electric field sustaining the **inductively coupled plasma (ICP) discharge**. Electric currents are also concentric with the coil elements and the *discharge itself is apparently electrodeless*. The ICP discharges can be very powerful and sustained at atmospheric and even higher pressures. The ICP electric field, according to the Maxwell equations, is proportional to frequency. To sustain the ICP discharge the RF frequencies 0.1–100 MHz are required. To avoid interference with radio communication systems the RF frequency commonly used in industrial ICP discharges is 13.6 MHz (wavelength 22 m).

Electrodynamics and thermal characteristics of the ICP discharges are described in the frameworks of **metallic cylinder model**. Let's consider a dielectric discharge tube of radius R inserted inside of solenoid coil, see Figure 12.1b. Plasma is sustained by Joule heating induced by alternating-current (AC) and stabilized by heat transfer to the walls of the tube. Radial temperature distribution is determined by energy balance (12.2):

$$-\frac{1}{r}\frac{d}{dr}rJ_r + \sigma <E_\varphi^2> = 0, \quad J_r = -\lambda\frac{dT}{dr}. \tag{12.6}$$

This equation is like that for positive column of arcs, but the electric field now is azimuthal and alternating. In MHz frequency range, the high-frequency conductivity coincides with that of the DC because $\omega^2 \ll v_{en}^2$; polarization and displacement currents can be neglected with respect to conductivity current; and the complex dielectric constant is mostly imaginary. Assuming $E, H \propto \exp(-i\omega t)$, the Maxwell equations become:

$$-\frac{dH_z}{dr} = \sigma E_\varphi, \quad \frac{1}{r}\frac{d}{dr}rE_\varphi = i\omega \mu_0 H_z. \tag{12.7}$$

Together with (12.6), the Maxwell equations complete the system describing the plasma column. The boundary conditions can be taken as: $J_r = 0, E_\varphi = 0$ at $r = 0$; $T = T_w \approx 0$ at $r = R$. Magnetic field in nonconductive gas near walls ($r = R$) is the same as inside of empty solenoid:

$$H_z(r = R) \equiv H_0 = I_0 n; \tag{12.8}$$

here I_0 is current in the solenoid coil, n is number of the coil turns per unit of its length. The metallic cylinder model, similarly to the arc channel model, considers plasma column as a metallic cylinder with the fixed conductivity corresponding to the maximum temperature T_m on the discharge axis. If plasma conductivity is high, the skin effect prevents penetration of electromagnetic fields deep into the discharge column. As a result, heat release related to the inductive currents is localized in thin skin layer. Thermal conductivity inside of the "metallic cylinder" provides the temperature plateau in the central part of the discharge cylinder where the inductive heating by itself is negligible. Radial distributions of plasma temperature, plasma conductivity, and the Joule heating are illustrated in Figure 12.2a.

The metallic cylinder model permits to separately consider electrodynamics and heat transfer aspects of the ICP discharges. Starting with the **electrodynamics of the ICP discharge**, plasma conductivity σ and radius r_0 can be taken as parameters. At the typical RF discharge frequency $f = 13.6$ MHz and high conductivity conditions, the skin layer δ is usually small ($\delta \ll r_0$) and the Maxwell relations can be expressed as:

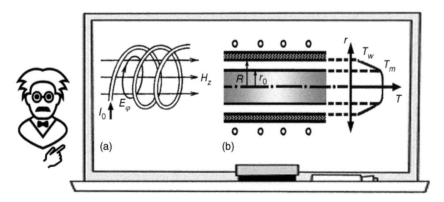

Figure 12.1 ICP discharge: (a) general schematic; (b) radial temperature distribution (R – discharge tube radius, r_0 – plasma column radius, T_m, T_w – maximum and wall temperatures).

$$\frac{dH_z}{dx} = \sigma E_y, \quad \frac{dE_y}{dx} = -i\omega\mu_0 H_z. \tag{12.9}$$

Coordinate "x" is positive for the direction inward to the plasma; coordinates "y" and "z" are tangential to the plasma surface. Boundary conditions: $H = H_0$ at $x = 0$, and E_y, $H_z \to 0$ at $x \to \infty$ lead to the solution: $H_z = H_0 \exp[-i(\omega t - x/\delta) - x/\delta]$, $\delta = \sqrt{2/\omega\mu_0\sigma}$; $E_y = H_0 \sqrt{\frac{\omega\mu_0}{\sigma}} \exp\left[-i\left(\omega t - \frac{x}{\delta} + \frac{\pi}{4}\right) - \frac{x}{\delta}\right]$. The electric and magnetic field amplitudes decrease exponentially in the skin layer, with the phase shift between them $\pi/4$. The electromagnetic energy flux is normal to the plasma surface and directed inward to plasma column:

$$< S > = S_0 \exp\left(-\frac{2x}{\delta}\right), \quad S_0 = H_0^2 \sqrt{\frac{\mu_0\omega}{4\sigma}}. \tag{12.10}$$

The flux S_0 is the total electromagnetic energy absorbed in the unit area of the skin layer per unit time. The total power w per unit length of the plasma column is: $w = 2\pi r_0 \cdot S_0$. For calculations, it is convenient to replace the magnetic field H_0 by current I_0 in a solenoid and the number n of turns per unit length of the coil:

$$S_0, \text{W cm}^{-2} = 9.94 \cdot 10^{-2} \cdot (I_0 n, \text{A} \cdot \text{turns cm}^{-1})^2 \sqrt{\frac{f, \text{MHz}}{\sigma, \text{Ohm}^{-1}\text{cm}^{-1}}}. \tag{12.11}$$

To describe **thermal characteristics of the ICP discharge**, (12.6) can be integrated between column and walls ($r_0 < r < R$, $\sigma = 0$) relating of temperature, radius r_0, and specific discharge power w per unit length:

$$\Theta_m(T_m) - \Theta_w(T_w) = \frac{w}{2\pi} \ln\frac{R}{r_0}. \tag{12.12}$$

Plasma temperature is expressed here in terms of the heat flux potential $\Theta(T)$; T_m is the maximum temperature in plasma column; T_w is the temperature of discharge tube walls. If distance Δr between plasma column and walls is short $\Delta r = R - r_0 \ll R$, the relation (12.12) can be simplified:

$$\Theta_m(T_m) - \Theta_w(T_w) \approx \frac{w}{2\pi r_0} \cdot \Delta r = S_0 \cdot \Delta r. \tag{12.13}$$

Plasma temperature in the central part of the ICP discharge is almost constant (see Figure 12.2a). The plasma temperature and conductivity decrease in the thin layer about $\delta/2$, where Joule heating is localized. The energy balance of the Joule heating induced by electromagnetic fields and thermal conductivity in this layer is:

$$\lambda_m \frac{\Delta T}{\delta/2} \approx S_0. \tag{12.14}$$

Considering relation between electromagnetic flux and discharge power per unit length: $w = 2\pi r_0 \cdot S_0$:

$$4\pi \lambda_m \Delta T \approx w \frac{\delta}{r_0}. \tag{12.15}$$

The thermal conductivity coefficient λ_m corresponds to the maximum plasma temperature T_m, and ΔT is the plasma temperature decrease, related to the exponential conductivity decrease. The temperature decrease ΔT at the boundary layer can be determined based on the Saha equation, then (12.14) can be expressed as:

$$2\sqrt{2} \lambda_m \frac{T_m^2}{I_i} \sigma_m = I_0^2 n^2, \tag{12.16}$$

This formula relates current I_0 and number of turns n per unit length in the solenoid coil with plasma conductivity σ_m and with ICP temperature T_m (here I_i is the ionization potential).

12.3 Plasma Temperature and Power of the Thermal ICP Discharges

Plasma conductivity σ_m strongly depends on temperature, while T_m and λ_m in ICP are changing only slightly. Therefore, the metallic cylinder model (12.16) predicts that $\sigma_m \propto (I_0 n)^2$; based on the Saha equation for conductivity $\sigma_m(T_m)$, the plasma temperature dependence on the solenoid current and number of turns per unit length as:

$$T_m = \frac{const}{const - \ln(I_0 n)}. \tag{12.17}$$

This dependence $T_m(I_0, n)$ is not strong (logarithmic) and corresponds to that derived for arc discharges. When the skin effect is strong, plasma temperature T_m does not depend on the electromagnetic field frequency. The following

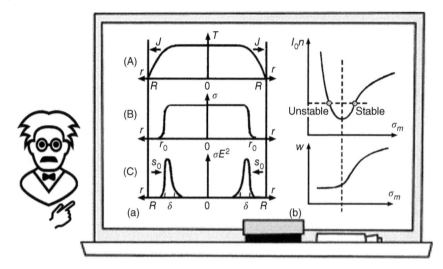

Figure 12.2 ICP discharge: (a) radial distributions of temperature (A), electric conductivity (B), and Joule heating (C); (b) plasma stability analysis.

relations can illustrate the dependence of the ICP power per unit length of discharge on the solenoid current and the other parameters:

$$w = 2\pi r_0 S_0 \propto H_0^2 \sqrt{\frac{\omega}{\sigma_m}} = (I_0 n)^2 \sqrt{\frac{\omega}{\sigma_m}} \propto I_0 n \sqrt{\omega}. \qquad (12.18)$$

Thus, the specific power grows not only with the solenoid current and the number of turns but also with the frequency of electromagnetic field (because of the increase of electric field with frequency. Considering $\sigma_m \propto (I_0 n)^2$, $w \propto \sqrt{\sigma_m}$, and even a small increase of temperature T_m (which according to the Saha leads to an exponential increase of electric conductivity) requires a significant increase of discharge power per unit length.

Using the above relations, typical characteristics of the thermal ICP in atmospheric pressure air at frequency $f = 13.6$ MHz can be estimated. For temperature about 10 000 K, discharge tube radius 3 cm, and a thermal conductivity $\lambda_m = 1.4 \cdot 10^{-2}$ W cm^{-1}·K, the electric conductivity of the thermal plasma is $\sigma_m \approx 25$ Ohm^{-1}cm^{-1} and the skin layer $\delta \approx 0.27$ cm. To sustain such plasma, the necessary electromagnetic flux is $S_0 \approx 250$ W cm^{-2}. Then the solenoid current and number of turns are: $I_0 n \approx 60$ A·turns cm^{-1}; magnetic field $H_0 \approx 6$ kA m^{-1}; the maximum electric field on the external boundary of the plasma column is 12 V cm^{-1}, the density of circular current is 300 A cm^{-2}, the total current per unit length of the column is 100 A cm^{-1}, thermal flux potential is about 0.15 kW cm^{-1}, the distance between the effective plasma surface and discharge tube is $\Delta r \approx 0.5$ cm, and finally the ICP discharge power per unit length is about 4 kW cm^{-1}.

A decrease of current in the solenoid coil leads to a growth of the skin layer δ ($\delta \gg r_0$, R), decrease temperature and electric conductivity. Then, the magnetic field is uniform $H = H_0$ in the absence of skin effect and the radial electric field distribution from the Maxwell equations is:

$$E(r) = \frac{1}{2}\omega\mu_0 H_0 \cdot r. \qquad (12.19)$$

The ICP power per unit length of the discharge with radius r_0 can be found by integrating Joule heating:

$$w = \int_0^{r_0} \sigma < E^2(r) > 2\pi r dr \approx \frac{1}{16}\pi\mu_0^2\omega^2\sigma_m H_0^2 r_0^4, w \approx \frac{1}{16}\pi\mu_0^2\omega^2\sigma_m I_0^2 n^2 r_0^4. \qquad (12.20)$$

In contrast to the case of strong skin-effect conditions where the temperature decrease takes place mainly in the boundary layer, here the temperature reduction is distributed over the entire plasma column radius. This leads to the following relation between specific power and maximum plasma temperature:

$$w \approx 4\pi r_0 \lambda_m \frac{\Delta T}{r_0} = 4\pi \lambda_m \Delta T \approx 8\pi \lambda_m \frac{T_m^2}{I_i}; \qquad (12.21)$$

here: ΔT is the temperature decrease across the plasma column; λ_m is the thermal conductivity corresponding to the maximum plasma temperature T_m; I_i is the effective value of ionization potential. Let's analyze what happens when the temperature and electric conductivity decreases in this weak skin-effect regime. The temperature cannot decrease significantly (because of the related exponential reduction of conductivity), which according to (12.21) makes the specific power w almost constant even when electric conductivity decreases. This effect of specific power stabilization is illustrated in Figure 12.2b. The plasma radius must also decrease with temperature reduction. Considering (12.20) leads to the interesting conclusion: current in the solenoid coil (factor $I_0 n$) in this regime is not decreasing but grows with the conductivity decrease in contrast to the strong skin-effect regime where $w \propto I_0 n \propto \sqrt{\sigma_m}$. This non-monotonic dependence of $I_0 n(\sigma_m)$ is also illustrated in Figure 12.2b.

Thus, the dependence $I_0 n(\sigma_m)$ has a minimum corresponding to the case when the skin layer is about the same size as the discharge tube radius $\delta \approx R$. This minimum value of the solenoid current $I_0 n$, which is necessary to sustain the quasi-equilibrium thermal ICP, can be found assuming electric conductivity σ_m corresponding to the critical condition $\delta \approx R$. In turn, this leads to the **minimum solenoid current $I_0 n$ to sustain thermal ICP discharge**:

$$(I_0 n)_{min} \approx \frac{2T_m}{R}\sqrt{\frac{\lambda_m \sqrt{2}}{I_i \mu_0 \omega}}. \tag{12.22}$$

As an example, the minimal current necessary to provide a thermal ICP discharge in a tube with $R = 3\,\text{cm}$ at frequency $f = 13.6\,\text{MHz}$ in air is $(I_0 n)_{min} \approx 10\,\text{A·turns cm}^{-1}$; corresponding minimum temperature is $T_{crit} \approx 7000 - 8000\,\text{K}$. As seen from Figure 12.2b, when the solenoid current exceeds the critical value, two steady states of the ICP discharge can be realized. One corresponds to high conductivity and strong skin effect and the other to low conductivity and no-skin-effect conditions. However only one of them, high conductivity regime is stable. The low conductivity regime (left branch on Figure 12.2b) is unstable. For example, if temperature (and hence conductivity) increases because of some fluctuation, then a current lower than actual one is sufficient to sustain the discharge. This leads to ICP plasma heating and further temperature increase until the stable high conductivity branch is reached.

12.4 Specific Configurations of Atmospheric Pressure RF ICP and RF CCP Discharges

A standard configuration of an ICP torch is illustrated in Figure 12.3a. The heating coil is water-cooled and is not in direct contact with plasma, which is therefore "clean." To initiate the discharge at low power levels about 1 kW, a graphite starting rod can be applied. At high power levels, it is convenient to use a pilot DC or RF small plasma generator to initiate the powerful ICP torch. Such **hybrid ICP plasma torches** operating at

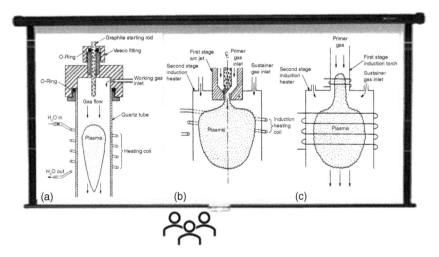

Figure 12.3 ICP discharge configurations: (a) kilowatt power level torch; (b) DC-RF hybrid plasma torch; (c) RF–RF hybrid plasma torch.

power levels of 50 – 100 kW are illustrated in Figure 12.3b,c. The ICP torches are often operated in transparent quartz tubes to avoid heat load on the discharge walls related to visible radiation. At power level exceeding 5 kW, the walls should be cooled by water. Plasma stabilization and the discharge walls insulation from direct plasma influence can be achieved in fast gas flows, including effective reverse-vortex stabilization, see Figure 11.20. More details on the subject can be found in Roth (1995, 2001), Fridman and Kennedy (2021), and Gutsol et al. (1999).

The **capacitively coupled plasmas (CCP)** are generated in capacitors and therefore supported by higher electric fields and can be far from thermodynamic equilibrium at moderate and even atmospheric pressures, the so-called **atmospheric pressure RF glow discharges**, see Roth (1995, 2001). The discharge can be confined by parallel electrodes across which a RF electric field is imposed. A bare metal screen can be located midway between the electrodes and grounded through a current choke. This median screen provides a substrate surface to support the material to be treated in the discharge. The electric field applied between the electrodes is on the level of kilovolts per centimeter; it should be sufficiently strong for electric breakdown and to sustain the discharge. In helium or argon, this electric field is obviously lower than in atmospheric air. Typical frequencies to sustain the uniform glow regime are about 1 – 20 kHz. At lower frequencies the discharge is difficult to initiate; at higher frequencies, the discharge is not uniform and has the filamentary structure. The atmospheric pressure uniform glow regime, corresponds to the specific RF frequencies sufficiently high to trap the ions between the median screen and an electrode, but not high to trap plasma electrons. This frequency range provides reduction of electron-ion recombination in the boundary layers, which promotes the ionization balance at lower electric fields. The power density of the discharge grows with voltage amplitude and the electric field frequency. However, even at relatively high values of the voltage amplitude and the electric field frequency, the discharge power density and total power are still not high relative to, for example, pulsed corona discharges. Maximum values of the power density in the uniform regime are approximately 100 mW cm^{-3}; maximum total power is about 100 W (see Roth 1995, 2001).

12.5 Microwave Sources of Thermal Plasma

Microwave plasma is sustained by electromagnetic waves in centimeters range comparable with the plasma size, and therefore is quasi-optical. Microwave generators like magnetrons are operating with power exceeding 1 kW in GHz frequency range and able to maintain thermal plasma at atmospheric pressure. Electromagnetic energy can be coupled with plasma in different ways, most conventionally through waveguides, see Figure 12.4a. The dielectric tube (usually quartz), which is transparent to the electromagnetic waves, crosses the rectangular waveguide. Plasma is ignited in the discharge tube by dissipation of the electromagnetic energy.

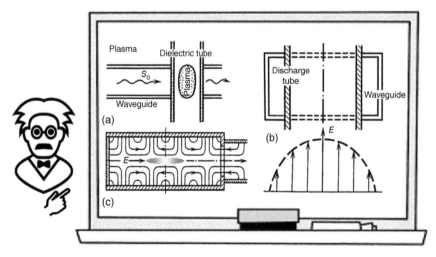

Figure 12.4 Microwave discharges: (a) discharge in a waveguide; (b) electric field distribution for H_{01} mode in a rectangular waveguide; (c) discharge in a resonator.

Different modes of electromagnetic waves formed in the rectangular **waveguide can be used to operate microwave discharges**. The most typical one is the H_{01} mode, illustrated in Figure 12.4b. The electric field in the H_{01}-mode is parallel to the narrow walls of the waveguide and is constant in this direction. Along the wide waveguide wall, the electric field is distributed as a sine function (in the absence of plasma) with a maximum in the center of the discharge tube and zero field on the narrow waveguide walls:

$$E_y = E_{\max} \sin\left(\frac{\pi}{a_w}x\right); \quad E_x = E_z = 0; \tag{12.23}$$

here E_x is the electric field component directed along the longer wall with length a_w; E_y is the electric field component directed along the shorter wall with length b_w; and E_x is the electric field component directed along the z. The maximum electric field $E_{\max}$ (in $kW\,cm^{-1}$) in the case of the mode H_{01} is related to the microwave power P_{MW} (expressed in kW) transmitted along the rectangular waveguide by the following numerical formula:

$$E_{\max}^2 = \frac{1.51 \cdot P_{MW}}{a_w b_w}\left[1 - \left(\frac{\lambda}{\lambda_{crit}}\right)^2\right]^{-1/2}; \tag{12.24}$$

here λ is the wavelength of the electromagnetic wave propagating in the waveguide; λ_{crit} is the maximum wavelength when the propagation is still possible ($\lambda_{crit} = 2a_w$); lengths of the waveguide walls a_w and b_w are expressed in cm. The H_{01} mode is most convenient for microwave plasma generation in the rectangular waveguides because the electric field in this case has a maximum in center of the discharge tube (see Figure 12.4a,b). Microwave discharges can be also generated in the cylindrical waveguides with round cross-section. The most typical oscillation mode in such waveguides is H_{11} (see Fridman and Kennedy 2021). In the case of the H_{01} mode in the rectangular waveguides, plasma is formed along the electric field and axis of the discharge tube, see Figure 12.4a,b. To provide plasma stabilization, gas flow can be supplied into the discharge tube tangentially. The waveguide dimensions are related to the frequency of the electromagnetic wave. Thus, for $f = 2.5\,GHz$ (wavelength in vacuum $\lambda = 12\,cm$), the wide waveguide wall should be longer than 6 cm and usually is equal to 7.2 cm. The narrow waveguide wall is 3.4 cm long and the dielectric discharge tube diameter is about 2 cm. The diameter of the generated microwave plasma column in such conditions is about 1 cm. The incident electromagnetic wave interacts with plasma, which results in partial dissipation and reflection of the electromagnetic wave. Typically, about half of the power of the incident wave can be directly dissipated in the high conductivity thermal plasma column, about a quarter of the wave is transmitted through the plasma, and the remaining is reflected. The powerful atmospheric pressure thermal **microwave plasma can be generated in resonator** with standing wave mode E_{01}, see Figure 12.4c. In this case, the electric field on the axis of the cylindrical resonator is directed along the axis and varies as the cosine function with the maximum in the center of the cylinder. The electric field then decreases further from the axis. Ignition of the discharge takes place in the central part of the resonator cylinder, where electric field has its maximum. Microwave plasma formed in the resonator appears as a filament located along the axis of the resonator cylinder.

Microwave plasma generation can be described by 1D model, which assumes that electromagnetic wave passes through a plane dielectric wall and then meets plasma, see Figure 12.5. Heat released in the plasma is subsequently transferred back to the dielectric wall, which is externally cooled, providing energy balance of the steady state. The model is based on the integral flux relation (10.1.5) with $n = 0$ and $const = 0$ because the temperature tends to the constant T_m deep inside the plasma ($x \to \infty$). As a result, the integral flux relation can be expressed as:

$$J + <S> = 0, \quad J = -\lambda\frac{dT}{dx}; \tag{12.25}$$

here λ is the coefficient of thermal conductivity, $<S>$ is the mean value of the electromagnetic energy flux density. Since microwave discharges are quasi-optical, the wave effects such as reflection and interference should be considered by means of the complex dielectric permittivity:

$$\frac{d^2 E_y}{dx^2} + \left(\varepsilon_\omega + i\frac{\sigma_\omega}{\varepsilon_0 \omega}\right)\frac{\omega^2}{c^2}E_y = 0; \tag{12.26}$$

here E_y is electric field along the plasma surface; σ_ω and ε_ω are high frequency conductivity and dielectric permittivity. Since plasma temperature and degree of ionization are not very high with respect to RF and arc discharges under similar conditions, the electron-ion collisions can be neglected in plasma conductivity. From the Saha equation, the quasi-equilibrium high frequency plasma conductivity and dielectric permittivity are: $\sigma_\omega \propto 1 - \varepsilon_\omega \propto n_e \propto \exp\left(\frac{I_i}{2T}\right)$,

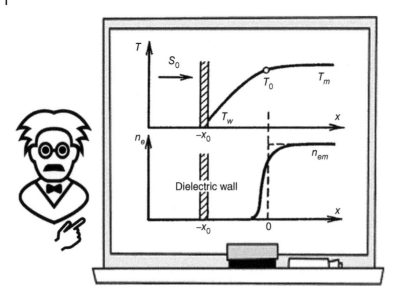

Figure 12.5 Microwave plasma: temperature and electron density distributions in a discharge sustained by electromagnetic wave (1D model).

where I_i is ionization potential. Boundary conditions for (10.25 and 10.26) are: electric field is assumed zero deep inside plasma ($E = 0$ at $x \to \infty$), temperature on the cooled dielectric wall is low ($T = T_w \approx 0$ at $x = -x_0$), and the electromagnetic energy flux density S_0 related to microwave power provided from generator is given. Solution of the system of (10.25 and 10.26) determines: the plasma temperature $T_m = T(x \to \infty)$; fraction of the electromagnetic energy flux density S_1, which is dissipated in the plasma; and the microwave reflection coefficient from the plasma $\rho = (S_0 - S_1)/S_0$. Analytical solution is possible assuming that plasma has a boundary at $x = 0$, where the temperature equals T_0 (see Figure 12.5). To the left from this point ($x \leq 0$, $T \leq T_0$) there is no plasma: $\sigma = 0$, $\varepsilon = 1$. To the right from the point ($x > 0$, $T_0 < T < T_m$) there is constant conductivity: $\sigma = \sigma_m, \varepsilon = \varepsilon_m$. This **constant microwave plasma conductivity model** is like the channel model of arc discharges and the metallic cylinder model of the thermal ICP discharges. The reflection coefficient of an incident electromagnetic wave normal to the sharp boundary of plasma $\rho = (S_0 - S_1)/S_0$ can be then expressed as:

$$\rho = \frac{(n-1)^2 + \kappa^2}{(n+1)^2 + \kappa^2};$$ (12.27)

here κ is the attenuation coefficient of electromagnetic wave; n is the refractive index. Both parameters n and κ are functions of the high frequency permittivity ε_ω and electric conductivity σ_ω. Microwave absorption and energy flux damping in plasma can be calculated from the Bouguer law with the absorption coefficient μ_ω mostly dependent on the electron density. According to the Bouguer law, dissipation of the electromagnetic wave takes place in a surface layer of about $l_\omega = 1/\mu_\omega$, where temperature grows from T_0 to T_m. The energy balance in this layer between the absorbed electromagnetic flux S_1 and the heat transfer to the wall can be expressed as: $S_1 \approx \lambda_m \Delta T/l_\omega$, $\Delta T = T_m - T_0$. As a result, the following equation determines the plasma temperature:

$$S_0[1 - \rho(T_m)] = \lambda(T_m) \cdot \frac{2T_m^2}{I_i} \cdot \mu_\omega(T_m).$$ (12.28)

Thermal microwave discharge parameters at $f = 10\,\text{GHz}$, ($\lambda = 3\,\text{cm}$) in air are presented in Table 12.1. At relatively low temperatures, the microwave plasma becomes transparent to electromagnetic microwave radiation, and hence to sustain such a plasma, high power must be provided. Further, low temperature and low conductivity regime is unstable in the same way as with RF discharges. Numerically, the minimum temperature is about 4200 K for the filament radius of 3 mm. Corresponding absorbed energy flux is $S_1 = 0.2\,\text{kW cm}^{-2}$; then the minimal microwave energy flux to sustain the thermal plasma is $S_0 = 0.25\,\text{kW cm}^{-2}$. Maximum temperature is limited by reflection of electromagnetic waves from plasma at high conductivities. This effect is seen in Table 12.1, where the reflection coefficient reaches the high value of $\rho = 81\%$ at $T_m = 6000\text{K}$. For this reason, the temperature of the microwave plasma in atmospheric pressure air does not exceed $5000 - 6000$ K.

Table 12.1 Atmospheric pressure microwave discharge in air at $f = 10\,\text{GHz}$, ($\lambda = 3\,\text{cm}$).

Plasma temperature (T_m)	4500 K	5000 K	5500 K	6000 K
Electron density (n_e, 10^{13}cm^{-3})	1.6	4.8	9.3	21
Plasma conductivity (σ_m, 10^{11}s^{-1})	0.33	0.99	1.9	4.1
Thermal conductivity (λ, $10^{-2}\text{W cm}^{-1}\cdot\text{K}$)	0.95	1.1	1.3	1.55
Refractive index n of plasma surface	1.3	2.1	2.8	4.3
Plasma attenuation coefficient (κ)	2.6	4.7	7.3	11
Microwave absorption layer ($l_\omega = \dfrac{1}{\mu_\omega}$, 10^{-2}cm)	9.1	5.0	3.2	2.2
Energy flux absorbed (S_1, kW cm^{-2})	0.23	0.35	0.56	1.06
Microwave reflection coefficient (ρ)	0.4	0.65	0.76	0.81
Microwave flux to sustain plasma (S_0, kW cm^{-2})	0.38	1.0	2.3	5.6

12.6 Thermal Plasma Generation in Continuous Optical Discharges

Thermal atmospheric pressure plasma can be generated by optical radiation similarly to the microwave discharges, see Figure 12.6a. It starts with **optical breakdown**, see Figure 12.6b. Usually, a high-power CO_2 laser (above 5 kW in air) is focused by a lens or mirror to sustain the discharge. To sustain the discharge in xenon at pressure of a couple of atmospheres, much lower power of approximately 150 W is sufficient. Light absorption coefficient in plasma significantly decreases with growth of the electromagnetic wave frequency. For this reason, application of visible radiation requires 100–1000 times more power than that of the CO_2 lasers. To provide effective absorption of laser radiation, the plasma density should be very high. As a result, plasma temperature in the continuous optical discharges is high, 15 000–20 000 K.

Let's analyze **laser radiation absorption in thermal plasma** depending on pressure and temperature. Absorption of CO_2 laser radiation quanta ($\hbar\omega = 0.117\,\text{eV}$), which is less than plasma temperature, is mostly provided by the bremsstrahlung absorption mechanism related to the electron-ion collisions. The absorption coefficient $\mu_\omega(CO_2)$ then can be calculated based on the Kramers formula:

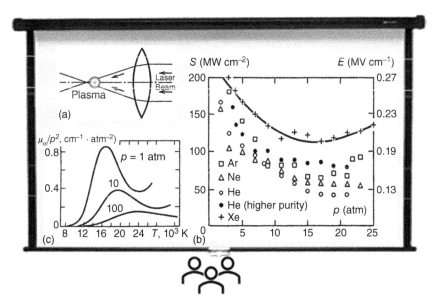

Figure 12.6 Continuous optical discharges: (a) schematic; (b) breakdown threshold under CO_2 laser radiation; (c) absorption coefficient for CO_2 laser radiation in air.

$$\mu_\omega(CO_2), cm^{-1} = \frac{2.82 \cdot 10^{-29} \cdot n_e(n_+ + 4n_{++})}{(T,K)^{3/2}} \cdot \lg\left(\frac{2700 \cdot T, K}{n_e^{1/3}}\right); \tag{12.29}$$

number densities are in cm^{-3}; n_e is the electron density; n_+ and n_{++} are densities of single-charged and doubly charged positive ions. The absorption coefficients for the CO_2 laser radiation in air is shown in Figure 12.6c as a function of temperature. It has a maximum at about 16 000 K at atmospheric pressure, which corresponds to complete single ionization. Further increase of temperature does not lead to additional ionization, while gas density and hence electron decrease at fixed pressures: $\mu_\omega \propto \frac{n_e^2}{T^{3/2}} \propto \frac{1}{T^{7/2}}$. The maximal absorption coefficient for the CO_2 lasers at atmospheric pressure is $\mu_\omega = 0.85\, cm^{-1}$, which corresponds to a light absorption length $l_\omega = 1.2\, cm$.

Energy balance of the continuous optical discharges determines plasma temperature. Plasma can be considered as a sphere of radius r_0 and constant temperature T_m and fixed absorption coefficient μ_ω. This plasma temperature is maintained by absorbing convergent spherically symmetric rays of total power P_0. If the plasma is transparent for the laser radiation, the fraction of the total radiation power absorbed in the plasma is: $P_1 = P_0\mu_\omega r_0$. At low laser power and not very high pressures, one can neglect radiation losses and balance absorption of the laser radiation with thermal conduction flux J_0:

$$P_1 = P_0\mu_\omega r_0 = 4\pi r_0^2 J_0 \approx 4\pi r_0^2 \frac{\Delta\Theta}{r_0} = 4\pi r_0 \Delta\Theta. \tag{12.30}$$

$\Delta\Theta = \Theta_m - \Theta_0$ is the drop of the heat flux potential in the plasma. Considering that gas is cold at infinity ($\Theta(r \to \infty) = 0$), the heat balance outside of the plasma sphere is:

$$P_1 = -4\pi r^2 \frac{d\Theta}{dr}, \quad \Theta(r) = \frac{P_1}{4\pi r}, \quad \Theta_0(r = r_0) = \frac{P_1}{4\pi r_0}. \tag{12.31}$$

Comparing (12.30 and 12.31), $\Delta\Theta = \Theta_0$, and hence, $\Theta_m = 2\Theta_0$. From this relation between maximum plasma temperature and one on the plasma surface: $P_1 = 2\pi r_0\Theta_m$. Recalling the above-mentioned formula: $P_1 = P_0\mu_\omega r_0$, the final energy balance relation, which determines the maximum plasma temperature T_m is:

$$P_0 = 2\pi \frac{\Theta_m(T_m)}{\mu_\omega(T_m)}. \tag{12.32}$$

Power required to maintain the optical discharge $P_0(T_m)$ has a minimum T_t because $\mu_\omega(T)$ has maximum (Figure 12.6c) while $\Theta(T)$ grows continuously. The temperature T_t, corresponding to **minimum power P_t to sustain the continuous optical discharge** is close to the temperature corresponding to the maximum of $\mu_\omega(T)$:

$$P_t = P_{0min} = 2\pi \frac{\Theta_m(T_m = T_t)}{\mu_\omega(T_m = T_t)}. \tag{12.33}$$

For example, CO_2 laser radiation in atmospheric air has the maximum of the absorption coefficient $\mu_{\omega max} \approx 0.85\, cm^{-1}$ at the threshold temperature $T_t = 18\,000\, K$. In this case, the heat flux potential is $\Theta(T_m = T_t) \approx 0.3\, kW\, cm^{-1}$, and the minimal power necessary to sustain the discharge is $P_t \approx 2.2\, kW$. Thermal plasma temperatures in continuous optical discharges are higher (about twice) than that of ICP and arc discharges, and significantly higher than in microwave discharges (about 3–4 times). This is related to the fact that $\mu_\omega \propto 1/\omega^2$ and plasma is transparent for the optical radiation. Only very high temperatures corresponding to almost complete ionization provide the level of absorption sufficient to sustain the discharges.

12.7 Radio-frequency (RF) Sources of Nonequilibrium Plasma, Capacitively Coupled Plasma (CCP), and Inductively Coupled Plasma (ICP) Discharges

RF plasma at lower pressures is far from equilibrium because electron-neutral collisions are less frequent and gas cooling is more intensive, which leads to electron temperatures much exceeding that one of the neutrals. The upper RF frequency limit is related here to wavelengths close to the system sizes. Lower frequency limit is related to characteristic frequency of ionization and ion transfer. Ion density in RF discharge plasmas and sheaths usually can be considered as constant during a period of electromagnetic field oscillation. Thus, typically RF frequencies exceed 1 MHz, though sometimes they can be smaller. *The nonthermal RF discharges can be subdivided into those of moderate and low pressures.* The discharges are referred to as those of moderate pressure (1–100 Torr) if they are nonequilibrium, but the electron energy relaxation length is small with respect to all characteristic sizes of the discharge system. In this case, the electron energy distribution function (EEDF) (and therefore ionization, excitation of neutrals, and other elementary processes) are determined by local electric fields. In opposite case of low pressures

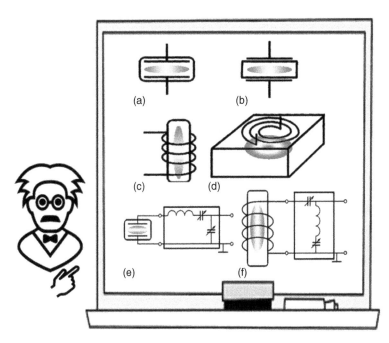

Figure 12.7 Nonthermal RF discharges: (a, b) CCP configuration; (c, d) ICP configuration; (e) coupling circuit for CCP; (f) coupling circuit for ICP.

p (p(Torr)·L(cm) < 1 in inert gases), the EEDF is determined by electric field distribution in the entire discharge zone L.

The nonthermal plasma of RF discharges can be either capacitively coupled (CCP) or inductively coupled (ICP), see Figure 12.7a–d. The CCP discharges provide electromagnetic field by electrodes located either inside (Figure 12.7a) or outside (Figure 12.7b) discharge chamber. They primarily stimulate electric field, which obviously facilitates their ignition. The inductive coil induces the electromagnetic field in ICP discharges: the discharge can be located either inside of a coil (Figure 12.7c) or aside of quasi-plane coil (Figure 12.7d). These discharges primarily provide magnetic fields, while corresponding nonconservative electric fields necessary for ionization are relatively low. For this reason, the nonthermal ICP plasma discharges are usually organized at low pressures, when the reduced electric field E/p is sufficient for ionization. Coupling between the inductive coil and plasma inside can be interpreted as a transformer where the coil represents the primary windings and the plasma the secondary windings. A coil as the primary winding consists of many turns while plasma has only one.

Therefore, the *ICP discharges can be considered as a voltage-decreasing transformer*. The ICP discharges are convenient to reach high currents, high electric conductivity, and high electron density. In contrast, the CCP discharges are more convenient to provide higher electric fields.

RF power supplies for plasma generation typically require active load of 50 or 75 ohm. To provide effective correlation between resistance of the leading line from the RF generator and the CCP/ICP discharge impedance, special coupling circuits should be applied, see Figure 12.7e,f. The RF generator should be not only correlated with the RF discharge during its continuous operation but should also provide sufficient voltage for breakdown. The coupling circuit should therefore form the AC-current resonance being in series with generator and the discharge system during its idle operation. The coupling circuit in the case of CCP discharge includes inductance in series with generator and discharge system (see Figure 12.7e). Variable capacitance is included there as well for adjustment because the design of variable inductances is more complicated. In the case of ICP discharge, the only variable capacitance, located in the coupling circuit in series with generator and idle discharge system provides necessary initial breakdown.

12.8 Fundamentals of Nonthermal Capacitively Coupled Plasma (CCP) Discharges

The fixed current amplitude regime is most conventional in CCP discharges. It can be represented by current density $j = -j_0 \sin(\omega t)$ flowing between two parallel plane electrodes. The RF plasma density $n_e = n_i$ can be assumed high enough providing electron conductivity current exceeding the displacement current: $\omega \ll \frac{1}{\varepsilon_0}|\sigma_e| \equiv \frac{1}{\tau_e}$. τ_e is time for electrons to shield electric field, v_e is electron collisional frequency, σ_e is the complex electron conductivity:

$$\sigma_e = \frac{n_e e^2}{m(v_e + i\omega)}. \tag{12.34}$$

The situation is opposite for ions because of their much higher mass $M \gg m$:

$$\omega \gg \frac{1}{\varepsilon_0}|\sigma_i| \equiv \frac{1}{\tau_i}, \quad \sigma_i = \frac{n_i e^2}{M(v_i + i\omega)}. \tag{12.35}$$

Thus, ion conductivity current can be neglected with respect to displacement current. In other words, the ion drift during an oscillation period can be neglected. The motion of charged particles is illustrated in Figure 12.8. Ions form a "skeleton" of plasma and can be considered being at rest, while electrons oscillate between electrodes. Electrons are present in the sheath of width L near an electrode only for a part of the oscillation period called the **plasma phase**. Another part of the oscillation period, when there are no electrons in the sheath, is called the **space charge phase**. The oscillating space charge creates an electric field, which forms displacement current and closes the circuit. Electric field of the space charge has a constant component in addition to an oscillating component, which is directed from plasma to the electrodes. The quasi-neutral plasma zone is also called the positive column. The constant component of the space charge field provides faster ion drift to the electrodes than in the case of ambipolar diffusion. As a result, ion density in the space charge layers near electrodes is lower with respect to their density in plasma (see Figure 12.8).

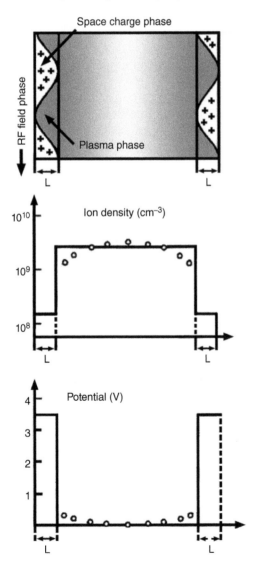

Figure 12.8 RF CCP discharge: ion density and potential distribution; boundary layer oscillation.

To describe *the discharge current and voltage*, let's assume that the RF discharge is symmetrical and ion densities in plasma and sheaths are fixed to n_p and n_s respectively (see Figure 12.8). Then the 1D- discharge zone can be divided each moment into three regions: (i) *plasma region* with thickness L_p, which is quasi-neutral throughout the oscillation period. Electric current here is provided by electrons and conductivity is active at relatively low frequencies $\omega < v_e$. (ii) *Sheath regions* in plasma phase, where the electric conductivity is also active and provided by electrons. (iii) *Sheath regions with space charge*. Considering the discharge symmetry and constancy of charge densities in the sheath, one can conclude that total thickness of the space charge region is always equal to the total thickness of the sheath region in plasma phase and equals to the sheath size L (see Figure 12.8). If total thickness of the sheath region in plasma phase is L, then active resistance of the region can be expressed as:

$$R_{ps} = \frac{Lmv_e}{n_s e^2 S};$$ (12.36)

here v_e is the electron collision frequency, and S is electrode area. Resistance of the quasi-neutral plasma region is:

$$R_p = \frac{L_p mv_e}{n_p e^2 S}.$$ (12.37)

The imaginary plasma and sheaths resistance in the plasma phase at all high frequencies $\omega > v_e$ is:

$$X_{p,ps} = iL_{p,s}\frac{m\omega}{n_{p,s}e^2 S} = i\omega L_{p,s}^{(e)}.$$ (12.38)

This impedance has the inductive nature, and factors $L_{p,s}^{(e)}$ are the effective inductance of plasma zone (p) and sheath (s) in plasma phase. The voltage drop on space charge sheath is proportional to the electric field near an electrode E and to the instantaneous size of the sheath $d(t)$. Assuming constancy of ion density in the sheath, the voltage drop is proportional to square of its instantaneous size:

$$U_r = \frac{1}{2}E_r d_r(t) = \frac{j_0}{4\varepsilon_0\omega}L\,[1 + \cos(\omega t)]^2.$$ (12.39)

Considering that the sheaths' half-size $L/2$ can be interpreted as amplitude of electron oscillation $\frac{j_0}{en_s}\cdot\frac{1}{\omega}$, the expression (12.39) for the voltage drop on the "right-hand" sheath (subscript "r") becomes:

$$U_r = \frac{j_0^2}{2\varepsilon_0\omega^2 en_s}[1 + \cos(\omega t)]^2.$$ (12.40)

The voltage drop on the sheath near the "left-hand" sheath (subscript "l") is in counter-phase with the voltage:

$$U_l = -\frac{j_0^2}{2\varepsilon_0\omega^2 en_s}[1 - \cos(\omega t)]^2.$$ (12.41)

The total voltage U_s related to the space charge sheaths is a sum of voltages on both electrodes:

$$U_s = U_r + U_l = \frac{2j_0^2}{\varepsilon_0\omega^2 en_s}\cos(\omega t) = \frac{j_0}{\varepsilon_0\omega}L\cos(\omega t).$$ (12.42)

Thus, total voltage drop on the space charge sheaths includes only the principal harmonic (ω) of the applied voltage, while the voltage drop on each sheath separately contains constant component and second harmonics (2ω). Considering that the electric current density as: $j = -j_0\sin(\omega t)$, one can see that the phase shift between voltage and current in the space charge sheath corresponds to the capacitive resistance.

From the relation (12.42), the space charge sheath represents the equivalent capacitance:

$$C_s = \frac{\varepsilon_0 S}{L},$$ (12.43)

which corresponds to the capacitance of vacuum gap with the width equal to the total size of the space charge sheaths L. If L does not depend on discharge current, the equivalent capacitance (12.43) is constant, and the discharge circuit can be considered as linear. Considering the impedance of a plasma (width L_p) has inductive component (12.38) and impedance of the space charge layers (width L) is capacitive; resonance in the discharge circuit is possible at frequency:

$$\omega = \omega_p\sqrt{L/L_p};$$ (12.44)

here ω_p is plasma frequency corresponding to the quasi-neutral plasma zone. The constant component of the space charge field is stronger than the electric field in plasma zone (see Figure 12.8). For this reason, even a small ionic

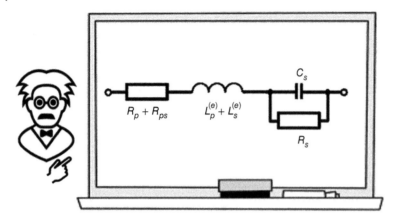

Figure 12.9 Equivalent circuit of nonthermal RF CCP discharge.

current can lead to energy release in the sheath exceeding that in plasma zone. Discharge power per unit electrode area, transferred to the ions, can be estimated as a product of the ion current density determined at the mean electric field in the sheath: $j_i = en_s v_i = en_s b_i \frac{j_0}{\varepsilon_0 \omega}$ and the constant component of the voltage (12.40). Here v_i and b_i are the ions drift velocity and mobility. Thus, the discharge power per unit electrode area related to ionic drift in a sheath can be as:

$$P_i = \frac{3}{4\varepsilon_0^2} \frac{j_0^3 b_i}{\omega^3} S. \tag{12.45}$$

The ionic power P_i is proportional to the cube of the discharge current, inversely proportional to the gas pressure (because of $b_i \propto 1/p$), proportional to the cube of frequency, and does not depend on the size of sheath. The power transfer to ions in a sheath can be presented in equivalent circuit by resistance R_s connected in parallel to the capacitance of sheaths. This resistance is determined by voltage drop on the capacitance and power released in sheaths:

$$R_s = \frac{U_s^2/2}{2P_i} = \frac{1}{3} \frac{\omega L^2}{j_0 b_i S}. \tag{12.46}$$

The **equivalent circuit of ICP discharge** is presented in Figure 12.9. It includes the resistance R_s and the capacitance C_s of sheaths connected in parallel to each other, and connected in series to the active resistance of sheaths in plasma phase R_{ps} and of plasma itself R_p.

12.9 Nonthermal RF Capacitively Coupled Plasma (CCP) Discharges of Moderate Pressure, α- and γ discharge Regimes

Moderate pressure CCP discharges are those, where the energy relaxation length λ_ε is less than size of plasma and sheaths: $\lambda_\varepsilon < L, L_p$. It which corresponds to the pressure interval $1 - 100$ Torr, where $\omega < \delta v_e$ (δ is the average fraction of electron energy lost per one electron collision, v_e is the frequency of the electron collisions). Electrons lose (and gain) energy here during a time interval shorter than the period of electromagnetic oscillations. EEDF and hence, ionization and excitation rates are determined in the moderate pressure discharges by local and instantaneous electric fields.

The moderate pressure CCP discharges operate in two qualitatively different forms, the so-called α- *and the γ- regimes, which differ by domination of either volumetric (α) or surface (γ) ionization mechanisms.* The α discharge has low luminosity in the plasma volume, see Figure 12.10a. Brighter layers are located closer to electrodes, but layers immediately adjacent to the electrodes are dark. The γ discharge occurs at higher values of current densities; layers adjacent to the electrodes are very bright, but relatively thin, see Figure 12.10a. The plasma zone is also luminous and separated from the bright electrode layers by dark layers like the Faraday dark space. The sheaths in γ discharges are like the cathode layers in glow discharges. The α- and γ-discharges operate at normal current density in the same manner as glow discharges. The current increase is provided by growth of the electrode area occupied by the discharge, while the current density remains constant. The normal

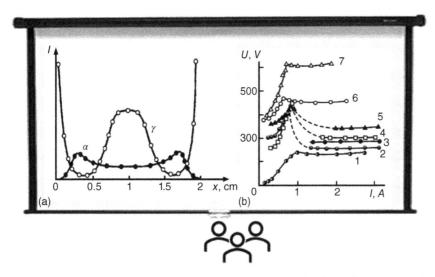

Figure 12.10 CCP discharges at moderate pressures: (a) emission intensity along the discharge gap (air, 10 Torr, 13.56 MHz, 2 cm); (b) current–voltage characteristics, 13.56 MHz: 1 – He ($p = 30$ Torr, $L_0 = 0.9$ cm); 2 – air (30 Torr, 0.9 cm); 3 – air (30 Torr, 3 cm); 4 – air (30 Torr, 0.9 cm); 5 – CO_2 (15 Torr, 3 cm); 6 – air (7.5 Torr, 1 cm); 7 – air (7.5 Torr, 1 cm).

current density in both regimes is proportional to the frequency of electromagnetic oscillations, and in the case of γ-discharges it exceeds the current density of α discharges more than 10 times.

Current–voltage characteristics of the moderate pressure CCP discharge are qualitatively different at low and high RF currents, see Figure 12.10b. Currents less than 1 A correspond to the α discharges, higher currents are related to the γ discharges. Normal α discharge can be observed on the curves 2,4, where there is no essential change of voltage at low currents. Most of the curves, however, show that an α discharge is abnormal just after breakdown. The current density is small in this regime and discharge occupies entire electrode immediately after breakdown, which leads to voltage growth together with current. Transition from α- to γ-regime occurs when the current density exceeds some critical value, which depends on pressure, frequency, electrode material, and type of gas. The $\alpha - \gamma$ transition is accompanied by discharge contraction and significant growth of the current density. Increase of current in the γ-regime after transition does not change the voltage and the discharge remains in normal regime.

Two mechanisms provide ionization in sheaths. One is related to the plasma phase when sheath is filled with electrons. This mechanism dominates in α discharges. The term "α discharge" influence of the first Townsend coefficient α on this ionization process. The second mechanism is related to the secondary electron emission from electrodes. This mechanism dominates in γ discharges. The term "γ discharge" shows the important contribution of the third Townsend coefficient γ on this ionization process.

Ionization rate in the α-regime can be calculated using the approximations for the first Townsend coefficient, and taking into account that the electric field and electron density are related in the plasma phase by Ohm's law: $j = n_e e b_e E$, where b_e is the electron mobility:

$$I_\alpha(x,t) = Ap\frac{j_0}{e}|\sin \omega t|\exp\left(-\frac{n_e}{n_0}\frac{1}{\sin \omega t}\right);\qquad(12.47)$$

here parameter $n_0 = j_0/eb_e Bp$ does not depend on pressure, because the electron mobility $b_e \propto 1/p$. This ionization rate grows exponentially with the instantaneous current and with decrease of the charge density, which leads to a monotonic decrease of ion density from plasma boundary to electrode. The exponential increase of $I_\alpha(x,t)$ at lower electron densities permits neglecting recombination in sheaths. The evolution of ionization rate in three different areas of the sheath can be summarized as follows: (i) near the plasma zone, the ionization rate of the α discharge I_α reaches a maximum when the current is in maximum; (ii) in the midpoint of the sheath, electrons appear only when the electric field in plasma already reached its maximum; (iii) closer to the electrode, electrons are absent in the moment of maximum current; the maximum of ionization is shifted. Considering the strong I_α dependence on electric field, the ionization rate significantly decreases near the electrode.

The **sheath parameters in the α-regime** can be estimated assuming that ion density in the sheath n_s is constant and ionization is absent near the electrode. In this case, the ion density in the sheath can be expressed as:

$$n_s = n_0 \ln \left[\frac{ABp^2\varepsilon_0}{e} \sqrt{\frac{2}{\pi} \frac{b_e}{b_i}} \frac{1}{n_0} \left(\frac{n_0}{n_s} \right)^{5/2} \right] = n_0 \ln \Lambda. \tag{12.48}$$

here A and B are Townsend parameters for the α-coefficient; n_0 is the pressure-independent parameter; p is gas pressure; b_e and b_i are electron and ion mobilities; $\ln \Lambda$ is the only slightly changing logarithmic factor. Considering $n_0 = j_0/eb_eBp$, the ion density in the sheath is proportional to the discharge current, does not depend on frequency, and only logarithmically depends on gas pressure. The sheath size can be expressed in turn as:

$$L = \frac{2j_0}{\omega n_s} = \frac{2eb_eBp}{\omega \ln \Lambda}; \tag{12.49}$$

it is inversely proportional to the electromagnetic oscillation frequency and does not depend on the current density.

The electric field of the α discharge in the space charge phase grows with current density. Also, the multiplication rate of the γ-electrons increases exponentially with the electric field in the sheath. As a result, reaching a critical current density leads to "the breakdown of the sheath" and **transition to the γ-regime of the CCP discharges**. The maximum ionization in the γ-regime of the CCP discharge occurs when electric field in the sheath has the maximum, which occurs when the oscillating electrons are located furthest from the electrode.

The effect of **normal current density in the γ-regime** can be explained keeping in mind that the normal current density j_n corresponds to the minimum of the current-voltage characteristics $U_0(j_0)$. At lower current density $j < j_n$ the current-voltage characteristics are "falling," which is generally an unstable condition. For example, small increase of current density in this case, results in lower voltages necessary to sustain the ionization. Then the actual voltage exceeds the required one, and current density grows until the normal regime $j = j_n$. The right branch ($j > j_n$) of the current-voltage characteristics $U_0(j_0)$ is stable in general, but not stable at the edge of the sheath on the boundary of current-conducting and currentless areas of the electrode. The normal current density and the sheath voltage drop in the γ CCP discharges can be expressed as:

$$j_n = \frac{Bp\omega\varepsilon_0}{2} \cdot \psi_1(\gamma); \tag{12.50}$$

here dimensionless functions $\psi_1(\gamma), \psi_2(\gamma)$ are about 10 and characterize influence of the secondary electron emission coefficient. The normal sheath voltage drop in the γ CCP discharges can be then expressed as:

$$U_n = \frac{B}{A} \psi_2(\gamma). \tag{12.51}$$

Similar to the case of a DC glow, the sheath voltage drop does not depend on pressure and is determined only by the gas and electrode material. The normal current density is proportional to the frequency of electromagnetic field and pressure. The pressure growth also leads to increase of the ion conductivity current:

$$j_i = \frac{b_i B^2 \varepsilon_0 A p^3}{4\psi_2 \psi_1^3}. \tag{12.52}$$

Let's discuss now the **α_γ Transition in moderate pressure CCP discharges**. The electric field in sheath is not very high at low current densities corresponding to the α-regime (see Figure 12.11a). For this reason, multiplication of electrons formed on the electrode surface due to secondary emission can be neglected. Growth of the current density and the related growth of the electric field lead to intense multiplication of the secondary (γ) electrons, which finally results in the phenomenon of the $\alpha - \gamma$ transition. At first, the increase of current density makes the multiplication of the secondary γ-electrons essential near the electrodes, where the electric field has maximum. The $\alpha - \gamma$ transition occurs when generation of ion flux due to multiplication of the secondary γ- electrons exceeds the ion flux produced in plasma phase. It requires current densities exceeding the critical value:

$$j_{crit} = Bp\omega\varepsilon_0 \ln^{-1} \left\{ \frac{ApL}{3\left[\ln\left(\left(1 + \frac{1}{\gamma}\right) \sqrt{\frac{\pi ApL}{4}} \right) - \frac{Bp\omega\varepsilon_0}{j_{crit}} \right]} \right\}. \tag{12.53}$$

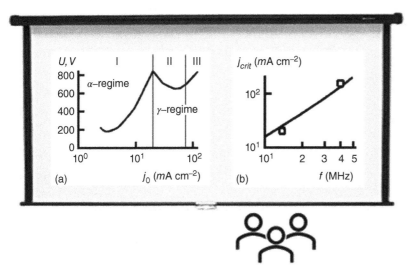

Figure 12.11 CCP discharges at moderate pressures, $\alpha-\beta$ transition: (a) current–voltage characteristic, nitrogen, 13.56 MHz, 15 Torr; (b) $\alpha-\beta$ transition critical current dependence on frequency: air, 0.75 cm, 15 Torr.

The critical current density (12.52) corresponding to the $\alpha-\gamma$ transition can be seen in Figure 12.11a. It grows with the oscillation frequency and pressure; see Figure 12.11b. More details on the subject can be found in Raizer et al. (1995) and Fridman and Kennedy (2021).

12.10 Low-pressure Capacitively Coupled Plasma (CCP) RF discharges

Low-pressure RF discharges (usually less than 0.1 Torr) are widely used in electronics for etching and chemical vapor deposition (CVD) processes, see Figure 12.12. Luminosity pattern of the low-pressure CCP-RF discharges is different from that of moderate pressure: the plasma zone is bright and separated from the electrodes by dark pre-electrode sheaths. The discharge is usually asymmetrical: a sheath located by the electrode where the RF voltage is applied is about 1 cm, while another sheath located by the grounded electrode is 0.3 cm thick. After breakdown, plasma fills out the entire low-pressure discharge gap, and normal current density does not take place.

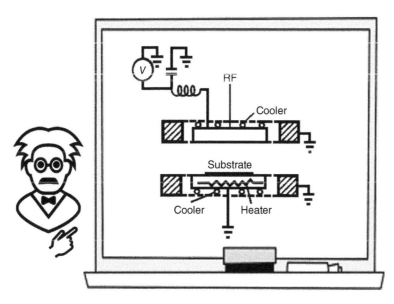

Figure 12.12 Schematic of a low-pressure RF CCP discharge.

Electron energy relaxation length λ_ε exceeds the sizes of the system: $\lambda_\varepsilon > L_p$, L, and *EEDF is not local here*; it is determined by the entire distribution of the electric fields in the zone of size λ_ε. Numerically, in the case of He, the low-pressure regime occurs when $p \cdot (L_p, L) \leq 1$ Torr·cm, which also corresponds to $\omega \gg \delta \cdot \nu_e$ permitting considering the EEDF as a stationary (δ is the average fraction of electron energy lost per one electron collision, ν_e is the frequency of the electron collisions). If $\omega < \nu_e$, then the "DC-analogy" is still valid and EEDF is the same as in the constant electric field $E_0 / \sqrt{2}$ (E_0 is amplitude of the RF oscillating field). The electric field has constant and oscillating components. The constant component provides balance of electron and ion fluxes for the quasi-neutrality. The oscillation component provides the electric current and heating of the plasma electrons. Distribution of plasma density and electric potential φ (corresponding to the constant component of electric field) are illustrated in Figure 12.8. The electric field in the space charge sheath exceeds that in the plasma, and the sharp change of potential on the boundary of plasma and sheath is vertical. Thus, the plasma electrons move between the sharp potential barriers satisfying the conservation of total energy $\varepsilon = \frac{1}{2} m v_e^2 + e\phi$ ($\phi = -\varphi$ is potential corresponding to the constant component of electric field).

 Plasma electrons in the low-pressure CCP can be divided into two groups. The first one includes electrons with kinetic energy below potential barrier on sheath boundary (see Figure 12.8). These electrons are trapped in the plasma zone and heated by the electric field determined by plasma density n_p:

$$E_p = \frac{j_0 m v_e}{n_p e^2}.$$ (12.54)

The second group of electrons has kinetic energy exceeding the potential barrier $e\phi$ on the sheath boundary. They spend part time inside sheath, where density n_s is lower, and the electric field in the plasma phase is higher:

$$E_s = \frac{j_0 m v_e}{n_s e^2} \gg E_p.$$ (12.55)

The energetic electrons of the second group are heated by the averaged electric field:

$$< E^2(\varepsilon) > = \frac{L_p}{L_0} E_p^2 + \frac{L}{L_0} \left(1 - \frac{e\phi}{\varepsilon}\right) E_s^2;$$ (12.56)

here L_p, L, L_0 are lengths of the plasma zone, sheaths, and total discharge gap respectively.

When gas pressure is very low ($\omega^2 \gg \nu_e^2$), the electric conductivity and Joule heating proportional to the frequency of electron-neutral collisions ν_e should be also very low. However, experimentally, the discharge power under such conditions can be significantly higher because of the contribution of the "collisionless" **stochastic heating effect** of fast electron in RF discharges (see Godyak 1971; Lieberman and Lichtenberg 2005). It can be explained considering so low pressures that the mean free path of electrons exceeds the size of sheaths. Then electrons entering the sheath are reflected by the space charge potential of the sheath boundary. If the sheath boundary moves from the electrode, then the reflected electron receives energy; conversely, if the sheath boundary moves toward the electrode and a fast electron is "catching up" to the sheath, then the reflected electron loses kinetic energy. The electron flux to the boundary moving from the electrode exceeds that for one moving in the opposite direction to the electrode, therefore energy is transferred to the fast electrons on average; which explains the stochastic heating.

Asymptotic solutions describing low-pressure capacitive RF discharges can be found for some extreme but practically important regimes. Consider the case of collisional heating ($\nu_e \gg \omega$) and assume proportionality of the ion velocity in plasma zone to the ambipolar electric field $v_p = \frac{e}{M\nu_i} \frac{2\phi}{L_p}$, which is acceptable for example in argon if $0.5 < pL_0 < 1$ Torr·cm. Let's analyze low- and high current regimes separately.

(1) ***Low-current regime of the low-pressure RF CCP discharge.*** Heating in the sheaths is negligible at low currents ($L_p E_p^2 \gg L_s E_s^2$), and the effective electric field is $E_{eff} = E_p / \sqrt{2}$. Then electric field in plasma is:

$$E_p = \sqrt{2} B^2 p \ln^{-2} \frac{E_p A p L_p e}{m \nu_e \sqrt{2\sqrt{2} e \lambda_i j_0 / \varepsilon_0 M \omega}}.$$ (12.57)

The electric field in plasma only logarithmically depends on the current density, which is similar to the positive column of the DC glow discharges. The potential barrier $e\phi$ on the plasma-sheath boundary is about equal to the lowest energy level ε_1 of electronic excitation and can be determined as:

$$e\phi = \varepsilon_1 \left[1 - \left(\frac{b_i \varepsilon_1}{3eL_p} \right)^{1/2} \left(\frac{\omega \varepsilon_0 M}{j_0 e \lambda_i} \right)^{1/4} \right]. \tag{12.58}$$

This potential barrier is almost constant and can be estimated as: $\phi = \frac{2}{3}\frac{\varepsilon_1}{e}$. Density of charges in plasma is proportional to current density $n_p = \frac{j_0 m v_e}{e^2 E_p} \propto j_0$. The ions flux is also proportional to the current density:

$$\Gamma_i = \frac{4}{3} \frac{j_0 \varepsilon_1 m v_e}{E_p L_0 e^2 M v_i} \propto j_0; \tag{12.59}$$

here v_e and v_i are the electron-neutral and ion-neutral collision frequencies. Sheaths do not make any contribution to electron heating in the low current regime. The sheath parameters are determined by the ion flux from plasma. Balancing ion flux in the sheath with that one in plasma, the sheath thickness, and the ion density in the sheath are:

$$L = \frac{3eE_p L_0 M v_i}{2\omega \varepsilon_1 m v_e} \sqrt{\frac{\sqrt{2}\lambda_i}{M} \frac{j_0}{\varepsilon_0 \omega}} \propto \sqrt{j_0}; \, n_s = \frac{4}{3} \frac{\varepsilon_1 m v_e}{E_p e^2 L_0 M v_i} \sqrt{\frac{M\omega \varepsilon_0 j_0}{\sqrt{2}\lambda_i}} \propto \sqrt{j_0}. \tag{12.60}$$

(2) **High-current regime of the low-pressure RF CCP discharge.** The main electron heating in the high current regime occurs in the sheaths. Low energy electrons are heated by the relatively low electric field in the plasma zone, and their effective "temperature" decreases. For this reason, the ambipolar potential maintaining the low-energy electrons in the plasma zone is less than the excitation energy. The effective electric field and sheath thickness are:

$$E_{eff} = \sqrt{\frac{L}{2L_0}} E_s = \sqrt{\frac{L}{2L_0}} \frac{m\omega v_e L}{2e}; L^{3/2} = \frac{2\sqrt{2L_0}eB^2 p}{m\omega v_e} \ln^{-2} \left(\frac{ApL^{3/2}L_p \omega}{2\sqrt{\frac{2\sqrt{2L_0}e\lambda_i}{M} \frac{j_0}{\varepsilon_0 \omega}}} \right). \tag{12.61}$$

The sheath thickness is determined by a balance of ionization and ion losses and does not depend on the discharge current. The potential barrier on the sheath boundary in the high current regime is:

$$e\phi = \varepsilon_1 \left(\frac{2\varepsilon_1}{M v_i L} \right)^2 \frac{M}{\sqrt{2}e\lambda_i} \frac{\omega \varepsilon_0}{j_0} \propto \frac{1}{j_0}. \tag{12.62}$$

In this case, the potential barrier of the sheath boundary is less than the excitation energy ε_1 and is inversely proportional to the discharge current. The charged particles' density in the plasma zone increases with current density:

$$n_p = \frac{2j_0}{\omega e L} \frac{L_p}{L} \left(\frac{\sqrt{2}e\lambda_i}{M} \frac{j_0}{\omega \varepsilon_0} \right)^{3/2} \left(\frac{LM v_i}{2\varepsilon_1} \right)^3 \propto j_0^{5/2}. \tag{12.63}$$

Ion concentration in sheath is proportional in this regime to current density $n_s = \frac{2j_0}{e\omega L} \propto j_0$.

Typical low-pressure CCP discharge parameters in argon are presented in Figure 12.13 as functions of current density in agreement with the above analytical relations. Thus, Figure 12.13a illustrates the plasma density dependence on current density. The low-current branch is linear, while at high currents this dependence is stronger. The average energy of plasma electrons $<\varepsilon>$ is related to potential barrier on the plasma-sheath boundary: $<\varepsilon> = \frac{2}{3}e\phi$. The average energy of plasma electrons is illustrated in Figure 12.13b. It grows slightly with j_0 at low current densities. In contrast, at current densities exceeding 1 mA cm^{-2}, the average electron energy decreases inversely proportional to the current density. The sheath thickness (Figure 12.13c) increases with current density as $\sqrt{j_0}$ at low currents and reaches the saturation level close to 1 cm at current densities exceeding 1 mA cm^{-2}. The discharge power grows with the current density (Figure 12.13d) and reaches about 1 W cm^{-3} at current densities 10 mA cm^{-2}.

More information on the subject, including organization of the low-pressure CCP discharges at lower frequencies can be found in Lieberman and Lichtenberg (2005), Fridman and Kennedy (2021), and Conti et al. (2001).

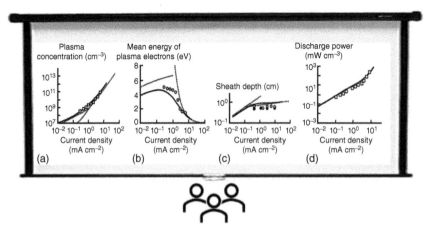

Figure 12.13 Low-pressure RF CCP discharge in Ar: plasma density (a), electron energy (b), sheath depth (c), and discharge power (d) as the function of current density. Solid curve – exact calculation, dotted line – asymptotic formulas, points – experiment ($f = 13.56$ MHz, $L_0 = 6.7$ cm, $p = 0.03$ Torr).

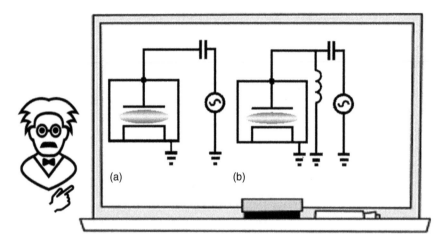

Figure 12.14 Low-pressure RF CCP discharges with disconnected (a) and DC-connected (b) electrodes.

12.11 Asymmetric and Magnetron RF CCP Discharges at Low Pressures

The asymmetric RF CCP discharges are organized in grounded metal chambers with one electrode connected to the chamber and other one powered. Plasma occupies large volume at low pressures and some fraction of the discharge current goes from the loaded electrode to the grounded walls. As a result, the current density in the sheath located near the powered electrode usually exceeds that in the sheath related to the grounded electrode. Lower current density in the sheath corresponds to lower voltages there. The constant component of the voltage, which is the plasma potential with respect to the electrode, is also lower at lower current density. For this reason, a constant potential difference occurs between the electrodes, if a dielectric layer covers them or if a special blocking capacitance is installed in the electric circuit (Figure 12.14) to avoid direct current between the electrodes. This potential difference is usually referred to as the auto-displacement voltage. The plasma is charged positively with respect to the electrodes. The voltage drop in the sheath near the powered electrode exceeds the drop near the grounded electrode. As a result, the loaded electrode has a negative potential with respect to the grounded electrode. Thus, the RF-loaded electrode is called sometimes a "cathode."

Consider the asymmetric low-pressure CCP-RF discharge between planar electrodes, illustrated in Figure 12.14. The ion flux to the powered electrode (cathode) can be expressed as:

$$\Gamma_i \approx n_s \sqrt{\frac{e\lambda_i}{M}\frac{j_0}{\varepsilon_0\omega}},$$

(12.64)

where n_s is the ion density near the loaded electrode, and j_e is the current density in this pre-electrode layer. In this case, the ion flux to the grounded electrode can be expressed by a similar formula:

$$\Gamma'_i \approx n'_s \sqrt{\frac{e\lambda_i}{M} \frac{j'_0}{\varepsilon_0 \omega}}, \tag{12.65}$$

where n'_s is ion density near the grounded electrode, and j'_e is the current density in this pre-electrode layer. The charged particle densities in the sheaths are lower than those in the plasma zone n_s, $n'_s \ll n_p$, and considering symmetry of the system, we can conclude that the fluxes ((12.63) and (12.64)) are equal. It leads to the inverse proportionality of the ion densities in the sheaths and the square roots of corresponding current densities:

$$\frac{n_s}{n'_s} = \sqrt{\frac{j'_0}{j_0}}. \tag{12.66}$$

Thus, charge densities near the grounded sheath only slightly exceed one in near the powered electrode. The electric field is determined by the space charge, therefore the ratio of current densities in the opposite sheaths is:

$$\frac{j'_0}{j_0} = \frac{n'_s L'}{n_s L}. \tag{12.67}$$

From ((12.65) and (12.66)), relation between the thickness of the opposite sheaths and current densities is:

$$\frac{L'}{L} = \left(\frac{j'_0}{j_0}\right)^{3/2}. \tag{12.68}$$

Thus, thickness of the sheath located near powered electrode is larger than that near grounded electrode. Considering that voltage drop in the sheaths U is proportional to the charge density and square of the sheath thickness:

$$\frac{U'}{U} = \left(\frac{j'_0}{j_0}\right)^{5/2}. \tag{12.69}$$

In asymmetric discharges, the voltage drop near grounded electrode can be neglected, and the auto-displacement voltage is close to the amplitude of the total applied voltage. The high charge density, low current density, and small sheath depth at the grounded electrode result in low intensity of electron heating in this layer. This is compensated by intensive heating of electrons in the "cathode" sheath.

The constant potentials of electrodes can be made equal by connecting (from the DC standpoint) the powered metal electrode with the ground through an inductance, see Figure 12.14b. Here direct current can flow between the electrodes, and the electron and ion fluxes to each electrode are not supposed to be equal. The average voltages on the sheaths are the same here, which results in equal thickness of the sheaths. The sheath related to the grounded electrode has a lower current density. For this reason, the amplitude of the plasma boundary displacement is shorter than the thickness of the grounded electrode sheath. The main part of the sheath stays for the entire period in the phase of space charge. The plasma does not touch the electrode and the ion current permanently flows to the grounded electrode. Here the average positive charge per period carried out from the discharge, is compensated by the electron current to the powered electrode. Thus, the direct current in this regime is determined by the ion current to the grounded electrode, as well as by the electron current to the powered one. The asymmetric RF discharge can be considered as a source of direct current. This phenomenon is called **the battery effect.**

The RF capacitive discharges have some disadvantages (high sheath voltages leading to low ion densities, and too high ion-bombarding energies), limiting their application. The **RF magnetron discharges**, see Figure 12.15, overcome these problems by applying weak DC-magnetic fields ($50-200$ G) to the powered electrode and hence, perpendicular to the RF electric field and current. The RF magnetrons permit increasing the degree of ionization at lower RF voltages and decreasing the energy of ions bombarding the powered electrode (or a sample placed on the electrode for treatment). Also, this discharge permits increasing the ion flux from plasma, which in turn leads to intensification of etching. The RF magnetrons can be sustained at very low pressures (down to 10^{-4} Torr) and ion beam penetrates the sheath without collisions. The ratio of ion flux to that of active neutral species grows, which also leads to higher-quality etching.

The **RF magnetron effect** assumes that electrons, oscillating together with the sheath boundary, additionally rotate around the horizontal magnetic field lines with the cyclotron frequency. When the magnetic field and the cyclotron frequency are sufficiently high, the amplitude of the electron oscillations along the RF electric field decreases significantly. The magnetic field "traps electrons." The cyclotron frequency plays the same role as the frequency of electron-neutral collisions: electrons become unable to reach the amplitude of their free oscillations in the RF electric field. The amplitude of electron oscillations determines the thickness of sheaths. Decrease of the amplitude of electron oscillations in magnetic field results in smaller sheaths and lower sheath voltage near

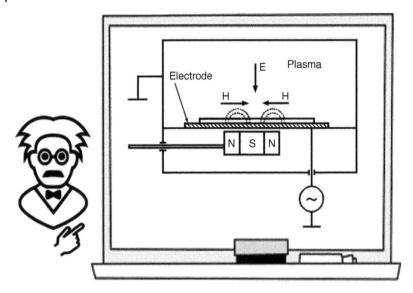

Figure 12.15 Typical configuration of RF – magnetron.

the powered electrode. This leads to lower values of the auto-displacement, lower ion energies, and lower voltages necessary to sustain the RF discharge.

To analyze physics of the RF magnetron, consider an electron motion in crossed electric and magnetic fields considering electron-neutral collisions. Assume electric field is directed along the "x"-axis perpendicular to the electrode surface $E \equiv E_x$, and the magnetic field is directed along the "z"-axis parallel to the electrode $B = B_z$. The electron motion equation can be then expressed as:

$$m\frac{d\vec{v}}{dt} = -\vec{E}_a e^{i\omega t} - e[\vec{v} \times \vec{B}] - mv_{en}\vec{v}; \tag{12.70}$$

here $\vec{v}$ is the electron velocity, $\vec{E}_a$ is the amplitude of oscillating electric field, v_{en} is the frequency of electron-neutral collisions. Projections of the motion equation to the "x" and "y" axis gives:

$$m\frac{d}{dt}v_x = -eE_a e^{i\omega t} - ev_y B - mv_{en}v_x, m\frac{d}{dt}v_y = ev_x B - mv_{en}v_y. \tag{12.71}$$

To solve this system with respect to velocities, it is convenient to transform variables to $v_x \pm iv_y$. The forced electron oscillations in the crossed electric and magnetic fields can be then described by the complex relation:

$$v_x \pm iv_y = \frac{eE_a e^{i\omega t}}{m(v_{en} + i(\omega \mp \omega_B))} \tag{12.72}$$

(ω_B is cyclotron frequency). It gives velocity of electron oscillations perpendicular to electrodes $v_x = d\xi/dt$:

$$v_x = \frac{eE_a e^{i\omega t}}{2m}\left[\frac{1}{v_{en} + i(\omega - \omega_B)} + \frac{1}{v_{en} + i(\omega + \omega_B)}\right]. \tag{12.73}$$

The coordinate $\xi(t)$ of the electron oscillations is determined as the integral of the normal velocity v_x:

$$\xi = -\frac{ieE_a e^{i\omega t}}{2m\omega}\left[\frac{1}{v_{en} + i(\omega - \omega_B)} + \frac{1}{v_{en} + i(\omega + \omega_B)}\right]. \tag{12.74}$$

At very low pressures $v_{en} \ll \omega$, and the electron oscillations can be considered collisionless. The amplitude of electron velocity in the collisionless conditions and in absence of magnetic field is $u_a = \frac{eE_a}{m\omega}$. The electron displacement amplitude (12.73) in the collisionless, nonmagnetized conditions is $a = \frac{eE_a}{m\omega^2}$. The objective of applying magnetic field is decreasing the amplitudes of the electron velocity and displacement, which results in smaller sheaths and lower sheath voltage near the powered electrode. It leads to lower auto-displacement, lower ion energies, and lower voltages necessary to sustain the RF discharge, which is desirable for RF magnetrons applications. If the magnetic field is sufficiently high ($\omega_B \gg \omega$), then the amplitude of electron velocity in the collisionless conditions is $u_{aB} = \frac{eE_a\omega}{m\omega_B^2}$. It is lower than the amplitude in the absence of magnetic field by the factor $(\omega_B/\omega)^2 \gg 1$. The amplitude of electron oscillation (12.73) in the magnetized collisionless conditions is $a_B = \frac{eE_a}{m\omega_B^2}$, which is also lower than the displacement under nonmagnetized conditions by the factor $(\omega_B/\omega)^2 \gg 1$. Ions motion in the RF magnetrons is not magnetized:

the ion cyclotron frequency is less than the oscillation frequency ($\omega_{Bi} \ll \omega$); and the trajectories of the heavy ions are not perturbed by magnetic field.

12.12 Nonthermal Radio-frequency (RF) Inductively Coupled Plasma (ICP) Discharges

The electromagnetic field in ICP discharges is induced by an inductive coil (Figure 12.7c,d), where the magnetic field is primary and nonconservative electric fields necessary for ionization is relatively low. Thus nonthermal ICP plasma discharges are usually configured at low pressures to provide the reduced electric field E/p sufficient for ionization. The coupling between the inductive coil and the plasma can be interpreted as a transformer, where the coil presents the primary windings and plasma represents the secondary ones. The coil as the primary winding consists of a lot of turns, while plasma has the only one. *Therefore, the ICP discharge can be considered as a voltage-decreasing and current-increasing transformer.* The coupling with the RF power supply requires a low plasma resistance. Thus, the ICP discharges are convenient to reach high currents, high electric conductivity, and high electron density at relatively low electric field and voltage. For example, the low-pressure ICP discharges effectively operate at electron densities $10^{11} - 10^{12} \text{cm}^{-3}$ (not more than 10^{13}cm^{-3}), which exceeds by an order of magnitude typical electron densities in the capacitively coupled RF discharges. Because of the relatively high electron densities in the discharges, they are sometimes called the **high-density plasma (HDP)** sources. Large-scale application of these discharges in electronics is due to their advantages with respect to CCP discharges, where sheath voltages and ion-bombarding energies are too high. Another advantage of the ICP discharges for high-precision surface treatment is the RF power coupling to the plasma across a dielectric window or wall, rather than by direct connection to an electrode in plasma. Such "non-capacitive" power transfer to plasma provides an opportunity to operate at low voltages across all sheaths at the electrode and wall surfaces. The DC plasma potential and energies of ions accelerated in the sheaths are typically 20–40 V, which is very convenient for surface treatment. The ion energies can be independently controlled by an additional capacitively coupled RF source called the RF bias, driving the electrode on which the substrate for material treatment is placed, see Figure 12.7c,d. ICP discharges provide independent control of the ion and radical fluxes by means of the main ICP source power, and the ion-bombarding energies by means of power of the bias electrode.

Consider the **inductive RF discharge in a cylindrical tube placed inside cylindrical coil**. The electric field $E(r)$ induced in the tube is $E(r) = -\frac{1}{2\pi r} \frac{d\Phi}{dt}$, where Φ is magnetic flux crossing the loop of radius r perpendicular to the axis of the discharge tube; r is distance from the discharge tube axis. The magnetic field is created by electric current in the coil $I = I_c e^{i\omega t}$ as well as by the electric current in the plasma (which mostly flows in the external plasma layers). Assuming plasma conductivity $\sigma(r) = $ constant, and considering harmonic current in plasma $j(r) = j_0(r)e^{i\omega t}$, equation for the current density distribution along the plasma radius is:

$$\frac{\partial^2 j_0}{\partial r^2} + \frac{1}{r}\frac{\partial j_0}{\partial r} - \frac{1}{r^2}j_0 = i\frac{\sigma\omega}{\varepsilon_0 c^2}j_0. \tag{12.75}$$

If pressure is low and plasma is collisionless, then its conductivity is inductive $\sigma = -i\varepsilon_0 \frac{\omega_p^2}{\omega}$, here ω_p is plasma frequency. The current density distribution in the ICP-plasma (finite at $r = 0$) can be then presented in form of the modified Bessel function $j_0(r) = j_b I_1\left(\frac{r}{\delta}\right)$; here $I_1(x)$ is the modified Bessel function, and δ is the skin-layer thickness. Current density on the plasma boundary in the vicinity of the discharge tube is determined by the non-perturbed electric field by the plasma conductivity on the boundary of the column $j_b = \sigma \frac{\omega a N}{2\varepsilon_0 c^2 l}I_c$, here a is the radius of the discharge tube; N is the number of turns in coil; l is the length of the coil; c is the speed of light; I_c is the amplitude of current in coil. Plasma conductivity usually grows with the electric field and the skin-layer thickness is smaller than discharge radius. Then most of electric current is in the relatively thin δ-layer on the discharge periphery. In this case, the Bessel function for the current density distribution can be simplified:

$$j_0(r) \approx j_b \exp\left(\frac{r-a}{\delta}\right). \tag{12.76}$$

Let's analyze the **equivalent circuit for the ICP RF discharge**. The total discharge current is related to the current in the coil I_c as $I_d = \sigma E_b \delta l = \sigma\delta\frac{a\omega N}{2\varepsilon_0 c^2 l}I_c$, where E_b is electric field in the thin skin layer on the plasma boundary. The complex ICP impedance can be then expressed as:

$$Z_p = \frac{U}{I_d} = \frac{2\pi a}{\sigma \delta l} = \frac{2\pi a m}{n_e e^2 \delta l}v_{en} + \frac{2\pi a m}{n_e e^2 \delta l}i\omega; \tag{12.77}$$

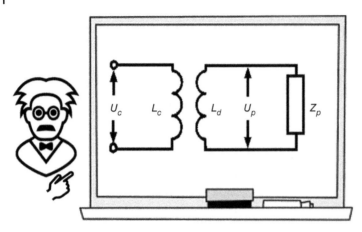

Figure 12.16 Equivalent circuit of an RF ICP discharge.

here n_e is electron density, v_{en} is the frequency of electron-neutral collisions. The plasma impedance has active (first term) and inductive (second term) components. In the circuit should be also a "geometrical" inductance attributed to the plasma considered as a conducting cylinder $L_d = \frac{\mu_0 \pi a^2}{l}$. If the skin layer is smaller than the discharge radius, the equivalent scheme of the ICP discharge can be represented as a transformer with a load in form of impedance (12.76), see Figure 12.16. The transformer can be characterized by plasma inductance L_d and coil inductance $L_c = \frac{\mu_0 \pi R^2}{l} N^2$, R is the coil radius, l is the coil length, and N is the number of its windings. The transformer is characterized by the mutual inductance $M = \frac{\mu_0 \pi a^2}{l} N$. Two equations for the equivalent circuit (Figure 12.16) describe amplitudes of the voltage applied to the inductor coil (c) and the voltage on the plasma loop (p):

$$U_c = i\omega L_c I_c + i\omega M I_d; \quad U_p = -I_d Z_p = i\omega M I_c + i\omega L_d I_d. \tag{12.78}$$

These equations lead to **analytical relations for the ICP discharge parameters**. The coil current I_c is determined by external circuit and can be considered as a given parameter. The electric field in plasma is related to the voltage on the plasma loop: $E_p = U_p / 2\pi a$, and only logarithmically depends on other plasma parameters. The electric field in plasma at high currents is smaller than in the idle regime: $E = \frac{\omega M I_c}{2\pi a}$, where M is the mutual inductance. The voltage drop U_p related to plasma can be neglected for high currents; therefore discharge current is:

$$I_d = -\frac{M}{L_d} I_c = -N I_c. \tag{12.79}$$

The current flows in direction opposite to the inductor current, and the plasma current exceeds that in the inductor. If the electric field is known, the electron density in plasma can be found from the current density in (12.78):

$$n_e = j \frac{m v_{en}}{e^2 E_p} = \frac{N I_c}{l \delta} \frac{m v_{en}}{e^2 E_p}. \tag{12.80}$$

Considering the Ohm's law for the current conducting plasma layer, the thickness of the ICP skin layer is:

$$\delta = \frac{2 E_p l}{\omega \mu_0 N I_c} \propto \frac{1}{I_c}. \tag{12.81}$$

Therefore, the ICP electron density (12.79) is proportional to square of electric current in inductor coil:

$$n_e = \left(\frac{N I_c}{e l E_p} \right)^2 \frac{\omega \mu_0 m v_{en}}{2} \propto I_c^2. \tag{12.82}$$

Similar to CCP, the ICP behavior especially plasma density and electric fields differ at moderate and low pressures, which is illustrated in Figure 12.17, see more details in Fridman and Kennedy (2021).

12.13 Planar Coil and Helical Resonator Configurations of the Low-pressure RF ICP discharges

The **ICP discharges in planar configuration** are widely used in electronics are illustrated in Figure 12.18a,b. They are geometrically like the RF CCP parallel plate reactors, but the RF power is applied here to a flat spiral inductive

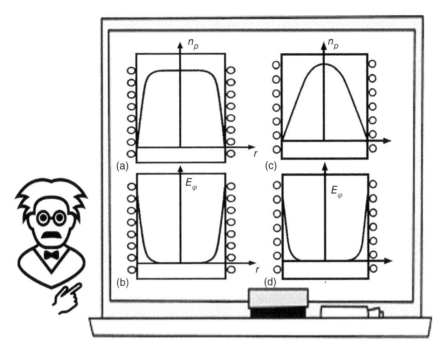

Figure 12.17 Radial distribution of plasma density (a, c) and electric field (b, d) in the RF ICP discharges at moderate (a, b) and low (c, d) pressures.

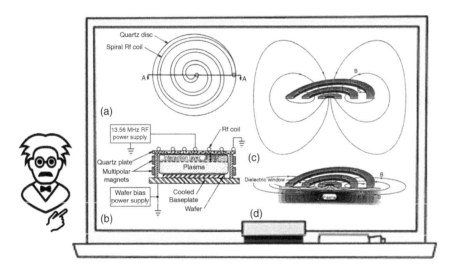

Figure 12.18 ICP parallel plate discharge reactor (a–d); RF magnetic field near the coil: without plasma (c), with plasma (d).

coil separated from plasma by dielectric insulating plate. The RF currents in the spiral coil induce image currents in the upper plasma surface corresponding to the skin layer providing the inductive coupling. The analytical relations derived for the low-pressure ICP discharge inside of an inductive coil can also be applied qualitatively for the planar coil configuration. The planar ICP discharges (Figure 12.18b) also include multi-polar permanent magnets (to improve plasma uniformity, confinement and to increase density) and a DC wafer bias (to control the energy of ions 30–400 eV impinging on the wafer). The magnetic field lines in the planar coil configuration are more complicated than in cylindrical inductive coil, see Figure 12.18c,d. They encircle the coil and are symmetric with respect to the plane of the coil. Deformation of the magnetic field in presence of plasma, formed below the coil is shown in Figure 12.18d. Conventional planar RF ICP discharges can produce uniform plasma for processing of wafers with diameters at least 20 cm. Power level here is about 2 kW, an order of magnitude greater than that of similar

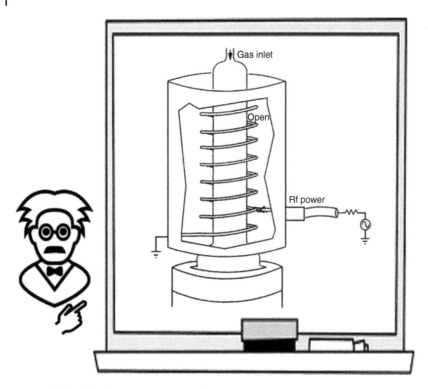

Figure 12.19 RF ICP helical resonator discharge.

CCP discharges. Typical frequency is 13.56 MHz, pressure from 1 to 20 mTorr, which is far lower than that of CCP discharges (several hundred mTorr).

Another special configuration of the low-pressure RF ICP discharges is **the helical resonator discharge**, see Figure 12.19. It consists of an inductive coil (helix) located inside of the cylindrical conductive screen and can be considered as a coaxial line with an internal helical electrode. Electromagnetic wave propagates in such a coaxial line with a phase velocity much lower than speed of light: $v_{ph} = \omega/k \ll c$, where k is the wavelength, and c is the speed of light. It allows the helical resonator to operate at the MHz frequencies, and permits generating low-pressure plasma. The coaxial line of the helical discharge becomes resonant, when an integral number of quarter waves of the RF field fit between the two ends of the system. The criterion of the simplest resonance is $2\pi r_h N = \frac{\lambda}{4}$, where r_h is the helix radius, N is number of turns in the coil, and λ is the electromagnetic wavelength in vacuum. Helical resonator discharges effectively operate at $3 - 30$ MHz, and do not require DC magnetic field. The resonators exhibit high Q-values, typically 600–1500 without plasma. In the absence of plasma, the electric fields are high facilitating the initial breakdown. The helical resonator discharges have high characteristic impedance and can be operated without matching network. Because of the resonance, large voltages necessarily appear between the open end of the helix and plasma. Hence the electric field is not exactly azimuthal in the helical resonator, and the discharge cannot be considered purely inductively coupled. The discharge sizes are close here to the electromagnetic wavelength, therefore the helical resonator discharge in some sense is like microwave discharges. More details on the subject can be found in the books of Lieberman and Lichtenberg (2005) and Roth (1995, 2001).

12.14 Nonthermal Wave-heated Plasma Sources: Electron–cyclotron Resonance (ECR) Microwave Discharges

The nonthermal discharges require high electric fields for heating of electrons and direct electron impact ionization. These electric fields can be achieved by high microwave power density, by applying resonators with high Q, or by application of magnetic field and effective electrons heating due to the electron cyclotron resonance (ECR). The ECR microwave discharges are the most conventional nonthermal low-pressure wave-heated plasma sources. In these discharges, a right circularly polarized waves (usually at microwave frequencies, e.g. 2.45 GHz) propagate along the strong DC-magnetic field (850 G at resonance) under the conditions of the ECR, which provides the wave absorption through a collisionless heating mechanism. The ECR-resonance between wave frequency ω and the electron cyclotron frequency $\omega_{Be} = eB/m$ (numerically $f_{Be}(\text{MHz}) = 2.8 \cdot B(\text{G})$) allows electron heating sufficient for ionization at low electric fields in the electromagnetic wave. Pressure in the ECR discharges system should be low $\nu_{en} \ll \omega_{Be}$ to provide the electron gyration sufficiently long to reach electron energy necessary for ionization. The microwave injection along the magnetic field ($\omega_{Be} > \omega$) allows the wave propagation to the absorption zone $\omega \approx \omega_{Be}$ even in dense plasma with $\omega_{pe} > \omega$, where the wave propagation in nonmagnetized plasma is impossible.

A typical configuration of the ECR-microwave discharge with microwave power injected along the axial nonuniform magnetic field is shown in Figure 12.20 (Lieberman and Lichtenberg 2005). Magnetic field profiles can provide multiple ECR-resonance positions as it is shown by the dashed line. A magnetic field coil at the wafer holder can be used to modify the uniformity of the etch or deposition. Typical ECR-microwave discharge parameters are: pressure 0.5–50 mTorr, power 0.1–5 kW, microwave frequency 2.45 MHz, volume 2–50 l, magnetic field about 1 kG, plasma density $10^{10} - 10^{12}\text{cm}^{-3}$, ionization degree $10^{-4} - 10^{-1}$ electron temperature 2–7 eV, ion acceleration energy 20–500 eV, typical source diameter is 15 cm.

Consider the magnetized electron heating in the ERC-microwave discharge sustained at low pressure by a linearly polarized electromagnetic wave, see Figure 12.21. This wave can be decomposed into the sum of two counter-rotating circularly polarized waves, right-hand-polarized and left-hand-polarized. The electric field of the right-hand-polarized wave rotates around the magnetic field at frequency ω, while an electron in uniform magnetic field also gyrates in the same "right-hand" direction at frequency ω_{Be}. Thus, at the ECR-resonance conditions $\omega = \omega_{Be}$, the electric field continuously transfers energy to the electron providing its heating. In contrast, the left-hand-polarized electromagnetic wave rotates in the direction opposite to the electron gyration.

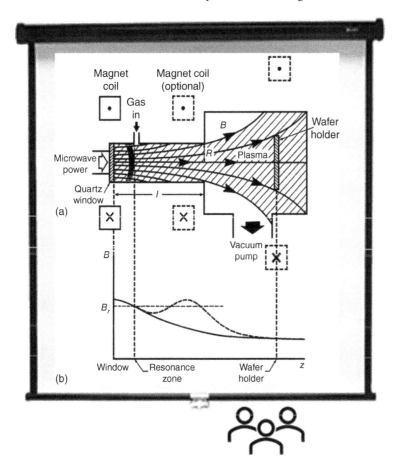

Figure 12.20 A high-profile ECR system: (a) configuration; (b) axial magnetic field distribution, showing one or more resonance zones.

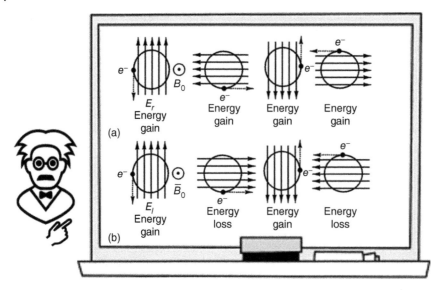

Figure 12.21 Mechanism of ECR electron heating: (a) continuous energy gain for right-hand polarization; (b) oscillating energy for left-hand polarization.

So, at the ECR-resonance $\omega = \omega_{Be}$, for a quarter of period the electric field accelerates the gyrating electrons, and for another quarter of period slows them down, resulting in no average energy gain. The described electron heating occurs only close to the ECR, which is necessary for the continuous energy transfer from microwave to an electron. Because the magnetic field and electron cyclotron frequency are not constant along the z-axis of the discharge, the electron heating occurs only locally, near the ECR-resonance, where the electron cyclotron frequency can be expressed as:

$$\omega_{Be}(z) = \omega \left(1 + \frac{1}{\omega} \frac{\partial \omega_{Be}}{\partial z} \Delta z \right) = \omega \left(1 + \frac{e}{\omega m} \frac{\partial B}{\partial z} \Delta z \right). \tag{12.83}$$

Then average electron energy gain per pass across the ECR-resonance zone in the collisionless regime is:

$$\varepsilon_{ECR} = \frac{\pi e E_r^2}{v_{res} |\partial B / \partial z|}; \tag{12.84}$$

here E_r is the amplitude of right-hand-polarized electromagnetic wave (which is a half of the linearly polarized microwave amplitude); $\partial B / \partial z$ is the gradient of magnetic field along the discharge axis near the resonance point; v_{res} is the component of electron velocity parallel to magnetic field also in the vicinity of resonance point. The width of the resonance zone along the discharge axis z, where most of the energy is transferred to the electron is:

$$\Delta z_{res} = \sqrt{\frac{2\pi m v_{res}}{e |\partial B / \partial z|}}. \tag{12.85}$$

The absorbed electromagnetic wave power per unit area (or microwave energy flux) is:

$$S_{ECR} = \frac{\pi n_e e^2 E_r^2}{e |\partial B / \partial z|}, \tag{12.86}$$

where n_e is the electron density in the ECR-resonance zone.

12.15 Helicon and Surface-wave High-density Plasma (HDP) Discharges

These nonthermal low-pressure HDP discharges are sustained by electromagnetic waves propagating in magnetized plasma in the so-called helicon modes. The driving frequency is in the RF range of $1 - 50$ MHz (13.56 MHz is commonly used for material processing). The helicon discharges can be considered as wave heated even though they operate in the RF range, which can be explained considering that the phase velocity of electromagnetic waves in magnetized plasma can be much lower than the speed of light. Magnetic field in the helicon discharges varies from 20 to 200 G (for fundamental plasma studies it reaches 1000 G) and is much lower than in the ECR-microwave discharges.

Plasma density is usually $10^{11} - 10^{12} \text{cm}^{-3}$ but can reach $10^{13} - 10^{14} \text{cm}^{-3}$. The helicon wave provided by RF antenna is coupled to the transverse mode structure across an insulating chamber wall. The wave then propagates along the plasma column in the magnetic field, and plasma electrons due to collisional or collisionless damping mechanisms absorb the mode energy. The plasma potentials in the helicon discharges are typically low, about 15–20 V, like in the ECR-microwave discharges. Important advantages of the helicon discharge with respect to ECR discharges are low magnetic fields and applied frequencies.

Let's consider first the helicon discharges analyzing propagation of the *helicon and whistler modes in magnetized plasma*. The **helicon waves** are propagating electromagnetic wave modes in an axially magnetized, finite diameter plasma column. The electric and magnetic fields have here radial, axial and, usually, azimuthal variations. They propagate in low frequency, high plasma density regime with relatively low magnetic fields: $\omega_{LH} \ll \omega \ll \omega_{Be}$, $\omega_{pe}^2 \gg \omega \omega_{Be}$. Here ω_{pe} is the electron plasma frequency; ω_{Be} is the electron cyclotron frequency; and ω_{LH} is the lower hybrid frequency in magnetized plasma:

$$\frac{1}{\omega_{LH}^2} \approx \frac{1}{\omega_{pi}^2} + \frac{1}{\omega_{Be}\omega_{Bi}}, \tag{12.87}$$

ω_{pi} is the plasma-ion frequency, and ω_{Bi} is the ion cyclotron frequency.

The right-hand polarized electromagnetic waves in magnetized plasma with frequencies between ion and electron cyclotron frequencies $\omega_{Bi} \ll \omega \ll \omega_{Be}$ are known as the **whistler waves**. For the right-hand polarized electromagnetic wave propagating along the magnetic field at frequencies below the frequency of the ECR, the dispersion equation is:

$$\frac{k^2 c^2}{\omega^2} = 1 + \frac{\omega_{pe}^2}{\omega \omega_{Be}}. \tag{12.88}$$

Propagation of the electromagnetic whistler waves is possible at frequencies below the plasma frequency $\omega < \omega_{pe}$. Considering the helicon frequency conditions and introducing the wave number $k_0 = \omega/c$ for the wave propagation without plasma, the dispersion equation for the whistler waves (12.87) becomes:

$$\omega = \frac{k_0^2 \omega_{pe}^2}{k^2 \omega_{Be}}. \tag{12.89}$$

In the helicon discharges $\omega_{pe}^2 \gg \omega_{Be}\omega$, which together with the dispersion equation gives $k^2 \gg k_0^2$. It means that wavelengths in these discharges with magnetized plasma are much lower than those without magnetic fields. Therefore, in contrast to RF CCP and RF ICP discharges, the helicon discharges are characterized by wavelengths comparable with the system size and are wave-heated, even though they operate at relatively low RF.

The electromagnetic modes in the helicon discharges superpose electromagnetic ($div\vec{E} \approx 0$) and quasi-static ($curl\vec{E} \approx 0$) fields: $\vec{E}, \vec{H} \propto \exp i(\omega t - kz - m\theta)$, where θ is the angle between the wave propagation vector and magnetic field, and the integer m specifies the azimuthal mode. Conventional helicon plasma sources have been developed based on excitation of the $m = 0$ and $m = 1$ modes. The $m = 0$ mode is axisymmetric and $m = 1$ mode has a helical variation, therefore both modes generate time-averaged axisymmetric field intensities. A schematic of RF antenna to excite the helicon $m = 1$ discharge mode is shown in Figure 12.22. The antenna generates B_x magnetic field over axial antenna length, which can be coupled to the transverse magnetic field of the helicon mode. The antenna also induces an electric current within the plasma column beneath each horizontal wire. This current produces a charge of opposite signs at the two ends of the antenna. These charges generate a transverse quasi-static RF electric field E_y, which can be coupled to the transverse quasi-static fields of the helicon mode. The helicon mode energy can be transferred to the plasma electron heating by collisional damping or collisionless Landau damping. The collisional damping mechanism transfers the electromagnetic wave energy to the thermal bulk electrons, while the collisionless Landau damping preferentially heats non-thermal electrons with energies much exceeding the bulk electron temperature. The collisional damping mechanism dominates at relatively higher pressures, while the Landau damping dominates at lower pressures (less than 10 mTorr in the case of argon).

In the **surface-wave discharges**, electromagnetic wave propagates along plasma surface and is absorbed by collisional heating of the plasma electrons near the surface. The heated electrons then diffuse from the surface into the bulk plasma. Surface wave discharges can be excited by either RF or microwave sources (the most conventional frequency however is $1 - 10$ GHz) and they do not require DC-magnetic field. The plasma potential with respect to all wall surfaces for the wave-heated discharges is relatively low (about five electron temperatures, $5T_e$) like in the case of ICP discharges. The surface wave discharges can generate the HDP with diameters as large as 15 cm. The electromagnetic surface wave can be arranged in different configurations. One of them is a planar configuration on the plasma-dielectric interface. In another configuration, plasma is separated from a conducting plane by a dielectric slab. This planar system also admits propagation of a surface wave that decays into the plasma region. Although this electromagnetic wave does not decay into the dielectric, it is confined within the dielectric layer by conducting plane. Finally, a surface wave can propagate in the cylindrical geometry on a nonmagnetized plasma column confined by a

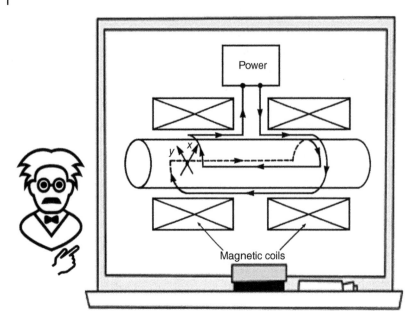

Figure 12.22 Helicon HDP discharge, antenna excitation of $m = 1$ helicon mode.

thick dielectric tube. More details on the subject can be found in the books of Lieberman and Lichtenberg (2005) and Fridman and Kennedy (2021).

12.16 Problems and Concept Questions

12.16.1 Microwave Discharge in H_{01}-mode of Rectangular Waveguide

Calculate the maximum electric field for an electromagnetic wave in the H_{01}-mode of a rectangular waveguide at the frequency $f = 2.5$ GHz (wavelength $\lambda = 12$ cm). Assume the microwave discharge power is 1 kW, the wide waveguide wall is equal to 7.2 cm, and the narrow one is 3.4 cm long.

12.16.2 Equivalent Circuit of RF CCP Discharge

Based on the equivalent scheme of the ICP discharges, find out the resonant frequency of electromagnetic oscillations for this circuit.

12.16.3 Critical Current of the $\alpha - \gamma$ Transition in moderate pressure CCP

Analyze the dependence of the critical current of the $\alpha - \gamma$ transition on RF frequency and gas pressure.

12.16.4 Stochastic Heating Effect

Analyze reflection of electrons from the sheath boundaries moving to and from electrodes and analyze the energy transfer to an electron.

12.16.5 Current Density Distribution in ICP Discharges

For the ICP discharge in inductor coil, analyze the current density distribution ((12.74) and (12.75)) along the discharge radius.

12.16.6 Equivalent Circuit of ICP Discharges

Analyze the equivalent scheme of the ICP discharge, shown in Figure 12.16, and explain why the discharge current exceeds the current in the inductor coil exactly N-times, where N is the number of windings in the coil.

12.16.7 Plasma Density in ICP Discharges

Why low-pressure ICP discharges operate at electron densities $10^{11} - 10^{12} \text{cm}^{-3}$ (even up to 10^{13}cm^{-3}), at least 10 times more than in CCP.

12.16.8 ECR-microwave Absorption Zone

Using formulas ((12.84) and (12.85)), estimate the width of the ECR-resonance zone and the absorbed microwave power per unit area.

Lecture 13

Atmospheric Pressure Cold Plasma Discharges: Corona, Dielectric Barrier Discharge (DBD), Atmospheric Pressure Glow (APG), Plasma Jet

13.1 Physics of Continuous Corona Discharges

Corona is the oldest known cold atmospheric pressure discharge. It is a weakly luminous discharge near the sharp points, edges, or thin wires where the electric field is sufficiently high. Coronas are always nonuniform: strong electric field, ionization, and luminosity are only in the vicinity of one electrode. Weak electric fields then drag charged particles to another electrode to close the circuit. No radiation appears from the "outer region" of the corona. Corona can be observed in air around high voltage transmission lines, around lightning rods, and even masts of ships, where they are called "Saint Elmo's fire." This is the origin of term corona, which means "crown."

 High voltage is required to ignite the corona, which occupies the region around one electrode. If the voltage grows even larger, the remaining part of the discharge gap breaks down and *the corona transfers into the spark,* which is a transitional quasi-thermal plasma discharge. The near-electrode electric field in coronas is much stronger than in the rest of the discharge gap because the characteristic size of an electrode r is much smaller than distance d between electrodes. For example, corona discharge in air between parallel wires occurs only if the inter-wire distance is sufficiently large $d/r > 5.85$. Otherwise, an increase of voltage results not in a corona, but in a spark.

 Neglecting plasma effects, the **electric field distribution in coronas** can be found analytically in the framework of electrostatics. For simple gaps (Figure 13.1), these field distributions without plasma are:

(a) The electric field in corona between **coaxial cylinders** of radii r (internal) and R (external) at distance x from the axis (Figure 13.1a) can be expressed depending on applied voltage V:

$$E = \frac{V}{x \ln(R/r)}, \quad E_{max}(x = r) = \frac{V}{x \ln(R/r)}. \tag{13.1}$$

(b) The electric field between **concentric spheres** (radii r and R, x is distance from the center, Figure 13.1b:

$$E = V \frac{rR}{x^2(R - r)}, \quad E_{max} \approx \frac{V}{r} \ (\text{if } R \gg r). \tag{13.2}$$

(c) The electric field in corona between **a sphere of radius r and a remote plane** $d/r \rightarrow \infty$ (Figure 13.1c) can be expressed as a function of distance x from the sphere center:

$$E \approx V \frac{r}{x^2}, \quad E_{max} = \frac{V}{r}. \tag{13.3}$$

(d) The electric field in corona between **a parabolic tip with curvature radius r and a plane perpendicular to it** at a distance d from the tip (Figure 13.1d) as a following function of distance x from the tip:

$$E = \frac{2V}{(r + 2x) \ln(2d/r + 1)}, \quad E_{max} \approx \frac{2V}{r \ln(2d/r)}. \tag{13.4}$$

(e) If corona is organized between **two parallel wires of radius r separated by distance d between each other, and b from the ground** (Figure 13.1e), the maximum electric field is near the surface of the wires and equals to:

$$E_{max} = \frac{V}{r \ln[d/\{r \cdot \sqrt{1 + (d/2b)^2}\}]}. \tag{13.5}$$

(f) Maximum field (13.5) for **a single wire and parallel plane** at distance b ($d \rightarrow \infty$), see Figure 13.1f, is:

$$E_{max} = \frac{V}{r \ln(2b/r)}. \tag{13.6}$$

Plasma Science and Technology: Lectures in Physics, Chemistry, Biology, and Engineering, First Edition. Alexander Fridman.
© 2024 WILEY-VCH GmbH. Published 2024 by WILEY-VCH GmbH.

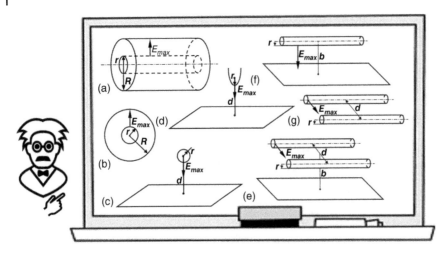

Figure 13.1 Different configurations of corona discharges, see (13.1)–(13.7).

(g) The maximum electric field between **two parallel wires** at a distance d ($b \to \infty$, see Figure 13.1g) is:

$$E_{max} = \frac{V}{r \ln(d/r)}. \tag{13.7}$$

Ionization mechanism in corona depends on polarity of the high electric field electrode. If the high electric field zone is around cathode, such discharge is referred to as the **negative corona**. Conversely, it is called **positive corona**. Ionization in the negative corona is due to multiplication of avalanches. Continuity of electric current from cathode to plasma is provided by secondary emission from cathode. Ignition of the negative corona has the Townsend breakdown mechanism considering nonuniformity and possible electron attachment processes:

$$\int_0^{x_{max}} [\alpha(x) - \beta(x)]dx = \ln\left(1 + \frac{1}{\gamma}\right); \tag{13.8}$$

here $\alpha(x)$, $\beta(x)$, and γ are the first, second, and third Townsend coefficients, describing respectively ionization, electron attachment, and secondary electron emission; x_{max} corresponds to the distance from the cathode, where the electric field becomes low enough and $\alpha(x_{max}) = \beta(x_{max})$, which determines not only ionization, but also the electronic excitation zone, and hence the zone of plasma luminosity (or an active corona volume). In the **positive corona**, the ionization processes are due to the cathode-directed streamers. The Meek breakdown criterion is a good approximation here, considering the corona nonuniformity and possible contributions of electron attachment:

$$\int_0^{x_{max}} [\alpha(x) - \beta(x)]dx \approx 18 - 20. \tag{13.9}$$

The critical electric field for ignition of positive and negative coronas are close although they are related to different breakdown mechanisms. It is due to strong exponential dependence of α on electric field. The Townsend coefficient α in air at $E/p < 150 \, \text{V cm}^{-1} \cdot \text{Torr}^{-1}$ can be approximated as:

$$\alpha, 1 \, \text{cm}^{-1} = 0.14 \cdot \delta \left[\left(\frac{E, \text{kV cm}^{-1}}{31\delta} \right)^2 - 1 \right]. \tag{13.10}$$

Then the igniting electric field for the coaxial electrodes in air can be calculated using the **Peek formula**:

$$E_{cr}, \text{kV cm}^{-1} = 31\delta \left(1 + \frac{0.308}{\sqrt{\delta \, r(\text{cm})}} \right); \tag{13.11}$$

here δ is ratio of air density to the standard one; r is radius of internal electrode. The formula can be applied for pressures $0.1 - 10$ atm, polished internal electrodes with radius $r \approx 0.01 - 1$ cm, with both direct current and AC with frequencies up to 1 kHz. The electrode's roughness decreases the critical electric field near electrode by $10 - 20\%$. The Peek formula for the critical electric field near electrode for two parallel wires is similar to (13.10):

$$E_{cr}, \text{kV cm}^{-1} = 30\delta \left(1 + \frac{0.301}{\sqrt{\delta \, r(\text{cm})}} \right). \tag{13.12}$$

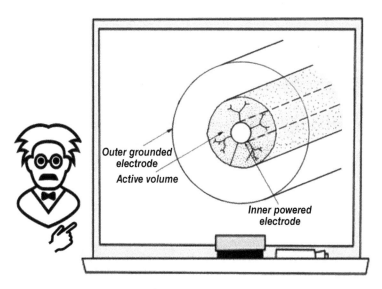

Figure 13.2 Active volume of corona discharge around a thin wire.

Generation of charged particles takes occurs in corona only in the electrode vicinity (called the **active corona volume**) where electric field is sufficiently high, see Figure 13.2. Radius of the active corona volume is determined by the breakdown electric field E_{break} on the boundary of the active volume, and in the case of the thin wire is:

$$r_{AC} = \frac{V}{E_{break}\ln(R/r)};\tag{13.13}$$

here V is the corona voltage. *The radius of active corona volume is about three times greater than the wire radius.*

To analyze the **space charge influence on the corona electric field distribution**, let's consider a corona generated between coaxial cylinders with radii R and r (e.g. corona around a thin wire). The electric current per unit length of the wire i is constant outside of active corona volume, where there is no charge multiplication:

$$i = 2\pi x \cdot en \cdot \mu E = const;\tag{13.14}$$

here x is distance from axis; n is the number density of charged particles providing the electric conductivity outside of the active volume; μ is the mobility of charged particles. The space charge perturbation of electric field is not strong, and the number density distribution $n(x)$ can be found from (13.14) and nonperturbed electric field (13.1):

$$n(x) = \frac{i}{2\pi e\mu Ex} = \frac{i\ln(R/r)}{2\pi e\mu V} = const.\tag{13.15}$$

The electric field distribution $E(x)$, can be found then using the Maxwell equation:

$$\frac{1}{x}\frac{d[xE(x)]}{dx} = \frac{1}{\varepsilon_0}en(x), \quad \frac{1}{x}\frac{d[xE(x)]}{dx} = \frac{i\ln(R/r)}{2\pi\varepsilon_0\mu V}.\tag{13.16}$$

It can be integrated, considering $E(x)$ following (13.1) with the critical voltage V_{cr} (corresponding to corona ignition) at the low current limit $i \to 0$. It gives the electric field distribution considering the space charge:

$$E(x) = \frac{V_{cr}\ln(R/r)}{x} + \frac{i\ln(R/r)}{2\pi\varepsilon_0\mu V}\cdot\frac{x^2 - r^2}{2x}.\tag{13.17}$$

Obviously, in this relation $\int_r^R Edx = V$.

13.2 Continuous Corona: Current–Voltage Characteristics and Discharge Power

Integration of the electric field (13.17) over the radius x considering $x^2 \gg r^2$, gives the relation between current (per unit length) and voltage of the discharge, which is the **current–voltage characteristic of corona** generated around a thin wire:

$$i = \frac{4\pi\varepsilon_0\mu V(V - V_{cr})}{R^2 \ln(R/r)}. \tag{13.18}$$

The corona current depends on the mobility of the charge particles providing conductivity outside of the active corona volume. Mobilities of positive and negative ions are nearly equal, therefore electric currents in positive and negative corona discharges are also close. Negative corona in gases without electron attachment (e.g. noble gases) provides much larger currents because electrons can rapidly leave the discharge gap without forming a significant space charge. The parabolic current–voltage characteristic (13.18) is valid not only for thin wires but for other corona configurations as well. Obviously, the coefficients before the quadratic form $V(V - V_{cr})$ are different for different configurations of corona discharges:

$$I = CV(V - V_{cr}); \tag{13.19}$$

here I is the total current in corona. For example, the current–voltage characteristic for the corona generated in atmospheric air between a sharp point cathode with radius $r = 3 - 50\,\mu m$ and a perpendicular flat anode located on the distance of $d = 4 - 16\,mm$ can be expressed as:

$$I, \mu A = \frac{52}{(d, mm)^2}(V, kV)(V - V_{cr}). \tag{13.20}$$

The corona ignition voltage V_{cr} is here $V_{cr} \approx 2.3\,kV$ and does not depend on distance d.

The **electric power released in the continuous corona discharge** around long thin wire (length L) can be expressed based on (13.18) as:

$$P = \frac{4\pi L\varepsilon_0\mu V(V - V_{cr})}{R^2 \ln(R/r)}. \tag{13.21}$$

In more general case of the corona discharges, the power can be determined as:

$$P = CV(V - V_{cr}). \tag{13.22}$$

For the corona generated in atmospheric air between the sharp pointed cathode with radius $r = 3 - 50\,\mu m$ and a perpendicular flat anode located at distance $d = 1\,cm$, the coefficient $C \approx 0.5$ if voltage is expressed in kV and power in mW. This power is about 0.4 W at the voltage of 30 kV. Corona generated in atmospheric air around thin wire ($r = 0.1\,cm$, $R = 10\,cm$, $V_{cr} = 30\,kV$) with voltage 40 kV releases $0.2\,W\,cm^{-1}$ of the discharge. Although, the corona power is very low, it can be significant in long wires. It occurs in the high voltage overland transmission lines, where coronal losses can exceed the resistive losses. The coronal power losses for a transmission line wire with 2.5 cm diameter and voltage of 300 kV are $0.8\,kW\,km^{-1}$ at fine and sunny weather. In rainy or snowy conditions, the critical voltage V_{cr} is lower, coronal losses grow significantly, and the losses of about $0.8\,kW\,km^{-1}$ correspond to a voltage about 200 kV.

13.3 Pulsed Corona Discharges

To increase the corona current and power, the voltage and therefore electric field should be increased. As the electric field increases, the active corona volume grows until it occupied the entire discharge gap. When streamers can reach the opposite electrode, spark channels occur resulting in local overheating and plasma nonuniformity. *Increasing of corona voltage and power without spark formation is possible using pulse-periodic voltages.* Streamer velocity is about $10^8\,cm\,s^{-1}$ and 10 times exceeds the electron drift velocity in avalanches. If the distance between electrodes is 1–3 cm, the total time necessary for the development of avalanches, avalanche-to-streamer transition, and streamer propagation between electrodes is 100–300 ns. The voltage pulses of this and shorter duration can sustain streamers and energy transfer into nonthermal plasma without streamer transformations into quasi-thermal sparks. The *pulsed coronas are more often organized as positive ones.* Typical ignition delay in positive corona is 30–300 ns. This interval is longer than streamer propagation time; which is related to the initial electrons formation time and propagation of initial avalanches.

Corona discharges can also operate as **periodic corona-current pulses at constant voltage**. Frequency of these pulses can reach 10^4 Hz in the case of positive corona and 10^6 Hz for negative corona. Such self-organized pulsed corona discharges are unable to overcome the current and power limitations of the continuous corona discharges because continuous high voltage still promotes the corona-to-spark transition. Consider at first a **self-pulsing positive corona discharge** formed between sharp point anode of radius 0.17 mm and flat cathode located 3.1 cm apart. Discharge ignition takes place at the critical voltage $V_{cr} \approx 5$ kV. This corona operates in the pulse-periodic regime starting from the ignition voltage to a voltage $V_1 \approx 9.3$ kV. Near the boundary voltages (V_{cr}, V_1), the frequency of the current pulses is low. The frequency of the pulses reaches the maximum 6.5 kHz in the middle of the interval (V_{cr}, V_1). This pulsing discharge is usually referred to as a **flashing corona**. The mean current reaches 1 μA at the voltage $V_1 \approx 9.3$ kV. The voltage increase from $V_1 \approx 9.3$ kV to $V_2 \approx 16$ kV stabilizes the corona. Further increase of voltage from $V_2 \approx 16$ kV up to the corona transition into a spark at $V_t \approx 29$ kV leads again to the pulse-periodic regime. Frequency grows up in this regime to about 4.5 kHz; mean value of current grows to 100 μA. The phenomenon can be explained by the positive space charge created when electrons formed in streamers fast disappear at the anode while slow positive ions remain in the discharge gap. The growing positive space charge decreases electric field near anode and prevents new streamers generation. Positive corona current is suppressed until the positive space charge goes to the cathode and clears up the gap. After that a new corona ignition takes place and the cycle can be repeated. The flashing coronas do not occur at the intermediate voltages $(V_1 < V < V_2)$, when the electric field outside of the active corona volume is sufficiently high for effective steady-state clearance of positive ions from the discharge gap but not high enough to provide intensive ionization.

Negative self-pulsing corona discharges sustained by continuous voltage also occur at low voltages close to ignition and called the **Trichel pulses**. Frequency of the pulses is much higher in this case ($10^5 - 10^6$ Hz); the pulse duration is short, about 100 ns. If the mean corona current is 20 μA, the peak pulse value current reaches 10 mA. Physical mechanism of the Trichel pulses is like that of the flashing corona and related to the space charge, which is due to the formation of negative ions. Therefore, the Trichel pulses occur only in electronegative gases.

The most practically important are **pulsed coronas sustained by nano-second pulse power supplies**, because of their ability to operate at higher powers without transition to sparks, see Figure 13.3a. The key element of these discharge systems is the nano-second pulse power supply generating short pulses (around 10–100 ns) to avoid the corona-to-spark transition. These power supplies provide very high voltage rise rate (up to about 100 kV ns^{-1}), which results in higher corona ignition voltage, and higher power.

The effect of increase of the voltage raise rate on the pulsed corona inception voltage is illustrated in Figure 13.3b, see more details in Fridman and Kennedy (2021). The most typical pulse corona configuration is based on using thin wires, which maximizes the active discharge volume (Figure 13.3a). This configuration is limited, however, by durability of the electrodes and not optimal interaction of the discharge volume with incoming gas flow. More

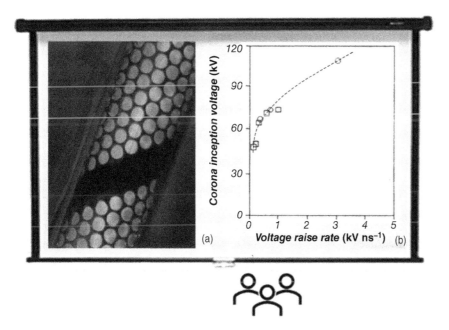

Figure 13.3 Pulsed corona discharge: (a) organization around a wire; (b) corona inception voltage as a function of voltage raise rate.

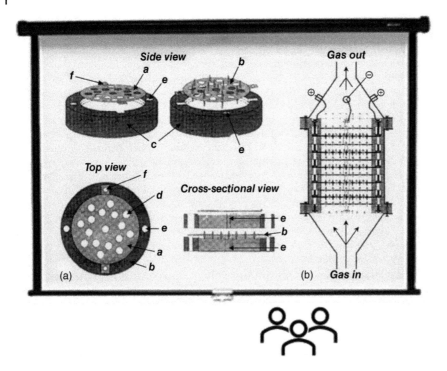

Figure 13.4 Pulsed corona in multi pin-to-plate configuration: schematic of electrodes and (a) and cross-sectional view of the assembled plasma reactor (b); a – anode plates; b – cathode plate; c – mounting block; d – holes for gas flow; holes for connecting post; f – connection wings.

practical from this perspective is corona configuration using multiple pin-to-plate electrodes, see Figure 13.4. Special pulse corona configurations permitted cold plasma combination with catalysts, water shower (called the **spray corona**), and thin liquid film on the walls (called **wet corona**). More details on the subject can be found, in particular, in Sobacchi et al. (2002) and Fridman and Kennedy (2021).

13.4 Dielectric-barrier Discharges (DBD): General Features and Configurations, Filamentary DBD Mode

Transition to sparks at high voltages and pressures can be achieved by placing a dielectric barrier in the discharge gap. This is the key idea of the **dielectric barrier discharges** (DBD) introduced by von Siemens in 1857. The dielectric barriers in the discharge gap surely require application of high frequency AC or pulsed voltage. Sometimes DBDs are also called **silent discharges** because of the absence of sparks, and therefore no local overheating, local shock waves, and noise. The DBD gap usually includes one or more dielectric layers (barriers), which are in the current path between metal electrodes. Two major DBD configurations, planar and cylindrical are illustrated in Figure 13.5. Typical clearance in the discharge gaps varies from 0.1 mm to several centimeters. Breakdown voltages are practically the same as those between metal electrodes without barriers. If the DBD gap is a few millimeters, the required AC driving voltage with frequency 500 Hz–100 kHz is about 10 kV in air. Typically, frequencies are 10–30 kHz to give possibility for ions to leave the gap before changing DBD polarity and starting a new cycle. The DBDs proceed in most of the conditions through many independent current filaments usually referred to as **DBD micro-discharges** or the **DBD filamentary mode**. From a physical point of view, these micro-discharges (filaments) are streamers, which are self-organized because of charge accumulation on the dielectric surface. Typical characteristics of the DBD micro-discharges in a 1-mm gap in atmospheric air are summarized in Table 13.1.

The snapshot of the filaments (micro-discharges) in a 1-mm DBD air gap photographed through a transparent electrode is shown in Figure 13.6a (original size 6 × 6 cm, exposure time 20 ms). The side view of the filamentary DBD is illustrated in Figure 13.6b. The micro-discharges are spread over the whole DBD zone quite uniformly. The extinguishing voltage of the micro-discharges is not far below the voltage of their ignition. Charge accumulation on surface of the dielectric barrier reduces the electric field at the location of a micro-discharge, which results in current termination within just several nanoseconds after breakdown. The short duration of micro-discharges leads to low

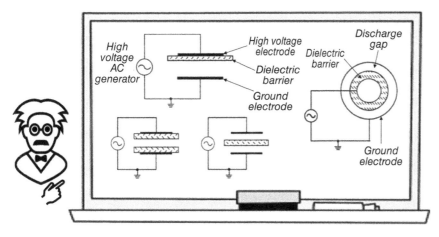

Figure 13.5 Conventional dielectric-barrier discharge (DBD) configurations.

Table 13.1 Typical parameters of DBD micro-discharge in a filament.

Lifetime	1–20 ns	Filament radius	$50 \div 100\,\mu m$
Peak current	0.1 A	Current density	$0.1 \div 1\,kA\,cm^{-2}$
Electron density	$10^{14} \div 10^{15}\,cm^{-3}$	Electron energy	1–10 eV
Charge per cycle	0.1–1 nC	Reduced electric field	$E/n = (1 \div 2)(E/n)_{Paschen}$
Energy per cycle	$5\,\mu J$	Gas temperature	300 K
Overheating per cycle	5 K		

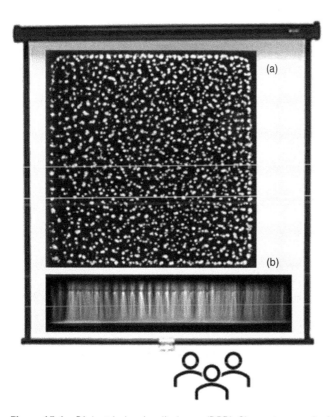

Figure 13.6 Dielectric-barrier discharge (DBD), filamentary mode: (a) end-on view micro-discharges; (b) side view.

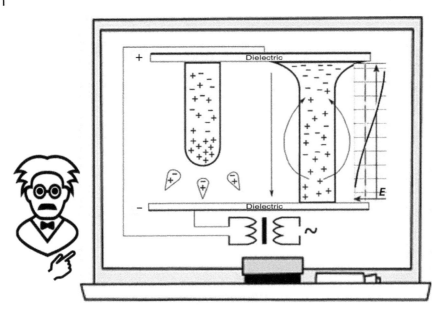

Figure 13.7 Repulsion of filaments (micro-discharges) in dielectric-barrier discharges (DBD).

overheating of the streamer channel, and the DBD plasma remains nonthermal. When the voltage is reversed, the next micro-discharges will be formed at the same locations because of the deposition of charges from the previous

micro-discharges. Thus, *the DBDs tend to reignite the old micro-discharge channels every half-period. These "family of micro-discharges striking in the same location we visually observe as filaments*. The principal micro-discharge properties (pulse energy about $5\,\mu J$ in air) do not depend much on the characteristics of the external circuit, but only on the gas composition, pressure, and electrode configuration. *Power of each filament is about the product of pulse energy ($5\,\mu J$ in air) and frequency. An increase of total power just leads to generation of a larger number of filaments.*

The DBD filaments interact with each other especially strongly when the DBD power is high and distance between the micro-discharges is short. Electrostatic repulsion of ions in the filaments as well as electric field E distribution along the streamer (decrease of E near anode and increase near cathode) results in preferential formation of new micro-discharges slightly further from each other, or in other words in **quasi-repulsion of the DBD filaments**, see Figure 13.7. The effect of filaments repulsion results in the **DBD filament patterning**, structuring, formation of the so-called "**2D Coulomb crystals**" of the DBD filaments. It is illustrated in Figure 13.8a,b by using the Voronoi Polyhedra characterizing level of the DBD pattern structuring, see Fridman et al. (2005a,b,c).

13.5 Nanosecond-pulsed Dielectric-barrier Discharges (DBD), Uniform DBD Mode

Similarly to the pulsed coronas (Section 13.3), the DBDs can be effectively organized in the micro- and nanosecond-pulsed regimes. Again, the key element of these systems are "pulsers" and especially the nano-second pulse power supply with pulses about 10–100 providing the voltage rise rate up to about $100\,kV\,ns^{-1}$. For comparison, the typical voltage rise rate in microsecond-pulsed DBD is about $5\,V\,ns^{-1}$, and in continuous wave sinusoidal DBD – about $1\,V\,ns^{-1}$). So *sharp increase of voltage provided by the nano-second pulsers permits achieving extremely high electric fields* up to 10 times exceeding those required for conventional breakdown, the so-called **overvoltage effect**, see similar tendency in Figure 13.3b. The reduced electric fields in air can achieve 1000 Td. The voltage grows so fast that it passes the critical one without breakdown simply because there is no time for the breakdown. *The significant overvoltage results, in particular, in fast multiplication of streamers and domination of the anode-directed-streamers over the conventional cathode-directed streamers*, see Section 9.3.

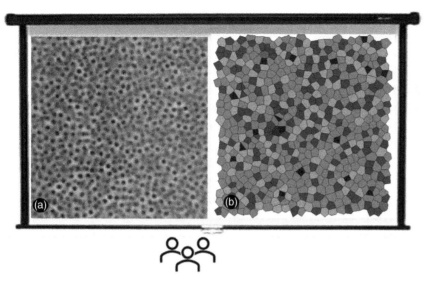

Figure 13.8 Patterning (structuring) of the DBD filaments: (a) DBD filaments pattern; (b) relevant Voronoi Polyhedra demonstrating high level of the DBD filaments self-organization.

 The significant overvoltage effect and domination of the cathode-directed over anode-directed streamers permits **DBD transition from filamentary to uniform mode** in the nanosecond-pulsed plasma. The anode-directed streamers developed at higher electric fields become "blind streamers" propagating from cathode to anode without much feedback between electrodes and therefore without necessary localization of breakdown and therefore without necessity of filamentation, see Figure 13.9. The nonuniformity-to-uniformity transition usually occurs at reduced electric fields about 400 Td, see Figure 13.10.

It is interesting that in nonuniform filamentary mode, the maximum electric field corresponding to the streamer heads significantly exceeds applied voltage, while in the uniform regime, these electric fields are close. Image comparison of the nonuniform filamentary structure of continuous-wave DBDs with intermediate uniformity in microsecond-pulsed DBDs, and high uniformity of the nanosecond-pulsed DBDs is illustrated in Figure 13.11. For more details on the subject see Fridman et al. (2008), Liu et al. (2014), and other publications of Professor Dobrynin.

Typical **initial evolution stages of the pulsed DBD** can be described in three steps illustrated in Figure 13.12 (Dobrynin et al. 2019a,b):

(1) During the first few hundreds of picoseconds, the discharge starts with the development of avalanches traveling from the negative (in this case – grounded) electrode toward the high-voltage positive electrode (anode) covered with a dielectric.

(2) As the avalanches reach the anode, electron concentration and local electric field are sufficient for the initiation of streamers. At this point, at about 1 ns, the presence of a dielectric surface facilitates the development of a surface discharge (surface-directed streamers) which shows up in experiments as a bright glow area near the positive dielectric-covered electrode. This **electrode glow–"pancake"** appears prior to the main volumetric discharge due to effects of the surface. In the equivalent-circuit language, the "pancake" can be described as an additional capacitor, which accumulates a portion of full discharge energy.

(3) In the third stage, the main volumetric discharge starts to develop with evolution of cathode-directed streamers. This stage corresponds to the most energetic phase of the DBD.

 The **uniform DBD mode at atmospheric pressure without nanosecond pulsing** can be also arranged in some special conditions. This so-called DBD glow mode can be operated at lower voltages (down to hundreds of volts). In this case, streamers are avoided because the electric fields are below the Meek criterion and discharge operates in the Townsend ionization regime. Secondary electron emission from dielectric surfaces, which sustains the Townsend ionization regime, relies on the uniform DBD glow upon adsorbed electrons (with binding energy only about 1 eV) that were deposited during previous DBD excitation (high voltage) cycle. If enough electrons "survive" voltage switching time without recombining, they can trigger transition to the homogeneous

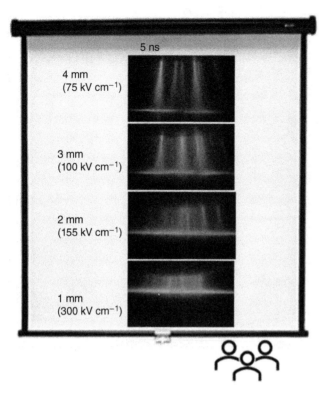

Figure 13.9 Transition of the pulsed DBD in air from filamentary to uniform mode by reducing gap and increasing electric field from 75 to 300 kV cm⁻¹.

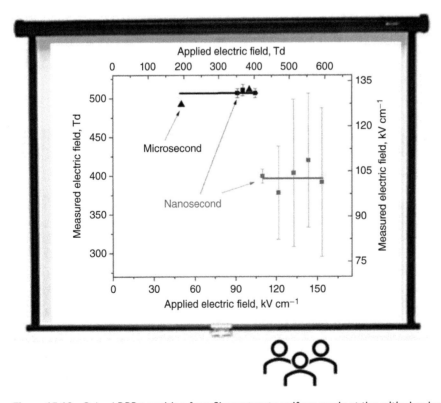

Figure 13.10 Pulsed DBD transition from filamentary to uniform mode at the critical reduced electric field 400 Td.

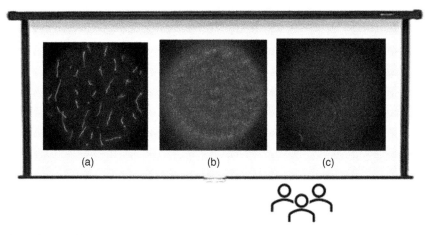

Figure 13.11 Comparison of DBD uniformity (typical end-on view): (a) continuous-wave; (b) microsecond-pulsed; (c) nanosecond-pulsed discharges.

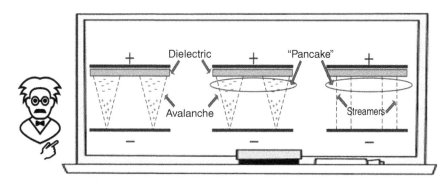

Figure 13.12 Three stages of DBD development: left, avalanches propagate from cathode (bottom) to anode (top); center, the "pancake" appears near anode; right: streamers develop from the "pancake" and start propagating toward cathode.

Townsend mode of DBD. "Survival" of electrons and crucial active species between cycles or the **DBD memory effect** is critical for organization of the **uniform atmospheric pressure DBD glow** and depends on properties of dielectric surface as well as operating gas. In electronegative gasses, the memory effect is weaker because of attachment losses of electrons. The avalanche-to-streamer transition depends here on the level of pre-ionization. Meek criterion is related to an isolated avalanche, while in the case of intensive pre-ionization, avalanches are produced close to each other and therefore interact. If two avalanches occur close enough, their transition to streamers can be electrostatically prevented and discharge remains uniform. A modified Meek criterion of the avalanche-to-streamer transition can be obtained considering two simultaneously starting avalanches with maximum radius R, separated by the distance L, (α is Townsend coefficient, d is distance between electrodes):

$$\alpha d - (R/L)^2 \approx const, \quad \alpha d \approx const + n_e^p R^2 d; \tag{13.23}$$

here distance between avalanches is approximated using the pre-ionization density n_e^p, the constant in these relations depends on gas (in air it is 20). According to the modified Meek criterion, the avalanche-to-streamer transition can be avoided by increasing avalanche radius and by providing sufficient pre-ionization. *The type of gas and special structuring of the dielectric barriers are important to DBD transition to the glow: helium gas is especially relevant for the purpose.* Helium has high-energy electronic excitation levels and no electron energy losses on vibrational excitation, resulting in higher electron temperatures at lower electric fields. Also, fast heat and mass transfer processes prevent in this case contraction and other instabilities at high pressures, see more in Kanazava et al. (1988) and Honda et al. (2001).

13.6 Time-evolution of the Short-pulsed DBD: Pulse Energy and Average Discharge Power, the "Maximum Power Principle"

To analyze energetics of the short-pulsed DBD, it can be presented by the equivalent circuit shown in Figure 13.13. In this circuit, the DBD dielectric barrier is represented by capacitor Cd, and the air (plasma) gap by capacitor Ca.

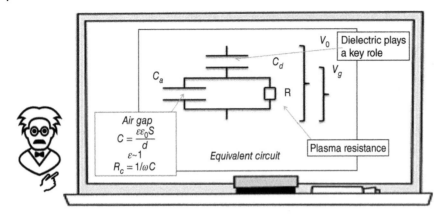

Figure 13.13 Equivalent electric circuit of the pulsed DBD discharge: C_d corresponds to a dielectric, C_a corresponds to an air gap, and R is a time-varying resistance of plasma in the gap.

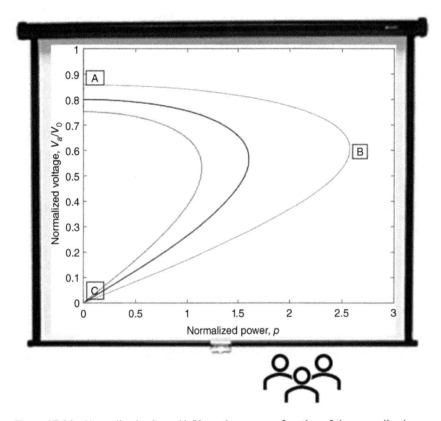

Figure 13.14 Normalized voltage V_a/V_0 on the gap as a function of the normalized power $p = 2P/(C_a\omega V^2_0)$, for $k = 1/3$ (left curve), 1/4 (middle curve), and 1/6 (right curve).

Characteristic capacitance ratio $k = Ca/Cd \sim 1/\varepsilon$, where $\varepsilon \approx 4$ is dielectric permittivity of the barrier; V_0 and V_g (V_a) are voltages on the total DBD and on gas (air) gap without barrier; voltage rise-time τ_r is characterized by effective frequency $\omega = 1/\tau_r$. When plasma appears in the gap, conductivity grows from zero to some finite value corresponding to the active resistance R. The Ohm's law in complex form (Dobrynin et al. 2019a,b) describes behavior of the pulsed-DBD equivalent circuit and explains evolution of voltage and power in the system, which is illustrated in Figure 13.14.

The discharge process starts at point (A) in the figure, as at the beginning there is no plasma and thus $R \to \infty$ and the normalized power $p = 0$. The evolution of the system can be viewed as a transition from (A) to (B), where the normalized power p reaches the maximum. The curve continues toward to origin point (C), where $R = 0$ (essentially a short circuit). The points on the curve from (A) to (B) to (C) are parameterized by decreasing value of $R(t)$. As the discharge starts developing, $R(t)$ drops. As a result, P increases, and V_a correspondingly drops. All along the curve, the plasma is in the nonthermal regime, and thus, the conductivity is defined by the ionization level, I. The higher the value of I is, the smaller is R. The value of I changes due to the power P released in plasma: the larger is P, the larger is I and, correspondingly, if P drops, so does I. There is a major difference between the top and bottom branches. On the top branch, a small drop in R (in other words, a shift to the right along the curve) causes P to increase. An increase in P causes I to increase, which, in turn, decreases R further. Thus, we have a self-sustained motion along the top branch. The situation is reversed on the lower branch. Now, a small drop in R shifts the system to the left along the curve, causing P to decrease. The decrease in P causes I to decrease, which increases R back. Thus, the system cannot propagate along the lower branch. The system comes to point (B) and is stuck there until streamers essentially bring the plasma resistance significantly down, upon which the systems jump straight to the origin. In other words, while the system can occupy the points on the top branch (and indeed passes through all of them), the bottom branch is unphysical. The time evolution of the voltage V and the power P along the top branch occurs with a varying rate. While in the beginning, both variables change fast, the rate of change of P vanishes at point (B) where $dP/dV = dP/dt = 0$.

In terms of the power release, *the system spends most of the time there at the point where the DBD power is at its maximum*, just like a pendulum spends most of the time near the turning points of maximum potential energy. At that moment of maximum DBD power, the resistance R effectively matches the impedance of the dielectric barrier capacitor. Detailed analysis of the equivalent circuit leads to the following vision of the pulsed DBD evolution:

(1) The pulsed plasma system evolution is governed by the decrease of the plasma resistance, until it reaches approximately the impedance of the dielectric barrier, at which state the discharge power reaches the maximum and the evolution drastically slows down. That is a clear manifestation of **the maximum power principle**.
(2) The maximum power release of the pulsed DBD discharge is proportional to the capacitance of the dielectric (because it corresponds to the minimum active plasma resistance) and the square of the applied voltage V (well expected from Ohm's law), and only weakly depends on the properties of the air gap in the discharge.

Principles of maximum (or minimum) power have been used for a theoretical description of major plasma discharges, including arc discharges, glow discharges, as well as gliding arcs. Application of these principles can be useful for analysis and quantitative description of different plasma systems. Although the maximum or minimum power principles can be viewed in the framework of the so-called "fourth principle of energetics in open system thermodynamics," in plasma physics systems they were explained based on the underlying physical laws.

The equivalent circuit determines the pulse energy by integrating the power P over the time τ of pulse duration (Dobrynin et al. 2019a,b):

$$E_t = \frac{C_d V^2}{4(k+1)}(\omega\tau) \tag{13.24}$$

Recalling that ω is defined by the reverse voltage rise-time, while τ is the total duration of the discharge pulse. Therefore, the overall average DBD power can be expressed as:

$$P_{av} = E_t f = \frac{C_d V^2}{4(k+1)}(\omega\tau)f. \tag{13.25}$$

where f is a pulse frequency. Concluding, *the DBD pulse energy and power are proportional to the square of the applied voltage, and significantly depend on the properties of the dielectric barrier.* Increase of the area or of the dielectric permittivity of dielectric or decrease of its width result in larger pulse energy and average power. The total energy is larger for steeper growing voltage profiles (higher ω). In this perspective, with other parameters fixed, nanosecond pulses release more power than the microsecond pulses.

13.7 Asymmetric, Packed-bed, Ferroelectric, and Other Dielectric-surface Discharges

The surface discharges are like DBDs but usually propagate along dielectric surfaces. *The dielectric surfaces decrease the breakdown voltage* because of nonuniformities of electric field and therefore local overvoltage. The surface discharges, like DBDs, can be supplied by AC or pulsed voltage. The effective decrease of breakdown voltage can be reached in the surface discharge configuration, where one electrode lays on the dielectric plate, with another one partially wrapped around as it is shown in Figure 13.15a. The component of electric field E_y normal to the dielectric surface plays an important role in generation of the pulse-periodic

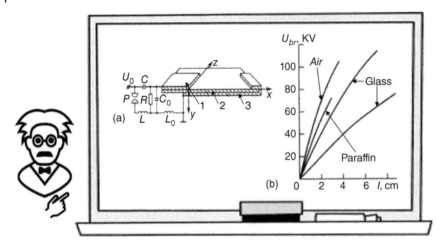

Figure 13.15 Pulsed asymmetric sliding surface discharge: (a) schematic, 1 – initiating electrode, 2 – dielectric, 3 – shielding electrode; (b) breakdown voltage in air along different insulation cylinders ($d = 50\,mm, f = 50\,Hz$).

sliding discharge that does not depend essentially on the distance l between electrodes along the dielectric (axis x in Figure 13.15a). For this reason, the breakdown voltages of the sliding discharge are lower and don't follow the Paschen's law, see Figure 13.15b. These plasma sources are called **sliding discharges or asymmetric DBDs** and can be uniform in some regimes on the dielectric plates of high surface area with linear sizes over 1 m at voltages not exceeding 20 kV. Two qualitatively different modes of the surface discharges can be arranged by changing the applied voltage amplitude: (i) complete one (**sliding surface spark**) and (ii) incomplete one (**sliding surface corona**). The sliding surface corona discharge takes place at voltages below the breakdown. Current in this regime is low and limited by charging the dielectric capacitance. Active volume and luminosity are localized near the igniting electrode and do not cover entire dielectric. The sliding surface sparks (or the complete surface discharges) take place at voltages above the breakdown, and plasma channels connect electrodes with multiple current channels of shapes called the **Lichtenberg figures**. Illustration of the DBD Lichtenberg figures is shown in Figure 13.16a (Ayan et al. 2008; you can also "Google" a lot of them). The number of the channels depends on the capacitance factor ε/d (ratio of dielectric permittivity over thickness of dielectric layer), which determines electric field on the sliding spark surface, see Figure 13.16b.

The **packed-bed corona** is a DBD combination with sliding surface discharge, where AC voltage 15–30 kV is applied to a packed bed of dielectric pellets creating nonequilibrium plasma in the voids between them, see Figure 13.17. The pellets effectively refract the electric field, making it nonuniform and 10–250 times stronger than the externally applied field depending on the shape, porosity, and dielectric constant of the pellet material. The packed bad corona and conventional DBDs can be organized using the ferroelectric ceramic materials of a high dielectric permittivity (ε above 1000) as the dielectric barriers. These plasma sources are called the **ferroelectric discharges,** see Opalinska and Szymanski (1996). Their mean power can be determined as:

$$P = 4f\, C_d V \left(V - V_{cr} \frac{C_d + C_g}{C_d} \right);$$ (13.26)

here f is the frequency of AC voltage; V is amplitude of the voltage; V_{cr} is the critical value of the voltage corresponding to breakdown; C_d and C_g are the electric capacities of a dielectric barrier and gaseous gap. When $C_d \gg C_g$, the relation for discharge power can be simplified:

$$P = 4f\, C_d V(V - V_{cr}).$$ (13.27)

High dielectric permittivity leads to reasonable power at low frequencies and applied voltages. The ferroelectric discharge based on $BaTiO_3$ which employs a ceramic barrier with a dielectric permittivity ε exceeding 3000, thickness 0.2–0.4 cm and gas discharge gap 0.02 cm operates effectively at AC frequency of 100 Hz, voltages below 1 kV, and electric power of about 1 W. Physical peculiarities of the ferroelectric discharges are related to the nature of the ferroelectric materials, which can be spontaneously polarized, and therefore have nonzero dipole moment even in the absence of external electric field. The long-range correlated orientation of dipole moments can be destroyed in

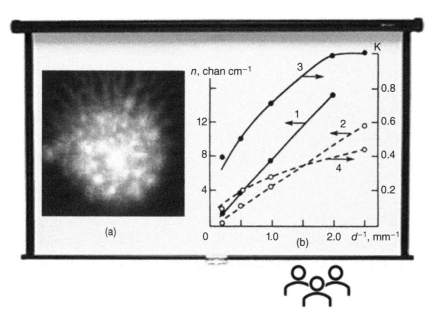

Figure 13.16 DBD Lichtenberg figures (a); linear density of channels n (1,2) and surface coverage by plasma K (3,4) as function of d^{-1} inverse dielectric thickness: He (1,3), air (2,4).

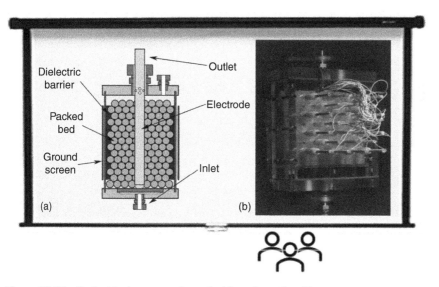

Figure 13.17 Packed-bed corona: schematic (a) configuration (b).

ferroelectrics by thermal motion. The temperature at which the spontaneous polarization vanishes is called the **ferroelectric Curie point**. Below the ferroelectric Curie point, the ferroelectric sample is divided into macroscopic polarized **ferroelectric domains.** Polarization vectors of individual domains in the equilibrium state are set up in the way to minimize internal energy of the crystal and to make polarization of the sample close to zero. External AC voltage leads to overpolarization of the ferroelectric material and reveals strong local electric fields on the surface, which can exceed $10^6 \mathrm{V} \, \mathrm{cm}^{-1}$ and stimulate the discharge.

13.8 Atmospheric Pressure Glow Discharges (APG)

The atmospheric pressure glow (APG) mode of DBD operating at low voltages was already discussed, in particular, in Section 13.5 in relation to using special gases (He, other noble gases, and their mixtures) and special barrier configurations and materials. Another APG configuration is related to the so-called **resistive barrier discharge (RBD)**, operating with DC or AC (60 Hz). It is based on DBD configuration, where dielectric barrier is replaced by highly resistive sheet (few $M\Omega\,cm^{-1}$) covering one or both electrodes. The system can consist of top wetted high resistance ceramic electrode and a bottom electrode. The highly resistive sheet plays a role of distributed resistive ballast, which prevents high currents and arching. If He-gas is used and the gap is not too large (below 5 cm), such spatially diffuse discharge can be maintained for several tens of minutes. If 1% of air is added to He, the discharge forms filaments. Even driven by DC voltage, the RBD current is pulsed with the pulse duration of few microseconds at a repetition rate of few tens of kHz. When the discharge current reaches a certain value, the voltage drop across the resistive layer becomes large and voltage across the gas became insufficient to sustain the discharge. The discharge extinguishes, and voltage across the gas increases to reinitiate the RBD. The **one-atmosphere uniform glow discharge plasma (OAUGDP)** OAUGDP is another configuration of the APG discharges (Roth 1995, 2001). It is like a conventional DBD but can be much more uniform because of the ion trapping. The **high γ stabilization of APG** was demonstrated by Van De Sanden and his team. Uniform plasma has been generated in argon DBD during the first cycles of voltage oscillations with relatively low amplitude (i.e. αd of about 3). Existence of the Townsend discharge at such low voltage requires an unusually high secondary electron emission coefficient (above 0.1). The high electron emission and the breakdown during the first low-voltage oscillations can be explained by considering low surface conductivity of most polymers applied as barriers. Surface charges occurring due to cosmic rays can be then detached by the applied electric field.

Assuming that the major cause of the DBD filamentation is instability leading to the glow-to-filaments transition, it has been also suggested to stabilize the glow mode using electronic feedback to fast current variations. The filaments are characterized by higher current densities and smaller RC constant. Therefore, in the **electronic APG stabilization**, the difference in RC constant can be used to "filter" the filaments because they react differently to a drop of the displacement current of different frequency and amplitude. Simple LC circuit, in which during the pulse generation the inductance is saturated, has been used to generate the displacement current pulses. The method of the electronic uniformity stabilization has been used for relatively high-power densities (in the range of $100\,W\,cm^{-3}$), and in a large variety of gases including Ar, N_2, O_2, and air. More details on the subject can be found in the book of Roth (1995, 2001).

13.9 Noble-gas-based RF Atmospheric Pressure Plasma Jets (APPJ)

The radio-frequency (RF) atmospheric pressure plasma jet (APPJ) can be organized as planar and co-axial system with the discharge gap or diameter of 1–2 mm, and frequency in the MHz range (13.56 MHz). The APPJ is an RF CCP discharge that can operate uniformly at atmospheric pressure in noble gases, typically in helium. In most APPJ configurations, electrodes are placed inside the chamber, and not covered by any dielectric in contrast to DBD. The discharge in pure helium has limited applications, therefore other gases like oxygen, nitrogen, nitrogen trifluoride, etc. are added. If the admixture of molecular gases exceeds a few percent, the discharge becomes unstable. The APPJ current is mostly due to electrons. The typical ionic drift velocity is about $3 \times 10^4\,cm\,s^{-1}$, and time needed for ions to cross the gap is about 3 µs which corresponds to a frequency of 0.3 MHz.

The RF frequency of the electric field is much higher, and thus ions in APPJ do not have enough time to move and they stay in the gap, while electrons move from one electrode to another as the polarity of the applied voltage changes. This is the key difference of APPJ discharges with respect to DBDs, where ions leave the gap between voltage pulses or changing polarity. Assuming the secondary emission coefficient $\gamma = 0.01$, the critical ion density $n_{p(crit)}$ in helium RF APPJ (like in RF CCP in general) before the α–γ transition is about $3 \times 10^{11}\,cm^{-3}$. Corresponding critical sheath voltage is about 300 V.

Typical helium **APPJ surface power density** is about $10\,W\,cm^{-2}$, and 10 times exceeds that of conventional DBDs. The power density that can be achieved in the uniform RF discharge is limited by two major instability mechanisms: thermal instability and α–γ transition instability. Critical power density for the thermal instability in APPJ is about $3\,W\,cm^{-2}$. Stable APPJ can be organized, however, with a power density exceeding this threshold due to stabilizing effect of the sheath capacitance. It can be described by the R parameter: square of the ratio of the plasma voltage to the sheath voltage. The smaller R, the more stable the discharge is with respect to the thermal instability. For the helium APPJ with $d_s = 0.3\,mm$, $V_s = 300\,V$, and $d = 1.524\,mm$, the parameter R is $R = (V_p/V_s)^2 = 0.36$, which corresponds to the critical discharge power density close to $100\,\mathrm{W\,cm^{-2}}$. *Major instability of the APPJ and loss of its uniformity, therefore, is mostly determined by the α–γ transition, or in simple words by breakdown of the sheath.* The α–γ transition in APPJ happens because of the Townsend breakdown of the sheath, which occurs when ion density and sheath voltage exceed the critical ones ($n_{p(crit)} = 3 \times 10^{11}\,\mathrm{cm^{-3}}$, $V_s = 300\,\mathrm{V}$). Even though the thermal stability of He discharge is better compared to the discharge with oxygen addition, a higher power is achieved with oxygen addition, which prevents the sheath breakdown. *It is easier to generate the uniform cold discharges in helium and argon.* It is due to not only high thermal conductivity of helium but also because of lower voltage, and therefore lower power density in these discharges, which helps to avoid thermal instabilities.

13.10 Atmospheric Pressure DBD-based Helium Plasma Jets, Plasma Bullets

The helium DBD jets transport cold plasma from the DBD source to a separate region for processing of different materials (including living tissue) without compromising plasma stability, see Figure 13.18 (Fridman and Friedman 2013). These jets operate at frequencies below $100\,\mathrm{kHz}$ (usually about $20\,\mathrm{kHz}$, like most DBDs, see Sections 13.4 and 13.5), which in contrast to RF jets permits ions to leave the gap between cycles or pulses. The DBD plasma jets can exist in form of the **plasma bullets**, a fast-moving train of highly luminous but discrete clusters along the jet when imaged on a nanosecond scale. The chaotic mode of the jet is observed after gas breakdown. With increasing input power, the discharge enters the **bullet mode**, see Figure 13.18b; and at high enough input power, the continuous mode is observed (Walsh et al. 2010). The Figure 13.18b shows a sequence of $10\,\mathrm{ns}$ exposure images taken throughout the positive and negative half cycles of the applied voltage while the plasma jet is operating in the bullet mode (Walsh et al. 2010; Fridman and Friedman 2013). Images (a) to (d) correspond to the negative half cycle, consequently the downstream ground electrode can be considered as the instantaneous anode and the powered electrode wrapped around the dielectric tube as the instantaneous cathode. The plasma bullets can propagate in long dielectric tubes

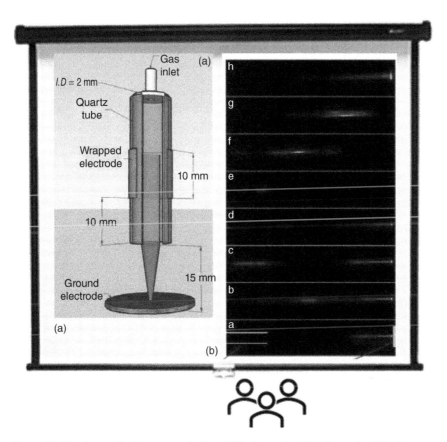

Figure 13.18 Atmospheric pressure helium DBD plasma jet: (a) schematic; (b) bullet propagation.

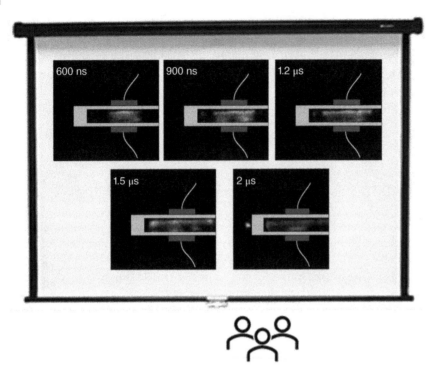

Figure 13.19 Nanosecond-pulsed atmospheric pressure DBD jet in helium: the "pancake" spreading out and formation of a bullet.

up to a few tens of centimeters away from the primary plasma-generation zone, which was widely investigated by Pouvesle and Robert and their group. The plasma bullet propagation can be effective in the tubes of very sophisticated configurations, including loops and splitting in T-shape connectors. It explains effectiveness of the multiple plasma jet formation from a single plasma source as well as formation of the 2D plasma jets.

 Energy balance and initial stages of the plasma bullet formation and evolution of the DBD atmospheric pressure plasma jets can be analyzed using the equivalent circuit approach discussed in Sections 13.5 and 13.6, see Dobrynin et al. (2019a,b). When avalanches reach electrode, they create plasma on the tube surface called "pancake" (see Figure 13.12). The equivalent circuit analysis determines fraction of the total energy released in a "pancake" as the ratio of air gap capacitance to that of a dielectric barrier.

The "pancakes" spread out along the tube stimulating the surface ionization waves forming the plasma bullets, which is illustrated in Figure 13.19 from Dobrynin et al. (2019a,b). Approximately half of the "pancake" energy is transferred to a bullet, which permits to estimate the fraction of the total *DBD* power released to the set of bullets creating the plasma jet as:

$$P_{jet}/P_{DBD} \approx \frac{1}{2\varepsilon + 1};$$

(13.28)

here ε is the barrier's dielectric permittivity. More details on the subject can be found in Laroussi et al. (2012), Fridman and Friedman (2013), and Sarron et al. (2011).

13.11 Problems and Concept Questions

13.11.1 Active Corona Volume

Why most of plasma-chemical processes occur in the active volume? Why the active volume cannot be increased much by increasing the voltage?

13.11.2   Power of Continuous Corona Discharges

Explain the physical limitations of the power increase of continuous coronas. Corona power can be increased, in principle, by diminishing the radius of external electrode. What is the limitation of this approach to increase power?

13.11.3   Voltage Rise Rate in Pulse Corona Discharges

Give your interpretation why not only pulse duration but also the voltage rise rates are important to reach high voltages.

13.11.4   Maximum vs Minimum Power Principles in Theory of Plasma Discharges

Why some plasma sources follow the maximum power principle, while others minimize the power?

13.11.5   Power Control of Nanosecond-pulsed Dielectric Barrier Discharges

Why the thinner is a dielectric barrier in DBD, the higher becomes average power of the pulsed DBD.

13.11.6   Evolution of the DBD Surface "Pancakes" vs DBD Streamers

Why are the "pancakes" developed and cover the electrode surface faster than propagation of the streamers?

13.11.7   Atmospheric Pressure RF vs DBD Plasma Jets

What is the qualitative difference of the ion behavior atmospheric pressure RF and DBD plasma jets? Why is it so crucial for parameters of these plasma sources?

13.11.8   Helium Plasma Jets

Why the cold atmospheric pressure RF and DBD plasma jets operate only in helium and some other noble gases?

13.11.9   Power of the DBD Plasma Jets

Estimate fraction of the total DBD power released to the set of bullets creating the plasma jet when dielectric barrier is made from quartz.

Lecture 14

Nonequilibrium Transitional "Warm" Discharges: Nonthermal Gliding Arc, Moderate-pressure Microwave Discharge, Different Types of Sparks and Microdischarges

14.1 Nonthermal Gliding Arc as an Example of the Nonequilibrium Atmospheric Pressure Transitional "Warm" Plasma Sources

 Ionization requires about 10 eV, and in both thermal and nonthermal plasma is mostly provided by electron impact. Therefore, *to provide effective ionization in majority of conventional plasma sources, typical electron temperature should be about 10% of the ionization potential, that is about 1 eV,* see Lecture 2. This **10% rule of thumb** is applicable by the way not only to plasma but for all high temperature chemical processes with small molecules. *While the electron temperature (about 1 eV) is determined by ionization requirements, gas temperature is simply controlled by energy and heat balance.* When cooling is fast (like in many low-pressure discharges) or energy release is low (like in coronas and dielectric-barrier discharges [DBDs]), gas temperature stays close to room temperature, which is the case of "cold" nonthermal plasmas. Vice versa, if energy release is significant and cooling is limited (like in conventional arcs), the gas temperature approaches equilibrium with electrons and reaches 5–10 000 K and above, which is the case of "hot" thermal plasmas. Thus, *in majority of plasma systems, the gas temperature goes to extremes: it is either room temperature cold or really very hot, 5–10 000 K and above. The transitional "warm" plasmas can stay due to different physical effects in the intermediate controllable gas temperature range,* which can be useful for some applications.

Clear representative example of the transitional plasmas is gliding arc, already considered at the end of Lecture 10. These auto-oscillating periodic discharges are illustrated in Figure 11.14. The gliding arc channel grows in different geometries until critical length determined by the power supply limits. Then the discharge extinguishes and reignites o start a new cycle. The gliding arc has transitional "warm" thermal or nonthermal properties depending on power input and flow rate. In the transitional regimes, the discharge starts with thermal plasma, but during the space and time evolution becoming the nonthermal one. Up to 70–80% of the total gliding arc power can be dissipated after the critical transition resulting in plasma where the gas temperature is approximately 1500–3000 K and the electron temperature is about 1 eV, see Kennedy et al. (1997). The "warm" plasma generation in gliding arcs is due to the time evolution of plasma from high to low gas temperature. Similarly, time evolution of sparks (in this case from low to high temperatures) also can lead to the generation of "warm" plasmas. The transitional plasmas can be also organized by space-evolution or space structure of the discharges, where high gas temperature zones coexist with cold ones. Good example in this regard is nonuniform microwave discharge sustained at moderate pressure.

14.2 Nonequilibrium Transitional "Warm" Microwave Discharges of Moderate Pressures

The general schematics of the moderate pressure microwave discharges are like those considered in Section 12.5 regarding atmospheric pressure microwave discharges. The waveguide- and resonator-based configurations of the microwave plasma generators (see Figure 12.4) can be applied to create nonthermal and strongly nonequilibrium plasma at moderate pressures, usually between 30 and 200 Torr. These plasmas in molecular gases, like CO_2, can be characterized by electron temperature T_e about 1 eV significantly exceeding the translational

Plasma Science and Technology: Lectures in Physics, Chemistry, Biology, and Engineering, First Edition. Alexander Fridman.
© 2024 WILEY-VCH GmbH. Published 2024 by WILEY-VCH GmbH.

one (≤ 1000 K); the degree of ionization and specific energy input (energy consumption per molecule) can be sufficiently high $n_e/n_0 \geq 10^{-6}$, $E_v \approx 1$ eV mol^{-1}, which is optimal for plasma-chemical processes stimulated by vibrational excitation of molecules (see Fridman 2008). Simultaneous achievement of these parameters is difficult in steady-state uniform discharges. For example, the conventional low-pressure nonthermal discharges are characterized by too large of specific energy inputs of at least 30–100 eV mol^{-1}; the streamer based atmospheric pressure discharges have low values of specific power and average energy input; powerful steady-state atmospheric pressure discharges usually operate close to quasi-equilibrium conditions.

This crucial advantage of the moderate pressure microwave discharges is due to the formation of an overheated plasma filament within the plasma zone, which sustains energy intensity but does not lead to the decrease of electric field because of the skin effect in the vicinity of the filament (Rusanov and Fridman 1984; Fridman and Kennedy 2021). Such peculiarity of the electrodynamic structure permits sustaining the strong nonequilibrium conditions $T_e > T_v \gg T_0$ and high level of vibrational excitation at relatively high specific energy inputs. Typical parameters of such plasma sources in molecular gases like CO_2 are frequency 2–3 GHz, power 1–3 kW (up to 1 MW), pressure 30–200 Torr, flow rate 0.1–3 sl s^{-1} (up to 1 sm^3 s^{-1}). Specific energy input is in the range of 0.1–3 eV mol^{-1}, and specific power is up to 500 W cm^{-3} (compare with conventional glow discharge values of $0.2 - 3$ W cm^{-3}). Vibrational temperature can be on the level of 3000–5000 K and significantly exceed the rotational and translational ones that are about 1000 K.

Let's first analyze the **discharge microstructure** (Rusanov and Fridman 1984; Fridman and Kennedy 2021) The discharge propagation in fast flows is mostly controlled by the processes on the plasma front. The velocity and front thickness of the ionization wave are determined by diffusion coefficient D of heavy particles and the characteristic time τ of the gas preparation for ionization (in particular, by vibrational excitation). Reactions of vibrationally excited molecules can be characterized by the same time interval $\tau_{chem} \approx 1/k_{eV} n_e \approx \tau$, where k_{eV} is the rate coefficient of vibrational excitation. If reactions are stimulated by nonequilibrium vibrational excitation, the total energy efficiency can be expressed as the function of initial gas temperature T_0^i, degree of ionization n_e/n_0, and specific energy input E_v (Fridman 2008):

$$\eta = \eta_{ex}\eta_{chem} \frac{E_v - k_{VT}\left(T_0^i\right) n_0 \hbar\omega \left(\tau_{eV} + \tau_p\right) - \varepsilon_v\left(T_v^{min}\right)}{E_v \left(1 - \dfrac{\varepsilon_v\left(T_v^{min}\right)}{\Delta H}\right)}; \tag{14.1}$$

here $\tau_{eV} = E_v/k_{eV} n_e \hbar\omega$ is time of vibrational excitation; $\tau_p = c_v^v \left(T_v^{min}\right)^2/k_{VT} n_0 E_a \hbar\omega$ is the reaction time in the passive phase of the discharge; η_{ex}, η_{chem} are the excitation and chemical components of the total energy efficiency; k_{VT} is the rate coefficient of vibrational VT- relaxation; $\varepsilon_v\left(T_v^{min}\right)$ is average vibrational energy of a molecule at the critical vibrational temperature T_v^{min} corresponding to equal rates of chemical reaction and vibrational relaxation; ΔH and E_a are the plasma-chemical reaction enthalpy and activation energy; c_v^v is the vibrational part of the specific heat per one molecule. Propagation of the nonequilibrium discharge is determined by the reduced electric field on the plasma front $(E/n_0)_f$, which depends on two external parameters, gas pressure p and electric field on the front E. Considering that the reduced electric field on the front is almost fixed by the ionization rate requirements, the initial gas temperature can be found by the following simple relation:

$$T_f = \left(\frac{E}{n_0}\right)_f \frac{p}{E}. \tag{14.2}$$

The electron concentration on the plasma front can be found from the energy balance as:

$$n_{ef} = \frac{D}{r_p^2} \frac{T_p}{T_p - T_g} \frac{p}{\mu_e e E^2}; \tag{14.3}$$

here r_p is the characteristic radius of the nonuniform microwave plasma zone, T_p, T_g are translational temperatures in the plasma zone and ambient gas respectively, μ_e is the electron mobility. The specific energy input E_v in the discharge, which determines the energy efficiency of the plasma-chemical process can be determined as:

$$E_v = P/Q_f; \tag{14.4}$$

here P is the microwave discharge power absorbed in the plasma, and Q_f is the flow rate.

The **macrostructure analysis** is focused on *transition between three major forms of the discharges: diffusive (homogeneous), contracted, and combined,* which occur at different pressures p and electric fields E related to the electromagnetic energy flux: $S = \varepsilon_0 c E^2$, see Figure 14.1. Critical pressures p and electric fields E, separating the three discharge forms, are shown in Figure 14.2. Area above the curve 1–1 corresponds to the microwave breakdown conditions; the microwave discharges are sustained below this curve. Critical curve 2–2 corresponds to the maximum E/p of the microwave discharges, which is sufficient to sustain ionization on the plasma front even

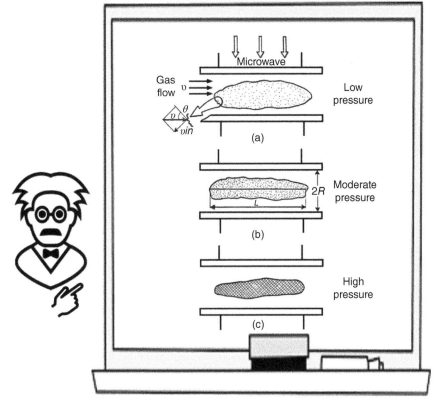

Figure 14.1 Macro-structure of the moderate-pressure microwave discharges: (a) diffusive regime; (b) combined regime; (c) contracted regime.

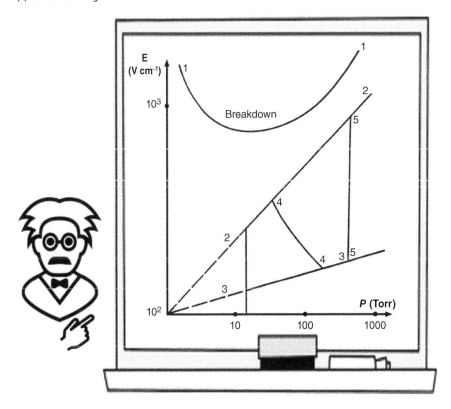

Figure 14.2 Moderate-pressure microwave discharges: electric field and pressure limits of the macrostructure regimes.

at room temperature. The **diffusive (homogeneous) regime**, see Figure 14.1a, occurs when $E/p < (E/p)_{max}$, but the moderate pressure is relatively low (close to 20–50 Torr). The discharge front in this regime is determined by the stabilization of the front in axial flow, which requires the normal velocity of the discharge propagation u_{in} to be equal to the component of the gas flow velocity $v\sin\theta$ perpendicular to the discharge front (see Figure 14.1a). The ratio E/p decreases at higher pressures and temperatures on the discharge front should increase to provide the necessary ionization rate. Critical curve 3–3 in Figure 14.2 determines the minimum reduced electric field $(E/p)_{min}$, when the microwave discharge is still nonthermal. The maximum temperature T^* on the discharge front, corresponding to $(E/p)_{min}$, is related to transition from nonthermal to thermal ionization mechanisms (Rusanov and Fridman 1984):

$$T^* = I \ln^{-1} \left\{ k_i \left[\left(\frac{E}{n_0} \right)_f \right] \times \frac{(4\pi\varepsilon_0)^5 \hbar^3 T^*}{m e^{10}} \frac{g_0}{g_i} \right\}. \tag{14.5}$$

here I is ionization potential; g_i, g_0 are statistical weights of ion and neutrals; $k_i[(E/n_0)_f]$ is the rate coefficient of ionization by direct electron impact at the reduced electric field on the discharge front. At an intermediate pressure range 70–200 Torr and $(E/p)_{min} < E/p < (E/p)_{max}$, **the combined microwave discharge regime** can be sustained, when the nonthermal ionization front co-exists with the thermal one, see Figure 14.1b.

In this case, *a hot thin filament of thermal plasma is formed inside the relatively large nonthermal plasma zone. Skin effect prevents penetration of electromagnetic waves into the hot filament and most of energy is absorbed in the nonequilibrium, nonthermal plasma surrounding.* The combined regime requires relatively high pressures and electric fields. To derive the criterion of existence for this regime, consider that T^* is the minimal temperature sufficient for the contraction of the filament. The energy balance of thermal filament is determined by the skin layer $\delta(T_m)$:

$$S = \varepsilon_0 c E^2 = \frac{4\lambda_m T_m^2}{I \delta(T_m)}; \tag{14.6}$$

here λ_m is the thermal conductivity coefficient at maximum plasma temperature T_m in the hot filament, I is the ionization potential. Then criterion of existence for the combined regime is:

$$E^2 \sqrt{p} \geq \frac{2\lambda_m e \cdot (T^*)^2}{\varepsilon_0 c I} \sqrt{\frac{2\omega\mu_0 T^* n_e(T^*)}{m k_{en}}}, \tag{14.7}$$

k_{en} is the rate coefficient of electron-neutral collisions. The corresponding critical curve 4–4 in Figure 14.2 separates the lower pressure regime of the homogeneous discharge (Figure 14.1a) from the higher-pressure regime of the combined discharge (Figure 14.1b). At high pressures, to the right from the critical curve 5–5, radiation heat transfer becomes significant. Because of reduction of the mean free path, the radiative front overheating becomes essential, and the **contracted microwave discharge** becomes completely thermal, see Figure 14.2c.

14.3 Vibrational–Translation Nonequilibrium ($T_v > T_0$) in Transitional "Warm" Microwave Discharges of Moderate Pressures

The microwave discharges in moderate-pressure molecular gases (for example, in CO_2; to be discussed in this section) can sustain significant vibrational–translational nonequilibrium ($T_v > T_0$) resulting in high efficiency of the relevant plasma-chemical processes. The radial profiles of temperatures $T_v(r)$, $T_0(r)$, and specific discharge power are qualitatively different in homogeneous discharge regime at lower pressures, contracted one at high pressures, and combined regime at intermediate pressures (see Figure 14.3). The radial profiles $T_v(r)$ and $T_0(r)$ are close in the quasi-equilibrium regime at higher pressures (Figure 14.3c). At relatively low pressures, electron density is low, the skin effect is weak, and the vibrational–translational nonequilibrium is significant on the axis of the discharge tube, where electron and power density have maximum (Figure 14.3a). *At intermediate pressures (Figure 14.3b), effective vibrational excitation ($T_v > T_0$) occurs mostly on the nonequilibrium front of the microwave discharge, which is located at some intermediate radii.* The vibrational excitation is not effective near the discharge axis because of too low electric field and high translational temperature (which accelerates vibrational relaxation). Near the walls of the discharge tube, the vibrational excitation is ineffective because of low

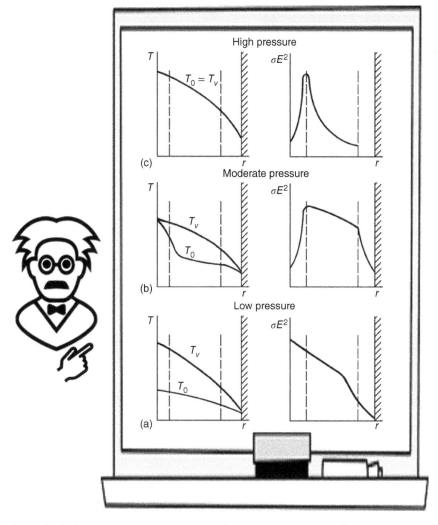

Figure 14.3 Moderate-pressure microwave discharges: temperatures T_v, T_0 and specific power profiles in (a) diffusive regime; (b) combined regime; (c) contracted regime.

electron density. *The intermediate pressure regime is the most effective for the plasma-chemical processes stimulated by vibrational excitation because most of the microwave energy is absorbed in the discharge zone with $T_v \gg T_0$*, see Figure 14.3b.

The maximum energy efficiency of the processes stimulated by vibrational excitation, like the dissociation of CO_2, can be achieved during transition from the homogeneous diffusive to contracted form of the moderate pressure microwave discharges. In this case, pressures are low enough to diminish VT relaxation losses, and high enough to provide sufficient temperature gradients to prevent penetration of the reaction products and electromagnetic energy into the thermal plasma zone. It is illustrated in Figure 14.4 for the case of CO_2 dissociation in microwave plasma, where the optimal pressure is about $p_{opt} \approx 120\,\text{Torr}$ (Rusanov and Fridman 1984).

Scaling up of the energy-effective plasma processes, like the dissociation of CO_2, can be based on the specific energy input E_v. Endothermic processes of gas conversion usually achieve the maximum energy efficiency at the specific energy inputs $1\,\text{eV mol}^{-1}$, which requires the fixed ratio of the discharge power to the gas flow rate close to $1\,\text{kWh}$ per standard m^3. This scaling rule should be corrected for the spatially nonuniform discharges (in particular, the microwave discharges), where portion of the flow can avoid contact with plasma. Consider a microwave discharge with the characteristic radial size of about the skin layer δ and length L, sustained in gas flow (see Figure 14.1). If the normal velocity of discharge propagation is u_{in} and axial gas velocity is v, then the angle θ between the vector of axial velocity and the quasi-plane of the discharge front is:

$$\sin \theta = \frac{u_{in}}{v}. \tag{14.8}$$

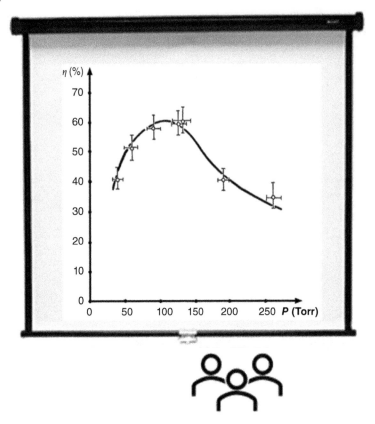

Figure 14.4 Moderate-pressure microwave discharges: pressure dependence of the energy efficiency of processes stimulated by vibrational excitation.

If the axial gas velocity v is low, the angle θ is large, and most of gas crosses the discharge front. In this case, if pressure is fixed (for example p_{opt}), the flow rate across the discharge front Q_f is close to the total gas flow rate Q and increases with the axial gas velocity v. In this regime the nonuniformity effects on the energy efficiency are small. At high gas velocities, influence of the nonuniformity becomes crucial. Further increase of the gas flow across the discharge front becomes impossible, when the axial gas velocity reaches the maximum:

$$v_{max} = \frac{u_{in}}{\sin \theta} \approx u_{in} \frac{L}{\delta}. \tag{14.9}$$

If the axial gas velocities exceed the critical one, $v > v_{max}$, the fast flow "compresses" the discharge, the effective discharge cross-section decreases as $1/v^2$, and the gas flow across the discharge front decreases as $1/v$. Thus, there is the maximum value of the gas flow rate across the discharge front, which can be expressed as:

$$Q_{fmax} \approx u_{in} \pi n_0 \delta L; \tag{14.10}$$

here n_0 is the gas density. Relation between the flow rate across the discharge front and total flow rate, assuming that at low velocities discharge radius is correlated with radius of the tube, is:

$$Q_f \approx Q, \quad \text{if} \quad Q < Q_{fmax}; \; Q_f \approx \frac{Q_{fmax}^2}{Q}, \quad \text{if} \quad Q \geq Q_{fmax}. \tag{14.11}$$

Scaling up of the process by proportional increase of the gas flow rate Q and discharge power P, keeping the specific energy input $E_v = P/Q$ constant, is possible only at flow rate values below the critical one $Q < Q_{fmax}$. Proportional increase of the gas flow rate and the discharge power at higher flow rates leads to an actual increase of the specific energy input proportionally to the square of power: $E_v \propto P^2$. In this case of higher flow rates $Q > Q_{fmax}$, the energy efficiency of plasma-chemical processes decreases as $\eta \propto 1/P^2$. The maximum power, where high energy efficiency is still possible in nonuniform microwave discharges of moderate pressure is:

$$P_{max} = \pi E_v u_{in} \delta L n_0. \tag{14.12}$$

If $E_v \approx 1 \, \text{eV mol}^{-1}$, $u_{in} \approx 300 \, \text{cm s}^{-1}$, $n_0 \approx 5 \cdot 10^{18} \text{cm}^{-3}$, $\delta \approx 1$ cm, $L \approx 10 \div 30$ cm, the power limit (14.12) gives $P_{max} = 10 - 30 \, \text{kW}$. Further increase of microwave power with high-energy efficiency requires special changes of the discharge geometry or by organizing the nonthermal microwave discharges in supersonic gas flows (Rusanov and Fridman 1984; Fridman and Kennedy 2021).

14.4 Spark Discharges

When streamers provide an electric connection between electrodes and neither a pulse power supply nor a dielectric barrier prevents further growth of current, it opens an opportunity for the development of a spark. *Spark can be considered as a pulsed transitional discharge with evolution from the nonequilibrium streamer-related discharge phase to the pulsed thermal plasma of the well-developed intensive spark.* The initial streamer channel does not have a very high conductivity and usually provides only a very low current of about 10 mA. The potential of the head of the cathode-directed streamer is close to the anode potential, and electric field around the streamer's head is strong. When the streamer approaches the cathode, this electric field is growing. It stimulates intensive formation of electrons on the cathode and subsequently their fast multiplication in the elevated electric field. New ionization waves more intense than the original streamer start propagating along the streamer channel but in opposite direction from the cathode to anode. This is referred to as the **back ionization wave**, which propagates back to the anode with high velocity about 10^9cm s^{-1}. This high velocity is not that of electron motion, but rather the phase velocity of the ionization wave. The back wave is accompanied by a front of intensive ionization and the formation of a plasma channel with sufficiently high conductivity to form a channel of the **intensive spark**.

The high-density current initially stimulated in the spark channel by the back ionization wave results in intensive Joule heating, growth of the plasma temperature, and contribution of thermal ionization. Gas temperatures in the intensive spark channel can reach 20 000 K, electron density rise to about 10^{17}cm^{-3}, which is already close to the complete ionization. Electric conductivity in the spark channel at such high degrees of ionization is determined by Coulomb collisions, does not depend on electron density, and reaches $10^2 \text{ Ohm}^{-1} \cdot \text{cm}^{-1}$. Further growth of the spark current is related not to increase of the ionization level and conductivity but to expansion of the channel and increase of its cross-section. The fast temperature increase in the spark channel leads to sharp pressure growth and to generation of a cylindrical shock wave. The amplitude of the shock wave is so high that the temperature after the wavefront can be sufficient for thermal ionization. The external boundary of the spark current channel grows first together with the front of the cylindrical wave. During the initial $0.1 - 1 \text{ μs}$ after the breakdown, the current channel expansion velocity is about 10^5cm s^{-1}. Subsequently, the cylindrical shock wave decreases in strength, and expansion of the current channel becomes slower than the shock wave velocity. Radius of the spark channel grows to about 1 cm, which corresponds to a spark current increase of $10^4 - 10^5 \text{A}$ at current densities of about 10^4A cm^{-2}. Plasma conductivity grows relatively high and a cathode spot can be formed on the electrode surface. Interelectrode voltage decreases below the initial one, and the electric field becomes about 100 V cm^{-1}. If voltage is supplied by a capacitor, the spark current obviously starts decreasing after reaching the maximum values.

The sparks can be effectively controlled by laser pulses, resulting in the so-called **laser-directed sparks** (Asinovsky and Vasilyak 2001), see Figure 14.5. Laser beams can direct spark discharges not only along straight lines but also along more complicated trajectories. Laser radiation is able to stabilize and direct the spark discharge channel in space through of three major effects: local preheating of the channel leading to higher levels of reduced electric field E/n_0, local photoionization, and optical breakdown of

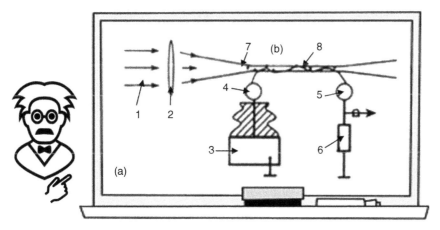

Figure 14.5 Laser (a) directed spark (b) discharge: 1, laser beam; 2, lens; 3, voltage pulse generator; 4,5, powered and grounded electrodes; 6, resistance; 7, optical breakdown zone; 8, discharge zone.

gas. The length of the laser-supported spark can be up to 1.5 m. The most intensive laser effect on spark generation can be provided by the optical breakdown of the gases. The length of such a laser spark can exceed 10 m. The laser spark in pure air requires power density of a Nd-laser ($\lambda = 1.06\,\mu m$) exceeding $10^{11} W\,cm^{-2}$. More information on the spark discharges can be found in books of Bazelyan and Raizer (2017) and Raizer (2011).

14.5 Spark Discharge in Nature: Lightning

Lightning is a large-scale natural spark discharge, occurring between a charged cloud and the earth, between two clouds or internally inside of a cloud. Lightning is caused by high electric fields related to the formation and space separation in the atmosphere of positive and negative electric charges. Formation of these charges is due mainly to ionization of molecules or micro-particles by cosmic rays. What is important for the interpretation of the lightning is that the negative charge in the thundercloud is in the bottom part of the cloud and positive charge mostly located in the upper part of the cloud. The polar water molecules on the water surface are mostly aligned with their positive ends oriented inward from the water surface due to the hydrogen bonds, see Figure 14.6a. Such orientation of the surface water molecules leads to the formation of a double electric layer on the surface of droplets with a voltage drop experimentally determined as $\Delta\varphi = 0.26$ V. This double layer traps negative ions and reflects positive ions until their charge Ne compensates the voltage drop $N \approx \frac{4\pi\varepsilon_0 r}{e}\Delta\varphi$. Droplets with radius $r = 10\,\mu m$ can absorb 2000 negative ions. The negative ions in clouds are trapped in droplets and descend, while positive ions remain in molecular or cluster form and remain in the upper areas of a cloud, see Figure 14.6b. This charge distribution in thunderclouds explains why most discharges occur inside the clouds.

Typical duration of the lightning is about 200 ms. The sequence of events in this natural modification of the long sparks is illustrated in Figure 14.7. The lightning consists of several pulses with duration about 10 ms each and intervals about 40 ms between pulses. Each pulse starts with propagation of a leader channel from the thundercloud to the ground. Current in the first negative leader is relatively low, about 100 A. This leader, called the **multi-step leader**, has a multi-step structure with a step length about 50 m and an average velocity $1 - 2 \cdot 10^7 cm\,s^{-1}$. The visible radius of the leader is about 1 m and the radius of the current-conducting channel is smaller. Leaders initiating sequential lightning pulses usually propagate along channels of the previous pulses. These leaders are called **dart-leaders**; they are more spatially uniform than the initial one and are characterized by a current of about 250 A and higher propagation velocities, $10^8 - 10^9 cm\,s^{-1}$. When the leader approaches the ground, the electric field increases, which results in the formation of a strong ionization wave moving in opposite direction back to the thundercloud. This extremely intensive ionization wave is usually referred to as the **return stroke**. The physical nature of the return stroke is like

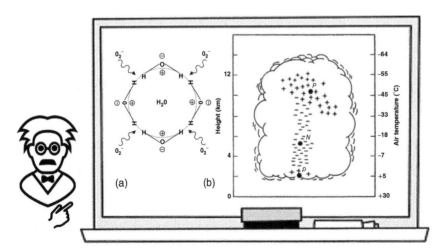

Figure 14.6 Physics of lightning: (a) water droplet trapping the negative ions; (b) charge distribution in thundercloud. Black dots mark centroids of charge clouds.

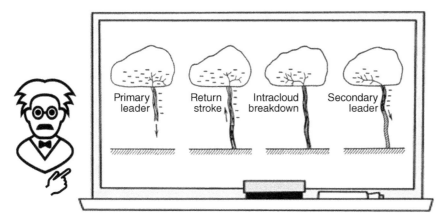

Figure 14.7 Illustration of a lightning evolution.

that of the back ionization wave in laboratory sparks. The return stroke is the main phase of the lightning discharge. Velocity of the ionization wave reaches gigantic almost relativistic values of 0.1–0.3 speed of light. Maximum current reaches 100 kA, which is the most dangerous effect of lightning. Temperature in the lightning channel reaches 25 000 K; electron density approaches $1 - 5 \cdot 10^{17} \mathrm{cm}^{-3}$, which corresponds to complete ionization. The electric field on the front of the return stroke is high, about 10 kV cm^{-1}. It leads to the gigantic specific powers on the front of the return stroke, $3 \cdot 10^5$ kW cm^{-1}. Intensive heat release leads to strong pressure increase in the current channel and to shock wave generation heard as thunder. Intensive heat release leads to fast expansion of the initial current channel during propagation of the return stroke and to the formation of a developed spark channel. Through this channel, some portion of the negative electric charge goes to the earth during about 40 ms. Electric currents through the spark channel during this 40 ms period is about 200 A. The lighting consists of several pulses. Each pulse results in transfer to the ground of only a part of the negative charge collected in the thundercloud. The positive charges located in the upper parts of the thundercloud mostly stay there, because distance to the ground is too long.

 Even a short discussion on lightning cannot be complete without mentioning the interesting and mysterious phenomenon of **ball lightning**. The luminous plasma sphere occurs in the atmosphere, moves in unpredictable directions (sometimes against direction of wind), and finally disappears sometimes with a strong explosion. The ball lightning has some special not trivial oddities, including ability of the plasma sphere to move through tiny holes, and strong explosions with heat release exceeding all reasonable estimations of energy contained inside of the ball. There are several hypotheses describing the ball lightning phenomenon. The most developed one is the so-called chemical model assuming production of some excited, ionized, or chemically active species in the channel of regular lightning followed by reaction between them with formation of luminous spheres of ball lightning. The typical ball lightning spherical shape is related in the framework of the chemical model to heat and mass balances of the process. Fuel generated by regular lightning is distributed over volumes much exceeding those of the ball lightning. Steady state exothermic reactions take place inside of the sphere. Fresh reagents diffuse into the sphere while heat transfer provides an energy flux from the sphere to sustain the steady-state process. This "steady-state combustion" sphere moves (in slightly nonuniform conditions) in the direction corresponding to the growth of temperature and "fuel" concentration. This explains possible motion of the ball lightning in the opposite direction to that of the wind. The high energy release during a ball lightning explosion can be explained noting that the explosion occupies much larger volume than the initial one related to the region of the steady-state exothermic reaction. More details on physics of the lightning, including models of the ball lightning can be found in Bazelyan and Raizer (2000), Boerner (2019), and Rusanov and Fridman (1984).

14.6 Pin-to-hole Discharge (PHD)

A special configuration of sparks called the pin-to-hole (PHD) discharge is illustrated in Figure 14.8 (Dobrynin et al. 2010a,b; Dobrynin et al. 2012b). The spark organized between the pin (needle) anode and the "hole" in cathode can be very energetic, short, and localized. The fast heated plasma with high density of active species is propelled through the whole with significant velocity and momentum, which makes these plasma sources interesting for multiple applications from propulsion to plasma medicine. In the specific example in Figure 14.8, a needle anode (usually about 1.5 mm diameter) is coaxially fixed in an insulator with gas inlet openings ($\sim$0.5 l min^{-1}) surrounded

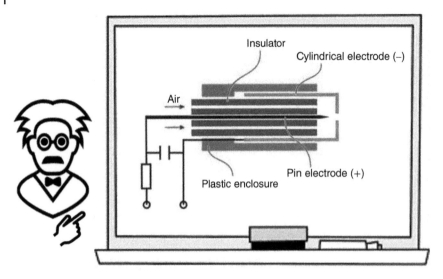

Figure 14.8 Schematic of the Pin-to-hole discharge (PHD).

by an outer cylindrical cathode (7 mm diameter) with an axial opening (2 mm diameter) for plasma outlet. The discharge is ignited by applying about 4 kV positive potential to the central electrode. To provide high discharge energy while keeping average gas temperature low, the electrode system is powered through a 0.33 μF capacitor. This forms a 35 μs dense energetic discharge pulse with average energy of ∼2 J/pulse. The plasma discharge appears as a series of microdischarges, in which the first one is the most energetic with 1 μs duration and 0.6 J energy. Due to the low frequency and short duration of the plasma microdischarges, the average plasma gas temperature is relatively low. It decreases from 75 °C at 2 mm distance from the plasma source down to ∼37 °C at 10 mm distance. By adding airflow, the maximum plasma gas temperature becomes ∼50 °C, and a temperature of 37 °C is reached by 6 mm away from the device. Although the average gas temperature is slightly above room temperature, the plasma temperature itself is high enough to produce a significant NO quantity, which is especially important for medical applications.

The PHD plasma is a local and powerful source of reactive oxygen species (ROS) and reactive nitrogen species (PNS). As an example, NO concentration in gas phase rapidly increases with plasma treatment and reached a stable level of 2000 ppm NO by around 50 s. In the liquid (phosphate buffered saline [PBS]), NO concentration increases linearly up to 240 plasma pulses. H_2O_2 concentration in PHD plasma-treated PBS increases up to 60 μM with 30 s of direct plasma treatment (210 pulses).

A time-averaged plasma excitation temperature in PHD plasma is about 9000 K. For this reason, it can also intensively radiate in the UV range. The total plasma UV radiation with and without airflow is 90 and 140 μW cm^{-2}, respectively. Another consequence of the high plasma excitation temperature is generation of a strong shock wave resulting in propulsion through the hole of a "plasma ball" characterized by significant plasma velocity and momentum.

14.7 Microdischarges: Atmospheric-pressure Micro-glow Discharge, Micro-hollow-cathode Discharge

Scaling down with a constant similarity parameter (*pd*) should not change the properties of discharges significantly, which is the **key idea of the microdischarges** in general. Conventional nonequilibrium discharges at low pressures are operated in the pd-range around 10 cm · Torr, for example, 3 cm and 3 Torr, to provide sufficient cooling through the walls. Therefore, *organization of the nonequilibrium microdischarges at atmospheric pressure should be effective in the sub-millimeter sizes* to provide similar effective cooling. Some specific new properties can be achieved by scaling down the plasma size to sub-millimeters: (i) Size reduction of nonequilibrium plasmas permits an increase of their power density to a level typical for the thermal discharges, because of intensive heat losses of the tiny systems. (ii) At high pressures, volumetric recombination and especially three-body processes can go faster than diffusion losses, which results in significant changes of plasma composition. For example, the high-pressure microdischarges can contain significant number of molecular ions in noble gases. (iii) Sheathes (about $10-30\,\mu m$ at atmospheric pressure) occupy a significant portion of the plasma volume. (iv) Plasma parameters move to the "left" side of the Paschen curve. The Paschen minimum is about $pd = 3\,cm \cdot Torr$ for some gases, therefore $30\,\mu m$ gap at 1 atm corresponds to the left side of the curve. (v) Differential resistance of the microdischarges can be positive, which allows supporting many discharges in parallel from a single power supply without using multiple ballast resistors.

A good example of the microdischarges is the **atmospheric pressure DC micro-glow discharge**, see (Staack et al. 2005, 2006; Farouk et al. 2006). The microplasma has been generated between a thin cylindrical anode and a flat cathode with the inter-electrode gap spacing from $20\,\mu m$ and above, see Figure 14.9. Current–voltage characteristics, visualization, and estimations of the current density indicate that the discharge operates in the normal glow regime, and plasma is in nonequilibrium. For 0.4 and 10 mA discharges, rotational temperatures are 700 and 1550 K, while vibrational temperatures are 5000 and 4500 K, respectively. It is possible to distinguish a negative glow, Faraday dark space, and positive column regions of the discharge. The radius of the column is about $50\,\mu m$ and remains relatively constant with changes in the electrode spacing and discharge current. Such radius permits balancing the heat generation and conductive cooling to help prevent thermal instability and the transition to an arc. For a medium-sized wire ($\sim 200\,\mu m$) as a cathode, width of the negative glow increases as the current increases until it covers the entire lower surface of the wire. If the current is further increased, the negative glow "spills over" the edge of the wire and begins to cover the side of the wire. This effect is like the transition from a normal glow to an abnormal glow in low-pressure glow discharges. For a normal glow discharge in air, the potential drop at the normal cathode sheath is around 270 V. A voltage drop above that occurs mostly in the positive column. For larger electrode spacing, the current–voltage characteristics have a negative differential resistance dV/dI. This is due to the discharge temperature increase with gap length resulting in growth of conductivity. A short discharge loses heat through the thermal conductivity of electrodes. A long discharge cooling is not efficient because the thermal conductivity of the gas is much lower than that of metal electrodes; therefore, the temperature of the long discharge is higher. Such behavior demonstrates a new property related to the size reduction to microscale. Diffusive

 heat losses can balance the increased power density only at elevated temperatures of a microdischarge and the traditionally cold glow discharge becomes "warm." In general, *the atmospheric pressure DC microdischarge is a normal glow discharge thermally stabilized by its size*, and able to maintain a high degree of vibrational–translational nonequilibrium. Table 14.1 summarizes the micro-glow discharge parameters corresponding to currents 0.4 and 10 mA.

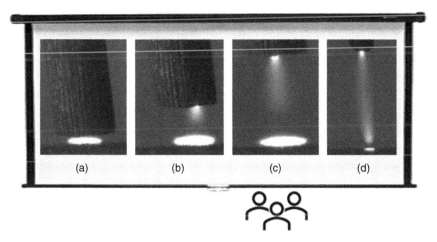

Figure 14.9 Micro-glow discharge in atmospheric-pressure air with electrode spacing: (a) 0.1 mm, (b) 0.5 mm, (c) 1 mm, and (d) 3 mm.

Table 14.1 Parameters of the atmospheric-pressure micro-glow discharge.

Micro-discharge and micro-plasma parameters	Micro-discharge current (mA)	
	0.4	**10**
Electrode spacing (mm)	0.05	0.5
Micro-discharge voltage (V)	340	380
Micro-discharge power (W)	0.136	3.8
Diameter of negative glow (μm)	39	470
Positive column diameter (μm)	—	110
Electric field in positive column (kV cm^{-1})	5.0	1.4
Translational gas temperature (K)	700	1550
Vibrational gas temperature (K)	5000	4500
Negative glow current density (A cm^{-2})	33.48	5.8
Positive column current density (A cm^{-2})	—	105
Reduced electric field E/n_0 (V $\cdot$ cm^2)	4.8×10^{-16}	3×10^{-16}
Electron temperature T_e (eV)	1.4	1.2

The atmospheric-pressure **micro hollow-cathode glow discharges** (micro-HCD) can be also organized in agreement with the (pd)-similarity. If (pd) is in the range of 0.1–10 Torr $\cdot$ cm, the discharge develops in stages. At low currents, a "pre-discharge" is observed, which is a glow discharge with the cathode fall outside the hollow cathode structure. As the current increases and the glow discharge starts its transformation into the abnormal glow with a positive differential resistance, a positive space charge region moves closer to the hollow cathode structure and can enter the cavity. After that, the positive space charge in the cavity acts as a virtual anode, resulting in the redistribution of the electric field inside the cavity. At the center of the cavity, a potential well for electrons appears, forming a cathode sheath along the cavity walls. At this transition from the axial pre-discharge to a radial discharge, the voltage drops. For more details, see for example, Kushner (2004, 2005).

14.8 Other DC, kHz-frequency, RF, Microwave Microdischarges, and Their Arrays

The microdischarges can be arranged in arrays, and plasma TV is a good example of that. The simple **microplasma array** consists of multiple identical microdischarges electrically connected in parallel. For stable operation, each microdischarge should have a positive differential resistance. One of the examples is the array consisting of microdischarges with inverted, square pyramidal cathodes, see Figure 14.10 (Park et al. 2001; modeled by Kushner 2004). In this 3 × 3 array, the microdischarges of 50 × 50 μm each are separated (center-to-center) by 75 μm. All the microdischarges (700 Torr of Ne) have common anode and cathode, i.e. the devices are connected in parallel. Ignition voltage and current for the array are 220 V and 0.35 μA. The array operates at a high-power loading (433 V and 21.4 μA); emission from each discharge is spatially uniform. Another example of microdischarge arrays is the **"fused" hollow cathode** (Barankova and Bardos 2002), which is based on the simultaneous generation of HCD plasmas in an integrated open structure with flowing gas. The resulting discharge is stable, homogeneous, luminous, and volume filling without streamers with power about one W cm^{-2} of the electrode structure.

KHz-frequency microdischarges are usually related to DBD. An integrated structure called the **coaxial-hollow micro dielectric-barrier discharge** (Sakai et al. 2005) has been made by stacking two metal meshes covered with a dielectric layer of alumina with thickness about 150 μm. The panel (diameter 50 mm) with hundreds of hollow structures (0.2 × 1.7 mm) has been assembled. He or N_2 have been used at pressures 20–100 kPa, and voltage below 2 kV even at the maximum pressure. Bipolar square-wave voltage pulses have been applied to a mesh electrode. The pulse duration is 3–14 μs; intermittent time 1 μs; repetition frequency 10 kHz. In each coaxial hole, the discharge occurs along the inner surface. Intensity of each microdischarge is uniform over the entire area. The glow with a length of some millimeters is observed in He. The electron density in He at 100 kPa is 3×10^{11} cm^{-3}. The array operates at

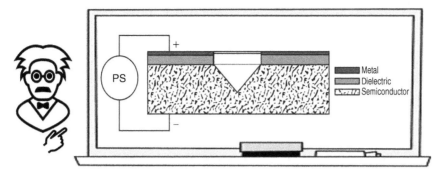

Figure 14.10 Single DC micro-discharge from the microplasma array.

low voltage (1–2 kV); the scaling parameter pd is 30–50 of Pa · m, corresponding to the Paschen minimum. Another kHz-range microdischarge is the **capillary plasma electrode (CPE)**, Becker et al. (2005). It employs dielectric capillaries that cover one or both electrodes. Although the CPE discharge looks like a conventional DBD, it exhibits a mode of operation that is not observed in DBD: the "capillary jet mode." The capillaries with diameter 0.01–1 mm and a length-to-diameter (L/D) ratio from 10 : 1 to 1 : 1 serve as plasma sources and produce plasma jets at high pressure. The jets emerge from the end of the capillary and form a "plasma electrode." When frequency of the applied voltage pulses is increased above a few kHz, one observes first a diffuse mode like the diffuse DBD. When the frequency reaches a critical value, the capillaries become "turned on," and bright intense plasma jets emerge from the capillaries. When many capillaries are placed near each other, the emerging plasma jets overlap, and the discharge appears uniform.

Between **RF microdischarges** (13.56 MHz), the **plasma needle** attracts special interest being one of the first applied in plasma medicine, see Figure 14.11 (Kieft et al. 2004). This discharge has a single-electrode configuration and is operating in helium. It operates near the room temperature. The plasma needle is confined in a plastic tube, through which helium flow is supplied. The discharge is entirely resistive with voltage 140–270 V; the electron density is about 10^{11} cm^{-3}. Conventional RF discharges, both inductively coupled (ICP) and capacitively coupled (CCP), have been also organized in micro-scale at atmospheric pressure. These plasmas are nonequilibrium because of their small sizes and effective cooling. Reduction in size requires reduction in wavelength and increase of frequency.

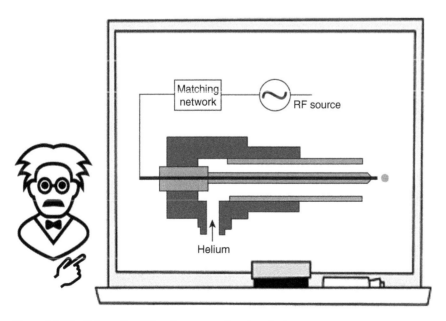

Figure 14.11 Schematic of the plasma-needle microdischarge.

The low-power **microwave micro-plasma source** based on a microstrip split-ring 900 MHz resonator operates at pressures 0.05 Torr – 1 atm has been developed by Iza and Hopwood (2003). Argon and air discharges can be self-started in the system with power less than 3 W. Ion density of 1.3×10^{11} cm^{-3} in argon at 400 mTorr can be produced with only 0.5 W power. Atmospheric discharges can be also sustained in argon with 0.5 W. The low power allows portable air-cooled operation of the system. More details about different microdischarges and their applications can be found in books of Fridman and Kennedy (2021) and Fridman (2008).

14.9 Problems and Concept Questions

14.9.1 Transitional "Warm" Discharges

Why majority of plasma sources operate either close to room temperature or vice versa at very high gas temperatures sufficient for thermal ionization, and only few "warm" discharges operate at intermediate temperatures?

14.9.2 Combined Regime of Moderate Pressure Microwave Discharges

Analyze formula (14.7) and Figure 14.2 estimate the critical value of $E^2 \sqrt{p}$ to form the hot filament. Why the filament formation requires relatively high pressures and electric fields.

14.9.3 Pressure Dependence of Energy Efficiency of Microwave Discharges

The energy efficiency of relevant plasma process reaches its maximum at the intermediate pressures, see Figure 14.4. Give your interpretation of the effect analyzing Figures 14.1–14.3.

14.9.4 Power and Flow Rate Scaling of Moderate Pressure Microwave Discharges

Explain why proportional increase of the gas flow Q and power P at high flow rates $Q > Q_{fmax}$ leads to increase of the specific energy input proportional to the square of the power: $E_v \propto P^2$.

14.9.5 Velocity of the Back Ionization Wave

Early stages of a spark generation are related to propagation of a back ionization wave, which provides a high degree of ionization in the spark channel. Interpret the experimental fact that the velocity of this wave reaches 10^9 cm s^{-1}.

14.9.6 Negative Ions Attachment to Water Droplets, Charge Separation in Thundercloud

Estimate the size of a water droplet (number of molecules in a cluster), which can provide effective trapping of at least one negative ion due to the surface polarization effect.

14.9.7 Propagation of Ball Lightning

Estimate the propagation velocity of the ball lightning in a steady state atmosphere with slightly nonuniform spatial distributions of temperature and fuel concentration. Can the ball lightning propagate faster than wind?

14.9.8 Pin-to-hole (PHD) Plasma Source

Estimate propagation velocity of the plasma ball generated by the PHD plasma source.

14.9.9 Atmospheric-pressure Micro-glow Discharge

Estimate cathode current density for the atmospheric-pressure micro-glow discharge in air operating in normal glow regime.

Lecture 15

Ionization and Discharges in Aerosols; Dusty Plasma Physics; Electron Beams and Plasma Radiolysis

15.1 Photoionization of Aerosols in Monochromatic and Continuous Spectrum Radiation

The photoionization of aerosols (radius r_a, density n_a) increases electron density n_e in dusty gases, especially if radiation quanta $E = \hbar\omega$ exceed the work function of macro-particles but is below the ionization potential of neutrals. The aerosol density n_m distribution over charges me should consider the **work function dependence on the charge**:

$$\varphi_m = \varphi_0 + \frac{me^2}{4\pi r_a}. \tag{15.1}$$

There are upper n_+ and lower n_- limits of the aerosol charge. The upper charge limit is due to the fixed photon energy $E = \hbar\omega$, and increasing the work function with a particle charge:

$$n_+ = \frac{4\pi\varepsilon_0(E - \varphi_0)r_a}{e^2}; \tag{15.2}$$

here φ_0 is work function of a noncharged macro-particle ($m = 0$), which slightly exceeds the work function A_0 of the same material having a flat surface: $\varphi_0 = A_0\left(1 + \frac{5}{2}\frac{x_0}{r_a}\right)$; $x_0 \approx 0.2$ nm. The lower aerosol charge limit n_e ($n_- < 0$) is related to the charges value, when the work function and hence, the electron affinity to a macro-particle become equal to zero:

$$n_- = -\frac{4\pi\varepsilon_0\varphi_0 r_a}{e^2}. \tag{15.3}$$

The equations for **aerosol density with different charges due to monochromatic photoionization** are:

$$\frac{dn_{n-}}{dt} = -\gamma p n_{n-} + \alpha n_e n_{n-+1}, \tag{15.4}$$

$$\frac{dn_i}{dt} = \gamma p n_{i-1} - \alpha n_e n_i - \gamma p n_i + \alpha n_e n_{i+1}, \quad n_- + 1 \le i \le n_+ - 1, \tag{15.5}$$

$$\frac{dn_{n+}}{dt} = \gamma p n_{n+-1} - \alpha n_e n_{n+}; \tag{15.6}$$

here, $\alpha \approx \sigma_a v_e$ is the coefficient of attachment of electrons with averaged thermal velocity v_e to the macro-particles of cross-section σ_a (constant at $i > n_-$); γ is the photo-ionization cross section of the macro-particles (constant at $i < n_+$); n_i is density of macro-particles carrying charge ie ($i < 0$ for negatively charged particles), p is the flux density of mono-chromatic photons. At the steady state $d/dt = 0$, the balance equation for the charged species is:

$$\alpha n_e n_{i+1} = \gamma p n_i, \quad q = \frac{n_{i+1}}{n_i} = \frac{\gamma p}{\alpha n_e}; \tag{15.7}$$

Considering the electro-neutrality and the mass balance, the equation for electron density is:

$$\frac{n_e}{n_a} = \frac{\sum_{k=n-}^{k=n+} kq^k}{\sum_{k=n-}^{k=n+} q^k} \approx \frac{\int_{n-}^{n+} xq^x dx}{\int_{n-}^{n+} q^x dx}. \tag{15.8}$$

Plasma Science and Technology: Lectures in Physics, Chemistry, Biology, and Engineering, First Edition. Alexander Fridman.
© 2024 WILEY-VCH GmbH. Published 2024 by WILEY-VCH GmbH.

$\lg(n_a/n_{max})$

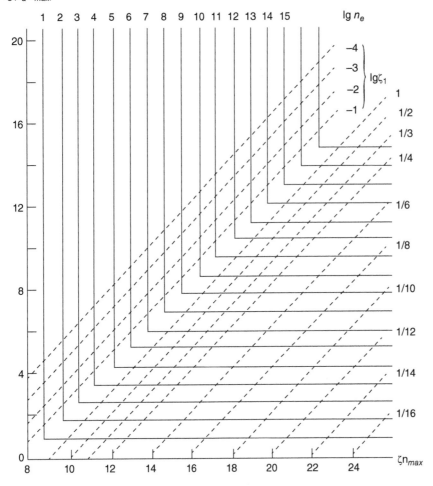

Figure 15.1 Chart to calculate the electron density due to the photoionization of aerosols.

Integrating (15.9) leads to the equation for density of electrons generated by photoionization:

$$\frac{n_e}{n_a} = \frac{n_+ q^{n+} - n_- q^{n-}}{q^{n+} - q^{n-}} - \frac{1}{\ln q}. \tag{15.9}$$

This general aerosol photo-ionization equation determines the relation between electron density n_e, photon energy $E = \hbar\omega$, flux p, and aerosol parameters: radius r_a, work function A_0, and concentration n_a. A chart for calculating the electron density n_e as a function of $\lg p$ and $\lg(n_a n_+)$ at monochromatic photoionization based on (15.9) is presented in Figure 15.1 assuming $\alpha/\gamma = 10^8 \text{cm s}^{-1}$. This chart also shows accuracy of the results. It is presented in a form of simple relative accuracy $\eta = \Delta n_e/n_e$, and accuracy of $\ln n_e$: $\eta_1 \approx \frac{\Delta n_e}{2.3 \cdot n_e \lg n_e}$.

Photoionization of aerosols by continuous spectrum radiation corresponds to $\gamma p_i n_i = \alpha n_e$, where p_i is the flux density of photons with energy sufficient for the $(n+1) - st$ photoionization. Using the energy distribution of photons in the form: $\xi_i = p_i/p_0$, $(\xi_0 = 1)$, the density of macro-particles with charge ke is:

$$n_1 = n_0 \left(\frac{\gamma p_0}{\alpha n_e}\right)\xi_0; n_2 = n_1 \left(\frac{\gamma p_0}{\alpha n_e}\right)\xi_1 = n_0 \left(\frac{\gamma p_0}{\alpha n_e}\right)^2 \xi_0 \xi_1; n_k = n_0 \left(\frac{\gamma p_0}{\alpha n_e}\right)^k \prod_{i=0}^{k-1}\xi_i. \tag{15.10}$$

Considering electro-neutrality, the electron density in continuous radiation spectrum can be expressed as:

$$\frac{n_e}{n_a} = \frac{\displaystyle\sum_{k=n-}^{\infty} kn_k}{\displaystyle\sum_{n-}^{\infty} n_k} = \frac{\displaystyle\sum_{k=n-}^{\infty} kq_0^k \prod_{i=0}^{k-1}\xi_i}{\displaystyle\sum_{k=n-}^{\infty} q_0^k \prod_{i=0}^{k-1}\xi_i}; \tag{15.11}$$

here the q_0 factor for the nonmonochromatic radiation is defined as: $q_0 = \gamma p_0 / \alpha n_e$. Assuming an exponential photon energy distribution: $p(E) = p \exp(-\delta E)$, the relation (15.11) with $\lambda = \delta e^2 / 4\pi\varepsilon_0 r_a$ becomes:

$$\frac{n_e}{n_a} = \frac{\sum\limits_{-\infty}^{\infty} k q_0^k \exp\left(-\lambda \frac{k(k-1)}{2}\right)}{\sum\limits_{-\infty}^{\infty} q_0^k \exp\left(-\lambda \frac{k(k-1)}{2}\right)}. \tag{15.12}$$

This equation can be solved by the elliptical functions: $\frac{n_e}{n_a} = y + \frac{\rho}{2\pi}\frac{d}{dy}\ln\theta_3(y,\rho)$; here $\theta_3(y,\rho)$ is the elliptical θ function of two variables $y = \frac{1}{\lambda}\ln q_0 + \frac{1}{2}$, $\rho = 2\pi/\lambda$. If $\rho \gg 1$:

$$\frac{n_e}{n_a} = \frac{1}{\lambda}\ln q_0 + \frac{1}{2}. \tag{15.13}$$

If $\rho \ll 1$, the electron density in dusty plasma is (see Fridman and Kennedy 2021):

$$\frac{n_e}{n_a} = \frac{q_0}{1+q_0}. \tag{15.14}$$

15.2 Thermal Ionization of Aerosols: Einbinder Formula, Langmuir Relation

The energy for ionization of macro-particles is related to their work function, which is lower than that of atoms and molecules. Considering that ionization potentials determine the exponential growth of ionization rate with temperature, one can conclude that thermal ionization of aerosols can be effective to provide high electron density and conductivity at relatively low temperatures. Let's assume that distance between macro-particles exceeds their radius: $n_a r_a^3 \ll 1$ (n_a is the aerosol density). If in quasi-equilibrium charge of macro-particles is small ($n_e \ll n_a$):

$$n_e = K(T) = \frac{2}{(2\pi\hbar)^3}(2\pi m T)^{3/2}\exp\left(-\frac{A_0}{T}\right). \tag{15.15}$$

When the macro-particle charge exceeds the elementary one, and electron density exceeds that of aerosols ($n_e/n_a \gg 1/2$), the electron density in quasi-equilibrium is determined by the **Einbinder formula** considering the work function dependence on the macro-particle charge:

$$\frac{n_e}{n_a} = \frac{4\pi\varepsilon_0 r_a T}{e^2}\ln\frac{K(T)}{n_e} + \frac{1}{2}. \tag{15.16}$$

When $n_a \approx 10^4 \div 10^8 \text{cm}^{-3}$ and work functions are about 3 eV, numerically:

$$n_e = 5\cdot10^{24}\text{cm}^{-3}\cdot\exp\left(-\frac{A_0}{T}\right), \quad \text{if} \quad T < 1000\,\text{K}, \tag{15.17}$$

$$n_e = n_a\frac{4\pi\varepsilon_0 r_a T}{e^2}\left(20 - \frac{A_0}{T}\right), \quad \text{if} \quad T > 1000\,\text{K}. \tag{15.18}$$

The Einbinder formula can be generalized for the nonequilibrium systems ($T_e \neq T_a$):

$$\frac{n_e}{n_a} = \frac{4\pi\varepsilon_0 r_a T_e}{e^2}\ln\frac{K(T_a)}{n_e} + \frac{1}{2}, \text{ or } n_e = n_a\frac{4\pi\varepsilon_0 r_a T_e}{e^2}\left(20 - \frac{A_0}{T_a}\right). \tag{15.19}$$

The electron density near particle surface (distance z) follows the **Langmuir relation**:

$$n_e(z) = \frac{K(T)}{\left(1 + \dfrac{z}{\sqrt{2}r_D^0}\right)^2}; \tag{15.20}$$

here $K(T)$ is the electron density (15.15), and $r_D^0 = \sqrt{\varepsilon_0 T/K(T)e^2}$ is the Debye radius corresponding to this density. Electric conductivity of the thermally ionized aerosols is determined by density of electrons n_e^* participating in the conductivity, and on the electron mobility b in heterogeneous medium. Dependence on the external electric field therefore is very sophisticated:

$$\sigma_0 = \frac{4\pi\varepsilon_0 r_a T}{e^2}\left(\frac{r_a}{r_D^0}\right)n_a eb \quad \text{if} \quad E_e \geq \frac{T}{er_D^0}; \sigma_0 = 4\pi\varepsilon_0 E_e r_a^2 n_a b \quad \text{if} \quad \frac{T}{er_a}\ln\frac{K(T)}{n_e{}^*} < E_e < \frac{T}{er_D^0}$$

$$\sigma_0 = \frac{4\pi\varepsilon_0 r_a T}{e^2}\left(\ln\frac{K(T)}{n_e{}^*}\right)n_a eb \quad \text{if} \quad E_e \leq \frac{T}{er_a}\ln\frac{K(T)}{n_e{}^*}. \tag{15.21}$$

The electric conductivity of aerosols begins depending on the external electric field explicitly, when $E_e > \frac{T}{er_a}\ln\frac{K(T)}{n_e{}^*}$; numerically at the external electric field exceeding $200\,\text{V cm}^{-1}$, if $T = 1700\,\text{K}$, $r_a = 10\,\mu\text{m}$, $A_0 = 3\,\text{eV}$. Maximum number of electrons participate in the conductivity if $E_e \geq \frac{T}{er_D^0}$, that is numerically above $1000\,\text{V cm}^{-1}$. The total electric conductivity of the heterogeneous medium can also depend on conductivity σ_a of the macro-particle material:

$$\sigma = \sigma_0\left(1 - \frac{4}{3}\pi r_a^3 n_a \frac{\sigma_0 - \sigma_a}{2\sigma_0 + \sigma_a}\right). \tag{15.22}$$

15.3 Electric Breakdown of Aerosols

Particles can be present in a discharge gap due to gas impurities, detachment of micro-pikes from the electrode surface, but also especially injected into a discharge gap to increase its breakdown resistance, or vice versa to initiate the breakdown. Three major physical factors determine the decrease of the breakdown resistance due to macro-particles in a discharge gap: (i) the particles like electrode surface irregularities can induce local intensification of electric field; (ii) the particle-surface collision can also lead to the field enhancement; (iii) particle-to-particle and particle-to-electrode micro-breakdowns can also decrease the overall breakdown resistance. These factors can decrease the breakdown voltage at atmospheric pressure air by a factor of two if the size of the aerosol particles is rather large (up to 1 mm), see Raizer (2011). In many cases, admixture of the aerosols leads vice versa to increase of the breakdown resistance mostly due to intensive attachment of electrons to the particles, which is to be discussed below.

The **kinetic equations for the electron and positive ion densities** in the aerosol can be expressed as:

$$\frac{dn_e}{dt} = An_e + Bn_i; \quad \frac{dn_i}{dt} = Cn_e - Dn_i \tag{15.23}$$

The effective frequencies A, B, C, D describing here the breakdown of the aerosol system are:

$$A = \alpha v_d^e - \beta v_d^e - \frac{D_e}{d^2} + K - \frac{v_d^e}{d}; B = \gamma_{ia}\sigma_{ia}v^i n_a + \gamma_{iw}\frac{v_d^i}{d}; C = k_i n_0$$

$$D = k_a n_a + \frac{D_i}{d^2} + \frac{v_d^i}{d}; K = \gamma_{pa}\mu v n_0 + \gamma_{pw}\mu(1 - v)n_0; \tag{15.24}$$

here, n_0, n_a are densities of neutral gas and aerosol particles; v_d^i, v^i, v_d^e, v^e are thermal and drift (with subscript "d") velocities of electrons (superscript "e") and ions (superscript "i"); α, β are the first and second Townsend coefficients; D_e, D_i are the diffusion coefficients of electrons and ions (free or ambipolar); d is the discharge gap; σ_{ia} is the cross-section of ion collisions with the aerosol particles; γ_{ia}, γ_{pa}, γ_{pw}, γ_{iw} are probabilities of an electron formation in an ion collision with a macro-particle, photon interaction with a macro-particle, photon interaction with a wall and ion collision with a wall; μ is the rate coefficient of a photon formation per electron-neutral collision; v is the probability for a photon to reach an aerosol particle; k_i is the rate coefficient of neutral gas ionization by electron impact; k_a is the rate coefficient of ion losses related to collisions with aerosol particles. The initial conditions: $n_e(t = 0) = n_e^0$ and $n_i(t = 0) = n_i^0$ then results in the electron density evolution formula:

$$n_e(t) = \frac{(An_e^0 + Bn_i^0) - \lambda_2 n_e^0}{\lambda_1 - \lambda_2}\exp(\lambda_1 t) + \frac{\lambda_1 n_e^0 - (An_e^0 + Bn_i^0)}{\lambda_1 - \lambda_2}\exp(\lambda_2 t). \tag{15.25}$$

Here $\lambda_{1,2}$ are solutions of the characteristic equation for the system of the differential equations (15.24):

$$\lambda_{1,2} = \frac{A - D \pm \sqrt{(A + D)^2 + 4BC}}{2}. \tag{15.26}$$

Let's analyze the **parameters of aerosol system related to breakdown**. The mean-free-paths of electrons with respect to all collisions, those with neutrals, and those with macro-particles can be respectively expressed as:

$$\lambda_\Sigma = \frac{1}{n_0\sigma_{e0} + n_a\sigma_a}, \quad \lambda_0 = \frac{1}{n_0\sigma_{e0}}, \quad \lambda_a = \frac{1}{n_a\sigma_a}; \tag{15.27}$$

here σ_{e0}, σ_a are cross-sections of collisions of electrons with neutral species and macro-particles. Then we can express electron temperature T_e, thermal and drift velocities of electrons v^e and v_d^e, electron diffusion coefficient D_e, the first and second Townsend coefficients α, β in aerosol systems, using the corresponding values of $T_e^0, v_0^e, v_{d0}^e, D_e^0, \alpha_0, \beta_0$ for the neutral gas without macro-particles. The electron temperature in the aerosol is:

$$T_e = \frac{eE}{\delta_1/\lambda_0 + \delta_2/\lambda_a} = \left[\frac{1}{T_e^0(E)} + \frac{\delta_2}{eE\lambda_a}\right]^{-1}; \tag{15.28}$$

δ_1, δ_2 the energy fractions lost in collisions with molecules or macro-particles. The electron thermal velocity is:

$$v^e = \sqrt{\frac{2eE}{m(\delta_1/\lambda_0 + \delta_2/\lambda_a)}} = \left[\left(\frac{1}{v_0^e}\right)^2 + \frac{\delta_2 m}{2\lambda_a eE}\right]^{-1/2}. \tag{15.29}$$

Then, diffusion coefficient and drift velocity of electrons in aerosols ($\xi = n_a\sigma_a/n_0\sigma_{e0}$) are:

$$D_e = \frac{\lambda_\Sigma}{3}\sqrt{\frac{2eE}{m(\delta_1/\lambda_0 + \delta_2/\lambda_a)}} = \frac{D_e^0}{1+\xi}\left(1 + \frac{\delta_2 T_e^0}{eE\lambda_a}\right)^{-1/2}; \tag{15.30}$$

$$v_d^e = \lambda_\Sigma\sqrt{\frac{eE}{2m}\left(\frac{\delta_1}{\lambda_0} + \frac{\delta_2}{\lambda_a}\right)} = \frac{v_{d0}^e}{1+\xi}\left(1 + \frac{\delta_2 T_e^0}{eE\lambda_a}\right)^{1/2}. \tag{15.31}$$

The first Townsend coefficient showing ionization per unit length of electron drift in the aerosols is:

$$\begin{aligned}\alpha &= \frac{2}{\lambda_\Sigma(\delta_1/\lambda_0 + \delta_2/\lambda_a)}\left[\frac{1}{\lambda_0}\exp\left(-\frac{I_i}{T_e}\right) + \frac{1}{\lambda_a}\exp\left(-\frac{I_a}{T_e}\right)\right] \\ &= \frac{2}{\lambda_\Sigma\left(eE/T_e^0 + \delta_2/\lambda_a\right)}\left[\frac{\alpha_0\lambda_0}{2}\frac{eE}{T_e^0} + \frac{1}{\lambda_a}\exp\left(-\frac{I_a}{T_e}\right)\right].\end{aligned} \tag{15.32}$$

I_i is ionization potential; I_a is electron energy when secondary electron emission coefficient from macro-particle reaches unity. The second Townsend coefficient describing electron attachment to molecules and aerosols is:

$$\begin{aligned}\beta &= \frac{2}{\lambda_\Sigma(\delta_1/\lambda_0 + \delta_2/\lambda_a)} \times \left[\frac{1}{\lambda_0}w^0 + \frac{1}{\lambda_a}w^a\right] \\ &= \frac{2}{\lambda_\Sigma\left(eE/T_e^0 + \delta_2/\lambda_a\right)} \times \left[\frac{\beta_0\lambda_0}{2}\frac{eE}{T_e^0} + \frac{1}{\lambda_a}w^a\right],\end{aligned} \tag{15.33}$$

here w^0, w^a are the electron attachment probabilities in a collision with a molecule or macro-particle. The developed approach permits to describe influence of aerosols on specific types of breakdown:

15.3.1 Pulse Breakdown of Aerosols

If τ is the pulse duration, c is the relevant speed of light, and the breakdown condition is $n_e(t = \tau) = n_b = 10^8 \times n_e^0$, the pulse breakdown criterion in aerosols can be expressed as:

$$\frac{\alpha_0\lambda_0 eE}{2T_e^0} + \frac{2\exp\left(-\dfrac{I_a}{T_e}\right) - 1}{\lambda_a} = \frac{eE\lambda_0\ln\dfrac{n_b}{n_0}}{2T_e^0 v_{d0}^e \tau}\sqrt{1 + \frac{\delta_2 T_e^0}{eE\lambda_a}}. \tag{15.34}$$

If $\tau = 10\,\text{ns}$, $p = 10\,\text{Torr}$, $d = 10\,\text{cm}$, $I_a = 100\,\text{eV}$, $\delta_2 = 1$, the pulse breakdown voltage in nitrogen without aerosol particles is $E_0 \approx 2\,\text{kV cm}^{-1}$ ($\xi = 0$). Addition of aerosol characterized by $\xi = n_a\sigma_a/n_0\sigma_{e0} = 0.1$ increases the breakdown voltage by factor of about 1.2 (20%).

15.3.2 Breakdown of Aerosols in High-frequency Electromagnetic Fields

The breakdown effective voltage increase due to electron attachment by aerosol particles in this case can be found from the criterion:

$$\lambda_0 \frac{eE}{T_e^0} \alpha_0 + \frac{2\xi}{\lambda_0} \left[2\exp\left(-\frac{I_a}{T_e}\right) - 1 \right] = \frac{\lambda_0}{d^2(1+\xi)}. \tag{15.35}$$

Assuming numerically: $p = 10\,\mathrm{Torr}$, $d = 10\,\mathrm{cm}$, $I_a = 100\,\mathrm{eV}$, $\delta_2 = 1$, with addition of aerosol particles, characterized by $\xi = n_a\sigma_a/n_0\sigma_{e0} = 0.1$ in nitrogen, gives a possibility to increase the breakdown voltage by factor of about 1.3 (30%) with respect to the same system without macro-particles ($\xi = 0$).

15.3.3 Townsend Breakdown of Aerosols

The breakdown voltage increase due to electron attachment by aerosol particles in this case can be found from the criterion

$$\frac{2d}{\lambda_\Sigma \left(eE/T_e^0 + \delta_2/\lambda_a\right)} \left[\frac{\alpha_0\lambda_0 eE}{2T_e^0} + \frac{2\exp(-I_a/T_e) - 1}{\lambda_a} \right] = n_a\sigma_{ia}d \frac{v^i}{2v_d^i} - \ln\gamma_{ia}. \tag{15.36}$$

Numerically, the Townsend breakdown voltage of nitrogen at 10 Torr with tantalum electrodes ($d = 1\,\mathrm{cm}$) can have 10% increase by adding particles with $\xi = n_a\sigma_a/n_0\sigma_{e0} = 0.1$, $\delta_2 = 1$.

15.3.4 Effect of Macro-particles on Vacuum Breakdown

In the above examples, the relatively low addition of macro-particles $\xi = n_a\sigma_a/n_0\sigma_{e0} < 1$ was assumed, resulting in maximum increase of the breakdown voltage about 10–30%. Much stronger effect can be achieved in effective Vacuum, when $\xi = n_a\sigma_a/n_0\sigma_{e0} > 1$ and most electron collisions are related to aerosol particles; for more details see Fridman and Kennedy (2021).

15.4 Steady-state DC Discharges in Heterogeneous Medium

Thermionic and field emission can be neglected in the steady-state cold heterogeneous DC discharges, and formation of electrons is provided by gas ionization through direct electron impact and by secondary electron emission from the aerosol particles. Electro-neutrality of the aerosol system including electrons, positive ions, and macro-particles can be expressed as:

$$n_e = n_i + Z_a n_a; \tag{15.37}$$

here n_e, n_i, n_a are densities of electrons, positive ions, and macro-particles, Z_a is the average charge of a macro-particle. If the density and radius of aerosol particles are relatively low, then formation of electrons is mostly due to neutral gas ionization. In this case usually $n_e \gg Z_a n_a$, and densities of electrons and positive ions are nearly equal $n_e \approx n_i$. This case is referred to as the **quasi-neutral regime**. Another heterogeneous discharge regime occurs if the macro-particle's density and sizes are sufficiently high and ionization is mostly provided by secondary electron emission. In this case: $n_i \ll n_e \approx Z_a n_a$, and the heterogeneous plasma consists mostly of electrons and positively charged macro-particles (instead of ions). This regime is usually referred to as **the electron-aerosol plasma** to emphasize the contrast with conventional electron–ion plasma.

Macro-particles in the *quasi-neutral regime* are negatively charged, and their potential is close to the floating value. The quasi-neutral regime of a steady-state DC discharge in aerosols can be described by the following system of equations:

$$\left(4\pi r_a^2 n_a + \frac{2}{R}\right)\sqrt{\frac{m}{M}} = n_0\sigma_{e0}\exp\left(-\frac{I_i}{T_e}\right); E = I_i n_S \left[1 + \frac{1}{n_S}(n_0\sigma_{e0} + n_e\sigma_{ei})\right] \cdot \Phi^2(T_e);$$

$$n_e = \frac{j_e}{e}\sqrt{\frac{m}{2eI_i}} \cdot \Phi(T_e); \Phi(T_e) = n_S\sqrt{\frac{m}{2M} \cdot \frac{T_e}{I_i}} \times \frac{\left(2 + \frac{1}{2}\ln\frac{M}{m}\right)\frac{T_e}{I_i} + 1}{n_0\sigma_{e0} + n_e\sigma_{ei}}; \tag{15.38}$$

here R is the radius of discharge tube; m, M are masses of an electron and a positive ion; n_0 is the gas density; T_e is the electron temperature; I_i is the ionization potential; E is the electric field; σ_{e0}, σ_{ei} are the characteristic cross-sections of electron-neutral and electron–ion collisions; j_e is the electron current density; n_S is "density of surfaces," which represents the surface area of macro-particles and discharge tube walls per unit volume. In the case of long cylindrical discharge tube $n_S = 4\pi r_a^2 n_a + \frac{2}{R}$. In the quasi-neutral regime, addition of aerosol particles at the fixed value of current density leads to an essential increase of electron temperature, while growth of electron density is relatively small. If the surface area of macro-particles is less than that of the discharge tube, (15.38) corresponds to the Langmuir–Klarfeld theory. Increase of the surface area density n_S leads to a logarithmic increase of electron temperature. If macro-particles make the major contribution to the total surface area $2/R \ll 4\pi r_a^2 n_a$, this dependence can be presented as:

$$T_e = I_i \ln^{-1} \frac{n_{a,cr}}{n_a}; \tag{15.39}$$

here $n_{a,cr}$ is the critical aerosol density when the electron temperature tends to infinity $T_e \rightarrow \infty$. The considered quasi-neutral regime occurs only at relatively low aerosol density:

$$n_a < n_{a,cr} = n_0 \frac{\sigma_{e0}}{4\pi r_a^2} \sqrt{\frac{M}{m}}. \tag{15.40}$$

At particle densities above critical (15.40), the **electron-aerosol plasma regime** occurs with significant electrons generation due to the secondary electron emission and $n_i \ll n_e \approx Z_a n_a$. Numerically, it takes place at $n_a = 10^8 \text{cm}^{-3}$, $r_a = 10\,\mu\text{m}$, if the molecular gas concentration is less than 10^{17}cm^{-3}. Then particle charge Z_a as a function of the electric field E is:

$$Z_a = Z_0 \frac{E - E_0}{E_0}; \tag{15.41}$$

E_0 is the breakdown field in the heterogeneous medium; and the aerosol charge parameter Z_0 can be numerically expressed as: $\lg Z_0 = 8 + \lg r_a(\text{cm})$. The electric field and electron density in the electron-aerosol plasma regime can be then expressed as a function of current density using a dimensionless factor $B = \frac{j_e}{Z_0 n_a e} \sqrt{\frac{2m}{\varepsilon_0}}$, which can be numerically presented as:

$$\lg B = 2 + \lg \frac{j_e(\text{A cm}^{-2})}{n_a(\text{cm}^{-3}) \cdot r_a(\text{cm})}. \tag{15.42}$$

At low current densities when the factor $B < 1$, the electric field in the heterogeneous discharge is slowly growing with current density as:

$$E = E_0(1 + B). \tag{15.43}$$

Electron density and particles charge at low currents ($B < 1$) grow proportionally to j_e:

$$n_e = n_a Z_0 B, \quad Z_a = Z_0 B. \tag{15.44}$$

At current densities ($B > 1$), n_e and Z_a reach the maximum values:

$$n_e = n_a Z_0, \quad Z_a = Z_0. \tag{15.45}$$

These saturation values of the electron density and particle charge are higher at larger particle radii. The electric field growth with j_e is stronger in the high current regime ($B > 1$):

$$E \approx E_0 B^2. \tag{15.46}$$

The dependence of the electric field $\lg \frac{E - E_0}{E_0}$ and electron density $\lg \frac{n_e}{n_a}$ on the current density $\lg \frac{j_e(\text{A cm}^{-2})}{n_a(\text{cm}^{-3}) \cdot r_a(\text{cm})}$ in the electron-aerosol plasma is illustrated in Figure 15.2. Thus, the average charge of aerosol particles of radius $10\,\mu\text{m}$ and density $n_a = 10^6\text{cm}^{-3}$ reaches its maximum at current densities exceeding $10\,\mu\text{A cm}^{-2}$. The maximum charge of the aerosol particles depends on their size, and can be quite large, about 10^5 for the $10\,\mu\text{m}$-macro-particles; for more details on the heterogeneous discharges see Fridman and Kennedy (2021).

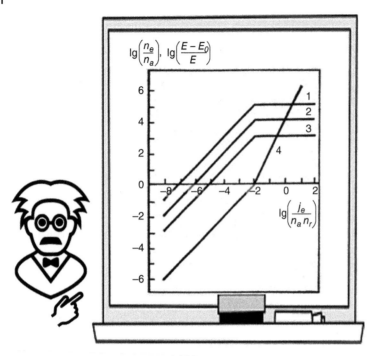

Figure 15.2 Characteristics of the electric discharge in aerosols: $1-3 - \lg(n_e/n_a)$; $4 - \lg[(E-E_0)]$. $1 - r_a = 10^{-3}$ cm; $2 - 10^{-4}$ cm; $3 - 10^{-3}$ cm; j_e (A cm^{-2}); n_a (cm^{-3}).

15.5 Dusty Plasma Structures: Coulomb Crystals and Phase Transitions

The $10 - 500$ nm dust particles can be significantly charged $Z_d e = 10^2 - 10^5 e$. Their Coulomb interaction can exceed thermal energy making **the dusty plasma** nonideal (7.1) and forming structures like those in liquids and solids. Such quasi-crystalline structures in dusty plasma are referred to as **Coulomb crystals.** The **nonideality of dusty plasma** is characterized by the **Coulomb coupling parameter**, which is ratio of the particle's interaction potential energy (density n_d, charge $Z_d e$, temperature T_d) and their kinetic energy:

$$\Gamma_d = \frac{Z_d^2 e^2 n_d^{1/3}}{4\pi\varepsilon_0 T_d}. \tag{15.47}$$

When interaction between the dust particles is strong and the nonideality parameter is high, the Coulomb coupling leads to quasi-crystallization. Change of the Coulomb coupling parameter result in the **quasi-phase transition in dusty plasma**. The dusty plasma is not structured and can be considered as "gas phase" at low values of the parameter $\Gamma \leq 4$. At higher nonidealities Γ, some coupling occurs, which can be interpreted as a quasi-"liquid phase." Finally, when the nonideality parameter exceeds the critical value $\Gamma \geq \Gamma_c = 171$, the 3D regular crystalline structure is formed according to the qualitative dusty plasma model.

A typical Coulomb crystal was observed in RF-CCP discharges (argon, 10 Torr, frequency 14 MHz, particle size about few μm, electron and ion densities $10^8 - 10^9$ cm^{-3}, Morfill and Thomas 1996). The observed structure can be characterized by the correlation function $g(r)$ showing the probability for two particles to be found at the distance r one from another. It substantiates the organized structure for at least five coordination spheres. Analysis of this structure shows that dust particles form a hexagonal 2D crystal grid in horizontal layers. In the vertical direction, dust particles position themselves exactly one under another and form a cubic grid between the crystal planes. "Melting" of the Coulomb crystals can be initiated by reducing the neutral gas pressure or by increasing the discharge power. The crystal-liquid phase transition in both cases is due to a decrease of the nonideality parameter. Similar Coulomb crystals were observed in dusty plasmas of DC glow discharges. Higher nonideality of dusty plasma and stronger coupling effects can be achieved by decreasing temperature, which was demonstrated in experiments with cryogenic plasmas of DC-glow and RF discharges.

15.6 Oscillations and Waves in Dusty Plasmas, Ionic Sound, and Dust Sound

Dust particles in plasma result in modification of the dispersion equations of plasma oscillations and waves even at low nonidealities and creates new modes specific only for dusty plasmas. To analyze these effects, the following dispersion equation is used:

$$1 = \frac{\omega_{pe}^2}{(\omega - ku_e)^2 - \gamma_e k^2 v_{Te}^2} + \frac{\omega_{pi}^2}{(\omega - ku_i)^2 - \gamma_i k^2 v_{Ti}^2} + \frac{\omega_{pd}^2}{(\omega - ku_d)^2 - \gamma_d k^2 v_{Td}^2}; \tag{15.48}$$

here, ω_{pe} is the plasma-electron frequency; ω_{pi} is the plasma-ion frequency; ω_{pd} is the plasma-dust frequency, taking into account the dust particles' mass m_d and charge $Z_d e$; u_e, u_i, u_d are the directed velocities of electron, ions, and dust particles; v_{Te}, v_{Ti}, v_{Td} are the average chaotic thermal velocities of electrons, ions, and dust particles; $\gamma_j(e, i, d)$ are factors related to the equation of state for electrons, ions and dust particles $p_j = const \cdot n_j^{\gamma_i}$, $\gamma_j = 1$ corresponds to isothermal oscillations of the j component, $\gamma_j = 5/3$ corresponds to the adiabatic ones. In the absence of direct motion of all three components, the **dispersion equation of dusty plasma** becomes:

$$1 = \frac{\omega_{pe}^2}{\omega^2 - \gamma_e k^2 v_{Te}^2} + \frac{\omega_{pi}^2}{\omega^2 - \gamma_i k^2 v_{Ti}^2} + \frac{\omega_{pd}^2}{\omega^2 - \gamma_d k^2 v_{Td}^2}. \tag{15.49}$$

This dispersion equation can be analyzed in different frequency ranges to describe the major oscillation branches of dusty plasma. For example, at high frequencies $\omega \gg k v_{Te} \gg k v_{Ti} \gg k v_{Td}$, the dispersion equation can be simplified:

$$1 = \frac{\omega_{pe}^2}{\omega^2} + \frac{\omega_{pi}^2}{\omega^2} + \frac{\omega_{pd}^2}{\omega^2}. \tag{15.50}$$

Considering that $\omega_{pe} \gg \omega_{pi} \gg \omega_{pd}$, the high frequency range is not affected by the dust particles and gives the conventional electrostatic Langmuir oscillations $\omega \approx \omega_{pe}$. At low frequencies $k v_{Te} \gg \omega \gg k v_{Ti} \gg k v_{Td}$, the dispersion can be expressed as ($\gamma_e = 1$):

$$\omega^2 \approx \omega_{pi}^2 \frac{k^2 r_{De}^2}{1 + k^2 r_{De}^2}; \tag{15.51}$$

here r_{De} is the electron Debye radius. When the wavelengths are shorter than the Debye radius ($k r_{De} \gg 1$), the dispersion (15.52) gives the conventional ion-plasma oscillations $\omega \approx \omega_{pi}$. When wavelengths are longer ($k r_{De} \ll 1$), the ionic sound $\omega \approx k c_{si}$ occurs and propagates with the speed:

$$c_{si} = \omega_{pi} r_{De} = \sqrt{\frac{T_e}{m_i} \cdot \frac{n_i}{n_e}} = v_{Ti} \sqrt{\tau(1 + zP)}; \tag{15.52}$$

here m_i is the ionic mass; factor $\tau = T_e/T_i$ is the ratio of electron and ion temperatures; factor $z = |Z_d|e^2/4\pi\varepsilon_0 r_d T_e$ is the dimensionless charge of a dust particle of radius r_d; factor $P = \frac{4\pi\varepsilon_0 r_d T_e}{e^2} \frac{n_d}{n_e}$ is the density parameter of dust particles and is approximately equal to the ratio of the charge density in the dust component to the charge density in electron component. The main influence of the dust particles on the ionic sound is due to the difference in densities of electrons and ions in the presence of charged dust particles when the dimensionless parameter $zP > 1$. At even lower frequencies of electrostatic plasma oscillations $k v_{Te} \gg k v_{Ti} \gg \omega \gg k v_{Td}$, the dispersion equation can be rewritten ($\gamma_e = 1$, $\gamma_i = 1$) as:

$$\omega^2 \approx \omega_{pd}^2 \frac{k^2 r_D^2}{1 + k^2 r_D^2}; \tag{15.53}$$

here the Debye radius is determined by electrons and ions: $r_D^{-2} = r_{De}^{-2} + r_{Di}^{-2}$. When the wavelengths of the oscillations are shorter than the Debye radius ($k r_D \gg 1$), the dispersion equation gives oscillations with plasma-dust frequency $\omega \approx \omega_{pd}$. In the opposite case of longer wavelengths ($k r_D \ll 1$), the dispersion equation brings us to a qualitatively new type of waves, the **dust sound** $\omega \approx k c_{sd}$. It propagates in the dusty plasmas with the velocity:

$$c_{sd} = \omega_{pd} r_D = v_{Td} \sqrt{\frac{|Z_d| T_i}{T_d}} \sqrt{\frac{\tau z P}{1 + \tau(1 + zP)}}, \tag{15.54}$$

which can exceed the thermal velocity of particles when their charges Z_d are high.

15.7 Electron Beam Plasmas: Generation, Propagation, Properties

The electron beam plasma is formed by injection into a neutral gas of a beam of electrons with energies usually from 10 keV to 1 MeV, see Figure 15.3. The beam is formed by electron gun usually located in a high-vacuum chamber and then transported through a foil or differential pumping window into a gas-filled discharge chamber. There, the high-energy electrons generate plasma and transfer energy into the gas by various mechanisms depending on pressure, beam current density, electron energy, and plasma parameters. If kinetic energy of electrons exceeds mc^2 (about 0.5 MeV), such beams are usually referred to as the **relativistic electron beam**. Total electron energy losses in dense gas due to interaction with neutrals along axis z are:

$$-\frac{1}{n_0}\frac{dE}{dz} = L(E);$$ (15.55)

n_0 is gas density; E is electron energy. $L(E)$ is the electron energy loss function (about 10^{-15} eVcm^2), which can be numerically expressed by the **Bethe formula** even for relativistic electrons:

$$L(E) = \frac{A}{E}\frac{(1+E/mc^2)}{(1+E/2mc^2)}\ln\frac{1.17\cdot E}{I_{ex}}.$$ (15.56)

In this relation, m is an electron mass; c is the speed of light; I_{ex} is the characteristic excitation energy (which is equal to 87 eV for air, 18 eV for hydrogen, 41 eV for helium, 87 eV for nitrogen, 102 eV for oxygen, 190 eV for argon, 85 eV for carbon dioxide, and 72 eV for water vapor). the factor A depends only on gas composition; $A = 1.3\cdot10^{-12}$eV^{2}cm^2 in air. The rate of ionization provided by an electron beam with current density j_b in a unit volume per unit time can be determined using the fast electron energy losses dE/dz as:

$$q_e = -\frac{dE}{dz}\frac{j_b}{eU_i};$$ (15.57)

here, U_i is the ionization energy cost. If the electron energy is in the range of 0.01–1 MeV, the ionization energy cost is always about 30 eV, see Table 15.1.

Different electron beam current densities and gas pressures result in four qualitatively different regimes of electron beam plasmas, which can be classified as follows:

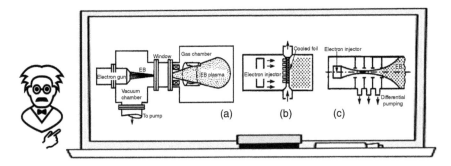

Figure 15.3 Electron beam generation: (a) schematic of a generator, (b) foil window, (c) differential pumping window.

Table 15.1 Ionization energy cost by electron beam with electron energies 0.1 − 1 MeV.

Gas	Ionization cost (eV)	Gas	Ionization cost (eV)
Air	32.3	Carbon dioxide (CO$_2$)	31.0
Nitrogen (N$_2$)	35.3	Oxygen (O$_2$)	32.0
Hydrogen (H$_2$)	36.0	Helium (He)	27.8
Argon (Ar)	25.4	Krypton (Kr)	22.8
Xenon (Xe)	20.8		

15.7.1 Powerful Electron Beam in Low-pressure Gas

The electron beam energy losses here are mostly due to the beam instability when the electron beam energy can be transferred by nonlinear beam-plasma interaction into the energy of Langmuir oscillations. Current and current density are high ($I_b > 1$ A, $j_b > 100$ A cm^{-2}), while pressure is below few Torr. This system is called **the plasma-beam discharge**, where T_e degree of ionization is high ($\alpha = n_e/n_0 \geq 10^{-3}$).

15.7.2 Low-current Electron Beam in Rarefied Gas

This regime is related to current densities $j_b < 0.1$ A cm^{-2} and pressures below 1 Torr. The beam instability is ineffective and the plasma-beam discharge cannot be sustained. The degree of ionization is low ($\alpha = n_e/n_0 < 10^{-7}$), and the temperature of plasma electrons is close to the neutral gas temperature.

15.7.3 Moderate-current Electron Beam in a Moderate Pressure Gas

Current density here is 0.1 A cm^{-2} and 100 A cm^{-2}, pressures 1–100 Torr. The degree of ionization depends on pressure and beam density and can vary in between 10^{-7} and 10^{-3}. The electron temperature is determined by the beam degradation. Gas temperature and specific energy are relatively low.

15.7.4 High Power Electron Beam in a High-pressure Gas

At high current densities $j_b > 100$ A cm^{-2} and pressures above 100 Torr, degree of ionization and electron temperatures are similar to those at moderate pressures and currents. High steady power and specific energy input result in intensive gas heating. The plasma is thermal, and T_0 can exceed 10 000 K.

Transportation length of the high-energy electron beams in dense gases is determined by energy losses due to ionization and excitation of neutrals and can be quite long. Numerically, the total length L (in m) for stopping high-energy electrons along their trajectory can be expressed as:

$$L = A E_{b0}^{1.7} \frac{T_0}{p}; \tag{15.58}$$

here T_0 is gas temperature in K, p is gas pressure in Torr, E_{b0} (eV) is initial electron energy, A is a constant, which depends only on gas composition, for air: $A = 1.1 \cdot 10^{-4}$. The electron beam transportation can also be limited by collisionless effects, related to beam instability, electrostatic repulsion of the beam electrons, and magnetic self-contraction. The first of these effects implies that the electron beam energy transfers to electrostatic Langmuir oscillations at the high-current – low-pressure regime. The second and third effects are related to the electric and magnetic fields of the beam. The total force pushing a periphery beam electron in a radial direction is:

$$F = \frac{n_{eb} r_b e^2}{2\varepsilon_0} (1 - \beta^2 - f_e); \tag{15.59}$$

here r_b is radius of the cylindrical beam, n_{eb} is the electron density of the beam plasma), $\beta = u/c$ is the ratio of beam velocity to the speed of sound, $f_e = n_i/n_{eb}$ is the beam space charge neutralization degree (n_i is the ion density). If $f_e > 1 - \beta^2$, which is usually valid in dense gases, the generated plasma neutralizes the electron beam space charge, and the magnetic field self-focuses the beam. If the high current electron beam is not neutralized, it can be destroyed by its own electric field unless electrostatic lenses are applied. When an electron beam is neutralized ($f_e = 1$), its current cannot exceed a critical value known as the **Alfven current**:

$$I_A = \frac{mc^3}{4\pi\varepsilon_0 e \beta\gamma}, \quad I_A(\text{kA}) = 17\beta\gamma; \tag{15.60}$$

here $\gamma = 1/\sqrt{1 - \beta^2}$ is the relativistic factor. The Alfven current provides the inherent magnetic field of the beam, which makes the Larmor radius less than half the beam radius and stops the electron beam propagation. For more details see Fridman and Rudakov (1973) and Mesyats et al. (2000).

15.8   Kinetics of Electron Beam Degradation Processes, Degradation Spectrum

The electron-beam energy degradation can be better described not in terms electron energy distribution function (EEDF), that is probability $f(E)dE$ for an electron to have energy E, but rather in terms of the **degradation spectrum**, which is the number of electrons $Z(E)dE$, having energy E during the degradation of one initial high-energy electron. The degradation process is determined by elastic and inelastic collisions of electrons with background particles, which can be considered at rest. Each of the processes "k" is characterized by the total cross-section $\sigma_k(E)$, and the differential cross-section $\sigma_k(E, \Delta E)$, where E is the electron energy during degradation, and ΔE is the electron energy loss during the collision. The differential cross-section $\sigma_k(E, \Delta E)$ can be normalized to the number of electrons formed after collision. The degradation, which does not change the number of electrons (for example, excitation processes), has the conventional normalization equation $\int_{-\infty}^{+\infty} \frac{\sigma_k(E, \Delta E)}{\sigma_k(E)} d(\Delta E) = 1$. Ionization, which doubles the number of electrons, has corrected normalization $\int_{-\infty}^{+\infty} \frac{\sigma_k(E, \Delta E)}{\sigma_k(E)} d(\Delta E) = 2$; and electron–ion recombination and electron attachment processes, where the electron "disappears," have the normalization $\int_{-\infty}^{+\infty} \frac{\sigma_k(E, \Delta E)}{\sigma_k(E)} d(\Delta E) = 0$. The differential probability for an electron with energy E to lose the energy portion ΔE in the collisional process "k" can be determined as $p_k(E, \Delta E) = \frac{\sigma_k(E, \Delta E)}{\sum \sigma_m(E)}$. The total probability for an electron with energy E to participate in the collisional process "k" can be then expressed as $p_k(E) = \frac{\sigma_k(E)}{\sum \sigma_m(E)}$. The degradation spectrum $Z(E)$ represents average number of electrons, which appears during the whole degradation process in the energy interval $E - E + dE$; therefore, the **kinetic equation for the degradation spectrum** in general is:

$$Z(E) = \int_0^\infty p(W, W - E) \cdot Z(W)dW + \chi(E),\tag{15.61}$$

with boundary condition $Z(E \to \infty) = 0$. Here, $\chi(E)$ is the source function of the high-energy electrons; if the degradation is started by a single electron with energy E_0, the source function is the delta function, $\chi(E) = \delta(E - E_0)$. Finally, $p(W, W - E)$ is the total probability of formation of an electron with energy E by collisions of an electron with initial energy W and the energy loss $\Delta E = W - E$: $p(W, W - E) = \sum p_m(W, \Delta E = W - E)$. At low energies (below the lowest excitation threshold), when the collisional cross-sections can be neglected, the degradation spectrum coincides with the change of the electron energy distribution at final and initial moments of the degradation process:

$$Z(E) = f(E, t \to \infty) - f(E, t \to 0).\tag{15.62}$$

At the absence of initial EEDF, $f(E, t \to 0) \to 0$, EEDF and degradation spectrum are equal under the excitation threshold. Integration of $Z(E)$ gives the total number of collisions "k" (ionization, excitation, dissociation, etc.) per one initial electron:

$$N_k = \int_0^\infty p_k(E) \cdot Z(E)dE = \int_0^\infty Z(E) \frac{\sigma_k(E) \cdot dE}{\sum \sigma_m(E)}.\tag{15.63}$$

If initial energy of a beam electron is E_0, the "energy cost of a process k" (like ionization energy cost, excitation energy cost, dissociation energy cost, etc.) can be calculated as:

$$U_k = \frac{E_0}{N_k} = E_0 / \int_0^\infty p_k(E)Z(E)dE.\tag{15.64}$$

The general degradation spectrum Eq. (15.61) can be simplified to a convenient formula known as the **Alkhazov's equation** (Nikerov and Sholin 1985):

$$Z(E) = \sum_k p_{ex,k}(E + U_k) \cdot Z(E + U_k) + \int_{E+I}^\infty p_i(W, W - E) \cdot Z(W)dW + \delta(E - E_0).\tag{15.65}$$

The boundary condition for the equation is again $Z(E \to \infty) = 0$. In the Alkhazov's equation, $p_{ex,k}(E + U_k)$ is the probability of excitation of an atom into the kth excited state in a collision with an electron having energy $E + U_k$; $p_i(W, W - E)$ is the differential probability of ionization of an atom in collision with an electron having initial kinetic energy W and during the collision losing energy, $\Delta E = W - E$; I and U_k are the ionization potential and the excitation energy of the kth state; $\delta(E - E_0)$ is the delta-function describing degradation of an initial electron with high energy

E_0. Simple solution of the Alkhazov's equation can be obtained in the Bethe–Born approximation considering only ionization by primary electrons:

$$Z(E) = \frac{\sigma(E)}{L(E)} = \frac{\aleph^2 \ln(cE)}{4R \ln\left(\sqrt{\dfrac{e}{2}\dfrac{E}{I^*}}\right)}. \tag{15.66}$$

$L(E) \approx (16R^2/E)\ln(\sqrt{e/2}E/I^*)$ describes electron energy losses; $\sigma(E) \approx (4R/E)\aleph^2 \ln(cE)$ is the total cross-section of inelastic collisions for an electron with energy E; $R = 13.595\,\text{eV}$ is the Rydberg constant; $\aleph^2 = 0.7525$ is square of the matrix element for the sum of transitions into continuous and discrete spectra; factor $c = 0.18\,\text{eV}$; $I^* = 42\,\text{eV}$ is the characteristic excitation potential (the numerical data are given for helium).

The Alkhazov's equation can be also solved analytically with sufficient accuracy resulting in the degradation spectrum above the ionization threshold in the form (for more information, see Nikerov and Sholin 1985):

$$Z(E) = \frac{0.6}{I} + \frac{0.5 \cdot E_0}{(E+I)^2}\left[1 + \frac{3I^2}{(E+I)^2}\ln\left(\frac{E}{I}+2\right)\right]. \tag{15.67}$$

Same approach for the degradation spectrum below the ionization threshold gives:

$$Z(E) = \frac{2}{(1+E/I)^4 I}\ln\left(\frac{E}{I}+2\right) + \frac{1}{I(1+E/I)^2}. \tag{15.68}$$

These analytical degradation spectra are illustrated in Figure 15.4 together with numerical solutions of the Alkhazov's equation for electron beam degradation in helium, hydrogen, and fluorine. Such degradation spectra can be applied for calculating the ionization and excitation energy costs and total number of ions and excited species produced by one high-energy electron (15.63 and 15.64); an example for the case of electron degradation in hydrogen is presented in Table 15.2.

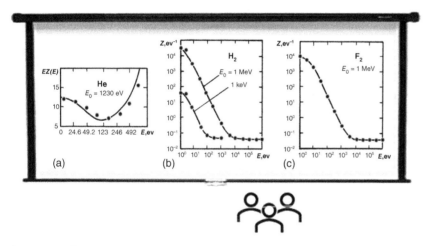

Figure 15.4 Degradation spectra in: (a) helium, (b) hydrogen, (c) fluorine. Stars – numerical calculations, curves – analytical calculations.

Table 15.2 Total production of ions and excited molecules in degradation of an electron with initial energies 1 MeV and 1 keV in molecular hydrogen.

State of molecule	$E_0 = 1\,\text{MeV}$	$E_0 = 1\,\text{keV}$
Ionized	$3.06 \cdot 10^4$	29.6
$H_2\left(\Sigma_u^+\right)$	$1.10 \cdot 10^4$	10.7
$H_2\left(B^1\Sigma_u^+\right)$	$1.43 \cdot 10^4$	13.9
$H_2^*(vibr.\ exc.)$	$1.84 \cdot 10^5$	187

15.9 Plasma-beam Discharge, Plasma-beam Centrifuge

Electron beam in the plasma-beam discharges does not directly interacts with neutral gas. At first, it interacts with Langmuir plasma oscillations, which subsequently transfer energy through electrons to neutrals. The electron beam can be considered as a source of microwave oscillations, which then sustain the "microwave discharge." In contrast to conventional microwave discharges, the source of electrostatic oscillations is present inside of plasma, which mitigates the skin-effect problems. The plasma-beam discharge requires high beam currents and current densities ($I_b > 1$ A, $j_b > 100$ A cm^{-2}), and pressures below few Torr ($10^{-4} - 3$ Torr). It usually operates at power $0.5 - 10$ kW. The degree of ionization varies widely $10^{-4} - 1$, as well as electron temperature $1 - 100$ eV. The principal operation condition (limiting the discharge to pressures below few Torr) requires beam energy dissipation into Langmuir oscillations to exceed energy losses in electron-neutral collisions, which in molecular gases means:

$$\frac{6\pi^2 \hbar \omega \nu_{eV}}{\overline{\varepsilon}_b \nu_{bn}} \left(\frac{\omega_{pb}}{\nu_{eV}} \right) > 1;$$

(15.69)

here ν_{eV} is the vibrational excitation frequency; $\hbar \omega$ is vibrational quantum; ν_{bn} is frequency of beam electrons' collisions with neutrals; $\overline{\varepsilon}_b$ is the average energy loss of a beam electron during its collision with a neutral particle; n_b is density of the beam electrons; ω_{pb} is plasma frequency corresponding to this density $\omega_{pb} = \sqrt{\frac{n_b e^2}{\varepsilon_0 m}}$. The second condition of effective plasma-beam discharge operation is related to energy transfer from Langmuir oscillations to plasma electrons through the modulation instability. This instability leads to a collapse of the Langmuir oscillations, and energy transfer along the spectrum into the range of short wavelengths. Energy of the short-wavelength oscillations can be then transferred to plasma electrons due to the Landau damping. Stimulation of chemical reactions through vibrational excitation of molecules requires sufficient power of the discharge $\sigma E^2 > \nu_{eV} n_e \hbar \omega$, and therefore very high level of the Langmuir noise W:

$$\frac{W}{n_e T_e} = \frac{\varepsilon_0 E^2}{2 n_e T_e} \geq \frac{\nu_{eV}}{\nu_{en}} \frac{\hbar \omega}{T_e};$$

(15.70)

it should exceed 10^{-3} for effective plasma-beam discharge operation.

A schematic of the plasma-beam discharge is shown in Figure 15.5a. The discharge chamber can be large: internal diameter 50 cm, length 150 cm. At 10 mTorr, the flow rate is about 30 cm^3 s^{-1}; pressure in the electron gun

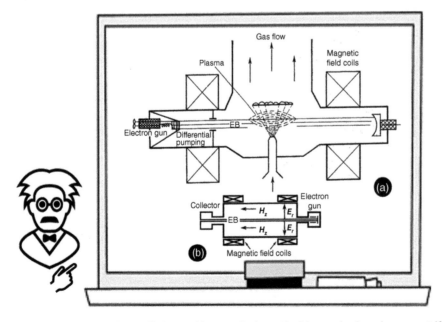

Figure 15.5 Plasma beam discharge: (a) general schematic, (b) organized as plasma centrifuge.

chamber is about 10^{-5} Torr and sustained by differential pumping. An electron beam with a maximum power 40 kW is transported into the discharge chamber along the magnetic field of 300 kA m^{-1}; beam electrons energy is 13 keV. The electron density is about $5 \cdot 10^{13}$cm^{-3} at the discharge power of 3 kW. The maximum fraction of the beam power absorbed in plasma exceeds 70%. The plasma-beam discharges can be used for organization of **plasma centrifuges** where the ionized gas is rotating fast in crossed electric and magnetic fields, which was discussed in Section 10.12. A specific feature of the plasma centrifuge, based on the plasma-beam discharge, is the radial electric field, which is provided directly by the electron beam, see Figure 15.5b. Detailed information regarding plasma-beam discharges and their applications can be found in the book of Ivanov and Soboleva (1978).

15.10 Non-self-sustained High-pressure Cold Discharges Supported by High-energy Electron Beams

The beam energy in these non-self-sustained discharges is relatively low and spent mostly for ionization and electronic excitation of neutrals, that is to create plasma. To provide necessary energy input, for example, to vibrational excitation of molecules, additional energy should be transferred to the plasma electrons from an external electric field. This electric field can be chosen sufficiently low to avoid ionization, but optimal for vibrational excitation or other processes requiring low electron temperatures. The non-self-sustained discharges are interesting for plasma-chemical and laser applications because they can provide a high level of nonequilibrium uniformly in very large volumes at high pressures (up to several atmospheres). An important advantage of these systems is the mutual independence of ionization and energy input in the molecular gas, which permits choosing the optimal reduced the electric field E/n_0 without caring to sustain the discharge. It is also important that ionization instabilities cannot destroy these discharges, simply because ionization is independent and cannot be affected by local overheating. Schematic of such discharges is illustrated in Figure 15.6. The electron beam is injected into the gas across the accelerator's window closed by a metallic foil. Voltage is applied to the further electrode through a capacitor. Thus, plasma generation is due to the electron beam ionization, and the electric current is provided by the applied external electric field. The discharge can be operated in stationary and in short-pulse regimes. In the short pulse regime: the pulse duration is usually about 1000 ns, the electron beam energy about 300 keV and the beam current density can exceed 100 A cm^{-2}. In atmospheric pressure air, the ionization rate provided by such electron beam reaches $q_e = 10^{22}$ 1/cm^{3}s. Losses of plasma electrons in this pulse-beam system are due to the electron–ion recombination and electron attachment to electronegative air components followed by the ion–ion recombination. These losses can be described by the effective recombination coefficient $k_r^{eff} = k_r^{ei} + \varsigma \, k_r^{ii}$, where k_r^{ei}, k_r^{ii} are rate coefficients of electron–ion and ion–ion recombination; $\varsigma = k_a/k_d$ is ratio of the attachment and detachment coefficients. In air, $k_r^{eff} \approx 10^{-6}$cm^3 s^{-1}. Plasma density n_e is then determined by electron beam ionization rate q_e:

$$n_e = \sqrt{q_e/k_r^{eff}}.$$ (15.71)

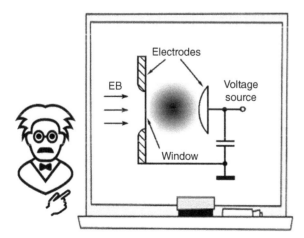

Figure 15.6 Non-self-sustained cold atmospheric discharge supported by high-energy electron beam.

In the above example, the plasma electron density is about $10^{14} \mathrm{cm}^{-3}$ and ionization degree about $3 \cdot 10^{-6}$. If the pulsed beam duration is 1000 ns, the specific energy input from the beam is $0.05 \, \mathrm{J \, cm}^{-3}$, and from the external electric field about $1 \, \mathrm{J \, cm}^{-3}$, which is 20 times higher.

The optimal specific energy inputs to stimulate endothermic plasma-chemical processes are usually $1 \, \mathrm{eV \, mol}^{-1}$ ($5 \, \mathrm{J \, cm}^{-3}$ at normal conditions). The minimum threshold values of the specific energy input in these systems are about $0.1 - 0.2 \mathrm{eV \, mol}^{-1}$ or $0.5 - 1 \, \mathrm{J \, cm}^{-3}$. Reaching these energy inputs in nonthermal homogeneous atmospheric pressure discharges is limited by numerous instabilities. Those can be suppressed in non-self-sustained discharges (since ionization is provided independently by electron beams), which makes these plasma sources so attractive.

15.11 Plasma in Tracks of Nuclear Fission Fragments, Plasma Radiolysis

The nuclear fission reaction between a slow neutron with uranium-235:

$$_{92}\mathrm{U}^{235} + {}_0 n^1 \rightarrow {}_{57}\mathrm{La}^{147} + {}_{35}\mathrm{Ba}^{87} + 2{}_0 n^1 \tag{15.72}$$

releases total fission energy 195 MeV. Most of this energy, 162 MeV, goes to kinetic energy of the fission fragments, barium and lanthanum. These fragments effectively produce plasma when transmitted into gas phase before thermalization. Such fragments appear in gas as multiply charged ions with $Z = 20 \div 22$ and initial energy of approximately $E_0 \approx 100$ MeV. Their energy degradation proceeds through formation of the so-called δ-**electrons** moving perpendicularly to the fission fragment trajectory with kinetic energy approximately $\varepsilon_0 \approx 100$ eV. These δ-electrons form a radial high-energy electron beam, which then generates plasma in the long cylinder surrounding the trajectory of the fission fragment. This plasma cylinder is referred to as the **track of the nuclear fission fragments.** If the plasma density is relatively low, then degradation of each δ-electron in the radial beam is quasi-independent and their interaction with gas is reduced to conventional radiolysis. Interesting physical and chemical effects occur if the plasma density in the tracks is relatively high, and collective effects of electron–electron interaction, vibrational excitation, etc. take place, which is referred to as the **plasma radiolysis**.

Let's analyze the **degree of ionization in the tracks of fragments** due degradation of δ-electrons with energy $\varepsilon_\delta \approx 100$ eV formed by the fragments. **The degradation length of the fragments** is:

$$\lambda_0 = \frac{4\pi \varepsilon_0^2 E_0 m v^2}{n_0 Z^2 Z_0 e^4} \ln^{-1}\left(\frac{2mv^2}{I}\right); \tag{15.73}$$

v, Ze, E_0 are velocity, charge, and kinetic energy of nuclear fission fragments; m, e are electron mass and charge; Z_0 is the sum of charge numbers of atoms forming molecules, which are irradiated by the nuclear fragments; n_0 is gas density; I is ionization potential. The number of δ-electrons formed by a fission fragment per unit length is:

$$N_\delta = \frac{n_0 Z^2 Z_0 e^4}{4\pi \varepsilon_0^2 m v^2 \varepsilon_\delta} \ln\left(\frac{2mv^2}{I}\right). \tag{15.74}$$

For example, generation of the δ-electrons in water vapor at atmospheric pressure is characterized by $N_\delta = 10^6 \mathrm{cm}^{-1}$, in carbon dioxide $N_\delta = 3 \cdot 10^6 \mathrm{cm}^{-1}$. The tracks start with very narrow channels of complete ionization. The radius of these initial completely ionized channels is:

$$r_0 = \frac{Ze^2}{2\pi \varepsilon_0 v \sqrt{m \varepsilon_\delta}} \ln\left(\frac{2mv^2}{I}\right) \approx 3 \cdot 10^{-8} \mathrm{cm}. \tag{15.75}$$

Expansion of this initial track is at first collisionless, when radius r of the channel is less than length λ_i necessary for ionization. Then the expansion mechanism becomes diffusive. During the **collisionless track expansion**, two regimes occur depending on the linear density of δ-electrons N_δ. If the linear density is sufficiently high:

$$N_\delta > 4\pi \varepsilon_0 \frac{\varepsilon_\delta}{e^2} \ln^{-1}\left(\frac{\lambda_i}{r_0}\right) \approx 10^8 \mathrm{cm}^{-1}, \tag{15.76}$$

then the plasma remains quasi-neutral during the collisionless expansion. In the opposite case of low linear density, when $N_\delta < 10^8 \mathrm{cm}^{-1}$, electrons and ions are independent during the expansion, and therefore the δ-electrons keep their energy. In the dense plasma regime $N_\delta > 10^8 \text{cm}^{-1}$, 2/3 of the δ-electrons energy (of about 100 eV) is transferred into radial motion of ions. The efficiency of these 70 eV ions in ionization processes and in chemical reactions is very low. Thus, this high-density regime is not effective in plasma radiolysis. Summarizing, the high energy efficiency can be achieved during the expansion of the initial track, only when the linear density of δ-electrons is relatively low ($N_\delta < 10^8 \text{cm}^{-1}$), and electrons do not draw ions with them. In this case, the radius of the tracks can be determined as $\lambda_i \approx 1/n_0\sigma_i$, where σ_i is cross-section of ionization provided by δ-electrons. The degree of ionization in the tracks of nuclear fragments can be then calculated as:

$$\frac{n_e}{n_0} = \left(\frac{3Z_0 e^4 \sigma_i^2}{4\pi\varepsilon_0^2 m\varepsilon_\delta} \right) \cdot \left(\frac{Z}{v} \right)^2 n_0^2. \tag{15.77}$$

The degree of ionization in the plasma channel of the nuclear tracks is proportional to the square of the neutral gas density and pressure. Numerically at atmospheric pressure, the degree of ionization in the tracks of nuclear fragments in water vapor is $n_e/n_0 \approx 10^{-6}$, in carbon dioxide, it is about $n_e/n_0 \approx 3 \cdot 10^{-6}$. The necessary degree of ionization for significant plasma effects in radiolysis (which can be estimated for water and carbon dioxide as $10^{-4} \div 10^{-3}$) cannot be achieved at atmospheric pressure. However, considering that $n_e/n_0 \propto n_0^2$, the sufficient degree of the ionization can be achieved by increasing pressure to 10–20 atm. Further pressure increase is limited by the criterion opposite to (15.76). At high pressures, the linear densities N_δ are also high, and the main portion of electron energy is ineffectively lost to ions. The ion heating and transfer to thermal plasma in the tracks takes place in water vapor at pressures of 100 atm, and in carbon dioxide at pressures about 30 atm. As a result, energy efficiencies of such plasma-radiolysis processes as H_2O and CO_2 decomposition have maximum at pressures 20–30 atm, see Figure 15.7. In principle, the highest energy efficiencies of water and carbon dioxide decomposition in the tracks of nuclear fission fragments can be achieved, if δ-electrons transfer their energy not directly to neutral gas but rather through intermediate transfer of energy to plasma formed in the tracks similar to the plasma-beam discharges described in Section 15.9. Then the energy of plasma electrons can be selectively transferred to the effective channels of stimulation of plasma-chemical reactions, for example to vibrational excitation. The nuclear fission fragments can transfer in this way most of their energy to formation of chemical products (like hydrogen) rather than to heating. Such chemo-nuclear reactor can be "cold," because only a small portion of its power would be released to heating.

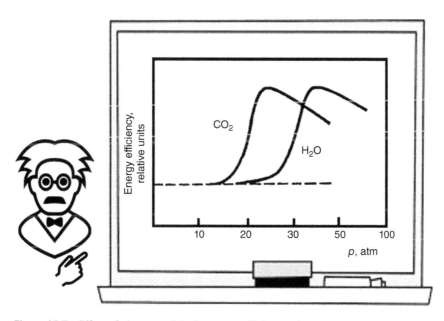

Figure 15.7 Effect of plasma radiolysis: energy efficiency of water vapor and carbon dioxide dissociation in tracks of nuclear fragments as a function of pressure.

15.12 G-Factors, Plasma-radiolytic Effects in Water Vapor and Carbon Dioxide

The energy efficiency of radiolysis is often characterized by the **G-factors**, showing the number of chemical processes (for example, number of dissociated water molecules) per 100 eV of radiation energy. The G-factor for **water vapor radiolysis** by electron beams can be expressed as:

$$\frac{G}{100} = \left[\frac{1}{U_i(H_2O)} + \frac{U_i(H_2O) - I_i(H_2O)}{U_i(H_2O) \cdot I_{ex}(H_2O)} \right] + \frac{\mu}{U_i(H_2O)}; \tag{15.78}$$

$U_i(H_2O) \approx 30$ eV is H_2O ionization energy cost; $I_i(H_2O) = 12.6$ eV, $I_{ex}(H_2O) = 7.5$ eV are energy thresholds of ionization and electronic excitation of water molecules; the factor μ shows how many times one electron with an energy below the threshold of electronic excitation $\varepsilon < I_{ex}(H_2O)$ can be used in the water molecule destruction. The first two terms in (15.78) are related to dissociation through electronic excitation and give the radiation yield $G \approx 12 \frac{mol}{100eV}$, which describes water dissociation by β-radiation and low current density electron beams. The third term in (12.5.2) describes the contribution H_2O destruction through dissociative attachment:

$$e + H_2O \rightarrow H^- + OH, \quad H^- + H_2O \rightarrow H_2 + OH. \tag{15.79}$$

Contribution of this mechanism is relatively low in conventional radiolysis is ($\mu = 0.1$ because of the resonance character of the dissociative attachment cross-section dependence on energy and low ionization degree. The factor μ can be significantly larger at higher degrees of ionization in plasma because of effective energy exchange (Maxwellization) between plasma electrons with energies below the threshold of electronic excitation $\varepsilon < I_{ex}(H_2O)$. The critical degree of ionization required for Maxwellization of plasma electrons and therefore for plasma effects in radiolysis can be found from the kinetic equation for electrons as:

$$\frac{n_e}{[H_2O]} \gg \sigma_a^{max} \frac{T_e^2 \varepsilon_0^2}{e^4 \Lambda}; \tag{15.80}$$

$\sigma_a^{max} \approx 6 \cdot 10^{-18} cm^2$ is the maximum dissociative attachment cross-section of electrons to H_2O; Λ is the Coulomb logarithm. Numerically, at $T_e = 2$ eV, this criterion requires $n_e/[H_2O] \gg 10^{-6}$, which can be achieved in tracks of nuclear fission fragments as well as in electron beam systems. The effective Maxwellization of electrons under the threshold of electronic excitation in plasma radiolysis permits the μ-factor to increase from about $\mu = 0.1$ (typical for conventional radiolysis) to higher values determined by the degradation spectrum $Z(\varepsilon)$:

$$\mu = \frac{\int_0^I \varepsilon \, Z(\varepsilon) \, d\varepsilon}{\varepsilon_a \int_0^I Z(\varepsilon) \, d\varepsilon}; \tag{15.81}$$

here I is the ionization potential; $Z(\varepsilon)$ is the degradation spectrum under the threshold of electronic excitation; ε_a is energy necessary for dissociative attachment. The integration (15.81) gives $\mu \approx 1$. Thus, the first two terms in (15.78) give the radiation yield $G \approx 12 \frac{mol}{100 \, eV}$, and the third term specific for plasma radiolysis gives an additional $G \approx 7 \frac{mol}{100 \, eV}$ due to the dissociative attachment. This brings the total radiation yield of H_2O dissociation to $G \approx 19 \frac{mol}{100 \, eV}$.

Radiation yield of CO_2**-dissociation**, similarly to the case of water vapor radiolysis, is determined by the conventional term G_{rad} typical for radiation chemistry, and by a special plasma radiolysis term related to the contribution of electrons under the threshold of electronic excitation. The first term, like in (15.78), is due to electronic excitation and dissociative recombination and can reach $G_{rad} = 8 \frac{mol}{100 \, eV}$. The plasma radiolysis effect here is due to contribution of the low-energy electrons to vibrational excitation of CO_2. At high degrees of ionization $n_e/n_0 \gg 10^{-3}$, the plasma radiolytic effect permits to double the radiation yield of CO_2 to reach $G = 16 \frac{mol}{100 \, eV}$, see Fridman and Kennedy (2021) for more details.

15.13 Dusty Plasma Generation by Relativistic Electrons, Radioactive Dusty Plasma

The high-energy electrons can ionize both neutral gas and aerosols, while the energies of photons are sufficient only for ionization of dust particles with relatively low work functions. The charge of macro-particles irradiated by high-energy electrons or electron beams is determined by two major effects. One is related to neutral gas ionization and plasma formation, which leads to negative charging of macro-particles due to electron mobility prevailing over that of ions. The second effect is related to secondary electron emission by electron impact from the aerosols, which tends toward positive charging of aerosols. Competition between these two ionization effects determines the composition and characteristic of the dusty plasma irradiated by electrons.

Special interest is directed toward interaction of β-radiation (β-radioactivity, which can be considered as a low current high-energy electron beam) with aerosols, and to plasma formation and charging related to β-radioactive "hot" aerosols. Considering the interaction of nuclear β-radiation with aerosols in air, one should consider fast attachment of plasma electrons to electronegative oxygen molecules and to electron conversion into negative ions, which makes the of negative charging effect not so strong. The β-radioactive hot aerosols present an example of dusty plasma where electron density is relatively low, and the macro-particle charge is usually positive. At the nuclear β-radiation rates exceeding one electron per second per particle, the total positive charge of the dust particles usually exceeds hundreds of elementary charges.

Generally, most aerosols are negatively charged if the electron bean flux is low, and secondary electron emission from their surfaces can be neglected. Assuming the Bohm flux of ions on the aerosol surfaces, and that electron–ion recombination mostly follows volume mechanisms, the relevant condition of low electron beam density and therefore negative aerosol charging is:

$$n_{eb} \ll n_0 \frac{\sigma_i}{\sigma_{rec}^{ei}} \frac{\bar{v}_e}{v_{eb}} \frac{1}{(\delta - 1)^2} \frac{m}{M}, \tag{15.82}$$

where m, M are respectively electron and ion masses. Numerically, it means $n_{eb} < 3 \cdot 10^7 \text{cm}^{-3}$, if electron thermal velocity is $\bar{v}_e = 10^8 \text{cm s}^{-1}$, ionization cross-section is $\sigma_i = 3 \cdot 10^{-18} \text{cm}^2$, beam electron velocity $v_{eb} = 3 \cdot 10^{10} \text{cm s}^{-1}$, $\sigma_{rec}^{ei} = 10^{-16} \text{cm}^2$, $\delta = 100$ (secondary electron emission coefficient, corresponding to a particle radius $r_a = 10 \,\mu\text{m}$), $m/M = 10^{-4}$, $n_0 = 3 \cdot 10^{19} \text{cm}^{-3}$. According to (15.82), the electron beam current density should be relatively low ($j_{eb} < 0.1 \text{ A cm}^{-2}$) to have mostly negative charging of the aerosol particles and to have the possibility to neglect the secondary emission from the macro-particle surfaces.

15.14 Problems and Concept Questions

15.14.1 Electron Density Due to Monochromatic Photoionization of Aerosols

Using Chart 15.1, calculate the electron density provided by monochromatic photoionization (photon flux $10^{14} \text{ 1/cm}^2\text{s}$, photon energy $E = 2 \text{ eV}$) of aerosol particles with radius 10 nm and work function $\phi_0 \approx A_0 = 1 \text{ eV}$. Consider particle densities $n_a = 10^3 \text{cm}^{-3}$ and $n_a = 10^7 \text{cm}^{-3}$.

15.14.2 Thermal Ionization of Aerosols

Consider thermal ionization of macro-particles of radius $10 \,\mu\text{m}$, work function 3 eV, and temperature 1500 K. Find the electron density near surface of the aerosol particle and corresponding value of the Debye radius r_D^0.

15.14.3 "Melting" of Coulomb Crystals

"Melting" of the 2D-Coulomb crystals in RF-CCP discharges can be initiated by reducing the neutral gas pressure or by increasing the discharge power. Explain how the change of pressure and power leads to a decrease of the nonideality parameter and to the Coulomb crystal-liquid phase transition.

15.14.4 Ionization Energy Cost Due to Irradiation by High-energy Electrons

Using Table 15.1, calculate the fraction of the electron beam energy going to excitation of neutrals in different gases. Why the ionization energy cost in molecular gases exceeds that for inert gases?

15.14.5 Degradation Spectrum vs EEDF

Why the electron energy distribution functions are less effective than degradation spectrum in description of stopping of electron beams.

15.14.6 Energy Cost of Ionization and Excitation by Electron Beams

Using Table 15.2, calculate energy cost of ionization and excitation of electronically and vibrational excited states in hydrogen irradiated by electron beam with an energy 1 MeV and 1 keV.

15.14.7 Initial Tracks of Nuclear Fission Fragments

Explain, why plasma in the initial tracks of in nuclear fission fragments is completely ionized?

15.14.8 Plasma Radiolysis of Carbon Dioxide

The plasma effect in CO_2 radiolysis is related to stimulation of the reaction $O + CO_2 \rightarrow CO + O_2$ by vibrational excitation of CO_2-molecules, which prevails over three-molecular recombination $O + O + M \rightarrow O_2 + M$. Determine the vibrational temperature of CO_2 molecules necessary for the plasma effect, considering that the necessary degrees of ionization should satisfy the requirement $n_e/n_0 \gg 10^{-3}$.

Lecture 16

Electric Discharges in Water and Other Liquids

16.1 Plasma Generation in Liquid Phase

Plasma is best known as a gas phase phenomenon, as an "ionized gas." At the same time, many efforts have been made in the past several decades to seek plasma in liquid phase. Most researchers, however, observed such plasmas not directly in liquids, but in low-density "pockets," gas phase bubbles or voids dispersed within fluids. While it looks like "liquid plasma," and surely it is plasma inside of liquid, there is no ionization of liquid phase itself in this case. These plasmas having a lot of applications, like plasma disinfection, sterilization, activation of liquids for washing, cleaning, plant growth stimulation, etc., are going to be discussed at the beginning of the lecture.

 Recent advances in pulsed power permitted application of very short voltage rise time (including the sub-nanosecond range) and revealed that *plasma-like phenomena can occur in liquid phase quasi-homogeneously without any bubbles as a nonequilibrium plasma in liquid phase. This can be interpreted as ionization of liquid itself*. Unique nonequilibrium properties of such discharges in homogeneous high-density medium, such as high densities of electrons and excited species, light and high energy radiation, and high electron energies together with low temperature of liquid may be associated with exclusive opportunities that may lead to fundamentally new effects and may have great impact in the fields of medicine, microelectronics, energy systems, and materials. These interesting new plasmas in different liquids, including liquid nitrogen, are going to be discussed in the second part of this lecture.

Let us start with consideration of plasma generation inside of liquids related to preliminary formation of bubbles, voids, etc. Electric breakdown of liquids, in general, is limited by their high density, short mean free path of electrons, and therefore requires very high electric fields E/n_0, see Paschen curves. Nevertheless, breakdown of liquids can be performed not at the extremely high electric fields required by Paschen curves but at those only slightly exceeding breakdown fields in atmospheric pressure molecular gases. This effect can be simply explained by different electrically induced mechanisms of formation of bubbles, macro- and micro-voids, and quasi- "cracks" inside the liquids. Inside those bubbles, voids, etc., plasma is formed in gas phase, which obviously requires not so high electric fields. Such discharge can be sustained, for example, in water by pulsed high-voltage power supplies and usually start from sharp electrodes. If the discharge does not reach the second electrode it can be interpreted as pulsed corona, branches of such a discharge are referred to as streamers. If a streamer reaches the opposite electrode a spark is usually formed. If the current through the spark is high (above 1 kA), this spark is usually called a pulsed arc. Various electrode geometries have been used for the plasma generation in water; the two simplest are point-to-plane and point-to-point configuration. The former is often used for pulsed corona discharges, whereas the latter is often used for pulsed arc systems. Concern in the use of pulsed discharges is the limitation posed by the electrical conductivity of water. In the case of a low electric conductivity (below $10\,\mu S\,cm^{-1}$), the range of applied voltage that can produce corona discharge without sparking is very narrow. In the case of a high electric conductivity of water (above $400\,\mu S\,cm^{-1}$), streamers become short and the efficiency of radical production decreases, and denser and cooler plasma is generated. Production of OH radical and O atoms is more efficient at conductivity below $100\,\mu S\,cm^{-1}$. For tap water, the bulk heating can be a problem for corona discharges. At frequencies about 200 Hz, the water temperature can rise from 20 to 55 °C in 20 min, indicating a significant power loss.

Plasma Science and Technology: Lectures in Physics, Chemistry, Biology, and Engineering, First Edition. Alexander Fridman.
© 2024 WILEY-VCH GmbH. Published 2024 by WILEY-VCH GmbH.

16.2 Major Conventional Breakdown Mechanisms and Discharge Characteristics in Water

Mechanisms of conventional (not assuming direct ionization of condensed phase) plasma discharges and breakdowns in liquids (specifically in water) can be classified into two groups: *the first approach presents the breakdown in water as a sequence of a bubble generation process and a following electronic process*; and *the second group divides the whole process into a partial discharge and a fully developed discharge such as arc or spark*.

In *the first group*, the bubble generation process starts from a micro-bubble which is formed by the vaporization of liquid in particular by a local heating in the strong electric field region at the tips of electrodes. The bubble grows, and an electrical breakdown takes place within the bubble. In this case, the cavitation mechanism can explain the slow bush-like streamers. The appearance of bright spots is delayed from the onset of the voltage, and the delay time tends to be greater for smaller voltages. The time lag to water breakdown increases with increasing pressure, supporting the bubble mechanism in a sub-microsecond discharge formation in water.

Time to form the bubbles is about 3–15 ns, depending on the electric field and pressure. The influence of water electrical conductivity on this regime of the discharges is small. Bulk heating via ionic current does not contribute to the initiation of the breakdown. The power necessary to evaporate the water during the streamer propagation can be estimated using the streamer velocity, the size of the streamer, and the heat of vaporization. Using a streamer radius of $31.6\,\mu m$, a power of 2.17 kW was estimated to be released into a single streamer to ensure its propagation in the form of vapor channels. In case of multiple streamers, the required power can be estimated by multiplying the number of visible streamers to the power calculated for a single streamer. In the electronic phase of the plasma generation process, electron injection and drift in liquid phase take place at the cathode, while hole injection through a resonance tunneling mechanism occurs at the anode. In the electronic process, electric breakdown occurs when an electron makes a required number of direct ionizing collisions during its transit across the breakdown gap.

In *the second group*, the discharges are divided into partial discharge, and arc and spark discharges, see Table 16.1. In the partial discharges, the current is mostly transferred by ions. In the case of high electrical conductivity water, a large discharge current flows, resulting in a shortening of the streamer length due to the faster compensation of the space charge electric fields on the head of the streamer. Subsequently, a higher power density in the channel is obtained, resulting in a higher plasma temperature, a higher UV radiation, and generation of acoustic waves. In the arc or spark discharge plasmas, the electric current is usually transferred by plasma electrons. The high current heats a relatively small volume of plasma in the gap between the two electrodes, generating quasi-thermal plasma with relatively high temperature. When a high voltage–high current discharge takes place between two submerged electrodes, a large part of the energy is consumed in the formation of a very energetic thermal plasma channel. This channel emits UV radiation and its expansion against the surrounding water generates intense shock waves. For the corona discharge in water in contrast to that, the shock waves are weak or moderate, whereas for the pulsed arc or spark the shock waves are strong.

To analyze major ***conventional mechanisms of water breakdown***, we must keep in mind that $30\,kV\,cm^{-1}$ is a typical breakdown voltage of air at 1 atm. When one attempted to produce direct plasma discharges in water, it could be expected that a much greater breakdown voltage and electric field in the order of $30\,000\,kV\,cm^{-1}$ might be required due to the density difference between air and water. A large body of experimental data on the breakdown voltage in water and water solutions shows, however, that this voltage is of the same magnitude as for gases. It can be explained fast formation of gas channels, bubbles, voids, etc. in the body of water under the influence of the applied high voltage. When formed, the gas channels give the space for the gas breakdown inside of the body of water. Therefore, the voltage required for water breakdown is of the same magnitude as for gases. The gas channels can be formed by development and expansion of gas bubbles already existing in water as well as by additional formation of the vapor channels by local heating and evaporation. Let us focus on the second mechanism, which is usually referred to as the ***thermal breakdown of liquids***. When a voltage pulse is applied to water, it induces a current and the redistribution of the electric field. Due to the dielectric nature of water, an electric double layer is formed near the electrode, which results in the localization of the applied electric field. This electric field can become high enough for the formation of a narrow conductive channel, which is heated up by electric current to temperatures of about 10 000 K. Thermal plasma generated in the channel is rapidly expanded and ejected from the narrow channel into water, forming a plasma bubble. The energy required to form and sustain the plasma bubble is provided by Joule heating in the narrow conductive channel in water. The physical nature of thermal breakdown can be related to thermal instability of local leakage currents through water with respect to the Joule overheating. If the leakage current is slightly higher at one point, the Joule heating and hence temperature also grows there. The temperature increase results in a significant growth of local conductivity and the leakage current. Exponential temperature growth to several thousand degrees at leads to formation of the narrow plasma channel in water, which determines the thermal breakdown.

The **thermal breakdown of liquids** occurs when the applied voltage exceeds a threshold value, and heat release in the conductive channel cannot be compensated by heat transfer. Thermal condition of water is constant during the breakdown; water stays liquid outside the discharge with the thermal conductivity about $0.68\,W\,mK^{-1}$. When the Joule heating between two electrodes is larger than a threshold value, the instability can occur, resulting in the instant evaporation and a subsequent thermal breakdown. When the Joule heating is smaller than the threshold value, nothing happens but electrolysis and the breakdown never take place unless release of gases during the electrolytic process results in formation of a bubble sufficient for breakdown. The thermal breakdown instability is characterized

Table 16.1 Parameters and characteristics of the pulsed discharges in water.

Pulsed corona in water	Pulsed arc in water	Pulsed spark in water
• Streamer-like channels	• Current is transferred by electrons	• Like pulsed arc, except for short pulse durations and lower temperature
• Streamer channels do not propagate across the entire electrode gap, i.e. partial electrical discharge	• Quasi-thermal plasma	• Pulsed spark is faster than pulsed arc, strong shock waves are produced
• Streamer length is about cm, channel width ~10–20 μm	• Arc generates shock waves within cavitation zone	• Plasma temperatures in spark are around a few thousand kelvin
• Electric current is transferred by ions	• High current filamentous channel bridges the electrode gap	
• Nonthermal plasma	• Gas in channel (bubble) is ionized	
• Weak to moderate UV	• High UV emission and radical density	
• Relatively weak shock waves	• Gap between electrodes (~5 mm) is smaller than that in pulsed corona	
• Water treatment area is limited at a narrow region near the corona	• Large discharges pulse energy (greater than 1 kJ per pulse)	
• Pulse energy is 1 J per pulse	• Large current (about 100 A), peak current greater than 1000 A	
• Frequency is 100–1000 Hz	• Electric field intensity at the tip of electrode is 0.1–10 kV cm^{-1}	
• Relatively low current, peak current is less than 100 A	• Voltage rise time is 1–10 μs	
• Electric field at the tip of electrode is 100–10 000 kV cm^{-1}	• Pulse duration ~20 ms	
• A fast-rising voltage, rise time from 1 to 100 ns	• Temperature can exceed 10 000 K	

by the instability increment showing frequency of its development:

$$\Omega = \left[\frac{\sigma_o E^2}{\rho C_p T_0} \right] \frac{E_a}{T_0} - D \frac{1}{R_0^2};$$

(16.1)

here, σ_o is water conductivity; E_a is activation energy for water conductivity; E is the electric field; ρC_p is the specific heat per unit volume; T_0 is temperature; R_0 is radius of the breakdown channel; $D \approx 1.5 \cdot 10^{-7} \, \text{m}^2 \, \text{s}^{-1}$ is thermal diffusivity of water. When the increment $\Omega > 0$, the perturbed temperature exponentially increases resulting in thermal explosion; when $\Omega < 0$, the perturbed temperature exponentially decreases sustaining the steady state. For the discharge in water, the minimum breakdown voltage in the channel with length L can be determined from (16.1) as:

$$V \geq \sqrt{\frac{D C_p \rho T_o^2}{\sigma_o E_a}} \cdot \frac{L}{R_0}.$$

(16.2)

The breakdown voltage increases with L/R_o. Assuming $L/R_0 = 1000$, it is about 30 kV. Detailed discussion on the subject can be found in publications of Prof. Young Cho, for example in the book Yang et al. (2012).

16.3 Nonequilibrium Nanosecond-pulsed Plasma in Water Without Bubbles

Although most of the plasmas inside of liquids discussed in the previous Section 16.2 were observed in gas bubbles, voids, etc., preliminarily created in these liquids, very interesting phenomenon of direct ionization of liquid phase without bubbles has been also observed recently. *Application of very fast rising voltage (like the sub-nanosecond rise) revealed that plasma-like phenomena can, in fact, occur in liquid phase quasi-homogeneously without any preliminary formation of bubbles, voids, etc.* This phenomenon can be

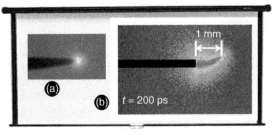

Figure 16.1 Pulsed discharges in water without bubbles: (a) micro-plasma, (b) macro-plasma.

interpreted as **direct ionization of liquid** itself. These plasmas can provide high density of electrons and excited species, high energy radiation, and high electron energies at low temperature of liquids. It provides interesting novel opportunities in the fields of medicine, microelectronics, energy systems materials, etc. One of the first observations of the nonthermal corona discharge plasma inside a liquid medium around electrodes with ultra-sharp tips or elongated nanoparticles have been presented by Staack et al. (2008), see more details in the book Fridman and Kennedy (2021). When electrode's tip is very sharp, for example on the level of nanometers, even not very high applied voltages are able to provide extremely high electric fields sufficient even for breakdown of liquids, see Figure 16.1a. Plasma shown in this figure is generated by negative corona with possible contribution of field emission. The active plasma radius of the corona discharges is usually only a couple of times greater than radius of curvature of the electrode's tip. It means that nano-corona's electrode tips should generate nano-size plasmas. Keeping in mind that ionization process requires presence of atoms inside of plasma (which is on nano-meters scale in this case), we should conclude that nano-plasma can be organized only in condensed phase (not in gases) where mean free path of electrons is also on the level of nanometers.

Cold macroscopic pulsed plasma in water without bubbles was observed by Starikovskiy et al. (2011) (more details can be found in the book Fridman and Kennedy 2021), and generated by high-voltage pulses with nanosecond duration, see Figure 16.1b. Important that this discharge has been propagating in this case from anode avoiding effects of field emission on electron generation. Applied voltages to generate such plasmas are usually from 30 to 220 kV, pulse duration from less than 1 to 30 ns, and voltage rise time from 0.2 to 1–2 ns. It was shown that the macroscopic cold strongly nonequilibrium discharge plasma can be developed on a picosecond time scale with extremely high propagation velocities up to 5000 km s^{-1} (5 mm ns^{-1}, which is about 15% of speed of light in vacuum). Such enormously high plasma propagation velocity was observed for a sub-nanosecond discharge (voltage 220 kV on the tip, 400 ps duration). The described macroscopic discharge in water has a different nature from the plasmas initiated by electrical pulses with a longer rise time as presented above. The macroscopic plasma propagating from a positive electrode is generated here in water without bubbles by direct streamer-like ionization of water without contribution of field emission from an electrode.

16.4 Nonequilibrium Nanosecond-pulsed Plasma Without Bubbles in Different Liquids: Comparison of Discharges in Water and PDMS

The macroscopic nanosecond-pulsed discharges can be organized without bubbles in different liquids characterized by different polarization properties. Let us compare with this regard discharges in water and silicon-based transformer oil. To analyze the effect of polarizability and dielectric properties of liquid on the nano-second pulsed discharge development, two types of liquids were considered: distilled deionized water (maximum conductivity 1.0 µS cm^{-1}), and polydimethylsiloxane (C$_2$ H$_6$ OSi)$_n$ transformer oil (PDMS)with dielectric constant $\varepsilon = 2.3$–2.8 (Dobrynin et al. 2013). The discharge initiation was provided by nanosecond pulses with +16 kV pulse amplitude in 50 Ω coaxial cable (32 kV on the high-voltage electrode tip due to pulse reflection), 10 ns pulse duration (90% amplitude), 0.3 ns rise time, and 3 ns fall time. Pulse frequency was 5 Hz. A 15 m long coaxial cable with a calibrated back current shunt mounted

Table 16.2 Discharge images with exposure time of 50 ns obtained for different discharge gaps and electrodes for the cases of water and silicon transformer oil (PDMS).

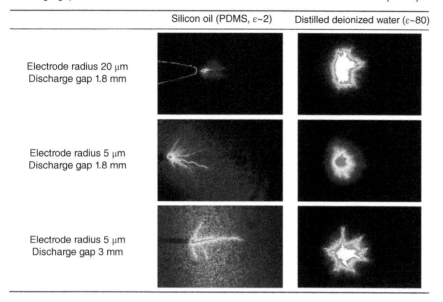

	Silicon oil (PDMS, ε~2)	Distilled deionized water (ε~80)
Electrode radius 20 μm Discharge gap 1.8 mm		
Electrode radius 5 μm Discharge gap 1.8 mm		
Electrode radius 5 μm Discharge gap 3 mm		

in a middle of it was used for control of applied voltage and power measurements. Discharge images with exposure time of 5 ns obtained for different discharge gaps and electrodes for the cases of water and silicon transformer oil (PDMS) are presented in Table 16.2. Twenty millimeter-needle image for PDMS shows in this table the artificially added electrode contour, and 5 mm-needle images for PDMS case are shadow images, image size 2 × 1.25 mm. As one can see from the Table 16.2, the size of the excited region near the tip of the high-voltage electrode is ~1 mm. The discharge has a complex multi-channel structure and this structure changed from pulse to pulse. The luminescence intensity and structure of the discharge are significantly different for the cases of water and PDMS: discharge in water is always more intense and has more "uniform" structure. At the same time, discharge size seems not to depend on the high-voltage electrode radius or inter-electrode distance, i.e. global (applied) electric field. This illustrates first image comparison of the liquid plasma in water and transformer oil.

To analyze the *spatial evolution of the nano-second pulsed discharge structure* a shorter camera gate (2 ns) has been used without signal accumulation (Dobrynin et al. 2013). Relevant nano-scale plasma structure evolution is shown in Figures 16.2 and 16.3 for water and PDMS respectively. In the case of water, when the voltage reaches maximum, the discharge propagation stops, and a "dark phase" appears ($t = 3$–9 ns in Figure 16.2). During this phase, the discharge cannot propagate because of space charge formation and decrease in the electric field. Presence of the "dark phase" proves that the liquid plasma is cold. Indeed, if the observed plasma is thermal, it could not be cooled down for a couple of nanoseconds and therefore the radiation could not disappear during the "dark phase." Voltage decrease in the "dark phase" leads to second stroke formation and the second emission phase (Figure 16.2, 11–13 ns). This means that the channels lose conductivity, and the trailing edge of the nanosecond pulse generates significant electric field and excitation of the media. It is comparable to the excitation corresponding to the leading edge of the pulse. When *the discharge is ignited in silicone oil* (PDMS, Figure 16.3), the "dark" phase is much less pronounced. This is due to lowering of the electric field in the discharge channels in PDMS compared to the discharge in water. Such decrease of the electric field can be interpreted assuming the pulsed plasma development in the so-called sub-micro pores and cracks, which are transversal. For the transversal pores $div\vec{D} = 0$ and $\varepsilon_{H_2O}E_{H_2O} = \varepsilon_{pore}E_{pore}$, therefore we can conclude that $E_{H_2O} \sim 40 \times E_{PDMS}$ since $\varepsilon_{H_2O} \sim 80$ and $\varepsilon_{PDMS} \sim 2$. In other words, local electric fields in water are significantly stronger than those in PDMS. Therefore, in water, the discharge has "thicker" channels and more pronounced "dark" phase. The space charge accumulates faster, and the screening effect is stronger.

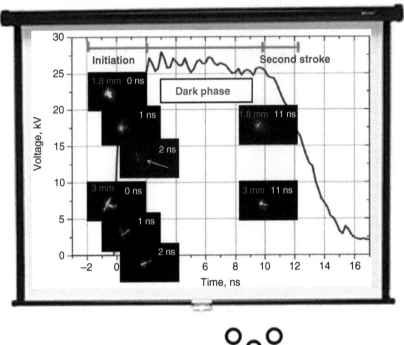

Figure 16.2 Pulsed discharge evolution in water without bubbles. Source: Dobrynin et al. (2013)/IOP Publishing.

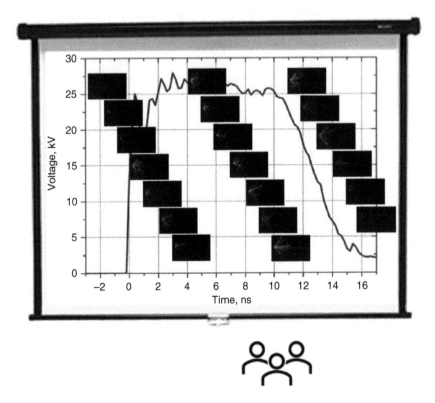

Figure 16.3 Pulsed discharge evolution in PDMS without bubbles. Source: Dobrynin et al. (2013)/IOP Publishing.

16.5 Characterization of the Nano-second Pulsed Discharges in Liquid: Shadow Imaging, Optical Emission Spectroscopy

To prove the homogeneous nature of the above-described liquid plasmas, the bubbles ~1 μm and other nonuniformities were analyzed in the liquid right before and after the discharge (Dobrynin et al. 2013) using a shadow imaging technique. The images for the PDMS oil are shown in Table 16.3.

The nanosecond plasma size here is ~1 mm (image size is 2 × 1.25 mm), but no density irregularities are observed right before and after the discharge, thus there are no bubbles related to the plasma. The discharge energy deposition ~5 mJ is determined by comparison of the first incident and reflected voltage pulses (back current shunt measurements). It results in only few degrees (~2 K for water and ~5 K for PDMS) increase in temperature. Thus, the liquid overheating in the discharge channel is not sufficient for generation of a shock-wave to create a void.

If the discharge imaging continues for a longer time, the second and other reflected pulses eventually should result in bubble formation due to electrostatic or energy dissipation, see Table 16.4 (Dobrynin et al. 2013). Time delay between the incident and the second reflected pulse here is ~110 ns. After the second and subsequent pulses,

Table 16.3 Nanosecond pulsed discharge plasma in PDMS oil: shadow imaging.

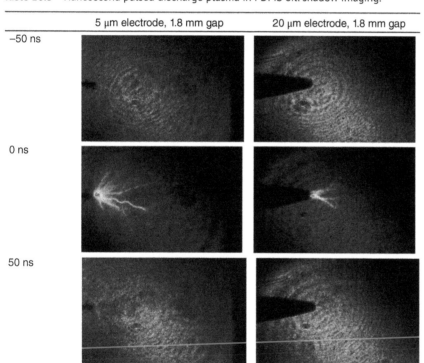

Table 16.4 Nanosecond discharge in PDMS: shadow imaging of bubble formation on longer time scale (reflected pulses).

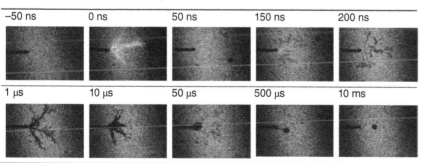

there is a formation of a tree-like gas void leading to a bubble. The bubble leaves the electrode and again no density perturbation can be observed before the next pulse (discharge here is 1 Hz).

The nanosecond-pulsed *liquid plasma spectrum* analyzed by Dobrynin et al. (2013) shows strong broadening of Balmer lines with almost continuum emission in the region 300–900 nm and weak broadened *OI* lines. Electron density and electron temperature can be then determined by analyses of the *Hα* and *I* (777 nm) profiles. The best fit was obtained in this case as a sum of two Lorentzian functions. Since Stark and Van Der Waals broadening can be treated independently and both have Lorentzian profile, one may suggest that complex *Hα* and *OI* profiles are primarily due to the combined contribution of these broadening mechanisms. It results in the electron density and electron temperature ~1.5×10^{17} cm^{-3} and ~3 eV, respectively. Corresponding Coulomb logarithm is ~6, and the Debye radius is ~50 nm.

16.6 Streamer Formation in Liquids and Nonequilibrium Nanosecond-pulsed Liquid Plasma Without Bubbles

There is currently no complete adequate theory of the nonequilibrium discharge initiation in liquids without bubbles. The liquid density is ~1000 times higher than that of the gas phase, which results in very short mean free path of electrons. Thus, an extremely high electric field ~30 MV cm^{-1} can be expected for effective direct formation of a streamer in a liquid phase without bubbles. In the conventional breakdown mechanisms (see Section 16.2), the micro-bubbles grow with the speed of sound in liquid, which is about ~1 km s^{-1}. Keeping in mind that plasmas described in the previous Section 16.5 were generated after about 1 ns, the micro-bubbles would be able to grow to maximum 1 micron, while size of the generated plasma was on millimeters scale. The observed speed of the liquid plasma propagation was up to 0.15 of speed of light in vacuum, which is obviously significantly exceeding the above-mentioned speed of sound in liquid. Also, the minimal size of a bubble that must be present inside of liquid for formation of an initial streamer is on the order of 20 μm. It was no bubbles of this size in the relevant discharges. Thus, special theories are expected to explain the intriguing phenomenon of direct ionization of liquids.

Analyzing images of the liquid plasma (see Table 16.2), one can see clear similarity of the nanosecond pulsed discharges in water with the discharges in long gaps, like long sparks (or lightning). We are surprised that electric fields below conventional one required by Townsend formulas and by Paschen curves are sufficient for ionization of liquids without bubbles, but we are not surprised that long sparks, and especially lightning can occur at electric fields way below those required by Townsend formulas and by Paschen curves. Understanding the similarity of lightning between clouds and ground with direct liquid ionization can be a key for understanding of the nanosecond pulsed plasma in liquids. The breakdown mechanism based on multiplication of avalanches (Townsend mechanism) is predominant at low pd < 200 Torr cm, while at pd > 4000 Torr cm, streamer formation dominates. The streamer mechanism works however only if the gap is not very long, the degree of electric field nonuniformity is not very high, and the attachment is absent. In very long gaps, although average electric field is low, discharge develops by leader mechanism discussed in Lecture 9. Once a plasma channel is formed, it can grow due to the very high electric field of its charged tip like in the leaders. Generally, propagation of a streamer in liquid phase without any voids or bubbles can occur when the avalanche head electric field reaches value of the applied external electric field. This electric field can be achieved in the avalanche head when $N_{emin} \sim 2 \times 10^8$ electrons are produced in the area with 10 μm – radius volume, which agrees with the observed electron density of ~2×10^{17} cm^{-3} mentioned in the previous Section 16.5.

Minimum number of electrons required for the avalanche-to-streamer transition corresponds to the Meek's criterion ($\alpha d = 20$) does not matter whether it is in liquid or gas phase:

$$\alpha \left(\frac{E_0}{p} \right) * d = \ln \frac{4\pi\varepsilon_0 E_0}{e\alpha^2} \approx 20, \quad N_e = \exp(\alpha d) \approx 3 \cdot 10^8. \tag{16.3}$$

The characteristic size here is about $d = 10$ μm, therefore $1/\alpha$ must be about 0.5 μm, which is very high for the condensed medium without any bubbles or voids. Some additional source of initial plasma electrons created inside of liquid to explain the direct liquid ionization can be related to nonuniformity of density, nonhomogeneity of liquid structure, specific transfer processes, etc.

One of the possible mechanisms of generation of initial electrons to start the streamer propagation is related to the pre-breakdown liquid structuring (Dobrynin et al. 2013). It can be introduced by an additional "ionization coefficient" μ representing, similarly to the Townsend coefficient α, effective electron multiplication inside the pre-breakdown low-density regions in liquid to be ionized. In this case, we may rewrite the condition for direct ionization of liquids in the form of the so-called *modified Meek's criterion*:

$$\left[\alpha\left(\frac{E_0}{p}\right) + \mu\right] * d \approx 20, \quad N_e = \exp((\alpha + \mu)d) \approx 3 \cdot 10^8. \tag{16.4}$$

To clarify the modified Meek's criterion let us imagine that highly inhomogeneous electric field in the vicinity of the high-voltage electrode creates several nano-sized pores (ellipsoids elongated along the electric field lines) due to the so-called ponderomotive electrostriction effect (see, for example, Shneider et al. 2012, and following publications of Sneider and his group). This **electrostriction effect** stimulates violation of the continuity of the liquid in the vicinity of the needle electrode due to the effective negative pressure, which is a result of the electrostrictive ponderomotive force pushing dielectric fluid to the regions with higher electric field. The electrostrictive effect is caused by rearrangement of the orientation of the elementary dipoles and occurs over a time much shorter than the characteristic time of the hydrodynamic processes. This effect can result in a possibility of formation of nanosecond discharge plasma in the liquids on time scales much shorter than the formation time for bubbles near the electrode. The liquid in the vicinity of the electrode ($r < R$, where r is the electrode radius and R is the radius of the region in which nanopores are created) is highly dispersed because of electrostriction. Thus, the liquid in the electrode's vicinity can be saturated by nanopores, and for the conditions of the experiments described above, it can be shown that $R \sim 1.9\, r_0$, where r_0 is the radius of the electrode. For the case of 5 μm radius electrode, we have a nanopore saturated region radius about $R \sim 10$ μm. Estimated pore size is about ~30 nm (Shneider et al. 2012), and corresponding energy that electron gains in the pore can be expected about ~100 eV. This energy is sufficient to cause ionization of liquid at the boundaries of nanopores, and to create several additional electrons. Effective "ionization coefficient" can be estimated in this case as $\mu \sim 2 \times 10^4$ cm^{-1} which corresponds to a distance between nanopores of about ~500 nm. Thus, about 20 nanopores per 10 μm distance would be sufficient for the formation of an initial streamer. Summarizing this liquid ionization mechanism, the breakdown scenario starts with strong inhomogeneous electric field at the needle electrode creating a region saturated by nanopores through electrostriction. In these pores, primary electrons are accelerated by electric field to the energies exceeding the potential of ionization of water molecule. This causes the formation of a primary streamer in the liquid phase if the modified Meek's criterion (Eq. (16.4)) is satisfied. After neutralization of electrons at the electrode, positive ions in the liquid phase form a virtual "needle electrode," and electrostrictive conditions for the appearance of the next set of cavities and density nonuniformities are fulfilled. This propagation of streamer continues until the drop in voltage on the moving streamer will not be an order of initial voltage on the needle electrode. In the process of propagation, the head of the streamer narrows, so that electrostrictive conditions for breakdown are reproduced in the vicinity of the streamer head.

More details on the subject can be found in Shneider et al. (2012), following publications of Sneider's group, Dobrynin et al. (2013), as well as in the book of Fridman and Kennedy (2021).

16.7 Cryogenic Liquid Plasma, Nanosecond-pulsed Discharge in Liquid Nitrogen

Compared to water and other dielectric liquids discussed in the previous Sections 16.1–16.6, very few studies are available for the homogeneous discharges in cryogenic liquids. The plasma imaging and spectroscopic characterization of the nanosecond-pulsed discharge in liquid nitrogen have been performed by Dobrynin et al. (2019a,b). The cryogenic conditions in the liquid nitrogen plasma permit to observe some unique plasma synthesis with effective stabilization of such products as polymeric nitrogen. Synthesis of polymeric nitrogen, which is expected to be highly energetic, has been successful, for the most part, at extreme conditions of high pressures (few to tens of GPa) but has shown to be unstable upon pressure release. Some polynitrogen species, like N^-_3, N_4, N^+_5, and N^-_5 are stable at ambient conditions only as salts and metal compounds.

For generation of the discharge, a sharp (75 μm radius of curvature) steel electrode was placed in liquid nitrogen contained in 450 ml double-walled glass flask (Dobrynin et al. 2019a,b). The flask was fixed inside an evacuated

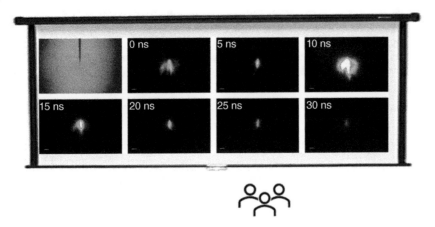

Figure 16.4 The high-voltage needle and ICCD images of the nanosecond-pulsed discharge in liquid nitrogen; exposure time 5 ns; time delay between images 5 ns; time corresponds to arrival of high-voltage pulse to the discharge gap. Source: Dobrynin et al. (2019a,b)/IOP Publishing.

metal chamber to decrease the liquid nitrogen evaporation rate and to screen electromagnetic noise. The vacuum flask was also closed from the top by a plastic lid with 3 mm diameter venting hole to minimize liquid nitrogen contamination by the surrounding air, especially oxygen. High-voltage electrode was powered by nanosecond-pulsed power supply capable of providing positive high-voltage pulses with maximum pulse amplitude of 120 kV, rise time at 10–90% amplitude less than 1 ns, and pulse duration at 90% amplitude 8–10 ns. The discharge energy was 100–130 mJ per pulse. Figure 16.4 shows long (5 ns) exposure images of the discharge in liquid nitrogen taken with 5 ns time step. The discharge is visible using an ICCD camera for about 30 ns, and high-voltage pulse reflections were absorbed by the power supply. Typical discharge size was on the order of a few mm and appears to be significantly larger than that one observed in the case of lower voltage ($\sim$30 kV) pulses. From these images, streamer propagation velocity in liquid nitrogen can be estimated as extremely high, on the level of at least 700–800 km s^{-1}. Similarly, to the nanosecond pulsed discharges in water and PDMS, the shadow imaging proved absence of gaseous bubbles in the experiments with liquid nitrogen as well. Only after about 15 ns, when the main plasma event starts to decay, visible gas voids start to appear at the location of the streamers. Optical emission in the 300–415 nm range was recorded with either 100 ns exposure time single accumulation or 3 ns exposure time and 50 accumulations. The emission in this range (300–415 nm) is mostly from molecular nitrogen, and specifically from the second positive system (SPS). Using rotational–vibrational emission spectrum of the SPS transition at around 337 nm and assuming equilibrium of the rotational temperature $T_r(C)$ of the C state and $T_r(X)$ of the ground state of nitrogen, the temperature of the discharge has been estimated about $\sim$110 K. The maximum discharge temperature obtained from the time-resolved spectra is around 140 K, which is 60 K above the liquid nitrogen temperature. These plasma characterization measurements prove the cryogenic nature of the nanosecond-pulsed discharge in liquid nitrogen.

The powder-like energetic polynitrogen material was generated at the liquid nitrogen treatment duration 30–60 min and pulse repetition frequency of 60 Hz (Dobrynin et al. 2019a,b). After 60 min of treatment, no erosion of the high-voltage electrode was registered. Liquid nitrogen after 1 h of treatment (about 50 ml) was heated in room air and in He atmosphere to increase the concentration of the produced material. With evaporation of the liquid nitrogen, the liquid darkened leaving a black powder-like material (see Figure 16.5), supposedly polymeric nitrogen. The powder then exploded within a second with significant generation of light and sound. No residue was left after the material decomposition. The observed material appears to be very similar to the polymeric nitrogen produced at the extremely high-pressure conditions. The Raman spectrum of the liquid nitrogen changes after treatment but cannot be reliably interpreted as polymeric nitrogen. Neither characteristic lines from azide groups (1360 cm^{-1}), nor any Raman peaks associated with ozone were registered, which further supports the hypothesis on liquid nitrogen-based plasma production of energetic nitrogen-rich material. More details on the subject can be found in Dobrynin et al. (2019a,b), and following publications of Dobrynin with co-authors.

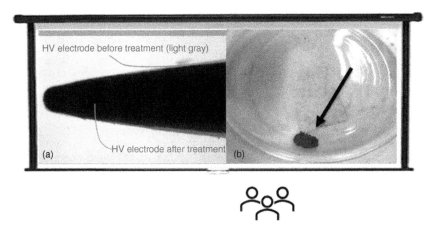

Figure 16.5 High-voltage electrode before and after 1 h treatment of liquid nitrogen at 60 Hz pulse repetition frequency (a); and polymeric nitrogen (to be exploded in about 1 s), produced as the results of 1 h treatment and after evaporation of excess liquid nitrogen (b).

16.8 Plasmas in Supercritical Fluids, Electric Breakdown of Supercritical CO$_2$

When temperature and pressure exceed some critical values, the dense fluid loses its "gaseous or liquid identity" and becomes supercritical, see CO$_2$ phase diagram in Figure 16.6 as an example. The supercritical fluids are characterized by very specific properties like strong solvents and attract interest to their applications. Plasmas in supercritical fluids, especially those generated in supercritical CO$_2$, have been investigated lately and have some common features with discharges in liquids, see for example Stauss et al. (2018), Kiyan et al. (2008), Furusato et al. (2012), Ihara et al. (2012). It was demonstrated that direct current (DC) corona discharges can be produced in supercritical CO$_2$ using a point-plane electrode configuration. The arc discharge is formed when the corona became large enough to bridge the gap between the electrodes. Rather than being homogenous, supercritical fluid has dense clusters and less dense regions, which simplifies its breakdown. An anode tip with a small radius of curvature (80 μm) can achieve electrical breakdown at a lower voltage than one with a larger radius of curvature (150 μm). The corona discharge with positive polarity precedes only in pulses right before the arc discharge in supercritical CO$_2$ is formed. A stable corona discharge can be produced using a negative polarity at lower voltages without arcing. The breakdown using a positive polarity (about 20 kV) is not dependent on fluid density. Whereas breakdown voltage (10–15 kV) is pressure dependent when a negative polarity is used. Images of the discharges in supercritical fluids are very similar to those in liquids.

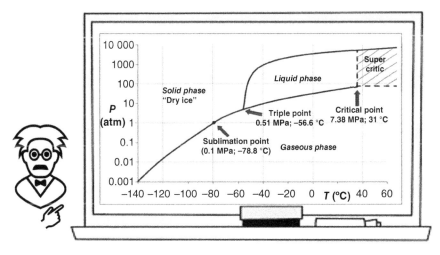

Figure 16.6 Thermodynamic phase diagram of CO$_2$ specifying the supercritical fluid conditions.

16.9 Problems and Concept Questions

16.9.1 Effect of Electric Conductivity on Breakdown of Water

Analyze, how electric breakdown of water and water solutions depends on their conductivity? Consider influence of applied voltage rise-time on the conductivity effect for breakdown of liquids.

16.9.2 Increment of the Thermal Breakdown Instability for Electric Breakdown of Water

Why increment of the thermal breakdown instability in water is growing with increase of activation energy-related temperature dependence of electric conductivity of water?

16.9.3 Breakdown Voltage of Water

The water breakdown voltage (Eq. (16.2)) is proportional to the channel's length, fixing the characteristic electric field in the discharge channel. Estimate this breakdown electric field depending on channel radius and water conductivity.

16.9.4 Generation of Nano-plasma by Nano-corona

Estimate voltage required to achieve electric field of $1\,\mathrm{MV\,cm^{-1}}$ if radius of curvature of an electrode tip is $20\,\mathrm{nm}$ and distance to the second electrode plate is 0.1 micron. Analyze contribution of field emission effect.

16.9.5 Comparison of Negative and Positive Pulsed Corona Discharges in Liquids Without Bubbles

Why generation of cold corona-like plasma in liquid without bubbles from anode is more challenging than that for a sharp-tip cathode?

16.9.6 The Dark Phase Effect During Evolution of the Nano-second Pulsed Discharge in Liquids

Why the dark phase effect during evolution of the nanosecond pulsed discharge plasmas in liquids proves the non-thermal nature of these plasmas?

16.9.7 Modified Meek's Criterion for Breakdown of Liquids

Why preliminary generation of plasma electrons with sufficient density is required for development of streamer breakdown of liquids? Discuss mechanisms of such initial plasma generation.

16.9.8 Nanosecond-pulsed Discharge in Liquid N_2, Synthesis of Polymeric Nitrogen

What other chemical compounds except polymeric nitrogen can be expected in the nanosecond pulsed discharges in liquid N_2? Keep in mind presence of metallic electrodes as well as possible presence of surrounding air during the cryogenic process.

16.9.9 Breakdown of Supercritical Fluids

Why the breakdown of supercritical fluids is less challenging than that of homogeneous liquids?

Part III

Plasma in Inorganic Material Treatment, Energy Systems, and Environmental Control

Lecture 17

Energy Balance and Energy Efficiency of Plasma-chemical Processes, Plasma Dissociation of CO_2

17.1 Energy Efficiency as a Key Requirement of Large-scale Plasma Processes: Comparison of Quasi-equilibrium and Nonequilibrium Plasmas

One of the crucial challenges of large-scale plasma-chemical technologies is minimization of energy consumption. Plasma-chemical processes consume electricity, which is presently a relatively expensive form of energy. It results in tough energy-efficiency requirements for the processes to be competitive with other technologies using just heat and chemical energy sources of energy. The energy efficiency η of a plasma-chemical process is determined as ratio of the thermodynamically minimal energy cost (which is usually the reaction enthalpy ΔH) to the actual energy consumption in plasma W_{pl}:

$$\eta = \Delta H / W_{pl}. \tag{17.1}$$

The energy efficiency is closely related to a process mechanism. The same processes, organized in different discharges at different conditions, consume very much different energy. For example, exhaust air cleaning from SO_2 using corona discharge requires 50–70 eV mol^{-1}. The same process organized in special plasma conditions provided by relativistic electron beams requires only 1 eV mol^{-1} (Baranchicov et al. 1992; Fridman 2008), almost hundred times less electric energy. While the energy-efficiency analysis is individual for each plasma process, there are some general energy balance features relevant for specific plasma-chemical mechanisms, which we are going to discuss now. First of all, *the energy efficiency of quasi-equilibrium plasma-chemical processes performed in thermal discharges is usually relatively low (less than 10–20%).* It is due to two major effects:

1. Thermal energy in quasi-equilibrium plasma is distributed over all components and all degrees of freedom, although most of them are useless in the stimulation of a required process.
2. If a high-temperature gas, generated in thermal plasma and containing the process products, is cooled down slowly, the products are converted back into initial substances via reverse reactions. Conservation of products of the quasi-equilibrium plasma-chemical processes requires **quenching** – very fast product cooling, which considerably limits energy efficiency.

The energy efficiency of nonequilibrium plasma-chemical systems can be much higher without contradicting the second law of thermodynamics. The discharge power can be selectively focused in this case on the chemical reaction of interest without heating the whole gas. Also, low gas temperature provides product stability with respect to reverse reactions without quenching. The energy efficiency in the nonequilibrium systems strongly depends on the process mechanism. The major mechanisms of nonequilibrium plasma processes are compared next regarding their energy efficiency. More details on the subject can be found in the books Fridman (2008), and Fridman and Kennedy (2021).

17.2 Energy Efficiency of Chemical Processes Stimulated in Plasma by Vibrational Excitation of Molecules, Electronic Excitation, and Dissociative Attachment

*The **vibrational excitation of molecules** can provide the highest energy efficiency of the endothermic plasma-chemical reactions in nonequilibrium discharge conditions* because of the following four factors:

(1) A major fraction (70–95%) of discharge power in several electropositive molecular gases (including N_2, H_2, CO, CO_2, CH_4) at $T_e \approx 1$ eV can be selectively transferred from plasma electrons to vibrational excitation of the molecules, see Section 3.5.

(2) vibrational–translational (VT)-relaxation is slow at low temperatures. Therefore, optimization of the ionization degree and energy input permits to utilize most of the vibrational energy in chemical reactions.

(3) Vibrational energy of molecules is the most effective in stimulation of endothermic reactions (see Sections 3.11 and 3.12).

(4) Vibrational energies required for endothermic reactions are close to their activation barriers and much lower than the energy threshold of processes proceeding through electronic excitation. For example, dissociation of H_2 through vibrational excitation requires 4.4 eV; and through excitation of an electronically excited state $^3\Sigma_u^+$ requires twice more energy: 8.8 eV.

Some chemical processes, stimulated by vibrational excitation, can consume most of the discharge energy. The best example here is dissociation of CO_2, which can be arranged in nonthermal plasma with energy efficiency up to 90% (see Section 17.6 in this Lecture). Almost all discharged power can be spent selectively in this case on organization of the chemical process. No other single mechanism can provide so high energy efficiency of a plasma-chemical process.

No one of the four kinetic factors mentioned above can be applied to the **stimulation of plasma-chemical reactions by electronic excitation** (see Section 3.4). Therefore, energy efficiency is relatively low in this case, usually below 20–30%. Plasma-chemical processes through electronic excitation can be energy effective only if they initiate chain reactions. Such a situation takes place, for example, in NO synthesis, where the Zeldovich mechanism can be effectively initiated by dissociation of molecular oxygen through electronic excitation.

The energy threshold of **dissociative attachment** is lower than the threshold of dissociation into neutrals. When the electron affinity of products is high, like for some halogens and their compounds, the dissociative attachment can even be exothermic. It means not only a low energy requirement for the reaction but also the transfer of a significant part of electron energy into dissociative attachment, which increases the energy efficiency of the dissociative attachment. The energy efficiency here is strongly limited, however, by energy cost of the production of an electron, which is lost during attachment and following ion–ion recombination. The energy cost of an electron, the ionization cost, is usually about 30–300 eV. Thus, the plasma-chemical process based on dissociative attachment can become more energy effective only if the same electron is able to participate in a reaction many times. In such a chain reaction, the detachment and liberation of the electron from a negative ion should be faster than the loss of the charged particle in ion–ion recombination. See as an example, water vapor decomposition in nonthermal plasma (Lecture 22) where the chain process can lead to the relatively high energy efficiency of 40–50%.

17.3 Energy Balance and Energy Efficiency of Plasma Processes Stimulated by Vibrational Excitation; Excitation, Relaxation, and Chemical Components of Total Energy Efficiency

As it was pointed out, energy efficiency of the plasma-chemical processes in electropositive gases stimulated by vibrational excitation of molecules can be very high in general when ionization degree and specific energy input exceed some critical values. To clarify this effect, let us consider the vibrational energy $\varepsilon(T_v)$ balance in a plasma-chemical system.

Such vibrational energy balance can be illustrated by a simplified one-component (quantum $\hbar\omega$) equation considering vibrational excitation by electron impact (rate coefficient $k_{eV}(T_e)$), VT relaxation (rate coefficient $k^{(0)}_{VT}(T_0)$), and chemical reaction (rate coefficient $k_R(T_v, T_0)$):

$$d\varepsilon_v(T_v)/dt = k_{eV}n_e\hbar\omega - k^{(0)}_{VT}n_0(\varepsilon_v - \varepsilon_{v0}) - k_R n_0 \Delta Q_\Sigma. \tag{17.2}$$

The left-hand side of (17.2) represents change in the average vibrational energy per molecule, which in one mode approximation (vibrational temperature T_v) can be expressed by the Planck formula $\varepsilon_v(T_v) = \frac{\hbar\omega}{\exp\left(\frac{\hbar\omega}{T_v}\right)-1}$. The first right-hand-side term describes vibrational excitation of molecules by electron impact (electron temperature T_e, density n_e), which stops when the vibrational energy input per single molecule reaches E_v. If most of the power is going to vibrational excitation, **the specific energy input E_v** per molecule is equal to the ratio of the discharge power and gas flow rate through the discharge. This convenient parameter is related to the energy efficiency (η) and conversion degree (χ) of a plasma-chemical process as:

$$\eta = \chi * \Delta H / E_v,\tag{17.3}$$

where ΔH is the enthalpy of formation of a product molecule. The second right-hand-side term in (17.2) describes VT relaxation of the lower vibrational levels; n_0 is gas density, T_0 is gas temperature; $\varepsilon_{v0} = \varepsilon_v(T_v = T_0)$. The third right-hand-side term is related to the vibrational energy losses in the chemical reaction and VV/VT relaxation channels from highly excited levels having rates corresponding to that of the chemical reaction (see Sections 5.8 and 5.9). In the weak excitation regime, the reaction rate strongly depends on vibrational temperature $k_R = A(T_0)\exp(-E_a/T_v)$, specifying the critical threshold value of vibrational temperature when the reaction and VT relaxation rates are equal:

$$T_v^{min} = E_a \ln^{-1}\frac{A(T_0)\Delta Q_\Sigma}{k_{VT}^{(0)}(T_0)\hbar\omega}.\tag{17.4}$$

Increase of vibrational temperature leads to much stronger acceleration of the chemical reaction than VT relaxation. Therefore, at high vibrational temperatures ($T_v > T_v^{min}$) almost all vibrational energy is going into chemical reaction. Conversely, at ($T_v < T_v^{min}$), almost all vibrational energy can be lost in vibrational relaxation.

The critical value of vibrational temperature (T_v^{min}) determines *threshold in the dependence $\eta(E_v)$ of energy efficiency of the plasma-chemical processes stimulated by vibrational excitation of molecules on the specific energy input E_v (ratio of the discharge power and gas flow rate through the discharge)*:

$$(E_v)_{threshold} = \varepsilon_v\left(T_v^{min}\right) = \frac{\hbar\omega}{\exp\left(\dfrac{\hbar\omega}{T_v^{min}}\right) - 1}.\tag{17.5}$$

If the specific energy input is lower than the threshold, the vibrational temperature cannot reach the critical value, therefore relaxational losses exponentially exceed the energy contribution to chemical reaction, and the energy efficiency is very low. Typical threshold values of energy input for reactions stimulated by vibrational excitation are about 0.1–$0.2\,eV\,mol^{-1}$ (0.5–$1\,J\,cm^{-3}$). Surely, the energy balance (17.2) requires vibrational excitation (the first term) to be faster than VT-relaxation (the second term) to reach the T_v^{min} and achieve high energy efficiency. It requires to exceed the **critical ionization degree** determined by the ratio of rate coefficients for VT relaxation and vibrational excitation:

$$\frac{n_e}{n_0} \gg \frac{k_{VT}^{(0)}(T_0)}{k_{eV}(T_e)}.\tag{17.6}$$

Numerically for the major plasma chemical processes stimulated by vibrational excitation the critical ionization degree is about 10^{-6}. Another important general feature of the processes stimulated by vibrational excitation is a **maximum in the dependence $\eta(E_v)$**. The optimal specific energy input E_v is usually about $1\,eV\,mol^{-1}$ (5–$10\,J\,cm^{-3}$). At larger E_v, the translational temperature becomes higher, which accelerates VT relaxation and decreases energy efficiency. Also at higher E_v, a significant part of the discharge energy is spent on the excitation of products. When the conversion degree χ is already high, the energy efficiency decreases hyperbolically $\eta \propto 1/E_v$.

The total energy efficiency η of any nonequilibrium plasma-chemical process stimulated by vibrational excitation can be subdivided into three main components: an excitation factor (η_{ex}), a relaxation factor (η_{rel}), and a chemical factor (η_{chem}): $\eta = \eta_{ex} * \eta_{rel} * \eta_{chem}$. The **excitation factor** gives the fraction of the discharge energy directed toward the production of the principal agent of the plasma-chemical reaction, in this case, vibrational excitation. The **relaxation factor** is related to conservation of the principal active species with respect to their losses, in this case, vibrational relaxation. The VT relaxation is slow if $T_v > T_v^{min}$); and near the E_v threshold:

$$\eta_{rel} = \frac{E_v - \varepsilon_v\left(T_v^{min}\right)}{E_v}.\tag{17.7}$$

It indicates the threshold of dependence, $\eta(E_v)$, and shows that when the vibrational temperature is lower than the critical one, most of the vibrational energy goes to VT relaxation. The **chemical factor** (η_{chem}) shows the efficiency of the principal active discharge species in the chemical reaction of interest. It restricts energy efficiency mostly because the activation energy often exceeds the reaction enthalpy: $E_a > \Delta H$.

17.4 Energy Efficiency of Quasi-equilibrium Chemical Processes in Thermal Plasmas: Absolute, Ideal, and Surer-ideal Quenching of Products

The processes in thermal plasmas can be subdivided into two phases. In the first one, reagents are heated to high temperatures required to shift equilibrium of endothermic processes in the direction of products. In the second phase called **quenching**, temperature decreases fast to protect the products produced in the first high temperature phase from reverse reactions. **Absolute quenching** means that the cooling process is sufficiently fast to save all products formed in the high temperature zone. In the high temperature zone, initial reagents are partially converted into the products of the process, but also into some unstable atoms and radicals. In the case of absolute quenching, stable products are saved, but atoms and radicals during the cooling process are converted back into the initial reagents. **Ideal quenching** means that the cooling process is very effective and able to maintain the total degree of conversion on the same level as was reached in the high temperature zone. The ideal quenching not only saves all the products formed in the high temperature zone but also provides conversion of all the relevant atoms and radicals into the process products. It is interesting that during the quenching phase, the total degree of conversion can be not only saved, but even increased. Such quenching is usually referred to as the super-ideal one. **Super-ideal quenching** permits increasing the degree of conversion during the cooling stage, using the chemical energy of atoms and radicals as well as the excitation energy accumulated in molecules. In particular, the super-ideal quenching can be organized when the gas cooling is faster than VT relaxation, and the VT, nonequilibrium $T_v > T_0$ can be achieved during the quenching. In this case, direct endothermic reactions are stimulated by vibrational excitation, while reverse exothermic reactions related to translational degrees of freedom proceed slower. Such unbalance between direct and reverse reactions provides additional conversion of the initial substances, or in other words provides super-ideal quenching.

As an example of the **ideal quenching of products**, let us consider dissociation of CO₂ in thermal plasma ($CO_2 \rightarrow CO + \frac{1}{2}O_2$, $\Delta H_{CO_2} = 2.9$ eV), see the equilibrium composition in Figure 17.1a. While main products are saturated molecules CO and O₂, the high temperature heating provides significant amount of atomic oxygen and even atomic carbon at very high T_0. If quenching is not sufficiently fast, the slow cooling in quasi-equilibrium would return the composition to the initial one, that is to pure CO₂. Fast cooling ($>10^8$ K s⁻¹) permits saving CO in the products. Atomic oxygen recombines $O + O + M \rightarrow O_2 + M$ faster than reacting with carbon monoxide $O + CO + M \rightarrow CO_2 + M$, which maintains the CO₂ degree of conversion providing the ideal quenching. In this case, the ideal and absolute quenching regimes are about the same.

 Energy efficiency of the quasi-equilibrium process in **thermal CO₂ plasma with ideal quenching** can be calculated as follows. First, the energy input per one initial CO₂ molecule required to heat the dissociating CO₂ at constant pressure to temperature T can be expressed as:

$$\Delta W_{CO_2} = \frac{\sum x_i I_i(T)}{x_{CO_2} + x_{CO}} - I_{CO_2}(T = 300 \text{ K}). \qquad (17.8)$$

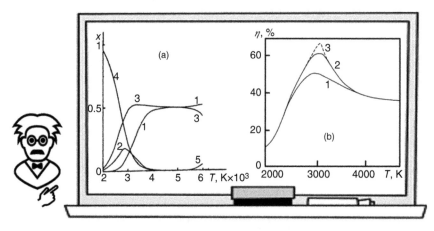

Figure 17.1 Dissociation of CO₂ in thermal plasma at $p = 0.16$ atm: (a) equilibrium composition: 1 – O, 2 – O₂, 3 – CO, 4 – CO₂, 5 – C; (b) energy efficiency: 1 – ideal quenching, 2 – super-ideal quenching, 3 – upper limit of the super-ideal quenching.

In this relation $x_i(p, T) = n_i/n$ is the quasi-equilibrium concentration of "i" component of the mixture, $I_i(T)$ is the total enthalpy of the component. If β is the number of CO molecules produced from one initial CO_2 molecule, then the total energy efficiency of the quasi-equilibrium plasma-chemical process can be expressed as:

$$\eta = \frac{\Delta H_{CO_2}}{(\Delta W_{CO_2}/\beta)}. \tag{17.9}$$

The conversion is equal to $\beta^0 = \frac{x_{CO}}{x_{CO_2}+x_{CO}}$ in the case of ideal quenching, thus energy efficiency is:

$$\eta = \frac{\Delta H_{CO_2} x_{CO}}{\sum x_i I_i(T) - (x_{CO_2} + x_{CO})I_{CO_2}(T = 300 \text{ K})}. \tag{17.10}$$

The energy efficiency of CO_2-dissociation in plasma with ideal quenching is presented in Figure 17.1b. Maximum energy efficiency 50% is reached at $T = 2900$ K. The energy cost of CO-production in thermal CO_2 plasma is shown in Figure 17.2 as a function of the cooling rate: clearly 10^8 K s^{-1} is required for the ideal quenching.

VT nonequilibrium effects during cooling can provide additional conversion. In CO_2 dissociation, this effect of **super-ideal quenching** is related to shifting equilibrium of the reaction $O + CO_2 \Leftrightarrow CO + O_2$, $\Delta H = 0.34$ eV during the cooling phase. The direct endothermic reaction of CO-formation can be effectively stimulated by vibrational excitation of CO_2 molecules, and not balanced by reverse exothermic reaction at $T_v > T_0$. The energy efficiency of CO_2-dissociation with super-ideal quenching is:

$$\eta = \frac{\Delta H_{CO_2}}{\Delta W_{CO_2}} \frac{x_{CO} + x_O}{x_{CO_2} + x_{CO}}, \quad \text{if } x_{CO_2} > x_O; \eta = \frac{\Delta H_{CO_2}}{\Delta W_{CO_2}}, \quad \text{if } x_{CO_2} < x_O \tag{17.11}$$

The energy efficiency of super-ideal quenching $\eta(T)$ as a function of temperature is shown in Figure 17.1b. The maximum efficiency $\eta = 64\%$ is reached at temperature in the hot zone about $T = 3000$ K. This efficiency is 14% higher than the maximum one for ideal quenching (but still less than that of the pure nonequilibrium process).

Difference between absolute and ideal quenching regimes can be illustrated using as an example **thermal decomposition of water vapor** $H_2O \rightarrow H_2 + \frac{1}{2}O_2$, $\Delta H_{H_2O} = 2.6$ eV, see the equilibrium composition in Figure 17.3a. The saturated products here are molecules H_2 and O_2, but in contrast to CO_2, thermal dissociation of water vapor results in a wider variety of atoms and radicals (concentrations of O, H, and OH are significant). Even when H_2 and O_2 initially formed in the high temperature zone are saved from reverse reactions, the active species O, H, and OH can be converted either into products (H_2 and O_2) or back into H_2O. This qualitatively different behavior of radicals determines the key difference between the absolute and ideal mechanisms of quenching. The absolute quenching, when the active species are converted back to H_2O, results in energy efficiency of hydrogen production:

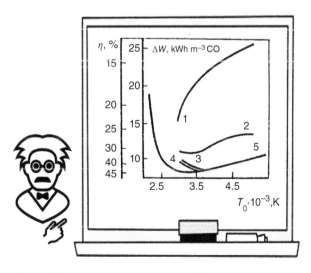

Figure 17.2 Dependence of energy efficiency η and energy cost ΔW of CO_2 thermal dissociation on heating temperature at $p = 1$ atm and different quenching rates: $1 - 10^6$ K s^{-1}, $2 - 10^7$ K s^{-1}, $3 - 10^8$ K s^{-1}, $4 - 10^9$ K s^{-1}, $5 -$ instantaneous cooling.

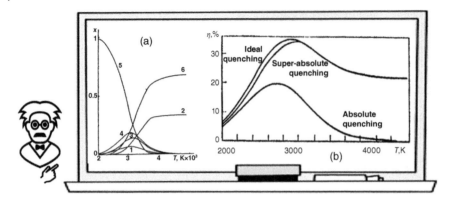

Figure 17.3 Dissociation of H$_2$O in thermal plasma at $p = 0.05$ atm: (a) equilibrium composition: 1 – O$_2$, 2 – O, 3 – OH, 4 – H$_2$, 5 – H$_2$O, 6 – H; (b) energy efficiency of the process.

$$\eta = \frac{\Delta H_{H_2O}}{\left(\Delta W_{H_2O}/\beta^0_{H_2}\right)};\tag{17.12}$$

here $\beta^0_{H_2}(T)$ is conversion of water into hydrogen. In other words, $\beta^0_{H_2}(T)$ is the number of hydrogen molecules formed in the quasi-equilibrium phase calculated per one initial H$_2$O molecule:

$$\beta^0_{H_2} = \frac{x_{H_2}}{x_{H_2O} + x_O + 2x_{O_2} + x_{OH}}.\tag{17.13}$$

The energy input per one initial H$_2$O molecule to heat the water at constant pressure to temperature T is:

$$\Delta W_{H_2O} = \frac{\sum x_i I_i(T)}{x_{H_2O} + x_O + 2x_{O_2} + x_{OH}} - I_{H_2O}(T = 300 \text{ K}).\tag{17.14}$$

Energy efficiency of water dissociation and hydrogen production with absolute quenching is then:

$$\eta = \frac{\Delta H_{H_2O} x_{H_2}}{\sum x_i I_i(T) - (x_{H_2O} + x_O + 2x_{O_2} + x_{OH}) I_{H_2O}(T = 300 \text{ K})}.\tag{17.15}$$

Energy efficiency of the thermal water dissociation in case of ideal quenching can be calculated in a similar way, considering the additional conversion of active species, H, OH, and O into product of the process (H$_2$ and O$_2$):

$$\eta = \Delta H_{H_2O} \left[\frac{\sum x_i I_i(T)}{x_{H_2} + 1/2(x_H + x_{OH})} - \frac{x_{H_2O} + x_O + 2x_{O_2} + x_{OH}}{x_{H_2} + 1/2(x_H + x_{OH})} I_{H_2O}(T = 300 \text{ K}) \right]^{-1}.\tag{17.16}$$

The energy efficiencies of water dissociation in thermal plasma for absolute and ideal quenching are presented in Figure 17.3b. From the figure, it is seen that complete usage of atoms and radicals to form products (ideal quenching) permit increasing the energy efficiency by almost a factor of two.

Slow cooling shifts the equilibrium between direct and reverse reactions in the exothermic direction of destruction of molecular hydrogen: O + H$_2$ → OH + H, OH + H$_2$ → H$_2$O + H; and the three-body recombination leads to: H + OH + M → H$_2$O + M (this reaction rate coefficient is 30 times larger than for alternative three-body recombination with formation of molecular hydrogen (H + H + M → H$_2$ + M). It explains the destruction of hydrogen during slow cooling. If the cooling rate is sufficiently high ($>10^7$ K s^{-1}), the reactions of O and OH with saturated molecules very soon become less effective. This provides conditions for absolute quenching. The active species OH and O react then to form mostly atomic hydrogen: OH + OH → O + H$_2$O, O + OH → O$_2$ + H. The atomic hydrogen (being in excess with respect to OH) recombines to molecular one providing at higher cooling rates ($>2 \cdot 10^7$ K s^{-1}) additional conversion of radicals into the stable process products. The "super-absolute" (almost ideal, see Figure 17.3b) quenching can be achieved at these cooling rates. Energy efficiency of thermal water dissociation and hydrogen production as a function of the cooling rate is presented in Figure 17.4. VT nonequilibrium during fast cooling of the water dissociation products can lead to the surer-ideal quenching effect in a similar way as was described CO$_2$ dissociation. The super-ideal quenching effect can be related in this case to the shift of quasi-equilibrium (at $T_v > T_0$) of the reactions:

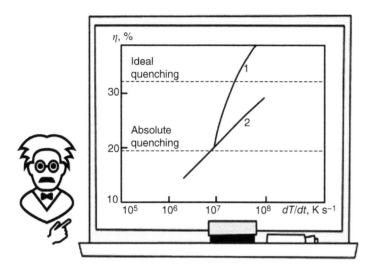

Figure 17.4 Energy efficiency of H_2O thermal dissociation as a function of quenching rate; heating temperature 2800 K, $p = 0.05$ atm: 1 – vibrational–translational nonequilibrium, 2 – vibrational–translational equilibrium.

$H + H_2O \Leftrightarrow H_2 + OH$, $\Delta H = 0.6$ eV, see Figure 17.4. The super-ideal quenching of the water dissociation products requires cooling rates exceeding $5 \cdot 10^7$ K s^{-1} and provides energy efficiency up to 45%.

17.5 Mass and Energy Transfer in Multi-component Thermal Plasmas, and its Effect on Energy Efficiency of Quasi-equilibrium Plasma-chemical Processes

The above consideration of the energy efficiency in different quenching regimes of thermal plasma processes assumed direct correspondence of the quasi-equilibrium chemical composition and local plasma enthalpy, which is not always the case. Let us analyze the influence of the transfer phenomena on energy efficiency of the quasi-equilibrium plasma-chemical processes.

The conservation equations, describing enthalpy (I) and mass transfer in multi-component (number of component N) quasi-equilibrium reacting gas are similar in thermal plasma and conventional combustion systems and can be started with the **conservation equation for total enthalpy**:

$$\rho \frac{dI}{dt} = -\nabla \vec{q} + \frac{dp}{dt} + \Pi_{ik} \frac{\partial v_i}{\partial x_k} + \rho \sum_{\alpha=1}^{N} Y_\alpha \vec{v_\alpha} \vec{f_\alpha}. \tag{17.17}$$

Then the **continuity equation for chemical components** can be expressed as:

$$\rho \frac{dY_\alpha}{dt} = -\nabla(\rho Y_\alpha \vec{v_\alpha}) + \omega_\alpha. \tag{17.18}$$

The thermal flux $\vec{q}$ neglecting radiation is: $\vec{q} = -\lambda \nabla T + \rho \sum_{\alpha=1}^{N} I_\alpha Y_\alpha \vec{v_\alpha}$; the diffusion velocity $\vec{v_\alpha}$ for α-chemical component can be found from the following relation:

$$\nabla x_\alpha = \sum_{\beta=1}^{N} \frac{x_\alpha x_\beta}{D_{\alpha\beta}} (\vec{v_\beta} - \vec{v_\alpha}) + (Y_\alpha - x_\alpha) \frac{\nabla p}{p} + \frac{\rho}{p} \sum_{\beta=1}^{N} Y_\alpha Y_\beta (\vec{f_\alpha} - \vec{f_\beta})$$

$$+ \sum_{\beta=1}^{N} \left[\frac{x_\alpha x_\beta}{D_{\alpha\beta}} \frac{1}{\rho} \left(\frac{D_{T,\beta}}{Y_\beta} - \frac{D_{T,\alpha}}{Y_\alpha} \right) \right] \frac{\nabla T}{T}. \tag{17.19}$$

Here: $I = \sum_{\alpha=1}^{N} Y_\alpha I_\alpha$ is the total enthalpy per unit mass of the mixture; Π_{ik} is the tensor of viscosity; $\vec{f_\alpha}$ is external force per unit mass of the component α; ω_α is the rate of mass change of a component α due to chemical reactions; λ, $D_{\alpha\beta}$, $D_{T,\alpha}$ are coefficients of thermal conductivity, binary diffusion, and thermo-diffusion respectively; $x_\alpha = n_\alpha/n$, $Y_\alpha = \rho_\alpha/\rho$

are the molar and mass fractions of the component α; v_i is a component of hydro-dynamic velocity; p is pressure. We can neglect the total enthalpy change due to viscosity; assume the forces $\vec{f_\alpha}$ the same for all components; and take the binary diffusion coefficients as $D_{\alpha\beta} = D(1 + \delta_{\alpha\beta})$, $\delta_{\alpha\beta} < 1$. After such simplification, the energy and mass transfer in addition to the continuity equation (17.18), can be rewritten as:

$$\rho \frac{dI}{dt} = -\nabla \vec{q} + \frac{dp}{dt}, \quad \vec{q} = -\frac{\lambda}{c_p} \nabla I + (1 - Le) \sum_{\alpha=1}^{N} I_\alpha D \rho_\alpha \nabla \ln Y_\alpha + \sum_{\alpha=1}^{N} \rho_\alpha I_\alpha \vec{v_\alpha^c} + \sum_{\alpha=1}^{N} \rho_\alpha I_\alpha \vec{v_\alpha^g},$$

$$\vec{v_\alpha} = -D\nabla \ln Y_\alpha + \vec{v_\alpha^c} + \vec{v_\alpha^g}.$$

(17.20)

Here $Le = \frac{\lambda}{\rho c_p D}$ is the Lewis number, and $c_p = \sum_{\alpha=1}^{N} Y_\alpha \frac{\partial I_\alpha}{\partial T}$ is the specific heat of the unit mass. The component of diffusion velocity $\vec{v_\alpha^c} = D \sum_{\beta=1}^{N} Y_\beta \left(\frac{\vec{F_\beta}}{x_\beta} - \frac{\vec{F_\alpha}}{x_\alpha} \right)$ is related to baro- and thermo-diffusion. Another component of the diffusion velocity $\vec{v_\alpha^g} = \sum_{\beta=1}^{N} x_\beta \delta_{\alpha\beta} \left(\vec{v_\beta^{(0)}} - \vec{v_\alpha^{(0)}} \right) + \sum_{\gamma=1}^{N} Y_\gamma \sum_{\beta=1}^{N} \delta_{\beta\gamma} \left(\vec{v_\beta^{(0)}} - \vec{v_\gamma^{(0)}} \right)$ is related to difference in the binary diffusion coefficients. The α-component force and the α-component velocities are:

$$\vec{F_\alpha} = (Y_\alpha - x_\alpha)\frac{\nabla p}{p} + \sum_{\gamma=1}^{N} \frac{x_\alpha x_\gamma}{\rho D_{\alpha\beta}} \left(\frac{D_{T,\gamma}}{Y_\gamma} - \frac{D_{T,\alpha}}{Y_\alpha} \right) \frac{\nabla T}{T}, \quad \vec{v_\alpha^{(0)}} = -D\nabla \ln Y_\alpha + D \sum_{\beta=1}^{N} Y_\beta \left(\frac{\vec{F_\beta}}{x_\beta} - \frac{\vec{F_\alpha}}{x_\alpha} \right).$$

(17.21)

These transfer equations prove the important rule, introduced by Potapkin et al. (1985): *transfer phenomena do not change the limits of energy efficiency of thermal plasma-chemical processes (absolute and ideal quenching) if the two following requirements are satisfied:*

(1) *The binary diffusion coefficients are fixed, equal to each other, and equal to the reduced coefficient of thermal conductivity (the Lewis number $Le = 1$, $\delta_{\alpha\beta} = 0$).*

(2) *The effective pressure in the system is constant implying the absence of the external forces and large velocity gradients.*

To interpret **the Potapkin's rule**, let us rewrite the enthalpy balance continuity equations corresponding to the above conditions for the steady-state one-dimensional case as:

$$\rho v \frac{\partial}{\partial x} I = \frac{\partial}{\partial x} \left(\rho D \frac{\partial}{\partial x} I \right), \quad \rho v \frac{\partial}{\partial x} Y_\alpha = \frac{\partial}{\partial x} \left(\rho D \frac{\partial}{\partial x} Y_\alpha \right) + \omega_\alpha;$$

(17.22)

and introduce new dimensionless functions $\xi = \frac{I - I^r}{I^l - I^r}$, $\eta = \frac{Y_\alpha - Y_\alpha^r}{Y_\alpha^l - Y_\alpha^r}$, where Y_α^l, I^l and Y_α^r, I^r are the mass fractions and enthalpy respectively on the left (l) and right (r) boundaries of the region under consideration. With these new functions, the enthalpy conservation and continuity equations can be rewritten as:

$$\rho v \frac{\partial \xi}{\partial x} = \frac{\partial}{\partial x} \rho D \frac{\partial \xi}{\partial x}, \quad \rho v \frac{\partial \eta_\alpha}{\partial x} = \frac{\partial}{\partial x} \rho D \frac{\partial \eta_\alpha}{\partial x} + \omega_\alpha \left(Y_\alpha^l - Y_\alpha^r \right);$$

(17.23)

the left and right sides boundary conditions are: $\xi(x = x_l) = \eta(x = x_l) = 1$, $\xi(x = x_r) = \eta(x = x_r) = 0$. To determine the maximum yield of the products, we can neglect the reaction rate during diffusion, assuming $\omega_\alpha = 0$ in the continuity equation. In this case, equations, and boundary conditions for η and ξ are completely identical, which means they are equal and gradients of total enthalpy I and mass fraction Y_α are related as:

$$\left(Y_\alpha^l - Y_\alpha^r \right)^{-1} \frac{\partial}{\partial x} Y_\alpha = (I^l - I^r)^{-1} \frac{\partial}{\partial x} I;$$

(17.24)

thus the minimum ratio of the enthalpy flux to the flux of products, taking into account chemical reactions equals to:

$$A = \frac{I^l - I^r}{Y_\alpha^l - Y_\alpha^r}.$$

(17.25)

The ratio (17.25) gives energy price of the plasma process product and completely correlates with the expressions for energy efficiency of quasi-equilibrium plasma-chemical processes considered in Section 17.4. Thus, *in absence of external forces and when the diffusion and reduced thermal conductivity coefficients are equal (the Lewis number $Le = 1$) the minimum energy cost of products of plasma-chemical reaction is determined by the product formation in the quasi-equilibrium high temperature zone. This*

minimum energy cost (maximum energy efficiency) corresponds to the limits of absolute and ideal quenching. This product energy cost considering the nonuniformity of heating is:

$$<A> = \frac{\int y(T)[I(T) - I_0]dT}{\int y(T)\chi(T)dT};$$ (17.26)

here $y(T)$ is the mass fraction of initial substance heated to temperature T; $I(T)$ and I_0 are the total enthalpy of the mixture at temperature T and initial temperature respectively; $\chi(T)$ is the degree of conversion of the initial substance (for ideal quenching) or degree of conversion into final product (for absolute quenching), achieved in the high temperature zone. Such possibility of using the thermodynamic relations for calculating the energy cost of chemical reactions in thermal plasma under conditions of intensive heat and mass transfer is a consequence of **the similarity principle of concentrations and temperature fields**. This principle is valid at the Lewis number $Le = 1$ in absence of external forces and is also applied in combustion. If the Lewis number differs from one ($Le = \frac{\lambda}{\rho c_p D} \neq 1$), the maximum energy efficiency can be different from that of the ideal quenching. If light and fast hydrogen atoms are products of dissociation, then the correspondent Lewis number is relatively small ($Le \ll 1$). In this case, diffusion of products is faster than the energy transfer, and energy cost of the products can be less with respect to ideal quenching.

17.6 CO₂ Dissociation in Plasma: Crucial Fundamental and Applied Aspects of the Process in Thermal, Nonthermal, and Transitional Discharges

The endothermic plasma-chemical process of CO_2 decomposition to CO and oxygen, staying in focus for more than half century, is illustrated in Figure 17.5, and can be summarized as:

$$CO_2 \rightarrow CO + \frac{1}{2}O_2, \quad \Delta H_{CO_2} = 2.9\,eV.$$ (17.27)

The enthalpy of the process is high and close to that of hydrogen production from water. Therefore, maximization of the process energy efficiency (17.1) is especially important in this case. The total decomposition (17.27) starts with and is limited by CO_2 dissociation ($CO_2 \rightarrow CO + O$, $\Delta H = 5.5\,eV\,mol^{-1}$) and ends with O conversion into O_2 by either recombination or reaction with another CO_2 molecule ($O + CO_2 \Leftrightarrow CO + O_2$, $\Delta H = 0.34\,eV$). The major reasons why the plasma CO_2 dissociation attracts so much attention of scientists and engineers are:

- It can solve the environmental challenge of CO_2 emission producing valuable products.
- It is stimulated by CO_2 vibrational excitation minimizing energy cost, with the highest energy efficiency between all other nonequilibrium endothermic plasma processes.
- It is an important process in kinetics of powerful CO_2 lasers.
- It is a crucial stage of H_2 production from H_2O in combination with catalytic shift ($CO + H_2O \rightarrow CO_2 + H_2$, $\Delta H = -0.4\,eV\,mol^{-1}$).
- It is involved in plasma-catalytic fuel conversion involving CO_2 and hydrocarbons.
- It can be used for fuel production on Mars from the CO_2 – reach atmosphere there.

Figure 17.5 Illustration of plasma-chemical CO_2 dissociation with production of CO and oxygen.

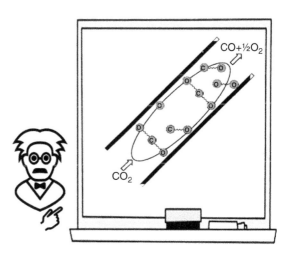

The CO_2 dissociation (17.27) was investigated in numerous thermal and nonthermal plasma systems (including arcs, gliding arcs, glow, hollow cathode and radiofrequency [RF] discharges, plasma beam discharges, plasma radiolysis, non-self-sustained discharges supported by electron beams and ultraviolet [UV] radiation). The major focus was on maximization of the energy efficiency, determined by the ratio of the process enthalpy $\Delta H = 2.9$ eV mol^{-1} to the actual energy cost. Major experimental and modeling results achieved in 1970s–1990s were reviewed in the books Rusanov and Fridman (1984), Givotov et al. (1985), and Fridman (2008). Relevant energy efficiency data are presented in Figure 17.6 (relevant references are in the above books). Recent additional modeling and experimental data are presented in Snoeckx and Bogaerts (2017) and Van Rooij et al. (2018). The highest energy efficiency achieved in quasi-equilibrium plasma is about 15%, which corresponds to a CO energy price of about 20 eV mol^{-1}. The theoretical maximum value of the energy efficiency of CO_2 dissociation in such plasmas is also not very high (about 43–48%, see Section 17.4). Nonequilibrium quenching effects (super-ideal quenching) permits to increase energy efficiency to about 60% (see Figure 17.1b), which can interpret recent experiments with moderate and higher pressure gliding arcs and microwave discharges described in Van Rooij et al. (2018) and Nunnally et al. (2011). Increased energy efficiency can be achieved in strongly nonequilibrium plasmas, see Sections 17.2 and 17.3. Low-pressure discharges, in particular glow discharges, are the simplest to achieve the stationary nonequilibrium conditions for CO_2 dissociation. The dissociation in these discharges with elevated reduced electric fields E/n, however, is mostly controlled by electronic excitation of CO_2, which is not a very energy-effective mechanism (see below), usually below 8%. Similar situation occurs in dielectric-barrier discharges (DBD) and corona discharges at atmospheric pressure where the reduced electric fields E/n are also elevated and not optimal for vibrational excitation. The most energy-efficient CO_2 dissociation through electronic excitation (up to 30%) has been achieved by plasma radiolysis in electron beam systems. Higher values of energy efficiency of CO_2 dissociation have been achieved in moderate-pressure nonequilibrium discharges. Experiments with pulsed microwave discharge at 300 Torr in magnetic field under electron cyclotron resonance (ECR) provide 60% energy efficiency. The same 60% energy efficiency was achieved at similar conditions in nonequilibrium RF discharges. *The highest energy efficiency of CO_2 dissociation was demonstrated in a nonequilibrium "transitional (warm)" microwave plasma at moderate pressures of 50–200 Torr, where process was effectively stimulated by vibrational excitation.* Performing the process in subsonic flow claims energy efficiency of 80% (Legasov et al. 1978a, b); in supersonic flow, it can reach 90%

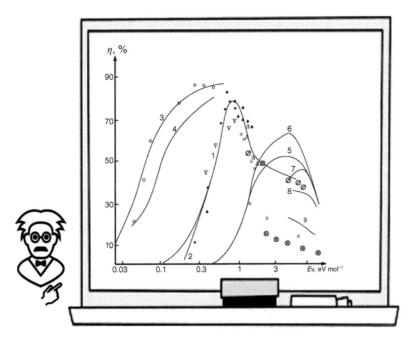

Figure 17.6 Energy efficiency of CO_2 dissociation as a function of specific energy input. (1, 2) Nonequilibrium calculations (one- and two-vibrational temperature approximations); nonequilibrium calculations for supersonic flows: (3) $M = 5$; (4) $M = 3.5$; thermal dissociation with (5) ideal and (6) super-ideal quenching; (7) thermal dissociation with quenching rates 10^9 K s^{-1}, (8) 10^8 K s^{-1}, (9) 10^7 K s^{-1}. Experiments: microwave discharges: o, ◆, Δ, X; supersonic microwave discharges: • ; RF-CCP discharges: o, ∇; RF-ICP discharges: Ø; arcs: ⊗, ⁚

(Asisov et al. 1983). Summarizing: although numerous chemical and relaxation processes occur simultaneously in CO_2 plasma, almost entire discharge energy (up to 90%) can be focused on dissociation.

17.7 About Mechanisms of CO₂ Dissociation in Plasma

Mechanisms of CO_2 dissociation in thermal plasma were discussed in Section 17.4 including absolute, ideal, and nonequilibrium super-ideal quenching regimes depending on cooling rates, see Figures 17.1 and 17.2. Let us focus now on the relevant mechanisms in nonthermal plasmas, where higher energy efficiency can be achieved, see Section 17.1. The most energy-efficient mechanism of plasma CO_2 dissociation in accordance with the general rule (Section 17.2) is related to **vibrational excitation of the molecules** by electron impact. The major portion of the discharge energy is transferred from plasma electrons to the CO_2 vibration at not very high reduced electric fields and electron temperatures about $T_e \approx 1$ eV, see Figure 17.7a. The rate coefficient of CO_2 vibrational excitation by electron impact in this case reaches $k_{eV} = 1-3 \times 10^{-8}$ cm³ s⁻¹. Vibrational energy losses through VT relaxation are slow at low gas temperatures T_0 (K) and mostly due to symmetric vibrational modes $k_{VT} \approx 10^{-10} \exp(-72/T_0^{1/3})$, cm³ s⁻¹. As a result, sufficiently high ionization degrees ($n_e/n_0 \geq 10^{-6}$) permit to reach significant VT nonequilibrium, see (17.6). Plasma electrons mostly provide excitation of low vibrational levels of the ground electronic state $^1\Sigma^+$. Population of highly excited vibrational levels providing dissociation is due to vibrational–vibrational (VV) relaxation, which determines the second kinetic order of the CO_2 dissociation. The elementary dissociation act can be illustrated by schematic of the CO_2 low-energy electronic terms shown in Figure 17.7b. The direct adiabatic dissociation of a vibrationally excited CO_2 molecule with total electron spin conservation:

$$CO_2^*(^1\Sigma^+) \rightarrow CO(^1\Sigma^+) + O(^1D), E_a > 7 \text{ eV}, \tag{17.28}$$

is related to the formation of an O atom in an electronically excited state ^1D and requires more than 7 eV. Nonadiabatic CO_2 transition $^1\Sigma^+ \rightarrow {}^3B^2$ in the terms crossing point with the spin change provides more effective dissociation process stimulated by the stepwise vibrational excitation:

$$CO_2^*(^1\Sigma^+) \rightarrow CO_2^*(^3B^2) \rightarrow CO(^1\Sigma^+) + O(^3P), E_a = 5.5 \text{ eV}. \tag{17.29}$$

The nonadiabatic dissociation (17.29) results in the formation of an oxygen atom in the electronically ground state $O(^3P)$. It requires the exact energy of OC—O bond to be spent (5.5 eV), and therefore this process is exponentially faster than adiabatic one (17.28). Atomic oxygen created in (17.29) is able then to participate in a secondary slightly endothermic reaction, also with a vibrationally excited CO_2 molecule, to produce a second CO molecule and molecular oxygen:

$$O(^3P) + CO_2^*(^1\Sigma^+) \rightarrow CO + O_2, \Delta H = 0.3 \text{ eV}, E_a \approx 0.5 - 1 \text{ eV}. \tag{17.30}$$

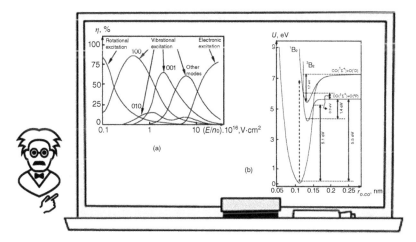

Figure 17.7 Mechanisms of CO_2 dissociation: (a) distribution of cold-plasma electrons energy between channels of CO_2 excitation; low electronic terms of CO_2 involved in dissociation; (b) schematic of the CO_2 low-energy electronic terms.

This reaction is faster than the three-body recombination $(O + O + M \rightarrow O_2 + M)$ and permits to produce a second CO molecule per one dissociation event, when vibrational temperature is not too low $(T_v \geq 0.1\,\text{eV})$.

The CO$_2$ dissociation stimulated by vibrational excitation in plasma (17.29 and 17.30) has the following three essential qualitative advantages in energy efficiency with respect to alternative nonequilibrium mechanisms of CO$_2$ dissociation:

- Not less than 95% of the total nonthermal discharge energy at electron temperature $T_e = 1\text{--}2\,\text{eV}$ can be transferred from plasma electrons to vibrational excitation of CO$_2$ molecules, mostly to their asymmetric vibrational mode, see Figure 17.7a.
- The vibrational energy is the most effective for stimulation of the endothermic reactions (17.29 and 17.30), see the Fridman–Macheret α-model, Section 3.12.
- Vibrational energy 5.5 eV required for (17.29) is less than that for dissociation through electronic excitation in direct electron impact, which due to the Frank–Condon principle (vertical transition $^1\Sigma^+ \rightarrow {}^1B^2$) requires at least 8 eV, see Figure 17.7b.

In nonthermal discharges with high reduced electric fields E/p (like low-pressure discharges, DBD, coronas) and when plasma is generated by energetic particles (high-energy electron beams or nuclear fission fragments), vibrational excitation is suppressed and **CO$_2$ dissociation through electronic excitation**, see Figures 17.7b and 17.8a, like for example

$$e + CO_2(^1\Sigma^+) \rightarrow CO(a^3\Pi) + O(^3P) \tag{17.31}$$

can be a dominant mechanism (Slovetsky 1980). Energy efficiency of this mechanism is relatively low, see Figure 17.8b. The maximum is about 25%, which corresponds to CO energy cost of ~11.5 eV. The energy threshold of dissociation through electronic excitation (and therefore the process energy cost) can be reduced by preliminary vibrational excitation. This effect can be understood from Figure 17.7b, considering the Frank–Condon principle of vertical transition between electronic terms induced by electron impact. Such a hybrid dissociation mechanism is nonlinear with respect to electron density and slightly improves the energy efficiency of the process.

Dissociative attachment of electrons to CO$_2$ molecules also contributes to CO$_2$ decomposition in nonthermal discharges with elevated reduced electric fields E/p:

$$e + CO_2 \rightarrow CO + O^-. \tag{17.32}$$

The process cross sections are illustrated in Figure 17.9 as a function of electron energy. The energy threshold here is lower than that of dissociation through electronic excitation. However, the maximum value of the cross section is not high $(\approx 10\text{--}18\,\text{cm}^2)$ and the contribution of the dissociative attachment toward the total dissociation kinetics is not significant, see Figure 17.8a. The energy efficiency of the CO$_2$ dissociation through dissociative attachment is limited by loss of an electron in the process (the energy price of an electron is quite high, usually 30–100 eV).

Plasma catalysis can enhance conversion of CO$_2$ by selectively accelerating nonendothermic stages of the process, see Britun and Silva (2018) and Snoeckx and Bogaerts (2017). Contribution of the catalytic effect to the most energy

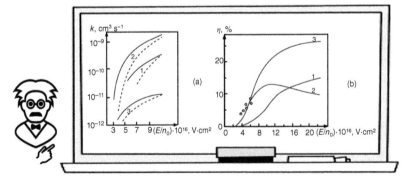

Figure 17.8 (a) Rate coefficients as function of reduced electric field: (1) CO$_2$ dissociation by direct electron impact; (2) CO$_2$ electronic excitation; (3) dissociative attachment of electrons to CO$_2$ (dashed lines reflect effect of CO admixture). (b) Energy efficiency of CO$_2$ dissociation through electronic excitation: (1) contribution of singlet states; (2) contribution of triplet states; (3) total; dots—validation data.

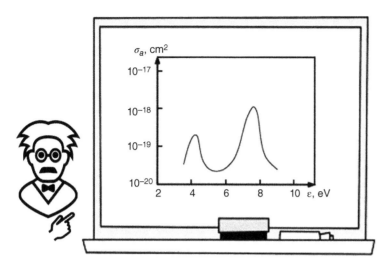

Figure 17.9 Dissociation of CO$_2$ molecules by dissociative attachment of electrons: the process cross-section as a function of electron energy.

effective dissociation regimes like (17.29 and 17.30) is however limited by relevant surface energy losses. Plasma catalysis can be more pronounced in CO$_2$ plasma in the presence of hydrocarbons or other hydrogen reach gases.

17.8 Physical Kinetics of CO$_2$ Dissociation in Nonthermal Plasma Stimulated by Vibrational Excitation of the Molecules

CO$_2$ vibrational excitation by electrons is the most effective mechanism of CO$_2$ dissociation in plasma. At $T_e = 1$–2 eV, the plasma electrons mostly provide excitation of low vibrational levels. Then VV-exchange leads to population of highly excited vibrational states with nonadiabatic transition $^1\Sigma^+ \rightarrow ^3B^2$ and dissociation (17.29), see Figure 17.7b. The VV exchange between vibrationally excited CO$_2$ molecules is a limiting stage of the dissociation kinetics. It is kinetically complicated process, which first proceeds along individual vibrational modes at low energies and then through mixed vibrations (vibrational quasi-continuum) at high energies.

CO$_2$ molecules are linear (symmetry group $D_{\infty h}$) and have three normal vibrational modes, see Figure 17.10. They are asymmetric valence vibration v_3 (energy quantum $\hbar\omega_3 = 0.30$ eV), symmetric valence vibration v_1 (energy quantum $\hbar\omega_1 = 0.17$ eV), and a double degenerated symmetric deformation vibration v_2 (energy quantum $\hbar\omega_2 = 0.085$ eV). The degenerated symmetric deformation vibrations v_2 are polarized in two perpendicular planes, which can result in quasi-rotation of the linear molecule around its principal axis. Angular momentum of this quasi-rotation is characterized by a special quantum number "l_2," which assumes the values $l_2 = v_2$, $v_2 - 2$, $v_2 - 4$, ... , 1 or 0, where v_2 is the number of quanta on the degenerated mode. The level of vibrational excitation can then be denoted as CO$_2\left(v_1, v_2^{l_2}, v_3\right)$ showing the number of quanta on each mode. There is a resonance between two types of symmetric vibrations – valence and deformational: $\hbar\omega_1 = 2\hbar\omega_2$. *The symmetric modes therefore are sometimes considered as one triple degenerate vibration* with vibrational quantum $\hbar\omega_s = \hbar\omega_2$ and generalized vibrational quantum number $v_s = 2v_1 + v_2$ to fit the simplified expression for the total vibrational energy of all symmetric modes in harmonic approximation: $E_{sym} = \hbar\omega_1 v_1 + \hbar\omega_2 v_2 = \hbar\omega_2(2v_1 + v_2) = \hbar\omega_s v_s$. The total vibrational energy then is:

$$E_v(v_a, v_s) = \hbar\omega_a v_a + \hbar\omega_s v_s - x_a v_a^2 - x_s v_s^2 - x_{as} v_a v_s, \tag{17.33}$$

where approximately: $x_a = 40$ cm^{-1}, $x_s = x_{as} = 12$ cm^{-1} are generalized coefficients of anharmonicity.

The VV-exchange process proceeds at low levels of vibrational excitation independently along symmetric and asymmetric CO$_2$ vibrational modes, see Section 5.7. These modes can be characterized therefore by individual vibrational temperatures T_{va} and T_{vs}, as well as by individual vibrational energy distribution functions. The major contribution to population of highly vibrationally excited states of CO$_2$ and, hence, to dissociation is related to the excitation of the asymmetric vibrational modes. It is due to: (i) the asymmetric mode is predominantly excited at $T_e = 1$–3 eV (see Figure 17.7); (ii) VT-relaxation from this mode is much slower than that of symmetric vibrations; (iii) VV exchange along the asymmetric mode is several orders of magnitude faster than along symmetric modes, which

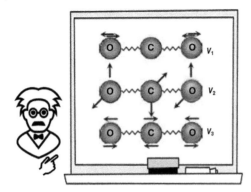

Figure 17.10 Vibrational modes of CO_2 molecules: v_1, symmetric valence vibrations; v_2, double degenerate symmetric deformation vibrations; v_3, asymmetric valence vibrations.

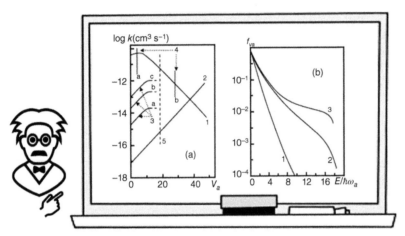

Figure 17.11 Vibrational kinetics of the CO_2 asymmetric mode: (a) rate coefficients of relaxation at $T_0 = 300$ K as functions of number of quanta on the mode: (1) VV relaxation; (2) VT relaxation; (3) intermodal VV relaxation (assuming [3a] $T_{vs} = 1000$ K, [3b] $T_{vs} = 2000$ K, [3c] $T_{vs} = 3000$ K); (4) intramolecular VV relaxation, transition to quasi-continuum (assuming [4a] equal excitation of all modes; [4b] predominant excitation of asymmetric vibrations); (5) CO_2 dissociation energy. (b) Population of the asymmetric mode of CO_2 vibrations: (1) $T_{va} = 3000$ K, $T_{vs} = 2000$ K, $T_0 = 1000$ K; (2) $T_{va} = 5000$ K, $T_{vs} = 3000$ K, $T_0 = 1000$ K; (3) $T_{va} = 5000$ K, $T_{vs} = 3000$ K, $T_0 = 500$ K.

intensify the population of highly excited states. Vibrational relaxation rate coefficients for the asymmetric mode are shown in Figure 17.11a. The domination of VV-relaxation results in the Treanor effect (see Section 5.7) and a overpopulation of highly excited levels of the asymmetric CO_2 vibrations in nonequilibrium plasma conditions, when vibrational temperatures T_{va} and T_{vs} exceed the translational temperature T_0, see Figure 17.11b.

As the level of excitation increases, the vibrations of different types are mixing collisionlessly due to the intermodal anharmonicity and the Coriolis interaction, see Figure 17.11a. The intramolecular VV quantum exchange results in creation of the **vibrational quasi-continuum of the highly excited states** providing dissociation (see Section 5.7). Energy of the asymmetric mode changes during vibrational beating corresponding to its interaction with symmetric modes ($v_3 \rightarrow v_1 + v_2$), and as a result the effective frequency of this oscillation mode changes as well:

$$\Delta\omega_a = (x_a\omega_a)^{\frac{1}{3}}(A_0\omega_a n_{sym})^{2/3}, \tag{17.34}$$

where $A_0 \approx 0.03$ is a parameter of interaction between modes ($v_3 \rightarrow v_1 + v_2$), and n_{sym} is the total number of quanta on the symmetric modes (v_1 and v_2). As the level of excitation, n_{sym}, increases, the of $\Delta\omega_a$ grows and at a certain critical number of quanta, covers the defect of resonance $\Delta\omega$ of the asymmetric-to-symmetric quantum transition:

$$n_{sym}^{cr} = \frac{1}{\sqrt{x_a A_0^2}}\left(\frac{\Delta\omega}{\omega_a}\right)^{2/3}. \tag{17.35}$$

When the number of quanta exceeds the critical value (17.35), the so-called **Chirikov stochasticity criterion**, the molecular motion becomes quasi-random, and the modes become mixed in the vibrational quasi-continuum. The critical excitation level n_{cr} decreases with number N of vibrational modes: $n_{cr} \approx 100/N^3$. For molecules with four or more atoms, the transition to the vibrational quasi-continuum occurs at quite low levels of vibrational excitation. CO_2, however, can maintain the individuality of vibrational modes up to higher levels of excitation.

When both asymmetric and symmetric modes are significantly excited and transition to vibrational quasi-continuum occurs at low excitation levels (17.35), vibrational kinetics of CO_2 is controlled by a single vibrational temperature. This case of **quasi-equilibrium of vibrational modes** is referred to sometimes as **one-temperature approximation of CO_2 dissociation kinetics** in nonthermal plasma. The CO_2 dissociation rate is limited not by elementary dissociation itself, but by energy transfer from a low to high vibrational excitation levels through VV-relaxation. The population of highly excited states with vibrational energy E depends here on the number of vibrational degrees of freedom "s" and is proportional to the density of the vibrational states $\rho(E) \propto E^{s-1}$. The CO_2 dissociation rate coefficient can be expressed in the quasi-equilibrium of vibrational modes as (Potapkin et al. 1980):

$$k_R(T_v) = \frac{k_{VV}^0}{\Gamma(s)} \frac{\hbar\omega}{T_v} \left(\frac{E_a}{T_v}\right)^s \Sigma\left(s, \frac{E_a}{T_v}\right) \exp\left(-\frac{E_a}{T_v}\right), \tag{17.36}$$

where $\Gamma(s)$ is the Γ-function, k_{VV}^0 and $\hbar\omega$ are the lowest vibrational quantum and VV-exchange rate coefficient, factor $\Sigma\left(s, \frac{E_a}{T_v}\right)$ is not a strong function of E_a/T_v and can be taken about 1.1–1.3 if $T_v = 1000$–4000 K. The rate coefficient (17.37) does not include any details related to the elementary act of dissociation itself, which reflects the nature of the fast reaction limit controlling the plasma-chemical CO_2 decomposition stimulated by vibrational excitation.

If temperatures of asymmetric and symmetric vibrational modes (T_{va}, T_{vs}) are different, transition to quasi-continuum occurs along the mode with the highest temperature (usually asymmetric vibrations). This kinetic approach is referred to as **two-temperature approximation.** The vibrational quantum number for symmetric modes can be fixed on its average value corresponding to Planck formula for these modes $<n_s>$. The vibrational energy on the asymmetric corresponding to transition to the quasi-continuum can be estimated based on (17.35) as:

$$E^*(<n_s>) = \frac{\Delta\omega}{x_{as}} \frac{\hbar\omega_a}{<n_s>}, \tag{17.37}$$

and numerically is about (15–20) $\hbar\omega_a$ at $T_{vs} = 1000$ K. It means that the transition to quasi-continuum occurs at rather high levels of excitation of the asymmetric mode, close to the dissociation energy (see Figure 17.11b). Thus, most of the vibrational distribution relevant to CO_2 dissociation in this case, in contrast to the one-temperature approach, is not continuous but discrete, significantly exceeds the Boltzmann distribution and corresponds to the Treanor function:

$$f_v(v_a, v_s, T_{va}, T_{vs}, T_0) \sim \exp\left(-\frac{\hbar\omega_a v_a}{T_{va}} - \frac{\hbar\omega_s v_s}{T_{vs}} + \frac{x_a v_a^2 + x_s v_s^2 + x_{as} v_a v_s}{T_0}\right). \tag{17.38}$$

More details on the kinetics of CO_2 dissociation through vibrational excitation, including transition from discrete distribution to the vibrational quasi-continuum and dissociation rates in the two-temperature approximation can be found in the books Fridman (2008) and Rusanov and Fridman (1984); recent details on the kinetic modeling are reviewed in Snoeckx and Bogaerts (2017).

17.9 Vibrational Kinetics and Energy Balance in Nonequilibrium CO_2 Plasma

The two-temperature approximation (T_{va}, T_{vs}) of CO_2 vibrational kinetics in nonequilibrium CO_2 plasma can be analyzed by energy balance equations describing major energy transfer processes, relaxation, and chemical reactions separately for different individual vibrational modes, including the dissociation products (Rusanov and Fridman 1984). Thus, time evolution of asymmetric and symmetric vibrational temperatures is shown in Figure 17.12 together with evolution of translational gas temperature at $E_v = 0.5$ eV mol^{-1} and different levels of electron density. The degree of nonequilibrium (difference between T_v and T_0) grows with ionization degree. Also, an interesting effect of *energy oscillations between asymmetric and symmetric vibrational modes*, related to the nonlinearity of interaction between the modes, can be observed in Figure 17.12 at relatively low degrees of ionization.

The energy efficiency of CO_2 dissociation in the two-temperature approximation as a function of E_v is shown in Figure 17.6 (at $n_e/n_0 = 3 \times 10^{-6}$); and as a function of ionization degree at the specific energy input $E_v = 0.5$ eV mol^{-1} in Figure 17.13. The dependence on ionization degree has a threshold at 3×10^{-7} and a tendency to saturate at high ionization degrees. The optimal specific energy input is about $E_v = 0.8$–1.0 eV mol^{-1}; the threshold is about 0.2 eV mol^{-1}

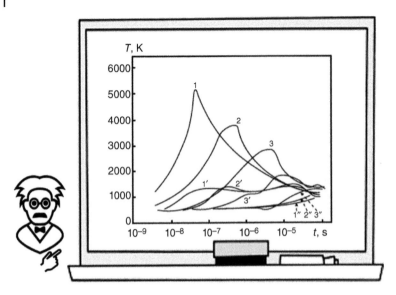

Figure 17.12 Time evolution of vibrational asymmetric T_{va} (curves 1–3), vibrational symmetric T_{vs} (curves 1′–3′), and translational T_0 (curves 1″–3″) CO$_2$ temperatures at specific energy input $E_v = 0.5$ eV mol^{-1} and ionization degrees 3×10^{-6} (curves 1), 10^{-6} (curves 2), 3×10^{-7} (curves 3).

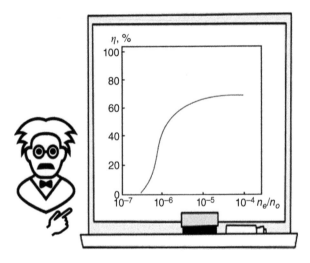

Figure 17.13 Energy efficiency of CO$_2$ dissociation in plasma at the specific energy input $E_v = 0.5$ eV mol^{-1} as a function of plasma ionization degree n_e/n_0.

for subsonic plasma (see Section 7.3). At high specific energy inputs $E_v > 1$ eV mol^{-1}, a significant part of the discharge energy is spent on the excitation of the dissociation products, especially CO. The conversion degree χ is already high at $E_v > 1$ eV mol^{-1}, and the energy efficiency decreases hyperbolically $\eta \propto 1/E_v$ following general tendency discussed in Section 17.3. More details on the subject are in Fridman (2008), Rusanov and Fridman (1984), and Snoeckx and Bogaerts (2017).

17.10 CO$_2$ Dissociation in Supersonic Cold Plasma Flows

The highest energy efficiency of CO$_2$ dissociation in plasma is achieved through the process stimulation by vibrational excitation, see Figure 17.6. The main cause of energy losses, in this case, is VT relaxation from low vibrational levels of CO$_2$ molecules, which decreases exponentially with reduction of gas temperature. *Significant decrease of gas temperatures (to the level of about 100 K; see Figure 17.14) and therefore*

further increase of the energy efficiency (see Figure 17.6) can be achieved by carrying out plasma-chemical CO_2 dissociation in a supersonic gas flow. Another reason for the supersonic process organization is due to limitations of productivity in subsonic discharges. The optimal energy input is about $1\,eV\,mol^{-1}$ (see Section 17.9) at moderate and high pressures, therefore high discharge power requires high flow rates through the discharge. Plasma instabilities restrict space sizes and pressures of the steady-state discharge systems, and velocities are usually limited by the speed of sound. Thus, flow rate and, hence, power are restricted by some critical value in subsonic plasma systems, which is about $100\,kW$ for nonequilibrium microwave discharges. Higher powers (up to 1–3 MW) and therefore higher process productivity can be achieved in supersonic gas flows. A simplified scheme and flow parameters of such discharge (power $500\,kW$, Mach number $M = 3$) are illustrated in Figure 17.14. Some gas-dynamic characteristics of the supersonic discharges are presented in Figure 17.15. Ignition of the nonequilibrium discharge takes place in this case after a supersonic nozzle in a relatively low-pressure zone, which is optimal for nonequilibrium discharges. Pressure in the after-discharge zone is restored in a diffuser, so initial and final pressures can be above atmospheric. Energy required for compression in the supersonic system is about 10% of the total energy cost.

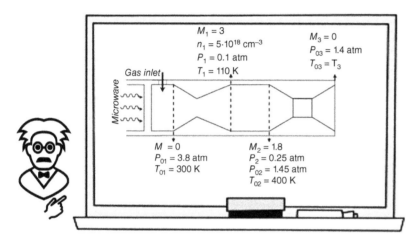

Figure 17.14 Typical parameters of a nozzle system for supersonic plasma CO_2 dissociation. Subscripts 1 and 2 are related to inlet and exit of a discharge zone; subscript 3 is related to exit from the nozzle system; subscript 0 is related to stagnation pressure and temperature.

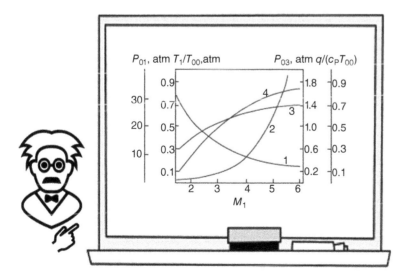

Figure 17.15 Supersonic plasma flow characteristics: (1) inlet temperature T_1; (2) initial tank pressure p_{01}; (3) exit pressure p_{03} at the critical heat release; (4) critical heat release q. Parameters (1–4) are shown as functions of Mach number M_1 in front of the discharge. Initial gas tank temperature $T_{00} = 300\,K$; static pressure in front of discharge $p_1 = 0.1\,atm$.

*Specific energy input (and therefore conversion degree) in supersonic plasma is restricted by **critical heat release** and choking of the flow.* Heating supersonic flow decreases its Mach number. The critical heat release q_{cr} corresponds to a drop of the initial Mach number from $M > 1$ before the discharge to $M = 1$ afterward leading to choking. The critical heat release for the supersonic flow reactor with constant cross section is equal to:

$$q_{cr} = c_p T_{00} \left[\frac{(1 + \gamma M^2)^2}{2(\gamma + 1)M^2 \left(1 + \frac{\gamma - 1}{2}M^2 \right)} - 1 \right]; \tag{17.39}$$

here T_{00} is the initial gas temperature in the tank before the supersonic nozzle, c_p and γ are specific heat and specific heat ratio. If the initial Mach number is not very close to one, the critical heat release is about $q_{cr} \approx c_p T_{00}$, see Figure 17.15. The heat release in plasma above the critical value leads to the formation of non-steady-state flow perturbations like shock waves and choke the flow. Even considering the high energy efficiency in supersonic flows, the critical heat release seriously restricts the specific energy input $E_v(1 - \eta) < q_{cr}$. As a result, the maximum conversion degree of CO$_2$ dissociation in supersonic discharge ($T_{00} = 300$ K and $M = 3$) does not exceed 15–20%. Special profiling of a supersonic discharge chamber partially suppresses this restriction.

Experiments with supersonic microwave discharge, operating at a frequency of 915 MHz at a power 10–100 kW (maximum 1 MW), confirm the possibility of achieving up to 90% energy efficiency of CO$_2$ dissociation (Asisov et al. 1983). The plasma reactor after a supersonic nozzle had diameter of 3.5 cm and effective length of 30 cm, crossing a waveguide. The critical supersonic nozzle had diameter 0.8 cm, and the expansion degree 20. The initial pressure in the tank before the nozzle varies from 1.8 to 7 atm, which corresponds to static pressure in the reactor from 0.05 atm (40 Torr) to 0.2 atm (150 Torr) and to flow rates of 5–50 l s^{-1}; $M \approx 2$–3. The electron density was 3–8×10^{12} cm^{-3}, electron temperature about 1 eV, vibrational temperature $T_v \approx 3500$ K, and translational gas temperature was $T_0 \approx 160$ K. The minimum energy cost of the CO$_2$ dissociation claimed in these experiments, was record low 3.2 eV mol^{-1}.

17.11 Gas-dynamic Stimulation of CO$_2$ Dissociation in Supersonic Flow Without Plasma, "Plasma Chemistry Without Electricity"

Stimulation of CO$_2$ dissociation can be achieved not only by means of an increase of gas temperature (better the vibrational one) but also by means of cooling the gas. It sounds nonrealistic but it is correct theoretically and proven experimentally. The explanation is simple: the dissociation rate in nonequilibrium systems depends not only on vibrational temperature (T_v) of CO$_2$ molecules but also on the degree of VT nonequilibrium (T_v/T_0) leading to the Treanor effect and overpopulation of the highly vibrationally excited states. The gas cooling, therefore, can accelerate the endothermic process almost as effectively as it's heating. It can be organized in a supersonic nozzle, similar to the supersonic plasma considered above. Most of the energy accumulated in CO$_2$ vibrational degrees of freedom before the supersonic nozzle can be localized through the Treanor effect in highly excited levels after the nozzle and transferred to dissociation. This effect, which can be observed not only in CO$_2$ dissociation, is sometimes referred to as **"plasma chemistry without electricity."** It is like gas-dynamic lasers, but nonequilibrium vibrational energy is transferred in this case into a chemical reaction instead of radiation.

The gas-dynamic stimulation effectively increases "kinetic temperature" of the gas. Due to the Treanor effect even at room temperature, the CO$_2$ dissociation rate is equivalent to that at more than 2000 K in quasi-equilibrium. The conversion degree, however, corresponds to the total energy accumulated in vibrational degrees of freedom at actual CO$_2$ temperature. Relevant experimental results in flow with Mach number $M \approx 6$ (Asisov et al. 1986) in comparison with kinetic simulations at $M = 4.5, 5.5, 6.5, 9.5$ ($T_v/T_0 = 5, 7, 10, 20$) are shown in Figure 17.16 presenting the CO$_2$ conversion degree α as a function of initial gas temperature before the supersonic nozzle. The conversion degree can reach 1–3% without direct without significant gas heating. At room temperature the CO$_2$ conversion degree is not high (about 0.2%) but quite visible.

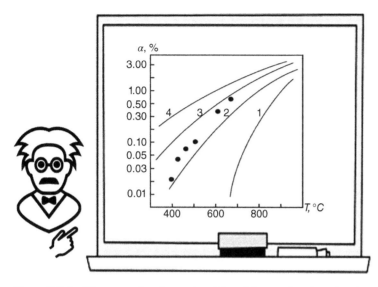

Figure 17.16 CO$_2$ conversion degree into CO and O$_2$ as a function of preheating temperature in supersonic system of gas-dynamic stimulation of CO$_2$ dissociation. Mach numbers: (1) $M = 4.5$ (corresponds to $T_v/T_0 = 5$); (2) $M = 5.5$ (corresponds to $T_v/T_0 = 7$); (3) $M = 6.5$ (corresponds to $T_v/T_0 = 10$); (4) $M = 9.5$ (corresponds to $T_v/T_0 = 20$). Solid lines correspond to theoretical calculations, dots correspond to experiments with $M = 6$.

17.12 Complete Plasma Dissociation of CO$_2$ to Carbon and Oxygen

Plasma parameters can be chosen to provide a predominantly complete dissociation of CO$_2$ with the production of solid carbon and gaseous oxygen:

$$CO_2 \rightarrow C(s) + O_2, \Delta H = 11.5 \text{ eV mol}^{-1}. \tag{17.40}$$

Plasma-chemical mechanism of the complete dissociation (17.40) can start with energy effective reactions ((17.29) and (17.30)) and then proceed with conversion of CO into elementary carbon, which is kinetically limited by record high value (11 eV) of CO bonding. The conversion of CO into carbon can be effectively performed by the reaction of disproportioning:

$$CO + CO \rightarrow CO_2 + C(g), \Delta H = 5.5 \text{ eV mol}^{-1}, E_a = 6 \text{ eV mol}^{-1}, \tag{17.41}$$

which can be effectively stimulated by vibrational excitation of CO molecules as well as proceed through electronically excited state CO($a^3\Pi$). Significant challenge of the complete dissociation (17.40) is related to the fast reaction reverse to (17.41) characterized by low activation energy $E_a = 0.5 \text{ eV mol}^{-1}$. Suppression of this reverse reaction requires heterogeneous stabilization of atomic carbon or its volumetric clusterization (Legasov et al. 1978a, b).

When the CO disproportioning (17.41) is stimulated by vibrational excitation, both molecules can be strongly excited $CO^*(v_1) + CO^*(v_2) \rightarrow CO_2 + C$. Relevant kinetics requires **generalization of the Fridman–Macheret α-formula** for efficiency of vibrational energy in overcoming activation energy of chemical reactions, see Section 3.12. Activation energy of the process ($E_a = 6 \text{ eV mol}^{-1}$) can be reduced here by the vibrational energy of both excited reagents:

$$\Delta E_a = \alpha_1 E_{v1} + \alpha_2 E_{v2}; \tag{17.42}$$

Here the subscript 1 corresponds to the molecule losing an atom and subscript 2 is related to another molecule accepting the atom; α_1 and α_2 are the relevant coefficients of efficiency for using vibrational energy. It is shown that $\alpha_1 = 1$ and $\alpha_2 \approx 0.2$, which means that *vibrational energy of the donating molecule is much more efficient in stimulation of endothermic exchange reactions than that of the accepting molecule.*

Between experiments focused on the complete dissociation of CO$_2$, we can point out those carried out in *microwave discharges in conditions of the ECR* and in the stationary plasma-beam discharges. The pulse-periodic microwave discharge in magnetic field under ECR conditions (Asisov et al. 1980) had wavelength 8 mm, power in each pulse 30 kW,

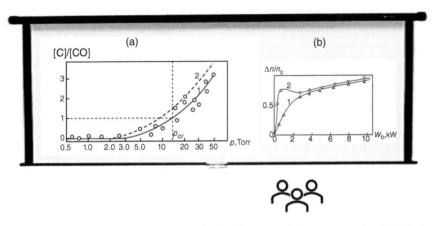

Figure 17.17 Complete dissociation of CO$_2$: (a) microwave discharge operating in ECR, density ratio [C]/[CO] as a function of pressure, measured by (1) pressure defect, (2) by mass-spectrometry; (b) plasma-beam discharge, conversion degree of (1) CO$_2$ and (2) CO as a function of electron beam power.

pulse duration 0.3 μs and frequency 0.5 kHz; reactor diameter 1.7 cm and length 10 cm; CO$_2$ at pressures 0.5–100 Torr. The C/CO ratio in products was measured from pressure analysis and using mass-spectroscopy, see Figure 17.17a. This ratio at a CO$_2$ pressure of 50 Torr reaches quite high values of C/CO ≈ 4, which means that complete dissociation can essentially dominate over incomplete dissociation. When the CO$_2$ temperature was preliminarily decreased below 300 K and the reverse reaction of carbon atoms was suppressed, the C/CO ratio grew significantly providing, almost exclusively, complete CO$_2$ dissociation.

The complete CO$_2$ dissociation has been also demonstrated in the *stationary plasma-beam discharges* (Atamanov et al. 1979), where electron energy in the beam was 30 keV, the beam current up to 2 A, gas pressure 10 mTorr, gas flow rate 40 cm^3 s^{-1}, power from 0.5 to 10 kW. During the electron beam transportation through plasma, 70% of the beam energy was transferred to plasma electrons. Plasma parameters were determined as T_e = 1–2 eV, $n_e/n_0 = 10^{-3}$–10^{-2}. Results of mass-spectrometric measurements are presented in Figure 17.17b as functions of electron beam power. At electron beam powers of 8–10 kW, most of the CO$_2$ was dissociated. A relatively low partial pressure of CO indicated the domination of complete CO$_2$ decomposition with the formation of elementary carbon, which was also confirmed by spectral diagnostics. The conversion degree of CO$_2$ reached 85%. The CO molecules were further decomposed to elementary carbon and oxygen with a conversion degree of up to 90%.

17.13 Problems and Concept Questions

17.13.1 Energy Efficiency of Quasi-equilibrium and Nonequilibrium Plasma Processes

Explain why the energy efficiency of nonequilibrium plasma processes can be higher than that of quasi-equilibrium ones. Is it a contradiction with the thermodynamic principles?

17.13.2 Plasma-chemical Processes Stimulated by Vibrational Excitation

Why the plasma-chemical reactions can be effectively stimulated by vibrational excitation only if the specific energy input exceeds the critical value, while reactions related to electronic excitation or dissociative attachment can proceed effectively at any levels of the specific energy input?

17.13.3 Absolute and Ideal Quenching of Products in Thermal Plasma

Explain why energy efficiencies of the absolute and ideal quenching of the CO$_2$-dissociation products are identical, while in the case of dissociation of water molecules, they are significantly different?

17.13.4 Super-ideal Quenching due to Vibrational–Translational Nonequilibrium

How can the VT nonequilibrium be achieved during the quenching, if degree of ionization is not sufficient to provide vibrational excitation faster than vibrational relaxation?

17.13.5 Super-ideal Quenching Effects Related to Selectivity of Transfer Processes

Compare conditions required for effective super-ideal quenching related to the following three different causes: (i) VT nonequilibrium during the cooling process; (ii) the Lewis number significantly differs from unity during separation and collection of products; (iii) cluster-products move relatively fast from the rotating high temperature discharge zone.

17.13.6 CO_2 Dissociation Through Electronic Excitation of Molecules in Cold Plasma

Using Figure 17.7b, calculate decrease in the energy threshold of CO_2 dissociation through electronic excitation related to preliminary vibrational excitation of the molecules. What is the kinetic order of the process with respect to the electron density?

17.13.7 Transition of Highly Vibrationally Excited CO_2 Molecules into Vibrational Quasi-continuum

Explain the mechanism for mixing different vibrational modes of CO_2 molecules when the oscillation amplitudes are relatively high. Consider the influence of vibrational beating on energy transfer between modes and the transition of a polyatomic molecule to a quasi-continuum of vibrational states. Analyze how the critical maximum number of quanta on an individual mode depends on the number of vibrational modes.

17.13.8 One-vibrational-temperature Approximation of CO_2 Dissociation Kinetics

Explain why the pre-exponential factor for the rate coefficient of CO_2 dissociation in one-temperature approximation is proportional to $(E_a/T_v)^s$. How does it reflect the fact that CO_2 dissociation kinetics through vibrational excitation in plasma is limited by VV exchange?

17.13.9 Plasma-stimulated Disproportioning of CO, and Complete Dissociation of CO_2 with Production of Elementary Carbon

C atoms generated in CO disproportioning in plasma can be lost in fast reverse reactions of the atoms with CO_2. Estimate the requirements for gas temperature and pressure to provide effective heterogeneous stabilization of the elementary carbon. Which temperature is more important for the stabilization: vibrational or translational?

Lecture 18

Synthesis of Nitrogen Oxides, Ozone, and Other Gas-phase Plasma Synthetic and Decomposition Processes

18.1 Plasma-chemical Synthesis of Nitrogen Oxides from Air: Fundamental and Applied Aspects of the Process in Thermal and Nonthermal Discharges

Plasmas have been applied both for NO generation from air (to be considered in this Lecture) for further production of fertilizers and explosives, and conversely for exhaust cleaning from NO_x emitted in combustion systems (to be considered later in this book). Interesting novel aspects regarding the NO synthesis applications to plasma agriculture are discussed in Pei et al. (2019), Hollevoet et al. (2020). The endothermic process of the nitrogen oxide synthesis in air plasma is one of the "old timers" in plasma technology:

$$\frac{1}{2}N_2 + \frac{1}{2}O_2 \rightarrow NO, \quad \Delta H = 1\,eV\,mol^{-1} \tag{18.1}$$

Henry Cavendish and Joseph Priestly were the first to investigate this plasma process in eighteenth century. Industrial implementation has been performed in 1900 by Kristian Birkeland and Samuel Eyde. They developed thermal arc furnaces for air conversion that produced about 1–2% of nitrogen oxides. The energy price for NO production in these systems was about $25\,eV\,mol^{-1}$, which corresponds to relatively low energy efficiency of about 4%. An alternative ammonia production technology developed shortly after by Fritz Haber and Karl Bosch appeared to be significantly less energy intensive (effective energy cost about $4\,eV\,mol^{-1}$, energy efficiency about 25%) and become the major approach for modern nitrogen fixation.

 Competition with the Haber-Bosch Process motivated more than century-long efforts to increase the yield and especially energy efficiency of the plasma-chemical NO synthesis from air. Modernization of thermal plasma systems, including operation at pressures of 20–30 atm, temperatures 3000–3500 K, and very high quenching rate up to 10^8 K/s permitted to decrease the energy cost to about $9\,eV\,mol^{-1}$, which corresponds to an energy efficiency of 11% (Polak et al. 1975). At slightly lower quenching rates, the energy cost of thermal plasma NO synthesis is about $20\,eV\,mol^{-1}$ (energy efficiency 5%), and the conversion degree is not higher than 4–5%. Thus, energy consumption of the thermal plasma (10–$20\,eV\,mol^{-1}$) much exceeds that of Haber-Bosch process ($4\,eV\,mol^{-1}$). It brings attention to nonthermal plasma, where the discharge energy can be selectively directed on the most productive channels of NO synthesis and fast quenching is not required.

In this regard, the high energy efficiency (14%, energy cost of $7\,eV\,mol^{-1}$) has been achieved in non-self-sustained strongly nonequilibrium discharges stimulated by relativistic electron beams (Basov et al. 1977, 1978). The threshold for the energy efficiency dependence on specific energy input $\eta(E_v)$ was observed at $E_v = 0.1$–$0.2\,eV\,mol^{-1}$, which indicates the crucial contribution of vibrationally excited molecules. The nonthermal pulsed microwave discharges (pulse duration 50–100 ns, repetition frequency 1000 Hz, pulse power 0.5–1.5 MW; Polak et al. 1975) gave similar results and were also attributed to vibrational excitation of nitrogen. Energy cost of the nonequilibrium NO synthesis is growing linearly with reduced electric field E/n_0 (Pei et al. 2019), which generalizes exceptional role of the N_2 vibrational excitation.

Plasma Science and Technology: Lectures in Physics, Chemistry, Biology, and Engineering, First Edition. Alexander Fridman.

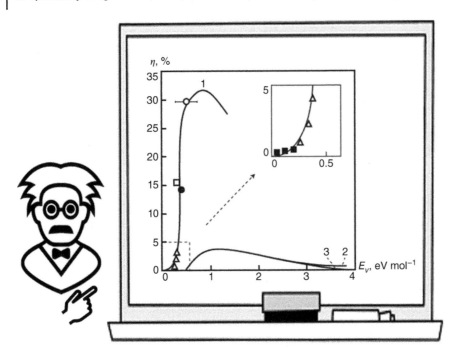

Figure 18.1 Energy efficiency of NO synthesis in air plasma as a function of energy input: (1) nonequilibrium process stimulated by vibrational excitation; (2,3) thermal synthesis with (2) ideal and (3) absolute quenching. Experiments with microwave discharges: °, •; discharges sustained by electron beams: •, ■, Δ.

The highest energy efficiency (about 30%) and lowest energy cost (about $3\,\text{eV mol}^{-1}$) were achieved in a nonthermal microwave discharge operating at low pressures in a magnetic field under electron-cyclotron resonance (ECR; Asisov 1980). *It is the only plasma system so far characterized by energy efficiency slightly better than that of the Haber-Bosch Process.* These exceptional results can be interpreted not only by ECR sustaining the optimal E/n_0 for vibrational excitation but especially by very effective wall-stabilization of NO at low pressures. Plasma conversion of air into NO is strongly limited by the product's reactions with N atoms, the same effect is also responsible for the relatively low energy efficiency of the process. At low pressures (ECR conditions; Asisov 1980), the products were trapped and stabilized in this system on the cold surface, which explains the mentioned highest process efficiency. In the same way, the highest conversion degree (about 20%) was achieved in experiments with low-pressure plasma beam discharges with products trapping and stabilization on the cold surface (Ivanov and Nikiforov 1978; Ivanov and Soboleva 1978). Obviously, the low-pressure conditions are not practical for large-scale industrial applications, but it shows feasibility of the nonthermal plasma approaches to become more energy effective than the Haber-Bosch Process.

The energy efficiency for NO synthesis in the described plasma systems as a function of the specific energy input $\eta(E_v)$ is summarized in Figure 18.1 (Rusanov and Fridman 1984; Fridman 2008) in comparison with modeling results. Recent data on experiments and simulations of the plasma NO synthesis coherent with the previous results can be found in Pei et al. (2019), Hollevoet et al. (2020).

18.2 Mechanisms and Energy Efficiencies of NO Synthesis from Air in Nonthermal and Thermal Plasmas, the Zeldovich Mechanism

The NO synthesis (18.1) proceeds in plasma by set of numerous elementary processes. The limiting stage is breaking of the strong N_2 bond (10 eV), which occurs efficiently in the **reactions of vibrationally excited N_2 with atomic oxygen in ground state**:

$$O(^3P) + N_2^*\left(^1\Sigma_g^+, v\right) \rightarrow NO(^2P) + N(^4S), \quad E_a \approx \Delta H \approx 3\,\text{eV mol}^{-1} \tag{18.2}$$

Formation of the highly vibrationally excited $N_2^*\left(^1\Sigma_g^+, v\right)$ molecules take place in VV exchange between the molecules at the lower vibrational states, excited by electron impact ($k_{eV} \approx 10^{-8}\,\text{cm}^3/\text{s}$ at $T_e = 1\,\text{eV}$). The key reaction

(18.2) is followed by a secondary exothermic process:

$$N + O_2 \rightarrow (k_2)\, NO + N, \qquad E_a \approx 3\,eV\,mol^{-1}, \quad \Delta H \approx -1\,eV\,mol^{-1} \tag{18.3}$$

which together with (18.2) creates a chain reaction of NO synthesis well known as the **Zeldovich mechanism**. Sum of the two reactions of chain propagation (18.2 and 18.3) gives the total NO synthesis (18.1). In contrast to the traditional Zeldovich mechanism, which is attributed to high-temperature oxidation of nitrogen during combustion and explosions, the chain reaction (18.2 and 18.3) occurs in nonequilibrium plasma at low temperatures. It is an example of a strongly endothermic chain reaction stimulated by nonequilibrium vibrational excitation.

Electronic excitation contribution to NO synthesis (18.1) can occur through the adiabatic channel of Zeldovich mechanism $O(^1D) + N_2^* \left(^1\Sigma_g^+, v \right) \rightarrow NO(^2P) + N(^2D)$ (see next Section's Figure 18.4) limited by fast relaxation of electronically excited atoms $O(^1D)$. NO synthesis through electronic excitation occurs with higher probability through dissociation of molecular nitrogen by direct electron impact. The rate coefficients here can reach $k_d = 10^{-11}\,cm^3/s$ at $T_e = 1\,eV$. After dissociation of N_2, NO synthesis occurs in in the exothermic reaction (18.3). Contribution of the electronic excitation becomes significant at high reduced electric fields (E/n_0) and electron temperatures and is less energy effective (below 3%) than the mechanism (18.2 and 18.3) related to vibrational excitation (see previous Section 18.1).

Contribution of charged particles to NO synthesis in limited to dissociated recombination $e + N_2^+ \rightarrow N + N$ (rate coefficient about $10^{-8}\,cm^3/s$ at electron temperature $T_e = 1\,eV$), ion–molecular reactions of positive atomic oxygen ions $O^+ + N_2 \rightarrow NO^+ + N$ (10^{-12}–$10^{-11}\,cm^3/s$), and molecular ions $O^+_2 + N_2 \rightarrow NO^+ + NO, N^+_2 + O_2 \rightarrow NO^+ + NO$ (low rate coefficients, about $10^{-16}\,cm^3/s$). Energy efficiency of these mechanisms is strongly limited below 3% by energy costs of the involved charged species usually much exceeding 30 eV.

NO synthesis in thermal air plasmas is somewhat like that in combustion systems and follows the quasi-equilibrium chain Zeldovich mechanism. The vibrational excitation of N_2 molecules is mostly responsible for the limiting endothermic reaction of NO synthesis ($O + N_2 \rightarrow NO + N$) in both quasi-equilibrium and strongly nonequilibrium systems. The difference is only in the selectivity of the discharge energy distribution over different reaction channels. Nonequilibrium discharges at $T_e = 1$–$3\,eV$ transfer most of their energy selectively into N_2 vibrational excitation (Figure 18.2), while thermal discharges distribute their energy uniformly over all degrees of freedom. The quasi-equilibrium composition of products in high-temperature plasma is shown in Figure 18.3 for atmospheric pressure and stoichiometric (molar 1 : 1) N_2–O_2 mixture. Maximum NO production requires high temperature of about 3300 K, but NO molar fraction is not high (below 5% at ideal quenching) even at these temperatures because of NO stability. The highest energy efficiency in thermal plasma doesn't exceed 5%, see Figure 18.1.

18.3 Elementary Zeldovich Reaction of NO Synthesis Stimulated by Vibrational Excitation of Nitrogen Molecules

The most energy-effective mechanism of NO synthesis in plasma (18.2 and 18.3) is related to the strongly nonequilibrium conditions and process stimulation through vibrational excitation of N_2 molecules by electron impact. Kinetics of this process follows the Zeldovich mechanism and is limited by the elementary endothermic reaction (18.2). This elementary reaction is limited and controlled not by VV relaxation and formation of molecules with sufficient energy (as in the case of CO_2 dissociation), but by the elementary process of the chemical reaction itself. The elementary process (18.2) is illustrated in Figure 18.4 by its reaction path profile including two pathways: through the electronically adiabatic channel aMbNc or through the electronically nonadiabatic channel aMbNc with transitions localized in points M and N. The adiabatic channels correspond to direct transfer of a nitrogen atom from N_2 to atomic oxygen, whereas the nonadiabatic channel proceeds through the formation of an intermediate long-lifetime vibrationally excited state $N_2O(^1\Sigma^+)$.

Although the probability of the nonadiabatic channel is limited by transition between different electronic states, it can be kinetically faster than the adiabatic one in nonequilibrium plasma conditions. The kinetic advantages of the nonadiabatic channel are due to the high efficiency (close to 100%) of vibrational energy in reactions proceeding through the formation of stable intermediate molecular complexes (Macheret et al. 1980a,b as well as Sections 3.12 and 3.13).

Probability of the electronically adiabatic channel aMbNc (Figure 18.4) $w_a(v) < 1$ of the Zeldovich reaction (18.2) at the translational gas temperature T_0, can be expressed depending on the vibrational quantum number v based on the model of energy distribution in products of reverse reaction (Macheret et al. 1980a,b):

$$w_a(v, T_0) = \exp\left[-\frac{1}{T_0}\left(E_a - \hbar\omega v \frac{\xi}{1-\xi}\right)\right] \tag{18.4}$$

here $\xi \approx 0.3$ is a fraction of energy released in the reverse reaction $N + NO \rightarrow N_2(v) + O$ to translational and rotational degrees of freedom. This probability is relatively small (10^{-3}–10^{-5} for $\hbar\omega v \geq E_a$ and $T_0 = 1000$ K) and exponentially decreases at lower translational temperatures.

The **electronically nonadiabatic channel** aMbNc of NO synthesis (Figure 18.4) proceeds through the formation and decay of an intermediate long-lifetime vibrationally excited complex:

$$O(^3P) + N_2^*\left(^1\Sigma_g^+, v\right) \leftrightarrow N_2O^*(^1\Sigma^+) \rightarrow NO(^2P) + N(^4S), \qquad E_a \approx \Delta H \approx 3 \text{ eV mol}^{-1} \tag{18.5}$$

This channel is kinetically more preferential than the electronically adiabatic one at low temperatures typical for nonequilibrium plasmas. It is mostly due to very high efficiency of vibrational energy (close to 100%) in overcoming energy barrier of $N_2O^*(^1\Sigma^+)$ dissociation exceeding that one for adiabatic direct exchange reaction (18.2). To illustrate the nonadiabatic interaction of atomic oxygen with an excited nitrogen molecule (O–NN), let us consider the cross section of the potential energy surface O–N–N corresponding to the angle $\gamma = 180°$ between N–N and N–O axis. The dependence of interaction energy between the O atom and N_2 molecule on the distance r between them can be presented by a set of intersecting **electronical-vibrational "vibronic terms"** (Nikitin and Umanskii 1984) illustrated in Figure 18.5a. The vibronic terms are formed by a vertical $\hbar\omega v$ displacement of electronic terms of linear N_2–O configuration. The potential energy difference related to intersection of the $v_1 + 1$ and v_1 singlet electronic terms with the v triplet term is equal to $\delta = 0.15$ eV, and it is independent of v and v_1 in the harmonic approximation of N_2 vibrations. The correlation diagram of electronic terms of N_2–O is shown in Figure 18.5b. Formation of the $N_2O^*(^1\Sigma^+)$ intermediate complex takes place because of nonadiabatic transitions between vibronic terms corresponding to adiabatic electronic terms $^3\Pi$ and $^1\Sigma^+$ on Figure 18.5b.

The **total probability of the nonadiabatic NO synthesis** is a product of probabilities related to formation and direct decay of an intermediate complex. It can be expressed at strongly nonequilibrium conditions ($T_v \gg T_0$) and translational temperatures below 1000 K as (Macheret et al. 1978):

$$w_n(v, T_0) = \frac{\pi V_{nn'}^2}{\hbar\Delta F}\sqrt{\frac{\mu}{T_0}}(\gamma^2 v)^{E_0/\delta}\left(\frac{\hbar\omega v - E_a}{\hbar\omega v - E_0}\right)^{s-1} \tag{18.6}$$

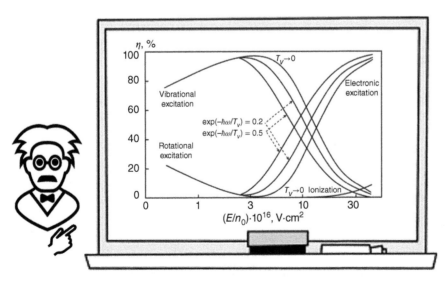

Figure 18.2 Distribution of electron energy in nonthermal discharges between different excitation and ionization channels in nitrogen as functions of reduced electric field at different vibrational temperatures.

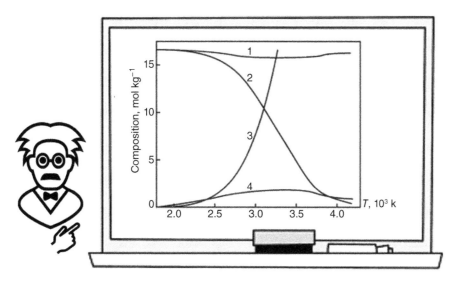

Figure 18.3 NO synthesis in atmospheric pressure thermal stoichiometric $N_2 + O_2$ plasma: composition of products: (1) N_2, (2) O_2, (3) O, (4) NO.

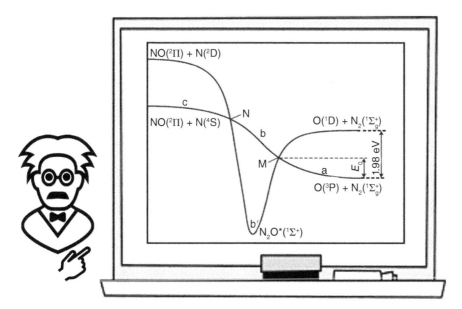

Figure 18.4 Reaction path profile (potential energy curve) for the elementary process of NO synthesis: $O + N_2 \rightarrow NO + N$, showing adiabatic and nonadiabatic reaction channels.

In this relation: $\frac{\pi V_{nn'}^2}{\hbar \Delta F}$ is the Landau–Zener factor reflecting transition between electronic terms, $V_{nn'} \approx 80 \text{ cm}^{-1}$ is the relevant matrix element of spin-orbital interaction between the electronic terms; ΔF is difference in slopes of the electronic terms at the point of their intersection, $\sqrt{\frac{\mu}{T_0}}$ characterizes the reverse relative O–N_2 velocity; $(\gamma^2 v)^{E_0/\delta}$ is the vibronic terms transition factor (see Figure 18.5a), γ is ratio of the N–N displacement during the nonadiabatic transition to the amplitude of N–N zero vibrations; $\left(\frac{\hbar \omega v - E_a}{\hbar \omega v - E_0}\right)^{s-1}$ is the statistical theory factor describing probability of the intermediate complex $N_2O^*(^1\Sigma^+)$ decay with the formation of NO, $s = 3$ is number of active vibrational degrees of freedom of the intermediate complex $N_2O^*(^1\Sigma^+)$.

The relation (18.6) shows that the nonadiabatic channel is very effectively stimulated by vibrational excitation [see the statistical theory factor $\left(\frac{\hbar \omega v - E_a}{\hbar \omega v - E_0}\right)^{s-1}$, as well as Sections 3.12 and 3.13]. The nonadiabatic channel probability is about 10^{-2}–10^{-4} and doesn't essentially depend on translational temperature, in contrast to that of adiabatic channel. When the vibrational energy slightly exceeds the activation energy, the nonadiabatic mechanism dominates at translational temperatures $T_0 < 1000$ K; the adiabatic channel dominates at higher temperatures.

18.4 Kinetics and Energy Balance of Plasma-chemical NO Synthesis in O_2–N_2 Mixtures Stimulated by Vibrational Excitation

The reaction $O + N_2 \rightarrow NO + N$ (18.2), limiting the Zeldovich mechanism of NO synthesis in air plasma, proceeds in nonequilibrium conditions mostly through the nonadiabatic channel (18.5), see Figure 18.4, with probability (18.6). To find out the **rate coefficient of the reaction under nonequilibrium conditions** $(T_v > T_0)$ as a function of vibrational (T_v) and translational (T_0) temperatures, the probability (18.6) should be averaged over the Treanor vibrational distribution function (Macheret et al. 1980a,b):

$$k_R(T_v, T_0) = k_0 \frac{\pi V_{nn'}^2}{\hbar \Delta F} \sqrt{\frac{\mu}{T_0}} \left(\frac{T_v}{E_a}\right)^{s-1} \left(\gamma^2 \frac{E_a}{\hbar \omega}\right)^{E_0/\delta} \exp\left(-\frac{E_a}{T_v} + \frac{x_e E_a^2}{T_0 \hbar \omega}\right) \tag{18.7}$$

In this relation: $k_0 \approx 10^{-10}$ cm^3/s is the rate coefficient of gas-kinetic collisions, $2x_e E_a T_v/\hbar \omega T_0$ is the Treanor factor, $\hbar \omega$ and x_e are vibrational quantum and anharmonicity of N_2 molecules. In contrast to CO_2 dissociation, the reaction $O + N_2 \rightarrow NO + N$ is classified in vibrational kinetics as a slow reaction and does not perturb the vibrational distribution function. It is due to the low density of atomic oxygen and low probability of the elementary chemical process with respect VV exchange between N_2 molecules. Numerically, at typical nonequilibrium plasma conditions $(T_v \approx 3000$ K, $T_0 \leq 1000$ K$)$, the pre-exponential factor in (18.7) is about $A_0 = 10^{-12}$–10^{-13} cm^3/s.

The N_2 **vibrational energy balance** for the nonequilibrium plasma-chemical NO synthesis stimulated by vibrational excitation can be qualitatively illustrated as:

$$d\varepsilon_v/dt = k_{eV} n_e \hbar \omega - k_R(T_v, T_0) [O] E_a - \left(k_{VT} n_0 + k'_{VT} [O]\right)(\varepsilon_v - \varepsilon_{v0}) \tag{18.8}$$

where ε_v is the average vibrational energy of N_2 related through the Planck formula to T_v; $\varepsilon_{v0} = \varepsilon_v(T_v = T_0)$; k_{VT} and k'_{VT} are the rate coefficients of VT relaxation of N_2 on molecules (mostly N_2 and O_2) and on atomic oxygen; $k_{eV} = 1$–$3 \cdot 10^{-8}$ cm^3/s is the rate coefficient of N_2 vibrational excitation by electron impact. Although the density of O atoms is not very high, its nonadiabatic relaxation is fast $(k_{VT} \approx 3 \cdot 10^{-13}$ cm^3/s$)$ and makes significant contribution. The density of atomic oxygen is typically proportional to the ionization degree; specifically $[O]_0 \approx 10^{16}$ cm^{-3} at electron temperature $T_e \approx 1$ eV and ionization degree $n_e/n_0 = 10^{-5}$.

Like all processes stimulated in plasma by vibrational excitation, NO synthesis also has the *critical vibrational temperature*, see Section 17.3. Based on energy the balance for NO synthesis (18.8), typically $T_v^{\min} = 0.2$–0.25 eV.

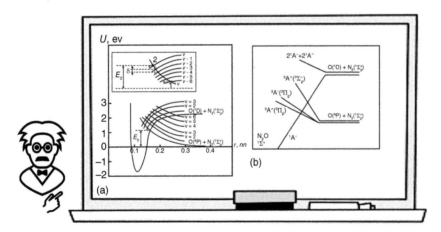

Figure 18.5 Atomic oxygen interaction with N_2: (a) vibronic terms for linear configuration of N_2–O interaction. Arrows (1,2) point out the most probable transitions leading to the formation of the N_2O complex: (1) lower temperatures, (2) higher temperatures; (b) correlation of electronic terms.

At lower vibrational temperatures, VT relaxation is faster than chemical reaction, and synthesis is ineffective. The critical vibrational temperature determines through the Planck formula the threshold $\varepsilon_v(T_v)$ of the dependence of energy efficiency, which is typically 0.5–0.8 J/cm^3 (0.1–0.15 eV mol^{-1}), see Figure 18.1. Also, plasma NO synthesis, like all other processes stimulated by vibrational excitation, requires ionization degree to be sufficient for eV processes to exceed VT relaxation. This *critical ionization degree for NO synthesis* is about 10^{-8}, which is less serious than that for CO_2 and H_2O dissociation due to low N_2 relaxation rate.

Total **energy efficiency (η) of the nonequilibrium plasma synthesis of NO** stimulated by vibrational excitation, can be subdivided into three factors: the excitation factor (η_{ex}), the relaxation factor (η_{rel}), and the chemical factor (η_{chem}), see Section 17.3: $\eta = \eta_{ex}* \eta_{rel}* \eta_{chem}$. The excitation factor (η_{ex}) is high in this case: $\eta_{ex} = 0.8$–0.9 at $T_e \approx 1$ eV, see Figure 18.2. When the ionization degree and specific energy input exceed the above-mentioned thresholds (10^{-8}, 0.1–0.15 eV mol^{-1}), the relaxation factor (η_{rel}) is also very high and close to 80–90%. *The chemical factor (η_{chem}) of energy efficiency is unfortunately low due to losses in the exothermic reactions and especially* (18.3). The exothermic reaction (6–3) of the Zeldovich chain $N + O_2 \rightarrow NO + N$ leads alone to losses of about 1 eV per single NO molecule formed, and to a 33% decrease in energy efficiency of NO synthesis in nonthermal plasma. It is one of the most important reasons why the total energy efficiency in this process is limited to 30–35% (Figure 18.1), which is significantly lower than in the case of CO_2 dissociation.

18.5 Stability of Products of Plasma-chemical NO Synthesis to Reverse Reactions, Effect of "Hot" Nitrogen Atoms and Surface Stabilization

Protection of products from active species in the active plasma zone seriously restricts the yield of the nonthermal plasma-chemical NO synthesis, which also crucially affects the total energy efficiency of the process, see Figure 18.1, and makes it less competitive with the Haber-Bosch Process (total energy efficiency about 25%). *The most important fast barrierless reverse reaction, leading to destruction of NO inside the active discharge zone, is $N + NO \rightarrow N_2 + O$, $k_N = 10^{-11}$ cm^3/s.* This reverse reaction competes with the propagation (18.3) of the Zeldovich chain $k_2 \approx 10^{-10} \exp(-4000/T_0, \text{K})$ cm^3/s, and can terminate the chain when NO density increases. To suppress the reverse reaction, atomic nitrogen should preferentially react not with NO but with O_2, propagating the Zeldovich chain. Unfortunately, the reverse reaction is fast and has no activation barrier whereas the direct Zeldovich reaction (18.3) has an activation barrier of about 0.3 eV. This kinetic competition restricts the limit NO yield at temperatures $T_0 < 1000$ K to about 3%. Chain propagation reaction (18.3) is exothermic and characterized by a low efficiency of vibrational energy in overcoming activation barriers ($\alpha \approx 0.2$). Therefore, this reaction cannot be effectively accelerated by vibrational excitation of oxygen until its vibrational temperature exceeds 5000 K. Even at these high vibrational O_2 temperatures, the chemical factor of energy efficiency for NO synthesis alone decreases to less than 40%. Higher NO yields can be reached without loss of energy efficiency under conditions of the effect of "hot atoms," or by using special NO surface stabilization approaches.

Generation of the **"hot atoms"** has been discussed in Section 5.13. When energetic atoms generated in exothermic reactions can participate in the following reactions before Maxwellization, they can significantly perturb the translational distribution function of the atoms. The "hot atoms" can then accelerate the exothermic chemical reactions because it is the translational (not vibrational) energy of reagents that stimulates overcoming the activation barriers of these reactions. This effect can significantly accelerate the exothermic Zeldovich reaction (18.3) and mitigate the restriction of NO yield. Detailed analysis of the N-atoms translational energy distribution function shows that the effect of "hot atoms" on NO yield and energy efficiency is stronger at higher oxygen fractions in the N_2–O_2 mixtures, see Figure 18.6.

Obviously, NO yield and therefore energy efficiency can be significantly improved by **NO stabilization at different surfaces**. Keeping in mind that typical NO lifetime is about 20 µs, such stabilization requires either low pressures (like in the ECR microwave and electron beam discharge plasmas, see Section 18.1) or short distances like in micro-plasma bubbles considered in Lecture 16. Plasma-chemical NO synthesis in the atmospheric pressure air microbubbles can be quite effective when the bubbles size is below 2 µm providing NO-related species stabilization time in water or water solutions below 20 µs. The NO energy cost in such systems is reduced to about 10 eV mol^{-1} and can be further improved by controlling the bubble size distribution. It should be mentioned that NO dissolving

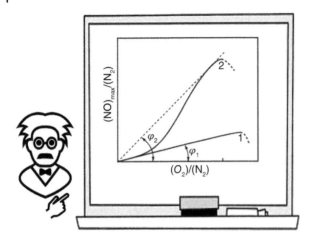

Figure 18.6 Maximum relative nitrogen oxide yield $[NO]_{max}/[N_2]$ as function of initial mixture composition ($[O_2]/[N_2]$ ratio) at relatively low translational temperatures ($T_0 < 1500\,K$): (1) without considering the effect of "hot atoms," $\tan \phi_1 = k_2/k_N \ll 1$; (2) taking into account the effect of "hot atoms," $\tan \phi_2 = k_2/k_N \approx 1$.

in water is very limited, which requires for its stabilization either preferential formation of HNO (fast solving in water) or using special solvents.

18.6 Plasma-chemical Synthesis of Ozone: Fundamental and Applied Aspects of the Process

Ozone generation from oxygen or air is one of the oldest plasma-chemical processes, with more than 150-year history of fundamental research and large-scale industrial application. Synthesis of ozone (O_3) is an endothermic process, which can be summarized as:

$$\frac{3}{2}O_2 \rightarrow O_3, \qquad \Delta H = 1.5\,eV\,mol^{-1} \tag{18.9}$$

The reverse process is obviously exothermic and can be explosive. Diluted ozone mixtures are relatively stable at low temperatures. However, even a relatively small heating leads to ozone decomposition, which can be explosive if the O_3 concentration is sufficiently high. Therefore, effective O_3 synthesis requires low temperatures, typically room temperature. Organization of the highly endothermic process (18.9) at room temperature requires application of strongly nonequilibrium methods of chemical synthesis. Nonequilibrium plasma of a dielectric barrier, pulsed corona, and other nonthermal atmospheric-pressure discharges are some of the most efficient methods for practical ozone production. The history of ozone generation and application has always been closely related to the history of plasma chemistry. It was Martinus van Marum who first observed in 1785 the formation of ozone in electric sparks generated by his electrostatic machine. Ozone was first identified as a new chemical compound in 1839 by Schonbein (1840). Soon after that, in 1857, Werner von Siemens made a breakthrough step proposing a reliable method of ozone generation by passing air or oxygen through an AC discharge bounded by at least one dielectric barrier (invention of DBD). The application of DBD in air or oxygen at pressures of 1–3 atm first proposed by Werner von Siemens remains today the major approach to large-scale ozone generation. The practical application of ozone is related to its very high ability to oxidize organic compounds (especially in the liquid phase) and to its ability to react with double bonds. Applications of ozone include treatment of drinking water, water in pools; cleaning of wastewater; bleaching processes; and oxidation steps in numerous chemical technologies. The largest ozone production is related to water cleaning. The first large-scale plasma installations in drinking water plants were built in Paris (1897), Nice (1904), and St. Petersburg (1910).

While industrial application of ozone is widespread, it is still limited by the energy cost of its generation. The thermodynamic minimum of energy cost for O_3 production (18.9) is 0.8 kWh/kg. However, conventional generators require about 15–20 kWh/kg for ozone synthesis from oxygen (energy efficiency 4–6%) and about 30–40 kWh/kg for ozone synthesis from air (energy efficiency 2–3%). Such low energy efficiencies are due to the high energy cost of O_2

Table 18.1 Types of industrial ozone generators, parameters, and electric energy costs.

Type & Configuration of Ozone Generator	Feed gas, Dew Point	Discharge Gap, Voltage, Frequency	Dielectric Barrier (mm)	Energy Cost (kWh/kg)
Otto (flat, glass DBD)	Air, −40 °C	3.1 mm, 7.5–20 kV, 50–500 Hz	3–5	10.2
Tubular reactor, DBD	Air, −60 °C	2.5 mm, 15–19 kV, 60 Hz	2.5	7.5–10
Tubular reactor, DBD	O_2, −60 °C	2.5 mm, 15–19 kV, 60 Hz	2.5	3.75–5
Lowther (flat, ceramic DBD)	Air, −60 °C	1.25 mm, 8–10 kV, 2 kHz	0.5	6.3–8.8
Lowther (flat, ceramic DBD)	O_2, −60 °C	1.25 mm, 8–10 kV, 2 kHz	0.5	2.5–5.5

dissociation, as well as to losses related to chemical admixtures and temperature effects. Some O_3 energy cost data achieved in different industrial ozone generators are presented in Table 18.1.

Especially low energy costs of ozone production were achieved in experiments with O_3 quenching by cooling to temperatures sufficient for its condensation. Thus, the energy cost in experiments with O_2 plasma was decreased to 3.3 kWh/kg (energy efficiency 25%) by cooling the plasma-chemical system to cryogenic temperature (−183 °C). Low energy cost of O_3 synthesis, 2.5 kWh/kg (energy efficiency about 30%), was claimed in experiments with nonuniform DBD in O_2, where the reactor walls were cooled for O_3 stabilization by liquid nitrogen. The economics of cryogenic O_3 stabilization is questionable, although deep cooling of the feed gas is also necessary to reach the low humidity required for effective synthesis in large-scale industrial systems, which is going to be discussed below. See more details on the subject in Fridman (2008).

18.7 Plasma-chemical Ozone Generation in Oxygen

Ozone generation in oxygen plasma requires about twice less electric energy than in air plasma (but consumes O_2, which is not free). When the oxygen plasma is produced in DBD, the strength of microdischarges influences kinetics and efficiency of ozone generation, see Kogelschatz (2003). The microdischarges in atmospheric-pressure oxygen last only for a few nanoseconds and transport electric charge less than 1 nC. The microdischarge current density is about 100 A/cm^2, local electron density reaches 10^{14} cm^{-3}, specific energy input is about 10 mJ/cm^3, and gas heating due to an individual microdischarge is negligible. The plasma-chemical kinetics of ozone formation in pure oxygen is determined by O_2 dissociation. The O_2 dissociation and ozone formation in strongly nonequilibrium discharges are mostly due to intermediate excitation of the excited state $A^3\Sigma_u^+$ with energy threshold of about 6 eV:

$$e + O_2 \rightarrow O_2^* \left(A^3\Sigma_u^+ \right) + e \rightarrow O(^3P) + O(^3P) + e \tag{18.10}$$

and intermediate excitation of the excited state $B^3\Sigma_u^-$ with energy threshold of about 8.4 eV:

$$e + O_2 \rightarrow O_2^* \left(B^3\Sigma_u^- \right) + e \rightarrow O(^3P) + O(^1D) + e \tag{18.11}$$

The O atoms formed in these reactions recombine with O_2 in the three-body process:

$$O + O_2 + O_2 \rightarrow O_3 + O_2, \quad k(\text{cm}^6/\text{s}) = 6.9 \cdot 10^{-34} (300/T_0, \text{K})^{1.25}. \tag{18.12}$$

Considering the rate coefficient (18.12), even though electrons disappear from the DBD microdischarges in less than 10 ns, the ozone formation takes longer, about a few microseconds.

The efficiency of O_3 generation in DBD depends on the strength of microdischarges. A strong one provides higher specific energy input and hence a relatively high density of atomic oxygen and ozone. It leads, however, to undesired side reactions of three-body recombination:

$$O + O + O_2 \rightarrow O_2 + O_2, \quad k = 2.45 \cdot 10^{-31} (T_0, \text{K})^{-0.63}, \text{cm}^3/\text{s} \tag{18.13}$$

as well as to the reverse reactions of destruction of ozone by atomic oxygen:

$$O + O_3 \rightarrow O_2 + O_2, \quad k = 2 \cdot 10^{-11} \exp(-2300 \text{ K}/T_0), \text{cm}^3/\text{s} \tag{18.14}$$

The relevant effectiveness of ozone synthesis decreases when DBD microdischarges are strong and relative density of atomic oxygen $[O]/[O_2]$ exceeds 0.3–0.5%. Lower limit of the microdischarge strength is due to losses related to ions, which dissipate energy even at low specific energy inputs. An optimal relative concentration of O atoms is about $[O]/[O_2] = 0.2\%$. Kinetic modeling predicts minimum energy cost of ozone production in oxygen-DBD at the level of 2.5 kWh/kg corresponding to energy efficiency about 30% (Kogelschatz 2003).

Why is the energy efficiency of O_3 synthesis so low (20–30%) even in optimal regimes and even in pure oxygen? Vibrational excitation initially collects most of the discharge energy in O_2 like in many other molecular gases at the typical T_e in atmospheric pressure cold discharges. In contrast to N_2 molecules considered above in this lecture, O_2 molecules lose their vibrational energy very fast in the so-called symmetrical exchange reactions:

$$O_2^*(v) + O \rightarrow O_3^*(v) \rightarrow O + O_2, \quad k \approx 10^{-11} cm^3/s \, (T_0 = 300 \text{ K}) \tag{18.15}$$

Thus, losses of vibrational energy and therefore energy efficiency of the ozone synthesis are mostly due to atomic oxygen, which is inevitable during the O_3 generation.

18.8 Plasma-chemical Ozone Generation in Air

Nonthermal discharges in air distribute energy between N_2 and O_2, which makes them less electric energy effective in O_3 synthesis than pure oxygen discharges. Discharge energy initially localized in N_2 is not completely lost, however, and can be partially used for O_3 synthesis. Therefore, ozone production from air requires only about twice more energy as needed in oxygen plasma. Considering the price of pure oxygen, smaller ozone generators are easier to make based on discharges in dry air. The mechanism of ozone synthesis in O_2 (18.10–18.12) still is still effective in air. In addition to that, significant contribution to O_3 generation in air is due to atomic and electronically excited nitrogen. Atomic nitrogen leads to ozone synthesis through intermediate formation of oxygen atoms in exothermic reactions:

$$N + O_2 \rightarrow NO + O, \quad E_a \approx 0.3 \text{ eV mol}^{-1}, \quad \Delta H \approx -1 \text{ eV mol}^{-1} \tag{18.16}$$

$$N + NO \rightarrow N_2 + O, \quad k = 10^{-11} cm^3/s, \quad \Delta H \approx -3 \text{ eV mol}^{-1} \tag{18.17}$$

These processes are limited by the strongly endothermic formation of N atoms. The energy cost of N atoms in DBD streamers in air is presented in Figure 18.7a in the form of **G-factors**, which characterize number of N atoms formed per 100 eV of streamer energy (Naidis 1997). For comparison, G-factors for the formation of O-atoms in the same conditions are shown in Figure 18.7b. The contribution of N atoms (18.16 and 18.17) can add about 10% to the total production of atomic oxygen and, therefore, ozone. More significant contribution to production of O atoms and ozone is related to reactions of electronically excited nitrogen molecules:

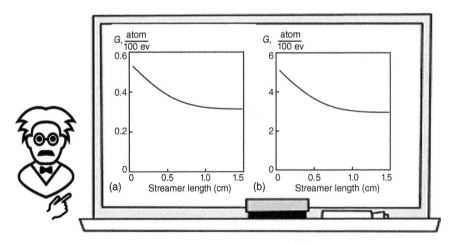

Figure 18.7 G-Factors of atoms generation in atmospheric air DBD microdischarges as a function of the streamer length: (a) G-factors for N-atoms, (b) G-factors for O-atoms.

$$N_2\left(A^3\Sigma_u^+\right) + O_2 \rightarrow N_2O + O \tag{18.18}$$

$$N_2\left(A^3\Sigma_u^+, B^3\Pi_g\right) + O_2 \rightarrow N_2 + O + O \tag{18.19}$$

In total, about 50% of O_3 generated in the nonthermal atmospheric-pressure air discharges can be attributed to atomic and electronically excited nitrogen. The characteristic time of ozone formation in atmospheric-pressure oxygen plasma is about $10\,\mu s$. The nitrogen-related reactions are longer, and ozone formation in air plasma takes about $100\,\mu s$.

An increase of specific energy input E_v in nonthermal air discharges leads to higher densities of NO and NO_2. The high NO_x densities stop the ozone generation, which is known as the **discharge poisoning effect** and related to fast reactions of O atoms with NO_x:

$$O + NO + M \rightarrow NO_2 + M, \quad k = 6{\cdot}10^{-32}\ cm^6/s \tag{18.20}$$

$$O + NO_2 \rightarrow NO + O_2, \quad k = 10^{-11}\ (T_0/1000\ K)^{0.18}\ cm^3/s. \tag{18.21}$$

When NO_x concentration exceeds the threshold of about 0.1%, the atomic oxygen reactions with NO_x (18.20 and 18.21) become faster than those with O_2 to form ozone, which explains the discharge poisoning effect. Even ozone molecules already formed under conditions of the discharge poisoning are converted back to O_2 in the fast reactions with nitrogen oxides:

$$O_3 + NO \rightarrow NO_2 + O_2, \quad k = 4.3{\cdot}10^{-12}\ \exp(-1560\ K/T_0)\ cm^3/s \tag{18.22}$$

Nitrogen oxides are not generally consumed in the reactions of discharge poisoning and can be interpreted as the catalytic recombination of oxygen atoms and catalytic destruction of ozone in the presence of NO_x. The maximum ozone concentration is typically about 0.3% and starts decreasing when the specific energy input exceeds $0.1\ eV\ mol^{-1}$ and the concentration of nitrogen oxides (NO_x) exceeds 0.03% (U. Kogelschatz, 2003). Similar effect occurs in the stratosphere, where nitrogen oxides at a level of 0.1% significantly decrease the ozone concentration.

 In general, *ozone and nitrogen oxides don't coexist well in plasma.* Nonthermal atmospheric-pressure discharges produce nitrogen oxides at elevated specific energy inputs and temperatures without any ozone formation. Conversely, at low specific energy inputs and temperatures close to room temperature, ozone is effectively produced but not NO_x, and the discharges can even be used in air purification. *Considering that ozone and nitrogen oxides have different smells, one can actually "smell" the temperature of the atmospheric-pressure discharges in air based on this effect.*

18.9 Stability of Plasma-generated Ozone: Negative Effects of Temperature, Water Vapor, Hydrogen, Hydrocarbons, and Other Admixtures

Elevated temperatures significantly reduce ozone production in plasma. On the one hand, it is due to the discharge poisoning: higher temperatures are related to higher specific energy inputs and higher concentrations of nitrogen oxides, which suppresses ozone synthesis. On the other hand, ozone itself even diluted in air is unstable at elevated temperatures. The rate coefficient of O_3 dissociation $(O_3 + M \Leftrightarrow O_2 + O + M)$ is

$$k_{O_3} = 1.3{\cdot}10^{-6}\ \exp\left(-11\,750\ K/T_0\right)\ cm^3/s \tag{18.23}$$

The thermal stability of ozone depends exponentially on temperature T_0. Ozone lifetime with respect to dissociation to O in pure dry air without catalytic admixtures at room temperature is quite long, about an hour. However, already at 100 °C, the ozone dissociation time is only 1 s.

The O_3 decomposition time to molecular oxygen (including O_3 dissociation and following O-atoms reaction with O_3 to form O_2) depends not only on temperature but also on the initial concentration of ozone. Diluted ozone is stable at low temperatures because O-atoms generated in O_3 dissociation stick back to O_2. For ozone relative concentration of 0.2%, the decomposition (to O_2) time is about 1 s at temperature of 145 °C. When the ozone relative concentration is 1%, the decomposition time is about 1 s at 125 °C. If temperature exceeds 125 °C and the O_3 density is above 1% the decomposition can be very fast and leads to thermal explosion.

Humidity (H_2O) significantly suppresses ozone synthesis in nonthermal atmospheric-pressure plasma, especially in DBDs. Therefore, ozone generators operate in dry air or dry oxygen (dew point $-60\,°C$ and below, which corresponds to water content of few ppm or less). The negative effect of water vapor is due to two major factors. The first one, specific to DBD, is related to increases of the surface conductivity of dielectrics at high humidity. It leads to stronger microdischarges with excessive density of atomic oxygen, higher temperature, and generation of NO_x poisoning the discharge. Another negative kinetic H_2O effect is due to the formation of OH and HO_2 radicals leading to catalytic ozone destruction in the chain mechanism:

$$OH + O_3 \rightarrow HO_2 + O_2, k = 5 \cdot 10^{-13}\ cm^3/s \tag{18.24}$$

$$HO_2 + O_3 \rightarrow OH + O_2 + O_2 \tag{18.25}$$

Typical dependence of the ozone yield (kgO_3/kWh) in oxygen DBD on humidity (dew point) is shown in Figure 18.8a. The curve is still far from saturation at dew point $-20\,°C$.

Small additions of molecular hydrogen have a negative effect on ozone synthesis also due to the formation in plasma of OH and HO_2 radicals and the chain mechanism (18.24 and 18.25) The relevant dependence of ozone yield (kgO_3/kWh) in DBD oxygen plasma on the volume fraction of hydrogen is illustrated in Figure 18.8b. Significant decrease in the ozone production efficiency occurs when the volume fraction of hydrogen exceeds 0.03% (300 ppm).

The *negative effect of hydrocarbons* on ozone synthesis (see Figure 18.8c) is also partially due to the formation of OH and HO_2 and the chain mechanism (18.24 and 18.25). The major effect of hydrocarbons, however, is related to trapping of oxygen atoms by the hydrocarbon molecules: $O + RH + M \rightarrow products + M$, which competes with O_3 production from atomic oxygen. The influence of different hydrocarbons of similar size on the yield of ozone synthesis is almost identical. The hydrocarbon volume fraction 1% suppresses ozone production almost completely.

Small *admixtures of halogens, like bromine and chlorine, also have negative effect* on production and stability of O_3 sometimes leading to ozone explosion. The halogens stimulate catalytic O_3 decomposition via the chain mechanism, which in the case of chlorine starts with formation of Cl atoms and ClO radicals ($Cl_2 + O_3 \rightarrow ClO + Cl + O_2$, $Cl_2 + O_3 \rightarrow ClO + ClO_2$), and then the chain propagation of the ozone destruction process (effect of Br_2 is similar):

$$OCl + O_3 \rightarrow Cl + O_2 + O_2, \quad Cl + O_3 \rightarrow ClO + O_2 \tag{18.26}$$

In contrast to the admixtures with a negative effect on ozone synthesis (water vapor, nitrogen oxides, hydrogen, hydrocarbons, halogens), some additives (N_2, CO_2) can stimulate O_3 generation in plasma, see Figure 18.9. The positive effect of these admixtures is due to related additional channels of oxygen dissociation (see Section 18.8); large additions of nitrogen and carbon oxide obviously suppress efficiency of O_3 synthesis. Thus, industrial ozone generators operating in oxygen, often have an additional 1% of nitrogen added to the feed gas.

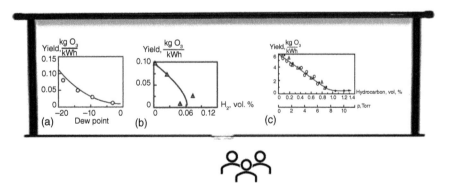

Figure 18.8 Effects suppressing plasma synthesis of ozone in room temperature oxygen-DBD: (a) effect of humidity (water vapor); (b) effect of molecular hydrogen; (c) influence of admixtures of different hydrocarbons on ozone yield (black circles correspond to cyclohexane, white circles correspond to heptane, and triangles correspond to hexane).

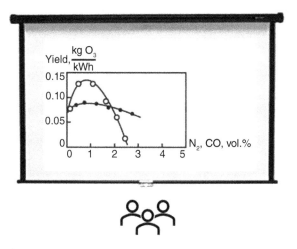

Figure 18.9 Influence of admixtures of N_2 and CO_2 on ozone yield in a DBD in room-temperature oxygen: black circles correspond to molecular nitrogen; white circles correspond to carbon monoxide.

18.10 Major Specific Configurations of Plasma Ozone Generators

Although O_3 synthesis can also be performed photo- and electro-chemically, most of the commercial ozone generation is based on plasma. Ozone generation requires low temperatures for product stability (surely below 100 °C) and elevated pressures for three-body attachment of O atoms to O_2, which makes nonthermal atmospheric-pressure discharges, and specifically DBDs and pulsed coronas the most interesting for industrial ozone generators. Werner von Siemens invented the DBD in 1857, which immediately became the champion for ozone synthesis. The DBD systems were installed in the first large-scale ozone generators for water purification plants in Paris (1897), Nice (1904), and St. Petersburg (1910). *Nowadays, DBD is still a champion in large-scale ozone production.* The energy efficiency of O_3 synthesis in DBD does not depend on applied frequency in the range between 10 and 50 kHz. The frequency increase leads to a larger number of DBD microdischarges, to an increase of total power, and total ozone production, keeping the O_3 energy cost at the same level. The use of higher frequencies, therefore, permits the same total discharge power to be sustained at lower operating voltages reducing the strain on dielectrics.

Most industrial DBD ozone generators are tubular, see Figure 18.10 and look somewhat like heat exchangers. They consist of 200–300 or more cylindrical discharge tubes with diameter 2–5 cm and length 1–3 m. Inner dielectric (usually borosilicate or other glass) tubes, closed at one side in most ozone generators, are inserted into outer metallic (steel) tubes. The two tubes form an annual discharge gap of 0.1–1 mm radial width. Some modern high-performance ozone generators use non-glass dielectric coatings on the steel tubes instead of glass dielectric tubes, which makes the tubes less fragile. Modern large-scale industrial ozone generators operate in a frequency range between 0.5 and 5 kHz, voltages about 5 kV (to reduce the failure of dielectric barriers), and power densities 0.1–1 W/cm². Large tubular DBD installations have not only high-power densities but also large total electrode surface areas allowing power level of several megawatts to produce about 100 kg of ozone per hour. Such systems produce ozone with concentrations up to 5 mass % using air as a feed gas, and up to 18 mass % using oxygen. *Flat or planar configurations of DBD-based ozone generators were also applied*, see Table 18.1. The dielectric barrier is arranged here as ceramic layers. Using ceramics allows thinner dielectric barriers and provides higher dielectric permittivity. *Flat surface DBD discharges* are the simplest but very small ozone generators. The discharge is maintained here between thin parallel electrode strips deposited on ceramic surface.

18.11 Ozone Generation in Pulsed Corona Discharges, Energy Efficiency of the Process

The pulsed corona discharges, and especially positive streamer coronas are alternative to the ozone generation in DBD. As an example, high-voltage pulses 25–40 kV with duration of about 100 ns can be applied in this case with

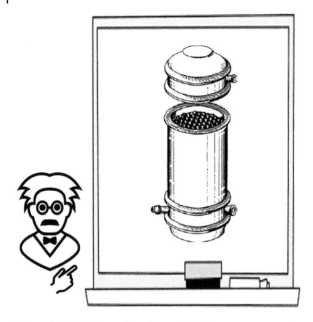

Figure 18.10   Illustration of a small tubular ozone generator.

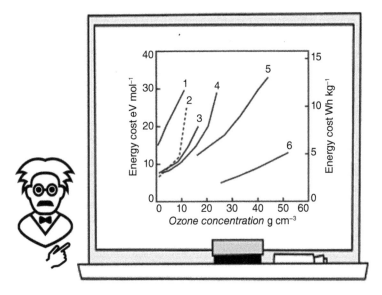

Figure 18.11   Energy cost of ozone production in pulsed corona discharge as function of concentration of produced ozone in different conditions (positive/negative corona, air/oxygen) and different discharge power (Watt per liter): (1) 40 W/L, air, negative corona; (2) 8 W/L, air, positive corona; (3) 20 W/L, air, positive corona; (4) 45 W/L, air, positive corona; (5) 70 W/L, air, positive corona; (6) 50 W/L, oxygen, positive corona.

repetition frequency 50 Hz to 2 kHz. While the short-pulsed discharges require more sophisticated power supplies, they have some important advantages in O_3 synthesis, in particular – the lowest achieved energy cost of the process, see Figure 18.11 (Knizhnik et al. 1999). The energy cost of ozone synthesis in the pulsed corona is shown in Figure 18.11 as a function of the produced ozone concentration, which is related to specific energy input. The negative pulsed corona is less effective than a positive one; the efficiency of ozone generation from oxygen is about twice higher than that from air; increase of specific power and, hence, gas temperature at the fixed level of produced O_3 concentration leads to decrease of ozone energy cost, which at its minimum is about 2.5 kWh/kg at room temperature.

Main features of the pulsed corona as ozone generator are quite different from those of DBDs. The most important is the discharge zone sizes. DBDs, operating in a quasi-uniform electric field, usually require very narrow discharge gaps. The pulsed corona operates in nonuniform electric field, which results in sustaining the discharge in much larger volumes and therefore at lower specific power of the discharge, lower temperature, and ozone concentration. The large discharge volume also leads to effective convective gas mixing in the discharge, which intensifies heat transfer about 10 times. Together with lower specific powers and related lower gas temperature, the intensive heat transfer results in no overheating of the corona and high stability of the synthesized ozone. The O$_3$ energy cost in a negative corona is 2–3 times higher than in a positive corona, see Figure 18.11. It is mostly due to the higher electron temperatures in the pulsed coronas with positive polarity, where the discharge is completely based on the formation of streamers. Higher electron temperatures increase the fraction of the discharge energy going into electronic excitation of molecular oxygen leading to ozone synthesis.

The high energy efficiency of ozone synthesis achieved in pulsed corona discharges in air and in oxygen can be explained by *contribution of vibrationally excited molecules to the ozone synthesis*. The *G*-factors (number of O$_3$ molecules synthesized per 100 eV) are presented in Figure 18.12a as a function of reduced electric field (1 Td = 10^{-17} V cm^2). The maximum *G*-factors for air and oxygen are, respectively, $G \approx 11$ mol/100 eV and $G \approx 22$ mol/100 eV, which correspond to minimum energy costs of 5 kWh/kg for air and 2.5 kWh/kg for oxygen. They agree with the pulsed corona results presented in Figure 18.11. Streamers are relatively short in DBD; therefore, the major contribution to O$_3$ synthesis is due to the streamer heads with relatively high electric fields. Pulsed corona discharges, which operate with much larger gaps, consist of longer streamers, where the contribution of the streamer channels with lower electric fields to O$_3$ synthesis can be much more significant. Relatively low electric fields typical for the streamer channels are actually optimal for vibrational excitation of both nitrogen and oxygen molecules. For this reason, vibrational excitation plays an important role in ozone synthesis, specifically in pulsed corona discharges.

The contribution of vibrational excitation of molecular nitrogen and oxygen into ozone synthesis in streamer channels is mostly due to the following kinetic effects:

1) *Effect of saturation of vibrational excitation.* The reduced electric field in a streamer channel (about 100 Td and less) corresponds to a regime when most of the discharge energy is transferred from electrons to vibrational excitation of N$_2$. Ozone synthesis is mostly provided by electronic excitation of molecular nitrogen and oxygen; therefore, vibrational excitation of the molecules is often considered to be an energy loss. Furthermore, vibrational excitation absorbs electron energy at $E/n_0 \leq 100$ Td so intensively that it results in a significant "drop" of the electron energy distribution function (EEDF) near the threshold of N$_2$ vibrational excitation. This kinetic effect exponentially slows down the electronic excitation of both nitrogen and oxygen molecules and, therefore, significantly decreases the energy efficiency of ozone production at the low reduced electric fields E/n_0 (see Figure 18.12a). However, the situation changes at higher specific energy inputs and higher levels of vibrational excitation. When most of the molecules are already vibrationally excited, they start to transfer energy back to electrons in the so-called superelastic collisions, which mitigates the EEDF "drop," exponentially accelerates electronic excitation, and finally decreases the energy cost of O$_3$ production; it is the effect of saturation of vibrational excitation on the electronic excitation rate and energy cost of O$_3$ synthesis, see Figure 18.12b. Thus, the *G*-factor of ozone production at $E/n_0 = 100$ Td increases by about 30% due to the saturation of vibrational excitation at the specific energy input to molecular vibration 0.75 eV mol^{-1}.

2) *Evolution of electronic excitation cross section with vibrational excitation.* The vibrational excitation of molecules also accelerates the electronic excitation due to changes in the elementary process cross sections. First, the process is related to an increase in the maximum of the cross sections, but more importantly, it is due to a shift of the threshold of electronic excitation for the vibrationally excited molecules to lower electron energies. The *G*-factor of ozone production from air at $E/n_0 = 100$ Td increases about 20% due to the cross-sectional effect at the specific energy input to molecular vibration, 0.75 eV mol^{-1}. The total effect of vibrational excitation including the saturation and cross-sectional effects is also presented in Figure 18.12b; the *G*-factor at $E/n_0 = 100$ Td and energy input 0.75 eV mol^{-1} increases total by about 50% due to vibrational excitation.

18.12 Plasma Synthesis of KrF$_2$ and Other Fluorine-based Compounds of Noble Gases

The synthesis of noble gas compounds is challenging because of their low stability and (even more importantly) low stability of the intermediate compounds for their synthesis. Nonthermal plasma is effective in such synthetic processes because high electron temperature intensifies the synthesis, and low gas and wall temperatures stabilize the products. Let us consider as an example the low endothermic process of **synthesis of krypton fluoride** (KrF$_2$):

$$Kr + F_2 \rightarrow KrF_2, \quad \Delta H = 0.6 \text{ eV mol}^{-1} \tag{18.27}$$

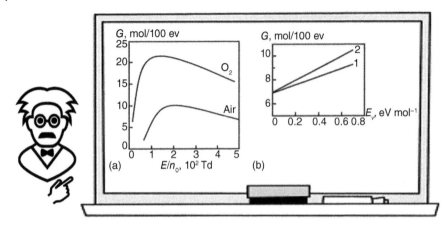

Figure 18.12 G-factors of ozone production: (a) as a function of reduced electric field E/n_0 in room-temperature atmospheric-pressure discharges in air and oxygen; (b) effect of vibrational excitation les in atmospheric pressure pulsed corona in air as function of specific energy input: (1) considering the effect of saturation of vibrational excitation, (2) total contribution of vibrational excitation including the cross-sectional effect.

Challenge of this synthesis is mostly due to low bonding energy (about 0.1 eV) of an intermediate compound, KrF. Synthesis of KrF_2 becomes possible in the nonequilibrium plasma conditions by intermediate formation of a stable electronically excited molecular compound, KrF^* (Legasov et al. 1978b, see also Fridman 2008):

$$Kr^* + F_2 \rightarrow KrF^* + F \tag{18.28}$$

The electronically excited KrF^* can then be converted into KrF_2 in reactions with F_2:

$$KrF^* + F_2 \rightarrow KrF_2 + F \tag{18.29}$$

Higher efficiency of KrF_2 production can be achieved if the intermediate product of the synthesis (KrF) is stabilized in a crystal structure of krypton (Legasov et al. 1978b). Mechanism of the synthesis includes four major steps: (i) reactor walls are cooled down to liquid nitrogen temperature providing condensation of Kr; (ii) F atoms are plasma generated from F_2 and transported across the gas phase to the surface with the condensed krypton matrix; (iii) interaction of F atoms with the Kr film leads to the formation of the intermediate compound KrF stabilized in the Kr crystal structure; (iv) reaction between KrF groups stabilized in the Kr matrix leading to the formation of KrF_2 molecules stable at normal conditions.

High efficiency of KrF_2 synthesis was achieved in nonequilibrium microwave discharge in a magnetic field with surface product stabilization (Asisov, 1980). The discharge was arranged in a reactor with a metallic wall cooled with liquid nitrogen. A layer of solid krypton was frozen on the wall. The saturated Kr vapor pressure is 2 Torr at the wall temperature (77 K), and the partial pressure of fluorine in the discharge was 10 Torr. The yield of KrF_2 is shown in Figure 18.13 as a function of total energy input. The energy cost of KrF_2 synthesis in this plasma-chemical system is 1.2 kWh/kg, which corresponds to yield of 800 g/kWh and to energy efficiency of 10%.

Plasma-chemical **synthesis of xenon fluorides** (XeF_2, XeF_4, and XeF_6) is like that of KrF_2, considered above. Experiments with glow discharges in Xe–F_2 (1 : 1) and Xe–CF_4 (1 : 1) mixtures were focused on the synthesis of XeF_2. The glow discharge current was 30–90 mA, wall temperature was −78 °C, and gas flow rate was 1–1.5 l/h. The yield of the product XeF_2 was 3.5 g/h in the experiments with Xe–F_2 (1 : 1) mixture, and 0.18 g/h in the experiments with Xe–CF_4 (1 : 1) mixture. XeF_2 synthesis can take place completely in the gas phase via the intermediate formation of XeF radicals $F + Xe + M \rightarrow XeF + M$ with further conversion into XeF_2 by recombination or disproportioning: $F + XeF + M \rightarrow XeF + M$, $XeF + XeF \rightarrow XeF_2 + Xe$. Synthesis of XeF_4 was performed in Xe–F_2 mixture (1 : 2) treated in a glow discharge with currents of 10–30 mA (flow rate 0.14 l/h and yield of XeF_4 0.43 g/h. Similar experiments with a Xe–F_2 mixture (1 : 2) and lower flow rate of 0.016 l/h resulted in effective generation of XeF_6. More details on the subject can be found in Legasov et al. (1978b).

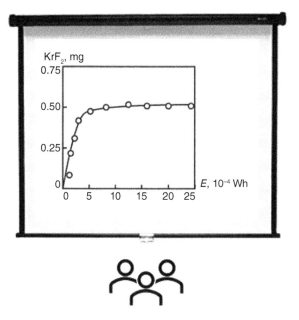

Figure 18.13 Yield of KrF$_2$ synthesis as a function of energy input in nonequilibrium microwave discharge with magnetic field at the ECR conditions.

18.13 Plasma F$_2$ Dissociation and Synthesis of Aggressive Fluorine-based Oxidizers

Atomic fluorine is a major starting agent in the synthesis of many fluorides, especially aggressive fluorine oxidizers. Plasma chemistry, therefore, plays an important role in these synthetic processes because application of nonthermal discharges is the most developed approach to F$_2$ dissociation and generation of fluorine atoms. Generation of fluorine atoms from CF$_4$, C$_2$F$_6$, C$_3$F$_8$, NF$_3$, etc. is an important step in plasma-chemical material processing in microelectronics and will be discussed separately. Generation of F-atoms from F$_2$, for example in glow discharge, typically has energy cost about 10 eV/atom (to compare, 12 eV/atom in electron beam plasmas). Reaction has the second kinetic order: the dissociation rate is proportional to the densities of electrons and fluorine molecules. The rate coefficient at E/n_0 typical for glow discharges at 10 Torr can be presented empirically as the function of T_0:

$$k_F(T_0), \text{cm}^3/\text{s} = 0.52\, T_0(\text{K})^{-2.155} \exp\left(-\frac{1910\,\text{K}}{T_0}\right) \tag{18.30}$$

Generation of atomic fluorine in nonthermal plasma permits, for example, effectively produce **O$_2$F$_2$ and other oxygen fluorides**. It usually starts with formation of an intermediate active radical O$_2$F stable at room temperature F + O$_2$ + M → O$_2$F + M. The major product of the synthesis, oxygen fluoride (O$_2$F$_2$), is then formed in the reactions: O$_2$F + F + M → O$_2$F$_2$ + M, O$_2$F + O$_2$F → O$_2$F$_2$ + O$_2$. The generation of O$_2$F$_2$ does not require dissociation of O$_2$ and can be neglected with respect to F$_2$ dissociation at relatively low electron temperatures in glow discharges, because F$_2$ bonding energy is significantly lower than that of O$_2$ molecules. Glow discharges with high oxygen concentrations (like O$_2$:F$_2$ = 2:1) and lower temperature of the discharge walls (−203 °C) permit to produce other aggressive oxygen fluorides (like O$_4$F$_2$) with a yield about 0.2 g/h. The formation of O$_4$F$_2$ is attributed to the recombination of O$_2$F radicals.

Synthesis of NF$_3$ from N$_2$ and F$_2$ was investigated in glow discharges with currents of 4–40 mA, pressures of 10–40 Torr, wall temperatures from −196° to −150 °C and gas flow rate was 0.09–1.2 l/h. The product yield in these experiments reached 30–70%. Also, N$_2$F$_4$ can be synthesized from NF$_3$ with high yield (50–65%) in some discharge regimes, as well as N$_2$F$_2$ with yield of 12–16%. Glow discharge in a mixture of NF$_3$ and O$_2$ (1 : 1) with current 30–50 mA, pressure 10–15 Torr, gas flow rate 1.7–3.4 l/h, and wall temperature −196 °C provides an effective synthesis of nitrogen oxyfluoride (NF3O). The yield of this aggressive fluorine oxidizer reaches 10–15%. Especially interesting is plasma-chemical synthesis in the mixture NF$_3$–AsF$_5$–F$_2$ (composition 1 : 1 : 2), which results in the formation of

NF_4ArF_6. The yield of the product in the glow discharge process was 0.025 g/h at pressure 80 Torr and wall temperature −78 °C.

Among other gas-phase synthetic processes performed in glow discharges and related to the production of fluorine oxidizers, we can mention generation of ClF_5 from a Cl_2–F_2 mixture (1 : 10) at 30 Torr, wall temperature −78 °C, and the gas flow rate 1.5 l/h. Finally, the synthesis of O_2BF_4 from a mixture of O_2–F_2–BF_3 was effectively performed in glow discharges. The yield here is high at temperature −196°C and characterized by the G-factor $G = 26$ mol/100 eV; energy cost of the synthesis is relatively low, only about 4 eV per molecule. More details on the subject can be found in Legasov et al. (1978b), and in Fridman (2008).

18.14 Plasma-chemical Synthesis of Hydrazine (N_2H_4) and Ammonia (NH_3)

Direct synthesis of hydrazine (N_2H_4) from N_2 and H_2 is a slightly endothermic process like the synthesis of ozone, krypton fluoride, and is very attractive for chemical technology:

$$N_2 + 2H_2 \rightarrow N_2H_4, \quad \Delta H = 1.1 \, eV \, mol^{-1} \qquad (18.31)$$

This synthesis is challenged by the low stability of anhydrous hydrazine. Electric sparking of hydrazine vapor at 100 °C results in explosive decomposition of N_2H_4 with the formation of ammonia, nitrogen, and hydrogen. Thus, similarly to the synthesis of ozone, compounds of noble gases, and aggressive fluorine oxidizers, the application of nonthermal plasma for direct N_2H_4 production is promising because it stimulates the process through high electron temperature and, at the same time, stabilizes products due to low gas temperature. An important positive feature of the synthesis (18.31) is high efficacy of vibrational excitation of N_2 and H_2 in plasma.

Effective **hydrazine and ammonia synthesis from N_2 and H_2** has been demonstrated in non-self-sustained electron beam (EB) supported discharges, where E/n_0 was optimized for vibrational excitation of the molecules. In these plasma systems (described in Fridman 2008; Rusanov and Fridman 1984; and Fridman and Kennedy 2021), the EB current density was 1–40 μA/cm², electron beam diameter 1 cm, reactor volume 3 cm³, applied voltage 6 kV, pressure 0.05–1 atm, gas velocity 0.5–40 m/s, and the non-self-sustained discharge current density was 1–100 mA/cm². Generated products were N_2H_4 and ammonia NH_3. The specific energy input in the discharge at atmospheric pressure was up to 0.1 J/cm³. The reduced electric field in the non-self-sustained plasma was relatively low ($E/p \leq 3$ V/cm Torr). A significant increase of discharge current and reduced electric field E/p occurs at lower gas pressures. The dependence of hydrazine yield on gas pressure at different gas velocities is presented in Figure 18.14a. The hydrazine yield

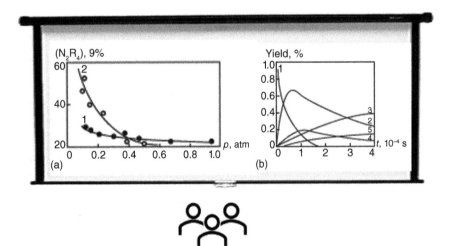

Figure 18.14 Hydrazine and ammonia plasma synthesis from N_2 and H_2: (a) relative yield of N_2H_4 with respect to total nitrogen hydrides (1 – discharge in the gas mixture, 2 – discharge in N_2 followed by mixing with H_2); (b) kinetic curves of the synthesis in N_2–H_2 mixture (1 : 1) at atmospheric pressure (concentrations of: (1) atomic nitrogen; (2) NH_2 radicals; (3) hydrazine (N_2H_4); (4) NH radicals multiplied by 10; (5) ammonia NH_3).

increases at lower pressures. Higher hydrazine yield is related to growth of the reduced electric field E/p and the specific energy input in the discharge at lower gas pressures. The hydrazine yield increases significantly when nitrogen alone is excited in the discharge and then subsequently mixed with hydrogen. Kinetics of the process starting with plasma dissociation of molecular nitrogen only is illustrated in Figure 18.14b. Atomic nitrogen is quickly converted in this case into NH$_2$ radicals after mixing with molecular hydrogen. When the NH$_2$ radicals are sufficiently accumulated, their recombination (NH$_2$ + NH$_2$ + M → N$_2$H$_4$ + M) and disproportioning (NH$_2$ + NH$_2$ → NH$_3$ + NH) lead to the formation of NH$_3$ and N$_2$H$_4$ with comparable production rates.

 Recently, significant interest was also focused on **plasma-catalytic synthesis of ammonia** from N$_2$–H$_2$ mixtures. For example, Shah et al. (2018) analyzed the NH$_3$ synthesis in ICP-RF plasma using Ga, In, and their alloys as catalysts to enhance plasma process. The Ga-In alloys (6 : 4 and 2 : 8) at power 50 W lead to relatively high yield 0.3 gNH$_3$/kWh and the process energy cost of about 200 MJ mol^{-1}. Suggested mechanism of the plasma-catalytic synthesis includes intermediate formation of the GaN groups on the Ga-In catalytic surface. More information on the subject can be found in the book ed. by Bogaerts (2019).

We should mention that the total energy cost of the plasma-chemical and plasma-catalytical processes is still very high in comparison to the conventional Haber-Bosch Process, see Section 18.1.

18.15 Plasma-chemical Synthesis of Polyphosphoric Nitrides (P$_6$N$_6$), Polymeric Nitrogen, Cyanides, and Some Other Inorganic Compounds

"Energetic but cold" nonthermal plasmas permit effective synthesis of some very interesting nitrogen compounds difficult or impossible to produce in the other way, see Bugaenko et al. (1992), Polak et al. (1975). The first example here is synthesis of **nitrides of phosphorus**. These valuable nitrogen compounds can be generated by surface reactions of vibrationally excited nitrogen molecules and nitrogen atoms with phosphorus. The process starts in glow discharge with N$_2$ dissociation followed by the formation of the PN-molecular groups on the surface of phosphorus:

$$N + P_4 \rightarrow PN + P_3, \quad \Delta H = 0.92 \, \text{eV mol}^{-1} \tag{18.32}$$

High energy efficiency of the synthesis of phosphorus nitrides was achieved in the non-self-sustained discharges supported by pulsed EBs Basov et al. (1977, 1978). The process was carried out at atmospheric pressure, gas temperature 418 °C, specific energy input up to 0.4 J/cm^3, pulse duration 10 µs, and reduced electric field E/p = 5–7 V/cm Torr optimal for vibrational excitation of nitrogen molecules and their reactions with P$_4$. The process energy cost, therefore, was relatively low: 6 eV per one PN group. The product was identified as polyphosphoric nitrides (P$_6$N$_6$) presenting significant practical interest.

The **cryogenic plasma synthesis of polymeric nitrogen** was discussed in Lecture 16 (Section 16.7) in relation with organization of cryogenic discharges in liquid nitrogen. The powder-like energetic polynitrogen material was generated at the liquid nitrogen treatment duration 30–60 min and pulse repetition frequency of 60 Hz (Dobrynin et al. 2019), see Figure 16.5. More details on the subject can be found in Dobrynin et al. (2019).

Another example of plasma-induced production of nitrogen compounds is **CN and NO synthesis in CO–N$_2$ discharges**, which can be presented as:

$$CO + N_2 \rightarrow CN + NO, \Delta H = 6 \, \text{eV mol}^{-1} \tag{18.33}$$

Both initial molecules, CO and N$_2$, have very high bonding energies (11 and 10 eV), which explains high temperatures (about 5000 K) required for their dissociation and initiation of the synthesis (18.33). The quasi-equilibrium concentration of products is low in the high-temperature plasma zone of thermal plasmas and most of the CN and NO production takes place during the quenching phase. Similarly, organization of the synthesis (18.33) in nonthermal discharges requires very high vibrational temperatures and has limited degree of conversion. The energy cost of cyanides (CN) in thermal plasma is shown in Figure 18.15 for absolute, ideal, and super-ideal quenching modes (Nester et al. 1988). The data are normalized to the production of a single CN molecule. Absolute quenching assumes conservation of CN molecules generated in the high-temperature zone; the minimal energy cost in this case is quite high (640 eV per CN molecule). Ideal quenching assumes additional CN formation due to recombination of carbon and nitrogen atoms during the cooling phase: the minimal energy cost in this case is lower: 27.6 eV per CN molecule.

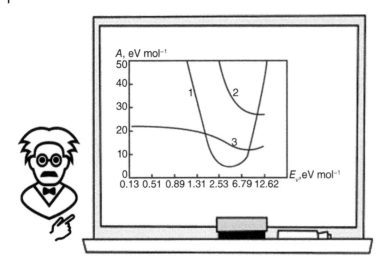

Figure 18.15 Synthesis of CN and NO in atmospheric-pressure thermal plasma. Energy cost of CN production as function of specific energy input: (1) absolute quenching (A × 100); (2) ideal quenching; (3) super-ideal quenching.

Super-ideal quenching takes place in the case of VT nonequilibrium during the cooling phase and permits the additional transfer of vibrational energy into the formation of CN molecules: the minimal energy cost in this case is the lowest: 12.9 eV per CN molecule.

Between the sulfur-related plasma-chemical synthetic processes, let us consider the endothermic **sulfur gasification by CO_2** leading to the formation of SO_2 and CO:

$$S(\text{solid}) + 2CO_2 \rightarrow 2CO + SO_2, \quad \Delta H = 1.4\,\text{eV/CO mol} \tag{18.34}$$

The energy efficiency of this process can be high because of its stimulation in nonequilibrium discharges by vibrational excitation of CO_2 molecules. A high process efficiency was achieved in nonthermal microwave discharge at moderate pressures (see Rusanov and Fridman 1984; Givotov et al. 1985; Fridman, 2008). Sulfur was supplied either as a powder or a vapor (in this case CO_2 was preheated to 450 °C), the gas pressure was 60–120 Torr, the CO_2 flow rate 0.1–0.5 l/s, the sulfur supply rate up to 0.2 g/s, and the power about 1 kW. The minimal energy cost of the process (18.34) with sulfur supplied as a vapor was 4.5 eV/CO mol. The optimal regime was achieved at a pressure of 100 Torr and a specific energy input of 6 J/cm³ (1.2 eV mol⁻¹). The CO_2 dissociation degree was 24%. Multiple other relevant examples of synthesis in both thermal and nonthermal plasmas can be found in Fridman (2008).

18.16 Gas-phase Plasma Decomposition of Inorganic Triatomic Molecules NH_3, SO_2, N_2O

The **dissociation of ammonia** with formation of N_2 and H_2 is slightly endothermic:

$$NH_3 \rightarrow \frac{1}{2}N_2 + \frac{3}{2}H_2, \quad \Delta H = 0.46\,\text{eV mol}^{-1} \tag{18.35}$$

This process was analyzed, in particular, in glow discharges at pressures (1–3 Torr; see Slovetsky 1980), where current varied in the range 5–60 mA, the gas temperature was kept below 470 K, the reduced electric field was $E/n_0 = 4$–$8 \cdot 10^{-16}$ V cm², and the electron density varied in the range $n_e = 10^9$–$2 \cdot 10^{10}$ cm⁻³. Main contribution to the ammonia decomposition was due to the direct dissociation by electron impact ($e + NH_3 \rightarrow NH_2 + H + e$) and by dissociative attachments ($e + NH_3 \rightarrow NH^-_2 + H$) with combined rate coefficient up to 10^{-10} cm³/s. NH_3 dissociation is also effective in collisions with sufficiently vibrationally excited N_2 molecules ($v > 20$), rate coefficient $k = 6 \times 10^{-13}$ cm³/s at room temperature.

Dissociation of sulfur oxide (SO_2) was analyzed both in thermal and nonthermal plasmas. Nonthermal plasma SO_2 dissociation leads to the production of elementary sulfur and oxygen, as a variety of lower sulfur oxides like SO and S_2O (Polak et al. 1975). Energy efficiency in the nonthermal plasma is low due to the wide variety of SO_2 dissociation products and, therefore, low process selectivity.

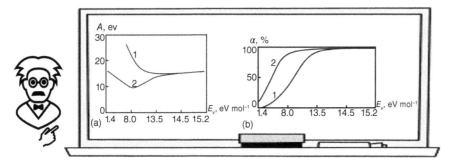

Figure 18.16 SO$_2$ dissociation in atmospheric pressure thermal plasma depending on specific energy input E_v: (a) energy cost of production of a sulfur atom in the case of (1) absolute quenching, (2) ideal quenching; (b) conversion degree of sulfur dioxide into molecular elemental sulfur in the case of (1) absolute quenching, (2) ideal quenching.

Thermal plasma dissociation of SO$_2$ was focused on the production of elementary sulfur:

$$SO_2 \rightarrow S(s) + O_2, \quad \Delta H = 3.1 \text{ eV mol}^{-1} \tag{18.36}$$

The energy cost and conversion degree of the process are presented in Figure 18.16 as functions of specific energy input (Nester et al. 1988). The minimal energy cost is 14 eV/S atom in the case of absolute quenching and 9.1 eV/S atom for ideal quenching. The energy efficiency of the thermal plasma SO$_2$ dissociation does not exceed 34%.

Destruction of nitrous oxide (N$_2$O) in nonthermal plasma is important to avoid its emission to atmosphere. Heating of N$_2$O in conventional chemical or thermal plasma systems easily leads to N$_2$O dissociation to N$_2$ and O$_2$. More interesting for applications is N$_2$O decomposition with selective production of NO. Significant anthropogenic source of N$_2$O emission is related to the production of adipic acid by nitric-based oxidation of cyclohexanol. Since the concentration of N$_2$O in adipic-acid-production off-gases is high, the effective way of reducing N$_2$O emission, according to the Rhone-Poulenc approach, is N$_2$O conversion back to NO (for further recirculation in the technology of production of adipic acid):

$$N_2O + \frac{1}{2}O_2 \rightarrow NO + NO, \quad \Delta H = 1 \text{ eV mol}^{-1} \tag{18.37}$$

The selective oxidation of N$_2$O to NO (18.37) becomes possible only in nonequilibrium plasma by dissociation of oxygen molecules with the following exothermic reactions of NO formation (O + N$_2$O $\rightarrow$ NO + NO). It was demonstrated in nonthermal gliding arc discharge (Czernichowski 1994) at flow rates from 22 to 82 slm with a molar fraction of N$_2$O 31.6–50.6%. The efficiency of N$_2$O oxidation to NO achieved in the nonthermal gliding arc was about 70%.

18.17 Dissociation of Hydrogen Halides, Hydrogen, Nitrogen, and Other Diatomic Molecules in Thermal and Nonthermal Plasmas

Dissociation of hydrogen halides in plasma with production of hydrogen and halogens can compete with electrolysis and in some cases can be very attractive for practical applications. The dissociative attachment (e + HHal $\rightarrow$ H + Hal$^-$) is very fast in nonthermal plasmas because of high electron affinity to Hal atoms. Relevant energy efficiency is limited by an electron loss (and cost of a "new electron" required for dissociative attachment is at least 30–50 eV). The energy efficiency of the HHal dissociation can be higher in thermal plasmas; however, requires effective quenching. It can be illustrated by the **plasma dissociation of hydrogen bromide**:

$$HBr \rightarrow \frac{1}{2}H_2 + \frac{1}{2}Br_2, \quad \Delta H = 0.38 \text{ eV mol}^{-1} \tag{18.38}$$

This endothermic plasma-chemical process is considered as an important step in the thermochemical calcium–bromine–water splitting cycle for hydrogen production (Doctor 2000). The HBr decomposition (13.38) leads to the formation of H$_2$ in gas phase and Br$_2$ in condensed phase, therefore assumes effective quenching and separation of products can be achieved by fast rotation of the quasi-thermal plasma. The process energy cost is presented in Figure 18.17a for the case of the absolute quenching mode also as a function of specific energy input in the thermal plasma. The minimal energy cost of HBr dissociation with the formation of molecular hydrogen and bromine is quite

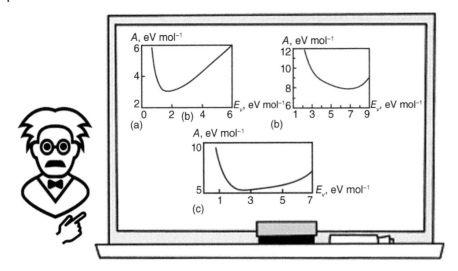

Figure 18.17 Dissociation of HBr (a), HF (b), and HCl (c) in atmospheric-pressure thermal plasma: the process energy cost as a function of specific energy input in the absolute quenching regime.

low, about $3\,\mathrm{eV\,mol^{-1}}$. Dissociation of HF (HF $\rightarrow \frac{1}{2}\mathrm{H_2} + \frac{1}{2}\mathrm{F_2}$, $\Delta H = 2.83\,\mathrm{eV\,mol^{-1}}$) and HCl (HCl $\rightarrow \frac{1}{2}\mathrm{H_2} + \frac{1}{2}\mathrm{Cl_2}$, $\Delta H = 0.96\,\mathrm{eV\,mol^{-1}}$, in contrast to HBr (18.38), result in two gas-phase products making the centrifugal product separation ineffective and decreases the process energy efficiency, see Figure 18.17b,c (Nester et al. 1988).

Plasma dissociation of fluorine was already discussed shortly in relation with synthesis of aggressive fluorine-based oxidizers (Section 18.13). It is similar to the dissociation of HHal: proceeding very fast in nonthermal plasmas through dissociative attachment but energy expensive due to losses of electrons. Same effect takes place in plasma dissociation of $\mathrm{Br_2}$, $\mathrm{Cl_2}$. Plasma dissociation of $\mathrm{O_2}$ was discussed in relation to ozone synthesis, see Section 18.7. Energy efficiency in this case is much higher because the process kinetics can be dominated here not by dissociative attachment (consuming an electron) but by dissociation through electronic excitation ((18.10), (18.11)).

The **dissociation of $\mathrm{H_2}$** ($\mathrm{H_2} \rightarrow \mathrm{H} + \mathrm{H}$, $\Delta H = 4.5\,\mathrm{eV\,mol^{-1}}$) in low-pressure nonthermal discharges is provided mostly by intermediate excitation of electronically excited states $\mathrm{H_2}(b^3\Sigma_u)$. The vibrational excitation of $\mathrm{H_2}$ by electron impact is very effective at electron temperatures around $T_e = 1\text{--}2\,\mathrm{eV}$. Therefore, the vibrational temperature in nonthermal $\mathrm{H_2}$ discharges can be high and can significantly affect the $\mathrm{H_2}$ dissociation rate. The major contribution of vibrational excitation into $\mathrm{H_2}$ dissociation in nonthermal discharges is related not directly to the dissociation of vibrationally excited molecules (as in the case of $\mathrm{CO_2}$ dissociation), but to acceleration of dissociation through the excited states $\mathrm{H_2}(b^3\Sigma_u)$ when an electron hits a vibrationally excited molecule. Molecular hydrogen can also be quite effectively dissociated in thermal plasma conditions; the relevant minimal energy cost is $6.2\,\mathrm{eV\,mol^{-1}}$.

Dissociation of molecular nitrogen is a strongly endothermic process ($\mathrm{N_2} \rightarrow \mathrm{N} + \mathrm{N}$, $\Delta H = 9.8\,\mathrm{eV\,mol^{-1}}$) because of very high bonding energy of the molecules. In low pressure glow discharges ($p < 1.5\,\mathrm{Torr}$), the dissociation is dominated by electronic excitation. However, at higher pressures ($p \geq 1.5\,\mathrm{Torr}$), the vibrational excitation of $\mathrm{N_2}$ leads to dissociation, not in combination with electronic excitation but directly through VV exchange and formation of molecules with vibrational energy sufficient for dissociation (Slovetsky 1980). The energy cost of $\mathrm{N_2}$ dissociation in thermal plasmas is, however, less than in nonthermal ones. The minimum energy cost of N-atoms is $6.3\,\mathrm{eV/atom}$ in thermal systems, which requires heating the gas to $6400\text{--}6600\,\mathrm{K}$ as well as absolute quenching. For more details on the subject see Fridman (2008).

18.18 Problems and Concept Questions

18.18.1 Rate Coefficient of Reaction $\mathrm{O} + \mathrm{N_2} \rightarrow \mathrm{NO} + \mathrm{N}$ Stimulated by Vibrational Excitation

Compare logarithmic sensitivity of the rate coefficient (18.7) to the vibrational ($\partial \ln k_R / \partial \ln T_v$) and translational ($\partial \ln k_R / \partial \ln T_0$) temperatures. Find out when the reaction stimulated by vibrational temperature is more sensitive to translational temperature?

18.18.2 Energy Efficiency of NO Synthesis in Plasma

Why the energy efficiency of NO synthesis in plasma is relatively low even when the process is stimulated in non-thermal plasma by very effective vibrational excitation of N_2 molecules?

18.18.3 Conversion Degree Limitations of NO Synthesis in Plasma

Why is the conversion degree of the plasma NO synthesis from atmospheric air so strongly limited both in thermal and nonthermal discharges?

18.18.4 Discharge Poisoning Effect During Ozone Synthesis

Comparing the rate of oxygen recombination (18.12) leading to O_3 with rates of oxygen reactions with NO and NO_2 ((18.20) and (18.21)), estimate concentration of nitrogen oxides resulting in the discharge poisoning. Which nitrogen oxide is more effective in suppression of ozone synthesis in nonthermal plasma?

18.18.5 Temperature Effect on Ozone Stability

Based on enthalpy released during the ozone decomposition, calculate the relative O_3 concentration in air doubling the initial room air temperature (heating to a temperature of about 600 K) provoking the ozone explosion.

18.18.6 Negative Effect of Humidity on Ozone Generation in DBD

Explain the negative effect of humidity on O_3 synthesis due to water enhancement of surface conductivity of dielectric barriers. Why does humidity influence the surface conductivity? Why does the elevated surface conductivity make microdischarges stronger? Why does making the microdischarges stronger lead to the reduction of O_3 synthesis efficiency?

18.18.7 Positive Effect of N_2 and CO Admixtures to Oxygen on Plasma-chemical Ozone Synthesis

Considering the mechanism of CO chain oxidation, explain the positive effect of CO addition on ozone synthesis. Why do small N_2 and CO admixtures have a positive effect on ozone generation, whereas more significant addition of these gases leads to a negative effect?

18.18.8 Plasma-chemical KrF_2 Synthesis with Product Stabilization in a Krypton Matrix

Using the KrF_2 synthesis as an example, identify the major challenge of production of compounds of the "inert noble" gases. Analyze the space distribution of fluorine intermediate groups KrF across the krypton film matrix.

18.18.9 Plasma Synthesis of NH_3 and N_2H_4

Why does plasma synthesis in N_2–H_2 mixtures usually leads to ammonia and not hydrazine? Which plasma approach is the most effective for fixing nitrogen: the synthesis of nitrogen hydrides or nitrogen oxides?

Lecture 19

Plasma Metallurgy: Production and Processing of Metals and their Compounds

19.1 Hydrogen-based Reduction of Iron Ore in Thermal Plasma: Using Hydrogen and Hydrocarbons, Plasma-chemical Steel Manufacturing

High plasma temperatures permit to intensify metallurgical processes, to accomplish them in a single stage, and make metallurgical equipment more compact. Plasma reduction of oxides is mostly carried out in thermal discharges. The conventional industrial **blast furnace process** of iron production from its oxides (Fe_2O_3, Fe_3O_4) with the manufacturing of steel includes multiple stages and uses large-scale equipment, see Figure 19.1a. Iron oxides are supplied in the blast furnace as ore, which requires preliminary agglomeration. Coke is burned as a fuel in a coke furnace to heat the blast furnace and produce CO, which reacts with metal oxides to produce iron:

$$Fe_2O_3 + 3CO \rightarrow 3CO_2 + 2Fe, \quad \Delta H = -0.28 \, eV \, mol^{-1}. \tag{19.1}$$

Disproportioning of carbon monoxide leads to the formation of carbon mixed with the iron. Iron produced in the blast furnace is called pig-iron and requires decarbonization in a special converter to become steel. Application of thermal plasma discharges permits the whole steel manufacturing to be accomplished directly in a single-stage process, see Figure 19.1b. The **plasma reduction of metal oxides** can be provided in this case by hydrogen or natural gas. *High plasma temperatures and use of hydrogen intensify the process, avoid carbon, and produces steel in a single stage* (Dembrovsky 1984; Mishra 2012; Paton et al. 2014; Tsvetkov and Panfilov 1980):

$$Fe_2O_3 + 3H_2 \rightarrow 2Fe + 3H_2O. \tag{19.2}$$

Plasma-metallurgical reduction of iron ore can also be organized in other configurations. Figure 19.2a illustrates the process, where thermal arc directly contacts the melt and reduction is provided by hydrogen or hydrocarbons. The scheme illustrated in Figure 19.2b uses iron ore preheating by the reduction process exhaust gases. In this case, the ore reduction is stimulated in slag by reduction gases heated up in arc. An arc, illustrated in Figure 19.2c, is organized between a cathode and a stationary ring anode. The arc moves along the ring anode and forms a plasma cone. Particulates of iron oxide and reduction gases cross the plasma cone, where the ore reduction takes place. Most such plasma systems applied for ore reduction operate at atmospheric pressure.

Electric power and, therefore, productivity of plasma-metallurgical furnaces can be high, 1 MW and more. Arc discharges in this case are usually equipped with copper anodes and tungsten or copper cathodes, and the arcs are stabilized by magnetic forces or by special gas flows. Pure hydrogen, natural gas, or their mixtures are used as reduction agents and heat carriers. The thermal efficiency of the plasma furnaces with respect to heat losses can be high – up to 90%. The theoretical minimum energy cost for iron production from magnetite (Fe_3O_4) and hematite (Fe_2O_3) is 2.21 kWh kg^{-1} Fe. As an example, 100 kW arc with hydrogen – natural gas (80–20%) mixture as a heat carrier and reduction agent provides single-stage pure iron production with energy cost of 4.8–5.9 kWh kg^{-1} Fe. Similar experiments with a 10-fold more powerful arc (1 MW) result in a decrease of the process energy cost under the same conditions to 4.2–5.2 kWh kg^{-1} Fe. Higher efficiency of the more powerful arcs is due to an increase of thermal efficiency of the reactor, better mixing, and decreased sizes of particulates of magnetite (Fe_3O_4) and hematite (Fe_2O_3).

Plasma Science and Technology: Lectures in Physics, Chemistry, Biology, and Engineering, First Edition. Alexander Fridman.
© 2024 WILEY-VCH GmbH. Published 2024 by WILEY-VCH GmbH.

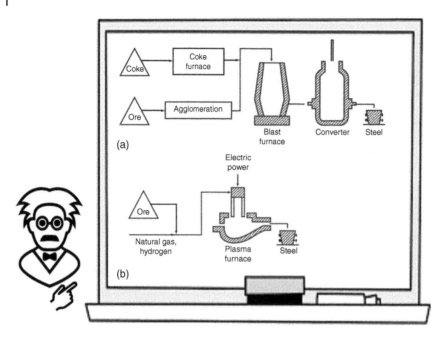

Figure 19.1 Iron production by reduction of its oxides: (a) conventional blast furnace; (b) plasma-metallurgical process.

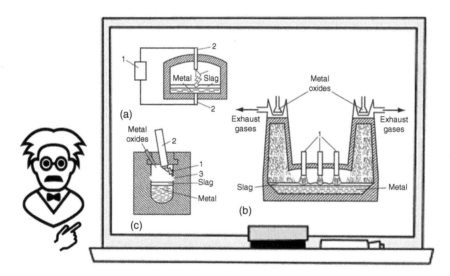

Figure 19.2 Plasma furnaces for iron reduction from its oxides: (a) reduction in slag zone by thermal arc electrically attached to the melt, (1) plasma generator, (2) electrodes, (3) arc; (b) treatment of the oxide melt by gas energy carrier generated in the three-phase thermal arc (1); (c) arc (1) rotating between cathode (2) and stationary circular anode (3).

19.2 Hydrogen-based Reduction of Refractory Metal Oxides in Thermal Plasma, Plasma Metallurgy of Tungsten and Molybdenum

Hydrogen reduction of tungsten oxide (as well as oxides of other refractory metals) proceeds in thermal plasma at high temperatures when molecular products are not stable. The reduction proceeds in this case through the so-called **dissociative reduction mechanism** (Tsvetkov and Panfilov 1980), which starts with dissociation of the oxides with the formation of metal atoms and oxygen atoms. Then, during the quenching stage, oxygen atoms recombine with hydrogen-based reduction agents, saving metal atoms from reverse reactions with oxygen and providing effective metal production. Production of metallic tungsten can be summarized in this case as:

$$WO_3 + 3H_2 \rightarrow W + 3H_2O. \tag{19.3}$$

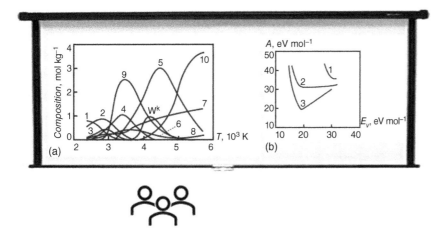

Figure 19.3 Metallic tungsten (W^k) production from WO_3 by decomposition in atmospheric pressure thermal plasma: (a) composition of products: (1) W_3O_9, (2) W_2O_6, (3) W_3O_8, (4) WO_3, (5) WO, (6) O_2, (7) [O] × 10, (8) W^+, (9) WO_2, (10) W; (b) energy cost of tungsten production as function of specific energy input: (1) absolute quenching, (2) ideal quenching, (3) super-ideal quenching.

The quasi-equilibrium composition of products of thermal WO_3 dissociation at atmospheric pressure is shown in Figure 19.3a as a function of temperature. The dissociative reduction mechanism means in this case that highly energy intensive dissociation of tungsten oxide (WO_3) and formation of atomic tungsten ($WO_3 \rightarrow W + 3/2\ O_2$, $\Delta H = 8.73\ \text{eV mol}^{-1}$) takes place at high temperatures (about 5000 K) where oxygen and water are completely atomized. So, water formation (19.3) occurs only at the quenching stage of the plasma-metallurgical process. The energy cost of tungsten production from tungsten oxide WO_3 in thermal plasma is shown in Figure 19.3b for atmospheric pressure as a function of specific energy input for different quenching modes. Minimal energy cost in the absolute quenching mode is 35 eV per W atom. Ideal quenching assumes in this process that all intermediate tungsten oxides participate in disproportioning reactions during the fast-cooling phase, forming atomic tungsten (W) and WO_3. The minimal energy cost of metal production in the case of ideal quenching is 30.6 eV per W atom. Super-ideal quenching in this process requires centrifugal separation of products and permits to decrease the energy cost to 19 eV per W atom at plasma temperature of 4500 K (Nester et al. 1988).

The **molybdenum plasma metallurgy** is another example of H_2 plasma reduction of oxides of refractory metals following the dissociative reduction mechanism:

$$MoO_3 + 3H_2 \rightarrow Mo + 3H_2O. \tag{19.4}$$

The mechanism and characteristics of the molybdenum production process are like those of tungsten. The dissociative reduction mechanism in this case means that dissociation of MoO_3 and formation of Mo atoms requires a high temperature exceeding 4500 K when oxygen and water are completely atomized. The minimal energy cost in the absolute quenching mode is 28 eV per Mo atom at a specific energy input of 25 eV mol^{-1}. Ideal quenching results in the minimal energy cost 25.4 eV per Mo atom and can be achieved at a specific energy input of 22.8 eV mol^{-1}. Super-ideal quenching with centrifugal separation of products permits to decrease the energy cost to 17.3 eV per Mo (Nester et al. 1988).

More information on the subject can be found in (Tsvetkov and Panfilov 1980; Tumanov 1981, 1989).

19.3 Thermal Plasma Reduction of Oxides of Aluminum and Other Inorganic Elements

Aluminum can be produced in thermal plasma from alumina (Al_2O_3) using CO, CH_4, or hydrogen as reduction agents. Plasma jets can be arranged in this case based on atmospheric-pressure arcs as well as thermal

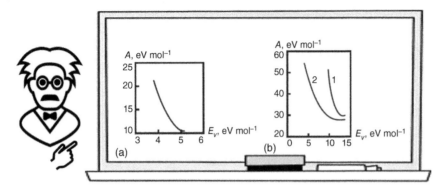

Figure 19.4 Aluminum reduction from Al_2O_3 in atmospheric pressure thermal plasma. Energy cost of aluminum production as a function of specific energy input: (a) CO as reducing agent, absolute quenching regime; (b) H_2 as reducing agent, (1) absolute quenching; (2) ideal quenching.

atmospheric-pressure radiofrequency (RF) discharges (Rykalin and Sorokin 1987). The reduction gaseous agents (CO, CH_4, or H_2) are in this case usually diluted by argon. The plasma-metallurgical process of reduction of alumina with carbon monoxide and production of aluminum is a strongly endothermic process:

$$Al_2O_3 + 3CO \rightarrow 2Al + 3CO_2, \quad \Delta H = 8.57\,eV\,mol^{-1}. \tag{19.5}$$

The process takes place at high temperatures, but lower than those required for reduction of tungsten and molybdenum. In contrast to the dissociative reduction mechanism, aluminum atoms are produced at temperatures when CO is not yet dissociated. The energy cost of aluminum production from Al_2O_3 in atmospheric-pressure thermal plasma is shown in Figure 19.4a as a function of specific energy input. The minimal energy cost of the process in absolute quenching mode is 10.6 eV per Al atom. This energy cost can be achieved at a specific energy input of 5.2 eV mol^{-1} with a conversion degree of 97%. The plasma-metallurgical process of reduction of alumina with hydrogen requires less energy:

$$Al_2O_3 + 3H_2 \rightarrow 2Al + 3H_2O, \quad \Delta H = 7.51\,eV\,mol^{-1}. \tag{19.6}$$

Alumina reduction requires here temperatures of 3500 K; relevant energy cost in aluminum production is shown in Figure 19.4b as a function of specific energy input in different quenching modes. The minimal energy cost in the absolute quenching mode is 29.4 eV per Al atom, which can be achieved at a specific energy input of 13.9 eV mol^{-1} with conversion degree 94%. The minimal energy cost of aluminum production in the ideal quenching mode is 27.9 eV per Al atom.

Thermal plasmas have been applied for oxide reduction of many other metals and nonmetallic inorganic elements, see Tsvetkov and Panfilov (1980); Tumanov (1981, 1989). Different reduction agents have been used in these processes. Thus, dispersed oxides of chromium (Cr), nickel (Ni), and silicon (Si) were reduced to produce the elements via the oxide treatment with reduction gases heated up in a thermal arc discharge. Plasma reduction of nickel (Ni), cobalt (Co), and tin (Sn) from their oxides was demonstrated using a hydrogen–nitrogen mixture heated up in thermal arcs. The plasma-metallurgical process of reducing tantalum (Ta) from its pentoxide (Ta_2O_5) was investigated in a plasma jet of helium with additions of hydrogen; conversion degree into pure metallic tantalum reached here 50%.

19.4 Reduction of Metal Oxides Using Nonthermal Hydrogen Plasma, Nonequilibrium Plasma Effect of Surface Heating and Evaporation

Although most of the hydrogen-based reduction of metal oxides were carried out in thermal plasmas, strongly nonequilibrium discharges were also investigated, see Legasov et al. (1978b). A major advantage of nonequilibrium plasma is the high energy efficiency for generation of atomic hydrogen, excited hydrogen molecules, and other active species stimulating the reduction process at low temperatures of gas and solid oxides. A major challenge of nonthermal plasma metallurgy is related to the kinetics of secondary surface reactions and the efficiency of desorption of gas-phase products to provide excess of active plasma species deeper in the solid oxide. The other challenge is the discharge power, and therefore, scaling up. Nonthermal plasma-generated atomic hydrogen was demonstrated to be effective in reduction of oxides of copper, bismuth, lead, silver, tin, mercury, titanium, vanadium, zirconium, hafnium, and titanium. Relevant reduction of thorium, beryllium, and aluminum oxides was not so effective.

Atomic hydrogen generated in nonthermal plasma reaches the surface of a solid metal oxide, stimulating the "cold" reduction process $Me_mO_k + 2k\,H \rightarrow m\,Me + k\,H_2O$. The reduction process starts from the surface. The front of metal formation propagates into the solid body, and the depth of the reduction layer is limited by the recombination of H atoms:

$$\delta = \sqrt[3]{\frac{D_0}{\Phi r_0}} \exp\left(-\frac{E_a^D}{3T_0}\right),$$

(19.7)

where D_0 (10^{-4}–$10^{-3}\,cm^2\,s^{-1}$) and $E_a^D \approx 1\,eV\,mol^{-1}$ are the pre-exponential factor and activation energy for diffusion of H-atoms in the metallic layer; Φ is the atomic flux; and r_0 is a critical distance between H atoms in the metal crystal structure sufficient for their recombination. The thickness of the reduction layer δ grows exponentially with surface temperature, which can be much higher than the gas temperature in nonequilibrium discharges ($T_e \gg T_0$). This effect of **nonequilibrium surface heating** can stimulate the surface oxide reduction processes, keeping the gas temperature low and efficiency of plasma H_2 dissociation high. If molecular plasma is in vibrational–translational (VT) nonequilibrium ($T_e > T_v \gg T_0$), the surface temperature (T_S) can significantly exceed the translational gas temperature (T_0) because surface VT relaxation is much faster than that in the gas phase. This overheating ($T_e > T_v > T_S \gg T_0$) can be accompanied by surface evaporation, desorption of chemisorbed complexes, and passivating layers.

The nonequilibrium surface heating and evaporation effect can be significant for plasma interaction with powders (to be considered below first) and flat surfaces (to be considered second). The preferential ***macro-particle surface heating*** without essential gas heating becomes possible in molecular gases when the total surface area of the macro-particles is so high that VT relaxation on the surface dominates over VT relaxation due to intermolecular collisions in plasma volume:

$$4\pi r_a^2 n_a > \frac{k_{VT}n_0}{P_{VT}^S v_T}.$$

(19.8)

In this relation, n_a and r_a are density and average radius of macro-particles; $k_{VT}(T_0)$ is the rate coefficient of gas-phase VT relaxation; n_0 is the gas density; v_T is the average thermal velocity of molecules; $P_{VT}^S \approx 3 \cdot 10^{-3}$ is the surface probability of VT relaxation called the accommodation coefficient. Assuming $k_{VT} = 10^{-17}\,cm^3\,s^{-1}$; $n_0 = 3$ $10^{19}\,cm^{-3}$ (normal conditions); $v_T = 10^5\,cm\,s^{-1}$, the amount of dispersed phase required for the nonequilibrium heating effect should satisfy the criterion $r_a^2 n_a > 0.1\,cm^{-3}$ (mining transparency at thickness above 10 cm). On the other hand, vibrational energy accumulated in gas is sufficient to evaporate a surface layer of the macro-particles with depth $r_S < r_a$ if the total volume fraction of the macro-particles is sufficiently small:

$$n_a r_a^3 < \frac{n_0 T_v}{n_k(T_S + r_S \varepsilon_k / r_a)}.$$

(19.9)

Here $n_k = 3 \cdot 10^{22}\,cm^{-3}$ is the number density of atoms in a macro-particle, and ε_k is energy required for evaporation of an atom. If the thickness of the reduced layer is relatively low $r_S < r_a T_S/\varepsilon_k$, then the major portion of vibrational energy input in (19.9) is related to heating the macro-particles up to temperature T_S. Numerically, in this case, the limitation (19.9) means $n_a r_a^3 < 10^{-3}$ for the conventional normal conditions of the heterogeneous nonequilibrium plasma, and $T_v = T_S = 0.3\,eV$. Summarized limitations of the volume fraction of particulates for effective nonequilibrium heating are illustrated in Figure 19.5. Typical parameters of a heterogeneous plasma system relevant for nonequilibrium surface heating and evaporation effect are $r_a = 10^{-4}\,cm$ (1 μm) and $n_a = 10^8\,cm^{-3}$ (Friedlander 2000). The nonequilibrium surface heating and evaporation effect in plasma treatment of powders play a significant role not only in metal reduction from oxides but in other technologies related to powder treatment as well (see Legasov et al. 1978b; Fridman 2008).

 The nonequilibrium flat-surface heating of a thin layer can be accomplished also by short nonthermal discharge pulses. Considering the low temperature of the surrounding gas, the short heating pulses can provide significant local temperature increase in a thin surface layer without heating up the whole system. To describe this transient process, we can assume the heat flux Q related to VT relaxation starting at the initial moment $t = 0$:

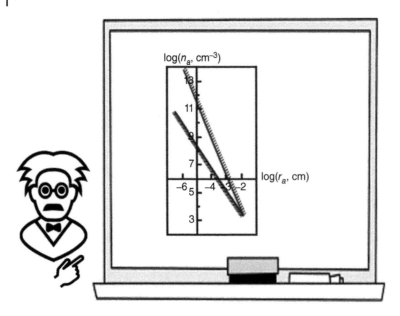

Figure 19.5 Parameters of a heterogeneous system (density of aerosols, n_a, and radius of particles, r_a) required for non-equilibrium heating of dispersed phase ($P_{VT}^S = 3 \cdot 10^{-3}, n_0 = 3 \cdot 10^{19} \text{cm}^{-3}$). The effective values of parameters are located between the two lines.

$$Q_0 = n_0 P_{VT}^S v_T \hbar \omega. \tag{19.10}$$

Solution of the heat transfer equation with the flux (19.10) from the surface ($x = 0$) gives the temperature profile in the plasma-treated layer:

$$T(x,t) = T_0 + 2\frac{Q_0}{\rho c}\sqrt{\frac{t}{\pi \chi}} \int_a^\infty \frac{\exp(-u^2)du}{u^2}. \tag{19.11}$$

In this relation ρ, c, χ are density, specific heat, and reduced heat transfer coefficient of the treated surface layer; the dimensionless depth a of this layer is $a = x/\sqrt{4\chi t}$. The heating of a thin layer by short nonthermal discharge pulses provides significant local temperature increase without heating up the whole system. The gas temperature also remains low because $T_e > T_v \gg T_0$). The heat flux Q_0 (19.10) required to provide a temperature increase ΔT in a layer with thickness h can be then expressed as:

$$Q_0 \approx \rho c \chi \frac{\Delta T}{h}. \tag{19.12}$$

19.5 Thermal Plasma Production of Metals by Carbothermic Reduction of Their Oxides: Pure Metallic Uranium, Niobium, Iron, Refractory, and Rare Metals

The carbothermic reduction of elements (El) is a reaction in the condensed phase with formation of the pure element also in condensed phase and carbon oxides:

$$(\text{El}_x\text{O}_y)_{cond} + \text{y(C)}_{cond} \rightarrow \text{x(El)}_{cond} + (\text{CO})_{gas}. \tag{19.13}$$

These processes are strongly endothermic and require high temperatures. Thus, *the carbothermic reduction of uranium (U), boron (B), zirconium (Zr), niobium (Nb), and tantalum (Ta) requires 2000–4000 K and, therefore, application of thermal plasma. In most plasma-chemical carbothermic reduction processes, an arc electrode is prepared from well-mixed and pressed oxide and carbon particles. The arc provides heating of* the mixture, stimulating the reduction process on the electrode. CO$_2$ leaves the electrode, finalizing the reduction process.

A good example here is **the carbothermic plasma reduction of metallic uranium** (Tumanov, 1981, 1989). The conventional nonplasma approach starts with uranium oxide (UO$_2$) and includes two stages: (i) exothermic conversion of the oxide into fluoride with HF: $(\text{UO}_2)_{solid} + 4(\text{HF})_{gas} \rightarrow (\text{UF}_4)_{solid} + 2(\text{H}_2\text{O})_{gas}$, $\Delta H = -1.8\,\text{eV mol}^{-1}$; (ii) solid-state exothermic UF$_4$ reduction with calcium, producing uranium and fluorite:

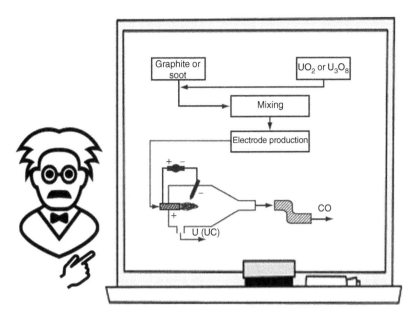

Figure 19.6 Schematic of the carbothermic reduction of uranium from oxides.

$(UF_4)_{solid} + 2(Ca)_{solid} \rightarrow (U)_{solid} + 2(CaF_2)_{solid}$, $\Delta H = -1.9 \, eV \, mol^{-1}$. In contrast to that, the carbothermic approach has a single stage:

$$(UO_2)_{solid} + 2(C)_{solid} \rightarrow (U)_{solid} + 2(CO)_{gas}, \quad \Delta H = 8.9 \, eV \, mol^{-1}. \tag{19.14}$$

This carbothermic process provides effective metallic uranium reduction not only from dioxide (UO_2) but also from a more accessible oxide U_3O_8. The schematic of the process is illustrated in Figure 19.6. The arc anode is prepared from the well-mixed and pressed uranium oxide and carbon particles. Temperature of the solid UO_2–C or U_3O_8–C mixtures on the electrode is 4000–5000 K. The described system also can be applied to produce uranium carbides. Similar approach was also effectively applied for **carbothermic reduction of niobium oxide Nb_2O_5**, where the condensed-phase Nb_2O_5–C or Nb_2O_5–NbC mixtures were heated up to temperatures exceeding 2900 K. Conversion degree of 99% is reached fast in these systems.

Production of high-purity metals often requires combination of the carbothermic plasma reduction with metal refining processes. As an example, we can consider the double-stage carbothermic plasma reduction of rare and refractory metals from their oxides (Tumanov 1981, 1989, see also Fridman 2008). The first stage of carbothermic reduction is conducted in a plasma furnace and results in production of so-called black metal, which can still contain some carbon or oxygen. The second stage is conducted in a high-current vacuum discharge with high-temperature hollow cathode, called an electron-plasma furnace, to provide refining and produce pure metals. Plasma reduction of metals from their oxides can also be effectively carried out in the **thermal plasma fluidized beds**, see the reactor schematic in Figure 19.7. Such technology is applied, as an example, for the carbothermic reduction of iron from FeO/TiO_2 mixture: $FeO/TiO_2 + C \, (H_2) \rightarrow Fe + TiO_2 + CO \, (H_2O)$. The reduction process is due to carbon mixed with FeO/TiO_2 as well as hydrogen used as a gas energy carrier in the discharge. The discharge electrodes also guide the hydrogen gas flow, which provides fluidization of the FeO/TiO_2 concentrate and carbon powders.

19.6 Direct Decomposition of Oxides to High-purity Elements in Thermal Plasma: Production of Aluminum, Vanadium, Indium, Germanium, and Silicon

Production of metals and other elements from their oxides is possible in high temperature conditions of thermal plasma even without the use of reduction agents like hydrogen, CO, or carbon. This approach can be attractive for production of high-purity products. Let us consider as an example **production of high-purity aluminum by alumina decomposition** ($Al_2O_3 \rightarrow 2Al + 3/2 \, O_2$) without any reduction agents in RF-ICP discharges, see Rykalin and Sorokin (1987). *While arc discharges have been applied in*

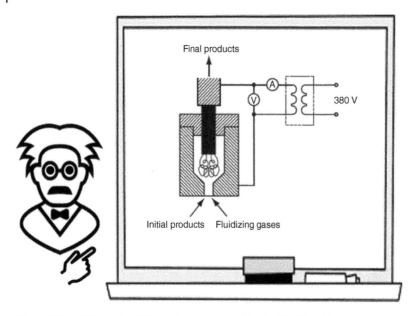

Figure 19.7 Schematic of thermal plasma-metallurgical fluidized bed.

many reduction technologies, requirements of high product purity restrict the use of arcs because of erosion of electrode materials contaminating the products. The high purity can be provided in RF inductively coupled plasma (ICP) which is thermal but electrodeless. Typically, the RF-ICP discharge is arranged in a flow of atmospheric pressure pure argon. Conversion of alumina in Ar plasma starts with the formation of monoxide $Al_2O_3 \rightarrow 2AlO + O$ with optimal temperature is about 4000 K. At higher temperatures closer to 10^4 K, which are reached in the Ar plasma, AlO is unstable and dissociates $AlO \rightarrow Al + O$. Relevant schematic of the RF-ICP plasma system for production of pure aluminum is shown in Figure 19.8. Most of the argon is supplied tangentially, but some is also supplied axially to intensify the interaction of alumina powder with plasma and to provide transportation for the powder. Product quenching is especially important in the absence of reduction agents; it was supported by additional cold Ar injection in a direction opposite

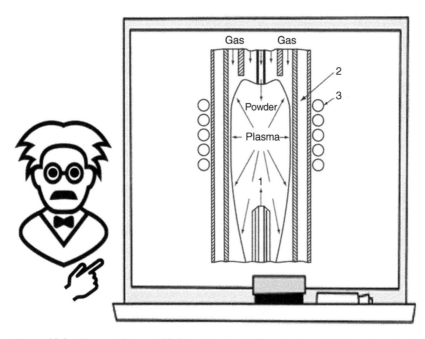

Figure 19.8 Thermal Plasma ICP-RF reactor for Al_2O_3 reduction to high-purity aluminum: (1) quenching zone; (2) water cooling system; (3) ICP-RF inductor.

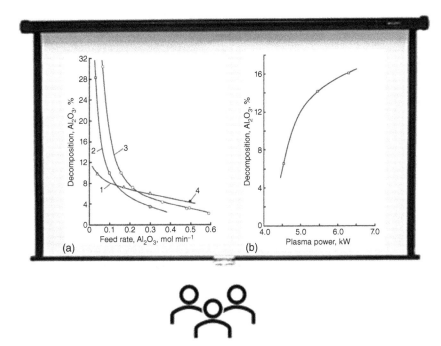

Figure 19.9 Decomposition of Al_2O_3 with production of aluminum without reducing agents: (a) as function of its feed rate and particles sizes (1) 45 μm, (2) 37 μm, (3) 26 μm, (4) 44 μm (argon plasma power 5 kW); (b) as function of the discharge power at particles size 26 μm, alumina feeding rate 0.23 g min^{-1}.

to the plasma flow. The main characteristics of the process are presented in Figure 19.9 illustrating the increase of reduction degree into aluminum at lower feeding rate of alumina, and by using powder with smaller sizes. Also, the alumina reduction degree increases with plasma power.

Similar approach has been analyzed for **direct production of vanadium from its oxides** (V_2O_5 and V_2O_3). Vanadium pentoxide V_2O_5 has melting temperature of 953 K and dissociates starting from about 1000 K. Decomposition of V_2O_5 is a strongly endothermic process:

$$V_2O_5 \rightarrow 2V + \frac{5}{2}O_2, \quad \Delta H = 16.1 \, \text{eV mol}^{-1}. \tag{19.15}$$

The energy cost of vanadium production from V_2O_5 is shown in Figure 19.10a as a function of specific energy input for absolute, ideal, and super-ideal quenching modes. The minimal energy cost for absolute quenching is 27.1 eV atom^{-1} and is reached at a specific energy input of 44.8 eV mol^{-1} and conversion degree of about 83%. The **direct production of indium from its oxide** is less endothermic than the reduction of vanadium:

$$In_2O_3 \rightarrow 2In + \frac{3}{2}O_2, \quad \Delta H = 9.6 \, \text{eV mol}^{-1}. \tag{19.16}$$

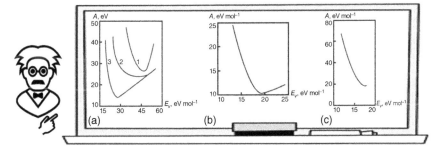

Figure 19.10 Energy cost of atmospheric pressure thermal plasma reduction of elements from their oxides as function of specific energy input: (a) vanadium from pentoxide V_2O_5 (1 – absolute quenching; 2 – ideal quenching; 3 – super-ideal quenching); (b) indium from In_2O_3, absolute quenching; (c) germanium from GeO_2, absolute quenching.

The melting temperature of indium oxide (In_2O_3) is 2183 K. Formation of In takes place at temperatures about 2300 K through decomposition of the intermediate oxide (In_2O). The energy cost of In production is shown in Figure 19.10b as a function of specific energy input for absolute quenching. The minimal energy cost is 10.3 eV atom^{-1} and can be achieved at a specific energy input of 19.1 eV mol^{-1} and conversion 93%. Also, the **direct production of germanium from its oxide** GeO_2 (melting temperature 1359 K) can proceed in thermal plasma as:

$$GeO_2 \rightarrow Ge + O_2, \quad \Delta H = 6.0 \text{ eV mol}^{-1}. \tag{19.17}$$

Germanium formation takes place at relatively high temperatures, exceeding 4000 K, through decomposition of the intermediate oxide (GeO). The energy cost of germanium production is shown in Figure 19.10c for the case of absolute quenching. The minimal energy cost is 19 eV atom^{-1} and can be achieved at a specific energy input of 18 eV mol^{-1} and conversion degree of about 95%.

A little different but relevant example here is **production of silicon monoxide (SiO) by SiO_2 decomposition in thermal plasma**. SiO is applied in electronics as a dielectric for capacitors, triodes, etc. Thermal plasma decomposition of SiO_2 at 1 atm and temperatures of 3500–5000 K leads to the conversion of 45–100% of dioxide into SiO. Electronics require production of very pure silicon monoxide. For this reason, plasma reduction of SiO_2 to SiO should be done without using any special reduction agents, which can contaminate the product. Silicon itself, however, can be used to intensify SiO_2 decomposition $SiO_2 + Si \rightarrow SiO + SiO$. The process is carried out at temperatures not exceeding the silicon melting point (1420 °C). However, the most attractive results were achieved in thermal plasma discharges at much higher temperatures (3000–6000 K), when all initial substances are gases. The most effective high-purity process has been organized in electrodeless RF-ICP discharges (Rykalin and Sorokin 1987).

19.7 Hydrogen-plasma Reduction of Metals, Metalloids, and Other Elements from Their Halides: Production of Boron, Niobium, Uranium, etc.

Reduction of pure metals, metalloids, and other elements from their oxides is applied only if the oxides are sufficiently pure. If this is not the case, *the oxides can be initially converted into halides (usually fluorides and chlorides), which are mostly volatile, even at relatively low temperatures, and therefore easier to purify.* Metals or other elements should then be derived from their halides, where plasma can be successfully applied. *The plasma reduction takes place in this case completely in the gas phase.* Reduction from halides (both fluorides and chlorides) can be organized with or without a special reduction agent, which is usually H_2. Avoiding hydrogen or other reduction additives permits one to produce uncontaminated elements but makes quenching more challenging. Thermal and nonthermal discharges can be applied for the halide reduction. *Reduction of oxides is mostly carried out in thermal plasmas more relevant to the treatment of solids. In contrast, treatment of gaseous halides can be more effective in cold plasmas.* Let us first consider as an example **production of boron by thermal plasma reduction of BCl_3 with hydrogen** in the slightly endothermic process:

$$(BCl_3)_{gas} + 3/2(H_2)_{gas} \rightarrow (B)_{solid} + 3(HCl)_{gas}, \quad \Delta H = 1.35 \text{ eV mol}^{-1}. \tag{19.18}$$

Performing the process in thermal RF-ICP discharges, see Figure 19.11, results in the most effective production of the high-purity fine boron powder, see Rykalin and Sorokin (1987). The quartz plasma reactor is surrounded by an inductor of an RF generator. Argon and BCl_3 are introduced tangentially, and H_2 is supplied in the opposite direction. Power of the RF generator varied from 2 to 30 kW at 2 MHz frequency. The purity of produced boron can be high (99.5%) without its deposition in the discharge area, at conversion 22–31%. **Production of niobium by thermal plasma reduction of gaseous pentachloride $NbCl_5$ with hydrogen** $(NbCl_5)_{gas} + 5/2(H_2)_{gas} \rightarrow (Nb)_{solid} + 5(HCl)_{gas}$ has been organized in a similar way in RF-ICP discharge. The degree of niobium reduction at atmospheric pressure is 98% at fivefold excess hydrogen over stoichiometry and the temperature in the reaction zone is about 1200 K.

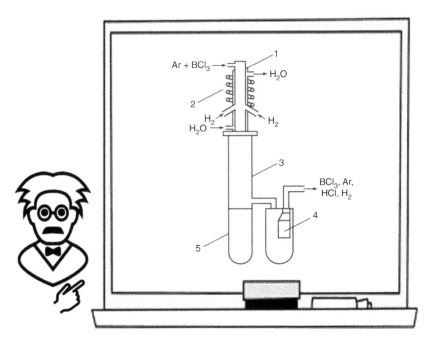

Figure 19.11 Schematic of the hydrogen-reduction of boron from BCl_3 in thermal plasma: (1) quartz plasmatron, (2) RF inductor, (3) quenching zone, (4) filter, (5) powder trapping system.

More challenging is one stage strongly endothermic process of **production of uranium by thermal plasma reduction of its gaseous hexafluoride UF_6 with hydrogen:**

$$(UF_6)_{gas} + 3(H_2)_{gas} \rightarrow (U)_{solid} + 6(HF)_{gas}, \quad \Delta H = 5.5\,\text{eV mol}^{-1}. \tag{19.19}$$

Conventional production of uranium from its hexafluoride is a two-stage process [first, production of $(UF_4)_{solid}$, and then $(UF_4)_{solid}$ reduction with calcium to metallic uranium in the solid-solid reaction]. The single-stage process (19.19) without significant losses of fluorine typical for conventional approach can be organized only in plasma through the step-by-step reduction from UF_6 to UF_5, from UF_5 to UF_4, from UF_4 to UF_3, and so forth, until it ends up with pure uranium. From all the intermediate fluorides, only one, UF_4, is a stable product and tends to remain in final products. Therefore, the single-stage production of metallic uranium requires very high temperatures (about 6000 K) and extremely fast quenching.

For more details on this important nuclear engineering subject see Tumanov (1981, 1989).

Thermal plasma hydrogen-reduction of tantalum (Ta), molybdenum (Mo), tungsten (W), zirconium (Zr), hafnium (Hf) titanium (Ti), germanium (Ge), and silicon (Si) from their chlorides has been demonstrated in arc and RF-ICP discharges in hydrogen with conversion up to 98%. Metals produced in these experiments are in powder form. Specific surface areas of the generated powder can be from 0.6 to 13 $m^2\,g^{-1}$. Hydrogen reduction of chlorides in thermal plasma can be applied not only to produce powder of individual metals but also to produce different intermetallic compounds. The composition of generated crystals copies that one of the gaseous chloride mixtures which permits production of the intermetallic compounds with required composition.

Although most investigations of the H_2 reduction of halides were carried out in thermal plasma, some results were obtained in nonthermal discharges. Hydrogen reduction of $TiCl_4$ in low-pressure glow discharge results in products consisting of 90% metallic titanium and 10% lower chlorides, $TiCl_2$ and TiCl. Following vacuum heating of the mixture results in the production of metallic titanium with relatively high purity (99.6%). Similar results were achieved in reduction of zirconium halides ($ZrCl_4$, $ZrBr_4$, ZrI_4, and ZrF_4), halides of boron, aluminum, scandium, arsenic, tin, and antimony. More detailed review on the subject can be found in Fridman (2008).

19.8 Direct Thermal Plasma Decomposition of Uranium Hexafluoride and Other Halides

The production of metals from their halides can be accomplished in plasma directly without using H_2 or other reducing agents, which is attractive due to avoiding hydrogen contamination of metals. On the other hand, the absence of H_2 makes the product quenching more challenging, because aggressive halogens are not bound with hydrogen and are supposed to coexist with produced metals. Let us consider the **thermal plasma decomposition of UF_6** as a relevant example:

$$(UF_6)_{gas} \rightarrow (U)_{solid} + 6(F_2)_{gas}, \quad \Delta H = 22.7\,\text{eV mol}^{-1}. \tag{19.20}$$

Uranium formation requires temperatures exceeding 5000 K. The energy cost of uranium production from UF_6 is shown in Figure 19.12a as a function of specific energy input for absolute and ideal quenching modes. The minimal energy cost in the absolute quenching mode is 49 eV atom^{-1} and achieved at specific energy input 46.7 eV mol^{-1} and conversion degree 95%. Ideal quenching permits disproportioning of lower fluorides into uranium and its tetrafluoride, UF_4. The minimal energy cost in the absolute quenching mode is 47 eV atom^{-1}. Figure 19.12a also shows for comparison of the energy cost of dissociation of UF_6, UF_5, and UF_4 in ideal quenching mode. The cooling rates required to quench products of direct plasma decomposition of UF_6 are presented in Figure 19.12b as a function of temperature in the quenching zone. One of the curves shows the minimal cooling rate for quenching the UF_4–F_2 mixture with initial temperature of 3200 K. Another curve shows the minimal cooling rates for quenching the U–F_2 mixture starting from plasma temperature of 6200 K and sufficient to produce metallic uranium. Effective condensation of UF_4 from the high-temperature reactive mixture requires the cooling rate to exceed $7 \cdot 10^7$ K s^{-1}. To quench metallic uranium, the cooling rates should be extremely high, 10^9 K s^{-1}, which is close to the maximum achieved quenching rates, see Ambrazevicius (1983), Tumanov (1981).

Between other examples of the direct thermal plasma decomposition of halides, we can point out reduction of halides of some alkali and alkaline earth metals (LiF, LiCl, LiBr, LiI, NaCl, KI, $BeCl_2$, MgI_2, $CaBr_2$, $BaCl_2$, etc.), halides of some elements of the groups 3–5 ($AlCl_3$, BF_3, $GaCl_3$, $SiCl_4$, GeF_4, $AsCl_3$, $SbCl_5$, etc.), halides of some transition metals (F_eCl_3, $CoCl_2$, $NiCl_2$, $CrCl_3$, MoF_6, WF_6, etc.), as well as halides of titanium ($TiCl_4$), zirconium ($ZrCl_4$), hafnium ($HfCl_4$), vanadium (VCl_4), and niobium ($NbCl_5$). More data on these plasma-metallurgical processes can be found in (Nester et al. 1988; Fridman, 2008).

19.9 Direct Decomposition of Halides and Reduction of Metals in Nonthermal Plasma

The effective direct decomposition of halides of metals, especially alkali metals, was first demonstrated in a strongly nonequilibrium microwave discharges, see McTaggart (1967). These plasma-metallurgical processes do not require any quenching and result in significant conversion degree. The most intensive dissociation takes place for gaseous **lithium and sodium halides**. The dissociation degree of **halides of potassium and cesium** is lower at similar discharge parameters. Specifically, the nonthermal plasma treatment of lithium iodide vapor (LiI) results in the production of metal with a conversion degree of 70%. The dissociation of NaCl vapor is characterized in these plasmas by a 30% conversion degree. Similar results were obtained in experiments with a direct nonthermal **microwave plasma decomposition of halides of alkaline earth metals** in nonthermal Products here are mostly mono-halides of the metals.

Nonthermal microwave and RF discharges of moderate pressure demonstrated effective **decomposition of higher metal halides, particularly ZrI_4 and $TiCl_4$**, see Moukhametshina et al. 1986). Application of nonthermal plasma provides effective stabilization of pure metals (specifically zirconium and titanium) without using H_2 and, therefore, without hydrogen contamination of produced metals. Both processes are of significant practical interest, especially the dissociation of ZrI_4, which is a crucial step in zirconium refinement in the nuclear industry. Energy cost of dissociation of ZrI_4 in nonthermal microwave and RF discharges at moderate pressure is summarized in Figure 19.13 (Moukhametshina et al. 1986). Complete conversion of the iodide and pure metal production without adding hydrogen has been achieved at an energy cost of about 40 kWh kg^{-1} of zirconium in the microwave discharge and at an energy cost of about 200 kWh kg^{-1} of zirconium in the RF discharge.

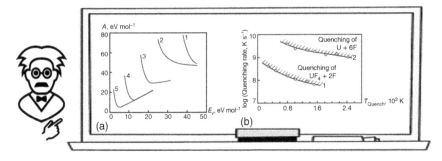

Figure 19.12 Direct decomposition of uranium fluorides in atmospheric-pressure thermal plasma: (a) energy cost at different quenching modes, (1) absolute quenching of uranium production from hexafluoride (UF_6), (2) ideal quenching of uranium from hexafluoride (UF_6), (3) ideal quenching of UF_3 production from tetrafluoride (UF_4); (4) ideal quenching of uranium tetrafluoride (UF_4) production from (UF_5), (5) ideal quenching of UF_5 production from hexafluoride (UF6); (b) threshold quenching rates required to avoid reverse reactions as functions of temperature in the quenching zone: (1) $UF_4 + 2F$, initial temperature 3200 K; (2) $U + 6F$, initial temperature 6200 K.

Figure 19.13 Energy cost of direct decomposition of zirconium tetraiodide as a function of specific energy input: triangles, experiments with microwave discharge; squares, experiments with radiofrequency discharge; solid line, kinetic modeling.

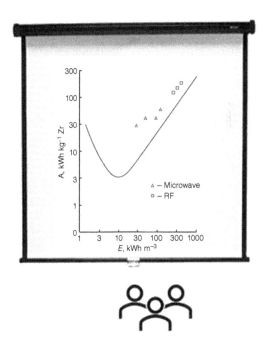

Complete dissociation of higher chlorides, which have higher bonding energies, is more challenging than that of iodides. The effective **one-step dissociation of $TiCl_4$ and production of metallic titanium** without adding hydrogen has been achieved in a strongly nonequilibrium stationary plasma-beam discharge (Atamanov et al. 1979). Very effective stabilization of products can be achieved in this case by product separation in plasma centrifuges, see Section 12.10, where plasma rotates in crossed electric and magnetic fields with extremely high rotation rates, see also Ivanov and Soboleva (1978) and Fridman and Kennedy (2021). Details on the physical and chemical kinetics of the processes can be found in Fridman (2008) and Ivanov and Soboleva (1978).

19.10 Synthesis of Nitrides and Carbides of Inorganic Materials in Thermal Plasmas

The most straightforward **synthesis of metal nitrides** can be organized by preliminary evaporation of metals in conjunction with gas-phase synthesis. Such gas-phase synthesis of nitrides of Ti, Zr, Hf, Al, Nb, and Ta has been demonstrated in thermal plasma of RF-ICP and arc discharges, see Polak et al. (1975), Rykalin and Sorokin (1987), Fridman (2008). Metal powder in these systems is injected for evaporation into high temperature N_2 plasma for further exothermic synthesis of nitrides. *More effective synthesis of nitrides can be organized by conversion of relevant gaseous chlorides.* Thus, the nitrides of elements, including titanium (Ti), silicon (Si), aluminum (Al), boron (B), zirconium (Zr), hafnium (Hf), niobium (Nb), and tantalum (Ta) were effectively produced by conversion of their chlorides in plasma in the presence of hydrogen and nitrogen, or ammonia NH_3. As an example, the yield of Si_3N_4 produced in this way is 80%. Consider as an example *synthesis of titanium nitride* from $TiCl_4$ interacting with H_2–N_2 is a slightly endothermic reaction:

$$(TiCl_4)_{gas} + 2(H_2)_{gas} + \tfrac{1}{2}(N_2)_{gas} \rightarrow (TiN)_{solid} + 4(HCl)_{gas}, \qquad \Delta H = 0.6\,eV\,mol^{-1}. \tag{19.21}$$

The maximum concentration of titanium nitride can be achieved at temperatures of 1000–1700 K. The yield of TiN depends on $TiCl_4$ dilution with nitrogen and hydrogen. Increasing the ratios H/Cl and N/Ti from 1 to 5 and from 1 to 100 results in a titanium nitride yield growth from 40% to 100%. The process has been demonstrated in microwave nitrogen plasma with injection of $TiCl_4$–H_2 mixture (see Fridman 2008). The nitride was synthesized in the form of a fine powder with a typical composition of $TiN_{0.8}$. The specific surface area of the powder was $45\,m^2\,g^{-1}$ and the particle size 10–50 nm. The energy cost of titanium nitride was $18\,kWh\,kg^{-1}$, which is close to the thermodynamic prediction of $15\,kWh\,kg^{-1}$. While most of the plasma production of nitrides is performed in thermal discharges, it can be also organized in cold plasmas. As an example of effective *nonthermal plasma production of nitrides*, we can mention the synthesis of silicon nitrides (Si_3N_4) from silane (SiH_4) and ammonia (NH_3) in low-pressure RF-ICP discharges.

Thermal plasma synthesis of carbides of metals and other elements can be organized in different ways, see Tumanov (1981), Vurzel (1970), Vurzel and Nazarov (2000), Mosse and Pechkovsky (1973), as well as Fridman (2008). First, it can be provided by heating dispersed or bricked mixture of the metal oxide and carbon in the form of graphite or soot (*solid-phase synthesis*) with formation of a carbide and CO using thermal discharges in inert gases. This method was applied to produce carbides of titanium, zirconium, lead, and bismuth. The efficiency here is limited by radiation heat losses and insufficient heat transfer inside of the brick, especially considering the decrease of density due to CO formation. Better in this regard is *condensed-phase synthesis of carbides from melt containing carbon compounds*. A relevant example here is the synthesis of carbides of uranium and plutonium from a melt containing their nitrites and carbon compounds (Tumanov 1981).

More effective is the *carbides synthesis by reaction of relevant oxides or halides with methane or other gaseous hydrocarbons*. As an example, titanium carbide can be produced from titanium oxide powder and methane $TiO_2 + CH_4 \rightarrow TiC + 2H_2O$. In this case, the TiO_2 powder carried by CH_4 is injected into argon thermal plasma generated by RF-ICP discharge. The production of carbide takes place at temperatures of 2000–4000 K. As an example of the gas-phase synthesis of carbides from gaseous halides and hydrocarbons, we can mention effective production of submicron boron carbide powder from gaseous boron trichloride and methane:

$$4(BCl_3)_{gas} + (CH_4)_{gas} \rightarrow (B_4C)_{solid} + 4(HCl)_{gas} + 4(Cl_2)_{gas}, \qquad \Delta H = 13.3\,eV\,mol^{-1}. \tag{19.22}$$

The process is stimulated by argon heated up inside of the RF-ICP discharge and takes place downstream from the thermal discharge, where reactive gases are injected into argon, see Figure 19.14 (Tumanov 1981). The yield of the carbide powder is 93–95%, composition of the boron carbide is close to the stoichiometric $B_{3.9}C$, size of the produced particles is 200–300 nm. We should note that the described approach can be applied not only for production of the carbide nano-powders but also for deposition of thin carbide films.

19.11 Plasma-chemical Production of Inorganic Oxides by Thermal Decomposition of Minerals, Aqueous Solutions, and Conversion Processes

Industrial production of multiple pure metals and other elements requires preliminary conversion of minerals or relevant solutions into oxides for further reduction and purification. It is especially challenging when the relevant minerals have very high melting temperatures, which attract interest to thermal plasma approaches. As an example of **production of inorganic oxides from minerals** with very high melting temperatures, let us consider the thermal plasma decomposition of zircon sand ($ZrSiO_4$) to

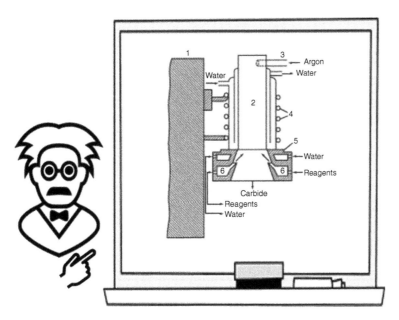

Figure 19.14 Thermal ICP reactor for boron carbide production by BCl_3 conversion in hydrocarbon plasma: (1) RF generator, (2) discharge chamber, (3) argon injection system, (4) RF inductor, (5) water-cooling system, (6) distributor of reagents.

produce zirconia (ZrO_2). Decomposition of the zircon sand $(ZrSiO_4)_{solid} \rightarrow (ZrO_2)_{solid} + (SiO_2)_{solid}$ requires temperatures exceeding 2050 K and results in the formation of a mixture of zirconia and silica. The heating of zircon sand is effectively provided by an energy-carrier gas, which is supplied to the reactor through arc discharges. A solid mixture of oxides produced in plasma is then treated by sodium hydroxide (NaOH), which leads to the formation of a soluble sodium salt extracting silica SiO_2 into solution ($SiO_2 + 2NaOH \rightarrow Na_2SiO_3 + H_2O$). Zirconia remains solid and can be easily separated as a product. General schematic of the technology is illustrated in Figure 19.15. Zirconia with purity of 99% is produced as hollow spheres with sizes of 100–200 nm; the energy cost of the zirconia is 1.32 kWh kg^{-1}. The relevant large-scale plasma technology has been accomplished with an arc power of 1 MW and productivity of 450 t year^{-1} of ZrO_2 (Tumanov 1981). Similar plasma approach has been demonstrated for production of manganese oxide (MnO) by decomposition of rhodonite ($MnSiO_3$) mineral, production of NiO by decomposition of serpentine minerals, and for production of different phosphorus oxides from tricalcium phosphate $Ca_3(PO_4)_2$ and fluorapatite $Ca_5F(PO_4)_2$, see Mosse and Pechkovsky (1973), as well as Fridman (2008).

As an example of **production of inorganic oxides from the relevant aqueous solutions**, let us consider thermal plasma decomposition of uranyl nitrate $[UO_2(NO_3)_2]$ solution, which is an important step in the nuclear fuel processing (Tumanov 1989):

$$[UO_2(NO_3)_2]_{solution} \rightarrow 1/3(U_3O_8)_{solid} + (NO)_{gas} + (NO_2)_{gas} + 7/6(O_2)_{gas} + x\,(H_2O)_{gas}. \qquad (19.23)$$

The aqueous solution is injected into an atmospheric air arc. Decomposition of the solution proceeds in three phases. The first one includes heating the droplets to the boiling temperature of the solution and partial evaporation of water. Complete evaporation of water occurs in the second phase. Further heating and decomposition of the residual salt take place in the third phase. A separator provides extraction of the solid product (U_3O_8); gases proceed to a condenser/absorber, where nitric acid is generated as a by-product. 300 kW arc in atmospheric-pressure air converts 80 kg h^{-1} of uranyl nitrate solution into U_3O_8 at temperature 4000 K. Conversion of uranyl nitrate into U_3O_8 is 99.8% in terms of uranium balance. The energy cost of the process is 2.2 kWh kg^{-1} of initial solution. Similar approach is effective for production of magnesium oxide (MgO) by thermal plasma decomposition of aqueous solution or melt of magnesium nitrate $Mg(NO_3)_2 \rightarrow MgO + NO + NO_2 + O_2$. Generally, the high-quality inorganic oxides of wide variety of metals can be effectively produced in plasma via thermal decomposition of aqueous nitrate solutions. The nitrate solutions are preliminarily produced in these technologies by treatment of relevant compounds with nitric acid. The high-quality ceramics produced in plasma are of special interest for the synthesis of high-temperature superconducting composites. This method was applied for production of such high-temperature superconducting composites as $YBa_2Cu_3O_x$, $YBa_2Cu_4O_x$, $Bi_2Sr_2Ca_2Cu_3O_x$, and $Bi_{1.7}Pb_{0.3}Sr_2Ca_2Cu_3O_x$. For more details see Fridman (2008).

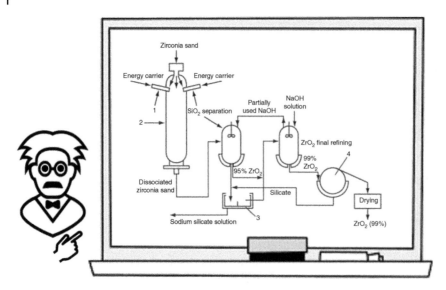

Figure 19.15 Zirconia (ZrO$_2$) production from zirconia sand (ZrSiO$_4$): (1) plasmatron; (2) plasma reactor; (3) centrifuge; (4) filter.

19.12 Production of Inorganic Oxides by Conversion of Relevant Halides with Water or Oxygen in Thermal Plasma

Considering the production of inorganic oxides by steam conversion of halides, we can use as an example *conversion of gaseous UF$_6$ into UO$_2$*, which is an important step in the nuclear fuel cycle. The process is effectively organized in arc discharges by injection of UF$_6$ in water vapor plasma (Tumanov 1989): UF$_6$ + 3H$_2$O → 1/3 U$_3$O$_8$ + 6HF + 1/6 O$_2$. The process temperatures 1050–1400 K at pressure 50 kPa are sufficient for almost complete conversion of uranium from UF$_6$ into U$_3$O$_8$, and fluorine into HF. At higher temperatures exceeding 1400 K, the production of uranium oxides is partially suppressed by the formation of gaseous uranyl fluoride (UO$_2$F$_2$). The thermal plasma process is conducted using a 200-kW arc discharge heating water vapor to the required temperatures; 150 kg h^{-1} of gaseous hexafluoride is supplied into the plasma reactor after the discharge. The molar ratio in the reactor is close to stoichiometric: UF$_6$:H$_2$O = 1 : 3. Uranium oxide is produced as powder with an average density of 4.5–5.7 g cm^{-3}; the specific surface area of the powder is 0.037–0.138 m^2 g^{-1}. The conversion degree of hexafluoride into oxide is very high (99%); the energy cost is 1.3–1.4 kWh kg^{-1} UF$_6$. The energy cost of U$_3$O$_8$ decreases at higher arc powers. The higher powers lead to the increased energy efficiency of the plasma-chemical reactor and power supply, and larger reactor volumes lead to the decrease of heat losses to the reactor walls. Similar organization of the thermal plasma process permits effective *steam conversion of silicon tetrafluoride (SiF$_4$) into silica (SiO$_2$) and HF* (SiF$_4$ + 2H$_2$O → SiO$_2$ + 4HF).

Thermal plasma *production of inorganic oxides is also effective using oxygen-conversion of the relevant halides*. A good example here is the *industrial-scale conversion of titanium tetrachloride into pigment titanium dioxide*: (TiCl$_4$)$_{gas}$ + (O$_2$)$_{gas}$ → (TiO$_2$)$_{solid}$ + 2(Cl$_2$)$_{gas}$, Rykalin and Sorokin (1987). Thermodynamic equilibrium is shifted in this case in the direction of products even at room temperature. The reaction, however, is kinetically limited and requires high temperatures, up to 1000–1500 K, when the process time decreases to 1–10 ms. Oxygen, air, or oxygen-containing gases are heated up to the required temperatures in thermal arc or RF-ICP discharges. TiCl$_4$ is injected into the oxygen plasma in the reactor chamber located after the discharge. Energy cost of the TiO$_2$ powder produced in the thermal air plasma is 2–3 kWh kg^{-1}. The energy cost can be reduced below 2 kWh kg^{-1} TiO$_2$ if enriched air (for example, 30% air and 70% oxygen) is applied as an oxygen-containing plasma gas.

19.13 Plasma-chemical Synthesis of Hydrides, Borides, and Carbonyls of Inorganic Materials

Plasma production of hydrides directly from elements can be organized in both thermal and nonthermal discharges. The relevant thermal technologies were effectively accomplished for the synthesis hydrides of tin (SnH$_4$)

and lead (PbH$_4$), see Shmakin and Marusin (1970). They were performed by heating H$_2$ in an atmospheric pressure RF-ICP discharge, and injection of tin and lead powders into the plasma. Nonthermal plasma approaches to production of hydrides were known for long time, see McTaggart (1967). Thus, hydrides of phosphorus, arsenic, and sulfur were produced in a glow discharge in hydrogen, when the elements were deposited on walls of the discharge tube. Different boron hydrides (B$_2$H$_6$, B$_{10}$H$_{16}$, B$_{20}$H$_{16}$) were also synthesized in a glow discharge in the mixture H$_2$:BCl$_3$ = 12 : 1 at pressure of 20 Torr with conversion up to 40%. Saturated germanium hydrides were synthesized in DBD at 300 Torr and discharge wall temperature of 78 °C. Hydrides of silicon were produced in DBD from silane at lower pressures with major products Si$_2$H$_6$ (66%), Si$_3$H$_8$ (23%), Si$_4$H$_{10}$ (11%), and even some Si$_8$H$_{18}$. Nonthermal plasmas can be especially effective in the hydride formation by hydrogen gasification of elements and by hydrogenation of thin films. Thus, vibrational excitation of H$_2$ in nonequilibrium plasma can effectively stimulate gasification of elements with the direct production of such hydrides as SiH$_4$, GeH$_4$, and B$_2$H$_6$ (Legasov et al. 1978b). Similar nonthermal plasma gasification stimulated by the dissociation of H$_2$ was also observed in production of hydrides of sulfur, arsenic, germanium, tin, tellurium, and selenium.

Nonthermal plasma synthesis of metal hydrides in thin surface films can be also effective. Consider as an example the solid-state direct synthesis of aluminum hydride (AlH$_3$) from the elements (Legasov et al. 1978a,b):

$$(Al)_{solid} + \frac{3}{2}H_2^* \left(X^1\Sigma_g^+, v\right) \rightarrow (Al)_{solid} + 3H \rightarrow (AlH_3)_{solid}. \tag{19.24}$$

This process proceeds through the intermediate formation of atomic hydrogen but requires low temperatures, because AlH$_3$ is stable only at temperatures below 400 K. The formation of atomic hydrogen stimulated by vibrational excitation takes place not only in the plasma volume (where dissociation through vibrational excitation requires 4.4 eV) but also on the surface of aluminum, where it requires only about 2 eV mol^{-1}.

The **plasma-chemical synthesis of borides** can be accomplished using different initial chemicals, not only in thermal but in nonthermal plasma as well. Borides of refractory metals were effectively produced in thermal Ar–H$_2$ RF-ICP discharges by simultaneous reduction of metal chlorides and BCl$_3$. Borides of vanadium, titanium, zirconium, and chromium were produced in thermal plasma by the interaction of their melts with boron hydrides (BH$_3$, B$_2$H$_6$). Different boron hydrides (B$_2$H$_6$, B$_{10}$H$_{16}$, B$_{20}$H$_{16}$) were synthesized in nonthermal plasma of a glow discharge at a pressure of 20 Torr. The process was organized under strongly nonequilibrium conditions in the mixture ratio H$_2$:BCl$_3$ = 12 : 1 (McTaggart 1967).

Direct plasma synthesis of metal carbonyls can be accomplished in nonthermal discharges by reactions of metals with excited CO molecules (Legasov et al. 1978a,b), which is especially important for the synthesis of carbonyls of chromium, molybdenum, manganese, and tungsten, which cannot be alternatively produced without solvents. The gaseous metal carbonyls are unstable at temperatures exceeding 600 K, which leads to specific restrictions of the nonthermal discharges.

The mechanism for plasma-chemical formation of metal carbonyls can be considered by taking as an example the synthesis of Cr(CO)$_6$ from chromium and vibrationally excited CO: (Cr)$_{solid}$ + 6(CO)$_{gas}$ → [Cr(CO)$_6$]$_{gas}$. This process includes the following four stages:

1. Chemisorption of the plasma-excited CO molecules on a chromium surface. The chromium crystal structure is characterized by a cubic lattice with valence 6 and coordination number 8. Therefore, each chromium surface atom can attach $k \leq 3$ carbonyl CO groups without the destruction of the lattice.
2. Formation of higher carbonyls ($k > 3$) requires taking a chromium atom with an attached CO group from the crystal lattice to a chemisorbed state.
3. Formation of chromium hexacarbonyl Cr(CO)$_6$ still adsorbed on the surface) takes place in the surface reactions of intermediate chemisorbed chromium carbonyl complexes.
4. The final stage is the desorption of volatile chromium hexacarbonyl Cr(CO)$_6$ into gas phase.

Summarizing the four stages for the synthesis of chromium hexacarbonyl, we can note that conducting the process in nonthermal plasma is supported by an interesting effect: intermediate carbonyls Cr(CO)$_m$ are stabilized on the surface due to chemisorption while vibrationally excited CO molecules stimulate the extraction of Cr atoms from the lattice. The final product, Cr(CO)$_6$, is not so strongly bound to the surface and can be effectively desorbed also by using energy accumulated in vibrational degrees of freedom of CO molecules. More details on these processes can be found in Legasov et al. (1978a,b), as well as in Fridman (2008).

19.14   Plasma-metallurgical High-temperature Material Processing Technologies: Plasma Cutting, Plasma Welding, and Plasma Melting

The principle of **plasma cutting** involves the constriction of an arc formed between an electrode (cathode) and a workpiece by a fine-bore copper nozzle. A schematic of the plasma cutting device is shown in Figure 19.16. Restricting the opening through which plasma gas passes increases the jet velocity and heat flux to the workpiece. Plasma temperatures in the system can exceed 20 000 K, providing fast melting. Gas velocities can approach the speed of sound, providing deep penetration of the plasma jet into a workpiece and removal of molten material, see Nemchinsky (1998, 2002, 2003). Plasma cutting is an alternative to the oxyfuel process, where the exothermic oxidation of metals provides heat for melting. Oxyfuel technology cannot be applied to cut metals such as stainless steel, aluminum, cast iron, and nonferrous alloys, which form refractory oxides. Plasma cutters can be applied in this case because their heat source is electric and not related to the formation of refractory metal oxides. A typical operating arc voltage is 50–60 V; the typical current is 50–80 A. The initiating voltage is as high as 400 V to start the arc between the cathode and the nozzle; then the arc is "transferred" by fast gas flow from the nozzle to the workpiece. Conventional plasma cutters use a tungsten electrode (cathode) and a nonoxidizing gas flow of argon, argon–hydrogen mixture, or nitrogen. To increase arc constriction and make "blowing away" of the "dross" more effective, a secondary gas shield can be introduced around the nozzle. The secondary gas is usually selected according to the metal being cut: air, O_2, or N_2 for cutting steel; N_2, Ar–H_2, or CO_2 for cutting stainless steel and aluminum. Additional arc constriction can be achieved by rotating the gas flow and applying a magnetic field surrounding the arc discharge. Instead of secondary gas, water can be injected radially into the arc around the nozzle. The injection of water additionally constricts the jet and considerably increases the temperature to about 30 000 K. The plasma system can also be operated with a water shroud, or even with a workpiece submerged 5–7.5 cm below the water surface. Water acts as a barrier, providing reduction of fume and noise and improving nozzle life.

Arc discharges for **plasma welding** are usually formed between a pointed tungsten electrode and a workpiece. Schematic of the plasma welding is shown in Figure 19.17. By positioning the electrode within the body of the torch, the arc can be separated from the shielding gas envelope. Plasma is then forced through a fine-bore copper nozzle, which constricts the arc. The electrode used for plasma welding is usually made from tungsten with 2% thoria; the plasma nozzle is made from copper. The normal combination of gases is argon for the plasma gas, and Ar plus 2–5% hydrogen for the shielding gas. By varying the bore diameter and plasma gas flow rate, we can distinguish three operating modes of plasma welding technology. The microplasma welding mode corresponds to low discharge currents of 0.1–15 A. This arc is stable even when the arc length is varied up to 20 mm. The microplasma mode is traditionally

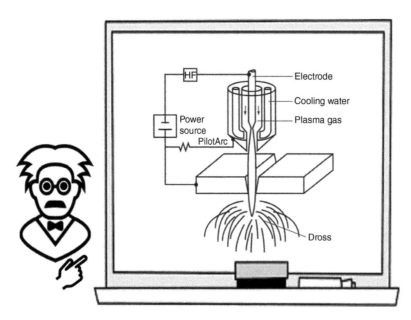

Figure 19.16   General schematic of a plasma cutting system.

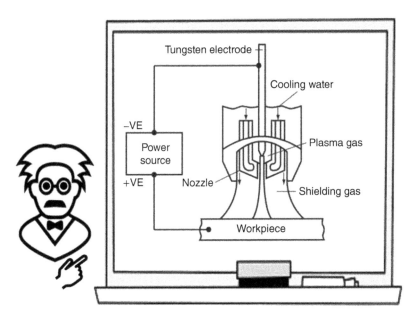

Figure 19.17 General schematic of a plasma welding system.

used for welding thin sheets (down to 0.1 mm thickness), as well as for wire and mesh sections. The medium current welding mode corresponds to higher currents (from 15 to 100 A) and like to what is known as tungsten inert gas conventional arc welding. The keyhole plasma welding mode corresponds to currents over 100 A and therefore to high gas flows. A very powerful plasma beam is created in this mode, which can achieve full penetration in a material, as in laser and electron beam welding. During welding, the hole progressively cuts through the metal with the molten weld pool flowing behind to form the weld bead under surface tension forces. This mode is characterized by high welding speed and can be used to weld thicker material (up to 10 mm of stainless steel) in a single pass. Plasma is also used for **melting and remelting of metals** as a source of process heat. Plasma melting makes use of the anode heat transfer characteristics of an arc between a cathode and metal (Heberlein 2002). The long process time (from 0.1 s to min) reduces the process instabilities.

19.15 Plasma Powder Metallurgy: Plasma Spheroidization and Densification of Powders

Plasma production of different powders in general and relevant plasma synthetic processes in powder metallurgy have been discussed earlier in this lecture. Let us now shortly discuss the plasma technologies for spheroidization and densification of different powders based on different thermal discharges, see for more details Boulos et al. (1994), Solonenko (1996). The powder densification and spheroidization process is "simple" with respect to other plasma processes. It takes place in thermal plasma by means of inflight heating and melting of feed material while passing through the discharge in the form of sintered or crushed powders. Molten spherical droplets formed in this way are gradually cooled down under free-fall conditions. The particle time of flight must be sufficient for complete solidification before reaching the bottom of reactor chamber. Finer particles are entrained by plasma gases and recovered in a filter. Different atmospheric pressure thermal plasmas are applied for such treatment of powders. Effective powder densification and spheroidization have been demonstrated in atmospheric-pressure RF-ICP discharges by Tekna Plasma Systems Inc. Their technique can melt relatively large particles and operates with a wide range of gases. Specifically, we can mention silica powder spheroidization performed in an open-air 100 kW ICP plasma. The thermal plasma treatment can produce dielectric powders with nano-sizes and to increase the specific surface area of the powder almost 1000-fold. Thermal plasma treatment of powders results not only in their spheroidization and densification but also in physical and chemical transformations, such as: (i) decreased powder porosity due to particle melting; (ii) surface oxidation or nitrification of particles; (iii) purification and refining of powders due to selective or reactive vaporization of specific impurities (for example Si and Zn impurities in tungsten powder can be decreased about threefold, Mn and Pb impurities in tungsten powder can be decreased 10 times); (iv) phase and chemical transitions (for example AlO_2 powder after plasma evaporation and fast quenching becomes $\gamma -Al_2O_3$); tungsten carbide (WC) powder loses carbon with increase of discharge power and plasma temperature forming new compounds: ($WC \rightarrow W_2C \rightarrow W$).

19.16   Problems and Concept Questions

19.16.1   Plasma Reduction of Metal Oxides with Hydrogen

What are the key advantages of using plasma approach for metal reduction from oxides with respect to conventional metallurgy? What is the key challenge of the plasma metallurgy?

19.16.2   Depth of Metal Oxide Reduction Layer in Nonthermal Hydrogen Plasma

Estimate the depth of metal oxide reduction layer using relation (19.7) at surface temperatures 300, 700, and 1100 K. Assume the interaction radius of hydrogen atoms in a metal structure is $r_0 \approx 0.4$ nm. To calculate the flux of atomic hydrogen from a discharge, assume a surface power of $1\,W\,cm^{-2}$, energy cost of atomic hydrogen in the discharge of 5 eV per atom, and probability for a hydrogen atom to reach the oxide surface 30%.

19.16.3   Nonequilibrium Surface Heating in Plasma Treatment of Thin Layers

Based on (19.11), derive the formula (19.12) for the heat flux Q_0 required to provide a temperature increase ΔT in a layer with thickness h. Calculate the discharge power per unit area required for $\Delta T = 700$ K in a layer with $h = 0.3$ cm. For treated materials, consider iron and iron oxide (Fe_2O_3).

19.16.4   Application of Arcs vs. RF-ICP Discharges in Plasma Metallurgy

What is the key advantage of using RF-ICP discharges in thermal plasma metallurgy with respect to conventional powerful arc discharges? What are the key challenges of the RF-ICP discharges?

19.16.5   Halides in Plasma Metallurgy

Why so many plasma-metallurgical processes involve processing of relevant metal halides (especially, fluorides and chlorides)?

19.16.6   Quenching Rate for Direct Plasma Reduction of Metallic Uranium from UF$_6$

Why the direct plasma reduction of uranium from its hexafluoride requires the record high quenching rates? Analyze approaches permitting to achieve cooling rates up to $10^9\,K\,s^{-1}$.

19.16.7   Decomposition of Titanium Tetrachloride TiCl$_4$ to Metallic Titanium in Nonthermal Plasmas

Is that possible to achieve the $TiCl_4$ dissociation to atomic titanium in a single direct electron impact even if electron energy is extremely high?

19.16.8   Production of Uranium Oxides by Decomposition of the Uranyl Nitrate [UO$_2$(NO$_3$)$_2$] Aqueous Solutions

Based on the data presented in Section 19.11, estimate the energy cost of plasma production of uranium from uranyl nitrate. Compare the result with the energy cost of uranium production by direct plasma dissociation of uranium hexafluoride (UF_6) discussed in Section 19.8.

Lecture 20

Plasma Powders, Micro- and Nano-technologies: Plasma Spraying, Deposition, Coating, Dusty Plasma-chemistry

20.1 Plasma Spraying of Powders as One of the Key Thermal Spray Technologies

 Thermal plasma application to produce powders of metals and other materials, to process them, to organize spheroidization and densification of powders has been discussed in the previous lecture. Let's focus now on the plasma spraying of powders. In this widely used thermal spray technology, the finely ground metallic and nonmetallic materials are deposited on a substrate in a molten or semi-molten state, see Kudinov et al. (1990), Zhukov and Solonenko (1990), Pawlowski (1995), Vurzel and Nazarov (2000), Fauchais (1997, 2004), Fauchais et al. (2001), Knight, et al. (1991), Knight and Smith (1998), Knight et al. (2005), Yoshida (2005), and Easter (2008). The thermal plasma heat source is usually based on direct-current (DC) arc or radio frequency inductively coupled plasma (RF-ICP) discharge.

 These thermal plasma systems provide very high temperatures, over 8000 K at atmospheric pressure, which allows melting of any material. This is a major distinctive feature of the plasma spraying. Although thermal plasma enables extremely high temperatures, the operational melting temperature in plasma spraying is usually kept at least 300 K below the vaporization or decomposition temperature to avoid losses in energy efficiency. Ground materials are either injected into plasma (RF discharges) or in the plasma jet (DC arcs), where they are heated, melted, softened, accelerated, and directed toward the surface or substrate being coated. Particles or droplets impacting the substrate are rapidly cooled, solidified, cooled further, contract, and build up incrementally to form a deposit. The basic coating building blocks, known as "splats," typically undergo cooling rates above 10^6 K s^{-1} in the case of metals. Individual splats are typically quite thin, about 1–20 μm; coating thicknesses range from 50 to 700 μm, which is thicker than most physical vapor deposits (PVDs) and chemical vapor deposits (CVDs).

The first industrial plasma spray torches based on DC arcs appeared in the 1960s. Later, in the 1980s, appeared those based on the thermal RF discharges. Today, thermal spraying effectively operates and consolidates virtually any material that exhibits a stable molten phase to produce coatings or deposits characterized by relatively homogeneous, fine-grained microstructures. Thermal spray is very versatile and is being used for several new applications, including rapid part prototyping, production of fiber-reinforced intermetallic matrix composites, coating of biomedical implants, and the production of "functionally gradient" materials. Plasma spraying, with the highest processing temperatures, is the most flexible in the thermal spray "family" of technologies with respect to materials and processing conditions. Plasma spray processes can operate in environments ranging from air to inert gases and to underwater conditions. They are commonly used especially for consolidating oxidation-sensitive and reactive materials.

20.2 DC-Arc Plasma Spray: Air Plasma Spray, VPS, LPPS, CAPS, SPS, UPS, and Other Specific Plasma Spray Approaches

A typical schematic of the DC-arc plasma spray is illustrated in Figure 20.1. These plasma spraying systems can be divided into three major categories: (i) atmospheric plasma sprays (APS); (ii) "vacuum" or low-pressure plasma sprays (VPS, LPPS); (iii) controlled-atmosphere plasma sprays (CAPS), which also include inert gas shrouded plasma spray (SPS) jets, and underwater plasma sprays (UPS). These distinctions are based on the level of interaction between the

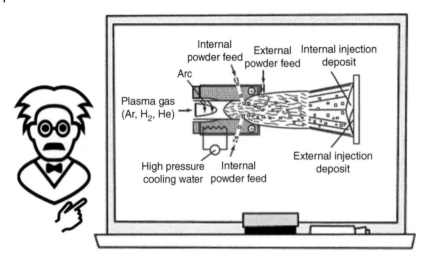

Figure 20.1 Schematic of DC arc thermal plasma spray system.

process jet and the materials being sprayed with the surrounding atmosphere and, thus, control the microstructure and properties of the sprayed materials. Let's discuss specifics of these DC arc plasma sprays.

The thermal plasma spray **APS systems** utilize high pressure vortex or axial gas-flow-stabilized non-transferred DC electric arcs. Arc currents range from 300 to 2000 A, and power from 20 to 200 kW. It results in plasma jets with temperatures of 3000–15 000 K. Such temperatures enable processing of any material that exhibits a stable melting point. A typical DC plasma spray torch, or gun, consists of a cylindrical, water-cooled, 1–2% thoriated tungsten cathode, which emits electrons thermionically when heated by an electric arc, located concentrically and coaxially inside and upstream of a cylindrical water-cooled copper nozzle. Electrons emitted from the cathode flow to the nozzle anode wall under the influence of the applied electric field, typically 30–80 V. The gas velocity may, in some cases, exceed $1500\,m\,s^{-1}$, corresponding particle velocities are as high as $500\,m\,s^{-1}$. A vortex flow is commonly used to spatially stabilize the arc and associated downstream plasma jet. Powders are injected into the plasma jet as fluidized streams of "carrier" gas and powder, either externally, via a tube directing the stream of powder particles radially into the hot gas jet immediately after it exits the nozzle, or internally, through powder feed ports through the nozzle wall itself. Internal injection produces longer particle heating or "dwell" times and improves melting efficiency, enabling higher-melting-point materials to be sprayed. Ar and N_2 are the most widely used carrier gases.

Oxidation during plasma spray is not a problem when spraying ceramics (Al_2O_3, Cr_2O_3, TiO_2, etc.); however, refractory metals such as W, Ta, Re, and "reactive" metals such as Ti and Al are readily oxidized during APS (especially in air), leading to inferior coating properties. Various approaches have been developed to overcome material oxidation during plasma spraying. Spraying at low pressures, in inert atmospheres, in enclosed pressure chambers, or using inert gas shrouds, can produce coatings with very low oxidation and porosity levels. This eliminates interaction between the molten particles and oxygen entrained into the jet from the surrounding air.

VPS/LPPS plasma spray systems operate with high-power non-transferred arc plasma jets at low pressures (50–200 Torr), which leads to significant extension of the jet. Spray distances during VPS are typically longer than in APS at the same power level and may be as high as 1 m. VPS processing also enables a process known as "reverse transferred arc" (RTA) to be used to "clean" the surfaces of metallic substrates prior to spraying. In the RTA systems, an auxiliary power supply is connected between the torch nozzle (anode) and the substrate, so a second arc is established between the torch and substrate. The presence of lower-work-function oxide films on the substrate surface favors preferential arc attachments, which quickly consume the oxide, leaving a metallurgically clean and active surface that promotes improved coating adhesion. VPS coatings generally exhibit low oxide contents, densities up to 99%, and bond strengths up to 70 MPa. Thus, the VPS-sprayed materials often have the same, or better, properties than relevant cast materials. It enables the VPS processes to be used to spray-form net shapes of "difficult-to-process" materials including ceramics, refractory metals, intermetallics, and composites.

Inert-gas SPS plasma spray systems achieve VPS-quality coatings under APS conditions (CAPS), thereby lowering the total processing cost. Several effective inert-gas shroud designs have been developed and reported, including porous metal nozzle inserts and discrete, multi-jet, parallel flow designs mounted downstream of the nozzle anode on conventional plasma spray torches. The **underwater plasma sprays (UPS)** have also been developed to overcome the oxidation issues associated with spraying in air without the need for costly and complex vacuum chambers, load locks, and so on. The UPS plasma systems can effectively operate in its own "bubble" of inert gas.

20.3 Radio Frequency (RF) Thermal Plasma Sprays

The RF plasma torches (Boulos et al. 1994), based on RF-ICP discharges, are electrodeless, in contrast to DC designs, and allow the use of a much wider range of plasma-forming gases, including air, O_2, H_2, and even hydrocarbons such as methane along with conventional Ar, Ar–H_2, Ar–He, and Ar–N_2. Continuous operation of RF discharges and torches is also unaffected by the electrode erosion and current carrying/cooling limitations of DC designs. Typical RF induction plasma spraying systems operate at pressures ranging from 50 Torr to atmospheric, power levels up to 150 kW, and frequencies 2–4 MHz; powders are injected axially.

RF induction plasma torches generate larger plasma volumes with gas velocities (10–$100\,\mathrm{m\,s^{-1}}$) an order of magnitude lower than in DC torches, resulting in residence times of 10–25 ms vs the ~1 ms typical for DC plasma spray. Coarser (>150 μm) powders can also be sprayed due to the larger volume of plasma region. Typical schematic of an RF plasma spray system is shown in Figure 20.2. RF plasmas are commonly "self-ignited" at reduced pressures but require pre-ionization at atmospheric pressure. Spraying is carried out usually at powers of 25–50 kW, pressures of 50–400 Torr, and spray distances of 150–300 mm; the latter is somewhat greater than in DC plasma spray.

Historically one of the major applications of RF plasma torches has been spheroidization of particulate materials due to larger and less-constricted volume of plasma as compared to DC systems and the relatively low gas velocities ($100\,\mathrm{m\,s^{-1}}$), which ensure efficient melting of injected materials. Effective RF-plasma-based vacuum spray (VPS) of various materials has been also demonstrated. It includes as examples the VPS the fiber-reinforced metal matrix composite "monotapes" as well as tungsten. Effective RF plasma spraying of near-net-shape structured Cr bipolar plates has been demonstrated for manufacturing solid oxide fuel cells and carbon fiber-reinforced aluminum coatings.

More details on the subject can be found in Boulos et al. (1994), Boulos (1985, 1992), and Jiang et al. (1993).

20.4 Thermal Plasma Spraying of Monolithic Materials

Three types of deposits can be thermally sprayed: (i) monolithic materials, including metals, alloys, intermetallics, ceramics, and polymers; (ii) composite particles, such as heat-resistant materials made of ceramic and sintered

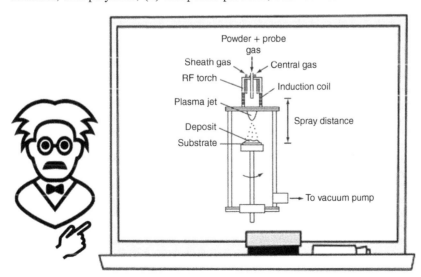

Figure 20.2 Schematic of RF thermal plasma coating system.

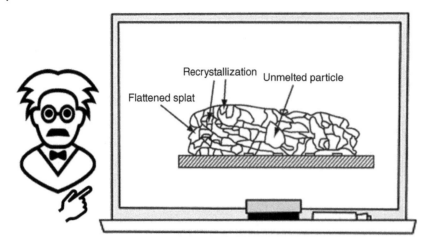

Figure 20.3 Illustration of microstructure of an "as-sprayed" metallic deposit.

metal (so-called cermets like WC/Co, Cr_3C_2/NiCr, WC/CoCr, NiCrAlY/Al_2O_3), reinforced metals (NiCr + TiC), and reinforced polymers; (iii) layered or functionally gradient materials (FGMs). Let's consider first the spraying of monolithic materials starting with the **thermal plasma spray of metals.** Tungsten, molybdenum, rhenium, niobium, titanium, superalloys (nickel, iron, and cobalt base), zinc, aluminum, bronze, cast iron, mild and stainless steels, NiCr and NiCrAl alloys, cobalt-based "stellites," Co/Ni-based tribaloys, and NiCrBSi "self-fluxing colmonoy" materials have all been thermally spray consolidated, either as coatings or free-standing deposits. A key feature of thermally sprayed deposits is their fine-grained structures and micro-columnar orientation. Plasma-sprayed metals have reported grain sizes of less than 1 μm prior to postdeposition heat treatment and associated grain growth/recrystallization. Grain structure across an individual splat thus normally ranges from 10 to 50 μm, with typical grain diameters of 0.25–0.5 μm, due to the high cooling rates (10^6 K s^{-1}). The "as-sprayed" microstructure of a typical metallic coating is shown in Figure 20.3.

The **thermal plasma spray of ceramics** is also very effective. Oxide ceramics such as Al_2O_3, ZrO_2 (stabilized with MgO, CeO, or Y_2O_3), TiO_2, Cr_2O_3, MgO, and hydroxyapatite; carbides such as Cr_3C_2, TiC, Mo_2C, SiC (with supporting metal matrix), and diamond; nitrides such as TiN and Si_3N_4; and spinels or perovskites such as mullite and superconducting oxides have all been deposited by plasma spray. Diamonds, SiC, and Si_3N_4 are especially suited to plasma spray consolidation due to high jet temperatures. Processing and materials flexibility as well as high temperatures give plasma spraying a leading role in spraying ceramic thermal barrier coatings (TBCs). APS is most suited to depositing ceramics and has found applications in TBCs, wear coatings on printing rolls (Cr_2O_3), and for electrical insulators (Al_2O_3). The microstructures of sprayed ceramics are like those of metals, with two important exceptions: grain orientation and microcracking. Inter- and intra-splat microcracking is widespread in plasma-sprayed ceramic coatings due to accumulation of highly localized, residual, quenching stresses. The microcracking of splats is a major contributor to the effectiveness of TBC, even under high-temperature gradients and moderate strains – conditions under which conventionally formed bulk ceramics would fail. Thermal spray consolidation and **forming of intermetallic powders** can be also very effective. Thermal spray's high heating and cooling rates reduce the segregation and residual stresses that ordinarily limit the formability of these brittle materials.

20.5 Thermal Plasma Spraying of Composite Materials

Difficult-to-process composites can be readily produced by the thermal spray. Also, the plasma spray is the process of choice for the most reactive matrix materials. Particulate-, fiber-, and whisker-reinforced composites have all been produced and used in various applications. Particulate-reinforced wear-resistant coatings such as WC/Co, Cr_3C_2/NiCr, and TiC/NiCr are the most common applications. Figure 20.4 illustrates the diverse forms of composites that can be thermally spray formed. Whisker particulates can be incorporated into sprayed deposits using so-called engineered powders, mechanical blending, or by co-injecting dissimilar materials into a single spray jet. Mechanical

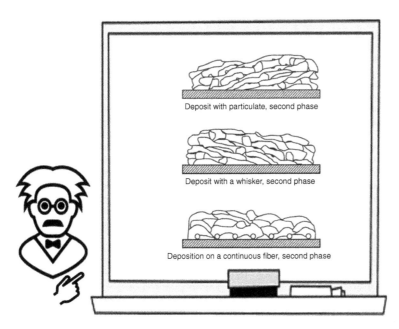

Deposit with particulate, second phase

Deposit with a whisker, second phase

Deposition on a continuous fiber, second phase

Figure 20.4 Illustration of different possible microstructures of the thermal-spray-formed composites.

blends and co-injection, although useful, have been found to result in segregation of the reinforcing phase, due to size and/or density differences and degradation of the second-phase whiskers. Thermal plasma spray composites can have reinforcing phase contents ranging from 10 to 90% by volume, where the metal matrix acts as a tough binder, supporting the reinforcing phase. The ability to consolidate such fine-grained, high-reinforcing-phase-content materials is a major advantage of thermal spray over other methods used to produce composites.

Thermal plasma spraying of composite materials with discontinuous reinforcements, such as particulates or short fibers, is accomplished by spraying powders or powder blends. Techniques to produce continuous fiber-reinforced materials which overcome the "line-of-sight" limitations of thermal spray processes were developed. This includes the "monotape" fabrication techniques, where continuous fibers are prewrapped around a mandrel, and a thin layer of a metal, ceramic, or intermetallic matrix material is sprayed. The fibers are thus encapsulated within thin monolayer tapes which are subsequently removed from the mandrel for consolidation to full density by hot pressing with preferred fiber orientations, producing continuously reinforced bulk composites. The discontinuously reinforced composites produced using thermal spray utilize either composite powders or direct reactive synthesis. Powders can be produced mechanically, chemically, thermos-mechanically, or by high-temperature synthesis. "Engineered" powders are defined as those in which different phases are incorporated to produce a "micro-composition" of the final desired structure. Blended powders are produced by mixing the required proportions of a binder phase (a metal, intermetallic, or ceramic) together with a reinforcing phase. The matrix and reinforcing phases may segregate over time during shipping, handling, and feeding into the thermal spray process and often yield poor "as-sprayed" deposit uniformities. Many mechanical blends can, however, be "agglomerated" and fixed using an organic binder, which reduces their sensitivity to handling. Spherical agglomerates have been shown to flow and feed better, and spray drying has been used to produce such materials. Agglomerated powders have enabled thermally sprayed deposits with improved uniformity to be produced, although the high viscosity and shear and thermal forces acting on injected agglomerated particles tend to break the agglomerates apart, leading to segregation and direct exposure of the second phase to the thermal spray jet. The mechanical stability of agglomerated composite particles can be improved by agglomeration and sintering, which is a solid or rapid "matrix" melting and cooling process. Metal matrix/ceramic hard-phase reinforced powders (for example, WC/Co, $Cr_3C_2/NiCr$, $NiCrAlY/Al_2O_3$, etc.) can be sintered in the solid state by heating in a protective atmosphere furnace, sometimes using a fluidized bed, followed by a gentle milling to break any weak inter-agglomerate bonds.

The retained porosity, although not a source of phase segregation in thermally sprayed deposits, can lead to higher as-sprayed porosity and increases the spray jet melting enthalpies required to melt the powders. Melting of metallic binding phases and retention of spherical particle morphologies can be achieved by processing the powders through a thermal plasma jet. Known as "plasma densification," this process produces essentially spherical powders exhibiting near 100% particle densities and uniform distributions of reinforcing phases. Plasma-densified powders produce the most uniform thermally spray consolidated deposits.

20.6 Thermal Plasma Spraying of Functionally Gradient Materials (FGMs), Reactive Plasma Spray Forming

The functionally gradient materials (FGMs) are important for systems and devices to be used in applications subject to large thermal gradients, requiring lower-cost clad materials for combinations of corrosion and strength or wear resistance, and perhaps improved electronic material structures for batteries, fuel cells, and thermoelectric energy conversion devices (Knight et al. 2005). The most immediate application for the FGMs is in thermally protective claddings, where large thermal stresses could be minimized, and component lifetime improved by "tailoring" the coefficients of thermal expansion, thermal conductivity, and oxidation resistance. These FGMs TBC are used in turbine components, rocket nozzles, chemical reactor tubes, incinerator burner nozzles, or other furnace components. Figure 20.5 illustrates an example of a plasma-sprayed FGM for the protection of copper using a layered ceramic FGM structure.

Thermal plasma spray forming can produce continuous gradations of metals, ceramics, and intermetallics, either in alternating stepwise layers or as micro-laminates. Materials could be sprayed to replicate the mechanical behavior of a material, for example, by making tungsten shear locally in high-impact applications. Normally, under high impact, tungsten, and its alloys behave in a ductile fashion, exhibiting significant plastic deformation, or "mushrooming." When localized shear behavior is required, very thin, alternating layers of intermetallic materials can be deposited concurrently while spray forming the bulk tungsten alloy. Thermal spray forming enables micro-laminate structures to be formed, which can be used to locally modify the overall mechanical behavior. The same approach could also be used to modify the electrical or thermal properties of materials. Thermally sprayed FGMs have also been proposed as strain-controlled coatings and/or structures and for electrical devices, as shown schematically in Figure 20.6 (Knight et al. 2005). The powder production route for materials permits oxides and/or brittle intermetallics with unique electrical and thermal properties to be produced. Figure 20.6 shows how the oxide-based fuel cells can be spray-formed by depositing alternating layers of oxide and metallic electrodes, and oxide electrolytes. Figure 20.6 also shows the spraying of alternating layers of thermoelectric materials (FeSi$_2$) and the concurrent deposition of metallic electrodes. This processing combines the grading capability of thermal spray together with its ability to spray materials with widely varying melting points, thus realizing the advantages of plasma spray forming.

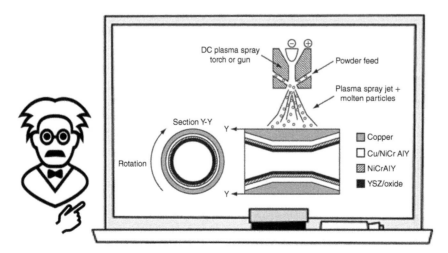

Figure 20.5 Illustration of plasma-sprayed functionally gradient material (FGM) for burner nozzle.

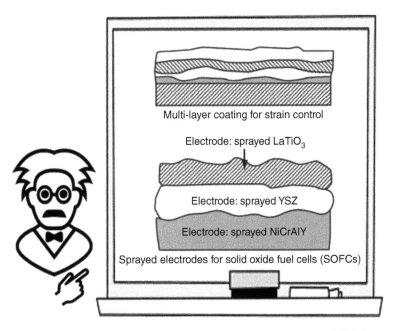

Figure 20.6 Schematic examples of functionally gradient material (FGM) thermally sprayed using plasma spraying system.

Let's now focus shortly on the concept of **reactive plasma spray forming.** The thermally sprayed deposits considered above are usually less than 500 μm thick. Thermal spray forming, however, can yield deposit thicknesses more than 25 mm. Free-standing shapes can be produced by spraying onto sacrificial mandrels, which are then mechanically or chemically removed after spraying. Plasma spray forming, a rapid particulate consolidation, has an inherent suitability for chemical synthesis. It is strong candidate for synthesizing multi-phase, advanced materials, which is usually referred to as reactive plasma spray forming (Knight et al. 2005). CAPS, adapted for reactive plasma spraying, has been developed to combine controlled dissociation and reactions in thermal plasma jets, for in situ formation of new materials, or to produce new phases in sprayed deposits. Reactive plasma spray forming is emerging as a viable method for producing a wide range of advanced materials. This process allows "reactive" precursors to be injected into the particulate and/or hot gas streams. These reactive precursors may be liquids, gases, or mixtures of solid reactants which, on the contact with the high-temperature plasma jet, decompose or dissociate to form highly reactive and ionic species which then react with other heated materials within the plasma jet to form new compounds. Reactive plasma spray applications include the synthesis of composite materials, shaped brittle intermetallic alloys, reinforced or toughened ceramics, and tribological coatings with hard or lubricating phases formed in situ. Reactive plasma spray forming enables a wide range of materials to be produced, for example, Al with AlN, Al_2O_3, or SiC reinforcements; NiCrTi alloys with TiC or TiN; intermetallics such as TiAl, Ti_3Al, $MoSi_2$, and other ceramics with oxides, nitrides, borides, and/or carbides. All of these have been produced in situ in reactive thermal plasma jets.

20.7 Microarc (Electrolytic Spark) Oxidation Coating: Aluminum Coating in Sulfuric Acid

The microarc oxidation coating is related to microdischarge plasma chemistry not in the gas phase but in the liquid phase, or to be more exact – in an electrolyte. That is why the coating process is often called electrolytic-spark oxidation. Such protective coatings can be arranged in on articles of certain metals like Al, Ti, Ta, Nb, Zr, Be, and their alloys. The main idea of electrolytic-spark oxidation is to sustain numerous electric sparks (or microarcs) at the surface of a workpiece placed in an electrolyte, which results in buildup of a protective oxide layer. Oxide coating deposition occurs in the microarc discharges, generated at the electrolyte–anode interface by applying pulsed (or sometimes DC) voltage exceeding 100 V to the electrolytic system, see Kharitonov et al. (1987, 1988), Yerokhin et al. (2003), Meyer et al. (2004), and Long et al. (2005). The method allows coatings to be formed with a set of properties that cannot be obtained by any other technique: uniform protective characteristics of oxide films with thickness 10–100 μm, high lifetime corrosion and wear resistance, adhesion over 5000 N cm^{-2}, and high thermal stability. Microarc oxidation coatings were initially designed especially for applications in nuclear reactors.

Let's consider an example of alumina (Al$_2$O$_3$) protective-coating formation on the surface of aluminum placed as an anode in a concentrated sulfuric acid electrolytic system with pulsed overvoltage exceeding 300 V (Kharitonov et al. 1987, 1988). Deposition of the Al$_2$O$_3$ oxide film on the surface of the aluminum anode takes place in a series of microdischarges (also called microarcs) at the anode–electrolyte interface. Major characteristics of the process can be summarized in the following ten statements: (i) light emission from the electrolyte–anode interface follows each voltage pulse with some delay and is well correlated with the current pulse; (ii) duration of the light emission pulses and related current pulses (1–3 mA) are 3–5 μs; (iii) when the amplitude of current pulses is stabilized, the required voltage V (in volts) grows with the film thickness d (in microns) as $V = 160 + \sqrt{1500d}$; (iv) specific resistance of the deposited coating ρ (in MOhm m) grows linearly with the film thickness d (in microns) as $\rho = 18.6 + 0.24d$; (v) the total electric charge crossing the oxide film corresponds to the electrochemical oxidation $3O^{-2} + 2Al^{+3} \rightarrow Al_2O_3$; (vi) the amount (and composition, 99% oxygen) of gas produced on the anode do not depend on the film thickness and is proportional to the total transferred charge; (vii) formation of the oxide film on the anode is accompanied by formation of passive colloid sulfur in the electrolyte; (viii) each current pulse corresponds to formation of a bright microdischarge (microarc) at the anode–electrolyte interface with the semispherical shape of radius about 100 μm; (ix) the microdischarge radiation is the superposition of two blackbody spectra: one corresponding to a relatively "cold" temperature of 1600–2200 K (the semispherical microdischarge), and another corresponding to relatively "hot" temperature 6800–9500 K (the hot nucleus, $2 \cdot 10^4$–$5 \cdot 10^5$ times smaller); (x) technological characteristics of the oxide films (adhesive, protective, and dielectric properties) are determined by the porosity of the films: decreasing porosity improvs film characteristics and can be achieved by optimization of the amplitude, duration, and frequency of the voltage pulses, as well as the current density.

20.8 Plasma Chemistry of the Microarc Oxidative Coating of Aluminum in Concentrated Sulfuric Acid Electrolyte

 The sophisticated plasma-chemical mechanism of the microarc oxidation in electrolyte includes the following stages. When a voltage pulse is applied, it induces current and a redistribution of electric field in the electrolyte. *A dielectric film already deposited on the anode leads to the formation of an electric double layer near the anode surface. Formation of the electric double layer results in localization of the major portion of the applied electric field in the anode vicinity. The electric field near the anode becomes high enough for thermal breakdown of the dielectric film and formation of a narrow conductive channel in the dielectric film.* The conductive channel is then heated up by electric current to temperatures of about 10 000 K. Thermal plasma generated in the channel is rapidly expanded and ejected from the narrow conductive channel into the electrolyte, forming a "plasma bubble" characterized by a very high temperature gradient and electric field. The energy required to form and sustain this bubble is provided by Joule heating in the narrow conductive channel in the dielectric film.

High current density in the channel is limited by conductivity in the relatively cold plasma bubble, where the temperature is about 2000 K. Electric conductivity in the bubble is determined not by electrons but by negative oxygen-containing ions. Further expansion of the plasma bubble leads to its cooling and to a decrease in the density of charged particles; it also results in a decrease of electric current and in a significant reduction of Joule heating in the conductive channel in the dielectric film, and finally in a cooling down of the channel itself. Product of the plasma-chemical process (γ-Al$_2$O$_3$) is condensed on the bottom and sidewalls of the channel. The oxide film grows mostly inside the substrate material. The described sequence of events takes place during a single voltage pulse. When the next voltage pulse is applied to the electrolytic system, new thermal breakdown and new microarcs occur at another surface spot where the dielectric film is less developed. This effect provides uniformity of the microarc oxidation coating.

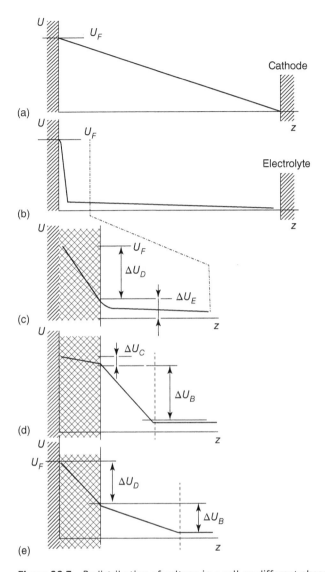

Figure 20.7 Redistribution of voltage in a cell on different phases of an elementary discharge: (a) at the initial moment of applying electric field; (b, c) at the end of the pre-breakdown stage (U_D is the voltage drop across the dielectric film; U_E is the voltage drop across the electrolyte); (d) at the moment of the developed discharge (U_E is the voltage drop in the channel; U_B is the voltage drop in the bubble); (e) at the final stage of the discharge and condensation of reaction products.

Let's shortly discuss the breakdown of the oxide film, forming the microarc discharge, and the following plasma chemistry. A voltage pulse applied to the electrolyte (concentrated sulfuric acid) stimulates the rearrangement of electric charges there. A layer of adsorbed HSO_4^- ions is formed on the Al_2O_3 dielectric barrier previously formed on the aluminum anode. Rearrangement of electric charges results in a redistribution of the electric field in the electrolyte, which is illustrated in Figure 20.7. Most of the voltage change and the highest electric field become concentrated at the oxide film. The voltage drop in the electrolyte decreases to about 30 V, which is due to the electrolyte resistance and leakage current across the thin film. Before breakdown of the thin dielectric film, the system can be represented by an equivalent electric circuit, shown in Figure 20.8a. The total current in the electrolytic cell includes a leakage current through the film as well as current charging of the electric double layer related to the adsorption of HSO_4^- ions on the anode. The evolution of current and voltage in the pre-breakdown stage of the process is shown in Figure 20.8b. Although the total voltage remains almost constant, the electric field in the thin film grows rapidly toward breakdown conditions. Thermal breakdown of the dielectric film occurs at electric fields of 10^5–$10^6 \, V \, cm^{-1}$. The physical nature of the thermal breakdown is related to the thermal instability of the local leakage currents through the dielectric film with respect to Joule overheating. If the leakage current is slightly higher at one point, Joule heating and, hence, temperature increases. The temperature increase results in

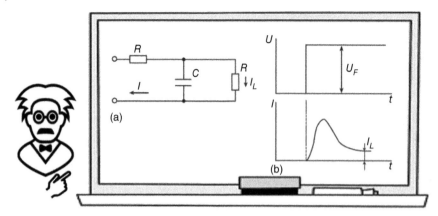

Figure 20.8 The pre-breakdown phase of the oxide film in microarc discharge: (a) equivalent electric circuit for the cell; (b) evolution of current and voltage.

a growth of local conductivity and the leakage current closing the loop of thermal instability. An exponential temperature growth to several thousand degrees at leads to the formation of a narrow plasma channel in the dielectric, which determines the breakdown. The phenomenon is usually referred to as **thermal breakdown of dielectrics**, see Section 16.2 (Eq. 16.1 and 16.2) regarding this type of breakdown in liquid dielectrics. More details on the thermal breakdown of solid oxide dielectrics and especially Al_2O_3 barrier coatings as a function of the film depth can be found in Fridman (2008).

The breakdown of the Al_2O_3 film in the microarc aluminum oxidation is a fast process requiring about 0.1 μs. The breakdown leads to the formation of a narrow plasma channel with a temperature of about 10 000 K in the coating material. The high temperatures result in intensive evaporation of sidewalls (Al_2O_3) and bottom (Al) of the channel. The metal evaporation from the bottom of the channel is much more significant. Pressure in the channel grows to thousands of atmospheres, which leads to ejection of the hot plasma into the electrolyte, forming a "plasma bubble" with a temperature of 1000–2000 K (Kharitonov et al. 1987, 1988). High electric conductivity in the plasma channel leads to shifting of the high electric fields from the channel to the "bubble," see Figure 20.7. These electric fields provide drift of negatively charged particles from the bubble into the channel. Considering that temperature in the plasma bubble is not sufficient for thermal ionization and that the electric field there ($2–5 \cdot 10^4 \, \mathrm{V \, cm^{-1}}$) is not sufficient for direct electron impact ionization, the negative oxygen-containing ions from the electrolyte make a major contribution in negative charge transfer from the bubble into the channel. Molar fractions of the major neutral components of the "bubble" plasma are H_2O, 40.7%; SO_2, 36.7%; O_2, 18.3%; SO_3, 4%; H_2SO_4, 0.12%; and OH, 0.08%. The ionization degree in the bubble plasma is quite low: $1–2 \cdot 10^{-8}$. The major negative ion is SO_2^-; its density exceeds that of free electrons by about 10^4. Major components in the thin and hot plasma channel are O atoms, 60.4%; Al^+ ions, 24.6%; and S atoms, 14.2%. Mass, electric charge, and energy transfer processes taking place in bubble and plasma microchannels are illustrated in Figure 20.9. Expansion of the plasma bubble leads to its cooling down and to a decrease of the flux of charged particles into the microchannel, which finally results in a temperature decrease in the channel. The bubble shrinks, reaction products move into the plasma channel, and alumina condenses at the bottom and walls of the channel. Because of fast quenching of products in the microchannel, the alumina is produced in the form of a metastable-phase γ-Al_2O_3.

20.9 Direct Micro-patterning, Micro-fabrication, Micro-deposition, Micro-etching, and Surface Modification in Atmospheric Pressure Nonequilibrium Plasma Microdischarges

The most widely used technique for creating microstructures in microelectromechanical systems (MEMS) and integrated circuits (ICs) is photolithography. Photolithography with typical pattern resolution from 80 nm to 10 μm is a multi-stepped and costly process, which involves application, masking, exposure, and development of photoresists to make polymer micropatterns. Then chemical, physical, or plasma deposition and etching techniques are applied to form metal, crystal, or ceramic microstructures.

It attracts attention to the *direct **microplasma micro-patterning** creating microstructures about 10 μm and below at elevated pressures in a single step.* These microdischarges can be arranged in a special array for simultaneous treatment of large surfaces. They can effectively operate at atmospheric pressure due to intensive cooling of the micron-size discharges. **Micropatterning and microfabrication** can be performed using different types of atmospheric pressure microdischarges, including the DC normal

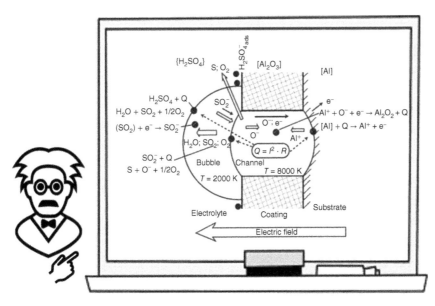

Figure 20.9 Major processes during the electrolytic spark microdischarges: dashed arrows, thermal fluxes; hollow arrows, mass transfer; solid arrows, charge transfer.

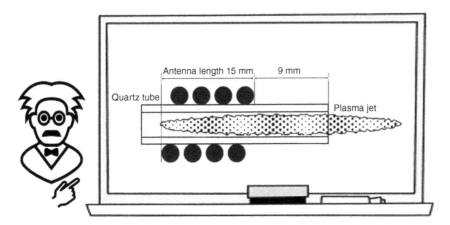

Figure 20.10 Schematic of the atmospheric pressure microplasma jet etcher.

micro-glow discharge, microwave microplasma, a miniaturized ICP source. These discharges, their parameters, and specifics for applications were discussed in Sections 14.7 and 14.8.

As an example, effective **micro-etching** of Si wafers was demonstrated using a microscale RF-ICP discharge. The miniaturized inductively coupled atmospheric pressure thermal plasma was generated at very high frequency (100 MHz) in a 1-mm diameter discharge tube, see Figure 20.10. Etch rates can be high in the micro-etcher: up to 4000 μm min^{-1} for Si wafers; and up to 14 μm min^{-1} for fused silica glass wafers. The **micro-deposition** can be illustrated by the processes organized in the DC normal micro-glow discharge at atmospheric pressure (Staack et al. 2005, 2006, 2008, see also Section 14.7). A micropattern was formed in this case on a silicon substrate by using microplasma in ambient air, which results in oxidation of the silicon. The resolution of the microplasma-generated patterns is approximately the size of the discharge. The same DC normal micro-glow discharge was also applied for PECVD of diamond-like carbon (DLC) coatings. In contrast to conventional plasma techniques of PECVD of DLC, application of a microplasma permits deposition at atmospheric pressure and allows deposition rates as fast as microns per minute. The process is accomplished in a micro-glow discharge in atmospheric-pressure hydrogen with a 2% admixture of methane. Higher yields of DLC are observed at higher substrate temperatures and lower methane concentrations in the mixture. The DC normal micro-glow discharge was also used to demonstrate application of microplasma for the **microscale differential wettability patterning**. Because nonthermal plasma is effective in improving of wettability of different surfaces, microplasma can be applied for differential wettability patterning at the microscale. The plasma-stimulated differential wettability patterning permits further micro-deposition using conventional liquid-phase techniques without masking and photolithography. For more details on the subject, see Staack et al. (2005, 2006, 2008), as well as Fridman (2008).

20.10 Nanoparticles in Cold Plasma, Physics and Kinetics of Dusty Plasma in Low-pressure RF Silane Discharges

Plasma physics and plasma chemistry are coupled in the so-called dusty plasma systems, see more details on this subject in Bouchoule (1993, 1999), Bouchoule and Boufendi (1993), Garscadden (1994), Watanabe et al. (1988), Kortshagen et al. (2016), Bhandarkar et al. (2003), and Melzer (2019). Basic plasma physics of the dusty plasmas has been discussed in Lecture 15. Applied aspects of the dusty plasmas are related to electronics, environment control, powder chemistry, and metallurgy. Investigation of particle formation and behavior has been stimulated by contamination phenomena of industrial plasma reactors used for etching, sputtering, and PECVD. For this reason, the process of particle formation in low pressure glow or RF discharges in silane (SiH_4, or mixture SiH_4 – Ar) has been extensively studied. Many efforts have dealt with the detection and dynamics of relatively large size (more than 10 nm) nanoparticles in silane discharge. Such particles are electrostatically trapped in the plasma bulk and significantly affect the discharge behavior. The most fundamentally interesting phenomena take place in the early phases of the process, starting from an initial dust-free silane discharge. During this period, small deviations in plasma and gas parameters (gas temperature, pressure, electron concentration) could alter cluster growth and the subsequent discharge behavior. Experimental data concerning super-small particle growth (diameter as small as 2 nm) and negative ion kinetics (up to 500 or 1000 amu) demonstrated the main role of negative ion- clusters in particle growth process.

Thus, the initial phase of formation and growth of the dust particles in a low-pressure silane discharge is a homogeneous process, stimulated by fast silane molecules and radicals' reactions with negative ion-clusters. The first particle generation appears as a monodispersed one with a crystallite size about 2 nm. Due to a selective trapping effect, the concentration of these crystallites increases up to a critical value, where a fast (like a phase transition) coagulation process takes place leading to large-size (50 nm) dust particles. During the coagulation step, when the aggregate size becomes larger than some specific value of approximately 6 nm, another critical phenomenon, the so-called "$\alpha - \gamma$ transition," takes place with a strong decrease of electron concentration and a significant increase of their energy. A small increase of the gas temperature increases the induction period required to achieve these critical phenomena of crystallites formation: agglomeration and the $\alpha - \gamma$ transition.

The dusty plasma formation has been observed in low pressure RF-discharge in the **Boufendi-Bouchoule experiments** (Bouchoule 1993, 1999; Bouchoule and Boufendi 1993) in a grounded cylindrical box (13 cm inner diameter) equipped with a shower type RF powered electrode. A grid was used as the bottom of the chamber to allow a vertical laminar flow in the discharge box; typical argon flow, 30 sccm; silane flow 1.2 sccm; total pressure 117 mTorr (so the total gas density is about $4\cdot10^{15}$ cm^{-3}, silane- $1{,}6\cdot10^{14}$ cm^{-3}); neutral gas residence time in the discharge is approximately 150 ms; RF power 10 W. The Boufendi-Bouchoule experiments demonstrated that the first particle size distribution is monodispersed with the diameter of about 2 nm and is practically independent of temperature. Appearance time of the first-generation particles is less than 5 ms. The particle growth proceeds through the successive steps of fast super-small 2 nm particle formation and the growth of their density up to the critical value of about 10^{10}–10^{11} cm^{-3}, when the new particles formation terminates and formation of aggregates with diameters of up to 50 nm begins by means of coagulation. During the initial discharge phase (until the $\alpha - \gamma$ transition, up to 0.5 s for room temperature and up to several seconds for 400 K) the electron temperature remains about 2 eV, the electron concentration is about $3 \cdot 10^9$ cm^{-3}, the positive ion concentration approximately 4×10^9 cm^{-3}, and the negative ion concentration about 10^9 cm^{-3}. After the $\alpha - \gamma$ transition, the electron temperature increases up to 8 eV while the electron concentration decreases 10 times and the positive ion concentration increases 2 times. These concentrations are correlated with the negative volume charge density of the charged particles through the plasma neutrality relation. The Boufendi-Bouchoule experiments show that the critical value of super-small particle concentration before coagulation practically is independent of temperature. The induction time observed before coagulation is a highly sensitive function of the temperature: ranging from about 150 ms at 300 K it increases more than 10 times when heated to only 400 K. For a temperature of 400 K, the time required to increase the super-small neutral particle concentration is much longer (10 times), than the gas residence time in the discharge, which can be explained by the neutral particle trapping phenomenon. The particle concentration during the coagulation period decreases, and their average radius grows. The total mass of dust in plasma remains almost constant during the coagulation.

20.11 Physical and Chemical Kinetics of Dust Nanoparticles Formation in Plasma: A Story of "Birth and Catastrophic Life"

The dust particle formation and growth in a SiH_4-Ar low-pressure radio-frequency (RF) discharge described above includes four steps: first, growth of super-small particles from molecular species and then three successive "catastrophic" events: selective trapping, fast coagulation and finally the strong modification of discharge parameters, the so-called $\alpha - \gamma$ transition (Fridman et al., 1996).

The first step, ***the super-small particle generation*** is illustrated in Figure 20.11a. It begins with SiH_3^- negative ion formation by dissociative electron attachment to silane. Non-dissociative three-body attachment to SiH_3 radicals could occur in a complementary way ($e + SiH_3 = (SiH_3^-)^*$, $(SiH_3^-)^* + M = SiH_3^- + M$). Then the negatively charged cluster growth is due to ion-molecular reactions such as $SiH_3^- + SiH_4 = Si_2H_5^- + H_2$, $Si_2H_5^- + SiH_4 = Si_3H_7^- + H_2$. This chain of reactions can be accelerated by silane molecules vibrational excitation. Typical reaction time is about

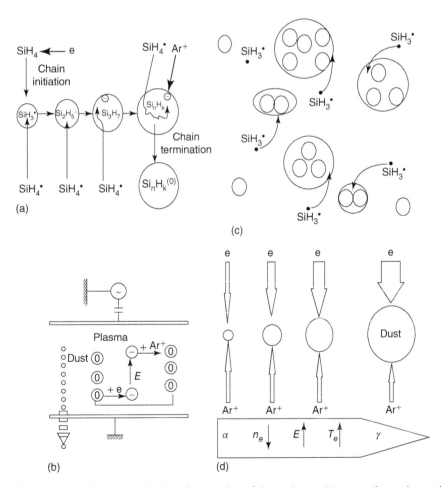

(a)

(b)

(c)

(d)

Figure 20.11 Illustrative kinetics of generation of dusty plasma: (a) generation and growth of first super-small particles; (b) electrostatic trapping of neutral nanoparticles; (c) Mechanism of particle coagulation of nanoparticles; (d) $\alpha-\gamma$ transition in RF dusty plasma.

0.1 ms and is much faster than ion-ion recombination (typical time 1–3 ms) terminating the cluster growth chain. As the negative cluster size increases, the probability of their reactions with vibrationally excited molecules decreases because of vibrational VT relaxation on the cluster surface. When the particle size reaches a critical value (about 2 nm at room temperature), the chain reaction of cluster growth becomes much slower, and is finally terminated by the ion–ion recombination process. The typical time of 2 nm particle formation by this mechanism is about 1 ms at room temperature. Vibrational temperatures of SiH_4 in the discharge are determined by VT relaxation in the plasma volume and on the walls and depend exponentially on the translational gas temperature according to the Landau-Teller effect. For this reason, even a small increase of gas temperature results in suppression of the vibrational excitation making the cluster growth reactions much slower.

The selective electrostatic trapping of neutral particles is illustrated in Figure 20.11b. Trapping of negatively charged particles is due to repelling forces exerted in the electrostatic sheaths when these particles reach the plasma boundary. For the super-small particles under consideration, the electron attachment time is about 100 ms and two orders of magnitude longer than the fast ion-ion recombination; therefore, most of the particles are neutral during this period. The effect of "trapping of neutral particles in electric field" allows the density of particles to reach the critical value sufficient for effective coagulation. For the 2 nm particles, the electron attachment time is shorter than the residence time and each neutral particle is charged at least once during the residence time in plasma and quickly becomes trapped by the strong electric field before recombination. The rate coefficient of two-body electron non-dissociative attachment grows strongly with the particle size. For this reason, *the attachment time for particles smaller than 2 nm is much longer than their residence time, and they can't be charged even once and therefore be trapped in plasma and survive. Only "large particles," exceeding the 2 nm – size, can survive: this is the size-selective trapping effect.* This "first catastrophe" associated with small particles explains why the first generation of particles has well-defined sizes of crystallites. It also explains the strong temperature effect on dust production. A small temperature increase results in reduction of the cluster growth rate and the initial cluster size. It leads to significant losses of the initial neutral particles with the gas flow and to long delay of coagulation, $\alpha - \gamma$ transition, and dust production process in general.

The fast nanoparticles coagulation phenomenon (see Figure 20.11c) occurs when the increasing density of the survived monodispersed 2 nm particles reaches a critical value of about 10^{10}–10^{11} cm^{-3}. At such densities, the attachment of small negative ions like SiH_3^- to 2 nm particles becomes faster than their chain reaction to generate new particles. New chains of dust formation become suppressed and the total particle mass remains almost constant. In this fast coagulation process, the mass increase by the "surface deposition" process is negligible. Moreover, when such high particle concentrations are reached, the probability of multi-body interaction increases. For this reason, the aggregate formation rate constant grows drastically, and the coagulation appears as a critical phenomenon of phase transition. The critical particle density does not depend much on temperature. Besides electrostatic effects, this is one of the main differences between the critical phenomenon under consideration and conventional gas condensation. The typical induction time before coagulation is about 200 ms for room temperature and much longer for 400 K due to the selective trapping effect. The trapped particle production rate for 400 K is much slower than at room temperature, but the critical particle density remains the same.

The critical phenomenon of fast changing of discharge parameters, $\alpha - \gamma$ ***transition in dusty plasma,*** occurs during the coagulation when the particle size increases and their density decreases. Before this critical moment, the electron temperature and other plasma parameters are mainly determined by the balance of volume ionization and electron losses on the walls. The $\alpha - \gamma$ transition occurs when the electron losses on the particles' surfaces become larger than those on the reactor walls. The electron temperature increases to support the plasma balance and the electron density dramatically diminishes. This fourth step of particle and dusty plasma formation story is the "plasma electron catastrophe" (Figure 20.11d). During the coagulation period, the total mass of the particles remains almost constant and so the overall particle surface in the plasma decreases during coagulation. The influence of particle surface becomes more significant when the specific surface value decreases. The probability of electron attachment to the particles grows exponentially with the particle size, and therefore the "effective particle surface" grows during the coagulation period. When the particle size exceeds a critical value of about 6 nm, an essential change of the heterogeneous discharge behavior takes place with significant reduction of the free electron density. Due to the increase of the electron attachment rate on these size-growing particles, most of them become negatively charged soon after the $\alpha - \gamma$ transition. The typical induction time before the $\alpha - \gamma$ transition is about 500 ms for room temperature and more than an order of magnitude longer for 400 K. This strong temperature effect is due to the threshold character of the $\alpha - \gamma$ transition, which takes place only when the particle size exceeds the critical value, and for this reason is determined by the strongly temperature-dependent time of the beginning of coagulation. See more on the "catastrophic" processes in Fridman et al. (1996); as well as in Fridman and Kennedy (2021).

20.12 Plasma Synthesis of Aluminum Nano-powders, Luminescent Silicon Quantum Dots, and Nano-composite Particles

Unipolar charging of nanoparticles in plasma suppresses their agglomeration, which makes plasma generation of nanoparticles especially attractive. Let's consider three examples of the relevant plasma nanotechnologies interesting for different special applications. First technology is related to the **synthesis of aluminum nanoparticles using DC thermal plasma** (see Zhang et al. 2005). Such nanoparticles are highly pyrophoric and very light. Being properly surface-passivated, the aluminum nanoparticles are of significant interest for solid fuel propulsion. The particles are synthesized in Ar–H$_2$ thermal DC-arc at near atmospheric pressure. The Ar–H$_2$ plasma is expanded in a subsonic flow through a ceramic nozzle to a pressure of about 50 kPa. AlCl$_3$ vapor is injected into the plasma at the upstream end of the nozzle. A cold argon counterflow opposes the plasma jet exiting the nozzle to provide dilution and cooling, which prevents coagulation and coalescence and maintains small particle sizes. At AlCl$_3$ flow rates of 20 sccm and carrier gas of 200 sccm, the produced aluminum particles have sizes of about 10 nm.

The second nanotechnology is related to high-yield **synthesis of luminescent silicon quantum dots in nonthermal RF discharge** (Mangolini et al. 2005). The nanosized quantum dots are of interest for a variety of applications, from solid-state lighting and optoelectronic devices to the use of fluorescent tagging agents. Conventional liquid-phase synthesis of the silicon quantum dots is characterized by limited process yields. Gas-phase approaches can provide higher yields; however, they are afflicted with problems of particle agglomeration. Nonthermal plasma synthesis offers an effective and fast gas-to-particle conversion common to gas-phase approaches and strongly reduces agglomeration at the same time since the particles in plasma are unipolar negatively charged. The silicon quantum dots have been produced in a continuous-flow Ar–SiH$_4$ nonthermal plasma reactor with 27 MHz RF-powered ring electrode at a pressure of 1.5 Torr. Typical flow rate is 40 sccm for Ar–SiH$_4$ (95 : 5) mixture. The residence time of particles in the plasma region is short (<5 ms), and the size of the produced particles is small (about 5 nm). The plasma-produced particles are collected on filters and washed with methanol. After the growth of a native oxide layer, the particles show bright photoluminescence in the red–orange region. The production rate of the nanoscale silicon quantum dots in these experiments is a few tens of milligrams per hour.

The third nanotechnological example is **plasma fabrication of the nanocomposites**, which are nanoparticles composed of one material and covered by a nanolayer of another. Relevant example is polymer-like C/H nanocoating of nano-powders of amorphous SiO$_2$ in nonthermal low-pressure RF-CCP discharge, see Kouprine et al. (2003). Nanoparticles of SiO$_2$ with diameter less than 100 nm are treated in 500–800 W discharge at 1–5 kPa in pure methane and ethane. Although non-saturated hydrocarbons are partially synthesized, their presence does not lead to soot nucleation. Instead, the SiO$_2$ particles passing through the nonthermal low-pressure RF-CCP discharge become individually coated by a polymer-like C/H film with thickness of about 10 nm. TEM images of such nanocomposites produced in CH$_4$ and C$_2$H$_6$ plasmas are shown in Figure 20.12. Thermal plasma treatment of CH$_4$ proceeds through generation of CH$_2$ and CH radicals and results in soot production. In contrast to that, the nonthermal plasma processes generate mostly CH$_3$ radicals and H atoms, which are not sufficient for soot production but very effective for deposition of thin films on the surface of already existing particles.

20.13 Plasma Synthesis of Highly Organized Carbon Nanostructures: Plasma Synthesis of Fullerenes

Plasma is effective in production of the highly organized nanostructures, such as fullerenes and carbon nanotubes. The fullerenes are closed-cage carbon molecules like C$_{36}$, C$_{60}$, C$_{70}$, C$_{76}$, and C$_{84}$, in which carbon atoms are in corners

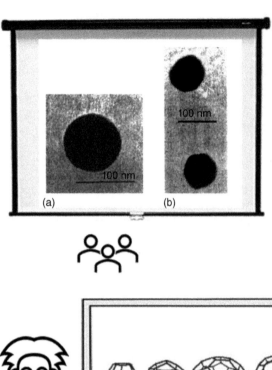

Figure 20.12 TEM images of single SiO_2 nanoparticles coated in the RF-CCP discharge by a nanolayer of polymer-like C/H film: (a) discharge in methane; (b) discharge in ethane.

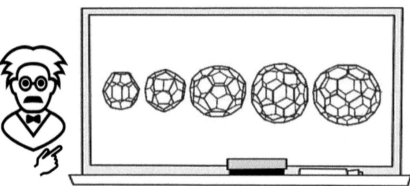

Figure 20.13 Structure of fullerenes like C_{36}, C_{60}, C_{70}, C_{76}, and C_{84}, constructed from pentagonal and hexagonal carbon rings.

of regular pentagons or hexagons covering the surface of a sphere or a spheroid (see Figure 20.13). The C_{60} molecule has the highest symmetry and, therefore, is the most stable. It is composed of 20 regular hexagons and 12 regular pentagons with a characteristic distance between carbon atoms of 0.142 nm. To some extent, one can say that fullerenes are spheroidal modifications of graphite-like sheets. Fullerenes have very specific physical and chemical properties; they are characterized by high oxidative potential and are able to simultaneously attach up to six electrons. Insertion of alkaline atoms in the crystal structure of C_{60} makes the crystal superconductive, with a transition to superconductivity at about 40 K. High pressure and high temperature modification of crystallite fullerenes leads to the formation of a new phase of carbon, which has a hardness like that of diamonds.

The most conventional approach to **plasma synthesis of fullerenes** uses thermal arcs (Farhat et al. 2002; Saidane et al. 2004). The process starts with the production of fullerene containing soot (see Figure 20.14). An AC arc is sustained between two graphite electrodes, with typical parameters: frequency 60 Hz, current 100–200 A, and voltage 10–20 V. The evaporation rate of a graphite rod is up to $10\,g\,h^{-1}$. As a result of the process, walls of the water-cooled reactor are covered with soot that contains up to 15% of fullerene. The productivity of the plasma system is high, about 1 kg of the fullerene-containing soot per day. Extraction of fullerenes from the plasma-produced soot is based on higher fullerene solubility in toluene, benzene, dimethyl-benzene (xylol), dichlorine-benzene, CS_2, and others.

Thermal arc systems like these shown in Figure 20.14 can be applied for **plasma production of endohedral fullerenes** (Eletsky 2000a,b; Lange et al. 2002). The endohedral fullerenes, or simply endohedrals ($M_m@C_n$), consist of a conventional fullerene shell C_n with an atom or molecule M_m encapsulated inside. Injection of metal vapor into the arc discharge leads to the formation of endohedrals with a relative concentration not exceeding a few percent from the total amount of fullerenes in the produced soot. The simplest method of metal injection into the thermal arc is based on the use of a composite electrode (anode), which is made of graphite with some small admixture of

Figure 20.14 Schematic of a plasma system of production of fullerene-containing soot: (1) graphite electrodes; (2) water-cooled copper element; (3) water-cooled surface for collection of the carbon condensate; (4) springs.

powder of the metal or its compound, oxide, or carbide. Consider, for example, formation of an endohedral fullerene La@C82 in the thermal arc discharge. The anode is made from a graphite rod (length 300 mm, diameter 15 mm) with a cylindrical hole (length 270 mm, diameter 10 mm). The cavity is filled with a mixture of La_2O_3 powder with graphite powder. The atomic fraction of lanthanum in the anode material is 1.6%. A pure graphite rod is used as a cathode. The thermal arc discharge is sustained in helium flow at a pressure of 500 Torr, and the DC is 250 A. Produced soot is transported by the helium flow and trapped in a filter in the absence of oxygen. Extraction of endohedral fullerenes from soot is done by using solvents such as toluene or CS_2. Metal injection in a thermal arc can be provided not only by adding metal powder to the electrode material but also by admixing gaseous compounds of the metal in the He flow. This approach was applied in plasma synthesis of $Fe@C_{60}$. Pentacarbonyl ($Fe(CO)_5$) was used as a gaseous admixture to helium in this case. Another approach to the production of endohedrals is based on ion bombardment of hollow fullerenes by ions of elements to be encapsulated. This approach is especially effective in the production of endohedrals containing highly chemically active atoms, specifically alkaline atoms, as well as $N@C_{60}$.

20.14 Plasma Nanotechnology: Synthesis of Nanotubes, and Nanotube Surface Modification

Plasma is also effective in production of such highly organized nanostructures as nanotubes. The graphite-like sheets form not only spheroidal nano-size configurations (fullerenes) but also high-aspect-ratio nano-cylindrical structures called nanotubes. The diameter of nanotubes can be as small as one or a few nanometers, while their length can exceed tens of microns. Depending on the number of graphite-like sheets forming the walls of the nanotubes, they are usually divided into two classes: *single-wall and multi-wall nanotubes*. **Single-wall carbon nanotubes** are formed by rolling the graphite sheet composed of regular carbon hexagons into a high-aspect-ratio cylinder. The properties of the nanotube, especially electrical conductivity, depend on the chirality of the structure, which is determined by the angle of orientation of the graphite sheet, α, with respect to the nanotube axis. Two specific chiralities permit rolling a graphite sheet into a nanotube without perturbation of the structure. One such chirality corresponds to the orientation angle $\alpha = 0$, which is called the *armchair configuration*. In armchair single-wall nanotubes, two C—C bonds of each regular hexagon are parallel to the axis of the nanotube ($\alpha = 0$). The armchair nanotubes have high metallic electric conductivity and high chemical stability. Another chirality, not perturbing the graphite sheet, corresponds to the orientation angle $\alpha = \pi/6$, which is referred to as having a *zigzag orientation*. An idealized model of

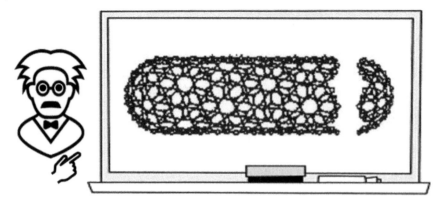

Figure 20.15   Idealized model of a single-wall nanotube.

a single-wall nanotube is shown in Figure 20.15. **Multi-wall carbon nanotubes** can be formed in a variety of configurations. Coaxial single-wall nanotubes inserted one into another form configuration called a *Russian doll*. Another variation of the Russian doll is composed of coaxial prisms; also, multi-wall nanotubes can look like a *scroll*. The distance between the graphite sheets in all three configurations is 0.34 nm corresponding to the interlayer distance in crystalline graphite.

The most effective plasma method to produce carbon nanotubes is **dispersion of graphite electrodes in thermal arc plasma** in helium, see Eletsky (2000a,b). Typical voltages are 15–25 V, currents are several tens of amperes, and the distance between electrodes is a few millimeters. Formation of nanotubes is due to dispersion of the graphite anode, which is characterized by temperature higher than that of the cathode. Carbon nanotubes are deposited on the discharge chamber walls and on the cathode. In contrast to the production of fullerenes, where the typical He pressure is 100–150 Torr, the production of nanotubes is optimal at higher pressures, around 500 Torr. Also, cathodes of larger diameters (exceeding 10 mm) are favorable for higher yields of nanotubes. To separate the nanotubes from the rest of deposit, ultrasonic dispersion can be effectively applied to the deposit preliminarily dissolved in methanol or other solvent. Thermal arc discharges with graphite electrodes produce mostly multi-wall nanotubes with mixed chirality. The production of single-wall nanotubes usually requires application of metallic catalysts of the platinum group (Ru, Rh, Pd, Os, Ir, Pt) and iron group (Co, Fe, Ni), as well as such metals as Mn, Sc, La, V, Ce, Zr, Y, or Ti. The presence of catalysts leads to the formation of a large variety of different nanostructures, which are determined by the type of the catalyst and by the parameters of the arc discharge. The iron group metals selectively promote the formation of single-wall nanotubes. Furthermore, the highest synthesis efficiency of single-wall nanotubes is achieved by application of mixed iron-group–metal catalysts, specifically Fe–Ni and Co–Ni.

The carbon nanotubes can be also produced by **thermal plasma dissociation of carbon compounds**, for example, by tetrachloroethylene (TCE), see Harbec et al. (2005). The arc plasma erodes the electrodes, which creates metal vapors transported toward the torch exit. Fast quenching in the torch nozzle leads to metal vapor nucleation into metal nanoparticles, which are catalysts for the formation of carbon nanotubes based on carbon atomized from TCE in the thermal plasma. Thermal arc electrodes made of tungsten, by means of generation of metallic tungsten particles, stimulate the production of multi-wall nanotubes having 10 concentric rolled-up graphite sheets. External and internal diameters of the multi-wall nanotubes produced in this system are, respectively, 10–30 nm and about 0.8 nm.

The surface of carbon nanotubes is usually non-reactive, which reduces their compatibility with other materials, especially organic solvents, and polymer matrices. **Surface modification of carbon nanotubes** can significantly increase their applicability. Thus, low-pressure RF-ICP plasma functionalization of the nanotubes with different gases, including O_2, NH_3, and CF_4 is especially interesting (Felten et al. 2005). Functionalization in RF oxygen plasma leads to the formation of C—O, C—O, and COO groups on the nanotube surface. Treatment in oxygen plasma results in a significant increase in the hydrophilicity of the carbon nanotubes. Treatment in CF_4 plasma, on the contrary, results in a significant growth of hydrophobicity of the carbon nanotubes. Another approach to surface modification of carbon nanotubes is related to plasma polymerization of monomers on their surfaces in low-pressure RF-ICP plasma. Thus, acrylic acid can be used as a monomer to produce water-soluble carbon nanotubes.

For more details on plasma nanotechnology see the books of Sankaran (2011), Keidar and Bellis (2018), Song et al. (2022), Ostrikov (2005), and Ostrikov et al. (2005).

20.15 Problems and Concept Questions

20.15.1 Thermal Plasma Spraying of Powders

What is the major distinctive feature of the plasma spraying with respect to other thermal spray technologies? Give example of powders, where the plasma spaying has the exclusive advantages.

20.15.2 Underwater Thermal Plasma Spraying (UPS)

It is claimed that UPS permits to avoid the oxidation issues associated with spraying in air without the need for costly and complex vacuum chambers, load locks, and so on. Explain how the powder oxidation with water vapor (forming oxides and hydrogen) can be suppressed in this technology?

20.15.3 Radio Frequency (RF) Thermal Plasma Sprays

What is the major advantage of application of the RF thermal plasma spraying with respect to spraying organized in way simpler conventional thermal arc discharges?

20.15.4 Microarc (Electrolytic Spark) Oxidation Coating

Calculate the required voltage to stabilize the amplitude of current pulses in the microarc oxidation coating of aluminum in sulfuric acid. Assume the oxide film thickness of $30\,\mu m$.

20.15.5 Trapping of Neutral Nanoparticles in Low-pressure Silane Plasma

Give your interpretation of how is it possible to "trap the non-charged neutral particles in the electrostatic field" of the low-pressure RF silane plasma?

20.15.6 The α–γ Transition During Coagulation of Nanoparticles in Silane Plasma

The α–γ transition, including a sharp growth of electron temperature and decrease of electron density, is related to a significant contribution of electron attachment to the particle surfaces and takes place during the coagulation. It is interesting that the total surface area of particles decreases during the coagulation process because their diameters grow but total mass remains almost fixed. Give your interpretation of the phenomenon of the "mystery."

20.15.7 Plasma Synthesis of Nano-powders

What is the most distinctive advantage of the plasma-based synthesis of nano-powders? What are the major challenges of this approach? Which discharges are the most suitable plasma technology?

Lecture 21

Plasma Processing in Microelectronics and Other Micro-technologies: Etching, Deposition, and Ion Implantation Processes

21.1 Plasma Etching as a Part of Integrated Circuit Fabrication: Etch Rate, Anisotropy, and Selectivity Requirements

The etching, removing material from surfaces in nonthermal plasma, is one of the key processes in microelectronic circuit fabrication. The etching can be chemically selective and anisotropic. Chemically selective etching means removing one material and leaving another one unaffected. Anisotropic etching involves removing material at the bottom of a trench and leaving the sidewalls unaffected. Today's plasma etching can create less than 0.2-μm-wide and more 5-μm-deep trenches in silicon films or substrates. *Application of plasma is the only viable approach today for anisotropic etching, which makes it a unique part of modern integrated circuit (IC) fabrication technology.* To analyze plasma applications for integrated circuit fabrication, let us consider as an example the six-step process for creation of a metal film patterned with submicron features on a wafer substrate, illustrated in Figure 21.1: (i) film deposition on a substrate; (ii) photoresist layer deposition over the film; (iii) selective exposition of the photoresist to light through a pattern; (iv) development of the photoresist and removal of the exposed resist regions (this leaves behind a patterned resist mask); (v) transferring the pattern into the film via the etching process (the mask protects the underlying film from being etched); (vi) removing of the remaining resist mask. Plasma processing is effectively used for film deposition (step 1) and etching (step 5). Plasma processing can also be used for photoresist development (step 4) and removal (step 6). The etching in step 5 is anisotropic, which is a unique feature of plasma processing that makes it so important in IC fabrication.

The **isotropic etching** can be illustrated by the etching of a silicon substrate. The process starts with the production of chemically aggressive atoms (e.g., F atoms) by dissociation of its nonactive compounds (e.g., CF_4) in a nonthermal, usually low-pressure RF discharge. The F atoms then react with the silicon substrate, yielding the volatile etch product SiF_4. The **anisotropic etching**, shown in Figure 21.1e, is due to substrate bombardment by high-energy ions, which mostly bombard the bottom of the trench, leaving the sidewalls untreated. The sidewalls of the trench can be additionally protected by special passivating film to achieve a stronger anisotropic effect from the plasma etching.

Considering the significant practical importance of the plasma etching, and deposition processes, numerous reviews and books have been published on the subject, including those of Powel (1984), Sugano (1985), Morgan (1985), Manos and Flamm (1989), Lieberman and Lichtenberg (2005), Roth (1995, 2001, 2003), Chen and Chang (2012), Nojiri (2014), Friedrich and Meichsner (2022).

The **etch rate requirements** can be analyzed keeping in mind the conventional set of films illustrated in Figure 21.2a and including about 1500 nm of resist, over 300 nm of polysilicon, and over 30 nm of gate oxide on an epitaxial silicon wafer. Each of these films should be etched for 2–3 min for a single wafer process. Therefore, the minimum etching rates for a photoresist layer are about $w_{pr} = 500\,\text{nm min}^{-1}$, for a gate oxide layer about $w_{ox} = 10\,\text{nm min}^{-1}$. For a polysilicon layer, the intermediate etch rate requirement is $w_{poly} = 500\,\text{nm min}^{-1}$.

The **etch selectivity requirements** can be analyzed by comparing the etching rates of polysilicon and photoresist used as a mask. To complete the polysilicon etch before eroding the resist, the required etching selectivity should satisfy the following criterion:

$$S = \frac{w_{poly}}{w_{pr}} \gg \frac{300\,\text{nm}}{1500\,\text{nm}} = 0.2. \tag{21.1}$$

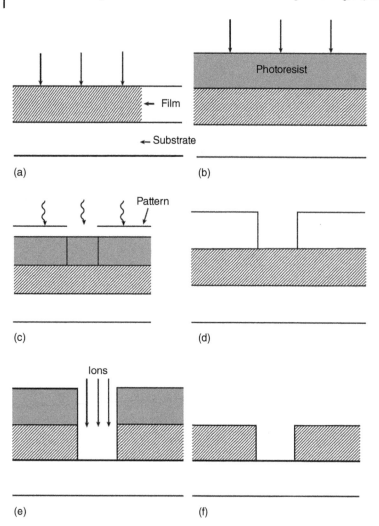

Figure 21.1 Six steps to create a thin metal film patterned with submicron features on a wafer substrate: (a) metal film deposition; (b) deposition of photoresist layer; (c) selective optical exposure of the photoresist through a pattern; (d) photoresist development, removing of the exposed resist regions leaving behind a patterned resist mask; (e) anisotropic plasma etch, is transferring pattern into the metal film; (f) removal of the remaining photoresist mask.

Due to non-uniformity of the polysilicon layer across the wafer, twice as much time may be required to clear polysilicon from all unmasked regions. To restrict destruction of the thin oxide layer during *the over-etching*, another selectivity criterion should be satisfied:

$$S = \frac{w_{poly}}{w_{ox}} \gg \frac{300\,\text{nm}}{30\,\text{nm}} = 10. \tag{21.2}$$

The **anisotropy requirements** are characterized by the anisotropy coefficient, which is the etching rate ratio in vertical and horizontal directions: $a_h = w_v/w_h$. Figure 21.2b illustrates the anisotropic etching of a deep trench of width l into a film of thickness d. Assuming that the mask is not eroded and horizontal over-etching under the mask is limited by δ, the anisotropy requirement can be expressed as $a_h \geq d/\delta$. If the mask width is l_m, the maximum width of the trench is $l = l_m + 2\delta$, see Figure 21.2b. Therefore, the anisotropy requirement can be presented as:

$$a_h \geq \frac{2d}{l - l_m}. \tag{21.3}$$

Typically: $l = 1\,\mu\text{m}$, $l_m = 0.5\,\mu\text{m}$, $d = 2\,\mu\text{m}$), therefore the required anisotropy coefficient should be at least 10. Small trench width can obviously be achieved by using a mask width that is as thin as permitted by limitations of lithography. The minimal trench width according to (21.3) is $l_{min} \approx 2d/a_h$. Fabrication of deep trenches ($d/l \gg 1$) requires significant anisotropy of the etching.

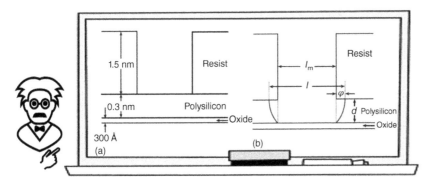

Figure 21.2 Plasma etching: (a) typical set of films; (b) anisotropy in polysilicon etching.

21.2 Basic Plasma Etch Processes: Sputtering, Pure Chemical Etching, Ion-energy Driven Etching, Ion-enhanced Inhibitor Etching

We can distinguish four basic types of plasma etching processes applied to remove materials from surfaces: sputtering, pure chemical etching, ion energy-driven etching, and ion inhibitor etching. The first process to discuss is **sputtering** (see Figure 21.3a), which involves the ejection of atoms from surfaces due to energetic ion bombardment. To provide an effective sputtering process, a low-pressure nonthermal discharge supplies energetic ions to the treated surface. The ion energies in practical sputtering systems are typically above a few hundred electron volts. Ions with energies exceeding 20–30 eV can sputter atoms from a surface. The sputtering yield γ_{sput}, which is the number of atoms sputtered per incident ion, rapidly increases with the ion energies up to few hundred electron volts, where the yield becomes sufficient for practical applications. Usual projectiles for practical physical sputtering are argon ions with energies of 500–1000 eV. In this energy range, a bombarding ion transfers energy to a large group of surface atoms. After redistribution of energy in the group, some atoms can overcome the surface binding energy, E_s, and escape the surface, enabling sputtering. The sputtering yield γ_{sput} is independent of the ion energy over a broad range of the energies and surface atom density and depends mostly on the surface binding energy E_s, and masses M_i and M_s of incident ions and surface atoms:

$$\gamma_{sput} \propto \frac{1}{E_s} \frac{M_i}{M_i + M_s}. \tag{21.4}$$

The sputtering yields are close to unity and not much different for different materials. Sputtering rates are generally low for this reason: about one atom per one incident ion. Because the ion fluxes from conventional low-pressure

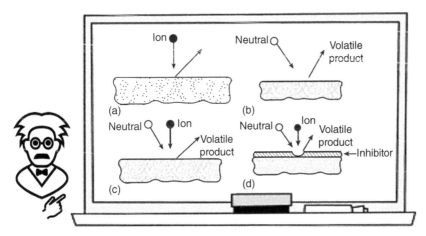

Figure 21.3 Basic plasma etch processes: (a) sputtering; (b) pure chemical etching; (c) ion-energy-driven etching; (d) ion-enhanced inhibitor etching.

etching discharges are relatively small, the sputtering rates are generally not sufficient for commercially significant material removal. The surface binding energy E_s (few electron volts) is much less than the incident ion energy (500–1000 eV), therefore energy efficiency of sputtering is only about 0.2–0.5%.

The **pure chemical etching**, see Figure 21.3b, involves a plasma discharge that supplies gas-phase etching atoms or molecules, which can react with the surface to form gas-phase products and remove the surface material. As examples, we can mention etching of silicon using fluorine atoms generated in SiF_4 plasma or etching of photoresist (which consists of hydrocarbons) using CO_2/H_2O plasma-generated oxygen atoms. Pure chemical etching is an isotropic process because the gas-phase neutral etchant atoms or molecules approach the substrate with near uniform angular distribution. While not anisotropic, pure chemical etching is characterized by high etching rates due to much higher fluxes of active chemicals than that of ions.

The **ion-energy-driven etching** is based on the synergetic effect of neutral chemically active etchants (for example, F atoms) and energetic ions, see Figure 21.3c. The synergetic etching effect occurs when neutral etchants and energetic ions together can provide much higher surface material removal rates than corresponding pure chemical etching alone or sputtering alone. For example, etch rates of silicon by XeF_2 etchant gas alone, as well as by Ar^+ ion beam sputtering alone, were less than 0.5 nm min^{-1}. The ion energy-driven etching, which combines these effects results in an etch rate exceeding 6 nm min^{-1}. Thus, ion energy-driven etching is chemical in nature but with a reaction rate determined by energetic ion bombardment. In contrast to sputtering, the etch rate increases in this case with growth of ion energy. Similarly, to pure chemical etching, the ion energy-driven etching requires products to be volatile.

The **ion-enhanced inhibitor etching**, illustrated in Figure 21.3d, involves application of inhibitor species. *This process permits the highly anisotropic etching with formation of vertical sidewalls, which is the key for the deep ditch etching.* Nonthermal discharge supplies here not only neutral chemically active etchants and energetic ions, as in ion energy-driven etching but also inhibitor precursor radicals and molecules. These inhibitor precursors can be absorbed or deposited anisotropically on the substrate to form a protective layer of polymer film, preventing etching. Anisotropic ion bombardment prevents formation of the inhibitor layer and, therefore, exposes the surface (e.g., bottom of the ditch) to chemical etching. Where there is no ion flux, the inhibitor is effectively formed and protects the surface (e.g., sidewalls of the ditch) from the chemical etchant. In particular, the inhibitor precursor radicals are CF_2 and CF_3, which are formed by plasma-chemical decomposition of CF_4, or CCl_2 and CCl_3, generated by plasma-chemical decomposition of CCl_4. These radicals, deposited on the substrate, can form fluoro- and chlorocarbon polymer films, protecting the substrate (sidewalls) from etching. A classic example of the ion-enhanced inhibitor etching process is anisotropic etching of aluminum trenches in nonthermal, low-pressure discharges in CCl_4–Cl_2 or $CHCl_3$–Cl_2 mixtures. In these systems, both atomic and molecular chlorine (Cl and Cl_2) chemically etch aluminum – intensively, but isotropically. Addition of carbon compounds to the feed gas leads to the formation of a protective chlorocarbon film on the surface. Ion bombardment normal to the substrate eliminates the protective film from the trench bottom, allowing the etch process to proceed there anisotropically. The chlorocarbon film created on the sidewalls of the trench protects them from chemical etching by Cl and Cl_2. Optimization of the process results in highly anisotropic etching with formation of vertical sidewalls.

21.3 Plasma Sources Applied for Etching and Other Material Processing: RF-CCP Discharges, RF-Diodes and Triodes, MERIE, Reactive Ion Etchers (RIE), High-density Plasma (HDP) Sources

The plasma etching discharges are mostly low-pressure, strongly nonequilibrium discharges. Most of them are radiofrequency (RF) based; however, some are supplied by microwave radiation or in some different manner. Low translational temperature in the discharges permits to avoid damaging the treated surfaces and to maintain high selectivity in the plasma processing. Low pressures help to sustain the discharge uniformity and nonequilibrium in sufficient volumes. Low pressures also provide effective contact of active species generated in the plasma volume with the treated surface, without product redeposition. Specifically, the most used etching discharges are RF capacitively coupled plasma (CCP) discharges, called RF diodes, and high-density plasma (HDP) discharges, in particular RF-ICP discharges.

The **RF diodes** used for anisotropic etching typically operate at pressures of 10–100 mTorr, at power densities of 0.1–1 W cm^{-2}, and at a driving frequency of 13.56 MHz. Multiple wafer treatment systems are common. Typical RF driving voltages are 100–1000 V, and plate separation is 2–10 cm. When a wafer is mounted for etching purposes on the powered electrode, the system is usually called a **reactive ion etcher (RIE)**. Typical plasma densities used for etching in the RF diodes are relatively low (10^9–10^{11} cm^{-3}); electron temperatures are about 3 eV. Sheath voltages, which determine ion acceleration energies, are relatively high and exceed 200 V. Dissociation degrees of molecules vary from 0.1% to almost 100%. Typical discharge volumes are 1–10 L, and typical cross-sectional areas are 300–2000 cm^2. A limiting feature of the RF diode application is that the ion bombarding flux in these discharges cannot be varied independently of the ion bombarding energy. Sheath voltages and hence bombarding ion energies should be high to achieve reasonable (although still not high) ion fluxes and sufficient dissociation degrees of the feed gas.

Control over the energy of ion bombardment can be achieved to some extent by using **RF triodes**. In these discharges, the wafer is located on the undriven electrode independently biased by a second RF source. Although the RF triodes can control the ion energy, processing rates and sputtering contamination are still issues in these discharges. Different magnetically enhanced modifications of the RF-CCP discharges have been developed to improve practical performance of the RF diodes and triodes. We can point out here the **magnetically enhanced reactive ion etchers (MERIEs)**. Magnetic field of 50–300 G is applied in these discharge systems parallel to the powered electrode, where the wafer is located. The magnetic field increases the efficiency of power transfer to the plasma and enhances plasma confinement, which results in reduction of sheath voltage and increase of plasma density.

The key feature of the low-pressure **HDP etchers**, which can effectively overcome the problems of RF diodes and triodes, is the coupling of RF or microwave power to the plasma across a dielectric window (in contrast to RF diodes and triodes, where electrodes have direct contact with the plasma). The non-capacitive coupling and power transfer allow operation at low voltages across sheaths. The ion acceleration energies are typically about 20–30 eV at all surfaces in this case. The electrode with wafer is biased by an independently driven capacitively coupled RF source, which permits independent control of ion energy. Thus, using HDP plasma sources makes possible the generation of plasma with a high density of charged and chemically active species, which leads to high processing rates, with independent control of the ion-bombarding energy. HDP sources significantly differ in how power is coupled to the plasma. It can be an **electron cyclotron resonance (ECR) plasma source** supported by microwave radiation; it can be a **helicon plasma source** applying an axial magnetic field with an RF-driven antenna; or it can be a **helical resonator plasma source** with the external helix and conducting cylinder surrounding the dielectric discharge chamber.

 The HDP etchers are very often organized with ICP sources. In these **ICP or transformer HDP etchers**, *plasma acts as a single-turn conductor, which is coupled to a multi-turn nonresonant RF coil across the dielectric discharge chamber. RF power is inductively coupled to the plasma, as in a transformer. Because plasma acts in a single turn, the transformer decreases voltage and increases current, which explains the generation of HDP in the discharges.* Although HDP sources are quite different, all of them have RF or microwave power coupled to the plasma across a dielectric window. HDP etching discharges are characterized in general by pressures of 0.5–50 mTorr, which is lower than that in RF diodes and triodes. HDP discharge power is in the range 0.1–5 kW, which is generally higher than that of RF diodes and triodes. Typical electron temperatures are 2–7 eV, which is higher than that in capacitively coupled discharges. In accordance with the term HDP, the plasma density in these discharges is high, 10^{10} to 10^{12} cm^{-3}, exceeding that of capacitively coupled discharges by about 10-fold. The low-pressure discharges described above (see also Lecture 12) are conventionally used not only for plasma etching but also for deposition and other material processing applications.

21.4 Kinetics of Etch Processes and Discharges: Surface Kinetics of Etching, Densities, and Fluxes of Ions and Neutral Etchants

Kinetics of the etching processes includes the *surface kinetics*, which determines the etch rate and anisotropy based on ion and neutral fluxes from a discharge, as well as the *discharge kinetics*, which describes the generation of etchant atoms and bombarding ions. These two subjects of the etching kinetics are shortly discussed below, more details can be found in the books of Lieberman and Lichtenberg (2005) and Roth (1995, 2001).

Consider first the **surface etching kinetics**, taking as an example the ion-energy-driven etching of carbon substrate by atomic oxygen. A simplified mechanism of the etching process starts with chemisorption of O atoms on the carbon substrate: O(gas) + C(solid) → C:O(chemisorbed), rate coefficient K_a. The chemisorbed complexes C:O can then be either simply desorbed from the surface C:O(chemisorbed) → CO(gas), rate coefficient K_d, or desorbed by ion impact, characterized by kinetic coefficients Y_i and K_i: C:O(chemisorbed) + ion → CO(gas). In the framework of **the Langmuir adsorption–desorption model**, all O atoms approaching the fraction $1 - \theta$ of the carbon surface not covered by C:O complexes immediately react, forming the complexes. The Langmuir kinetic equation for the fraction θ can be then presented as:

$$\frac{d\theta}{dt} = K_a n_O (1 - \theta) - K_d \theta - Y_i K_i n_i \theta. \tag{21.5}$$

The limiting step of the etching is CO desorption into the gas phase, therefore the steady-state carbon surface coverage with the C:O groups can be found by assuming $d\theta/dt = 0$. Here, n_O and n_i are O and ion densities near the carbon surface (at the plasma-sheath edge), and K_a is the O adsorption rate coefficient, ($K_a = v_{tO}/4n_S$), v_{tO} is the velocity of O atoms; n_S is the surface density of the sites; Y_i is the ion energy-driven yield of CO molecules desorbed per ion incident on a fully covered surface, which is greater than unity and can be estimated as $Y_i = \eta E_i/E_b$; E_i is ion energy; E_b is the C:O group-surface binding energy ($E_i \gg E_b$); $\eta \leq 1$ is the efficiency of ion energy in desorption of the C:O complex from the carbon; K_i is the ion flux coefficient ($K_i = u_B/n_S$), u_B is the Bohm velocity. The steady-state carbon surface coverage with the C:O groups, θ, can be found from (21.5) as: $\theta = K_a n_O/(K_a n_O + K_d + Y_i K_i n_i)$. The flux of CO molecules leaving the surface because of the etching equals to $(K_d + Y_i K_i n_i) \theta n_S$. The vertical etch rate is determined by the above etching flux of CO and the C-atom density, n_C, and can be expressed as:

$$w_v = \frac{n_S}{n_C} \left((K_d + Y_i K_i n_i)^{-1} + (K_a n_O)^{-1} \right)^{-1}. \tag{21.6}$$

To calculate the horizontal etch rate, it is reasonable to assume that bombarding ions strike the substrate surface at normal incidence. The ion flux incident on a vertical trench sidewall in this case is equal to zero, and:

$$w_h = \frac{n_S}{n_C} \left((K_d^{-1} + (K_a n_O)^{-1} \right)^{-1}. \tag{21.7}$$

The etch anisotropy coefficient in practical conditions can be then expressed from ((21.6) and (21.7)) as:

$$a_h = \frac{w_v}{w_h} = \frac{Y_i K_i n_i}{K_d} \left(1 + \frac{Y_i K_i n_i}{K_a n_O} \right)^{-1}. \tag{21.8}$$

The maximum anisotropy coefficient for the ion energy-driven etching then is equal to:

$$a_h = \frac{w_v}{w_h} = \frac{Y_i K_i n_i}{K_d}. \tag{21.9}$$

The maximum anisotropy can be achieved when the ion flux is not too high, and the surface is covered by chemisorbed complexes ($\theta \to 1$). In this ion flux-limited regime, the etching anisotropy increases with an increase of ion energies and fluxes, as well as with a decrease of substrate temperature, which reduces the desorption, K_d.

Thus, the surface kinetics of etching is controlled by densities of ions and active neutrals near the surface. Determination of these parameters requires a detailed consideration of etching discharges. Some useful relations, however, can be derived from the general **kinetics of low-pressure discharges applied for etching**. The balance of charged particles in plasma between electrodes with area A (characteristic radius R) and narrow gap l between them ($l \ll R$), controlled by ionization and losses to the electrodes, can be estimated as:

$$2 n_i u_B A \approx k_{ion}(T_e) n_0 n_e l A \tag{21.10}$$

where n_e, n_i, and n_0 are the number densities of bulk plasma, ions at the plasma-sheath edge, and neutrals; $u_B = \sqrt{T_e/M_i}$ is the Bohm velocity; T_e is the electron temperature; M_i is an ion mass; and $k_{ion}(T_e)$ is the rate coefficient of ionization by direct electron impact, which strongly depends on electron temperature. Relation between ion density n_i near the surface at plasma-sheath edge and the bulk plasma density n_e can be expressed at low-pressures as:

$$n_i = h_l n_e, \quad h_l = 0.86 \left(3 + \frac{l}{2\lambda_i} \right)^{-1/2}. \tag{21.11}$$

where $\lambda_i = 1/n_0 \sigma_{i0}$ is the ion–neutral mean free path; $\sigma_{i0} \approx 10^{-14}$ cm² is a typical value of the ion–neutral collision cross section. Thus, the electron temperature in the discharge can be determined from $k_{ion}(T_e)$ as a function of $n_0 l$:

$$k_{ion}(T_e) = \frac{2 u_B h_l}{n_0 l} = 1.72 \frac{\sqrt{T_e/M_i}}{\sqrt{3 + (n_0 l)\sigma_{i0}/2}} \frac{1}{n_0 l}. \tag{21.12}$$

The discharge power can be expressed through the generation rate of charged particles and energy cost to the creation of a single electron–ion pair, E_{ei} (which depends on electron temperature, and numerically is about 30–300 eV): $P \approx 2AE_{ei}n_i u_B$. This way, we can calculate the near-surface ion density at the plasma-sheath edge:

$$n_i = \frac{P}{2AE_{ei}u_B} = \frac{P}{2AE_{ei}}\sqrt{\frac{M_i}{T_e}}. \tag{21.13}$$

The ion flux at the etching surface can also be expressed as a function of the total discharge power:

$$\Gamma_i = n_i u_B = P/2AE_{ei}. \tag{21.14}$$

Then, maximum anisotropy of the ion energy-driven etching as a function of the total discharge power is:

$$a_h = \frac{w_v}{w_h} = P\frac{Y_i}{2AE_{ei}n_S K_d}. \tag{21.15}$$

Similar relations can be derived for density and fluxes of neutral etchants. For example, atomic oxygen losses at the electrode during the etching lead to the decrease of the etching rate called the **loading effect**, see Fridman (2008).

21.5 Gas-phase Composition in Plasma Etching, Etchants-to-unsaturates Flux Ratio

Consider the role of different neutral chemical compounds applied in etching. Gases used for plasma etching can have, in general, quite complex compositions because they essentially affect the balance between etch rate, etching selectivity, and anisotropy. The gas-phase compounds participating in the etching can be classified into the following six groups:

1. *Saturates like CF_4, CCl_4, CF_3Cl, COF_4, SF_6, and NF_3.* The saturates are chemically not very active and unable themselves either to etch or forma surface film. They dissociate in plasma through collisions with electrons, producing chemically aggressive etchants and unsaturates.
2. *Unsaturates like radicals CF_3, CF_2, and CCl_3, and molecules such as C_2F_4, C_2F_6, and C_3F_8.* The unsaturates can react with substrates, forming surface films. Sometimes, however, the unsaturates can etch a substrate (e.g., SiO_2), producing volatile products.
3. *Etchants like atoms F, Cl, Br, and O (applied for resist) and molecules such as F_2, Cl_2, and Br_2.* Etchants are obviously the major players in etching processes; they are very aggressive, reacting with substrates and producing volatile products.
4. *Oxidants like O, O_2, and so forth.* The oxidants added to feed gas can react with unsaturates, converting them into etchants and volatile products.
5. *Reductants like H, H_2, and others.* The reductants added to the feed gas can react with etchants and neutralize those producing passive volatile products.
6. *Nonreactive gases like N_2, Ar, He, and so forth.* The nonreactive additives are used to control electrical properties of etching discharges and thermal properties of a substrate.

An important parameter of etching is the **etchants-to-unsaturates flux ratio** at the substrate. When the ratio is high, pure chemical etching can dominate, resulting in mostly isotropic etching. On the other hand, a low etchants-to-unsaturates ratio leads to the formation of surface films, which increases the contribution of the ion energy-driven etching and makes the etching anisotropic. There can also be an intermediate etchants-to-unsaturates flux ratio when the inhibitor film is deposited on sidewalls but is cleared from the bottom of trenches by ion bombardment. This regime corresponds to the anisotropic ion-enhanced inhibitor etching. The etchants-to-unsaturates flux ratios can be increased (for more isotropic etching) by adding to the feed gas such etchants as Cl_2 or Br_2, which do not produce unsaturates. Another way to increase the ratio is by adding oxidants, which can convert unsaturates into etchants ($CF_3 + O \rightarrow COF_2 + F$). To decrease the etchants-to-unsaturates flux ratio and increase the sidewalls' protection (for more anisotropic etching), unsaturated gases like C_3F_8 or C_2F_4 can be added to the feed gas. The effect can also be achieved by adding H_2 or other reductants to the feed gas, which react with etchants and neutralize them.

21.6 Atomic Fluorine and Chlorine-based Plasma Etching of Silicon, Flamm Formulas, and Doping Effect

Etching of silicon by F-atoms is the most practically important and the best characterized etch process. Let us consider first the **pure chemical isotropic silicon etching by F-atoms**. The F atoms intensively react with silicon, producing volatile silicon fluorides and, hence, providing effective surface etching. The reaction is of first kinetic order with respect to F-atom density, n_F. For undoped silicon and for thermally grown silicon dioxide, the F-atom etching rates can be expressed by the numerical **Flamm formulas**:

$$w_{Si} \text{ nm min}^{-1} = 2.86 \; 10^{-13} \; n_F(\text{cm}^{-3})\sqrt{T(K)} \; \exp(-1248 \, K/T). \tag{21.16}$$

$$w_{SiO_2} \text{ nm min}^{-1} = 0.61 \; 10^{-13} \; n_F(\text{cm}^{-3})\sqrt{T(K)} \; \exp(-1892 \, K/T). \tag{21.17}$$

The Flamm formulas lead to the expression for the Si/SiO_2 etching selectivity:

$$S = w_{Si}/w_{SiO_2} = 4.66 \; \exp(644 \, K/T). \tag{21.18}$$

Thus, the silicon-to-silicon dioxide etching selectivity at room temperature is quite good ($S \approx 40$). The F-atom silicon etching also has a good selectivity over Si_3N_4 etching and reasonable selectivity over resist etching. The mechanism of pure chemical F-atom etching of silicon includes the formation 2–5 intermediate silicon fluoride monolayers with a composition corresponding to SiF_3 on the surface. The main silicon etching product at room temperature is gaseous SiF_4 (about 65%); the rest is composed of other volatile silicon fluorides, Si_2F_6 and Si_3F_8. The halogen-atom etch rate depends on silicon doping: n-type silicon etching is faster than p-type, which is known as the **doping effect**. The doping effect in the case of F-atom etching gives only a factor-of-2 difference in the etch rates, whereas in the case of Cl-atom etching, the difference in etch rates can reach many orders of magnitude. The doping effect is related to the formation of negative halogen ions on the surface and their contribution to the etching. **Ion-energy-driven F-atom etching of silicon** follows the general features presented in Section 21.2 (Figure 21.3c). High fluxes of energetic ions increase the etch rates of silicon 5- to 10-fold with respect to the pure chemical etching described by Flamm formulas. A single 1 KeV Ar$^+$ ion can cause the removal of 25 Si atoms and 100 F atoms (to compare, such an ion removes only one atom in the case of sputtering). Although the anisotropy of Si etching can be about 5–10, pure chemical etching on the trench sidewalls remains a limiting aspect of fluorine-based anisotropic silicon etching.

Silicon etching by Cl atoms has many features in common with the etching provided by F atoms. Between major differences in pure chemical etching, the most important are related to strong crystallographic and doping effects in the case of Cl-atom etching. The silicon etch rate by Cl atoms can be described in Arrhenius form like the Flamm formulas. The **chlorine-based ion-assisted etching of silicon** can be performed with Cl_2, even without plasma volumetric dissociation of the molecules. For example, 1 keV Ar$^+$ ions with an adequate flux of Cl_2 provide the removal of three to five silicon atoms per incident ion (etching yield 3–5). Dissociation of Cl_2 does not result in a significant increase of the etching yield. To compare, dissociation of F_2 in a similar situation results in a 5- to 10-fold increase in yield of ion energy-driven silicon etching. To conclude our considerations of the halogen-atom based etching of silicon, we can make some remarks regarding bromine abilities in etching. Br atoms are in general less reactive than chlorine atoms, and much less reactive than F atoms. Pure chemical etching of even heavily n-doped silicon is negligible at room temperature. Pure chemical etching by Br atoms is observed at higher temperatures with a very strong doping effect.

21.7 Plasma Etching of Silicon in CF_4 Discharges: Competition Between Etching and Carbon Deposition

Plasma-generated F atoms are widely used for etching processes. Molecular fluorine (F_2) is not applied, however, as a feed gas in silicon etching or source of F atoms, because F_2 itself etches Si, leaving a rough and pitted surface as a result. Major feed gases for plasma generation of F atoms in this case are CF_4, SF_6, and NF_3 (obviously, unsaturates and other special admixtures can be added to the feed gases). The best studied plasma etching feed gas is CF_4, which can be used as an example. Plasma generated F-atoms and CF_x radicals play in this case an important role in anisotropic etching and forming protective films on the treated surfaces. While major contribution of F-atoms is etching itself (see Section 21.6), an important contribution of the CF_x radicals is related to their ability to dissociate and react on active surfaces, including silicon and the layers of intermediate silicon fluorides, SiF_x. The surface dissociation with formation of chemisorbed carbon leads to a buildup of carbon or polymer film on the surface, which prevents etching and is unlikely to be removed without ion bombardment. To characterize the reduction of the Si etch rate due to the flux of CF_x radicals ($x < 4$), the balance of fluorine on the surface can be presented as a balance of adsorbed and desorbed fluxes. The balance of the fluxes of CF_x radicals (Γ_{CF_x}) and F atoms (Γ_F) absorbed on the surface and the fluxes of volatile molecular gases CF_4 (Γ_{CF_4}) and SiF_4 (Γ_{SiF_4}) desorbed from the surface can be expressed as:

$$x\Gamma_{CF_x} + \Gamma_F = 4\Gamma_{CF_4} + 4\Gamma_{SiF_4}. \tag{21.19}$$

Steady-state conditions for atomic carbon on the treated surface also require a balance of the adsorption–desorption fluxes $\Gamma_{CF_4} = \Gamma_{CF_x}$, which results in the fluorine balance equation:

$$\Gamma_F = (4 - x)\Gamma_{CF_x} + 4\Gamma_{SiF_4}. \tag{21.20}$$

The rate of Si removal from the surface is related to the desorption of volatile tetrafluoride, $w_{Si} = \Gamma_{SiF_4}/n_{Si}$ (where n_{Si} is silicon number density). Then based on (21.20), the etch rate can be presented by the formula showing competition between fluxes of F atoms and CF_x radicals:

$$w_{Si} = \frac{\Gamma_{SiF_4}}{n_{Si}} = \frac{\Gamma_F - (4 - x)\Gamma_{CF_x}}{4\,n_{Si}}. \tag{21.21}$$

Assuming $x = 3$, etching is possible only if $\Gamma_F > \Gamma_{CF_x}$. In the opposite case, carbon deposition dominates and the etch rate is zero. Ion bombardment can shift the balance in the direction of etching, which is surely used for anisotropic etching. The effect of ion bombardment is related to increase of the ratio of the net fluxes absorbed, Γ_F/Γ_{CF_x}, and the effective physical sputtering of CF_y ($y < 4$). When sputtering of the CF_y groups dominates the desorption process, expression (21.21) for the etch rate can be modified for the case of ion energy-driven etching:

$$w_{Si} = \frac{\Gamma_{SiF_4}}{n_{Si}} = \frac{\Gamma_F - (y - x)\Gamma_{CF_x}}{4\,n_{Si}}. \tag{21.22}$$

 Silicon etching dominates over polymerization in this regime if $y \le x$. Increase of bias applied to the surface leads to higher ion bombardment energies, lower values of "y," and further shifting of the treatment in the direction of etching. The **F/C ratio of fluorine atoms to CF_x radicals** in the discharge strongly affects the etching deposition balance ((21.21) and (21.22)). When the F/C ratio is high (F/C > 3), etching dominates at any applied bias voltages and both trench sidewalls and bottoms are etched. When the F/C ratio is low (F/C < 2), film deposition dominates, stopping etching of both trench sidewalls and bottoms. *Highly anisotropic etching takes place in the intermediate case (2 < F/C < 3), when the trench bottom is etched at sufficient bias voltages (about 200 V) but the sidewalls are protected by the deposited film.* Correction of the F/C ratio can be achieved in the CF_4 discharges by additions of unsaturates, O_2, and H_2. Addition of oxygen up to about 16% converts CF_3 and other unsaturates into F atoms, removes carbon from the surface (by O atoms), increases the F/C ratio, and promotes etching. Addition of hydrogen and unsaturates decreases the F/C ratio and stimulates polymerization.

21.8 Plasma Etching of Silicon Oxide (SiO₂) and Nitride (Si₃N₄), Aluminum, Photoresists, and Other Materials

The rate of pure chemical **plasma etching of SiO₂ by F atoms** is determined by the Flamm formula (21.17). For a typical density of F atoms ($3 \cdot 10^{14}$ cm^{-3}), the etch rate is 0.59 nm min^{-1} at room temperature. The etch rate of silicon dioxide is low, about 40 times slower than that of silicon; CF_x radicals are even less effective in etching SiO_2. In contrast to the case of silicon, CF_x radicals are unable even to dissociate on the SiO_2 surface. Thus, effective SiO_2 etching in fluorocarbon plasmas is usually ion energy driven. When ion energies are 500 eV and higher, F and CF_x ion energy-driven etching of SiO_2 is characterized by high rates exceeding 200 nm min^{-1}. The etching is obviously anisotropic: the etch rates are correlated with ion bombarding energy and are independent of substrate temperature. The loading effect is smaller than in silicon etching. F atoms have no etching selectivity for SiO_2 over Si, whereas

CF_x radicals do; therefore, unsaturated fluorocarbon feedstocks, or CH_4–H_2 mixtures, are used to achieve selective SiO_2/Si etching.

Silicon nitride (Si_3N_4) etching is important due to application of this material as a mask for patterned oxidation of silicon and as a dielectric. Pure chemical etching of Si_3N_4 by F atoms is isotropic and is characterized by Arrhenius kinetics with an activation energy of about 0.17 eV. Si_3N_4 etching has a selectivity of 5–10 over that of etching of SiO_2. Pure chemical etching of Si_3N_4 by F atoms is not selective, however, over the etching of Si. Anisotropic etching of silicon nitride can be performed by ion enhancement of the etching process. Ion-energy-driven anisotropic Si_3N_4 etching is usually carried out using unsaturated fluorocarbon feed gases. It is characterized by low selectivity over silicon dioxide (SiO_2), but high selectivity over silicon and resist.

Plasma etching of aluminum is mostly due to its application as an interconnect material in ICs. Fluorine atoms cannot be applied for aluminum etching because the process product AlF_3 is not volatile. For this reason, other halogens (Cl_2 and Br_2) are used as feed gases for plasma etching of aluminum. In the absence of ion bombardment, chlorine, and bromine provide isotropic etching. Molecular chlorine can etch clean aluminum even without plasma. Etching of aluminum in Cl_2 discharges at low temperatures ($T \leq 200\,°C$) can be described as $Al_{solid} + 3/2$ $(Cl_2)_{gas} \to (AlCl_3)$gas. At higher substrate temperatures, the main product of pure chemical etching is Al_2Cl_6, which is also volatile. To provide anisotropic etching of aluminum, ion energy-driven performance of the processes with an inhibitor is required. In this case, special additives such as CCl_4, $CHCl_3$, $SiCl_4$, and BCl_3 are added to the main feed gas, which is molecular chlorine (Cl_2).

Photoresist masks are important in manufacturing ICs for electronics. **Plasma etching of photoresist** is key for two processes: the first one is isotropic etching of the resist mask materials from wafers (called ***stripping***), and the second is anisotropic pattern transfer into the resist (in the ***surface-imaging dry development***). The photoresist masks are long-chain organic hydrocarbon polymers, so oxygen atoms can be effectively used as etchants for both applications. Thus, O_2 discharges are usually used for plasma etching of the photoresist mask materials. Pure chemical etching of photoresists by O atoms is isotropic and highly selective over silicon and SiO_2. When the kinetics of the process is presented in Arrhenius form, the activation energy is 0.2–0.6 eV.

Plasma can also be effectively applied for etching layers of silicon doped with refractory or rare metals, semiconductors, or via interconnects made with rare or refractory metals. These processes normally use fluorine-containing gases as a feedstock and have hexafluorides as volatile etching products. However, chlorine is preferred in some specific cases.

21.9 Plasma Cleaning of Chemical Vapor Deposition (CVD) and Etching Reactors, In-situ Cleaning in Micro-electronics, and Other Active and Passive Cleaning Processes

Plasma etching processes considered above are directly focused on the IC fabrication. Besides direct treatment of wafers, the microelectronics also requires chamber cleaning, where plasma is effectively applied as well. CVD or etching chamber cleaning from deposited silicon materials is an isotropic F-atom etching process, which has, however, several individual special features. In the production of ICs, the cleaning of treatment chambers is a very time-consuming operation because deposits of silicon oxides are difficult to remove from surfaces of the treatment chamber. Cleaning is usually achieved by etching the chamber surfaces with active particles, among which atomic fluorine is the most effective. Atomic fluorine can be conveniently produced from stock gases such as NF_3, CF_4, C_2F_6, and SF_6 in low-temperature discharge plasmas. Plasma cleaning does not require ion bombardment and has fewer limitations related to process anisotropy and selectivity. As a result, operational discharge pressures can be essentially increased with respect to etching discharges to the level of several Torr. Higher pressures applied in this pure chemical etching permit significant acceleration of the cleaning process, see the Flamm formulas. To avoid ion-related chamber damages during elevated pressure plasma chamber cleaning, **remote plasma sources (RPSs)** are applied, where F atoms are generated in a special discharge and then delivered to the chamber through a special transport tube, to be discussed in the next section. The high cleaning rates can lead to significant emissions of **perfluoro-compounds (PFCs)**, which are products of conventional cleaning and create environmental control problems in semiconductor fabrication plants. PFC emission restrictions stimulated changes in feed gases applied as a source of F atoms for plasma chamber cleaning, leading to a wider application of NF_3 as feed gas. Plasma dissociation of NF_3 to N_2 and F is effective and not accompanied by PFC emission. However, high NF_3 prices dictate strong requirements for utilization efficiency of F atoms, especially the efficiency of F-atom transportation from the RPS to the chamber. All those factors determine features of in situ plasma cleaning, distinguishing this process from conventional etching.

The **in-situ plasma cleaning** performed after CVD of dielectric thin films is one of the major emitters of PFCs in semiconductor manufacturing; it can represent 50–70% of the total PFC emission in a semiconductor fabrication plant. To clean the process chambers of deposited by-products, conventional cleaning methods use CF_4 or C_2F_6 gases activated by RF CCP (usually 13.56 MHz) inside the process chamber. These PFC gases achieve a relatively low degree of dissociation and unreacted molecules are emitted in the process exhaust. To overcome these limitations, a new plasma cleaning technology is applied, which uses a remote RF-ICP source (see Section 21.10) to completely break down NF_3 gas into an effective cleaning chemical.

Plasma can be also effective in other cleaning processes. Thus, plasma can effectively remove adherent monolayers of hydrocarbons, thin layers of chemical contaminations like surface oxides, radioactive contaminations, and more. Such *plasma cleaning systems can be subdivided into two classes: passive and active plasma cleaning*. This classification, based on differences in the collection of electric current on the workpiece to be cleaned, is rather artificial but nevertheless differentiates between types of nonequilibrium discharges applied for cleaning.

Passive plasma cleaning is provided by plasma-generated active species but does not introduce the treated material as a current-collecting electrode; hence, the treated workpiece does not draw any "real" electric current from the plasma. For example, effective passive plasma cleaning of stainless-steel samples using intermediate-pressure argon and dry-air discharges provides high lap shear strength for adhesive bonding. Another example of effective passive plasma cleaning is related to removal of thin layers ($\leq 4\,\mu m$) of hydrocarbon vacuum pump oils and poly-a-olefins from the surface of stainless steel and other metals. It was performed using RF discharge in oxygen at gas pressures of 15–400 mTorr. The removal rate of the hydrocarbon oil film is 1–$2\,\mu m\,min^{-1}$. Surface cleaning is achieved here without drawing any essential electric current from the RF plasma, so the process can be qualified as passive plasma cleaning.

Active plasma cleaning, in contrast to passive cleaning, is characterized by applying electric current directly to the treated material. The workpiece to be cleaned is placed on an electrode, for example on the cathode of an abnormal glow discharge. Active plasma cleaning can also be carried out by placing the workpiece on a powered RF electrode that is DC biased to draw current, or on a hollow cathode electrode. Although active plasma cleaning provides a flux of energetic charged particles to the surface in the form of a "real" electric current, the cleaning rate can sometimes be even lower than that of passive plasma cleaning. Active plasma cleaning can be represented by cleaning Al surfaces from carbon and oxygen impurities by using a DC 60 Hz abnormal glow discharge. The rate of removal of the impurities is 1–10 monolayers min^{-1}. Another example of active plasma cleaning involves the application of DC glow discharge to clean the interior of stainless-steel accelerator beam tubes.

 Active and passive cleaning of metallic surfaces can result in **significant improvement of wettability**. Wettability is one of the most important surface characteristics related to surface energy and can be significantly improved by plasma treatment. Wettability is directly related to the effectiveness of printing, dyeing, and so forth on surfaces. Wettability can be determined as the solid surface's ability to adsorb liquid or to absorb the liquid in the bulk of fibrous materials. Wettability implies a high surface energy on the level of 50–$70\,dyn\,cm^{-1}$ and low contact angles. *Plasma is very effective in increasing the surface energy, not only by chemical conversion of the surface but often simply through plasma cleaning. Thus, plasma cleaning can significantly improve the wettability of the surfaces.* Whereas plasma treatment of polymers surfaces and other organic materials often leads to an increase in surface energy and wettability by means of special chemical activation, improvements in the wettability of metals are usually due to simple plasma surface cleaning. Both active and passive plasma cleaning can be effectively applied to metallic surfaces. Plasma cleaning can make these surfaces so clean that their energies approach $70\,dyn\,cm^{-1}$, which is sufficient for good wettability.

21.10 Remote Plasma Cleaning in Microelectronics, Choice of Cleaning Feedstock Gases

Plasma cleaning is more effective at elevated pressures when pure chemical etching is faster, see the Flamm formulas and discussion in the previous section. To avoid ion-related chamber damages at elevated pressure, the **RPSs** can be effectively applied in the plasma cleaning of CVD and etching chambers in microelectronics. A fluorine-containing gas (in particular, NF_3) is introduced in a remote discharge chamber, where plasma is sustained by microwave or RF energy. The cleaning feed gas is plasma-dissociated into charged and neutral species (F, F_2, N, N_2, NF_x, electron, ions, and excited species). Because plasma is confined inside the applicator, and since ions have a very short lifetime, mainly neutral species are injected through a transport tube and a special shower head into the main deposition chamber for cleaning purposes. The fluorine radicals react in the main chamber to be cleaned with the deposition residues (SiO_2, Si_3N_4, etc.) to form non-global-warming volatile by-products (SiF_4, HF, F_2, N_2, O_2). The volatile by-products are pumped through the exhaust and can be removed from the stream using conventional scrubbing technologies. Due to the high efficiency of the microwave/RF excitation, NF_3 gas utilization removal efficiency can be as high as 99% in standard operating conditions. With this "remote" technique, no plasma is sustained in the main microelectronics processing chamber and the cleaning is much "softer" on the chamber components, compared to the in-situ plasma cleaning approach, see Raoux et al. (1997, 1999).

Let us discuss the **choice of cleaning feed gas.** The PFC molecules have very long atmospheric lifetimes (50 000 years for CF_4; 10 000 years for C_2F_6; 7000 years for C_3F_8; 740 years for NF_3); which is a direct measure of their chemical stability. It is not surprising then that high plasma power level is required to achieve near-complete destruction efficiencies of these gases. Among the commonly used fluorinated gas sources, nitrogen trifluoride (NF_3) has been determined to be the best source of gas. NF_3 has similar *global warming potential (GWP)* compared to most other PFCs (with a 100-year integrated time horizon). However, when considering the GWP of NF_3 over the life of the molecule, it is much lower than that of CF_4, C_2F_6, and C_3F_8 due to shorter lifetime. This should be taken into consideration for estimating the long-term impacts of PFC gas usage. Another reason why NF_3 is well suited to this application is the weaker nitrogen–fluorine bond as compared to the carbon–fluorine bonds in CF_4 or C_2F_6. The relative ease of destruction of NF_3 results in a high usage efficiency of this source gas for plasma cleaning. Another advantage of NF_3 plasma chemistry is that it is a noncorrosive carbon-free source of fluorine. The use of fluorocarbon molecules such as CF_4, C_2F_6, and so forth requires dilution with an oxidizer (O_2, N_2O, etc.) to prevent the formation of polymeric residues during the cleaning process. Dilution of NF_3 with oxygen enhances the etch rate but leads to the formation of NO_x by-products.

More information regarding kinetics of F-atom generation from NF_3, CF_4, C_2F_6, and transportation in and from the RPS, kinetics of surface and volume recombination of F-atoms can be found in Conti et al. (1999); Iskenderova et al. (2001, 2003); Stueber et al. (2003); as well as in Fridman (2008).

The crucial aspect of the RPS efficacy is transportation of the plasma generated F-atoms from the remote source to CVD or etching chamber along the **transportation tube**, see typical parameters in Figure 21.4a. Typically, gas with molar composition F 60%, F_2 10%, and Ar 30% flows into the tube at a rate of about 250 sccm. The temperature of the tube walls is 300 K and the gas pressure varies from 1 to 8 Torr. Figure 21.4b shows losses of F atoms along the tube at different pressures. Losses of F atoms are more significant at higher pressure, which indicates a contribution of volume recombination. The ratio of F-atom recombination losses attributed to surface and volume processes integrated over the whole tube, $f(p)$, is shown in Figure 21.5a. Volume recombination (rate proportional to p^3) dominates at higher pressures; contributions of volume and surface mechanisms are equal at a pressure of 4.5 Torr (Iskenderova, 2003). A related question involves the optimization of pressure in the transport tube to maximize the F-atom density at the tube exit. If the pressure is too low, the initial density of fluorine atoms is low; if the pressure is too high, recombination losses are significant. The F-atom concentration at the exit of the transport tube is shown in Figure 21.5b as a function of pressure. The optimal pressure at the conditions under consideration is about 2 Torr (Iskenderova, 2003; Stueber et al., 2003).

21.11 Plasma-enhanced Chemical Vapor Deposition (PECVD), Amorphous Si-film Deposition

Plasma plays an important role in the production and modification of thin films by means of plasma deposition, including plasma-enhanced chemical vapor deposition (PECVD) and sputtering deposition, implantation, and surface modification processes. Production of a thin film from a feed gas through a set of gas-phase and surface chemical reactions is called chemical vapor deposition (CVD). If the gas-phase and surface reactions, starting from dissociation of the feed gas, are stimulated by plasma, the CVD process is called **PECVD**. Like plasma etching, PECVD also

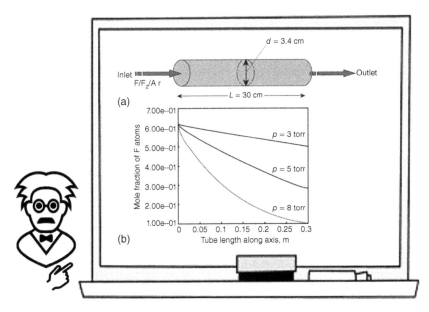

Figure 21.4 F-atoms transportation from RPS for etch and CVD chamber cleaning: (a) transport tube; (b) mole fraction of F atoms along the transport tube length at pressures: 3, 5, and 8 Torr.

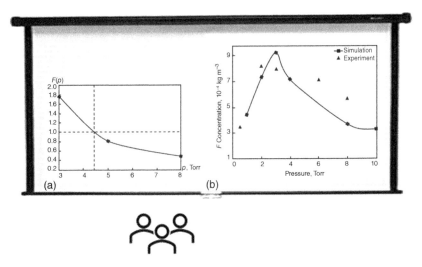

Figure 21.5 F-atoms transportation from RPS for etch and CVD chamber cleaning: (a) ratio of surface and volume recombination rates of F-atoms at different pressures (critical pressure is 4.5 Torr); (b) F-atoms concentration after the transport zone as function of pressure.

plays a key role in microelectronics for fabrication of ICs. As an example, PECVD is applied for deposition of the final insulating silicon nitride (Si_3N_4) layer in many electronic devices. This PECVD requires a surface temperature near 300 °C, whereas the non-plasma CVD activated thermally requires 900 °C, which is unacceptable. The PECVD often requires reactions between gas-phase precursor components; therefore, gas pressure in the PECVD discharges cannot be too low. Usually, the operational pressures in these systems are 0.1–10 Torr, which is considerably higher than the pressure in etching discharges typical plasma density in PECVD discharges is 10^9 to 10^{11} cm^{-3}, and ionization degrees are 10^{-7} to 10^{-4}. Deposition rates are usually not sensitive to the temperature of the substrate. The substrate temperature does, however, determine such properties of the deposited film as morphology, composition, stress, and so on. A critical issue of PECVD is uniformity of the deposited thin film. Relatively high gas pressures and flow rates in the discharge make the uniformity requirements quite challenging. The uniformity requirements also dictate careful control of discharge power deposition per unit area. For this reason, RF-CCP discharges with a parallel-plate configuration are especially relevant for PECVD applications. We should note, however, that several effective plasma deposition processes have been performed in cylindrical HDP discharges, including ECR, helicons, and RF-ICP transformer discharges.

First and the most relevant example is the **PECVD production of thin films of amorphous silicon**, which has a wide range of applications, including thin-film transistors for flat-panel displays, solar cells, and exposure drums for xerography. PECVD amorphous silicon produced from SiH_4 usually incorporates 5–20% of H atoms in lattice. Its density is $2.2\,g\,cm^{-3}$ – lower than that of epitaxial crystalline silicon ($2.33\,g\,cm^{-3}$). PECVD amorphous silicon is inexpensive and can be deposited over large areas of different substrates such as glasses, metals, polymers, and ceramics. Silane (SiH_4) is a typical feed gas for the PECVD process in RF discharges at pressures of 0.2–10 Torr. At somewhat higher pressures, H_2 and Ar can be admixed to the silane. To grow p-type material, B_2H_6 is usually added; PH_3 addition is used to produce *n*-type material. Deposition rates of the process are typically $5–50\,nm\,min^{-1}$, which requires an RF discharge power of $0.01–0.1\,W\,cm^{-2}$. The substrate temperatures for the PECVD process are 25–400 °C depending on the application.

Kinetics of the PECVD of amorphous silicon films has been investigated and reviewed in several publications, see Kushner (1988); McCaughey and Kushner (1989); Lieberman and Lichtenberg (2005), see also Fridman (2008). Let us shortly discuss some key kinetic features of the process. SiH_4^+ positive ions are unstable; therefore, ionization of silane is dissociative with formation of SiH_3^+ and SiH_2^+. Surface bombardment by these positive ions plays a critical role in the growth of amorphous silicon films. The silane radicals SiH_3 and SiH_2 have positive electron affinity; therefore, the silane discharge is essentially electronegative, and dissociative attachment processes make contribution in the balance of charged particles and production of negative silane ions SiH_3^- and SiH_2^-. The most important precursors for amorphous silicon thin-film growth are silane radicals, SiH_3 and SiH_2, produced by direct electron impact dissociation of the silane molecules (SiH_4). The non-dissociated SiH_4 molecules also participate in the surface reactions, but mostly as deactivating chemical agents. We should remind also that the silane discharge plasma chemistry has been discussed in Sections 20.10 and 20.11 regarding formation of the dusty plasma nanoparticles.

Amorphous silicon deposition rates from silane in the PECVD process are typically in the range $5–50\,nm\,min^{-1}$. To analyze the kinetics of the deposition process, consider a simple surface model illustrated in Figure 21.6. The surface of amorphous silicon consists of a combination of active (a) and passive (p) sites. The active sites contain at least one open free bond; the fraction of the surface covered by active sites is θ_a. In the passive sites, all bonds are occupied by silicon or hydrogen atoms; the fraction of the surface covered by passive sites is θ_p ($\theta_a + \theta_p = 1$). Creation of new active sites is due to ion bombardment, which also removes hydrogen atoms from the surface with the yield Y_i (per one ion):

$$SiH_3^+ + \theta_p \rightarrow \theta_a + Y_i H(g), \quad K_i = \frac{u_B}{n_S}, \tag{21.23}$$

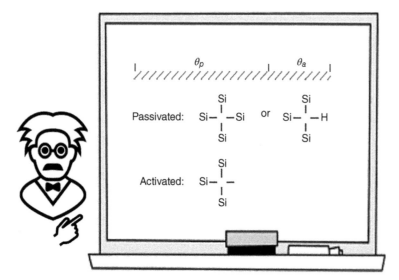

Figure 21.6 Surface coverage during amorphous silicon deposition in plasma: θ_a and θ_p represent the active and passive surface site fractions.

where K_i is the rate coefficient of the process, u_B is Bohm velocity, and n_S is the area density of the surface sites. Collisions of SiH_2 radicals with both active and passive sites lead to their insertion into the lattice. The probability of their insertion into the passive silicon surface is lower than into the active one. The SiH_2 radicals' insertion does not change the type of the surface sites; that is, the passive sites remain passive, and the active sites remain active:

$$SiH_2 + \theta_a \rightarrow \theta_a, \quad K_2 = s_2 <v_2> \frac{1}{4n_S}; \quad SiH_2 + \theta_p \rightarrow p, \quad K_{2p} = s_{2p} <v_2> \frac{1}{4n_S}. \tag{21.24}$$

Here K_2 and K_{2p} are rate coefficients of the surface processes on the active and passive sites; s_2 and s_{2p} are sticking coefficients for an SiH_2 radical on the active and passive sites; and $<v_2>$ is the average thermal velocity of the radicals. *Film growth related to the insertion of SiH_2 radicals is characterized by poor quality, voids, surface roughness, and other surface defects. Growth of a smooth high-quality film is provided by SiH_3 radicals, which are adsorbed on the surface, diffuse along the surface, and finally can be inserted into the lattice selectively at the active sites only.* Insertion of SiH_3 radicals into the active lattice sites converts them into passive sites, fills in the surface roughness, and results in growth of a smooth, high-quality film:

$$SiH_3 + \theta_a \rightarrow \theta_p, \quad K_3 = s_3 <v_3> <M> \frac{1}{4n_S}, \tag{21.25}$$

where K_3 is the rate coefficient of the surface process, s_3 is the sticking coefficient for SiH_3 radicals on the active surface, $<v_3>$ is the thermal velocity of the radicals, and factor $<M>$ characterizes the mean number of sites visited by a surface-diffusing adsorbed SiH_3 radical before desorption. Non-dissociated silane molecules (SiH_4) can also be adsorbed upon direct impact at active sites. The saturated SiH_4 molecule can lose a hydrogen atom in this case, passivate the site, and release an SiH_3 radical back to the gas phase:

$$SiH_4 + \theta_a \rightarrow \theta_p + SiH_3(g), \quad K_4 = s_3 <v_4> \frac{1}{4n_S}, \tag{21.26}$$

where K_4 is the rate coefficient of the surface process, s_4 is the sticking coefficient for SiH_4 on the active surface site, and $<v_4>$ is the thermal velocity of SiH_4. The balance equation for the fraction of the active surface sites is:

$$d\theta_a/dt = Y_i K_i n_i (1 - \theta_a) - K_3 n_3 \theta_a - K_4 n_4 \theta_a, \tag{21.27}$$

where n_i is the ion density at the plasma-sheath edge, and n_3 and n_4 are near the surface number densities of SiH_3 and SiH_4. In steady-state conditions ($d/dt = 0$), fraction of the active surface sites, θ_a from the above balance equation is:

$$\theta_a = \frac{Y_i K_i n_i}{Y_i K_i n_i + K_3 n_3 + K_4 n_4}. \tag{21.28}$$

In typical PECVD conditions, deactivation of the active surface sites by SiH_4 molecules dominates the kinetics ($K_4 n_4 \gg Y_i K_i n_i + K_3 n_3$), and relation (21.28) becomes $\theta_a = Y_i K_i n_i / K_4 n_4$. It results in the following rate of amorphous silicon film deposition from SiH_4 in the PECVD process:

$$w_{D,Si} \approx (K_3 n_3 \theta_a + K_2 n_2) \frac{n_S}{n_{Si}} = \left(\frac{K_3 n_3 Y_i K_i n_i}{K_4 n_4} + K_2 n_2 \right) \frac{n_S}{n_{Si}}, \tag{21.29}$$

where n_{Si} is the density of Si atoms in the lattice. Conventionally: $Y_i = 5$–10, $<M> \approx 10$, $n_i/n_4 \approx 10^{-4}$, and all sticking coefficients s are of order unity ($s \approx 1$). Then, the fraction of the active surface sites is $\theta_a \approx 10^{-2}$. Thus, most of the surface remains passive during the PECVD process, and only about 1% of sites are active for the growth of silicon film. Compare the first and second terms of (21.29), which represent high-quality (related to SiH_3) and low-quality (related to SiH_2) deposition mechanisms. The ratio of these terms, ξ representing the film quality is:

$$\xi = \frac{K_3 n_3 \theta_a}{K_2 n_2} \approx <M> \theta_a \frac{n_3}{n_2}. \tag{21.30}$$

In the conventional SiH_4 PECVD conditions, the concentration of SiH_3 exceeds that of SiH_2: $n_3/n_2 \approx 100$. It results in $\xi \approx 10$, which means that the contribution of the high-quality film deposition mechanism usually dominates the process. Improvement in film quality requires higher ion fluxes (i.e., higher n_i) and energies (i.e., higher Y_i), higher SiH_3/SiH_2 ratios, and finally better SiH_3 surface diffusion (i.e., higher $<M>$).

21.12 PECVD of Silicon Oxide (SiO₂) and Silicon Nitride (Si₃N₄) Films, Conformal and Non-conformal Deposition in Trenches, Atomic Layer Deposition (ALD)

Production of SiO_2 thin films is of considerable importance in microelectronics. It can be performed using many approaches, including direct oxidation of silicon and different types of CVD. All these processes can be stimulated by nonthermal plasma. The **SiO₂ film growth by direct silicon oxidation** in O_2 or H_2O can be achieved thermally at 850–1100 °C. The process can be enhanced by application of nonthermal plasma, which accelerates the deposition and decreases the required substrate temperature. High-quality thin SiO_2 films have been grown on single-crystal silicon at substrate temperatures of 250–400 °C using a nonthermal O_2 discharge. This direct plasma oxidation process is usually referred to as **plasma anodization**. The substrate is biased positively with respect to plasma, which draws a net DC through the film while it grows. The mechanism of direct oxidation is related to the drift of negative ions (O^-) from the surface across the Si–SiO₂ interface into silicon. The O^- transport limits the kinetics of SiO_2 film growth.

More effective, from this perspective, are deposition processes, which do not require oxygen transportation through the interface to build up the SiO_2 film. **PECVD growth of SiO₂ films** can be organized at 100–300 °C using either *silane–oxygen mixtures*, or a special *tetraethyl orthosilicate (TEOS)–oxygen mixture* as feed gases. To compare, the thermal CVD approach in this case requires much higher temperatures (600–800 °C).

Use of different feed gases leads to different **conformality characteristics of the PECVD on topographical features** *like trenches.* **Silane–oxygen PECVD of SiO₂ films is non-conformal** and can be performed in such gas mixtures as SiH_4–Ar–N_2O, SiH_4–Ar–NO, and SiH_4–Ar–O_2. The molecular gases N_2O, NO, and O_2 are used here as a source of atomic oxygen, which determines the SiO_2 film growth. The most conventional source of oxygen, for this purpose, is N_2O. The SiO_2 deposition precursors are SiH_3, SiH_2 radicals, and O atoms, which are created by plasma dissociation of feed gases. Initial surface reactions of the film growth include surface formation of such intermediate molecular oxides as $(SiH_3)_2O$ and SiH_3OH. Further oxidation removes hydrogen from the surface in form of water vapor, usually leaving about 2–9% of hydrogen atoms in the final SiO_2 film. The SiO_2 film deposition rate in the described silane–oxygen discharges is quite high, up to 200 nm min⁻¹. The high deposition rates are due to high sticking coefficients of SiH_3 and SiH_2 radicals (the sticking probabilities are about $s \approx 0.35$). On the other hand, *the high sticking coefficients of SiH_3 and SiH_2 radicals result in a non-conformal deposition of SiO_2 films within the trenches.* This means that deposition rates at different points of the trench are different, which limits the use of silane–oxygen discharges for PECVD.

Let us consider the **conformality** using a simple model of SiO_2 deposition on the sidewalls and bottom of a trench, which is provided by a uniform isotropic source of precursors at the top of the trench (see Figure 21.7a). Assume the sticking coefficient equals unity, and the precursor's mean free path is much greater than the characteristic sizes of the trench. Then the deposition flux of SiO_2 on the sidewalls can be expressed as $\Gamma_{SiO_2} \propto 1 - \cos\theta_s$, where θ_s is the angle subtended by the trench opening as seen at a position along the sidewall (see Figure 21.7a). Thus, the maximum deposition rate value corresponds to near the top area of the trench, which can lead to the formation of undesirable keyhole non-conformality within the trench (see Figure 21.7b). Two following effects help to make the PECVD process within the trenches uniform and conformal, to avoid formation of the "keyholes," and to permit complete filling of the trench with insulating dielectric material (SiO_2): (i) low sticking probabilities of active precursors and numerous reflections within the trench before the final random sticking to the surface lead to uniform deposition within a trench; (ii) high surface diffusion rates of the adsorbed active precursors in the trench also make the deposition uniform and conformal. The silane-containing feed gases have high sticking coefficients (they are also characterized by insignificant rates of surface diffusion), which provide high deposition rates on the one hand, but poor deposition conformality on the other hand. When deposition conformality is an important issue, another feed-gas mixture, TEOS–O_2, is usually applied for PECVD of silicon dioxide films.

The **tetraethyl orthosilicate (TEOS, Si(OC₂H₅)₄)** is used as an SiO_2 deposition precursor in the **conformal PECVD processes**. *TEOS produces deposition with low sticking probabilities, which permits excellent conformality of the process.* It is also important for practical application that TEOS is a relatively inert liquid at room temperature, in contrast to silane (SiH_4), which is an explosive gas at room temperature. A feedstock mixture, which is typically used for conformal PECVD deposition, is 1% TEOS and 99% O_2. The discharge kinetics is dominated in this case by molecular oxygen. SiO_2 deposition from the TEOS–O_2 mixture is usually organized in nonthermal plasma at pressures of 0.2–0.5 Torr and at substrate temperatures of 200–300 °C. Deposition rates are usually not higher than 50 nm min⁻¹ (to compare, those in SiH_4-based mixtures are considerably higher, up to 200 nm min⁻¹). However, the sticking coefficients for the TEOS precursor deposition ($s \approx 0.045$) are

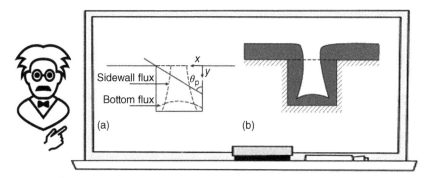

Figure 21.7 Illustration of a non-conformal plasma deposition within a trench: (a) trench before deposition (the dashed lines present the deposition flux incident on the sidewall and bottom); (b) intermediate moment of the deposition.

about 10-fold lower than those for the SiH_4 precursor deposition. This results in good deposition conformality and indicates the effectiveness of TEOS application for PECVD on topographical features like trenches. Ion bombardment of the substrate anisotropically stimulates the vertical deposition rate of SiO_2, providing directionality to the PECVD process, see Stout and Kushner (1993).

PECVD produced amorphous silicon nitride (Si_3N_4) films are resistant to water vapor, salts, etc., and, therefore, are applied as a final encapsulating layer for ICs. A typical feed-gas mixture for PECVD is SiH_4–NH_3. The process is performed in plasma at pressures of 0.25–3 Torr; conventional substrate temperatures are in the range 250–500 °C. Deposition rates of silicon nitride films under such conditions are about 20–50 nm min^{-1}. Major film precursors are SiH_3, SiH_2, and NH radicals, generated in plasma by dissociation of a feed gas by direct electron impact.

Extremely thin film deposition can be achieved by the so-called **atomic layer deposition (ALD) and plasma-assisted atomic layer deposition (PA-ALD)** based on the sequential processes using two gaseous precursors. These precursors react with the surface one at a time in a sequential, self-limiting, manner. In contrast to conventional chemical vapor deposition, the precursors are never present simultaneously in the reactor, but they are inserted as a series of sequential, nonoverlapping pulses. In each of these pulses, the precursor molecules react with the surface in a self-limiting way, so that the reaction terminates once all the reactive sites on the surface are consumed. By varying the number of cycles, it is possible to grow materials uniformly and with high precision on arbitrarily complex and large substrates. ALD produces very thin, conformal films with control of the thickness and composition at the atomic level. PA-ALD allows to reduce the deposition temperature without compromising the film quality; also, a wider range of precursors can be used and thus a wider range of materials can be deposited as compared to thermal ALD, see for more details Oviroh et al. (2019).

21.13 Sputter Deposition Processes: Physical and Reactive Sputtering

Sputtering is a process of ejection of atoms from surfaces due to energetic ion bombardment. The sputtering process can be applied not only for material removal from surfaces (etching, discussed above in this lecture) but also as a source of atoms for following deposition. This method is usually referred to as **sputter deposition** and can be divided into two approaches: physical sputtering and reactive sputtering. *Physical sputtering* implies that atoms sputtered from the target material are directly transported without any special conversion to a substrate, where they are deposited. An important issue of physical sputtering is process uniformity. The sputtered atoms' energy distribution is also very important, because it determines mixing of the sputtered atoms with substrate atoms and influences the properties of the deposited film. In the case of *reactive sputtering*, a special feed gas is also supplied to the plasma deposition system. Molecules of the feed gas dissociate in plasma, producing chemically active species, which then react with the target during ion bombardment and sputtering. As a result, the film deposited on a substrate includes not only material of the ion-bombarded target but also compounds produced due to the reactive gases. Chemical reactions between the target material and reactive gas species take place in this type of sputter deposition process at both target and substrate surfaces. Nonthermal plasma sources used for the sputtering deposition of thin conductive films are generally low-pressure DC discharges and especially DC planar magnetrons. RF-CCP discharges or RF-driven planar magnetrons are usually used in the case of sputtering deposition of insulating films.

The **physical sputter deposition** implies ion bombardment of a target, which results in sputtering of the target atoms. The sputtered atoms are then ballistically flowing to the substrate and finally are deposited on the substrate. The ion bombardment is typically provided by Ar^+ ions with energies in the range 0.5–1 keV. Sputter yields for different target materials are of order unity. Therefore, a wide variety of materials, including metals, alloys, and insulators, can be deposited this way over large areas with reasonable deposition rates, excellent uniformity, and good surface smoothness and adhesion. The sputter deposition is not conformal; the conformality can be improved in this case only by ion bombardment of the deposited film and redeposition. Typical physical sputter deposition rate is 75 nm min^{-1}. Properties of the deposited film are sensitive to the sputtered-atom energy distribution. It can be explained considering that the energy of the atoms striking the substrate determines mixing and diffusion processes between the incoming sputtered atoms and substrate material. These processes are crucial for good bonding and adhesion, resulting in better quality of the sputter-deposited film.

The **reactive sputter deposition** uses active species from the gas phase for chemical reactions with the target material during ion bombardment. The chemical reactions take place at both the target and the substrate. The gas-phase chemically active species are produced by dissociation of the special feed gas introduced into the discharge. Thus, the deposited film in this case is a compound formed from the target material and the feed gas. The sputtering deposition of SiO_2 explains the advantages of reactive sputter deposition. After SiO_2 targets physical sputtering, Si and O atoms approach the substrate. Silicon atoms stick to the substrate surface, while O atoms partially form molecular oxygen, leaving the deposited film with a deficiency of oxygen. To restore the SiO_2 stoichiometry of the deposited film, O_2 is added as a feed gas to the discharge, which leads to the additional formation of O atoms and their incorporation into the growing film. This is the main idea of the reactive sputter deposition approach. The feed gas can control oxygen concentration in the film; therefore, even pure silicon can be used for reactive sputter deposition of SiO_2 films in this case. Considering the advantages of the approach when one component has a high vapor pressure, reactive sputter deposition is widely applied in the production of thin films of oxides, nitrides, carbides, and silicides. Feed gases commonly used as a source of O atoms are O_2 and H_2O; N_2 and NH_3 are sources of N atoms; CH_4 and C_2H_2 are sources of C atoms; and silane (SiH_4) is a source of Si atoms. Reactive sputter deposition is used for the production of thin films of different metal compounds like titanium nitride (TiN). *The deposition processes with a metal target and a compound film can be arranged in two modes: "metallic" mode and "covered" mode.* The covered mode takes place when the ion flux is low and the gas flux is high; hence, the target is covered by the compound during the deposition process. The metallic mode, on the other hand, takes place when the ion flux is high and the gas flux is low; hence, the target remains metallic during the deposition. Deposition rates are generally higher in the metallic mode. If the ion flux is fixed, increased gas flow

leads to a transition from metallic to covered mode. The transition exhibits hysteresis: the "increasing" transitional gas flux corresponding to the metallic to covered mode transition is larger than the "decreasing" transitional gas flux corresponding to the covered-to-metallic mode transition. More details on the subject, especially on kinetics of the sputter deposition and the hysteresis effect can be found in Berg et al. (1989); Lieberman and Lichtenberg (2005), see also Fridman (2008).

21.14 Ion Implantation Processes, Ion-beam Implantation

Ion implantation is based on directing high-energy ions to surfaces and penetration of the energetic ions into the atomic structure of a material over many atomic layers. Ion energies typically used for the implantation processes are in the range 10–300 keV and sometimes higher. The penetration depth of the energetic ions can be up to a micron, which is sufficient for many applications. Ion implantation is an important process for semiconductor doping but also has other practical applications, especially in surface hardening of materials, especially metals. Silicon doping can be provided by ion implantation of boron, phosphorus, and arsenic. Surface hardening of metals can be provided by ion implantation of nitrogen or carbon. The ion implantation technology can be divided into two approaches: conventional ion-beam technology, and plasma ion implantation to discuss in the next section. General schematics of the ion-beam implantation systems are shown in Figure 21.8. Ion-beam implantation can be accomplished by two types of systems, depending on energy of the ions. The first type (Figure 21.8b) applies lower-energy ions (10–300 keV) and is mostly used in microelectronics for wafer doping. Current densities of these ion sources are limited by space charge. The second type (Figure 21.8a) uses higher-energy ions (above 300 kV). Acceleration of

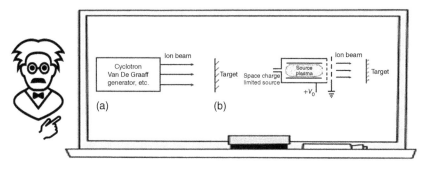

Figure 21.8 Illustration of ion-beam implantation systems: (a) schematic of a high-energy (above 300 keV) implantation system; (b) schematic of an intermediate-energy (between 10 and 300 keV) implantation system.

ions is provided in this case by cyclotrons, Van De Graaff generators, or by other accelerators, which makes the method more expensive. The second type of ion-beam implantation makes it possible to implant atoms to greater depths, which determines its application niches. Advantages of ion-beam implantation are related to the possibilities of effectively adjusting ion energy and therefore the implantation depth and adjusting implantation dose by controlling the exposure time. The biggest issue of ion-beam implantation is that the technology is expensive. As a result, ion-beam implantation is mostly used for high-value products in microelectronics. An important challenge of the conventional ion-beam implantation is the requirement of a normal incidence of ions to the workpiece, which limits its application for three-dimensional objects. Implantation on three-dimensional objects requires in this case application of masking and movable fixturing, which makes the method quite complicated and expensive. In other words, between the initial step of ion generation and the final step of ion implantation, three intermediate costly steps are required: beam extraction, beam focusing, and beam scanning. Relatively low pressure in the system requires vacuum or batch processing, which is also an issue for implantation technology. Plasma-immersion ion implantation (PIII), to be considered next, allows to overcome the challenges by avoiding masking and fixturing of the workpiece.

21.15 Plasma-immersion Ion Implantation (PIII)

In PIII technology, the target is immersed in plasma. Positive ions are extracted directly from the plasma and accelerated toward the target for implantation by a series of negative high-voltage pulses applied to the target. Comparing PIII with conventional ion-beam implantation, the PIII skips intermediate stages of beam extraction, beam focusing, and beam scanning. PIII technology is effective, for example, in doping of semiconductors in microelectronics, in surface hardening of metallurgical components, and in hardening of medically implantable hip joints. A schematic of the PIII system is illustrated in Figure 21.9. Relevant uniform plasma can be generated by microwave, DC, or RF discharges; for example, 3 kW microwave discharge at frequency 2.45 GHz and gas pressures 20–100 mTorr. Plasma volumes in PIII systems are typically large, hundreds of liters or more. To improve efficiency of plasma sources and to increase plasma densities, a magnetic field can be applied to magnetize plasma electrons. Typical electron densities vary in the range 10^8 to 10^{11} cm^{-3}. Electrically conducting workpieces are positioned in the plasma for implantation. Pulsed negative DC voltage (10–200 kV) is applied to the workpieces by a fast-switching circuit to attract and accelerate ions. Depending on the plasma density and the workpiece area, current pulses are 1–500 A during the implantation. To control the implantation dose and avoid surface overheating, the pulse repetition rates are varied by the switching circuit in the range 200–500 Hz (pulse duration 10–30 μs); the pulse rise time is about 1 μs.

Sheath evolution in the PIII process is illustrated in Figure 21.10 for a planar target in plasma with density n_p. It is assumed that a negative voltage pulse with amplitude $-V_0$ and duration t_p is applied to the target. After a short period of time (about the inverse electron plasma frequency, $1/\omega_{pe}$), electrons are driven away from the target surface and create the matrix sheath, which consists of uniformly distributed positive ions (Figure 21.10a). The matrix sheath thickness s_0 is determined by the applied voltage and plasma density. After a longer period (about inverse ion plasma frequency, $1/\omega_{pi}$), ions are accelerated in the sheath and implanted, and the sheath edge recedes,

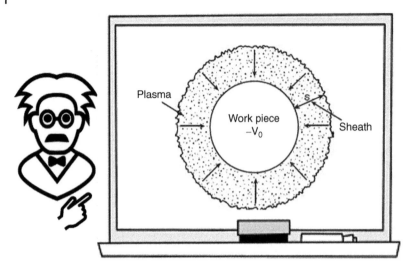

Figure 21.9 Illustration of the plasma ion implantation on a workpiece by negative biasing and the attraction of ions from plasma.

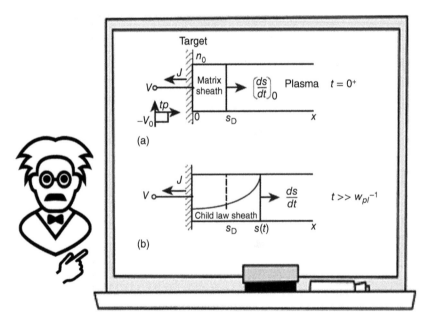

Figure 21.10 Illustration of the planar PIII geometry evolution: (a) density distribution just after formation of the matrix sheath; (b) density distribution after evolution of the quasi-static Child law sheath.

exposing new ions to be extracted. The process can be interpreted as sheath propagation. The steady-state Child law sheath is finally developed (Figure 21.10b), which is about $(eV_0/T_e)^{1/4}$ times larger than the matrix sheath. The dynamics of sheath evolution determines the implantation current and energy distribution of the ions to be implanted.

Microwave discharges (2.45 GHz) operating under ECR conditions can be effectively used for the PIII processing of semiconductor materials. These discharges can operate at pressures as low as 0.2 mTorr, providing the relatively high ion densities (10^{10} to 10^{11} cm^{-3}) required for sufficiently high implantation current densities. To accelerate the ions, the substrate is biased by pulsed negative voltages (amplitude 2–30 kV, pulse duration 1–3 μs). Two PIII configurations are applied for the semiconductor processing: *diode configurations and triode configurations.*

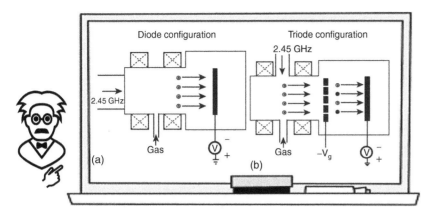

Figure 21.11 Illustration of PIII systems for semiconductor implantation: (a) diode configuration; (b) triode configuration.

The **diode configuration** is shown in Figure 21.11a. Gases like argon, nitrogen, oxygen, water vapor, and BF_3 are introduced into the diode systems to be converted into ions for direct implantation. The diode configuration can be applied when the dopant gaseous sources are available. In particular, the PIII diode configuration is successfully applied for such semiconductor doping applications as shallow junction formation and conformal doping of nonplanar device structures. The **triode configuration** (Figure 21.11b) adds an additional negatively biased target controlled by a separate power supply. Atoms from the target are sputtered into the plasma by the carrier-gas plasma ions. Some of the atoms emitted from the intermediate target are ionized in the plasma and, therefore, can be implanted into the major substrate. A main advantage of the triode system is possibility of implanting components when dopant gaseous sources are not available.

Between different **specific PIII applications for processing of semiconductor materials**, we can point out application of boron implantation in formation of sub-100-nm $p+/n$ junctions. PIII technology, characterized by high ion fluxes and low implantation energies, is very suitable to form ultra-shallow junctions. Prior to the 2 kV BF_3 implantation, the silicon is pre-amorphized with 4 kV SiF_4 implantation. The technology permits to fabricate the junctions with a total leakage current density lower than $30\,nA\,cm^{-2}$ at a reverse bias of $-5\,V$. Selective metal plating is another PIII application in IC fabrication. PIII has been used for selective and planarized plating of copper interconnects using palladium seeding. The process is conducted in the triode configuration. A palladium sputtering target is immersed in plasma and independently negatively biased. The sputtered palladium atoms then provide deposition, while Ar^+ and Pd^+ ions stimulate penetration of the deposited palladium into the substrate. The third specific PIII application in semiconductor processes is related to conformal doping of silicon trenches. Conformal BF_3 doping of high-aspect-ratio silicon trenches has been achieved using angular divergence of implanting ions at relatively elevated pressures. Silicon trenches (1 μm wide, 5 μm deep) were implanted at a pressure of 5 mTorr and voltage of 10 kV. A $p+/n$ junction with relatively uniform depth on the top, bottom, and sidewalls of the trench was produced using this approach.

Plasma-immersion ion implantation technology, applied for the **PIII treatment of metallurgical surfaces**, is usually referred to as **plasma source ion implantation (PSII)**. The PSII objective is to improve wear, hardness, and corrosion resistance of metal surfaces. As with PIII, the PSII approach can be used to treat nonplanar targets such as tools. In contrast to the PIII systems in microelectronics, metallurgical PSII technology can use hot-filament plasma sources because contamination is not so much serious issue in this case. A typical PSII process is done by immersion of a treated target in nitrogen plasma with a density of about $5 \cdot 10^9\,cm^{-3}$. A series of negative voltage pulses (50 kV, duration 10 μs, pulse frequency 100 kHz) are applied to the target under treatment. The total PSII treatment time of a target varies depending on application, from minutes to hours. PSII for metallurgy can be much less sophisticated than that for semiconductor processing applications. Plasma can be generated by a hot tungsten filament inserted into a chamber with background gas at pressure of about 0.1 mTorr and negatively biased at $-(100–300)\,V$. The hot filament emits electrons, which are accelerated in the filament's sheath and then ionize the background gas. Multipole magnets are required on the surface of the implantation chamber to confine the primary electrons, see Lieberman and Lichtenberg (2005) for more details.

21.16 Problems and Concept Questions

21.16.1 Anisotropy Requirements for Plasma Etching

Using relation (21.3), make a graph demonstrating the effect of mask feature size l_m on requirements for etching anisotropy. Discuss the minimization of the mask feature size, considering the size limitations for lithography.

21.16.2 Sputtering

Explain why the sputtering yield (21.4) is low when the mass of an incident atom is much less than that of surface atoms?

21.16.3 Etching Anisotropy Analysis in the Framework of Surface Kinetics of Plasma Etching

Based on relation (21.8), analyze the substrate temperature effect on etching anisotropy. Which step of the ion energy-driven etching is most affected by substrate temperature? Compare effects of translational gas temperature and electron temperature on the plasma etching process.

21.16.4 Etching Anisotropy as a Function of Discharge Power

Analyze dependence of the etching anisotropy coefficient on discharge power (21.15). Estimate the anisotropy coefficient for carbon etching in oxygen assuming a discharge power 100 W, pressure of 30 mTorr, cross-sectional area of $400 \, \text{cm}^2$, and $E_{ei} = 300 \, \text{eV}$. Assume the substrate temperature $T_0 = 400 \, \text{K}$.

21.16.5 Flamm Formulas for F-atom Silicon Etching

Using the Flamm formulas ((21.16) and (21.17)), calculate the F-atom pure chemical etching rates of undoped silicon and thermally grown silicon dioxide for room-temperature substrates. Assume a typical F-atom density in the etching discharges, $3 \cdot 10^{14} \, \text{cm}^{-3}$. Based on the found reaction rates, determine the Si-to-SiO$_2$ etching selectivity at room temperature.

21.16.6 Competition Between Silicon Etching and Carbon Film Deposition in CF$_4$ Discharges

Generalize relation (21.21) for the silicon etch rate in CF$_4$ discharges additionally including physical sputtering of CF$_y$ groups from the surface by ion bombardment. To derive the generalized relation, modify equations for the surface balance of F atoms (21.19) and surface balance of carbon atoms (21.20), considering Γ_{CF_y}, the desorption flux provided by sputtering. Compare the generalized relation with the simplified one (21.22) assuming sputter-dominated desorption.

21.16.7 Rate of Amorphous Silicon Film Deposition in Silane (SiH$_4$) Discharges

Amorphous silicon deposition rates from SiH$_4$ PECVD process are in the range 5–50 nm min^{-1}. Compare these deposition rates with theoretical ones, which you can calculate using (21.29).

21.16.8 Non-conformal Deposition of SiO$_2$ Within Trenches During PECVD Process in SiH$_4$–O$_2$ Mixture

Calculate the nonuniform (non-conformal) deposition flux Γ_{SiO_2} on the sidewalls within a trench as a function of the angle θ_s subtended by the trench opening as seen at a position along the sidewall (see Figure 21.7). Assume a uniform isotropic source of precursors at the top of the trench, a sticking coefficient equal to unity, and a precursor mean free path that is much greater than the trench.

21.16.9 Conformal and Non-conformal Deposition in Trenches

Give your interpretation why the silane–oxygen PECVD of SiO_2 films is non-conformal, while the TEOS-oxygen PECVD of SiO_2 films is conformal?

21.16.10 Plasma-assisted Atomic Layer Deposition (PA-ALD)

Give your interpretation why the PA-ALD process permits in principle conformal deposition of the extremely thin films up to a single atomic layer?

Lecture 22

Plasma Fuel Conversion and Hydrogen Production, Plasma Catalysis

22.1 Plasma-assisted Production of Hydrogen from Gaseous Hydrocarbons: Partial Oxidation, Water-vapor Conversion, Dry (CO$_2$) Reforming, Direct Pyrolysis, Two Concepts of Plasma Catalysis

Demand for H$_2$ and H$_2$-rich gases is growing, resulting in development of the **hydrogen energy concept** in energy production, metallurgy, and transportation, including fuel cell technology, conversion of natural gas into syngas (CO–H$_2$ mixture) for liquefaction and numerous processes of organic synthesis. Conventional thermo-catalytic technology for H$_2$ production is limited by relatively low specific productivity, high metal capacity, and the large equipment size, which becomes especially important in the case of small- and medium-scale H$_2$-generation systems. The application of plasma in the production of hydrogen and syngas creates an attractive alternative to conventional thermo-catalytic technologies, see Deminsky et al. (2002). Plasma stimulation of hydrogen-rich gas production results in a significant increase of specific productivity and reduces the capital and operational costs of the process. The use of traditional catalysis is limited by a time delay due to heating the catalyst to the required high temperature, which is especially critical in transportation applications such as hydrogen production from hydrocarbons on board a vehicle. The plasma hydrogen production has almost no inertia, which makes the plasma approach very attractive to the automotive industry.

 The application of nonthermal plasma for fuel conversion and hydrogen production is especially effective because plasma is used here not as a source of energy but as a source of radicals, charged and excited particles leading to long-chain reactions. *The energy required for fuel conversion and hydrogen production can be provided mostly by chemical energy of reagents and low-temperature heat in nonthermal plasma. The plasma-generated active species just stimulate this process and contribute only a very small fraction (on the level of only a couple percent) of the total process energy. This effect is usually referred to as* **plasma catalysis**. Different kinetic mechanisms can be responsible for the plasma catalysis, including generation of atoms, radicals, and excited and charged species. Charged and excited particles can stimulate chain processes, which cannot be accomplished in conventional chemistry. *Plasma charged particles are especially important in the plasma catalysis considering that exothermic ion–molecular reactions proceed without activation barriers*, which has been discussed in the Section 2.14. This mechanism of plasma catalysis should not be confused with the *other plasma-catalytic approach (also referred to as* **plasma catalysis***) when plasma and catalytic technologies are combined* to increase a process efficacy. It was already discussed in the Section 18.14 and will be additionally discussed in this lecture. Plasma production of H$_2$-rich gases can be divided into three classes: (i) pyrolysis (direct decomposition) processes with the production of carbon and hydrogen; (ii) steam (or CO$_2$, called dry) reforming of hydrocarbons with production of syngas, and (iii) partial oxidation of hydrocarbons with O$_2$ also with syngas production. These classes are characterized in Table 22.1.

Fuel conversion and hydrogen production are characterized by high conversion degrees especially for the partial oxidation both in thermal and nonthermal plasmas. Energy cost, however, is higher in thermal plasmas: usually not lower than 0.15–0.25 eV mol^{-1} of syngas (CO/H$_2$), see Bromberg et al. (2001); Deminsky et al. (2002). These energy costs are still higher than that of conventional thermo-catalytic processes, and to become competitive with conventional systems require heat recuperation. A significant decrease in CH$_4$ partial oxidation energy cost becomes possible

Table 22.1 Thermodynamic and thermal temperature characteristics of the different classes of fuel conversion and hydrogen production processes.

Fuel conversion, H_2 production process	Enthalpy, ΔH, eV mol^{-1}	Temperature, T, K	Combustion enthalpy of initial fuel, eV mol^{-1}
Pyrolysis	Endo-therm.		
$CH_4 \rightarrow C(solid) + 2H_2$	0.9	1000–1100	−8
$C_3H_8 \rightarrow 3C(solid) + 4H_2$	1.3	1000–1100	−20.4
Steam reforming	Endo-therm.		
$C(solid) + H_2O \rightarrow CO + H_2$	1.3	900–1000	−3.9
$CH_4 + H_2O \rightarrow CO + 3H_2$	2.2	1000–1100	−8
$C_3H_8 + 3H_2O \rightarrow 3CO + 7H_2$	5.4	1000–1100	−20.4
Partial oxidation	Exo-therm.		
$C(solid) + \frac{1}{2}O_2 \rightarrow CO$	−1.1	—	−3.9
$CH_4 + \frac{1}{2}O_2 \rightarrow CO + 2H_2$	−0.2	1,100	−8
$C_3H_8 + \frac{3}{2}O_2 \rightarrow 3CO + 4H_2$	−2.1	1,100	−20.4

with application of nonthermal plasma catalysis, for example in the "warm" pulsed microwave discharges or gliding arcs. Relevant energy efficiency in the cold discharges (dielectric-barrier discharge [DBD], coronas, etc.) is usually lower even in the presence of plasma-catalysis because of relatively high energy costs of the active species in this case and temperatures too low for the long-chain reactions.

22.2 Plasma-catalytic Syngas Production from Methane and Other Gaseous Hydrocarbons by Partial Oxidation in Nonthermal "Tornado" Gliding Arcs and Other Discharges

Partial oxidation of methane is a slightly exothermic process ($\Delta H = -0.2$ eV/mol), see Table 10.1, where air is often used as a source of oxygen. The oxygen-to-carbon ratio in the stoichiometric mixture is [O]/[C] = 1. Considering the exothermicity of the partial oxidation, the process can be organized using plasma catalysis, that is, with very low energy input through stimulation of the low-temperature long-chain reactions by selective

 generation of radicals and charged and excited particles. High pressure nonequilibrium discharges can be applied for this purpose, including pulsed corona discharges, DBD, RF, microwave discharges, and gliding arc discharges. The subject was presented and reviewed by multiple researchers, including A. Bogaerts (2019); Bogaerts et al. (2020); Nozaki and Okazaki (2013); Mutaf-Yardimci et al. (1999a,b); Deminsky et al. (2002); Kalra et al. (2005); Fridman (2008).

While the high-energy-efficiency plasma catalysis of partial oxidation is relevant to the nonthermal discharges, industrial applications of the technology require relatively high productivity of syngas and, therefore, high power of the applied discharges. The gliding arc discharges are well suited for this due to their ability to remain strongly nonequilibrium and mostly nonthermal even at relatively high power, see Lectures 11 and 14. The most practically attractive yield and energy efficiency of the CH_4 partial oxidation and syngas production have been achieved in these "warm" discharges stabilized at atmospheric pressure in the reverse vortex "tornado flow" (see Section 11.12, Kalra et al. 2005). Gases (methane, air) enter the system at room temperature. After preheating in the heat exchanger, gases flow into the plasma reactor, typically at 700 K, and leave it at 1050 K. The methane-to-syngas conversion as a function of the [O]:[C] ratio in the inflow is presented in Figure 22.1a. The conversion rises at first with the [O]:[C] ratio due to an increase of the process temperature, but at [O]:[C] > 1.35 the conversion starts decreasing because of CO_2 and water formation. Maximum methane-to-syngas conversion in the nonequilibrium plasma system is quite high, 80–85%, compared to 60%, which is the maximum for the thermal process without plasma.

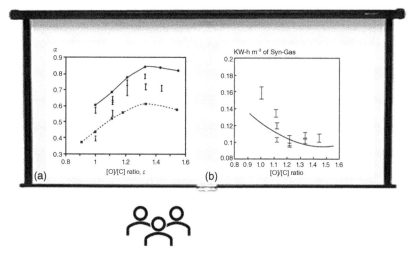

Figure 22.1 Plasma-stimulated methane-air partial oxidation process: (a) CH_4 conversion to syngas as a function of [O]:[C] ratio; (b) electric energy cost as a function of [O]:[C] ratio. Solid lines represent the kinetic simulations; solid bars represent the experimental results; dotted line and bar represent kinetic simulation and experimental results without plasma.

Electric power of the nonequilibrium gliding arc "Tornado" (GAT), about 200 W, is much lower than chemical energy flux of the fuel (more than 6–10 kW), indicating low electric energy cost of the partial-oxidation process and its plasma-catalytic nature. The electric energy cost of syngas production as function of the [O]:[C] ratio is shown in Figure 22.1b. Like the methane-to-syngas conversion (Figure 22.1a), electric energy efficiency is also optimal at [O]:[C] = 1.3–1.4. *The minimum electric energy cost is quite low, only about 0.09 kWh m^{-3} of syngas. Considering that the total chemical energy of 1 m^3 of syngas is about 3 kWh, not more than 3% of the energy is consumed in the form of electricity*; everything else is chemical or thermal energy of the initial products. Nonequilibrium GAT plasma mainly stimulates the process kinetically by operating as a catalyst and does not contribute significant energy to partial oxidation. Therefore, the process can be qualified as plasma catalysis. Partial oxidation of methane in the nonequilibrium gliding arc completely avoids soot formation in the relatively rich mixtures presented in Figure 22.1.

Kinetic analysis shows that syngas production in the system can be divided into two stages: first, the short combustion stage, and second, the longer reforming stage. During the combustion stage, densities of atomic oxygen and OH radicals are relatively large, which results in the kinetics resembling that of combustion. This phase is exothermic and, therefore, gas temperature grows. The first stage ends when no more molecular oxygen is available. The second stage proceeds in the absence of oxygen, where the rest of the methane should be oxidized and converted into syngas (H_2–CO mixture) by water vapor and CO_2 formed during the first phase. The reforming reactions are endothermic, causing the temperature decrease and slowing down the partial-oxidation process. The methane partial-oxidation process produces syngas with H_2:CO ratio 2 : 1. If for a specific practical application (i.e., fuel cells) it is preferable to produce mostly hydrogen, water vapor can be added to stimulate the shift reaction converting CO into hydrogen.

Plasma-chemical conversion of gaseous hydrocarbons into syngas and hydrogen-rich mixtures is not limited to methane, natural gas, or ethane. *Effective conversion and plasma-catalytic effects have been observed in the conversion of other gaseous hydrocarbons.* Thus, partial oxidation of propane has been investigated in atmospheric pressure gliding and transitional arcs, see Czernichowski (1994); Bromberg et al. (2001). The experiments with a 0.1 kW gliding arc were performed with a propane–air mixture. Based on the gas input/output analysis and mass balance, the process can be presented as follows: $C_3H_8 + 1.85O_2 \rightarrow 2.7CO + 3.6H_2 + 0.3CO_2 + 0.4H_2O$. The O/C ratio in the process is 1.23, which is close to the ideal partial oxidation. The process produces nitrogen-diluted syngas containing up to 45% of H_2 + CO. The output syngas flow rate is 2.7 standard m^3 h^{-1}, which is equivalent to 8.6 kW of output power. In terms of plasma catalysis, discharge power in this case (0.1 kW) is only 1.2% of the total power of the produced syngas. The gliding arc discharge has also been applied for partial oxidation of cyclohexane (C_6H_{12}), heptane (C_7H_{17}), and toluene (C_7H_8).

22.3 Plasma-catalytic Syngas Production in Mixtures of CH_4/H_2O (Steam Reforming) and CH_4/H_2O (Dry Reforming)

Syngas can be produced in plasma by CH_4 oxidation with water vapor referred to as the **steam reforming of methane** $CH_4 + H_2O \rightarrow CO + 3H_2$, $\Delta H = 2.2\,eV\,mol^{-1}$. In contrast to CH_4 partial oxidation, the steam reforming is endothermic and, therefore, requires significant energy consumption, see Table 22.1. The energy cost of the steam reforming in atmospheric-pressure thermal plasma is presented in Figure 22.2 as a function of specific energy input in different quenching modes (Nester et al. 1988). The minimum energy cost, $0.59\,eV\,mol^{-1}$ H_2–CO (ideal quenching) and $0.69\,eV\,mol^{-1}$ H_2–CO (absolute quenching), are relatively high assuming the energy should be provided in the form of electricity.

The situation can be improved by using the plasma catalysis. In the case of endothermic reactions, it means that most of the energy required for the process is provided in form of the reasonably low-temperature heat and not as electric energy going through the electric discharge.

As an example, the plasma catalysis of endothermic processes supported by a combination of low-temperature heat and plasma energy has been demonstrated using pulsed microwave discharge (Givotov et al. 2005). The initial products with flow rates of 50–$200\,cm^3\,s^{-1}$ at atmospheric pressure are preheated to 360–$570\,°C$ and then treated in the pulsed microwave discharge (frequency of 9 GHz, power in the pulses up to 100 kW, pulse duration 0.3–1 µs, pulse repetition frequency of 1 kHz, average power up to 30–100 W). The ratio of microwave and heating power is $P_p/P_T = 5$–10%, and the ratio of flow rates of CH_4:H_2O varies from 1:1 to 1:2. Gas-phase products in this case are syngas (H_2, CO) and some carbon dioxide (CO_2). The added microwave discharge energy, which is only a small fraction of the total energy (5–10%), increases energy efficiency of the steam reforming from 30% to 60%, which proves the plasma-catalytic nature of the process. Similar plasma-catalytic approaches to steam reforming of methane have been developed using nonequilibrium pulsed corona, sparks, and gliding arc discharges.

Syngas can be also effectively produced in plasma by CH_4 oxidation with carbon dioxide (CO_2) usually referred to as the **dry reforming of methane**:

$$CH_4 + CO_2 \rightarrow 2CO + 2H_2, \qquad \Delta H = 2.6\,eV\,mol^{-1}. \tag{22.1}$$

The process is kinetically like the steam reforming of methane; however, it is more endothermic. In contrast to the steam reforming, the CO_2-based dry reforming can be more effectively stimulated by vibrational excitation of the reagents, considering the completely gas-phase nature of the reaction and effectiveness at sustaining vibrational–translational nonequilibrium in CO_2. Although the energy efficiency of the reformation exceeds 40%, the total consumption of electric energy is still high. Plasma catalysis allows here also like in steam reforming a significant decrease in the consumption of electric energy. The plasma-chemical process (22.1) consumes CO_2 during the fuel conversion,

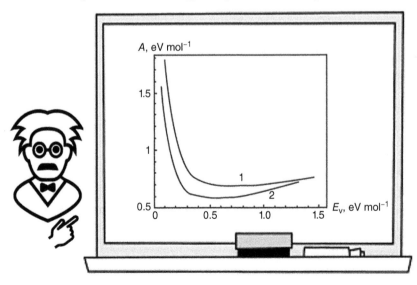

Figure 22.2 Energy cost of production of a syngas molecule (H_2–CO mixture) in the thermal plasma process of steam reforming of methane, $CH_4 + H_2O \rightarrow 3H_2 + CO$: (1) absolute quenching; (2) ideal quenching.

which can be attractive from the perspective of environmental control. Produced syngas is not supposed to be burned in this case but rather used for further chemical synthesis.

Significant reduction in the CH_4-reforming energy cost (both in steam- and dry reforming processes) can be achieved by adding air (or oxygen) to the initial mixture. The reaction enthalpy can be close to zero in such regimes, usually called the **auto-thermal regimes of fuel reforming**. Production of syngas in this case would be due to the combination of endothermic steam or dry reformation with exothermic partial oxidation in nonequilibrium plasma. These processes can be also effectively stimulated by combination of plasma processes with conventional or especially configured catalysis, see Bogaerts (2019); Bogaerts et al. (2020); Nozaki and Okazaki (2013).

22.4 Direct Decomposition (Pyrolysis) of Methane and Other Gaseous Hydrocarbons, Plasma-catalytic Effects in the Pyrolysis, the Winchester Mechanism

Direct decomposition (pyrolysis) of hydrocarbons with **production of soot (or black carbon) and hydrogen** is another type of fuel conversion processes. Whereas hydrocarbon combustion results in the production and exhaust of CO_2, pyrolysis produces hydrogen leading to CO_2-free clean combustion. The carbon produced in this process should be not burned in this case, but rather used for chemical purposes. When plasma pyrolysis is conducted without special quenching, the hydrocarbons can be mostly converted into soot and hydrogen. The quality of the soot is better than that of the soot generated in furnaces. The soot produced in plasma is highly dispersed and characterized by a developed secondary structure. The process has been performed in arc discharges and thermal RF inductively coupled plasma (ICP) discharges directly, as well as in plasma jets of argon, nitrogen, and hydrogen at temperatures from 1500 to 5000–6000 K. Different hydrocarbons were used as a feedstock, including methane, natural gas, propane, toluene, naphthalene, xylol, gasoline, and special mixtures of aromatic compounds. The maximum yield of soot in these plasma systems (45–70 mass%) is achieved by pyrolysis of aromatic compounds. The energy cost of soot is 7–$13\,kWh\,kg^{-1}$ in this case. Thermal plasma pyrolysis of paraphenes and naphthenes leads to a lower yield of soot (18 mass%) and a higher energy cost ($70\,kWh\,kg^{-1}$).

Strong **plasma-catalytic effect in the pyrolysis of hydrocarbons** has been observed in "warm" nonequilibrium atmospheric pressure microwave discharges in methane, see Givotov et al. (2005); as well as Fridman (2008). Application of the nonequilibrium plasma results in a significant decrease of temperature required to achieve certain level of conversion. Small amount of plasma energy ($<20\%$) initiates intensive contribution of CH_4 thermal energy into the endothermic process of methane pyrolysis, see Table 22.1. Nonequilibrium plasma shifts the conversion curves closer to thermodynamic equilibrium, demonstrating the effect of plasma catalysis. The energy cost of hydrogen production in the pyrolysis $CH_4 \rightarrow C(solid) + 2H_2$ is shown in Figure 22.3 as a function of specific energy input in plasma. As one can see from the upper curve of the figure, adding only 15% plasma energy (with respect to thermal energy) allows for a twofold decrease of the energy cost of hydrogen production. The lower curve in Figure 22.4 represents the electric (plasma) energy cost of hydrogen production, which is lower than the enthalpy of the process, see Table 22.1. When the fraction of plasma energy is 10%, the temperature during the process decreases from 550 to 480 °C, and the electric (plasma) energy cost of hydrogen production is only $0.2\,eV\,mol^{-1}$. It proves that most of the required energy is provided by the thermal reservoir, while the effect of plasma is mostly catalytic.

Thus, temperature of the preheated methane decreases during the plasma-catalytic process because a fraction of the thermal energy is spent to increase CH_4 conversion to H_2 in the endothermic reaction of pyrolysis. Different **mechanisms of the nonequilibrium plasma-catalytic effects in CH_4 pyrolysis** have been analyzed by Givotov et al. (2005), including process stimulation by radicals and excited atoms and molecules, nonuniform temperature distribution in the discharge, stimulation of CH_4 decomposition on the surface of carbon clusters, and finally process stimulation by the **ion-molecular Winchester mechanism**, which is proved to be the dominating one in the CH_4 pyrolysis.

The Winchester mechanism is based on the thermodynamic advantage of the ion-cluster growth processes, for example, in the sequence of negative ion clusterization $A_1^- \rightarrow A_2^- \rightarrow A_3^- \rightarrow \cdots \rightarrow A_n^-$, where electron affinities $E_{A1}, E_{A2}, E_{A3}, \ldots, E_{An}$, are increasing in the sequence to finally reach the value of the work function (electron extraction energy), which is generally higher than the electron affinity for small molecules. As a result, each elementary reaction of the cluster growth $A_n^- \rightarrow A_{n+1}^-$ tends to be exothermic. Each elementary step $A_n^- \rightarrow A_{n+1}^-$ includes the cluster rearrangements with an electron usually going to the farthest end of the complex, which explains the term "Winchester." The Winchester mechanism enables positive ion-cluster growth as well:

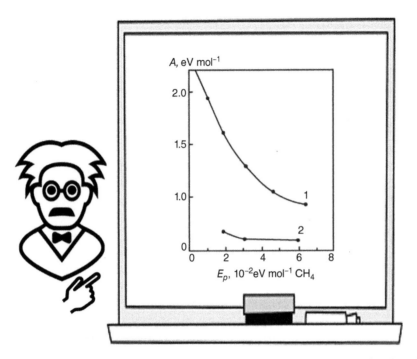

Figure 22.3 Energy cost of the plasma-catalytic pyrolysis of methane as a function of specific plasma energy input: upper curve, total energy cost of the process with respect to the total produced hydrogen; lower curve, electrical energy cost with respect to additional hydrogen produced by the discharge.

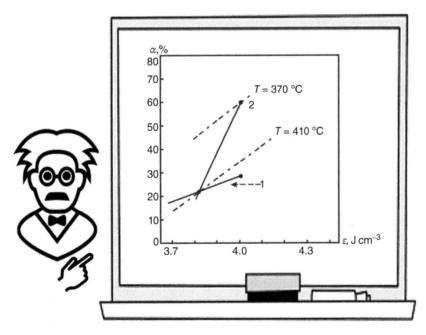

Figure 22.4 Plasma-catalytic reforming of ethanol: conversion degree of ethanol as a function of specific energy input. Curve 1 corresponds to only thermal energy input; curve 2 corresponds to application of the microwave discharge; dashed lines correspond to the isothermal conditions.

$A_1^+ \to A_2^+ \to A_3^+ \to \cdots \to A_n^+$, because ionization energies $I(A_1)$, $I(A_2)$, $I(A_3)$, ... , $I(A_n)$ are decreasing in the sequence to finally reach the value of the work function, which is generally lower than the ionization energy of small molecules. It explains the thermodynamic advantage of the ion-cluster growth for positive ions as well.

The Winchester mechanism makes the biggest contribution to the plasma catalysis of CH_4 pyrolysis with hydrogen production. Although the role of the ion–molecular reactions of negative ions has been discussed, it is concluded that the CH_3^+ ion-radical stimulates the most effective chain of CH_4 conversion into carbon clusters and H_2:

$$C_nH_{2n+1}^+ + CH_4 \to \left[C_{n+1}H_{2n+5}^+\right] \text{ (excited), } k \approx 6 \cdot 10^{-9} \text{ cm}^3 \text{ s}^{-1}. \tag{22.2}$$

The excited complexes $\left[C_{n+1}H_{2n+5}^+\right]$ (excited) are unstable with respect to decomposition, rearrangements, and hydrogen desorption. The Winchester mechanism of methane conversion into hydrogen with growth of the carbon clusters can be interpreted as a chain reaction with multiple uses of a positive ion. Decrease of the ionization potentials in the sequence $C_nH_{2n+1}^+$ makes the chain propagation exothermic and fast because the ion–molecular reactions have no significant activation barriers. It determines mechanisms of the nonequilibrium plasma-catalytic CH_4 pyrolysis.

22.5 Plasma Partial Oxidation and Steam-reforming of Liquid Fuels: On-board Generation of Hydrogen-rich Gases, Reforming of Kerosene, Ethanol, Aviation and Diesel Fuels, Gasoline, Renewable Biomass, Waste-to-energy Processes

Liquid fuels, such as gasoline, diesel, biodiesel, diesel oil, kerosene, ethanol, and so forth can be effectively converted into syngas (CO–H_2) by nonequilibrium plasma-assisted partial oxidation or steam–air conversion. These processes are like the gas conversion ones considered above, especially because in most of cases the liquid fuels are vaporized in plasma before the reforming. Such technologies of hydrogen-rich gas production can be applied in different industrial schemes, but they are especially interesting for transportation systems. The applications of plasma for transportation systems focus on either development of hydrogen filling stations or development of devices for on-board generation of hydrogen-rich gases. The **plasma systems for on-board generation of hydrogen-rich gases** *should meet two crucial requirements: the first is compactness and low weight of the equipment; the second is no inertia of the process or, in other words, the possibility of fast start*. Both requirements can be met by using plasma stimulation of the fuel conversion processes.

 Plasma systems are characterized by specific productivity exceeding that of conventional catalytic systems by up to 1000-fold (Rusanov and Fridman 1984) and have no technological inertia related to preheating. The plasma-chemical fuel-conversion processes are sulfur resistant (in contrast to catalysis), which is important considering the composition of the automotive fuels. Plasma-stimulated onboard generation of hydrogen-rich gases can be used in several applications to improve environmental quality and reduce total fuel, especially petroleum fuel consumption in internal combustion engine vehicles. Four major applications can be specifically mentioned in this regard: (i) production of relevant hydrogen-rich fuel for fuel cells providing on-board electricity for automotive vehicles; (ii) generated syngas can be directly used in internal combustion engines as an admixture to the conventional fuel–air mixture to reduce toxic exhausts and improve the major engine characteristics; (iii) on-board syngas plasma generated from liquid automotive fuel can be effectively applied for after treatment of particulates and NO_x emissions; (iv) the syngas produced on board the diesel vehicles can be effectively used for regeneration of the diesel particulate filter. The effective conversion of liquid fuels (gasoline, diesel, biodiesel, diesel oil, kerosene, ethanol, etc.) into syngas (CO–H_2) by plasma-assisted partial oxidation or steam–air conversion becomes possible because of the plasma catalysis: *consumption of the electric (plasma) energy in plasma is only a small fraction of the total energy of the produced syngas*. Detailed review of specific plasma-stimulated liquid fuel reforming processes can be found in Fridman (2008). Key plasma-chemical features of some of those processes are going to be shortly discussed below.

Plasma-catalytic **steam reforming and partial oxidation of kerosene** into syngas have been demonstrated in the nonequilibrium atmospheric pressure pulsed microwave discharge (Givotov et al. 2005). The endothermic process of kerosene steam reformation can be represented as $C_{11}H_{22} + 11H_2O \to 11CO + 22H_2$; the exothermic partial oxidation of kerosene is $C_{11}H_{22} + 5.5O_2(+N_2) \to 11CO + 11H_2(+N_2)$. The minimum quasi-equilibrium temperature for steam–air reforming of kerosene is about 1170 K. The kerosene steam–air reforming stoichiometry corresponding to the minimum temperature providing syngas production without soot formation is $28.05C_{11}H_{22} + 112.4O_2 + 414N_2 + 83.78H_2O \to 308.57CO + 392.35H_2 + 414N_2$. Preheating and evaporation of the liquid fuel can be incorporated into either the plasma system or in a special recuperative heat exchange system. The lowest achieved electric energy cost of the process is about $0.2\,eV\,mol^{-1}$ CO–H_2.

The effect of plasma catalysis has been demonstrated for the **conversion of ethanol (C_2H_5OH) into syngas** using nonequilibrium atmospheric pressure pulsed microwave discharge and atmospheric-pressure transitional arc discharge. The process was investigated in different regimes including direct conversion, steam reforming, and partial oxidation. The plasma-catalytic effect on the ethanol reforming is illustrated in Figure 22.4 by the dependence of the conversion degree on specific energy input in the pulsed microwave discharge (Givotov et al. 2005). Although electric (plasma) energy in this case is about 5% or less, the plasma-catalytic effect is strong. The conversion degree of ethanol increases from 23% to 62%, while the thermal reservoir temperature decreases from 410 to 370 °C. Curve 1 in the figure shows (for comparison) the thermal effect of preheating without application of nonequilibrium plasma, which clearly demonstrates the effect of plasma catalysis.

Effective plasma-stimulated **partial oxidation of diesel fuel and diesel oils** has been first demonstrated in atmospheric-pressure transitional arcs and gliding arcs. The minimal electric energy cost for production of syngas (CO–H_2) from the dirty diesel fuel demonstrated in gliding arcs is $0.12\,kW\,m^{-3}$ (Czernichowski 1994). The electric (plasma) energy spent in the process is less than 4% with respect to the total process energy. Thus, this reforming process can also be qualified as plasma catalytic. Sulfur contained in the diesel fuel is converted here into H_2S and can be removed by conventional methods, for example, by a ZnO bed.

Similar results were achieved in plasma-stimulated **partial oxidation of gasoline 95 and aviation fuels (like JP8)**. The plasma-stimulated JP8 aviation fuel conversion has been integrated with the solid-oxide fuel cells (SOFC). In the integrated system, the fuel has been converted into an H2–CO–CH_4 mixture with a quality sufficient for effective operation of the SOFCs. See for more details Gallagher et al. (2010).

Plasma **reforming of renewable biomass** is important, not only for the generation of syngas for further production of effective liquid fuels and for electricity production in fuel cells but also for the onboard production of hydrogen-rich gases for reducing petroleum consumption in automotive transportation. The most abundant renewable fuels are biodiesel and ethanol (already discussed above). Atmospheric-pressure transitional arc plasma has been effectively used for conversion of biodiesel, different bio-oils, including corn oil and such vegetable oils as soy and canola, into syngas. Plasma stimulation of the reforming can be combined with a conventional catalyst, which has been demonstrated for the partial oxidation of unrefined soybean oil. The catalyst in this case doubles the production of H_2 and significantly increases the production of CO.

The plasma reforming of biomass is closely related to the **plasma-stimulated waste-to-energy processes**, which can be organized as steam-air conversion of both organic liquid waste (as landfill leachate) and organic solid waste (like municipal waste). These plasma processes are usually result in production of syngas. In the case of application at the landfills, the syngas produced this way can be effectively combined with the conventionally collected landfill gases to be further used as feed gases for electricity production using special engines or turbines.

22.6 Combination of Plasma and Catalysis in Hydrogen Production from Hydrocarbons: Plasma Pre-processing and Post-processing, Plasma Treatment of Catalysts

The plasma-catalytic approach to fuel conversion and hydrogen production considered above were mostly focused on "plasma replacement" of conventional catalysts to stimulate the process without providing major energy input. The **combined plasma-catalytic approach** is quite different method of plasma catalysis. It implies application of both plasma and conventional catalytic systems to convert hydrocarbons to hydrogen-rich gases. The combined plasma–catalysis can be arranged in different configurations, which are always intended to achieve synergy of the plasma and catalysis. One configuration, which can be referred to as **plasma preprocessing**, means that incoming gas is first processed by the plasma before treatment in the catalytic reactor. Plasma preprocessing aims to enrich the gas with reactive species and then process it by the catalyst, thus enhancing the kinetics of hydrogen production. Another configuration, which can be referred to as **plasma postprocessing**, means that plasma is used for treating the exhaust gases from a catalytic unit. The postprocessing operation is intended to complete the reformation process and to destroy unwanted by-products generated during traditional catalytic reactions.

Both plasma preprocessing and plasma postprocessing effects can be obviously combined in one plasma-catalytic reactor system, where also plasma stimulation of the conventional catalyst itself can also make additional contribution.

Comparison of contribution of the plasma preprocessing and postprocessing effects in the combined plasma catalysis has been analyzed by investigation of partial oxidation of isooctane (trimethylpentane, C_8H_{18}, liquid used to represent gasoline): $C_8H_{18} + 3.7(O_2 + 3.76N_2) \rightarrow$ products (H_2, CO, and other), see Sobacchi et al. (2002, 2003). Different separate (stand-alone) and combined methods were applied for the reformation: (i) a stand-alone catalytic reactor, (ii) a stand-alone nonequilibrium atmospheric-pressure plasma (pulsed corona) reactor, (iii) a combined plasma-catalytic reactor in the plasma preprocessing configuration, (iv) a combined plasma-catalytic reactor in the plasma postprocessing configuration. Water was added for the shift reaction ($CO + H_2O \rightarrow H_2 + CO_2$) to replace CO with hydrogen. It was demonstrated that *plasma preprocessing seems to be the optimal configuration of the combined plasma-catalytic system.* For all tested temperatures, this configuration resulted in larger hydrogen production than plasma postprocessing or the catalytic treatment alone. Moreover, at higher temperatures when the production of hydrogen is highest, plasma preprocessing reduces the concentration of CO, which is oxidized to CO_2. The final concentration of CO at these high temperatures is lower than that for the other configurations. Better conversion of isooctane into hydrogen in the preprocessing configuration with respect to the stand-alone operation of the catalytic unit suggests that the catalyst acts more effectively on the intermediate oxidized compounds produced by the pulsed corona discharge than on the initial gas mixture. For the postprocessing configuration of the combined system, plasma is formed in the gases produced by the catalytic conversion process, when the oxidation processes are mostly completed. On the other hand, in the case of the preprocessing configuration, the gas composition in the nonequilibrium plasma reactor is closer to the initial composition, with only some preliminary level of fuel oxidation into CO and CO_2. Generally, it can be concluded that *plasma stimulates the kinetically suppressed initial stages of oxidation of isooctane (and probably other hydrocarbons as well), whereas the catalyst supports the kinetically "easier" reactions of further oxidation of the intermediate oxidized compounds to syngas.* It generalizes the major idea of the plasma catalysis as the synergistic combination of plasma and catalysis. Not to forget, that *plasma can also stimulate the catalytic abilities of the conventional catalysts themselves through formation and activation of the partially charged and active centers on their surface*, see Section 2.14.

22.7 Plasma-chemical Conversion of Coal, Plasma Coal Pyrolysis, Coal Conversion in Thermal Plasma Jets

Coal is the fuel most able to cover world deficiencies in oil and natural gas. Wide application of coal is, however, somewhat limited by its transportation and ecological "nonpurity" when used directly as a fuel, which is related to its significant formation of ashes, sulfur compounds, nitrogen oxides, and so forth. It motivates development of new plasma technologies for coal conversion into other fuels, its gasification, and direct liquefaction. Mostly thermal plasma sources, however also the nonequilibrium "warm" discharges, were used to convert coals (including those of low quality) into other fuels with high selectivity, simplicity of the process control, and minimal production of ashes, sulfur, and nitrogen oxides. *Plasma conversion of coals can be subdivided into pyrolysis, when plasma effect is mostly limited to heating of coal particles (to be discussed in this and next sections), and plasma-chemical transformation of coals (especially gasification) via air, steam, or dry CO_2-based reforming (to be considered later in this Lecture).*

Details on these plasma coal conversion technologies, both purely pyrolytic ones as well as those focused on plasma-chemical gasification can be found in publications Xie et al. (2002); He et al. (2004); Messerle (2004); Polak et al. (1975); Fridman (2008).

Coal has quite complicated composition and structure, which are to be discussed first. Major chemical elements that form coal are carbon, hydrogen, oxygen, sulfur, and nitrogen. Coal also contains moisture and mineral components. The age of coal determines degree of the **coal metamorphism**: the oldest coals have the highest degree of metamorphism and usually have the highest quality. Low-quality (low-metamorphism-degree) coals have about 65% carbon, whereas the high-quality (high-metamorphism-degree) coals can have up to 91% carbon. The mass fraction of hydrogen in the same sequence of growing degree of metamorphism decreases from 8% to 4%. The mass fraction of oxygen decreases in the metamorphism sequence from 30% to 2%. All types of coal contain 0.5–2% nitrogen and 0.5–3% sulfur. The concentration of mineral components varies in a wide range, from 1–3% to

10–30%. Most organic mass of coal is a rigid irregular three-dimensional polymer. The remaining organic mass is made of mobile monomolecular or slightly polymerized substances which are immobilized in pores of the rigid skeleton. Oxygen contained in coal is mostly bound in hydroxyl, carbonyl, and carboxyl groups. The concentration of phenol-hydroxyl groups decreases with higher concentrations of carbon in coal and almost disappears when the carbon mass fraction reaches 89%. Sulfur is contained in coal in both organic (30–50%) and inorganic (50–70%) compounds in the form of disulfides and sulfates. Disulfides are mostly represented in coal by pyrite (FeS_2); sulfates are mostly represented by gypsum ($CaSO_4$) and ferrous sulfate ($FeSO_4$). The total amount of sulfates in coal is low and usually does not exceed 0.1 mass %. Nitrogen is incorporated only in the organic fraction of coal, mostly in heterocyclic compounds or as a bridge between two carbocyclic compounds. Coal is characterized by high porosity. The pores can be divided into three classes: nanopores with sizes below 30 nm, intermediate pores with size 30–300 nm, and macropores with size 0.3–3 µm. The nanopores have a very large surface area. In these pores, most of the coal destruction products are generated during thermal treatment. Diffusion of chemicals during adsorption and reactions also takes place through the nanopores. The intermediate pores provide major channels for the transport of reagents and products in and out of the coal particles. The fraction of macropores in the coal is relatively low.

Thermal plasma pyrolysis of coal is mostly due to heating, usually in inert atmosphere of the thermal plasma discharges. Heating of coal in an inert atmosphere to temperatures of 620–670 K already leads to its decomposition with formation of primary volatile products and coke. Formation and release of the volatile products continue until the temperature is about 1220 K. The **volatile compounds of thermal plasma pyrolysis** include water vapor, hydrogen, methane, carbon oxides, hydrocarbon gases, and heavy organic products (tars). The fraction of the volatile compounds varies in different types of coal and generally is about 15–40%. The amount of the volatile compounds is highest in the younger coals with the lowest degree of metamorphism. Anthracite, the highest-quality hard coal with the highest degree of metamorphism, releases the minimum amount of volatile compounds. The fraction of heavy organic products (tars) in volatile compounds is high (40–80%) and increases with the rate of coal heating. Coal heating in an inert atmosphere also significantly changes the residual solid fraction. First, its plasticity grows and the microparticles agglomerate. Then the solid microparticles become rigid again, forming the porous structure of the coke. Formation of the volatile compounds, as well as formation of coke, is the result of multiple chemical reactions taking place during the coal heating. For this reason, the terms "formation of volatile compounds", "cracking of coal," and "pyrolysis of coal," which describe the thermal plasma conversion of coal, are synonymous.

Release during the thermal plasma coal pyrolysis of the functional groups and products of their decomposition can be described by zero kinetic order. This is applicable when the characteristic size of the coal particles exceeds some critical value that depends on the heating rate. The critical diameter of the coal particles changes from 2 to 0.2 mm, when heating rate grows from 10^2 to 10^4 K s^{-1}. When the coal particles are large enough (greater than the critical value), the relevant chemical reactions take place mostly on the surface and are controlled by heat transfer, which results in the zero kinetic order of the pyrolytic processes.

The **coal conversion in thermal plasma jets** assumes that the coal particles are not introduced directly in the thermal plasma discharge (like in electric arc) but introduced in the high temperature jet produced by the thermal discharge. Coal conversion in the high-temperature plasma jets is characterized by high selectivity. Products contain only substances stable at high temperatures, such as H_2, C_2H_2, HCN, CO, and CO_2. Higher hydrocarbons and tar are practically absent in the products. The specific product composition and yield are determined by the type of coal, by plasma gas, and by degree of heating of the coal particles. The degree of heating of the particles depends on their size, the type of the plasma-chemical reactor, and the degree of mixing of the solid particles with the hot plasma gas. The most conventional of this group of the coal plasma conversion processes is the thermal arc jet pyrolysis of coal in relatively inert gases like argon, hydrogen, and nitrogen to be discussed below.

22.8 Thermal Plasma Jet Pyrolysis of Coal in Relatively Inert Gases (Ar, N₂, H₂): Production of Acetylene (C₂H₂), Hydrogen Cyanide (HCN), Transformation of Sulfur- and Nitrogen-compounds of Coal

The Ar–H₂ plasma jet pyrolysis of coals with a high degree of metamorphism leads to ***formation of a single major gas-phase product: acetylene (C₂H₂).*** The addition of hydrogen does not change many characteristics of the pyrolysis; it only increases the yield of C_2H_2. For the coal pyrolysis in an argon plasma jet, an increase of the coal feeding rate leads to decrease of the conversion degree of carbon from coal to acetylene and to decrease of the energy cost for the formation of acetylene. Acetylene yield is proportional to the mass fraction of the volatile compounds in coal. When coal particles are supplied along the radius, the maximum yield of acetylene from coals containing 50% volatiles is 20%. The release of volatile compounds starts in the H_2 plasma when the temperature of the coal particles reaches about 1220 K. This critical temperature of volatile formation is almost independent of the plasma-gas temperature. The temperature of the coal particles remains fixed during the intensive generation of volatile compounds, which prevents effective heat exchange between the particles and the plasma gas. Coal-to-carbon conversion also remains almost fixed at different plasma-gas temperatures.

Acetylene is formed during plasma jet pyrolysis from volatile compounds released from solid coal particles. Therefore, the acetylene yield is determined by the completeness of volatile release, which is related to the rate and depth of heating the particles, effectiveness of mixing the particles with the plasma gas, size distribution of particles, temperature, and time of the reaction.

Heating the coal particles is a major factor determining the completeness of releasing volatile compounds from coal. The release of volatile compounds in plasma jets is often incomplete, although the plasma-gas temperatures significantly exceed the required values of about 1000–1200 K. For example, pyrolysis of coal in a hydrogen plasma jet at temperature of 1560–2180 K leaves 7–8% of volatiles in coke. Coal pyrolysis often leads to intensive formation of small and mostly spherical particles with sizes of 60–90 nm. They originate from fast plasma heating of the coal particles and intensive formation of volatile compounds inside of them. This leads to an explosive decay of the coal particles into very small sizes.

The thermal pyrolysis of coal is also effective in **plasma jets of nitrogen or Ar–N$_2$ mixture** (Ar + 10% N$_2$). The *major products of pyrolysis in this case are acetylene, hydrogen cyanide (HCN).* When the coal pyrolysis occurs in a plasma jet of pure nitrogen, the major product of pyrolysis is HCN, whereas acetylene dominates if the amount of nitrogen in the Ar–N$_2$ mixture is relatively small. The yield of HCN in the pyrolysis in the pure nitrogen plasma jet is close to the amount of volatile compounds in the coal. Acetylene yield in this case is about 3–5%. Generally, the yields of acetylene and HCN in the coal pyrolysis in nitrogen-based plasma jets are proportional to the amount of volatile compounds in the treated coal.

Let us discuss now **transformations of the sulfur-containing compounds during plasma pyrolysis of coal**. Sulfur is mostly contained in coal as pyrite (FeS$_2$) and organic compounds (30–50%). *Thermal treatment of coal leads to the decomposition of pyrite and organic sulfur compounds and the formation of mostly H$_2$S, which then in turn reacts with the organic mass of coal and ash components.* A general schematic of the transformations of sulfur-containing compounds during pyrolysis in an inert or reducing atmosphere is illustrated in Figure 22.5. Decomposition of pyrite (FeS$_2$) starts at temperatures exceeding 820 K and leads to the formation of elemental sulfur and iron sulfide (FeS). In the presence of hydrogen (H$_2$) formed during coal pyrolysis, the pyrite is more intensively converted into iron sulfide and hydrogen sulfide at temperatures of about 770 K: FeS$_2$ + H$_2$ → FeS + H$_2$S. Pyrite and FeS also react with carbon oxides, CO, and CO$_2$ (formed by the pyrolysis), at temperatures around 1100 K, which results in the formation of COS. Organic sulfur in coal is contained mostly in thiophenes (40–70%) and sulfides (60–30%). Thiophenes are thermally stable, and their decomposition requires temperatures at about 1100 K. For this reason, thiophenes are the major

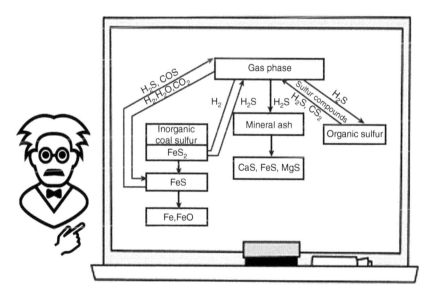

Figure 22.5 Schematic of transformations of sulfur-containing compounds during thermal plasma conversion of coal in inert and reduction medium.

sulfur-containing products of coal pyrolysis. Sulfides are presented in coal as aryl sulfides (about 50%), cyclic sulfides (about 30%), and aliphatic sulfides (about 20%). Their thermal decomposition leads mostly to gas-phase production of H_2S and CS_2, which then reacts with the organic and inorganic fraction of coal.

Consider, finally, the **transformations of the nitrogen-containing compounds during plasma pyrolysis of coal.** Nitrogen is contained almost completely in organic compounds of coal, mostly in aromatic rings. Release of nitrogen in the gas phase during pyrolysis does not start immediately; it begins after the preliminary formation of about 10% of volatile compounds not containing nitrogen. The volatile nitrogen-containing products of the plasma pyrolysis are NH_3, HCN, N_2, and such tar components as pyridine, pyrol, nitryl, carbosol, chinoline, and indole. The amount of nitrogen released from coal as volatile compounds depends on the pyrolysis temperature, heating rate, and type of coal. Temperature increase from 770 to 1170 K leads to growth of the nitrogen fraction released into the gas phase from 20% to 80%. The rest of the nitrogen is fixed in the rigid coal skeleton. It remains in coke even at very high temperatures. Conversion of the nitrogen, strongly bound in the coke skeleton, into the gas phase becomes possible only in the processes of complete gasification of coke using, for example, steam and air.

22.9 Coal Gasification Using Thermal Plasma Discharges: Partial Oxidation, Steam, and Dry Reforming Processes

The most effective plasma-based coal gasification can be organized based on the air-steam reforming with production of syngas or more generally the hydrogen-reach gases. The **plasma-activated steam reforming of coal** $C(s) + H_2O \rightarrow CO + H_2$ is endothermic (see Table 22.1) and requires high temperatures for reasonable reaction rates. To provide necessary heat into the system, a portion of the carbon should be burnt. Heat in this case is released by exothermic combustion processes $C(s) + O_2 \rightarrow CO_2$, $2\,C(s) + O_2 \rightarrow 2\,CO$. To provide the required energy balance, the amount of CO_2 in the products of conventional gasification is relatively high. The conventional industrial coal gasification processes produce 15–30% of CO_2 (Zhukov and Solonenko 1990). The application of plasma jets provides an independent source of energy, thereby opening wide possibilities for process control and significantly diminishing the production of CO_2. Coal conversion in a plasma jet of water vapor results in the production in the gas phase of mostly (about 95%) syngas, the H_2–CO mixture. Usually, the syngas produced this way contains more hydrogen than carbon monoxide (H_2:CO > 1). The conversion products also contain CO_2, usually about 3%, and some traces of methane and hydrogen sulfide. The yield of CO_2 is much smaller in plasma than in conventional technologies. Also, the plasma conversion of coal does not produce tars, phenols, or polycyclic hydrocarbons, which decreases the costs of syngas purification and makes plasma technology more preferential from an ecological standpoint. Coal gasification with water vapor in atmospheric-pressure thermal plasma jets is usually based on one of the following two approaches. In the first approach, the cold coal powder is injected into the atmospheric pressure plasma jet of water vapor. In the second approach, both the coal particles and the water vapor are heated in plasma jets of some other gases. Energy consumption during the coal gasification in atmospheric-pressure plasma jets is significantly decreased by adding oxygen (or air) to the water vapor, which is usually referred to as the **partial oxidation of coal in thermal plasma**. The degree of conversion of the carbon from the coal is typically about 96% in the steam–oxygen plasma gasification system. The composition of the produced gas is 39 vol% H_2, 54 vol% CO, and 7 vol% CO_2. The consumption of electric energy in this not-strongly-endothermic process is relatively low, not more than 7% from the total consumed process energy.

An interesting opportunity is related to the plasma reforming of coal using CO_2 plasma, which is usually referred to as the **plasma-activated dry reforming of coal** $C(s) + CO_2 \rightarrow 2\,CO$. While this process is still very endothermic like the plasma-activated steam reforming, it can be effectively stimulated by vibrational excitation of CO_2 molecules, which can significantly increase energy efficiency of the reforming process (see Lecture 17). In contrast to majority of the above-considered processes, the plasma-activated dry reforming of coal $C(s) + CO_2 \rightarrow 2\,CO$ is much more effective not in thermal plasma but in the nonequilibrium "warm" discharges (like gliding arcs and microwave discharges), permitting effective vibrational excitation of CO_2 molecules and controllable gas temperature required for effective surface reactions between CO_2 molecules and solid carbon matrix. More on the subject can be found in Zhukov and Solonenko (1990); Rusanov and Fridman (1984); Fridman (2008).

22.10 Energy and Hydrogen Production from Hydrocarbons with Carbon Bonding in Solid Suboxides without CO_2 Emission

One of the key challenges of the modern world is possible to extract energy from hydrocarbon feedstock without CO_2 emission into the atmosphere. Considerable efforts are focused on answering this challenge by CO_2 sequestration, dissolving CO_2 into ocean water, or pumping CO_2 into the salt caverns or old gas and oil fields. The application of nonequilibrium plasma gives a completely different "out of the box" option for energy generation, using coal for example, without CO_2 production. As a background, there exist carbon oxides other than CO and CO_2. These are carbon suboxides, such as C_3O_2, which are stable and can be polymerized. Very significant amounts of carbon on the Earth exist as such polymer substance, the humic acids, which are natural organic fertilizers and major organic components of soils. *Oxidation of hydrocarbons (fossil fuels) to CO_2 obviously releases high energy and is a key modern energy source, but it exhausts CO_2, which is unacceptable for our environment. In contrast to that, **oxidation of hydrocarbons (fossil fuels) to carbon suboxides** (C_3O_2-based materials) obviously releases a little bit less energy but with fertilizers as products, and surely without any CO_2 emission.* The oxidation of hydrocarbons (fossil fuels) to the C_3O_2-based materials with relevant energy release surely proceeds in nature usually provided by microorganisms (digestion, fermentation, landfill processes). These natural processes are, however, very slow and unacceptable for industrial energy systems like vehicles, power plants, etc. *Nonthermal plasma-chemistry can sustain these processes very fast, in milliseconds starting from different gaseous (methane, ethane, butane, etc.) and condensed phase hydrocarbons (liquid fuels, biomass, coal, etc.).* The oxidation of hydrocarbons (fossil fuels) to the C_3O_2-based materials can produce energy in form of heat or, instead, generate hydrogen. The processes were demonstrated using the nonequilibrium "warm" discharges (especially, the low-current nonequilibrium gliding arcs and microwave discharges), see more details in Fridman et al. (2006a,b); Odeyemi et al. (2012).

The crucial question regarding practical implementation of the oxidation of hydrocarbons (fossil fuels) to the C_3O_2-based materials is comparison of the energy release in this technology with respect to conventional combustion. Because if the energy losses are significant, the implementation of the technology is questionable even though the environmental effect is impressive. Answer to this question surely depends on type of initial hydrocarbons (fossil fuels) and composition of the process products. For example, if we start with high quality coal characterized by high carbon content, the energy losses related to not complete oxidation are maximal. If we start with low quality coal characterized by low carbon content and high content of hydrogen, the energy losses related to not complete oxidation are minimal. Relevant analysis will be discussed in the next section and requires detailed knowledge of thermodynamic and other properties of the suboxides. These detailed data are available in Odeyemi et al. (2012), and shortly described below.

Carbon suboxide (C_3O_2) is a foul-smelling lachrymatory gas. Its molecules are linear and symmetric with structure that can be represented as O=C=C=C=O. The suboxide is stable, but at 25 °C it polymerizes to form a highly colored solid material with a polycyclic six-membered lactone structure, see Figure 22.6. The carbon suboxide is the acid anhydride of malonic acid, and it slowly

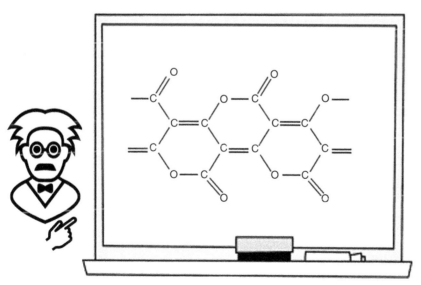

Figure 22.6 Basic chemical structure of the polymeric carbon suboxide $(C_3O_2)_n$.

reacts with water to produce that acid. It can be stored at a pressure of a few Torr, but under standard conditions, C_3O_2 forms a yellow, red, or brown polymer $(C_3O_2)_n$ (ruby-red above 100 °C, violet at 400 °C, and it decomposes into carbon at 500 °C). The formation energy for the gas-phase carbon suboxide C_3O_2 is $-(96-97)$ kJ mol^{-1}, or -95.4 kJ mol^{-1}. The formation energy for the liquid polymerized carbon suboxide $(C_3O_2)_n$ is -59 kJ per one mole of carbon, and the formation energy for the solid polymerized carbon suboxide $(C_3O_2)_n$ is about -112 kJ per one mole of carbon.

22.11 Plasma-chemical Conversion of Coal and Methane into Suboxides for Production of Hydrogen and Energy without CO_2 Emission

Under the natural conditions of weathering, slow oxidation of coal occurs with the formation of carbon suboxide and no, or very little, carbon dioxide is generated. The heat release accompanying this **coal oxidation process to the polymerized substance**, and surely the same for plasma-stimulated oxidation, can be described as:

$$C(s) + (1/3)\,O_2 \rightarrow 1/3n\,(C_3O_2)_n(\text{solid}), \quad \Delta H = -112\,\text{kJ mol}^{-1} \approx -1.1\,\text{eV mol}^{-1}. \tag{22.3}$$

Thus, the carbon oxidation to a suboxide phase still provides considerable heat release but, in contrast to burning, it occurs without any CO_2 emission. Considering the essential hydrogen content in coal, water will also be produced during the process, and the heat release of the oxidation can, therefore, be significantly higher than the enthalpy just shown. It should also be emphasized that suboxides can be further utilized as fertilizers and/or in biochemical technologies. The natural oxidation of coal to suboxide (22.3) is a very slow process at low temperatures. It takes many days, which prevents effective use of the process in the industrial energy generation. Acceleration of the process by temperature increase is impossible because temperature increase above 300 °C results in the formation of gas-phase carbon oxides, CO, and CO_2. The process is effective in the nonequilibrium nonthermal plasma conditions (especially in the "warm" plasma discharges). The energy of carbon oxidation to suboxide can also be extracted in the chemical form. For example, water vapor can be added to the oxygen, or air, to make the oxidation process thermoneutral:

$$2.16\,C(s) + 0.22\,O_2 + H_2O \rightarrow 0.72/n(C_3O_2)_n(\text{solid}) + H_2, \quad \Delta H \approx 0. \tag{22.4}$$

Coal contains hydrogen (typically 0.5–1 H-atoms per C-atom), so the process (22.4) with coal results in a higher hydrogen yield. Additional high-quality energy can be generated if a part of the oxygen in the process (22.4) is substituted with water vapor and low-temperature waste heat.

Changing from coal to natural gas as a feedstock, the **plasma partial oxidation of CH_4 with formation of polymerized solid suboxides** $C_3O_2)_n$ without CO_2 emission proceeds as:

$$CH_4 + 1/3\,O_2 \rightarrow (1/3n)\,(C_3O_2)_n(\text{solid}) + 2H_2, \quad \Delta H = -37.4\,\text{kJ/mol}. \tag{22.5}$$

Addition of a small amount of water vapor leads to the thermoneutral plasma process:

$$1.3\,CH_4 + 1/3\,O_2 + 0.2\,H_2O \rightarrow (1.3/3n)\,(C_3O_2)_n(\text{solid}) + 2.8\,H_2, \quad \Delta H \approx 0. \tag{22.6}$$

The LCV of the produced hydrogen is about 520.8 kJ per mole of CH_4. Using the LCV of methane, 802.5 kJ mol^{-1}, which can be released in combustion, the conversion of CH_4 into hydrogen without CO_2 emission has 65% energy efficiency. This energy efficiency is higher than that of the coal conversion and is due to the high H_2 content of methane. The rest of the energy (in this case about 35%) stays in the form of suboxides, which can then be used as organic fertilizer. Additional high-quality energy in form of hydrogen can be produced if part of the oxygen in reaction is substituted by a low-temperature waste heat. It is interesting to compare CH_4 conversion into polymerized carbon suboxides with direct thermal plasma pyrolytic CH_4 conversion into carbon and H_2 ($CH_4 \rightarrow C(s) + 2H_2$, $\Delta H = 74.6$ kJ mol^{-1}), see Steinberg (2000). Although this plasma process leads also to CH_4 conversion to H_2 with no CO_2 emission, the thermodynamic efficiency here is 50% that is lower than in the production of suboxides. Experimental data on the plasma oxidation of hydrocarbons to carbon suboxides can be found in Fridman et al. (2006a,b); Odeyemi et al. (2012); D'Amico and Smith (1977); Kalra et al. (2005). We should mention that dissociation of CO_2 in nonthermal plasma (see Lecture 17) can also be organized with carbon suboxides as the process products ($CO_2 \rightarrow (1/3n)\,(C_3O_2)_n(\text{solid}) + 4/3\,O_2$). Thermodynamically minimal energy cost of the process in this case is obviously lower than that of the CO_2 dissociation with CO/O_2 as the products.

22.12 Plasma-assisted Liquefaction of Natural Gas, Direct CH₄ Incorporation into Nonsaturated Liquid Hydrocarbon Fuels

Transformation of natural gas (that is mostly CH_4) into valuable liquid hydrocarbons including liquid fuels is in focus from beginning of the twentieth century, and especially important today due to opening of great number of new shale gas sources all over the world. Natural gas has a potential to become the key feedstocks for production of valuable chemicals and liquid fuels, which explains the importance of the effective CH_4 liquefaction. One of the most developed CH_4 liquefaction approach is an indirect one, first involving CH_4 conversion into syngas (see Sections 22.2 and 22.3), following by the Fischer–Tropsch synthesis to obtain the desired liquid products. Although, successful from a technical perspective, these methods are marked by low overall yields and high energy inputs, which triggers interest to alternative nonconventional reforming technologies, and especially to plasma-chemical processes.

Figure 22.7 overviews of the different **plasma-based natural gas liquefaction approaches**, pointing out their major drawbacks and challenges. The "direct gas phase plasma methane liquefaction" is the pyrolytic conversion of pure methane into hydrogen gas and liquid hydrocarbon chains originating from the remaining CH_2 blocks, see Section 22.4. The "oxidative plasma liquefaction" methods, can be subdivided in an indirect and a direct approach. The indirect approach converts methane into syngas together with an oxidant and then processed into liquids using the Fischer–Tropsch technology, see Sections 22.3 and 22.4. The "direct oxidative approach", on the other hand, tries to convert methane (usually also with oxidants) into oxygenated liquid products, such as alcohols and aldehydes, in one step (see Lecture 25). Finally, recently a new approach the "direct two-phase plasma-assisted liquefaction" has emerged, which aspires the direct liquefaction of methane through its incorporation into a second phase, namely existing liquid hydrocarbons. Looking at the minimal energy cost in eV/molecule and subsequently the resultant operational expenditure (OPEX) cost for these different approaches, shown in Figure 22.7, the two most promising techniques to date are the "direct oxidative plasma liquefaction" and the "direct two-phase plasma-assisted liquefaction." The former still needs a lot of research to increase of the selectivity. The latter, although looks the most attractive, still requires significant research and development, and is going to be shortly discussed below in some more detail.

 The **direct two-phase plasma-assisted CH₄ liquefaction** *not only permits incorporation of natural gas into fuel increasing its volume, which is illustrated in Figure 22.8a, but also permits improvement of the low-quality liquid fuel, for example by saturating the double bonds and converting the heavy crude fractions to lighter ones.* Plasma-excited species provide here the cracking allowing operation at ambient temperature and atmospheric pressure unlike the traditional noncatalytic cracking processes. The **CH₄ molecules incorporation into the liquid hydrocarbons** follows two possible pathways: (i) saturation of double bonds in the non-saturated hydrocarbons $CH_4 + R_1 = R_2 \rightarrow HR_1 - R_2CH_3$, $\Delta H \approx -0.5$ eV/mol, (ii) polymerization of the saturated hydrocarbons $CH_4 + R_1 - R_2H \rightarrow R_1 - R_2CH_3 + H_2$, $\Delta H = +0.5$ eV/mol. The saturation of double bonds is the preferred process since it is exothermic reaction with typical activation energy and eventual plasma energy cost of only 0.3 eV/molecule of CH_4. The other hydrocarbon reactions in nonthermal plasmas, such as polymerization and

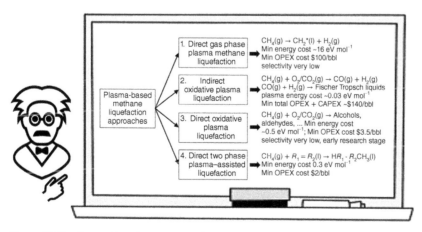

Figure 22.7 Plasma-based natural gas liquefaction approaches.

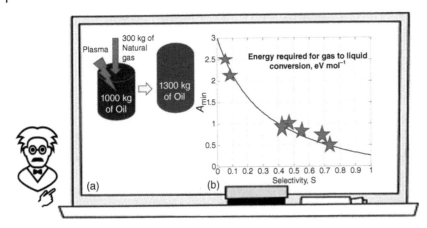

Figure 22.8 Plasma-assisted direct liquefaction of CH_4 by its incorporation into non-saturated liquid hydrocarbon fuels: (a) technology illustration; (b) energy cost of the process as a function of selectivity of incorporation into double bonds; stars correspond to experimental results, curve to modeling (initial liquid composition: 30% aromatic molecules, e.g. anthracene $C_{14}H_{10}$, methyl naphthalene; 70% saturated molecules, e.g. dodecane $C_{12}H_{26}$).

dissociation, require much higher energy inputs. To promote the saturation process (and hence its energy efficiency), local heating needs to be avoided, by working in the nonequilibrium discharge region. The key to this approach lies in the selective vibrational excitation of CH_4, which allows for an exothermic incorporation with low energy cost. Theoretically, this allows the energy cost to remain beneath 0.3 eV/molecule, which would be a significant improvement over conventional gas-to-liquid Fischer–Tropsch synthesis. For two plasma systems, the successful incorporation of CH_4 into liquid hydrocarbons at low energy cost has already been achieved: an atmospheric pressure nanosecond pulsed DBD and an atmospheric pressure glow discharge. Both discharges were ignited inside gaseous bubbles fed through liquid hydrocarbons; this bubbling is used to increase the reaction surface area as well as the mixing. Relevant process energy costs are shown in Figure 22.8b and demonstrate advantages of the approach over conventional gas-to-liquid Fischer–Tropsch synthesis. More information on the subject can be found in Snoeckx et al. (2016); Liu et al. (2017).

22.13 H$_2$S Decomposition in Plasma with Production of Hydrogen and Elemental Sulfur: Fundamental and Technological Aspects, Energy Efficiency in Different Plasma Systems

Hydrogen sulfide dissociation to produce hydrogen and elemental sulfur is slightly endothermic, almost thermoneutral process:

$$H_2S \rightarrow H_2 + S_{solid}, \ \Delta H = 0.2\,\text{eV mol}^{-1}. \tag{22.7}$$

Although the enthalpy required for H_2S decomposition is theoretically very low, a thermal process (22.7) with low energy cost is not possible. When the temperature is increased to stimulate the dissociation, sulfur is produced not in the condensed phase, but as a gas (S_8), which requires more energy: $H_2S \rightarrow H_2 + 1/8\,S_8$, $\Delta H = 0.3\,\text{eV mol}^{-1}$. Because the enthalpy of the gas-phase sulfur process exceeds that of the solid phase one, the required temperature to stimulate H_2S dissociation is higher. Continuing along these lines, the higher temperature required for the decomposition results in the production of sulfur in the form of the dimer S_2 (not S_8), increasing the enthalpy to 0.9 eV. As a result, the quasi-equilibrium thermal H_2S decomposition requires very high temperatures in the thermal plasma and even the minimal energy cost at the ideal quenching is 1.8 eV mol^{-1} (corresponds to a maximum energy efficiency of only 22%).

The minimal energy cost can be significantly decreased below 1.8 eV mol^{-1}, and energy efficiency significantly increased above the thermal plasma limit in **nonequilibrium plasma-chemical H$_2$S dissociation**, *where the essential deviation from equilibrium is due to forced selectivity of the transfer processes (or plasma catalytic effects).* The selective transfer of the sulfur clusters to the discharge periphery produces hydrogen and sulfur with a minimal energy cost of 0.5 eV mol^{-1}, corresponding to the energy efficiency 40%. Major

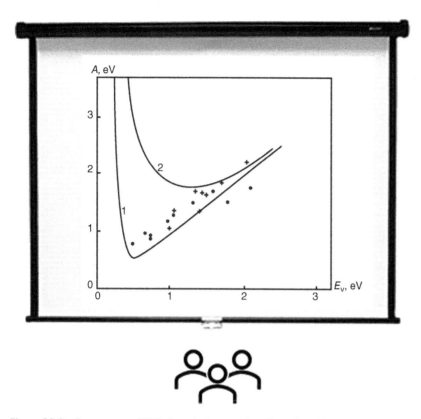

Figure 22.9 Energy cost of H_2S dissociation as a function of specific energy input: (1) non-equilibrium process, modeling; (2) minimal energy cost in quasi-equilibrium process, modeling. Experiments: •, microwave discharge, 50 kW; +, RF discharge, 4 kW.

experimental and modeling results illustrating this plasma-chemical process are summarized in Figure 22.9. Thus, experiments in the nonequilibrium microwave and RF discharges stabilized by fast gas rotation providing strong centrifugal effect at moderate and atmospheric pressures give the minimal energy cost $0.7-0.8 \, \text{eV} \, \text{mol}^{-1}$. The optimal reaction temperature in these nonequilibrium systems is about 1150 K, and the effective clusterization temperature is 850 K. Advantages of the high-frequency, and especially microwave nonequilibrium discharges, for the high energy efficiency H_2S dissociation is due to the ability of this discharges to absorb most of electromagnetic energy in their low-temperature periphery see section, where gas rotation velocity can be close to speed of sound providing the required centrifugal effect.

More information on the subject, including both the two-phase process fundamentals and industrial implementation of the technology, can be found in Rusanov and Fridman (1984); Balebanov et al. (1985a,b, 1989); Krasheninnikov et al. (1986); Harkness and Doctor (1993); Gutsol et al. (2012), Fridman (2008).

The low energy cost and high-conversion characteristics of the H_2S decomposition process achieved in the nonequilibrium, high-pressure plasma attracts significant industrial interest, especially in natural gas processing and in oil refining. Some natural gas from major sources contains 25% or more of H_2S, which is usually separated from the natural gas and converted into sulfur and H_2O (losing hydrogen) in the Claus process $H_2S + \frac{1}{2} O_2 \rightarrow S(s) + H_2O$.

Industrial application of the plasma technology of H_2S dissociation in natural gas processing produces sulfur and saves the hydrogen with minimal energy expenditures. The application of this plasma technology in oil refineries relates to hydrogen recycling in the hydrogen desulfurization process of crude oil. Hydrogen used for oil desulfurization is conventionally produced in refineries by partial oxidation of the oil. The hydrogen desulfurization of oil leads to H_2 conversion into H_2S, which is then conventionally transferred into sulfur in the Claus process. New hydrogen is continuously produced in conventional refineries to purify the crude oil. Plasma technology allows for recycling the hydrogen, making the hydrogen desulfurization of crude oil more effective and ecological (Harkness and Doctor 1993).

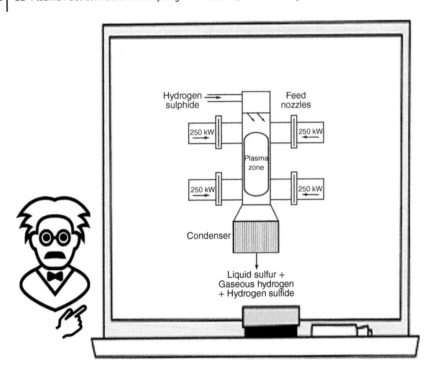

Figure 22.10 Schematic of the 1 MW plasma-chemical non-equilibrium microwave plasma reactor for H_2S treatment in Orenburg Gas Plant.

A **large-scale industrial-level plasma-chemical facility** was built in the Orenburg Gas Plant (Orenburg, Russia) for treatment of H_2S from natural gas (Balebanov et al. 1989; Harkness and Doctor 1993). The four-port microwave discharge configuration used in the facility is illustrated in Figure 22.10. The plasma-chemical reactor is a short section of circular waveguide with four ports for injecting the microwave energy. *The microwave power is produced by four 250 kW magnetron generators and combined in the plasma-chemical reactor to yield a total of 1 MW microwave power.* Orenburg natural gas contains a large fraction of sour gas (H_2S–CO_2 mixture), which is extracted from crude natural gas and then treated in the plasma-chemical reactor. The gas flow through the plasma reactor reaches several thousands of standard $m^3\,h^{-1}$, and the pressure is as high as 1.3 atm. The end of the reactor volume is defined by the sulfur condenser (Figure 22.10), which brings the temperature of the reactor effluent to between 130 and 135 °C. The plasma-chemical reactor is a major component, but not the sole part of the complicated technological scheme required to produce pure hydrogen and sulfur from hydrogen sulfide. A simple flow process diagram for the Orenburg experimental facility, including sulfur recovery and polymer membrane gas separation, is shown in Figure 22.11. Polymer membrane technologies were effectively applied in this case for separation of the H_2S-containing gases. An important advantage of the polymer membrane technologies is their compatibility with the high specific productivity of the plasma-chemical reactor. Effective **plasma dissociation of H_2S into H_2 and sulfur in oil refin**ing was demonstrated at the industrial scale using the microwave plasma system at a power level of 50 kW with productivity of 50 standard m^3 of H_2 per hour at a refinery near L'viv in the Ukraine (Harkness and Doctor 1993). The plasma unit was effectively applied for hydrogen recycling in the hydrogen desulfurization of crude oil.

22.14 Nonequilibrium Kinetics of H_2S Decomposition in Plasma: Nonequilibrium Clusterization in Centrifugal Field, Effect of Additives

The most effective plasma-chemical decomposition of hydrogen sulfide occurs when the process starts with thermal decomposition followed by sulfur clusterization and product separation in a centrifugal field. Consider the **kinetics of the H_2S decomposition process** beginning with thermal decomposition $H_2S + M \rightarrow H + HS + M$. Atomic hydrogen formed in the dissociation process then reacts with and decomposes another H_2S molecule and forms molecular hydrogen in the fast exothermic process $H + H_2S \rightarrow H_2 + HS$ ($k = 1.3 \cdot 10^{-11}\,cm^3\,s^{-1}\,exp(-860\,K\,T^{-1})$. Active

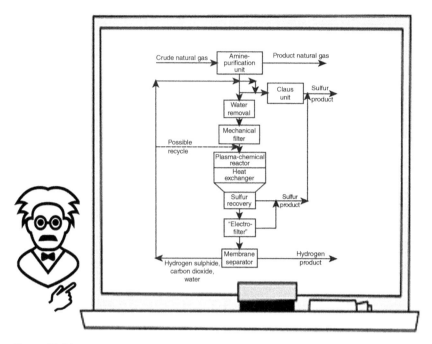

Figure 22.11 Flow sheet for the H_2S treatment Orenburg Gas Plant plasma facility.

radical HS then react with each other and form molecular hydrogen and sulfur in the fast exothermic reaction: $HS + HS \rightarrow H_2 + S_2$, $k = 2.1 \cdot 10^{-10}$ cm³ s⁻¹. The decomposition time is about 5 ms, if it is assumed that the initial temperature in the thermal plasma system is 1500 K. The final composition of the products corresponds to the thermodynamic equilibrium of the mixture. Complete quenching of H_2S dissociation products is achieved at relatively low cooling rates about 10^4 K s⁻¹. The secondary reactions converting H and HS into H_2 and sulfur, are strongly exothermic. As a result, the reverse reactions to secondary processes are essentially endothermic and, therefore, relatively slow at the quenching temperatures. The relatively slow kinetics of the reverse reactions explains the low requirements for the quenching rates (10^4 K s⁻¹) for the H_2S decomposition. The energy cost of hydrogen production in the thermal decomposition of H_2S is presented in Figure 22.12 as a function of the process temperature (or the specific energy input) at different quenching rates and in the case of ideal quenching. The relatively low quenching rates ($\geq 10^4$ K s⁻¹) are sufficient to reach the minimum energy cost for the thermal plasma (about 1.8 eV mol⁻¹), like it was mentioned above. It should be mentioned that the minimum energy cost for the thermal plasma can be slightly decreased through the **plasma-catalytic mechanism of H_2S dissociation**, involving chain reactions through hydrogen disulfide (H_2S_2) to (H_2S_{2n}) initiated through nonequilibrium excited and charged plasma species (see Section 2.14, and Gutsol et al. 2012).

The sulfur **clusterization process accompanying the H_2S dissociation** can be strongly affected by fast plasma rotation (in the RF and microwave discharges, the tangential velocities v_ϕ can be close to the speed of sound). Centrifugal forces proportional to the cluster mass push the particles to the discharge periphery faster than the heat transfer occurs. This can result in a significant shift of chemical equilibrium toward product formation. This effect provides product separation in the processes and decreases its energy cost significantly below the minimum of the quasi-equilibrium thermal process. Also, the relatively large cluster formation in the fast-rotating plasma can occur at temperatures exceeding the quasi-equilibrium condensation point. These effects, reviewed in Fridman (2008) and Fridman and Kennedy (2021), are significant when the centrifugal forces are sufficiently strong, and the centrifugal factor is high:

$$\frac{mv_\varphi^2}{T} \alpha(T)\, n_m^2 > 1. \tag{22.8}$$

Here v_φ is tangential velocity, m is a sulfur monomer mass, n_m is the "magic" cluster size (for sulfur at lower temperatures $n_m = 8$), $\alpha(T)$ is the fraction of these clusters with respect to total sulfur at given temperature. The criterion (22.8) determines temperature when there are more large clusters than molecules moving to the discharge periphery.

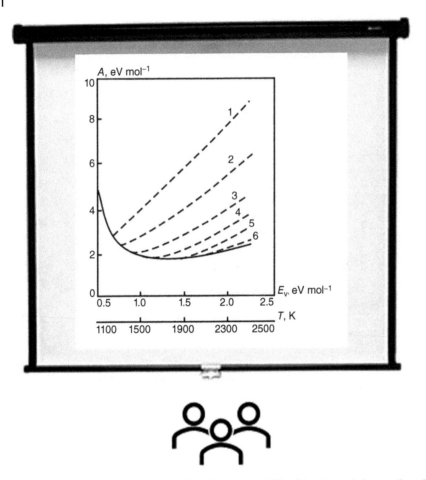

Figure 22.12 Energy cost of production of hydrogen (eV/mol) by thermal plasma dissociation of H_2S as a function of specific energy input (and heating temperature) at different quenching rates: (1) 10 K/s; (2) 100 K/s; (3) 10^3 K/s; (4) 10^4 K/s; (5) 10^5 K/s; (6) 10^6 K/s; (7) 10^7 K/s. Solid line represents the regime of the ideal quenching.

The higher centrifugal factor is the more this critical temperature exceeds the condensation point. For example, the condensation temperature of sulfur is $T_c \approx 550$ K at 0.1 atm. At high centrifugal factors, when tangential velocity is close to the speed of sound, the effective clusterization temperature reaches 850 K. The "magic clusters" for this case are sulfur compounds S_6 and S_8. Transfer phenomena do not affect the maximum energy efficiency of plasma processes when there are no external forces, and the Lewis number is close to unity see Section 17.5. A strong increase of energy efficiency is achieved in the centrifugal field if the molecular mass of the products essentially exceeds that of other components. Then the fraction of products moving from the discharge can exceed the relevant fraction of heat, which results in a decrease of product energy cost with respect to the minimal value for quasi-equilibrium (1.8 eV mol^{-1}; see Figure 22.9). If plasma rotation is sufficiently fast and criterion (22.8) is satisfied, selective transfer of the sulfur clusters to the discharge periphery produces hydrogen and sulfur with a minimal energy cost of 0.5 eV mol^{-1}.

Let us start the short discussion of the **effect of additives on the plasma H_2S dissociation** with the ***plasma-chemical processes in H_2S–CO_2 mixture***. The hydrogen sulfide is present in natural gas as sour or acid–gas mixture (H_2S–CO_2). The previously considered technologies mostly assumed preliminary separation of the acid gas before plasma-chemical dissociation of H_2S and the production of hydrogen and sulfur. In some cases, however, it can be preferable to treat acid gas in plasma directly without the preliminary separation: $H_2S + 2CO_2 \rightarrow H_2 + 2CO + SO_2$, $\Delta H = 3.2$ eV mol^{-1}. The enthalpy of production of a syngas molecule in this case is about 1 eV mol^{-1}. Energy cost of the process in the nonequilibrium moderate-pressure microwave discharges is also at least 3 times higher than that of dissociation of pure H_2S. ***Small addition of oxygen***, vice versa, can decrease the energy cost of the H_2S decomposition. Addition of only 3% of oxygen (and therefore burning 6% of the produced hydrogen) fourfold reduces the required process enthalpy. Final remark is about the ***plasma dissociation of H₂S with admixture of hydrocarbons***. The hydrogen sulfide extracted from natural gas and from crude oil desulfurization gases sometimes contains up to 10% of hydrocarbons in industrial conditions. These hydrocarbons in this case react with the sulfur produced during the plasma H_2S dissociation process and form CS_2. In other words, the presence of hydrocarbons can replace sulfur with CS_2 in the products of H_2S dissociation. More details on the effect of additives on the plasma H_2S dissociation can be found in Balebanov et al. (1989); Nunnally et al. (2014).

22.15 Dissociation of Water Vapor and Direct H₂ Production in Plasma: Fundamental and Applied Aspects, Mechanisms and Energy Efficiency of the Process

The direct decomposition of water vapor is the most straightforward approach to the plasma-chemical hydrogen production from H_2O:

$$H_2O \rightarrow H_2 + \frac{1}{2}O_2, \quad \Delta H = 2.6\,\text{eV mol}^{-1}. \tag{22.9}$$

The principal advantage of the plasma-chemical approach to direct hydrogen production from water is very high specific productivity (in other words, very high hydrogen productivity per unit volume or per unit mass of the industrial device). Because of its volumetric nature, the plasma-chemical process can produce up to 1000 times more hydrogen than electrolytic or thermos-catalytic systems of the same size. The major disadvantage and challenge of the direct hydrogen production from water in plasma is relatively low energy efficiency of the process. The cumulative dependence of the energy efficiency of H_2O dissociation on energy input for different specific mechanisms of the process is presented in Figure 22.13 together with collection of the relevant experimental results. Comparing with plasma dissociation of CO_2 (see Figure 17.6), the energy efficiency of H_2O dissociation is about 3 times lower.

To explain the relatively **low energy efficiency of the process**, we should mention that major contribution to the energy-efficient H_2O dissociation in nonthermal plasmas is provided by two mechanisms: (i) vibrational excitation of water molecules, and (ii) dissociative attachment of electrons to water molecules. The ***H₂O dissociation in plasma stimulated by vibrational excitation*** is strongly limited by fast VT relaxation of water molecules ($10^{-12}\,\text{cm}^3\,\text{s}^{-1}$, 2–3 orders of magnitude faster than that of CO_2). Thus, effective process requires ionization degree above 10^{-4}–10^{-3}

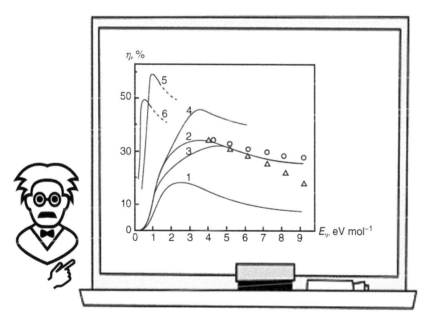

Figure 22.13 Energy efficiency of water vapor dissociation as function of specific energy input: (1, 2) thermal dissociation with absolute and ideal quenching; (3, 4) super-absolute and super-ideal quenching mechanisms; (5) non-equilibrium dissociation stimulated by vibrational excitation; (6) non-equilibrium decomposition due to dissociative attachment. Circles and triangles represent different experiments with microwave discharges.

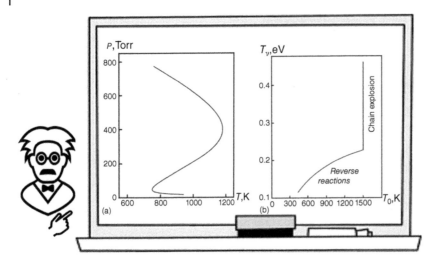

Figure 22.14 Stability of products of plasma-chemical H_2O dissociation ($H_2 - O_2$): (a) explosion limits (explosion takes place to the right of the curve); (b) T_v/T_0 temperature restrictions for stability of the products of H_2O dissociation in plasma stimulated by vibrational excitation (reverse reactions take place to the right and down from the curve, stability is to the left and up).

(see Lecture 17), which is difficult to achieve in moderate and high pressure nonthermal discharges. The **dissociative attachment process** proceeds mostly as $e + H_2O \rightarrow H^- + OH$. The energy efficiency of the dissociation is limited here by the loss of an electron in the process, and the energy price of an electron is not less than 30 eV. The negative ions, formed in the dissociative attachment are subjects of very fast ion–ion recombination and therefore charge neutralization. The only way for the dissociative attachment to become energy effective is via recuperation of electrons from negative ions by means of detachment, like $H^- + e \rightarrow H + e + e$. For the electron detachment to suppress the ion–ion recombination, the ionization degree again should exceed $10^{-4}–10^{-3}$, which is difficult to achieve in the moderate and high pressure nonthermal discharges.

Another challenge of the direct plasma decomposition of water vapor (22.9) is **co-production of H_2 and O_2 in the plasma process,** which should be protected from any kind of reverse reactions including explosion of the mixture, see Figure 22.14a. The domain of vibrational and translational temperatures of water vapor, where the product stability with respect to explosion and reverse reactions can be achieved, are shown in Figure 22.14b. Also, standard restrictions for dilution of the $H_2 - O_2$ mixture are required. An interesting approach for **stabilization of the products as H_2 and H_2O_2** has been demonstrated by Gutsol et al. (1990, 1992). The direct H_2O dissociation stimulated by vibrational excitation in cold plasma usually proceeds as a chain reaction ($H + H_2O^* \rightarrow H_2 + O$, $E_a = 0.9$ eV mol^{-1}, $OH + H_2O^* \rightarrow H_2O_2 + H$, $E_a = 3.0$ eV mol^{-1}) with intermediate formation of H_2 and hydrogen peroxide H_2O_2, which further dissociates to water and oxygen. Because of H_2O_2 instability at elevated plasma temperatures, final H_2O dissociation products usually contain oxygen instead of H_2O_2. Stabilization of H_2O_2 has been achieved at low temperatures when the plasma process was organized in supersonic flow where the gas temperature after the nozzle can be very low. Such experiments in supersonic microwave discharge with Mach number $M = 2$ and vapor temperature before nozzle 400 K results in 1% conversion of H_2O in H_2O_2 at specific energy input 4 J cm^{-3}. The energy efficiency of the supersonic plasma process $H_2O + H_2O \rightarrow H_2 + H_2O_2$ ($\Delta H = 3.2$ eV mol^{-1}) in this regime is about 6%.

22.16 Hydrogen Production from Water in Double-step Plasma-chemical Cycles, Plasma Chemistry of $CO_2 - H_2O$ Mixture

Plasma technologies of fuel conversion and hydrogen production are not limited to the plasma sources itself. Products should be separated, additionally purified, and so forth to integrate plasma-chemical processes into industrial technologies. Considering the high specific productivities of plasma devices (productivity per unit volume of equipment), the system integration can be especially challenging. It can be very challenging to find conventional chemical engineering equipment (product separators, catalytic reactors,

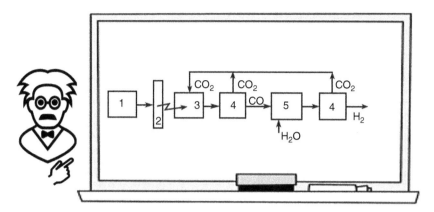

Figure 22.15 Technological scheme of the plasma-chemical double-stage process of hydrogen production from water: (1) nuclear energy power source; (2) electric power supply system; (3) plasma-chemical reactor; (4) gas separation system; (5) steam conversion shift reactor.

heat exchangers, etc.) consistent with the process intensity and specific productivity of plasma reactors. *If the system integration and technological compatibility problems are not solved, the plasma-chemical technology of fuel conversion and hydrogen production could look like a dinosaur with a small head symbolizing the compact and highly productive plasma unit itself, and a massive tail representing the auxiliary (tail) slow chemical processes.* This challenge can be illustrated by energy-effective, two-step, **technological cycle of H₂ production from water** using a nuclear power as the electric energy source, see Figure 22.15 (Rusanov and Fridman 1984). Electricity produced during the night-time at a nuclear power plant (off-peak electricity) is the source of electric energy for the plasma-chemical reactor, where CO_2 is dissociated in strongly nonequilibrium conditions with the production of oxygen and carbon monoxide $CO_2 \rightarrow CO + \frac{1}{2} O_2$.

This plasma-chemical process in nonequilibrium plasma conditions provides energy efficiency up to 90%, and low energy cost of CO production 3.2 kWh m⁻³. Specific productivity (productivity per unit volume of the reactor) in this case is very high, 0.03–0.1 standard L (s⁻¹ cm⁻³). The produced $CO–O_2–CO_2$ mixture is then separated, with the CO_2 recirculating back to the plasma-chemical reactor. Oxygen is extracted at this stage from the mixture as a product. Special gas-separating polymer membranes are used to achieve specific productivity comparable with that of a plasma reactor. In the second chemical stage, the produced carbon monoxide is converted into hydrogen in the exothermic shift reaction $CO + H_2O \rightarrow CO_2 + H_2$. Conventional catalysis using the shift reaction is characterized by a relatively low specific productivity. Better specific productivity was achieved by an electrochemical method applied to the shift reaction and H₂ production. The final stage of the technology is based again on using polymer membranes for separation of gas products and recirculation of CO_2. More details on the technology can be found in Legasov et al. (1988).

Another example of the plasma-chemical cycle of hydrogen production from water is based on **plasma-chemical decomposition of HBr with the production of hydrogen and bromine**: $HBr \rightarrow \frac{1}{2} H_2 + \frac{1}{2} Br_2$, see Doctor (2000), as well as Lecture 18. This endothermic plasma-chemical process requires the super-ideal quenching and the product separation by the fast plasma rotation like in the case of plasma-chemical H₂S dissociation, see Section 22.13. The above plasma-chemical stage using off-peak electricity is a key step in the HBr cycle of hydrogen production from water. The second stage of the cycle is conventional bromine conversion to HBr in reaction with water $Br_2 + H_2O \rightarrow 2HBr + O_2$.

Let us also shortly discuss the processes of **H₂ production in plasma-chemical CO₂–H₂O mixture**. The effectiveness of plasma H₂O dissociation is restricted by requirements of high ionization degree and by reverse reactions of OH radicals, see Section 22.15. Both limitations can be mitigated by the addition of CO_2 (and its dissociation product CO) to water vapor. The presence of CO decreases the concentration of OH radicals in the fast exothermic reaction $OH + CO \rightarrow CO_2 + H$. On the other hand, CO_2 and CO have larger cross sections of vibrational excitation and much lower probabilities of VT relaxation, which makes the ionization degree requirements less strong. CO_2 molecules can play the role of a "physical catalyst" of water decomposition. Restrictions of the CO_2–H₂O mixture composition and the vibrational temperature, required for selective hydrogen production, are shown in Figure 22.16 together with energy efficiency of H₂ production. The energy efficiency in optimal conditions (essentially an excess of CO_2) can be close to 60%, which corresponds to the process energy cost of 4.3 eV mol⁻¹ H₂.

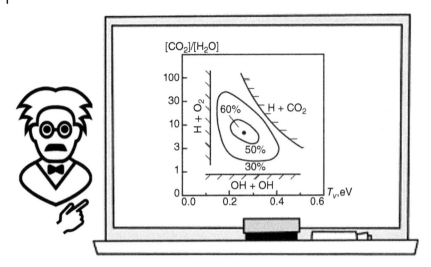

Figure 22.16 Energy efficiency of hydrogen production and restrictions of H_2O-CO_2 mixture composition for effective water dissociation at different values of vibrational temperature in non-equilibrium plasma conditions: closed curves represent lines of constant value of energy efficiency.

We should mention that plasma-chemical and especially plasma-catalytic processes in the CO_2-H_2O mixture also results in only **slightly endothermic nonequilibrium production of relatively large organic molecule** (like oxalic acid and different modifications of formic acid) with partial release of oxygen, see Rusanov and Fridman (1984):

$$2CO_2 + 3H_2O \rightarrow (COOH)_2 * 2H_2O + \frac{1}{2}O_2. \tag{22.10}$$

This process characterized by low energy consumption can be applied to destroy CO_2 and move it to liquid exhaust for environmental purposes. Also, it can "mimic the photosynthesis" because of the release of oxygen without significant consumption of plasma energy. The nonequilibrium plasma process (22.10) can be also the first step of hydrogen production from water. The second step is then the low-temperature thermal decomposition of the organic acids (like $(COOH)_2 * 2H_2O$). Hydrogen is produced in this case from water and the total energy consumption is surely determined by the enthalpy of the H_2O dissociation (22.9). Most of this energy, however, can be applied as a low-temperature heat, which makes this approach attractive.

22.17 Problems and Concept Questions

22.17.1 Plasma Catalysis of Hydrogen Production by Direct Decomposition (Pyrolysis) of Ethane

Interpreting the plasma-catalytic effect of decomposition of hydrocarbons and hydrogen production (see Section 22.5), explain why the application of thermal plasma results in an increase of gas temperature, while application of nonequilibrium plasma results in gas cooling and additional hydrogen production. Compare the thermodynamics of these systems with that of refrigerators and heat pumps.

22.17.2 Plasma-stimulated Partial Oxidation of Liquid Fuel into Syngas (CO–H_2)

Explain why the total energy efficiency of syngas production by partial oxidation of liquid fuels is always less than 100%, even in the ideal plasma process (for example, the efficiency is 65% for diesel fuel conversion). A fraction of energy is always converted into heat in the processes.

22.17.3 Gasification of Coal by Water Vapor in Thermal Plasma Jet

Explain why the thermal plasma treatment of coal with water vapor provides not only gasification of the volatile compounds of the coal but also gasification of coke?

22.17.4   Plasma Conversion of Coal into Carbon Suboxides without CO₂ Emission

Why the energy effectiveness of this process is higher for low quality coals containing significant amount of hydrogen than for high quality carbon-rich coals?

22.17.5   Plasma Dissociation of H₂S with production of Hydrogen and Elemental Sulfur

Why this only slightly endothermic process (enthalpy about $0.2\,\mathrm{eV\,mol^{-1}}$) requires about 10 times more energy even in the regime of the ideal quenching of products?

22.17.6   Reverse Reactions and Explosion of Products of Plasma Dissociation of Water Vapor

Interpret the explosion limitation of translational gas temperature, as shown in Figure 22.14b. Why does this restriction not depend on vibrational temperature in contrast to the limitation related to reverse reactions?

22.17.7   Contribution of Dissociative Attachment to H₂O Dissociation in Nonthermal Plasma

Explain the $\eta(Ev)$ curve for H_2O dissociation stimulated by dissociative attachment presented in Figure 22.13. What is the physical cause of the threshold? What is the physical cause of the decrease in energy efficiency at high values of specific energy input?

22.17.8   H₂O Dissociation in Nonequilibrium Supersonic Plasma with Formation of H₂ and Stabilization of Peroxide in Products

Would you expect condensation of water molecules and H_2O_2 molecules in the supersonic flow of the plasma-chemical reactor?

Lecture 23

Plasma Energy Systems: Ignition and Combustion, Thrusters, High-speed Aerodynamics, Power Electronics, Lasers, and Light Sources

23.1 Plasma-assisted Ignition and Stabilization of Flames: Ignition of Fast Transonic and Supersonic Flows, Sustaining Stable Combustion in Low-speed Flows

Spark ignition is one of the oldest applications of plasma, known and successfully applied for thousands of years. Even in the automotive industry, spark ignition has been applied for more than a hundred years. Discharges other than sparks, especially nonthermal discharges, are attracting attention now for special and more efficient ignition and stabilization of flames. The classical effect of discharges with different currents on the explosion limits of an $H_2–O_2$ mixture is summarized in Figure 23.1. The curve marked "0 A" (no Amps, no current) in the figure corresponds to the conventional $H_2–O_2$ explosion limit (the so-called Z-curve, see Figure 22.14a). High current discharges significantly shift the Z-curve to lower temperatures and pressures.

Two qualitatively different mechanisms, thermal and nonthermal, make contributions to plasma-assisted ignition and combustion. The **thermal ignition mechanism** implies that plasma as an energy source provides a temperature increase (usually local), which results in exponential acceleration of the elementary reactions of dissociation, chain propagation of fuel oxidation, and, finally, ignition. The locally ignited mixture then propagates, establishing a stable combustion process. Application of thermal plasma discharges like sparks and arcs leads usually to thermal ignition. Strongly nonequilibrium nonthermal discharges (like dielectric-barrier discharge [DBD], corona, gliding arc, etc.) also can increase gas temperature and stimulate the thermal ignition mechanism. Nonequilibrium discharges, however, can also trigger a much more effective **nonthermal ignition mechanism**. The nonthermal ignition mechanism implies that specific plasma-generated active species – radicals, excited atoms, molecules, charged particles, and so on – stimulate the chain reactions of fuel oxidation leading to the ignition of the fuel–oxidizer mixture. Even relatively small concentrations of radicals and other active species (molar fraction $10^{-5}–10^{-3}$) can be sufficient to initiate combustion. The temperature increase in the nonthermal discharges corresponding to this amount of active species is usually not more than 300 K. The nonthermal plasma ignition has significant advantages over the thermal one because it requires much less energy and can provide more uniform and well-controlled ignition and stabilization of flames. The nonthermal mechanism uses plasma energy selectively, only in the active plasma species relevant to the flame ignition and stabilization, whereas the thermal ignition distributes plasma energy over all degrees of freedom, and most of those are irrelevant to flame ignition and stabilization. Also, the relative uniformity of nonthermal plasma ignition keeps it far from the detonation condition.

Plasma is effective in **ignition and combustion control in fast and especially supersonic flows**. It is used in important applications for aircraft engines, where problems of reduction in ignition delay time, high-altitude (low-pressure) flame holding, and extension of the flame blow-off limits are of crucial importance. It is especially important for use in supersonic combustion ram (SCRAM) engines. *The plasma ignition of supersonic flows is challenging because normal flame propagation velocity is less than the speed of sound.* Plasma ignition of the supersonic hydrocarbon–air mixture has been demonstrated using RF, microwave, and pulsed discharges. The effective plasma ignition of the supersonic flows with stagnation pressures about 1 atm was shown in this case at Mach numbers M about 2. Nonequilibrium plasma is also effective in **sustaining combustion in turbulent or laminar atmospheric-pressure slow flows** ($3–100\,\mathrm{m\,s^{-1}}$) employed in industrial burners. For example, application of

Plasma Science and Technology: Lectures in Physics, Chemistry, Biology, and Engineering, First Edition. Alexander Fridman.
© 2024 WILEY-VCH GmbH. Published 2024 by WILEY-VCH GmbH.

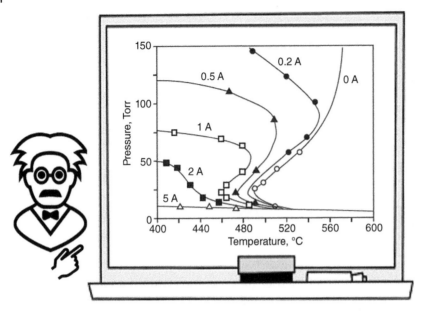

Figure 23.1 Illustration of extension of explosion limits of the H_2-O_2 mixture due to the plasma stimulation of ignition. Different curves correspond to different currents (Amps) in a primary electric circuit.

pulsed nanosecond DBD allows doubling of the blow-off velocity of the premixed propane–air flame. Application of the pulsed nanosecond DBD also allows for stable burning of lean propane–air (and other hydrocarbon–air) mixtures without flame deceleration. Three different types of discharges demonstrated stabilization of the non-premixed lifted flame, namely, a single-electrode corona discharge, a DBD, and a nanosecond pulsed discharge between two electrodes. The nanosecond pulsed discharge can increase the flame stability limit 10-fold, whereas the effect of the DBD discharge is less pronounced. The power consumption of all three discharges is less than or about 0.1% of the chemical power.

A certain level of translational gas temperature is helpful even in strongly nonthermal stimulation of ignition and combustion. Anyway, some increase of translational temperature is inherent to the nonequilibrium discharges. It attracts interest to the application of **combined (hybrid) nonthermal/quasi-thermal discharges** with significantly super-equilibrium generation of active plasma species and at the same time with a controlled level of translational temperature. The nonequilibrium gliding arcs are especially promising in this regard (Omrello et al. 2006a,b; Fridman et al. 2007a–e). These discharges are also interesting for flame ignition and stabilization because of its periodic structure in space and time, see Sections 11.11, 11.12, and 14.1. Homogeneous nonequilibrium plasma activation of flames leads to power requirements that can become extremely high, sometimes more than several $MJ\,m^{-3}$. The nonequilibrium gliding arcs provide excitation of not entire combustible gas flow but only some well-organized time-space structured array, which decreases energy consumption but still avoids unstable detonation-related effects on the fast gas flow during combustion. Such time-space structured array of plasma-assisted ignition, referred to as **"zebra" ignition** permits a significant decrease of the nonequilibrium plasma power required for ignition of large-volume flows, particularly in the case of scramjet engines.

Significant number of publications and reviews are focused on plasma-assisted combustion, see Anikin et al. (2003, 2004, 2005); Starikovskaia (2006, 2014); Starikovskaia et al. (2004, 2006); Matveev et al. (2005); Brovkin and Kolesnichenko (1995); Pancheshnyi et al. (2005); Bozhenkov et al. (2003); Starikovskii (2003); Lou et al. (2006); Leonov et al. (2006); Esakov et al. (2006); Klimov et al. (2006); Shibkov et al. (2006); Ombrello et al. (2006a,b); Fridman et al. (2007a–e); Matveev (2013, 2015); Starikovskiy and Alexandrov (2013); Ju and Sun (2015).

23.2 Mechanisms and Kinetics of Nonequilibrium Plasma-stimulated Combustion, Ignition Below the Auto-ignition Limit

Various species generated in nonequilibrium plasma make contributions to the stimulation of combustion. In discharges with relatively low reduced electric fields E/n_0, low-energy excitation of molecules such as vibrational excitation and excitation of lower electronic states can be essential in the stimulation of combustion. Thus, hydrogen–air mixture with a small (<1%) addition of ozone molecules, vibrationally excited by CO_2 laser at 9.7 µm, has 2–3 orders of magnitude increase of ignition efficiency in comparison to equivalent local gas heating by infrared (IR) radiation. At high reduced electric fields, the electronic excitation to high energy levels results in ignition by O_2 dissociation either through direct electron impact or by the collisions with highly electronically excited nitrogen. The role of nonequilibrium discharges in ignition of hydrocarbon–air, H_2-air, and CO–air mixtures can be often explained by a branching mechanism that multiplies the primary plasma generated radicals. It is also known that plasma-generated ions can play an important role in ignition kinetics through plasma catalysis, see Section 2.14. *At very lower temperatures, the main plasma effect leading to ignition acceleration is the fast gas heating (mostly due to the recombination of radicals). The injection of O atoms becomes more and more effective itself starting from several hundreds of kelvins when the atoms not only lead to heating but also initiate chemical chains.*

Interesting and important that the nonthermal plasma-stimulated ignition in H_2, CH_4, and C_2H_4 can occur at temperatures apparently below their auto-ignition thresholds. It means that *nonthermal plasma can accelerate the ignition not only where it is possible according to traditional chemical kinetics, but also below the auto-ignition limit where production of new radicals is suppressed by their recombination and ignition is considered impossible in the framework of conventional combustion kinetics.* Kinetics of this plasma effect, called **"the ignition below the auto-ignition limit"**, is going to be discussed in the next section; but let us first discuss chemistry of the explosion limits for fuel–oxidizer mixtures without plasma.

A large number of fuel–oxidizer mixtures share qualitatively similar ignition characteristics with respect to branched-chain ignition. The characteristic Z-shaped curve, the auto-ignition or explosion limit, divides the "pressure versus temperature plane" a region of explosion and no explosion, see Figure 23.2. Most hydrocarbons exhibit more complicated behavior (e.g., lobes of multi-stage ignition, zones for single and multiple cool flames not leading to explosion) because of their oxidation kinetics; however, the general Z shape remains. The explosion limit demarcates the locus of initial temperature and pressure conditions where termination and branching reaction rates are balanced, leading to a nonexplosive reaction. Since the rate of termination of radical chains exceeds that of chain branching below and/or to the left of the limit, it is generally held in combustion science that the

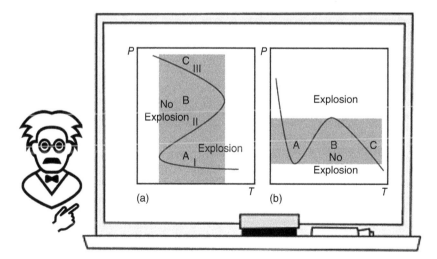

Figure 23.2 Explosion limits diagram for (a) hydrogen-oxygen mixture, (b) hydrocarbons-oxygen mixture.

mixture cannot ignite at these conditions without thermal stimulus to raise at least local temperatures to the explosion limit. To nonthermally ignite a mixture at a temperature below the threshold requires that either the termination rate must be reduced, or the branching rate must be accelerated to extend the region of auto-ignition in the temperature–pressure plane to lower temperatures.

The **auto-ignition of H_2–O_2** is governed by three explosion limits (I–III in Figure 23.2a). Below the first explosion limit (I), chain-carrying radicals and heat diffuse to vessel walls where their chains terminate faster than branching occurs, inhibiting the overall reaction rate explosion. Above I, in region A, radical chains branch explosively for initial pressures up to the second limit, II, at which point effective termination reaction rates again exceed branching rates. In this case, the rate of the pressure-dependent three-body reaction $H + O_2 + M \rightarrow HO_2 + M$ eclipses the rate of the branching reactions: $H + O_2 \rightarrow OH + O$, $O + H_2 \rightarrow H + OH$. The hydroperoxyl-radical (HO_2) terminates the atomic H branching chain since it is a relatively unreactive radical at the temperatures corresponding to limit II. The plasma extension of the auto-ignition limits in the vicinity of limit II appears to depend on either recovering active radicals (H, O, or OH) from HO_2 or preventing HO_2 formation. At higher pressures corresponding to limit III, this is the case; the reaction $HO_2 + H_2 \rightarrow H_2O_2 + H$ recovers chain carriers from HO_2, once again enabling explosion.

The **hydrocarbon auto-ignition limits** differ from those for hydrogen because the reactions competing with chain branching are in this case function of temperature rather than pressure. The region B in Figure 23.2b can be attributed to competition between reactions: $R + O_2 \rightarrow olefin + HO_2$, and $R + O_2 \rightarrow RO_2$; where R is any alkyl radical (except methyl). The former reaction only serves to propagate the chain, whereas the latter contributes to branching after subsequent steps $RO_2 + R^1H \rightarrow ROOH + R^1$, $ROOH \rightarrow RO + OH$, or $RO_2 \rightarrow R^1CHO + R^2O$, $R^1CHO + O_2 \rightarrow R^1CO + HO_2$. At the lowest temperatures, chain branching pathway predominates, but termination rates exceed branching rates, at least up to region A. At temperatures corresponding to region B, the non-branching step predominates, preventing explosion; however, reaction self-heating causes explosive runaway to occur at temperatures corresponding to that in region C.

Thus, according to traditional combustion kinetics, auto-ignition is a critical (threshold-type) phenomenon. It becomes possible only at such pressures and temperatures when multiplication of radicals overcomes recombination and quenching, leading to explosive growth of their density. From this point of view, the nonthermal plasma generation of radicals without influencing temperature should not result in ignition, because the radicals would be quenched anyway. Nonthermal plasma can overrule this requirement due to contribution of specific nonthermal plasma species including charged and excited atoms and molecules to be discussed next.

23.3 Subthreshold Plasma Ignition, Kinetics of Plasma "Ignition Below the Auto-ignition Limit"

The **subthreshold plasma-stimulated ignition** has been demonstrated in H_2–O_2 and hydrocarbon–air mixtures using different nonequilibrium discharges. Let us point out some examples of observation of the effect. The **subthreshold auto-ignition of a stoichiometric hydrogen–oxygen mixture** ($12H_2$:$6O_2$:$82He$) has been observed by Bozhenkov et al. (2003), using shockwaves to bring the mixture to near-threshold temperature and the ***fast ionization wave (FIW)*** as single-pulse plasma stimulation of the gas mixture. None of the measured ignition delay times were faster than 10^{-4} s, and many ignition results suggested ignition below the explosion limit for stoichiometric H_2–O_2 mixtures. The theoretical explosion limit corresponds to an infinite ignition delay time, although practically it is determined as corresponding to ignition delays of many seconds or minutes. Each order-of-magnitude change in ignition delay time corresponds to an increase of many tens of Kelvins.

RF plasma-stimulated **subthreshold volumetric ignition of premixed methane–air and ethylene–air flows** in a supersonic combustion has been demonstrated by Chintala et al. (2005, 2006). The plasma-stimulated subthreshold ignition of ethylene–air flow has been achieved in nanosecond pulsed DBD discharge (Lou et al. 2006, 2007).

 To summarize, ignition from tens to hundreds of degrees below the auto-ignition temperature threshold, as well as a substantial degree of flameless oxidation for some conditions can be plasma triggered. The nonthermal chain branching (explosion) occurs under plasma stimulation in the temperature–pressure (T–P) range where conventional combustion chemistry, dealing with temperature and radicals, predicts only slow reaction (i.e., branching radical termination rate exceeds the branching rate). Let us analyze the major effects explaining the phenomenon.

23.3.1 Subthreshold Plasma Ignition Initiated Thermally: the "Bootstrap" Effect

For a potentially explosive mixture at a point in T–P space initially below the explosion limit, an increase in temperature up to or beyond the limit will result in a rate of branching exceeding that of termination and an ensuing explosion. Such temperature increase can be due to not only direct plasma heating (even in nonthermal discharges) but also due to recombination and relaxation of the plasma generated radicals and excited species. In addition, *the chain reactions of the plasma-generated radicals can release significant heat (more than simply heat of recombination) in the exothermic chain processes even before the branching of this chain dominates and lead to explosion.* This, the so-called "**bootstrap effect**", can provide up to several hundred degrees of heating and explain in some cases the ignition Below the Auto-Ignition Limit, see Chintala et al. (2005, 2006).

23.3.2 Subthreshold Ignition Initiated by Plasma-generated Excited Species

A good example of this effect is related to suppression of HO_2-chain termination by the plasma generated excited species. As it was clarified in the previous section, the auto-ignition of both hydrogen and hydrocarbons is significantly due to the formation of the passive radical HO_2 terminating the explosive chain reaction (in H_2–O_2 mixture, it is determined by competition of the chain branching $H + O_2 \rightarrow OH + O$, and recombination $H + O_2 + M \rightarrow HO$). Both latter reactions stem from the same process, illustrated in Figure 23.3. Reactants associate to a metastable state, HO_{2*}, and temporarily store 205 kJ mol^{-1} of energy of the H–O_2 bond in mostly vibrational modes. In hydroperoxyl-radical (HO_2) formation, the third body M collisionally stabilizes the HO_{2*} complex into the stable HO_2 potential energy well. Failing stabilization, the HO_{2*} may decompose back to H and O_2, while at higher temperatures, the energy barrier about 70 kJ mol^{-1} for the $O + OH$ formation channel is increasingly likely to be overcome, leading to chain branching.

Two plasma-stimulated reaction schemes recover the active radicals OH and O from HO_2. In the first scheme, an active plasma particle, either vibrationally or electronically excited (for example, N_{2*}), collides with stable HO_2 providing roughly 2.87 eV per HO_2 to lift the system out of the potential energy well and produce OH and O. In the alternative, energetically less costly scheme (0.73 eV), the vibrationally or electronically excited species (e.g., N_{2*}) decompose the intermediate complex HO_{2*} to OH and O. Combination of the effect can decrease the required ignition temperature up to several hundred degrees assuming significant N_2-vibrational excitation in the plasma-activated mixture, see Chintala et al. (2005, 2006), as well as Fridman (2008).

23.3.3 Subthreshold Ignition Initiated by Plasma-generated Neutrals like NO and CH₂O

Not only excited but plasma-generated active molecules can convert the passive HO_2 radicals into active ones stimulating ignition and combustion processes. This can be illustrated by effect of the plasma-generated NO, which can effectively convert HO_2 into OH in the fast chain process $NO + HO_2 \rightarrow NO_2 + OH$, $H + NO_2 \rightarrow NO + OH$. This NO effect can significantly decrease the ignition limit, especially when concentration of the plasma-generated NO is significant.

The above-considered "bootstrap effect" (Section 23.3.1) was focused on temperature increase during the initial induction period of ignition, which promotes the explosion. Very similar effect is related to production of the relatively active neutrals, like formaldehyde (CH_2O) and some peroxy compounds during the initial induction period of ignition. This **active neutrals related "bootstrap effect"** can effectively decrease the ignition limit.

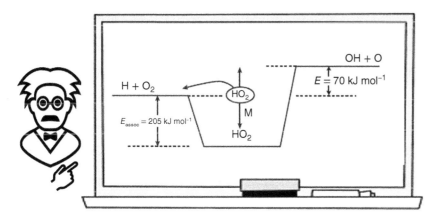

Figure 23.3 Potential energy diagram for the chemical reaction path profile $H + O_2 \rightarrow$ products.

23.3.4 Contribution of Ions to the Subthreshold Ignition

Conversion of the passive HO_2 radicals into active ones stimulating ignition and combustion can be also stimulated by ions. The relevant chain process can be illustrated as $M^+ + H_2 \rightarrow MH_2^+$, $HO_2 + MH_2^+ \rightarrow H_2O + OH + M^+$ and was considered in application to auto-ignition by Shibkov and Konstantinovskij (2005). It should be mentioned that corresponding neutral–neutral reaction $HO_2 + H_2 \rightarrow H_2O + OH$ is exothermic ($-192.7\,kJ\,mol^{-1}$) and releases a very active OH radical, which could serve to propagate the hydrogen oxidation chain reaction. This neutral–neutral reaction does not, however, make any significant contribution into the hydrogen oxidation kinetics because of its high activation energy $E_a \sim 100\,kJ\,mol^{-1}$. This activation energy can be effectively eliminated if process is organized in the framework of ion catalysis as an ion–molecular reaction. Another interesting ion-molecular mechanism of hydrogen oxidation in nonequilibrium plasma proceeding through negative ions and mitigating the explosion limit has been proposed by Starikovskii (2000).

23.4 Plasma-assisted Ignition and Combustion in Ram/Scram Jet Engines, Energy Efficiency of Transonic and Supersonic Ignition

Plasma stimulation is promising for application in the ramjet and scramjet (supersonic combustion ramjet) engines, especially for hypersonic the airbreathing vehicles associated with several challenges, including ignition, flameholding, and spreading. In Mach 4–8 flights with dynamic pressures of 0.5–1 atm, the combustor inlet conditions are static pressures 0.5–2 atm, static temperatures 400–1000 K, and flow velocity 1200–2400 m s^{-1}. Hydrocarbon fuel is injected at Mach numbers around 4 as a liquid, and at higher Mach numbers as a gas. Mixing in the scramjets is helped somewhat by the flow turbulence. The auto-ignition delay exceeds 10 ms for hydrocarbon fuels at pressure of 1 atm and temperature 600–1200 K, which results in unacceptably long ignition region. Thus, forced ignition is required, which can be provided by plasmas.

Efforts have been focused on the volumetric (space-uniform) plasma ignition to solve the flame spread problem. Gas is preheated in the ram/scramjet combustors due to its compression facilitating plasma ignition. Ignition at temperatures of 1000–1400 K depends on plasma generation of the pool active radicals. The molar fraction of active radicals $\alpha = 1 - 3 \cdot 10^{-3}$ permits the ignition delay time to be shortened to about 10 µs and thus the ignition length to about a few centimeters. The specific energy input E_v in the discharge required to achieve the initial fraction α of the active radical can be calculated as (Macheret et al. 2005):

$$E_v = \alpha W = \frac{\alpha e^2}{m k_{en} k_{rad}} (E/n_0)^2, \tag{23.1}$$

where W is the energy cost of production of one radical; E/n_0 is the reduced electric field; k_{en} is the gas-kinetic rate coefficient for the electron–neutral collisions; k_{rad} is the effective rate coefficient for production of radicals (for example, through dissociation) in the electron–neutral collisions; and e and m are an electron charge and mass. In the conventional low-electric-field nonequilibrium discharges in molecular gases under consideration, electron temperature is $T_e \approx 1\,eV$ and reduced electric field is $E/n_0 \approx 30\,Td$. Most of the discharge energy in such conditions goes into the vibrational excitation of molecules, and the energy cost of a radical is relatively high, at least $W \approx 100\,eV$. Then, according to (23.1), the specific energy input required to achieve the molar fraction of plasma-generated active radicals ($\alpha = 1-3 \cdot 10^{-3}$) and to shorten the ignition delay time to about 10 µs is rather high and equals 0.1–0.3 eV mol^{-1}. This specific energy input per molecule corresponds to very high macroscopic energy input of 0.3–1 MJ kg^{-1}, which is comparable with the total flow enthalpy of 1.5–2.5 MJ kg^{-1} at Mach numbers of 5–7. In power requirements, this translates to more than 100 MW per square meter of a combustor cross section, which is unacceptable, see Macheret et al. (2005).

The energy requirements for ignition in the ram- and scramjet engines can be reduced by application of plasma systems with higher electron energies, including electron beams and electric discharges such as high-voltage pulse discharges with high E/n_0. Using electron beams or the high-voltage nanosecond pulse discharges with megahertz repetition rate operating at $E/n_0 \approx 300$–1000 Td permits the effective energy of a radical production to be reduced to about $W \approx 15\,eV$. Then, the specific energy input required to achieve the molar fraction $\alpha = 1-3 \cdot 10^{-3}$ and to shorten the ignition delay time to about 10 µs decreases to 0.015–0.05 eV mol^{-1}. It corresponds to the macroscopic energy input 0.05–0.15 MJ kg^{-1}. This energy is still very high and is equivalent to at least several percent of the total flow enthalpy.

If 50–100 µs ignition delay time (0.1–0.2 m ignition length) is still acceptable for the ram- and scramjet engines, and the static temperature is 1000–1400 K, then a mole fraction $\alpha \approx 10-4$ of the plasma-generated radical can be sufficient. Then the minimum required specific energy input during application of the high-electron-energy plasma system can be reduced to about 10^{-3} eV mol^{-1}, which corresponds to the power budget of several megawatts per square meter. Further increase of the energy efficiency is possible in the space- and time-periodic plasma ignition systems including the zebra igniters discussed in Section 13.1, and the multi-spot igniters proposed by Macheret et al. (2005). These volumetric but not completely homogeneous plasma systems require much less energy and power budget because they do not cover all the flow uniformly, but their volumetric nature still helps avoid significant flow perturbations typical for local igniters.

23.5 Plasma Ignition and Stabilization of Combustion of Pulverized Coal: Application for Boiler Furnaces

Plasma is effectively applied for ignition and stabilization of combustion of not only gaseous or evaporated fuels but also solid fuels. This application is especially important for the ignition of low-quality coals in power plant boilers. The conventional combustion of the low-quality coals requires addition of up to 21–25% natural gas and heavy oils, which is effective neither economically nor technologically. Although intensive spark discharges can be used for the ignition of the coal powder, the thermal arcs prove to be more effective. Air heated in a thermal arc discharge has a very high temperature and active species density, which are sufficient to provide ignition of coal powder more effectively than the conventionally used pilots burning natural gas and heavy oil. This plasma process was first demonstrated at industrial scale in a power plant boiler in Marietta, Ohio, involving the burning of bituminous coal and producing a total of 200 MW of electric power. The energy efficiency of the thermal arc igniter varies in this case in the range 45–68%. Stable ignition requires arc power of 100 kW and exit plasma jet temperature of 5700 K. A single arc discharge provides stable ignition of 1.5 tons of pulverized coal per hour. The fraction of plasma power applied for ignition is only 1% of the combustion power. To compare, application of heavy oil requires about 6% of the total combustion power.

The most effective industrial-scale plasma ignition of coals can be achieved if fraction of volatiles in the coal is high and their release simplifies the ignition, see Messerle et al. (1985); Messerle and Sakipov (1988); Messerle (2004). The thermal arc applied for ignition usually has power 20 kW and increases air temperature before it mixes with pulverized coal. Such plasma is effective in ignition of 300–600 kg coal h^{-1}. The effectiveness of the igniter can be characterized by the ratio of the arc power to the power of the coal burner, which is 0.75–1%. Even better efficiency (0.3%) can be achieved when the arc is organized directly in the air–coal mixture.

More challenging is the *plasma ignition of coals containing low amount of volatiles,* especially anthracites with a percentage of volatiles below 5%. Effective ignition can be achieved in this case by special thermochemical coal preparation before combustion, see Solonenko (1986, 1996). It assumes thermal plasma heating of the pulverized coal premixed with some fraction of air to temperatures sufficient for release of volatiles and partial gasification of the coke. It results in total generation of combustible gases on the level of 30%, which corresponds to the characteristics of highly reactive coals. This approach permits the conversion of low-reactive coals into a double-component coke-combustible gas mixture, which can then be conventionally ignited without any additional chemical admixtures. Such precombustion plasma treatment also decreases the combustion temperature and, therefore, reduces the amount of nitrogen oxides in the combustion exhaust gases by about 20%. Additional information on the subject can be found in the review of Karpenko and Messerle (2000) and Messerle and Sakipov (1988).

23.6 Ion and Plasma Thrusters: Electric Propulsion, Specific Impulse of Electric Rocket Engines

Launching of a spacecraft to an orbit requires powerful high-thrust rocket engines based on the use of liquid or solid fuel. Attitude control, precision spacecraft control, and low-thrust maneuvers can be effectively provided by ion and plasma thrusters, which are high-specific-impulse, low-power electric propulsion systems. The useful load of a spacecraft is very expensive and should be spent carefully. *The electric propulsion systems can decrease the required mass of the spacecraft because they are characterized by very*

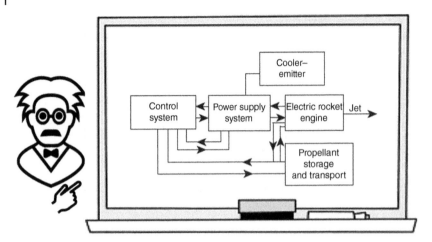

Figure 23.4 Schematic and general components of an electric rocket engine.

high specific impulse, I_{sp}, which is the impulse (change of momentum) per unit mass of propellant. The specific impulse is determined by exit velocity of the propellant and measured in meters per second. The I_{sp} is often presented not per unit mass but per unit weight of propellant; then it is $g = 9.8 \, m \, s^{-2}$ times less and measured in seconds. The specific impulse I_{sp} is used as a measure of the efficiency of the propulsion system. In practice, the specific impulses of real engines, vary with both altitude and thrust; nevertheless, I_{sp} is a useful to compare engines, like "miles per gallon" is used for cars. *The specific impulse of rocket engines based on liquid and solid fuels does not exceed 4500–5000 m s^{-1} (450–500 s), which is limited by neutral gas velocities at high temperatures.*

Much higher exit velocities can be achieved by forced electromagnetic acceleration of the propellant, using accelerators of charged particles or plasma. Such electric propulsion principles lead to the creation of ion and plasma thrusters with very high values of the specific impulse. A schematic of the electric rocket engine is presented in Fig. 23.4. A distinctive feature of the electric rocket engine is a special onboard electric energy generator, which determines the limit for the total thrust of the engines. Thrusts exceeding 50–100 N already require an unacceptable mass of onboard electric energy generators. As an example, thrust of 100 N and specific impulse $I_{sp} = 40 \, km \, s^{-1}$ correspond to jet power of an electric rocket engine of 2 MW. The typical required power of the onboard electric energy generator in this case is 2.8 MW, and the mass of the generator is 86 tons. The electric rocket engines cannot be applied for launching spacecraft and can be used only in zero gravity conditions. The acceleration provided by electric rocket engines is usually in the range 10^{-4}–$10^{-2} \, m \, s^{-2}$. To reach velocities of tens of kilometers per second, conventional for spacecrafts, the electric rocket engines would require several thousand hours.

Each flight of a spacecraft with an electric rocket engine (ion or plasma thruster) has an optimal specific impulse, which minimizes the initial mass of the spacecraft and cost of its launching to an orbit. It distinguishes the electric propulsion based on the ion or plasma thrusters from the rocket engines using liquid or solid fuel, for which efficiency always increases with the specific impulse. To find the optimal specific impulse for an electric rocket engine, consider a mass model of a spacecraft with an electric rocket engine. Initial mass M_0 of such a spacecraft at low(near-Earth) orbit includes the useful load M_{use}, mass of the onboard electric generator and the electric rocket engine $M_{electric}$, mass of the propellant (working fluid) M_{wf}, mass of the fuel reservoirs M_{fr}, and mass of the construction elements M_{const}: $M_0 = M_{use} + M_{electric} + M_{wf} + M_{fr} + M_{const}$. The mass of an electric rocket engine increases linearly with the total consumed power of the engine, $P_{electric}$, where the coefficient of proportionality γ electric can be estimated today as $0.03 \, kg \, W^{-1}$: $M_{electric} = \gamma_{electric} P_{electric}$. The mass of the propellant (working fluid) can be calculated based on the thrust F, operation time of the engine, τ, and the specific impulse I_{sp}: $M_{wf} = F \tau / I_{sp}$. The jet power of the propellant, P_j, leaving the engine is determined by thrust and specific impulse: $P_j = F I_{sp}/2$. Also, the jet power of the propellant, P_j, is related to the total consumed power of the engine, $P_{electric}$, through the thrust efficiency η_{thrust}, which can be numerically estimated as $\eta_{thrust} = 0.7$: $P_{elecrtic} - P_j = \eta_{thrust} P_{electric}$. The relation between mass of the fuel reservoir and mass of the propellant (working fluid) is conventionally described using the coefficient γ_{fr} characterizing relative mass of the fuel reservoir: $M_{wf} - M_{fr} = \gamma_{fr} M_{wf}$. A typical numerical value of the relative fuel reservoir mass is $\gamma_{fr} = 0.2$. Combination of the above equations results in the following relation between the initial mass M_0 of a spacecraft at low (near-Earth) orbit and its specific impulse I_{sp}:

$$M_0 = M_{use} + \frac{\gamma_{electric} F I_{sp}}{2\eta_{thrust}} + \frac{F\tau}{I_{sp}}(1 + \gamma_{fr}) + M_{const}. \tag{23.2}$$

This relation shows that high specific impulses permit, on one hand, decreasing initial spacecraft mass through reduction of the mass of propellant (working fluid). On the other hand, however, the higher specific impulses lead to an increase of the initial spacecraft mass because of the required larger mass of the onboard electric generator and the electric rocket engine. It leads to the optimal specific impulse minimizing the initial mass of a spacecraft:

$$I_{sp}^{opt} = \sqrt{2\eta_{thrust}\tau(1 + \gamma_{fr})\gamma_{electric}}. \tag{23.3}$$

The optimal specific impulse does not depend on the thrust and is mostly determined by operation time of the electric rocket engine, or simply on the flight time of the spacecraft. For example, the flight to Mars (when $\tau \approx 1$ year) requires an optimum $I_{sp} \approx 40\,km\,s^{-1}$. The parameters of electric rocket engines are expected to be improved soon to the values $\gamma_{electric} = 0.02\,kg\,W^{-1}$, $\gamma_{fr} = 0.1$, and $\eta_{thrust} = 0.8$. For the improved parameters, the same flight to Mars requires an optimum of $I_{sp} \approx 50\,km\,s^{-1}$. Flights to Jupiter and Saturn ($\tau \approx 3$–4 years) require an optimum of $I_{sp} \approx 90$–$100\,km\,s^{-1}$. Optimal specific impulse of ion or plasma thrusters for short near-Earth flights and for corrections of orbits can be estimated as $I_{sp}^{opt} = 15$–$30\,km\,s^{-1}$. The propellant (working fluid) velocities in electric propulsion systems (ion and plasma thrusters) typically expected today are in the range 15–100 $km\,s^{-1}$.

More information on the subject can be found in Grishin (2000); Grishin et al. (1975); Morozov (1978); Tajmar (2004); Jahn (2006); Keidar and Bellis (2018); Herrera (2021); Diaz and Seedhouse (2017).

23.7 Electric Rocket Engines Based on Ion and Plasma Thrusters: Operation of Ion Thrusters, Ion Acceleration Mechanisms, Classification of Major Plasma Thrusters

An **ion thruster** is an electric rocket engine where the propellant (working fluid) is represented by positive ions accelerated in a constant electric field. Ion thrusters were developed historically before plasma thrusters mostly because of the relative simplicity of generation and acceleration of ions in these engines. A schematic of a simple ion thruster is shown in Fig. 23.5. An external electrode and a neutralizer (cathode compensator) are grounded on the spacecraft body. An ion source is sustained under positive potential. The accelerating electrode is sustained under negative potential, and the external electrode has a zero potential (grounded). Such distribution of electric potential prevents the neutralizer electrons from getting to the accelerating electrode and the ion source. The accelerated ions leave the thruster and go out into the environment, entraining electrons from the neutralizer along with them. Therefore, although the accelerated ions leave the thruster, the electric potential of the spacecraft does not grow, and the space charge of the ion beams is neutralized. The ion thrusters can be also based on different methods of ion generation, like gas-discharge ion sources, where ionization occurs due to electron impact, as well as ion generation based on surface ionization. Also, electric rocket engines like the ion thrusters can be organized by accelerating not ions but nanometer-size charged liquid droplets. Such electric engines are referred to as the **colloidal thrusters**.

The acceleration zone of ion thrusters lies between the ion source and the external electrodes and is filled with ions of the same polarity, which creates a significant space charge limiting the operational characteristics of the electric rocket engines. Better characteristics can be achieved in the **plasma thrusters** using quasi-neutral plasma with a high ionization degree (usually 90–95%) as a propellant. Electrons neutralize the positive space charge of ions in the accelerating channel and thus improve the characteristics of the thrusters. Both electrons and ions are accelerated in plasma thrusters to velocities of around 100 $km\,s^{-1}$. Most of the energy in this case is accumulated obviously in ions; thus, the major task of the plasma thrusters, like that of ion thrusters, is ion acceleration. The motion of an ion in the completely ionized rarefied plasma can be described (see Fridman and Kennedy 2021) as:

$$M\frac{d\vec{v}_i}{dt} = e\vec{E} - \frac{\nabla p_i}{n_{e,i}} - \frac{1}{\sigma}\vec{j} + e[\vec{v}_i \times \vec{B}], \tag{23.4}$$

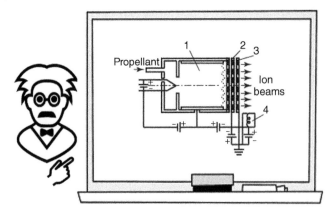

Figure 23.5 General schematic of a simple ion thruster: (1) source of ions; (2) accelerating electrode; (3) external electrode; (4) neutralizer or cathode compensator.

where, $\vec{v}_i$, e, and M are the velocity, charge, and mass of an ion; $n_{e,i}$ and σ are the number density and conductivity of plasma; $\vec{E}$ and $\vec{B}$ are the electric and magnetic fields; ∇p_i is the gradient of the ionic pressure; and $\vec{j}$ is the current density. The first three forces on the right-hand side of the ion-dynamic equation can increase the ion energy and, in these terms, to accelerate the ion. Acceleration of the ions by the force $e\vec{E}$ is called "electrostatic"; by the force $\nabla p_i / n_{e,i}$ is called "gasdynamic" or "thermal"; and by the force corresponding to $\frac{1}{\sigma}\vec{j}$ is called "ohmic" or the acceleration provided by the "electronic wind". Considering the different mechanisms of ion acceleration, the plasma thrusters are usually subdivided into three major categories: electrothermal plasma thrusters, electrostatic plasma thrusters, and magneto-plasma-dynamic plasma thrusters, which are going to be discussed below.

23.8 Electrothermal, Electrostatic, Magneto-plasma-dynamic, and Pulsed Plasma Thrusters

In the **electrothermal plasma thrusters**, plasma as a propellant is generated in a thermal gas discharge (usually either thermal arc discharge or thermal RF inductively coupled plasma discharge) and then accelerated in Laval nozzle similarly to the case of conventional rocket engines using liquid fuel. An advantage of electrothermal plasma thrusters with respect to conventional liquid fuel propulsion systems is due to the possibility of using the light gas, hydrogen, as a propellant (hydrogen velocity is higher at the same temperature), and due to the possibility of heating gas in thermal discharges to extremely high temperatures, which leads to high exit velocities of the propellant and therefore high values of the specific impulse.

Ion acceleration in the **electrostatic plasma thrusters** is provided by an electric field generated in plasma in a circular gap between poles of a magnet due to the sharp decrease of the transverse electron mobility in the magnetic field (see Fig. 23.6a). The magnetic field in the thrusters is chosen to magnetize plasma electrons but leave the ions non-magnetized. When the anode has a typical potential (0.3–3 kV), and the magnet and external cathode compensator are sustained under zero potential, a symmetrical discharge occurs in the circular gap with strong axial electric field. Drift of the magnetized electrons is in the tangential directions, while ions are not constrained by the magnetic field and are accelerated in the axial directions. Because electrons are trapped in the layer of ion acceleration and they are not able to exit the propulsion system together with ions, a cathode compensator should be placed at the exit from the ion acceleration channel. In the cathode compensator, some fraction of electrons compensates the ion beam similar to the case of ion thrusters. Another fraction of electrons goes to the ion-acceleration channel and sustains the electric discharge. Neutral atoms of the propellant are injected into the accelerating channel and are ionized there by the rotating magnetized electrons. Collisions of electrons with heavy particles break their tangential drift and result in a shift of electrons in the direction toward the anode providing the axial electron current to the anode. Ions formed by ionization of the propellant (working fluid) are accelerated in the electric field and exit the propulsion system. Thus, electrostatic plasma thrusters have a combined layer for both ionization and acceleration of the propellant. In contrast to the ion thrusters, ion acceleration in the electrostatic plasma thrusters takes place inside the cloud

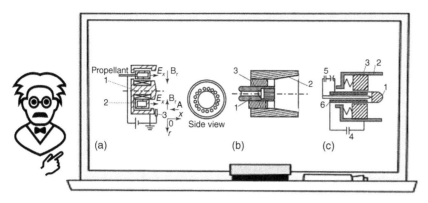

Figure 23.6 Plasma thrusters: (a) electrostatic thruster with closed drift of electrons: (1) magnet; (2) anode/vapor-distributor; (3) cathode-compensator (B magnetic field; E electric field); (b) high-current magneto-plasma-dynamic thruster: (1) thermionic cathode; (2) ring-anode; (3) insulator; (c) erosion pulsed plasma thruster: (1) cathode; (2) anode; (3) propellant; (4) main battery of capacitors; (5) battery of capacitors of ignition system; (6) igniting needle.

of electrons. It compensates space charge of the ions and improves the characteristics of the engine. Electrostatic plasma thrusters can be constructionally organized in two different ways. The so-called stationary thruster has a long ionization and acceleration layer. Another configuration, the so-called thruster with an anode layer, has a narrow ionization and acceleration layer adjacent to the anode.

In **the magneto-plasma-dynamic thrusters**, ions are accelerated by electrons through electron ion friction. The magnetic field in such a thruster can be induced either by its own discharge current or by external magnetic systems. The effectiveness of ion acceleration in the thruster with a magnetic field induced by its own discharge current is higher at higher levels of electron current, which is sustained in this case usually in the range 1–10 kA. Such thrusters, for this reason, are usually referred to as **high-current thrusters**. Electrons in the thrusters leave the system together with ions; therefore, a special cathode neutralizer is not required in these systems. A schematic of a high-current thruster is shown in Fig. 23.6b. The thruster consists of a thermionic cathode, through which vapor propellant comes to the engine partially ionized, an external ring anode with the shape of the de Laval nozzle, and an insulator. A cathode in the high-current thrusters is shorter than an anode. Therefore, plasma confinement to the axis of a thruster can be effectively achieved in these kinds of electric engines. We should also mention a little more sophisticated but deeply theoretically and experimentally investigated **Hall-effect thrusters (HET)**, usually characterized by moderate specific impulse (1600 s). In these thrusters, based on the Hall effect, the propellant is accelerated by electric field. Magnetic field is applied to limit the axial motion of electrons, used then to ionize the propellant, efficiently accelerate the ions, to produce thrust, as well as neutralize the ions in the plume. See for more details Morozov (1978); Keidar and Bellis (2018).

Propellant and electric energy are introduced in the **pulsed plasma thrusters**, not continuously but by separate pulses with a duration in the millisecond range. The pulsed thrusters include capacitors to collect sufficient energy and a system for periodic ignition of the discharge. The most effective of the pulsed thrusters are so-called **erosion thrusters**, see Fig. 23.6c. Propellant is in the thruster between the anode and the cathode in the form of solid or paste-like dielectric and works also as an insulator. Operation of the pulsed plasma thruster is initiated by an additional capacitor, generating initial plasma going into the accelerating space between the major cathode and anode, where the voltage and energy of the major bank of capacitors are ready to be discharged. Plasma generation in the vicinity of the dielectric initiates surface discharge and evaporation of the dielectric. Interaction of the discharge current with its own tangential magnetic field leads to plasma acceleration in the coaxial system. When the major bank of capacitors is discharged, formation of the vaporized propellant stops automatically, which terminates a pulse.

23.9 Plasma Aerodynamics, Plasma Interaction with High-Speed Flows and Shocks

Plasma aerodynamic effects, high speed plasma flows, and shock waves in plasma have been already discussed in Lecture 8. In addition to that, *weak ionization of gases can result in large changes in the standoff distance ahead of a blunt body in ballistic tunnels, in reduced drag, and in modifications of traveling shocks.* These interesting plasma effects are of great practical importance for high-speed aerodynamics and flight control. Energy addition to the flow results in an increase in the local sound speed that leads to modifications of the flow and changes to the pressure distribution around a vehicle due to the decrease in local Mach number. Intensive research has been focused on determining strong plasma effects influencing high-speed flows and shocks, which are the most attractive for applications. It has been demonstrated that, although the heating in many cases is global, experiments with positive columns, DBDs, and focused microwave plasmas can produce localized special energy deposition effects attractive for energy efficiency in flow control. Numerous schemes have been proposed for modifying and controlling the flow around a hypersonic vehicle. These schemes include novel approaches for plasma generation, magnetohydrodynamic (MHD) flow control, and power generation. Two major physical effects of heat addition to the flow field lead to drag reduction at supersonic speeds: (i) reduction of density in front of the body due to the temperature increase (that assumes either constant heating or a pulsed heating source having time to equilibrate in pressure before impinging on the surface) and (ii) coupling of the low-density wake from the heated zone with the flow field around the body, which can lead to a dramatically different flow field (the effect is very large for blunt bodies, changing their flow field into something more akin to the conical flow). Efficiency of the effect grows significantly with Mach number.

 Klimov et al. (1982) were first to investigate the **propagation of shockwaves in low-pressure plasmas**, and they found that the observations could not completely be explained via plasma induced thermal gradients. A pulsed electric discharge was used to launch the shockwave at one end; a fast-response piezoelectric transducer for measuring the shock pressure was placed at the other end. Shock velocity in the system with a neutral gas was $740\,\mathrm{m\,s^{-1}}$ and reached $1200–1300\,\mathrm{m\,s^{-1}}$ velocity $1–3\,\mu s$ after entering the plasma. The shock front broadened from a temporal width in the neutral gas of $2\,\mu s$ to a width of $8–10\,\mu s$ in the plasma. The calculated increased velocity due to the temperature gradient at the plasma boundary was only $900\,\mathrm{m\,s^{-1}}$ versus the measured $1200–1300\,\mathrm{m\,s^{-1}}$. The effect can be explained by the release of stored vibrational energy. Further experiments (Klimov and Mishin 1990) provide additional evidence that the *plasma influence on the shockwave structure and velocity is essentially stronger than expected if plasma affects only the thermal gradients.*

Significant **plasma aerodynamic effects in the ballistic range tests** were also observed, see for example Mishin (1997). Thus, the flight of simple-shaped gun launch projectiles (spheres, cones, and stepped cylinders) through steady DC discharges has been analyzed. The shock standoff distance significantly increases when the sphere is flown through a steady DC discharge region. For projectile velocities in the range of $1200–2400\,\mathrm{m\,s^{-1}}$, the shock standoff distance in plasma was observed to increase about twice. Anomalous shock standoff distances were observed for spheres within the discharge zone at distances far from the thermal non-uniformities near the entrance to the plasma. Time-of-flight measurements have been made to determine the drag coefficient of the sphere. The relevant measurements indicate a substantial decrease in the drag coefficient at subsonic speeds, with a small increase at supersonic speeds. Analysis of the plasma aerodynamic effects, observed in ballistic range tests, and especially on the mechanisms leading to the anomalous shock standoff distances can be found, for example, in Serov and Yavor (1995); Lowry et al. (1999); Candler et al. (2002). The gas discharges used in ballistic ranges result in significant energy tied up in the vibrational modes; therefore, the flow field around the sphere can be altered significantly if the energy relaxes back into translational and rotational modes within the shock layer. The role of vibrational energy relaxation in the shock layer has been analyzed, and it has been shown that its contribution to the anomalous shock standoff distances is usually less significant than that of thermal non-uniformities.

Shockwave propagation induces polarization of plasma and leads to specific electric effects. Some flow kinetic energy can be converted this way into electrical energy through the charge separation. A traveling strong double layer can be formed near the shock front in the case of shock propagations through a strongly nonequilibrium ($T_e \gg T_0$) plasmas including those of glow and other cold discharge. The existence of a strong double layer with a significant potential jump leads to local electron heating, excitation, ionization, and local gas heating. The last effect causes shock front broadening and velocity changes in addition to the effect of overall discharge heating. More details on the relevant aerodynamic effects in specific discharges including filamentary and non-filamentary RF, microwave, electron-beam, and pulsed discharges are presented in several reviews including Fridman (2008).

The **reduction of drag on a body at supersonic speeds** can also be achieved by injection of plasma from an onboard generator (see, for example, Ganiev et al. 2000). Plasma in this case can be generated by a thermal arc providing plasma heating to about $6000\,\mathrm{K}$ prior to its injection through a Laval nozzle at the leading edge of a $30°$ cone cylinder model. Decreases in the drag on the model between two- and fourfold have been achieved at speeds between

Mach 0.4 and 5.0. The reduction of drag coefficient is due to a modification of the flow field over the cone with the plasma jet injection resulting in an effectively lower-angle cone, see review of Bletzinger et al. (2005). More information on the subject can be found in a special issue of the Journal of Propulsion and Power, vol. 22, edited by Macheret (2006).

23.10 Plasma Effects on Boundary Layers, Aerodynamic Plasma Actuators, Plasma Flow Control

Surface discharges permit rapid and selective heating of a boundary layer at or slightly off the surface, as well as creating plasma zones for effective MHD interactions, see for example, Shibkov et al. (2001); Leonov et al. (2001); Grossman et al. (2003). The surface discharges for supersonic and hypersonic aerodynamic applications have been effectively organized using surface microwave discharges, DC, and pulsed discharges between surface-mounted electrodes and internal volumes and sliding discharges, see Section 13.7. Leonov et al. (2001) investigated the use of a surface discharge to reduce the friction drag force of a flat plate. At transonic flow speeds with a local static pressure of 500 Torr, application of the 15 kW discharge results in gas heating sufficient to produce a temperature of about 1500 K. A portion of the flat plate downstream of the discharge was isolated and mounted on a force balance. The drag measurements provided evidence that the friction drag downstream of the discharge is significantly reduced 20–30%. Thus, the surface discharges are effective for adding thermal energy to a boundary layer. For speeds below approximately Mach 6, the creation of a hot layer stabilizes the boundary layer. Above Mach 6, alternative transition modes become dominant.

 Asymmetric DBDs (see Section 13.7), with one electrode located inside (beneath) the barrier and another mounted on top of the dielectric barrier and shifted aside, stimulate intensive **"ion wind" along the barrier surface, influence boundary layers**, and can be used as aerodynamic actuators (see for example, Boeuf et al. 2007; Likhanskii et al. 2007). Pioneering experiments in this direction were carried out by R. Roth and his group. Aerodynamic plasma actuators have lately attracted interest in numerous applications for flow control above different surfaces (aircraft wings, turbine blades, etc.) The ion wind pushes neutral gas with not very high velocities (typically 5–10 m s^{-1} at atmospheric pressure). However, even those DBD-stimulated gas velocities near the surface can make significant changes in the flow near the surface, especially when the flow speed is not fast. The major contribution to the ion-wind drag in DBD plasma actuators is mostly due not to streamers themselves but to the ion motion between the streamers when ions move along the surface from one electrode to another, similarly to the case of corona discharges. During the phase when the electrode mounted on the barrier is positive, the positive ions create the wind and drag the flow. During the following phase when the electrode mounted on the barrier is negative, the negative ions formed by electron attachment to oxygen (in air) make a significant contribution to total ion wind and drag the flow in the same direction (Likhanskii et al. 2007). The nano-second pulse DBD permitted A. Starikousky and his group to achieve *strong effect at very high speeds (M = 0.7). The effect is mostly due in this case not to ion wind but to fast heating of the boundary layer formation of a shock wave.*

Plasma aerodynamic flow control can be applied for controlling shock positions within an inlet to reduce or eliminate the complexity of a variable geometry inlet. The plasma flow-control techniques are usually based on the use of Lorentz force while minimizing the role of heat addition.

MHD inlet flow control has been suggested for use in control of the captured flow rate, shock positioning, compression ratio, and local shockwave/boundary layer interactions for scramjet inlets. Generation of adequate electric conductivity is a requirement for the development of an MHD inlet flow control. For blunt vehicles, significant electron number densities can occur naturally in the viscous shock layer, but the scramjet inlets are long and slender with small flow deflection angles; therefore, the electric conductivity for MHD interaction requires an additional source of ionization. The most energy-efficient technique for this purpose is the use of an electron beam. The electron beam can be operated at moderate energy fluxes with injection through foil windows or at high energy fluxes with injection through aerodynamic windows. Analysis of the inlet performance shows that MHD flow control can be used to modify the shock structure in an inlet, although operation in the flow expansion mode is not efficient. Macheret et al. (2000) have examined the feasibility of using an external MHD system with electron-beam ionization for control of a scramjet inlet. They show that the power required for ionization is less than the power generated within the MHD system for inlet operation at Mach 8 and 30 km altitude operating conditions. Plasma-initiated MHD effects can also be used to reduce heat transfer. Analysis of the hypersonic flow over a hemisphere shows that the stagnation-point heat transfer could be reduced by approximately 25% at Mach 5 with a magnetic interaction parameter of 6. Reduction in heat transfer at a blunt leading edge can be achieved using current flows along the leading edge to produce a circumferential magnetic field. The resulting Lorentz force leads to a deceleration of the flow approaching the leading edge.

Another concept is **onboard power extraction from a scramjet flow**, which has been considered both as energy bypass and as auxiliary power generation, see Macheret et al. (2000, 2001a,b). The energy bypass involves extraction of flow energy upstream of the combustion using an MHD power extraction system with re-introduction of the power into the propulsive stream downstream of the combustor using an MHD flow acceleration system. Energy bypass results in the possibility of organizing the combustion process at a lower effective velocity. Auxiliary MHD power generation can be used for plasma aerodynamic applications, plasma-assisted combustion, or nonequilibrium ionization systems.

23.11 Plasma Power Electronics: Magneto-hydrodynamic (MHD) Generators, Plasma Thermionic Converters

Low-temperature plasma can be effectively applied to produce electricity from high temperature heat energy "directly," without any special mechanical devices, and without any rotating or linearly moving mechanical parts such as rotors, pistons, and so forth. The major plasma power electronic systems of the direct production of electricity are MHD generators and thermionic converters. The MHD generators use fossil fuels as their primary energy source and create thermal plasma through combustion. In the future, nuclear energy is to be used to produce plasma in MHD generators. Plasma flow moving in an MHD channel works against the magnetic field generating electricity. Thermionic converters (or thermo-electronic converters) are based on electric current generation by means of the thermionic emission from a hot cathode (called an emitter) through a vacuum gap to a colder electrode (called a collector). An interest to thermionic converters is partially due to their application in nuclear power systems for autonomous electricity production without any mechanical devices and moving parts.

Application of **plasma as a working fluid in MHD generators** permits conversion of both thermal and kinetic energy of the working fluid into electric energy. Plasma motion with velocity u in magnetic field B results in the formation of electromotive force $\vec{u} \times \vec{B}$. When electrodes connected to an external electric circuit are in the plasma flow, the electromotive force $\vec{u} \times \vec{B}$ leads to electric current density $\vec{j}$. A schematic of a channel of MHD generator is shown in Fig. 23.7. In contrast to a conventional turbine-based electric generator, the MHD generator has

Figure 23.7 Schematic of an MHD generator channel: (1) electrode wall; (2) insulator wall; (3) magnetic coil; (4) load.

no moving parts, which permits a significant increase of the working fluid temperature. Depending on the type of working fluid, MHD generators can be divided into two classes: open cycle and closed cycle. Open-cycle MHD generators use high-temperature combustion products as the working fluid. Closed-cycle MHD generators operate with inert gases (for example, in the case of nuclear energy sources). With respect to the method of induction of electric field, MHD generators can be divided into two other classes: conductive and inductive MHD generators. Conductive MHD generators are characterized by potential electric field; and electric current generated in a working fluid goes through an external load (see Fig. 23.7). Although conductive MHD generators can operate with AC and DC, they always include electrodes. The electric field in inductive MHD generators is induced by alternating magnetic field. The electric current generated in the system can be closed inside of the working fluid.

The **electric conductivity of plasma as the MHD working fluid** is one of the critical parameters of MHD generators. To increase the electric conductivity, small amounts of alkaline metals (K, Cs, etc.), characterized by low ionization potentials, are usually added to the working fluid. Electric conductivity of thermal plasma exponentially depends on temperature. In the case of molecular gases, plasma in MHD generators is not far from the thermodynamic quasi-equilibrium, which imposes strong limitations on operational temperature in the MHD generators to reach a necessary level of electric conductivity. For example, MHD generators, using the products of combustion of natural gas with alkaline-metal additives as working fluid, operate with exit temperatures always exceeding 2000 K. The operating temperature of MHD generators can be decreased by using plasma of inert gases as a working fluid. Such a situation takes place in closed-cycle MHD generators (when heating of the working fluid is provided, for example, from a nuclear power source). A large contribution to the ionization of inert gases in this case is made by nonequilibrium ionization by electric fields present in the system.

The **thermionic converters** are other systems for direct conversion of heat energy to electricity. They can be combined with nuclear and solar power sources, which attracts particular interest to their application in spacecraft. A schematic of thermionic converter is shown in Fig. 23.8. A thermionic converter is a flat or cylindrical diode consisting of a hot cathode to emit electrons and a cold anode to collect electrons. If the work function of a cathode, U_C, exceeds that of an anode, U_A, then connecting the electrodes (and equalizing the Fermi levels of cathode and anode) leads to the formation of an electric field in the interelectrode gap, which pushes electrons from cathode to anode (see Fig. 23.9a). In a vacuum thermionic converter, the electron current remains constant until the load voltage V_L does not exceed the contact difference of potentials: $V_L \leq U_C - U_r$. At larger load voltages, the external circuit current decreases, because a fraction of the emitted electrons returns via the electric field back to the cathode. Ideal current-voltage characteristics of a vacuum thermionic converter are shown in Fig. 23.9b. Generated power is larger at the highest values of cathode work function and the lowest values of anode work function. To provide high thermionic current from a cathode with high work function, the cathode temperature should be sufficiently high. Usually, such metals as W, Mo, Re, Nb, and their alloys, characterized by high melting temperatures, are used for manufacturing the high-temperature cathodes. Work functions of the cathode are usually in the range 4.4–5.1 eV. They operate at temperatures exceeding 2800 K to provide a thermionic current density on the level of 10 A/cm². Anodes are usually coated by special thin films that reduce their work function.

An important role in the thermionic converters is played by the addition of cesium, which is characterized by very low work function (1.8 eV) and ionization potential (3.89 eV); it makes a thermionic converter works as a plasma-chemical reactor. Cesium plays a multi-functional role in thermionic converters. First, adsorption of cesium

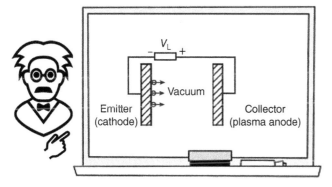

Figure 23.8 Simple illustration of a thermionic convertor.

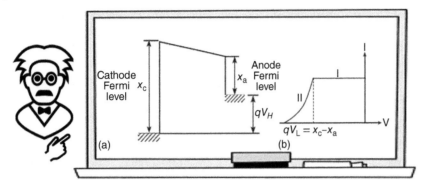

Figure 23.9 Physics of thermionic convertors: (a) electron energy diagram; (b) ideal current-voltage characteristic.

on the high-melting-temperature cathode surface changes operation of the cathode in the so-called film-cathode regime. In this regime, the work function of the cathode can be controlled by flux of the Cs atoms to the surface and the surface temperature. The cathode work function in the film-cathode regime can be varied from the highest value, corresponding to the work function of the initial cathode material, to the lowest value, corresponding to the work function of cesium. A second effect of Cs addition is related to a decrease of anode work function to the low values required for high effectiveness of thermionic converters. The reduction of the anode work function occurs due to the formation of a Cs film on the surface of a cold anode. Finally, the third positive effect of cesium is related to the generation of a significant concentration of Cs^+ ions in the gap, which results in neutralization of the electron space charge there and growth of the converter effectiveness. Cesium atoms have very low ionization potential (3.89 eV) and are easily ionized in the gap by different ionization mechanisms.

23.12 Gas-discharge Communication and Special Devices, Plasma Metamaterials

The application of electric discharges as **switches and communication devices** has already been briefly discussed in Section 11.12 (see Fig. 11.19) in relation to gliding-arc discharges. Gas discharges are also widely applied as **pulsed commutation devices**, specifically for the generation of pulses with current amplitude exceeding 100 A and pulse voltages from several to hundreds of kilovolts, see, for example, Bochkov and Korolev (2000). The gas-discharge commutation devices can be classified, either according to gas pressure or to their position on the Paschen curve. Thyratrons, tasitrons with hot or cold cathodes, and pseudo-spark discharges operate at pressures of 1–100 Pa, which corresponds to the left branch of the Paschen curve. Vacuum commutating discharges with plasma generation due to erosion of electrodes and discharges with liquid (usually Hg) cathode can also be classified in the lower-pressure group. In contrast to that, commutating spark discharges can be classified as high-pressure gas-discharge commutation devices. These commutation devices operate at pressures from 100 Pa to several atmospheres in conditions corresponding to a minimum of the Paschen curve or its right-hand branch.

 Very interesting applications are related to **plasma behavior as a metamaterial**. Generally, the metamaterials are those having properties not found in nature. They can be made from assemblies of multiple elements from composite materials such as metals and plastics, as well as plasmas. The metamaterials are usually arranged in repeating patterns, at scales that are smaller than the wavelength of the phenomena they influence. *The plasma metamaterials derive their properties not from the properties of the base materials, but from their newly designed structures. Their precise geometry and arrangement give them their smart properties capable of manipulating, for example, electromagnetic waves. Those that exhibit a negative refraction index for specific wavelengths have been the focus of a large amount of research. These materials known as the* **negative-index metamaterials** *often including plasma elements open possibilities for the* **metamaterial cloaking**, *which is the use of metamaterials as the "invisibility cloak".* When we form a structure of plasmas distributed in a certain space in which electromagnetic waves propagate, such plasma structure serves as a medium very different from a homogeneous bulk plasma. It is also possible to enhance or generate novel functions of the plasmas when we add other structural materials such as functional components. Relevant material properties of such a medium including permittivity, permeability, and conductivity demonstrate extraordinary functional effects that arise from the synthesis of the structure. Details on this subject including fundamental of electromagnetic wave propagation in such specific plasma structures, new functions of the plasmas as metamaterials, including a photonic-crystal-like behavior, a negative refractive index state, and a nonlinear bifurcated electric response can be found in the review of Sakai and Tachibana (2012).

23.13 Plasma in Lasers: Classification of Lasers, Inversion Mechanisms, Lasers on Self-limited Transitions, Ionic Gas-discharge Ar and He–Ne Lasers

Strongly nonequilibrium gas-discharge plasmas ($T_e \gg T_0$) overpopulate highly excited states of atoms, molecules, and ions and, therefore, can be used as an active medium for lasers. The lower working level, however, is usually also excited in the plasma system; therefore, lasing requires special plasma-kinetic mechanisms to provide effective inversion in the population of upper and lower working levels. Lasers in gas discharges can be classified in different ways. For example, lasers can be classified according to the type of working transition into those on transitions between electronically excited states and those on transitions between vibrationally excited states. When laser generation takes place during ionization (in an active discharge zone), they are usually called gas lasers. If generation occurs in the recombining plasma (that is not in an active discharge phase), the lasers are usually referred to as plasma lasers. Let us first discuss the lasers in strongly nonequilibrium gas discharges on the transitions between electronically excited states of atoms, molecules, and ions. The most successful organization of the inversion in lasers on electronic transitions has been achieved in the so-called **plasma-gas lasers on self-limited transitions**. The self-limited transitions are those to a metastable state "a," which is operating as a lower working level of the laser system. The upper working level, "b," is usually a resonant one that can be easily excited in a strongly nonequilibrium plasma ($T_e \gg T_0$). The lasers on self-limited transition efficiently operate on the front of current pulses, in pulse-periodic regimes. The optimal time of the pulse front should be shorter than the inversion time, that is, less than 10^{-8}–10^{-6} s. The pulsed self-limited lasers only provide intensive excitation of upper levels, whereas deactivation of the lower level is not as important because of the pulsed-periodic operation. The average power of these lasers is limited by the pulse-periodic mode of their operation.

Gas-discharge lasers using metal vapors can be considered an example of lasers on self-limited transitions (between different electronically excited states of atoms). The highest average and pulse power have been achieved for the case of copper vapors (lasing wavelengths $\lambda = 510.5$ nm and $\lambda = 578.2$ nm). These lasers can provide an average radiation power exceeding 100 W from one active element. Between other gas-discharge lasers on metal vapors, we can point out those in lead vapor ($\lambda = 722.9$ nm), barium vapor ($\lambda = 1499.9$ nm), europium vapor ($\lambda = 1759.6$ nm), and gold vapor ($\lambda = 627.8$ nm and $\lambda = 312.3$ nm). A gas-discharge laser operating on the vapor of a single metal can generate radiation at two or more wavelengths. The operating temperatures of metal vapor lasers are high. The lasers using copper vapors require temperature around 1500 °C. Lasers based on self-limited transitions of vapors of copper, lead, and gold are widely applied because of their high average power and stable operation in the pulse-periodic regime with pulse frequencies exceeding 10 kHz.

Quasi-stationary inversion mechanisms permit laser generation to be longer than the lifetime of the working levels. It requires not only population of the upper lasing level but also deactivation of the lower working level. To decrease the excitation of the lower working level, the upper level can be excited by resonant energy-exchange collisions with particles of a buffer gas, which are plasma-excited to long-lifetime states. The molar fraction of the buffer gas usually exceeds that of the working gas in order to initially transfer most of the energy of plasma electrons to the buffer. Excitation transfer from the buffer to the working gas can be organized either through energy-transfer relaxation collision of neutral particles or through charge exchange processes. For example, the He–Ne laser is based on excitation of upper levels of Ne in energy-exchange relaxation collisions with metastable electronically excited levels of helium (buffer). Inversion and lasing can also be achieved by the so-called radiative deactivation if radiative decay time of a lower working level is shorter than that of an upper working level. Such type of inversion is not common for most atoms and ions with simple configurations, because the rapidly decaying lower working levels are usually also very effective for excitation. The inversion mechanism based on the radiative deactivation can be achieved, however, in the case of ions with complex configuration.

The radiative mechanism of deactivation of the lower level determines the inversion and operation of the low-pressure **ionic gas-discharge lasers**. Effective lasing has been achieved in numerous ionic transitions of such atoms as Ne, Ar, Kr, Xe, Cl, I, O, N, Hg, C, Si, S, Cd, Zn, and P. The most powerful low-pressure laser today is the argon laser, operating on the radiative mechanism of deactivation of the lower working level. The lifetime of the upper working levels in the Ar laser is about 10^{-8} s; the lifetime of the lower levels having radiative transition to the ground state is about 10^{-9} s, which leads to population inversion and lasing. The Ar lasers usually are organized in low-pressure discharge tubes with current density of $1–2 \, kA \, cm^{-2}$, and energy efficiency of about 0.1%. The highest power of the argon laser in the blue-green range is 0.5 kW. The widely used He–Ne laser operates based on the population inversion sustained by excitation of the upper working levels of neon atoms Ne (4s, 5s) in collisions with helium atoms (buffer) preliminarily excited in plasma to the metastable states He(2^3S) and He(2^1S) by plasma electrons. Inversion in He–Ne lasers is also essentially supported by radiative deactivation of the lower working levels. Lasing is achieved in the system on more than 200 spectral lines; the most used from those are $\lambda_1 = 3.39 \, \mu m$, $\lambda_2 = 0.633 \, \mu m$, and $\lambda_3 = 1.15 \, \mu m$. The excitation of the laser is usually organized in the low-pressure positive column of a glow discharge with typical current densities $100–200 \, mA \, cm^{-2}$. Optimal electron density in the gas-discharge lasers is about $10^{11} \, cm^{-3}$; the typical density of neon is $2 \cdot 10^{15} \, cm^{-3}$ and the typical density of helium is $2 \cdot 10^{16} \, cm^{-3}$.

23.14 Plasma Lasers: Inversion in Plasma Recombination, He–Cd, Penning, and Other Lasers

In most of the gas lasers using transition between electronically excited states considered earlier, lasing takes place during ionization in the active discharge zone. In contrast to that, generation in the so-called **plasma lasers** takes place during plasma recombination. Inversion can be sustained as a quasi-stationary state in the overcooled recombining plasma, when a significant fraction of electrons appear on the upper working level after recombination and sequence of gradual relaxation processes. Effective lasing in this case also requires deactivation of the lower working levels, which can be achieved by means of radiative deactivation, or deactivation in collisions with electrons or heavy particles. Radiative deactivation of the lower working levels takes place in many metal vapor and inert gas lasers. This mechanism is important in sustaining inversion in short-wavelength lasers using transitions of multi-charged ions. When electron density is relatively high ($n_e > 10^{12}–10^{13} \, cm^{-3}$), effective deactivation of the lower working levels can be provided via electron impact. This deactivation mechanism is typical for conditions when plasma recombination processes dominate over ionization. The deactivation of the lower working levels in collisions with electrons is most effective for the levels located close to the ground state. Deactivation of lower working levels and inversion population in dense gases can be provided by collisions with heavy neutral particles. A significant role can be played in such laser systems by Penning deactivation. Electronically excited particles in these systems, corresponding to the lower working level, lose their activation in ionizing collisions with particles of a specially added gas characterized by low ionization potential. The Penning deactivation effect is limited in the active phases of conventional gas discharges because significant presence of additives with low ionization potential leads to its intensive ionization and restriction of ionization of the major gas. Effective Penning deactivation is organized in lasers using transitions of neon and helium.

Plasma He–Cd lasers operate at high pressures with inversion based on radiative deactivation of the lower working level. When excitation of plasma laser is provided by electron beams or by charged products of nuclear reactions, effective lasing has been achieved on the following electronic transitions of the singly charged cadmium ions: $4d^9 5s^2 \, ^2D_{5/2} \rightarrow 4d^{10} 5p^2 P_{3/2} (\lambda = 441.6 \, nm)$, $4d^{10} 4f \, ^2F_{5/2,7/2} \rightarrow 4d^{10} d^2 D_{3/2,5/2} (\lambda = 533.7 \text{ and } 537.8 \, nm)$. The recombination regime lasing (radiative deactivation of the lower working level) on the cadmium ion transition $4d^9 5s^2 \, ^2D_{3/2} \rightarrow 4d^{10} 5p^2 P_{1/2} (\lambda = 325.0 \, nm)$ in the He–Cd mixture of high pressure has been achieved only by nanosecond electron-beam excitation and on the front of the microsecond electron-beam excitation. Quasi-stationary generation is achieved in the system only by adding to the mixture a small amount of electronegative CCl_4 gas. Another recombination-regime lasing (radiative deactivation of lower working level) on the cadmium ion transitions $\lambda = 325.0 \, nm$ and $\lambda = 853.1 \, nm$ in He–Cd mixture at high pressure has been achieved by nuclear excitation with fragments from nuclear fission of ^{235}U. The generation power here is not high (36 W), and energy efficiency is also low (0.02%). The highest generation power (1 kW) and energy efficiency (0.4–0.7%) have been achieved by nuclear excitation in the quasi-stationary regime at wavelength $\lambda = 441.6 \, nm$.

Penning plasma lasers, operating at high pressures and based on the Penning deactivation of lower working levels, have been organized in mixtures of inert gases. The lasers provide high power generation in the visible range with energy efficiency below 1%. The Ne Penning laser is generated in He–Ne–Ar, He–Ne–Ar–H$_2$, Ne–Ar, He–Ne–Kr, He–Ne–Xe, Ne–Kr, He–Ne–H$_2$, and Ne–H$_2$ mixtures using electronic transition with wavelengths 585.3, 626.7, 633.4, 703.2, 724.5, and 743.9 nm. Excitation of the mixtures in the Ne Penning laser can be organized for use in electron beams, special electric discharges, and nuclear excitation. A penning plasma laser using electronic transition of He ($\lambda = 706.5$ nm) has been organized in the afterglow of high-voltage (50–170 kV) pulsed (10–20 ns) discharge in an He–H$_2$ mixture with pressure 15 Torr. The lasing duration in the system is around 1 μs. Using helium and neon excited in a discharge with admixtures of NF$_3$ results in laser generation of an electronic transition in the helium ($\lambda = 706.5$ nm) and neon ($\lambda = 585.3$ nm). The addition of electronegative NF$_3$ gas to the plasma laser system provides effective deactivation of the lower working level and increases the effectiveness of the excitation.

Among the **dense gas plasma lasers**, lasers using electronic transitions of xenon atoms in Ar–Xe, He–Ar–Xe, and He–Xe mixtures have the lowest excitation thresholds and highest energy efficiency. The energy of a radiation pulse in this case reaches several hundreds of Joules when the lasing is initiated by an electron beam. The high-pressure Xe plasma laser intensively radiates on wavelengths 1.73, 2.03, 2.63, 2.65, 3.37, and 3.51 μm. Maximum energy efficiency of the Xe laser is achieved at atmospheric pressure by electron-beam excitation in the mixture; Ar:Xe = 200:1 is 4.5%. The interest in **plasma lasers using electronic transitions of multi-charged ions** is due to the high energy of the transitions and possibilities of achieving effective lasing at very low wavelengths. Effective lasing in the short-wavelength range is very challenging because spontaneous radiation from the upper working levels significantly grows with frequency, whereas cross sections of the induced photo-transitions decrease with frequency (see Fridman and Kennedy 2021). Examining the lasing based on the transitions of multi-charged ions, we can point out the powerful amplification of the Ne-like ion SeXXV (wavelengths 20.63 and 20.93 nm). The radiation energy in the pulse is about 3 mJ, and radiation power is several megawatts.

23.15 Molecular Lasers on Vibrational-rotational Transitions, CO$_2$, and CO Lasers, Excimer Lasers, Chemical Lasers

Lasing based on the vibrational–rotational transitions in the IR and submillimeter range has been achieved using gas-discharge excitation of different molecular gases, including CO$_2$, CO, N$_2$O, SO$_2$, CS$_2$, COS, NH$_3$, HCN, ICN, H$_2$O, and C$_2$H$_2$. An advantage of such lasers is since for many molecules most of the discharge energy can be transferred to molecular vibration (see Section 5.1), which leads to their high energy efficiency and high lasing power. The **gas-discharge CO$_2$ laser** is particularly advantageous from this point of view; its efficiency can exceed 10%, and its power in a continuous regime is the record high and can exceed 100 kW. CO$_2$ lasers, both in continuous and pulsed regimes, are usually operated in CO$_2$–N$_2$–He mixture (typical composition 1:1:8). Lasing occurs on the transition $00^01 \rightarrow 10^00$ from more energetic asymmetric vibrations of CO$_2$ to a lower-lying energy level corresponding to the deformational oscillations of CO$_2$. Nitrogen molecules (as well as CO$_2$ molecules) are very effectively vibrationally excited by direct electron impact in strongly nonequilibrium discharges. The vibration of N$_2$ molecules is close to resonance with the asymmetric oscillation mode of CO$_2$ and effectively transfers energy there through vibrational–vibrational (VV) exchange. The N$_2$ molecules accumulate most energy and then transfer it to the lasing asymmetric oscillations of CO$_2$. At optimal plasma parameters, more than 80% of the total discharge energy is transferred specifically to vibration of N$_2$ and the asymmetric mode of CO$_2$, which results in the high energy efficiency and power of the CO$_2$ lasers. Atomic helium increases heat conductivity, decreases gas temperature, and deactivates the deformational CO$_2$ oscillations, which increases the lasing efficiency. Conventional gas-discharge CO$_2$ lasers are performed in low-pressure (3–10 Torr) DC discharges; but also, can operate in pulsed regime at high pressures, up to 100 atm. Gas temperature in optimal lasing conditions should not exceed 600 K. *The energy efficiency of CO$_2$ lasers can be very high, up to 30%.*

CO lasers operate in CO–He or CO–N$_2$–He mixtures using the vibrational–rotational transitions. The Treanor distribution with overpopulation of highly excited vibrational states can be organized in strongly nonequilibrium CO gas discharges $T_e \gg T_v \gg T_0$ as well as by supersonic expansion (**gas-dynamic lasers**, see also Section 17.11). The Treanor effect is the major kinetic mechanism of the vibrational population inversion and lasing in the CO–He and CO–N$_2$–He mixtures. Generation occurs usually from several vibrational transitions, $v \rightarrow v-1$ ($v = 6$–15), with wavelength $\lambda = 5$–$5.8\,\mu m$. Continuous CO lasers usually work in low-pressure strongly nonequilibrium discharges in mixtures of CO:He = 1:10/1:30 with parameters close to those of CO_2 lasers. *The energy efficiencies of CO lasers can be even higher than those of the CO_2 lasers.*

The **excimer lasers** are based on chemical compounds stable in an electronically excited state and easily dissociating in a ground state called "exciplex" (short for "excited complex"). An exciplex built from identical atoms or fragments is referred to as an excimer. The term excimer is often generalized to any kind of dimers stable only in electronically excited states. The exciplex molecules used in lasers usually contain atoms of inert gases, which do not form stable compounds in a ground state (with rare exceptions like XeF_2). The excimer lasers radiate using photo-dissociative electronic transitions from a stable electronically excited state to a dissociating ground state. The spectral width of photo-dissociative transitions is large, which causes major difficulties with lasing in these systems. Electron-beam excitation of liquid xenon results in the amplification of a photo-dissociative transition, which demonstrates the excimer lasing effect. Electron beams also have been used for laser excitation in compressed inert gases. Effective lasing has been achieved this way for such dimers as Xe_{2*} ($\lambda = 172\,nm$), Kr_{2*} ($\lambda = 146\,nm$), and Ar_{2*} ($\lambda = 126\,nm$). The most effective are the excimer lasers on halides of inert gases: KrF^* ($\lambda = 250\,nm$), XeF^* ($\lambda = 350\,nm$), $XeCl^*$ ($\lambda = 308\,nm$), and ArF^* ($\lambda = 193\,nm$). *Special interest to excimer lasers is due to the high energy density of their radiation (up to $40\,kJ\,m^{-3}$), high lasing energy efficiency (up to 9–15%), and ability to generate radiation over a wide range of wavelengths, from the visible to vacuum ultraviolet (VUV)*, see more in the book Basting and Marowsky (2005).

Population inversion in chemical lasers is due to fast exothermic chemical processes releasing energy into internal degrees of freedom, usually in molecular vibration. For effective lasing, the rate of the chemical reactions should exceed that of the corresponding relaxation processes. The most relevant reactions for this type of lasers are those leading to the formation of hydrogen halides because they release 30–70% of the energy into vibrational energy of the products and create absolute inversion in the population of several vibrational levels. The first chemical laser was organized in an H_2–Cl_2 mixture, where the chain reaction generating vibrationally excited HCl^* molecules ($Cl + H_2 \rightarrow HCl + H$, $H + Cl_2 \rightarrow HCl^* + Cl$) was initiated using pulsed discharges or pulsed photolysis. Most chemical lasers are organized in pulse regimes, although some of those are effective in continuous operation. The most used of the chemical lasers is the **HF chemical laser** ($F + H_2 \rightarrow HF + H$, $H + F_2 \rightarrow HF^* + F$). The chemical lasers are usually organized in a preliminary prepared mixture with total pressure 1–100 Torr (in some cases at much higher pressures). External energy sources, such as pulsed discharges, electron beams, or energetic photon sources, create active chemical species, which are able then to react with a laser-mixture component, producing a vibrationally excited molecular product. Inversion in the vibrational population of the molecules results in lasing. Because the lasing releases in this case "energy of the chemical explosion", *pulse energy and power of the chemical lasers can be extremely high.*

23.16 Plasma Sources of Radiation with High Spectral Brightness, Plasma Lighting: Mercury-containing and Mercury-free Lamps

High spectral brightness radiation is usually defined as radiation characterized by brightness exceeding that of an absolute blackbody with temperature 20,000 K. An absolute blackbody with temperature 20,000 K generates about 85% radiation in the UV and VUV ranges. Thus, radiation sources of high spectral brightness are supposed to radiate in the UV and VUV ranges, usually with high power ($\geq 1\,MW\,cm^{-2}$). The radiation sources of high spectral brightness are usually divided into the following four groups (see Fridman and Kennedy 2021):

1) *Laser sources* are based on the stimulated radiation mechanism and generate monochromatic radiation with low angular divergence, which results in extremely high brightness corresponding to the absolute blackbody temperatures exceeding 10^6 K.
2) *Luminescent sources* are based on the nonequilibrium spontaneous radiation of atoms, ions, and molecules, which are excited by nonequilibrium discharges, electron beams, and so forth (as an example, excimer VUV lamps on the dimers of inert gases excited by electron beams).
3) *Synchrotron (ondulator) sources* are based on the bremsstrahlung radiation of the relativistic electrons in continuous magnetic fields (synchrotron sources) and space-time modulated magnetic fields (ondulator sources).
4) *Thermal plasma sources* are based on the radiation of thermal plasma heated to very high temperatures. Considering their wide area of application, let us consider these high-spectral-brightness radiation sources in more detail.

The **thermal plasma light sources** are effective radiation devices in generating short wavelength (UV and VUV) radiation. Thermal plasma radiation cannot exceed that of the absolute blackbody at the plasma temperature. Therefore, fixing the required spectral emission interval determines the required values of the thermal plasma temperature. Specifically, for effective emission in UV and VUV spectral ranges, the thermal plasma temperature should be in the interval from 2 to 15 eV. Plasma in high-brightness light sources is usually characterized by high electron density (10^{18}–10^{20} cm^{-3}). The following three methods are applied for generation and heating of thermal plasma for high-brightness radiation sources:

1) *Joule (ohmic) heating* is provided by electric current of high amplitude (10^4–10^5 A) and high density (10^4–10^6 A cm^{-2}). This method is the mostly used in the high-brightness radiation sources. To reach the brightness temperature in the range 20,000–40,000 K, the power of thermal discharges with Joule heating is very high, 4–60 MW cm^{-3}.

2) *Plasma-dynamic heating* is based on the thermalization of kinetic energy of fast-moving gas or plasma flow. Advantage of this method is energy transfer between heavy plasma particles, which is more effective at high temperatures than electron–ion energy transfer in the Joule heating.

3) **Radiative heating** is provided by the interaction of highly concentrated radiation fluxes with materials. Plasma, generated in such interaction, can be a powerful source of thermal radiation in the range of VUV and X-rays. Conversion of the powerful coherent laser radiation into high-temperature thermal radiation can reach 50–70% in such systems.

Plasma is crucial today in light sources. It is widely used in both **mercury-containing and mercury-free light sources**, electric lamps. The search for efficient and environmentally friendly substitutes for mercury today is a key issue in many plasma lighting technologies such as automotive lamps and fluorescent lighting systems. As an example, the use of Ne–Xe-based low-pressure discharges is an interesting solution for the excitation of phosphors. The development of a pulse power supply, generating high-repetition current pulses in the microsecond time range, allows a three- to fourfold enhancement of both the luminous flux and the efficiency of mercury-free plasma lamps to be achieved (Robert et al. 2005). Thus, the development of mercury-free plasma lamps is quite successful.

Overwhelming environmental effects from the use of electric lighting are from the power generation over the life of the lamps. These effects depend on the power source and include atmospheric emissions of mercury, carbon dioxide, nitrous oxide and sulfur dioxide, noise, and radioactive waste. Efficient lighting systems that use mercury-containing lamps will typically reduce these effects by at least 75% because of their lower power requirements. Power generation effects dwarf any environmental emissions which might occur during the manufacture or proper disposal of mercury-containing lamps. Generally, mercury-free electric lamps cannot be substituted for mercury-containing lamps because of incompatibilities of light output, shape, color, life, electrical characteristics, and excessive heat, or because their increased energy consumption may violate energy codes and overload electrical circuits. Let us shortly discuss some mercury-containing and mercury-free specific lighting sources:

1) *Fluorescent lamps.* No viable replacement has yet been discovered for mercury in general-purpose fluorescent lamps. Mercury-free xenon-based fluorescent discharges are available in a flat-panel format, suitable for backlighting of liquid crystal displays, but the efficiency is approximately 30% of a normal mercury-based fluorescent lamp.

2) *High-intensity discharge (HID) lamps.* HID lamps are used in street lighting, floodlighting, and industrial and some commercial applications. There are better prospects for the mercury-free HID lamps. Mercury-free high-pressure sodium lamps are available at powers of 150 W and more. This has been achieved by reengineering the arc tube geometry and fill pressure. Metal halide lamps without mercury present a greater challenge. These lamps may not be a "screw-in replacement" for existing types. The high-pressure sulfur lamp is fundamentally mercury-free but is unstable and requires forced cooling. Lamps which have been marketed so far are high wattage (at least 1 kW), and they require coupling to a lighting distribution system such as a light pipe. The overall system efficiency is lower than an equivalent fluorescent or HID system.

3) *Low-pressure sodium lamps.* These types of lamps are mercury-free plasma light sources. Although very efficient in photometric terms, their visual efficiency in typical outdoor (street lighting) applications is less than that of other lamps. All colors are rendered in shades of brown or gray, making recognition of people and vehicles very difficult. The lamp contains sodium in sufficient quantities to fail tests for reactivity and ignitability.

4) **Light-emitting diodes (LEDs)**. LEDs are mercury-free somewhat plasma related lamps that have long lifetimes and high energy efficiency. They are very bright and lend themselves to very wide use, see more on the subject in the books of Held (2016); Li and Zhang (2019).

23.17 Plasma Display Panels and Plasma TV

The Plasma TV is not very large plasma application today, but surely the most publicly known. The flat plasma display is a competitor among several flat-panel display technologies, competing for the high-definition monitor's market. The strongly nonequilibrium plasma in the display panels is generated by tiny surface DBDs. The plasma display panel (PDP) is essentially a collection of numerous very small DBD lamps, each a few tenths of a millimeter in size. If we look closely, it is easy to distinguish the individual PDP cells – the tiny color elements of red, green, and blue light that together form what is called a pixel. The DBD arranged on the dielectric surface in a PDP cell is illustrated in Fig. 23.10. Like fluorescent lamps, the light we see from the PDP does not come directly from plasma, but from the phosphor coatings on the inside walls of the cells when they are exposed to UV radiation emitted by the plasma. Because each cell emits its own light, a PDP is called an "emissive display". It contrasts with the liquid crystal display (LCD), a type of flat display in which the light comes from a plasma lamp behind the liquid crystal, which has arrays of small switches controlling where light is allowed to pass through. It is interesting that LCD TV can also be called plasma TV according to the type of light source used in that system. The DBD surface discharge, illustrated in Fig. 23.10, operates in xenon mixed with a buffer gas, which consists usually of neon or helium to optimize UV emission. The operating conditions of the display (gas composition, pressure, voltage, geometry, etc.) represent a compromise, considering performance requirements such as low-voltage operation, long life, high brightness, and high contrast.

The plasma display consists of two parallel glass plates separated by a precise spacing of some tenths of a millimeter and sealed around the edges. The space between the plates is filled with a mixture of rare gases at a pressure somewhat less than one atmosphere. Parallel stripes of transparent conducting material (electrodes) with a width of about a tenth of a millimeter are deposited on each plate, with the stripes on one plate perpendicular to those on the other. Thus, the stripes operate as "electrodes" to which voltage is applied. The intersections of the rows of electrodes on one side and the columns of electrodes on the opposite glass plate define the individual color elements, or cells, of the PDP. For high-quality color images, it is important to keep the UV radiation from passing between cells. To isolate the individual cells, barriers are created on the inside surface of one of the plates before sealing. Honeycomb-like structures and other shapes have been used. The red, green, and blue phosphors are deposited inside these structures. The plasma in each individual cell can be turned on and off rapidly enough to produce a high-quality moving picture. To help turn the individual cells on and off, there are two electrodes on one side and a third electrode on the opposite side of each cell. PDPs in modern plasma TVs consist of several million DBD cells which are switched at a rate that creates at least 60 TV picture frames per second. More details on the PDP physics can be found in Boeuf (2003).

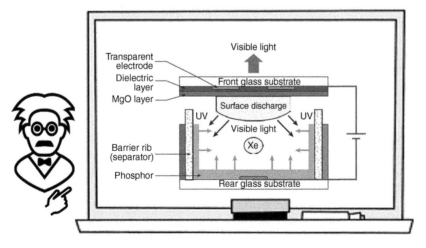

Figure 23.10 Plasma display panels: plasma cell cross section of a typical PDP.

23.18   Problems and Concept Questions

23.18.1   Radical-thermal "Bootstrap" Effect in the Subthreshold Plasma-ignition of Hydrogen

Estimate how long a non-branching chain of radical-stimulated oxidation in stoichiometric H_2–O_2 mixture should be to provide the radical-thermal "bootstrap" ignition 200 K below the temperature threshold of auto-ignition. Assume that the initial concentration of the plasma-generated radicals is low, molar fraction 10^{-4}, and neglect direct plasma heating during the nonequilibrium plasma ignition.

23.18.2   Contribution of Vibrationally and Electronically Excited Molecules into Plasma Stimulated Ignition of H_2–Air Mixtures

Estimate the relative contribution of nitrogen molecules, vibrationally and electronically excited in nonequilibrium plasma, in the stimulation of ignition of H_2–air mixtures. Compare the possible contribution to the ignition of electronically and vibrationally excited nitrogen molecules with that of oxygen molecules.

23.18.3   Energy Requirements for Plasma Ignition in Ram/Scramjet Engines

Analyze equation (23.1) to explain why the specific energy input E_v required for a nonequilibrium plasma ignition in ram- and scramjet engines decreases when using discharge systems operating at higher values of the reduced electric field E/n_0.

23.18.4   The Electric Propulsion Systems Decrease the Required Mass of the Spacecraft

This effect is due to very high specific impulse, I_{sp} of the electric propulsion systems. Why the specific impulse, I_{sp}, of the ion and plasma thrusters is so high?

23.18.5   Optimal Specific Impulse of Electric Propulsion Systems (Ion and Plasma Thrusters)

Based on relation between the initial mass M_0 of a spacecraft at low (near-Earth) orbit and its specific impulse I_{sp}, derive formula (23.3) for the optimal specific impulse.

23.18.6   Optimal Specific Impulse Dependence on the Trust

Explain why the optimal value of the specific impulse of the ion and plasma thrusters does not depend on the value of their thrust and is mostly determined by the duration of space flight.

23.18.7   Plasma Aerodynamics

Give your interpretation why the plasma-induced ionic wind can suppress separation of the boundary layers from the wings in the very high speed flights?

23.18.8   Plasma as a Metamaterial

Give your interpretation why plasma can be a crucial element of the negative-index metamaterials opening possibilities for the metamaterial cloaking, which is the use of metamaterials as the "invisibility cloak"?

Lecture 24

Plasma in Environmental Control: Cleaning of Air, Exhaust Gases, Water, and Soil

24.1 Exhaust Gas Cleaning from SO_2: Fundamental and Applied Aspects, Application of Relativistic Electron Beams and Coronas

Industrial SO_2 emissions cause acid rain and result in serious environmental disasters. The world's most severe acid rain regions today are Northern America, Europe, and China. The major sources of SO_2 emissions are coal-burning power plants, steelworks, non-ferrous metallurgical plants, as well as oil refineries, and natural gas purification plants. The SO_2 emissions in air are usually high-volume and low concentration: the SO_2 fraction is usually on the level of hundreds of ppm, while total flow of polluted air in one system can reach a million cubic meters per hour. Oxidation of SO_2 in air to SO_3 results in the rapid formation of sulfuric acid (H_2SO_4). The kinetics of this process is limited by very low rates of natural oxidation of SO_2 at low-temperature air conditions. Therefore, the formation of SO_3 and sulfuric acid from the exhausted SO_2 occurs not immediately in the stacks of the industrial or power plants (burning sulfur-containing fuel) but later in the clouds, which ultimately results in acid rain. Non-thermal plasma can be used to stimulate oxidation of SO_2 into SO_3 inside the stack. This permits collection of the sulfur oxides in the form of sulfates, for example, in the form of a fertilizer, $(NH_4)_2SO_4$, if ammonia (NH_3) is admixed to the plasma-assisted oxidation products. An important advantage of the nonthermal plasma method is possibility of simultaneous air and exhaust gas cleaning from SO_2 and NO_x. Considering the very large volumes to be cleaned from SO_2, the nonthermal atmospheric pressure plasma systems mostly applied for this purpose are often based on electron beams and pulsed corona discharges, see for example Chae et al. (1996); Jiandong et al. (1996); Mattachini et al. (1996).

The crucial question is **energy cost of the plasma cleaning processes**, which strongly depends on the treatment dose rate (current density in the case of electron beams), air humidity, temperature, and other factors. Nonthermal plasma just stimulates exothermic oxidation of SO_2 to SO_3, which then is removed chemically. Nevertheless, the process can be quite energy expensive. At low current densities ($j < 10^{-5}\,A\,cm^{-2}$) conventional for many experimental systems, the energy cost of the desulfurization process is usually high, on the level of 10 eV per SO_2 molecule. In this case, cleaning exhaust of a 300 MW power plant containing 0.1% (1000 ppm) of SO_2 requires about 12 MW of plasma power. Even when such an energy cost of plasma exhaust cleaning is acceptable; the use of such powerful electron beams or pulsed coronas is questionable. Decreasing the energy cost requires the use of very specific plasma parameters, particularly dose rates, and current densities in the case of electron beams (Baranchicov et al. 1990a,b, 1992).

The **application of relativistic electron beams** (electron energies ≥ 300–500 keV) is attractive for exhaust gas cleaning in large ducts, because the propagation length of high-energy electrons in atmospheric air can be estimated as 1 m per 1 MeV (see Section 15.7). Also, the high-energy electrons are characterized by the lowest energy cost of formation of charged particles (about 30 eV per an electron–ion pair), which permits low energy cost of SO_2 oxidation in air stimulated by the chain ion–molecular reactions. Electron-beam-generated plasmas provide effective SO_2 oxidation to SO_3, which is removed afterward using conventional chemical methods like addition of NH_3 and filtration of $(NH_4)_2SO_4$, see Baranchicov et al. (1990a,b, 1992).

The energy costs of SO_2 oxidation into SO_3 in different electron-beam experiments are presented in Figure 24.1 as a function of the electron-beam current density. In the high-current density range ($j_{eb} \geq 0.1\,A\,cm^{-2}$), the energy cost of SO_2 oxidation is proportional to the square root of the relativistic electron-beam current density: $A_{SO_2} \propto j_{eb}^{1/2}$. The minimum achieved energy cost, A_{SO_2}, without considering oxidation in droplets is 2–3 eV mol^{-1} SO_2. Including oxidation in the droplets the minimum achieved oxidation energy cost is 0.8 eV mol^{-1} SO_2 (to be discussed in the next

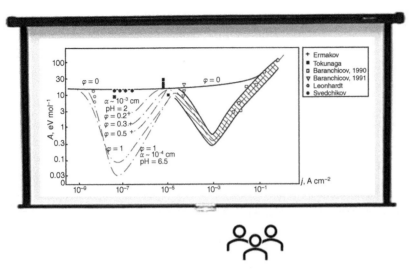

Figure 24.1 Energy cost of SO_2 oxidation as a function of electron beam current density. Dashed line shows energy cost decrease by the cluster effect; dotted line shows energy cost decrease by the effect of droplets.

section). The low oxidation energy costs are due to plasma stimulation of chain oxidation mechanisms at relatively high values of electron-beam current density. Electron beams with lower current densities ($\leq 10^{-4}\,A\,cm^{-2}$) are unable to initiate chain oxidation, which results in an oxidation energy cost of about $10\,eV\,mol^{-1}$; see Figure 24.1.

Corona and pulsed corona discharges are also effective in SO_2 oxidation to SO_3 and therefore in the exhaust gas cleaning of sulfur oxides. Baranchicov et al. (1992) investigated SO_2 oxidation in air using both continuous and pulsed corona discharges. The minimum energy cost for the pulsed coronas is $3–5\,eV\,mol^{-1}\,SO_2$, and $10\,eV\,mol^{-1}$ SO_2 for the continuous coronas. Thus, the minimum SO_2 oxidation energy costs are close for the pulsed corona discharges and electron beams. At low current densities ($\leq 10^{-4}\,A\,cm^{-2}$), the energy cost of SO_2 oxidation to SO_3 remains almost constant at the level of $10\,eV\,mol^{-1}\,SO_2$. At high current densities ($0.1\,A\,cm^{-2}$ and higher), the corona energy cost follows the same tendency $A_{SO_2} \propto j_{eb}^{1/2}$ as in the case of electron beams with an expected minimum at $10–100\,mA\,cm^{-2}$. The low plasma-stimulated SO_2 oxidation energy costs in air at current densities $10–100\,mA\,cm^{-2}$ are due to the chain ion-molecular mechanism of SO_2 oxidation, which is unavailable in conventional gas-phase chemistry.

24.2 Kinetics and Energy Balance of Plasma-catalytic Ion-molecular Chain Oxidation of SO_2 to SO_3 in Airflow

Atomic oxygen and other reactive oxygen species generated in nonthermal air plasma with an energy cost of about $10\,eV$ per particle can oxidize SO_2 to SO_3 with the same energy cost, which explains experiments with low current densities (see Figure 24.1). *The SO_2 oxidation energy costs of $1–3\,eV\,mol^{-1}$ and below, observed in plasma at higher current densities, can be attributed to plasma-induced chain oxidation processes.* There is no chain mechanism of SO_2 oxidation to SO_3 in conventional gas-phase chemistry (without a catalyst). The plasma-catalytic ion-molecular chain process occurs in gas phase, and even more effectively in clusters and droplets, formed by water condensation around H_2SO_4, ions, and other active species. When the cold plasma is generated in atmospheric air containing sulfur compounds, the active species and especially sulfuric acid generated in the system immediately lead to water clusterization/condensation and formation of mist. Kinetics of the plasma-catalytic ion-molecular chain processes of SO_2 oxidation in liquids, droplets, and clusters is reviewed in Fridman (2008).

To analyze **energy balance of the plasma cleaning in exhaust gases**, it is sufficient to consider the kinetics of simple ions, neutrals, and excited molecules. The chain oxidation starts with conventional formation of the negative ions (O_2^-) in air through the three-body electron attachment process. Then the SO_2^- ions are formed due to a fast exothermic charge exchange process $O_2^- + SO_2 \rightarrow O_2 + SO_2^-$. The SO_2^- ions are then converted in collisions with O$_2$ into its chemically active peroxide-configuration SO$_4^-$. Formation of SO$_3$, the oxidation products, occurs in the SO$_4^-$ reactions with oxygen SO$_4^- + O_2 \rightarrow SO_3 + O_3^-$. Restoration of electrons and chain propagation is provided by ion–molecular reactions and associative detachment, producing additional sulfuric acid: O$_3^- + SO_2 \rightarrow SO_3^- + O_2$, SO$_3^- + H_2O \rightarrow H_2SO_4 + e$. The minimum SO$_2$ oxidation energy cost can be calculated for this chain reaction mechanism as:

$$A_{\min} = \frac{G_i W_i}{2G_v} = \frac{100\,\text{eV}}{G_v}. \tag{24.1}$$

In this relation, G_i and G_v are the radiative yields (*G*-factors) of ions and vibrationally excited molecules, and W_i is the energy cost of generation of an ion. This process energy cost can be significantly lower than the energy cost of an ion (W_i) because of the long oxidation chain length. Considering that the G_v factor in air is 100–300, the minimal values of the energy cost of the plasma-stimulated SO$_2$ oxidation to SO$_3$ is 0.3–1 eV mol^{-1} SO$_2$ (see Figure 24.1).

24.3 Plasma-stimulated Combined Oxidation of NO$_x$ and SO$_2$ in Air: Simultaneous Industrial Exhaust Gas Cleaning from Nitrogen and Sulfur Oxides

An important advantage of plasma-assisted exhaust gas cleaning, especially in the case of power plant exhaust, is due to the possibility of **simultaneous oxidation of SO$_2$ and NO$_x$ to sulfuric and nitric acids**. The products can then be collected in the form of non-soluble sulfates and nitrates, for example, in the form of the fertilizers (NH$_4$)$_2$SO$_4$ and (NH$_4$)NO$_3$ if ammonia (NH$_3$) is added to the plasma-assisted oxidation products. *Simultaneous NO and SO$_2$ oxidation processes appear to be more effective than their plasma-assisted individual oxidation in air. In the presence of SO$_2$, the radiative yield (G-factor, yield per 100 eV) of NO oxidation to NO$_2$ and nitric acid can exceed 30 (energy cost below 3 eV mol^{-1} NO).* The low-energy-cost effect of simultaneous NO and SO$_2$ oxidation is due to the chain co-oxidation mechanism propagating through the radicals OH and HO$_2$:

$$HO_2 + NO \rightarrow OH + NO_2, \tag{24.2}$$

$$OH + SO_2 \rightarrow SO_3 + H, H + O_2 + M \rightarrow HO_2 + M. \tag{24.3}$$

The formation of high concentrations of NO$_2$ in this chain can be limited by plasma-generated atomic nitrogen (N + NO$_2$ → 2NO, N + NO$_2$ → N$_2$O + O). The product quenching can be provided by the plasma-generated droplets. In such droplets at certain pH conditions, it is possible not only to stabilize NO$_2$ by the formation of nitric acid in the solution but also to reduce NO$_2$ to molecular nitrogen (N$_2$) with simultaneous oxidation of sulfur from S(IV) to S(VI): NO$_2$ + 2HSO$_3^- \rightarrow 1/2N_2 + 2HSO_4^-$. More kinetic details regarding the simultaneous plasma-chemical cleaning of industrial exhaust gases of sulfur and nitrogen oxides using relativistic electron beams can be found, in Potapkin et al. (1993, 1995).

Regarding the nonthermal plasma treatment of large-volume exhausts, we should point out the **concept of "radical shower"**, applied in particular to NO$_x$-removal flow processing (Yan et al. 2001; Chang 2003; Wu et al. 2005). The radical shower approach implies that plasma treats directly only a portion of the total flow (or even separate gas) producing active species (in particular, radicals), which then treat the total gas flow as a "shower." Although the approach is limited by mixing efficiency of the plasma-treated and nonplasma treated gases, radical shower gas cleaning can be energy effective especially in the case of very high flow rates of the exhaust gas streams. Because the radical shower approach is often organized using corona discharges, it is sometimes referred to as the **corona radical shower**.

24.4 Plasma-assisted After-treatment of Automotive Exhaust, Double-stage Plasma-catalytic NO$_x$, and Hydrocarbon Remediation

Lean-burn car engines are in wide use because of their higher efficiency (more miles per gallon) and lower relative CO$_2$ emission. Combustion of the lean mixtures results, on the other hand, in larger nitrogen oxide emissions, which then require their effective remediation. Such after-treatment of automotive exhaust can be effectively organized using the strongly nonequilibrium plasma, often pulsed corona discharges. The lean-burn engines operate under oxidizing conditions when conventional three-way catalysts are ineffective in controlling emissions of NO_x through stimulating its reduction by hydrocarbons. Concentrations of hydrocarbons are low in these conditions, which makes the three-way catalyst ineffective in controlling NO_x emission. Application of plasma provides effective NO_x-removal aftertreatment even under oxidizing conditions, simultaneously with burning out the residual hydrocarbons.

Although plasma-generated atomic nitrogen can reduce NO to molecular nitrogen in the barrierless fast elementary reaction $N + NO \rightarrow N_2 + O$, plasma in lean-combustion exhaust (containing a significant amount of oxygen) mostly stimulates not reduction but oxidation of NO to NO_2. The plasma-stimulated NO-to-NO_2 conversion becomes especially effective in the presence of hydrocarbons. *Complete NO-to-NO_2 conversion permits effective NO_2 reduction to molecular nitrogen (N_2) subsequently by using special catalysts. Thus, plasma-catalytic NO-reduction aftertreatment means in this case the actual combination of plasma and catalytic technologies to provide the* **double-stage automotive exhaust cleaning**. The plasma-chemical phase of the double-stage exhaust treatment process is usually organized in the pulsed corona or similar strongly nonequilibrium atmospheric-pressure discharges in an automotive exhaust mixture containing conventional combustion products as well as NO, hydrocarbons, and residual oxygen (O_2). The major purpose of the plasma-chemical phase is the complete conversion of NO to NO_2 with simultaneous partial destruction of hydrocarbons.

The **oxidation process mechanism** starts with plasma dissociation of molecular oxygen (O_2) and generation of atomic oxygen through electronic excitation and dissociative attachment $e + O_2 \rightarrow O + O + e$, $e + O_2 \rightarrow O + O^-$. Atomic oxygen then quickly reacts with the hydrocarbons (which can be represented as propylene, C_3H_6), stimulating initial destruction of the hydrocarbons with the production of hydrogen atoms, methyl radicals, and HO_2 radicals: $O + C_3H_6 \rightarrow CH_2CO + CH_3 + H$, $O + C_3H_6 \rightarrow CH_3CHCO + H + H$, $H + O_2 + M \rightarrow HO_2 + M$. Intermediate hydrocarbons produced by the propylene destruction further react with molecular oxygen, producing additional HO_2 radicals: $CH_2OH + O_2 \rightarrow CH_2O + HO_2$, $CH_3O + O_2 \rightarrow CH_2O + HO_2$, $HCO + O_2 \rightarrow CO + HO_2$. Plasma-generated HO_2 radicals are the major particles responsible for effective plasma-chemical conversion of NO into NO_2 in the systems under consideration: $NO + HO_2 \rightarrow NO_2 + OH$. The OH radicals produced in the system are effective in both destruction of hydrocarbons ($OH + C_3H_6 \rightarrow C_3H_6OH$, $OH + C_3H_6 \rightarrow C_3H_5 + H_2O$) and in NO_x conversion into acids relevant for further catalytic reduction in the second (catalytic) phase of the double-stage automotive exhaust aftertreatment: $NO + OH + M \rightarrow HNO_2 + M$, $NO_2 + OH \rightarrow HNO_3 + M$. After plasma conversion of NO into NO_2 and the acids, the double-stage aftertreatment is completed by NO_2 reduction to nitrogen and burning out the residual hydrocarbons in the catalytic stage of the process: $NO_2 + Hydrocarbons + Catalyst \rightarrow N_2 + CO_2 + H_2O + Catalyst$. Metal oxides or metal-covered zeolites (such as Na-ZSM-5, Cu-ZSM-5, Ce-ZSM-5, H-ZSM-5, Al_2O_3, ZrO_2, and Ga_2O_3) can be applied as catalysts for the process.

The double-stage system for **plasma-catalytic NO_x reduction in automotive exhaust with a pulsed corona discharge** is illustrated in Figure 24.2. Typical parameters of the pulsed coronas used for this application are: maximum pulse voltage 20 kV; pulse duration 100 ns; pulse repetition frequency, 10 kHz; voltage rise rate, $1-2\,kV\,ns^{-1}$; and average power of a pulsed power supply, about 1 kW. The effectiveness of NO oxidation to NO_2 at the relevant temperatures about 300 °C significantly increases in the presence of hydrocarbons. Without hydrocarbons, plasma-generated atomic oxygen intensively reacts with NO_2 in the reverse process ($O + NO_2 \rightarrow NO + O_2$). The presence of hydrocarbons makes the atomic oxygen mostly reacting with them preventing the destruction of the product (NO_2) in a wide temperature range. The NO_x reduction strongly increases at higher specific energy inputs ($J\,L^{-1}$). Complete plasma-catalytic cleaning of the automotive exhausts containing 200–300 ppm of NO_x typically requires about $30\,J\,L^{-1}$ of pulsed corona energy input, which corresponds to an energy cost of about $30\,eV\,mol^{-1}$ NO. The typical exhaust gas flow rate per unit engine power can be estimated as $10\,L\,min^{-1}$ per 1 kW of engine power. A car with a power of about 200 hp, or about 150 kW, exhausts gas with a maximum flow rate of about $1500\,L\,min^{-1}$. It corresponds to minimum power of the pulsed corona discharge on the level of 750 W, required for the effective plasma-catalytic exhaust treatment from nitrogen oxides. More information on the subject can be found in Pu and Woskov (1996); Hammer and Broer (1998); Hoard and Servati (1998); Penetrante et al. (1998); Puchkarev et al. (1998); Slone et al. (1998).

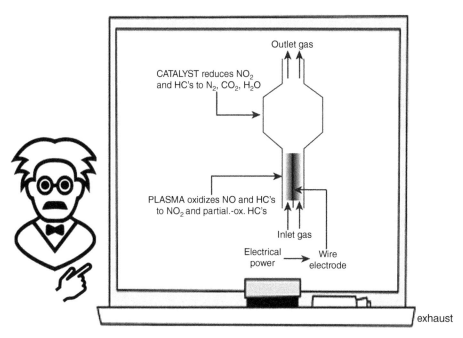

Figure 24.2 A schematic of the plasma-assisted catalytic reduction process of NO_x in automotive exhaust.

24.5 Nonthermal Plasma Abatement of Volatile Organic Compounds (VOC) in Air, Plasma Cleaning of Emission from Paper Mills and Wood Processing Plants

Volatile organic compounds (VOCs) form a class of air pollutants that has been addressed by environmental regulations due to the pollutants' toxicity and contribution to global warming mechanisms. The control of air pollution from dilute large-volume sources, such as paint spray booths, paper mills, pharmaceutical, and food and wood processing plants, is a challenging problem. Conventional technologies, such as carbon adsorption/solvent recovery or catalytic/thermal oxidation, widely used in industry, and regenerative thermal oxidation (RTO) systems, have too high annual cost and are not economical for large gas flow rates (50,000–250,000 scfm) and low VOC concentrations (<100 ppm). Among the emerging low-temperature VOC treatment technologies is nonthermal plasmas (like pulsed coronas, DBDs, and electron beams), which solves most of the problems typical for alternative VOC treatment methods. Plasma approaches usually become more energy efficient and competitive with other cleaning methods when stream volumes are large, and concentration of pollutants is small.

The highest energy efficiency of the cleaning of large-volume air exhaust from low concentration VOCs has been achieved in plasma systems with the highest electron energies (such as electron beams, pulsed coronas, and pulsed DBD), where the energy cost of generation of oxidation active species (such as OH) is lowest. Nonthermal plasma systems have been successfully applied for cleaning of high-volume low-concentration (HVLC) VOC exhausts, particularly from volatile hydrocarbons (acetone, methanol, pinene, etc.), sulfur-containing compounds (dimethyl sulfide, H_2S, etc.), and chlorine-containing compounds (vinyl chloride, trichloroethylene, trichloroethane, carbon tetrachloride, etc.).

The **mechanism of plasma oxidation of VOC hydrocarbons in air** *can be generally interpreted as the low temperature burning out of the VOCs in air to preferentially CO_2 and H_2O using different plasma-generated active oxidizers* (OH, atomic oxygen, electronically excited oxygen, ozone, etc.). It can be generally outlined by the following illustrative kinetic mechanism. Between different plasma-generated active oxidizers responsible for the cold burning out of VOC, OH radicals play a major role. Formation of the OH radicals in air plasma follows numerous channels, from which it is interesting to point out a selective one starting with the charge exchange from any positive air ions M^+ to water ions, H_2O^+: $M^+ + H_2O \rightarrow M + H_2O^+$. The H_2O^+ ions then react with water molecules to produce active OH radicals in the fast ion–molecular reaction $H_2O^+ + H_2O \rightarrow H_3O^+ + OH$.

Considering that the ionization potential of the water molecules is relatively low, most of the positive ions initially formed in air have a tendency for charge exchange, formation of H_2O^+, H_3O^+, and OH. Therefore, the discharge energy initially distributed over different air components is selectively localized on the ionization of water and the selective production of OH radicals. It leads to relatively low OH energy cost in air: 10–30 eV/radical, although the fraction of water molecules in air is not large.

Representing a hydrocarbon VOC molecule as RH, where R is an organic group, the OH-based oxidation of the molecule starts with fast barrierless reaction of its dehydrogenization:

$$OH + RH \rightarrow R + H_2O. \tag{24.4}$$

Almost immediate attachment of molecular oxygen to the active organic radical R results in the formation of an organic peroxide radical:

$$R + O_2 \rightarrow RO_2. \tag{24.5}$$

The peroxide radical RO_2 reacts with another saturated VOC molecule RH, forming saturated organic peroxide (RO_2H), propagating the chain mechanism of the RH oxidation in air:

$$RO_2 + RH \rightarrow RO_2H + R. \tag{24.6}$$

Depending on the temperature and chemical composition, peroxides RO_2 and RO_2H are further oxidized up to CO_2 and H_2O with or without additional consumption of OH.

Exhaust gases of paper mills (especially the brownstock washer ventilation gases) and **wood processing plants** (especially strandboard press and dryer ventilation gases) are good examples of HVLC VOC streams, that can be effectively cleaned by nonthermal plasma following the above-described mechanism. Summarizing the mechanism, the energy cost corresponding to the VOC treatment process is on the level of 10–30 eV per molecule of the pollutant RH. This energy (10–30 eV mol^{-1}) is relatively high, but the total energy consumption can be small because of the very low concentration of pollutants. A major conventional approach to VOC control, RTO (regenerative thermal oxidation), consumes about 0.1 eV mol^{-1}, but this energy is calculated per molecule of air. Therefore, energy consumption in plasma becomes lower than that of conventional RTO, when the VOC concentration in air is below 0.3–1% (3,000–10,000 ppm). In the case of plasma treatment of large-volume exhausts, the "radical shower" approach can be effective (see Section 24.3) like in the NO_x reduction and air desulfurization processes. More details on the subject can be found in Sobacchi et al. (2003); Gutsol et al. (2005); Harkness and Fridman (1999); Paur (1999); Hunter and Oyama (2000); Yamamoto et al. (1992, 1993); Hsiao et al. (1995); Evans et al. (1993); Rosocha et al. (1993); Neely et al. (1993); Masuda (1993); Mutaf-Yardimci et al. (1998); Czernichowski (1994); Fridman et al. (1999); Penetrante et al. (1996a,b).

24.6 Nonthermal Plasma Air Cleaning from Acetone, Methanol, Dimethyl Sulfide (DMS), α-Pinene, and Chlorine-containing VOC

Let us consider some specific examples of plasma air cleaning from VOC starting with **removal of acetone and methanol from air using pulsed corona** (wire-in-cylinder coaxial electrode configuration) with power 1–20 W, voltage pulse amplitude 9–12 kV, pulse repetition rate 0.2–2 kHz, pulse duration 100 ns, and rise time 10 ns (Sobacchi et al. 2003). Typical atmospheric-pressure gas flow through this plasma system in 2 slm; corresponding residence time about 13 s. The initial VOC concentration in air between 5 and 1000 ppm, temperature 200 °C, and the air humidity corresponds to saturation at about 55 °C. Destruction and removal efficiency (DRE, the mole percentage of VOC removed with respect to initial amount) of acetone and methanol depends strongly on the initial composition and on the corona power. Higher power levels result in higher DRE; increase of the initial VOC concentration requires higher power to reach the same level of DRE. Summarizing the results, the relatively low energy cost of treatment (about 20 Wh m^{-3} of air) is sufficient for 98% of DRE both for acetone and methanol, which is well within acceptable limits for the industrial applications.

Plasma air cleaning from sulfur-containing VOCs can be illustrated by **removal of dimethyl sulfide, DMS,** $S(CH_3)_2$, using pulsed corona like described above (Sobacchi et al. 2003). DMS is an air pollutant of major concern in different industrial exhausts, particularly including the brownstock ventilation gases of paper mills. DMS is highly toxic, flammable, and emits a foul odor (odor threshold is ppb). The products of nonthermal plasma DMS removal are CO_2, H_2O, and SO_2. The process is highly energy effective. The SEI of about 50 Wh m^{-3} is sufficient for processing even the highest DMS concentrations with DRE exceeding 98%.

VOC emissions of paper mills and wood processing plants usually contain significant amount of terpenes, especially α-*pinene* ($C_{10}H_{16}$). **Effective plasma removal of α-pinene** has been demonstrated also in pulsed corona discharges at temperatures of 70–200 °C. The major initial by-product of α-pinene plasma removal is acetone. The α-pinene plasma removal tests demonstrate that 98% DRE for the typical initial VOC concentration of 200 ppm can be reached at quite low SEI values into the air at 25 Wh m^{-3}. Treatment efficiency and especially energy cost of this exhaust treatment can be enhanced at higher humidity, see next section.

Nonthermal atmospheric plasma systems, including those based on electron beams, are effectively applied for the **destruction of the diluted large volume chlorine-containing VOCs** including vinyl chloride, trichloroethylene, trichloroethane, and carbon tetrachloride. The major specific plasma systems used for the application are electron beams, pulsed corona, and DBD. Let us consider as an example the **destruction of carbon tetrachloride (CCl_4) diluted in air using the electron beam plasma.** The products of the plasma processing of CCl_4 in air are Cl_2, $COCl_2$, and HCl, which can be easily removed from the gas stream by dissolving in aqueous scrubber solutions with $NaHCO_3$ to form NaCl. Major contribution to the process kinetics is provided by direct electron-impact destruction through dissociative attachment (Penetrante et al. 1996a,b). Therefore, energy cost of the CCl_4 decomposition is determined by the energy cost of an electron consumed in the dissociative attachment. The cost of an electron, the ionization energy cost, is lower at higher electron energies in a plasma system. The lowest ionization energy cost can be achieved in plasma generated by electron beams (about 30 eV per anion–electron pair, and about the same for the CCl_4 decomposition). To compare, the ionization energy cost in pulsed corona and DBD is usually on the level of hundreds of electron volts, which explains the higher energy efficiency of electron beams in the destruction of carbon tetrachloride. The advantages of electron beams versus pulsed corona discharges have also been demonstrated (Penetrante et al. 1996a,b) in the plasma treatment of 100 ppm of methylene chloride ($C_2H_2Cl_2$) in dry air at 25 °C, and in the plasma treatment of 100 ppm of TCE (trichloroethylene, C_2HCl_3) in dry air at 25 °C.

24.7 Treatment of Large-scale Exhaust Gases from Paper Mill and Wood Processing Plants by Wet Pulsed Corona: Combination of Plasma VOC Cleaning with Wet Scrubbing

The significant reduction of VOC-removal energy cost as well as the complete elimination of by-products can be achieved by application of the so-called **wet or spray pulsed coronas**, which is a combination of pulsed corona discharge with a scrubber (Gutsol et al. 2005). When the system is organized as shower, one says "spray corona", if water is supplied as a thin film on the walls, one says "wet corona". General schematics of the wet and spray configurations of the pulsed corona discharges (which are both often referred to the **wet pulsed corona**) are shown in Figure 24.3. First, water droplets (or film) absorb the soluble VOCs (such as methanol) present in the exhaust gases, which obviously simplifies the task for plasma treatment. Non-soluble organic compounds (in particular, nonpolar hydrocarbons) cannot be directly removed by water scrubbing. Their plasma treatment, however, leads to the formation of soluble peroxides (ROOH) and peroxide radicals (RO_2) which can be effectively removed by water scrubbing. Thus, plasma cleaning of VOC emissions is provided in this plasma approach not by complete oxidation of the organic compounds to CO_2 and H_2O, but only by plasma-induced conversion of the nonpolar non-soluble compounds into soluble ones, which are then removed by water scrubbing.

 Complete VOC oxidation to CO_2 and H_2O requires much more plasma-generated oxidizers (such as OH) than conversion of the nonpolar non-soluble compounds (RH) into soluble ones (ROOH or ROO), where a single OH radical can be sufficient. It results in much lower energy requirements for VOC treatment in wet pulsed corona discharges. The same level of DRE (especially without the treatment by-products) can be reached in wet pulsed coronas at about order of magnitude lower energy inputs (SEI) than in conventional coronas (Gutsol et al. 2005). The by-products of plasma VOC treatment are usually soluble; therefore, application of wet pulsed corona permits removal of all of them from the gas flow. For example, in plasma cleaning of sulfur-containing exhausts (such as DMS), sulfur oxides and polar by-products of oxidation are effectively removed from the airflow in

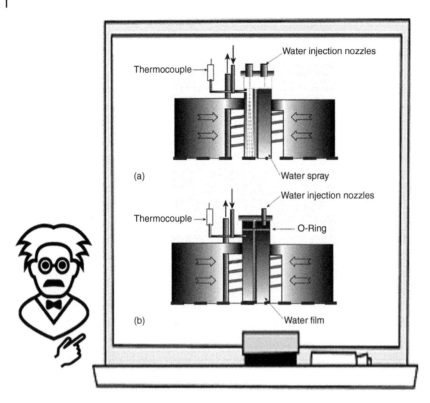

Figure 24.3 Schematics of (a) wet and (b) spray configurations of the pulsed corona discharges.

the form of relevant acidic solutions. Plasma-stimulated oxidation of organic compounds continues when they are already absorbed in water droplets, which increases the absorbing capacity of the water droplets. Therefore, water consumption for the VOC exhaust cleaning approach is very low.

A mobile plasma laboratory for VOC removal from the exhaust streams of paper mills and wood processing plants has been built at Drexel University (Figure 24.4, see Gutsol et al. 2005) based on 10 kW wet (spray) pulsed corona discharge. The discharge covered big volume to be observed from the six windows, see Figure 24.5. The pneumatic and hydraulic schematic of the system is shown in Figure 24.6. Exhaust gas in the pilot plant first goes to a scrubber where it is mostly washed out from soluble VOCs. After that, the exhaust gas is directed to the wet pulsed corona discharge, where the non-soluble VOCs are converted into soluble compounds to be scrubbed out by water spray in the chamber. Final air refining is provided by the mist separator, finalizing the exhaust cleaning process. The pilot plant is mounted on a trailer providing opportunity for industrial field experiments. The mobile plasma laboratory for VOC removal includes not only the gas cleaning system but also, special exhaust gas characterization capability for gas diagnostics before and after plasma treatment. The DRE and SEI characteristics of the VOC exhaust treatment were like those in the small laboratory-scale wet corona systems.

24.8 Nonthermal Plasma Removal of Elemental Mercury from Coal-fired Power Plant Emissions, and Other Industrial Off-gases

Elemental mercury and mercury compounds are extremely toxic, volatile, and difficult to remove from the off-gases. Hg removal can be effectively organized, however, using nonthermal plasma. Major sources of mercury are combustion facilities (coal-fired power plants), municipal solid waste incinerators, hazardous waste incinerators, and medical waste incinerators, which altogether account for 87% of total Hg emissions. The U.S. Environmental Protection Agency established a standard regulating Hg concentration to $40 \, \mu g \, m^{-3}$ (about 4.5 ppb). Typical off-gas from incineration or a coal-fired boiler contains from 100 to $1000 \, \mu g \, m^{-3}$ (0.1–1 ppm) of mercury in both elemental and oxidized form. While oxidized forms of mercury (Hg (II), primarily $HgCl_2$ and HgO) can be removed by wet scrubbers,

Figure 24.4 Mobile Environmental Laboratory carrying exhaust gas characterization facility as well as 10 kW wet (spray) pulsed corona discharge for treatment of VOC exhaust gases.

Figure 24.5 Mobile Environmental Laboratory with six access windows to the 10-kW wet pulsed corona discharge for VOC treatment.

elemental mercury is chemically inert, insoluble in water solutions, and therefore difficult to remove. A significant portion (10–30%) of the total mercury load in coal-fired power plant off-gases stays in the elemental form. Conventional Hg removal technologies based on adsorption, scrubbing, and electrostatic precipitation (ESP) are not effective and generate secondary wastes. Mercury capture in conventional control devices is limited to 27% in the cold-side ESP, 4% in the hot-side ESP, and 58% in the fabric filters.

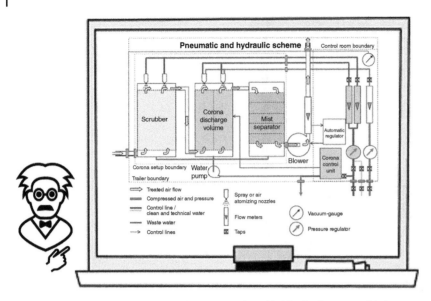

Figure 24.6 Pneumatic and hydraulic scheme of the Mobile Environmental Laboratory.

Application of nonthermal plasma permits significant improvement in the effectiveness of mercury capture, especially in the case of elemental mercury. The effective Hg removal from flue gas has been demonstrated using the **pulsed-corona-enhanced dry ESP**. The experiments show high (much over 90%) removal efficiency of mercury in pulsed and DC coronas of both polarities. The removal efficiency increased with the decrease of gas temperature (from about 350 to 100 °C). Negative polarity discharges performed better. A removal efficiency close to 99% was achieved in the negative pulsed corona regime at an energy input level of an ordinary ESP (about 0.3 kJ m^{-3}), see Helfritch et al. (1996).

The effective mercury capture can be also achieved using **DBD discharges**, see Alix et al. (1997). The process is capable of oxidizing and removing 80–90% of mercury at initial concentrations in the off-gas of 26 µg cm^{-3} (out of which ∼6 µg cm^{-3} is in elemental form). The DBD reactor is operated at a power 18 W scfm^{-1} (∼37 kJ m^{-3}) at flue gas flow rate of 1500 scfm. Removal of mercury from air using a **capillary corona reactor** has been demonstrated by Babko-Malyi et al. (2000). At initial concentrations of elemental mercury (30–500 µg cm^{-3}), removal efficiencies of up to 97% have been achieved at energy input of 6 kJ m^{-3}.

Analyzing **mechanism of plasma-induced removal of mercury** we should keep in mind that mercury is present in off-gases in both elemental (Hg^0) and oxidized (Hg(II)) forms, which require different capture approaches. Oxidized mercury (HgO, built from Hg(II)) can be transferred into water solution and separated from the off-gas by using conventional scrubber or WESP technologies. Elemental mercury, on the other hand, it is not soluble in water and therefore easily escapes scrubber systems. Plasma stimulation of mercury removal is mostly based on plasma's ability to oxidize elemental mercury (Hg^0) into Hg (II), which then can be effectively removed using the conventional technologies. Oxidation of elemental mercury (Hg^0) by molecular oxygen has very high activation barrier (above 3 eV mol^{-1}) and, therefore, can be completely neglected not only at room temperature but even at elevated temperatures. Thus, oxidation of mercury requires the participation of stronger oxidizers. The mechanism of plasma-stimulated mercury oxidation starts with the generation in air of such strong oxidizers as atomic oxygen and electronically excited molecular oxygen, ozone, and OH radicals, see Babko-Malyi et al. (2000). The energy cost of mercury oxidation and removal through the nonthermal plasma mechanisms is determined by the energy cost of plasma generation of the strong oxidizers (electronically excited oxygen, OH radicals, etc.). Thus, DBD, for example, requires 10 Wh m^{-3} of combustion off-gases.

24.9 Plasma Decomposition of Freons (Chlorofluorocarbons) and Other Gaseous Waste Treatment Processes in Thermal and Transitional "Warm" Discharges

Thermal and transitional (quasi-thermal) plasma systems are effectively applied for decomposition, incineration, and reforming of different concentrated industrial and municipal wastes. The decomposition of freons (chlorofluorocarbons) [CFCs], can be considered an example of such waste treatment processes. Freons have been widely used in cooling systems, in the production of insulation foams and aerosols, and as solvents, especially in the electronic industry. The total amount of freons industrially produced and applied is very large. Exhaust from freons can penetrate to the upper atmosphere, destroying the ozone layer. Solar UV radiation dissociates CFC-producing Cl atoms, which stimulates a chain reaction of ozone destruction (a single chlorine atom can destroy up to 10^6 ozone molecules). For these reasons, CFC production has been prohibited and remaining stocks of freons by now have been destroyed.

Decomposition of the freons in thermal plasma jets and under direct action of the electric discharges is an effective practical approach to CFC destruction, see Murphy (1997); Murphy et al. (2001, 2002); Wang et al. (1999); Jasinski et al. (2000, 2002a,b). Most of plasma CFC destruction processes are carried out in the presence of oxygen, air, and/or water vapor. As an example, the CFC decomposition with a destruction degree of 99.99% in argon plasma jets in the presence of oxygen and water vapor, or water vapor alone has been demonstrated by Murphy (1997); Murphy et al. (2001). The major decomposition products are CO_2, HCl, and HF. Presence of water vapor significantly reduces the production of other "unwanted" products such as other freons, dioxins, and furans. Although CFC destruction effectiveness in thermal plasma jets is good, this technology is characterized by significant heat losses and relatively high electric energy cost. The electric energy cost of plasma CFC destruction can be reduced by performing the process not in a thermal plasma jet but directly in the active discharge zone (Wang et al. 1999; Jasinski et al. 2000, 2002a,b). The process has been organized at atmospheric pressure as well as at reduced pressures. Concentration of freons varied from very low (several dozen ppm) in diluted mixtures to very high (several dozen percent) in non-diluted exhausts.

Effective decomposition of $CFCl_3$, CF_2Cl_2, and CHF_2Cl was achieved in the **transitional regimes of gliding arc discharges** in the presence of water vapor (Czernichowski 1994) and oxygen (Opalska et al. 2002a,b). An admixture of H_2 to freons during their treatment in gliding arc discharges limits products to simple aliphatic hydrocarbons, completely avoiding formation of the chlorinated dioxins and furans or phosgene, which can be generated in oxidative or redox media (Czernichowski 1994). Effective CFC destruction by steam plasma generated in an atmospheric-pressure DC discharge has also been demonstrated by Watanabe et al. (2005). Like freons, plasmas can be effectively applied for reducing of gaseous perfluorinated compound (PFC; for example, CF_4, SF_6) emissions from chamber cleaning and dielectric etch tools in electronics (Kabouzi, et al. 2005). Although the plasma PFC abatement is mostly organized today at relatively low pressures, more and more attention is being given now to the use of atmospheric pressure discharges for this application.

24.10 Plasma Decontamination of Water: Fundamental and Applied Aspects, Suppression of Hazardous Organic Compounds, Challenges of Energy Cost

Way more than billion people are unable to acquire today safe drinking water, which determines the need for improved methods of water treatment. *Contaminated water can be attributed to several factors, including chemical fouling, inadequate treatment, and deficient or failing of the water treatment and distribution systems. The water decontamination challenge and relevant technologies can be subdivided in two groups: removal of hazardous chemicals from water, and water disinfection and sterilization.* The first group, the **chemical decontamination of water**, is going to be considered in this Lecture. The second group, water disinfection and sterilization, will be considered in Lecture 29 together with analysis of physics, chemistry, and biology of plasma-activated water and its applications in plasma agriculture and food processing. To provide water cleaning in general, the commercial applications of chemical treatments, ultraviolet radiation, and ozone injection units have been developed and implemented into potable water delivery systems. The commercialization of these water treatment technologies has, however, some deficiencies. Regarding human consumption, chemical treatments such as chlorination can render potable water toxic. UV-radiation and ozone injection have also been proven to be two practical methods of chemical cleaning and bacterial inactivation in water, but the effectiveness of such methods largely depends upon adherence to regimented maintenance schedules. Plasma methods effectively combining contribution of UV-radiation, active chemicals, high electric fields etc. are considered therefore as very effective approach to water treatment, see Fridman et al. (2007a–e), Locke et al. (2006).

The effective **plasma degradation of organic contaminants** in water and decolorization of a dye in liquid has been first shown by Clements et al. (1987). After that, plasma effectiveness has been demonstrated in treatment and abatement of several contaminants of environmental importance including pharmaceuticals, herbicides, pesticides, warfare agents, bacteria, yeasts, and viruses, see as an example Ajo et al. (2018). Different reactor configurations have been used for this purpose, including direct-in-liquid discharges with and without bubbles, and discharges in a gas over and contacting the surface of a liquid. Different plasma sources generating nonthermal and thermal plasmas have been used, including AC, nanosecond pulsed and DC voltage sources, pulsed coronas, glow discharges, sparks, arcs, gliding arcs, etc.

Plasma-induced mechanisms of destruction of the organic compounds in water has some similarities with that in gas phase considered in Section 24.5 but also has significant specifics (as an example those discussed in the Section 24.2). Like in the gas phase, many reactive oxygen and nitrogen species (RONS) contribute to the destruction of the organics. But *the major oxidant both in gas and liquid phase is OH radical, see equations (24.4-6), which can, however, recombine in water creating hydrogen peroxide* H_2O_2. *The oxidation process in the liquid phase is strongly supported by acidity of the plasma-activated water (pH down to 1.5–2.5), helping to convert H_2O_2 back to the chemically aggressive OH.* Like, it was discussed in the Section 24.5, formation of the OH radicals in plasma is partially due to the selective charge exchange from any positive air ions M^+ to water ions, H_2O^+: $M^+ + H_2O \rightarrow M + H_2O^+$. The H_2O^+ ions then react with water molecules to produce active OH radicals in the fast ion–molecular reaction $H_2O^+ + H_2O \rightarrow H_3O^+ + OH$ together with the H_3O^+ responsible for the acidity of the plasma-activated water (the so-called **plasma acid effect**, see Shainsky et al. 2012). Acidity in the plasma-activated water is, obviously, also due to nitrogen oxides coming to the water from plasma-generated nitrogen oxides (often through the HNO radicals easily dissolved in H_2O). The combination of the acidity and hydrogen peroxide leads to the formation of the peroxynitrite (NOO_2^-): $H^+ + H_2O_2 + NO_2^- \rightarrow H^+ (NOO_2^-) + H_2O$. It is the peroxynitrite that dissociates forming back OH from peroxide and, therefore, plays a crucial role in the plasma oxidation of organics in water. Specifics of this oxidation mechanism of organic compounds in water are going to be discussed in detail in Lecture 29 regarding water disinfection and sterilization.

The **energy cost of plasma decontamination of water** is a critical issue of the technology, which is generally considered a highly energy intensive. While the energy efficiency of the process can differ depending on details of the specific approaches, the typical level of the energy cost can be estimated in general based on the above-described oxidation mechanism. Concentration of the RONS required for effective abatement of organic compounds in water is on the level of at least $10\,mg\,L^{-1}$, or about 10 ppm (Yang et al. 2012). G-factor of the required reactive species is about $G = 1/100\,eV$. Thus, the plasma energy per one water molecule should be about $10^{-3}\,eV$ to generate necessary amount of active species. This energy per water molecule corresponds to *the required energy cost of the RONS-induced water decontamination process of at least $30\,kJ\,L^{-1}$*. Such energy cost of the water decontamination is on the borderline of being acceptable for the large-scale environmental control application. Some relevant improvements in the energy cost by, for example, surfactancy-based concentration of the contaminants on the plasma/water surface would be preferential, see later in the Lecture.

24.11 Plasma-induced Water Softening, Mechanisms of Removal of Hydrocarbonates from Water

Thermoelectric generation is accounted for about 40% of all fresh-water withdrawals in the US, second only to irrigation. Each kWh of thermoelectric generation requires withdrawal of 25 gallons of water, which is primarily consumed for cooling purposes. Since heat removal from condenser tubes requires evaporation of pure water, the concentration of mineral ions such as calcium and magnesium in the circulating cooling water increases with time. Even though the makeup water is relatively soft, the continuous circulation eventually increases the hardness of the water due to pure water evaporation. These mineral ions, when transported through piping in ordinary plumbing system, can cause various problems, including the loss of heat transfer efficiencies in condensers and pipe clogging due to scale formation. Thus, to maintain a certain calcium hardness level in the cooling water, one must discharge a fraction of water through blowdown and replace it with makeup water. In a typical cooling tower application, the cycle of concentration (COC)in cooling water is often maintained at 3.5. That means if the calcium carbonate hardness of the makeup water is $100\,mg\,L^{-1}$, the hardness in the circulating cooling water is maintained at $350\,mg\,L^{-1}$. If the COC can be increased through continuous precipitation and removal of calcium ions, one can significantly reduce the amounts of makeup and blow-down water, resulting in the conservation of freshwater.

Various chemical and nonchemical methods are used to prevent scaling and thus increase the COC. The scale-inhibiting chemicals like chlorine and brominated compounds are conventional for the control of mineral fouling, but leading to several problems related to their storage, handling, disposal, leaks, and negative impact on environment and human health. To avoid these problems, **the nonchemical plasma** removal of hydrocarbonates and the **plasma-induced water softening** has been demonstrated using different plasmas and for softening of different types of water. This water cleaning technology is mostly based on application of the transitional "warm" plasma sources (spark discharges, gliding arcs, etc.) due to their abilities to combine effects of reactive plasma species with controllable local heating as well as their relatively high power suitable for treatment of large volumes of water. For more details $CaCO_3$, see Yang et al. (2010, 2012); Wright et al. (2014); Yang et al. (2011).

Let us shortly discuss the **mechanisms of plasma water softening**. Calcium carbonate is one of the most common scale-forming minerals occurring in water systems. It is generally the first mineral to precipitate out when water is heated due to its inverse solubility. Thus, the control of calcium carbonate scale is often the limiting factor in most industrial water applications. The carbonates can be present in the form of two negative ions: the carbonate ion CO_3^{2-} ($CaCO_3$ is non-soluble and precipitates combining Ca^{2+} and CO_3^{2-}) and the hydrocarbonate ion HCO_3^- ($Ca(HCO_3)_2$ is soluble therefore HCO_3^- stays in water as a free ion). The hard water contains high concentration of the hydrocarbonate ions HCO_3^-, which at elevated water temperatures and no acidity (typical for the cooling water heat exchangers, etc.) start dissociating: $HCO_3^- \rightarrow H^+ + CO_3^{2-}$ with further scaling on the surfaces ($CO_3^{2-} + Ca^{2+} \rightarrow CaCO_3$, solid deposit), which obviously damages the surfaces.

Mechanism of the plasma water softening is related to two major effects: acidic neutralization and "stochastic heating". First, if the plasma-treated water is sufficiently acidic (see the previous section), the hydrocarbonate ions HCO_3^- are effectively neutralized by H^+ ions in the fast recombination reducing the acidity:

$$HCO_3^- + H^+ \rightarrow H_2O + CO_2. \tag{24.7}$$

Second, the quite strong softening effect is due to local intensive pulsed water heating, the so-called **"stochastic heating"**, provided in sparks and gliding arcs and leading to local stochastic dissociation of the hydrocarbonate ions HCO_3^- with formation of CO_3^{2-}:

$$HCO_3^- \rightarrow H^+ + CO_3^{2-}; HCO_3^- \rightarrow OH^- + CO_2; OH^- + HCO_3^- \rightarrow H2O + CO_3^{2-}. \tag{24.8}$$

The stochastic local intensive heating ends up with precipitation $CO_3^{2-} + Ca^{2+} \rightarrow CaCO_3$, that is formation of the solid calcium carbonate $CaCO_3$. In contrast to the conventional scaling related to precipitation on the wall surfaces, the plasma-induced precipitation results in the formation of small particulates easy to remove by filtration. The two mechanism of plasma-induced water softening have different energy efficiencies. The acidic neutralization is provided by plasma-generated active species, and therefore requires more energy, at least $30\,kJ\,L^{-1}$, see previous section. The local intensive stochastic heating is a global effect requiring an order of magnitude less energy, see Yang et al. (2012).

24.12 Plasma-induced Cleaning of Produced and Flowback Fracking Water

Natural gas production through hydraulic fracturing (fracking) of shale formations is a rapidly accelerating field. During the widespread optimism towards the shale gas, certain environmental concerns have arisen that center on seismic effects of hydraulic fracturing (fracking) as well as the aggressive consumption of water and, more importantly, possible pollution of local water resources by produced and flowback waters. Conventional methods of water purification (for example, activated carbon, various forms of filtration, organic clay adsorbers, chemical oxidation, etc.) are challenged by extremely high organic load of the fracking water, which looks more like "dirty mineralized yogurt" than water. The problem of the fracking water cleaning (both produced and flowback) could be ultimately solved, in principle, by such methods as reverse osmosis or distillation. These approaches can be reasonably costly if operating with not hard water and with limited amount of organic load to avoid any kind of precipitation and scaling. As an example, while water evaporation energy is quite high (about $2000\,kJ\,L^{-1}$), energy requirements of the reverse osmosis and distillation can be reasonable (below $100\,kJ\,L^{-1}$, when energy efficiency of the equipment is high). This high efficiency can be achieved, however, if water to be cleaned is soft and with only limited amount of "sticky" organic load.

This challenge can be met by plasma precleaning of the fracking water. Like it was discussed in the Sections 24.10, Section 24.11, plasma treatment of water can effectively suppress both organic load and carbonates responsible for water hardness. Energy cost of both processes is on the level of about $30 \, kJ \, L^{-1}$ and is still less than that of reverse osmosis and distillation. It means that combination of the plasma pretreatment with the reverse osmosis and distillation technologies increase the energy cost only about 30%, which makes it attractive for the ultimate cleaning of the produced and backflow fracking water. Like in the general case of water softening (see Section 24.11), the most suitable discharges for the fracking water cleaning are the transitional "warm" plasma sources, especially spark discharges, gliding arcs, and in some cases, nonequilibrium microwave discharges, see Lecture 14. Possibility of simultaneous removal of the water hardness and organic load without significant increase of the process energy cost has been demonstrated using spark and gliding arid discharges submerged in the liquid. It should be mentioned that suppression of the water hardness and organic load is especially challenging in the case of fracking water because of very high concentration of the impurities. For example, the produced fracking water to be cleaned contained up to $4000 \, mg \, L^{-1}$ of bicarbonate, and concentration of calcium ions up to, 30,000 ppm. More details on the technology can be found in Wright et al. (2014); as well as in the book of Yang et al. (2012).

24.13 Plasma Water Cleaning from PFOS/PFOA and Other Perfluoro-alkyl-substances (PFAS, the "Forever Chemicals")

The **poly- and perfluoroalkyl substances (PFAS)** are a class of organofluorine compounds that are persistent in the environment, bioaccumulative and can be soluble in aqueous matrices. PFAS are thermally stable, lipid and water repelling, and have been deemed **"forever chemicals"**. PFAS have been liberally used over the past 60 years in aqueous film forming foam (AFFF), household products such as carpets, paper, and nonstick cookware, and even coated cardboard takeout containers. Extensive groundwater and surface water contamination of PFAS originates from a variety of sources, including industrial and commercial manufacturing plants that produce or use PFAS contaminated biosolids application, contaminated landfill leachate, wastewater, and water treatment plants discharges, and AFFF training facilities. Drinking water supplies for millions of U.S. residents have been contaminated with various PFAS, with both perfluorooctanesulfonic acid (PFOS) and perfluorooctanoic acid (PFOA) often surpassing the US EPA's lifetime health advisory ($70 \, ng \, L^{-1}$). Since this advisory was set for these two compounds (PFOS & PFOA), most research done to date focuses on PFOS and PFOA, but a growing number of studies are investigating a wider range of PFAS. Due to PFAS contamination in groundwater, surface water, agriculture, and drinking water and their associated health risks, there has been a great focus on developing the relevant water treatment technologies.

The **PFAS treatment technologies** developed so far have included adsorptive and destructive methods. Adsorptive methods to date include granular activated carbon (GAC), ion exchange polymers, and protein addition. These methods show promise in removing some of the compounds from contaminated water but can be expensive and do not degrade the PFAS compound, leading to the generation of PFAS contaminated residues and concentrates. Due to the recalcitrant nature of PFAS and the C–F bond, destructive methods are strongly restricted. In contrast to hydrocarbons RH, which can be destroyed by OH, see (24.4–24.6), the RF compounds have no simple exothermic oxidative pathways, see Figure 24.7. It focuses attention to the **plasma technologies of the PFAS abatement** providing combined effect of the RONS and temperature as well as contribution of electrons triggering the dissociative attachment to the strongly electronegative PFAS compounds. The plasma process is usually organized either in DBD or coronas of noble gases (like argon) to provide stronger effect of electrons (because these gases are not electronegative), or in transitional "warm" discharges like gliding arcs, sparks, or streamer coronas to have adaptationally the controlled contribution of temperature. Discharges are usually located above the water surface or submerged in water (often with bubbles), with general tendency to maximize the plasma/surface contact. Maximizing the plasma/surface contact is especially important for plasma treatment because the PFAS compounds are surfactants and can be to some extent concentrated on the water surface. As an example, Figure 24.8 illustrates the submerged gliding arc discharge with bubbling effectively applied to water cleaning from PFOS and PFOA, see Lewis et al. (2020), as well as Lectures 11, 14. Analyzing the PFOS/PFOA destruction in water, we should note that often this process is limited to only their conversion to other types of PFAS or related radicals. While it can be profitable in some cases (for example to

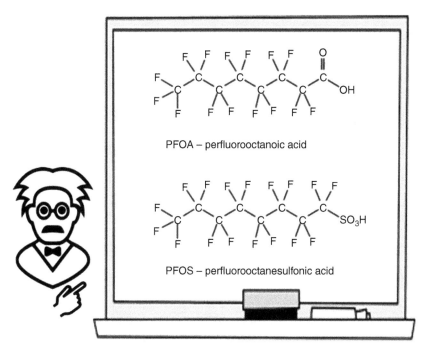

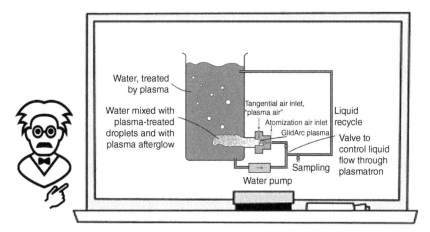

Figure 24.7 Chemical structure of PFOA and PFOS as examples of the PFAS.

Figure 24.8 Schematic of submerged gliding arc water bubbling system for PFAS abatement.

stimulate further filtration), the ultimate objective of the plasma cleaning is obviously "mineralization", that is PFAS conversion to CO_2 and HF. It requires special organization of the plasma processes effectively combining effects of the RONS, electrons, and controllable level of temperature at the plasma/water surfaces, which is going to be shortly discussed in the next section.

For more details on the experimental organization of the PFAS abatement technology see Nau-Hix et al. (2021); Nzeribe et al. (2019); Lewis et al. (2020).

24.14 Plasma Chemistry of PFAS Mineralization in Water: Mechanisms and Energy Balance

As it was explained in the previous section, the complete mineralization (conversion to carbon oxides and HF) is the optimal goal of the plasma PFAS abatement techniques because PFAS are known to transform during the water treatment into recalcitrant by-products of concern. *Abilities of the complete PFAS mineralization in plasma and relevant optimizations of the plasma technology require first understanding the mineralization reaction enthalpies and especially their endothermic level to determine whether a treatment technique is providing sufficient energy to propel mineralization to completion.* The thermodynamic properties of individual PFAS components are not completely known but still can be used to characterize enthalpy of at least the complete mineralization processes in aqueous systems, see Snitsiriwat et al. (2022); Surace et al. (2022). Let us first focus on **thermodynamics of the PFOA mineralization in water**, without reactive species or additional oxidizing agents:

$$C_8HF_{15}O_2(PFOA, aq) + 7\,H_2O(l) \rightarrow 15\,HF(aq) + 7\,CO(g) + CO_2(g). \tag{24.9}$$

This reaction is slightly endothermic with $\Delta H \approx 120$ kcal mol^{-1} (slightly below 5 eV). The endothermic level here is low keeping in mind that the PFOA molecule is quite big, and even per carbon atom, the process enthalpy is only 15 kcal mol^{-1} (about 0.6 eV per carbon atom). Although the enthalpy of the reaction is not high, it is still positive. It means that *the complete mineralization process is although slightly but endothermic and require energy and some level of temperature to be accomplished. If a process is exothermic (like oxidation of hydrocarbons), its acceleration requires only plasma active species to be catalyzed. The endothermic process (24.9) requires for its acceleration not only plasma active species but also some energy input and some level of temperature to be stimulated.*

The understanding of the specific energy input and temperature requirement for the complete mineralization in plasma can be achieved by analyzing the thermodynamic composition of the mixture and relevant total enthalpy as a function of temperature. Then energy cost of the plasma-chemical process can be found assuming the absolute quenching mechanism, see Section 17.4. Example of such dependence of the degree of the complete PFOA defluorination (mineralization) on the specific energy input in plasma is presented in Figure 24.9 in comparison with the experimental results achieved using the submerged gliding arc, see Figure 24.8 (Surace et al. 2022; Lewis et al. 2020). It should be mentioned that this figure shows the modeling results corresponding to different values of the PFOA standard entropy and specific heat (the best fit is achieved at higher relevant values of the standard entropy and specific heat.

The Figure 24.9 shows that *achieving significant level of mineralization requires specific energy inputs on the high but reasonable level of 500–1000 kJ L^{-1}. Relevant temperature requirement for the plasma-water surface area, where most of the process occur, is about 500 K.* Decrease of the process energy cost is possible by increasing the plasma-water contact surface, for example by bubbling, as well as concentrating of the PFAS on the plasma-water surface using, for example, the surfactancy effect.

Analysis of the **plasma-chemical kinetics of PFAS destruction in water** indicates the dominating role of the plasma-generated OH radicals and electrons (hydrated in water), see Stratton et al. (2017); Zhang et al. (2021a,b). Surace et al. (2022) estimated the activation energy barriers of the major OH-induced advanced oxidation reaction of

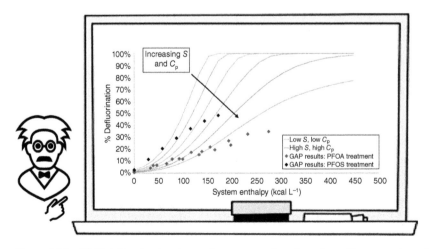

Figure 24.9 PFAS defluorination degree in water as a function of specific energy input in plasma: curves show modeling results (left to the right corresponds to increase of PFOA entropy and specific heat); dots correspond to experiments with submerged gliding arc discharge (PFAS, higher; PFOA lower).

PFAS (represented as RF) OH + RF → HF + RO, as well as electron-induced defluorination e + RF → F$^-$ + RF. Both activation energies are about 1 eV mol^{-1}, while the activation energy of the dissociative attachment is slightly lower than that of oxidation. These activation energies bring us back again to temperatures about 500 K and required energies on the level of 500–1000 kJ L^{-1} (to be improved by increase of the effective plasma-water contact surface, and stimulation of the relevant surface processes).

24.15 Plasma Environmental Cleaning of Soil: Destruction of PFAS Compounds, Vitrification of Contaminated Radioactive Soil, and Other Related Solid Waste Treatment Technologies

Plasma abatement of PFAS in water has been discussed above pointing out the relevant big challenge of the tiny concentration of these hazardous compounds still requiring significant energy deposition in the entire water (500–1000 kJ L^{-1}), if there is no PFAS pre-concentration. The **plasma cleaning of soil from PFAS** does not have this problem in principle, because the PFAS compounds are mostly located on the surface of the soil particles from the beginning. PFAS abatement in soil has, however, another challenge, namely plasma access to the surface of the soil particles. Zhan et al. (2020) claim of 70% PFAS degradation (with 20% mineralization) in soil using the pulsed streamer coronas requires about 50,000 kJ kg^{-1} of soil, which is two orders of magnitude higher than in the case of water treatment. Surely, this technology requires better plasma contact with the soil particles by using for example the gliding arc fluidized bed, see Lecture 11. Similar approach demonstrated effective PFAS suppression to regenerate GAC after its use in filtration of the polluted water, see Section 24.13. Organization of the processes not in cold plasma but the transitional "warm" discharges, especially gliding arcs permits significant increase of the process energy efficiency above that of the plasma PFAS abatement in water.

Thermal and transitional "warm" plasmas can be also quite effective in environmental compaction of solid hazardous wastes, especially including treatment of radioactive wastes by their plasma vitrification for final disposal, see Watanabe (2003); Tzeng et al. (1998). Vitrification is a proven and reliable technology used at waste processing facilities. The process converts the contaminated soil as well as liquid radioactive and chemical waste into a solid, stable glass, eliminating environmental risks. It should be mentioned in this regard that the thermal plasma process in this case can emit some amount of the waste into gas phase as a gaseous exhaust, tar, or relevant particulates. It can require additional gaseous exhaust treatment, which can be also effectively organized in plasma especially using the gliding arc discharges, see Rusanov and Fridman (1984) and Fridman (2008), for more details.

24.16 Problems and Concept Questions

24.16.1 Application of Relativistic Electron Beams in Exhaust Gas Cleaning

Why plasma generated by the relativistic electron with the electron energies on the level of 0.5–1.5 MeV are more often used for large scale air and exhaust gas cleaning than in other relevant environmental plasma applications?

24.16.2 Plasma-catalytic Ion-molecular Chain Oxidation of SO$_2$ to SO$_3$ in Airflow

Why the ion-molecular chain oxidation mechanism in contaminated airflow, see Section 24.2, occurs only at elevated current densities, but energy efficiency of the process decreases with further intensification of plasma?

24.16.3 Plasma-stimulated Combined Oxidation of NO$_x$ and SO$_2$ in Air

Explain why the simultaneous NO and SO$_2$ oxidation processes appear to be more effective than their plasma-assisted individual oxidation processes in air?

24.16.4 VOC Removal from Exhaust Gases Using Wet Pulsed Corona Discharge

Explain why the combination of nonthermal plasma treatment with water scrubbing permits a significant decrease of energy cost of VOC removal from polluted gas streams? Why a single OH radical can be sufficient to convert a non-soluble organic molecule into a soluble one?

24.16.5 Plasma-induced Mechanisms of Destruction of the Organic Compounds in Water

Plasma-generated OH radicals very effectively stimulate oxidation and destruction of VOC, the organic compounds in the gas phase. Why organization of the same plasma-induced oxidation process in liquid phase becomes more effective in acidic environment, especially in the plasma produced acids?

24.16.6 Plasma-induced Water Softening, Removal of Hydrocarbonates from Water

Why heating of hard water leads to the dangerous effect of the scale deposition on the walls of heat exchangers, but the plasma initiated "stochastic" heating of the same hard water results in its softening and cleaning?

24.16.7 Plasma Chemistry of PFAS Mineralization in Water

Why the plasma-stimulated mineralization of PFAS in water requires not only active and charge species (like OH and hydrated electrons) but also requires some level of heating or energy deposition?

24.16.8 Plasma Environmental Cleaning of PFAS Contaminated Soil

Why energy efficiency of the plasma PFAS abatement and mineralization in contaminated soil can be higher than that of plasma treatment of the contaminated water? What is the key challenge of this plasma environmental technology?

Part IV

Organic and Polymer Plasma Chemistry, Plasma Medicine, and Agriculture

Lecture 25

Organic Plasma Chemistry: Synthesis and Conversion of Organic Materials and Their Compounds, Synthesis of Diamonds and Diamond Films

25.1 Thermal Plasma Pyrolysis of Methane: The Kassel Mechanism, the Westinghouse Process, Co-production of Acetylene and Ethylene

Among numerous plasma processes related to the conversion of hydrocarbons, special attention is always focused on the endothermic process of **methane conversion into acetylene**:

$$CH_4 + CH_4 \rightarrow C_2H_2 + 3H_2, \Delta H = 3.8 \, \text{eV mol}^{-1}. \tag{25.1}$$

The process is of great practical interest related to the direct organic synthesis based on conversion of natural gas for further use in brazing, soldering metals, glass industry, production of acetic acid and acrylonitrile, and especially for manufacturing of synthetic rubber to make tires, etc. It is mainly important when the oil and the oil accompanying gases are not available, and the organic synthesis should be based on natural gas and coal. This plasma technology has been organized at large industrial scale using thermal arcs with an optimal energy cost of 8.3 eV mol^{-1} (energy efficiency 46%). An alternative carbide method of acetylene production requires 10–11 eV mol^{-1} (energy efficiency 35–40%, product yield 90–100 g C$_2$H$_2$ per kWh), which explains the advantage of the plasma-chemical approach with respect to applying calcium carbide.

Plasma pyrolysis of hydrocarbons proceeds through a long sequence of chemical processes that generally have condensed-phase carbon and hydrogen as final products. Formation of the condensed-phase products at 1100–1300 K is mostly due to the contribution of radicals; soot formation at temperatures exceeding 1300 K is mostly due to molecular processes on the carbon surfaces. Thermal decomposition of methane generally follows the phenomenological (reduced) scheme suggested by Kassel in 1932. According to the **Kassel mechanism**, the thermal pyrolysis of methane is determined by the initial endothermic formation of CH$_2$ radicals:

$$CH_4 \rightarrow CH_2 + H_2. \tag{25.2}$$

Other radicals, like CH$_3$, make the principal contribution in the real plasma systems to the initial kinetics of the plasma pyrolysis. The Kassel mechanism is reduced one and focuses only on CH$_2$, which is sufficient for the qualitative analysis. The CH$_2$ radicals, according to the Kassel mechanism, lead to methane conversion into ethane:

$$CH_4 + CH_2 \rightarrow C_2H_6. \tag{25.3}$$

Further dehydrogenization sequence results in a gradual conversion of ethane into ethylene, ethylene into acetylene, and finally acetylene into soot:

$$C_2H_6 \rightarrow C_2H_4 + H_2, C_2H_4 \rightarrow C_2H_2 + H_2, C_2H_2 \rightarrow 2C(\text{solid}) + H_2. \tag{25.4}$$

The major reactions of the Kassel mechanism (25.2)–(25.4) follow the first kinetic order, their reaction rate coefficients are discussed below in Section 25.5.

 *Although the acetylene conversion into soot includes numerous stages, this exothermic reaction proceeds relatively quickly with respect to other steps of the Kassel mechanism. For this reason, the major intermediate components of methane pyrolysis are **ethylene and acetylene**. Fixation of a significant amount of higher hydrocarbons preceding the soot formation has low probability. During the thermal plasma pyrolysis of methane, ethylene is usually formed after 10^{-6}–10^{-5} s, and acetylene after 10^{-4}–10^{-3} s. Then fast*

quenching should be performed to avoid significant soot formation. The **conversion of natural gas into acetylene, ethylene, and hydrogen** is the most developed industrial plasma process of hydrocarbon pyrolysis. The process takes place directly in thermal plasma discharges and, therefore, is called **electric cracking**. Industrial electric cracking units usually operate at power levels of 5–10 MW; the length of the thermal arc is about 1 m. The total conversion degree of natural gas is 70%; conversion into acetylene is 50%; volume fraction of acetylene in products is 14.5%; and consumption of electric energy is 13 kWh per 1 m^3 of natural gas. A general challenge of the plasma-chemical process is the considerable amount of soot in the reaction products.

The energy-effective **Westinghouse plasma process of natural gas conversion** is conducted in two steps. The first step is the electric cracking of natural gas, and the second step is the product quenching by heavy hydrocarbons with their simultaneous pyrolysis. This approach results in high product yields: the conversion of methane into acetylene reaches 80%, the volume fraction of acetylene in products is 20%, and the consumption of electric energy is 1 about 10 kWh per 1 kg of acetylene. These data characterize large-scale industrial plasma systems producing 25 000 tonnes of acetylene per year. Industrial-scale electric cracking is used also for **co-production of ethylene**, which was demonstrated for example by Huls, Chem. W., Germany. The relative yield of acetylene and ethylene depends on temperature and duration of the process, the type of quenching, the composition of initial products, and so on. The maximum production of acetylene usually corresponds to temperatures of 1800–2000 K (for CH$_4$ and C$_3$H$_8$) and temperatures of 1600–1880 K (for heavier hydrocarbons); the process duration is about 10^{-5}–10^{-4} s. In such regimes, the combined yield of acetylene and ethylene can be up to 80 mass %. The energy cost of the co-production is 4.6 kWh kg^{-1} in the case of pyrolysis of gasoline or crude oil, and 9.5 kWh kg^{-1} in the case of pyrolysis of methane. At lower temperatures (1300–1600 K) and similar process duration (10^{-5}–10^{-4} s), pyrolysis of hydrocarbons with more than three carbon atoms results in higher concentration of ethylene. Pyrolysis of gasoline, for example, leads in these conditions to the production of 13 vol% of ethylene in products and corresponds to equal production of acetylene and ethylene. In general, the double-step pyrolysis leads to a larger production of ethylene. Changing the starting moment of the quenching-focused injection of hydrocarbons affects the acetylene-to-ethylene ratio in products of the pyrolysis. Earlier injection results in a larger amount of ethylene in products, see more details in Fridman (2008).

25.2 Thermal Plasma Pyrolysis of Higher Hydrocarbons

Thermal plasma pyrolysis is surely not limited to natural gas. The hydrocarbon mixtures often used for pyrolysis are crude oil compounds and gas condensates, which consist of not only alkanes but also significant amount of aromatic compounds. A typical composition of a gas condensate, for example, is normal alkanes (represented as C$_8$H$_{18}$), 21%; naphtheno-isoparaphenes (represented as C$_7$H$_{14}$), 58%; aromatic compounds (represented by C$_9$H$_{14}$), 10%; and multi-nuclei aromatics (represented by C$_{12}$H$_{10}$), 11%. The **kinetics of plasma-chemical pyrolysis of complicated mixtures** is discussed by Polak et al. (1975). Kinetic curves illustrating the plasma pyrolysis of gas condensate are shown in Figure 25.1 (those for gasoline are similar) and are not much different qualitatively from those for the methane pyrolysis. However, the time required to reach maximum concentrations of ethylene and acetylene is shorter in the case of the gas condensate pyrolysis. The ethylene concentration reaches maximum after 10^{-6} s, and acetylene after 10^{-5}–10^{-4} s. It is due to lower enthalpy of the conversion reaction in the case of gas condensate and gasoline than in the case of methane (25.1). The plasma jet temperature first decreases during the initial 10^{-4} s due to endothermic processes of the CH$_4$ and C$_2$H$_4$ decomposition. Then the temperature increases during exothermic C$_2$H$_2$ conversion into soot (see Figure 25.1). To save acetylene and other valuable intermediate products, quenching should start after no later than 10^{-4} s with cooling rate of 5 × 10^6 K s^{-1}. The cooling processes significantly affect the kinetic curves shown in Figure 25.1 starting from about 10^{-5} s: the acetylene concentration decreases, while the ethylene concentration increases with respect to those values without cooling. Starting from about 3 × 10^{-5} s, the concentrations of products stop changing almost completely. Thus, cooling rates of 10^6–10^7 K s^{-1} provide effective quenching of the products of the plasma pyrolysis.

Thermal pyrolysis of numerous hydrocarbons has been investigated in thermal plasma systems, mostly in arc jets over a wide range of power, from 10–30 kW to 10 MW. Typical energy carriers in plasma jets in this case are hydrogen or H$_2$–CH$_4$ mixture. In the case of **plasma pyrolysis of hydrocarbons with more than three carbon atoms** (C > 3), the composition of the initial products (alkanes) does not much affect the process yield. However, an increase in the fraction of hydrocarbons with higher boiling temperatures leads to decrease of acetylene and ethylene yield.

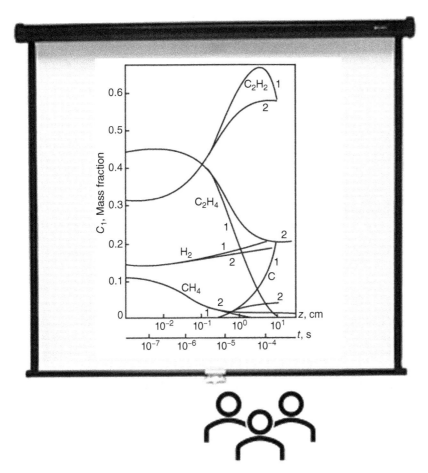

Figure 25.1 Kinetics of the plasma-chemical pyrolysis of gas condensate: (1) without cooling losses in the reactor; (2) with cooling losses in the reactor.

Also, aromatic compounds in the mixture undergoing plasma pyrolysis lead to higher yields of soot. An increase of gas pressure affects the plasma pyrolysis of hydrocarbons in a negative manner. The pressure increase from 1 to 6 atm results in lower yields of acetylene and higher energy costs of its production. An increase of temperature from 1300 to 1700 K leads at first to an increase in the yields of vinyl acetylene (C_4H_4) from 0.18 to 0.30 vol% at 1500 K; then the yield decreases to 0.24 vol%. The concentration of diacetylene in this case does not exceed 0.1 vol%; the concentration of other compounds decreases from 1.44 to 0.4 vol%.

The plasma pyrolysis can be conducted with a relatively low yield of products (60–80%) in the gas phase, with the rest formed mostly in the liquid phase. The liquid products then include significant amount of aromatic compounds. For example, the paraphene/naphthene fraction of gas condensate does not initially include any aromatic compounds. Pyrolysis of this fraction can result in formation of liquid products containing 20% aromatic compounds, specifically toluene, benzene, and xylols. More details on the subjects can be found in Fridman (2008).

25.3 Technologies Based on Thermal Plasma Pyrolysis of Hydrocarbons: Production of Vinyl Chloride and Production of Acetylene by Carbon Reactions with Hydrogen and Natural Gas

Acetylene and ethylene produced by plasma pyrolysis of hydrocarbons can be used for further synthesis of different monomers, including vinyl chloride, acryl-nitryl, and chloroprene. The introduction of the plasma stages leads in this case to a fundamental reorganization of the entire technology. As a specific technological example, let's consider the **plasma-based industrial process for the production of vinyl chloride**, which is a monomer applied for

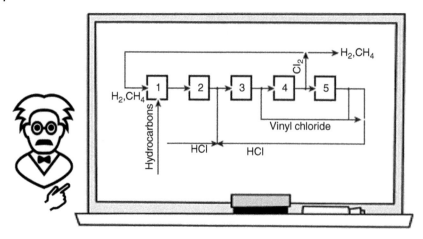

Figure 25.2 Technological scheme of production of the vinyl chloride monomer: (1) plasma-chemical pyrolysis; (2) cleaning from higher unsaturated hydrocarbons; (3) hydrochlorination of acetylene; (4) chlorination of ethylene; (5) thermal pyrolysis of dichloroethane.

large-scale production of polyvinyl chloride (PVC). The technology is based on plasma pyrolysis of hydrocarbons with production of acetylene and ethylene. Vinyl chloride is produced by hydrochlorination of acetylene and by chlorination of ethylene with a subsequent pyrolysis of dichloroethane. Hydrochlorination of acetylene and chlorination of ethylene is completed without extracting them from the plasma pyrolysis gases. A schematic of the vinyl chloride production technology is illustrated in Figure 25.2. The first step of the technology is pyrolysis of hydrocarbons in the plasma jet containing 80–90% hydrogen and 10–20% methane (see Section 25.1). Quenching is accomplished by using either water or gasoline jets. The gas composition after separation of the quenching liquid is hydrogen (H_2) 60.8–66.9 vol%, acetylene (C_2H_2) 12.1–20.0 vol%, ethylene (C_2H_4) 5.0–12.9 vol%, C_2H_6 up to 0.3 vol%, CH_4 7.5–10.5 vol%, C_3H_6 0.1–1.9 vol%, and higher nonsaturated compounds (vinyl acetylene, methyl acetylene, diacetylene, propadiene, and butadiene) 0.5–1.5 vol%. The second step is separation of the higher nonsaturated hydrocarbons (vinyl acetylene, methyl acetylene, diacetylene, propadiene, and butadiene) by kerosene absorption (pressure 6–10 atm). Hydrogen chloride (HCl) is injected in the plasma pyrolysis gases in stoichiometry with acetylene for further hydrochlorination (step 3). HCl applied in this stage is recycled from the dichloroethane pyrolysis (step 4). Catalytic hydrochlorination of acetylene and production of vinyl chloride takes place in the third step without C_2H_2 extraction from the pyrolysis gases: $C_2H_2 + HCl \rightarrow C_2H_3Cl$. The gas composition after this stage is hydrogen (H_2) 63.1–67.3 vol%, vinyl chloride (C_2H_3Cl) 12.5–20 vol%, ethylene (C_2H_4) 5.0–13.3 vol%, C_2H_6 0–0.3 vol%, and CH_4 7.6–10.8 vol%. Vinyl chloride is extracted from the mixture by dichloroethane absorption at 10 atm. The fourth step of the technology is chlorination of ethylene in the pyrolysis gases by injection of chlorine (Cl_2): $C_2H_4 + Cl_2 \rightarrow C_2H_2Cl_2$. The chlorination of ethylene proceeds in liquid dichloroethane in the presence of a catalyst (iron chloride) at a pressure of 5 atm without separation of the pyrolysis gas mixture. The composition of the residual plasma pyrolysis gas mixture after the liquid-phase chlorination is hydrogen (H_2) 85.5–90 vol% and methane (CH_4) 10–14.5 vol%. This mixture is then applied as an energy carrier gas for plasma pyrolysis of hydrocarbons in the first stage. Dichloroethane formed in liquid phase then undergoes thermal pyrolysis in the fifth step of the technology. The pyrolysis of dichloroethane is performed at temperatures of 450–550 °C and pressures of 7 atm and results in production of vinyl chloride and hydrogen chloride: $C_2H_4Cl_2 \rightarrow C_2H_3Cl + HCl$. Vinyl chloride produced at this stage is a product of the entire technology together with the vinyl chloride produced in the third step. HCl produced in the fifth step is recycled back to the third step for hydrochlorination of acetylene.

Another relevant technological example is the **thermal plasma production of acetylene by carbon reactions with hydrogen or methane.** High-intensity arc discharges operating at atmospheric pressure and temperatures sufficient for sublimation of graphite have been applied to produce acetylene by reactions of the carbon vapor with hydrogen or methane. Reaction with hydrogen provides the formation of acetylene (19.3 vol% of C_2H_2) in products. Using a plasma jet in a mixture of helium and hydrogen ($He:H_2 = 2:1$), the concentration of acetylene in products is 24 vol%, attracting technological interest. This value exactly corresponds to the thermodynamic equilibrium. Replacement of hydrogen by methane as the gas energy carrier in plasma jets leads to an increase of acetylene concentration in products reaching 52 vol%, again in good agreement with the thermodynamic equilibrium.

25.4 Thermal Plasma Technology of Pyrolysis of Hydrocarbons with Production of Soot and Hydrogen

The fundamental scientific aspects of the plasma pyrolysis of methane and other hydrocarbons with formation of soot, especially the kinetics and energy balance of the plasma-catalytic effects related to this process has been discussed in Section 22.4. Here, we are going to consider shortly the **technological aspects of the plasma soot production technology**. Generally, when plasma pyrolysis is conducted without special quenching, the hydrocarbons can be mostly converted into soot and hydrogen. The quality of the soot is better than that of the soot generated in furnaces. The soot produced in plasma is highly dispersed and characterized by a developed secondary structure.

 The mechanism and detailed kinetics of the soot formation in the thermal plasma technology can be found, for example, in Rykalin et al. (1970). It shows that *the most intensive soot production takes place during pyrolysis of aromatic compounds*. The process has been performed in the arc discharges and thermal radiofrequency (RF) inductively coupled plasma (ICP) discharges directly, as well as in plasma jets of argon, nitrogen, and hydrogen at temperatures from 1500 to 5000–6000 K and residence times of 10^{-3}–10^{-2} s. Different hydrocarbons were used as a feedstock, including methane, natural gas, propane, toluene, naphthalene, xylol, gasoline, and special mixtures of aromatic compounds. The maximum yield of soot in these plasma systems (45–70 mass%) is achieved by pyrolysis of aromatic compounds. The energy cost of soot is 7–13 kWh kg^{-1} in this case. Thermal plasma pyrolysis of paraphenes and naphthenes leads to a lower yield of soot (18 mass%) and a higher energy cost (70 kWh kg^{-1}).

The structure and quality of soot are usually characterized by the so-called **soot oil number**, which is usually not higher than 2.4 cm^3 g^{-1} for conventional soot. The oil number for plasma produced soot reaches 4.0 cm^3 g^{-1}, and the specific surface exceeds 150 m^2 g^{-1}. High-quality soot with fixed crystal structure can be also produced by **thermal plasma pyrolysis of low-quality soot** or other cheap carbon-containing materials. Initial products in this case are injected into a plasma jet, evaporated, and condensed after the quenching process to form special high-quality soot and carbon structures. Plasma pyrolysis of hydrocarbons leads to the formation of a condensed phase not only in the volume but also on the walls of the reactor or specially designated surfaces. The reaction of condensed phase carbon with hydrocarbons usually results in the formation of **pyro-graphite** on the reactor walls. The process takes place in thermal plasma conditions at temperatures of 1800–2100 K and atmospheric pressure. We should mention that the thermal plasma produced by soot can be used for further extraction of the special carbon-based structural materials including fullerenes, carbon nanotubes, graphene, etc. Some of this process has been already discussed in Lecture 20, some will be discussed later in this lecture in connection to the diamond film technologies.

25.5 Nonthermal Plasma Conversion of Methane into Acetylene: Contribution of Vibrational Excitation, Energy Efficiency of the Process

One of the key challenges for the plasma conversion of methane into acetylene (25.1) is improvement of its energy efficiency or, in other words, reduction of the **energy cost of acetylene production**. The lowest energy cost achieved in the technological thermal plasma systems so far is 8.5 eV mol^{-1} C$_2$H$_2$ corresponding to the energy efficiency of less than 45%. Lower energy costs can be achieved in nonthermal plasma conditions if the process is stimulated by vibrational excitation of CH$_4$ molecules. Conducting the methane-to-acetylene conversion in a nonequilibrium microwave discharge at moderate pressures (10–80 Torr) in conditions optimal for vibrational excitation permits an energy cost of acetylene production of 6 eV mol^{-1} C$_2$H$_2$ (energy efficiency 63%), and a methane-to-acetylene conversion degree of 80% (Babaritsky et al. 1991, see also Fridman 2008). When the pressure increases from 10 to 80 Torr, the methane-to-acetylene conversion degree in these systems rises and reaches a maximum of 80% at the specific energy input $E_v = 2.6$ eV mol^{-1} CH$_4$. The gas product composition in this regime is H$_2$, 73 vol%; CH$_4$, 5 vol%; and C$_2$H$_2$, 22 vol%. The concentration of ethylene, ethane, and higher hydrocarbons is less than 1%; soot production is not significant. Nonequilibrium plasma conversion of methane into acetylene does not require special quenching because the translational gas temperature remains relatively low.

The total endothermic process (25.1) of acetylene production from methane under the quasi-equilibrium conditions corresponds to the first kinetic order and can be characterized by the reaction rate coefficient $k = 10^{12}$ exp $(-39\,000\,K/T)$, $1\,s^{-1}$. The **detailed Kassel mechanism kinetics** can be represented as the sequence of the quasi-equilibrium transformations of methane to ethane (endothermic), ethane to ethylene (endothermic), ethylene to acetylene (endothermic), and acetylene to soot (exothermic). The reaction rate coefficients of the Kassel mechanism are all the first kinetic order:

$$2\,CH_4 \rightarrow C_2H_6 + H_2, \quad \Delta H = 0.7\,eV\,mol^{-1}, k = 4.5 \times 10^{13}\,\exp(-46\,000\,K/T), 1/s; \tag{25.5}$$

$$C_2H_6 \rightarrow C_2H_4 + H_2, \quad \Delta H = 1.4\,eV\,mol^{-1}, k = 9.0 \times 10^{13}\,\exp(-35\,000\,K/T), 1/s; \tag{25.6}$$

$$C_2H_4 \rightarrow C_2H_2 + H_2, \quad \Delta H = 1.8\,eV\,mol^{-1}, k = 2.6 \times 10^{8}\,\exp(-20\,500\,K/T), 1/s; \tag{25.7}$$

$$C_2H_2 \rightarrow 2\,C_{solid} + H_2, \quad \Delta H = -2.34\,eV\,mol^{-1}, k = 1.7 \times 10^{6}\,\exp(-15\,500\,K/T), 1/s. \tag{25.8}$$

To produce acetylene, the sequence of reactions (25.5)–(25.8) should be interrupted after stage (25.7) of C_2H_2 formation is completed and before the fast exothermic reaction (25.8) converts a significant amount of acetylene into soot. It can be achieved when the gas residence time in the discharge is 10^{-3} s and the quenching rate is $10^6\,K\,s^{-1}$.

The quasi-equilibrium kinetics (25.5)–(25.8) determines the **lower limit of the energy cost for acetylene production from methane in thermal plasma**, which is $8.5\,eV\,mol^{-1}$ C_2H_2, which can be achieved at a specific energy input $E_v = 2.5–3\,eV\,mol^{-1}$. The energy costs achieved in the nonthermal microwave discharge at pressures about 80 Torr are significantly below the limit of the quasi-equilibrium systems and reach in the minimum $6\,eV\,mol^{-1}$ C_2H_2, which can be interpreted by the nonequilibrium contribution of the vibrational excitation of CH_4 molecules.

VT nonequilibrium $(T_v > T_0)$ *can increase the energy efficiency of methane conversion* in accordance with general principles discussed in Lecture 17. Specifically for the case of methane conversion, the first three steps in the Kassel scheme (25.5)–(25.7) are endothermic and are, therefore, effectively stimulated by vibrational excitation, while acetylene conversion into soot (25.8) is exothermic, and therefore is controlled by translational temperature. Thus, the $T_v > T_0$ nonequilibrium promotes the formation of acetylene without significant soot production. A limiting elementary reaction of the methane conversion mechanism is initial dissociation of CH_4 molecules, characterized by relatively high activation energy: $CH_4 + M \rightarrow CH_3 + H + M$, $E_a = 4.5\,eV\,mol^{-1}$. This elementary reaction proceeds mostly through vibrational excitation of CH_4 molecules even in quasi-equilibrium systems. Therefore, providing the nonequilibrium discharge energy selectively to the vibrational degrees of freedom of CH_4 molecules permits significantly lower the process energy cost. In contrast to CO_2 dissociation, however, the level of the VT nonequilibrium in CH_4 plasma of moderate pressure is not so high. It is mostly due to fast VT relaxation of CH_4 molecules in collisions with multiple intermediate products of the conversion process, which explains why the maximum energy efficiency of methane conversion into acetylene achieved in nonequilibrium conditions (63%) is lower than that of CO_2 dissociation (up to 90%).

The **vibrational excitation of methane**, however, still improves the energy efficiency with respect to quasi-equilibrium systems, even when the parameter of nonequilibrium, $\gamma = (T_v - T_0)/T_0$, is not very high (usually $\gamma = 0.1–0.5$). For these not very high values of the nonequilibrium parameter, $\gamma = (T_v - T_0)/T_0$, the acceleration of the methane dissociation $CH_4 + M \rightarrow CH_3 + H + M$ can be determined as: $\ln k_R(T_0, T_v) = \ln k_{R0}(T_0) + A\,\gamma$, where the parameter $A = E^*(T_0)/T_v$ is about unity and only slowly changing with temperature $(E^*(T_0)$ is the vibrational energy of a CH_4 molecule when the VT relaxation rate becomes equal to that of the vibrational–vibrational exchange). Relevant kinetic and energy balance modeling explains the minimum energy cost of the acetylene production about $6\,eV\,mol^{-1}$ C_2H_2, in a good agreement with the nonequilibrium microwave plasma experiments.

25.6 Other Processes of Decomposition, Elimination, and Isomerization of Hydrocarbons in the Nonequilibrium Plasma Chemistry

The decomposition of numerous hydrocarbons has been investigated in nonequilibrium glow, microwave, and RF discharges mostly at low pressures (below 10 Torr). Some nonequilibrium hydrocarbon decomposition processes were analyzed at higher pressures (including atmospheric pressure), using corona and dielectric barrier discharges (DBDs). Specifically, we can mention nonthermal plasma decomposition of linear alkanes such as CH_4, C_2H_6, n-C_4H_{10}, n-C_5H_{12}, n-C_6H_{14}, and n-C_8H_{18}; cyclic alkanes and their derivatives, such as cyclopentane, cyclohexane, and methylcyclohexane; alkenes such as C_2H_4 and C_4H_8; and aromatic compounds such as C_6H_6, $C_6H_5CH_3$, and $C_6H_4(CH_3)_2$, see Slovetsky (1981) for more details.

Nonequilibrium plasma has been effectively applied for several hydrocarbon isomerization processes, see Suhr (1973, 1974). The conversion degree of isomerization can be very high, reaching 80% in the treatment of 2-naphtyl-methyl-ester. The selectivity of the reactions can also be very high. For example, from all conversion products of *trans*-stylbene, 95% is *cis*-stylbene. Similar characteristics are achieved in nonequilibrium plasma stimulation of elimination reactions. In the elimination processes, plasma provides selective cutting of some specific functional groups from initial hydrocarbon molecules. For example, the nonthermal plasma treatment of phthalic anhydride is characterized by a high conversion degree (70%) and selectivity, which means that almost 90% of the products are di- and triphenylene. Such conversion and selectivity characteristics of the nonequilibrium plasma systems allow them to be used in preparative organic chemistry.

The combination of plasma with catalysts, usually referred to as plasma catalysis, can be helpful in improving yield characteristics of some processes of plasma conversion of hydrocarbons. The term "plasma catalysis," like it has been discussed in Lecture 22 is applied to two very different situations: combination of plasma chemistry with conventional catalysts and replacing of conventional catalysts by plasma. Here we consider the simple combination of plasma-chemical conversion of methane with application of catalysts to increase the process yield. As an example, the combination of atmospheric-pressure pulsed corona discharge with Ce/CaO–Al$_2$O$_3$ catalyst results at room temperature in a 14% yield of acetylene from methane with selectivity of C$_2$H$_2$ production exceeding 70% (Gong et al. 1997). A combination of atmospheric pressure DBD with Ni catalyst increases the energy efficiency of the process by a factor of 40 (Nozaki 2002; Kado et al. 2003). See Lecture 22 for more examples.

25.7 Thermal Plasma Synthesis and Conversion of Nitrogen-organic Compounds: Production of C$_2$N$_2$ from Carbon and Nitrogen; Co-production of HCN and C$_2$H$_2$ from Methane and Nitrogen

Dicyanogen (C$_2$N$_2$) can be effectively formed in thermal N$_2$ discharges by the plasma-chemical reaction $2C + N_2 \rightarrow C_2N_2$. Carbon can be introduced into the system either from evaporation of a carbon electrode or by special soot injection into the thermal plasma jet. According to the thermodynamics of the C–N mixture (composition C:N = 1:1, pressure 1 atm), the maximum concentration of the dicyanogen (about 0.5 vol%) is formed at 3500 K. The CN concentration reaches 40–55% at 3500–5000 K. Assuming complete CN recombination into dicyanogen results in an energy cost of 10.2–11.2 kWh per 1 kg of C$_2$N$_2$. A stationary concentration of CN radicals is achieved after about 5×10^{-4} s. Formation of the dicyanogen takes place during the quenching stage and reaches 12%.

Thermal plasma permits also effective co-production of hydrogen cyanide (HCN) and acetylene (C$_2$H$_2$) from methane and nitrogen. Methane injection into a nitrogen plasma jet leads to the reaction: a CH$_4$ + N$_2$ → 2 HCN + (a/2−1) C$_2$H$_2$ + 3a/2 H$_2$. Considering that H:C = 1:1 in both C$_2$H$_2$ and HCN, the conversion of methane in N$_2$ plasma is always accompanied by hydrogen production. According to the high temperature thermodynamics, optimal conditions of C$_2$H$_2$ and HCN co-production from the mixture CH$_4$:N$_2$ = 1:1 correspond to a temperature of about 3000 K at atmospheric pressure. The concentration of HCN in the products is 5–6 vol% in this case; the concentration of acetylene is about 1%. Better characteristics can be achieved when the H-to-C ratio is less than 4. Such composition can be achieved by considering the contribution of carbon evaporated from the arc discharge electrodes. For example, an optimal synthesis in atmospheric-pressure mixtures C:H:N = 1:2:2 and C:H:N = 1:1:4 takes place at lower temperatures (1500–2500 K) and leads to the formation of 22.1 vol% of C$_2$H$_2$ and 20.6 vol% of HCN for the first mixture, and 13.9 vol% of C$_2$H$_2$ and 12 vol% of HCN for the second mixture. The energy cost of the C$_2$H$_2$–HCN co-production in the case of the second mixture is 4.8–5.3 kWh kg^{-1}.

Kinetic analysis of the thermal plasma-chemical process shows that production of acetylene in the system proceeds mostly via the Kassel mechanism, whereas a significant amount of HCN is formed in reactions of atomic nitrogen with acetylene, ethylene, and methyl radicals: N + C$_2$H$_4$ → HCN + CH$_3$, N + C$_2$H$_2$ → HCN + CH, N + CH$_3$ → HCN + H$_2$. Quenching starts at maximum HCN concentration; the cooling rate is supposed to be equal to 10^6 K s^{-1}.

25.8 Nonthermal Plasma Production of Hydrogen Cyanide (HCN) from Methane and Nitrogen

The production of HCN from methane and nitrogen is an endothermic process:

$$CH_4 + \tfrac{1}{2} N_2 \rightarrow HCN + 3/2 \, H_2, \, \Delta H = 2.6 \, eV \, mol^{-1}. \tag{25.9}$$

HCN production in thermal plasma arc discharges requires about 10 eV mol^{-1} HCN, which corresponds to an energy efficiency of 26%. Low energy cost and higher energy efficiency can be achieved by conducting HCN synthesis

(25.9) in nonequilibrium plasma with the process stimulation through vibrational excitation of N_2 molecules, see Rusanov and Fridman (1984). HCN synthesis stimulated in nonthermal plasma by vibrational excitation of molecular nitrogen, $N_2^*(^1\Sigma_g^+, v)$, proceeds through the chain mechanism:

$$H + N_2^* \left(^1\Sigma_g^+, v\right) \rightarrow NH + N, \qquad E_a \approx \Delta H = 7\,eV, \tag{25.10}$$

$$N + CH_4 \rightarrow HCN + H_2, \quad \Delta H = -0.5\,eV, \quad E_a = 0.5\,eV. \tag{25.11}$$

The NH is then consumed mostly in the barrier-less radical–radical reaction of hydrogen production: $NH + NH \rightarrow N_2 + H_2$. The limiting stage (25.10) of the chain process is effectively stimulated by vibrational excitation of nitrogen, $N_2^*(^1\Sigma_g^+, v)$. The energy efficiency of the process is limited by losses in the exothermic reactions and reaches a maximum of 40%. It corresponds to an energy cost of about $6.5\,eV\,mol^{-1}$ HCN, which is 1.5 times lower than that of HCN production in thermal plasma systems. The nonequilibrium plasma experiments with HCN synthesis stimulated by vibrational excitation were carried out using the non-self-sustained discharge supported by a relativistic electron beam. The process has been performed directly in the methane–nitrogen mixture at atmospheric pressure, as well as by preliminary N_2 excitation in the atmospheric-pressure discharge followed by mixing with CH_4.

The nonequilibrium mechanism of HCN production does not include direct plasma dissociation of CH_4 molecules; it is due to much faster excitation of N_2 to the highly excited and highly reactive vibrational levels. As a result, although CH_4 bonding energies are weaker with respect to nitrogen, vibrational excitation leads preferentially to effective dissociation and reactions of N_2 molecules. Nonequilibrium plasma systems with higher electron temperatures, especially low-pressure discharges, are characterized by a lower contribution of vibrational excitation and preferential electronic excitation and direct dissociation of reacting molecules. In such plasma systems, dissociation, and reactions of CH_4 molecules mostly through electronic excitation dominate over the mechanism based on the vibrational excitation of nitrogen. These regimes are characterized by preferential production of acetylene (C_2H_2) from methane, while the HCN production is suppressed. The relevant example is the process for HCN/C_2H_2 co-production, which has been carried out in the nonthermal plasma of a glow discharge in a CH_4–N_2 mixture at pressures of 10–15 Torr. The maximum methane conversion into HCN and C_2H_2 (about 80%) has been achieved when the molar fraction of methane in the initial mixture was 15%. The HCN concentration in the product mixture was up to 10%. The energy cost of the co-production process in low-pressure glow discharges is higher than that in the atmospheric-pressure plasma processes stimulated by vibrational excitation. It is even higher than the energy cost achieved in the thermal plasma systems.

25.9 Other Thermal and Nonthermal Plasma Processes of Synthesis and Conversion of the Organic Nitrogen Compounds

Plasma production of HCN and H_2 from CH_4–NH_3 mixture is an endothermic process:

$$CH_4 + NH_3 \rightarrow HCN + 3H_2, \Delta H = 2.6\,eV\,mol^{-1}. \tag{25.12}$$

The product composition of the atmospheric-pressure thermal plasma process in the temperature range 1000–3500 K, which corresponds to a specific energy input range of 0.95–$10.5\,eV\,mol^{-1}$, is presented in Figure 25.3 (Nester et al. 1988). At temperatures below 2000 K, dissociation of ammonia mostly leads to the formation of molecular nitrogen. The maximum yield of the HCN corresponds to a temperature of 3200 K. That temperature is also optimal for direct production of acetylene. The energy cost of HCN production is presented in Figure 25.4 as a function of specific energy input for the cases of (i) absolute and (ii) ideal quenching mechanisms (Nester et al. 1988). The minimum energy cost in the absolute quenching mode is quite high: about $50\,eV\,mol^{-1}$ HCN (energy efficiency only about 5%). A much lower energy cost (about $3\,eV\,mol^{-1}$ HCN) can be achieved in the ideal quenching mode, when radical–radical reactions during the cooling stage make a significant contribution to HCN production. Production of HCN and H_2 from CH_4 and NH_3 (25.12) can also be accomplished in nonthermal plasma. Conducting the process in a glow discharge, for example, results in a high conversion (70%) of methane to hydrogen. The energy efficiency of the process (25.12) in nonequilibrium conditions is low with respect to that in the thermal plasma.

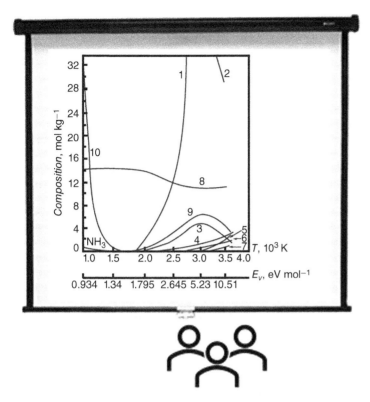

Figure 25.3 Thermodynamic quasi-equilibrium composition of the mixture CH_4–NH_3 at atmospheric pressure and different temperatures corresponding to different values of the specific energy input: (1) H; (2) H_2; (3) C_2H_2; (4) C_2H; (5) C; (6) CN; (7) C_2; (8) N_2; (9) HCN; (10) CH_4.

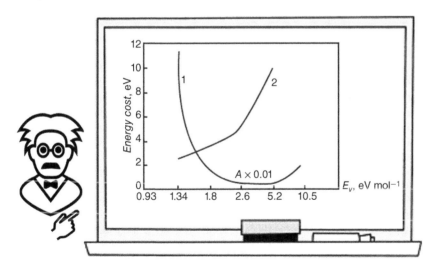

Figure 25.4 Energy cost A (eV mol^{-1}) of production of HCN in the process $CH_4 + NH_3 \rightarrow HCN + 3H_2$ at different quenching modes: (1) absolute quenching, $A10^{-2}$, $A_{min} = 50$ eV mol^{-1}; (2) ideal quenching.

Co-production of NO and CN in the CO–N_2 mixture can be also accomplished in both thermal and nonthermal plasma conditions. This endothermic process can be presented as:

$$CO + N_2 \rightarrow CN + NO, \quad \Delta H = 6 \text{ eV mol}^{-1}. \tag{25.13}$$

Intensive conversion starts in thermal plasma at relatively high temperatures, exceeding 4500 K, due to the very large binding energies of both reagents. The energy efficiency is relatively low in thermal plasma conditions. The minimum energy cost of CN production in the absolute quenching mode is very high (640 eV mol^{-1}) because of the low concentration of the CN radicals at the high temperatures required for the process. Even ideal quenching and super-ideal quenching modes assuming VT nonequilibrium during the fast cooling, decrease the minimum energy cost only to 28 eV CN^{-1} and 13 eV NO^{-1}, respectively.

Higher energy efficiency of the process (25.13) can be achieved in nonequilibrium plasma with the process stimulation through vibrational excitation because of the excellent characteristics of vibrational excitation for both N$_2$ and CO in nonequilibrium discharges. Plasma co-production of CN and NO proceeds in this case by the chain-reaction mechanism:

$$O + N_2^* \rightarrow NO + N, \quad \Delta H \approx E_a = 3\,eV, \tag{25.14}$$

$$N + CO^* \rightarrow CN + O, \quad \Delta H \approx E_a = 3\,eV. \tag{25.15}$$

The first reaction in the chain is the limiting reaction of the Zeldovich mechanism and its stimulation by vibrational excitation of N$_2$ molecules, discussed in Lecture 18. The second reaction is stimulated by vibrational excitation of CO molecules due to the relatively high energy barrier of the reaction and the absence of activation energy of the reverse reaction.

Complex organic synthesis of nitrogen compounds has been performed in a low-pressure glow discharge, with an example of the process:

$$N_2 + CO + 3H_2 \rightarrow NH_3 + HCN + H_2O \tag{25.16}$$

This nonthermal plasma process leads to the formation of two nitrogen compounds and can be accomplished at specific discharge parameters with relatively high selectivity. The conversion degree in this synthetic process reaches 80–90%. The plasma-chemical reaction of ethylene with HCN resulting in the formation of ethyl cyanide has been demonstrated in the similar nonthermal conditions with relatively high selectivity and conversion degree: $C_2H_4 + HCN \rightarrow C_2H_5CN$. Nonthermal plasma organic synthesis of nitrogen aromatic compounds has also been demonstrated with an example of production of aniline ($C_6H_5NH_2$) from benzene and ammonia: $C_6H_6 + NH_3 \rightarrow C_6H_5NH_2 + H_2$. Details on the subject can be found in Fridman (2008).

25.10 Organic Plasma Chemistry of Chlorine Compounds

Let's start with **thermal plasma synthesis of reactive mixtures for production of vinyl chloride** (C_2H_3Cl), a monomer for further production of a widely used polymer PVC. Vinyl chloride is conventionally produced by catalytic hydrochlorination of acetylene ($C_2H_2 + HCl \rightarrow C_2H_3Cl$). The stoichiometric acetylene–hydrogen chloride mixture (C_2H_2: HCl = 1 : 1), which is applied for vinyl chloride production, can be effectively formed in thermal plasmas either by pyrolysis of hydrocarbons in plasma jets containing H$_2$, HCl, Cl$_2$, or CH$_4$, or by pyrolysis in H$_2$ plasma jet of the organic chlorine industrial wastes. In particular, the stoichiometric C_2H_2–HCl mixture has been produced by pyrolysis of gasoline (with final evaporation temperature 165 °C) in atmospheric-pressure plasma jets containing H$_2$, HCl, Cl$_2$, and CH$_4$ in different proportions. The process usually requires temperature 1300–1700 K and effective contact time 10^{-5}–10^{-4} s. A temperature increase leads to acetylene concentration growth from 10 to 20 vol%. The maximum concentration of ethylene is 14 vol% and can be reached at 1400 K. Equal concentrations of C_2H_2 and HCl (17–20 vol%), required to produce vinyl chloride, are achieved in the temperature range 1550–1700 K at effective contact times of 0.5–1.5 × 10^{-4} s. The energy cost of the C_2H_2 and HCl co-production in this regime is 6.3 kWh kg^{-1} C_2H_2. The energy cost is lower when the plasma jet contains Cl$_2$, which is due to heat release in the reaction of HCl production. The use of gasoline for quenching also decreases the energy cost by about 15–20%; the optimal ratio of gasoline used for pyrolysis with respect to gasoline used for quenching is 1 : 0.7.

 Plasma pyrolysis of organic chlorine industrial wastes is also applied for the co-production of acetylene and hydrogen chloride for further synthesis of vinyl chloride and PVC (Volodin et al. 1970a,b, 1971a,b). The process is carried out by interaction of thermal plasma jets of hydrogen (or H$_2$–CH$_4$, or H$_2$–C$_2$H$_4$ mixtures) with the organic chlorine compounds, which can initially be in different phases. *The most effective in this regard is the plasma pyrolysis of dichloroethane, butyl chloride, hexachlorane, and organic chlorine industrial waste mixture. Products of the pyrolysis are applied for the synthesis of vinyl chloride.* The initial composition of the organic chlorine industrial waste mixture used as a feedstock for the pyrolysis is trichloroethane, 15 vol%; dichloroethane, 15 vol%; butyl chloride, 15 vol%; ethyl chloride, 10 vol%; dichloroisobutane, 5 vol%; trichloroisobutylene, 5 vol%; monochlorobutane, 5 vol%; and other organic compounds, 30 vol%. The pyrolysis is carried out in the plasma jet of hydrogen (or H_2–CH_4, or H_2–C_2H_4 mixtures) at 1400–1700 K and at an effective contact time of about 10^{-4} s. In the case of thermal pyrolysis of hexachlorane, the plasma jet is submerged in a tank containing the dry initial product. Temperature increase of the plasma pyrolysis leads to an increase in the concentration of acetylene (C_2H_2) in the products and to a decrease of concentrations of ethylene (C_2H_4), C_3H_6, iso-C_4H_8, and vinyl chloride (C_2H_3Cl). Quenching of the plasma pyrolysis products can be provided by injection of gasoline or the mixture of organic chlorine compounds, which increases the energy efficiency of the process. Typical pyrolysis-to-quenching feeding rate ratios of the organic chlorine compounds are about $1:2$. Kinetic analysis of the thermal plasma pyrolysis of the organic chlorine compounds at temperatures of 1600–2000 K shows that almost all chlorine from the initial products is converted into HCl. The composition of hydrocarbons in the products of pyrolysis of the organic chlorine compounds is like that of the pyrolysis of pure hydrocarbons at the same C:H ratio and the same temperature.

25.11 Organic Plasma Chemistry of Fluorine Compounds

High-intensity thermal arc discharges with carbon electrodes have been effectively applied for **plasma conversion of tetrafluoromethane (CF_4) into tetrafluoroethylene (C_2F_4)** and fluorine, see for example Bronfine (1970):

$$CF_4 + CF_4 \rightarrow C_2F_4 + 2F_2. \tag{25.17}$$

The yield of the tetrafluoroethylene C_2F_4 increases at higher discharge powers and lower pressures. The maximum yield of C_2F_4 (75 vol%) is achieved at the discharge power of 20 kW and pressure of 0.1 atm. In addition to CF_4 and C_2F_4, products of the plasma-chemical process also contain C_2F_6 and C_3F_8.

Quasi-equilibrium heating of organic fluorine compounds to temperatures of 2000–4000 K leads mostly to the formation of C_2F_2 and different kinds of carbon clusters. During the quenching phase, significant amounts of C_2F_4 and soot can be generated. The formation of the tetrafluoroethylene C_2F_4 during the product quenching phase is due to attachment of atomic fluorine to C_2F_2 molecules: $C_2F_2 + F \rightarrow C_2F_3$, $C_2F_3 + F \rightarrow C_2F_4$. The formation of C_2F_4 from C_2F_2 during cooling is accompanied by soot formation: $C_2F_2 + C_2F_2 \rightarrow C_2F_4 + 2C_{(solid)}$. Plasma decomposition of CF_4 can also be optimized for the **production of soot and fluorine** in the strongly endothermic process:

$$CF_4 \rightarrow C(solid) + 2F_2, \quad \Delta H = 9.5 \, eV \, mol^{-1}. \tag{25.18}$$

The energy cost of the process in atmospheric-pressure thermal plasma is shown in Figure 25.5 as a function of specific energy input for different quenching modes (Nester et al. 1988). The minimum energy cost in the case of absolute quenching of products is 25.8 eV C^{-1} atom, which can be achieved at the specific energy input of 25.1 eV mol^{-1} and CF_4 conversion degree of 97%. The minimal energy cost in the ideal quenching mode is 21.6 eV C^{-1} atom and that in the super-ideal quenching mode is 17.4 eV C^{-1} atom.

The **pyrolysis of CF_4, CHF_3, and other organic fluorine compounds in a nitrogen plasma** leads to synthesis of different nitrogenated fluorocarbons, see for example Polak et al. (1975). In the case of pyrolysis of the carbon tetrafluoride CF_4 in the N_2 plasma jet with a quenching rate of 10^6 K s^{-1}, the major nitrogen-containing products are NF_3 and trifluotomethylamine CF_3NH_2. Smaller amounts of N_2F_4 and N_2F_2 are also present in the products. The plasma-induced conversion of nitrogen into different fluoride compounds is relatively low, about 1%. Yields of nitrogen fluorides increase at higher discharge powers and larger F:N ratios. Pyrolysis of CHF_3 in the N_2 plasma jet leads to the formation of such fluorocarbons as CF_4 (57.7%), C_2F_6 (18.2%), C_2F_4 (12.9%), and C_3F_8 (3.4%). The major nitrogen-containing organic fluorine compound in this case is the trifluoroacetonytrile, CF_3CN–7.8%.

Thermodynamic quasi-equilibrium in the mixture C:N:F $= 1:4:4$ at atmospheric pressure and temperature of 4000 K gives major fluorocarbons: CF_2 radicals and C_2F_2; the major nitrogen-containing compounds at 2500–4200 K are FCN and CN radicals. The maximum concentration of FCN is 1 vol%, and it is reached at 3500 K. The maximum concentration of CN radicals is 5 vol%, and it is reached at temperature 4500 K.

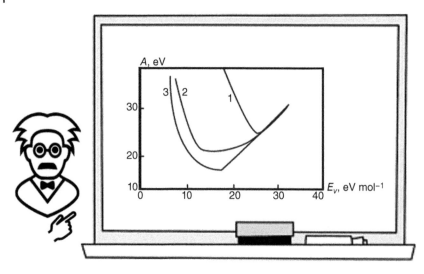

Figure 25.5 Energy cost A (eV mol^{-1}) of production of a carbon molecule in the thermal plasma process of dissociation of CF$_4$ at different quenching modes: (1) absolute quenching; (2) ideal quenching; (3) super-ideal quenching.

25.12 Thermal and Nonthermal Plasma Processing of Chlorofluorocarbons (CFCs)

Chlorofluorocarbons (CFCs) are associated with high ozone-destroying and global warming potential. It determines interest to their destruction in plasma, which can be quite effective. Although some of these processes are accomplished in nonthermal discharges, most of the **CFC decomposition processes** are effectively performed in thermal plasmas with addition of hydrogen and oxygen into the plasma gas. Detailed information on this practically important subject can be found in Badie et al. (1997), Brozek et al. (1997), Coulibaly et al. (1997), Park and Cha (1997), Rutberg et al. (1997), Murphy et al. (2002), Ponelis and Van der Walt (2003), Watanabe et al. (2005); see also review in Fridman (2008) and Watanabe et al. (2005).

When the process is carried out in Ar plasma jets at powers of 2–10 kW without adding O$_2$, the main products of pyrolysis of chlorofluoromethane are chlorofluoroethane compounds, benzene derivatives, and so on. It is interesting to discuss the effect of hydrogen and oxygen on the plasma-chemical process of CFC pyrolysis. Introducing hydrogen improves thermal conductivity of the plasma, which increases the efficiency of the process. The most important function of hydrogen is its involvement in the irreversible reactions with fluorine and chlorine-containing radicals and formation of HF and HCl, which are products of the pyrolysis. Introducing oxygen helps avoid soot formation during pyrolysis. Relatively small amounts of oxygen lead mostly to the formation of CO, and larger oxygen admixtures result in the formation of CO$_2$. When oxygen admixtures are relatively high, the concentration of CO in products can be increased with respect to CO$_2$ by increasing the flow rate of hydrogen. Generally, it can be concluded that the degree of perfluoro compound (PFC) destruction and product composition are essentially controlled by the amounts of H$_2$ and O$_2$ added to the plasma gas. Hydrogen and oxygen can be introduced into the thermal plasma in the form of steam. In this case, CFCs are injected in a direct-current (DC) steam plasma torch operating at a power of about 1.5 kW. The system provides the complete destruction of CFC compounds with a high degree of fluorine recovery – about 95%. The destruction degree of the PFCs in thermal plasma can be very high (99–99.99%).

Let's shortly analyze now, the **nonthermal plasma conversion of CFCs** and other plasma processes with halogen-containing organic compounds. An advantage of the nonthermal plasma approaches is possibility of higher energy efficiency in destruction and conversion of CFC compounds diluted in other gases, primarily in air. Electrons and active species of the nonequilibrium plasma systems can destroy the highly diluted CFC compounds without heating the whole gas, which makes the energy cost of the process lower. For example, application of a DBD to dry atmospheric air with 100–300 ppm of trichloroethylene (C$_2$HCl$_3$) requires energy input of only about 60 J l^{-1} for a destruction efficiency of C$_2$HCl$_3$ up to 80%. Different types of nonthermal discharges have been applied for the destruction of CFC compounds, specifically low-pressure RF capacitively coupled plasma (CCP) and glow discharges, moderate-pressure (about 100 Torr) microwave and atmospheric-pressure sliding surface discharges, as well as pulsed electron beams.

A challenge of the nonequilibrium systems is the **wide variety of by-products**. While nonequilibrium plasma stimulates some elementary conversion processes, many intermediate products are not completely converted into desirable final products because of low gas temperature in the system. For example, decomposition of CF_4 in the presence of oxygen leads not only to the formation of CO and CO_2 but also to the production of highly toxic phosgenes, COF_2 and COF. The phosgene yields can be reduced, for example, by applying calcium oxide particles that trap fluorine in the form of CaF_2. Among other nonthermal plasma processes with halogen-containing organic compounds, there are also dehydration reactions of organic acids, $Cl_2 + HCOOH \rightarrow 2HCl + CO_2$, chlorine–bromine exchange reactions in conversion of organic bromine and chlorine compounds, $Cl_2 + CCl_3Br \rightarrow BrCl + CCl_4$, and halogen attachment to hydrocarbons, specifically the formation of dibromoethylene from C_2H_2: $C_2H_2 + Br_2 \rightarrow C_2H_2Br_2$, see Baddur and Timmins (1970) and Rusanov and Fridman (1984).

25.13 Direct Plasma Synthesis of Methanol and Formaldehyde by Oxidation of Methane

Let's start with the **plasma-induced methane oxidation to methanol by carbon dioxide**, CO_2. The reaction of methane with carbon dioxide in thermal plasma conditions leads mostly to the production of syngas. While nonthermal plasma can also be applied for the conversion of the CH_4–CO_2 mixture into syngas, the nonequilibrium plasma conditions can also lead to the direct (without additional use of a catalyst) formation of methanol (Rusanov and Fridman 1984):

$$CH_4 + CO_2 \rightarrow CH_3OH + CO, \quad \Delta H = 1.6 \, \text{eV mol}^{-1}. \tag{25.19}$$

This process is promoted in strongly nonequilibrium conditions ($T_v \gg T_0$), when the vibrational excitation of CO_2 molecules stimulates the chain reaction of methanol production:

$$CH_3 + CO_2^*(v) \rightarrow CO + CH_3O, \quad \Delta H \approx E_a = 1.5 \, \text{eV mol}^{-1}, \tag{25.20}$$

$$CH_3O + CH_4 \rightarrow CH_3OH + CH_3, \quad \Delta H = 0.1 \, \text{eV mol}^{-1}, E_a \approx 0.3 \, \text{eV mol}^{-1}. \tag{25.21}$$

Reactions between CH_3O radicals also contribute to synthesis in the CH_4–CO_2 mixture:

$$CH_3O + CH_3O \rightarrow CH_3OH + CH_2O. \tag{25.22}$$

Although this fast elementary reaction produces valuable products, methanol, and formaldehyde, it decreases the total yield of methanol because of termination of the chain (25.20) and (25.21). Direct production of methanol and formaldehyde in the CH_4–CO_2 mixture (25.19) is possible only at very low gas temperatures. A temperature increase leads to CH_3OH and CH_2O decomposition in the system and to formation of the syngas (CO, H_2) as major reaction products. Generally, the conversion degree of the direct production of methanol and formaldehyde in the CH_4–CO_2 mixture is limited without a catalyst, see Lecture 22.

The reaction of methane and water vapor in thermal plasma leads to the production of syngas (H_2, CO). Nonthermal plasma is also applied to produce the H_2–CO mixture. However, as it was demonstrated by Okazaki and Kishida (1999), application of **nonthermal plasma permits the direct production of methanol from methane and water**:

$$CH_4 + H_2O \rightarrow CH_3OH + H_2. \tag{25.23}$$

These experiments were carried out in an atmospheric-pressure DBD supported by short pulses (duration 400 ns, repetition frequency 500 Hz). Typical plasma-chemical parameters of the system were: peak voltage, 3 kV; residence time, 50 s; gas temperature, 393 K; and specific energy input, 11 J cm^{-3}. The methanol yield was about 1% at a water vapor concentration of about 50%. The energy cost of methanol production from methane was relatively high, about 200 eV mol^{-1} CH_3OH. The reaction proceeds via dissociation of water and methane molecules with formation of CH_3 and OH radicals, which leads then to methanol production. In contrast to the process with CO_2 molecules (25.19), the reactions (25.23) with H_2O molecules cannot be effectively stimulated by vibrational excitation, which results in a lower energy efficiency for methanol production. An increase of the methanol yield from the CH_4–H_2O mixture can be achieved by combination of DBD with an Ni catalyst (Nozaki 2002; Kado et al. 2003).

Although partial oxidation of CH_4 in thermal plasma leads mostly to the generation of syngas, special quenching can also result in a significant **direct thermal plasma production of formaldehyde (CH_2O) by oxygen-based partial oxidation of methane**:

$$CH_4 + O_2 \rightarrow CH_2O + H_2O. \tag{25.24}$$

The major products of the plasma-chemical process are H_2, CO, CH_2O, CO_2, and H_2O. The highest yield of CH_2O is 3.5% with respect to the initial methane concentration. This yield has been achieved at reaction temperatures of 1600–1700 K. The plasma-chemical process (25.24) can be also stimulated in nonequilibrium discharges by vibrational excitation of oxygen molecules (Rusanov and Fridman 1984, also see Fridman 2008), but the energy efficiency is strongly limited in this case by fast nonadiabatic VT relaxation of molecular oxygen. Nonthermal plasma in CH_4–O_2 mixture can be more effectively optimized to the **production of methanol**: $CH_4 + \frac{1}{2} O_2 \rightarrow CH_3OH$. This process has been demonstrated in atmospheric pressure pulsed DBD by Okazaki et al. (1995). The yield of methanol reached 2.4%; the selectivity of CH_3OH production from CH_4 (percentage of methanol in converted products) was as high as 33%. Higher methanol yields can be achieved by combining nonthermal plasma with catalysts and recirculation of products. Generally, the different ways of combination of plasma-chemical and catalytical processes can make a significant impact on the organic synthesis. For more details on this interesting subject see Lecture 22, as well as relevant reviews and publications, especially (Bogaerts 2019; Bogaerts et al. 2020; Nozaki and Okazaki 2013; Deminsky et al. 2002).

25.14 Nonthermal Plasma Synthesis of Aldehydes, Alcohols, Organic Acids, and Other Organic Compounds in Mixtures of Carbon Oxides with Hydrogen and Water

A hydrogen mixture with carbon monoxide is widely used for conventional catalytic synthesis; the mixture is called syngas. Nonthermal plasma processing of the CO–H_2 mixture leads to direct thermoneutral synthesis of formaldehyde:

$$H_2 + CO \rightarrow CH_2O, \quad \Delta H \approx 1 \, \text{kcal mol}^{-1}. \tag{25.25}$$

This process has been demonstrated in different nonthermal discharges. Although the reaction (25.25) is thermoneutral, the energy cost of CH_2O production in these systems is quite high, about 70 eV mol^{-1}. The maximum yield of formaldehyde is 7%, and it is reached in the initially hydrogen-rich mixtures (up to 75% of H_2). Lower energy costs of the process are possible when synthesis of formaldehyde is stimulated by vibrational excitation of CO and H_2 molecules (Givotov et al. 1984). When the syngas CO–H_2 mixture is treated in nonequilibrium plasma with effective vibrational excitation of both molecules (CO, H_2), a sufficient increase in the specific energy input leads to partial conversion of formaldehyde into methanol (Givotov et al. 1984):

$$H_2 + CH_2O \rightarrow CH_3OH, \Delta H = -1 \, \text{eV mol}^{-1}. \tag{25.26}$$

Starting not from the syngas CO–H_2 but from higher initially oxidized mixtures, CO–H_2O or CO_2–H_2, plasma-chemical synthesis leads to the formation of formic acid:

$$CO + H_2O \rightarrow HCOOH, CO_2 + H_2 \rightarrow HCOOH. \tag{25.27}$$

These direct synthetic processes were demonstrated in different nonequilibrium plasma-chemical systems but with limited level of conversion, and therefore low energy efficiency, see Rusanov and Fridman (1984).

An interesting nonequilibrium process for plasma-chemical synthesis of formic acid from a mixture of CO_2 and H_2O was first demonstrated 130 years ago by Losanitch and Jovitschitsch:

$$CO_2 + H_2O \rightarrow HCOOH + \frac{1}{2} O_2. \tag{25.28}$$

This process can be also considered as the first plasma-chemical stage of a two-step process for hydrogen production from water. The second stage in this case is formic acid decomposition with formation of hydrogen and carbon dioxide:

$$HCOOH \rightarrow H_2 + CO_2. \tag{25.29}$$

The carbon dioxide produced this way from formic acid is then recycled back to the plasma chemical synthesis (25.28). This two-step cycle (25.28) and (25.29) of hydrogen production from water is especially interesting because the plasma-chemical stage (9–61) does not require thermodynamically significant energy input. Most of the energy required for hydrogen production from water should be spent in this cycle during decomposition of formic acid in the second stage (25.29), which proceeds at a relatively low temperature. Therefore, *the two-step cycle (25.28) and (25.29) of hydrogen production from water can be performed using mostly low-temperature heat, like in the natural process of photosynthesis. CO_2 and H_2O molecules are first converted into a relatively large organic molecule, HCOOH, with the production of oxygen without consuming significant energy but using selective and catalytic properties of nonthermal plasma. Then the relatively large molecule dissociates, consuming significant energy but at low temperatures.* Significant improvements in the process selectivity are related to combination of the plasma synthesis with relevant catalysis. Similar plasma-chemical and plasma-catalytic effects have been observed in the formation of oxalic acid in CO_2–H_2O mixture $2CO_2 + 3H_2O \rightarrow (COOH)_2 * 2H_2O + \frac{1}{2}O_2$, which was discussed in Lecture 22, Section 22.16, regarding hydrogen production. More on this interesting subject of the **plasma mimicking photosynthesis** can be found in Novikov et al. (1978) and Rusanov and Fridman (1984).

Methane and acetylene can also be effectively produced from syngas (CO–H_2) in nonthermal plasma conditions. Syngas-based production of methane is an exothermic process:

$$CO + 3H_2 \rightarrow CH_4 + H_2O, \quad \Delta H = -2\,eV\,mol^{-1}. \tag{25.30}$$

The total plasma-chemical process of methane synthesis can be considered a continuation of the carbon monoxide hydrogenation sequence, which starts with the production of formaldehyde from syngas (25.25), then proceeds to production of methanol from formaldehyde (25.26), and finally leads to the production of methane (25.30) in the reaction:

$$CH_3OH + H_2 \rightarrow CH_4 + H_2O, \quad \Delta H = -1\,eV\,mol^{-1}. \tag{25.31}$$

The entire carbon monoxide hydrogenation sequence from CO to methane can be effectively stimulated by vibrational excitation of hydrogen molecules in nonequilibrium plasma conditions (Rusanov and Fridman 1984). Methane formed in the process (25.25) can be partially converted into acetylene, which determines CH_4 and C_2H_2 as the major products of the nonthermal plasma treatment of syngas (CO, H_2). To reach higher conversion degree in the synthetic process (25.30), water produced in the reaction can be frozen out. As a result, 80% of carbon monoxide can be converted in the nonthermal plasma into methane with some additional small amount of acetylene.

25.15 Plasma-chemical Synthesis of Diamonds and Diamond Films

Nonequilibrium plasma is effectively used in the production of synthetic diamonds, and especially diamond films, which are widely used in modern technology. Diamond is a metastable form of carbon at normal conditions; therefore, it can be synthesized in thermodynamic quasi-equilibrium conditions only at very high pressures (150 kbar) and temperatures (about 3000 °C). The application of special catalysts decreases diamond production pressures to less than 70 kbar and temperatures less than 1500 °C, which are still quite extreme conditions. Plasma provides nonequilibrium conditions that permit effective diamond synthesis at atmospheric and lower pressures from different carbon compounds including CH_4, C_2H_2, CO, CO_2, C_3H_8, CH_3OH, C_2H_5OH, CH_3Cl, CF_4, CCl_4, and others. Detailed information on the subject can be found, for example, in Prelas et al. (1997), Asmussen and Reinhard (2002), Robertson (1993a,b, 1994a,b, 1997, 2002, 2004), Mankelevich and Suetin (2000), He et al. (1999), and Conway et al. (2000).

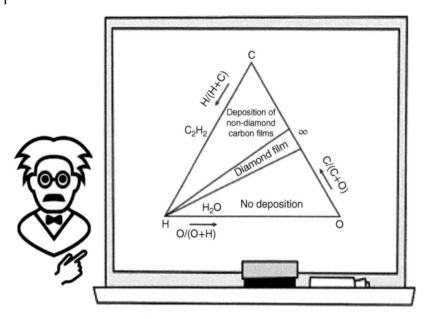

Figure 25.6 The Bachmann triangle.

High electron temperature and relatively low gas temperature in nonequilibrium plasma stimulate intensive dissociation of the carbon compounds and provide fast carbon-atom deposition on different special substrates. The nonequilibrium plasma systems also provide a high concentration of hydrogen atoms, usually by dissociation of specially added molecular hydrogen.

 The hydrogen atoms selectively etch graphite from the deposited film, leaving diamond intact. To provide better selective deposition of diamond films, the substrates (made, for example, from silicon, metals, or ceramics) are especially treated to create diamond nucleation centers. Typical substrate temperatures for effective deposition of diamond films are 1000–1300 K. The influence of the initial C/H/O composition of the carbon compounds used for the nonequilibrium diamond synthesis on the composition of the deposited films can be illustrated by the so-called **Bachmann triangle**, shown in Figure 25.6. The Bachmann triangle summarizes experimental data and indicates the C/H/O compositions corresponding to deposition of nondiamond films, diamond films, or no deposition at all (which corresponds to domination of etching over deposition). Deposition of diamond films from the gas phase is typically performed at pressures of 10–100 Torr in a mixture of hydrogen with about 1–5% methane.

The plasma-chemical conversion of carbon compounds can lead not only to the formation of diamond films but also to production of the so-called **diamond-like carbon (DLC)**. DLC films are hard amorphous films with a significant fraction of sp^3-hybridized carbon atoms, which can contain some hydrogen. Depending on deposition conditions, DLC films can be fully amorphous or contain diamond crystallites. The DLC films are not called "diamond" unless a full three-dimensional crystalline lattice of diamond is proven. The physical and chemical properties of DLC films are often not as attractive as those of diamond films. On the other hand, the production of DLC films is generally easier and requires lower substrate temperature, gas pressure, hydrogen fraction in the initial mixture, and a lower level of gas-phase activation, which attracts interest to their large-scale industrial application as protective, anticorrosive, and other special coatings

The most conventional plasma-chemical systems for synthesizing diamonds and diamond films from gas-phase carbon compounds are based on the application of microwave discharges. The important advantage of these systems is purity of the produced films due to absence of electrodes and sustaining plasma only near the substrate. Deposition of diamond films from the gas phase is performed in this case at pressures of 10–100 Torr in a mixture of hydrogen with about 1–5% methane. The growth rate of the diamond films is about 10 μm h^{-1}. The price of diamond films produced in microwave discharge plasma is about $200 per carat. The diamond films can also be effectively synthesized in low-pressure DC glow discharges, where the film growth rate can be relatively high, exceeding 30 μm h^{-1}. The substrate in this case is located on the grounded anode and heated up to the required temperature by the discharge itself. Hydrogen with 1–10% methane is injected into the low-pressure plasma reactor through the cathode. *The most "popular" laboratory deposition method, especially for the case of DLC, is RF discharge plasma-enhanced chemical vapor deposition.* The reactor usually consists of two electrodes of a different area; the RF power is capacitively coupled to the smaller electrode, on which the substrate is mounted. The other electrode, typically including the reactor walls, is grounded. For DLC deposition, the plasma should be operated at lower pressures to maximize the ion-to-radical ratio.

The highest deposition rates of the diamond films (1 mm h^{-1}) are achieved in reactors based on thermal arc discharges. In contrast to the nonequilibrium plasmas, the discharge is clearly thermal and close to the quasi-equilibrium conditions; gas temperature is about 3000–5000 K. The deposition uniformity is not as good as in the previously nonthermal discharge systems. To improve the deposition uniformity, several arc discharges can be applied simultaneously. Typical gas pressures in the arc discharges applied for diamond film synthesis are in the range 250–760 Torr, while pressure in the reactor is about ten times lower (20–50 Torr). A plasma jet with velocities of 100–300 m s^{-1} formed in a narrow nozzle is directed from the arc discharge zone into the reactor, where the substrate is located. The reacting H_2–CH_4 mixture is usually diluted in the discharge zone with 90% argon or other inert gas. Effective deposition of diamond films requires in this case fast gas cooling and quenching in the substrate area. Considering that the thickness of the boundary layer near the substrate is inversely proportional to the square root of the jet velocity, the transonic jet velocities are usually applied to provide effective deposition of diamond films in thermal plasma systems.

25.16 Mechanisms of Major Plasma-volume and Surface Processes Leading to the Plasma-chemical Diamond-film Growth

Nonequilibrium microwave and glow discharges applied to produce diamond films have similar plasma parameters: the gas temperature is relatively high 2000–3000 K and equal to the ion temperature, the average electron energy is 1–2 eV with a strongly non-Maxwellian electron energy distribution function, the electron density is about 10^{11}–10^{12} cm^{-3}, the gas pressure is moderate 50–150 Torr, and the reduced electric field is $E/n_0 = 35$–50 Td. DC glow discharges are usually stratified in this case. The plasma gas applied in the diamond film deposition systems is a mixture of molecular hydrogen with a small admixture of hydrocarbons. Nonequilibrium gas-phase processes lead to the formation of acetylene, ethylene, and such atoms and radicals as H, CH_3, CH_2, CH, C, and so forth. The active species formed in the gas phase are deposited on the surface, where they participate in a sophisticated sequence of such processes as adsorption/desorption, recombination, surface diffusion, and surface chemical reactions, which result in the formation of the diamond film. Atomic hydrogen predominantly etches nondiamond carbon structures, promoting selective production of the diamond films. The discharge is usually first initiated in pure hydrogen to clean the substrate.

Atomic hydrogen is generated by thermal dissociation as well as by electron impact through excitation of electronically excited states. A major contribution is provided by the electronically excited states of hydrogen molecules with relatively low excitation thresholds: 8.9, 11.75, and 11.8 eV. Thermal dissociation dominates the only at very high translational gas temperatures exceeding 2750–2850 K. Atomic hydrogen formed in the thermal and nonthermal processes initiates conversion of hydrocarbons and formation of radicals (CH_x, $x < 4$), which are the gas-phase precursors of the diamond film. The major active gas-phase species that provide the diamond film growth are atomic hydrogen (H), methyl radical (CH_3), and acetylene (C_2H_2). The temperature of the substrate is sustained 1000–1300 K to provide the effective diamond synthesis.

Mechanism of the surface reactions leading to crystal growth in the diamond film can be considered by the analysis of growth of the (100) diamond plane from the plasma generated CH_x ($x = 0$–3) components, which is illustrated in Figure 25.7. A monohydrated diamond surface is thermodynamically more stable than a dehydrated surface. On such surfaces, at temperatures conventional for diamond film growth, the fraction of the surface centers (C^-) with a hydrogen free bond is about 10–20%, while the fraction of the hydrogen-covered (C–H) centers is about 80–90%. Losses of hydrogen atoms due to their recombination through surface are characterized by a relatively high probability, in the range 10–20%. The major active carbon compound, generated in nonequilibrium H_2–CH_4 plasma, is the methyl radical. Its concentration far exceeds that of other CH_x compounds and that of acetylene (C_2H_2). When the CH_3 radical from the gas phase reacts with the surface, it attaches to the surface in the form of CH_2, releasing a hydrogen atom. Attachment of the CH_2 group to the diamond surface occurs at the (C^-) center of the dimer carbon pair, which

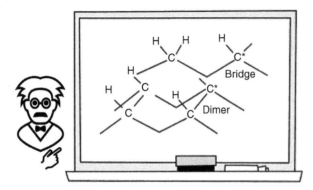

Figure 25.7 Mechanisms of the diamond growth, different surface centers on the (100) diamond plane.

is illustrated in Figure 25.7. The attachment process proceeds through breaking of the dimer bond and incorporation of the CH_2 group in the form of a "bridge" between the two carbon atoms of the dimer pair. Formation of the bridges results in the creation of a new layer of the diamond lattice. Formation of the bridges above the dimer carbon pair proceeds very effectively and is characterized by a low-energy barrier. However, when vacancies above the dimer carbon pairs are occupied, growth of the new diamond layer slows down since further incorporation of the CH_x groups between the dimers in the through site centers and between the bridge sites is kinetically limited by high-energy barriers. A significant contribution to the completion of the building of the new diamond layer is made by surface migration of the CH_2 groups, which leads to effective filling of the intermediate empty spots between the bridges.

 The attachment of acetylene (C_2H_2) to the diamond film surface makes an essential contribution to the growth of the (100) diamond plane. The process is dominated by surface migration of the $C=CH_2$ groups. Although the rate coefficient for C_2H_2 attachment to the diamond surface about 20 times lower than that of the methyl radicals, the total contribution of acetylene in the diamond film growth can be significant because its density in plasma volume can exceed that of CH_3 by about 10–100-fold. The key property of the diamond (and DLC) films is sp^3 bonding. The plasma deposition process which promotes the sp^3 bonding and, therefore, diamond synthesis is the ion bombardment. The highest sp^3 fractions are formed by energetic C^+ ions with energies of around 100 eV. The sp^3-bonded graphite occupies 50% more volume than sp^3-bonded diamond. Therefore, energetic ions, penetrating the film, create compressive stress and stabilize the diamond phase. The process is essentially a subsurface one, and usually is referred to as low-energy subsurface implantation or simply "subplantation." The "subplantation" creates a metastable increase in density of the film, which tends to cause the bonding to change to sp^3 and promote synthesis of diamond crystal structure.

25.17 Problems and Concept Questions

25.17.1 Plasma Synthesis of Organic Compounds from Methane

Explain why the thermal plasma conversion of methane predominantly leads to the formation of acetylene (sometimes ethylene) or soot, while the nonthermal plasma process in CH_4 has tendency to be limited mostly to production of ethane (and sometimes other saturated compounds)?

25.17.2 The Kassel Mechanism of Methane Conversion in Thermal Plasma

Why the Kassel mechanism describing the sequence of plasma conversion processes from methane to soot doesn't cover kinetics of transformation of the hydrocarbons containing more than two carbon atoms and directly jumps from acetylene to soot?

25.17.3 Westinghouse Thermal Plasma Process of Natural Gas Conversion

Why the Westinghouse processes permit significantly increase the yield of the synthesis of organic compounds (especially production of acetylene and ethylene) as well as significantly decrease the energy cost of the plasma technology.

25.17.4 Mechanism of the Thermal Plasma Production of Soot from Hydrocarbons

Give your interpretation, why the most intensive soot production in thermal plasma takes place during pyrolysis of aromatic compounds?

25.17.5 Direct Plasma Synthesis of Methanol and Formaldehyde by Oxidation of Methane

Why these direct synthetic processes while feasible have limited applications, and only using the different ways of combination of plasma-chemical and catalytical processes can make a significant impact on this kind of organic synthesis?

25.17.6 Nonthermal Plasma Synthesis of Formic Acid in CO_2-H_2O Mixture

Determine the minimum energy efficiency of the plasma-chemical HCOOH synthesis in the CO_2-H_2O mixture required for effective hydrogen production in the double-step cycle. Assume that thermodynamically about 70% of the total energy required for hydrogen production from water should be consumed in this case for decomposition of formic acid to form hydrogen and to recycle carbon dioxide back to the plasma process.

25.17.7 Plasma-chemical Synthesis of Diamonds and Diamond Films

Why the atomic hydrogen plays so important role in the plasma-chemical synthesis of diamonds and diamond films? Use the Bachmann triangle to support your opinion.

Lecture 26

Plasma Polymerization, Processing of Polymers, Treatment of Polymer Membranes

26.1 Plasma-chemical Polymerization of Hydrocarbons: Formation of Thin Polymer Films, Mechanisms of Plasma Polymerization

 The formation of high-molecular products (polymers) from initial low-molecular substances (monomers) in nonthermal, mostly low-pressure (0.01–10 Torr), discharges is referred to as **plasma polymerization**. *The formation of high-molecular-weight products (polymerization) proceeds either on solid surfaces contacting with plasma, which results in growth of polymer films, or in the plasma volume, which results in production of polymer powders or some forms of polymer macroparticles. In contrast to conventional polymerization which requires the use of specific monomers, the application of plasma permits the polymerization to start from practically any organic compound.* Plasma-stimulated deposition of polymer films is of significant interest due to the applications of these films to surface processing and formation of thin dielectric and protective films, which is especially important in microelectronics as well as in biology and medicine.

 Numerous research efforts have been focused on the plasma polymerization for a century, thus Linder and Davis described the plasma synthesis of 57 different polymers already in 1931. Reviews on plasma polymerization and related topics can be found, for example, in Yasuda (2012), Inagaki (2014), D'Agostino (2012), Park (2022), Hollahan and Bell (1974), Vinogradov and Ivanov (1977), Vinogradov (1986), Biederman and Osada (1992), Biederman (2004), Oehr (2005), Sardella et al. (2005), and D'Agostino et al. (2010).

Characteristics of plasma polymerization processes essentially depend on the specific power of nonequilibrium discharges, gas pressure, and dilution degree of the initial hydrocarbons in noble gases. At low values of the specific power (about $0.1\,\mathrm{W\,cm^{-3}}$), which are typical for discharges in mixtures highly diluted with noble gases, the translational gas temperature is usually relatively low ($T_0 < 450\,\mathrm{K}$). In this case, the initial volume dissociation of hydrocarbons is mainly due to nonequilibrium processes stimulated by direct electron impact. At higher levels of specific power, which is typical for discharges in nondiluted hydrocarbons, translational gas temperatures can exceed $500\,\mathrm{K}$. In this case, thermal decomposition of hydrocarbons and their reactions with atomic hydrogen make larger contribution to the polymerization kinetics. Also, the higher temperatures lead to higher volume concentration of heavier gas-phase hydrocarbons and to the acceleration of plasma polymerization. An increase of pressure above 3–10 Torr in the discharge slows down diffusion of the hydrocarbon radicals to the reactor walls and stimulates their volume reactions and recombination. Numerous experiments have been focused on the kinetics of plasma polymerization in low-pressure nonequilibrium plasma. For example, plasma polymerization of benzene, toluene, ethyl benzene, and styrene in glow discharges shows an exponential decrease of the polymer deposition rate with temperature, which indicates the contribution of absorption kinetics to the polymerization process. It shows also significant contribution of charged particles in the deposition rate, while contribution of ultraviolet (UV) radiation, as well as excited and chemically active neutrals, is also essential.

Which plasma components, ions, or radicals, mostly dominate polymerization is still an open question. Consider as an example the glow discharge plasma polymerization of cyclohexane. In this system, the film deposition rate linearly grows with cyclohexane concentration in the initial mixture with noble gases, whereas all plasma parameters including the charged particles flux to the surface remain the same. It shows that gradual cyclohexane ion attachment does not contribute to the polymer film deposition. First, cyclohexane dissociates in the plasma volume to form a cyclohexane radical and atomic hydrogen. The radicals diffuse to the walls and recombine there with open bonds formed on the polymer film surface by plasma particle bombardment. Further bombardment leads to more bond breaking and reorganization of the C—C and C—H bonds on the polymer surface. Hydrogen is then desorbed from the surface, while free surface bonds recombine between one another and with gas-phase radicals, creating a highly cross-linked polymer structure. Development of the general plasma polymerization model is restricted by an essentially nonlinear contribution of different plasma components in the process. Plasma particles can simultaneously build and destroy thin polymer films.

The **Osipov–Folmanis plasma polymerization model** assumes charged particles activating adsorbed molecules, and then allowing the cross-link with the polymer film. The rate of polymer film deposition from a nonequilibrium low-pressure discharge (number of deposited molecules per unit time) can be calculated in this case as:

$$w_d = \Phi \Big/ \left[1 + \frac{1}{\sigma j}\left(\frac{1}{\tau} + a\Phi\right)\right] \tag{26.1}$$

where Φ is the flux of the polymer creating molecules to the surface, j is the flux of the charged particles to the surface, σ is the cross-section for activation of the adsorbed molecules by charged particles, τ is the effective lifetime of the adsorbed molecule, and a is the surface area occupied by an absorbed molecule. When fluxes to the surface of neutral and charged particles are not very high, $\sigma j \tau \ll 1$, $a\Phi\tau \ll 1$, the rate of polymer film deposition (26.1) can be rewritten, considering the adsorption time dependence on surface temperature:

$$w_d \propto \Phi \sigma j \exp\left(\frac{E_a^a + \Delta H_a}{T_s}\right). \tag{26.2}$$

Here E_a^a and ΔH_a are activation energy and enthalpy of the adsorption relevant to the plasma polymerization and T_s is the surface temperature. Formula (26.2) interprets the plasma polymerization kinetics, when the film deposition rate is proportional to fluxes of both neutral and charged particles. It explains the exponential acceleration of the polymer film deposition rate in plasma with reduction of surface temperature. This kinetic effect is due to the surface stabilization of intermediate products at lower temperatures, which accelerate the polymerization rate. Polymer film growth rates in nonthermal plasma vary over a large range, between $1\,\text{nm s}^{-1}$ and $1\,\mu\text{s}^{-1}$. These values are time-averaged; instantaneous film growth rates can be higher.

26.2 Plasma Polymerization Kinetics: Initiation of Polymerization by Dissociation of Hydrocarbons in Plasma Volume, Heterogeneous Polymerization of C_1/C_2 Hydrocarbons

Consider the kinetics of the plasma process starting with **decomposition of hydrocarbons in low-pressure glow discharges**, which initiates polymerization. If the initial concentration of hydrocarbons mixed with an inert gas exceeds 10–20 vol%, dissociation of hydrocarbons is mostly due to electronic excitation by direct electron impact. When the concentration of hydrocarbons is less that 3 vol%, electronically excited atoms of inert gases and hydrogen atoms start making an essential contribution to the dissociation of hydrocarbons. Ion–molecular processes also can contribute to the dissociation at very low hydrocarbon concentrations. If gas temperature in the nonthermal discharges exceeds 500 K, additional decomposition of hydrocarbons is due to their reactions with atomic hydrogen, which can double the plasma dissociation rate of hydrocarbons. Vibrational excitation of hydrocarbons stimulates dissociation if the ionization degree is sufficiently high. For low-pressure glow discharges in hydrocarbons, the criterion of significant contribution of vibrational excitation in dissociation can be presented as:

$$\frac{AP_v}{p^2} \gg 1, \tag{26.3}$$

where P_v is the specific power in the discharge, in W cm^{-3}; p is the gas pressure, in Pa; and A is a numeric factor, which equals 10^4 for discharges in CH_4 and decreases for discharges in higher hydrocarbons. When the pressure is about 1000 Pa, a significant contribution of vibrational excitation requires very high values of specific power exceeding $100\,\text{W cm}^{-3}$. At lower pressures, power requirements are not as strong, but essential losses of vibrational excitation can be related to heterogeneous relaxation.

Plasma polymerization, in contrast to conventional polymerization, permits a polymer to be built up starting from any organic compound. Building up a polymer in plasma does not require any specific chemical behavior of monomers, like opening of double bonds, but is based on the attachment of chemically active species (radicals, atoms, non-saturated molecules, etc.) generated in plasma beforehand. Such a mechanism of plasma polymerization based on the preliminary "plasma activation of monomers" is sometimes referred to as "stepwise" mechanism. The **heterogeneous mechanism of polymerization** is quite complicated in this case. In the specific case of plasma polymerization of C_1/C_2 hydrocarbons, mechanisms of heterogeneous stages have been well investigated. The key heterogeneous process initializing the plasma polymerization is the formation of free bonds near the polymer film surface (depth up to several monolayers), which are called centers of polymer growth. Center formation is mostly due to recombination of electrons with ions of hydrocarbons, which proceeds by electron tunneling through the potential barrier of the dielectric polymer film. The center of polymer growth is usually related to a broken C—H bond; the probability of breaking C—C and C=C bonds is much lower. Major growth of the polymer films is due to the attachment of radicals and non-saturated hydrocarbons produced in the plasma volume to the centers of polymer growth. Not only radicals and non-saturated molecular hydrocarbons but also hydrocarbon ions and ion radicals produced in plasma can be attached to the centers of polymer growth (and even to the polymer macromolecules). UV radiation from nonthermal plasma also leads to breaking bonds in the macromolecules of hydrocarbons. Analysis of the heterogeneous kinetics also indicates that plasma polymerization includes two competing processes: plasma-stimulated polymer film growth and plasma-stimulated etching of the polymer film. Generally, "more active" plasma interaction with the surface leads to domination of etching.

26.3 Plasma Initiated Chain-polymerization, Mechanism and Kinetics of Plasma Polymerization of Methyl Methacrylate

Nonthermal plasma can be applied not only for the stepwise polymerization discussed above but also for effective stimulation of more conventional chain polymerization processes. **Plasma-initiated polymerization of methyl methacrylate (MMA)** with production of practically important polymer, the polymethyl methacrylate (PMMA) is a good example of such processes (Ponomarev 1996, 2000). MMA is a quite large organic molecule: $CH_2=C(CH_3)—C(=O)—O—CH_3$, creating polymer (PMMA) by opening of the C=C double bond. Conventional chain propagation reactions of MMA polymerization can be represented by the elementary MMA attachment to a radical R($\bullet$):

$$R(\bullet) + CH_2=C(CH_3)—C(=O)—O—CH_3 \rightarrow (RH_2)C—C(\bullet, CH_3)—C(=O)—O—CH_3 \qquad (26.4)$$

Plasma initiation of the chain polymerization is due to formation of a primary free radical R($\bullet$), starting the traditional scheme (26.4), and by formation of positive or negative ion radicals, which are also capable of initiating the MMA polymerization. The primary free radical R($\bullet$) as well as the charged centers of polymer growth are formed from the absorbed monomers by electron/ion bombardment and UV radiation from plasma. Formation of a positive ion radical from an adsorbed MMA molecule on the surface under electron/ion bombardment and UV radiation can be schematically shown as the ionization process:

$$CH_2=C(R_1)—R_2 \rightarrow CH_2(\bullet)—C^+(R_1)—R_2 + e. \qquad (26.5)$$

The positive ion radical then initiates a sequence of attachments of further and further MMA molecules in the ion–molecular chain propagation reactions:

$$CH_2(\bullet)—C^+(R_1)—R_2 + CH_2=C(R_1)—R_2 \rightarrow CH_2(\bullet)—(R_1)C(R_2)—CH_2—C^+(R_1)—R_2. \qquad (26.6)$$

Formation of a negative ion radical (which is also a center of polymer growth) from an adsorbed MMA molecule on the surface is due to direct electron attachment:

$$CH_2=C(R_1)—R_2 + e \rightarrow CH_2(\bullet)—C^-(R_1)—R_2. \qquad (26.7)$$

Similarly to the process (26.6), the negative ion radical operating as a center of polymer growth also initiates a sequence of attachment processes of further MMA molecules in the ion–molecular chain-propagation reactions involving the negative ions:

$$CH_2(\bullet)—C^-(R_1)—R_2 + CH_2=C(R_1)—R_2 \rightarrow CH_2(\bullet)—(R_1)C(R_2)—CH_2—C^-(R_1)—R_2. \qquad (26.8)$$

To deposit the PMMA thin film on a substrate of interest, the partial pressure of MMA should be high enough to provide sufficient MMA concentration in the absorbed surface layer. Such a requirement is common for all plasma-initiated chain polymerization processes. Discharge power should not be too high in the plasma-initiated chain polymerization to minimize conversion of MMA (or other monomers) in the plasma volume. The role of plasma should be limited in this case to the generation on the surface of relatively low concentrations of active centers initiating chain polymerization in the layer of absorbed monomers.

26.4 Plasma-initiated Graft Polymerization

Polymer film deposition on polymer substrates or other surfaces can be accomplished by the so-called **graft polymerization**. *In this case, chain polymerization of a monomer proceeds on a polymer (or other material) substrate preliminarily treated in nonthermal plasma.* A common approach to such graft polymerization is based on the preliminary polymer substrate treatment in O_2-containing plasma, which forms organic radicals $R(\bullet)$ on the surface and converts them into the organic peroxides. Formation of the organic peroxide compounds is the chain process, which starts with the direct attachment of molecular oxygen and formation of an organic peroxide radical:

$$R(\bullet) + O_2 \rightarrow R\text{—}O\text{—}O(\bullet). \tag{26.9}$$

Further propagation of the plasma-initiated chain leads to production of the organic peroxides and restoration of organic radicals:

$$R\text{—}O\text{—}O(\bullet) + RH \rightarrow ROOH + R(\bullet), R\text{—}O\text{—}O(\bullet) + RH \rightarrow ROOR_1 + R_2(\bullet). \tag{26.10}$$

The polymer substrate activated this way in plasma (by attachment of the surface peroxide groups) can initiate the graft polymerization of gas-phase or liquid-phase monomers. To initiate the graft polymerization process, the substrate should first be heated up to dissociate the organic peroxide on the surface (26.10) and to form active organic radicals:

$$ROOH \rightarrow RO(\bullet) + (\bullet)OH, ROOR_1 \rightarrow RO(\bullet) + R_1O(\bullet) \tag{26.11}$$

The RO radicals then function as the centers of polymer growth. They initiate the graft polymerization, which in the specific case of MMA polymerization into PMMA is the sequence of attachment reactions of the monomer to the center of polymer growth:

$$RO(\bullet) + CH_2{=}C(CH_3)\text{—}C({=}O)\text{—}O\text{—}CH_3 \rightarrow RO\text{—}CH_2\text{—}C(\bullet, CH_3)\text{—}C({=}O)\text{—}O\text{—}CH_3 \tag{26.12}$$

Grafting of monomers with special functional groups significantly changes the surface properties of initial polymer substrates and allows the creation of new special compounds. Such an approach has been effectively applied to create new types of immobilized catalysts on a polymer base. Polyethylene powder with specific surface area $2\,m^2\,g^{-1}$ has been activated in oxygen-containing plasma for further graft polymerization of acrylic and methacrylic acids from the gas phase. Then vanadium (V), titanium (Ti), and cobalt were chemically deposited on the powder to produce highly effective immobilized metal catalysts. Such plasma-activated powders were successfully used as catalysts for ethylene polymerization into polyethylene (PE), as bifunctional catalysts of C_2H_4 dimerization and C_2H_4—C_4H_8 co-polymerization. Also, they were used for the nonthermal plasma-initiated graft polymerization of vinyl monomers on the surface of polytetrafluoroethylene (PTFE), which significantly changes the PTFE surface properties. In this case, the PTFE film is first activated in nonthermal plasma of a low-pressure discharge and then treated in a special liquid monomer that significantly enhances its adhesion to different materials and especially to steel, see Ponomarev (1996, 2000).

26.5 Formation of Polymer Macroparticles in Volume of Nonthermal Plasma of Hydrocarbons

Sometimes, when the residence time of hydrocarbons in the discharge volume is relatively long, the polymer film growth on substrates, reactor walls, and the electrode is accompanied by the formation of polymer powder in the plasma volume. Polymer macroparticles of the powder are deposited on different surfaces but also can be incorporated into the growing polymer film. The typical size of spheroidal macroparticles is 0.1 μm and larger. Although mechanisms of growth of the polymer powder in the plasma volume are like those of polymer films on surfaces, some specific features are related in this case to the formation of precursors of the macroparticles. The precursors of macroparticles are usually sufficiently large hydrocarbon molecules formed in plasma. The minimal size of a macromolecule to be considered a macroparticle is determined by the ability of the particle to be effectively negatively charged in plasma. In the case of large hydrocarbon molecules, the ability to be charged usually requires the macroparticle sizes to exceed 5 nm. While the growth rate of the plasma polymer film on substrate surfaces usually varies between 1 nm s^{-1} and 1 μ s^{-1}, the growth rate of the polymer macroparticles can be significantly faster (10^{-8}–10^{-4} cm s^{-1}). Therefore, formation of a macroparticle larger than 1 μm typically requires more than 1 s of residence time. The polymer macroparticles leave the discharge zone through different channels. Relatively large macroparticles (5–10 μm) simply fall to the bottom of the plasma-chemical reactor due to gravity. While gas flow drags the polymer powder, stagnation zones lead to longer residence time and to an increase of size and concentration of the macroparticles. The polymer macroparticles are usually charged in plasma, which essentially influences their evolution and their dynamics. Factors that prevent significant formation of the polymer macroparticles include the high flowrate of hydrocarbons through the discharge active zone, the absence of stagnation zones there, and the limitation of the film growth rate.

26.6 General Properties of Plasma-polymerized Thin Films

Configurations of reactors applied for plasma polymerization can be quite different. They can be based on low-pressure DC glow discharge and discharges over a wide variety of frequencies from industrial (50–60 Hz) to RF and microwave frequencies. The DC and industrial-frequency alternating-current (AC) discharges use special electrodes located inside the plasma-chemical reactor, whereas RF and microwave discharges are usually electrodeless. A conventional reactor for plasma deposition of the polymer films in a low-pressure DC glow discharge has typical voltage of hundreds of volts and the typical distance between parallel planar electrodes 3–5 cm. The substrate is located between the electrodes. Typical pressure in the discharge varies in the range 30–300 Pa. A typical plasma polymerization reactor based on the electrodeless low-pressure RF ICP discharge has similar parameters. While the properties of plasma-polymerized thin films obviously depend on the configurations of reactors and relevant plasma parameters, some of these properties to be discussed below are common.

 The most specific property of the plasma-polymerized films is a high concentration of free radicals in the films and many cross-links between macromolecules. The concentration of free radicals can be very high, up to 10^{19}–10^{20} spin g^{-1}. The cross-linkage immobilizes the free radicals, significantly slowing down their recombination and chemical reactions. The slowly developing chemical processes with the free radicals result in "slow but sure" changing of gas permeability, electric characteristics, and other physical and chemical properties of the plasma-polymerized films, which is usually referred to as the **aging effect**. Plasma-polymerized films are often characterized by high internal stresses, up to 5–7 × 10^7 N m^{-2}. In contrast to conventional polymerization, internal stresses in the plasma-polymerized films are related to extension of the material. The effect is due to intensive insertion of free-radical fragments between already deposited macromolecules with simultaneous cross-linking during the polymer film growth. The **solubility of plasma-polymerized films** in water and organic solvents is usually very low because of very strong cross-linkage between macromolecules. Most plasma-polymerized films are characterized by **high thermal stability**. A thin polymer film deposited from a nonthermal low-pressure discharge in methane remains stable and does not lose weight after being treated in argon at temperatures up to 1100 K and in air at temperatures up to 800 K.

The **wettability of plasma-polymerized** films is related to their surface energy and depends on the type of plasma gas used for polymerization. For example, polymer films formed from fluorocarbon and organic silicon compounds are characterized by low surface energy and low wettability. Plasma-polymerized hydrocarbon films usually have high wettability. It is important that their wettability is higher than that of their counterparts formed without plasma. The effect is due to oxygen-containing groups usually incorporated into the plasma-polymerized films. These groups usually make a significant contribution to increased surface energy and wettability. Thin plasma-polymerized films are also specific in **selective permeability for different gases**. Many cross-links make this film somewhat like a molecular sieve, creating interest for applications of such films as membranes for gas separation. Ultrathin plasma-polymerized films with thickness of 0.1 μm and smaller are characterized by very **strong adhesion to a substrate**.

The **electric properties of plasma-polymerized films** are especially important in connection with their applications as dielectrics in microelectronics. The **dielectric permittivity** ε is slightly higher and the **dielectric loss tangent**, tan δ, is significantly higher in the case of plasma-polymerized films than in conventional polymer films, which is due to very high concentration of the polar groups. **Resistance of plasma-polymerized films to electric breakdown** is high. As an example, plasma-polymerized fluorocyclobutane film is characterized by breakdown electric fields: 5×10^6 V cm^{-1} for a film thickness of 150 nm, 6×10^6 V cm^{-1} for a film thickness of 100 nm, and 7–8×10^6 V cm^{-1} for a film thickness of 75 nm. To compare, breakdown of a conventional PTFE film with thickness 0.1–0.2 mm requires only 0.4–0.8×10^6 V cm^{-1}. **Electric conductivity of the plasma-polymerized films** is low and strongly depends on temperature with typical activation energy of about 1 eV. The unique properties of the plasma-polymerized films determine their application niches in microelectronics, gas separation, catalysis, and protective coating, see Ponomarev (1996, 2000).

26.7 Plasma Treatment of Polymer Surfaces: Initial Surface Products, Treatment of Polyethylene

Nonthermal plasma treatment of polymers leads to significant changes of their surface properties, in particular surface energy, wettability, adhesion, surface electric resistance, dielectric loss tangent, dielectric permittivity, catalytic activity, tribological parameters, gas absorption, and permeability characteristics. Plasma treatment and modification of polymer surfaces are widely used today in numerous applications, from painting of textiles and printing on synthetic wrapping materials to treatment of photographic materials and in microelectronic fabrication. Plasma treatment of polymers can be performed in nonthermal plasma at both low and high pressures. Low-pressure nonthermal discharges are usually applied for polymer surface chemical functionalization and for "specific and accurate" chemical modification, as in the application for functionalization of photographic materials and in microelectronics. At the same time, high-pressure nonthermal discharges are usually sufficient for less specific treatment of polymers directed to surface cleaning, changing wettability, and so on.

 Considering the significant practical interest, numerous research efforts have been focused on investigation of plasma treatment of polymer surfaces, like in the case of plasma polymerization, see for example Boenig (1988), Kramer et al. (1989), Liston (1993), Liston et al. (1994), Gilman (2000), Gilman and Potapov (1995), Garbassi et al. (1994), Ebdon (1995), Ratner (1995), Chan (1996), Yasuda (2012), Inagaki (2014), D'Agostino (2012), Park (2022), Hollahan and Bell (1974), Vinogradov and Ivanov (1977), Vinogradov (1986), Biederman and Osada (1992), Biederman (2004), Oehr (2005), Sardella et al. (2005), D'Agostino et al. (2010), Maximov et al. (1997), Arefi-Khonsari et al. (2001), Hocker (1995, 2002), Wertheimer et al. (2003), and Tatoulian et al. (2004).

Chemical processes in the thin polymer surface layers are stimulated by all major plasma components, especially by electrons, ions, excited particles, atoms, radicals, and UV radiation. The major primary products of plasma polymer treatment are free radicals, non-saturated organic compounds, cross-links between polymer macromolecules, products of destruction of the polymer chains, and gas-phase products (mostly molecular hydrogen). Processes for the formation of radicals on the polymer surface under plasma treatment, which are due to electron impact and UV radiation, are related to breaking of R—H and C—C bonds in polymer macromolecules $RH \rightarrow R(\bullet) + H$, $RH \rightarrow R_1(\bullet) + R_2(\bullet)$. Direct formation of non-saturated organic compounds with the double bonds on the surface of a polymer, which are treated by nonthermal plasma, can be illustrated as $RH \rightarrow R_1$—CH=CH—R_2. Secondary reactions of atomic hydrogen usually lead to the formation of molecular hydrogen through different mechanisms, including recombination and hydrogen transfer with the polymer macromolecule: $H + H \rightarrow H_2$, $H + RH \rightarrow R(\bullet) + H_2$. The secondary reactions of atomic hydrogen with organic radical R can result not only in recombination but also in simultaneous formation of molecular hydrogen and a double bond in the organic macromolecule $H + R(\bullet) \rightarrow R_1$—CH=CH—$R_2$.

When the plasma gas contains oxygen, the free organic radical R generated by nonthermal plasma treatment on polymer surfaces very effectively attaches molecular oxygen from the gas phase, forming active organic peroxide radicals $R(\bullet) + O_2 \rightarrow R—O—O—$. The RO_2 peroxide radicals formed on the polymer surface by treatment in non-thermal plasma systems can initiate different important chemical surface processes. The simplest processes started by the RO_2 radicals are related to the formation of hydro-organic peroxide and other peroxide compounds on the surface of polymers: $R—O—O— + RH \rightarrow R—O—O—H + R(\bullet)$, $R—O—O— + RH \rightarrow R—O—O—R_1 + R_2(\bullet)$. These reactions together with attachment process create the chain reaction for the formation of hydro-organic peroxide and other peroxide compounds on the surface of polymers.

The plasma-induced chemical transformation occurs in relatively thin surface layers of the polymer because energies of plasma electrons and ions are limited and coefficients of extinction of UV radiation in polymers are high. Electrons make significant contribution to the modification of polymer materials; this has been demonstrated in the specific cases of plasma treatment of polystyrene films and biological objects. The nonthermal plasma electrons usually penetrate the polymers to a depth of about $x_0 \approx 2\,\mu m$. The characteristic depth of modification of polymer material and formation of cross-links between macromolecules can be estimated as [see Rusanov and Fridman (1978a,b)]:

$$x^2 \approx x_0^2 (\ln \sigma_{i0} N - \ln \ln K), \tag{26.13}$$

where σ_{i0} is the effective cross-section of modification of the macromolecules (in the case of bio-macromolecules it can be estimated as $\sigma_{i0} \approx 10^{-9}\,cm^2$); N (cm^{-2}) is the total time-integrated flux of electrons to the polymer surface; and $K \gg 1$ is degree of polymer modification. Assuming, for example, $N = 10^{13}\,cm^{-2}$ and $K = e$, the characteristic depth for modification of a polymer material according to (26.13) is about $6\,\mu m$.

Kinetic investigations of the **chemical modification of the polyethylene (PE) surfaces** in the low-pressure nonthermal pulsed RF discharges show that at very low pressures (10–100 mTorr), the depth of the plasma polymer treatment is 20–30 μm. At higher pressures (≥ 5 Torr), the treatment depth is shorter, below 2 μm, see Ponomarev and Vasilets (2000) and Vasilets (2005). The kinetics of free radical formation in PE is determined in this case by relative contribution of electrons and UV radiation. The saturation level of the free radical concentration treated by only UV radiation ($\lambda > 160$ nm) from the plasma does not depend on discharge pressure in the range 1–20 Torr and does not depend on thickness of the polymer film in the range 2–110 μm. The saturation value of the free radical concentration is about 2.2×10^{17} radicals g^{-1}, which is much below the value corresponding to total plasma treatment at very low pressures (10–100 mTorr). It leads to the conclusion that free radical formation in PE under pulsed RF discharge treatment is dominated by the contribution of UV radiation at higher pressures, whereas at very low pressures (10–100 mTorr) it is mostly due to the contribution of plasma electrons.

26.8 Nonthermal Plasma Etch of Polymers, Contribution of Charged Species, Atoms, Radicals, and UV-radiation in Polymer Treatment and Etching

Plasma etching of polymer materials, like etching in general, is provided by two mechanisms: physical sputtering and chemical etching. Physical sputtering is due to simple ion bombardment, while chemical etching is due to surface reactions and gasification of polymers, particularly provided by atomic oxygen, fluorine, ozone, and electronically excited oxygen molecules $O_2(^1\Delta_g)$. Typical examples of plasma gases applied for polymer etching are CF_4 and $CF_4—O_2$ mixture, which are widely used in microelectronics for cleaning substrates of organic deposits and for etching of photoresist layers. Treatment of polymers in oxygen-containing plasma initially leads to the formation on the surface of the oxygen-containing chemical groups: $—C=O$, $—C(R)—O—H$, $—C(R)—O—O—H$, $—C(R_1)—O—O—C(R_2)—$. Further interaction of the polymer with nonthermal oxygen-containing plasma can result in further oxidation, formation of CO_2 and H_2O, and their transition to the gas phase. Such process can be interpreted as chemical etching of polymer oxygen-containing plasma. Etching rates of different polymers are very different. Fluorocarbon polymers, for example PTFE, are characterized by the slowest etching rates in plasma. Etching of hydrocarbon polymers is much faster. Heteroatoms and especially oxygen-containing groups are the least resistive to plasma etching. Branching of polymer chain usually leads to higher etching rates, whereas cross-links between macromolecules usually decrease the etching rate. Etching rates are different for amorphous and crystal regions of a polymer, which results in changing the polymer surface morphology and can lead to the formation of porous surface structure after long plasma treatment. Etching of the amorphous regions is usually faster, which means the crystal phase can dominate after plasma treatment. This etching effect is applied to increase the durability of polymer fibers.

The formation of chemical products of plasma polymer treatment – radicals, double bonds, cross-links, hydrogen, and so forth – or, in other words, the chemical effect of plasma treatment is due to several plasma components. *Usually, the primary plasma components active in high-depth polymer treatment are divided into two groups, electrons, and UV radiation, which can penetrate relatively far into the surface of polymer material.* The term "contribution of electrons" also assumes the contribution of secondary factors related to the primary plasma electrons such as gas-phase radicals, excited atoms, molecules, and so on. Interaction of the chemically active heavy particles with polymers is usually limited, however, to very thin surface layers (except the effect of ozone and other relatively stable neutrals). The "UV radiation" includes contribution of vacuum-UV (VUV) radiation as well as softer UV radiation. VUV radiation is characterized by short wavelengths (110–180 nm) and very high photon energies sufficient for electronic excitation and direct dissociation of chemical bonds. The high chemical activity of VUV radiation leads, on the other hand, to short absorption lengths. Effective transfer of this type of radiation across the relatively long distance is possible only in low-pressure gases, which explains the name "vacuum ultraviolet." Softer UV radiation is characterized by longer wavelengths, exceeding 200 nm, and therefore lower photon energies. Softer UV radiation is less chemically active but can penetrate deeper into the polymer film. As an example, treatment of PE by VUV radiation of a low-pressure nonthermal RF discharge results in the significant formation of intermolecular cross-links but not very deep into PE. Softer UV radiation from a mercury lamp (typical wavelength 253.7 nm) does not produce the intermolecular cross-links in PE.

Formation of molecular hydrogen (H_2), *trans*-vinylene bonds, and free radicals in PE powder (with specific surface $2\,m^2\,g^{-1}$) due to its treatment in nonthermal pulsed RF discharge in an inert gas has been analyzed by Vasilets (2005). It has been demonstrated that the formation of H_2 and *trans*-vinylene bonds at the depth 1–2 μm after more than 60 min of the plasma–polymer interaction is dominated by VUV in the range 140–160 nm. Plasma electrons at the same time make a significant contribution toward the formation of free radicals in the PE powder. The effect of plasma-generated active heavy particles can be neglected when considering depths of 1–2 μm. The nonthermal low-pressure plasma treatment of PTFE leads to the formation of two types of radicals: those formed by detachment of an F atom, —CF_2—$CF(\bullet)$—CF_2—, and those formed by breaking a macromolecule, —CF_2—$CF_2(\bullet)$. During the initial period of the treatment ($t < 1$ min), formation of radicals is mostly due to the contribution of plasma electrons; later the partial contribution of VUV (140–155 nm) becomes significant. Quantum yields of the free radical formation in this case are 3×10^{-3} for VUV with wavelength 147 nm, and 5×10^{-3} for VUV with wavelength 123 nm. The depth of the effective formation of free radicals in PTFE is about 0.6 μm.

There is significant **synergistic effect of plasma-generated active species and UV radiation in the plasma interaction with polymers**. Therefore, answer to a simple question which plasma component dominates the plasma–polymer treatment often cannot be so simple and unambiguous. Two or more plasma components can make a synergistic contribution into the total process. Generally, the etching of polymers in low-pressure oxygen plasma is mostly due to atomic oxygen, electronically excited molecular oxygen, and UV radiation. For example, experiments with these plasma components individually give total etching rates after summation about three times lower than the combined contribution of the components applied together. Ponomarev and Vasilets (2000) claim that PE and PVC etching is a synergetic effect of atomic particles and UV radiation. Same effect occurs in the plasma treatment of biopolymers and specifically in plasma sterilization processes, where UV radiation, radicals, and chemically active and electronically excited molecules make essentially nonlinear synergistic contributions.

26.9 Plasma-chemical Oxidation, Nitrogenation, and Fluorination of Polymer Surfaces

The interaction of plasma-activated heavy particles with polymers plays an important role in plasma–polymer treatment but is usually localized on the surface in only few molecular monolayers. Probably only ozone is an exception because of its ability to effectively diffuse and penetrate deeper into the polymer. Molecular and atomic particles provide etching of polymer surface layers as well as form new functional groups, significantly changing the surface characteristics of the polymer. The volume properties of the polymer materials remain the same. Let us start with **plasma-chemical oxidation of polymer surfaces**, which is widely used today in different industrial areas. Interaction with oxygen-containing plasma results in the formation of polar groups on the polymer surfaces, which leads to growth of the polymer surface energy and significant increase of the polymers' wettability and adhesion to metals and different organic compounds. Photo-electronic spectra of plasma-treated PE show that the plasma-chemical oxidation leads first to formation on the polymer surfaces of —C(−)—O— bonds. These bonds correspond to such specific groups as peroxides, alcohols, ethers, and epoxies. Second in the row are —C(−)=O bonds, which are typical for aldehydes and ketones. The least probable are —C(—O—)=O bonds, corresponding to the carboxyl-acidic groups. Plasma-chemical oxidation always includes the simultaneous formation of oxygen-containing surface groups and surface etching. As a result, the polymer surface oxidation degree essentially depends on the polymer composition and structure and can be significantly varied by discharge power, plasma parameters, and treatment time. It is possible to avoid the etching of oxygen-containing groups by preliminary treatment of the polymer in plasma of inert gases and following contact of the activated surface with oxygen-containing gases outside of the discharge.

Plasma-chemical nitrogenation of polymer surfaces is provided by plasma-activated nitrogen and nitrogen compounds (especially NH_3) and results in the formation of amine groups ($R—NH_2$), amide groups ($R_1—NH—R_2$), and imine groups ((R_1,R_2)C=N—H) on the polymer surfaces. These groups promote surface metallization or adhesion of different materials to the polymers. The nonthermal plasma nitrogenation approach has been used, for example, for promoting silver adhesion to polyethylene terephthalate (PET). Another example is plasma nitrogenation of polyester webs to promote adhesion of gelatin-containing layers, which were used in the production of photographic materials. Nonthermal nitrogen plasma is also effective in promoting adhesion on the surface of polyethylene-2,6-naphthalate. In these examples, effective nitrogenation is achieved in N_2 plasma, and nitrogen has been incorporated into the polymers in the form of amine and amide groups.

The types of nitrogen-containing groups incorporated into the polymer surface depend on the plasma gas. For example, NH_3 plasma treatment of polystyrene leads mostly to the formation of amine groups (NH_2) on the polymer surface, whereas N_2 plasma in similar conditions does not produce amine groups. Treatment of polymers in N_2 plasma results in a more significant formation of imine groups (C=N—H) on the surface. Experiments with low-pressure nonequilibrium microwave discharges in molecular nitrogen (N_2) indicate that the total surface concentration of nitrogen-containing groups can be very high, reaching up to 40 at%. Ammonia (NH_3) plasma in similar conditions results in a lower surface concentration of the nitrogen-containing groups. Fluorine-containing polymers are the most resistive to nitrogenation and oxidation in nonthermal plasma. For example, treatment of PTFE in N_2 or NH_3 plasma usually leads to the formation of not more than 6 at% of the nitrogen-containing groups in a thin 10 nm surface layer.

Plasma nitrogenation of polymer surfaces strongly promotes further oxidation of the surfaces in atmospheric air. For example, nitrogenation of PE in N_2 plasma and following contact of the surface with atmospheric air results in an oxygen concentration in PE of about 8 at% in the 10 nm layer. A similar procedure in NH_3 plasma leads to an oxygen concentration in PE of about 4–6 at% in the 10 nm surface layer. *Surface modification of polymers in nitrogen-containing plasmas is widely used to improve the biocompatibility of polymer materials.* For example, amine groups formed in plasma on polymer surfaces provide effective immobilization of heparin and albumin on the surfaces. Biocompatibility and adhesion of different cells to the polystyrene surface are significantly enhanced by the formation of amine groups on the surface during treatment in NH_3 plasma. Similarly, the amine groups created in NH_3 plasma on the surface of PTFE provide adhesion of collagen.

Plasma-chemical fluorination of polymer surfaces *leads to a decrease of surface energy for hydrocarbon-based polymers and makes these polymer surfaces hydrophobic, which is widely used for practical applications.* The interaction of hydrocarbon polymer materials with fluorine-containing plasmas results in the formation of different surface groups, especially C—F, CF_2, CF_3, and C—CF. The interaction of CF_4 plasma with hydrocarbon-based polymer materials leads mostly to the formation of C—F and CF_2 groups on the polymer surface, while interaction with CF_3H plasma mostly leads to the formation on the surface of C—CF and CF_3 groups. Treatment of polymer surfaces with fluorine-containing plasmas stimulates three groups of processes simultaneously: formation of the fluorine-containing groups (C—F, CF_2, CF_3, and C—CF), polymer etching, and plasma polymerization. The relative contribution of these three processes strongly depends on the relative concentration in plasma of CF and CF_2 radicals on the one hand, and F atoms on the other hand. The CF and CF_2 radicals are building blocks for plasma polymerization, whereas atomic fluorine is responsible for etching and formation of the fluorine-containing surface groups. The relative concentration in plasma of CF and CF_2 radicals and F atoms depends on the type of applied plasma gas. Typical fluorine-containing gases applied for polymer treatment are CF_4, C_2F_6, C_2F_4, C_3F_8, and CF_3Cl. *Generally, higher C:F ratios in the initial plasma gases lead to an increase of relative concentration of CF and CF_2 radicals in volume with respect to F atoms. Therefore, higher C:F ratios are favorable for plasma polymerization, whereas lower C:F ratios are favorable for polymer etching and formation of fluorine-containing surface groups.* An example of the application of plasma fluorination of polymers is the plasma treatment of PMMA-based contact lenses to minimize their interaction with the eye's tissues. Plasma fluorination of polymers is also used to increase their durability and to decrease the friction coefficient.

26.10 Aging Effect in Plasma-treatment of Polymers

The composition and space distribution of products of plasma treatment of polymer materials can keep changing long after the plasma treatment process is finished. This phenomenon is referred to as **the aging effect in plasma-treated polymers**. Four major mechanisms of the aging effect can be pointed out: (i) re-orientation and shift of the polar groups formed on the polymer surface inside of the polymer material due to thermodynamic relaxation; (ii) diffusion of the low-molecular-mass admixtures and oligomers from the volume of the polymer material to the polymer surface; (iii) diffusion to the polymer surface of the low-molecular-mass products formed during the plasma treatment in the relatively thick surface layer; and (iv) post-plasma treatment reactions of free radicals and other plasma-generated active species and groups between themselves and with the environment.

Aging of hydrocarbon-based polymer materials treated in oxygen plasma is mostly due to the reorientation and shift of the polar peroxide groups formed on the polymer surface inside of the polymer, which is related to thermodynamic relaxation. The wettability contact angle in H_2O decreases several times immediately after treatment because of formation of the polar peroxide groups. Then the wettability contact angle starts increasing and because of the aging effect can almost return to the initial value after several days of storage in atmospheric air. Aging of hydrocarbon-based polymers treated in nitrogen is mostly due to the postprocessing reactions of nitrogen-containing surface groups with the environment. As an example, the major effect of plasma nitrogenation of PE in N_2 discharges is related to the formation of imine groups ($(R_1, R_2)C=N—H$) on the polymer surface. Storage of the plasma-treated polymer in atmospheric air results mostly in the hydrolysis of the imine groups:

$$(R_1, R_2)C=N—H + H_2O \rightarrow (R_1, R_2)C=O + NH_3. \qquad (26.14)$$

Longer storage in atmospheric air results in additional reactions of nitrogen incorporated into polymer with atmospheric water:

$$R_1—CH=N—R_2 + H_2O \rightarrow R_1—CH=O + H_2N—R_1. \qquad (26.15)$$

Polypropylene is characterized by the strongest aging effect. PET is affected by the aging effect a little less. The most durable with respect to aging are PE and polyimide. Generally, a higher level of crystallinity of polymers leads to their stronger durability with respect to the aging after plasma treatment.

26.11 Plasma Modification of Wettability of Polymer Surfaces

One of the most important results of the plasma treatment of polymers, which are produced on the industrial scale, is the change in their wettability and adhesion characteristics. As was discussed earlier, plasma treatment can make polymers more hydrophilic as well as more hydrophobic. Both effects are widely used for practical applications. The change of wettability is usually characterized by the **contact angle** θ, *which is formed on the solid surface along the linear solid–liquid borderline of air (see Figure 26.1).* An increase of wettability or making a polymer more hydrophilic leads to a decrease of the contact angle. An example illustrating improvement of wettability and decrease of the contact angle because of plasma treatment of a polymer is shown in Figure 26.2, see Fridman (2008), for more details. The wettability increase effect is related to plasma-stimulated formation of polar peroxide groups on the polymer surface. Changes in contact angles for plasma-modified polymer

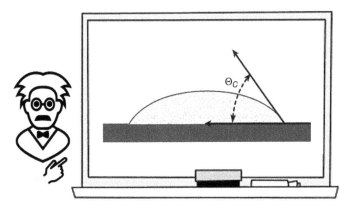

Figure 26.1 Plasma control of surface wettability, contact angle θ_c of a liquid sample.

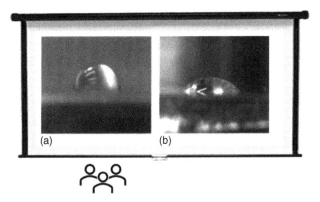

(a)   (b)

Figure 26.2 Increase of wettability and relevant change (decrease) of the contact angle because of surface treatment by nonthermal atmospheric plasma: (a) before, (b) after the treatment.

surfaces depend on the applied plasma gas and conditions of the plasma treatment. The application of discharges in air, oxygen, nitrogen, and ammonia transforms initially hydrophobic surfaces into hydrophilic surfaces. Application of discharges in fluorine-containing gases such as tetrafluoroethylene, perfluoropropane, and octafluorocyclobutane provides significant enhancement of hydrophobic properties of polymer surfaces. An increase of gas pressure, discharge current, and plasma treatment time leads to reduction of the contact angle θ and to an enhancement of wettability. Major enhancements of hydrophilic properties usually occur during the initial treatment time of 30–120 s. The plasma-enhanced hydrophilic properties become less strong with time (aging effect).

Maintaining the enhanced hydrophilic properties over a long period of time is important in practical applications. The nonthermal plasma treatment of polypropylene permits high wettability maintenance (contact angle $\theta < 60°$) for 30 days, and plasma treatment of polyimide permits high wettability maintenance (contact angle $\theta < 50°$) for 12 months. Generally, some level of restoration of the wettability contact angle θ takes place mostly during the first 10 days, and then the aging process slows down.

26.12 Plasma Enhancement of Polymer Surface Adhesion, Metallization of Polymer Surfaces

Plasma treatment cleans the polymer surface and makes it highly hydrophilic, which results in enhancement of adhesion properties of the polymer materials and, as a result, in a wide variety of different relevant applications. The process is widely used in practice for gluing polymers in different combinations and fabrication of composite materials with special mechanical and chemical properties. Thus, plasma treatment-enhanced adhesion of fibers to binding materials significantly improves the characteristics of composites. Plasma technology can be applied for fabrication of high-temperature-resistant electric insulation. Contact properties of polymer surfaces are often improved in practical applications by plasma deposition of special polymer films characterized by strong contact properties with respect to the adhesive material as well as to the major polymer. Plasma enhancement of polymer surface adhesion can be considered as a generalization of the plasma stimulation of wettability. Plasma enhancement of adhesion is widely used in various largescale industrial applications.

An increase of the adhesion energy of polymer materials by their treatment in oxygen-containing plasma is related to an increase of surface energy for the plasma-treated polymers. Plasma modification changes **two components of the surface energy: polar and dispersion**. The polar component of the surface energy characterizes polar interactions between the surface of the polymer material and the working fluid or surface film. This component is determined by the presence of polar groups, electric charges, and free radicals on the polymer surface. In contrast, the dispersion component of the surface energy characterizes the dispersion interaction between the surface and working liquid or surface film. This component is determined by the roughness, unevenness, and branching level of the polymer surface. The polar component of surface energy grows during treatment in the oxygen containing plasma. This effect is due to the formation of polar groups, especially peroxide groups, on the polymer surface. In contrast, the dispersion component of the surface energy, which is relatively large before plasma treatment, can be decreased after plasma modification.

The plasma-enhanced adhesion of polymer surfaces permits effective **metallization of surfaces** using such conventional methods as vacuum thermal and magnetron spraying, deposition by decomposition of organic metal compounds, and so on. Plasma treatment of industrial polymer materials leads to a significant improvement of adhesion to the vacuum thermally sprayed thin aluminum (Al) films and to a decrease of the wettability contact angles, see Gilman (2000), for more details on the subject. Polymers are used in composite materials as a dispersed phase (fibers or powders) as well as a matrix phase. In both cases, adhesion between the phases can be significantly improved by treatment in nonthermal plasma. Plasma modification of porous and nonporous polymer substrates allows the production of composite membranes for gas separation, pervaporation (separation of liquids by evaporation through the membrane), and water cleaning by reverse osmosis.

Plasma modification approaches differ for porous and nonporous substrates. Treatment of porous substrates is focused on the porous size reduction due to cross-linkage of the polymer surface in air, O_2, or inert gas discharges; activation of the substrate surface followed by grafting; and deposition of thin polymer film ($<1\,\mu m$) on the porous substrate surface or on the special adhesive sublayer. Plasma treatment of nonporous membranes can be focused on functionalization, hydrophilization, and cross-linking of the polymer surfaces in plasma of air, O_2, N_2, NH_3, and so on, on plasma deposition of thin polymer films with preliminary surface activation; and on grafting on a preliminarily plasma-activated membrane surface.

26.13 Plasma Treatment of Textiles: Processing of Wool

The plasma approach plays an important and multi-functional role in the treatment of natural as well as man-made textile material, see Maximov (2000) and Hocker (2002). The contribution of plasma technology is not limited to the well-known and widely used plasma effect on dyeing and printing of textiles. Plasma is also effectively used for more specific treatment of natural fibers, including enhancement of shrink-resistance of wool and selective oxidation of lignin in cellulose, which transforms the lignin in a water-soluble form for further extraction.

To discuss **plasma treatment of wool** it should be first mentioned that the cuticle cells in wool overlap each other to create a directional frictional coefficient, and the very surface is highly hydrophobic. The hydrophobic behavior leads in aqueous medium to aggregation of the fibers, which move to their root, resulting in felting and shrinkage of the wool. A significant effect of wool treatment in oxygen-containing (in particular, air) plasma is due to the oxidation and partial removal of the hydrophobic lipid layer on the very surface of wool. This applies both to the adhering external lipids and to the covalently bound 18-methyl-eicosanoic acid. Thus, the plasma-induced decrease of the hydrophobic behavior results in the **shrink resistance**. The effect can be achieved in both low-pressure and atmospheric-pressure nonthermal plasma discharges.

Another nonthermal plasma effect, which can also be achieved in both low-pressure and atmospheric-pressure nonthermal discharges, is related to a significant **reduction of the cross-link density of the exocuticle layer in wool**. The exocuticle is the layer located below the fatty acid layer of the very surface, which is called the epicuticle. The exocuticle layer is highly cross-linked via disulfide bridges. Treatment of the wool in oxygen-containing plasma leads to oxidation and breaking of the disulfide bonds, which results in a significant reduction of the cross-link density and improvement of the wool properties. Protein loss after even intensive plasma treatment and extraction is very low (about 0.05%) because of the surface-oriented nature of the plasma treatment. The specific surface area of the wool is significantly increased because of plasma treatment from about 0.1 to $0.35\,\mathrm{m}^2\,\mathrm{g}^{-1}$. Due to the surface-directed nature of the plasma treatment, the tenacity of the fibers is only slightly influenced. Thus, plasma modification of the wool surface leads to a decrease in the shrinkage behavior of the wool top. The felting density of the wool top before spinning decreases from more than $0.2\,\mathrm{g\,cm}^{-3}$ to less than $0.1\,\mathrm{g\,cm}^{-3}$. Especially strong shrinkage resistance can be achieved by additional resin coverage of the plasma-treated fiber surface. This combined plasma–resin procedure leads to the formation of a smooth surface with reduced scale height and shrinkage of about 1% after 50 simulated washing cycles. To compare, the area felting shrinkage of untreated wool is 69%, and of plasma-treated wool without resin is 21%. As the wool is oxidized during treatment in oxygen-containing nonthermal plasma, the hydrophobic behavior of its surface is changed to become increasingly hydrophilic. It results in additional advantages of plasma treatment related to improved dyeing kinetics, enhanced depth of shade, and improved bath exhaustion.

26.14 Plasma Treatment of Textiles: Processing of Cotton, and Synthetic Textiles, the Lotus Effect

Treatment of cotton and synthetic textile fibers can be performed in nonthermal plasma at atmospheric pressure as well as reduced pressures depending on the specific modification needs of the materials. First, consider some physical and chemical features of plasma treatment of cotton fibers. Like wool treatment, the nonthermal plasma treatment of cotton in oxygen-containing gases leads to a significant increase of specific surface area of cotton. On the other hand, treatment of cotton using hemamehtyldisiloxane (HMDSO) plasma results in a strong **hydrophobization effect**. It leads to formation of smooth surfaces with increased contact angle, about 130° with respect to water. The strong hydrophobization effect of the cotton fiber can be also achieved using nonthermal hexafluoroethane plasma, whereby fluorine is effectively incorporated into the fiber.

 Plasma-induced hydrophobization of cotton fabric in conjunction with increased specific surface area leads to an interesting and practically important effect. Water droplets can effectively remove dirt particles from the surface of the cotton fabric. This phenomenon is illustrated in Figure 26.3 for the case of HMDSO-plasma-treated cotton fabric and is usually referred to as **the Lotus effect**, see Fridman (2008), for more details. Thus, the highly hydrophobic plasma-treated surface of cotton with specific plasma-modified surface topography is extremely dust- and dirt-repellant in contact with water. *As an important consequence, the plasma-treated surface also becomes repellant to bacteria and fungi.* The effect is relevant not only

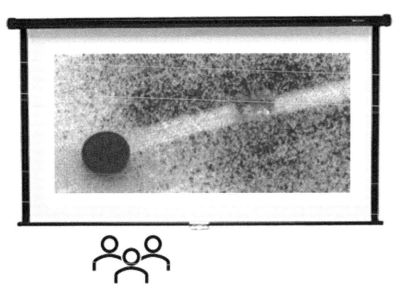

Figure 26.3 Water drop–induced dye removal from an HMDSO-plasma-treated cotton fabric.

to cotton fiber but to other materials as well. Effective treatment of synthetic fibers can be achieved by using atmospheric-pressure plasma of DBD or corona.

Treatment of polypropylene (PP) can significantly increase the hydrophilicity of the surface. In particular, the contact angle with respect to water can be decreased from 90° to 55°; even after two weeks, the contact angle remains about 60°. The hydrophilicity increase is due to a plasma-induced high oxygen-to-carbon ratio, which is significant even at the tenth surface layer of the polymer. The nonequilibrium plasma intensification of the PP surface hydrophilicity can be additionally enhanced by using maleic acid anhydride as an assisting reagent. Incorporation of oxygen into the polymer fiber surface is permanent in this case, and a contact angle with respect to water can be decreased to 42°. PET fibers are used for enforcing the PE matrix in the process of fabrication of polymer composite materials. In this case, nonequilibrium ethylene plasma treatment of the PET fibers significantly increases their adhesion strength to the PE matrix.

Another example of effective plasma treatment of synthetic textiles is treatment of polyaramid textile fibers, such as Nomex, which are high-performance fibers but are prone to hydrolysis. Treatment of the fiber in hexafluoroethane–hydrogen plasma creates a diffusion barrier layer on the surface that is resistant to hydrolysis. The fiber becomes resistant even to 85% H_2SO_4. Contact of the treated fibers with sulfuric acid for 20 h at room temperature leaves the fibers completely intact whereas conventional fluorocarbon finishing under the given conditions causes significant damage.

Industrial plasma-chemical reactors used in final fabric treatment process large amount of materials and sustain stable operation of uniform plasma in large volumes. The reactors can treat fabric widths of about 2 m on a roll of diameter 1–1.4 m. Discharge powers are 100–200 kW, treatment rates are 10–100 m min^{-1}, current densities are 1–3 mA cm^{-2}, and pressures are 50–150 Pa. The plasma-chemical reactors are used for different applications related to final fabric treatment. Specifically, we can mention treatment of wool fabric before printing. The fabric is treated in this technological process in air plasma after smoothing and drying. Subsequent printing has better quality than in the traditional wet chemical approach and does not require chlorination, treatment in tin salt solution, and washing. Another application is related to pre-treatment of rough cotton and mixed fabric (cotton/polyether, cotton/nylon) before dying. The water absorption ability of the plasma-treated fabric is sufficiently high to exclude the otherwise required process of alkaline boiling of the fabric. This feature is especially important for cotton/polyether and cotton/nylon mixed fabrics. Plasma treatment permits the production of fabrics that cannot be manufactured using other technologies. An example is the production of cotton/polyether lace fabric; the semitransparency of the lace fabric is achieved by special plasma modification of the cotton component of the mixed fabric with subsequent partial elimination of the material during alkaline boiling and bleaching.

26.15 Plasma-chemical Treatment of Plastics, Rubber Materials, and Special Polymer Films

Industrial equipment has been developed for the plasma-chemical surface modification of rolled materials, different films, and special products from plastics, rubber, paper, metal, textiles, different fabrics, and so forth for protection of the materials and preparation for dyeing, varnishing, gluing, antifriction treatment, and so on, see Maximov (2000). Plasma-chemical equipment can be operated in this case either continuously or periodically and is used either for plasma activation and coating or just for plasma activation of surfaces as a pretreatment phase of the whole technology. Nonthermal discharges applied in the equipment are DBDs and corona discharges operating at atmospheric-pressure, or glow discharges operating at low pressures. A DBD system for surface modification of polymer films shown in Figure 26.4 can be considered an example of the atmospheric-pressure plasma systems. The system can activate about 10 m min^{-1} of polymer band with width up to 1.6 m. It is applied for treatment of PE, PP, and PET. A glow discharge system for surface modification of polymer films shown in Figure 26.5 can be considered an example of low-pressure plasma systems. The discharge power in the system is 50 kW, voltage is 1.4–1.6 kV, current is 0.3–0.35 A, and gas pressure is about 30 Pa. The system can activate about 3 m min^{-1} of polymer band with width up to 60 mm. It has been specifically designed for activation of high-density PE applied as a gliding ski surface. A similar glow discharge system has been applied for 7 m min^{-1} pretreatment of fluorine-based film before gluing; the width of the glow discharge–treated films can be 0.6 m or more. Such discharge system has also been used for activation of fluoroplastic insulation of 1–4 mm wires with treatment rate 5 m min^{-1}.

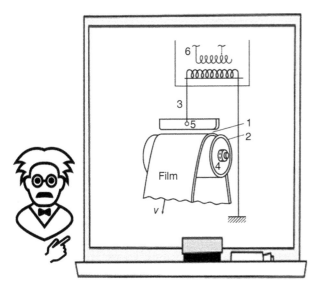

Figure 26.4 General schematic of a system for surface modification of polymer films based on a surface barrier (DBD) discharge: (1) air gap; (2) metallic roller covered with a dielectric; (3) high-voltage wiring; (4) ground system; (5) electrode; (6) high-voltage power supply.

Figure 26.5 General schematic of a reactor for activation of polymer films by an industrial frequency AC glow discharge: (1) gas control system; (2) treatment zone; (3) film-moving system.

Plasma-chemical surface modification of rolled materials and different films from plastics, rubber, paper, metal, textiles, and different fabrics has been also performed using industrial-scale low-pressure RF and microwave discharges. The power of the plasma systems is as high as 300 kW, treatment rate is 60 m min^{-1}, and the width of the treated band is as much as 1.55 m. Such systems have been also applied for the treatment of different shaped products and powders.

26.16 Plasma Modification of Gas-separation Polymer Membranes: Enhancement and Control of Selectivity and Permeability

Polymer membranes and specifically those treated in plasma are used in separation of gases, see for example, Biederman (2004). These membranes are characterized by the **dimensionless selectivity S** determined as a ratio of the membrane permeabilities for different gases. Increase in the selectivity of a polymer material is related to a decrease in its permeability. Creation of the polymer membranes with high permeability and simultaneously high selectivity for gas separation becomes possible via fabrication of the asymmetric multi-layer polymer membranes. Such asymmetric membranes consist of a high-permeability and low-selectivity mechanically strong substrate layer, which is covered by thin high-selectivity and low-permeability gas separating layer. Considering that the mechanical strength of the asymmetric membrane is provided by the substrate layer, the gas separating layer can be made very thin to maintain high multi-layer membrane permeability.

Plasma polymerization is used in this case for deposition of the thin and highly selective polymer layer of the asymmetric membranes. The application of low-pressure RF discharges allows the deposition of layers to be uniform with thicknesses from 30 to 300 nm. The plasma-polymerized films have sufficient adhesion with the high-permeability substrate layer and permit the fabrication of gas-separation asymmetric membranes with sufficiently good characteristics. *A more effective approach to fabrication of asymmetric polymer membranes for gas separation is related not to plasma polymerization but to plasma surface modification of the polymer membranes. A high-permeability and low-selectivity polymer membrane is used in this case as a substrate. Plasma modification of a surface of the membrane leads to high selectivity of its thin surface layer, creating an asymmetric high-permeability membrane effective for gas separation.* The permeability change during plasma modification of polymer membrane surfaces is different for different gases, which leads to selectivity of the polymer membranes and to effective separation of gases. Plasma modification of polymer surfaces permits the fabrication of asymmetric polymer membranes with high permeability and simultaneous high selectivity for gas separation (Arbatsky et al. 1988, 1990). Values of the polymer membrane selectivity achieved by plasma treatment exceed those of any other polymer materials and approach the values typical for inorganic materials.

Microwave plasma modification of the surfaces of siloxane membranes (particularly lestosil and polycarbosil membranes) as well as acetate cellulose membranes is an example of such plasma fabrication of asymmetric highly selective gas-separating polymer membranes. A magnetron with frequency 2.45 GHz and power 2.5 kW is applied as a microwave source in this case (Arbatsky et al. 1988, 1990). Air as well as different atomic and molecular plasma gases have been used with a flowrate of about 40 cm^3 s^{-1}, and the typical gas pressure was above 0.5 Torr. Usually, the initial membranes have high permeability coefficients but low selectivity for separation of practically important gases. It should be mentioned that siloxane polymer materials are generally characterized by the highest permeability coefficients between polymer materials applied for gas separation. The plasma treatment can lead to either decrease or increase of permeability of the gas-separating polymer membranes. For example, plasma treatment of asymmetric membranes from acetate cellulose leads to the destruction of a thin dense surface layer, which results in a significant permeability increase. Siloxane polymers are not etched during microwave plasma treatment; therefore, surface modification results in the densification of a surface layer due to cross-linking and other effects. This effect leads to a decrease of permeability of the siloxane membranes after plasma treatment. Longer treatment times of siloxane membranes saturate the decrease of their permeability, which is due to the presence of pores in the membranes. While plasma treatment reduces the permeability of the dense siloxane, it cannot affect gas penetration through the pores in the material. As a result, longer treatment does not decrease permeability below some level. In agreement with common sense, the equality of fluxes through the pores and plasma-treated dense polymer layers occurs faster (after shorter plasma treatment time) for larger penetrating molecules (CO_2, CH_4, N_2, O_2) than for smaller ones (He, H_2).

The **selectivity of gas-separating polymer membranes** can be significantly increased by treatment in a microwave plasma, which is illustrated in Figure 26.6a,b. Figure 26.6a shows the dependence of selectivity of a lestosil membrane with respect to different pairs of gases as a function of plasma treatment time. Similar dependences for acetate cellulose and polycarbosil are presented in Figure 26.6b (Arbatsky et al. 1988, 1990; Fridman 2008). The selectivity increases for separation of He—CH_4 and H_2—CO_2 mixtures, permeabilities of the plasma-treated membranes with respect to helium and hydrogen remain almost as high as before the plasma treatment. Because of the presence of micropores, the maximum selectivity of membranes is determined not only by the plasma treatment process but also by the quality of the initial polymer material, specifically by the concentration and sizes of the pores. A decrease of selectivity for higher treatment times (Figure 26.6a,b) is due to the fact that permeability for smaller molecules (like hydrogen) keeps reducing, while permeability for bigger molecules (like methane) is already determined by the gas penetration through pores and remains almost constant.

Although plasma treatment of polymer membranes is generally affected by the aging effect, the plasma enhancement of gas-separating membrane selectivity remains stable for a long period of time, at least 100 days of storage in air. The stability to aging is due to the chemical and structural stability of compounds formed on the polymer surface during plasma treatment which enhances the gas-separating properties of the membranes. The permeability of the plasma-modified membranes and, hence, their selectivity depends not on the discharge power and treatment time separately but on their product, which is the discharge energy input. This so-called **dose effect** is illustrated in Figure 26.7, where He—CH_4 selectivity of the acetate cellulose membrane is shown as a function of dose at different values of average power and treatment time, see Fridman (2008). The dose effect is due to the permeability of the

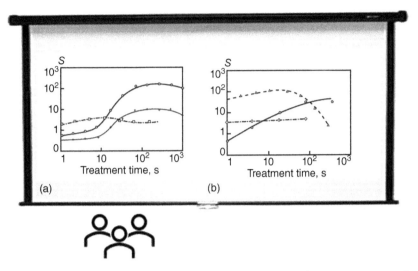

Figure 26.6 (a) Selectivity of the lestosil polymer membrane with respect to separation of gases as a function of microwave plasma treatment time: °, S(He/CO$_2$); +, S(H$_2$/CO$_2$); squares, S(O$_2$/N$_2$); (b) selectivity of the acetate cellulose polymer membrane: triangles, S(He/CH$_4$); ♦, S(H$_2$/CO$_2$) and polycarbosil polymer membrane:(°, S(He/CH$_4$) as function of microwave plasma treatment time.

Figure 26.7 Gas-separation selectivity of the acetate cellulose polymer membrane S(He/CH$_4$) as function of the microwave discharge energy (dose) deposited during total duration of treatment: microwave pulse power 2 kW; pulse duration 100 μs; gas pressure in the microwave discharge chamber 2 Torr; nitrogen/oxygen ratio in the plasma gas N$_2$:O$_2$ =4:1; flow rate of the plasma gas 40 cm^3 s^{-1}. The same dose can be achieved at different values of average microwave power: +, 10 W (pulsing period 20 ms); °, 20 W (pulsing period 10 ms); cubes, 100 W (pulsing period 2 ms).

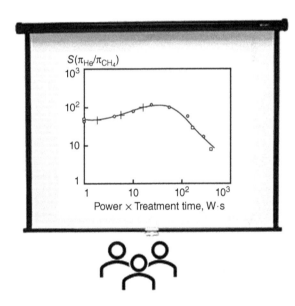

plasma-modified membranes is determined by surface reactions and the density of active oxygen species, which is proportional to the specific energy input in the microwave discharge.

26.17 Mechanisms of Plasma Modification of Gas-separating Polymer Membranes, Lame Equation

The plasma-induced selectivity enhancement of polymer membranes is mostly due to cross-linking of linear macromolecular chains into the polymer network. The cross-linking, in turn, is mostly due to the excited and atomic oxygen penetrating the membrane. The high density of the cross-links leads to decrease in membrane permeability, especially with respect to larger penetrating molecules, which results in selectivity enhancement.

A **theoretical model describing the selectivity enhancement of polymer membranes** due to the polymer cross-linking induced by plasma considers a membrane as a potential field of mechanical forces with respect to the penetrating molecules. The model is based on the three assumptions:

1. Change in selectivity is due to the formation of cross-links or other local quasi-rigid areas in the polymer matrix. The surface density of the cross-links or other local quasi-rigid areas is proportional to the energy input in the discharge during the entire period of plasma treatment.
2. The polymer membrane material is considered to be a uniform elastic medium with Young modulus E. The cross-links and other local quasi-rigid areas are presented as immobile solid globules dispersed in the elastic medium.
3. There are through pores in the membrane, but most of the gas penetration across the non-treated membrane is due to diffusion through the polymer material. As permeability is reduced by plasma treatment and formation of cross-links, the gas flow through the pores becomes a major fraction of the total permeability.

The elastic medium representing the polymer material in the absence of shearing can be described by the **Lame equation**:

$$\lambda \Delta \vec{u} + \vec{f} = 0, \tag{26.16}$$

where λ is coefficient of elasticity, $\vec{u}$ is the displacement of medium from its equilibrium position, and $\vec{f}$ is external force acting on the unit volume. The Lame vector equation (26.16) resembles the scalar Poisson equation in electrostatics, which allows to use the analogy with electrostatics to describe the perturbed elastic medium.

To describe the influence of nonequilibrium plasma-treatment and cross-links on the permeability and selectivity of the gas-separating polymer membranes, let us first calculate the energy required for formation in an unbounded elastic medium of a hollow sphere with radius R corresponding to the radius of a penetrating molecule. Solving the Lame equation in the absence of external forces ($f = 0$) for radial displacement $u_r = R$, the compression energy of the elastic medium is:

$$W_0 = \frac{E}{2} \int_R^\infty \left(\frac{\partial u_r}{\partial r} \right)^2 4\pi r^2 dr = 2\pi ER^3. \tag{26.17}$$

This relation corresponds in electrostatics to the electric field energy of a charged conductive sphere. The displacement of elastic medium in this case is an analog of electrostatic potential, while electric field and electric charge correspond, respectively, to $\sqrt{E} \frac{\partial u_r}{\partial r}$ and $4\pi \sqrt{E} R^2$; here the Gaussian system of units is chosen. Developing the elasticity/electrostatics similarity, the interaction energy of repulsion between two molecules with radii R_1 and R_2 penetrating in the elastic medium can be expressed as:

$$W_1 \approx E \frac{R_1^2 R_2^2}{r}, \tag{26.18}$$

where r is the distance between two penetrating molecules. Similarly, the interaction energy of repulsion inside the polymer matrix between a penetrating molecule of radius R and a rigid globule of radius R_c, representing a cross-link or another local quasi-rigid area, can be calculated as:

$$W_1 \approx E \frac{R^4 R_c}{r^2 \left(1 - \frac{R_c^2}{r^2} \right)}. \tag{26.19}$$

In this relation, r is the distance between the penetrating molecule and the center of the sphere. The presence of a cross-link (effective size R_c) significantly affects the motion of a penetrating molecule and, therefore, the permeability of a polymer membrane when the penetrating molecule moves sufficiently close to the cross-link and the energy of their repulsion exceeds temperature T. The critical distance r_D between the penetrating gas molecule and the cross-link (when the energy of their repulsion exceeds temperature T) is like the Debye radius in the framework of the elasticity/electrostatics analogy. This critical distance can be calculated by assuming $R \approx R_c \ll r$ as:

$$r_D \approx \sqrt{ER^5/T}. \tag{26.20}$$

The relation between the density of the cross-links, n, and average distance L between them can be estimated as $L = 1/n^{1/3}$. If the quasi-Debye radius is short with respect to distance between the cross-links ($L \gg r_D$), the motion of the penetrating molecules is only slightly perturbed by the cross-links. The density of the cross-links should be relatively low in this case $n \ll (T/ER^5)^{3/2}$, and the permeability P of the membrane is decreased only slightly:

$$P = P_0 \left[1 - (r_D/L)^2\right], \tag{26.21}$$

where P_0 is the permeability of the initial membrane before plasma treatment and without any cross-links ($n = 0$). At a higher density n of the cross-links, $r_D > L$, and the energy of interaction between penetrating molecules and the cross-links exceeds temperature everywhere in the volume of the polymer membrane. The permeability P of the polymer membrane is significantly decreased in this case by plasma treatment and formation of the cross-links and can be expressed by the following exponential relation:

$$P = P_0 \exp\left(-ER^4 R_c n^{\frac{2}{3}}/T\right). \tag{26.22}$$

Thus, the size of the penetrating molecules R very strongly affects the permeability, which explains the strong effect of plasma treatment on the selectivity of the gas-separating membranes. The cross-link's mobility and clusterization can also significantly affect the permeability of plasma-treated polymer membranes. More details on the subject can be found in Fridman (2008).

26.18 Modeling of Selectivity of the Plasma-treated Gas-separating Polymer Membranes

The strong exponential dependence of permeability of the plasma-treated membranes on the size of the penetrating molecules determines plasma effect on their selectivity for gas separation. Penetration of larger molecules is usually strongly suppressed by nonthermal plasma treatment of the membrane surface, which leads to significant plasma enhancement of the membrane selectivity, determined as the ratio of permeabilities with respect to two gases characterized by molecular sizes R_1 and R_2. The effect of plasma treatment is determined by specific energy input, or dose, which is proportional to the factor γ – the volume fraction of the elastic medium occupied by the quasi-rigid globules, representing the cross-links. Modeling of the selectivity of the plasma-treated gas-separating polymer membranes has been developed based on the membrane permeability mechanisms described above in Section 26.17. In the case of relatively low doses and low values of factor γ ($\gamma < \gamma_{cr} = 0.16$), when clusters of cross-links are not interconnected, the model predicts:

$$S = S_0 \exp\left[2\pi E \left(R_2^3 - R_1^3\right) \gamma^{\frac{4}{3}}/T\right], \tag{26.23}$$

where S_0 is the selectivity of the polymer membrane before plasma treatment. This relation shows that the selectivity starts increasing exponentially when the specific energy input and factor γ exceed the threshold value:

$$\gamma_{th} = \left[T/2\pi E \left(R_2^3 - R_1^3\right)\right]^{3/4}. \tag{26.24}$$

Thus, the plasma treatment makes the thin surface layer of the polymer membrane "stiffer," which significantly decreases the permeability with respect to larger molecules and results in an enhancement of selectivity in gas separation. The increase in stiffness of the polymer surface after plasma treatment can be characterized by an effective increase of the Young modulus:

$$E_{eff} = E \left(1 + \gamma^{\frac{4}{3}} \frac{R}{R_{cl}}\right). \tag{26.25}$$

These modeling predictions (26.23)–(26.25) are in good agreement with experimental data on microwave plasma treatment of lestosil membranes, which is illustrated in Figure 26.8 for the case of He—CH$_4$ gas separation (Arbatsky et al. 1988, 1990; Fridman 2008). The polymer membrane parameters used for the modeling are $E = 5 \times 10^6$ N m^{-2}, $T = 300$ K, R(CH$_4$) = 0.41 nm, R(He) = 0.215 nm, $R_{cl} = 1$ nm. Based on experimental measurements, the time evolution of the cross-link density can be characterized as $\gamma \propto t^k$ ($k \approx 0.6$).

Thus, the maximum lestosil selectivity for He—CH$_4$ separation achieved in the microwave plasma is 120. Generally, the presence of a maximum in the dependence of selectivity on plasma treatment time is due to porosity of the initial membranes. Although plasma treatment reduces the permeability of the dense polymer, it cannot affect gas penetration through the pores in the material. As a result, longer plasma treatment does not decrease permeability

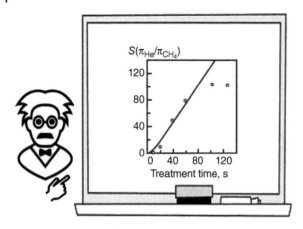

Figure 26.8 Gas-separation selectivity of the lestosil polymer membrane S(He/CH$_4$) as function of microwave plasma treatment duration: °, experimental results; solid line, theory. The membrane treatment regime: microwave pulse power 2 kW; pulsing period 2 ms; pulse duration 100 µs; gas pressure in the microwave discharge chamber 2 Torr; nitrogen/oxygen ratio in the plasma gas N$_2$:O$_2$ = 4 : 1; flow rate of the plasma gas 40 cm^3 s^{-1}.

below some level, which explains the presence of the maximum in the dependence of selectivity on plasma treatment time. The initial porosity of the polymer membranes essentially limits the maximum value of their selectivity to gas separation. When the lestosil membrane is especially prepared without pores, its selectivity with respect to He—CH$_4$ separation exceeds 1000 after microwave plasma treatment. More details on the subject can be found in Arbatsky et al. (1988, 1990), as well as in Fridman (2008).

26.19 Problems and Concept Questions

26.19.1 Mechanisms and Kinetics of Plasma Polymerization

Using kinetic relation (26.1), find the plasma polymerization regimes (derive criteria for the regimes) when the polymer film deposition rate depends on and is proportional to (i) only the flux of charged particles, (ii) only the flux of relevant neutral species, and (iii) fluxes of both active neutral and charged particles.

26.19.2 Temperature Dependence of the Plasma Polymerization Rate

Analyze formula (26.2) and interpret the plasma polymerization kinetics, when the film deposition rate is proportional to fluxes of both neutral and charged particles. Explains the exponential acceleration of the polymer film deposition rate in plasma with reduction of surface temperature.

26.19.3 Temperature Dependence of Electric Conductivity of Plasma-polymerized Films

The electric conductivity of plasma-polymerized films is low and strongly depends on temperature following the Arrhenius formula $\sigma = \sigma_0 \exp(-E_a/T_0)$. Considering the high concentration of radicals in the films, give your interpretation of why different polymers have relatively close values of activation energy (around 1 eV).

26.19.4 Depth of Plasma Modification of Polymer Surfaces

Analyze the double logarithmic dependence of the modification depth x on the polymer modification degree (26.13). Based on the analysis, explain the sharp front of plasma modification of polymer materials. Compare for illustration the depth of polymer treatment corresponding to its slight modification ($K = e$) and very significant modification ($K = 1000$). Assume a total flux of electrons $N = 10^{13}$ cm^{-2} and effective cross-section of modification of the macromolecules, $\sigma_{i0} \approx 10^{-9}$ cm^2.

26.19.5 Primary Plasma Components Active in High-depth Polymer Treatment

Why the high-depth plasma polymer treatment is mostly provided by electrons, and UV radiation, which can penetrate relatively far into the surface of polymer material. Which effects limit the relevant contribution of active heavy particles?

26.19.6 Plasma-chemical Nitrogenation and Fluorination of Polymer Surfaces

Why the plasma-chemical nitrogenation of polymer surfaces (and especially formation of the amine-group NH_2) is effectively used for the biocompatibility of these surface, while the plasma fluorination of polymer surfaces leads to effective surface repulsion of the biomaterials including the surface protection of the bacteria?

26.19.7 The Lotus Effect

Give your interpretation why the plasma-induced hydrophobization of cotton fabric in conjunction with increased specific surface area leads to the water droplet's ability of effective removing dirt particles from the surface of the cotton fabric? Why the HMDSO plasma treatment of cotton fabric is especially useful in the Lotus effect?

26.19.8 Permeability of Plasma-treated Gas-separating Polymer Membranes

Analyzing the relation (26.22), compare the influence of sizes of penetrating molecules and cross-links on the permeability of plasma-treated gas-separating polymer membranes. Why do these two sizes appear in relation (26.22) for gas permeability through the membrane in a nonsymmetric way with significantly different powers.

26.19.9 Threshold Effect of Plasma Treatment on Selectivity of Gas-separating Polymer Membranes

Using relation (26.24), calculate the threshold of the γ-factor (which characterizes the volume fraction of the elastic medium occupied by the quasi-rigid globules, representing the cross-links) corresponding to the beginning of exponential growth of selectivity of a gas-separating membrane. Take numerical values of parameters specifically for lestosil, which can be found in the end of Section 26.18.

Lecture 27

Plasma Biology, Nonthermal Plasma Interaction with Cells

27.1 Plasma Biology as a Fundamental Basis of Plasma Medicine, Plasma Agriculture, and Plasma Food Processing

The following last six lectures are covering plasma medicine, plasma agriculture, and plasma food processing, which are all fundamentally based on the **plasma biology**, science focused on plasma interaction with living tissues. New ideas bring new hopes: plasma medicine is one of those. Recent developments in physics and engineering have resulted in many important medical advances. Among such various medical technologies are applications of ionizing radiation, lasers, ultrasound, magnetism, and others, see for example Fridman and Friedman (2013) and Baura (2020). Plasma technology is a relative newcomer to the field of biology and medicine. Very recent exponential developments in physical electronics and pulsed power engineering promoted consequent significant developments in the non-thermal atmospheric pressure plasma science and engineering. Space-uniform and well-controlled cold atmospheric pressure plasma sources become a reality. All that created an opportunity to apply plasma "safely and controllably" to animal and human bodies.

The nonthermal plasma can provide breakthrough solutions of challenging biomedical problems. It is effective in sterilization of different surfaces including living tissues, disinfects large-volume air and water streams, deactivates dangerous pathogens including those in food and drinks, and is able to stop serious bleeding without damaging healthy tissue. Nonthermal plasma can be directly used to promote wound healing and to treat multiple diseases including skin, gastrointestinal, cardiovascular, and dental diseases, as well as different forms of cancer. It is also proved to be effective in treatment of blood to control the blood properties. Nonthermal discharges are also proved to be very useful in treatment of different biomaterials and in tissue engineering, in tissue analysis and diagnostics of diseases, and even in pharmacology by changing properties of existing drugs and creating new ones.

 Nonthermal plasma is far from the thermodynamic equilibrium and can be therefore very "creative" in interactions with biomolecules. *As it was first demonstrated in 1950s by Stanley Miller and his colleagues from University of Chicago, plasma is even able to generate amino acids from methane and inorganics. It is very much possible that plasma as strongly nonequilibrium and strongly multi-parametric medium can be even responsible for creation of life itself.* Recent experiments prove controllable changes of DNA after the nonthermal plasma treatment are very sensitive to plasma parameters. *It explains the great importance of the "controllability" of plasma parameters and deep understanding of mechanisms for successful progress of the plasma-medical science. Success of plasma medicine, agriculture, and food processing requires detailed understanding of physical, chemical, and bio-medical mechanisms of the strongly nonequilibrium-plasma interaction with cells and living tissues,* which is usually referred to as the **plasma biology**.

Complexity of larger multicellular organisms such as the human body and other macroscopic living tissues makes it difficult to study fundamentally their interaction with various plasmas. It is somewhat easier to study interactions of cells with different plasmas experimentally. Initial steps in deciphering possible mechanisms of interactions between cells and plasma have been made over the last decade or so, but many parts of these mechanisms remain unclear, and the entire subject remains "work in progress," which is going to be reviewed in this lecture starting with some general description of different types of cells and relevant biochemical processes.

Plasma Science and Technology: Lectures in Physics, Chemistry, Biology, and Engineering, First Edition. Alexander Fridman.
© 2024 WILEY-VCH GmbH. Published 2024 by WILEY-VCH GmbH.

27.2 Types of Cells and Primary Cell Components Involved in Interaction with Plasma

Life exhibits hierarchical organization. Atoms are organized into molecules, molecules into organelles, organelles into cells, and so on. All living things are composed of one or more cells as the most basic units of life, and the functions of a multicellular organism are a consequence of the types of cells it contains and how these cells are arranged and work together. Cells fall into two broad groups: prokaryotes and eukaryotes. *Prokaryotic cells (usually bacterial) are smaller and lack much of the internal compartmentalization and complexity of **eukaryotic cells** (typical for humans and animals). Thus, plasma disinfection and sterilization are mostly focused on selective suppression of the prokaryotic cells, while plasma medicine is more related to interaction with the eukaryotic cells. No matter which type of cell we are considering, all known cells have certain features in common such as a cell membrane, DNA and RNA, cytoplasm, and ribosomes.* The natural shapes of cells vary. For example, neurons can grow parts called axons that are often many centimeters long. Skeletal muscle cells can also be several centimeters long. Others such as parenchyma (a common type of plant cell) and erythrocytes (red blood cells) are much more equidimensional. In general, cells range in size from small bacteria (about 1 μm) to unfertilized eggs produced by birds and fish. Let's analyze shortly the major cell components involved in the plasma biology.

27.2.1 The Cell Envelope: Membranes and Walls

The cell membrane functions as a semi-permeable mechanically flexible barrier, allowing very few molecules across it while fencing the majority of organically produced chemicals inside the cell. The cell membrane can be represented as a **lipid bilayer**, which is illustrated in Figure 27.1 (Fridman and Friedman 2013). The most common molecule in the lipid bilayer is the phospholipid, which has a polar (hydrophilic) head and two nonpolar (hydrophobic) tails. These phospholipids are aligned tail to tail so the nonpolar areas form a hydrophobic region between the hydrophilic heads on the membrane surfaces, facing toward the inside and outside of the membrane. The bilayer membrane is fluid-like and various molecular structures in it can move when these structures are not anchored in some way to other molecular structures within cells. In mammalian cells, these anchors and structural support are partly provided by the cytoskeleton filaments. Cell membrane proteins are typically suspended within the bilayer, as illustrated in Figure 27.1, although the more hydrophilic areas of these proteins "stick out" into the cell interior as well as outside the cell. These proteins function as gateways that allow certain molecules to cross into and out of the cell by moving through open areas of the protein channel. The outer surface of the membrane will tend to be rich in glycolipids, which have their hydrophobic tails embedded in the hydrophobic region of the membrane and their heads exposed

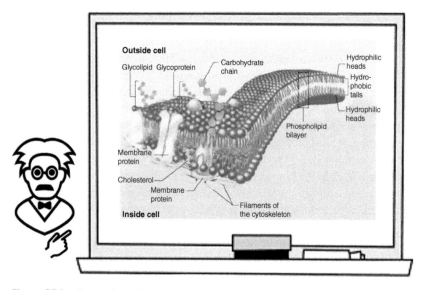

Figure 27.1 Illustration of a cell membrane and associated molecular structures.

outside the cell. These, along with carbohydrates attached to the integral proteins, are thought to function in the recognition of self, a sort of cellular identification system.

Most animal and animal-like cells are enveloped only by cell membranes. Fluidity and flexibility differentiate these cell membranes from cell walls around many nonmammalian cell types. Many bacteria, fungi, and plant cells are enveloped by **cell walls** that are complex semi-rigid structures. In gram-positive bacteria, for example, the cell wall envelopes the inner cytoplasmic lipid membrane. In gram-negative bacteria, the cell wall and the inner membrane are surrounded by an outer lipid-based membrane. The chemical composition of the cell wall differs among various bacteria but is substantially based on peptidoglycans. Peptidoglycan is essentially a polymer formed by repeating disaccharides interconnected by polypeptides. Some species of bacteria have a third protective covering, a capsule made up of polysaccharides.

27.2.2 The Nucleus, DNA, and Chromosomes

The **nucleus** is found only in eukaryotic cells and is the location for DNA and RNA. DNA is the molecular carrier of inheritance and, is mostly restricted to the nucleus. RNA is formed in the nucleus using the DNA base sequence as a template and moves out into the cytoplasm where it functions in the assembly of proteins. The nucleolus is an area of the nucleus (usually two nucleoli per nucleus) where ribosomes are constructed. The nuclear envelope is a double-membrane structure. Numerous pores occur in the nuclear envelope allowing RNA and other molecules to pass, but not DNA. Chromatin is the combination of DNA and proteins that make up the contents of the nucleus. The primary protein components of chromatin are histones. Chromatin is found mostly in eukaryotic cells, while DNA-associated structure in prokaryotes is different and referred to as genophore. In eukaryotic cells, chromatin is typically divided into separate **chromosomes**, which have linear strands of DNA containing anywhere from 105 to 109 nucleotides. In prokaryotic cells, the separate pieces of genetic material (which are also called **chromosomes**) are typically connected in a circle. Genetic information contained in chromosomes is segregated between chromosome sets. In humans, a single set consists of 23 chromosomes and most cells have a total of 46 chromosomes.

27.2.3 Cytoplasm and Cytosol

The **cytoplasm** is the material between the plasma membrane (cell membrane) and the nuclear envelope. The part of the cytoplasm that is outside all the organelles is called the cytosol. The **cytosol** is a gel-like complex mixture of cytoskeleton filaments (fibrous proteins), dissolved molecules, and water that fills much of the volume of a cell. Cytoskeleton filaments maintain the shape of the cell, serve as organelle anchors, and control movement of the cell and its internal reorganization. Three primary types of filaments are usually found in eukaryotic cells: **microtubules, actin filaments (microfilaments) and intermediate filaments.** Microtubules are rope-like polymers of tubulin that can grow up to 25 μm in length. The outer diameter of a microtubule is c. 25 nm. Microtubules are important for maintaining cell structure, providing platforms for intracellular transport, and forming the spindle during mitosis as well as other cellular processes. There are many proteins that bind to the microtubules including motor proteins such as kinesin and dynein, severing proteins like katanin, and other proteins important for regulating microtubule dynamics. Actin filaments are the thinnest filaments of the cytoskeleton. Linear polymers of actin subunits are flexible and relatively strong. Microfilaments are highly versatile, playing the key role in cell movement and changes in cell shape. Intermediate filaments (IFs) are a family of related proteins that share common structural and sequence features. They have an average diameter of 10 nm, which is between that of 7 nm actin filaments and 25 nm microtubules. Most types of IFs are cytoplasmic, but lamins are nuclear. IFs function as tension-bearing elements to help maintain cell shape and rigidity and to anchor in place several organelles, including the nucleus and desmosomes.

27.2.4 Vacuoles and Vesicles

Vacuoles are single-membrane organelles that essentially form a cellular exterior within the boundaries of the cell membrane. Many cells will use vacuoles as storage compartments. **Vesicles** are much smaller than vacuoles and function as material transport vehicles both within and to the outside of the cell.

27.2.5 Endoplasmic Reticulum

Endoplasmic reticulum is a mesh of interconnected membranes that perform functions involving protein synthesis and transport. Rough endoplasmic reticulum (rough ER) is so named because of its rough appearance due to the numerous ribosomes (see the following section) that occur along the ER. Rough ER connects to the nuclear envelope through which the mRNA (the blueprint for proteins) travels to the ribosomes. Smooth ER lacks the ribosomes characteristic of rough ER and is thought to be involved in transport and a variety of other functions.

27.2.6 Ribosomes, Golgi Apparatus

Ribosomes are the sites of protein synthesis within the cytoplasm. They are not cell membrane bound and occur in both prokaryotes and eukaryotes. Eukaryotic ribosomes are slightly larger than prokaryotic ribosomes. Structurally, the ribosome consists of a small and larger subunit. Biochemically, the ribosome consists of rRNA and some 50 structural proteins. In eukaryotic cells, ribosomes often cluster on the endoplasmic reticulum, in which case they resemble a series of factories adjoining a railroad line. **Golgi apparatus** complexes are flattened stacks of membrane-bound sacs. Golgi function as a packaging plant, modifying vesicles produced by the rough ER. New membrane material is assembled in various cisternae (layers) of the Golgi.

27.2.7 Lysosomes

Lysosomes are relatively large vesicles formed by the Golgi. They are cellular organelles that contain acid hydrolase enzymes to breakdown waste materials and cellular debris and are found in animal cells; in yeast and plants, the same roles are performed by lytic vacuoles. Lysosomes digest excess or worn-out organelles and food particles and engulf viruses or bacteria that might end up within the cell. They are frequently nicknamed "suicide-bags" or "suicide-sacs" by cell biologists due to their autolysis. The size of lysosomes varies over the range 0.1–1.2 μm. At pH 4.8, the interior of the lysosomes is acidic compared to the slightly alkaline cytosol (pH 7.2). The lysosomal membrane protects the cytosol, and therefore the rest of the cell, from the enzymes within the lysosome.

27.2.8 Mitochondria

A mitochondrion (plural mitochondria) is a membrane-enclosed organelle found in most eukaryotic cells. These organelles range from 0.5 to 1.0 μm in size. Mitochondria are sometimes described as "cellular power plants" because they generate most of the cell's supply of ATP used for chemical energy exchange (just like electricity generated by power plants is used in all kinds of devices requiring energy exchange). In addition to supplying cellular energy-exchange currency, mitochondria are involved in a range of other processes such as signaling, cellular differentiation, cell death as well as the control of the cell cycle and cell growth. In humans, 615 distinct types of proteins have been identified from cardiac mitochondria; 940 proteins encoded by distinct genes have been reported in rats. Although most of the DNA of a cell is contained in the cell nucleus, the mitochondrion has its own independent genome. A mitochondrion contains DNA that is organized as several copies of a single, circular chromosome. The mitochondrial genome provides the codes for RNAs of ribosomes and the tRNAs necessary for the translation of messenger RNAs into protein. The circular structure is also found in prokaryotes, and the similarity is extended by the fact that mitochondrial DNA is organized with a variant genetic code like that of proteobacteria. Mitochondria are also bounded by two membranes, similarly to some bacteria. The inner membrane folds into a series of cristae, which are the surfaces on which ATP is generated. The matrix is the area of the mitochondrion surrounded by the inner mitochondrial membrane. Ribosomes and mitochondrial DNA are found in the matrix. One important mechanism of regulating mitochondrial activity in cells is calcium (Ca^{2+}) signaling (mitochondria is one of many possible targets of calcium signals in cells). In general, calcium performs a signaling role through binding to various organic molecules, which modulates their function. Accumulation of Ca^{2+} in mitochondria regulates mitochondrial metabolism and causes a transient depolarization of mitochondrial membrane potential. Mitochondria may act as a spatial Ca^{2+} buffer in many cells, regulating the local Ca^{2+} concentration in cellular microdomains. Mitochondrial Ca^{2+} uptake plays a substantial role in shaping Ca^{2+} signals in many cell types. Under pathological conditions of cellular Ca^{2+} overload, particularly in association with oxidative stress, mitochondrial Ca^{2+} uptake may trigger pathological states that lead to cell death. *Observations of **intracellular calcium concentration variations because of nonthermal plasma treatment** may suggest one mechanism by which nonthermal plasma modulates mitochondrial activity, resulting in increased production of intracellular reactive oxygen species (ROS).*

27.2.9 Plastids

Plastids are also membrane-bound organelles that only occur in plants and photosynthetic eukaryotes. Leucoplasts store starch, as well as some protein or oils. Chromoplasts store pigments associated with the bright colors of flowers and/or fruits. Like mitochondria, chloroplasts have their own DNA, called cpDNA. Chloroplasts of green algae (protista) and plants (descendants of some of the green algae) are thought to have originated by endosymbiosis of a prokaryotic alga like living prochloron.

27.3 Transport Processes Across Cell Membranes and Their Relevance to Plasma Treatment of Cells and Its Selectivity

Transfer of plasma species into a cell is a complicated process crucial for plasma interaction with living tissues. Transport of molecules across the cell membranes can be strongly affected by plasma treatment. Knowledge of cell membrane transport mechanisms can therefore be important for developing an understanding of plasma-medical effects. The cell membranes act as barriers to most molecules. Development of a cell membrane that could allow some materials to pass while constraining the movement of other molecules was a major step in the evolution of the cell. Most of the transport mechanisms can be divided into passive and active transport processes. **Passive transport** processes do not require cells to expend energy for the transport process, while **active transport** requires cellular energy expenditure.

Water (H_2O), carbon dioxide (CO_2), oxygen (O_2), hydrogen peroxide (H_2O_2), and nitric oxide (NO) are among the few molecules that can cross the cell membrane by diffusion (that is **passive transport**), a flow process driven by concentration gradients. The ability of these molecules to diffuse is primarily due to their small size, lack of net charge in solution, and absence of strong chemical reactivity with organic molecules. Animal cells typically produce carbon dioxide and consume oxygen through cellular metabolic processes. Carbon dioxide gradient (higher within cell, lower outside) and oxygen gradient (higher outside cell, lower inside) maintain the flow of these gases out and into the cell. By contrast, plant cells use CO_2 as a carbon source through photosynthesis while producing oxygen. Concentration gradients of these gases are consequently different across plant cell walls to across animal cell membranes.

Flow of water across the cell membrane often occurs by **osmosis**, a phenomenon closely related to diffusion. In osmosis, a different concentration of some solute across the membrane drives the water preferentially in a direction that would tend to equalize solute concentration. For example, cells containing less salty water inside than outside will tend to lose water and shrink, while cells surrounded with less salty water will tend to uptake water and burst. *Plasma generates initially significant amount of charged and active species in water outside of the cells, therefore the **osmotic pressure** can intensively pump water from the cells. This effect can lead to cell shrinking and death and can be crucial in plasma sterilization (somewhat like the salt-induced food sterilization and long-term storage).*

Larger molecules or even small ions are generally much less likely to diffuse across a lipid bilayer. However, limited diffusion even of these species can occur through "**flickering**" **pores** in the lipid membrane. The pore flickering in a lipid bilayer can occur occurs because of thermal fluctuation processes. Pores below certain sizes related to the Debye screening length in the cellular medium tend to close quickly, while pores above this critical size could be more stable. Also, phenomenon of **electroporation,** which can be viewed as increased probability of larger pores under external electric-field-induced cell membrane polarization, has been shown to reliably permit the transfer of genetic material into cells that would otherwise not occur.

Facilitated diffusion is a passive form of the transport process that occurs in the direction opposite to the concentration gradient. In contrast to simple diffusion, *facilitated diffusion requires the assistance of proteins integrated into various cell membranes* (including intracellular organelle membranes). Facilitated diffusion is a process particularly important for transport ions, polar molecules, and larger nonpolar molecules. Two types of transmembrane proteins which facilitate diffusion can be distinguished: (i) those that restrict diffusion to species by size, shape, charge, and other properties of the channel they form; and (ii) those that change their conformation and work like a rotating doorway when they recognize the species they are meant to transport. **Channel-forming proteins** are mostly used to facilitate ion transport (for this reason they are often called ion channels) and are often "ligand-gated" (require

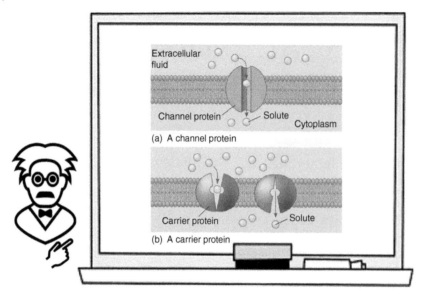

Figure 27.2 Illustration of facilitated transport processes across lipid membranes: (a) effect of channel protein, (b) effect of carrier protein.

binding of a small signaling molecule to open the protein channel), see Figure 27.2 (Fridman and Friedman 2013). Diffusion of larger polar and nonpolar molecules typically occurs with the help of proteins that change their confirmation. These are called carrier, transporting, or permeases proteins. Although carrier proteins are designed by nature to change their confirmation when binding to a target species, they can also restore their initial state (initial confirmation prior to binding) through thermal fluctuation.

 Aquaporins (AQPs), also called water channels, should be especially mentioned in specifically regarding the selectivity of plasma effect on cells. The AQPs are channel proteins from a larger family of major intrinsic proteins that form pores in the membrane of biological cells, mainly facilitating transport of water and other molecules including H_2O_2 between cells. *The plasma effect on the AQPs can be potential interpretation of the **selective anticancer behavior of plasma**, explaining why the cancer cells respond to plasma treatment with a greater rise in ROS than homologous normal cells. Cancer cells tend to express more AQPs on their cytoplasmic membranes, which may cause the H_2O_2 uptake speed in cancer cells to be faster than in normal cells.*

 As a result, plasma treatment kills cancer cells more easily than normal cells. Preliminary observations indicate that glioblastoma cells consumed H_2O_2 much faster than did astrocytes in either the plasma-treated or H_2O_2-rich media, which supports the selectivity effect of plasma on AQPs and related selective plasma ability in cancer treatment. More information about the AQPs selective effect on plasma cancer treatment can be found in publications of Keidar and his group, reviewed by Yan et al. (2015), Keidar and Bellis (2018), and Keidar (2020).

 Active transport of ions and molecules often occurs in the direction of the gradient (that is, against diffusion direction) and requires the cell to spend chemical energy, usually through ATP molecules, to carry out the transport. The problem is that cellular events such as production by the cell of appropriate molecules that are integrated into membranes to assist in the transport are also indirectly associated with any transport mechanism. Protein channels associated with facilitated diffusion may need to be renewed over time for example, and this requires cellular energy. One example where ATP molecules are directly involved in transport is protein pump, a transmembrane protein. In a protein pump, an ATP molecule donates energy to the transmembrane protein usually by letting it "take" one phosphate group and becoming an ADP molecule to change the transmembrane protein confirmation after it binds some ions or molecules within it. This conformational change allows the bound species to be moved across a membrane from lower concentration to higher. Protein pumps are employed by bacteria to pump out anti-biotic molecules. A good example of a protein pump in mammalian cells is the sodium-potassium pump, where Na^+ is maintained at low concentrations inside the cell and K^+ is at higher concentrations. The reverse is the case on the outside of the cell. Nerve cells employ the sodium-potassium pumping to establish conditions necessary for propagation of electrical impulses.

Another interesting mechanism of what is often viewed as active transport is a phosphotransferase system that bacteria such as *Salmonella* and *Escherichia Coli* use to bring glucose and other forms of sugars inside the cell. Beginning with ATP, it involves a sequence of phosphate group transfers that eventually transform glucose (or another sugar) into glucose-6-phosphate while the glucose is being transferred through a transmembrane protein (Enzyme II C, for example). Once glucose is converted into glucose-6-phosphate, it cannot go back out of the cell using the same route. The phosphotransferase system is an example of where molecules are modified in the transport process, circumventing the need to work against the concentration-gradient-driven flow. Although the energy for this transport does come from ATP-donated phosphate groups, it can be argued that glucose is a source of cell-produced ATP and is therefore the energy source for its own transport. However, this energy is delivered through the cell rather than directly. For this reason, this mechanism is appropriately classified as active. One more mechanism of transport often viewed as active is vesicle-mediated endocytosis-exocytosis. Endocytosis is a process by which cells absorb molecules (such as proteins) by engulfing them. It is used primarily for large molecules. The process opposite to endocytosis is exocytosis (removal of large molecules from the cells). Endocytosis can be subdivided into several different types, depending on the molecular mechanism by which it is mediated. Clathrin-mediated endocytosis occurs through small (approximately 100 nm in diameter) vesicles that have a morphologically characteristic crystalline coat made up of a complex of proteins that are mainly associated with the cytosolic protein clathrin. Clathrin-coated vesicles (CCVs) are found in virtually all cells and form domains of the plasma membrane referred to as clathrin-coated pits. Coated pits can concentrate large extracellular molecules that have different receptors responsible for the receptor-mediated endocytosis of ligands, for example, low-density lipoprotein, transferrin, growth factors, antibodies, and many others.

27.4 Cell Cycle, Cell Division, Cellular Metabolism as a Possible Base of Treatment Selectivity in Plasma Biology

Effect of plasma treatment of cell cultures depends on the state of the culture, on the attribution of the cells to specific point of the cell cycle. The cells at different states show different susceptibility to plasma treatment. Following the cell cycle and cell division mechanisms is therefore important for the plasma biology. Despite differences between prokaryotes and eukaryotes, there are several common features in their **cell division processes**. Replication of the DNA must occur. Segregation of the "original" and its "replica" must follow. **Cytokinesis**, the process that occurs after cell division where one cell splits off from its sister cell, ends the cell division process. The **cell cycle** is the sequence of growth, DNA replication, growth, and cell division that all cells go through. Beginning after cytokinesis, the daughter cells are quite small and low on ATP (energy-exchange molecule). They acquire ATP and increase in size during the G1 phase of Interphase. Most cells are observed in Interphase, the longest part of the cell cycle. After acquiring a sufficient size and amount of ATP, the cells then undergo DNA synthesis (replication of the original DNA molecules, making identical copies including one "new molecule" eventually destined for each new cell), which occurs during the S phase. Since the formation of new DNA is an energy-draining process, the cell undergoes a second growth and energy (ATP) acquisition stage, the G2 phase. The energy acquired during G2 is used in cell division. Some cells divide rapidly (beans, for example, take about 20 h for the complete cycle, while bacteria take about 20 min). Others such as nerve cells lose their capability to divide once they reach maturity. Cancer cells are those which undergo a series of rapid divisions such that the daughter cells divide before they have reached "functional maturity." Environmental factors such as changes in temperature and pH and declining nutrient levels lead to declining cell division rates. These effects can be surely used for plasma control on cells.

Cellular metabolism can be also used to provide selective plasma control of cells. Living cells constantly consume material and energy. They also dispose of various materials, perform mechanical work, and generate heat. From a thermodynamic point of view, these cellular metabolic processes are what keep cells from thermal equilibrium and from the resulting death. From a biochemical point of view, cellular metabolism is the collection of chemical transformations that take place within a cell through which energy and molecular components are provided for all essential cellular processes, including the synthesis of new molecules and the breakdown and removal of others. Different forms of life and different cells use different nutrients as sources of carbon, nitrogen, phosphorus, sulfur, and various metals. **Autotrophs** such as plants and cyanobacteria, for example, obtain carbon from inorganic sources such as carbon dioxide (CO_2), while **heterotrophs** such as animals take carbon from organic molecules, often consuming other living forms for this purpose.

 The differences in cellular metabolism, its intensity, and "specifics of tastes" of different type of cells can be a good base for selectivity of plasma treatment. It can provide, for example, predominant poisoning of pathogens while keeping "good cells in good shape." Also, this effect can contribute to the selective suppression of cancer cells without damaging normal cells.

Sources of energy for different life forms and cells can be light, as in the case of plants that carry out photosynthesis or chemical compounds (inorganic and organic). All sequences of metabolic chemical reactions (metabolic pathways)

are redox processes that involve transfer of electrons between different atoms. Organisms can be classified according to primary electron donors in these pathways and electron acceptors. When the final electron acceptor is oxygen, the metabolic pathway is said to be aerobic; it is anaerobic otherwise. Given the large variety of nutrients and sources of energy, it is natural that many metabolic pathways exist. In fact, each cell type often has many metabolic pathways available to it. The choice of a particular pathway depends on abundance of available nutrients in the environment and on other environmental factors. In some cases, once an organism chooses a particular pathway it may continue to employ it even when this pathway becomes sub-optimal (produces less energy and needed chemical compounds for a given amount of nutrients). This results in existence of multiple possible stable metabolic phenotypes for the same set of environmental conditions. The choice of a particular metabolic pathway can even be inherited (this is a form of so-called epigenetic inheritance or inheritance that is over and above genetic). Metabolic pathways consist of catabolism and anabolism. The catabolic part of a pathway breaks down organic matter to harvest energy and basic construction materials. The anabolic part of a pathway uses net energy and materials provided through catabolism and the environment as input to construct components of cells such as lipids, proteins, and nucleic acids.

27.5 Reactive Species in Cells and Their Similarity with Plasma-generated Reactive Species, Reactive Oxygen Species (ROS)

Several highly reactive or unstable molecules and ions play extremely important role in cellular processes. These are mostly oxygen, nitrogen, hydrogen, and carbon (sometimes also chlorine) containing compounds that are jointly referred to as the **reactive species**. Some of these reactive species can be classified as radicals, while others fall apart to produce radicals or are in a state of electronic excitation. Radicals are traditionally defined in chemistry as molecules or ions whose electrons have a nonzero total spin. Important example of an uncharged highly reactive radical in organic chemistry and biology is the hydroxyl radical OH. Due to its high reactivity with organic molecules, it has a typical lifetime of the order nanoseconds in organic solutions. Some free radicals involving oxygen and hydrogen as well as some nonradical species are depicted in Figure 27.3. The oxygen-based active species in cells are usually referred in biochemistry as the ROS. A less reactive electrically neutral radical that can diffuse across cell membrane is nitric oxide (NO). As a result of its lower reactivity than hydroxyl, nitric oxide is often involved in transmitting signals across cellular membranes (e.g. acting as a neurotransmitter). The nitrogen-based active species in cells are usually referred in biochemistry as the **reactive nitrogen species** (RNS).

It is interesting and intriguing that the ROS and RNS in cells are quite like those generated in air plasma and especially in air plasma interaction with water, see Graves (2012). It hints again on the role of plasma and plasma interaction with water in the creation of life.

Let's consider first the **ROS and their transformations**. Organic radicals often play a role in various biochemical chain oxidation reactions. Peroxidation reactions, for example, can be self-propagating reaction chains. Once they are initiated, as for example by $RH + OH \rightarrow R + H_2O$, peroxidation will propagate by the reactions: $R + O_2 \rightarrow ROO$ and $ROO + RH \rightarrow R + ROOH$. Radicals such as ROO are called **peroxy-radicals** and molecules such as ROOH are called **organic hydroperoxides**. Ionic radicals can form by accepting or donating an electron. A common example of such a radical is **superoxide O_2^-**. This ROS ion-radical is much less reactive with organic

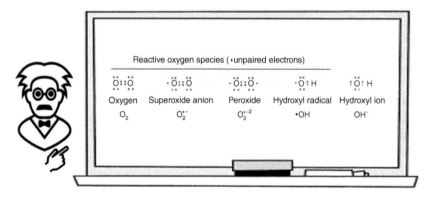

Figure 27.3 Radicals containing only oxygen and hydrogen in comparison with nonradicals.

molecules than hydroxyl, although it does react with organic molecules such as quinones and metalloproteins. However, superoxide is directly or indirectly responsible for production of other more reactive species. **Singlet oxygen** $^1O_2$ is a di-oxygen molecule where the bond-forming electrons have opposite spin (singlet state) and zero net spin. The energy of the singlet oxygen bonding is $94\,kJ\,mol^{-1}$ (roughly 1 eV) greater than the bond energy of the ground state (triplet) oxygen. Reactivity of singlet oxygen is significantly higher than that of superoxide or triplet state oxygen, partly since most bonds in organic molecules are in a singlet state and no longer require transition to the triplet state to react. Singlet oxygen will often react directly with lipids, for example, breaking the carbon–carbon double bonds (particularly when they are weakened by the adjacent alkyl groups. The reaction rates of singlet oxygen with amino acids mostly depend on the presence of double bonds or the presence of electron-rich sulfur atom in the amino acid. In practice, therefore, singlet oxygen reacts primarily with five amino acids: tryptophan, histidine, tyrosine, methionine, and cysteine. The category of reactive species also includes molecules that can split apart spontaneously or with the assistance of catalysts to produce radicals. Peroxides are examples of such molecules, R—O—O—R, where R and R are some atoms or organic molecular groups. When both R and R are hydrogen atoms, the peroxide is the **hydrogen peroxide H—O—O—H or H_2O_2**. The peroxide bond O—O is rather weak and, when it breaks, the resulting two electrically neutral molecules are radicals. As an example, of the ROS reaction, we can mention the **lipid peroxidation**, which is a major mechanism of toxicity in living systems. Abstraction of hydrogen by radicals and attachment of the peroxy-group occurs most easily in polyunsaturated fatty acid chains at carbon locations that are adjacent to carbon double bonds. The lipid peroxidation process essentially breaks down longer polyunsaturated fatty acids, producing aldehydes.

 Let's shortly discuss **the cellular sources of ROS**, which typically include superoxide O_2, OH, singlet oxygen $^1O_2$, and H_2O_2. *Most of the natural pathways for generation of intracellular ROS begin with the capture of an electron by molecular oxygen and formation of the superoxide O_2^-. Electron transport chains such as those occurring in mitochondria are responsible for the greatest amount of the superoxide produced under normal circumstances in cells converting of the order 1–2% of oxygen used by mitochondria into this radical.*
The superoxide radicals can then generate much more reactive hydroxyl radicals and singlet oxygen. This occurs during the spontaneous dismutation reaction (reaction where some of the oxygen is oxidized and some is reduced) of superoxide with production of hydrogen peroxide, and subsequent reaction of the superoxide with hydrogen peroxide:

$$2\,O_2^- + 2\,H^+ \rightarrow H_2O_2 + {}^1O_2, \quad O_2^- + H_2O_2 \rightarrow {}^1O_2 + OH + OH^- \tag{27.1}$$

In addition, the presence of some multivalent metals in cells, such as Cu and Fe, catalyzes further production of hydroxyl radicals from hydrogen peroxide in the **Fenton reaction**:

$$Fe^{2+} + H_2O_2 \rightarrow Fe^{3+} + OH + OH^- \tag{27.2}$$

The Fenton mechanism assumes restoration of Fe^{2+} back from Fe^{3+} in the reaction $Fe^{3+} + H_2O_2 \rightarrow Fe^{2+} + OOH + H^+$. Given the high reactivity of the hydroxyl radicals and of the singlet oxygen, it is not surprising that several enzymes exist, allowing cells to bypass the generation of these ROS from superoxide (at least to some extent). One such enzyme is the **superoxide dismutase** (SOD), which catalyzes the superoxide dismutation to hydrogen peroxide:

$$2\,O_2^- + 2\,H^+ \rightarrow H_2O_2 + O_2 \tag{27.3}$$

The catalase enzymes also help dispose of the hydrogen peroxide, avoiding the overproduction of the hydroxyl radicals by catalyzing the reaction of H_2O_2 conversion to water and oxygen $2H_2O_2 \rightarrow 2H_2O + O_2$.

27.6 Cellular Sources of Reactive Nitrogen Species (RNS), Some NO-based Plasma-biological and Cell Processes

 *The RNS are those reactive molecules which originate from the nitric oxide radical NO, just as ROS typically originates from superoxide. The most active among these species are nitrogen dioxide NO_2 radical and **peroxynitrite $ONOO^-$**. The primary known mechanism of nitric oxide production in cells is oxidation of amino acid L-arginine catalyzed by an enzyme from a family of enzymes known as NOS or nitric oxide synthase. This catalytic reaction is:*

$$2\,(\text{L-arginine}) + 3NADPH + 2H^+ + 4O_2 \rightarrow 2(\text{Citrulline}) \rightarrow 3NADP^+ + NO. \tag{27.4}$$

The above reaction is a two-stage process with each stage oxidizing a different amount of nicotinamide adenine dinucleotide phosphate (NADPH). Nitric oxide is not a highly reactive radical, particularly when compared to hydroxyl, but also with singlet oxygen and many organic peroxy-radicals. Relatively weak reactivity permits diffusion of NO through cytosol and often across cell membranes, which is critical to the signaling functionality of this molecule in biology. Given that nitrogen's electronegativity on the Pauling scale is 3.04 and oxygen is 3.44, which is not a large difference, NO can be considered hydrophobic. *That makes* **NO** *diffusion through cell membrane lipids often faster than diffusion of oxygen. This effect can play very significant role in NO-generating plasma (for example, air gliding arcs) treatment of living tissues, when the plasma-generated NO effectively penetrates the cells. Like molecular oxygen (triplet), nitric oxide also reacts with radicals quickly. However, in contrast to the di-oxygen molecule which has two unpaired spins and whose reaction with other radicals often results in radical chains or unstable molecules, NO often terminates a chain reaction when it reacts with other radicals. For example, it can convert thiol radicals into nitrosothiols by acting as a chain-terminating agent: $RS + NO \rightarrow RS - NO$, where R is some organic group and S is the sulfur atom. Generally, the* **nitrosylation thiols** *play significant role in the nonoxidative sterilization in plasma agriculture and food processing.*

The two most active RNS produced from nitri–oxide are **peroxynitrite ONOO⁻ and nitrogen dioxide NO$_2$ radical,** both of which are the major reactive products of nitric oxide with oxygen. Peroxynitrite, which is a structural isomer of the nitrate NO_3^- ion (in the nitrate all oxygen atoms are bound to nitrogen, while in peroxynitrite two oxygen atoms are nitrogen bound and one is oxygen bound), is produced in cells primarily in the reaction with superoxide:

$$NO + O_2^- \rightarrow ONOO^- \qquad (27.5)$$

The rate of the above reaction is high, 3×10^9 M⁻¹ s⁻¹, and is comparable to the rate of superoxide dismutation by SOD, 2×10^9 M⁻¹ s⁻¹. This means that, at sufficiently high concentrations, nitric oxide can compete effectively for superoxide with SOD. Peroxynitrite can react directly with many important biomolecules, particularly those containing sulfhydryl (SH) groups and transition metal centers (iron, copper, etc.). Peroxynitrite also causes direct DNA strand cleavage by oxidizing deoxyribose. Indirect oxidative effects of peroxynitrite occur through several mechanisms. One is its decomposition resulting in the production of hydroxyl radicals, a moderately reactive nitrogen dioxide radical and nitric acid:

$$2\,ONOO^- + 2H^+ \leftrightarrow 2\,ONOOH \rightarrow NO_2 + OH + HNO_3 \qquad (27.6)$$

The hydroxyl radicals can in some cases be produced more effectively by this process than by the Fenton mechanism. Lipid peroxidation and DNA damage associated with peroxynitrite are likely to occur through this mechanism of hydroxyl radical generation. It reflects again the significant similarity between the RNS biochemistry in cells and the plasma chemistry of air plasma and its interaction with water (Graves 2012). Nitrogen dioxide radical NO$_2$ can abstract a hydrogen from some of the weaker C—H bonds (allylic hydrogen that is bound to a carbon which is adjacent to a carbon–carbon double bond). Protein tyrosine is a major target of NO$_2$ in biological systems. Tyrosine nitration involves covalent protein modification resulting from the addition of a nitro ($-NO_2$) group adjacent to the hydroxyl group on the aromatic ring of tyrosine residues. Reactions with NO$_2$ also result in nitration of lipids.

27.7 Cell Signaling Functions and Their Role in Plasma Biology, Contribution of Reactive Species in Cell Signaling

Cell signaling systems can be viewed as communication networks that control cellular functions and coordinate activities of different cells. Molecular carriers of signals between cells are often called "first messengers," while intracellular signal carriers can be referred to as "second messengers." In contrast to telephony systems that establish communication channels between specific users or end points, cellular communications (of the biological kind) resemble a set of broadcasting systems where the broadcast messages can be targeted to a specific set of receivers, accepted by these receivers, interpreted and then re-broadcast, possibly using a different set of chemical carriers. An intracellular broadcast signal might be initiated, for example, at the cell membrane when a specific membrane receptor (usually a transmembrane protein) binds to a specific extracellular ligand (this could be a protein, hormone, a small molecule, or an ion). For example, a ligand called epidermal growth factor (EGF) binding to the EGF receptor

(EGFR) on the extracellular side of the cell membrane usually leads to phosphorylation of the EGFR on the intracellular side, activating binding to the intracellular adaptor protein (GRB2 protein helps mediate protein–protein interactions). This couples to other proteins propagating the signal further inside the cell along several different pathways to different eventual targets. One of these pathways is called the mitogen-activated protein kinase (MAPK) pathway. The MAPK protein is a protein kinase (an enzyme) that phosphorylates various other proteins such as the transcription factor MYC, which leads to an altered gene transcription and, ultimately, cell cycle progression.

The cell signaling plays very important role in plasma treatment of living tissue, plasma biology, and plasma medicine in general because the plasma species directly affect only surface of the tissue, and further penetration in depth significantly relies on the cell signaling and cell-to-cell communication, which is going to be discussed in Lecture 30.

Intracellular signals are usually passed by the mechanism of protein–protein interactions, often involving phosphorylation and other redox-related events and through diffusion of small molecules and ions. It is natural for reactive species to participate in regulatory cell signaling. Generation of superoxide during cellular metabolism, for example, can be viewed as a signal of aerobic metabolic activity. Reduction of the amount of superoxide produced may signal, for example, reduction in the availability of oxygen or in the availability of glucose. Either way, cells will likely need to adjust their behavior, possibly switching to a different metabolic pathway or by enhancing glucose transport across cell membrane. Given that all cellular activities require energy and materials generated through metabolism, it might be expected that the ROS, whose production is initiated through metabolic activity, might play an important signaling role in the cells.

There are various examples of growth factors, cytokines, or other ligands that trigger ROS production in non-phagocytic cells through their corresponding membrane receptors. It has also been confirmed that intracellular ROS production can be triggered by extracellular messages using membrane-bound NADPH oxidase isoforms as the mechanism. ROS generation by the cardiovascular NADPH oxidase isoforms can be induced by hormones or hemodynamic forces, for example. Angiotensin II, a peptide hormone usually associated with blood vessel constriction, increases NADPH-driven superoxide production in cultured vascular smooth muscle cells and fibroblasts. Thrombin, platelet-derived growth factor (PDGF), and tumor necrosis factor-α (TNF-α) stimulate NADPH oxidase-dependent superoxide production in vascular smooth muscle cells. Interleukin-1, TNF-α, and platelet-activating factor all increase NADPH-dependent O_2^- generation in fibroblasts.

It would be difficult to detect small variations in the concentration of reactive species as a signal if their normal or "background" concentrations were too high. Maintenance of reactive species homeostasis is therefore essential not only to minimize the damaging effects of the reactive species but also to achieve a good signal-to-noise ratio. *This homeostasis is usually maintained through scavenging of intracellular reactive species by various molecules called **antioxidants**.* Antioxidants come in several varieties. Some are water soluble, while others are lipophilic and tend to reside in lipid membranes. Some antioxidants are molecules that can become radicals relatively easily when exposed to reactive species but do not easily convert other organic molecules into radicals. Radicals of such antioxidants readily combine with other organic radicals to terminate radical chain reactions. A well-known example of such antioxidant includes Vitamin E, which is in a class of lipophilic molecules. Although intracellular amino acids are not as easily oxidized by reactive species, their large concentration within cells may also buffer cells against high levels of reactive species. Consistent with this hypothesis, the proteases (enzymes which degrade proteins into amino acids) appear to be over-expressed in cells during periods of increased oxidative stress.

An important class of antioxidants is based on oxidation of thiol groups (S—H) and generation of disulfide bond (S—S, sometimes called bridges). The bond of sulfur to hydrogen is weaker than C—H or O—H bonds and is relatively easily broken when sulfur donates electrons in redox reactions with reactive species. At the same time, the disulfide bond is also relatively weak, making disulfide-containing compounds good candidates for reduction. This type of redox cycling is exactly what occurs in many cysteine residues containing proteins. A common and important redox cycling antioxidant mechanism in many eukaryotic cells is based on glutathione (most prokaryotes lack this antioxidant mechanism). Glutathione (GSH) is a tripeptide that contains an unusual peptide linkage between the amine group of cysteine (which is attached by normal peptide linkage to a glycine) and the carboxyl group of the glutamate sidechain. One of the glutathione functions is to reduce disulfide bonds formed within cytoplasmic proteins to cysteines by serving as an electron donor. In donating an electron, glutathione itself becomes reactive, but readily reacts with another reactive glutathione to form glutathione disulfide (GSSG, also called L-glutathione). Such a reaction is possible due to the relatively high concentration of glutathione in cells (up to 5 mM in the liver, for example). In the process, GSH is converted to its oxidized form, GSSG. Once oxidized, glutathione can be reduced back by an enzyme glutathione reductase, using NADPH as an electron donor. Provision of cysteine is the rate-limiting factor in glutathione synthesis by the cells, since this amino acid is relatively rare in foodstuffs. Furthermore, if released as the free amino acid, cysteine appears to be toxic and spontaneously catabolized in the gastrointestinal tract and blood plasma. Raising GSH levels through direct supplementation of glutathione is difficult as it does not readily enter most cells from outside. Intracellular GSH concentrations can be raised by administration of certain supplements that serve as GSH precursors. N-acetylcysteine, commonly referred to as NAC, is the most bioavailable precursor of glutathione as it enters cells and provides the needed cysteine.

Many enzymes appear to be direct receivers of ROS signals. Some examples include guanylyl cyclase, phospholipase C, phospholipase A2, and phospholipase D. Ion channels also appear to be the ROS signaling targets, including calcium channels. Sulfurhydryl groups (R—S—H) and transition-metal-containing proteins appear to be the primary targets (receivers) of ROS signals. There are two sulfur-containing amino acids – cysteine and methionine – but only cysteine contains thiols. Through thiols, cysteine plays an important role in controlling protein confirmation by oxidation of thiols and formation of disulfate bridges (S—S bond) between different parts of a protein. Overall oxidative stress state (concentration of various ROS) can therefore modulate thiol/disulfate redox state of many proteins. This has been suggested as the mechanism by which levels of ROS affect human insulin receptor kinase activity and p38 MAPK signaling pathways described above, for example. ROS and a pro-oxidative shift of the intracellular thiol/disulfide redox state appear to increase the overall signal strength in cellular broadcasting networks. There has been a growing body of evidence suggesting that this occurs by modulation of protein kinases and protein phosphatases (enzymes that phosphorylate and de-phosphorylate proteins) activities directly. Recall that protein phosphorylation–dephosphorylation processes play critical roles in regulating many cellular metabolic processes in eukaryotes. They also govern many signal transduction pathways. Among the extracellular signals, growth-factor-dependent protein tyrosine kinases and protein tyrosine phosphatases are of critical importance to mitogenesis, cell adhesion, cell differentiation, oncogenic transformation, and apoptosis, to name a few. Target proteins are typically phosphorylated at specific transduction sites (usually at serine/threonine or tyrosine residues) by one or more protein kinases and the phosphates are removed by specific protein phosphatases. Changing the activity of protein kinases, phosphatases or both can regulate the likelihood of phosphorylation at a particular site. The overall effect of reactive species on the signaling networks exerted by activation of protein kinases and phosphatases might be expected because all protein tyrosine phosphatases, for example, have a conserved (as for many cells and species) cysteine residue in their catalytic domain; for complete activity, this catalytic domain needs to be in the fully reduced form. Conditions of oxidative stress will lead to changes in the activity of several signaling enzymes.

27.8 Mechanisms of Plasma Interaction with Cells: Direct vs Indirect Plasma Effects, Main Stages, and Key Players

The primary stages and the key players involved in interaction between plasma and cells are illustrated in Figure 27.4, see Fridman and Friedman (2013) and Dobrynin et al. (2009). Different active agents can be more or less important for different types of plasma treatment. In this regard, it's crucial to separate **direct vs indirect plasma effects**. When **plasma is applied directly** to the medium or to the substrate containing the cells, charges, and local electric field as well as UV can have important effects together with the chemically active species. In contrast to that, the **indirect plasma applications** (through preliminary activation of different liquids, including the special biological

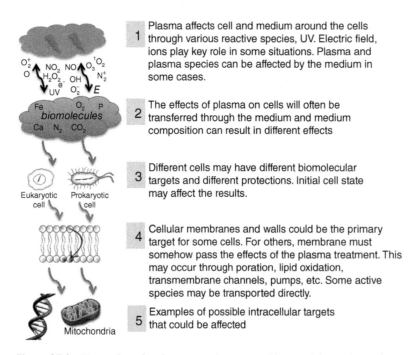

1. Plasma affects cell and medium around the cells through various reactive species, UV. Electric field, ions play key role in some situations. Plasma and plasma species can be affected by the medium in some cases.

2. The effects of plasma on cells will often be transferred through the medium and medium composition can result in different effects

3. Different cells may have different biomolecular targets and different protections. Initial cell state may affect the results.

4. Cellular membranes and walls could be the primary target for some cells. For others, membrane must somehow pass the effects of the plasma treatment. This may occur through poration, lipid oxidation, transmembrane channels, pumps, etc. Some active species may be transported directly.

5. Examples of possible intracellular targets that could be affected

Figure 27.4 Illustration of various general stages and key participants in the interaction between plasma and cells.

media) provide mostly chemically active species generated in plasma, and then converted to numerous ROS and RNS in the liquids and media further interacting with cells and tissues. The efficacy comparison of the direct and indirect plasma effects (sometimes called **plasma pharmacology**) on different cells and tissues is going to be discussed for specific plasma applications in medicine and agriculture.

Generally, ozone is known to occur in higher concentration at lower gas temperature in discharges, while concentrations of nitrogen oxides in air discharges tend to increase at higher gas temperatures. Different discharges produce different UV intensities in different wavelength ranges, which may be differently absorbed depending on the gas composition and the medium above the cells. Ability of the electric field to reach the cells and affect them in some way depends on the amount, conductivity, and dielectric properties of the medium above the cells.

*With some exceptions, such as sporulated bacteria, cells are rarely in a dry environment. The **liquid layer surrounding the cell** creates a barrier significantly influencing the plasma-cell interaction. Even if mammalian cells were capable of surviving for a short time without water around cell membrane, it would be difficult to remove this water completely due to hydrophilic nature of the outside surface of the mammalian membranes.* The actual amount of water, however, may vary. Plasma jets, for example, having their flow directed toward the surface of cell medium may remove some of this medium from the cells.

One may estimate how much medium is left over the cells very roughly by setting the surface tension forces of water equal to the shear force created tension due to the air velocity U at the surface of the water film. Assuming the airflow at a high velocity of $1\,\mathrm{m\,s^{-1}}$ is maintained roughly over the planar substrate length L of about 1 cm, the thickness of the water film on this substrate should be approximately

$$\delta \approx \frac{\mu U}{\gamma}L = \frac{1\times10^{-3}(\mathrm{N\cdot s\,m^{-2}})1\,\mathrm{m\,s^{-1}}}{7\times10^{2}\,\mathrm{N\,m^{-1}}}0.01(\mathrm{m}) = 10\,(\mathrm{nm}), \tag{27.7}$$

where μ is the dynamic viscosity of water, γ is its surface tension and δ is the thickness of the water film. In this situation a shear force $F \approx \mu U\frac{D_{cell}^2}{\delta} \approx 10\,(\mu\mathrm{N})$ would be applied to an individual cell whose characteristic size is $D_{cell} = 10\,\mu\mathrm{m}$ when this cell is attached to the substrate. This is a very large shear force for a single cell handle, much larger than what an endothelial cell might experience due to blood flow, for example. This illustrates the difficulty of removing water from the cells by mechanical means (such as gas flow). Thus, it is reasonable to assume that during the plasma treatment, there is a water layer greater than 10 nm remaining on the mammalian cells.

Given this water layer over the mammalian cells, one can wonder to what extent a charge delivered through plasma to the surface of this water layer can influence processes at the cell membrane. The electrostatic influence can be estimated roughly by comparing the energy of electrostatic interactions of two electrons or ions (having a single electron charge) and the energy of thermal fluctuations. The effective distance at which two-point charges in water have interaction energy equal to thermal fluctuation is known as the **Bjerrum length**:

$$\lambda_B = \frac{e^2}{4\pi\varepsilon_r\varepsilon_0 k_B T}, \tag{27.8}$$

where e is the electron charge, k_B is the Boltzmann constant, T is the absolute temperature, ε_0 is the free space permittivity and ε_r is the relative dielectric constant of water equal to about 80. The Bjerrum length is about 0.7 nm in water. The presence of free ions in water provides another mechanism for screening of cells from the direct electrostatic effects of charges that arrive to water from plasma. The effective screening length in this case is the Debye length for electrolytes given by $\lambda_D = \sqrt{\frac{\varepsilon_r\varepsilon_0 k_B T}{2N_A e^2 I}}$, where I is the ionic strength of the solution and N_A is the Avogadro's number. At typical intracellular salt concentrations (150–200 mM), the Debye length is $\lambda_D \approx 1\,\mathrm{nm}$.

*Thus, there is at least about 10 nm of water on top of mammalian cells and that charges are effectively screened over the distance smaller than about 1 nm, we can conclude that charges arriving from plasma will have negligible electrostatic influence on membranes of mammalian cells. Significant effects on cells can occur due to **chemical modification of the medium surrounding the cells**.* In this case, the exact chemical composition of the cell medium can play an important role. The nature of chemical species produced in plasma and their concentration will be critical as well in such mechanisms. Charges from plasma can play an important role in this chemistry in several ways. On the one hand, their presence can influence peroxidation and oxidation processes catalytically. On the other, they can change the medium acidity and, though this, various reactions, and their rates.

Cellular membrane as the outermost barrier separating the cell interior from its environment is likely the first and most important biological object in the mechanism of interaction between plasma and cell. Various processes can occur at the cell membrane because of plasma treatment. Electric field that penetrates the cell medium for a short time can induce or enhance formation of pores, a phenomenon known as **electroporation**. Besides the obvious possibility of membrane damage, this can influence transport of ions and other biological molecules leading to a variety of signaling effects. Chemically active species created in the medium through its interaction with plasma can lead to oxidation or peroxidation of various molecules in the lipid membrane. Byproducts of lipid peroxidation, such as aldehydes, can end up within the cell. Transmembrane proteins can also be affected by the chemically active species in the medium.

The interaction of plasma and cells does not end with the cell membrane. Through biological signaling, electroporation, lipid peroxidation, or other mechanisms, the membrane can transfer the effects resulting from plasma treatment toward various intracellular targets. This transfer seems to occur through or be accompanied by intracellular oxidative stress. It is possible that this oxidative stress occurs through changes in the functionality of mitochondria. One of the targets of this oxidative stress appears to be the intracellular DNA.

27.9 Contribution of Plasma-generated Charged Species to Plasma Interaction with Cells

Analysis of the contribution of plasma-generated charged species to plasma Interaction with cells is directly related to comparison of indirect and direct plasma effects where plasma charges are clearly present. The first investigations comparing effects of direct and indirect modes of plasma were carried out using the FE-DBD discharge (Fridman et al. 2007a–e). These initial investigations were focused on bacteria as the biological target and employed agar as the supporting surface. It was shown that direct application of plasma yields roughly two orders of magnitude faster rate of bacteria inactivation on agar as compared to indirect plasma application. These investigations also demonstrated that effects of active species without charges were negligible compared to the effects observed with the charges when bacteria were treated on agar. In short, leaving only UV radiation (placing quartz window for UV transmission or magnesium fluoride window for Vacuum UV transmission) removes the ability of dielectric-barrier discharges (DBD) plasma to cause 6 log reduction of bacterial population. Average gas temperature and the applied electric fields have also been shown to have a negligible effect on bacteria. Effects of neutral active species produced in DBD, on the other hand, could not be ignored. It has been shown that neutrals by themselves are able to sterilize as well as direct plasma treatment, although this takes an order of magnitude more time. Therefore, the role of charges in killing bacteria on agar appeared to be significant and catalytic, in that charges increased the rate of bactericidal effects of reactive species.

More detailed DBD experiments comparing direct and indirect treatment controlled the flux of charged particles (Dobrynin et al. 2009, 2010a,b). In case of direct plasma treatment, the discharge is ignited on the treated surface with 1.5 mm discharge gap. For indirect plasma treatment, a grounded metal mesh is used as a second electrode (22 wires per cm, 0.1 mm wire diameter, 0.35 mm openings, 60% open area, weaved mesh). The gap between the mesh and the quartz dielectric insulator on the electrode is set to 1.5 mm, just as in the case of the direct treatment setup. Agar placed into metal dishes was employed as the substrate on which bacteria were treated. These dishes are then biased with unipolar potential to extract charges from the discharge. The plasma was ignited in the same electrode-mesh configuration as in the indirect plasma treatment experiment. Distances between the powered electrode and the mesh, and between the mesh and the agar (note that both the dish and the agar are conductors), were both set to 1.5 mm. When *E. coli* is treated with plasma directly, they are exposed simultaneously to charged particles, UV, and all active plasma components such as ozone (O_3), hydroxyl radicals (OH), and other excited molecular and atomic species, and thus maximum inactivation effect is obtained in this case.

The results of indirect plasma treatment show that it is significantly less effective, due to absence of charged species in plasma afterglow. Applying bias potential to the agar leads to an increase in inactivation efficiency. In general, presence of charged species may lead to a significant increase of plasma treatment efficiency.

Gas composition can certainly affect concentration of various ions in plasma and, therefore, affect the plasma effect on cells. The effect was observed in the *E. coli*, skin flora (mix of streptococci, staphylococci, and yeast obtained from human patient samples), and *Bacillus subtilis* spores treated by the DBD plasma on agar in various gases followed by a 24-h incubation and colony counts (Dobrynin et al. 2010a,b). Gases tested included air, O_2, N_2, Ar, He, and N_2/NO mixture (700 ppm NO). Complete inactivation (>7 log reduction in colony forming units) of *E. coli* was achieved in direct plasma treatment in air and oxygen at 2 J cm^{-2}; in other gases tested dose of over 12 J cm^{-2} was required to achieve any visible effect and much higher doses were needed to achieve complete inactivation (no significant effect, compared to O$_2$ and air, was observed in Ar or He at all, even at >600 J cm^{-2}). Although one can conclude that the presence of oxygen makes significantly enhances bacterial inactivation with DBD, no clear conclusion could be drawn regarding the specific ions playing the key roles, particularly given the fact that some bacterial inactivation could be observed using DBD in nitrogen.

DC corona discharge was used to separate the **effects of positive and negative ions** without significant flux of chemically active neutral species (Dobrynin et al. 2009, 2010a,b). It was shown that ions of both polarities can inactivate bacteria. The positive ions show slightly higher efficiency; however, the difference between positive and negative ions is no more than ~10–15%.

27.10 Contribution of UV, Electric Field, Hydrogen Peroxide, Acidity, Ozone, NO$_x$, Other ROS, and RNS to Direct Plasma Cell Treatment, Effect of Presence of Water and Media

Energetic UV radiation (UV-C and VUV) has sufficient quantum energy to break organic bonds and thus significantly affect large organic molecules, which was discussed in Lecture 26. Therefore, the UV radiation (especially UV-C and VUV) has also very strong effect on cells, for example in sterilization processes in water (see Lecture 28). *In non thermal plasma systems, however, such as DBD, corona, and cold plasma jets, which are in focus of this lecture, the UV intensity is usually low and not sufficient for significant damage of cells. In the thermal plasma systems, like air sparks and arcs, contribution of UV in plasma cell interaction and especially plasma sterilization can be very significant.* One way to separate the UV effect of plasma is to protect cells from everything that is generated in plasma, except for the UV photons. It can be achieved by placing a quartz glass on top of the treated surface. Quartz, used in the relevant DBD experiments, was transparent to UV photons of >200 nm wavelengths, and MgF$_2$, which is transparent to vacuum UV (VUV) photons of >140 nm wavelengths (Dobrynin et al. 2009, 2010a,b). This way only UV/VUV photons generated in plasma reach bacteria. Bacteria that were protected from direct discharge by MgF$_2$ slide stayed unaffected (highest plasma dose used was 600 J cm^{-2} with no observed difference between untreated bacteria and MgF$_2$-protected bacteria). Thus, the action of ultraviolet radiation in DBD can be neglected in this case.

Effect of electric field on cell membranes has been already discussed in Section 27.8 regarding effectiveness of its penetration across the liquid layer as well the major biological response related to **electroporation**. In relation to the direct plasma effect of DBD discharges, we should point out that electric field in the streamer heads building up the DBD plasma can be significantly (3–10 times) higher than the applied electric field, see Sections 9.3 and 9.4, as well as Lecture 13. Physics and biochemistry of the electroporation is a special subject mostly related to the effect of high electric field on the cell membranes without involvement of plasma. Plasma can provide in this case, however, a significant synergistic effect. More details on the subject can be found, for example, in Davalos et al. (2005), Li et al. (2020), Rubinsky (2009), Kee et al. (2011), and Pakhomov et al. (2010).

The product of recombination of two directly plasma-generated OH molecules, **hydrogen peroxide** molecule, is much less reactive than OH and may pass through the cell membrane relatively easily causing a variety of downstream lethal effects, for example, fatal DNA damage. The biochemical effects of H$_2$O$_2$ have been already discussed in Section 27.5. The ability of hydrogen peroxide to sterilize is widely known and well-studied. Experiments were carried out to analyze the role of hydrogen peroxide specifically in DBD plasma-based inactivation of bacteria (Dobrynin et al. 2009, 2010a,b). Dependence of peroxide concentration on the treated volume is almost linear and increases approximately three times when amount of treated liquid is decreased 10 times. The estimated amount of liquid under the electrode (surface area of ~5 cm^2) in the case of "moist" agar is a few microliters; therefore, one can expect the concentration of peroxide to be on the order of a few tens of mmol l^{-1} for plasma dose of several J cm^{-2}. To determine the concentration of H$_2$O$_2$ that causes the same sterilization effect on agar in direct plasma treatment, 50% by volume of water solution of H$_2$O$_2$ was used and further diluted with distilled de-ionized water. The 0.1 ml droplet of the solution is poured onto the bacteria that are placed on agar and spread over the whole agar surface. The results show that concentration of H$_2$O$_2$ that corresponds to about 0.5 J cm^{-2} of direct DBD plasma is more than 200 mmol l^{-1}. The direct DBD plasma produces about 6.5 mmol l^{-1} H$_2$O$_2$ at plasma dose more than an order of magnitude higher than that required for inactivation by plasma. The strength of the H_2O_2 effect strongly depends on the **plasma-generated acidity**, see for more details Lecture 29 regarding efficacy of the plasma-activated water. This synergistic effect is related to the acidity-based H_2O_2 conversion to OH (often through peroxynitrite), see also Sections 27.5 and 27.6.

Bactericidal effect of plasma-generated ozone is well known and has already been utilized in industry for some time, see Lecture 18. In room air, DBD generates ozone, which is partly responsible for the observed bacterial inactivation in direct plasma treatment (Dobrynin et al. 2009, 2010a,b). The antimicrobial effect of ozone generated by the DBD has been evaluated in two ways: (i) by comparing the results to those found with an ozone generator and by scavenging ozone in DBD. The DBD in room air at ~60% relative humidity produces about 28 ppm of ozone as measured outside of the discharge zone. Ozone generator (~500 ppm max output) was used to produce same concentration of ozone in room air without plasma. No significant inactivation effect was observed on *E. coli* cultures on agar and on skin flora (mix of streptococci, staphylococci, and yeast) in as much as 30 min of treatment. DBD plasma in gas produced by diluting a stock 700 ppm NO in N_2 with oxygen leads to the same inactivation efficiency as when pure N_2 with O_2 mixture is used, while the ozone concentration is zero in the first case and 28 ppm in the second. Thus, it can be concluded that, as with UV and H_2O_2, ozone does not play a major role in the direct DBD inactivation of bacteria.

 Effect of thin water layer on direct plasma interaction with bacteria is significant and depends on amount of water. To analyze this possibility (Dobrynin et al. 2009, 2010a,b), the DBD treatment conditions were separated into three groups: (i) *dry* treatment, when a droplet with bacteria is placed on a glass slide and then dried (~1 h); (ii) *moist* treatment, when droplet with bacteria is placed on agar surface and left to dry until agar appears dry (~1 h); and (iii) *wet* treatment, when a droplet with bacteria is placed in a cavity on a glass hanging drop slide and treated immediately before any water has a chance to evaporate (although some evaporation likely occurs during the treatment). DBD plasma treatment is then performed in direct or indirect conditions. Significant qualitative difference has been found between these three cases. While on agar bacteria are covered with extremely thin (the exact amount is unknown, hypothesized to be anywhere from few nanometers to microns) layer of free water, in solution bacteria are submerged in water, and therefore the effect of charged particles may be significantly diluted. In addition, if in the case of bacteria on agar, the water layer is on the order of nanometers, bacteria may experience short, but very strong electric field pulses associated with streamers. In case of bacteria dried on glass slides, when there may be only a few molecules of water bound to bacteria, one might expect almost entirely gas-phase plasma effects such as those that are observed in the treatment of polymers by plasma. These may include plasma chemistry and ion bombardment, rather than typical oxidative processes that occur in the presence of water. In general, the direct plasma treatment achieves inactivation at significantly lower doses than indirect in the case of "moist" treatment on agar.

The ratio of plasma doses in direct and indirect plasma treatment decreases significantly as the amount of water around the bacteria increases in the "wet" plasma treatment case. Overall, this suggests that importance of charges directly in contact with the treated sample diminishes as the amount of water increases, while neutral species from plasma continue to play a significant role. *Summarizing, (i) presence of water is required for effective interaction of plasma with bacteria and (ii) that charges play a much more important role when there is a relatively small amount of water, while the relative importance of neutral species in the interaction of plasma with bacteria increases as the amount of water increases.*

27.11 Biological Mechanisms of Direct DBD Plasma Interaction with Mammalian Cells: Key Role of Intracellular ROS, Plasma-induced DNA Damage

The biological effects of plasma treatment of mammalian cells are strongly mediated by ROS. As it has been demonstrated using the direct DBD plasma application, the **plasma-induced intracellular ROS** is the key mediates of direct plasma interaction with mammalian cells, see Kalghatgi et al. (2011). Breast epithelial cells (MCF10A) that adhered to the surface of glass slides were employed and covered by a relatively small amount of cell medium (100 μl, about 0.3 mm fluid layer) during DBD treatment (doses ranged from 0.13 to 7.8 J cm^{-2}). DNA damage as detected by phosphorylation of a histone protein called H2AX was used as the primary readout of the plasma interaction with the cells. Using DNA as the readout of the effects of interaction is convenient for several reasons. On the one hand, knowing what happens to DNA is important in determining long and short-term toxic effects of plasma treatment. On the other, DNA damage, repair, and regulation are involved in many intracellular processes. Phosphorylated form of H2AX, so-called γ-H2AX, can be detected by attachment to it of highly specific fluorescently labeled antibody. This can be done directly in the treated cells. The results indicates that phosphorylation of H2AX and, therefore, DNA damage does increase with the increasing dose of the DBD treatment. This is the first important conclusion that is not entirely intuitive, since DBD plasma does not generate any significant amount of penetrating radiation or charged particles that can penetrate through cell medium, through the cell membrane, through nuclear membrane, and into the nuclei to affect the DNA. This result clearly suggests that biological chain reaction/diffusion process takes place through the cell medium and across the cell membrane to create the observed effects.

Following direct DBD plasma treatment, ROS were detected inside cells by using fluorescent indicators 5-(and- 6)-chlroromethyl-2′, 7′ dichloro-dihydro-fluorescein diacetate, acetyl ester (CM-H$_2$DCFDA, Molecular Probes). To test the hypothesis that the intracellular ROS are directly responsible for the observed DNA damage, the cells were incubated with a well-known modification of cysteine protein called *N*-acetyl cysteine (NAC). NAC effectively acts as a scavenger of intracellular ROS because by increasing levels of intracellular glutathione. The resulting levels of DNA damage were compared at different plasma doses for cells with and without NAC using immunoblotting-based measurements of γ-H2AX. *It was clearly demonstrated that cells in which NAC scavenged intracellular ROS showed no detectable level of DNA damage in contrast to cell that were not incubated with NAC. This provided definitive confirmation that intracellular ROS are directly responsible for mediating interactions between cells and plasma, at least any cellular responses that involve DNA.*

DNA damage is a relatively common occurrence in cells. There are several different types of DNA damage due to different causes. Intracellular ROS that occur as a byproduct of cellular respiration is one cause of occasional DNA damage. Ultraviolet radiation (UV) from the environment is another typical cause. Medical therapies such as ionizing radiation (a common cancer therapy) and more recently developed photodynamic therapy cause substantial DNA damage primarily through generation of OH radicals and singlet oxygen, respectively, directly within cells. The probability of any given type of DNA damage depends on the underlying cause. The most common type of DNA damage due to UV, for example, is formation of thymine dimers in the DNA molecule. Cellular respiration produced ROS most often cause **single-stranded DNA breaks**. These breaks are usually repaired by cellular processes that employ the second undamaged DNA strand as the source of information of the appropriate DNA sequence. Ionizing radiation creates one of the most difficult-to-repair DNA lesions called **double-strand breaks (DSB)** where both complementary DNA strands are both broken. These types of DNA breaks are more difficult to repair when the reliable source of information, the second DNA strand that acts as a repair blueprint, is missing and this is what happens due to ionizing radiation. As a result, ionizing radiation is more likely to cause DNA mutations. Hydrogen peroxide creates DNA damage like the ionizing radiation and is often employed to simulate the effects of ionizing radiation. It is important to note that DSBs can also occur in the process of other type of DNA damage repair. These DSBs do not increase the likelihood of mutation because a reliable repair blueprint exists. Phosphorylation of H2AX used to detect DNA as described above does occur in the process of repair that involves DSB. However, this may be the type of DSB that occurs in the process of repair, rather than a DSB created directly by an external influence. Therefore, it would not be correct to conclude that plasma treatment causes double-stranded DNA breaks just because the damage is detected using γ-H2AX. Different types of DNA damage are associated with different repair pathways. All these pathways proceed through phosphorylation of different proteins and involve different kinases (recall that kinases are enzymes that help phosphorylate proteins). Three most common kinases involved in DNA repair are called ATM, DNA-PK, and ATR. Roughly speaking, ATM and DNA-PK kinases are activated because of double-stranded breaks caused by some external agent, while ATR is typically activated in association with repair of single-stranded DNA damage or UV damage.

Identifying the type of kinase that is activated following the plasma treatment provides important information on the **type of DNA damage that occurs in cells**, see Kalghatgi et al. (2011). It was observed that phosphorylation of H2AX in response to DBD plasma (as well as H$_2$O$_2$) was markedly reduced in cells pretreated with 100 mM Wortmannin, which at 100 mM is known to inhibit ATM, ATR, and DNA-PK. In contrast, one-hour pretreatment with 10 mM KU55933, an ATM-specific inhibitor, did not significantly reduce the phosphorylation of H2AX in response to plasma treatment, whereas it significantly reduced it in response to H$_2$O$_2$. It indicates that ATR and/or DNA-PK is required for the phosphorylation of H2AX in response to DBD plasma. In summary, the intracellular ROS generation following the DBD plasma treatment results in DNA damage that is very different from the types of DNA damage that typically occurs in response to ionizing radiation, hydrogen peroxide, or UV radiation. Not only does this suggest that plasma is a novel tool that differs from ionizing radiation, hydrogen peroxide, and UV in its mechanisms of interaction with cells, but also that this novel tool is probably less dangerous than tools like ionizing radiation (ionizing radiation is more likely to lead to dangerous mutations).

27.12 Effect of the Cell Medium on Plasma Interaction with Mammalian Cells, Plasma-induced Factors Crossing the Cell Membrane

The direct plasma application produces a much stronger effect when the cells have some water around them compared to the effect of indirect treatment where neutral species play the dominant role. Direct and indirect treatments do not differ in the inactivation of bacteria as much when the bacteria are surrounded by greater amount of water. Similar conclusions can be made for the treatment of mammalian cells. Direct and indirect (with a grounded mesh placed between the cell sample and the insulated high voltage DBD electrode) treatments produce similar results as measured by phosphorylation of H2AX, with only a slightly stronger effect due to the direct treatment. This clearly confirms that plasma charges play a critical role when the amount of water around the cells is smaller, while their role is diminished when the amount of water around the cells is much larger. Consistent with the result that intracellular ROS mediated the DBD plasma-induced DNA damage, it has been demonstrated that longer incubation in the original 100 µl of medium, in which cells were treated (before dilution into 2 ml of medium), resulted in higher levels of γ-H2AX, see Kalghatgi et al. (2011). Additionally, when the treated medium was subjected to different dilutions one minute after treatment, the amount of damage correlated with the dilution, i.e. damage was greater at the lowest dilution. It suggests that the generation of intracellular ROS and the induction of DNA damage are the results of plasma's interaction with the extracellular medium, and that the effects of plasma depend strongly on the concentration of active species in the medium as well as the length of exposure of cells to these active species. The effects of DBD plasma can be due to modification of the cell medium by the plasma treatment as opposed to a direct effect on the cells. To demonstrate this effect, medium was treated separately and added to cells. The term "separated treatment" was used to describe the form of treatment where medium (100 µl) on a coverslip (without cells) was treated with the DBD plasma and then transferred to a fresh coverslip with cells already on it. The effect of medium separately treated with the DBD plasma and added to cells turned out to be not significantly different from the effect of direct treatment of cells overlaid with the same type of medium.

The question arises: **which components of the medium are responsible for the "storage" and transfer of the plasma treatment**? Is the organic content of the medium important or can one attribute the observed behavior to water and other inorganic medium content? The second question was answered by comparing the effects of separated medium treatment to separated treatment of medium reduced only to its inorganic content (phosphate buffered saline or PBS). Little/no DNA damage in cells exposed to the separately treated PBS was observed, whereas separately treated medium induced the DNA damage as anticipated. *This suggests that quasi-stable organic components in the medium, such as* **organic peroxides, are probably the main solution mediators** *of the plasma treatment.*

Cell culture medium is composed of amino acids, glucose, vitamins, growth factors, and inorganic salts, as well as serum. The γ-radiation induces formation of amino acid and protein hydroperoxides in aqueous solutions containing BSA or individual amino acids. Equivalent levels of H2AX phosphorylation were plasma-induced in cells subjected to separately treated serum-containing medium, serum-free medium, or PBS with BSA, but not PBS alone, suggesting that amino acid peroxidation may be involved. Peroxidation efficiency is widely variable among different amino acids, see Kalghatgi et al. (2011). To determine whether the observed results are related to the peroxidation efficiency of organic components in cell culture medium, 11 different amino acids with a range of peroxidation efficiencies were dissolved individually in PBS and separately treated with plasma and then added to cells. Phosphorylation of H2AX was directly proportional to the peroxidation efficiency of the amino acids, with valine producing the most significant level of damage and serine and methionine producing no detectable DNA damage. There is a direct correlation between the peroxidation efficiency of 11 different amino acids and the level of DNA damage, providing strong support for the hypothesis that *organic peroxides are produced in the plasma-treated medium and are responsible for the observed effects on DNA.*

The nonthermal plasma-generated outside cell medium can initiate a sequence of events that ends up inducing intracellular ROS and various other associated biological effects including repairable DNA damage. Chemical modifications of organic molecules in the cell medium play a key role in transferring the observed effects from the plasma phase into the cell. *One of such mechanisms of the plasma effect transfer across a membrane is* **nonspecific lipid peroxidation**. *Lipid peroxidation is a relatively common outcome of the presence of oxidizing agents in the extracellular space. Various by-products of lipid peroxidation, such as* **malondialdehyde (MDA),** *have been known to create bulky adducts on DNA which is a form of damage requiring repair. MDA is one of the most abundant carbonyl products of lipid peroxidation. It reacts with DNA to form adducts to deoxyguanosine and deoxyadenosine. One of the adducts to DNA is a pyrimidopurinone called M_1G. Site-specific mutagenesis experiments indicate that M_1G is mutagenic in bacteria and is repaired by the nucleotide excision repair pathway.*

The MDA production due to the DBD treatment depends on the presence of organic components of the medium and does not occur in PBS treatment.

27.13 Problems and Concept Questions

27.13.1 Osmosis, Flow of Water Across the Cell Membrane

Give your interpretation of the cell suppression mechanism by significant increase of concentration of the active and charged hydrophilic plasma-generated species in the cell environment. Is this process due to increase or decrease of the relevant osmotic pressure in the cell environment? What is the similarity and difference of this process with respect to disinfection and sterilization of food by simple NaCl solution known for thousands of years?

27.13.2 Aquaporins (AQPs) and Selective Anticancer Behavior of Plasma

Explain why the AQPs can provide the selectivity of the plasma treatment of cancer cells with respect to the corresponding normal cells.

27.13.3 Cellular Metabolism and Selective Anticancer Behavior of Plasma

Explain possible mechanisms dominating the cellular metabolism-related selectivity of the plasma treatment of cancer cells with respect to the corresponding normal cells. Which selectivity effect do you expect more significant contribution to the plasma cancer treatment, AQPs or cellular metabolism related?

27.13.4 Intracellular ROS Transformations

Usually, superoxide is considered as the primarily generated intracellular ROS. Which are the major transformation pathways converting the superoxide into other ROS, like singlet oxygen, hydrogen peroxide, and hydroxyl radicals? What are the similarities and differences in the ROS chemistry in cells and in air plasma?

27.13.5 Intracellular RNS Transformations

Usually, the nitrogen oxide NO is considered as the primarily generated intracellular RNS. Which are the major transformation pathways converting the superoxide into other major RNS species? What are the similarities and differences in the RNS chemistry in cells and in air plasma?

27.13.6 Liquid Medium Film Covering Cells

Using Eq. (27.7), estimate how much medium is left over the cells assuming the jet gas at velocity of $10\,\text{m s}^{-1}$ is maintained roughly over the planar substrate length L of about 1 cm. What are the major effects of this liquid medium film on the plasma interaction with cell?

27.13.7 Propagation of the Electric Charge Effect Across Liquid Medium Film Covering Cells, the Bjerrum Length

Using the Bjerrum length formula (27.8), estimate to what extent a charge delivered through plasma to the surface of this water layer can influence processes at the cell membrane.

27.13.8 Plasma-induced Electroporation

Electric field that penetrates the cell medium for a short time can induce or enhance formation of pores in the cell membranes, which is known as electroporation. Estimate electric field required to be applied to the lipid layer to achieve the electroporation effect. Compare from this perspective effectiveness of plasma jets and dielectric-barrier discharges (DBD).

Lecture 28

Plasma Disinfection and Sterilization of Different Surfaces, Air, and Water Streams

28.1 Nonthermal Plasma Surface Sterilization, Microorganism Survival Curves, *D*-value of the Microorganism Deactivation

Killing bacteria is probably the first known biological application of cold plasmas. Disinfection and sterilization are both decontamination processes, which can be effectively induced by plasma. Terminologically, bio-cleaning simply reduces the number of contaminants present and, in doing so, removes a proportion of organisms present. Disinfection removes most pathogenic organisms. Sterilization is the killing or removal of all organisms. *Nonthermal plasma is an effective source of active species and factors (see the previous lecture), which can deactivate, kill, or even completely disintegrate bacteria, viruses, and other microorganisms without any significant temperature effects. It attracts interest to plasma applications in sterilization and disinfection of different surfaces (especially subtle and temperature-sensitive), living tissues (including skin, wounds, and ulcers), liquids (including water, blood, and other bio-liquids), and air streams.* **Disinfection** usually implies couple of orders of magnitude reduction of population of microorganisms, while **sterilization** usually requires at least 10^4–10^5 times reduction in the number of microorganisms. Earlier nonthermal plasma sterilization experiments have been carried out mostly at low gas pressures, and usually in gas mixtures containing components with germicidal properties such as H_2O_2 and aldehydes. Advantages of plasma sterilization are revealed, however, when it uses gases (such as air, He-air, He—O_2, N_2—O_2) without germicidal properties on their own, which become biocidal only when plasma is ignited. Such studies motivated by decontamination of interplanetary space probes and sterilization of medical tools have been performed using RF and microwave low-pressure discharges in oxygen and O_2—N_2 mixtures, see Bol'shakov et al. (2004), Moreau et al. (2000), and Moisan et al. (2001). **Effect of the low-pressure RF oxygen plasma** on different microorganisms has been investigated in both inductively coupled (ICP) and capacitively coupled (CCP) discharges. The ICP provided better efficiency in destroying biological matter due to higher electron and ion densities in this mode. High densities of atomic and electronically excited oxygen in synergy with ultraviolet (UV) photons induced chemical degradation of the biological materials followed by volatilization of the decomposition products (CO_2, CO, etc.). DNA degradation was evaluated for both ICP and CCP modes of the low-pressure nonthermal RF plasma. It was found that at the same power, the ICP discharge destroyed over 70% of supercoiled DNA in 5 s, while only 50% of DNA was destroyed in the same conditions in the CCP discharge.

Effect on bacteria of the low-pressure N_2/O_2 afterglow plasma generated using a surfatron source driven by microwave power with frequencies 0.915 and 2.450 GHz has been investigated by Moreau et al. (2000) and Moisan et al. (2001). *The **survival curves** (**colony forming units**, CFUs, versus treatment time) in the experiments with Bacillus subtilis spores exhibit three distinctive inactivation phases, characterized by different D-values are illustrated in Figure 28.1, see Philip et al. (2002); as well as Fridman and Friedman (2013). The first phase in this figure, which exhibited the shortest **D-value** (decimal value, time (or energy) required to reduce an original concentration of microorganisms by 90%, one log reduction), corresponds to the **action of UV radiation** on isolated spores or on the first layer of stacked spores. The second phase, which is characterized by the slowest kinetics, can be attributed to a slow erosion by active species (such as atomic oxygen, O). The third phase starts after the outer spore debris are cleared during the phase 2, hence allowing UV to hit the genetic material of the still-living spores. The D-value of this phase is close to that of the first phase.*

Plasma Science and Technology: Lectures in Physics, Chemistry, Biology, and Engineering, First Edition. Alexander Fridman.
© 2024 WILEY-VCH GmbH. Published 2024 by WILEY-VCH GmbH.

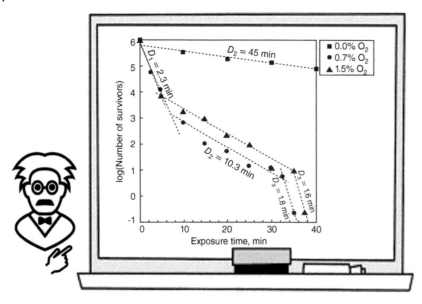

Figure 28.1 Illustration of three phases of bacteria (*B. subtilis*) deactivation with different *D*-values using the survival curves (N_2–O_2 low-pressure, 5 Torr, microwave discharge afterglow.

28.2 Cold Atmospheric Pressure Plasma Inactivation of Microorganisms on the Surfaces, Mechanisms and Kinetics of Plasma Sterilization

Nonthermal atmospheric pressure discharges, particularly in air, are very effective and convenient in deactivation of microorganisms. *Atmospheric pressure nonthermal discharges in air are characterized by a high density of strongly reactive oxidizing species and are able not only to kill microorganisms without any notable heating of the substrate, but also destroy and decompose the microorganism. This ability of complete but cold destruction of microorganisms in atmospheric pressure plasmas is attractive for complete steril-*ization of spacecrafts in the framework of the planetary protection program. Atmospheric pressure cold DBD plasma has been demonstrated as an effective tool in destruction of not only bacteria but also such resistive microorganisms as B. subtilis (spores), Bacillus anthracis (anthrax spores), and Deinococcus radiodurans (microorganisms surviving strong radiation of nuclear materials). Effective sterilization (5-log reduction) and significant destruction (cell wall fracture and leakage as well as complete disintegration) of *B. subtilis* spores in DBD air plasma (120 s treatment, *D*-value 24 s) is illustrated in Figure 28.2, Fridman (2003), Fridman and Friedman (2013). Therefore, range of applications of plasma sterilization at atmospheric pressure is wide, from medical instruments and spacecrafts to different food products.

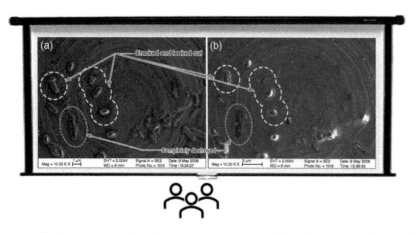

Figure 28.2 DBD plasma inactivation of microorganisms: *B. subtilis* before (left) and after (right) treatment with DBD plasma (120 s, 0.8 W cm^{-2}).

Mechanisms and kinetics of the sterilization processes are more sophisticated in the high-pressure strongly nonequilibrium plasmas, because of significant contribution of collisional gas-phase processes at higher pressures and therefore wider variety of active species involved in the sterilization kinetics. In particular, the germicidal effect of nonthermal atmospheric pressure plasma is characterized by different shapes of the survival curves (see Section 28.1) depending on the type of microorganism, the type of the medium supporting the microorganisms, and the method of exposure (Laroussi 2005). The "single slope" survival curves (one-line curves) for the inactivation of the bacteria strains have been observed in atmospheric pressure plasma sterilization by Herrmann et al. (1999), Laroussi et al. (1999), and Yamamoto et al. (2001). The D-values range in this case is from 4 s to 5 min. Two-slope survival curves (two consecutive lines with different slopes) have been observed in atmospheric pressure plasma by Kelly-Wintenberg et al. (1998), Kelly-Wintenberg (2000), and Laroussi et al. (2000). The D-value of the second line (D_2) is smaller (shorter time) in these systems than the D-value of the first line (D_1). The D_1 value is dependent on the species being treated, and the D_2-value is dependent on the type of surface supporting the microorganisms. The two-slope survival curve can be explained considering that the active plasma species during the first phase react with the outer membrane of the cells, inducing damaging alterations. After this process has sufficiently advanced, the reactive species can quickly cause cell death, resulting in a rapid second phase. Multi-slope survivor curves (3 kinetic phases or more) were also observed in sterilization by nonthermal atmospheric pressure plasmas. It occurs for example in plasma deactivation of Pseudomonas aeuroginosa on filter exposed to He-air DBD plasma (Laroussi et al. 2000). Interpretation of the multi-slope kinetic effect can be similar in this case to that one described in the previous section regarding sterilization by low-pressure plasma, see Figure 28.1.

28.3 Contribution of Different Plasma Species and Factors in Cold Atmospheric Pressure Plasma Sterilization of Surfaces

The bacteria inactivation kinetics reveals complexity of mechanisms of sterilization by nonequilibrium high-pressure plasmas. Several factors can impact the killing process: type of bacteria, type of medium in/on which the cells are seeded, number of cell layers, the type of exposure, contribution of UV, operating gas mixture, etc. Generally, if UV plays an important or dominant role, the survival curves tend to exhibit a first rapid phase (small D-value) followed by a second slower phase (outer layer of bacteria is deactivated by UV during the fast first phase; then slower erosion of the bacteria proceeds during the longer second phase). When UV does not play a significant role, such as in the case of air plasma (where contribution of active oxidizing species is essential), the single-phase survival curves are mainly observed. Generally, the plasma sterilization kinetics at high pressures can be quite sophisticated and, obviously, depends on specifics of mechanisms of plasma sterilization in specific atmospheric pressure discharges.

Plasma sterilization is quite complicated process determined by multiple plasma species and factors, including charged and excited species, reactive neutrals, UV radiation, etc. *Contribution of different plasma species and factors to plasma sterilization differs in different types and regimes of nonequilibrium plasma discharges; also, it differs from the point of view of the induced biological pathways. The synergistic nature of interaction between the plasma factors is important in sterilization similarly to the case of plasma treatment of polymers (see Lecture 26). While the contribution of different plasma species and factors to sterilization has some specifics and peculiarities, they follow most general tendencies of the plasma interaction with cells described in Sections 27.8–27.10.* Specifically, these general tendencies can be referred to the biochemical sterilization effect of electrons and other charged species, UV radiation, and active neutrals (ROS, RNS). Some interesting specifics of plasma sterilization are due to the sterilization effect of energetic ion bombardment and electric fields related to charged plasma particles.

Deactivation of micro-organisms can be due to destruction of the lipid layer of their membranes or other **membrane damages by ion bombardment** from plasma. This effect is especially important in the case of direct plasma treatment of the micro-organisms. Although the ion bombardment makes bigger contribution in the low-pressure plasmas when the ion energies are quite high, effect of ion bombardment in atmospheric pressure discharges with relatively low ion energies can be also significant. Average energy of the ions $<\varepsilon_i>$ in the ion bombardment process in the high-pressure nonequilibrium ($T_e \gg T_0$) plasmas can be estimated as energy received by the ions in electric field E during their displacement on the mean free path $eE\lambda$ (Fridman and Kennedy 2021):

$$<\varepsilon_i> \approx T_0 + T_e \sqrt{\delta} \qquad (28.1)$$

Here, T_e, T_0 are electron and gas temperatures in plasma; δ is the average fraction of energy lost by electrons in the electron-neutral collisions. Considering that $\delta \approx 0.1$ in molecular gases, average energies of the ion bombardment even in high-pressure plasma systems can reach 0.3–0.7 eV. Although these energies are not sufficient to break strong chemical bonds, they are high enough to break hydrogen bonds responsible for integrity of lipid layers forming cellular membranes. Thus, in addition to chemical effects, ions can provide deactivation of micro-organisms and sterilization through mechanical effects of bombardment (and possibly related osmotic pressure effects).

Direct effect of externally applied electric fields on sterilization and deactivation of micro-organisms (electroporation) in nonthermal plasma systems (as well as those thermal) can be usually neglected, see Neumann et al. (2001). At the same time, **effects of electric fields related to collective motion and deposition of charged particles** can be significant, see Laroussi et al. (2003). The deposited charged particles can play significant role in rupture of the outer membrane of bacterial cells. Electrostatic forces caused by charge accumulation on the outer surface of the cell membrane could overcome the tensile strength of the membrane and cause its rupture. Charged bacterial cells with μm sizes experience large electrostatic repulsive forces proportional to the square of the charging potential Φ and inversely proportional to the square of the radius of the cell's curvature r. The charging (floating) potential Φ is negative and depends on the ratio of the ion and electron mass. Condition for disruption of the bacterial cell membrane can be expressed in this case as (Laroussi et al. 2003):

$$|\Phi| > 0.2 \cdot \sqrt{r\Delta F_t} \tag{28.2}$$

where Δ is the thickness of the membrane, and F_t is its tensile strength. This mechanism can be relevant for gram-negative bacteria with irregular surface membranes. These irregularities with small radii of curvatures can result in localized high electrostatic forces (Laroussi et al. 2003).

High strengths of electric fields and electroporation effect can be achieved not only because of the charge deposition on irregular surfaces of gram-negative bacteria but also due to collective and strongly localized motion of the charged particles especially in streamers of dielectric barrier discharges (DBD). Strengths of electric fields in the DBD streamers significantly exceed those of the externally applied electric fields. Therefore, while the externally applied electric fields in the atmospheric pressure discharges are not sufficient for electroporation and sterilization, electric fields in the streamers are able, in principle, to contribute to plasma deactivation of micro-organisms. It should be mentioned that the biological effect of very high electric fields of streamer heads, especially of those in short-pulsed DBD plasma systems, is not limited to electroporation but can be also related to triggering of the intracellular signaling, which is especially important in the case of plasma interaction with cells.

This discussion as well as detailed consideration of the plasma sterilization kinetics, see Fridman et al. (2007a–e), confirms *the general conclusion stated in Sections 27.8–27.12 that at least in FE-DBDs direct plasma sterilization provides stronger effect than the indirect one. To some extent, this conclusion can be generalized to other nonthermal discharges applied for plasma sterilization.* Simply, plasma that comes in direct contact with bacteria can sterilize significantly faster than afterglow or plasma-activated medium. Only 5 s of direct plasma treatment results in significant inactivation of microorganisms. Complete sterilization occurs within 15 s when direct treatment is employed. Over 5 min of indirect treatment is usually required to achieve sterilization results like direct treatment obtained within 15 s. Therefore, even with substantial UV radiation, indirect plasma treatment is substantially weaker than direct one provided by charged particles. It is important that effective plasma treatment of bacteria, cells, and other biomaterials is possible not only if they are in contact with plasma but even when they are separated from plasma and protected by a layer of intermediate substance, for example, submerged into a culture medium. It is especially important in plasma suppression of biofilms, see Fridman and Friedman (2013).

28.4 Nonthermal Plasma Sterilization of Spores and Viruses: Inactivation of *Bacillus cereus*, *Bacillus anthracis* Spores, SARS-CoV-2 Coronavirus

Nonthermal plasma is proved to be effective for inactivation of spores. Bacillus species, which are ubiquitous in the environment, are aerobic or facultative anaerobic gram-positive bacteria. The genus *Bacillus* is divided into three broad groups, depending, among other characteristics, on the morphology of the spore. *Bacillus cereus*, *B. anthracis* (anthrax)), and *Bacillus thuringiensis* belong to the *B. cereus* group. Moreover, morphological, and chromosomal similarities between these species have prompted the view that *B. anthracis*, *B. thuringiensis*, and *B. cereus* are all varieties of a single species. Bacilli can produce a **dormant cell type called a spore** in response to nutrient-poor conditions. Bacterial spores have little or no metabolic activity and can withstand a wide range of environmental assaults including heat, UV, and solvents. It especially attracts attention to their plasma sterilization. ***B. anthracis* spores**, as opposed to vegetative cells, are the infectious form and cause anthrax. The spores of *B. anthracis* represent a noteworthy bioterrorism agent and can be easily distributed in dry form in parcels and letters via postal service (as what occurred in 2001, when anthrax-contaminated letters sent through the U.S. postal service killed 5 people and sickened 23 others), in aerosols, or in contaminated water, for instance. In response to these social challenges, an effective, low-energy, and cost-effective method of spore inactivation or sterilization is required. Clearly, the direct plasma approach is one of them.

The **Bacillus spores have been treated in DBD plasma** at room temperature in air, see Dobrynin et al. (2010a,b). Inactivation of bacteria in spore form both in liquid or air-dried on surface requires higher doses of DBD plasma treatment, and up to 5 log reduction can be achieved within a minute of exposure to plasma. The discharge gap was kept at 1.5 mm. The microsecond-pulsed DBD discharge was ignited by applying an AC-pulsed high voltage of 30-kV magnitude (peak to peak) and 1.3-kHz frequency between the electrodes. Current peak duration was 1.2 μs, and the corresponding plasma surface power density was 0.3 W cm^{-2}. The spores were treated in either dry form or in aqueous suspension on glass slides. The slides were placed on top of grounded metal, and the plasma was ignited directly between the powered electrode and the treated spores. Another set of experiments was carried out using indirect plasma treatment. In this case, a grounded metal mesh (22 wires cm^{-1}, 0.1-mm wire diameter, 0.35-mm openings, 60% open area, and weaved mesh) was used as a second electrode. The gap between the mesh and the second electrode was kept at 1.5 mm, which is the same as that in the direct treatment setup. The distance between the mesh and the treated surface was also kept at 1.5 mm. Ten microliters of *B. cereus* or *B. anthracis* spores in distilled water at concentrations of 10^7, 10^6, and 10^5 spores/ml were placed in hanging drop glass slide wells and treated for 5–45 s with direct DBD plasma. After treatment, bacteria were appropriately diluted and plated.

Also, spores were placed inside a plastic chamber (to mimic the envelopes), covered with a glass cover slide, and dried for about 30 min with a constant airflow of about 0.1 l/min (control experiments with only gas flow through the chamber showed no loss of collected spores). Ten microliters of *B. cereus* at 10^8, 10^7, and 10^6 spores/ml in distilled water were used. Dried spores were treated with direct dielectric-barrier discharge DBD plasma for 5–45 s and then washed out of the chamber with 30 ml of distilled water, appropriately diluted and plated. Similar experiments were done using *B. anthracis* spores, except that 10 μl of *B. anthracis spores were placed inside either plastic or paper envelopes,* dried in room air for one hour and treated with DBD plasma. In these experiments, the discharge was ignited in the volume of either the plastic chamber or the paper or plastic envelope (envelopes were slightly inflated to assure that walls are in contact with powered and grounded electrodes, see Figure 28.3, Dobrynin et al. (2010a,b), as well as Fridman and Friedman (2013).

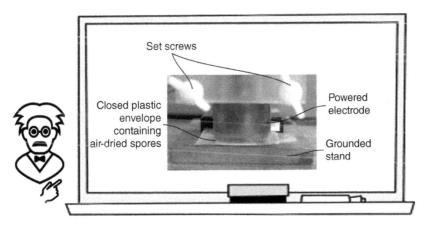

Figure 28.3 Atmospheric DBD plasma treatment of air-dried spores, *B. cereus*, *B. anthracis* (anthrax), contained inside a closed envelope.

In the series of experiments, carried by Dobrynin et al. (2010a,b), the spores of two Bacillus species, namely, **B. cereus** and **B. anthracis**, were treated in water droplets on the surface of glass slides using atmospheric-pressure DBD plasma in room air. Although the spores of both species were effectively inactivated (up to 5 log reduction in less than a minute of treatment), log N, i.e. killing kinetics, was quite different for *B. cereus* versus *B. anthracis* spores—two-step versus linear, respectively. In addition, anthrax spores appeared to be slightly more resistant to plasma treatment. In contrast to the inactivation of spores suspended in distilled water, the inactivation by the plasma treatment of air-dried *B. cereus* and *B. anthracis* spores both appear to be linear. *The sterilization of 10^7 ml^{-1} anthrax spores inside a closed paper envelope has the DBD plasma dose of 10–15 J cm^{-2}. These results are especially interesting, as they show the high efficiency of DBD-plasma-based systems to sterilize within (inside of) temperature-sensitive materials (envelopes).*

To see if morphological differences occurred, the spore morphology was analyzed before and after DBD treatment at the doses of 10–15 J cm^{-2} using the scanning electron microscopy (SEM). The SEM photomicrographs show that the inactivation mechanism is not the erosion of the spores since their protective coats appear to be undamaged. Therefore, it is believed that the diffusion of chemically active oxygen species into spores followed by the damage of internal macromolecules or molecular systems may be the primary mechanism of the spore inactivation.

Nonthermal plasmas in different modifications (mostly atmospheric pressure DBD and plasma jets) have been also applied to suppress different types of viruses. Relevant plasma parameters and the treatment dose requirements are like those described above, see Fridman and Friedman (2013). Recently, relevant experiments were accomplished demonstrating efficacy of the plasma abatement of the SARS-CoV-2 coronavirus on different surfaces, including masks and food packages, as well as in airflow (see below in this lecture). Details on this topic can be found, for example, in Cheng et al. (2020) and Kaganovich and Keidar (2020); as well as in publications of the Matteo Gerardi and Vittorio Colombo group, Capelli et al. (2021) and Bisag et al. (2022).

28.5 Decontamination of Surfaces from Extremophile Organisms and Prion Proteins Using Nonthermal Atmospheric-Pressure Plasma

*Nonthermal plasma has an ability to kill any microorganisms, including those which are extremely resistant to other sterilization methods, the so-called **extremophiles**. A good example in this regard, is the nonthermal DBD plasma killing of the **D. radiodurans**, an extremophile organism, by compromising the integrity of its cell membrane*, see Cooper et al. (2009, 2010). In samples of *D. radiodurans*, which were dried in a laminar flow hood, it was observed that DBD plasma exposure resulted in a 6-log reduction in CFU (colony-forming unit) count after 30 min of treatment. When the *D. radiodurans* cells were suspended in distilled water and treated, it took only 15 s to achieve a 4-log reduction of CFU count. It is especially interesting in the planetary-protection arena, where the proliferation of terrestrial bacteria beyond Earth is known as forward contamination. It is important, when sending spacecraft and probes to other planets and moons, to prevent the contamination of the environment, particularly when searching for native life. If bacteria from the location in which a sample was collected returns to Earth and proliferates, then reverse contamination has taken place.

The goal with this regard is to achieve surface sterilization of spacecraft materials with complete disintegration of spores and bacteria. This effect was successfully demonstrated by cold ambient-air plasma, which leads to sterilization, lysing, and disintegration of microorganisms (Cooper et al. 2009, 2010). D. radiodurans has been chosen for such plasma sterilization experiments considering that if we can show that cold plasma can destroy one of the toughest microorganisms found on Earth, then we can claim that, probably, we can destroy less robust organisms as well. D. radiodurans is very resistive to radiation, temperature change, reactive oxygenated species, and vacuum. It can withstand an instantaneous radiation dose of 5000 Gy with no loss of viability (60 Gy sterilizes a culture of *Escherichia coli*), an instantaneous dose of up to 15 000 Gy with 37% viability loss, and exposure to space vacuum (~10−6 Pa) for three days with decreased cell survival by four orders of magnitude. Before treatment, the number of viable *D. radiodurans* was approximately 10^6 CFU. After 30 min of DBD treatment, there was a 6-log reduction in the number of *D. radiodurans* cells. Temperature measurements show that during plasma treatment,

Figure 28.4 Plasma decontamination of surfaces from extremophile organisms: SEM images of *D. radiodurans* on blue steel (a) before and (b) after 30 min of DBD plasma treatment.

the average temperature was 26 °C. Imaging techniques were applied to visualize the effect of treatment with cold plasma on the morphology of microorganisms. The viability measurements were supported by SEM images taken before and after 30 min of DBD treatment of *D. radiodurans* on blue steel at 1 W cm^{-2}, see Figure 28.4 [Cooper et al. (2009, 2010), as well as Fridman and Friedman (2013)]. Comparison of Figure 28.4a (before DBD plasma treatment) and Figure 28.4b (after 30 min DBD plasma treatment) clearly shows that DBD plasma causes significant morphological changes and "physical" destruction of *D. radiodurans* on the sterilization surface at 30 °C.

Another critical sterilization challenge, which can be solved based on application of nonthermal plasma, stems from the fact that most of the medical contamination is not only provided by microorganisms but also by biomolecules such as proteins, DNA, and lipids. The example in this regard is the **contamination by the misfolded prion proteins**, widely regarded as the etiologic agent of spongiform neurodegenerative pathologies such as bovine spongiform encephalopathy (BSE), scrapie, and Creutzfeldt-Jakob diseases (CJD). Prion proteins are resistant to all current conventional sterilization strategies including autoclaving, ionizing radiation, ethylene oxide, and formaldehyde, and as a result have forced the unsustainably expensive option of single-use surgical instruments. The nonthermal plasmas, and in particular micro-plasma jets, developed by Kong and his group [see for example, Iza et al. (2008)] are capable of the prion protein destruction. A helium-oxygen atmospheric pressure plasma jet has been successfully applied by this group to inactivate proteins deposited on stainless-steel surfaces. Using a laser-induced fluorescence technique for surface protein measurement, a maximum protein reduction of 4.5 logs has been achieved by varying the amount of the oxygen admixture into the background helium gas. This corresponds to a minimum surface protein of 0.36 femtomole mm^{-2}.

It is found that plasma reduction of surface-borne protein is through protein destruction and degradation and that its typically biphasic reduction kinetics is influenced largely by the thickness profile of the surface protein. By interplaying the protein inactivation kinetics with optical emission spectroscopy, it has been shown in this research that the main protein-destructing agents are excited atomic oxygen (via the 777 and 844 nm emission channels) and excited nitride oxide (via the 226, 236, and 246 nm emission channels). It is also demonstrated that the most effective protein reduction is achieved possibly through a synergistic effect between atomic oxygen and nitride oxide. The studies of Michael Kong and his group represent an important step toward a full confirmation of the efficacy of nonthermal plasma as a sterilization technology for surgical instruments contaminated by prion proteins.

28.6 Sub-lethal Plasma Effect on Bacterial Cells, Apoptosis vs Necrosis in Plasma Treatment of Cells

The **sub-lethal plasma effect** on bacterial cells has been demonstrated by Laroussi et al. (2002). When a sub-lethal plasma exposure is administered to bacterial cells, a change in their metabolic behavior could occur. To demonstrate these effects, Laroussi et al. (2002) addressed the biochemical impacts of plasma on *E. coli* with sole carbon substrate utilization (SCSU) experiments. It clarifies if exposure to plasma altered the heterotrophic pathways of the bacteria. It was presumed that any changes in metabolism would be indicative of plasma-induced changes in cell function. The Biolog GN2 plate was comprised of a control well and 95 other wells, each containing a different carbon substrate. Color development of a redox dye present in each well-indicated utilization of that substrate by the inoculated bacteria. The 95 substrates were dominated by amino acids, carbohydrates, and carboxylic acids. The plasma exposure caused an increase of utilization of some substrates and a decrease of utilization in others, indicating noticeable changes in the corresponding enzyme activities without causing any lethal impact on the cells.

 Nonthermal plasma can stimulate both apoptosis and necrosis in treated cells. **Apoptosis, or programmed cell death,** *is a complex biochemical process of controlled self-destruction of a cell in a multicellular organism. This process plays an important role in maintaining tissue homeostasis, fetal development, immune cell "education," development, and aging. Examples of apoptosis that occur during normal body processes include the formation of the outer layer of skin, the inner mucosal lining of the intestine, and the endometrial lining of the uterus, which is sloughed off during menstruation.* During apoptosis, cellular macromolecules are digested into smaller fragments in a controlled fashion, and ultimately the cell collapses into smaller intact fragments that can be removed by phagocytosis without damaging the surrounding cells or causing inflammation. In contrast, during **necrosis, also termed "accidental cell death,"** the cell bursts, and the cellular contents spill out into the extracellular space, which can cause inflammation. Necrosis is induced by cellular injury, for example, extreme changes in osmotic pressure or heat, that lead to adenosine triphosphate (ATP) depletion of the cell.

Plasma-induced apoptosis and necrosis in mammalian cells have been initially studied by Stoffels (2003, 2006). Chinese hamster ovarian cells were used as a model. The cells were exposed to an RF small-volume cold plasma generated around the tip of a needle-shaped electrode (plasma needle). Necrosis (cell death due to catastrophic injury) occurs in the plasma needle treatment for powers above 0.2 W and exposure times longer than 10 s. The cell membranes are damaged, and the cytoplasm is released. Lower doses of exposure lead to apoptosis. If the power level and exposure time were reduced significantly to about 50 mW and 1 s, the Chinese hamster ovarian cells partly detach from the sample, take a more rounded shape, and did not undergo apoptosis, see Stoffels (2003, 2006). Plasma-stimulated apoptotic behavior in melanoma skin cancer cell lines has been demonstrated by Fridman et al. (2007a–e). This effect has important potential in cancer treatment because the cancer cells frequently acquire the ability to block apoptosis and thus are more resistant to chemotherapeutic drugs.

28.7 Different Levels of Deactivation/Destruction of Microorganisms Due to Plasma Sterilization: Are They Dead or Just Scared to Death? Concept of VBNC

Plasma physicists involved in plasma sterilization research often ask the "tricky" question: we see that the microorganisms are not culturable (not proliferating, not multiplying) after plasma treatment, does it mean that they are dead? I have a good friend of mine, he has no kids, but he is very much alive (he is not a bacterium though). Also, physicists often are curious about possibility of "resurrection" after the bacterial death during the plasma sterilization. While physicists are very excited about this kind of arguments, microbiologists usually keep quiet knowing that the concepts of "death" and "resurrection" can be very sophisticated for micro-organisms.

Special attempt to answer these questions has been tried my microbiological group of Joshi. This study (Cooper et al. 2009, 2010) was focused on biological responses to plasma of one of the robust organisms, *Bacillus stratosphericus*. DBD plasma was applied over various durations to *B. stratosphericus* either surface-dried or suspended in de-ionized water, and viability, culturability, and viable but nonculturable (VBNC) were assayed using standard techniques. *Depending upon the exposure of B. stratosphericus to DBD plasma resulted in three viability states,* **viable and culturable** *at low plasma doses and VBNC* **(viable-but-not-culturable)** *or disintegrated* **bacteria** *at higher plasma doses.* Although organism's respiration levels in the case of VBNC were at relatively low levels immediately after plasma treatment; over the course of 24 hours, respiratory activity was increased eight times (and it was found still nonculturable during colony assays). The loss of culturability in the case of VBNC is hypothesized to be induced as one of the responses to oxidative stress and it remains to be unclear if the response is temporary or indefinite. Appropriate plasma powers should be used obviously to avoid the VBNC-like status. It was also shown that *B. stratosphericus* exhibits similar phenomena of etching and disintegration of the membrane, and damage to DNA with plasma treatment in dry air, and furthermore, when treated in liquid, the oxygenated species can inactivate bacteria to cause lethal and sublethal damages. Oxidative stress can induce a VBNC state in bacteria, and it remains consistent with bacteria whose membrane is not peroxidized beyond the repairable limit by plasma. It should be mentioned that *B. stratosphericus* tolerates 17% NaCl and is resistant to UV as well as selected antibiotics such as penicillin, vancomycin, and erythromycin. Its resistance characteristics make it an interesting choice for understanding the effect of plasma exposure. The organism is multi-resistant and originally isolated from high altitude where radiation, UV, and other rays are likely affecting it. Studies of Cooper et al. (2010) has been focused on a comparing culturability, membrane integrity, bacterial morphology, and respiration capabilities to analyze the effects of FE-DBD exposure at doses which are lethal to other bacteria and to show that robust bacteria are able to adapt and survive these treatments.

An exposure to DBD plasma at otherwise "lethal" doses for most of microorganisms induces a VBNC state in *B. stratosphericus*, which are not completely inactivated/disintegrated by such treatment (Cooper et al. 2009, 2010). Such doses might be leading to activation of genetic mechanisms of switch from viable to VBNC state in *B. stratosphericus*. The presence of VBNC cells poses a major public health hazard. These cells cannot be detected by traditional culture methods, and the cells may remain potentially pathogenic upon favorable conditions. It can be hypothesized that the ROS produced by plasma may be inducing this state. Also, the mechanisms of inactivation in *B. stratosphericus* may be different under dry environment (directly ionic interactions with cell envelop) and wet environment (probably via ROS). To ensure the death of such bacteria, relatively longer plasma treatment time is advisable. *Thus, when plasma physicists involved in plasma sterilization research*

ask the "tricky" question: when the micro-organisms after sufficient plasma treatment are not culturable (not proliferating, not multiplying), is that possible that they are not dead but only not culturable? The answer should be probably that "yes" the microorganisms after not-sufficient plasma treatment can be in the so-cold VBNC state, meaning "viable-but-not-culturable." More information about the biological responses of the *B. stratosphericus* to FE-DBD plasma treatment including the VBNC phenomenon can be found in Cooper et al. (2009, 2010), as well as in the book Fridman and Friedman (2013).

28.8 Nonthermal Plasma Sterilization of Air Streams, Direct Air Sterilization vs Application of Filters, Pathogen Detection and Remediation System (PDRF)

Most of the first plasma systems for air sterilization were coupled with the high-efficiency particulate air (HEPA) filters to provide both trapping and killing of microorganisms, see Gadri (2000), Kelly-Wintenberg (2000), and Jaisinghani (1999). Usually, the micro-organisms are initially trapped with the filter with following plasma sterilization of the filter. The downside of relying on HEPA filters is that they have limited efficiency at trapping submicron-sized airborne microorganisms, and they also cause significant pressure losses in heating, ventilation, and air conditioning (HVAC) systems giving rise to higher energy and maintenance costs.

A large-scale experimental facility, named the Pathogen Detection and Remediation Facility (PDRF), was designed by Gallagher et al. (2005, 2007) and build in the Nyheim Plasma Institute of Drexel University to perform direct plasma air decontamination using a Dielectric Barrier Grating Discharge (DBGD), see Figure 28.5. The PDRF combines plasma device with a ventilation system and bioaerosol sampling capabilities. Experiments with PDRF show that *direct contact of the bio-aerosol and the DBGD plasma with very short duration can cause an approximate 2-log reduction (97%) in culturable E. coli with a ~5-log reduction (99.999%) measured in the 2 min following exposure. Fast treatment times within plasma are due to a high airflow rate and high velocity of the bio-aerosol particles in flight and a small discharge length of the DBGD device, which results in a residence time of treatment of approximately 1 ms. Thus, direct plasma treatment leads to extremely fast and effective deactivation of the airborne bacteria.* The bioaerosol can be treated in the PDRF with repeated passes through the same plasma discharge. It is a plug flow reactor, where airflow is turbulent so that radial variation of the bacterial concentration is minimized. The PDRF system has a quite large total volume of 250 l and operates at high airflow rates (~25 l s^{-1} or greater), which are typical for indoor ventilation systems. The PDRF system residence time, that is, time for one bio-aerosol particle to make one complete revolution through the system, is 10 s.

Special DBD configuration, called the **dielectric barrier grating discharge DBGD**, has been applied in the PDRF to provide effective direct sterilization of airflow. The DBGD consists of a thin plane of wires with equally spaced air

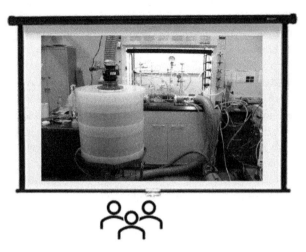

Figure 28.5 Plasma sterilization of air stream, Pathogen Detection and Remediation Facility (PDRF).

Figure 28.6 The PDRF, Dielectric Barrier Grating Discharge (DBGD) with air sterilization chamber.

gaps of 1.5 mm. The high voltage electrodes are 1 mm diameter copper wire shielded with a quartz capillary dielectric that has an approximate wall thickness of 0.5 mm. The total area of the DBGD discharge including electrodes is 214.5 cm^2 and without electrodes is 91.5 cm^2. Figure 28.6 shows an image of the DBGD device, which has two air sample ports located at 10 cm from each side of the discharge area so that bioaerosol can be sampled right before and after it enters the plasma discharge. When the PDRF system is operated at a flow rate of 25 l s^{-1}, the air velocity inside the DBGD discharge chamber is 2.74 m s^{-1} and the residence time of treatment, that is, the duration of one bio-aerosol particle (containing one *E. coli* bacterium) passing through the DBGD device is approximately 0.73 ms. The DBGD device is operated using a quasi-pulsed power supply that delivers a distinct sinusoidal current–voltage waveform with a very fast rise time that nearly simulates a true square wave pulse. The period between pulses is approximately 600 µs, peak-to-peak voltage is 28 kV, and pulsed current is nearly 50 amps (peak-to-peak value). The average power of the discharge is approximately 330 W and considering the discharge area of 91 cm^2, the power density is 3.6 W cm^{-2}. The typical concentration of bio-aerosol is 5×10^5 bacteria per liter of air, which translates to approximately 9×10^3 bacteria within the cross section of discharge area at any given time (in each 2 mm wide cross section of flow passing through the DBGD).

The culture test results from four replicate trials of DBGD-treated *E. coli* bioaerosol in the PDRF system are shown in Figure 28.7, Gallagher et al. (2005, 2007), as well as Fridman and Friedman (2013). In the DBGD-treated trials, an approximate 1.5-log reduction (97%) in the surviving fraction of *E. coli* was measured between samples 1 and 2 (before and after the plasma treatment). Plasma was only ignited in a short period during each set of samples, which is noted

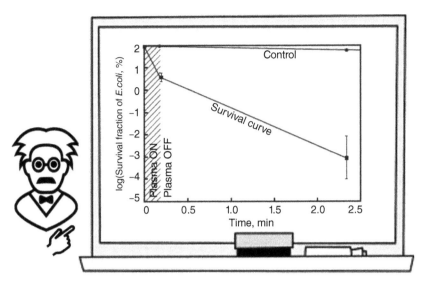

Figure 28.7 Plasma sterilization of air stream in PDRF, culture tests results of DBD-treated *E. coli*; the grey shaded area indicates the time when the DBD discharge is ignited.

by the grey shaded areas in Figure 28.7. Secondary decrease in the surviving fraction of *E. coli* shown in the figure occurred in the time between plasma treatments. In this second decline, the rest of culturable bacteria (3% from initial number) decreased by an additional 99.95% (3.5 logs) when the plasma discharge was switched off. Additionally, after second and third passes through plasma, the number of culturable *E. coli* was less than the detection limit (10^{-6}).

Flow cytometry analysis of air samples taken during the trials indicates that the total number of bacteria (both active & inactive) remains almost constant; therefore, the DBGD device is not acting as an electrostatic precipitator and the concentration of bioaerosol particles remains undisturbed for the duration of each experiment. Flow cytometry results also showed that bacterial outer membranes of *E. coli* were not disintegrated with up to three passes of direct exposure in the DBGD plasma device. However, culture test results demonstrated a 97% reduction in culturable *E. coli* with a millisecond exposure time in DBGD plasma (one pass through discharge) and a subsequent 3.5-log reduction in the two minutes following treatment.

The direct plasma exposure time of 0.73 ms (per pass) allows enough time for bacteria to be attacked by all chemically active components of plasma: charged particles, UV radiation, OH radicals, atomic oxygen, and ozone, which is one explanation for the initial 97% reduction in culturability. Subsequent remote exposure to ozone in the two minutes following direct plasma treatment, as well as the high sensitivity of airborne bacteria to slight atmospheric changes may account for the remaining 3.5-log reduction. Thus, direct plasma treatment of airflow permits its rapid sterilization (much faster than corresponding indirect plasma treatment), similarly to the case of surface sterilization. Similar results were obtained using the PDRF and other nonthermal plasma systems for sterilization of airflow infected by the SARS-CoV-2 coronavirus, see for example publication of the Gherardi and Colombo group (Bisag et al. 2022).

28.9 Phenomenological Kinetics of Nonthermal Plasma Sterilization of Air Streams

Sterilization of airborne micro-organisms by plasma-generated UV and active chemical species (atoms, radicals, excited and charged particles, as well as active stable molecules like O_3, etc.) can be described in terms of conventional gas-phase chemical kinetics. The sterilization process is considered in this case as an elementary reaction, which occurs with some probability during a "collision" of the micro-organisms and relevant plasma-generated active specie. Rate and effectiveness of the "elementary sterilization reaction" can be described in such chemical kinetic approach in terms of conventional reaction rate coefficients. Such approach to plasma sterilization kinetics, proposed by Gangoli et al. (2005), is a phenomenological, because it cannot describe biochemical pathways of the complicated process of deactivation of micro-organisms. However, the phenomenological plasma sterilization kinetics permits

Table 28.1 Phenomenological reaction rate coefficients for the nonequilibrium plasma sterilization of bacteria (*E. coli*) using ozone O_3, hydroxyl radicals OH and UV radiation.

Effect of ozone (O_3)	1.5×10^{-16} (cm^3 s^{-1})
Effect of hydroxyl radical (OH)	3.6×10^{-13} (cm^3 s^{-1})
Effect of UV radiation	3.8×10^{-3} (cm^2 µJ^{-1})

generalization of the experimental kinetic data obtained in different discharge systems (with different concentrations of different relevant active plasma components) applied to deactivation of different types of micro-organisms.

Relevant reaction rate coefficients for plasma sterilization (focused on the plasma-generated ozone, OH-radicals, and ultra-violet) have been calculated by Gangoli et al. (2005) through analysis of empirical data (available in the food and water decontamination literature) that resulted from many experiments in which ozone and ultraviolet radiation were used to destroy a variety of bacteria, viruses, and spores. The reaction rate coefficients k have been calculated assuming second kinetic order of the micro-organism deactivation:

$$\frac{d[M]}{dt} = -k \cdot [M] \cdot [A] \tag{28.3}$$

where d[M]/dt is the change of concentration of viable microorganisms with time, [M] is the concentration of microorganisms, [A] is the concentration of active chemical species A. The reaction rate coefficient for interaction of A with the microorganism, k_A, is defined in this case as:

$$k_A = \frac{\ln\left(\frac{1}{S}\right)}{[A] \cdot t}$$

where, S is the surviving fraction of microorganism M, and t is treatment time. Using the same approach, the model accounts for the destructive effects of the plasma-generated UV on each individual microorganism by assigning the reaction rate coefficient k_{UV} for each microorganism. The density of active species [A] are replaced in the case of UV-radiation by energy flux [J] usually measured in µJ/(s · cm^2), the sterilization reaction rate coefficient is measured in the case of UV-radiation not in cm^{3}s^{-1} (which is conventional for the second order reactions) but in cm^2 µJ^{-1}. All described the plasma sterilization rate coefficients are presented in Table 28.1.

Summarizing the data, the inactivation ability of OH exceeds that of O_3 by 10^3–10^4 times. In the frameworks of the above-introduced kinetic model of the plasma sterilization, the following formula describes the rate of inactivation of airborne microorganisms inside a DBD discharge from the combination of effects of ozone, hydroxyl, and UV-radiation:

$$\frac{d[M]}{dt} = -k_{O_3} \cdot [M] \cdot [O_3] - k_{UV} \cdot [M] \cdot [I] - k_{OH} \cdot [M] \cdot [OH] \tag{28.4}$$

In this kinetic equation: [M] is the concentration of micro-organisms; [O_3] and [OH] are concentrations of the active chemical species; [I] is energy flux of UV-radiation; k_{O_3}, k_{OH}, and k_{UV} are corresponding reaction rate coefficients (see Table 28.1). Since the DBD is operated in air, typical concentration of ozone ranges from 100 ppm, which corresponds to about 2.7×10^{15} cm^{-3} at standard conditions. The G-factor for OH formation (number of OH radicals produced per 100 eV energy of the discharge) is about 0.5, which corresponds in the PDRF conditions (DBD power 330 W, 25 l s^{-1} flow rate) to OH concentration about 15 ppm (4×10^{14} cm^{-3}). The UV radiation intensity in the wavelength range 200–300 nm is below 50 µW cm^{-2}. Relevant modeling results are in reasonable agreement with experimental results shown in Figure 28.7. Better agreement requires accounting for an additional contribution of other plasma-generated, not considered in (28.4), especially charged species (ions and electrons) and other radicals (such as atomic oxygen and nitric oxides) present in the DBGD plasma.

Summarizing the kinetics of the killing of the airborne microorganisms in the PDRF system, we can conclude that the process is extremely fast (shorter that 1 ms). Between different effects accounting for the so fast DBD plasma sterilization of airborne microorganism, the most "intriguing" is related to the steamers "following the airborne microorganisms" like missiles able to follow a moving target. This hypothesized effect is due to a significant polarization induced in the airborne microorganisms by strong electric field in and around the DBD streamer heads.

More details on the kinetics of the plasma killing of the airborne microorganisms, contribution of different plasma species and factors in the effect, comparison of direct and indirect effects, can be found in Gangoli et al. (2005) and Gallagher et al. (2005, 2007), in publications of the Matteo Gerardi and Vittorio Colombo group (for example, Bisag et al. 2022), as well as in the book of Fridman and Friedman (2013).

28.10 Plasma-provided Water Disinfection and Sterilization Using Ozone, UV, and Pulsed Electric Fields

Plasma-induced water disinfection and sterilization has a lot of common features with water decontamination from hazardous chemicals already considered in the Lecture 24 (see Sections 24.10–24.14), but also has some important specifics. *The most important specific feature of the plasma water disinfection and sterilization is ability of some plasma-generated factors (especially ozone and UV radiation) to kill microorganisms in water indirectly, being separated from plasma itself. Plasma-generated ozone can be produced in DBD or pulsed coronas in one location (see Lecture 18) and then transported quite far for water disinfection and sterilization in other location. In a similar way, plasma-generated UV radiation (see Lecture 23, for example Section 23.16) is generated inside the plasma UV lamps, and then leave the lamps for water disinfection and sterilization outside of plasma itself*. These approaches to water disinfection and sterilization are in a large-scale industrial use for more than a century, see Lectures 18 and 23. Direct application of plasma has way shorter history.

More than 1 billion people are unable to acquire today safe drinking water, which determines the need for improved methods of water treatment. Contaminated water can be attributed to several factors, including chemical fouling, inadequate treatment, and deficient or failing water treatment and distribution systems. An additional important cause of contamination is the presence of untreated bacteria and viruses within the water. As estimated by the Environmental Protection Agency (EPA), nearly 35% of all deaths in developing countries are related directly to contaminated water. The increased presence of *E. coli*, along with various other bacteria within some areas of the United States, has been a cause for national concern. To inactivate these bacteria, successful experiments and commercial applications of chemical treatments, ultraviolet radiation, and ozone injection units have been developed and implemented into potable water delivery systems. The experimental success and commercialization of these water treatment methods are not, however, without deficiencies. Regarding human consumption, chemical treatments such as chlorination can render potable water toxic. UV radiation and ozone injection have also been proven to be two practical methods of bacterial inactivation in water, but the effectiveness of such methods largely depends upon adherence to regimented maintenance schedules. Plasma methods effectively combining contribution of UV-radiation, active chemicals, high electric fields etc. are considered therefore as very effective approach to water treatment (Fridman et al. 2007a–e; Locke et al. 2006; Fridman and Friedman 2013). Before considering direct application of plasma to water treatment (which is a major goal of the section), consider shortly independent application of UV-radiation, active chemicals, and high electric fields into deactivation of micro-organisms in water.

Ozonation (obviously, O_3 is plasma-generated) is one of the powerful and growing method of water treatment. Ozone gas, generated in plasma aside, is bubbled into a contaminated solution and dissolves in it. The ozone is chemically active and is capable of efficiently inactivating microorganisms at a level comparable to chlorine. Achieving a 4-log reduction at 20 °C with an ozone concentration of 0.16 mg l^{-1} requires an exposure time of 0.1 min (Anpilov et al. 2001). At higher temperatures and pH levels, ozone tends to rapidly decay and requires more exposure time. Plasma discharges, especially DBD has been used to produce ozone in the past several decades to kill microorganisms in water. Ozone has a lifetime of approximately 10–60 min, which varies depending on pressure, temperature, and humidity of surrounding conditions. Because of the relatively long O_3 lifetime, ozone gas can be produced in air or oxygen plasma, and then stored in a tank, and finally injected to water. Bactericide effect of O_3 in water is usually characterized by *Ct*-factor, defined as the product of ozone concentration C [mg l^{-1}] and the required time *t* [min] to disinfect a microorganism in water. For example, for *Ditylum brightwelli*, important ballast water species, the *Ct* value was 50 mg-min l^{-1}. In other words, if the ozone concentration is 2 mg l^{-1}, it takes 25 min of contact time to disinfect this organism in ballast water (Dragsund et al. 2001). Energy efficiency of ozonation is limited by O_3 losses during storage and transportation.

UV radiation (obviously also plasma-generated) has proven to be effective in decontamination processes and is gaining popularity particularly in European countries because chlorination leaves undesirable by-products in water. Most bacteria and viruses require relatively low UV dosages for inactivation, which is usually in a range of 2000–$6000\,\mu W\,s\,cm^{-2}$ for 90% kill. For example, *E. coli* requires a dosage of $3000\,\mu W\,s\,cm^{-2}$ for a 90% reduction. Cryptosporidium, which shows an extreme resistance to chlorine requires a UV dosage greater than $82,000\,\mu W\,s\,cm^{-2}$. The criteria for the acceptability of UV disinfecting units include a minimum dosage of $16,000\,\mu W\,s\,cm^{-2}$ and a maximum water penetration depth of approximately 7.5 cm. UV radiation in the wavelength range from 240 to 280 nm causes irreparable damage to the nucleic acid of microorganisms. The most potent wavelength of UV radiation for DNA damage is approximately 260 nm. Total energy cost of the UV water treatment is also quite large.

Pulsed electric fields can be also mentioned as a novel approach to water treatment, which can be applied as a part of direct plasma treatment or separately. The electric field associated with this technology is not necessarily strong enough (membrane potential of more than 1 V can sometimes kill a bacteria) to initiate electrical breakdown in water and deactivation of micro-organisms is due to electroporation, which is creation of holes in the cell membranes. It means, by the way, that plasma-originated electric fields (for example, those in DBD streamers) can be sufficient for electroporation. At nominal conditions, the energy expense for the 2-log reduction of number of viable cells is relatively high, about $30,000\,J\,l^{-1}$ (Katsuki et al. 2002a,b).

Summarizing, ozone, and UV radiation generated in remote plasma sources are effective means of water cleaning and sterilization. In a similar way, electric field can be also mentioned. If plasma is organized not remotely but directly in water, effectiveness of the treatment due to plasma-generated UV-radiation, active chemicals, and electric field can be much higher. Organization of plasma inside of water also leads to additional significant contribution of short-living active species (electronically excited molecules, active radicals like OH, O, etc.), charged particles, and plasma-related strong electric fields into cleaning and sterilization. All separate plasma-related species and factors provide significant synergistic effect when applied all together in direct plasma sterilization (Fridman et al. 2007a–e; Locke et al. 2006; Fridman and Friedman 2013; Sun et al. 1997, 1999). Direct water treatment by plasma organized inside of the water, therefore, can be much more effective than individual applications of the separate plasma-generated factors.

28.11 Application of Different Types of Direct Pulsed-plasmas for Water Disinfection and Sterilization

Application of electrical pulses in the microsecond range to different biological cells in water has been investigated for example by Schoenbach et al. (1997, 2000), Joshi et al. (2002a,b), Katsuki et al. (2002a,b), Abou-Ghazala et al. (2002). They used a point-to-plane geometry to generate pulsed-corona discharges for bacterial (*E. coli* or *B. subtilis*) decontamination of water with a 600 ns, 120 kV square wave pulses. Concentration of *E. coli* could be reduced by three orders of magnitude after applying 8 corona pulses to the contaminated water with the corresponding energy expenditure of $10\,J\,cm^{-3}$ ($10\,kJ\,l^{-1}$). Plasma pulses cause the accumulation of electrical charges at the cell membrane, shielding the interior of the cell from the external electrical fields. Since typical charging times for the mammalian cell membrane are on the order of 1 μs, these microsecond pulses do not penetrate cells. Hence, shorter pulses in the nanosecond range can penetrate the entire cell, nucleus, and organelles, and affect cell functions, thus disinfecting them. High-voltage pulse generators have been used to apply nanosecond pulses as high as 40 kV to small test chambers. Biological cells held in liquid suspension in the cuvettes were placed between two electrodes for pulsing. The power density was up to $10^9\,W\,cm^{-3}$, but the energy density was rather low (less than $10\,J\,cm^{-3}$, a value that could slightly increase the temperature of the suspension by approximately 2 °C).

Heesch et al. (2000) also applied pulsed electric fields and pulsed corona discharges to inactivate microorganisms in water. They used four different types of plasma treatment configurations: a perpendicular water flow over two wire electrodes, a parallel water flow along two electrodes, air-bubbling through a hollow needle electrode toward a ring electrode, and wire cylinder. They used 100 kV pulses (producing a maximum of $70\,kV\,cm^{-1}$ electric field) with a 10 ns rising time with 150 ns pulse duration at a maximum rate of 1000 pulses s. The pulse energy in these systems varied between 0.5–$3\,J\,pulse^{-1}$ and an average pulse power was 1.5 kW with 80% efficiency. Inactivation of microorganism was found to be $85\,kJ\,l^{-1}$ per 1-log reduction for Pseudomonas fluorescents and $500\,kJ\,l^{-1}L$ per 1-log reduction for spores of *Bacillus cereus*. It was demonstrated that corona directly applied to water was more efficient than pulsed electric fields. With direct corona, Heesch et al. (2000) achieved $25\,kJ\,L^{-1}$ per 1-log reduction for both gram-positive and gram-negative bacteria.

Pulsed plasma discharges in water have been applied for sterilization and removal of organic compounds such as dyes by Sato et al. (1996), Sugiarto et al. (2003), and Sun et al. (1997, 1999). The streamer discharge was produced from a point-to-plane electrode, where a platinum wire in a range of 0.2–1 mm in diameter was used for the point electrode, which was positioned 1–5 cm from the ground plane electrode. Lisitsyn et al. (1999) used dense medium plasma reactor for the disinfection of various waters. The plasma reactor consisted of a rotating upper electrode at a range of 500–5000 rpm and a hollow conical cross-sectional end piece. Advantage of the rotating the electrode was that the rotating action spatially homogenized the multiple micro-arcs, activating a larger effective volume of water. In addition, spinning the upper electrode also simultaneously pumped fresh water and vapors into the discharge zone. Water sterilization in the electrohydraulic (arc) discharge reactor was investigated by Ching et al. (2001, 2003). A typical operational condition included a discharge of 135 mF capacitor bank stored energy at 5–10 kV through a 4 mm electrode gap within 40 µs with a peak current of 90 kA. They studied the survival of *E. coli* in aqueous media exposed to the above electrohydraulic discharges. They reported the disinfection of 3 l of a 4×10^7 CFU ml^{-1} *E-coli* suspension in 0.01 M PBS at pH 7.4 by 50 consecutive electrohydraulic discharges. It was demonstrated that UV-radiation emitted from the electrohydraulic discharge was the lethal agent that inactivated *E. coli* colonies rather than the thermal/pressure shocks or the active chemical species. Detailed specific data regarding the efficiency of different thermal and nonthermal discharges for water cleaning and sterilization can be found in reviews: Fridman et al. (2007a–e), Locke et al. (2006); as well as in the special book on this subject of Yang et al. (2012).

28.12 Challenge of Electric Energy Cost of Plasma-induced Water Disinfection and Sterilization, Most Energy-effective Technologies Based on Application of Sparks and Other "Quasi-thermal" Warm Discharges

Energy cost of water treatment is one the most critical parameters of the plasma technologies focused on water disinfection and sterilization. While the conventional technology of water chlorination is not ecologically friendly, it is very cheap, which creates significant challenge for all relevant plasma-based technologies. Especially important is operational cost of the plasma-based technologies, because of significant consumption of electric energy in the related processes. To be competitive in cleaning of large volumes of water, electric energy consumption is expected not to exceed 3–10 kJ l^{-1} for treatment of strongly contaminated nontransparent water, and 30–100 J l^{-1} for disinfection of slightly biologically contaminated, chemically clean, transparent water (where plasma-generated UV is effective). In general, energy cost of the plasma-induced water disinfection and sterilization strongly depends on quality of water to be cleaned. The plasma-generated UV-radiation can provide the energy-effective disinfection and sterilization if water is sufficiently UV-transparent, which permits to meet the above-mentioned electric energy cost requirement. If water is strongly contaminated and not UV-transparent, the decontamination requires significant contribution of the plasma-generated active species (ROS, RNS) resulting in about 100-times higher energy consumption, which can be still acceptable, see Fridman et al. (2007a–e), Locke et al. (2006); as well as Yang et al. (2012).

In this perspective, when water is UV transparent, the low energy cost of disinfection and sterilization can be achieved when plasma can effectively generate the UV emission in the range of wavelengths optimal for killing relevant microorganisms. Such effective UV radiation is usually provided by thermal and quasi-thermal plasmas, especially those in air bubbles with nitrogen oxides emitting the UV. Keeping in mind the restrictions related to electric energy cost and electrode erosion, the energetic thermal plasma sources (like high current thermal arcs) have significant problems and disadvantages, see for example, publications of Philip Rutberg and his group (Kolikov and Rutberg 2016). In contrast to that the "warm" discharges, like sparks and gliding arcs are preferential (see Lectures 14 and 16) for water disinfection and sterilization.

The highest energy efficiency of water sterilization has been achieved using the pulsed spark discharge (Campbell et al. 2006; Fridman et al. 2007a–e), organized in the point-to-plane electrode configuration (see Figure 28.8). Variance in the generated plasma from corona discharge to spark discharge was observed in these experiments to be dependent on the gap distance measured from the anode to the grounded cathode. A stainless-steel electrode (0.18 mm) was encased in silicon residing in a hollow Teflon tube, providing the necessary insulation for the electrodes. The electrode

Figure 28.8 Energy-effective water disinfection and sterilization: point-to-plane spark discharge.

extended approximately 1.6 mm beyond the bottom of the glass tube, providing a region for spark discharge initiation. The critical distance between which spark discharge and corona discharge exist was observed to be approximately 50 mm between electrodes. Inter-electrode distance greater than 50 mm resulted in the corona discharge, whereas the distance less than 50 mm resulted in the spark discharge. The initial steep rise in the voltage indicates the time moment of breakdown in the spark gap, after which the voltage linearly decreased with time over the next 17 μs due to a long delay time while the corona was formed and transferred to a spark. The rate of the voltage drop on the capacitance. The current and power profiles show the corresponding histories which show initially sharp peaks and then very gradual changes over the next 17 μs. Duration of the initial peak was measured to be approximately 70 ns. At $t \cong 17$ μs, there was a sudden drop in the voltage, indicating the onset of a spark or the moment of channel appearance, which was accompanied by sharp changes in both the current and power profiles. The duration of the spark was approximately 2 μs, which was much longer than the duration of the corona.

Plasma sterilization data collected in the pulsed spark discharge by treatment of *E. coli* (Campbell et al. 2006; Fridman et al. 2007a–e) are different for two different initial conditions. When the initial cell count is high (1.8×10^8 cells ml^{-1}), the spark discharge produces a 4-log reduction at 100 pulses and 2-log reduction at about 65–70 pulses. When the initial cell count was on intermediate level (2×10^6 cells ml^{-1}), the spark discharge produced a 2-log reduction at 50 pulses.

 Thus, energy per 1 l of water for 1-log reduction in E. coli concentration is in minimum as low as about 70 J l^{-1}. Similar but more detailed experiments with spark discharge treatment of E. coli in water at lower bacterial concentrations (Arjunan et al. 2007) resulted in similar very low values of the plasma sterilization energy cost (about 100 J l^{-1}).

28.13 Dominating Role of UV-radiation in the Highly Energy-effective Inactivation of Microorganisms in Water Using the Pulsed Spark Discharge System

To analyze the regime with the highest energy efficiency of water disinfection and sterilization (Arjunan et al. 2007), water, contaminated with *E. coli*, was treated with a pulsed spark discharge. Three relevant samples were taken at regular intervals. Samples for determining the initial population were taken before applying the plasma discharge. To analyze the treatment results, serial dilutions of the samples were prepared using sterile water and enumerated using a spread plate counting method. 100 μl aliquots of diluted samples were spread on Brain Heart Infusion (BHI) agar plates and incubated at 37 °C for 12–18 hours and the CFU were counted. The effectiveness of the spark discharge in inactivating *E. coli* has been expressed in terms of *D*-value, which is identified as the energy required to achieve 1 $\log_{10}$ reduction in bacterial concentration at the specific plasma treatment condition, see Section 28.1.

Inactivation experiments with the sparks were conducted for *E. coli* concentrations of 10^8 and 10^6 CFU ml^{-1} of water. Two suspensions of different concentrations were chosen to check any influence of bacterial concentration on *D*-value. For an *E. coli* concentration of 10^8 CFU ml^{-1}, the *D*-value was found to be 174 J l^{-1} of water, while for a lower concentration of 10^6 CFU ml^{-1}, a low *D*-value of 14 J l^{-1} was obtained. This indicates some dependence of *D*-value on initial bacterial concentration. *As the initial bacterial concentration increased, the D-value also increased. This **load effect** may be due to the inability of ultraviolet radiation, produced by spark discharge, to reach E. coli through water lacking transparency. Also at high concentration, E. coli can aggregate to each other partially shielding one another and thus, preventing the active species from effectively attacking them.*

To optimize the spark discharge system, the energy per pulse was varied by varying the capacitance of the capacitor source. This was achieved by varying the number of capacitors in the capacitor bank. For energy per pulse of 1.7 J, the *D*-value obtained was about 150 J l^{-1}. A low *D*-value of 98 J l^{-1} was obtained for energy per pulse of 1 J. It may be assumed that only a portion of the energy input into the water contributes to the inactivation of microorganisms, while the rest of the energy is dissipated into the water. For energies per pulse of 0.68 J and 0.34 J, *D*-values were 140 J l^{-1} and 366 J l^{-1} respectively. Thus, *the **optimized treatment system corresponds to a minimum D-value of about and below 100 J l^{-1} and energy per pulse of 1 J.***

Let us now analyze the **role of ultraviolet radiation in the energy-efficient inactivation of microorganisms.** A special UV absorber can be used for this purpose. To absorb UV radiation, 2, 2′-dihydroxy-4, 4′-dimethoxybenzophenone-5, 5′-disulfonic acid (Benzophenone-9, BASF), a sunscreen agent of the *o*-hydroxybenzophenone class, can be effectively used. It is a nontoxic ingredient commonly found in commercial water-soluble sunscreens which dissipates the absorbed energy without radiation. Various degrees of UV absorption in water can be achieved by varying the concentration of benzophenone-9 (BP-9). BP-9 solutions of concentrations 1, 3, 10, 30, 100, 300, 1000, and 3000 mg l^{-1} of water were prepared, and absorbance was measured using a Perkin-Elmer Lambda 40 UV–Vis Spectrometer. The transmission spectrum for wavelengths between 200–800 nm was obtained for BP-9 solutions of concentrations 3, 30, 300, and 3000 mg l^{-1}. The BP-9 solution of concentration 3 mg l^{-1} transmitted a major portion of UV-A, B, and C. BP-9 solution of concentration 30 mg l^{-1} absorbed a significant portion of UV-C, UV-B, and some UV-A. A concentration of 3000 mg BP-9 l^{-1} completely absorbed UV-A, B, and C. The concentrations 3, 30, and 3000 mg l^{-1} can be, hence, selected for conducting the inactivation experiments. As a control, the sensitivity of *E. coli* to BP-9 was tested by exposing 1.44×10^7 CFU ml^{-1} to 7.5 g BP-9 l^{-1} of water for two hours. No change in bacterial concentration was observed.

Figure 28.9 shows the survival curves for various concentrations of BP-9. It could be observed that disinfection of E. coli in water was almost completely suppressed by adding 30 mg of BP-9 l^{-1} of water. Thus, it is the effect of ultraviolet radiation which make the major contribution to the most energy-efficient inactivation of microorganisms in water (about 100 J l^{-1} per 1 log reduction). Similar conclusions have been made by other researchers analyzing the role of ultraviolet radiation in the inactivation of microorganisms in water by pulsed discharges (Ching et al. 2001, 2003). Like it was discussed above, the ultraviolet radiation is known for effective killing of microorganisms in water. DNA has an important absorption peak near 254 nm. The UV photons can cause irreparable damage to the bacterial DNA thus inactivating them. These results indicate that ultraviolet radiation produced by spark discharge in water plays a major role in inactivating microorganisms. Generally, the spark discharge in water is very effective in sterilizing water. Energy efficiency of the sterilization in this case is high compared not only to other plasma approaches but also to conventional sterilization methods, see Section 28.12. Another, similar technology of direct plasma-induced water disinfection and sterilization is based on application of transitional "warm" plasma of gliding arcs. This technology also can provide possibility of treatment of large amount of water, which is going to be discussed in the next lecture regarding wash water disinfection activation for agricultural applications.

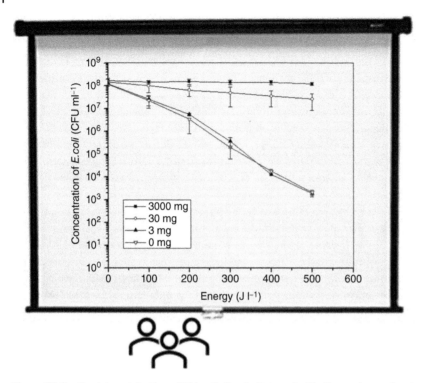

Figure 28.9 Crucial contribution of UV radiation in the most effective regimes of water disinfection and sterilization: survival plot of *E. coli* for various concentrations of BP9 UV absorber.

28.14 Problems and Concept Questions

28.14.1 Multi-slope Survival Curves for Plasma-induced Killing of Microorganisms

Give your interpretation of the possible manifestation of three different slopes on the survival curves. Which specific plasma inactivation mechanisms can correspond to these different slopes characterized by different *D*-values? Would you expect different plasma sources and different types of microorganisms to have their distinguished number of slopes?

28.14.2 Effect of Membrane Damages by Ion Bombardment During Plasma Sterilization of Surfaces

Estimate protective and destructive effect of humidity and thin layers of water (or biological and water solutions) on the cells on relative contribution of ion bombardment to the plasma-induced sterilization of surfaces.

28.14.3 Plasma-Induced Sterilization of Anthrax Spores Inside a Closed Paper Envelope Using DBD Plasma

Direct contact with plasma provides stronger sterilization effect than indirect plasma treatment. How to interpret the high efficiency of DBD plasma inactivation of the anthrax spores inside of the paper envelope, which obviously creates a significant barrier between plasma and the microorganisms.

28.14.4 Plasma-Induced Killing of Microorganisms vs Transferring them to the VBNC (viable-but-not-culturable) State

Explain challenge and danger of confusing killing of the microorganisms with transferring them into the VBNC state. What are the typical discharge parameters and plasma treatment doses corresponding to these two modes of plasma inactivation of microorganisms?

28.14.5 Nonthermal Plasma Sterilization of Air Streams, Plasma Suppression of the Airborne Microorganisms

What is your interpretation of the very short time required in PDRF to inactivate the airborne microorganisms? Which specific features of DBD in the DBGD plasma sources are accounted for in the plasma sterilization of air streams in these systems?

28.14.6 Phenomenological Kinetics of Nonthermal Plasma Sterilization of Air Streams

Using the sterilization rate coefficients presented in Table 28.1, and kinetic Equation (28.4), calculate typical time of inactivation of *E. coli* by ozone in airflow. Consider numerical value of the number density of ozone discussed in Section 28.9.

28.14.7 Application of Different Types of Direct Pulsed-Plasmas for Water Disinfection and Sterilization

Why the combined effect of multiple plasma-generated factors (ozone, UV, etc.) applied together in the direct plasma treatment provide significantly stronger synergistic effect in water disinfection and sterilization than simple summation of these factors? Compare this effect with relevant synergistic effect in plasma treatment of polymers discussed in Lecture 26.

28.14.8 Dominating Role of UV-radiation in the Highly Energy-Effective Inactivation of Microorganisms in Water Using the Spark Discharge System

Explain the load effect in sterilization of water using spark discharges, meaning the dirtier is water to be biologically cleaned the higher is energy cost of the technology.

Lecture 29

Plasma Agriculture and Food Processing, Chemical and Physical Properties of Plasma-activated Water, Fundamentals and Applications to Wash and Disinfect Produce

29.1 Plasma Agriculture and Food Processing: A Rapidly Emerging Field of Plasma Science and Technology, Direct Application of Plasma vs Use of Plasma-activated Water and Solutions

Plasma inactivation of microorganisms discussed in the previous lecture can be effectively accomplished on the surfaces of plants, fresh produce, and other foods without damaging and reducing their quality. Plasma technologies of disinfection and sterilization of large volumes of water discussed in the previous lecture can be also effectively applied for fresh produce washing water and for hydroponic water. In addition to that, plasma can be applied for acceleration and increasing efficacy of seed germination and crop growth, see illustration of the plasma effect on plant growth in Figure 29.1. All these plasma-biological effects and technologies define the **plasma agriculture and food processing** as an emerging novel and rapidly growing field of plasma science and technology.

Plasma bio-processing methods possess many advantages, such as a low-temperature treatment, short processing time, and absence of harmful chemicals. In agricultural fields and plant protection stations, pesticides are sprayed to protect crops from various insects and viruses. Fungi, such as Aspergillus or Penicillium, contaminate foods, such as cereals, fruits, vegetables, and meats. Residual agricultural chemicals (e.g. thiabendazole, imazalil, and *ortho*-phenylphenol) can be in this case harmful to the human body and the environment. Methyl bromide is an effective and widely used pesticide. In 2005, however, it was prohibited under the "Montreal Protocol on Substances that Deplete the Ozone Layer" because of its high ozone depletion potential. Nonequilibrium low-temperature atmospheric-pressure plasmas permit to avoid the pesticides and showed a promise as a very effective disinfection and sterilization system with minimal damage to crops, foods, seeds, humans, and the environment.

 Like in all other branches of plasma biology, discussed in the Lectures 27 and 28, *plasma can be applied for the agricultural and food processing applications both directly and indirectly, usually through plasma-activated water and solutions, which is in plasma medicine sometimes referred to as **plasma pharmacology**. Direct application of plasma while being usually more "intensive" and effective in most of plasma-biological applications (see Lectures 27 and 28), can be limited to local, specific, and not very large-scale plasma agricultural and food processing technologies. Direct plasma can be especially effective in this regard, for example, in disinfection and sterilization of the individually packaged food produce, in specific treatment of individual plants to control biology of their development (like affecting the hemp buds to control further CBD extraction and refining into delta-8 THC, THC-O, and other novel cannabinoids), as well as in stimulation of seed germination. In contrast, application of plasma-activated water and solutions is focused on large-scale large volume technologies like stimulation of plant growth, fertilization and sanitization of the hydroponic water, disinfection, and activation of fresh produce wash water, and sanitization of the growing plants. Depending on specific application, the plasma-activated water and solutions can be used not only as a "bulk liquid," but also through spraying and especially misting, which is going to be discussed below in this lecture.*

Figure 29.1 Illustration of the plasma-induced effect on a plant growth used at the IWOPA-1: control plant (left), plant watered by plasma-activated water (right).

 Plasma agriculture and food processing have been "energized" relatively recently as a spinoff from Plasma Medicine with starting the International Workshops on Plasma Agriculture (IWOPA). The first IWOPA meeting has been organized and held in May of 2016 by C&J Nyheim Plasma Institute at Drexel University, Philadelphia. Figure 29.1 illustrating plasma-induced plant growth has been used as a symbol of the IWOPA-1. Since then, the scope of research and number of publications grew up significantly. Between several recent reviews on the subject, we can recommend Ranieri et al. (2021), Attri et al. (2020), Simec and Homola (2021), Ito et al. (2012), Puac et al. (2018).

29.2 Where Plasma Technologies Can be Specifically Applied in the Life Cycle of Fresh Produce?

When we discuss the plasma agriculture and food processing, we should keep in mind that significant interest and expectations are related to stimulation of plant growth, enhancement of produce safety, and especially enhancement of the fresh produce safety. Analyzing relevant abilities of plasma technologies, we need to realize that the lifecycle of the fresh produce is complex and includes a lot of specific steps (Figure 29.2, see Park et al. 2013a,b). Plasma may be effectively used to all of them, in all steps of the lifecycle of fresh produce. These major steps are listed below, and most of them are especially discussed in different specific lectures as well as in the following sections of the current lecture.

- **Sterilization of seeds** while in storage; these technologies are closely related to the surface sterilization and disinfection processes discussed in the previous lecture (some specifics are due to use of the fluidized beds).

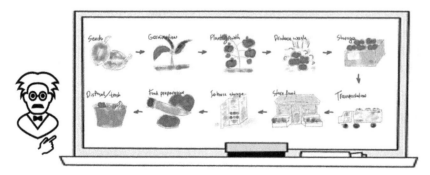

Figure 29.2 Plasma-assisted treatment of fresh produce: lifecycle of the fresh produce includes multiple steps and plasma technologies can be beneficial to most of them.

- Enhancement of **seed germination and growth** of seedlings; these processes were effectively organized using low- and atmospheric pressure plasmas, as well as applying plasma-activated water, see Sections 29.3 and 29.4.
- Plasma generation of NO and RNS in air and **capture atmospheric nitrogen in water to be used as fertilizer and stimulate plant growth**, see Lectures 18 and 24, as well as current one, below regarding stimulation of plant growth.
- Plasma generation of reactive oxygen species (ROS) and other oxidizers, combined with controlled (lowered) pH, to **reduce pathogen invasion of soils** and hydroponic water, see Lecture 24 and the current lecture, below.
- **Air cleaning, disinfection/sterilization of air and plant leaves, removal of volatile organic compounds (VOC) in greenhouse facilities**; these technologies were effectively organized by application of nonthermal plasma in atmospheric pressure air (mostly DBD), see Lectures 24 and 28.
- Activating plasma wash water treatment, **sterilization, and cleaning of water used for the produce wash after harvest; plasma enhanced washability of fresh produce**, see Lecture 28, as well as and the current lecture, below.
- **Disinfection of the fresh produce before packaging; treatment of the fresh produce conveyers to suppress counter-contamination**; these technologies are effectively organized using nonthermal plasma in atmospheric pressure air (mostly DBD). The processes are closely related to the surface sterilization and disinfection processes discussed in the previous lecture, see also the current lecture, below.
- **Sterilization/disinfection of food in closed packages;** these technologies are organized using mostly DBD discharges, which can effectively generate plasma across the barriers, see Lecture 13, as well as the current lecture, below.
- **Air cleaning, sterilization, and removal of VOC in the packaged produce storage facility and transportation vehicles**; see Lecture 28, as well as the current lecture, below.
- **Control of pests and pathogens at the in-store display cases and in-store storage**; can be effectively organized using the plasma-mist sanitization systems sustained by nonthermal air discharges, see the current lecture below.
- **Removal of ethylene from air to reduce rate of the fresh produce aging and deterioration** (significantly controlled by amount of ethylene in air responsible for the produce counter contamination aging and deterioration), as well as prolonged storage time in the storage facilities and stores, see Lecture 18.
- **Sterilization of cutting boards, knives, and other food processing equipment both at home and in food processing facilities or grocery stores**; see Lecture 28.
- **Plasma-assisted destruction of hazardous waste and/or waste-to-energy conversion of the nonhazardous food wastes,** see Lecture 22.

29.3 Direct Plasma-stimulated Seed Germination and Growth: Application of Low, Medium, and Atmospheric Pressure Discharges

The **low- and medium-pressure direct plasma treatment of various seeds** in the presence of different feeding gases were reviewed and analyzed by Attri et al. (2020). The low/medium-pressure plasmas were used, for example, to treat the seeds of radish sprouts, wheat, ajwain, black gram, poppy, oilseed rape, garlic, sweet basil, and bean. Radish Sprouts (*Raphanus sativus*) seeds were treated by low-pressure (100 Pa) radiofrequency (RF, 13.56 MHz) plasma with O_2 and N_2 as feed gases. While effect on the seed germination itself was not strong, the average length of sprouts was 60% higher. Plasma treatment on *Arabidopsis thaliana* seeds results in increased length of stems by 1.5 times and area of leaves by 2 times as compared with control (without plasma treatment). Wheat is an essential strategic crop; therefore, many researchers used plasma to treat the wheat seeds. The wheat seeds (*Triticum aestivum*), for example, were treated in glow discharge plasma with a mixture of air/O_2 gases at 1333 Pa pressure. 6 min of treatment in glow discharge plasma could result in 95–100% seed germination and a 20% increase in wheat yield. In a recent study, RF (13.56 MHz) low-pressure air plasma was applied for 180 s treatment of the same wheat seeds. After this plasma treatment, the grain and spike yield were enhanced to 58% and 75%, respectively, compared to control in the presence of haze stress (Saberi et al. 2020). Dubinov et al. (2000) revealed 27% increase in the quantity of germination of Oat seeds treated with glow discharge plasma at pressure 13 Pa with respect to reference seeds. Hosseini et al. (2018), treated the artichoke seeds (*Cynara cardunculus* var. *scolymus* L.) with capacitively coupled RF plasma at pressure of 1.8 Pa. The length of root increased by 28.5% and 50% after 10- and 15-min plasma treatment, respectively. The dry weight of roots was increased by 13% and 53% after 10- and 15-min plasma treatment, respectively. Recent study by Singh et al. (2019), showed the increased in germination percent of sweet basil (*Ocimum basilicum* L.) seeds when treated with RF plasma at 13.56 MHz frequency, 40 Pa pressure with variable power 30–270 W in the mixture of O_2 (80%) and Ar (20%) gases. The germination percentage increased by 16.3% and 20.5% than control at power 90 and 150 W, respectively.

While the low-pressure plasma-induced germination results described above are quite promising, the recent efforts in seed germination and growth enhancement are more focused on the more practical **atmospheric pressure plasma sources**, see Attri et al. (2020). This is due to the difference in the treatment cost of the devices as well as the user-friendly operation of atmospheric pressure plasma devices. The various seeds, such as radish sprouts, wheat, sunflower, pea, bean, maize, rice, pumpkin, cucumber, pepper, barley, spinach, basil, black pine, etc., were effectively treated in these systems. As an example, **dielectric-barrier discharge (DBD)** treated the *R. sativus*, *Oryza sativa*, *A. thaliana*, *Plumeria*, and *Zinnia* seeds at 9.2 kV discharge voltage and 0.2 A discharge current and 1.5 W cm^{-2} of discharge power density. The growth enhancement was 250%, 80%, 60%, 30%, and 20% for *R. sativus*, *O. sativa*, *A. thaliana*, *Plumeria*, and *Zinnia*, respectively, after DBD treatment (Sarinont et al. 2014, 2015). Kitazaki et al. (2014), used the DBD plasma to analyze the growth of radish sprouts (*R. sativus* L.). They observed 250% growth enhancement when seeds were placed at $x = 5$ and $y = 3$ mm. In a separate study, this DBD treated the radish sprouts (*R. sativus*) seeds for 3 min in the presence of different feed gases like He, Ar, N_2, Air, O_2, and NO (10%) + N_2. For He, N_2, and Ar feeding gases, plasma treatment showed a limited influence on plant growth; however, for O_2, Air, and NO (10%) + N_2 gases plasma had significant effect on growth enhancement. Amnuaysin et al. (2018), treated the rice (*O. sativa* L.) seeds with air-DBD operated at 5.5 kHz frequency for 60 s. The germination rate was 93% after plasma treatment, while only 85% for control. Additionally, the vigor index, germination rate, seedling growth, fresh weight (root and shoot), and dry weight (root and shoot) showed significant improvement after the DBD plasma treatment.

Thuringian Mallow (*Lavatera thuringiaca*) seeds have been treated with **atmospheric pressure gliding arc plasma** in dry nitrogen as working gas at 50 Hz discharge frequency, and 40 W of mean power (Pawlat et al. 2018). The germination was 60% for both 2- and 5-min plasma-treated seeds, while only 36% for control seeds. In another study, pre-treatment of hemp with gliding arc working at 50 Hz power frequency at a flow rate of 10 l min^{-1} of humid air, resulted in increased length of seedlings, seedling accretion, and weight of seedling (Sera et al. 2017). *Concluding, very different types of plasmas from strongly nonthermal to warm transitional once, operating in very wide range of pressures, from low to atmospheric one, all demonstrated significant enhancement of seed germination and further growth. The atmospheric pressure plasma sources (especially, DBD and gliding arc discharges) proved to be more practical in this regard due to lower treatment cost as well as the user-friendly operation of atmospheric pressure, see Attri et al. (2020).*

29.4 Some Biological Effects of Direct Plasma-induced Enhancement of Seed Germination and Growth of Seedlings

Mechanisms of the plasma effects on seed germination and growth of seedlings are not sufficiently clear at this point. Some plasma-induced physical effects on germination can be related to damaging and activation of the bran, the seeds skin. More interestingly, some observations, reviewed by Attri et al. (2020), indicate **biological and biochemical effects accompanying the plasma treatment of seeds**, which can clarify mechanisms of the plasma-induced enhancement of seed germination and growth of seedlings. In this regard, for example, Sarinont et al. (2014, 2015), revealed an increase in chlorophyll and carotenoid concentration in Radish sprout seeds after DBD treatment. Saberi et al. (2020), showed improvement in the photosynthesis rate, stomatal conductance, and chlorophyll content in low-pressure plasma-treated wheat seeds. Ji et al. (2016), showed an increased level of chlorophyll and total polyphenols contents in spinach seedlings after air DBD treatment. Also, they observed an elevated level of GA3 hormone in spinach seedlings after plasma treatment. Li et al. (2017), showed an elevated level of proline, soluble sugar contents, and osmotic-adjustment products after DBD treatment on wheat seeds. Another study showed that low-pressure plasma treatment on the soybean seeds results in increased protein and soluble sugar content Li et al. (2014). Similarly, Guo et al. (2017), observed improvement in proline and soluble sugar contents and decreased level of the malondialdehyde content in wheat seeds treated with DBD.

Hosseini et al. (2018), showed **improved seed's water uptake** in artichoke seeds after low-pressure N_2 plasma treatment. Meng et al. (2017), showed that DBD treatment on wheat seeds results in an improvement in the capacity for water absorption and activation of several physiological reactions. Bormashenko et al. (2015), showed that low-pressure plasma treatment noticeably increased the water imbibition in bean seeds. Alves et al. (2016), observed the change in hydrophilicity and water absorption of Mulungu seeds after plasma jet treatment. Volkov et al. (2019), showed that pumpkin seeds treated with plasma jet showed structural deformations of seeds such as surface defects, and hydrophilic pores that results in enhanced water uptake. Fadhlalmawla et al. (2019), reported that Fenugreek seeds treatment with cold atmospheric pressure plasma jet showed increased water imbibition and absorption that was due to etching on seed coat surfaces. Additionally, Li et al. (2017) showed that improvement in the water absorption capacity of wheat seed after DBD treatment is due to the etching effect. Saberi et al. (2020), demonstrated the improved tolerance of wheat plants against the haze and increased relative water content after low-pressure treatment.

Further, it was shown that the **production of reduction-type thiol compound** changes significantly in radish sprout seeds after treated with plasma torch; these thiol compounds are responsible for growth regulation mechanisms in plants (Hayashi et al. 2015). Hayashi et al. (2015), also mention this effect confirming that thiol content in seeds with plasma irradiation may be associated with plant growth. The increased amount of thiol is due to the reduction of cystine by active hydrogen. The increased amount of cysteic acid is due to oxidative modification of cysteine during plasma irradiation. Recently, Song et al. (2020), demonstrated the enhanced contents of the primary metabolites, especially the free amino acids and soluble sugars, as well as secondary metabolites like phytochemicals, e.g. saponarin, GABA, and policosanols in barley seeds after treatment with the surface DBD plasma. Kyzek et al. (2019), noticed the DNA damage in Pea seeds due to DCSBD plasma treatment. Hosseini et al. (2018) and Tong et al. (2014), observed the increase in catalase activity after DBD and low-pressure treatment.

Summarizing, *multiple biological and biochemical effects are accompanying the plasma treatment of seeds. They include significant changes in concentrations of some crucial biochemicals, modification of the thiol compounds responsible for growth regulation in plants, strong changes in water uptake abilities, etc. These effects can clarify mechanisms of the plasma-induced enhancement of seed germination and growth of seedlings.*

29.5 Indirect Plasma Effects on Seeds, Effect of Plasma-treated Water on Seed Germination and Enhancement of Seedling Growth

Application of the plasma-treated water (with high concentration of the plasma-generated ROS and RNS, especially those with long lifetime, like H_2O_2, NO_2^-, NO_3^-, etc.) in agriculture has been in focus in last couple of decades, particularly for stimulation of seed germination and enhancement of seedling growth, see Bruggeman et al. (2016), Locke et al. (2006), Bourke et al. (2018), Schnabel et al. (2019), Yong (2018). Thus, *Vigna mungo* seeds, for example, have been treated with H_2O–O_2 discharge plasma (3–6 kV, 3–10 kHz) generated water for 3, 6, 9, 12, and 15 min. This treatment revealed significant improvement in seedling growth of black gram seeds. The plasma-activated water generated by 6 min plasma treatment showed the foremost cumulative germination, while 3 min plasma treatment produced activated water providing the higher vigor index. Seeds treated with plasma-activated water created showed longer shoot and root as compared to untreated samples. Additionally, the plasma-activated water produced after plasma treatment for 12 min displayed the highest dry weights of shoots among other treated samples.

Sarinont et al. (2017), demonstrated the effect of **plasma-activated water generated using DBD plasma** in various gases like air, O_2, N_2, He, and Ar on the Radish sprout (*R. sativus* L.) seeds. The power was 10, 9, 6, 5, and 4 W for air, O_2, N_2, He, and Ar working gases, respectively. The plasma-activated water was kept for 1 h and 1 day after DBD treatment at room temperature to minimize the effect of short-lived reactive species. The plasma-activated water produced from Air, O_2, He, N_2, and Ar plasmas kept for 1 h and showed 1.62-, 1.38-, 1.13-, 1.12-, and 1.04-times increase in seedling length of sprouts as compared to control. This plasma-activated water (generated using the air, O_2, He, N_2, and Ar plasmas) kept for 1 day showed 1.52, 1.28, 1.13, 1.10-, and 1.08-times increased length of sprouts, respectively, more than control. Sivachandiran and Khacef (2017) treated the Radish sprouts (*R. sativus* L.) seeds with plasma-activated water from dielectric barrier discharge with 15- and 30-min treatment. This treatment showed enhance of seeds germination rate and seedling growth.

Zinnia annual (*Zinnia elegance*) seeds were treated with **plasma-activated water produced by underwater discharge** (5 min treatment, operated at a voltage not more than 1.5 kV, and discharge current 50–70 mA). Germinability rate has been increased by 50%, and plant roots rose to 1.5- to 2-fold as compared to the control after the plasma-activated water treatment (Naumova et al. 2011). *Brassica rapa* var. *perviridis* seeds were treated with plasma-activated water generated similarly by the underwater discharge with a repetition rate of 250 pps, peak voltage of 30 kV, and pulsed power generator for 10- and 20-min. The dry weight of the plant increased in this case 3.9 and 6.6 times; the leaf length increased 2.1 and 2.5 times after treatment with plasma-activated water generated by plasma during 10- and 20-min, respectively, as compared to control (Takaki et al. 2013).

Zhang et al. (2017) treated the Lentils seeds with **plasma-activated water created with He-plasma jets.** They demonstrated 80% germination rates after the indirect plasma treatment (control rate was 30%). Additionally, they also noticed the higher stem elongation rates and bigger final stem lengths in the plasma-water treated samples than those produced using commercial fertilizer (control). Recently, Adhikari et al. (2019), treated the Tomato (*Solanum lycopersicum* L.) seeds with plasma-activated water generated also using the plasma jet. The frequency, current, and applied voltage of the plasma jet discharge were 83.5 kHz, 70.39 mA, and 0.66 kV, respectively. Plasma-activated water generated by plasma jet is 15- and 30-min treatment showed better morphological growth compared to control seedlings. Also, this plasma-activated water is produced. In the 15- and 30-min plasma jet treatment showed higher shoot and root length, but no significant change was observed with plasma-activated water generated in 60 min plasma treatment.

Summarizing, the indirect plasma effects on seeds, which is mostly the positive effect of plasma-treated water on seed germination and enhancement of seedling growth, is quite significant and comparable with the relevant direct plasma effect. Challenge of the indirect plasma treatment of seeds is sometimes due to low tolerance of some seeds to water and even humidity. Advantage of the indirect plasma treatment of seeds through activation of water is related to possibility to treat large volumes of the agricultural products using this approach.

29.6 Plasma Stimulation of Plant Growth Using Plasma-activated Water: Major Mechanisms, Comparison of Different Plasma Sources, Plasma Hydroponics

Watering different plants with plasma-activated water can accelerate the plant growth, as well as increase size of the plants, and their mass. This impressive effect known for quite long time has been observed with watering different plants from flowers and grasses to sweet basil and even hemp (both in soil and in hydroponic systems). Therefore, it is not a surprise that photos illustrating this effect have been used as a symbol of the 1st IWOPA organized and held in May of 2016 by C&J Nyheim Plasma Institute in Drexel University, Philadelphia. While the effect of plasma-water stimulation of plant growth has been demonstrated using several different plasma sources, the detailed understanding of biology and biochemistry of the relevant processes is still very far from understanding. It is clear, however, that *nature of the plasma-water stimulation of plant growth is related to plasma ability to generate nitrates and nitrites in water in the range of about 50–300 ppm, which is optimal for fertilization of several plants, plasma's ability to produce different peroxides in range of concentrations 0.1–10 ppm also optimal for the plant sanitization, plasma compatibility with conventional plant nutrition solutions, and plasma ability to control water pH in the optimal range of 4.5–6.5 (sometimes with the use of special buffers).*

The detailed analysis comparing different types of plasma-activated water (using different discharges for water activation, different water feedstock, and different plants) for stimulation of plant growth has been accomplished by Park et al. (2013a,b). The researchers of Nyheim Plasma Institute of Drexel University, Park et al. (2013a,b), have studied the effects of three different types of plasmas: underwater (submerged) spark discharge (2 kV, 600 mA, 5 Hz), transferred arc discharge (1.6 kV, 300 mA), and gliding arc discharge (0.8 kV, 300 mA) on plant growth. Each plasma-treated water for 2 min. Schematics of these three types of transitional "warm" plasma sources are shown in Figure 29.3. Choice of the transitional "warm" plasma sources is due to the possibility of such discharges to synthesize maximum amount of nitrogen compounds in air, and therefore in water, see Lecture 18. These plasmas were used to treat either tap, spring, or distilled water to clarify influence of the water feedstock on the stimulation of plant growth. The plasma-activated water has been then applied for watering of watermelon (*Citrullus lanatus*), zinnia (*Zinnia peruviana*), alfalfa (*Medicago sativa*), pole beans (*Phaseolus coccineus*), and shade champ grass. All plant samples were given their respective

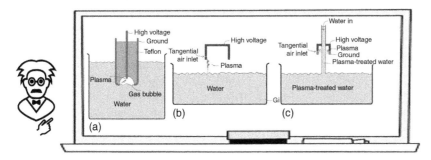

Figure 29.3 Transitional "warm" discharges applied to activate water for further watering plants with the purpose of stimulation of their growth: (a) submerged spark discharge, (b) transferred arc discharge, (c) gliding arc discharge (plasmatron).

water types 30 ml every Monday and Wednesday and given 40 ml every Friday to compensate for the weekend. Lights were placed above all the plants on an automatic schedule from 8 a.m. to 6 p.m. The plant samples were cut, measured, and weighed after approximately 3 weeks. Plant samples were taken by removing the plant from its starter pot. Excess soil was shaken off the roots and rinsed to ensure dirt removal. Roots were then patted dry with paper towels. The plants were cut to separate the roots from the stems. The root and stem lengths were measured, and the weight of both root and stems were taken. Pesticide (Fungicide, Garden Safe) and fertilizer (Flower and Vegetable 10-10-10, Scotts) were used to compare against the sterilization and fertilizer-like properties of plasma-treated water.

Changes in water chemistry following the plasma treatment were analyzed. In most of the cases, measured concentrations of nitrates, nitrites, peroxides, and pH were in optimal range mentioned above (starting pH was slightly basic, 7–8). Gliding arc discharge, especially in in the case of treatment of spring water provides some stronger chemical effect on the treated water. Park et al. (2013a,b), present detailed results covering all investigated **plasma-induced growing plant** samples. The measured variables are the length and weight of the stem (from the start of the root, upward, marked "top") and the length and weight of the roots (marked "bottom"). The data were normalized to the values of the control plants. The data for zinnia plants show that the plasmatron system achieved the best results in the categories for plasmatron (tap water), weight of the stem and length of the root, and plasmatron (spring water) length of the stem. Less but still significant results were found for plasmatron (distilled) length of the stem. The best results for the alfalfa weight of the stem and roots were achieved with the plasmatron system while used with spring water, and less but still significant results were found for plasmatron fed with tap water (weight and length of the root), see Figure 29.4 and with distilled water (weight of the stem and roots).

Summarizing, the plasma water generated using all types of transitional "warm" discharges (effectively producing nitrogen compounds and peroxides) consistently provide significant effect on growth of all analyzed

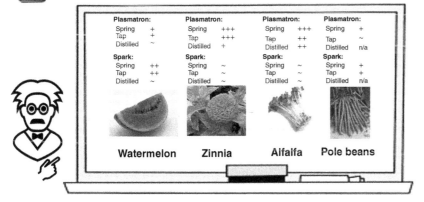

Figure 29.4 Illustrative comparison of effectiveness in stimulation of different plant growth by plasma-water generated using the gliding arc plasmatron and submerged sparks treatment of spring, tap, and distilled water (number of + illustrate the effectiveness).

plants (the strongest effect on alfalfa and zinnia). Increase of the plant mass is about 3–5 times for alfalfa and zinnia and 1.5–2 times for pole beans and watermelons. While being quite close, the best fertilizing effect has been achieved using gliding arc discharge treatment of spring water.

Thus, based on the obtained results, the gliding arc plasmatron was shown to provide most promising effect on plants' growth and development. Therefore, the plasmatron system was used to treat spring water (which proved to be most acceptable for the plasma treatment in preliminary experiments). The plasma-activated water generated in this case has been applied for watering of radishes (*R. sativus*), tomatoes (*S. lycopersicum*), and banana peppers (*Capsicum annuum*). The major objective of these experiments was comparison of the plasma-activated water with water treated with conventional pesticides (Fungicide, Garden Safe) and fertilizers (Flower and Vegetable 10-10-10, Scotts). The results show that plasma, being ecologically more friendly, provides effect like the conventional pesticide/fertilizer combination. These results also demonstrate that combination of plasma with conventional pesticides/fertilizers can provide even stronger effect. It is especially important for plasma application in the so-called **plasma hydroponics**. Similar effects, including the gliding arc plasma activation of plant nutrition solution in the larger scale hydroponic systems were reviewed in Fridman (2023). More information on the subject and relevant data can be found in Park et al. (2013a,b).

Finally, we should mention that application of the plasma-activated water (and especially water activated with transitional "warm" gliding arcs) not only stimulates the plant growth but also permits to save water (Fridman 2020). There is an increased effort to improve plant irrigation without compromising plant yield, and plasma water meets this challenge. 20–30% less water can be required for watering plants if the water has been preliminarily plasma-activated. The effect has been demonstrated on *A. thaliana* plants, see Brar et al. (2016). The study also monitored molecular changes in the plasma watered plants increasing their tolerance to drought. This **plasma effect of conservation of the agricultural water** can be due to plasma-induced changing of physical properties of water (surface tension, capillarity, wettability, viscosity, evaporation rate, see Section 29.19 of this lecture).

29.7 Direct Application of Plasma for Disinfection of Agricultural Products

Sterilization and disinfection of different surfaces have been discussed in Lecture 28 (Sections 28.1–28.7). While plasma disinfection of agricultural products, illustrated in Lecture 1 by Figure 1.11, follows many general principles discussed over there, it has some important specific peculiarities. The major specifics of plasma disinfection of agricultural products include:

- Often, very **sophisticated surfaces of the agricultural products** containing multiple deep wrinkles, pores, etc., provide possibilities for dangerous pathogens to "hide." If the relevant contaminated produce surfaces are flat (see Figure 1.11), effectiveness of plasma disinfection and sterilization is way higher.
- Sometimes, undesirable microorganisms and dangerous pathogens can be located **not on the surface but inside of the agricultural products or their conglomeration** like in the case of powder products (coffee, child food, bakery products, meats, etc.). Penetration of plasma sanitization effects deep inside of the products is possible but challenging, which is going to be discussed in Lecture 30 regarding plasma medicine.
- Sterilization and disinfection of any surfaces always require **selectivity of the treatment** (not damaging substrates, etc.), but it becomes crucial in the case of plasma treatment of the agricultural products. Plasma **disinfection of the produce should not damage them**, should not change their taste, and surely not making them poisonous. Such selectivity is easier to achieve if the products are well protected by cuticles, bran, and "skin" of the produce like in the case of apples, tomatoes, some seeds, etc. (see Figure 1.11). It is not the case in most of the agricultural products, however.
- Not all the specifics of the plasma disinfection of agricultural products are negative, some of them are positive. Good example here is **preventing of the produce counter contamination** aging and deterioration, which is often due to ethylene transferring the deterioration between the fresh produce. Removal of the ethylene from air to reduce rate of the fresh produce aging and deterioration, as well as prolong storage time in the storage facilities and stores be effectively provided by plasma. Such kind of suppression of microorganisms is not applicable for not-agricultural sterilization (electronics, spacecrafts, furniture, etc.).

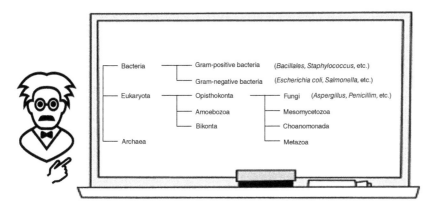

Figure 29.5 Disinfection of agricultural products, classification of relevant microorganisms.

The major dangerous pathogens crucial in the food safety challenges are *Escherichia coli*, *Listeria*, and *Salmonella*. The practical list of food poisoning microorganisms is, however, much longer. Not all the microorganisms are equal with respect to sanitization agents, like bleaching chemicals, ozone, UV, etc. To compare, the relative resistance of different microorganisms to the sanitization agents, let us analyze the simple classification of microorganisms presented in Figure 29.5. Thus, *Bacillus* and *Escherichia* are classified as bacteria. *Penicillium digitatum* is classified as eukaryota. Spores of *P. digitatum* are different from other microorganisms because of resistant structure, composition, and the function of their cell wall. The doses of the ultraviolet UV–C irradiation (between 100 and 280 nm) required for 90% inactivation of the microorganism are: (i) *Bacillus subtilis* (nonspores) – 7.1 mJ cm^{-2}; (ii) *B. subtilis* (spores) – 12.0 mJ cm^{-2}; (iii) *E. coli* – 1.3–3.2 mJ cm^{-2}; *P. digitatum* spores – 44 mJ cm^{-2}; *Aspergillus niger* spores – 132 mJ cm^{-2}. Thus, larger dose for inactivation of mold spores such as *Penicillium* and *Aspergillus* is required compared with bacteria such as *Bacillus* and *E. coli*. These bacteria and eukaryota on crops and foods are targets for inactivation using plasmas. Standard procedures for qualifying and testing sterilizers require the inactivation of 10^6 (6 log) microorganisms in medical applications while minimal damage, no toxic residue, and high-inactivation rate are expected for the disinfection of crops and foods.

Plasma disinfection or sterilization of agricultural products can be organized either directly (direct contact with plasma), which is going to be discussed in the following sections, or indirectly through washing, spraying, or misting with plasma-activated water, which is going to be considered later in this lecture. Let us start with short analysis of the direct disinfection of crops and seeds provided by direct contact with different nonthermal plasma.

29.8 Direct Plasma-induced Disinfection of Crops and Seeds

Different types of nonthermal plasma can be effectively applied to directly disinfect different crops and seeds, which was reviewed by Ito et al. (2012). For example, disinfection of apples, cantaloupes, and lettuce using **atmospheric uniform glow discharges** was reported by Critzer et al. (2007). The remote chamber technique was applied in this case to pathogens on apples, cantaloupes, and lettuce. *E. coli*, *Salmonella*, and *Listeria monocytogenes* populations were reduced by >1 log after a 1-min exposure, and >2 log after a 2-min exposure.

Perni et al. (2008a,b), inactivated one pathogenic (*E. coli*) and three spoilage organisms (*Saccharomyces cerevisiae*, *Pantoea agglomerans*, and *Gluconacetobacter liquefaciens*) on the pericarps of mangoes and melons by using **cold atmospheric-pressure plasma** and suggested that the production of oxygen atoms was related to the level of bacterial inactivation achieved from optical emission spectroscopy. Furthermore, the *E. coli*, *S. cerevisiae*, *P. agglomerans*, and *G. liquefaciens* were inactivated on the inoculated membrane filters and inoculated fruit surfaces. The inactivation efficiency was markedly reduced for microorganisms on the cut surfaces against there on the surfaces of the filter membrane. They concluded that the lack of effect was not the result of quenching of the reactive plasma species responsible for microbial inactivation but principally the result of the migration of microorganisms from the exterior of the fruit tissue to its interior. Similar strong effect of atmospheric pressure plasma has been demonstrated on apples a biofilm layer using the **nonthermal regime of the gliding arc discharge**, see Fridman and Friedman (2013), as well as illustration in Figure 1.11.

Basaran et al. (2008), Selcuk et al. (2008), developed a **low-pressure cold plasma** (LPCP) system using air gases and sulfur hexafluoride (SF_6) and tested it for anti-fungal efficacy against *Aspergillus parasiticus* on various nut samples, such as hazelnuts, peanuts, and pistachio nuts. Air-gas plasma treatment for 5 min resulted in a 1-log reduction of *A. parasiticus* while SF_6 plasma application was more effective, resulting in approximately a 5-log decrease in fungal population. The effectiveness of plasma treatment against aflatoxin was also tested. A 20-min air-gas plasma treatment resulted in a 50% reduction in total aflatoxin while only a 20% reduction in total aflatoxin was observed after a 20-min SF_6 plasma treatment. The same direct plasma system (LPCP) also demonstrated the inactivation efficacy using air gases or SF_6 for suppression of two pathogenic fungi (*Aspergillus* and *Penicillium*) on seed surfaces. The plasma decontamination was performed by using a batch process in a vacuum chamber, with gas injection being followed by plasma discharge for a duration of 5–20 min. The plasma treatment reduced the fungal attachment to seeds to below 1% of the initial load, depending on the initial contamination level, while preserving germination quality of the seed. A reduction of 3-log for both species was achieved within 15 min of SF_6 plasma treatment. In these studies, a rapid, functional clean-up method for the elimination of aflatoxin-producing fungus from shelled and unshelled nuts and of pathogenic fungi on the seed surface was demonstrated as a suitable decontamination method.

29.9 Plasma-induced Disinfection of Meats, Cheeses, Other Foods, and Relevant Food Containers and Storage Areas

Direct plasma disinfection of meats (and other similar products) represents additional challenges related to more sophisticated positioning of microorganisms to be treated, see Section 29.7, as well as review of Ito et al. (2012). As an example, Noriega et al. (2011), examined the efficacy of cold atmospheric gas plasmas for **decontaminating chicken skin and muscle inoculated with *Listeria innocua*.** A 10 s treatment gave >3 log reductions of *L. innocua* on membrane filters, an 8 min treatment gave a 1-log reduction on skin, and a 4-min treatment gave >3 log reductions on muscle. These results show that the efficacy of gas plasma treatment is greatly affected by the surface topography. Scanning electron microscopy (SEM) images of chicken muscle and skin revealed surface features wherein bacteria could effectively be protected from the chemical species generated within the gas plasma, see Section 29.7.

Kim et al. (2011), investigated the effect of nonequilibrium atmospheric pressure plasma (NEAPP) with He or He/O_2 on the **inactivation of pathogens (*L. monocytogenes*, *E. coli*, and *Salmonella*) on bacon.** Plasma with helium could only reduce the number of inoculated pathogens by about 1–2 log cycles. On the other hand, the helium/oxygen gas mixture was able to achieve microbial reduction of about 2–3 log cycles. The number of total aerobic bacteria showed 1.89 and 4.58 decimal reductions after plasma treatment with helium and the helium/oxygen mixture, respectively. Microscopic observation of the bacon after plasma treatment did not show any significant changes. As one can see, the inactivation effect in the case of application of NEAPP discharge is higher than in the case of use of the cold plasma described above (Noriega et al. 2011). This is due to the transitional "warm" nature of the NEAPP discharge, providing stronger effect, like in the case of gliding arc and spark discharges (Lecture 14). See more details regarding strong disinfection effect of the NEAPP in Ito et al. (2012).

Group of J.J. Quinlan of Drexel University demonstrated effectiveness of DBD plasma on *Salmonella enterica* and *Campylobacter jejuni* inoculated onto the surface of **boneless skinless chicken breast and chicken thigh with skin.** Chicken samples were inoculated with antibiotic-resistant strains of *S. enterica* and *C. jejuni* at levels of 10^1–10^4 CFU and exposed to plasma for a range of time points (0–180 s in 15-s intervals). Surviving antibiotic-resistant pathogens were recovered and counted on appropriate agar. To determine the effect of plasma on background microflora, noninoculated skinless chicken breast and thighs with skin were exposed to air plasma at ambient pressure. Treatment with plasma resulted in elimination of low levels (10 CFU) of both *S. enterica* and *C. jejuni* on chicken breasts and *C. jejuni* from chicken skin, but viable *S. enterica* cells remained on chicken skin even after 20 s of exposure to plasma. Inoculum levels of 10^2, 10^3, and 10^4 CFU of *S. enterica* on chicken breast and chicken skin resulted in maximum reduction levels of 1.85, 2.61, and 2.54 log, respectively, on chicken breast and 1.25, 1.08, and 1.31 log, respectively, on chicken skin following 3 min of plasma exposure. Inoculum levels of 10^2, 10^3, and 10^4 CFU of *C. jejuni* on chicken breast and chicken skin resulted in maximum reduction levels of 1.65, 2.45, and 2.45 log, respectively, on chicken breast and 1.42, 1.87, and 3.11 log, respectively, on chicken skin following 3 min of plasma exposure. Plasma exposure for 30 s reduced background microflora on breast and skin by an average of 0.85 and 0.21 log, respectively. This research demonstrates the feasibility of nonthermal dielectric barrier discharge plasma as an intervention to help reduce foodborne pathogens on the surface of raw poultry.

Song et al. (2009), examined the inactivation efficacy of the NEAPP applied to **sliced cheese and ham inoculated by *Listeria monocytogenes* (LMC)**. Calculated D values from the survival curves of 75, 100, 125, and 150 W of NEAPP treatments were 71, 62, 20, and 17 s for LMC in sliced cheese, respectively, and those in sliced ham were 476, 88, 71, and 64 s. These results indicate that the inactivation effects of NEAPP on LMC strongly depend on the type of food, which is obviously understandable, even keeping in mind energetic nature of this discharge generating "warm" transitional plasma, see Ito et al. (2012), as well as Lecture 14.

Atmospheric pressure micro-second pulsed DBD plasma has been effectively applied to **extend the storage time and shelf life** of banana exemplifying storage of fresh fruit and vegetables, see Trivedi et al. (2019). Bananas were exposed to plasma-treated air for 1 week at room temperature and normal pressure and humidity. It was proved that plasma treatment caused no significant change in weight, color, surface morphology, or sugar content of bananas. The mold growth and deterioration were observed in untreated samples after storage but were absent in plasma-treated samples. This study demonstrated that the atmospheric pressure DBD technique has the potential to prolong the storage and shelf life of bananas compared with conventional methods by inhibiting pathogen growth in post-harvest storage conditions as well as suppressing the negative effect of ethylene on the produce counter contamination, see Section 29.7.

Yun et al. (2010), used the nonequilibrium atmospheric pressure discharge (NEAPP, generating energetic transitional "warm" plasma) discharge to decontaminate *L. monocytogenes* inoculated onto **disposable food containers**, including disposable plastic trays, aluminum foil, and paper cups. A strong influence of the material on the inactivation kinetics was observed, demonstrating again influence of the surface structure on disinfection of the food-related bio and storage products, see Section 29.7.

29.10 Direct Plasma-induced Disinfection of Foods Inside of Closed Packages

Plasma and especially DBDs provide an interesting and practically important opportunity for direct disinfection of food in closed packages without any heating. These technologies are organized using mostly DBD discharges, because they can naturally generate plasma across the barriers. Thus, Chiper et al. (2011), reported generation and evolution of a DBD produced inside a closed package made of a commercially available packaging film and filled with gas mixtures of Ar/CO_2 at atmospheric pressure. The $Ar + 7\%$ CO_2 cold-plasma treatment reduced the concentrations of *Photobacterium phosphoreum* in **cold-smoked salmon** by $\sim$3 log CFU g^{-1} in 60 s whereas the Ar-cold plasma reduced concentrations of this fish spoilage bacterium in cold-smoked salmon by $\sim$3 log CFU g^{-1} in 120 s. In contrast, a 120 s-treatment with an Ar or an Ar/CO_2 cold plasma did not significantly reduce the concentrations of *L. monocytogenes* or *Lactobacillus sakei* in cold-smoked salmon.

Schwabedissen et al. (2007), investigated the disinfection in a similar way of **cherry tomatoes and strawberries in a closed package**. In this plasma system, ozone as a disinfecting agent has been produced inside a closed package equipped with disposable labeled electrodes. Using *B. subtilis* as a test microorganism, a viable count reduction on plate count agar of approximately five orders of magnitude was observed following a 10-min treatment, and an increase in the shelf life of cherry tomatoes and strawberries was evident.

Fast and effective **disinfection of bread in a closed package** (infected by *Penicillium* spp., important in spoilage of bread) has been demonstrated using microsecond pulsed DBD plasma (Fridman 2020; Ranieri et al. 2018), see illustration in Figure 29.6. While the bread surface is quite sophisticated, effectiveness of the disinfection is sufficient for practical interest. The most serious challenge of the technology at this point is rate of the plasma treatment, which is not sufficiently fast so far. The packaged bread moves on the conveyer providing for the treatment not more than 1 s, which is not sufficient so far for effective suppression of the practically relevant microorganisms, see Fridman (2020). We should mention that the described plasma disinfection methods inside of closed packages have a significant additional advantage because they decontaminate the contents of a pre-sealed container without the risks associated with handling reactive species such as ozone in open atmosphere, see Ito et al. (2012).

Figure 29.6   DBD plasma disinfection of bread inside of closed package.

29.11   Plasma-water, and How Does It Work in Agriculture for Disinfection and Washing of Fresh Produce and Other Foods? Plasma Chemistry in Water, Plasma-water Spraying and Misting

 The indirect plasma effect (through activation of water and relevant solutions) on germination of seeds and growth of plants is proved to be effective, which has been discussed above in this lecture. Even stronger impact on the food processing industry is related to the plasma-water application for disinfection and washing of fresh produce and other foods. Plasma-water works in food processing due to plasma affecting physical and chemical properties of water. Let us start with **plasma-chemical effects in water**, leaving the plasma-induced changes of physical properties of water to Section 29.18. Significant research efforts are focused recently on the plasma-chemical effects in water. They were already discussed in Lectures 27 and 28, and especially in Sections 27.5 and 27.10. More information on this wide-spread subject can be found in Brandenburg et al. (2019), Bruggeman and Leys (2009), Bruggeman et al. (2016), Lukes et al. (2014), and Graves (2012).

 Summarizing, the plasma-chemically activated water effects in disinfection and washing of fresh produce and other foods can be usually reduced to the effects of relatively long-living factors, while short-living species are already consumed in primary reactions. *These **long-living factors in plasma water** are limited in most of cases to: (i) peroxides (and especially, hydrogen peroxide H_2O_2), and (ii) acidity factors (nitric, nitrous, and specific "plasma" acids). It is important that the peroxides and acids are not strong sterilizers by themselves. The strong oxidative sterilizing effect is mostly due to extremely aggressive OH radical which they can generate.* The conversion of the hydrogen peroxide H_2O_2 into hydroxyl radicals OH usually occurs in biochemistry through the Fenton mechanism discussed in Section 27.5 and requiring the presence of Fe_2^+ and Fe_3^+ ions. It is not the case in the plasma water, where the conversion of the hydrogen peroxide H_2O_2 into hydroxyl radicals OH occurs through intermediate formation of peroxynitrite $ONOO^-$ (or it's conjugate peroxynitrous acid H^+ONOO^-):

$$H^+ + H_2O_2 + NO_2^- \rightarrow H^+ \left(NOO_2^-\right) + H_2O \tag{29.1}$$

The peroxynitrite $ONOO^-$ (or it's conjugate peroxynitrous acid H^+ONOO^-) then dissociates at low temperatures (including room temperature) producing the aggressive hydroxyl radicals OH responsible for the strong oxidative sterilizing effect:

$$H^+ \left(NOO_2^-\right) \rightarrow OH + NO_2 \tag{29.2}$$

Thus, the combination of the acidity and hydrogen peroxide provides production of OH and the strong sterilization effect of plasma-activated water. Effect of temperature on this sterilization pathway and possibilities of nonoxidative sterilization are going to be discussed later in this lecture regarding application of "warm" discharges to activate water. *It is important to point out that the hydroxyl radicals OH leading to the strong oxidative sterilization effect in plasma-water are not directly generated in plasma, but first converted to peroxides and only inside of water converted back to OH proving the antimicrobial effect.*

Administration of the plasma-activated water to agricultural products for their disinfection and washing can be organized using three approaches: (i) washing in bulk water, (ii) water spraying, and (iii) water misting. While washing in bulk water does not require special explanation, spraying and misting approaches do. Both spraying and misting are usually applied to minimize amount of water required for disinfection and washing. The **plasma water spraying approach** (illustrated in Figure 29.7) usually assumes preliminary preparation of the plasma-activated water, and then it's spraying to wash and disinfect the food. Typical droplet size in the case of spraying surely exceeds 20 μm, which results in falling of the droplets on the treated surface (gravity much exceeds air viscosity). Figure 29.7 illustrates the relevant plasma disinfection of raw ground beef inoculated with *E. coli* (10^8 CFU) in comparison with acidified sodium chlorite, and pure Peptone water, as control. The DBD plasma-water demonstrated in this case better disinfection than not only pure control water but even better results than conventional acidified sodium chlorite solution (Fridman 2020). In general, the plasma water spraying is like washing and disinfecting of products with the bulk plasma water. As an advantage, the spraying approach uses less water contributing to water conservation.

Even more significant water conservation can be achieved in the framework of the **plasma-water misting approach**, illustrated in Figure 29.8, see Fridman (2020). The mist water droplets are generated by a special nebulizer and usually have sizes smaller than 20 μm, therefore they are not so much affected by gravity and "go with a flow." The mist crosses the plasma zone (DBD, in the case illustrated in Figure 29.8), and the mist particulates are negatively charged and plasma-chemically activated. The electric charge of the mist particulates prevents their coagulation and stimulates their electrostatic explosion when electrostatic forces dominate over surface tension, see

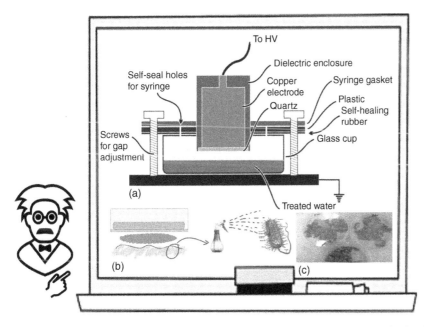

Figure 29.7 Plasma-water spraying for food sterilization: (a) plasma-water production (using DBD in this case); (b) spraying the plasma-water; (c) sterilization of raw ground beef (plasma-water, PTW, comparison with acidified sodium chlorite, ASC, and pure Peptone water, PW, control).

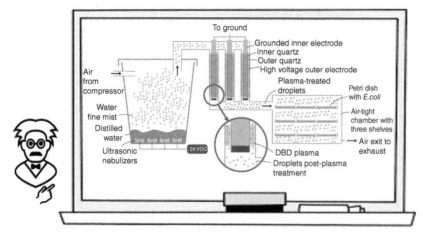

Figure 29.8 "Plasma mister" for disinfection of fresh produce using three DBDs to activate mist generated in ultrasonic nebulizer.

below in this lecture. The mist of the supermall (often submicron) particulates is "suspended" in the agricultural products storage space and can provide strong disinfection effect due to very high specific surface area of the plasma-activated water. Physics and microbiological effects of the plasma mist are going to be considered below (Section 29.13), after preliminary discussion of the metastable nature of the plasma-activated water, and metastable plasma acids.

29.12 Metastable Nature of Chemical and Biological Activity of Plasma-activated Water, Metastable Plasma Acids

A question is often asked: if the effect of plasma-activated water is simply limited to a combination of an acid and hydrogen peroxide (see the previous Section 29.11), why not to take (buy) these inexpensive on-shelf products and mix them instead of more complicated plasma-chemical approach. The answer is in the metastable nature of plasma-water. As one can see from the above discussion, *plasma-treated water contains significant amount of the **metastable active species** (unstable acids, non-stabilized peroxides, peroxynitrous compounds, etc.) with minutes/hours lifetime. It makes the plasma-water-based approaches so much unique in agriculture, food processing, as well as in medicine. In other words, the best is to use plasma-activated water when it's fresh.* The **metastable nature of plasma-water** is obvious when considering contribution of hydrogen peroxide, ozone, etc. These compounds have limited lifetime, and disappear after a while, leaving water clean without residuals.

It raises an interesting question about acidity, are plasma-generated acids stable or also metastable? To answer this question, let us first remind that plasma treatment of water can make water quite acidic. As an example of the **plasma-water acidity**, Figure 29.9 shows the pH reduction of deionized water (black squares) after application of the atmospheric-pressure DBD, ignited in air, to the surface of the water, like in the Figure 29.7b (Shainsky et al. 2010, 2012; Fridman and Friedman 2013). As it was discussed in the current and previous lectures, acidity of the plasma-water generated by air plasma can be significantly due to the formation of nitric and nitrous acids (corresponding to the conjugate bases: nitrites NO_2^-, and nitrates NO_3^-). The plasma-generated nitrogen oxides are not strongly soluble in water themselves, so formation of NO_2^-, and NO_3^- is often determined by dissolving in water of the plasma-generated radical HNO.

It is interesting, however, that DBD discharges in pure oxygen also can produce acidic plasma-water (Figure 29.9). There are no gas-phase nitrogen compounds in such plasmas, which indicates possibility of *production of the so-called **metastable plasma acid** generated from the charged plasma species, positive and negative ions. As it has been already discussed in Lectures 24 and 27 regarding the air purification by nonthermal plasmas, the "nonchemical" plasma acidification starts with the fast charge exchange from different positive ions to water ion H_2O^+ (Section 2.13,*

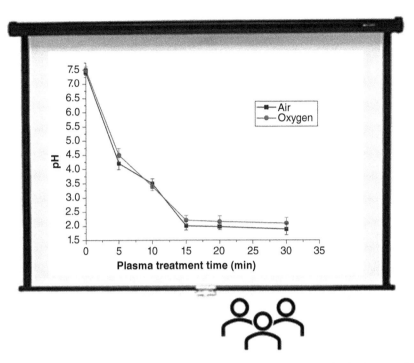

Figure 29.9 Acidity of plasma-water: variations of pH of degassed deionized water after DBD plasma treatment in air and oxygen (no pH variation after treatment in Ar was observed).

elementary process [2.85]) occurring because the ionization potential of water molecules both in gas and liquid phase are low. These water ions H_2O^+ further react very fast with water molecules forming H_3O^+-ions (that's plasma acid) and OH-radicals:

$$H_2O^+ + H_2O \rightarrow H_3O^+ + OH\Delta H = -12\,kcal\,mol^{-1}, k(300\,K) = 0.5{*}10^{-9}\,cm^3\,s^{-1}. \tag{29.3}$$

Important contribution of the metastable plasma acid to the total acidity of the plasma-activated water is supported by comparison of the acidification provided by direct (charges contact water) and indirect (charges are blocked by metal mesh) DBD discharge mode, see Shainsky et al. (2010, 2012), Fridman and Friedman (2013). Direct DBD where ions reach water leads to low pH values (Figure 29.10a). But what is the **conjugate base of the metastable plasma acid**? High efficiency of oxygen plasma in water acidification (Figure 29.9) suggests that the negative oxygen-based ions, superoxide O_2^- and even O_3^- can be the conjugate base in this case. This hypothesis is supported by experiments with addition of superoxide dismutase (SODe) converting O_2^- into H_2O_2 suppress the plasma-water acidity (Lee et al. 2011). Also, if the DBD is organized in Ar and water is degassed (no N_2 and O_2) neither in plasma nor in water, there is no visible acidification, see Figure 29.10b. While H_3O^+-ions are effectively formed through the charge exchange with Ar^+, the hydrated electrons are very fast recombining with H_3O^+ suppressing the acidity.

 While temporary concentration of the plasma acid can be relatively high, the plasma-ions generated super-equilibrium concentrations of H_3O^+ and O_2^- (or O_3^-) cannot stay long at room temperature, resulting in the possible metastable behavior of the plasma acids. The ions simply recombine forming week HO_2 or HO_3 acidic radicals. Way stronger nitrogen-based acids (especially, strong nitric acid) create stable effect, *which is illustrated in Figure* 29.11, *see Shainsky et al. (2010, 2012), Fridman and Friedman (2013). We* should note that ability of the oxygen-based plasma-water to increase pH and suppress the acidification can be also due to the presence of chlorides Cl^- (for example from NaCl), which reacts in water with plasma-generated O atoms (either directly or through dissociation of ozone) forming ClO^-, the conjugate base of the weak hypochlorous acid. Because the weakness of the hypochlorous acid, formation of ClO^- in water, results in its recombination with H_3O^+ and making the solution more basic, or even the **plasma alkaline solution**.

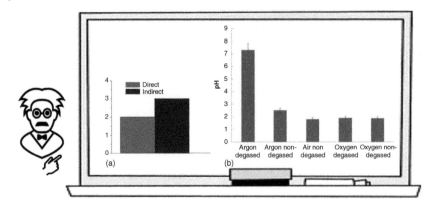

Figure 29.10 Metastable plasma acids: (a) Comparison of pH for deionized water after direct activation (charges enter water) and indirect activation (charges do not enter water) after 15 min of DBD treatment; (b) acidity of deionized water after plasma degassed and non-degassed water treatment for 15 min in argon, oxygen, and air plasma.

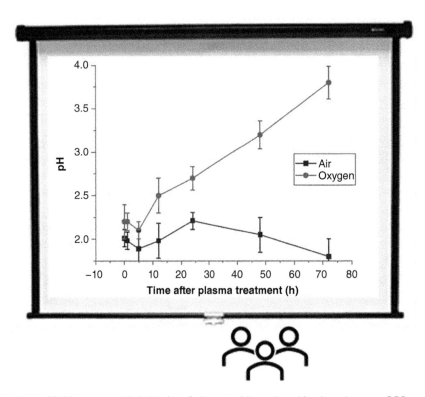

Figure 29.11 Metastable behavior of plasma acids produced in air and oxygen DBD: variation of pH of plasma-treated deionized water as a function of time after the treatment (0 indicates measurement immediately after the treatment).

Summarizing, the plasma-activated water, water solutions, and plasma-activated non-water-based solutions, all are quite metastable. It concerns not only peroxides, peroxynitrous, and nitrosylated (see later in this lecture) compounds, but even acidity, which can also be temporary. If the plasma-activated water is very stable, most probably it is already converted into nitric acid. The metastability and aging effect of the plasma-treated liquids can be stabilized, however, by lowering storage temperature and using some special stabilizers. Going to other extreme, one can imagine the plasma-activated water extremely aggressive initially due to peroxide, ozone, and acidity, but after some sufficient time completely self-neutralized to be returned safely back to the nature without any residuals.

29.13   Plasma Misting Fundamentals: Mist Droplet Size Distribution, Mist Particulates Explosion and Evaporation-condensation Effects

Conventional fresh produce technologies utilize bulk washing of produce in large volumes of water with the addition of active chemicals such as hydrogen peroxide and sodium hypochlorite. As the microbial burden in this water increases so does the incidence of cross-contamination. To minimize the generation of large reservoirs of contaminated liquids, the industry is transitioning toward usage of misting systems that contain high concentrations of antimicrobial agents, but they utilize significantly smaller volumes of liquid to coat the surface of produce in bulk, see Walia et al. (2017), McCartney and Lefsrud (2015). The standard misting nozzles tend to generate droplet sizes smaller than 20 μm to make them fill the treatment area without falling due to gravity and uniformly wash and disinfect the produce, see Section 29.11.

Plasma misting, illustrated in Figure 29.8 and mentioned in Section 29.11, assumes that the droplets cross the plasma zone where they are charged and plasma-chemically activated. The electric charges of the plasma mist particulates prevent their coagulation and stimulate their electrostatic explosion when electrostatic forces dominate over surface tension. It leads to the extra-small submicron sizes of the plasma mist droplets and therefore to their extremely high chemical and biological activity. Additionally, the plasma-treated mist has been shown to penetrate multiple layers of storage fresh produce much more effectively than other relevant methods.

There are two major **physical processes affecting the droplet size distribution** occurring during the plasma treatment of mist: particles explosion, and particles evaporation/condensation. The **electrostatic mist explosion** happens when a particle size r becomes smaller than the critical one r_{cr}, and the pressure of surface tension (γ) squeezing the droplet ($\sim 1/r$), becomes lower than negative electrostatic pressure ($\sim 1/r^2$) induced by negative charge of the particle under the floating potential (which is independent on radius, see Lecture 7). The critical radius of this Coulombic fission known as the Rayleigh instability is:

$$r_{cr} = \frac{\varepsilon_0 T_e^2}{4\gamma e^2} \ln^2 \frac{M_i}{2\pi m_e} \tag{29.4}$$

For an electron temperature of 3 eV and surface tension of 72.8 mN·m^{-1}, the critical radius of a water droplet is about 2 μm. Thus, smaller particles are expected to explode. This is in a good agreement with experimental results observed in the DBD mister illustrated in Figure 29.8, see Ranieri et al. (2019), where the mist particle size distribution has been measured using the laser diffraction droplet size analyzer.

Figure 29.12 illustrates that *while the initial distribution has a single peak (without plasma, droplet sizes 3–5 μm), the DBD plasma charging of the mist results in the two-picks distribution. The explosion of droplets reduces their size about 10 times, to the submicron sizes about 0.2–0.3 μm, 200–300 nm relevant volumes of individual droplets become about 1000 smaller, while total volume of the particulates does not change significantly. The droplet size reduction due to the explosion is prominent on smaller droplets* as shown in the normalized distribution in Figure 29.12. Here, the onset diameter of Coulombic fission is between 2.9 and 3.4 μm, corresponding to radius sizes of 1.45–1.7 μm, respectively. Therefore, droplets up to approximately 1.7 μm in diameter are susceptible to Coulombic fission during the DBD plasma treatment in this system. The size distribution of the submicron charged droplets is relatively uniform (about 0.2–0.3 μm, 200–300 nm) at the outlet of the discharge. This indicates that these smaller submicron droplets (the explosion products) are not sufficiently charged and therefore stable. It is interesting to note that if the charging of mist particles during the DBD plasma treatment is blocked by a biased grid, the explosion effect is obviously suppressed, and the droplet size stays non-perturbed by DBD plasma.

The second major physical process affecting the droplet size distribution occurring during the plasma treatment of mist is related to the **mist particles evaporation and condensation**. Plasma treatment of mist particles stimulates their evaporation and relevant decrease of their sizes and volumes. In contrast to the water particle explosion, the total liquid volume significantly decreases in this case with corresponding increase of humidity which can approach 100% and stimulate reverse process of vapor condensation on plasma ions and hydrophilic ROS and RNS. Two interesting physical effects accompany this plasma-induced evaporation/condensation process: (i) droplets shrink during the water evaporation but not completely disappear because of growing concentration of the water-dissolved nonvolatile plasma active species (charges, ROS, RNS); the droplet evaporative shrinking stabilizes at the chemically active cluster sizes about 20–30 nm; (ii) vapor condensation on plasma ions and hydrophilic ROS and RNS leads also to formation

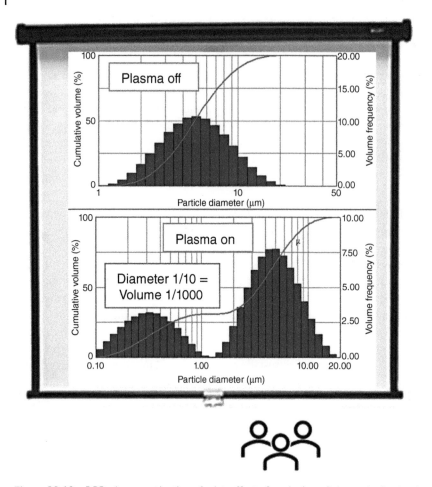

Figure 29.12 DBD plasma activation of mist: effect of explosion of charged mist droplets.

of the chemically active clusters (hydrated ions and active plasma species) with the same sizes of about 20–30 nm (if humidity stays below 100%). Thus, both plasma-induced evaporation-condensation effects lead to formation at the chemically active clusters with sizes about 20–30 nm, observed in mist particles size distribution functions and significantly contribution to the chemical and biological activity of plasma-activated mist (He et al. 2022; Fridman 2023).

 Thus, summarizing, see illustration in Figure 29.13, *DBD plasma (as well as other cold plasmas) can significantly change the mist particle size distribution, which (due to the droplet explosion and evaporation-condensation effects) results in formation of two types of plasma mist particles, microdroplets (with sizes approximately 0.2–0.3 μm) and molecular clusters (with typical sizes being 20–30 nm), both contributing to the fresh produce washing and disinfection, as well as to other applications.*

29.14 Application of Plasma Misting to Fresh Produce Disinfection and Other Sanitization Technologies

The microsecond-pulse DBD-based plasma mister described above (see schematic in Figure 29.8) has been effectively used for inactivation of microorganisms in different applications. It is important that plasma mist not only operates as an effective sterilizer but also significantly absorbs ozone generated by nonthermal plasmas. Due to evaporation and therefore self-concentration of the mist microdroplets, their acidity can be even higher than that of plasma-treated bulk water (see Section 29.12) reaching pH = 1. High water acidity increases ability of the microdroplets to absorb ozone, which reduces technological risk factors, and very favorite most technological processes.

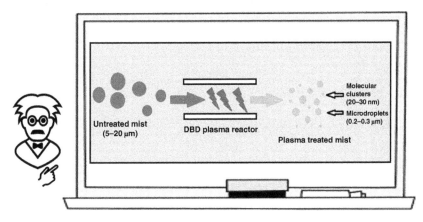

Figure 29.13 Plasma misting, evolution of droplet sizes due electrostatic explosion of particulates and droplet evaporation-condensation effect.

For demonstration of efficacy of the DBD plasma mister, this **100-l system was used for growth inhibition of *E. coli*** when exposed to the plasma-treated mist for less than 30 min, which is quite competitive, see Ranieri (2019). 26-kV pulses have been used at 3500 Hz with 100% duty cycle to three concentric electrodes. These pulses have a 5-kV ns^{-1} rise time, 2-μs pulse duration, and are single-polarity positive. Approximately 2.5 cm of tap water covered 14 ultrasonic nebulizers for droplet generation inside of the bucket. Compressed air, unfiltered, carried the droplets from the bucket to the plasma discharge at a flowrate of 25–50 SLPM. The droplets were elevated vertically prior to entering the electrodes, to remove sufficiently large droplets affected by gravity (and maximize the droplet explosion effect, see Section 29.13). Three identical DBD electrodes were used with the total plasma treatment cross-sectional area roughly equal to the tube used to deliver the mist (approximately 80 cm^2). Thus, there should be no pressure drop or mass flow restriction between the tubes and the discharge zone. Once in plasma, droplets were treated in a 12.5-cm discharge zone and enter the refrigerator through the top inlet. It takes 2–4 min for the plasma-treated mist to fill the 100-l refrigerator at 50 and 25 SLPM, respectively, see illustration in Figure 29.14. The approximate power per square centimeter of the DBDs is 1 W cm^{-2}. For the discharge volume, the total discharge power was about 50 W. For treatment at 30 min and approximately 700 ml (the nebulizers need 100 ml h^{-1} each), this equates to about 0.036 kWh l^{-1}. *E. coli* plated on TSA plates was treated for 30 min at 50 SLPM and compared to untreated control

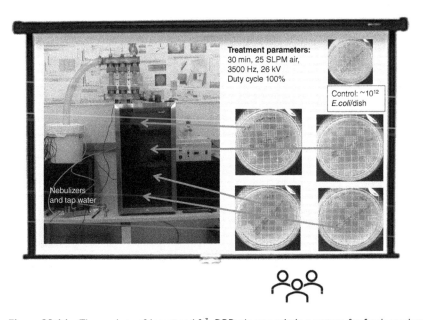

Figure 29.14 Three-pipe refrigerator 4 ft^3, DBD plasma misting system for fresh produce processing.

Figure 29.15 DBD plasma mister sterilization of fresh produce: strawberries, kale, spinach, and lettuce; 6-log reduction of *E. coli* in 10 min.

cultures. *E. coli* plates were placed on different shelves (see Figure 29.14); as seen from the figure, plasma-treated mist inhibited *E. coli* growth on all shelves when compared to the untreated control.

The same 100-l microsecond pulsed DBD plasma mister demonstrated very high efficiency of **sterilization of fresh produce, specifically of strawberries, kale, spinach, and lettuce**. The fresh strawberries, kale, spinach, and lettuce were inoculated with 5, 10 µl drops of 10^6 *E. coli* and placed in the DBD plasma mister (10 SLPM, 9 µs DBD pulsed plasma with frequency 3500 Hz), see Figure 29.15. As a post-treatment procedure, the fresh produce was placed in individual stomacher bags and stomached for 120 s; 500 µl was taken from each stomacher bag and plated for analysis. The impressive 6-log reduction of *E. coli* has been achieved after 10 min of the fresh produce treatment with mist. Efficacy of the plasma-mist-based disinfection technology has been demonstrated not only for storage of the fresh produce but also for disinfection of growing plants, including those growing in the green houses (Fridman 2023).

The DBD misting approach has been also effectively applied for **disinfection and regeneration of the personal masks used against the COVID-19 pandemic**. Due to the shortage of personal protective equipment (PPE) during the COVID-19 pandemic, the interest and demand for sterilization devices to reuse PPE has increased. For reuse of face masks, they must be effectively decontaminated of potential infectious agents without compromising their filtration ability during sterilization. It was effectively accomplished by using DBD plasma-mister like the one described above but engineered slightly differently to meet the purpose, see Figure 29.16 (He et al. 2022). MS2 and T4 bacteriophages were used to conduct the decontamination tests on two types of N95 respirators. Results showed at least a 2-log reduction of MS2 and T4 on N95 respirators treated in one cycle with 7.8% hydrogen peroxide plasma-activated mist (PAM) and at least a 3-log reduction treated in 10% hydrogen peroxide PAM. In addition, it was found that there was no significant degradation in filtration efficiency of N95 respirators (3M 1860 and 1804) treated in 10% hydrogen peroxide PAM found after 20 cycles. In terms of re-useability of masks after treatment as determined, it was shown

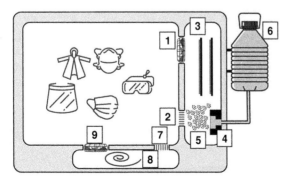

Figure 29.16 53-L Plasma mister for mask regeneration: 1. Fan forcing the air into the plasma chamber. 2. Exit from the plasma chamber activated droplets are introduced to PPE. 3. Plasma is generated with a 20 kHz, 3000 V DBD. 4. Ultrasonic nebulizer ~ 5 µm. 5. Water microdroplets dense fog. 6. Each cycle consumes 1 ml of DI water or other solutions, 750 ml container. 7. For the heated air cycle, the air is sucked in from the chamber with the PPE. 8. The air moisture is heated to 50 °C to facilitate removal of the residual reactive chemistry and drying of the PPE. 9. Warm air is forced into the chamber with the PPE by a fan.

that the elastic straps of 3M 1804 were fragmented after 20 treatment cycles rendering them unusable, while the straps of 3M 1860 were not negatively affected even after 20 disinfection cycles. Without any doubt, the plasma-activated mist can be effectively applied for multiple different applications.

29.15 Large-volume Produce Washing with Plasma-activated Water, Effect of Treatment Temperature

While disinfecting and water saving characteristics of the plasma-activated mist are impressive, this technology is not applicable (so far) for washing and sanitization of large volume of produce, where required flowrates of water are at least from hundreds to thousands of gallons per minute (GPM). Keeping in mind that wash water is not chemically clean and sufficiently transparent, electric energy cost of the required water activation is at minimum $3-10\,kJ\,l^{-1}$, see Section 28.12. For flow rates of 100–1000 GPM, it requires electric power about at least 100 kW. These discharge powers are too high for conventional DBD, corona, and other similar cold atmospheric pressure plasma sources, but achievable for the transitional "warm" discharges, like gliding arcs, etc., see Lecture 14. Application of high-power gliding arcs and other transitional "warm" discharges for effective oxidative water activation is supposed, however, to overcome the **negative temperature effect on oxidative water activation**, illustrated in Figure 29.17.

The negative temperature effect on oxidative water activation is due to the kinetically competitive pathways of transformation of the plasma-induced peroxynitrite in water. As it was discussed in Section 29.11, the hydroxyl radicals OH leading to the strong oxidative sterilization effect in plasma water are not directly generated in plasma, but first converted to peroxides and only inside of water converted back to OH proving the antimicrobial effect. This conversion of H_2O_2 to OH inside plasma water occurs through formation (29.1) and decomposition (29.2) of the peroxynitrite. The peroxynitrite decomposition pathway (29.2) dominates at low room temperature and leads to the required formation of OH. The peroxynitrite $ONOO^-$ (conjugate base of peroxynitrous acid HONOO), however, is an isomer of the nitrate ion NO_3^- (conjugate base of strong nitric acid HNO_3). Conversion rate coefficient of peroxynitrite $ONOO^-$ to nitrate NO_3^- has high pre-exponential factor, but also high activation energy (Lukes et al. 2014), see Figure 29.17. Therefore, low room temperature is preferable to the formation of OH (and strong oxidative sterilization effect), while elevated temperature results in formation of NO_3^- and nitric acid without strong direct oxidative sterilization effect.

Summarizing, the plasma activation of water with significant production of OH (through peroxynitrite) for oxidative sterilization is very effective at room temperature, corresponding to cold discharges (like DBD, corona, etc.), but requires some special controls (to avoid the peroxynitrite conversion to nitric acid) when using the transitional "warm" discharges (like gliding arcs, etc.) where the treated water temperature near plasma can be elevated.

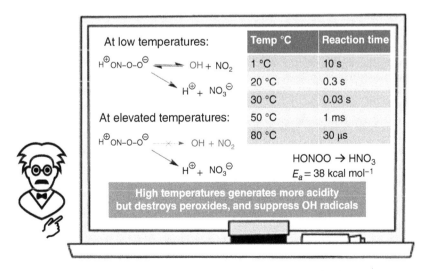

Figure 29.17 Negative temperature effect on oxidative water activation.

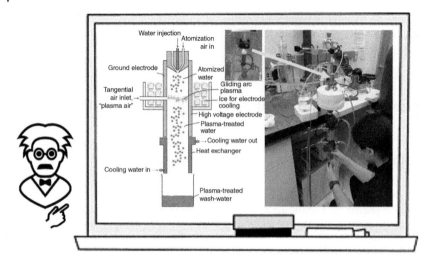

Figure 29.18 Controlling the negative temperature effect on oxidative water activation with gliding arc discharges.

The negative temperature effect on oxidative water activation for suppression of *E. coli* has been demonstrated using gliding arc activation of water with controlled dry-ice-cooling of an electrode, see Figure 29.18 (Fridman 2020). While no *E. coli* deactivation has been observed without cooling, the dry-ice-cooling permitted to achieve significant *E. coli* suppression effect, illustrated in Figure 29.19. Obviously, application of the dry-ice-cooling is not practical for industrial implementation of the wash water activation (it was done only for analysis of the effect). The relevant water temperature control has been achieved by relevant regulation of the gliding arc parameters (decreasing current, and going to stronger nonequilibrium regimes, see Lecture 14). The high efficacy fresh produce washing with the wash water activated using the temperature-controlled nonequilibrium gliding arc is going to be presented below, in Section 29.17, after preliminary consideration of the organic load effect on produce washing.

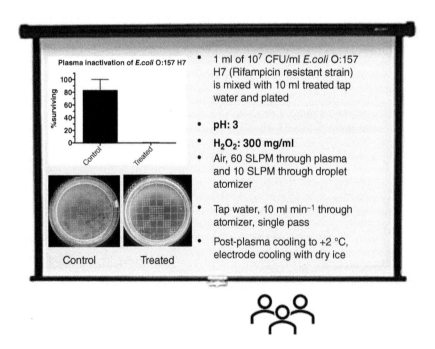

Figure 29.19 Disinfection of *E. coli* using the gliding arc-activated water with temperature control.

29.16 Produce Washing with Plasma-activated Water, Effect of Organic Load, Nonoxidative Disinfection Processes

The presence of organic compounds compromises the antimicrobial efficacy of conventional oxidation-based disinfectants, because instead of reacting with microorganisms they mostly react with the organic load. In the fresh produce washing, the organic compounds come from the product itself during washing steps. Thus, it is relatively easy to effectively clean and activate the wash water if it's not dirty, which is a kind of nonsense. Obviously, the **effect of organic load** can be "overregulated" by applying excessive amount of disinfectants, which is usually the case in conventional systems leading to chemical pollution of the wash water and produce. While it occurs with the plasma-activated wash water as well to some extent, *the antimicrobial effect of plasma-water is based not only on oxidative ROS-based sterilization but also on nonoxidative suppression of microorganisms induced by nitrogen compounds, RNS, which are less sensitive to organic load*. This **nonoxidative plasma sterilization effect** *is mostly attributed to plasma-generated NO with following nitrosylation and transnitrosylation to transfer the plasma-generated NO to bacteria suspended in a solution with organic compounds. This effect is determined by NO concentration in plasma, therefore sensitive to plasma temperature, and predominantly occurs in the transitional "warm" discharges (like gliding arcs, see Lectures 14 and 18).* **Nitrosylation, the covalent bonding of NO, to thiol groups** *is known to occur in cells, see Lecture 27. Thiols and thiol-containing proteins are naturally found in the cellular antioxidant system. While known to reduce oxidants, low molecular weight thiols such as glutathione are susceptible to nitrosylation when the concentration of nitrosative compounds is relatively high. Nitrosylated thiols, or nitrosothiols (general structure RSNO) are known to be antimicrobial through the release of NO. Thus, nitrosothiols are viewed as NO donor molecules, which prolongate the plasma-generated NO in its antimicrobial activity.* The two methods by which nitrosothiols transport NO to bacteria are through diffusion or transnitrosylation. Transnitrosylation, which can be described as:

$$RSH + R'SNO \rightarrow RSNO + R'SH \tag{29.5}$$

is the transfer of NO from one nitrosylated thiol to a free thiol, see Ercan et al. (2013). For transport of the plasma-generated NO through organic load, this may be a transfer mechanism where RSNO exchanges NO with thiols that come from the washed produce. Any low molecular weight thiol may react with a nitrosothiol through transnitrosylation or form a disulfide bridge (RSSR) that releases the bound NO to further transportation and antimicrobial activity. The tendency for disulfide formation or transnitrosylation is dependent on the steric and electrostatic factors of the NO donor. The pathway of the nitrosothiols transporting NO to bacteria is illustrated in Figure 29.20, see Ercan et al. (2013), Ranieri (2018) for more details.

The stabilization of gliding arc plasma-generated NO in plasma-treated water and solutions (including those with significant organic load) and subsequent NO delivery to and for inactivation of *E. coli* suspended in solution containing organic compounds has been demonstrated in Ranieri (2018), see Figure 29.21. Normal acetylcysteine (NAC) was used to exemplify thiols due to its low molecular weight, making it more favorable for nitrosylation. The evidence of nitrosothiol formation in plasma-treated samples was shown by absorption spectroscopy. Comparison to the standard curve generated by the mixture of NAC and sodium nitrite showed that *S*-nitrosylated NAC (SNAC) formation increased with plasma treatment time. When *E. coli* suspended in distilled water was exposed to the plasma-treated NAC, it exhibited comparable antimicrobial efficacy with plasma water. Interestingly, NAC did not reduce the efficacy or species concentrations in plasma-treated samples when compared to the plasma water. *When the plasma-treated NAC was mixed with E. coli suspended in organic load solution OLS, it inhibited microbial growth compared with its activity in distilled water alone,* see illustration in Figure 29.21. Thus, the presence of nitrosothiols in plasma-treated NAC contributed to the retention of the antimicrobial effects. Furthermore, the measured H_2O_2, NO_2^- and NO_3^- concentrations, and pH were similar between the solutions, which alluded to the role of nitrosothiols for the observed antimicrobial effects. The applicability of the plasma-treated NAC solutions (representing wash water with thiols) to address the produce cross contamination during washing was evaluated in comparison to pure plasma-treated water, see Ranieri (2018). Nitrosothiols formed from plasma treatment were demonstrated to stabilize and deliver NO to bacteria. These solutions retained antimicrobial activity for up to 7 days independent of the storage temperature. In contrast to pure plasma-treated water, this effect is not dependent upon the decay of solvated ROS/RNS, such as NO_2^-. Once the thiol is modified, the nitrosothiols extended the storage capabilities of PTW by delivering the stabilized NO

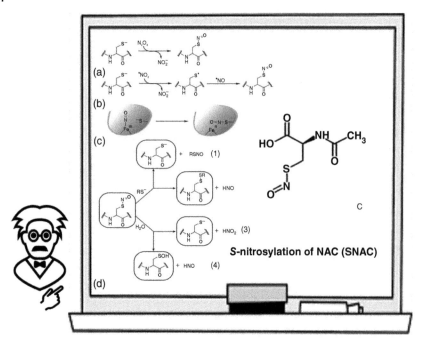

Figure 29.20 The pathway of the nitrosothiols transporting NO to bacteria for inactivation.

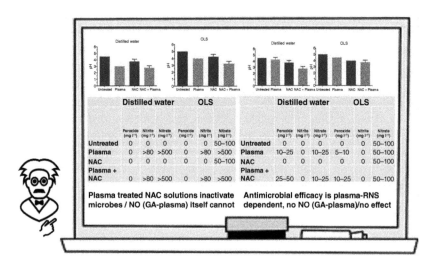

Figure 29.21 Nonoxidative plasma disinfection (suppression of *E. coli*) through plasma-NO stimulated *S*-nitrosylation of thiols.

to the suspended microorganisms. The formation of nitrosothiols is plasma-generated RNS-dependent. When nitrogen is removed from the gliding arc discharge, the solution does not exhibit significant antimicrobial activity. This indicated nitrogen was necessary to generate nitrosative species, see Figure 29.21.

Summarizing, plasma disinfection of the "dirty" wash water with organic load can be quite effective in contrast to conventional oxidative approaches (provided by strong oxidants mitigated by the organic load). Effectiveness of the plasma-induced disinfection of the wash water is due to the nonoxidative mechanisms induced by plasma-generated NO with following nitrosylation of thiols present in the fresh produce wash water. The transfer of NO in the organic load solutions OLS is determined by diffusion or transnitrosylation between SNAC and the free thiols, see illustration in Figure 29.22. There are three major pathways of inactivation once the nitrosothiol reaches the membrane of the bacteria: (i) NO is released on the outside of the membrane, which can lead to the formation of peroxynitrite and peroxidation of the membrane through OH; (ii) the nitrosothiol can be transferred intracellularly by reacting with cell-penetrating peptides, where they release NO eventually leading to

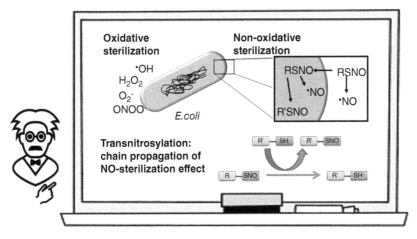

Figure 29.22   Stabilization of the plasma-generated NO sterilization effect by thiol nitrosylation.

the formation of OH; (iii) the nitrosothiol can be transferred intracellularly and undergoes a transnitrosylation with a protein required for reproduction, inactivating the microbe through a nonoxidative process. See more on the subject in Ercan et al. (2013), Ranieri (2018).

29.17 Plasma Treatment of Water Leads Not Only to Disinfection but additionally, to Enhancement of Washability of Fresh Produce

Plasma treatment of washing water is a multi-faced phenomenon. Not only it increases the sterilization ability of water by killing pathogens (like it was discussed above), but it also can significantly enhance washability, that is the ability to remove pathogens from the surface of the fresh produce. He et al. (2022) demonstrated that the plasma-activated water stimulates removing (washing out) of the microorganisms from the produce surface with further killing them already in liquid, see Kim et al. (2013). The main difficulties of inactivation on leafy produce are due to two factors. Complicated surfaces with folds make it hard for water to reach the microorganisms that hide in pores. *In the everyday life, there are two standard approaches to reduce surface tension and therefore stimulate washability: increasing the temperature and/or adding a surfactant (such as soap). Plasma treatment provides similar effect of surface tension decrease and therefore enhancement of washability without any soap and temperature increase (kind of "plasma shampooing effect").* Mechanisms leading to significant changes of physical properties of water (including the surface tension and therefore washability) are going to be discussed in the Section 29.18, while here we'll shortly focus on washability of fresh produce.

He et al. (2022) used the low-current nonequilibrium gliding arc plasma (operational arc voltage of 1–2 kV, frequency of 1–2 kHz, power 1–3 kW) to treat the wash water, see illustration in Figure 29.23. The gliding arc plasmatron is connected to a tank filled with tap water. Air is injected tangentially into the gap between two cylindrical electrodes and creates a vortex. Water is injected by a water pump into the plasmatron, passes through the plasma zone, and collected at the exit of the plasma system. Rifampicin-resistant *E. coli* O157: H7 has been used to contaminate Romain Lettuce. Plasma-activated water (PAW) was collected and cooled in a chiller to 4 °C before washing. 10 g of lettuce that was inoculated with *E. coli*. The decontamination efficiency was determined by comparing the plate count of washed and unwashed samples. Therefore, the decontamination of lettuce washed in water includes inactivation of *E. coli* on the surface of lettuce and wash-out to the water. The washed lettuce showed 0.7-log reduction in plasma-activated water compared to unwashed lettuce, while the decontamination efficiency was 0.5-log reduction when washed in tap water, see Table 29.1, so the plasma-activated water has a stronger ability to decontaminate *E. coli* on lettuce.

The total amount of DNA of *E. coli* that remained on the lettuce after washing in the gliding arc plasma-activated water or tap water was measured by qPCR. These numbers included both alive and dead *E. coli* on the surface of lettuce. More total DNA was found after washing in tap water (105, 490) than after washing in plasma-activated water (69, 757). It means that 50% more *E. coli* (either alive or dead) was washed out from the surface of lettuce when the PAW was used.

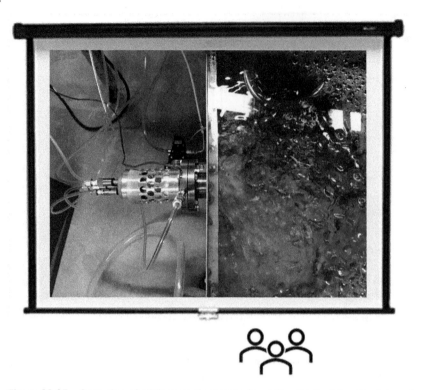

Figure 29.23 Operation of gliding arc plasmatron in wash water tank to stimulate water disinfection and enhance washability.

Table 29.1 The effectiveness of gliding arc plasma-activated water in bacterial removal.

Type of water	Washing out efficiency (log)	*E. coli* remained (%)
Plasma-activated water	0.67 ± 0.21	$24\% \pm 11\%$
Plasma-activated water with organic load	0.70 ± 0.33	$26\% \pm 19\%$
Tap water	0.55 ± 0.32	$36\% \pm 25\%$

Therefore, it was demonstrated that plasma-activated water not only kills pathogens both on the produce surface and washing water but also enhance washing ability of the water. The total remaining amount of *E. coli* (dead or alive) is significantly lower for the plasma-activated water compared with tap water, which cannot be explained by the mere presence of ROS or RNS. Explanation of this interesting effect related to plasma affecting of physical properties of water, especially, surface tension and wettability are going to be discussed below.

29.18 Plasma Treatment Effect on Surface Tension, Viscosity, and Other Physical Properties of Water

Plasma treatment effect on produce washability, and therefore enhancement of water surfactancy and decrease of surface tension, discussed above, triggered interest to plasma affecting the physical properties of water in general, see Shaji et al. (2023). It has been demonstrated that *atmospheric-pressure air plasma treatment can decrease the surface tension of water at room temperature, increases the viscosity of water at high temperatures, and lowers the contact angle of droplets on glass surfaces at room temperatures. Factors influencing these changes include plasma alteration of the mesoscopic structure of water at low temperatures and plasma additives acting as foreign particles in water at higher temperatures.* Shaji et al. (2023), produced the plasma-activated water using the nonequilibrium reverse-vortex-stabilized gliding arc system like one described in the Section 29.17.

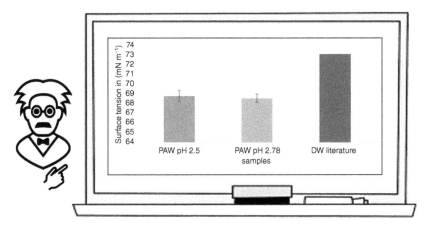

Figure 29.24 Surface tension of plasma-activated water at 18 °C (two left columns); the surface tension of distilled water (DW) at this temperature is shown for reference.

The high accuracy pendant drop tensiometry using the Open Drop software has been used to measure surface tension and the Cannon-Fenske capillary viscometer was chosen for viscosity measurements, and the contact angles were measured by the telescope–goniometer (also using the Open Drop image processing software).

Shaji et al. (2023), measured the **surface tension** of two PAW samples with pH 2.5 and pH 2.78 (characterizing the treatment dose) at a room temperature of 18 °C, see Figure 29.24. The PAW at pH 2.5 had a surface tension of 68.7 mN m^{-1}, while the PAW at pH 2.78 had a surface tension of 68.6 mN m^{-1} – both lower than the surface tension of distilled water at 18 °C, which is 73.1 mN m^{-1}. On average, the PAW displayed a viscosity 6.1% lower than that of distilled water. The relative accuracy of the measurements was determined to be 3.6%. According to He et al. (2022), the surface tension of water is lowered by the plasma-induced transition of the crystalline mesoscopic structure of water to an amorphous mesoscopic structure. Reduction in surface tension increases the surfactant behavior exhibited by a liquid, making it more suitable for cleaning and removal of particles. The mentioned transition in the water's structure is aided by the plasma lowering the mesoscopic transition temperature, which is going to be discussed in the Section 29.19.

The mesoscopic structural changes and foreign plasma additives in PAW also influence **viscosity of plasma-activated water**. The structure change at low temperatures should reduce the viscosity of PAW, while the foreign plasma additive should increase its viscosity. The effect of foreign plasma additives in increasing the viscosity of water is like how gel added to water can affect its viscosity; foreign additives can lead to increased friction in the liquid flow, resulting in higher viscosities. Since the effect of plasma results in mesoscopic structural changes in water at low temperatures, PAW should display lower to almost identical viscosity relative to water at low temperatures. This is because the viscosity-reducing effect of mesoscopic structural change at low temperatures can possibly be countered by the viscosity-increasing effect of foreign plasma additives. If the proposed mesoscopic structural change does not occur, PAW should have a higher viscosity at lower temperatures because of the foreign plasma additives. The foreign plasma additives should also cause PAW to have higher viscosity at higher temperatures. At temperatures above 35 °C, normal water has an amorphous structure, and the only influence differentiating the rheological behaviors of PAW and normal water is that of the plasma foreign additives.

The viscosity of PAW at pH 2.78 was measured by Shaji et al. (2023), from 10 to 40 °C using the Cannon-Fenske viscometer, see Figure 29.25. The kinematic viscosity of PAW at 10 °C was 1.28 mm^2 s^{-1}, compared to the 1.30 mm^2 s^{-1} kinematic viscosity of nontreated water at the same temperature. The viscosity of PAW was 1.3% lower than that of distilled water at 10 °C. The slightly lower viscosity of PAW at low temperatures supports the mesoscopic structural changes in water proposed by He et al. (2022). Under normal conditions, PAW should have exhibited a higher viscosity since the foreign plasma additives have the natural effect of increasing viscosity. The amorphous structure of PAW at low temperature might have countered the viscosity-increasing effect of the plasma additives, resulting in a lower viscosity than that of distilled water. With the increase in temperature from 10 °C, the high viscosity of normal water, caused by its crystalline structure, started to decrease due to the transition of its structure to amorphous. As the crystalline structure in water began to weaken, the relatively lower viscosity of PAW, due to its amorphous structure, became less pronounced and began to match the viscosity of normal water, as demonstrated at about 15 °C

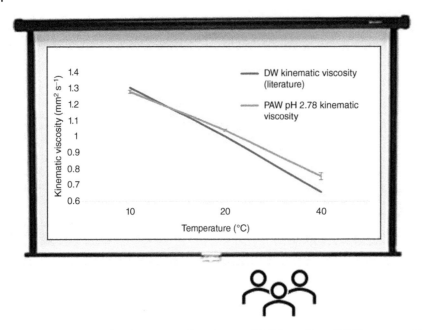

Figure 29.25 Viscosity of plasma-activated water (PAW) from 10 to 40 °C is presented by upper-right curve; the viscosity of distilled water (DW) is shown for reference.

in Figure 29.25. With further increase in temperature from 15 °C, the higher viscosity of normal water due to its crystalline structure began to decrease as its structure transitioned further into amorphous, resulting in PAW having higher viscosity than normal water. The viscosity-reducing effect of the amorphous structure in PAW no longer countered the viscosity-increasing effect of its foreign plasma additives. The kinematic viscosity of PAW at 20 °C was 1.04 mm^2 s^{-1}, while that of normal water was 1.00 mm^2 s^{-1}; at 20 °C, the viscosity of PAW was 2.19% higher than that of normal water. Following this trend of increased viscosity in PAW at higher temperatures, PAW at 40 °C had a kinematic viscosity of 0.76 mm^2 s^{-1}, while that of normal water was 0.66 mm^2 s^{-1}, meaning that PAW had a 12.8% higher viscosity than normal water at 40 °C. At 40 °C, the high viscosity caused by the crystalline structure was minimal in normal water, as its mesoscopic structure might have transitioned very close to amorphous; therefore, the higher viscosity demonstrated for PAW, which also has an amorphous structure, could be attributed to the foreign plasma additives.

The **contact angles made by PAW** at pH values of 2.47, 2.68, and 2.85 on a glass microscope slide are shown in Figure 29.26 (Shaji et al. 2023). The contact angle made by a distilled water droplet of the same volume in the same

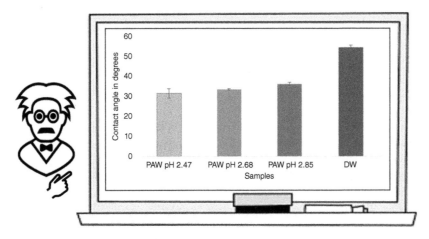

Figure 29.26 Gliding arc plasma water treatment effect on contact angle; contact angle made by PAW and distilled water (DW) on glass slides.

setup is shown for accuracy indication. The measurements were conducted at a room temperature of 18 °C. PAW droplets make smaller contact angles on glass surfaces than water droplets, by an average of 20°, or 36%. At lower pH or higher plasma production power, PAW makes smaller contact angles. Therefore, the plasma treatment increases the surface energy during interaction between the glass surface and the water. The contact angle is an indication of the adhesive and cohesive forces exhibited by the liquid. If the adhesive force of a liquid is high relative to its cohesive force, the liquid will wet the surface more, resulting in a lower contact angle. If the cohesive force of the liquid is high relative to its adhesive force, the liquid will wet the surface less, resulting in a higher contact angle formed on the surface. Since plasma activation resulted in water forming smaller contact angles, it increased the adhesive force of water during contact with glass. Interesting that similar effect can be achieved not by plasma treatment of water but by plasma treatment of substrate, see Section 26.11.

29.19 About Mechanisms of Plasma-induced Changes of Physical Properties of Water

The mesoscopic structure of normal water is nano-crystalline at temperatures below 35 °C, and it transitions to being amorphous at temperatures between 35 and 60 °C. The nano-crystalline structure is characterized by high surface tension and high viscosity, while the amorphous structure is characterized by lower surface tension and viscosity relative to the nano-crystalline structure. One example of differences in the physical properties of water with different mesoscopic structures is that hot water is more effective in cleaning applications than cold water, due to the lower surface tension and higher surfactancy of hot water's amorphous structure compared to cold water's nano-crystalline structure. Plasma water treatment can lower the temperature required for mesoscopic structural changes in water.

According to He et al. (2022), Shaji et al. (2023), plasma activation and therefore adding ROS and RNS lowers the free energy F of water:

$$F = U - T\sigma + q(qV_s + (1 - q)V_r)$$
$$= qE_s + (1 - q)E_r + (qV_s + (1 - q)V_r)P + k_B T \left(q \ln \frac{q}{g_s} + (1 - q) \ln \frac{1 - q}{g_r} \right) \quad (29.6)$$

where, U is the internal energy of water, σ is the entropy, q is the percentage of structured state, $E_{s,r}$ are the specific energies of the structured and random states, respectively ($E_s < E_r$), $V_{s,r}$ are the specific volumes, $g_{s,r}$ are the statistical degeneracy values ($g_s \ll g_r$), T is the temperature, P is the pressure, and k_B is the Boltzmann constant. When ions are added, it changes the fluid's free energy, which in the presence of ions contains an additional term, the Debye–Huckel term (U_{DH}):

$$U_{DH} = -\sum_{i=1}^{M} \frac{N_i}{2} \frac{z_i^2}{4\pi\varepsilon_r\varepsilon_0} \frac{e^2 K}{1 + K\alpha_i}; K^2 = \frac{2e^2}{\varepsilon_r\varepsilon_0 k_B T} \sum_{i=1}^{M} z_i^2 n_i, \quad (29.7)$$

where i denotes the type of ion species, N_i is the ion's concentration, α_i is the ion's radius, z_i is the charge of an ionic species, e is the charge of the electron, M is the total number of ionic species, ε_r is the dielectric permittivity of the medium, ε_0 is the dielectric permittivity of a vacuum, and K is the inverse of the Debye screening length. The ions are unlikely to penetrate the clusters; thus, in the first approximation, their impact is proportional to the percentage of water in the amorphous phase. From this, we can derive the free energy of plasma-activated water (F_{DH}), modified with the Debye–Huckel term as follows:

$$F_{DH} = F + (1 - q)u_{DH}. \quad (29.8)$$

From these equations, plasma thermodynamically "favors" the amorphous state of water; thus, the transition temperature decreases. Importantly, the plasma-activated water presents interesting "nonideal" behavior as a solution. An **ideal solution** is a solution whose properties change in proportion to the concentration of solute added to it. As PAW displays changes in properties (i.e. 6.1% for surface tension, 1.3–12.6% for viscosity, 36% for contact angle) that are significantly higher than the concentration of active species added to it (0.01%), we can conclude that **PAW behaves as nonideal solution**.

Summarizing the plasma effect on water's physical properties, we can especially point out the decrease of surface tension and controllable change of water viscosity and contact angle. While controlling the viscosity and contact angle has very diverse range of applications, the low surface tension of PAW indicates its surfactant behavior, especially interesting for the agricultural, food processing, and general washing applications.

Surfactants are required for numerous industrial processes, including but not limited to detergents, paints, food emulsions, biotechnological processes, biosciences, pharmaceuticals, and cosmetic products. PAW can be an eco-friendly and cost-effective alternative to current products used for these applications. As PAW is antibacterial, antifungal, and has demonstrated its ability to disinfect bacteria from fresh produce, it can be an excellent washing-out agent – an industrial process that prevents disease breakouts due to microbes on produce. Surfactants are also important ingredients for the preparation of detergents and cleaning agents. PAW with surfactant properties can potentially be used in these processes as a biodegradable and eco-friendly ingredient.

29.20 Problems and Concept Questions

29.20.1 Plasma Stimulation of Plant Growth Using Plasma-activated Water

Plasma-activated water (especially produced using gliding arcs and other transitional "warm" plasma sources favorable for generation of nitrogen compounds) provide nitrates and nitrites concentration close to the optimal one for fertilization of plant growth. This plasma water can be quite acidic at the same time. How to keep the acidity in the optimal range using the plasma water for stimulation of plant growth?

29.20.2 Direct Plasma-induced Disinfection of Foods Inside of Closed Packages

How can plasma be used to provide effective disinfection inside of close packages? Which types of the nonthermal discharges are optimal for the in-package plasma disinfection? How to intensify these processes to meet the technological requirements for such conveyer-based technologies?

29.20.3 Plasma Chemistry of Agricultural Water, Oxidative Disinfection

Inactivation of microorganisms in water is usually directly provided by hydroxyl OH radicals, which react with majority of organic molecules without activation energy. Why these OH radicals providing the disinfection in plasma water are not directly generated in plasma itself but usually produced in the secondary processes already in liquid phase?

29.20.4 Metastable Nature of Chemical and Biological Activity of Plasma-Activated Water

Is that possible to consider possibility of safely drinking any plasma-activated water after sufficiently long time demonstrating the nonchemical nature of this water? Can the plasma-generated water acidity be temporary? Is that possible to generate plasma-alkaline water?

29.20.5 Plasma Misting for Fresh Produce Disinfection

What are major advantages of plasma misting application in food processing with respect to plasma spraying and plasma washing? Which are the major challenges of this technology?

29.20.6 Mist Droplet Size Distribution, Mist Particulates Explosion and Evaporation-condensation Effects

Using Eq. (29.4), calculate the critical size for explosion of the plasma-charged droplets at electron temperatures 1, 2, and 3 eV. Why the charged droplets explosion effect leads to bigger sub-micron particulates after plasma treatment than those generated due to the evaporation-condensation processes?

29.20.7 Effect of Treatment Temperature on Produce Washing with Plasma-activated Water

Explain why even relatively small water temperature increase can significantly suppress effectivity of plasma washing of fresh produce? Using the kinetic data presented in Figure 29.17, estimate the maximum water temperature increase allowed to sustain the high efficiency of washing and disinfection.

29.20.8 Effect of Organic Load on Produce Washing, Plasma-induced Nonoxidative Disinfection

Give your interpretation of the plasma-water-induced nonoxidative disinfection of fresh produce. How does this approach suppress the negative effect of organic load on disinfection of fresh produce? Which types of organic load are favorable for the plasma-water-induced nonoxidative disinfection of fresh produce? Why the transitional "warm" plasma discharges are the most suitable for this washing technology?

29.20.9 Plasma-induced Enhancement of Washability of Fresh Produce vs Stimulation of Direct Disinfection with PAW

Explain the qualitative difference between the plasma-Induced enhancement of washability of fresh produce and stimulation of their direct disinfection with plasma-activated water. Which of these plasma washing technologies is expected to be more ecologically friendly? Which of these technologies is primarily based on plasma-induced chemical composition of the plasma-activated water, and which one is primarily based on the plasma-induced changing of physical properties of the wash water.

29.20.10 Plasma Treatment Effect on Surface Tension of Water, It's Surfactancy, and Washability

Using the statistical thermodynamics Eqs. (29.6–29.8), give your interpretation of decrease of surface tension because of plasma treatment of water. Why the relative decrease of the surface tension can be significantly higher than relative concentration of plasma-generated species in water, which can be expected only for the nonideal water solutions? Why the decrease of surface tension results in enhancement of surfactancy and washability characterizing the plasma-activated water?

Lecture 30

Plasma Medicine: Safety, Selectivity, and Efficacy; Penetration Depth of Plasma-Medical Effects; Standardization and Dosimetry

30.1 Safety, Selectivity, and Efficacy are Key Factors in Direct Plasma Treatment of Wounds and Diseases: Prehistory, History, and Nowadays of Plasma Medicine

Long ago, in the **"prehistory" of plasma medicine**, it was noted that electric discharges in atmosphere result in "unusual smells" influencing living species, and especially human health. These ideas and observations, which are way older than *plasma medicine* and even *concept of plasma* itself, can be attributed to the experiments of Dutch chemist Martinus van Marum in 1785 with electric discharges above water. German-Swiss chemist Christian Friedrich Schönbein observed generation of the same "unusual pungent smell" and recognized it as one often following a bolt of lightning. He isolated in 1839 the gaseous chemical responsible for this "smell" and named it "ozone" (from the Greek word "to smell"). Air plasma applications in biology have been significantly strengthened in the 1850s when Werner von Siemens first used DBD plasma to generate ozone for bio-decontamination of water, see details in Lecture 18. Industrial scale of such plasma-based bio-decontamination has been achieved half-century later, when the first drinking water plasma plant was built in Nice, France. At that time, interestingly, ozone was considered as an electric discharge-generated healthy component: the Beaumont-city in California even had as its official slogan "Beaumont: Zone of Ozone."

First direct plasma application to treat diseases can be attributed to experimental works of French physicist and physiologist Jacques-Arsène d'Arsonval in the early 1900s. The relevant devices had been made feasible at that time by Nikola Tesla, who worked with extremely high-frequency currents at high voltage, creating "impressive light phenomena which proved harmless to humans in the case of direct contact with the effusions." In early 1900s Germany, the devices were further developed for the caloric treatment of patients (diathermy). Rumpf developed a device which differed from the French ones by implementing a capacitively coupled electrode consisting of a Leydener bottle which was directly applied to the patient's skin. This device can be considered the **first DBD plasma source in medicine**. Starting from the 1920s, **thermal electrosurgical devices and then argon plasma coagulator and tissue ablation devices** became an important step in medical applications of plasma. Important initial development in this direction is related to works of American scientist and inventor William Bovie in the 1920s.

 See more information on the subject "thermal plasma medicine," namely thermal electrosurgical devices, argon plasma coagulator and tissue ablation devices, in Peng et al. (2018), Stalder and Woloshko (2012), Shimizu and Ikehara (2018); as well as Fridman and Friedman (2013).

Starting from 1960s first patents and publications started to appear all over the world with a focus on cold plasma sterilization, plasma control of biological activity of microorganisms, as well as plasma regulation of biological properties of medical polymer materials. Number of relevant patents and publications started growing in the 1990s, which can be seen from that time materials of the IEEE International Conferences on Plasma Science (ICOPS), the Gaseous Electronics Conference (GEC), the International Conference on Phenomena in Ionized Gases (ICPIG), and the International Symposium on Plasma Chemistry (ISPC). Based on these initial successes, as well as breakthrough development of novel devices able to be directly applied to human body for medical treatment, the new field of PLASMA MEDICINE was born in the early 2000s on the nexus of plasma physics and modern medicine.

The key distinctive focus of the novel field of plasma medicine is focus on the direct application of nonthermal plasma to treat wounds and heal diseases. It becomes possible due to the breakthrough development in the early 2000s novel controllable and sufficiently uniform atmospheric pressure plasma discharges (especially short-pulsed

Plasma Science and Technology: Lectures in Physics, Chemistry, Biology, and Engineering, First Edition. Alexander Fridman.
© 2024 WILEY-VCH GmbH. Published 2024 by WILEY-VCH GmbH.

DBD and plasma jets, see Lecture 13). These novel discharges have opened opportunities for plasma to electrically safe interact with living tissue at sufficiently high level of plasma power required to treat wounds and heal diseases. The breakthrough development of the newborn plasma medicine happened at this point by exponential growth of involvement of medical doctors and clinical researchers. The key challenges of the newborn plasma medicine became safety, selectivity, and efficacy of the nonthermal plasma treatment of patients. At this point, in October of 2007, author of this book, Alex Fridman together with Gary Friedman from the Nyheim Plasma Institute of Drexel University organized the first International Conference on Plasma Medicine, ICPM. More than 60 physicist, chemists, biologists, and most importantly medical doctors came to ICPM-1, which were held during one week in Corpus Christi, Texas, USA. Participants were very excited about the newly born field, discussions continued till late night, see pictures in Figure 30.1. All agreed to continue these conferences biannually, number of involved researchers and medical practitioners continued the exponential growth provoking proportional growth of publications. The second conference ICPM-2 was also organized by Alex Fridman and Gary Friedman and held in San Antonio, Texas, USA in March of 2009. Number of participants exceeded 110. At the ICPM-2, a new scientific society, "The International Society for Plasma Medicine (ISPM)" has been formed to support the activities and interests of the fast-growing plasma medicine scientific community. Author of this book, Alex Fridman, has been elected the Founding President of the Society. The Founding Members of the Board of the International Society for Plasma Medicine were Jean-Michel Pouvesle (France), Farzaneh Arefi-Khonsari (France), Pietro Favia (Italy), Michael Wertheimer (Canada), Michael Kong (UK), Klaus-Dieter Weltmann (Germany), Gregor Morfill (Germany), Alex Fridman (USA), David Graves (USA), Mounir Laroussi (USA), Geoff Lloyd (UK), Satoshi Hamagouchi (Japan), Alex Dolgopolski (USA), Gary Friedman (USA), Richard Satava (USA), see Figure 30.2.

 Nowadays, after 20 years, 9 International Conferences (ICPM-10 is expected to be in 2024, in Slovenia), and countless original publications, we are glad and proud that Plasma Medicine made it strongly to the hospitals and saved already so many lives. While we surely have so many opportunities in front of us, we'll always remember the first human life saved in the nonthermal plasma surgery by a student at that time, Danil Dobrynin, now Professor of Drexel University, Nyheim Plasma Institute, see Figure 30.3 (Dobrynin et al. 2010a,b). The deadly phlegmonous eyelid defeat has been successfully treated (with a special individual permission) using the Pin-to-Hole spark discharge (PHD, see Section 14.6), the first human life saved by this plasma-medical treatment. While number of original plasma-medical publications is today countless, still for deeper review perspectives the following books focused on Plasma Medicine can be recommended: Fridman (2008), M. Laroussi et al. (2012), Fridman and Friedman (2013), Metelmann et al. (2018a,b), Toyokuni et al. (2018), Lu et al. (2019), Kuo (2019), Keidar et al. (2019), Laroussi (2020), Keidar (2020), Martinez (2020), Chen and Wirz (2021), and Metelmann et al. (2022).

Figure 30.1 1st International Conference on Plasma Medicine, Texas, USA, 2007.

Figure 30.2 The Founding Members of the Board of the International Society for Plasma Medicine, Texas, USA, 2009.

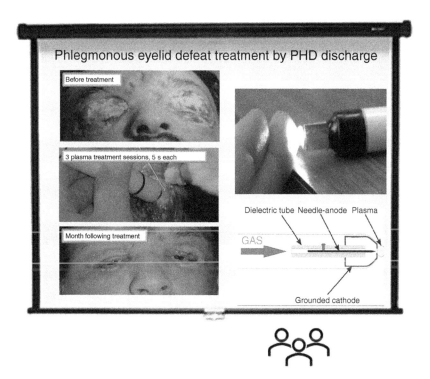

Figure 30.3 Phlegmonous eyelid defeat treatment by Pin-to-Hole spark discharge (PHD), the first human life saved by the plasma-medical treatment.

30.2 Major Discharges in Plasma Medicine: Floating-electrode Dielectric Barrier Discharge (FE-DBD), Plasma Jets, Pin-to-hole (PHD) Discharge, and Electrical Safety Issues

*Between different **nonthermal atmospheric pressure discharges applied in plasma medicine**, the most convenient and widely spread are dielectric-barrier discharges (DBD), plasma jets, and pin-to-hole (PHD) discharges, mentioned above regarding the first human life saved by plasma medicine (Figure 30.3), see Laroussi et al. (2022) and Fridman and Friedman (2013). These discharges are convenient for the plasma-medical treatment because of their good controllability, atmospheric pressure operation with temperature at the contact* with living tissue close to room temperature, and ability to operate without dangerous side effects like excessive UV-radiation, ozone production, leak current to the tissue, and so on, see Sakakita and Ikehara (2018) and Shimizu and Urayama (2018). Obviously, each plasma-medical discharge has its own advantages, challenges, and therefore specific areas of medical applications. Thus, floating-electrode dielectric barrier discharge (FE-DBD) discharges (and other DBD modifications, like surface DBDs) can treat larger surfaces, which justify their nonlocal applications in dermatology, wound healing, cosmetics, etc. Plasma jets, which can be vice versa effectively focused to a point are optimal for local highly controllable medical treatments. PHD plasma sources (as well as other transitional "warm" plasma sources like microwave discharges) generating higher concentrations of NO are more suitable for wound healing and other medical treatment where NO therapies are effective.

Physics and engineering of all these plasma-medical discharges have been discussed in detail in Lectures 13 and 14 (Sections 13.4–13.10, and 14.6). Here we should focus on the plasma safety and treatment dose-related issues, which are easier to present analyzing a specific DBD configuration, called the **FE-DBD** actively used in plasma medicine, see Fridman et al. (2007a–e, 2008).

Developed at the Nyheim Plasma Institute of Drexel University, the FE-DBD is one of the most applied cold plasma sources in plasma medicine (see Lecture 1, Figure 1.4). The FE-DBD plasma is ignited between two electrodes: one electrode under high voltage is covered by a dielectric barrier, and the living tissue itself is used as a second electrode. Let us first focus on the **electrical safety of the FE-DBD plasma source**, as an example, keeping in mind that other major plasma-medical sources are less sensitive to these issues. The principle of operation of the FE-DBD plasma source can be explained with the help of a relatively simple model. Consider the insulated DBD electrode as a sphere of diameter D_{el} while the "biological" object whose surface is being treated is modeled as a sphere of diameter D_{ob}. In the absence of the object, the electrode capacitance with respect to the far away (located at infinity) ground is $C_{el} = 2\pi\varepsilon_0 D_{el}$. If the biological object being treated has a relatively high dielectric constant (such as water), it effectively expels most of the electric field from its interior when it is brought close to the electrode. From that point of view, this object behaves like a good conductor and its capacitance with respect to the far away ground can therefore also be modeled as $C_{ob} = 2\pi\varepsilon_0 D_{ob}$. The region between the object and the electrode can be modeled roughly as a parallel plate capacitor $C_{gap} = \pi\varepsilon_0 D_{el}^2/2g$ (where g is gap distance) if the gap is significantly smaller than the electrode diameter. We should note that for typical characteristic sizes:

$$\frac{C_{gap}}{C_{ob}} = \frac{D_{el}^2}{4gD_{ob}} \ll 1 \tag{30.1}$$

In the absence of any conduction current, the electrical models of the electrode by itself and the electrode near the treated object are well approximated by the circuits shown in Figure. 30.4a,b. When the electrode is removed from the ground, the magnitude of the applied voltage V is insufficient to create an electric field strong enough to cause the breakdown and discharge. However, when the object with a high dielectric constant is sufficiently close to the electrode, most of the applied voltage appears across the gap. This is because the capacitance of the object with respect to ground is much larger than the gap capacitance, and the voltage divides across these capacitors proportionally to the inverse of their size. This results in a strong electric field in the gap, which can then lead to breakdown and discharge. The electrical circuit model can be further refined by considering the nonlinear resistance and the capacitance of the plasma created in the gap; the resulting circuit refinement is shown in Figure 30.4c. The refined circuit does not change the main conclusion that most of the applied voltage appears across the plasma gap. At about 10 kHz, the

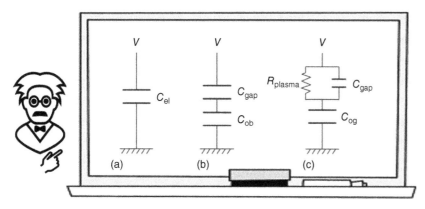

Figure 30.4 Illustration of the FE-DBD equivalent schematic: (a) electrode itself, (b) electrode near the treated biological object, and (c) plasma discharge on the treated object.

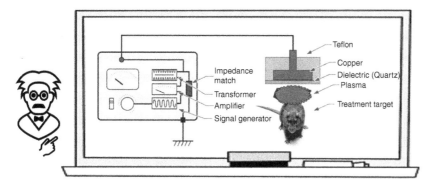

Figure 30.5 General schematic of FE-DBD plasma-medical treatment of living tissue.

following circuit parameters typical for FE-DBD can be estimated assuming that body diameter is roughly 1 m, the electrode diameter is about 25 mm and the gap is about 1 mm: the gap capacitance is 4 pF; capacitive resistance (impedance) of the gap is 4.2 MΩ; plasma resistance is 5–10 MΩ; the biological object capacitance is 50 pF; and the capacitive resistance (impedance) of the object is 300 kΩ. Electrical safety of the biological object being directly treated by FE-DBD plasma is ensured because the current which the power supply delivers is less than 5 mA. Although such currents may cause some mild discomfort, they do not cause muscle or cardiovascular malfunction in a human and are therefore deemed safe by the US Occupational Safety and Health Administration.

The simplest FE-DBD system is based on a conventional dielectric barrier discharge and is basically a system driven by AC high voltage applied between two conductors where one or both are covered with a dielectric to limit the current and to prevent transition to an arc. Amplitude and waveform of the high-voltage signal are obviously quite important, which is discussed in the following section. A simplified schematic of the plasma-medical treatment setup is shown in Figure 30.5. Here, the signal of any frequency, amplitude, and waveform is current-amplified and voltage is then stepped up in the transformer, which is then connected to the "powered electrode." The electrode is basically a conductor covered by a dielectric. Plasma is then generated between the surface of the dielectric and the treatment target. As we discussed above, the electric safety issues related to the FE-DBD plasma sources, and other crucial DBD safety issues related to plasma uniformity and plasma treatment dose are going to be discussed below.

30.3 FE-DBD Plasma Uniformity and Plasma Controllability for Safe Treatment of Living Tissue

One of the key issues for the safe plasma-medical application of FE-DBD in air is its uniformity. Voltages applied in DBD are very high, however, the dielectric barriers surely limit the total current to the values safe for plasma-medical applications, see Lecture 13. While the total DBD currents are safe, the discharge nonuniformities (in filamentary mode) result in high current densities making the plasma-medical treatment not comfortable for patients. Only sufficiently uniform-plasma modes of the FE-DBD in air meet the plasma-medical safety requirements. The most promising method of uniform plasma generation in DBD is the application of very short pulses, shorter than the time required to create the DBD filaments, see Section 13.5. The way in which the voltage is delivered to the electrode can be quite important in FE-DBD; the main issue is the rise time of the voltage. In dielectric barrier discharges at atmospheric pressure, breakdown occurs following the spark breakdown mechanism. The initial electron avalanche transitions to streamer occur in approximately 20–60 ns. If the voltage is increased slowly (as is the case with sinusoidal excitation wave, about 1 V ns^{-1}), the number of streamers occurring during a single cycle can be quite high. Since the initial streamer forms a pre-ionized channel, the probability of the next streamer striking in the same position is increased. In the case of a sinusoidal excitation wave, multiple streamers strike at the same position creating a rather energetic channel called a microfilament. The temperature in this channel can locally approach couple hundred degrees, which is non-acceptable for plasma-medical treatment.

 *To solve this problem of FE-DBD filamentation, it is helpful to employ pulses rather than a sinusoidal wave, allowing for a voltage rise time of at least 5–10 V ns^{-1} and a pulse duration of at maximum few μs. These discharges are usually referred to as the **microsecond-pulsed FE-DBD**. The time between pulses is sufficient for the gas to return to the initial conditions so that, during the next pulse, streamers again strike randomly.* The relevant power supplies operate based on voltage amplification: a low voltage pulse or a sinusoidal wave is generated, and the current is amplified and passed through a transformer with a high turns' ratio. A high-voltage signal is then taken from the transformer's secondary coil and collected to the load. In filamentary DBD modes, the temperature in the filament can rise much above room temperature and potentially damage the surface being treated, particularly in the case of sensitive surfaces such as polymer surface treatment or in biological applications. During a microsecond pulse, the gas does not heat up in DBD. However, there is sufficient time for streamer formation and development into filaments. This way the discharge is more uniform than in the case of a sinusoidal excitation wave, but a nonuniformity is still observed as can be seen in Figure 13.11 and discussed in Lecture 13. This way the treated surface is covered with reasonable uniformity in as little as a third of a second or less.

For some applications a more uniform plasma treatment might be needed, however. For this, a different power supply with a shorter pulse duration (high-voltage pulses with short rise time of 5–20 ns and less) and a high pulse repetition frequency (up to 2000 Hz) should be applied.

 *These uniform atmospheric pressure DBD discharges in air are usually referred to as the **nanosecond-pulsed FE-DBD**. The image of a single pulse of the nanosecond-pulsed FE-DBD plasma in air is shown in Figure 13.11 (Lecture 13) and demonstrates high plasma uniformity, absence of any filaments, and therefore safety of the relevant plasma-medical treatment. Detailed discussion and explanation of this high-level uniformity of the nanosecond-pulsed FE-DBD related to the strong discharge overvoltage and overlapping of streamers usually achieved in this regime are provided in Section 13.5.*

While the nanosecond-pulsed DBD is uniform between two parallel electrodes, there is a specific plasma-medical question of how this plasma uniformity depends on "nonuniformity" (nonuniform morphology) of the "second" electrode, which is a living human or animal tissue. Figure 30.6 illustrates that plasma uniformity is only slightly perturbed when the second electrode is not parallel to the first one. This is called sometimes the ***effect of "blind" streamers***, *assuming that the self-propagating streamers are not sensitive (not seeing, being "blind") to the second electrode before they approach him very closely. It explains the possibility of uniform FE-DBD plasma-medical treatment of relatively large areas of living tissue of humans and animals. Also, it opens possibility to make sophisticated shapes of the "first" (high-voltage) FE-DBD electrode.* Figure 30.7 illustrate, for example, generation of uniform plasma using a semispherical FE-DBD electrode. As a reminder from Lecture 13 (see Section 13.6), *DBD power and therefore plasma-medical treatment intensity can be controlled by thickness of dielectric barrier,* ***the thinner barrier, the more powerful discharge, and higher treatment intensity***. *Thus, application of the semispherical FE-DBD soft electrode permits also controlling the plasma-medical treatment intensity by simply changing position of the electrode with respect to the treated tissue.*

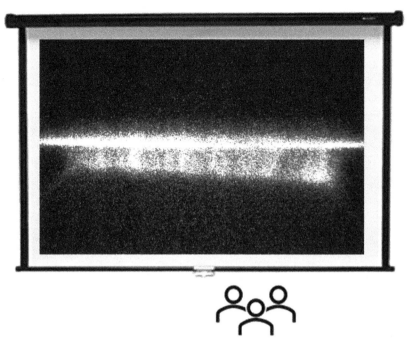

Figure 30.6 Effect of "blind" streamers in nanosecond-pulsed FE-DBD; influence of the second electrode morphology on the nanosecond-pulsed DBD plasma uniformity.

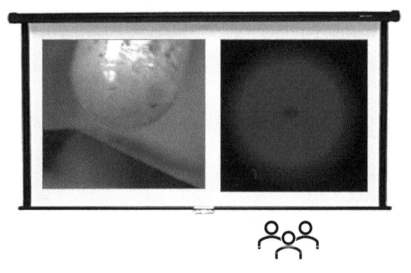

Figure 30.7 Nanosecond-pulsed FE-DBD with semispherical electrode: controllability and uniformity of the discharge.

30.4 Indirect Plasma Application for Medical Purposes: Plasma Pharmacology, Use of Plasma Activated Aqueous Media

The of plasma-medical treatments varies from killing bacteria or some malignant cells to diminishing inflammatory reactions or enhancing cell proliferation and tissue healing. In many cases, like it has been discussed in Lecture 27, the organic medium surrounding the cells plays an important role in mediating the observed effects of the plasma treatment of cells and living tissues. This is unsurprising as it is well-known that organic molecules in aqueous phase can be modified (primarily oxidized, but also nitrosylated) by plasma created in the gas phase above or within the fluid. It has also been widely observed that water treated by plasma undergoes a change in acidity and composition (Lecture 29).

These facts suggest that *plasma can be used not only for direct treatment of living systems but also to modify the medium which could be used separately for treatment of cells and tissues. In this sense, the medium treated by plasma can be viewed as a pharmacological substance. Such treatment is referred sometimes as **plasma pharmacology**,* see Fridman and Friedman (2013). The key issue in this regard is the stability of the plasma-treated medium and its effects on living systems. The ability to store the effects of plasma treatment in the form of plasma-treated material may have advantages in certain medical applications. In some cases, it may be difficult, for example, to apply plasma to certain surfaces (e.g. intestinal surfaces). On the other hand, applying plasma-treated fluids to such surfaces can be substantially easier. Moreover, controlling the material that delivers plasma treatment may provide an additional degree of control over the effect of the treatment.

Let us start with **biomedical effect of the plasma-treated aqueous media**, its stability, and the stability of its effects on living systems. As it was discussed in Lecture 29, various types of plasma treatment of water result in oxidation of organic molecules in water. In plasma pharmacology, the goal may be to find organic molecules that can enhance the effect of plasma-treated water alone or can allow longer-term storage of the plasma treatment effects. We can consider various types of organic molecules in the search for enhancement of desired effects. As a very interesting example in this regard, we can consider the effects of treated deionized water in comparison with phosphate-buffered saline (PBS) and with PBS plus *N*-acetylcysteine (NAC) solution. The specifics of plasma activation of NAC for biomedical purposes have been already discussed regarding the effect of non-oxidative sterilization in treatment of wash water due to *S*-nitrosylation of NAC (NO-based conversion of NAC to SNAC, see Section 29.16). In the frameworks of plasma medicine, these processes are important not only for sterilization purposes (in wounds, and medical equipment), but also for wound healing because prolongation of NO lifetime and deep penetration in tissue lead to intensive angiogenesis.

The consistent analysis of these effects has been accomplished by the group of S. Joshi in Drexel University, see Ercan et al. (2013, 2016). The DBD plasma treater of aqueous media in this case is like the one illustrated in Figure 29.7a. Briefly, the bottom of the insulated electrode is positioned 1.5 mm above the surface of liquid sample, and the discharge is generated by applying alternating polarity 1.7 kHz pulsed voltage of about 17 kV, pulse width 1.5 μs, and rise time 5 V ns^{-1}. The volume of the treated liquid is 1 ml, and the sample holder has the area such that the liquid is always about 0.6 mm in thickness. It was demonstrated that all three plasma-treated liquids (NAC solution, PBS, and deionized water) have strong antimicrobial properties. Less than 3 min of the plasma treatment generated sufficient reactive species in the liquids for complete inactivation of planktonic *Escherichia coli*. Even by the end of 2 min, there was a significant inactivation of either by NAC solution ($p < 0.05$), PBS ($p < 0.05$) or de-ionized water ($p < 0.05$) when compared to respective controls (0 min or no treatment). Plasma-treated NAC solution was the most powerful of all three in carrying antimicrobial effect. After the treatment for 1 min (14.5 J cm^{-2}), it inactivated a highly significant amount of *E. coli* ($p < 0.05$) compared to the same time points for DIW or PBS alone. During exposure to plasma-treated liquid, it was found that bacteria (107 CFU ml^{-1}) were inactivated in their free-floating planktonic form. Experiments were therefore carried out where *E. coli* cell suspensions of various CFUs were exposed to the plasma-treated liquids. A range of NAC concentrations (1–20 mM) with or without plasma treatments were tested to determine whether NAC solution upon plasma treatment inactivates common pathogens such as *Staphylococcus aureus*, *Staphylococcus epidermidis*, *Acinetobacter baumannii*, *Candida albicans*, and *Candida glabrata* in addition to *E. coli*. The observations demonstrated that 5 mM is sufficient for inactivation and demonstrated a linear relationship during inactivation studies in colony assays. Planktonic form studies showed that 3 min of plasma treatment makes the NAC solution completely inactivate all the pathogens tested. Most of the pathogens in their biofilm form were almost equally sensitive to the biocidal effect of treated NAC solution.

Summarizing, Ercan et al. (2013, 2016), demonstrated the existence of a plasma pharmacological effect of 5 mM and higher concentrations of NAC. This amino acid derivative significantly intensified the antimicrobial effect of the plasma-treated deionized water and PBS solutions. In addition, the plasma-treated NAC solution demonstrated that its effect does not diminish with storage time of several months.

30.5 Plasma Pharmacology: Use of Plasma Activated Biological Media (PAM), Plasma Activated Lactate Solution (PAL), and Gels for Medical Purposes

As it was discussed in Lecture 29 and in the previous section, plasma-activated water, and solutions are effective in disinfections of different surfaces including biological surfaces like for example fresh produce and living tissues.

Even cancer cells are not the exception and can be considered somewhat as "infection" to be inactivated. Thus, Sato et al. (2011), have shown that plasma-treated medium produced hydrogen peroxide sufficient to kill cervical cancer, HeLa cells. Tanaka et al. (2011, 2014), demonstrated that not only direct treatment of cells with plasma but also **plasma-activated medium (PAM)** induced apoptosis on glioblastoma brain tumor cells. Plasma-treated water generally shows a decrease in pH; however, plasma-activated biological medium (PAM) is buffered in pH, and PAM did not experience the decrease in pH, which is positive for the plasma-medical treatment. When plasma treatment of cells relevant to plasma medicine has been discussed in Lecture 27, it was pointed out that the cells in most of cases are cultivated and stored in the special biological media. For this reason, treatment of cells using the preliminarily activated PAM can be seen as a very logical approach. The natural next step after PAM treatment of cells is treating this way animal and human living tissues, which is going to be considered in next two lectures regarding the plasma-medical treatment and healing of specific diseases, in particular cancers.

 The biological medium (or growth medium) is the solid or liquid substratum on which, or in which, cells or organ explants can be made to grow; a medium may include well-defined factors such as salts, amino acids, and sugars, as well as less well-defined factors such as serum or blood. *Medium contains around 30 components, including nutrients, amino acids, glucose, and inorganic salts. Most of these compounds are subject of intensive peroxidation in PAM, and some of them are also subject of nitrosylation* (see *Section 29.16*) *induced by plasma treatment. Tanaka et al. (2011, 2014), demonstrated that PAM is killing glioblastoma not because plasma destructed nutrients in medium but because plasma produced toxic materials to cancer cells in medium.* When medium was treated with plasma for 1 min, the PAM killed 1000 glioblastoma cells. When the medium was treated with plasma for more than 3 min, the PAM killed even 10 000 cells. The duration of the effectiveness in PAM was estimated more than eight hours and less than 18 hours. When medium was treated by plasma for 3 min, the PAM induced apoptosis on glioblastoma brain tumor cells, while the same PAM did not affect astrocyte normal cells. Selective killing of cancer cells by the same PAM might be explained by the differences between cancer cells and normal cells in intracellular signaling network structures. PAM also showed antitumor effects in vivo experiments, and PAM killed antitumor drug-resistant ovarian cancer cells in mice, see Utsumi et al. (2013). These results suggest that PAM is a promising tool for the future cancer chemotherapy and possibly other medical procedures.

According to Tanaka et al. (2011, 2014), the **antitumor and other medical effects of PAM** are probably based on hydrogen peroxide, and more complicated organic peroxides are produced in medium. However, other factors might be important for selective killing of cancer cells as well. The whole molecular mechanisms for selective killing cancer cells by PAM remain to be elucidated. We should note that free radicals generated in PAM are generally short lifetime species, so it is not completely clear why PAM can have long time duration of effectiveness. It can be due to lower chemical activity of the organic peroxides or by contribution of the nitrosylation effects discussed in Section 29.16.

PAM composition is determined by composition of the conventional growth medium and is quite complicated and difficult to analyze, which creates some challenges with certification of PAM for use in hospitals. For this reason, it's more favorable to achieve treatment results like those with application of PAM but using plasma treatment of biological medium with less sophisticated composition. It has been achieved by Tanaka et al. (2016), by plasma treatment of Ringer's solutions for clinical applications. The Ringer's solutions are based on only four components: lactate, NaCl, KCl, and $CaCl_2$ and permitted to injections into human body for medical purposes. In vitro and in vivo experiments demonstrated that **plasma-activated Ringer's lactate solution, so-called PAL**, also has anti-tumor effects. Tanaka et al. (2016), proved that of the four components in Ringer's lactate solution, only lactate exhibits anti-tumor effects through activation by nonthermal plasma. Nuclear magnetic resonance analyses indicate that plasma irradiation generates acetyl and pyruvic acid-like groups in Ringer's lactate solution, which can be involved in healing mechanism. Overall, the plasma-activated Ringer's lactate (PAL) solution can be considered promising for chemotherapy.

 *In addition to PAM and PAL, the indirect plasma treatment of living tissue can be provided through oils, creams, and especially different kind of hydrogels. A good example of such **plasma-activated oils, creams, and gels** is plasma activation of alginate hydrogels for cold atmospheric plasma-based therapies, suggested by Labay et al. (2019). While the PAM or PAL is expected to be injected into the body, it is not a case for administration of the plasma activation of oils, creams, and gels. Generally, injection of a liquid in the body results in fast diffusion due to extracellular fluids and blood flow. Therefore, the development of efficient vehicles like alginate hydrogels which allow local confinement and delivery of ROS and RNS to the diseased site can be very effective.* Labay et al. (2019), investigated the generation of ROS/RNS (H_2O_2, NO_2^-, short-lived RONS) in alginate hydrogels using two atmospheric pressure plasma jets: kINPen and a helium needle, at a range of plasma treatment conditions (time, gas flow, distance to the sample). The physical and chemical properties of the hydrogels remain unchanged by the plasma treatment, while the hydrogel shows several-fold larger capacity for generation of ROS/RNS than a typical isotonic saline solution. Part of the ROS/RNS is quickly released to a receptor media, so special attention must be put on the design of hydrogels with in-situ crosslinking. Remarkably, the hydrogels show capacity for sustained release of the RONS. The plasma-treated hydrogels remain fully biocompatible (due to the fact that the species generated by plasma are previously washed away), indicating that no cytotoxic modifications have occurred on the polymer. Moreover, the ROS/RNS generated in alginate solutions showed cytotoxic potential toward bone cancer cells. These results open the door for the use of hydrogel-based biomaterials in plasma therapies. We can mention as a final remark that special plasma-activated media and gels can be also applied for the immune system modulation, including the vaccination effect, which is going to discuss in Lecture 32.

30.6 FE-DBD Plasma in Direct Living Tissue Sterilization

Direct plasma treatment is very effective in sterilization and other aspects of tissue treatment. The direct plasma treatment using FE-DBD (Sections 30.2 and 30.3) implies that living tissue itself is used as one of the electrodes and directly participates in the active plasma discharge processes. Figure 30.8 illustrates direct plasma sterilization of skin of a live mouse, see Fridman and Friedman (2013), and for more details Fridman et al. (2005a–c, 2006a,b, 2007a–e, 2008). FE-DBD plasma is generated in this case between the quartz-covered high-voltage electrode and the mouse as a second electrode. Direct application of the high-voltage (10–40 kV) nonthermal plasma discharges in atmospheric air to treat live animals and people requires a high level of safety precautions. Discharge current should be limited to below the values permitted for treatment of living tissue. Moreover, discharge itself should be homogeneous enough to avoid local damage and discomfort, see Sections 30.2 and 30.3. As soon as this atmospheric discharge is safe, it can be applied directly to the human body as illustrated in Figure 1.4, Lecture 1.

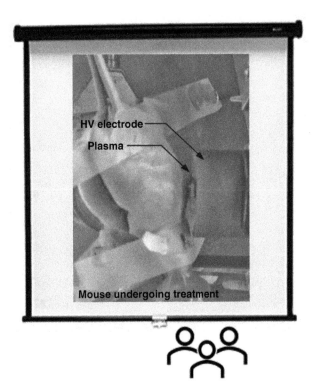

Figure 30.8 FE-DBD sterilization of living tissue: animal directly treated in plasma for up to 10 min remains healthy and no tissue damage is observed immediately and to two weeks after treatment.

Table 30.1 FE-DBD human flora sterilization.

Original concentration	FE-DBD, 5 s treatment	FE-DBD, 10 s treatment	FE-DBD, 15 s treatment
10^9	850 ± 180	9 ± 3	4 ± 4
10^8	22 ± 5	5 ± 5	0 ± 0
10^7	6 ± 6	0 ± 0	0 ± 0

Plasma sterilization of different fragile surface has been analyzed in Lecture 28. The significant level of inactivation of different microorganisms has been achieved without damaging the surfaces. The living human and animal tissues can be considered as one of these fragile sensitive surfaces but obviously require extreme attention to toxicity and safety issues. Generally, the sterilization of living animal or human tissue with minimal or no damage is of importance in a hospital setting. Chemical sterilization does not always offer a solution. For example, transporting chemicals for sterilization becomes a major logistics problem in a military setting. Also, the use of chemicals for sterilization of open wounds, ulcers, or burns is not possible due to the extent of damage they cause to punctured tissues and organs. Nonthermal atmospheric-pressure plasma is non-damaging to the animal and human skin, and it is a potent disinfecting and sterilizing agent, which opens a pathway to its medical applications.

Human tissue sterilization *has been initially investigated by Fridman et al. (2005a–c, 2006a,b, 2007a–c, 2008). Bacteria in this case were the "skin flora" (a mix of bacteria collected from cadaver skin containing Staphylococcus, Streptococcus, and yeast). Direct FE-DBD plasma sterilization leads roughly to a 6-log reduction in bacterial load in 5 s of treatment, as illustrated in Table 30.1.*

A similar level of the skin flora sterilization using the indirect DBD approach requires 120 s and longer of plasma treatment at the same level of the discharge power. Sterilization of the mix of bacteria collected from cadaver skin generally occurred in after 4 s of treatment in most cases and 6 s in a few cases. Nonthermal atmospheric plasma, especially when applied directly, is therefore an effective tool for sterilization of living tissue. It opens possibilities for pre-surgical patient treatment, sterilization of catheters (at points of contact with human body), sterilization of wounds and burns as well as treatment of internal organs in gastroenterology. Specific medical examples of plasma sterilization in relation to wounds and surgical conditions are going to be discussed in the next lectures focused on specific plasma-medical applications.

30.7 Analysis of Toxicity (Nondamaging) in Direct FE-DBD Plasma Treatment of Living Tissues

Plasma is an excellent sterilization tool for different surfaces as it has been discussed in Lecture 28. The key aim of direct plasma sterilization of skin and other human and animal tissues is for the skin in the example of skin sterilization to remain intact after the plasma treatment. The challenge of non-damaging is the key issue of the entire plasma medicine and specific plasma-medical techniques. *In this regard, the magnitude of the energy or* **dose of direct plasma treatment** *is very important. Correspondence of the treatment dose with treatment energy is somewhat specific to DBD discharges with some level of generalization to other discharges, which is going to be discussed in the end of this lecture. Comparing the direct DBD plasma effect with radiation therapy (or radiation biology) it can be concluded that, kinetically, they are quite similar, including possible double strand breaks in DNA (see Lecture 27). However, typical DBD plasma doses after seconds of treatment are about 10^4–10^5 Gy (J kg^{-1}). Similar treatment effects in the case of gamma-radiolysis require only 0.5–1 Gy. Thus, typical direct nonthermal plasma is about 10^4–10^5 times more "energetic" than penetrating radiation. Alternatively, the plasma effect is less "penetrating" than the ionizing radiation by the same factor.* See more on this subject below in Section 30.14.

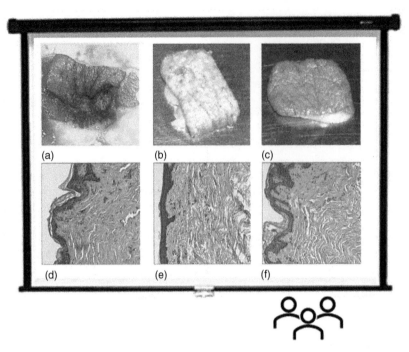

Figure 30.9 FE-DBD toxicity analysis. Photos (top) and tissue histology (bottom) of cadaver skin samples after FE-DBD treatment: (a, d) control; (b, e) after 15 s of treatment; and (c, f) after 5 min of treatment – no visible damage is detected.

A topical plasma-medical treatment which damages the living tissue surface would not be obviously acceptable to the medical community, and so cadaver tissue was first tested in the experiments followed by escalating skin toxicity trials on SKH1 hairless mice in the FE-DBD experiments of Fridman et al. (2005a–c, 2006a,b, 2007a–e, 2008). **Cadaver tissue treatment by FE-DBD plasma** has been accomplished for up to 5 min without any visible or microscopic change in the tissue, as verified with tissue sectioning and staining via the hematoxylin and eosin (H&E) procedure illustrated in Figure 30.9, see Fridman and Friedman (2013).

Based on the knowledge that FE-DBD plasma has non-damaging regimes, the **plasma-toxicity animal model test** has been carried out using the SKH1 mouse model. The skin treatment was carried out at varying doses to locate damaging power/time (dose) combination (Fridman et al. 2005a–c, 2006a,b, 2007a–e, 2008). First, the animal was treated at what was deemed to be a toxic (damaging) dose based on trials with cadaver skin tissue. Once the dose where the damage was visible was determined, a new animal was treated at a lower dose. If no damage was observed at that dose, two more animals were treated. If no damage was observed in all three, the dose was deemed "maximum acceptable dose." Once the maximum dose was located, three animals were treated at that dose and left alive under close observation for two weeks.

 Based on the experimental matrix, a dose at 0.6 W cm^{-2} for 10 min was deemed maximum acceptable prolonged treatment, and a dose of 2.3 W cm^{-2} for 40 s was deemed maximum acceptable high-power treatment. Histological (microscopic) comparison of control SKH1 skin sample with toxic and nontoxic plasma doses shows regions where the plasma dose is high while the animal remains unaffected (see animal after the treatment in Figure 30.10, and histological samples in Figure 30.11, see more details in Fridman and Friedman 2013). Note that sterilization was achieved at 2–4 s at high-power treatment of 0.8 W cm^{-2} and at 10 s at half that power. Following the investigation on mice, an investigation on pigs was carried out achieving the same results. The toxicity due to the FE-DBD treatment of living tissue not only depends on the treatment dose (discharge power and treatment duration) but also strongly depends on the shape of voltage applied to the discharge. Pulsing of the DBD discharges can significantly decrease its damaging ability. Application of nanosecond pulses completely prevents the formation of streamers and therefore DBD microdischarges, which helps to reduce the toxicity of the direct plasma-medical treatment of living tissue (Ayan et al. 2007, 2008).

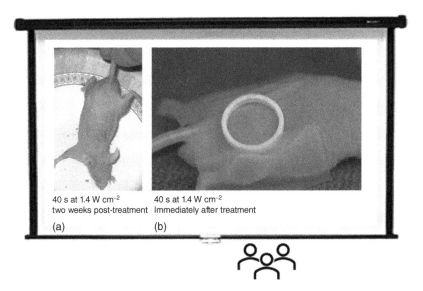

Figure 30.10 Animal after a reasonably high FE-DBD plasma dose (40 s at 1.4 W cm^{-2}; more than 10 times higher than needed for skin sterilization): (a) immediately after treatment and (b) two weeks after treatment.

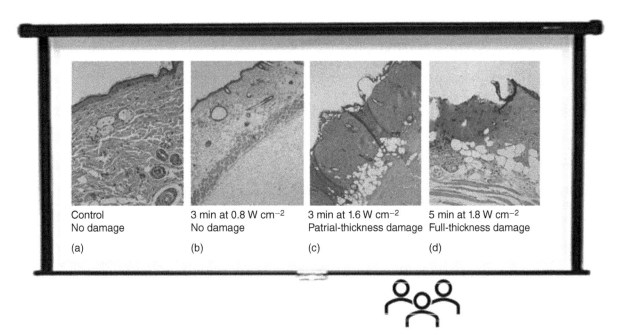

Figure 30.11 Histology of toxic and nontoxic to SKH1 skin plasma doses compared to untreated skin: (a) control; (b) 3 min at 0.8 W cm^{-2} (no damage); (c) 3 min at 1.6 W cm^{-2} (partial-thickness damage); and (d) 5 min at 1.8 W cm^{-2} (full-thickness damage).

30.8 Direct Plasma Interaction with Living Tissue: Not only Plasma Affects Tissue, but Tissue Affects Plasma as Well; In-vitro Model Mimicking the Living Tissue

One of the major mechanisms of direct plasma-medical treatment of tissues is related to the plasma-catalysis of the reduction/oxidation reactions occurring in the biological system. The action of specific charged or neutral active species or radiation is frequently associated with the corresponding specific effect (e.g. anti-inflammatory effect of NO, highly oxidative hydroxyl radical and other ROS, etc.). *It is surely proved that plasmas can cause some significant, perhaps positive, clinical effects in patients. It is*

therefore crucial to analyze what happens to plasma itself when it interacts with living tissue, not only plasma effects on the organism. Clearly, plasma in contact with dirty oily bloody tissue is not the same as that generated between two perfect electrodes. To clarify this effect, Dobrynin et al. (2012a,b), have developed an in vitro model of the FE-DBD plasma treatment of living tissue, where the second electrode for plasma generation is a specially prepared agarose gel. Dobrynin et al. (2012a,b), compared measurements of plasma-generated species on agarose gel to those on a store-bought chicken meat and to those on a live rat wound. Surprisingly, the findings suggest that properly prepared agarose gel may serve as a very good model of penetration of plasma-generated species into the tissue. This provides an opportunity to analyze plasma (perform spectroscopic, microwave, and other measurements) with a second electrode of a simple agarose gel model and obtain results which are a sufficiently accurate representation of the real environment of the DBD treatment of living tissue.

FE-DBD plasma was generated in this case by applying alternating polarity pulsed (1 kHz) voltage of about 20 kV magnitude (peak-to-peak) and a rise time of $5\,V\,ns^{-1}$ between the insulated high-voltage electrode and the sample undergoing treatment. The powered electrode comprised a 1.5-cm-diameter solid copper disk covered by a 1.9-cm-diameter 1-mm-thick quartz dielectric. The discharge gap was kept at 1.5 mm. Current peak duration was 1.2 μs, and the corresponding plasma surface power density was $0.3\,W\,cm^{-2}$. In the case of ex vivo measurement in rat tissue, a special pen-size electrode was used. A 1-mm-thick polished clear fused quartz was used as an insulating dielectric barrier and a handheld pen-like device with the quartz tip was used for treatment. The average power density for the active area of the high-voltage electrode was kept at a level of about 0.74 W for 6 mm electrode diameter. Agarose gels were prepared using the standard procedure with pure agar powder in either distilled water or PBS. To determine the best concentration of agarose gels which would closely represent tissue, the agar has been used at concentrations of 0.6%, 1.5%, and 3% weight percentage, see Dobrynin et al. (2012a,b).

 Agarose gels at 0.6% concentration closely resemble in vivo brain tissue with respect to several physical characteristics. To compare 4% agar phantoms are widely used as a tissue model for radiology studies. To note, the 1.5% concentration of agar was chosen as a median point which is often used as a microbiological substrate.

It has been demonstrated that composition, state, and morphology of the agar phantoms representing living tissue can significantly affect characteristics and behavior of the DBD plasma applied to treat the tissue. Relevant changes of surface conductivity (typical for the living tissue under the plasma treatment) and surface nonuniformities influence sizes and mobility of filaments in the DBD filamentary regimes. Higher surface conductivity results in spreading out the charges deposited on the surface, and therefore thicker filaments. The thicker DBD filaments are typically characterized by higher temperatures because of the suppressed heat exchange inside of the filaments, which finally significantly affects the tissue treatment. Fortunately, transition to the nanosecond-pulse DBD discharges with sufficiently high reduced electric fields (see Lecture 13, Figure 13.10) suppress filamentation of the DBD discharge and provides plasma uniformity required for safe tissue treatment. Same short-pulse discharge parameters permit the effect of "blind streamers" (see Section 30.3) suppressing significant treatment nonuniformities due to morphological peculiarities of the treated tissue.

30.9 Depth of Penetration of Plasma-generated Active Species into Living Tissue

Experiments with the especially developed **agar phantoms representing living tissue** (Dobrynin et al. 2012a,b) permitted to analyze the depth of penetration of plasma-generated active species into living tissue. Measurements of H_2O_2 and pH penetration into agarose gels (0.6%, 1.5%, and 3% wt) and tissues were made using Amplex UltraRed reagent and Fluorescein fluorescent dyes, respectively. In the case of H_2O_2, 75 μL PBS containing 100 μM Amplex UltraRed with horseradish peroxidase (MP Biomedicals) was placed between 1-mm thick agar slices and incubated for about 15 min before the treatment, to provide the presence of the dye in the agar volume. For the pH measurement, the agarose gels were prepared by adding fluorescein dye before it solidified. To measure the H_2O_2 and pH in tissue, dyes were injected using a syringe into a 1-cm-thick skinless chicken breast tissue samples at various points up to a depth of 1 cm.

Ex vivo measurements were made in rat tissue using an animal (hairless Sprague–Dawley male rat) which had been euthanized just before the procedure. A 200 μL dye solution (Amplex UltraRed) was injected subcutaneously using a sterile syringe, and the animal skin was treated with FE-DBD plasma at various time intervals after 5 min

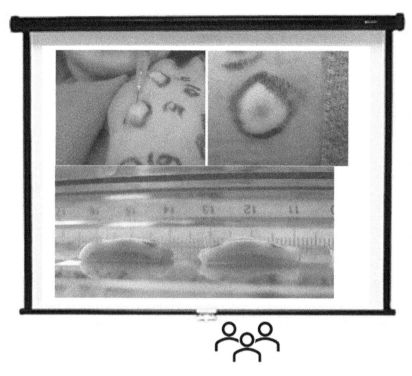

Figure 30.12 Deep tissue penetration of plasma-generated active species: Subcutaneous injection of the fluorescent dye (euthanized rat, top left), the rat skin after the plasma treatment (top right), and the skin sample cross-section just before measurement (bottom).

incubation period [Figure 30.12, see Fridman and Friedman (2013), for more details]. Immediately after the treatment, skin tissue samples were extracted and analyzed as follows. Treated samples were sliced in a vertical direction with thickness of 1 mm, and fluorescence was measured using a fluorescent spectrometer equipped with XY reader accessory. To obtain calibration curves for hydrogen peroxide in the plasma-treated samples, a standard stabilized 3% H_2O_2 water solution properly diluted to obtain various concentrations was used.

With longer treatment time, the depth of penetration as well as concentration of hydrogen peroxide increases. In general, several millimoles of H_2O_2 are produced in tissue after plasma treatment, while it diffuses 1.5–3.5 mm deep. The same tendency is observed in the case of tissue acidity change. However, the effect of pH lowering leads to deeper penetration of up to 4.5–5 mm. To investigate depth of reactive species penetration, three concentrations of agarose media were used. Hydrogen peroxide concentration on the agar gel surface was different for different agar densities: 0.5 mM for 0.6%, 0.7 mM for 3%, and 1.9 mM for 1.5% gels after 1 min of plasma treatment. Compared to the dead chicken breast tissue, the depth of penetration appears to be very similar: up to 4 mm after 2 min treatment (although concentration is almost ten times higher due to corresponding higher power of the discharge by a factor of 8). The depth of H_2O_2 penetration for all types of agarose media was about the same. In contrast to hydrogen peroxide, the dynamics of acidity change inside the agar gels were significantly different for different agar media compositions. Tissue appeared to have better buffer properties compared to non-buffered agarose gels. However, the addition of PBS to agar produces required similarities in terms of pH changes.

 In the case of real tissue, active species produced by plasma on the surface may travel in tissue volume to depths of up to millimeters. The depth of penetration is determined by diffusion and reaction rates, which depend highly on the type of tissue and its biochemical characteristics. Measurement of these parameters is an extremely complicated task, but the creation of a simple model that closely represents certain tissue is possible. For the case of hydrogen peroxide, depth of diffusion is about the same for all types of agars. The concentration of H_2O_2 varies however, being the closest for the 1.5% agar weight percent case and about 3 times lower for both 0.6% and 3% wt agar gels. This behavior may possibly be explained by both diffusion properties and reaction rates of H_2O_2 in agar. Acidity of tissue was measured to be consistently increased (lowering of pH) because of exposure to the discharge. The tissue compared to agarose gels prepared in distilled water acts as a significantly better buffer, and the depth of pH changes inside such gels is much greater. In fact, a significant drop in the whole volume of a phantom was observed – up to 1 cm thickness of agar. This problem may be addressed simply by adding a buffer into the agarose media in the form of 1.5% wt agar prepared in PBS. Summarizing, generally, plasma-generated active species can directly penetrate the living tissue about one or couple of millimeters in depth. Further penetration, so much important to stimulate immune system and treat internal diseases (see Lecture 32), requires relevant contribution of the cell-to-cell communication, which is going to be discussed below.

30.10 Plasma Effects Propagating into Living Tissue Deeper than Plasma Generated ROS/RNS, Effects of Radiation and Electric Field, Contribution of Cell-to-cell Signaling

Plasma ROS/RNS penetration inside of human or animal tissue is controlled by reactive diffusion as well as by protective effect of skin, especially stratum corneum (typical thickness 10–30 μm). These two effects limit the effective direct penetration to about one to couple of millimeters in depth, like it has been discussed in Section 30.10. ROS/RNS penetration through the stratum corneum, which is a thin layer of dead skin, has similarities with penetration through conventional polymer membranes considered in Sections 26.16–26.18. At some treatment doses discussed in Sections 26.16–26.18, typical for plasma-medical treatment, plasma stimulates cross-linking of this biopolymer and induces significant penetration without damage. These effects will be additionally discussed in the next lecture regarding plasma dermatology and cosmetics. Special analysis has been focused on depth of penetration of electric field and radiation during the plasma-medical treatment. While the penetration of electric field is determined by geometrical sizes of the charged areas, the depth in this case also usually does not exceed millimeters. Regarding radiation, UVB is almost completely absorbed by the epidermis (about 0.1 mm), with comparatively little reaching the dermis. Penetration of UVC is also usually not more than 0.1 mm.

Direct effect of energetic electrons becomes stronger and deeper when their energies approach MeV level of energy (typical depth of penetration in this case is about 1 MeV per 1 mm).

 Thus, summarizing, *depth of penetration of all direct plasma effects inside the human and animal bodies usually does not exceed millimeters unless the cell-to-cell communication is triggered. The* **cell-to-cell communication (sometimes referred to as the bystander effect),** *from the other hand, can lead to prolongation of the plasma effects up to about a centimeter. If this depth of penetration is sufficient to trigger the immune system (see Lecture 32), the plasma becomes able to affect the whole human or animal body.* DBD treatment of cancer cells can be a good example of this important plasma-medical depth of penetration effect. It has been accomplished by plasma treatment of cancer cells *in-vitro* and tumors in-vivo induced the immunogenic cell death, see Lin et al. (2018). According to Lin et al. (2018), the immunogenic cell death marker, calreticulin (CRT), was observed 7 mm into the depth of the tumor after plasma treatment, see Figure 30.13. Details on this subject are going to be considered in Lecture 32 regarding the plasma-induced immunomodulation for cancer treatment.

To analyze the plasma-induced cell-to-cell communication in propagation of the immunogenic cell death (Lin et al. 2018), and observation of the CRT marker 7 mm into the depth of tumor, special experiments have been carried out by Ranieri et al. (2017) and Ranieri (2018). To investigate the depth of penetration of plasma effect in this case, the 3-D cell cultures were placed on a monolayer of target cancer cells, see Ranieri et al. (2017) and Ranieri (2018). After DBD plasma was applied atop the barrier, the target cells were measured for immunogenic cell death markers to assess if the treatment affected the cells. The major conclusion of this work was confirmation that the cells in the 3-D culture transferred the plasma-induced effect of immunogenic cell death to the target cells. When plasma was applied to a 3-D without those cells, there was no significant increase in immunogenic cell death measured in the target cells. The cells within the 3-D culture were affected by DBD plasma components and transferred the plasma-induced effect to reach the target cells through cellular communication. Since the immunogenic cell death marker CRT is the major calcium-binding protein in the cell, it was suggested that the transfer may be due to calcium release. The results of this analysis are going to be discussed below.

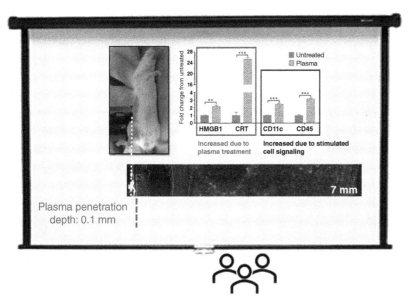

Figure 30.13 Cell-to-cell communication: DBD treatment induces CRT expression in tumor cells up to 7 mm in the solid tumor, while ROS/RNS penetration is about 0.1 mm.

30.11 Transferring Nanosecond Pulsed DBD Plasma Effects Across Barrier of Cells

To investigate the effect, Ranieri et al. (2017) and Ranieri (2018), treated a monolayer of CT26 cells with the nanosecond pulsed DBD plasma in the presence or absence of a barrier to evaluate the depth of penetration of plasma treatment. The expression of ecto-calreticulin (CRT), a marker for the initiation of ICD, in the monolayer was used to indicate if plasma effects reach the cells below the barrier. Plasma interacts with cells in the barrier through the direct diffusion of plasma components (UV radiation, reactive oxygen and nitrogen species, electric field, and neutral species), induces oxidative stress, and stimulates redox-based signaling responses in both cancerous and healthy cells. Plasma-generated reactive species have limited half-times and diffusion distances, which limit the depth plasma treatment effect. To distinguish between tumor cell responses to plasma components or to a combination of plasma-stimulated intercellular signaling and components, the barriers were acellular or loaded with cells, respectively. A baseline ecto-CRT expression in CT26 colorectal cancer cells was established in cells grown and treated in the absence of a barrier. The cells were collected 24 h post-plasma treatment and stained for CRT. Plasma treatment at 300 mJ increased the externalization of CRT on cells from 7.5% to 16.8%. This result, a twofold increase in the expression of ecto-CRT, is comparable with previous studies of Lin et al. (2018) at the same energy. It can serve as a positive control to determine if plasma treatment penetrated the tissue model to affect cells underneath.

To investigate if plasma components penetrated to affect ecto-CRT levels on tumor cells, Ranieri et al. (2017) and Ranieri (2018), utilized a tissue model as a barrier and evaluated the cellular response in the monolayer below. Treatment at the same energy as the positive control resulted in a negligible increase to 7.8% from 7.5% ecto-CRT positive population. This suggests that plasma components generated at 300 mJ are unable to pass through the acellular barrier to influence the monolayer of cells but possibly, higher energy plasma could. To assess if the CRT expression increased with increasing energy, the experiment was repeated with treatments at 700, 900, and 2000 mJ which resulted in partial recovery even at the highest energy tested. While this shows plasma components penetrate further with higher energy, the energy required is too high to be safe for biological tissue.

 While plasma components may not penetrate deep, the biological signaling may facilitate the deep transfer of plasma effect. Therefore, Ranieri et al. (2017) and Ranieri (2018), populated the barrier with cells to examine the influence of plasma-stimulated intercellular signaling on CRT induction in the target cells. The addition of cells to the barrier is expected to cause the plasma components to interact with the embedded cells. Any effect on the target cells below would presumably be because of cell-to-cell communication.

Treatment atop the cell-laden barrier increased the percentage of cells with ecto-CRT expressed from 7.5% to 9.95% at 300 mJ. This indicates that the presence of cells enabled the plasma effect to travel to the target cells albeit the effect was weak. The influence of this signaling is more evident with higher energy plasma as the percent CRT positive cells is higher at both 900 mJ (19.27%) and 2000 mJ (22.82%) than at the positive control (16.8%). In comparison to the acellular barrier treated at the same plasma energies, the addition of cells to the barrier allowing for intercellular signaling resulted in higher ecto-CRT expression in the monolayer. In conclusion, plasma-stimulated intercellular signaling can increase the anti-tumor effects of plasma treatment. The typical reaction time for the expression of ecto-CRT in response to ICD inducers is in a few hours after treatment, which is orders of magnitude longer than the lifetime of the plasma components. Thus, to study the immediate direct effects of plasma treatment, the extracellular ATP concentration can be measured. DBD plasma causes cancer cells to release ATP within 10 min post treatment. The concentration of extracellular ATP released in response to plasma treatment at 300 mJ was established in the monolayer. Plasma treatment increased measured ATP in the extracellular fluid from 14 nM in the untreated cells to 3.28 μM. With the addition of the acellular barrier and treatment at 2000 mJ, the measured ATP was 1.84 μM suggesting that the barrier impedes the plasma components from penetrating through to the target cells. Plasma treatment atop the cell-laden barrier at 2000 mJ resulted in 2.99 μM, comparable to the ATP released from treatment of the monolayer. While this seems to indicate that plasma-stimulated intercellular signaling can stimulate the target cells to release more ATP comparable to the positive control, this is not the case. When the cell-laden barrier was treated at 2000 mJ without the target cells below, the measured ATP was 1.34 μM. This indicates that the cells embedded in the barrier contribute to the total measured ATP (2.99 μM) when evaluated in the presence of the basal target cells. This shows that both, the cells embedded in the barrier, and the target cells react to plasma faster than the CRT response would indicate. The described experiments permitted clarify some mechanisms of the plasma-induced cell-to-cell communication and especially role of calcium ions in this process, which is going to be discussed below.

30.12 Mechanisms of Plasma-induced Cell-to-cell Communication Leading to Deep Propagation of Plasma-medical Effects, Calcium Ion Wave Propagation

Like it has been discussed above, for the in-vivo xenograft models, the propagation distance of anti-tumor effects from the skin to the tumor below is further than the penetration depth of plasma-generated species. This observation suggests the contribution of intercellular signaling. Signal transfer between two cells happens through the release of factors from one cell to their neighbors, called **paracrine signaling**, and/or through **direct cell-to-cell contact signaling, such as gap junctions**. The cell-laden barrier data suggests the plasma components penetrate the tissue model and interact with the cells within the barrier to stimulate subsequent signaling events. To assess if paracrine signaling was the dominant signaling mechanism for inducing ecto-CRT expression in the target cells, the cell-laden barrier was treated in a separate well prior to transfer to prevent any immediate, direct contact with plasma components affecting the target cells (Ranieri et al. 2017; Ranieri 2018). After the barrier was transferred to the monolayer and incubated for 24 h, CRT expression was measured. The observed response in the CT26 cells was 11.49% at 2000 mJ, which is comparable to the control without plasma treatment (10.65%).

 *This indicates that paracrine signaling alone cannot sustain the transfer of effect and requires a proximity between cells. For the **transfer of signals through direct contact, such as through gap junctions**, the embedded cells should form connections with their cytoskeleton. To examine if this was the case, the cell-laden barriers were fixed and stained for nucleic acid and the cytoskeleton to identify the nuclei and the cytoskeletal junctions respectively. As shown in Figure 30.14, the cytoskeleton surrounding the nuclei of the cells connect forming a network capable of signal propagation. After plasma treatment, cell-laden barriers were fixed in 4% PFA for one hour and washed with PBS without Ca, Mg. The barriers were permeabilized, blocked, and stained for nucleic acid and the cytoskeleton with Hoechst stain and Phalloidin respectively. The nuclei of the cells (bright white) were connected to each other through junctions of the cytoskeleton (gray). The formation of junctions between cells in the barrier creates an intercellular network allowing cells to signal through direct contact, such as gap junctions. Summarizing, signal transfer through gap junctions could be a likely pathway (Ranieri et al. 2017; Ranieri 2018).*

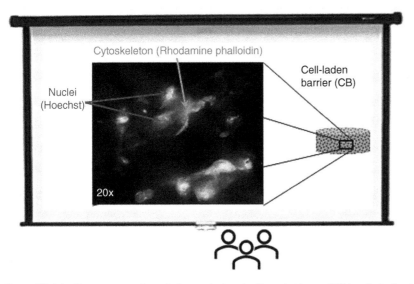

Figure 30.14 Deep penetration of plasma-induced effects in tissue: CT26 cells in the barrier form junctions enabling direct cell-to-cell signaling.

The propagation of anti-tumor effects of plasma through cell-to-cell communication can be interpreted in the frameworks of the **Evans mechanism of calcium ion wave propagation**, see Evans (2015). This mechanism is based on the calcium-induced calcium release (CICR), which requires both paracrine and direct contact signaling to propagate a wave of intracellular calcium between cells. Plasma treatment of cancer cells increases the intracellular calcium concentration (Dobrynin et al. 2009). Calcium can signal to nearby cells through inducing the release of ATP or itself be transferred through junctions to adjacent cells, see illustration of the mechanisms in Figure 30.15 (Ranieri et al. 2017; Ranieri 2018). Plasma-induced calcium increase can be propagated from cell to cell via paracrine and direct contact signaling. The released ATP in response to plasma can signal to nearby cells by binding to surface purinergic receptors raising the calcium concentration. However, extracellular ATP rapidly degrades limiting the propagation distance due to paracrine signaling. Calcium increase triggered this way originates from the ER, where the depletion of ER calcium influences the expression of surface CRT.

Calcium can travel as a wave between adjacent cells through direct contact signaling, reaching cells that cannot be influenced by ATP released due to plasma treatment. If the concentration is sufficient, these cells can further propagate the wave by calcium-induced release of ATP, thus regenerating the paracrine signaling propagation. Therefore, CICR propagation may be the mechanism of the propagation of plasma-stimulated effects. Released ATP degrades rapidly; thus, it will only signal to nearby cells. ATP binds to purinergic receptors on the cell membrane that induces the release of calcium from the endoplasmic reticulum (ER), which increases the intracellular calcium concentration. This supports the model, as the induction of surface CRT expression is favored when calcium in the ER is depleted. When the concentration of extracellular ATP is not sufficient to trigger ER calcium release, propagation due to paracrine signaling ceases. Excess calcium will also propagate as a wave between cells via gap junctions. If the concentration of intracellular calcium is adequate, calcium can signal the release of ATP from cells that are not directly treated with plasma. Therefore, plasma treatment induces stress that may cause cells to propagate a gradient of intracellular calcium through CICR that results in expression of surface CRT and the release of ATP.

Summarizing, enhanced propagation of plasma-stimulated anti-tumor effects through a tissue model through intercellular signaling has been demonstrated. The indicators of plasma treatment affecting the target tumor cells, the expression of ecto-CRT, and the release of ATP, were higher in the cancer cells beneath a cell-laden barrier than those under an acellular barrier. This suggests that cells in the barrier responded to plasma treatment by enhancing the propagation of ICD-inducing factors to affect the target cancer cells. Additionally, increasing the plasma energy caused more of these factors, both the components and the intercellular signals, to reach the cells beneath the barriers, indicating that higher energies increase the depth of penetration of plasma treatment.

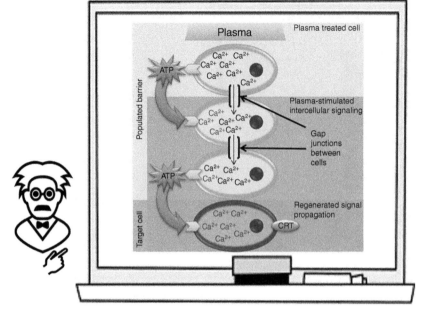

Figure 30.15 Plasma-stimulated cell-to-cell communication: calcium-induced calcium release (CICR) propagation of surface CRT expression.

More details on the subject can be found in Ranieri et al. (2017), Ranieri (2018), and Von Woedtke et al. (2022). We should note that the described Evans mechanism of calcium ion wave propagation to about 1 cm explains propagation of the plasma effects deep enough to stimulate immune system and therefore affect the whole animal or human body, which is going to be discussed in Lecture 32 regarding the plasma treatment of cancers.

30.13 Standardization of Plasma-medical Devices and Approaches

While efficacy of the specific plasma-medical treatment procedures (to be discussed in Lectures 31 and 32) are widely demonstrated, the crucial challenge becomes related to regulatory issues, standardization and certification of medical devices, official approvals of the relevant approaches and protocols. Regulations of medical devices are intended to guarantee their safety and are rooted in the analysis of the risks associated with the technology under consideration. Even at the early stage of the development of plasma technologies for medical applications, the need to consider safety aspects was already well understood, see Sections 30.1 and 30.2, as well as Fridman and Friedman (2013). The awareness of these aspects grew over the years, with an increasing number of papers referring to the need to consider the compliancy with regulations in the development of plasma devices and citing several documents for the analysis of associated risks. The matter was complicated by the local nature of regulations, often different from country to country, and the exponential increase of activity in the field, resulting in many different plasma sources intended for different medical applications.

A special DIN SPEC (2014), a specification registered by the German Institute of Standardization (DIN) and indicating the criteria for the characterization of different plasma devices intended for medical applications, has been introduced in 2014 by Weltmann and Von Woedtke and their group in INP, Greifswald, Germany. The specification was meant to complement the safety evaluation requirements indicated in the DIN EN by introducing standards for the characterization of plasma sources to (i) obtain systematic and comparable results from researchers all over the world and (ii) produce results that could be checked against the safety limits imposed by regulations not specifically developed for plasma technology and that could vary from country to country. The test procedures proposed in the DINSPEC, in accordance with already available standards, encompass the measurements of temperature, UV irradiance, emitted gas species, chemical species in liquids, leakage current, antimicrobial activity, and cytotoxicity. This specification has been proven as a useful stepping stone for the regulatory approval of plasma sources as medical devices and is, overall, an accepted instrument with several publications reporting the characterization of plasma devices according to its standard procedures. Despite all these positive aspects, this specification is rooted in risk analysis performed for plasma devices intended for dermatological applications; therefore, the indicated methodologies do not specifically address plasma devices intended for intrabody medical procedures. It should nonetheless be noted that, considered its evolutive nature, the DIN SPEC could be adapted in the future to account for specificities of intrabody plasma delivery and the associated clinical applications. Also, the DIN SPEC does not cover the regulatory aspects of plasma-treated liquids. This aspect is particularly thorny since plasma-treated liquids intended for indirect medical applications can arguably be considered as pharmaceutical compounds, which are associated with harsh regulations, strict manufacturing procedures, and the need to univocally identify active principles.

The DIN SPEC (2014) specification has been registered by the German Institute of Standardization (DIN). Similar specification has been registered in South Korea by Prof. Eun Ha Choi and his group in Kwangwoon University of Seoul. The South Korea regulations have been accomplished coherently with those developed in Germany. The South Korean and German groups used their regulatory specification as the basis to create the relevant international policy under umbrella of International Electrotechnical Commission (IEC). The IEC is an international standards organization that prepares and publishes international standards for all electrical, electronic, and related technologies. Especially selected groups of experts from multiple countries are in process of preparation of these regulatory standards. In the United States, significant efforts in addition to that are focused on the relevant Food and Drug Administration (FDA) approvals of the plasma-medical technologies.

We should especially mention that several prospective medical applications of plasmas, especially the oncological applications, envision intrabody plasma delivery. Plasma devices intended for this scope of applications face the peculiar constraints. Besides the requirement of biological efficacy, which is a challenge itself given the variability of conditions at the site of application (see Lecture 32), plasma devices must be flexible and composed of materials with a limited rate of erosion when contacting the plasma. Other aspects that require even more attention for the case of intrabody plasma application with respect to cutaneous application are electromagnetic compatibility, leakage currents, gas delivery, and plasma stability. While these constraints have been faced before and were overcome by the plasma community during the development of endoscopic plasma coagulators and ablators, the nondestructive nature of medical applications of nonthermal plasma creates additional challenges. More information on this subject can be found in the review by Von Woedtke et al. (2022).

30.14 Dosimetry of Plasma Medical Treatments and Procedures, Absorbed Energy Related Physical Approach to Dosimetry

Dosimetry is one of the crucial challenges of the clinical plasma medicine absolutely required for wide application of the plasma devices in practical medical environment. Development of the unified plasma-medical dosimetry is not an easy task, however, keeping in mind so much wide diversity of the plasma-medical devices and relevant treatment protocols. The term "dose" refers to the quantity of a therapeutic agent or of radiation, in the case of radiotherapy, administered to the patient during treatment. Medical doctors rely on this concept when considering the therapeutic window of a certain drug or treatment, whose boundaries are defined by the minimum quantity required to exert the desired biological effects (minimum effective dose) and the maximum quantity not eliciting unacceptable toxic effects (maximum tolerated dose). Moreover, the dose offers a standard to compare different treatments and is a suitable control parameter, since it connects the inputs (the administered quantity of therapeutic agent) with the outputs (the biological effects) of the process.

In the case of dielectric-barrier discharges, the physical definition of dose has been already considered in this lecture (see Sections 30.7 and 30.8) and was related to total energy of the treatment (discharge power integrated over the treatment time). The physical energy units of the dose determined this way are $J\,cm^{-2}$. As it was pointed out that direct relation between the medical treatment dose and the treatment energy is somewhat specific to DBD discharges with only some level of generalization to other discharges. It can be explained keeping in mind the **concept of G-factors** widely applied in radiation chemistry and biology and effectively applied in plasma chemistry as well, see Sections 15.12, 18.7, and 18.8, as well as Figure 18.7.

The G-factors show the number of the primarily plasma-generated or radiation-generated species (like electrons, ions, atoms, radicals, excited atoms, and molecules) per 100 eV of discharge plasma or radiation energy. G-factors are functions of gas composition and reduced electric field, E/n_0. Therefore, for specific discharges (for example, DBDs), the G-factors are coefficients of proportionality between discharge power and total generation rate of species, between discharge energy and total amount of generated species, and between energy per unit treatment area ("physical dose," $J\,cm^{-2}$) with relevant flux of excited species. For example, the G-factors of production of reactive species in DBD are about 1 specie/100 eV (a little more for ROS and less for RNS, see Figure 18.7). Therefore, if the "physical treatment dose" is $D = 1\,J\,cm^{-2}$, in corresponds to the to the total integrated flux of reactive species $(D \times G)$ about 6×10^{16} reactive species per cm^2. Assuming that the "biological" treatment dose is determined by the total integrated absorbed flux of reactive species $(D \times G)$, we can conclude (completely like in the case of the radiation treatment) that plasma-medical treatment dose can be characterized by the integrated absorbed plasma energy $(J\,cm^{-2})$.

Based on this approach of the "absorbed DBD energy doses," Fridman and Friedman (2013), combined the efficacy and toxicity data on the FE-DBD plasma sterilization of living tissue (see Sections 30.6 and 30.7 of this lecture) and FE-DBD treatment of mammalian cells (see Lecture 27) to determine the safe vs damaging "absorbed FE-DBD energy doses, which is illustrated in Figure 30.16. It can be summarized that below $1\,J\,cm^{-2}$, there is no significant sterilization and no cell-treatment effects; $1–2\,J\,cm^{-2}$ is sufficient for effective sterilization; $4–5\,J\,cm^{-2}$, effective non-damaging treatment of cells; above $10\,J\,cm^{-2}$, significant biological damage of cells; above $100\,J\,cm^{-2}$, significant necrosis. Thus, **effective, and safe absorbed energy doses of the plasma medical DBD treatment** *stay in the range from 1 to $10\,J\,cm^{-2}$.*

Comparing the direct DBD plasma effect with radiation therapy (or radiation biology) it can be concluded that, kinetically, they have some similarities, but significantly differ in the depth of penetration. The radiation biology and medicine are based upon the highly energetic "penetrating radiation," while the direct plasma depth of penetration is quite limited to about millimeters, see Section 30.9. Thus, the above-mentioned effective, and safely absorbed energy doses of the plasma medical DBD treatment from 1 to $10\,J\,cm^{-2}$ corresponds in this terminology to radiation dose of $10^4–10^5$ J kg^{-1} ($10^4–10^5$ Gy). Even this level of absorbed energy leads to pulsed temperature increase of only about 10 K. The unit of absorbed radiation in radiation chemistry and biology is 1 Gray (1 Gy = 1 J kg^{-1}), and typical values of effective and safe doses are about 1 Gy. Thus, typical direct nonthermal plasma is $10^4–10^5$ times more "energetic" than penetrating radiation. Alternatively, the plasma effect is less "penetrating" than the ionizing radiation by the same factor and therefore less damaging.

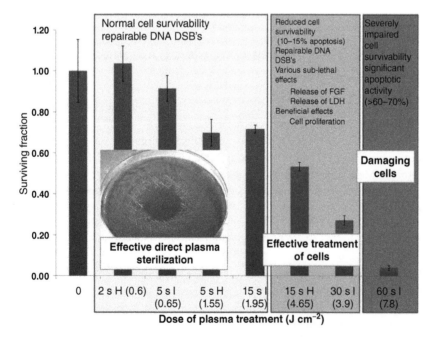

Figure 30.16 Safe vs damaging "absorbed DBD energy doses": below $1\,J\,cm^{-2}$, there is no sterilization and no cell-treatment effects; $1–2\,J\,cm^{-2}$ is sufficient for effective sterilization; $4–5\,J\,cm^{-2}$, effective non-damaging treatment of cells; above $10\,J\,cm^{-2}$, significant biological damage of cells; above $100\,J\,cm^{-2}$, significant necrosis.

Simplicity of the integrated absorbed energy approach to the plasma-medical dosimetry makes it attractive wherever it can be effectively applied like for example in FE-DBD. Application of this approach to plasma jets is quite limited, however, because, of the more complicated relation between the discharge energy and integrated flux of reactive species to the living tissue to be treated. The G-factor approach, described above, is still applicable for plasma jets but only regarding the initial production of the "primary" plasma-generated ROS/RNS in the discharge itself and in the ionization wave propagating from the discharge to the treated object. Which percentage of these "primary" reactive species are transferred to the tissue is not clear and significantly depends on configuration of the plasma jet. Also, in the case of helium jet significant portion of the reactive species are not "primary" but generated in the plasma jet boundary with the surrounding air, which makes correlation between the integrated absorbed energy and the flux of reactive species to the tissue even more complicated. Thus, development of the plasma-medical dosimetry unifying the direct DBD, and plasma jet treatment require different approaches based not on the absorbed energy in plasma but on the chemical, biological, or biochemical response of the treated tissue. The joke between physicists is that dose should be defined not by how much energy is in plasma but "how loud a patient yell or cry" because of the treatment. Such plasma-medical dosimetry approaches based on the chemical, biological, or biochemical responses of the treated tissue are going to be discussed below.

30.15 Dosimetry of Plasma Medical Treatments and Procedures Based on Biological Responses of Tissues

As the understanding of the fundamental mechanisms driving plasma medical applications in the so different plasma-medical systems (especially DBDs and plasma jets) progressed over the years, interesting biology-related concepts of the plasma-medical dosimetry has been proposed. As example of these approaches, we can first mention those relying on the equivalent total oxidation potential (ETOP) and the bacterial reduction factor, see Cheng et al. (2020a), Von Woedtke et al. (2022). The ETOP accounts for the oxidation potential of (i) the reactive species produced in the discharge and flowing toward the target; (ii) the other agents associated with the plasma discharge, such as UV radiation and electric fields; and (iii) the synergistic effects of the previous two factors. Nonetheless, a comprehensive description of dose through the ETOP parameters faces many obstacles. First, determining the oxidation potential of the reactive species requires the measurement of their flow toward the substrate in real time, posing significant challenges in terms of process monitoring. Second, quantitatively determining the oxidation potential of agents other than reactive species is complicated. Third, determining the oxidation potential associated with the synergistic effects would require a complete mechanistic comprehension of the plasma biomedical processes. Furthermore, the proposed parameter does not account for the oxidation potential of the reactive species produced in the liquid environment surrounding the tissues and has yet to be shown to be suitable for therapeutical applications different from disinfection. More details on this approach can be found in Cheng et al. (2020a), Von Woedtke et al. (2022).

 An interesting approach to the dose based not on the plasma/device but on biological response to plasma treatment (let us count killed rabbits not number of bullets used) has been proposed by Mazur and his colleague from INP, Greifswald, Germany in collaboration with Nyheim Plasma Institute. *This approach is based on introduction of the* **Plasma Treatment Unit (PTU)** *based on the biological readout: 50% decrease in metabolic activity (IC50) of specific cancer cells (CT26) in specified medium and volume using the Alamar Blue Assay, see details of this approach in Bekeschus et al. (2018a–c). The definition of the dose of 1 PTU (one plasma treatment unit) is independent of plasma devices and plasma conditions and therefore can be applied to different plasma sources.* When 1 PTU is determined this way, 2 PTU simply means application of such a treatment twice, 3 PTU, 4 PTU, etc. are defined likewise. For practical dose measurements, it is not necessary each time to check 50% decrease in metabolic activity (IC50) of specific cancer cells (CT26) in specified medium and volume using the Alamar Blue Assay. The practical dose measurements can be provided for example, by linearly related chemical measurements like H_2O_2, proportional to the biological readout (calibration procedure). Surely such calibration could be plasma/device specific. Application of calibrated chemical, electronic conductivity, and heat flux "strips" are also possible in clinical conditions.

For example, 1 PTU in the case of uniform nanosecond pulsed FE-DBD treatment at voltage 29 kV, frequency 75 Hz, gap distance 1 mm, requires 10 s of treatment. FE-DBD treatment energy corresponding to 1 PTU, in this case, is 350 mJ cm^{-2}; when the total treated volume is 700 mJ, the absorbed radiation dose is 140 J kg^{-1} (Gy). Equivalent production of H_2O_2 in deionized water is ∼ 4.5 μM. Such calibration helps to provide simple PTU measurements in clinical environment. Bekeschus et al. (2018a–c), also performed a relevant comparative study of two plasma sources (floating-electrode dielectric barrier discharge, DBD, Drexel University; atmospheric pressure argon plasma jet, kINPen, INP Greifswald) on cancer cell toxicity. Cell culture protocols, cytotoxicity assays, and procedures for assessment of hydrogen peroxide (H_2O_2) were standardized between both labs. The inhibitory concentration 50 (IC50) and its corresponding H_2O_2 deposition were determined for both devices. For the DBD, IC50 and H_2O_2 generation were largely dependent on the total energy input but not pulsing frequency, treatment time, or total number of cells. DBD cytotoxicity could not be replicated by addition of H_2O_2 alone and was inhibited by larger amounts of liquid present during the treatment. Jet plasma toxicity depended on peroxide generation as well as total cell number and amount of liquid. Thus, the amount of liquid present during plasma treatment in vitro is key in attenuating short-lived species or other physical effects from plasmas. These in vitro results suggest a role of liquids in or on tissues during plasma treatment in a clinical setting. More details on the subject can be found in Bekeschus et al. (2018a–c).

30.16 Problems and Concept Questions

30.16.1 Electrical Safety of the FE-DBD Plasma Source for Direct Medical Treatment of Humans and Animals

Using relation (30.1), explain the electrical safety of the FE-DBD to living tissue and show that current which the power supply delivers in this case is less than 5 mA.

30.16.2 Safety of the FE-DBD Plasma in Treatment of Nonuniform Living Tissues

Analyzing the effect of "blind" streamers, explain the possibility of safe and uniform plasma treatment of living tissue, which can be by itself significantly nonuniform both morphologically and in its composition and electric conductivity.

30.16.3 Medical Applications of the Plasma-Activated Ringer's Lactate Solution (PAL)

What are the major advantages of medical applications of the plasma-activated Ringer's lactate solution (PAL) with respect to the more conventional PAM.

30.16.4 Mechanisms of Plasma-Induced Cell-To-Cell Communication Leading to Deep Propagation of Plasma-Medical Effects

Give your interpretation of the calcium ion wave propagation effect in cell-to-cell signaling. What is the role played by ATP in the calcium ion wave propagation.

30.16.5 Direct Relation Between the Medical Treatment Dose and the Treatment Energy in FE-DBD Plasma

Use the concept of G-factors in kinetics of generation of ROS/RNS active species to explain the proportionality of the medical treatment effect and the integrated absorbed discharge energy.

30.16.6 Effective, and Safe Integrated Absorbed Energy Doses of the Plasma Medical DBD Treatment

Keeping in mind that the effective, and safe integrated absorbed energy doses of the plasma medical DBD treatment stay in the range from 1 to 10 J cm^{-2}, estimate the relevant temperature increase of the treated tissue.

30.16.7 Relation Between Integrated Absorbed Energy Doses in DBD with Absorbed Radiation Doses

Keeping in mind that the effective, and safe integrated absorbed energy doses of the plasma medical FE-DBD treatment of tissues stay in the range from 1 to 10 J cm^{-2}, calculate the corresponding value of the absorbed radiation dose in (Gy, J kg^{-1}). Give your interpretation why the corresponding value of the absorbed radiation dose is so much higher than typical values of the absorbed radiation doses leading to similar biochemical outcomes, which are about 1 J kg^{-1} (1 Gy).

30.16.8 Integrated Absorbed Energy Doses in DBD Leading to Necrosis of Treated Tissue

Keeping in mind that the integrated absorbed energy doses of the FE-DBD plasma leading to necrosis of the tissue is about 100 J cm^{-2}, estimate the corresponding value of the absorbed radiation dose and relevant temperature increase of the treated tissue.

30.16.9 Plasma Treatment Units (PTU), Dosimetry Approach Based on Biological Readout

Explain why using the *50% decrease in metabolic activity (IC50) of specific cancer cells (CT26) in specified medium and volume using the* Alamar Blue Assay as the plasma treatment unit permits the plasma-medical dosimetry independent on type of plasma device and specific discharge regimes and parameters.

Lecture 31

Plasma Medicine: Healing of Wounds and Ulcerations, Blood Coagulation

31.1 Plasma Wound Healing, Types of Wounds, and Relevant Healing Processes

The most straightforward way of using plasma to treat human body is to apply the FE-DBD, plasma jets, or other plasma-medical devices superficially to the skin, wounds, teeth, etc., see for example Figure 1.4. That is why from the beginning of plasma medicine (see Section 30.1), a central focus of the application-oriented research in plasma medicine was the nonthermal plasma use in wound healing, see Fridman et al. (2008), Laroussi (2009), Lloyd et al. (2010), and Fridman and Friedman (2013). Above all, chronic wounds are important challenge for patients, health care professionals, and health care systems worldwide. Plasma was estimated to have great potential in this field because of its early-predicted effectivity, in two ways: inactivation of wound-contaminating microorganisms (significantly due to plasma-generated reactive oxygen species [ROS]) and direct stimulation of tissue regeneration [significantly due to plasma-generated nitric oxide (NO) and other reactive nitrogen species (RNS)]. Before focusing on plasma effects in wound healing let's shortly discuss the major types of wounds and relevant healing processes, which can be supported by application of plasma.

Nonhealing (refractory) or slowly healing cutaneous wounds represent a considerable burden for healthcare, markedly reducing the quality of life of patients and increasing cost of care. Although wounds and wound healing pathologies can be relatively complex, to some extent the persistence of refractory wounds is a consequence of inaccurate diagnosis and wound assessment as well as suboptimal management of both wounds and underlying pathologies. For this reason, development of more convenient wound care tools, such as those based on plasma treatment, can provide substantial value for patients and for the healthcare system. Let's start with **skin damage and wound healing.** When injured, the skin displays a remarkable ability to repair or regenerate itself through a regulated biochemical cascade. The wound healing process can be categorized into four distinct phases based on a complex array of dynamic mechanisms including cellular migration, adhesion/de-adhesion, proliferation, differentiation, and apoptosis: (i) hemostatic phase: immediately following injury, damaged vessels constrict to reduce blood flow and the coagulation cascade is initiated, leading to thrombus formation; (ii) inflammatory phase: within six to eight hours of injury, tissue damage induces release of biochemicals that attract white blood cells, resulting in phagocytosis of invading bacteria and other organisms; (iii) proliferative (repair) phase: two to three days post-injury, angiogenesis, collagen deposition and granulation tissue development occur together with wound epithelialization and wound contracture; (iv) maturation phase (epithelialization and remodeling): this may occur for a considerable period of time following injury and is primarily a process of connective tissue remodeling that results in scar formation and increased wound tensile strength.

Several cell types play a major role in normal wound healing processes including macrophages, fibroblasts, and keratinocytes. Within about 24 h of injury, growth factors released by platelets and other cells attract macrophages to the site of injury (second phase). *These **immune system cells (macrophages)** perform a number of critical functions in normal wound healing, including direct phagocytosis of bacteria; generation of antimicrobial proteins and reactive oxygen and nitrogen species; secretion of elastase and collagenase (enzymes that break up elastin and collagen fibers) affecting debridement of devitalized tissue; and release of various growth factors, cytokines, and enzymes that control cellular proliferation and angiogenesis (plasma stimulation of these processes often controlled by plasma RNS to be discussed below). In this way, the macrophages not only affect a broad range of actions in the early stages of wound healing but also push the wound into the next phase of the healing*

Plasma Science and Technology: Lectures in Physics, Chemistry, Biology, and Engineering, First Edition. Alexander Fridman.
© 2024 WILEY-VCH GmbH. Published 2024 by WILEY-VCH GmbH.

process. It is important that the macrophages can be effectively stimulated by plasma making significant contribution to the wound healing, see Lecture 32. The third phase involves proliferation of fibroblasts that constitute the dominant cell type in the wound by the end of the first week or so. The fibroblasts deposit ground substance into the wound bed and lay down collagen fibers, leading to the formation of a rudimentary granulation tissue and of the extra-cellular matrix (ECM). Protease enzymes released by fibroblasts later aid in the process of ECM remodeling. The fibroblasts also secrete chemo attractants and growth factors, which aid angiogenesis. The fourth and final phase of healing is dependent on the formation of granulation tissue since the keratinocytes responsible for re-epithelialization require a viable tissue surface across which to migrate. The keratinocytes detach themselves from the basement membrane and migrate laterally from the wound edges over the ECM, regenerating the basement membrane. Epithelial stem cells in hair follicles in the deep layers of the dermis also proliferate and migrate over the wound bed. Again, keratinocytes (stimulated by NO, which is one of the major RNS compounds generated in air plasma especially at elevated vibrational temperatures, see Lecture 18, Sections 18.1–18.3) produce a cocktail of growth factors, cytokines, and protease enzymes. The growth factors and cytokines promote angiogenesis, while the proteases help dissolve nonviable tissue.

Let's shortly discuss the **factors affecting the wound healing.** As a rule, normal healing follows the sequence of events described above. In many ways, the initial inflammatory response can be regarded as the key process in this sequence responsible not only for the defense against bacterial contaminants but also for initiating cell proliferation and migration. Prolongation or stasis of the inflammatory stage prevents progression into the proliferative phase and may lead to persistent impairment of healing, destruction of the ECM, and development of necrotic tissue. Such prolongation is influenced primarily by (i) wound infection/colonization and (ii) hypoxia (low oxygenation and blood circulation). **Hypoxia** can be related to a variety of underlying causes such as diabetes, obesity, malignant disease, corticosteroid therapy, chemotherapy treatment, or poor nutrition. Wound infection remains a major cause of serious illness and death, however, and is little impacted by antibiotic use. The consequences of wound infections, particularly in nonhealing chronic wounds, can persist for years and significantly reduce the quality of life enjoyed by patients. For example, most deaths in severely burn-injured patients are still due to burn wound sepsis or complications due to inhalation injury. Approximately 2–10% of post-operative patients get infected wounds (equivalent to some 3.6 million patients in the United States and some 4 million patients in Europe), requiring up to 10 days extra of in-hospital treatment. Diabetic patients with infected ulcers have a significantly higher overall mortality rate compared to diabetic ulcer patients without infections. Pressure ulcer bacteremia results in more than 50% mortality. The types of bacteria found colonizing a wound depend on the type of wound and the phase of wound healing. Initial endogenous and exogenous seeding of the wound results in the creation of microenvironments that ironically encourage later colonization by completely different microbial species. Therefore, chronic wounds will invariably be subject to a polymicrobial colonization in which typically benign commensals, such as *Staphylococcus aureus* and *Pseudomonas aeruginosa*, may become pathogenic. A variety of anaerobes including *Bacteroides* spp., *Prevotella* spp., and *Peptostreptococcus* spp. are also sometimes isolated from wound samples, although their contribution to delaying/preventing wound healing is probably underestimated due to the difficulty associated in culturing and identifying them. When combined with changes in the wound milieu, bacterial proliferation also increases the probability of biofilm formation. This is a combination of cellular components and an extracellular polysaccharide matrix secreted by micro-colonies attached to the wound bed. Bacteria found within biofilms tend to be of higher virulence and display an increased resistance to both immunological defense mechanisms and antimicrobial agents. Biofilm structures have been identified in around 60% of biopsies from chronic wounds. Biofilms have also been implicated in prolonging the inflammatory response and delaying wound healing.

31.2 Specifics of Acute and Chronic Wounds Important for Their Plasma Healing

Wounds are categorized in medicine as acute (abrasions, scalds, burns, and post-operative incisions) or chronic (long-term wounds such as diabetic ulcers, venous ulcers, arterial ulcers, and pressure sores). Acute wounds can develop into a nonhealing state and/or become infected, which limits their process through the phases of healing and so can also become chronic in nature. Let's start with the **acute wounds**. **Abrasions**, as example of acute wounds, are largely uncomplicated and easily managed wounds that usually do not require tools beyond those already existing. Surgical wounds and burns are however much more prone to bacterial colonization, which often develops into overt wound infection. Current prevention and treatment of these wound infections rely on the use of topical antimicrobials, which have significant side effects and are of variable clinical efficacy. Significant concerns remain regarding proliferation of antibiotic-resistant microorganisms.

Surgical wounds also considered as acute ones are often associated with a range of local and systemic changes, not least of which are those that affect the immune system. Surgical trauma, for example, initially causes an early hyper-inflammatory response followed by cell-mediated immunosuppression. Similarly, general anesthetics also produce a short-lived immunosuppression by decreasing T-lymphocyte production, monocyte activity, and cytokine secretion and inhibiting a respiratory burst by neutrophils. Factors complicating surgical wound healing include hypothermia; anesthesia (and therefore immunosuppression); poor aseptic technique; and changes in blood glucose level, oxygenation, and perfusion of tissue (affected by many factors such as cardiac output and perioperative fluid management). All of these have an impact on the risk of wound infection and prolongation of wound healing.

Burns, are considered acute wounds, and may be either partial thickness (first and second degree) or full thickness (third degree). Partial thickness burns involve the epidermis or the epidermis and a portion of the dermis, whereas full-thickness burns involve the entire epidermis and dermis. Full-thickness or extensive partial-thickness burns are associated with significant levels of fluid and protein loss (which may lead to hypovolemia and shock), a hypermetabolic state, and a degree of immunosuppression (possibly because of T-cell dysfunction with failure of interleukin-2 production). Immediately following injury, the surface of a burn is aseptic, but endogenous wound colonization occurs typically within 48 h. Unfortunately, the necrotic tissue and protein-rich wound exudates found in a burn represent an ideal growth medium for bacteria and burn wound infection is a common finding. Burns typically have three zones of involvement, as follows: (i) zone of coagulation, which usually occurs in the areas of maximal damage (in this zone there is irreversible tissue loss due to coagulation of the constituent proteins); (ii) zone of stasis, typically surrounding the zone of coagulation and characterized by decreased tissue perfusion and sluggish capillary flow (the tissue in this zone is potentially salvageable); (iii) zone of hyperemia is the outermost zone in which tissue perfusion is increased (the tissue here will invariably recover unless there is severe sepsis or prolonged hypoperfusion).

Very important focus of plasma treatment is the **chronic wounds**. Normal wound healing proceeds in an orderly sequential manner with four discrete stages: hemostasis, inflammation, proliferation, and maturation. Interruption or prolongation of any of these processes may lead to sub-optimal healing and a failure of the wound to return to functional and anatomical integrity in a timely manner. That said, most chronic wounds are due to persistence of the inflammatory phase with a resulting build-up of ROS, continuing tissue damage, and perpetuation of inflammation. Cytokines (particularly interferon gamma) may play a pivotal role in this pro-inflammatory response. The major chronic wound subtypes are venous, arterial, diabetic, and pressure ulcers; some examples are shown in Figure 31.1, see Fridman and Friedman (2013).

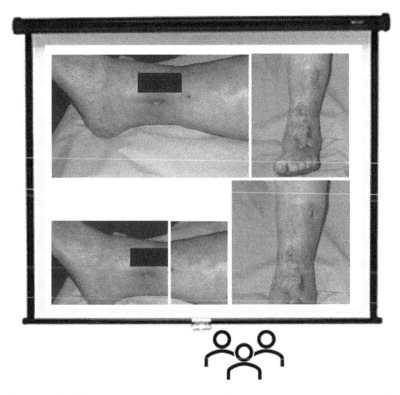

Figure 31.1 Different chronic wounds: top left is venous leg ulcer, top right is arterial leg ulcer, bottom left is ischemic diabetic foot ulcer, bottom right is grade 3 pressure ulcer.

It is generally accepted that **venous ulceration** as a chronic wound is the result of a functional failure of venous valves in the lower limbs, which leads to increased backflow and venous hypertension. This high pressure is transmitted back to the venules and results in distension and increased permeability of the capillary beds. As a result, fibrinogen leaks into the dermis of the skin and (in the presence of reduced fibrinolysis) forms a fibrin cuff around vessels. This prevents diffusion of oxygen and nutrients, contributes to local tissue hypoxia and ischemia, and traps growth factors essential for wound healing in the extravascular space. Venous insufficiency also causes white blood cells (leukocytes) to become trapped in the venules where they release inflammatory factors, proteolytic enzymes, and ROS, which together with the physical plugging of small vessels increase local ischemia and capillary pressures. This in turn increases capillary permeability and leads to further extravasation. Substances that extravasate out of the vessel include red blood cells, which release hemoglobin into the extra vascular space. This is digested and the breakdown products are responsible for the characteristic brown staining and thickening of skin seen in chronic venous ulcer patients.

Arterial ulcers, another type of chronic wounds, are usually caused by progressive atherosclerosis (hardening of the arteries), in which cholesterol plaques gradually narrow and eventually reduce blood flow in large- and medium-sized arteries. Similar changes are found in smaller arteries and arterioles and combine to produce global or regional arterial hypo-perfusion (insufficient perfusion). The accompanying local hypoxia and lack of nutrients result in focal tissue breakdown, particularly in locations most distant from the left side of the heart.

Foot ulcers are a common complication of diabetes and often precede lower-extremity amputation. They have many different causes but, due to a wide number of coexisting metabolic abnormalities, diabetics have a higher incidence of vascular abnormalities including thickening of capillary basement membranes, thickening, and calcification of the arterial tunica media (recall that this is the middle cellular layer in arteries) and arteriolar hyalinosis (accumulation of amorphous material in the walls of small arteries and arterioles, a common degenerative change in the elderly). They are also more prone to developing atherosclerotic disease below the knee. This leads to a generalized reduction in arterial inflow, which prevents adequate oxygenation of tissues and predisposes the patient to developing chronic wounds. Also important is the frequent presence of neuropathy (damage to nerves), which inhibits the perception of pain. As a result, patients may not initially notice small wounds to legs and feet and may therefore develop infections or exacerbate the original injury.

Finally, **pressure ulcers** (bed sores) are caused by ischemia that occurs when external pressure on the tissue leads to capillary occlusion and restriction of blood flow into the area, causing skin and tissue necrosis. After the above classification and simple description of the wounds, we can consider specific examples of the plasma-assisted wound healing.

31.3 Effects of Plasma-generated NO on Wound Healing and Other Medical Treatments

As it was demonstrated in Lecture 18, Sections 18.1–18.3, atmospheric air plasmas are generating significant amount of NO especially when vibrational temperature T_v of molecular nitrogen is high and the Zeldovich mechanism (plasma-stimulated chain process of formation of NO from air, see Sections 18.1–18.3) dominates the nitrogen fixation from air. While more energy effective is NO generation in nonequilibrium plasmas where $T_v \gg T_0$, the more conventional thermal plasmas (where general gas temperature T_0 is also high) are also intensive generators of NO for medical treatment. NO has received particular and significant attention regarding healing wounds and other medical procedures. The importance of NO as a biological molecule was especially recognized relatively recently (1980s). Since Furchgott, Ignarro, and Murad received the Nobel Prize for investigating the function of NO as a signaling molecule in 1998, the idea that a molecule of gas produced by some cells can dissolve in biological fluid, penetrate lipid membranes, and transmit regulatory signals to other cells became much better established. Three such molecules seem to play physiological roles in living systems are: carbon monoxide (CO), NO, and hydrogen sulfide (H_2S). All these molecules can pass through cell membranes, although to a somewhat different extent. All these molecules can serve as electron donors to varying extents and therefore behave as antioxidants (serving as scavengers of multiple oxidizers).

*In humans, NO, and particularly the plasma-generated NO appears to serve a multitude of essential biological functions. It offers antimicrobial and antitumor defense and regulates blood vessel tone, blood coagulation, some immune system activity, apoptosis, neural communication, flat bronchial and gastrointestinal muscles, and hormonal and sex functions. NO also plays an important role in adaptation, stress, tumor growth, immunodeficiency, cardiovascular, liver, and gastrointestinal tract diseases. **NO role in wounds has been particularly noted**. For example, inducible NO synthase (iNOS) grows substantially in traumatic wounds, burn wound tissues, and bone fracture site tissues in the inflammatory and proliferation phases of the healing process.* Activation of iNOS was discovered in cultivation of wound fibroblasts. Macrophage activation in a wound, cytokine synthesis, proliferation of fibroblasts, epithelialization, and wound healing processes are all linked to the activity levels of iNOS. In animal models, injection of iNOS inhibitors disrupts all these processes and especially the synthesis of collagen, while NO synthesis promoters accelerate these processes. Animals with iNOS deficiency demonstrate a significant decrease in wound healing rate; however, this can be reversed by injection of iNOS gene. In complicated wound models (e.g., in experimentally induced diabetes) and in patients with tropic ulcers, lowered activity of iNOS is often found to correlate with slowed healing processes. NO in tissues can be regulated either through drugs that control release of endogenous NO (generated internally to the body by tissues) or administration of exogenous NO (produced externally to the body). Use of exogenous NO in infection and inflammation processes is well studied. It has been linked either to direct antimicrobial effects whereby NO (partly through interaction with ROS) kills bacteria by reacting with various organic molecules, or to stimulation of the immune system responses such as activation of macrophages and T-lymphocytes as well as induction of cytokine and antibody production. The influence of NO on increasing microcirculation also promotes delivery of immune system components to the site of infection. Exogenous NO was demonstrated to be important in traumatic wound processes. Its delivery through NO-donors (nitrogen-containing compounds) to the wound promotes healing processes in animals with complicated wounds and in animals with inhibited iNOS. The various effects of NO can be complex. In some environments NO acts as an antioxidant, while in others it acts as an oxidizer. The ultimate effects may depend on the NO concentration. Understanding the effects of different NO concentrations, coupled with theoretical and experimental data on NO generation in air plasmas, can serve as a basis for a series of biomedical experiments focused on the use of the plasma-generated exogenous NO delivered directly to the pathologic site for control of inflammatory processes and increase in the rate of wound healing.

Given the importance of NO, there has been a tendency in some discussions of plasma treatment to focus on NO as the key molecule generated by plasma. If that were the case, however, we could question the use of plasma as a source of NO in healthcare setting when bottled NO is often available. One argument in favor of plasma is that it requires only electrical energy, which today is widely accessible in various environments. Bottled NO, on the other hand, can be associated with various logistical challenges. Beyond this argument, there could be more compelling reasons to employ plasma in various treatments. Plasma can be a source of NO as well as various other reactive species. In some situations, reviewed in the following section, NO may have been the key factor. In other situations, however, it is not entirely clear what active species play what roles in the treatment.

31.4 Intensive-NO-generating Plasma Sources for Wound Treatment

One of the most investigated plasma-medical devices whose therapeutic effects are based primarily on NO effects is probably the so-called **Plazon system** based on the jet of hot air plasma rapidly cooled upon exit from the plasma generation region (Shekhter et al. 1998; Pekshev 2001). This plasma device has been investigated for: sterilization of wound surfaces; destruction and desiccation of dead tissue and pathologic growths; dissection of biological tissues with the plasma jet; and for stimulation of regenerative processes and wound healing by the gas flow with temperature of 20–40 °C. Plazon generators (Shekhter et al. 1998; Pekshev 2001) are DC arcs with different configurations of the exit channels corresponding to the different applications (blood coagulation, tissue destruction, therapeutic manipulation/stimulation). The main elements of the system construction are the liquid-cooled cathode, intra-electrode insert, and anode. Atmospheric air enters the manipulator through the built-in micro-compressor, passes through the plasma arc, heats up and accelerates, and exits through the hole in the anode of the plasma generating module. Plasma temperature at the anode exit varies for different configurations of the device, corresponding to different medical applications. Temperature drops rapidly away from the anode: at 30–50 mm from the anode, the flow is composed simply of the warm gas and the plasma-generated NO. NO content in the gas flow is mainly determined by the quenching rate. The necessary quenching rate for effective operation of the medical device is about 10^7–10^8 K s^{-1}. Commonly, the cooling rate of plasma jets is of the order 10^6 K s^{-1}. To achieve the cooling rate of 10^7–10^8 K s^{-1}, it is necessary to utilize additional cooling of the plasma jet which has been achieved by special construction of the plasma nozzles. The therapeutic manipulator–stimulator configuration of the Plazon discharge system is used solely for therapeutic treatment by exogenous NO. The principal difference of this manipulator is that the air-plasma jet does not freely exit into the atmosphere. Instead, it exits the anode into a two-step cooling system in which gas channels are created in a maze scheme to force-cool the jet by the liquid circulating from the cooling system. This construction permits NO-containing gas flow (NO-CGF) with sufficiently low temperature, which makes it possible to apply this manipulator for the treatment of external body surfaces by using the cooling hose of 150 mm length (temperature of NO-CGF at the exit is about 36 °C). NO content in the gas flow also depends on the distance from the exit channel. For laparoscopic operations, a special manipulator of 350 mm length and 10 mm diameter is utilized. The possible operating regimes of the apparatus are defined by the characteristics of the gas flow exiting from the manipulator, the main parameters of which are its temperature and the nitrogen oxide content. The first group of regimes is that of free-flowing plasma afterglow exiting the manipulator, while the second group of regimes is the treatment of bio-tissues by completely cooled (20 °C) NO-CGF. The second set of regimes not only allows the tissues to be directly treated by NO but also allows the NO to be delivered to a pathologic location through drainage tubes, puncture needles, or any endoscopic device (gastroscope, bronchoscope, cystoscope, rectoscope, etc.).

Another example of the intensive nitric-oxide-producing discharge for plasma-medical applications is a special DC spark microplasma called **pin-to-hole (PHD) discharge system**. Engineering aspects of this plasma source have been discussed in Section 14.6. This system has been developed for medical applications and investigated in situations where access to the affected site might be limited, such as in corneal infections (Gostev and Dobrynin 2006). The device allows generation of plasma afterglow with average gas temperature not exceeding 30–40 °C. It consists of a tube cathode with a small hole at the end that functions as a nozzle and a pin-like anode positioned coaxially inside the cathode tube. The discharge gap is formed essentially between the tip of pin and the edges of the hole at the end of the cathode tube. The gas is fed through the tube from the back to the discharge gap. The voltage is applied between the pin anode and the tube cathode as a capacitor discharges through the plasma gap. Different gasses, including moist and dry air and Xe, have been employed to form the plasma. Typical discharge voltage decays start from 1 to 3 kV, while the voltage pulse duration is around 50 μs and total power is of the order 1–2 W. The typical diameter of the afterglow sphere outside the nozzle is about 3 mm. This microplasma operated in Xe radiates intensively in the UV range; operated in air it generates excited oxygen species, ozone, nitrogen oxides (primarily NO, but also NO_2), and OH radicals (Gostev and Dobrynin 2006). Both regimes have bactericidal effects and air plasma is also able to aid in **tissue regeneration via NO therapy.** UV radiation of Xe plasma in this case is: UVA (315–400 nm) 180 μW cm^{-2}, UVB (280–315 nm) 180 μW cm^{-2}, and UVC (200–280 nm) 330 μW cm^{-2}. UV-radiation of air plasma is: UVA (315–400 nm) 53 μW cm^{-2}, UVB (280–315 nm) 25 μW cm^{-2}, and UVC (200–280 nm) 90 μW cm^{-2}. The ability of the above microplasma system to sterilize surfaces has been demonstrated by Misyn et al. (2000). *Staphylococcus* culture in liquid media (2×10^6 CFU ml^{-1}) have been treated by the air plasma plume of 3 mm diameter, incubated for 24 h, and counted. A 6-log reduction in viable bacteria is achieved in 25 s of treatment; however, the sterilization efficiency drops off with increasing volume of liquid, which inhibits UV penetration and diffusion of active species generated in plasma. During the investigation of plasma treatment of ulcerous dermatitis of rabbit cornea, two important observations were made: (i) plasma treatment has a pronounced and immediate bactericidal effect and (ii) the treatment influences wound pathology and the rate of tissue regeneration and wound healing process.

31.5 Nitric Oxide Effects in Plasma-medical Plazon Treatment: Cell Cultures, Wound Tissues, Animal Models

To analyze the biochemistry of the plasma wound healing, Shekhter et al. (2005), investigated the dynamics of levels of **endogenic NO in wound tissues and in organs** in an animal model (70 rats). Electron paramagnetic resonance or EPR spectroscopy, a method commonly employed to detect free radicals, was utilized to investigate the endogenic NO. Given the relatively short lifetime of many radicals in organic solutions and tissues, special molecules called "traps" are employed. These traps react preferentially with a designated radical and can be relatively long lived. Shekhter et al. (2005) used diethylthiocarbamate (DETC) as a NO trap. The DETC was injected into rats with full-thickness flat wound of $300\,mm^2$ area five days prior to EPR analysis. Following euthanasia, samples were collected from the animals' blood, granular tissue from the bottom of the wound, and from internal organs (heart, liver, kidney, and the small intestine). For a portion of the animals, five days after initial wound introduction the wound surface was treated by the NO-CGF (500 ppm).

Without the NO treatment, the results indicated a high content of endogenic NO in wound tissues ($10.3\,\mu M$). The liver of the animals with the wound contained $2.3\,\mu M$ of DETC-iron-mononitrosyl complex (IMNC, a complex of the trap, NO, and Fe), while the control group (without the wound) had much lower levels of only $0.06\,\mu M$. Animals without the wound were used to investigate capability of gaseous exogenous NO to penetrate through undamaged tissues of the abdominal wall. Treatment by the NO-CGF was performed for 60 and 180 s. A linear dependence of the amount of DETC-IMNC produced in the liver and blood of the animal on the NO-CGF treatment time was observed. When the animal was euthanized 2 min after the 180 s treatment, a maximum signal was registered in the bowels of the animal which was 2.6 times higher than the control. In the heart, liver, and kidney the difference was a factor of 1.7. These results clearly indicate the ability of the exogenous NO molecules to penetrate undamaged tissues and potentially trigger endogenic response.

A more complex relationship was observed in **treatment by exogenous NO of the wound tissues.** If the animal was euthanized 30–40 min following the 180 s treatment, the NO content in wound tissue and blood was observed to increase by a factor of 9–11 over that observed in the case of the 2 min interval. Several explanations of this effect are possible. One is that an activation of the first cascade of antioxidant defense following exogenous NO treatment leads to a significant decrease in the levels of superoxide. This may have considerably decreased the influence of ROS on DETC-IMNC and the nitrosyl complexes of the hemoproteins (complexes where NO– ion binds to iron forming Fe—NO). Another possible explanation is activation of iNOS and resulting endogenic NO generation. This partially explains the discovered phenomena of stimulation of wound development processes via the influence of exogenous NO, when there is a deficiency of endogenic NO or excess of free radicals including superoxide.

In experiments on rabbit cornea, mucous membranes of hamster mouths and meninx membrane in rats via lifetime microscopy, it was found that the effect of the expansion of the opening of the micro-vessels under the influence of exogenous NO (500 ppm) lasts with varying intensity up to 10–12 h, while the lifetime of NO molecules is much shorter (Shekhter et al. 1998; Shekhter et al. 2005). This provides additional evidence that a single application of plasma-generated exogenous NO initiates a cycle of cascade reactions, including biosynthesis of endogenic NO, which leads to a long-lasting effect and explains the success of the NO therapy.

The **effect of the exogenous NO on the cellular cultures of the human fibroblasts and rat nervous cells** was studied by Shekhter et al. (1998), Shekhter et al. (2005), and Ghaffari et al. (2005). Single treatment by the plasma-generated NO of the cell cultures significantly increases (2.5 times) the cell proliferation rate via the increase of DNA synthesis (tested by inclusion of C14 thymidine) and, to a lesser extent (1.5 times), the increase of protein synthesis by the cells (tested by inclusion of C14 amino acids). As expected, the stimulating effect is dose dependent. The action of exogenous NO on the phagocytic activity of the cultured wound macrophages from the washings of the trophic human ulcers, studied by photochemiluminescence (Krotovskii et al. 2002) revealed that a maximum increase in the luminous intensity (a factor of 1.95 greater than control) confirms the activation of the proteolytic enzymes of macrophages under the effect of NO-CGF. A statistically significant increase in fluorescence of macrophages was observed in less than 24 h following a 30 s treatment.

31.6 Clinical Tests of Plasma-induced NO-therapy of Wounds: Plasma-medical Plazon Healing of Ulcers

*Clinical tests of the plasma-induced NO-therapy of wounds and especially ulcers using the Plazon system provide quite promising results. Application of air-plasma-generated exogenous NO in the **treatment of venous and arterial trophic ulcers** of lower extremities with an area of 6–200 cm² in 318 patients showed high efficiency,* see Shekhter et al. (1998), Shekhter et al. (2005), Fridman et al. (2008), and Fridman and Friedman (2013). For assessment of the effectiveness of the plasma NO-therapy, clinical and planimetric indices were analyzed during the sanitation and epithelialization of ulcers; a bacteriological study of discharge from

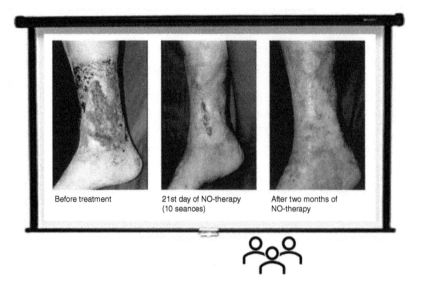

Before treatment

21st day of NO-therapy
(10 seances)

After two months of
NO-therapy

Figure 31.2 Plasma-medical wound healing, clinical test: reduction of ulcer size as a function of time during the Plazon-generator-based plasma NO therapy.

the ulcer; cytological study of exudates; a histopathological study of biopsies from the boundary of a trophic ulcer; the indices of microcirculation (according to the data obtained by laser doppler flowmetry or LDF); and transcutaneous partial pressure of oxygen. In the main groups of observations, trophic ulcers were processed in the regime of NO therapy (500 and 300 ppm). Prior to beginning the therapy, the ulcer surface was treated in the regime of coagulation until the evaporation of necrotic debris. Following initial treatment, the wound was treated for 10–30 days in the NO-therapy regime. Proteolytic and antimicrobial drugs were used in the control group in the exudation and necrosis phases and wound coatings in the tissue regeneration and epithelialization phases.

Planimetric observation of the **dynamics of decrease of the trophic ulcer area** showed that, on average, traditional treatment methods applied to the control group led to 0.7% per day decrease, while in the experimental group demonstrated a decrease of 1.7% per day. Cleansing of ulcers from necrosis and exudate and the appearance of granulation and boundary epithelialization were accelerated with NO-therapy by a factor of 2.5. The time to final healing was reduced by a factor of 2.5–4, depending on the initial ulcer size, see Figure 31.2 (Shekhter et al. 1998; Shekhter et al. 2005; Fridman et al. 2008; Fridman and Friedman 2013). Larger ulcers tended to close faster than smaller ulcers. LDF investigation of microcirculation in the tissues of trophic ulcers showed that following the NO-therapy, pathologic changes in the amplitude-frequency characteristics of the microvasculature were normalized and regulatory mechanisms were activated. By 14–18 days, the average index of microcirculation, value of root-mean-square deviation, coefficient of variation, and index of fluctuation of microcirculation approached those of the symmetrical sections of healthy skin. In the control group, the disturbances of microcirculation remained. Against the background of treatment, normalization of the level of transcutaneous partial pressure of oxygen (TpO_2) happened at a higher rate in the experimental group than in the control group, especially at the NO concentration of 500 ppm, see Figure 31.3, see more details in Shekhter et al. (1998), Shekhter et al. (2005), Fridman et al. (2008), and Fridman and Friedman (2013).

A bacteriological study of **wound discharge from the trophic ulcers** showed that for the experimental group, the plasma NO-therapy (especially in combination with the preliminary coagulation of ulcerous surface) reduced the degree of bacterial seeding (microbial associations). The level of bacterial seeding fell below the critical level necessary for maintaining the infectious process in the wound by days 7–14, see Figure 31.4 (Shekhter et al. 1998; Shekhter et al. 2005; Fridman et al. 2008; Fridman and Friedman 2013).

Use of the plasma-generated NO for **local treatment of ulcerous and necrotic tissues** in patients with diabetes (diabetic foot ulcer) was demonstrated by Shulutko et al. (2004), see review in Fridman and Friedman (2013). Patients were selected for this study following two months of unsuccessful treatment by state-of-the-art techniques. The improved success was evident from the first few sessions: inflammatory reaction was clearly reduced; patients reported a reduction in pain and cleansing of the ulcer surface was clearly visible. Following 10 sessions, most patients

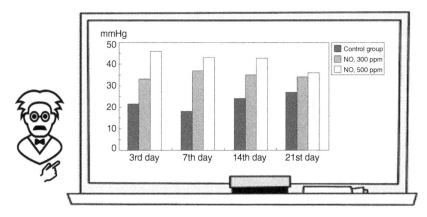

Figure 31.3 Plasma-medical wound healing, clinical test: normalization of the level of transcutaneous partial pressure of oxygen in trophic ulcer tissues for different NO concentrations and for the negative control.

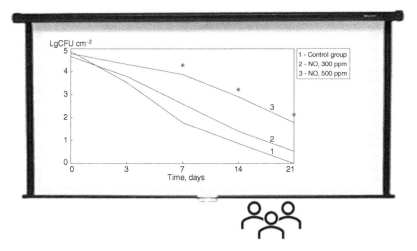

Figure 31.4 Plasma-medical wound healing: bacterial load reduction in trophic ulcer tissue as a function of time for different NO concentrations versus negative control.

expressed positive healing dynamics and the ulcer size decreased to between one-third and one-quarter of the original size. LDF markers, pO_2, and bacteriological investigation all showed a positive dynamic. In patients with relatively small-sized ulcers (initial diameter less than 1 cm), full epithelization occurred after six to eight plasma NO treatment sessions. The period of stationary treatment and full clinical recovery of patients was noticeably shortened (on average by 2.3 times). In the cases of large ulcerating wounds, the necessity for amputation was reduced by a factor of 1.9, see Figure 31.5 (Shekhter et al. 1998; Shekhter et al. 2005; Fridman et al. 2008; Fridman and Friedman 2013).

31.7 Other Plazon Clinical Tests of Plasma-induced NO-therapy of Wounds

The plasma-induced NO-therapy using the Plazon thermal plasma system demonstrated high efficacy of treatment multiple types of wounds. Thus, effectiveness of the exogenic NO and air plasma on **healing of pyoinflammatory diseases of soft tissues** has been demonstrated by studying 520 patients with purulent wounds of different etiology and 104 patients with the phlegmonous-necrotic form of the erysipelatous inflammation, see Lipatov et al. (2002). By the fifth day of therapy, wounds on most of the patients in the experimental group (90%) were clear of necrotic tissue (contrary to the control group) and were beginning to be covered by bright spots of granular tissue. Microbial infestation of the wound tissue had lowered from 10^6 to 10^8 colony-forming units (CFU) per gram of tissue to 10–100. Data from complex analysis of microcirculation (LDF, pO2) showed significant repair of the microvasculature and blood flow in the wound tissues in most of the patients in the experimental group. The predominant types

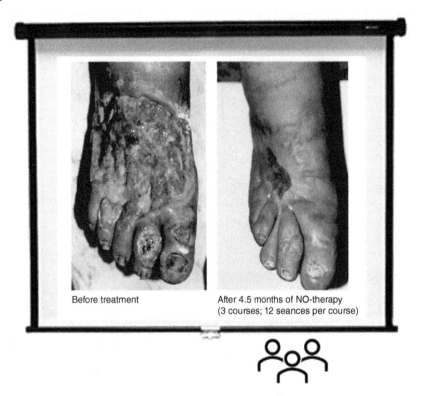

Before treatment

After 4.5 months of NO-therapy
(3 courses; 12 seances per course)

Figure 31.5 Plasma-medical wound healing, clinical test: example of a large, ulcerated wound before and after 4.5-month duration of NO treatment.

of cytograms were regenerative and regenerative-inflammatory with a notable increase in fibroblast proliferation (on average, 18.5%). By day 7–10 of treatment, large, suppurated wounds, for example suppurated burn wounds were clear of the pyonecrotic exudate and were beginning to be covered by granular tissue; in other words, these wounds were ready for dermatoplasty.

 *Effectiveness of plasma NO-therapy is most apparent in the treatment of the **pyonecrotic form of erysipelatous inflammation**, that is, the most severe cases of purulent surgery departments,* see Lipatov et al. (2002). The combination of surgical preparation of extensive pyonecrotic centers and local NO-therapy allowed in most of the patients with phlegmonous-necrotic erysipelas during 12–14 days of treatment to liquidate heavy pyonecrotic process and to create conditions for completion of reparative procedures. In **maxillofacial surgery**, plasma NO-therapy was used to accelerate the healing of post-operative wounds and in preventive maintenance of the formation of hypertrophic and keloid scars, treatment of the formed scars, treatment of pyonecrotic processes (abscesses, phlegmon, etc.). In the latter case, preliminary coagulation of purulent centers is sometimes utilized, see Shekhter et al. (2005) and Fridman et al. (2008).

Plasma NO treatment has been also successfully applied to **surgical oncology** (Kabisov et al. 2000; Reshetov et al. 2000). Inter-operative treatment in the coagulation regime of the Plazon system ensures ablation; it considerably decreases blood plasma and whole blood losses from extensive wound surfaces because of thin film formation over the wound surface consisting of coagulative necrotic tissue. As a result of **plasma-induced NO-therapy of post-operative wounds**, a significant decrease in inflammation is observed along with stimulated proliferation of granular tissue and epithelization. The effect is observed independently of the location of the wound on the body and of the plastic material used. An additional benefit is the prophylactic treatment of the local relapses of the tumor, which enables this method to be widely applied in oncoplastic surgeries. The effectiveness of **NO-therapy in treatment of early and late radiation reactions allows surgeons** to carry out a full course of radiation therapy in 88% of the patients. Treatment of radiation tissue fibrosis also yields a statistically significant improvement, confirmed in morphological investigation of these tissues. Plasma NO-therapy is successfully used for the prevention of the formation of **post-operative hypertrophic and keloid scars**, and for treatment of already formed scars by softening the scar tissue, decreasing fibrosis, and preventing their relapse with the surgical removal.

In **ophthalmology**, treatment by the plasma-generated NO NO-CGF (300 ppm) does not induce a toxic reaction and does not cause intraocular pressure changes or morphological changes in the tissues of eye, but considerably accelerates the healing of wounds and burns of cornea. Such therapy has been used in clinics for the effective treatment of cornea burns, erosions, injuries, and burn ischemia of conjunctiva (Chesnokova et al. 2003). In **gynecology**, the effectiveness of plasma NO-therapy has been demonstrated for patients with purulent inflammation of appendages of the womb (Davydov et al. 2002, 2004; Kuchukhidze et al. 2004). Where the abdominal cavity was opened in surgery, the purulent wound was treated by air plasma in the coagulation regime. Later, plasma NO-therapy was applied remotely through the front abdominal wall and vagina. With operational laparoscopy after dissection and sanitation of the purulent center, the region of surgical incision and the organs of the small basin were treated by NO-CGF delivered locally through the aspiration tube. Plasma NO-therapy was also continued in the post-operation period. The use of NO-CGF in surgical and therapeutic regimes aided the rapid reduction in the microbial load and in swelling, lowered the risk of post-operative bleeding, encouraged rapid development of reparative processes, and decreased the overall time that patients remained in hospital by six to eight days on average. NO-CGF was also used in organ-saving surgical operations on the womb, the uterine pipes, and the ovary. More information on the subject has been reviewed by Fridman et al. (2008) and Fridman and Friedman (2013).

31.8 Wound Healing Using Pin-to-hole (PHD) and Microwave Transitional "Warm" Plasma Discharges

Similarly to the Plazon plasma-medical treatment, other transitional "warm" and thermal plasma sources, especially the PHD and microwave discharges, have been effectively used for wound healing. The results of **PHD microdischarge**, described in Sections 14.6 and 31.4, on ulcerous dermatitis of rabbit cornea, demonstrated that treatment has a pronounced and immediate bactericidal effect and clear benefits for wound pathology, the rate of tissue regeneration and the wound healing process. These results provide strong support for the clinical application of the transitional "warm" PHD microplasma system for treatment of human patients with complicated ulcerous eyelid wounds. A before-and-after PHD discharge treatment example is shown in Figure 30.3 (Misyn and Gostev 2000; Gostev and Dobrynin 2006). Necrotic phlegm on the surface of the upper eyelid was treated by an air plasma plume of 3 mm diameter for 5 s once every few days. By the fifth day of treatment (two 5 s plasma treatment sessions) the eyelid edema and inflammation were reduced. By the sixth day (third session), the treated area was free of edema and inflammation and a rose granular tissue appeared. Three more plasma treatments were administered (six in total), and the patient was discharged from the hospital six days after the last treatment. PHD microplasma treatment is being further developed for stimulation of reparative processes in various topical wounds, tropic ulcers, chronic inflammatory complications, and other diseases of soft tissues and mucous membrane (Misyn and Gostev 2000; Gostev and Dobrynin 2006).

A microwave plasma-medical transitional **"warm" plasma device for indirect (argon plasma afterglow jet) treatment**, called MicroPlaSter, has been developed in 2005 by the Max Planck Institute for Extraterrestrial Physics in Garching, Germany, group led by Gregor Morfill. This plasma-medical has been built for clinical treatment by ADTEC Plasma Technology Co. Ltd., Hiroshima, Japan and London, UK. The device creates plasma using microwave electromagnetic field. Flow of gas (e.g., argon or air) through the discharge region delivers the cooled afterglow to the skin. The typical torch-skin distance is about 20 mm. The system allows the treatment of relatively large inhomogeneous and topographically uneven areas (about 5 cm in diameter) below the threshold of thermal damage in contact-free mode. The active agents are carried with inert argon gas from production in the torch to the desired region. Typical flow delivers up to 10^{10} active species per cm^2 per second. A mixture of hydrogen peroxide, NO, and nitrogen dioxide has been reported in argon afterglow of this plasma. Various nitrogen species including NO and nitrogen dioxide have also been reported when air is employed as the plasma gas. Clinical trials conducted using the MicroPlaSter focused on lower leg ulcers have been performed by Dr. Isbary and his group, see Isbary et al. (2010, 2012, 2013). Chronic ulcers of the lower leg are associated with considerable patient morbidity and account for an estimated 1–2% of the annual healthcare budget in European countries, for example. Bacterial colonization of such wounds is a well-recognized factor contributing to impaired wound healing.

Having carried out Phase I of the trial, focusing on safety and bactericidal dosing, Isbary a group of researchers from the Max Planck Institute for Extraterrestrial Physics and the Department of Dermatology, University of Regensburg, Germany, reported promising results (Isbary et al. 2010, 2012, 2013) in a Phase II study of 38 chronic infected wound treatments in 36 patients. Patients with chronic infected skin wounds who attended the outpatient and inpatient clinics of the Department of Dermatology, Allergology and Environmental Medicine of Hospital Munich Schwabing in Germany were invited to participate in this trial. Patients could be included in the trial if they had at least one colonized wound large enough for plasma treatment and a control area of 3 cm². In addition to standard wound care, patients received a daily 5 min cold plasma treatment to the randomized wound(s) using the MicroPlaSter plasma torch operating at 2.46 GHz, 86 W and having Ar gas flow of 2.2 slm (standard liters per minute) at a distance of 2 cm. Control wounds remained undressed during plasma treatment. The same standard wound care was given to both plasma-treated and control areas. Because of the uneven surface of most of the wounds, high-pressure water jet or scalpel was used to clean all wounds of debris before the initial plasma therapy. Bacterial load on the wounds was assessed in several ways. Once each week, two standard bacterial swabs were taken from all control and plasma-treated areas immediately after dressing removal and before redressing to detect the types of bacteria present. On the other days of the week, nitrocellulose filters were used to detect changes in bacterial load. These filters were applied to the wounds with gentle pressure before and after the treatment and then placed on Columbia blood agar plates and incubated for 12 h at 36 °C. Semi-quantitative assessment of the plates was carried out by a manual count. An important benefit of the filter technique is that it displays where the bacteria are primarily situated within the wound. Plasma treatment in this trial resulted in a highly significant (c. 34%) reduction in bacterial count in plasma-treated areas compared to nontreated areas. The corresponding bootstrap test confirmed the high significance level of the results ($n = 500$). This is probably the first clinical evidence of a significant bactericidal effect in patients of plasma treatment when used in addition to standard wound care for chronic infected wounds, see Isbary et al. (2010, 2012, 2013) for more details.

31.9 Treatment of Skin and Infected Wounds Using FE-DBD Plasma

Fridman et al. (2005a–c, 2006a,b, 2007a–e, 2008), analyzed human tissue sterilization including skin sterilization suppressing the "skin flora" (a mix of bacteria collected from cadaver skin containing *Staphylococcus*, *Streptococcus*, and yeast), which has been discussed in the Section 30.6. It has been demonstrated that direct FE-DBD plasma sterilization leads roughly to a 6-log reduction in bacterial load in 5 s of treatment, which is illustrated in Table 30.1. The challenge in this regard is high sensitivity of the FE-DBD sterilization effectiveness to the superficial specifics of the tissue. Dry live skin has complicated morphology with a lot of possibilities for microorganisms to "hide" from the direct plasma treatment. Special laser scanning confocal study performed in the frameworks of the FE-DBD wound sterilization and toxicity analysis in the porcine animal model (Figure 31.6) showed that microorganisms (*S. aureus*) penetrate the skin quite deep, about 1 mm. This is probably a reason why even direct FE-DBD plasma treatment of live dry skin results usually in not more than 2 log reduction of bacterial load. It looks like the bacteria are hiding in the skin. This level of deactivation of microorganisms can be achieved at quite low nondamaging treatment doses, see Figure 31.6. It is interesting that when the skin is even slightly wounded effectiveness of the bacterial load suppressing is usually higher and achieved even at lower FE-DBD treatment doses, which is illustrated in this figure. The wound wetness and bleeding help direct plasma effects to penetrate deeper in the wounded skin, see Lecture 30, more information in Dobrynin (2011).

 *This effect becomes even more pronounced in the case of direct FE-DBD treatment of heavier bleeding and infected wounds, see Figure 31.7. Due to the intensive penetration of the direct plasma-generated active species in the wet and bleeding tissue of the wounds, the **effectiveness of suppressing the bacterial load on the infected wounds** can be orders of magnitude higher than that of the intact dry skin, which stays on the level of about 2 log reduction at the FE-DBD plasma treatment doses not damaging the skin.* Plasma treatment of the infected wounds in the live rat model demonstrating this effect is illustrated in Figure 31.7 (Dobrynin 2011). The infected wound treatment protocol included 2 cm incision of the animal followed by a ring insertion and wound closure, see Figure 31.7. Then *S. aureus* bacteria (at least 10⁴ CFU) were applied to the wound followed by a special dressing of the infected wound for internal development of the infection. Then the infected wounds were treated

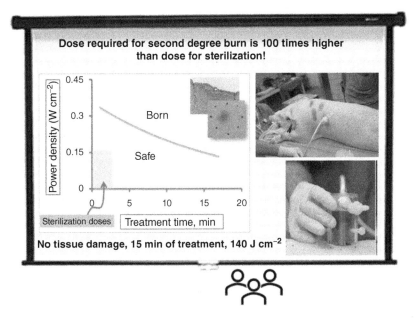

Figure 31.6 FE-DBD wound sterilization and toxicity analysis; plasma treatment of wounded skin: porcine animal model; wound is easier to sterilize than skin.

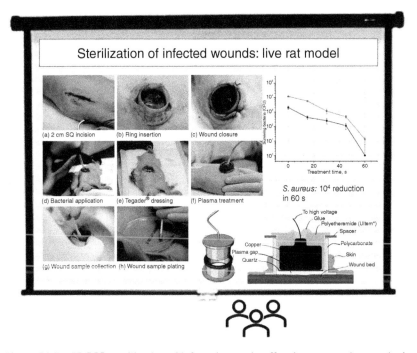

Figure 31.7 FE-DBD sterilization of infected wounds, effect is stronger than on the intact skin.

for 60 s using the FE-DBD plasma at the treatment doses way below the damaging level, see Lecture 30. The wound samples have been finally collected, plated, and analyzed. These animal tests demonstrated 4 log reduction of the *S. aureus* bacterial load on the infected wound because of 60 s direct FE-DBD plasma treatment, which is significantly stronger effect than relevant result in suppressing same microorganisms on the dry skin at the similar plasma and treatment conditions (Dobrynin 2011). More information on the FE-DBD treatment of infected wounds in different models can be found in Fridman et al. (2005a–c, 2006a,2006b, 2007a–e, 2008) and Dobrynin (2011), as well as in the book of Fridman and Friedman (2013).

31.10 Plasma Wound Healing in Clinical Studies

Being the first in focus of plasma medicine, it is not surprising that plasma wound healing is the first to go through significant clinical studies. There are several clinical trials at this moment proving the plasma wound healing safety and efficacy, summarized for example in such recent books and review papers as Metelmann et al. (2018a,b, 2022), Boeckmann et al. (2020), Bernhardt et al. (2019), Gan et al. (2021), and Von Woedtke et al. (2022), where one can find a lot of details on this subject. The first clinical trials were mainly focused on the safe application of the nonthermal plasma devices and the reduction of the microbial load of chronic wounds, see Isbary et al. (2010, 2012), Brehmer et al. (2015), and Ulrich et al. (2015). In these studies, the nonthermal plasma application was used in addition to regular wound care, where antibiotic and/or antiseptic therapy was not excluded generally. Moreover, in some studies, plasma was tested against local antiseptics. Any statements applying to wound closure or other indications of healing processes were avoided or mentioned marginally. Consequently, a first meta-analysis considering these studies concluded that the use of cold plasma "in wound care is safe, but the retrieved evidence and meta-analysis show that there is no clinical benefit … in chronic open wounds" (Assadian et al. 2019). This was not surprising because the effectivity of well-proven local antiseptics or systemic antibiotic therapy to reduce the microbial load of chronic wounds is undisputed. Consequently, it was crucial to demonstrate that nonthermal plasma is more than a "local physical antiseptic." Several clinical studies and case reports in human volunteers or patients with well-defined artificial acute wounds without microbial contamination or infection, respectively, could reproduce the results from animal trials, that nonthermal plasma application is also effective in stimulating tissue regeneration directly, see for example Metelmann et al. (2012), Heinlin et al. (2013), Vandersee et al. (2014), and Von Woedtke et al. (2022).

Meanwhile, there are meaningful randomized controlled clinical trials available where stimulation of chronic wound healing by nonthermal plasma was the first objective. In a clinical trial by Stratmann et al. (2020), 62 diabetic foot ulcers were nonthermal plasma-treated over 14 days using the argon-driven plasma jet kINPen MED. A significant reduction of the wound area in the plasma group compared to a placebo group was found, whereas both nonthermal plasma and placebo treatment reduced the bacterial load of the wounds. It must be noted that all wounds received standard wound care including local disinfection and systemic antibiotic treatment if indicated. Therefore, this study demonstrated beneficial nonthermal kINPen plasma jet effects on wound healing based on direct stimulation of tissue regeneration independent from antiseptic plasma effects in a randomized, placebo-controlled clinical trial in a population-based, representative cohort of patients.

These nonthermal kINPen plasma jet effects were confirmed in another clinical study including 44 diabetic foot ulcer patients. In this study, a helium-driven high-frequency (HF) plasma jet was used by Mirpour et al. (2020). In a further clinical trial, 37 patients suffering from chronic ulcers of different etiology were nonthermal plasma treated one time or three times per week over twelve weeks maximum with the argon-driven plasma torch SteriPlas. Here again, a significant reduction of wound area was found in the nonthermal plasma-treated groups. Moreover, the nonthermal plasma treatment three times a week was not superior to the once-a-week treatment (Moelleken et al. 2020). In a similar study, another argon-driven HF plasma jet treated 50 patients with pressure ulcers once a week over eight consecutive weeks additionally to standard wound care. Both a significantly improved wound healing and a reduction in bacterial load were found (Chuangsuwanich et al. 2016). In a case series with 32 patients suffering from venous and mixed leg ulcers, a daily treatment with a needle-type helium-driven plasma device improved wound healing and reduced pain (González-Mendoza et al. 2019). Consequently, the support of chronic wound healing by nonthermal plasma is proven according to the current state of clinical research. This is particularly remarkable against the background that generally in the field of wound healing few randomized, controlled trials of wound dressings, bandages, etc., which are classified as medical devices, are available because wounds present as part of such a complex presentation that any generalization within the scope of a clinical trial is estimated to be very difficult (Cutting et al. 2017).

The described wide range of successful clinical trials focused on application of plasma for wound healing has opened the pathway for the "scientifically based plasma-medical devices" to the medical market, see Metelmann et al. (2018a,b, 2022). It is interesting that this wound-healing-focused pathway permits application of the relevant certified devices to the wider range of medical use. Similar situation occurs also in the case of the ICE standardization and certification of plasma-medical devices discussed in Section 30.13.

31.11 "Scientifically Based" Plasma-medical Devices Healing Wounds and Ulcerations in Medical Market

While the thermal plasma-based medical devices, especially blood cauterization devices, are not newcomers in the medical market, see next section, the nonthermal plasma devices are. Based on the relatively long-term experimental and clinical research, the first nonthermal plasma devices are on the market that are CE-certified as medical devices class II a, see Metelmann et al. (2018a,b, 2022) and Von Woedtke et al. (2022). Between these devices, we should especially mention the argon-driven HF plasma jet kINPenMED (Neoplas Med, Greifswald, Germany), the argon-driven microwave plasma torch SteriPlas (ADTEC, Hunslow, UK), and the DBD-based devices PlasmaDerm (CINOGY, Duderstadt, Germany) and Plasma Care (Terraplasma Medical, Garching, Germany). The latter two devices use atmospheric air as working gas. Currently, these devices are mainly used in reconstitution or re-stimulation of healing (especially wound healing) processes that are defective caused by other primary reasons resulting in chronic wounds. The direct (initial) stimulation of tissue regeneration is a unique characteristic of nonthermal plasma with a supportive role of antimicrobial/antiseptic plasma effect.

First devices are on the medical marked since 2013 with increasing successful application in clinical practice. Unfortunately, comprehensible, and systematic documentation of this practical experience beyond clinical studies is currently rare. Reported experimental and clinical results on direct stimulation of tissue regeneration by nonthermal plasma make it also applicable in the field of acute wound healing, even if acute surgical but also traumatic wounds typically do not need additional stimulation because they are healing regularly under physiological conditions. However, in cases where the risk of retarded healing is enhanced because of the patient's health status, nonthermal plasma application may be promising, which has been proven in the first case reports, see Hartwig et al. (2017). Moreover, the use of nonthermal plasma for the disinfection of primary and secondary wounds to avoid postoperative wound infection is repeatedly discussed, see Metelmann et al. (2018a,b, 2022) and Von Woedtke et al. (2022). However, it will be difficult to prove the clinical benefit of such preventive nonthermal plasma application directly. The same is with the question if any initial acceleration of wound closure by CAP treatment as demonstrated experimentally in vivo, has any positive impact on scar formation, see Cutting et al. (2017). Nevertheless, together with possible preventive effects on wound infection, nonthermal plasma treatment is mentioned as a promising option to support several interventions also in the field of plastic surgery and aesthetic medicine, see Metelmann et al. (2018a,b, 2022) and Von Woedtke et al. (2022).

A specific nonthermal plasma application close to wound healing, which is extremely helpful for the patients concerned, is the treatment of left ventricular assist devices (LVAD) driveline infections. The certified plasma-medical devices have been successfully used for this application. LVAD are mechanical support devices for patients with severe acute and chronic heart failure. These implanted devices are connected to an external power supply by a so-called driveline. The transcutaneous driveline exit site is a conduit for bacterial entrance with the risk of creating a bacterial biofilm growing along the driveline into the body. This can become very dangerous for the patient because the removal and replacement at another site of these driveline catheters are impossible or intricate, respectively. Here, the nonthermal plasma and, above all, the argon-driven jet devices kINPen and SteriPlas demonstrated repeatedly to be very effective tools to successfully fight such driveline infections, see Metelmann et al. (2018a,b, 2022), Von Woedtke et al. (2022), Hilker et al. (2017), Rotering et al. (2020), and Kremer et al. (2020, 2021). We should point out that above discussed plasma-medical devices applied for healing wounds and ulcerations in the medical market are "scientifically based" (Metelmann et al. 2018a,b, 2022) in contrast to some plasma devices distributed and applied without sufficient studies and making negative impact on the development of plasma medicine in general, see Metelmann et al. (2018a,b, 2022) and Fridman and Friedman (2013).

31.12 Preventing and Suppressing Bleeding Using Thermal Plasma Cauterization Devices, Argon Plasma Coagulators (APS)

Blood coagulation is an important issue in medicine in surgeries, wound treatment, etc. Thermal plasma has been traditionally used for this application in form of the so-called cauterization devices such as **argon plasma coagulators (APC)** and **argon beam coagulators**. Widely used in surgery and wound treatment, plasma in these devices is simply a source of local high-temperature heating with high temperature gradients typical for thermal plasma sources.

The high temperature thermal plasma "beam" cauterizes (actually, "cooks" or carbonize) the blood. APC involves the use of a jet of ionized argon gas (plasma) directed through a probe passed through the endoscope. The choice of argon in these blood coagulation devices is dictated by no-chemical-reactivity of argon, and possibility to sustain the discharge at relatively low voltages. The probe is placed at some distance from the bleeding lesion (depending on specific medical protocols), and argon plasma gas is emitted from the probe resulting in coagulation of the bleeding lesion. As no physical contact is made with the lesion, the procedure is safe. The depth of coagulation is usually only a few millimeters, which is effectively controlled by high-temperature-gradient boundary layer of the thermal plasma beam jet.

The APC is used to treat very different types of wounds for blood coagulation and healing purposes. Specifically, we can mention the following conditions: (i) colonic polyps, after polypectomy, (ii) Barrett's esophagus, (iii) esophageal cancer, (iv) rectal bleeding post-radiation (radiation proctitis), (v) gastric antral vascular ectasia, or "watermelon stomach." The APC can be based on different types of thermal plasma sources. For example, the thermal HF discharge (conventional frequency about 350 kHz, voltage up to 6 kV) in the APC generates high-temperature (10 000 K) plasma in flowing argon, which leads to rapid cauterization and tissue desiccation.

APC devices are widely used today in surgery and represent a good example of clinical FDA-approved application of plasma for coagulation of the bleeding lesions. It is important to note that the APC devices can effectively combine plasma-induced blood coagulation and surgical the thermal plasma-based tissue cutting. A good relevant example in this regard is an FDA-approved APC developed by Professor Jerome Cannady called Canady Vieira Hybrid Plasma Scalpel. While APC are very effective and FDA-approved plasma-medical devices, they also have some disadvantages and challenges related to thermal damage of the living tissue, which stimulates interest to the non-thermal-plasma-induced blood coagulation to be discussed below.

31.13 Nonthermal Atmospheric-pressure Plasma-assisted Blood Coagulation

 While the thermal plasma-based APC are today FDA approved and widely used in hospitals in multiple surgeries and procedures for coagulation of the bleeding lesions, the challenge of thermal tissue damage stimulates interest to plasma blood coagulation without very high temperatures. The recent development of effective nonthermal plasma-medical systems enables effective blood coagulation to be achieved without any negative thermal effect. In such systems, the **coagulation is achieved through nonthermal plasma stimulation of specific natural mechanisms** *in blood without any "cooking" or damage to surrounding tissues (Fridman et al. 2006a,b).* We should note that both coagulating the blood and preventing the coagulation could be needed, depending on the specific medical application. For example, in wound treatment, the surgeon would want to close the wound and sterilize the surface around it. Flowing blood, in that case, would prevent wound closure and create the possibility of re-introduction of bacteria into the wound. Where blood coagulation would be detrimental is, for example, in sterilization of stored blood in blood banks. A potential exists for blood to contain or to have somehow acquired bacterial, fungal, or viral infection which needs to be removed for this blood to be usable. In this case, the treatment cannot coagulate the blood. An understanding of the mechanisms of blood coagulation by nonthermal plasma is therefore needed. Starting with the first experiments with FE-DBD, plasma was confirmed to significantly hasten blood coagulation in vitro (Fridman et al. 2006a,b). Visually, a drop of blood drawn from a healthy donor and left on a stainless-steel surface coagulates on its own in about 15 min; a similar drop treated for 15 s by FE-DBD plasma coagulates in about 1 min, see Figure 31.8 (Fridman et al. 2006a,b; Fridman and Friedman 2013).

FE-DBD treatment of cuts on organs leads to similar results where blood is coagulated without any visible or microscopic tissue damage. Figure 31.9 shows a human spleen treated by FE-DBD for 30 s (Fridman et al. 2006a,b; Fridman and Friedman 2013). Blood is coagulated and tissue surrounding the treatment area looks "cooked"; the temperature of the cut however remains at room temperature (even after 5 min of FE-DBD treatment) and the wound remains wet, which could potentially decrease healing time. Additionally, a significant change in blood plasma protein concentration is observed after treatment by plasma of blood plasma samples in healthy patients, patients with hemophilia, and blood samples with various anticoagulants. Anticoagulants, like sodium heparin or sodium citrate, are designed to bind various ions or molecules in the coagulation cascade, thus controlling coagulation rate or preventing it. Analysis of changes in concentration of blood proteins and clotting factors indicates that FE-DBD aids in promoting the

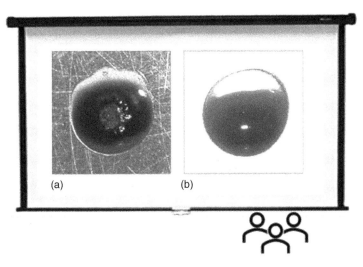

Figure 31.8 Plasma-induced blood coagulation, in-vitro: blood drop treated by FE-DBD: (a) after 15 s of FE-DBD and (b) control; photo was taken 1 min after the drops were placed on brushed stainless steel substrate.

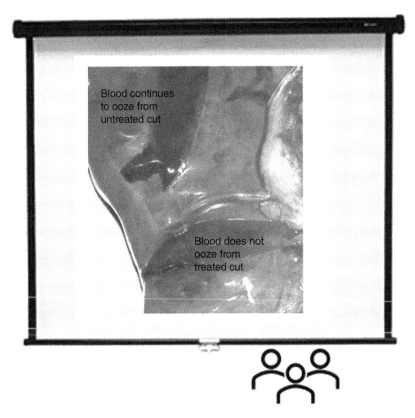

Figure 31.9 Plasma-treatment of human tissue: after 30 s of FE-DBD treatment of human spleen, blood coagulates without tissue damage. Top cut: blood continues to ooze from an untreated area; bottom cut: blood coagulates while the wound remains wet.

advancement of blood coagulation or, in other words, plasma can catalyze the biochemical processes taking place during blood coagulation, see Fridman et al. (2006a,b).

FE-DBD plasma stimulation of in vivo blood coagulation has been demonstrated by Fridman et al. (2007a–e), in experiments with live hairless SKH1 mice. Fifteen seconds of FE-DBD plasma treatment have been proved to be able to coagulate blood at the surface of a cut Saphenous vein (Figure 31.10, see book of Fridman and

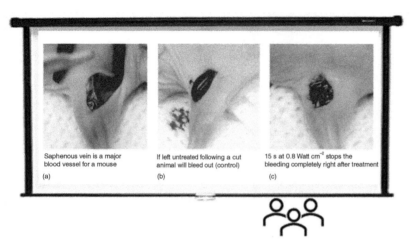

Figure 31.10 FE-DBD plasma-induced blood coagulation, live animal model. (a) Saphenous vein is a major blood vessel for a mouse; (b) if a cut is left untreated, the animal will bleed out; (c) after 15 s at 0.8 W cm^{-2}, the bleeding is stopped.

Friedman 2013) and vein of a mouse. Only the ability of direct nonthermal plasma treatment to coagulate blood was tested in these experiments and the animal was not left alive to test improvement in healing times. Full in vivo investigation of ability of plasma to hasten wound healing through sterilization and blood coagulation has been also discussed in Fridman et al. (2007a–e). We should mention that significant number of publications on plasma-induced blood coagulation using nonthermal plasma jet has been presented recently by Prof. Sakakita and his colleagues, see for example Shimizu et al. (2022). These nonthermal plasma jet results have a lot of similarities with the above-considered results obtained in the nonthermal FE-DBD plasma.

31.14 Biochemical Mechanisms of Nonthermal Plasma-induced Blood Coagulation, Contribution of Ca^{2+} Ionic Factor to Blood Coagulation Cascade

Detailed biochemical pathways of nonthermal plasma-stimulated blood coagulation remain largely unclear. Possible mechanisms have however been investigated, see for example, Fridman et al. (2006a,b, 2007a–e), Kalghatgi et al. (2007a,b) and, Fridman and Friedman (2013). *Most importantly, it was demonstrated that direct nonthermal **plasma can trigger natural, rather than thermally induced, coagulation processes**. The evidence points to selective action of direct nonthermal plasma on blood proteins involved in natural coagulation processes, which is going to be discussed in the next section.*

The mechanisms of plasma interaction with blood can be deduced from the following facts observed in experiments with FE-DBD plasma (Fridman et al. 2006a,b, 2007a–e; Kalghatgi et al. 2007a,b; Fridman and Friedman 2013):

1. plasma can coagulate both normal and anticoagulated blood, but the rate of coagulation depends on the anticoagulant used;
2. plasma can alter the ionic strength of the solution and change its pH, but normal and anticoagulated blood buffers these changes even after long treatment time;
3. plasma changes the natural concentration of clotting factors significantly, thus promoting coagulation;
4. plasma effects are nonthermal and are not related to gas temperature or the temperature at the surface of blood;
5. plasma can promote platelet activation and formation of fibrin filaments, even in anticoagulated blood.

Observations of the fast blood coagulation effect stimulated by FE-DBD plasma clarify the process mechanism. For example, in vivo, the pH of blood is maintained within a very narrow range of 7.35–7.45 by various physiological

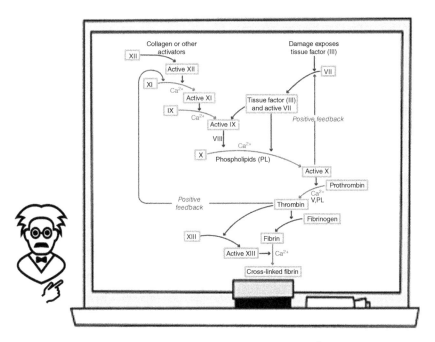

Figure 31.11   Natural blood coagulation cascade, the role of Ca^{2+} ionic factor.

processes. The change in pH by FE-DBD plasma treatment (about 0.1 after 30 s) is less than the natural variation of pH, which indicates that the coagulation is not due to pH change in blood. Anticoagulants such as sodium heparin bind thrombin in the coagulation cascade, thus slowing coagulation. Sodium citrate or ethylene diamine tetra-acetic acid (EDTA) are however designed to bind calcium, an important factor in the cascade, thereby preventing coagulation altogether. Plasma treatment promotes visible coagulation in blood with the above anticoagulants. Let's shortly discuss the plasma-stimulated mechanism of natural blood coagulation by controlling the Ca^{2+} ionic factors.

Plasma-induced Ca^{2+}-based mechanism of natural blood coagulation is the hypothesis focused on plasma controlling of Ca^{2+}, which is an important factor in the natural blood coagulation cascade. The major element and factors of the natural blood coagulation cascade are illustrated in Figure 31.11. The coagulation cascade indicates that the Ca^{2+} ion-related factors are crucial in conversion of prothrombin to thrombin, which then stimulates conversion of fibrinogen into fibrin and then cross-linked fibrin. The cross-linked fibrin represents microfilaments able to connect blood platelets and therefore coagulate blood. This simplified sequence explains contribution of the plasma-released Ca^{2+} ions in stimulation of blood coagulation. It is suggested that plasma stimulates generation of Ca^{2+} through the redox mechanism: $[Ca^{2+}R^{2-}] + H^+_{H_2O} \Leftrightarrow [H^+R^{2-}]_{H_2O} + Ca^{2+}_{H_2O}$ provided by hydrogen ions produced in blood in a sequence of ion-molecular processes induced by plasma ions. As it was discussed in Section 29.12, ionic charge exchange leads to significant conversion in liquid phase of several positive plasma ions to hydrated hydrogen ions (hydronium), which can replace the chemically bounded Ca^{2+} ions releasing hydrated Ca^{2+} ions into the blood flow. Biochemical kinetics of this process influencing the natural blood coagulation cascade has been simulated by Fridman et al. (2006a,b, 2008), explaining significant FE-DBD plasma-induced prothrombin-to-thrombin conversion and blood coagulation, see Figure 31.12 (Fridman et al. 2006a,b, 2008; Fridman and Friedman 2013). This hypothesis was also tested experimentally by measuring Ca^{2+} concentration in the plasma-treated anticoagulated whole blood using a calcium-selective microelectrode. Calcium ion Ca^{2+} concentration was measured immediately after plasma treatment and changed only slightly for up to 30 s of treatment; it then increased for prolonged treatment times of 60 and 120 s. Further research in this direction is required for better understanding of a quite strong effect of FE-DBD plasma stimulation of natural blood coagulation cascade.

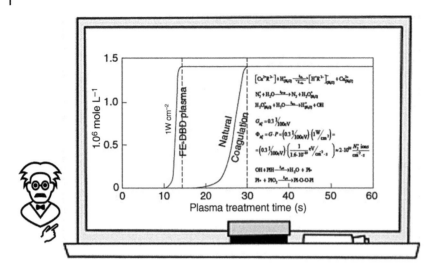

Figure 31.12 Plasma-induced prothrombin-to-thrombin biochemical kinetic simulation.

31.15 Plasma-induced Blood Coagulation: Plasma Influence of Protein Activity, Natural-soft, Intermediate, and Strong Regimes of Stopping Bleeding

 FE-DBD treatment of whole blood sample changes concentrations of proteins participating in the coagulation cascade. Plasma treatment is shown to "consume" coagulation factors (proteins and enzymes) and a visible film is formed on the surface of the treated samples. An increase in the sample volume and keeping the surface area fixed decreases the effect, indicating that plasma treatment initiates clot formation at the surface and not in the volume. *When the surface of blood is protected by small thin aluminum foil, which prevents contact between blood and FE-DBD plasma but transfers all the heat generated by plasma, no influence of blood is observed. This proves nonthermal rather natural mechanism of plasma-stimulated blood coagulation.*

 The final step in the natural biological process of blood coagulation is the production of thrombin which converts fibrinogen into fibrin monomers that polymerize to form fibrin microfilaments. FE-DBD plasma treatment of fibrinogen solution in physiological medium coagulates it, which is confirmed visually through a change in the color of the solution (from clear to milky white) and through dynamic light scattering. Note that plasma does not influence fibrinogen through a pH or temperature change. *FE-DBD treatment, however, does not usually polymerize albumin (directly not participating in coagulation cascade) as no change in its behavior is observed either visually or through dynamic light scattering (DLS). Nonthermal FE-DBD plasma therefore selectively affects proteins (specifically, fibrinogen) participating in the natural coagulation mechanism.*

To assess the **influence of plasma on protein activity**, compared to plasma influence on the protein itself, trypsin (pretreated with L-1-tosylamido-2-phenylethyl chloromethyl ketone or TPCK to inhibit contaminating chymotrypsin activity without affecting trypsin activity) was treated by plasma for up to 2 min and its total protein weight and protein activity analyzed via fluorescence spectroscopy. Total protein weight, or the amount of protein in the treated solution, remains practically intact after up to 90 s of treatment while the enzymatic (catalytic) activity of this protein drops to nearly zero after 10–15 s of treatment. Similar behavior is also observed for albumin. This proves that the plasma effect on proteins is not just destructive but quite selective and "natural." Morphological examination of the clot layer by scanning electron microscopy (SEM) further proves that FE-DBD plasma does not "cook" blood but initiates and enhances the natural sequences of blood coagulation processes. Activation followed by aggregation of platelets is the initial step in the coagulation cascade, and conversion of fibrinogen into fibrin is the final step. Figure 31.13 shows in this regard extensive platelet activation, platelet aggregation, and fibrin formation following FE-DBD plasma treatment, see Fridman and Friedman (2013).

Figure 31.13 Untreated (a, b) and FE-DBD treated (c, d) anticoagulated whole blood: (a) whole blood (control) showing single activated platelet (white arrow) on a red blood cell (black arrow); (b) whole blood (control) showing many nonactivated platelets (black arrows) and intact red blood cells (white arrows); (c) whole blood (treated) showing extensive platelet activation (pseudopodia formation) and platelet aggregation (white arrows); and (d) whole blood (treated) showing platelet aggregation and fibrin filament formation (white arrows). Source: Vasilets et al. (2009)/Pleiades Publishing, Ltd.

Thus summarizing the plasma-induced blood coagulation effects considered in the last four sections (Sections 31.12–31.15), we can distinguish three qualitatively different types of the plasma blood treatment: a natural-soft regime related to the natural coagulation cascade, an intermediate regime related to the intensive nonspecific cross-linking of proteins, and strong regime cauterizing the blood and tissue due to intensive heating. The **strong cauterizing regime**, discussed in Section 31.12, is provided by thermal plasma sources. The strong plasma-induced high-temperature effect leads to carbonization ("cooking") of blood and tissues. While this regime can result in tissue damage, the significant medical advantage of this approach is related to possibility to stop major bleedings. The **natural-soft coagulation regime** is related to plasma stimulation of the natural coagulation cascade, as discussed in Sections 31.13–31.15. This "soft" treatment regime observed mostly using the FE-DBD plasma systems doesn't lead to any thermal tissue damages and even does not result in "nonnatural" cross-linking of proteins not directly related to the natural blood coagulation cascade. While this regime is "natural" and nondamaging, it cannot provide significant stimulation of blood clogging to stop major bleedings. Finally, the **intermediate blood coagulation regime** doesn't lead to thermal tissue damage, while in addition to stimulation of the natural cascade results in significant nonspecific cross-linking and aggregation of the major blood proteins like albumin not directly related to the natural blood coagulation cascade. This intermediate regime combining advantages (and disadvantages) of the previous two "extreme" approaches has been investigated and effectively applied by Prof. Sakakita and his colleagues, see for example Shimizu et al. (2022).

31.16 Problems and Concept Questions

31.16.1 Intensive-NO-generating Plasma Sources for Wound Treatment

Why the major intensive-NO-generating plasma sources like the Plazon system and pin-to-hole discharge (PHD) applied for wound healing are based on thermal plasma? What types of reactive nitrogen species are expected to be generated in these discharges? Calculate the expected total NO flux generated by these plasma-medical discharges, assuming their power of 1 kW, and G-factor of NO production about 1–3 NO-particles per 100 eV.

31.16.2 Direct FE-DBD Treatment of Heavy Bleeding and Infected Wounds

What is your interpretation of the medical test results indicating that the FE-DBD plasma sterilization of the heavy bleeding and infected wounds is more effective than FE-DBD plasma sterilization of the intact dry skin?

31.16.3 Thermal Plasma Cauterization Devices, Argon Plasma Coagulators (APC)

Why the thermal plasma jet-based cauterization devices are more attractive for medical application than similar devices based on combustion jets? What are the advantages of plasma cauterization devices with respect to laser beam cauterization? Why the APC devices use argon as a plasma gas?

31.16.4 Biochemical Mechanisms of Nonthermal Plasma-induced Blood Coagulation

Explain possible mechanism of the nonthermal plasma stimulation of natural blood coagulation cascade. Keeping in mind possible contribution of Ca^{2+} ionic factors in the plasma-induced natural blood coagulation cascade, compare effectiveness in stimulation of the "soft natural" regime of blood coagulation by direct plasmas (like FE-DBD) vs indirect approaches like plasma jet afterglow.

31.16.5 Natural-soft, Intermediate, and Strong Regimes of Stopping Bleeding

Analyzing temperature effects on kinetics of plasma interaction with biopolymers (see Lecture 26), explain possibility of the intermediate regime of the plasma-induced stopping bleeding when there is no thermal tissue damage, while in addition the natural cascade there is significant nonspecific cross-linking and aggregation of the major blood proteins.

Lecture 32

Plasma Medicine: Dermatology and Cosmetics, Dentistry, Inflammatory Dysfunctions, Gastroenterology, Cardiovascular, and Other Diseases, Bioengineering and Regenerative Medicine, Cancer Treatment and Immunotherapy

32.1 Plasma Dermatology, Clinical Trials

Beyond wound healing as the most investigated application of plasma in medicine, there is strong potential in the use of plasma devices in a wider context of dermatology, see for example, Bernhardt et al. (2019), Friedman et al. (2020), Friedman et al. (2019), Liu et al. (2020), Klebes et al. (2014), Von Woedtke et al. (2022), Fridman and Friedman (2013), Metelmann (2018a,b, 2022).

First effective plasma dermatological treatments have been initially demonstrated using thermal plasma-generated NO, especially with the Plazon system. The high efficacy has been observed in treatment of psoriasis, eczemas, dermatitis, ulcerous injuries with local and systemic angiitises, scleroderma, red flat lishchaya, and several other skin illnesses, see Zaitsev (2003) and Fridman et al. (2006a,b).

Further relevant success in the **cold atmospheric pressure plasma dermatology**, including clinical studies, has been demonstrated using nonthermal plasma discharges, especially dielectric barrier discharges (DBD)-based (like floating-electrode dielectric barrier discharge [FE-DBD]) and nonthermal plasma jets. Relevant single medical case reports, clinical case series, and clinical trials are available today on the plasma-medical treatment of several infective and inflammatory skin diseases. A randomized placebo-controlled clinical trial is reported including 37 patients suffering from **herpes zoster, a viral skin disease** characterized by blister formation and painful skin rash in a localized area. Weekday 5-min treatments with the argon-driven microwave plasma torch MicroPlaSter (a predecessor device of SteriPlas, see Lecture 31) resulted in a more rapid clinical improvement in the first one to two days in the plasma-treated group, including acute pain reduction and initial healing of the herpes zoster lesions, see Isbary et al. (2014). In a randomized two-sided placebo-controlled study with 46 patients with different pruritic (itchy) skin diseases, a 2 min daily treatment with the MicroPlaSter argon plasma torch additional to standard treatment did not result in higher pruritus reduction compared to placebo treatment with argon gas only, see Heinlin et al. (2013). Another relatively widespread skin disease is **psoriasis**, an autoimmune disease that is noncontagious and characterized by sometimes-large areas of angry, dry, itchy, and scaly skin. In a case series with six patients, a three-times-a-week treatment of psoriatic plaques with the argon-driven atmospheric pressure plasma jet kINPen (see Lecture 31) showed results comparable with conventional therapies (Klebes et al. 2014).

Mainly based on the antimicrobial and antiviral effectivity of nonthermal plasma, clinical investigations have demonstrated the successful plasma effect on **onychomycosis**, fungal infection of the nail, see for example, Lipner et al. (2017). These clinical results have been achieved using FE-DBD discharge. Especially interesting that the microorganisms responsible for the onychomycosis are located below a nail while the DBD plasma is surely located above the nail. The nature of DBD permits them to sustain plasma below the nail which operates in this case as an additional dielectric barrier. Plasma sterilization in this case is somewhat like sterilization of food inside of closed packages, which has been discussed in Section 29.10, see also Figure 29.6. This is a good example when specifics of plasma sources open possibilities to treat very specifically located tissues.

Another example of effective clinical application of FE-DBD in dermatology is **plasma treatment of warts**. Results of treatment of two patients are illustrated in Figure 32.1 (Friedman 2018). In this figure, the first patient presented by photos (a–e). Partially subungual warts located on the thumb tip before the first treatment are shown in (a); (b) represents status five months after the second treatment; (c–e) wart on index finger (c) before treatment,

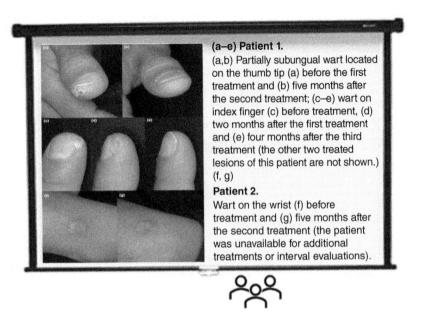

(a–e) **Patient 1.**
(a,b) Partially subungual wart located on the thumb tip (a) before the first treatment and (b) five months after the second treatment; (c–e) wart on index finger (c) before treatment, (d) two months after the first treatment and (e) four months after the third treatment (the other two treated lesions of this patient are not shown.) (f, g)

Patient 2.
Wart on the wrist (f) before treatment and (g) five months after the second treatment (the patient was unavailable for additional treatments or interval evaluations).

Figure 32.1 Clinical plasma dermatology: micro-sec pulsed DBD treatment of warts.

(d) two months after the first treatment and (e) four months after the third treatment (the other two treated lesions of this patient are not shown). The second patient is presented by photos (f, g). Wart on the wrist is shown: (f) before treatment and (g) five months after the second treatment (the patient was unavailable for additional treatments or interval evaluations). More details on the subject can be found in Friedman et al. (2020) and Friedman et al. (2019). Other results of Friedman and colleagues on FE-DBD plasma dermatology will be discussed below regarding plasma oncology and treatment precancerous lesions.

32.2 Plasma Cosmetics Closely Related to Dermatological Studies

Close to therapeutic plasma dermatology is the field of **plasma applications in cosmetics**. It must be stated that there is not always a clear distinction possible between therapeutic use as it is intended with plasma medicine devices and improvement of appearance, as it must be intended with cosmetics as well as aesthetic medicine. For instance, the nonthermal plasma application in **acne treatment** might be focused on both the healing of a skin disease and aesthetic skin improvement (Chutsirimongkol et al. 2014). On the World Wide Web, there are several proposals that offer promising effects of plasma application for corrective treatments and skin improvement, some of them explicitly referring to plasma medicine. Cosmetic indications of plasma treatment are focused on **skin tightening, wrinkle removal, face and body lifting, skin rejuvenation, or blepharoplasty** (tightening of eyelids). The techniques are often based on needle-like devices where an electrical spark is generated between the needle-tip and the skin, resulting in dark spots on the skin because of tissue carbonization that resolves during the following days; search for example the World Wide Web using a keyword like "plasma cosmetics." Nevertheless, these applications are often described as "noninvasive" and "nonthermal." They are surely based on **thermal or transitional "warm" plasmas**.

One of the best described plasma devices of this kind for cosmetic and aesthetic applications is the nitrogen plasma-based jet-like Portrait PSR3 system, which is mainly used for skin regeneration. However, its effectivity in acne treatment is also reported. Its activity is based on instantaneous skin heating in a controlled, uniform manner without an explosive effect on tissue or epidermal removal; see for example Bogle et al. (2007), Foster et al. (2008), Gonzalez et al. (2008), and Bentkover (2012). Because most of the known applications of plasma for cosmetic purposes are based on intensive thermal effects, any reference to plasma medicine in terms of this foundation paper, above all, with respect to safety, is illegitimate because nearly all recent investigations into plasma medicine are based on NTP–cell and plasma–tissue interactions.

Generally, **acne is one of the "bread and butter" diseases of dermatology**. It is an inflammatory disease of the sebaceous follicles, having many factors playing a role in its development. Just like in the case of leg ulcers, the role bacteria play in the pathogenesis of acne makes this condition a feasible target for cold plasma. Effective relevant treatment has been demonstrated using FE-DBD plasma, see Friedman (2020). A semi-spheric DBD electrode to treat acne was reported to reduce the number of acne lesions and propionibacterium colonization of the skin. A recent case report of two patients treated with an argon-jet plasma torch device once a week for several weeks, also found that cold plasma was an effective and well-tolerated treatment for acne, based on both clinical assessment and related metrics, such as sebum production. Both reports are limited by the very small number of cases, but they are encouraging as acne treatment is another area in dermatology, which lacks definitive solutions. One of the obstacles for cold plasma to treat acne is that the treatment must be applied to a large, incongruous surface (the face) evenly, and in a time efficient manner. DBD electrodes and single-jet plasmas reach a relatively smaller surface, but both can be moved around continually to cover larger areas. This, however, introduces a variable to the treatment. It is difficult to standardize the speed of moving, and the way of evenly covering all areas, especially concave surfaces, and this can lead to operator-dependent varying outcome. One way to avoid this is the use of devices, which can treat a larger surface without being moved and the exposures may be timed. One approach is a plasma torch, such as the above referenced case report. Another is a variation of the DBD electrode concept, the surface micro-discharge plasma, which can generate substantial amount of plasma on a larger target surface without using an external gas supply. Many "classical" acne treatments are antibiotics: topical or systemic. In line with our overall goal in medicine to reduce the use of antibiotics, dermatology is always in search of nonantibiotic acne treatments. It remains to be seen if cold plasma devices can replace some of our current acne treatments. For more information on the subject see Friedman (2020).

There are several approaches recently indicating that nonthermal plasma might also be a promising tool for the stimulation and/or regeneration of skin and, therefore, useful for cosmetic applications. A forward-looking review on the potential of **nonthermal plasma for skin treatments** is given in Busco et al. (2020). This group also organized the 1st International Meeting on **Plasma Cosmetic Science** in Orleans, France, where many nonthermal plasma-cosmetic and dermatological-cosmetic treatments have been discussed. In particular, the repeatedly reported nonthermal plasma effect to enhance skin permeability and to enhance the transcutaneous penetration of substances may be promising both for cosmetic purposes and for transcutaneous drug release, see for example Liu et al. (2020), Lademann et al. (2011), Choi et al. (2014), Shimizu et al. (2015), Gelker et al. (2018), and Kristof et al. (2019).

The **FE-DBD plasma** discussed in the previous section has been recently effectively applied also for the plasma-cosmetic and dermatological-cosmetic treatments by Friedman and the team. Especially, we can mention the indirect DBD **plasma treatment to stimulate hair growth,** see Khan et al. (2020). In the center of this limited clinical study was the androgenetic alopecia (AGA), which is a chronic form of hair loss characterized by the miniaturization of hair follicles. There are different prescription and over-the-counter treatments for AGA, which have various advantages and drawbacks, but the search for the ideal AGA treatment continues. In the framework of this pilot study, four patients with AGA have been treated for three months and additional 10 patients have been treated for six months. In all the cases, treatment has been well tolerated. Most patients as well as medical investigator reported improvement of the AGA conditions. These investigations open a pathway for the plasma stimulation of hair growth, especially in the indirect mode of treatment. In the cosmetic applications, the indirect plasma treatment (see Section 30.4.5) has especial advantages, especially when plasma activates gel and creams safe and effective in cosmetic treatment. For example, activation of gels and creams with the intensive NO-generating discharges (see Section 31.4) adds the NO therapeutic factors (like healing of wounds, lesions, stimulation of angiogenesis, etc.) to conventional positive cosmetic effects of the treated gels and creams (Fridman 2023).

Direct FE-DBD plasma-cosmetic and dermatological-cosmetic treatments have been also recently suggested and accomplished by Friedman and the team. They include healing of the **molluscum contagiosum (MC)**, a disease of viral etiology, like warts, which typically occurs in children, but also in adults Although "self-limited," MC has a significant impact on quality of life. Physical removal, cryotherapy, and topical pharmacotherapies have been tried with varying success. Application of the direct FE-DBD plasma to a 12-year-old patient presented with several months' history of multiple, untreated lesions of MC on the face, neck, shoulder, and thighs was successful. Details on this study can be found in Friedman et al. (2022).

One more similar example of the successful direct FE-DBD dermatological-cosmetic treatments has been demonstrated by Friedman and the team in **treatment of rosacea**. The Demodex mites have been implicated in the pathogenesis of skin disorders, including rosacea, for which a topical formulation of the anti-parasitic drug ivermectin is used as a treatment. It has been demonstrated on three patients that direct DBD plasma effectively kills Demodex in vivo, without damaging the skin. Details on this study can be found in Khan et al. (2020). We should finally mention that very interesting application of nonthermal plasma in cosmetics can be related to the **skin and more general rejuvenation** due to plasma-induced stimulation of stem cells, which is going to be discussed later in this lecture.

32.3 Plasma Dentistry: Structure of Teeth, Challenges for Plasma Application

From the first years of plasma medicine, three primary applications of plasma in dentistry appear to have been widely discussed. One application deals with the killing of bacteria on exposed teeth surfaces and inside tooth root canals as well as providing some anti-inflammatory activity. The second is related to improved adhesion and incorporation of materials such as those contained in tooth fillings. The third is whitening of the tooth enamel, which is mostly a cosmetic application.

Prior to discussing these and other aspects of plasma dentistry, it may be useful to review some facts about tooth with respect to capabilities of plasma approaches. The typical structure of a tooth is illustrated in Figure 32.2, and its various components are described below.

Enamel is the hardest and most highly mineralized substance of the body. It is normally visible and must be supported by underlying dentin; 96% of enamel consists of hydroxyapatite mineral. The normal color of enamel varies from light yellow to grayish white. At the edges of teeth where there is no dentin underlying the enamel, the color sometimes has a slightly blue tone. Since enamel is semi-translucent, the color of dentin and any restorative dental material underneath the enamel strongly affects the appearance of a tooth. Enamel varies in thickness over the surface of the tooth and is often thickest at the cusp (up to 2.5 mm) and thinnest at its border.

Dentin is a mineralized connective tissue with an organic matrix of collagenous proteins. It has microscopic channels called dentinal tubules which radiate outward through the dentin from the pulp cavity to the exterior cementum or enamel border. The diameter of these tubules ranges from 2.5 μm near the pulp to 1.2 μm in the middle portion and 900 nm near the dentin/enamel junction. As a result, dentin has a degree of permeability which can increase the sensation of pain and the rate of tooth decay. By weight, 70% of dentin consists of the mineral hydroxyapatite, 20% is organic material and 10% is water. Primary dentin, the most prominent dentin in the tooth, lies between the enamel and the pulp chamber. The outer layer closest to enamel is known as mantle dentin. This layer is unique to the rest of the primary dentin. Mantle dentin is a layer approximately 150 μm thick formed by newly differentiated odontoblast cells. Unlike primary dentin, mantle dentin lacks phosphoryn, has loosely packed collagen fibrils, and is less mineralized. Below it lies the circumpulpal dentin, a more mineralized dentin which makes up most of the dentin layer and is secreted by the odontoblasts after the mantle dentin.

Cementum is a specialized bone-like substance covering the root of a tooth. It is approximately 45% inorganic material (mainly hydroxyapatite), 33% organic material (mainly collagen), and 22% water. Cementum is excreted by cementoblast cells within the

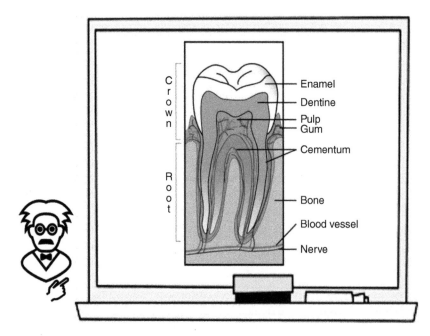

Figure 32.2 Structure of a tooth.

root of the tooth and is thickest at the root apex. Its coloration is yellowish, and it is softer than either dentin or enamel. The principal role of cementum is to serve as a medium by which the periodontal ligaments can attach to the tooth for stability. At the cement–enamel junction, the cementum is acellular (free of cells). This acellular type of cementum covers at least two-thirds of the root. The more permeable form of cementum, cellular cementum, covers about one-third of the root apex.

The dental pulp is the central part of the tooth filled with soft connective tissue. This tissue contains blood vessels and nerves that enter the tooth from a hole at the apex of the root. Along the border between the dentin and the pulp are odontoblasts, which initiate the formation of dentin. Other cells in the pulp include fibroblasts, preodontoblasts, macrophages, and T lymphocytes. The pulp is commonly called the "nerve" of the tooth.

The most common dental problems are probably the result of bacterial activity. One of the reasons bacteria on the teeth can be difficult to fight is formation of bacterial biofilm, known as plaque. If not removed regularly, plaque buildup can lead to dental cavities (caries) and/or periodontal problems such as gingivitis (inflammation of the gum). Given time, plaque can mineralize along the gingiva, forming tartar. The microorganisms that form the biofilm are almost entirely bacteria, with the composition varying by location in the mouth. *Streptococcus mutans* is probably the most important bacterium associated with dental caries (cavities). Certain bacteria in the mouth live off the remains of foods, especially sugars and starches. With insufficient oxygen, they produce lactic acid, which dissolves the calcium and phosphorus in the enamel. This process, known as "demineralization," leads to tooth destruction. Enamel begins to demineralize at a pH of 5.5. Saliva can gradually neutralize the acids raising pH above critical in some locations. However, saliva is often unable to penetrate through plaque to neutralize the acid produced by the bacteria at the foundation of the biofilm.

Tooth restoration typically involves cleaning out infected cavities in teeth and placement of dental filling material in place of removed parts of the teeth. Complete removal of bacteria has been problematic, particularly during root canal procedures. Root canal usually needs to be cleaned out when a tooth's nerve tissue or pulp is damaged, and bacteria begin to multiply within the pulp chamber. The bacteria and other decayed debris can cause an infection or abscessed tooth. An abscess is a pus-filled pocket that forms at the end of the roots of the tooth when the infection spreads all the way past the ends of the roots of the tooth. In addition to an abscess, an infection in the root canal of a tooth can cause swelling that may spread to other areas and cause bone loss around the tip of the root. A hole can occur through the side of the tooth with drainage into the gums or through the cheek with drainage into the skin. During a root canal procedure, the nerve and pulp are removed, and the inside of the tooth is cleaned and sealed. A tooth's nerve is not considered to be vitally important to a tooth's health and function after the tooth has emerged through the gums. Its only function appears to be sensory, providing the sensation of hot or cold. Removing all bacteria from a root canal is essential because the procedure removes not only pulp and nerves but also blood vessels which could supply immune cells capable of fighting infection.

Different dental filling materials have been employed in restorative dental procedures. The main difficulties involved are usually poor adhesion and/or material incompatibilities such as shrinkage of the filling material. In the past, dentists created special structures by removing some healthy parts of a tooth to help hold the filling in place. To create a porous surface that the adhesive can infiltrate, current preparation techniques etch and demineralize dentin. The culprit that foils mechanical bonding is a protein layer, the so-called "smear layer," which is primarily composed of type I collagen that develops at the dentin/adhesive junction. Interactions between demineralized dentin and adhesive give rise to the smear layer, which inhibits adhesive diffusion throughout the prepared dentin surface. This protein layer may be partly responsible for causing premature failure of the composite restoration. It contributes to inadequate bonding that can leave exposed, unprotected collagen at the dentin adhesive interface, allowing bacterial enzymes to enter and further degrade the interface and the tissue. Let us start consideration of plasma applications in dentistry with those first accomplished in the frameworks of plasma medicine.

32.4 First Effective Plasma Applications to Dental Health

One of the first effective use of plasma in dentistry is related to **healing of gingivitis**, which is an inflammation/infection of the gum. Effectiveness of the plasma-induced NO-therapy has been demonstrated on chronic gingivitis using the Plazon plasma source, see Section 31.4. After the first session of the therapy, gum bleeding ceased. After one to two weeks, normalization of tissue and regional blood flow in the tissues of periodontium occurred (Grigorian et al. 2001). Normalization of cytological signs was observed in two to three months, while in the control group normalization was not observed at all. Utilization of NO-CGF from plasma after surgical intervention for periodontal disease showed that normalization of clinical and cytological signs occurs by day 7 in the experimental group, but not until day 14 in the control group. Complications were not observed in the experimental group but were observed in the control group.

Also, recent years have seen a significant rise in interest in the **use of ozone for dentistry**. Ozone is obviously generated in nonthermal plasmas, like DBD or pulsed corona (see Section 18.6), although it is usually not reflected in the dentistry-related literature. Since inhalation of ozone is considered undesirable, different ozone delivery fittings are available to expose teeth and areas surrounding them, while minimizing inhalation. The effect of ozonated water on oral microorganisms and plaque has been investigated by several researchers (Nagayoshi et al. 2004; Arita et al.

2005). At 2 and 4 mg l^{-1} of ozone, the water was able to reduce the bacteria and *Candida albicans* dramatically. Other studies pointed out that the effect of ozone gas directly exposed to dentures was even stronger (Oizumi et al. 1998). Irrigation with ozonated water and sonication (Arita et al. 2005; Nagayoshi et al. 2004; Huth et al. 2007, 2009) had antimicrobial activity like 2.5% NaOCl (sodium hypochlorite, a standard root canal rinse used to remove dentin chips and kill bacteria in 3 min). Ozonated oil was also shown to be similar in its effect to calcium hydroxide paste (with camphorated paramonochlorophenol and glycerin) in endodontic treatment of teeth (root canal; Silveira et al. 2007).

Key to all these studies is appropriate dozing of ozone. Ozone works best with reduced organic debris present; the use of either ozonated water or ozone gas at the end of the cleaning and shaping process is therefore recommended. Ozone is effective when it is used in sufficient concentration for an adequate time but is not effective when the dose is insufficient. It was shown, for example, that by increasing the contact time from 10 to 20 s, the bacterial kill rate changed in some cases from ozone being a disinfectant to acquiring sterilizing effect (6-log reduction). This killing was reduced in the presence of saliva, although increasing the ozone application time to 60 s overcame the neutralizing effect of saliva (Johansson et al. 2009) which tends to decompose ozone faster and offers additional organic targets. One study on primary root caries lesion (PRCL) found that ozone application for either 10 or 20 s dramatically reduced most of the microorganisms in PRCLs without any side effects recorded at recall intervals between 3 and 5.5 months (Baysan and Lynch 2004). The antimicrobial properties of ozone may not be the only reason for the interest in it. Ozone and ozonated water apparently have strong anti-inflammatory properties. Some researchers (Huth et al. 2007, 2009) have examined the effect of ozone on the influence on the host immune response. They considered the NF-kappa B system, a well-known mechanism for inflammation-associated signaling/transcription. Their results demonstrated that NF-kappaB activity in oral cells in periodontal ligament tissue from root surfaces of periodontally damaged teeth was inhibited following incubation with ozonized medium.

One of the first groups to propose **applying plasma directly to teeth** was probably Eva Stoffels. She and her group at the Eindhoven Institute of Technology developed a RF-powered helium plasma jet called a plasma needle and investigated it as a tool for fighting dental cavities (Sladek et al. 2004; Sladek and Stoffels 2005; Sladek et al. 2006). This plasma needle demonstrated the ability to kill relevant bacteria. Substantial work has been performed then on development and characterization of various discharges in the form of small jets, many of which have also been referred to as plasma needles. These include discharges operating in various gases and excited by various means using RF sources as well as high-voltage AC sources. Many of these discharges have been shown to be effective against various bacteria in vitro (cultures on dishes). Some recent studies started looking for antimicrobial activity not only on agar but also on targets more relevant for dentistry. A microwave-powered nonthermal atmospheric-pressure helium plasma jet was evaluated (Rupf et al. 2010) for its antimicrobial efficacy against adherent oral microorganisms. Agar plates, as well as dentin slices, were inoculated with 106 CFU cm^{-2} of *Lactobacillus casei*, *S. mutans*, and *C. albicans*, with *Escherichia coli* as a control. Dentin slices were rinsed in liquid media and suspensions were placed on agar plates. The plasma jet treatment reduced the CFU by three to four orders of magnitude on the dentin slices in comparison to untreated controls, demonstrating that plasma can have a strong antimicrobial efficacy on relevant dental targets.

Other research groups considered the application of plasma jets to **disinfection of root canals**. Jiang et al. (2009a,b), at the University of California employed a helium/oxygen plasma jet triggered by fast-rising (around 10 ns) high-voltage pulses and demonstrated that bacteria can be killed inside the root canal, although the disinfecting action did not extend through the entire root canal. Another approach where the plasma probe is placed into the tooth (Lu et al. 2009) was able to achieve deeper penetration of plasma into the root canal. It was found that this device efficiently kills *Enterococcus faecalis* (an important root canal pathogen) within several minutes of plasma treatment.

Between first plasma applications to dental health, we can also point out **improving filling adhesion**. Preliminary data has shown that plasma treatment increases the bonding strength at the dentin/composite interface by roughly 60%; that interface-bonding enhancement can then significantly improve composite performance, durability, and longevity. Ritts et al. (2010), from University of Missouri (Kansas City) investigated the plasma treatment effects on dental composite restoration for improved interface properties. Atmospheric cold plasma brush (ACPB) treatment can modify the dentin surface and thus increase the dentin/adhesive interfacial bonding. The solution is to introduce bonds that depend on surface chemistry rather than surface porosity. The effects of plasma treatment on the shear bond strength between fiber-reinforced composite posts and resin composite for core buildup were also studied (Yavirach et al. 2009). It was concluded that plasma treatment appeared to increase the tensile-shear bond strength between post and composite.

32.5 Modification of Implant Surface, Enhancing Adhesive Qualities, Polymerization, Surface Coating, and Other Material-related Plasma Applications in Dentistry

The implant surface hydrophilicity or wettability has recently received considerable attention in dentistry. Studies of the effect of **plasma on titanium implant surfaces** indicated that this treatment could improve cell adhesion by changing surface roughness and wettability, which decreases after plasma exposure (Kawai et al. 2004; Shibata et al. 2002). These studies used glow discharge treatment at low pressure during the manufacturing process. Similar effect of nonthermal atmospheric plasma immediately prior to implant placement was also reported (Giro et al. 2013; Duske et al. 2012, which stated that plasma treatment reduced the contact angle and supported the spread of osteoblastic cells. One of the advantages of plasma treatment is that it leaves no residues after treatment. Several studies reported the **plasma treatment of zirconia implants**, which is increasing as an alternative to the conventional titanium implant due to its superior esthetic properties (Park et al. 2013a,b). They also demonstrated the increase in hydrophilicity and enhanced osseointegration in in vitro as well as in vivo experiments.

Enhancing adhesive qualities is another key in the dental health care. For most dental joints, one set of adherents is usually composed of any dental substrates or previous restorations while the other set consists of restorative ones such as composite, amalgam, or ceramics. Optimal adhesion could be achieved when the adhesive material is spread impulsively across the entire adherend surface. In other words, optimal wettability of the substrate is achieved with reference to that adhesive. Conventional adhesive systems employed several methods to improve wettability, to elevate the surface energy, and to increase the roughness through techniques involving etch-and-rinse, acid primer, Hydroxy-ethyl-methacrylate (HEMA) primers, and laser irradiation. In the same respect, plasma treatment has been introduced as an alternative or additional procedure, especially in the bonding of ceramic restorations, which is more difficult to achieve. As an alternative for adhesion enhancement in dental ceramic bonding, atmospheric pressure plasma treatment has been suggested (Cho et al. 2011). It enhances adhesion by producing carboxyl groups on the ceramic surface and improves the surface hydrophilicity as a result (Han et al. 2012). The nonreactive surface of zirconia, sometimes described as "ceramic steel," presents a consistent issue of poor adhesion strength to other substrates (Piascik et al. 2011a,b). Silane treatment used for silica-based substrates is not applicable. In addition, zirconia itself is known to be hydrophobic and possesses very low surface concentrations of OH groups. For this more complicated bonding, several other methods have been proposed. Plasma treatment was also tested, and the results showed that a significant increase in the microtensile bond strength to zirconia surfaces was observed when nonthermal plasma was applied alone or in combination with resin.

For enamel, dentin, and composite, Chen et al. (2013) reported that a super-hydrophilic surface could be easily obtained by plasma brush treatment without affecting the bulk properties regardless of the original hydrophilicity. Another study revealed that plasma treatment of the peripheral dentin surface resulted in an increase in the interfacial bonding strength, while over 100 s of prolonged treatment resulted in a decrease in the interfacial bonding strength (Ritts et al. 2010). As for the post surface, Costa Dantas et al. (2012), showed that plasma treatment favored the wettability of the post, however, real adhesion improvement was not observed after argon plasma. The ethylene-diamine plasma treatment resulted in significant chemical modification as indicated by the high roughness.

Polymers synthesized by plasma exposure demonstrate high cross-linking and high degrees of polymerization (see Lecture 26), which is so much important for applications in dental care. For composite resin, plasma arc curing units are popular because of their short curing time in comparison to conventional ones, see Friedman and Hassan (1984) and Rueggeberg et al. (1993). The nonthermal plasma brush was also reported to be effective in the polymerization of self-etch adhesives with no negative effects of water on the degree of conversion of plasma-cured samples (Chen et al. 2012a,b). Plasma approaches were demonstrated to be effective in **surface coating.** Variation in surface texture or nanoscale topography features can affect the cell response for a titanium implant. **Plasma spraying of implants**, which used to be one of most popular coating techniques for implants, can be considered as one type of plasma treatment. Several other types of thin coating techniques using plasma to enhance osseointegration, have been introduced and have shown some promise such as plasma nitriding, titanium nitride oxide coatings, plasma polymerized hexa-methyl-di-siloxane, plasma polymerized allylamine, and plasma polymerized acrylic acid, see Da Silva et al. (2011), Durual et al. (2011), Hayakawa et al. (2004), Walschus et al. (2012), and Schroder et al. (2010).

In addition to the disinfective role of plasma cleaning on elastomeric impression materials, plasma treatment increases the surface wettability. The wettability of impression materials is an important requirement for the accurate reproduction of intraoral structures, since it is directly related to the quality of die stone casts, and, therefore, the castability of prostheses. The surface properties of several set elastomeric impression materials contaminated with saliva were inspected after plasma cleaning and exhibited an increase in the critical surface tension and an improvement in the castability of materials (Fernandes et al. 1992; Vassilakos et al. 1993; Yamazaki et al. 2011). More details on the material-related plasma applications in clinical dentistry can be found in review of Cha and Park (2014).

32.6 Direct Application of Nonthermal Plasma in Dentistry

First, efficacy of the direct plasma application in dentistry has been demonstrated in **microbicidal treatments.** Methods for the decontamination and conditioning of intraoral surfaces are of great interest in the field of dentistry. Cold plasmas are of particular interest in dentistry in this regard because it prevents heat and chemical damage to dental pulp, see Rupf et al. (2010). Thus, the substantial reduction of oral microorganisms adherent to dentin slices was also reported with direct use of plasma, where the sterilization effect was attributed to reactive oxygen species, see Rupf et al. (2010).

In the real oral environment, micro-organisms exist in the form of a biofilm and not in a planktonic state. The established and matured oral biofilm is a three-dimensionally structured community of many microbial species and is relevant to the development of caries and periodontal disease. For example, dental plaque, a biofilm on the tooth surface, consists of complex communities of oral bacteria with hundreds of species present. Furthermore, biofilms are also present on artificial surfaces in the oral cavity such as dentures or implants. Therefore, there was a shift in research model from planktonic to biofilm, and the effects of nonthermal plasma were evaluated on biofilm models. Several direct plasma treatment studies in this direction exhibited highly promising results, see Yang et al. (2011) and Sladek et al. (2007). Research has succeeded in improving the biological acceptance and osseointegration of dental implants. However, its long-term success is still challenging because of peri-implant diseases caused by the formation of biofilms. The decontamination of implant surfaces represents a basic procedure in the management of peri-implant diseases but remains a challenge. Biofilms play a major role in the pathogenesis of various oral diseases, especially peri-implant mucositis. In recent studies, the use of nonthermal plasma has been suggested for the removal of biofilms in general, and for the treatment of both peri-implant mucositis and peri-implantitis (Koban et al. 2011a,b; Rupf et al. 2011; Idlibi et al. 2013). The mechanisms are not clearly elucidated but can be hypothesized generally as two: One is the generation of reactive species such as oxygen, nitrogen, or nitrogen oxide radicals and the other is the etching effect of the plasma on biofilms during the chemical process of oxidation.

Plasma root canal disinfection can be considered as a special type of plasma decontamination in general. It is especially important in the endodontic procedures for successful outcome and difficult nature of eradicating the root canal micro-organisms due to its biological and geometrical characteristics. The tooth root canal system has complicated structures, such as isthmuses, ramifications, deltas, irregularities, and in particular dentinal tubules. It has been reported that bacteria can enter dentinal tubules as deep as 500–1000 µm. A variety of methods have been performed such as mechanic cleaning, irrigation, laser irradiation, ultrasound, and application of hypochlorite and other antibacterial compounds. However, eliminating the residual micro-organisms especially within the biofilm is still a challenging task, and clinical investigations showed that there are around 10% of treatment failures when conventional disinfections were performed. Since persistent endodontic infections are frequently caused by *E. faecalis,* numerous in vitro experiments have been reported on this issue using this micro-organism, see Jablonowski et al. (2013), Chen et al. (2012a,b), Zhou et al. (2010), Cao et al. (2011), Xiong et al. (2011), and Bussiahn et al. (2010). They have shown quite promising results after using nonthermal plasma alone or with conventional approaches pointing out that plasma as a gas phase has a capability of reaching deep into the complex canal. Therefore, there is a unique advantage of direct contact with bacteria when using plasma which is compulsory but impossible with conventional methods. The effective inactivation of *E. faecalis* has been attributed to several mechanisms, such as excited species, charged particles, and ultraviolet radiations.

Between **other applications of plasma dentistry**, we can mention results reported by Koban et al. (2011a,b), regarding direct plasma application to reduce the contact angle of untreated dentin surface, which caused a superior spreading of osteoblasts on the dentin. These results may be utilized to optimize periodontal regeneration. According to Miletic et al. (2013), the plasma interaction with the human periodontal ligament mesenchymal stem cells (MSCs) demonstrated that nonthermal plasma inhibited the migration of the cells and induced some detachment, without affecting their viability. In addition, the plasma significantly attenuated the cells' proliferation but promoted their osteogenic differentiation. Also, an important issue as to how deep can plasma penetrate a biofilm was addressed by several researchers. Among them, Pei et al. (2012), showed that 25.5 μm thick *E. faecalis* biofilm layer was penetrated by nonthermal plasma.

Tooth bleaching has become a popular esthetic service in dentistry where nonthermal plasma can be effectively used. Conventionally, hydrogen peroxide (H_2O_2) is a widely used bleaching material that is effective and safe. In-office bleaching systems usually use a 30–44% H_2O_2 bleaching gel and a high-intensity light source. The light source may enhance bleaching by heating the H_2O_2 and consequently accelerating bleaching, but this mechanism has yet to be confirmed. Since human dental tissues are sensitive to heat, many studies have investigated thermal damage to the pulp of teeth induced by light sources, especially plasma-related light sources like PAC (plasma arc lamp, bleaching mode). Temperatures more than 42.5 °C may induce degradation of the pulp tissues. The PAC and diode laser increased the temperature of the tooth surface to approximately 43 and 42.5 °C, respectively, after 20 min. The nonthermal helium jet plasma applied for teeth bleaching, however, took about 10 min to reach a steady-state temperature, and thereafter the tooth surface temperature is maintained around 37 °C, see Nam et al. (2013). The fact that the temperature of the nonthermal plasma treatment is like that of the human body means that a patient could not feel a sense of pain from cold or heat. Furthermore, it does not induce thermal damage in the tooth pulp or the tooth around the tissue. Tooth bleaching using the nonthermal helium jet plasma resulted in a better bleaching effect than the use of the PAC and diode laser with a 15% CP. In addition, the nonthermal helium jet plasma treatment did not increase the temperature of the tooth surface above 37 °C, indicating that the nonthermal plasma does not cause any thermal damage to the tooth. Therefore, the nonthermal helium jet plasma can be a useful tooth bleaching device that provides a high bleaching effect with a low concentration of hydrogen peroxide. As such, it could replace conventional light sources that have limitations, such as questionable bleaching efficacy and high temperatures. Plasma treatment was suggested to be complementary to the conventional method because it provides effective bleaching without thermal damage.

More details on the plasma tooth bleaching and other dental treatments can be found in Nam et al. (2013), Cha and Park (2014), Lee et al. (2009), and Park et al. (2011); as well as on Internet, where multiple companies suggest their tooth bleaching devices although without clear information on their safety and efficacy.

32.7 Plasma Treatment of Inflammatory Dysfunctions and Infections

Let us start with the treatment of inflammatory dysfunctions and infections **using plasma-generated NO**. The possibility of using thermal nitric-oxide-producing plasma for treatment of inflammatory and erosive processes in tissues within the pleural and abdominal cavities, lungs, stomach and bowels, and ear-nose-throat (ENT) organs has been noted in several different studies. In some cases, plasma-generated NO was inhaled or directed through puncture needles, vent lines, and endoscopic instruments. In other cases, plasma was placed in direct proximity to the affected tissues using endoscope-like delivery. Several examples of the thermal nitric-oxide-producing plasma application for treatment of inflammatory, erosive, and infection-related processes have been demonstrated by using Plazon system, see Shekhter et al. (1998, 2005), Fridman et al. (2008), and Fridman and Friedman (2013).

In **pulmonology**, the strong effect of plasma-generated NO was demonstrated in the treatment of pleural empyema via insufflation (inhalation) from the Plazon system into the cavity of the pleura through the vent lines (Shulutko et al. 2004). Therapy in treatment of 60 patients with pleural empyema showed regulative influence on the development of the wound tissues and stimulated healing. Acceleration of the purification of pleural cavity from the microorganisms and the debris, stimulation of phagocytosis, and normalization of microcirculation accelerate the passage of the phase of inflammation during wound regeneration, which leads to a significant decrease in the drainage time for all patient categories in the experimental group (compared to control) and to the reduction in the hospitalization time. The inhalation application in treatment of patients with complex chronic unspecific inflammatory lung diseases led to the clearly expressed positive dynamics of the endoscopic picture of the tracheobronchial tree: decrease of the degree of inflammatory changes in the mucous membrane of bronchi and the reduction in the quantity and the normalization of the nature of contents of the respiratory tract. Through biopsies of mucosa of bronchi, it was verified that for all cases the liquidation or the considerable decrease of inflammatory changes occurred in addition to a complete or partial restoration of the morphological structure of the bronchi. Plasma NO-therapy was also employed as an adjunct treatment in patients with infiltrative and fibrous cavernous pulmonary tuberculosis via NO insufflation through the bronchoscope or cavernostomy for cavernous tuberculosis, through the vent line with tubercular pleurisy or empyema. A significant acceleration of healing of the cavities, tubercular bronchitis, and pleurisy was achieved (Seeger 2005) in 8–10 therapy sessions.

In **gastroenterology**, plasma-generated NO therapy was delivered through endoscopic instruments for the treatment of chronic ulcers, erosions of stomach and duodenum, and blowholes of small intestine (Chernekhovskaia et al. 2004). Stomach ulcers healed twice as fast as in the control group. The proliferating activity of the epithelium, according to the data of the immunomorphology of biopsies, was increased by a factor of 8.

In **purulent peritonitis**, caused by diseases of the abdominal cavity organs, a positive effect was achieved by the direct treatment of peritoneum first and in the post-operative period by cooled plasma-generated NO delivered through the vent lines (Efimenko et al. 2005). It was argued that NO carries bactericidal action, stimulates microcirculation and lympho-drainage, normalizes the indices of cellular and humoral immunity, dilutes inflammatory processes, and serves as a factor of the preventive maintenance of adhesions in the abdominal cavity.

Suppression of infections can be demonstrated by using the floating-electrode dielectric barrier discharge **FE-DBD in ophthalmology**, which is illustrated in Figures 32.3 and 32.4. Figure 32.3 illustrates FE-DBD plasma treatment of corneal infection in live rabbit model. Studies were focused on eye sterilization with simultaneous control of toxicity. 5-log inactivation of *S. aureus* has been demonstrated in 15 s of plasma treatment without any observed toxicity (Fridman 2020, collaboration with the Wills Eye Hospital). These studies were further developed by carrying out in vivo pig eye sterilization before and during surgery using FE-DBD plasma. Complete sterilization of cornea (>5-log reduction in *S. aureus*) has been demonstrated in these live porcine tests without any damage to cornea or sclera, which is illustrated in Figure 32.4 (Fridman 2020, collaboration with the Wills Eye Hospital). Regarding suppression of corneal infection in animal and human tests using PHD discharge see Section 31.4 and Figure 30.3.

Effective **nonthermal plasma suppression of infection during surgery** has been demonstrated also using FE-DBD (including its combination with a spay), which is illustrated in (Figure 32.5) (Fridman 2020, collaboration with Johnson & Johnson). The studies were focused on spinal surgery of live rabbits. Plasma sterilization of musculature, ligaments, and bone has been demonstrated. Animals survived the plasma surgery. Complete sterilization for all *S. aureus* counts (10^7, 10^5, 10^3) has been achieved. Complete healing has been also achieved with no evidence of procedure damage except for scar on day 7 (Fridman 2020, collaboration with Johnson & Johnson). Effective treatment in

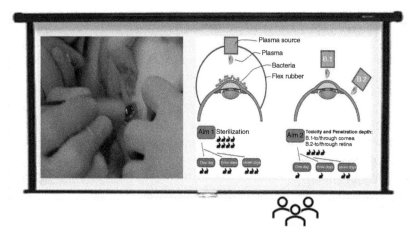

Figure 32.3 FE-DBD plasma ophthalmology: treatment of corneal infections, live rabbit model, no observed toxicity; 5-log inactivation of *S. aureus* in 15 s.

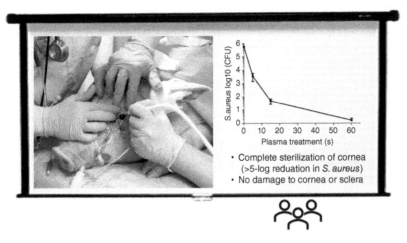

Figure 32.4 FE-DBD plasma ophthalmology: in vivo pig eye sterilization before and during surgery without damage to cornea and sclera.

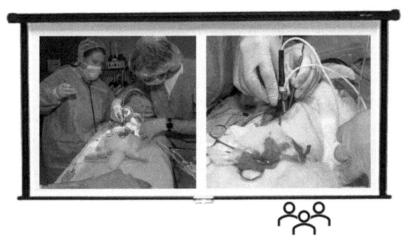

Figure 32.5 Plasma suppression of infections during spinal surgery on live rabbits. Plasma sterilization of musculature, ligaments, and bone.

the similar approach has been also demonstrated with suppression of **cutaneous Leishmaniasis**, see (Figure 32.6) (Fridman 2020).

32.8 Plasma Gastroenterology: Inflammatory Bowel Diseases (IBD)

32.8.1 Pin-to-hole Microdischarge Plasma Treatment of Ulcerative Colitis

Inflammatory bowel diseases (IBD) consist of two major chronic, relapsing, and debilitative forms of diseases known as **ulcerative colitis** and **Crohn's disease**. The etiology (causes) of these diseases remains a mystery although genetic, environmental, and immunological factors are found to play a major role in the induction, chronicity, and relapses of these diseases. Crohn's disease may appear in any part of the gastrointestinal tract from the mouth to anus and affect' the entire thickness of the bowel wall. On the contrary, ulcerative colitis is an inflammatory disorder affecting colonic mucosa and submucosa. There are no known curative therapies for these diseases; however, recent advances in IBD therapeutics have shown that certain biological therapies have been successful in maintaining remission, particularly in Crohn's disease. A study of the use of pin-to-hole discharge treatment of both healthy colon tissue and experimentally induced ulcerative colitis disease in a live animal model was carried out by Chakravarthy et al. (2011). The goals of the study were to examine whether this plasma treatment adversely affects the healthy

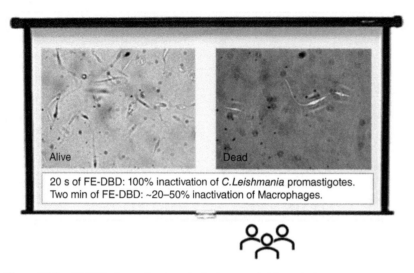

20 s of FE-DBD: 100% inactivation of *C.Leishmania* promastigotes.
Two min of FE-DBD: ~20–50% inactivation of Macrophages.

Figure 32.6 FE-DBD plasma treatment of cutaneous Leishmaniasis.

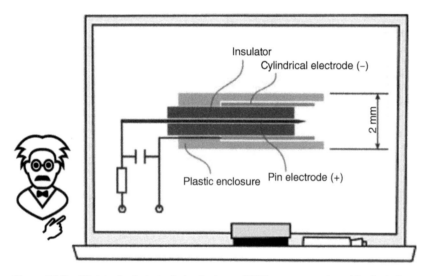

Figure 32.7 Miniaturized pin-to-hole discharge (PHD) system employed in the inflammatory bowel disease (IBD) treatment demonstration on live mice.

mucosa, and to evaluate if the plasma treatment results in acceleration or worsening of the disease during its induction phase. Additional pilot experiments were conducted to study whether cold plasma discharges provide therapeutic effects, and whether these effects are comparable to a standard therapy or enhance the beneficial effect of a standard therapy.

A **miniaturized pin-to-hole discharge** (see Section 14.6) was employed in the study. It consisted of a central copper needle covered by dielectric inserted into a grounded stainless steel cylindrical electrode as shown in Figure 32.7. To cause minimal mechanical damage to colon tissues, the external electrode was covered by a polyethylene sleeve. Each PHD pulse existed for 3 μs, with a peak voltage of 3.2 kV and an energy per pulse of 0.1 J. Due to the low repetition frequency of about 7 Hz and short pulse duration, the average gas temperature did not exceed room temperature. Plasma temperature was around 7200 K. This plasma was found to radiate in the UV range at a total power density of 5 μW cm^{-2}. The outer tubing was 1.5 mm longer than electrode system, covering the discharge from the sides to significantly reduce the mucosa UV exposure. The NO production varied from 900 to 1200 ppm while plasma was applied for 15–60 s.

The Dextran sodium sulfate (DSS) model was used in this study to produce ulcerative colitis in mice, which is representative of human ulcerative colitis. Female Swiss Webster mice 25–30 g in weight aged approximately 6–8 weeks were used. The disease is induced in an animal through a daily oral administration of 2.5% DSS dissolved in drinking water at a concentration of 2.5%. The animal develops an acute form of inflammation beginning on the third day and typically has a full-blown colitis two days later. The primary characteristics of acute inflammation are the increased number of neutrophils in the mucosal layer, shortening of the epithelial crypts, and hyalinization in the lamina propria, accompanied by severe weight loss, diarrhea, and blood in the stool. This form of the disease provides the opportunity to study efficacy of drugs and compounds. The most striking feature of this model is that it works using a very simple pathway to produce the disease as DSS overcomes the barrier of the epithelium to expose the mucosa to the flora present in the intestine, resulting in an inflammatory response; this in turn leads to activation of macrophages and monocytes. The model also shows close links to the disease in human beings and is simple to induce and reproduce. To quantify the disease induced by the DSS model, a disease activity index (DAI) was used. This index has the scale of 0–4 with four being the lethal stage of the disease. DAI is scored on the parameters of weight loss, consistency of the stool and presence of blood in the stool. This index is linearly correlated to the histology score based on changes in the architecture of the crypt.

One of the objectives of the study was to test for any damage caused to the colon tissue or to the animal itself due to the plasma treatment. For this purpose, 12 mice were divided into 4 groups of 3 animals in each group. The first group was a control group which did not receive any treatment, and the other three groups remained as experimental groups. In the experimental groups, a laparotomy was performed, and the colon was exposed and kept moist covered with saline gauze.

The plasma probe was introduced through the anal verge of a live animal up to 4 cm into the colon (see Figure 32.8, Fridman and Friedman 2013). Plasma treatment was administered for 0, 4, 30, and 60 s in the respective group (for "0" time, the probe was inserted into the colon with no plasma ignited). To check colon tissue damage, the mice were intravenously injected with 30 mg kg^{-1} of Evans Blue (EB); 10 min later the administration colon was washed with 1 ml physiological saline and EB presence was analyzed spectrophotometrically. Animals were then euthanized with an overdose of Nembutal. Colon tissue samples were surgically removed and were preserved in formalin for further histopathological analysis. Spectro-photometrical analysis of the saline fluid collected from the colon showed no traces of EB, indicating that plasma did not affect the tissue integrity. Histology analysis also showed that no macroscopic damage was induced to the colon tissues by plasma treatment or probe manipulation. To check the response of the disease progression to plasma treatment, 24 animals were divided into four groups each receiving 0, 4, 30, or 60 s of plasma treatment every alternate day for seven days.

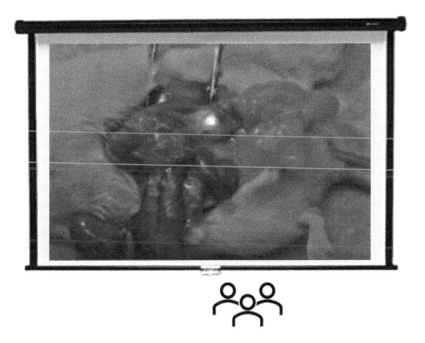

Figure 32.8 The IBD surgery, live mouse model. The miniaturized pin-to-hole discharge in operation, the tip was inserted through the anal verge up to 4 cm into the mouse colon. The light due to plasma was observed through the intestinal wall when the abdomen was open.

All animals were fed 2.5% DSS for seven days in parallel with plasma treatment. DAI was scored every day to see which dosage of plasma was most effective in controlling the progression of the disease. For the plasma probe to be inserted and to go through the colon, the colon has been cleansed of stool specks. To summarize the results of the plasma treatment study, it was found that 30 s of treatment produced the best results after seven days. Thus plasma treatment significantly slows down progression of the Crohn's disease, see for more details Chakravarthy et al. (2011), as well as Fridman and Friedman (2013).

The last stage of the study investigated the **effectiveness of plasma treatment as an adjuvant to conventional antioxidant drug** (5-amino salicylic acid or 5-ASA) treatment. Based on previous results, where it was shown that 30 s of plasma treatment gives the best results in controlling and reducing the disease progression compared to all other groups, this treatment dose was selected for the next step. In this set of experiments, 24 animals were divided into four groups: (i) control group, where no plasma or drug treatment was performed and groups where mice were treated either (ii) with plasma alone; (iii) 5-ASA alone; or (iv) drug and plasma together.

All four groups of animals were fed with DSS for six days. On days 2, 4, and 6 they received 2.5% DSS dissolved in water and for days 1, 3, and 5 (one day before plasma treatment) they received 2.5% DSS dissolved in 15% polyethylene-glycol-based laxative to clean the colon. The plasma probe was introduced into the colon 4 cm from anal verge in groups (ii) and (iv). Group (ii) received a 30 s dose of plasma treatment only, while group (iv) received the same dose of plasma treatment together with 0.1 ml of 5-ASA treatment. Animals in groups (iii) and (iv) were treated with 0.1 ml of 5-ASA. The DAI was scored every day during the tenure of the study. On the seventh day of DSS treatment and final plasma treatment, DAI was measured, and animals were euthanized with an overdose of Nembutal. The control group showed a steady increase through the course of seven days with the DAI reaching 2.7 on day 7. The plasma treatment group was administered 30 s of plasma treatment; obtained data are in direct correlation with the data acquired in the previous study and show a constant increase of disease progression. However, the disease curtailed to near 1.8 on the DAI scale. Group (iv) showed very positive results; the treatments in combination proved to be effective in controlling the disease and keeping the DAI at a level of 1.2. In summary, this study assessed the possibility that pin-to-hole discharge could be used to treat IBD. The plasma treatment did not adversely affect the animals and did not increase the progress of the disease. In fact, it reduced the disease from progressing rapidly as compared to the control. Combination therapy of 5-ASA and cold plasma showed that there is a significant therapeutic relevance when it comes to adding plasma to the drug in controlling colitis in the DSS model during the induction phase. The exact mechanism by which cold plasma induced its beneficial effect remains unknown, although the authors of this study suggest that the observations can be partly attributed to the effects of NO. However, other effects including oxidation of cytokines responsible for activation of the immune system are possible.

32.9 Plasma in Cardiovascular Diseases: Plasma Effect on Whole Blood Viscosity (WBV)

One of the important aspects of plasma treatment of cardiovascular diseases is related to the plasma effect on rheological properties of blood (blood viscosity), and plasma control of low-density-lipoprotein (LDL) cholesterol, see Jung et al. (2012), Yang et al. (2012), and Fridman and Friedman (2013). Plasma is not only able to control blood coagulation but can also control and regulate multiple physical and biochemical properties of blood. One of these properties, **blood viscosity,** has been analyzed by Jung et al. (2012).

The viscosity of a fluid represents the frictional resistance between a moving fluid and stationary wall. Blood viscosity is the inherent resistance of blood to flow and represents the thickness and stickiness of blood. Since about 45% of blood volume is made up of suspended cellular particles, primarily red blood cells, the blood behaves as a non-Newtonian fluid where its viscosity varies with shear rate (i.e. the ratio of flow velocity to lumen diameter). The dynamic range of the **whole blood viscosity** (WBV) is relatively large, that is, 40–450 millipoise (mP) (or 4–45 centipoise). This highlights the potential utility of this parameter as a biomarker, to the degree that viscosity provides additional incremental prediction of clinical outcomes and is modifiable by therapeutic modalities.

WBV has been strongly associated with cardiovascular disease, stroke, and peripheral arterial disease. Mechanical interaction between blood and blood vessels mediated by the increase of WBV has a crucial pathogenic role in the release of endothelium-derived mediators (NO and endothelia), thus causing subsequent vascular remodeling by activation of endothelium, initiation of inflammation, alteration of lipid metabolism, and finally progression of atherosclerotic vascular disease. There are several variables for the WBV, which include hematocrit, plasma proteins (i.e. fibrinogen, immunoglobulin, and albumin), total cholesterol, low-density lipoprotein (LDL) cholesterol, high-density lipoprotein (HDL) cholesterol, and triglyceride. In addition, the aggregation and deformability of erythrocytes critically affect the WBV. Since both the plasma proteins and LDL molecules influence the aggregation of erythrocytes, it is important to keep both levels within the respective normal ranges. For example, lipid-lowering statin drugs have been widely used to keep the LDL cholesterol within the normal range (i.e. 62–130 mg dl^{-1}). Statins are powerful cholesterol-lowering drugs in clinical practice and have a life-saving potential in properly selected patients,

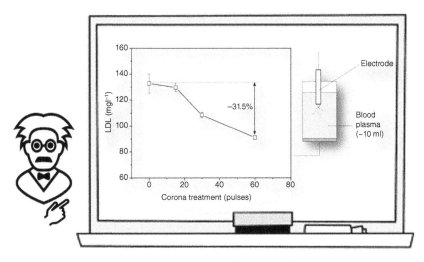

Figure 32.9 Pulsed corona plasma-induced LDL removal from blood plasma.

particularly those with severe hyperlipidemia and atherosclerotic disease. Results from randomized clinical trials have demonstrated a decrease in congested heart diseases (CHD) and total mortality, reductions in myocardial infarctions, revascularization procedures, stroke, and peripheral vascular disease. However, statins are prescribed for less than half of the patients who should receive this therapy, unfortunately, because of the side effect which remains a major impediment to the appropriate use of these drugs: liver and muscle toxicity.

Nonthermal plasma can be a powerful tool to solve the above-described issues. It was demonstrated that selective coagulation of fibrinogen, which is one of the major viscosity determinants in blood, could be induced by the application of DBD in air (Kalghatgi et al. 2007a,b). It is therefore hypothesized that the hemorheological properties of blood may be improved by lowering the levels of fibrinogen and LDL molecules in blood plasma through DBD-assisted coagulation and the subsequent filtration of the coagulated particles from the treated blood plasma (Figure 32.9, see more details in the Section 32.10).

The WBV of blood samples measured using the Hemathix Analyzer at five different shear rates of 1000, 300, 100, 10, $1\,s^{-1}$ before DBD treatment have been analyzed. The normalized WBV of untreated blood varied from 35.38 to 301.01 mP when shear rate decreased from 1000 to $1\,s^{-1}$. When blood plasma was treated with DBD, a white layer formation was found on the top surface of the treated blood plasma samples and at the interfacing edge with a copper surface, whereas the blood plasma without DBD treatment did not exhibit any white particle formation on the sample surface after being exposed to open atmosphere for the same duration. The viscosity of blood plasma showed irregular values in accordance with the DBD treatment time for all shear rates. Over the range of shear rates, the normalized WBV increased with increasing DBD treatment time. WBV measured at a high shear rate of $300\,s^{-1}$ is usually referred to as **systolic whole blood viscosity (SBV)**, whereas WBV measured at a low shear rate of $1\,s^{-1}$ is referred to as **diastolic whole blood viscosity (DBV)** as WBV varies during a cardiac cycle in the same way as blood pressure. DBV values are significantly higher in DBD-treated samples, i.e. 368 mP for 8 min of DBD treatment, than untreated control samples (301 mP). SBV values were also elevated from 38.0 mP for control to 49.3 mP for 8 min of DBD treatment.

WBVs of blood samples made with DBD-treated blood plasma, which was filtered before being mixed with the original red blood cells have been analyzed. The systolic blood viscosity (SBW) after 8 min of DBD treatment without filtration increased by 5.5% from the baseline value of 35.4 mP. When the DBD-treated plasma was filtered before being mixing with red blood cells, the SBW decreased by 9.1% from the baseline SBV of 35.4 mP. The DBV of blood treated by DBD but without filtration was elevated with increasing DBD treatment time. For example, the DBV of blood increased by 29.9% from the baseline value of 301.0 mP. When the DBD-treated plasma was filtered before being mixing with red blood cells, the DBV of blood for 8 min of treatment decreased by 17.7% from the baseline value of 301 mP.

These DBD plasma effects on WBV clarify pathways to control the rheological blood properties and therefore treatment of the cardiovascular diseases. For more details see Jung et al. (2012) and Yang et al. (2012).

32.10 Plasma in Cardiovascular Diseases: Control of Blood Rheological Properties and Low-density-lipoprotein (LDL) Cholesterol, Stimulation of Angiogenesis

As it was discussed in the previous section, when blood plasma is treated by DBD a white clot layer is formed at the top of blood plasma sample (Jung et al. 2012). It suggests that the plasma proteins and lipids in blood plasma might have precipitated into coagulated particles through DBD treatment. Accordingly, *if these coagulated particles are removed, improvement in the rheological properties of whole blood may be anticipated. Note that the size of fibrinogen (diameter of 5–7 nm and length of 48 nm) and LDL molecules (approximately 22 nm) are too small to be removed by filter. The application of DBD treatment to blood plasma helps the excess fibrinogen and LDL molecules in blood plasma to be removed with a regular filter via DBD-assisted oxidation and coagulation.*

Although DBD discharge employed in this study is nonthermal there could be some amount of thermal energy to be transferred to the blood plasma sample, causing the creation of the white layer of coagulated particles. To test this hypothesis, temperature was measured before and after the treatment of the samples with DBD. No significant change in temperature was observed, suggesting that the coagulated particles must have come from the oxidation and subsequent coagulation by active species generated by DBD. In addition, there was no change in hematocrit before and after DBD treatment, indicating no significant evaporation in blood plasma during the DBD treatment procedure.

The blood mixed with the blood plasma treated by DBD without filtration showed a significant increase in WBV. This can be attributed to the fact that plasma proteins and lipids formed large groups of coagulated particles of micron size, increasing frictional resistance to flow over the entire range of shear rates. Since the coagulated particles are visible (i.e. large in size) they could be relatively easily removed with filtration, a process which helped to reduce WBV as the plasma proteins and lipids are the key determinants of WBV.

SBV of blood with DBD treatment and filtration decreased by c. 9.1% from the baseline value, whereas the DBV dropped by 17.7% from the respective baseline value. Both represent significant improvements after the DBD treatment. The reason why the DBD treatment improved the DBV more than the SBV is that the SBV is usually affected by erythrocyte deformability while DBV is affected by the erythrocyte aggregation. Since the plasma proteins (i.e. fibrinogen and immunoglobulins) and LDL molecules are instrumental in the red blood cells (RBC) aggregation, their removal by the DBD treatment and filtration should mitigate the erythrocyte aggregation, subsequently reducing the DBV.

In terms of the mechanism of DBD treatment in the **precipitation of plasma proteins and lipids from blood plasma**, Kalghatgi et al. (2007a,b) reported that DBD treatment might have activated some of the protein coagulation process which resulted in rapid fibrinogen aggregation. It was also mentioned that the selective coagulation of proteins was observed – not of albumin but of fibrinogen only, see Lecture 31. It can be hypothesized that this coagulated fibrinogen and subsequent removal might have affected the WBV of blood with DBD-treated blood plasma. Jung et al. (2012), utilized DBD treatment and filtration to reduce blood viscosity. The results indicate that the nonthermal DBD treatment could precipitate and coagulate plasma proteins (i.e. fibrinogen) and LDL molecules in blood plasma. The formation of a white layer on the surface of blood plasma after DBD treatment confirmed the precipitation and coagulation of plasma proteins and lipids in blood plasma. It has been demonstrated that WBV could be significantly reduced by 9.1% and 17.7% for SBV and DBV, respectively, from the respective baseline values when DBD-treated blood plasma was filtered prior to mixing.

In addition, **oxidized LDL molecules** *are known to adhere to arterial wall surfaces, playing a key role in the* **progression of atherosclerosis**. *In a similar manner, when LDL molecules in blood plasma are oxidized by DBD treatment, the oxidized LDL molecules tend to precipitate and coagulate. Since LDL molecules are one of the important variables in WBV, their removal with DBD treatment and filtration should not only help to reduce WBV but also to reduce the progression of atherosclerosis.* This effect has been demonstrated by the pulsed corona plasma-induced LDL removal from blood plasma. It shows the effective plasma-stimulated control of blood composition and critical biophysical and biochemical properties of blood. For more details see Jung et al. (2012) and Yang et al. (2012).

Another process related to the cardiovascular health is plasma control of the vasculature and **stimulation of angiogenesis**. Blood vessels are affected during plasma treatment of many tissues and may be an important potential target for clinical plasma therapy. Porcine aortic endothelial cells were treated in vitro with a DBD-based nonthermal plasma device by Clyne and her group in Drexel University, see Kalghatgi et al. (2010) and Arjunan et al. (2012). Low dose plasma (up to 30 s or 4 J cm^{-2}) was relatively nontoxic to endothelial cells in these studies while treatment at longer exposures (60 s and higher or 8 J cm^{-2}) led to cell death. Endothelial cells treated with plasma for 30 s demonstrated twice as much proliferation as untreated cells five days after plasma treatment. Endothelial cell release of fibroblast growth factor-2 (FGF2) peaked three hours after plasma treatment. The plasma proliferative effect was abrogated by an FGF2 neutralizing antibody, and FGF2 release was blocked by reactive oxygen species scavengers. These data demonstrated that low dose nonthermal plasma enhances endothelial cell proliferation due to reactive oxygen species-mediated FGF2 release. Thus, nonthermal plasma may be a novel therapy for dose-dependent promotion or inhibition of endothelial cell-mediated angiogenesis. More information on this subject can be found in Kalghatgi et al. (2010) and Arjunan (2012).

32.11 Plasma-assisted Tissue Engineering: Regulation of Bio-properties of Polymers, Bioactive Micro-xerography, Cell Attachment and Proliferation on Polymer Scaffolds

Plasma methods are widely applied today in different aspects of tissue engineering. First consider plasma control of the biological properties of medical polymers and biocompatibility. The main requirement imposed on all polymer biomaterials applied in medicine is a combination of their desired physicochemical and physicomechanical characteristics with biocompatibility.

 Depending on applications, the biocompatibility of polymers can include various requirements which can sometimes be contradictory to each other. In the case of artificial vessels, drainages, intraocular lenses, biosensors, or catheters, the interaction of the polymer with a biological medium should be minimized for the reliable operation of the corresponding device after implantation. In contrast, in most of orthopedic applications, the active interaction and fusion of an implant with a tissue is required. General requirements imposed on all medical polymers consist in nontoxicity and stability.

The body's response to a polymer implant mainly depends on its surface properties: chemical composition, structure, and morphology. Physical techniques are required for regulating the biological properties of polymer materials. These techniques should vary the physicochemical, structural, and functional properties of surfaces over a wide range without affecting bulk characteristics such as strength, elasticity, transmission factor, refractive index, and electrophysical parameters.

Treatment in a low-temperature gas-discharge plasma, which is widely used for modifying the surface characteristics of polymers, is such a multipurpose and multifunctional technique, see Lecture 26. The cross-linking effect under the action of inert gas plasmas can be used for the **immobilization of biocompatible low-molecular-weight compounds and various functional groups on polymer surfaces**. For example, the treatment in argon plasma was used for grafting the Pluronic 120 copolymer, which is highly biocompatible, onto the surface of polytetrafluoroethylene (PTFE), Vasilets (2005). The Pluronic120 triblock copolymer (polyethylene oxide/polypropylene polyethylene oxide or PEO/PP/PEO) was initially supported on the polymer surface by physisorption from solution and then grafted by treatment in low-pressure argon plasma. A similar procedure was used for the covalent immobilization of sulfur groups on the surface of polyethylene (PE). Plasma etching changes the polymer surface morphology, which plays an important role in biocompatibility. For example, surface smoothing due to plasma etching upon treatment in oxygen- and fluorine-containing plasmas positively affects hemocompatibility because it decreases the probability of thrombosis at surface irregularities in a bloodstream. The surface smoothing of a polymethylmethacrylate (PMMA) lens without changes in its bulk optical characteristics is a positive consequence of plasma treatment in a CF_4 discharge.

Since most biological fluids are aqueous solutions or contain water, the wettability or **hydrophilicity of polymer surfaces is of paramount importance for biocompatibility**. According to the **hypothesis of complementarity**, to improve hemocompatibility it is necessary to minimize not only the average interfacial surface energy at the material–blood interface but also at every point in the surface; that is, the character of free energy distribution at the interface of the biomaterial with an adsorbed protein layer should be taken into consideration. In other words, the hydrophilic and hydrophobic regions of a surface that contacts with blood and analogous regions of an adsorbed protein molecule should be complementary for the product to be highly hemocompatible. The hydrophilicity and surface energy of biomaterials can be varied over wide ranges by generating various surface polar groups with the use of treatment of medical polymers in oxygen plasma, oxygen-containing CO_2, H_2O, SO_2 plasmas, or nitrogen-containing (N_2, NH_3) plasmas. The processes of protein adsorption and cell adhesion can thereby be regulated. Plasma treatment is therefore effective in regulating adhesion of medical polymers (making their surfaces adhesive or repulsive to certain biological agents), as well as controlling their biocompatibility. See Vasilets (2005), Fridman and Friedman (2013) for more details.

Biochemical patterning (bio-printing) which allows for micro-scale resolution on nonplanar substrates has been demonstrated by Fridman et al. (2003, 2005a–c). Utilizing this method, biomolecules (including DNA, proteins, and enzymes) can be delivered to charged locations on surfaces by charged water buffer droplets. Charging of water droplets has been accomplished in using

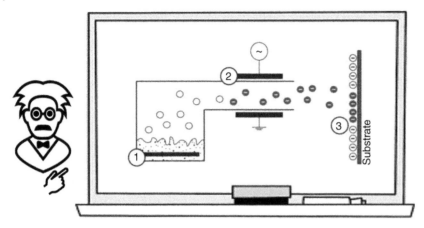

Figure 32.10 Plasma bio-printer. 1: droplet atomizer; 2: DBD plasma reactor; and 3: charge droplet deposition substrate.

atmospheric DBD stabilized in the presence of a high concentration of micron-size water droplets. Patterning of biomolecules on surfaces has many applications ranging from biosensors used in genetic discovery and monitoring of dangerous toxins to tissue engineering constructs where surfaces control tissue assembly or adhesion of cells. Most methods of biochemical patterning are only suitable for planar surfaces. In addition, micro- and nanoscale patterning often relies on complex sequences of lithography-based process steps. Plasma provides the biochemical substance patterning or printing allowing micro- and even nanoscale resolution on planar and nonplanar substrates. It implies printing droplets of buffer containing biomolecules (including DNA), peptides, and cells. The method involves the creation of droplets with relevant molecules or cells in their respective buffer solutions and then DBD charging of the droplets. Finally, the droplets are delivered onto substrate with charge pre-written via conventional xerography (micron-resolution) or via charge stamping (nanometer-resolution); see Figure 32.10. The key issue of the plasma-assisted bio-printing is protection of the biological agents contained in the buffer droplets from their plasma-based deactivation. Survival of the biological agents in plasma can be achieved by choosing relatively low specific DBD power and a special composition of the protective buffer solutions.

A significant **effect of nonthermal plasma treatment of polymer scaffolds on both attachment of bone cells and on their further biological activity**, specifically on their proliferation, has been demonstrated by Yildirim et al. (2008). The experiment in the atmospheric-pressure DBD plasma addressed attachment and proliferation of osteoblasts cultured over poly-ε-caprolactone (PCL) scaffolds. Traditional bone grafting procedures due to pathological conditions, trauma, or congenital deformities have disadvantages such as graft rejection, donor site morbidity, and disease transmission. Tissue engineering of bone is increasingly becoming the treatment of choice among surgeons to alleviate these problems. Favorable cell-substrate interaction during the early stage of cell seeding is one of the most desirable features of tissue engineering. The ability of bone cells to produce an osteoid matrix on the scaffold can be affected by the quality of the cell–scaffold interaction. Plasma treatment of the scaffolds significantly improves this interaction. Plasma makes the PCL surface hydrophilic, which leads to successful formation of tissue constructs and cell proliferation, differentiation, and new tissue ingrowth. The influence of oxygen-pulsed DBD plasma-treated PCL on early-time points osteoblastic cell adhesion and proliferation has been demonstrated. The PCL scaffold has been treated with oxygen-based microsecond-pulsed DBD plasma with different exposure time, then mouse osteoblast cells (7F2) were cultured on treated PCL scaffolds for 15 h.

Yildirim et al. (2008), first consider the effect of atmospheric-pressure oxygen DBD plasma on **surface hydrophilicity and surface energy of the PCL scaffold** by contact angle measurements. The total surface energy of PCL increased with prolonged oxygen plasma treatment time from $95\,\mathrm{mN\,m^{-1}}$ for untreated to $115\,\mathrm{mN\,m^{-1}}$ for 5 min oxygen-plasma-treated scaffold. While the dispersive component was relatively unchanged, the polar component increased significantly. This demonstrates that the major contribution of total solid surface energy to the increment is due to the formation of polar (mostly peroxide) groups on the polymer surface during its treatment in atmospheric-pressure nonthermal oxygen plasma. The effect of oxygen plasma treatment on PCL microstructure was characterized using AFM. The average roughness of untreated film was 13 nm and the surface pattern was relatively smooth. As a result of 1 and 3 min DBD plasma treatment, the roughness was increased to 20 and 26 nm, respectively. For the 5-min treatment time, the change in surface morphology became obvious by large peaks in nanoscale with the roughness increase to 39 nm. The results therefore show that with the prolonged treatment time the mean surface roughness is increased by a factor of almost 3.

The **metabolic activity and the morphology of mouse osteoblast cells** on plasma-treated and untreated PCL scaffolds were examined by the live/dead cell vitality assay and the Alamar Blue assay. On the untreated PCL scaffolds, cell proliferation, cell attachment, and cell spreading with culturing time were significantly lower as compared to oxygen-plasma-treated PCL scaffolds. The mouse osteoblasts on untreated PCL scaffolds did not show significant metabolic activity during the incubation time. In contrast, a higher degree of cell attachment and proliferation were seen on the plasma-treated surface. The viable cell count was also increased with prolonged oxygen-plasma treatment. The highest improvement for proliferation rate was observed for the 5-min treated sample. Normalized fluorescence intensity data shows that after a 15-h incubation period, the cell proliferation rate on the 5-min treated samples increased by 90% from the beginning of the experiment. From the fluorescence images, it was observed that the mouse osteoblast cells are hardly attached on untreated scaffolds. After a 15-h incubation period, cells were hardly spread out on the 1-min treated scaffold. With the increased plasma treatment time from 1 to 5 min, osteoblastic adhesions improved. On the 5-min oxygen-plasma-treated samples, a confluent mouse osteoblast cell layer was observed. A similar trend was found for cell proliferation. Oxygen-plasma-treated PCL scaffolds are therefore much more favorable than untreated scaffolds for cell attachment, due to the higher hydrophilicity and increased roughness. More details on the subject can be found in Yildirim et al. (2008), as well as in Fridman and Friedman (2013).

32.12 Plasma Control of Stem Cells and Tissue Regeneration, Differentiation of Mesenchymal Stem Cells (MSC) into Bone and Cartilage Cells, Plasma Orthopedics

Plasma orthopedics is a good example of **plasma control of stem cells and tissue regeneration** demonstrated by Freeman and her group in Jefferson University Hospital, Philadelphia, see for more details Freeman et al. (2012), Steinbeck et al. (2013), and Chernets et al. (2015). Fracture repair is a dynamic process requiring the mobilization and activation of stem cells to achieve new bone formation and increase resorption of old bone. Research into stem cell use, molecular signaling cascades, and application of growth factors is still in the initial phase with minimum impact on patients. At the same time, biophysical applications through either electrical, electromagnetic, or ultrasound stimulation have been in use for many years with reasonable success. The application of nonthermal plasma in this regard provides new methods to exploit these phenomena to advance the rapidly growing field of regenerative medicine.

From the very first experiments of Freeman and her group, a high efficiency of *nonthermal plasma on* **MSC** *differentiation and related bone tissue regeneration has been successfully demonstrated in animal models. They investigated the potential of FE-DBD-generated reactive oxygen species (ROS) and RNS to promote mesenchymal cell (MC) proliferation, commitment, and differentiation along skeletal lineages. It was shown that DBD-generated ROS can be used to enhance skeletal cell differentiation by increasing intracellular ROS, which leads to the activation of kinases and transcription factors known to influence genes associated with differentiation and skeletal development. Freeman et al. They also investigated the effect of plasma on activation of cell signaling promoting differentiation. It was shown that DBD plasma significantly enhanced both early and late osteoblast differentiation gene expression.* It was also observed that DBD plasma alone is not sufficient to initiate significant changes in osteogenic differentiation. However, once differentiation has been initiated, FE-DBD plasma significantly enhances the differentiation.

The enhanced differentiation of MSCs into chondrocytes or osteoblasts is of paramount importance in tissue engineering and regenerative therapies. The appendage regeneration is dependent on ROS production and signaling. *The MSC stimulation by nonthermal plasma, which produces and induces ROS (i) promote skeletal cell differentiation and (ii) limb autopod development. Stimulation with a* **single treatment of the FE-DBD plasma enhanced survival, growth, and elongation of mouse limb autopods** *in an in vitro organ culture system. Noticeable changes included enhanced development of digit length and definition of digit separation. These changes were coordinated with enhanced Wnt signaling in the distal apical epidermal ridge (AER) and presumptive joint regions. Autopod development continued to advance for approximately 144 h in culture, seemingly overcoming the negative culture environment usually observed in this in vitro system. Real-time quantitative polymerase chain reaction analysis confirmed the up-regulation of chondrogenic transcripts.* Mechanistically, plasma increased the number of ROS positive cells in the dorsal epithelium, mesenchyme, and the distal tip of each phalange behind the AER, determined using dihydrorhodamine. The importance of ROS production/signaling during development was further demonstrated by the stunting of digital outgrowth when antioxidants were applied. Results of this study show that plasma initiated and amplified ROS intracellular signaling to enhance development of the autopod. Parallels between development and regeneration suggest that the potential use of plasma could extend to both tissue engineering and clinical applications to enhance fracture healing, trauma repair, and bone fusion. More details on this interesting subject can be found in Freeman et al. (2012), Steinbeck et al. (2013), and Chernets et al. (2015).

In addition, plasma treatment of polymers can control sophisticated behavior of biological systems on treated surfaces. An example of such an application is the **selective inhibition of type X collagen expression in human MSC differentiation on polymer substrate** surface modified by low-pressure RF CCP discharge plasma. Such plasma-assisted tissue engineering experiments were carried out by Nelea et al. (2005). A major drawback of current cartilage- and disc-tissue engineering is that human MSCs rapidly express type X collagen, which is a marker of chondrocyte hypertrophy associated with endochondral ossification. Some studies have attempted to use growth factors to inhibit type X collagen expression, but none to date have addressed the possible effect of the substratum on chondrocyte hypertrophy. Nelea et al. (2005) examined the growth and differentiation potential of human MSCs cultured on two polymer types (PP and nylon-6), both of which have been surface modified by low-pressure RF CCP discharge plasma treatment in ammonia gas. Cultures were performed for up to 14 days in Dulbecco's modified Eagle medium plus 10% fetal bovine serum. Commercial polystyrene culture dishes were used as control. Reverse transcriptase-polymerase chain reaction was used to assess the expression of types I, II, and X collagens and aggrecan using gene-specific primers. Glyceraldehyde-3-phosphate dehydrogenase was used as a housekeeping gene. Types I and X collagens, as well as aggrecan, were found to be constitutively expressed by human MSCs on polystyrene culture dishes. Whereas both untreated and treated nylon-6 partially inhibited type X collagen expression, treated PP almost completely inhibited its expression. These results indicate that plasma-treated polypropylene or nylon-6 may be a suitable surface for inducing MSCs to a disc-like phenotype for tissue engineering of intervertebral discs in which hypertrophy is suppressed. The Nelea et al. (2005), experiments demonstrate the effectiveness of plasma polymer treatment in controlling not only simple surface properties but also very specific biological behavior of cells on the treated surfaces.

32.13 Plasma Treatment of Cancer, Direct and Indirect (Plasma-treated Solutions) Approaches

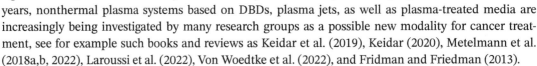

Depending on the type of cancer, conventional treatment options may include surgery often followed by chemotherapy and/or radiation. Photodynamic therapy and immunotherapy are more recent addition to the toolset of cancer treatment options. *Interestingly, all these treatments one way or another involve ROS. It is probably for this reason that plasma treatment can provide effective alternative cancer treatment approach. The novelty of the plasma approach lies in its multi-factorial effects that include reactive oxygen and nitrogen species (ROS/RNS) produced in the plasma, physical factors like emitted electromagnetic waves, and the electric fields that are formed when plasma impinges on tissue.* In recent years, nonthermal plasma systems based on DBDs, plasma jets, as well as plasma-treated media are increasingly being investigated by many research groups as a possible new modality for cancer treatment, see for example such books and reviews as Keidar et al. (2019), Keidar (2020), Metelmann et al. (2018a,b, 2022), Laroussi et al. (2022), Von Woedtke et al. (2022), and Fridman and Friedman (2013).

Cancer is often described as a disease where some cells have increased metabolism and escape the processes of "normal" regulated growth and cell death for various reasons and, as a result, can accumulate in the body faster than other cells of similar origin. When accumulated in sufficient numbers, these cells can then "hijack" or divert various processes in the body such as formation of blood vessels for their own benefit leading to various dysfunctions and ultimately death. Cancers are typically classified by the type of cells that the malignant cells tend to resemble and are, therefore, often presumed to be the origin of the tumor. The different types commonly include the following. **Carcinoma**: cancers derived from epithelial cells. This group includes many of the most common cancers, particularly in older individuals, and include nearly all those developing in the breast, prostate, lung, pancreas, and colon. **Sarcoma**: cancers arising from some connective tissue (i.e. bone, cartilage, fat), each of which develops from cells originating in mesenchymal cells outside the bone marrow. **Lymphoma and leukemia**: these two classes of cancer arise from hematopoietic (blood-forming) cells that leave the marrow and tend to mature in the lymph nodes and blood, respectively. These cancer cells do not form solid tumors. **Germ cell tumor**: cancers derived from pluripotent cells, most often presenting in the testicle or the ovary. **Blastoma**: cancers derived from immature "precursor" cells or embryonic tissue. These are common in children. Some cancer cells may acquire the ability to penetrate and infiltrate surrounding normal tissues in the local area, forming a new tumor. The newly formed "daughter" tumor in the adjacent site within the tissue is called a **local metastasis**. Other cancer cells acquire the ability to penetrate the walls of lymphatic and/or blood vessels, after which they can circulate through the bloodstream (circulating tumor cells) metastasizing to other nonadjacent sites and tissues in the body.

Like in plasma medicine in general, plasma cancer treatment can be direct and indirect. The experimental approaches of **direct plasma treatment to cancer** encompass not only direct **plasma killing of cancer cells** but also possible **stimulation of immune responses** by inducing controlled oxidative stress in cancer cells, which is going to be discussed at the end of this lecture. The direct plasma treatment is performed when cells, samples, animal, or human tissues are directly accessible to bulk plasma. A layer of physiological solution typically covers the cells during direct in vitro plasma treatments. In vivo plasma treatment also proceeds through some level of "wetness." Such layers facilitate the diffusion and formation of reactive species from the gas phase to the liquid or "wet" phase. Direct plasma treatment has already been widely used in many in vivo experiments and clinical studies, which is going to be discussed below.

The basic strategy of the **indirect plasma treatment of cancer** is the application of plasma-treated solutions (PTS) such as cell culture medium or aqueous solutions like lactate to kill cancer cells or destroy tumor tissues. The anticancer effects are achieved by the chemical components, including reactive species and other chemically modified substances in these solutions Tanaka et al. (2021), Yan et al. (2001, 2018), and Tanaka et al. (2016). Typically, PTS are produced through direct contact of plasma with the solution during treatment. However, PTS can also be produced by the gas discharge within the solution. PTS has been shown to selectively kill many cancer cell lines in vitro or have anti-tumor effects when directly injected in subcutaneous or intraperitoneal tumor models in vivo. Once produced, PTS is independent of plasma source, which is an advantage over direct plasma treatment. Treatment by PTS eliminates the local effects of electric fields, photons, and temperature. Direct and indirect treatment share many similarities: many cellular responses have been widely observed in both treatment strategies, including perturbations in the cell membrane, changes in mitochondrial structure and function (increase in intracellular ROS), and damage to the endoplasmic reticulum and DNA, see Malyavko et al. (2020). Let us start with consideration of plasma abatement of cultured malignant cells.

32.14 Plasma Abatement of Malignant Cells, Apoptosis vs Necrosis of Cancer Cells

The first reports of the apoptotic effects of nonthermal plasma on cancer cells were probably from the Drexel University, where FE-DBD plasma has been employed to treat melanoma cells, see Fridman et al. (2007a–e). Melanoma cancer cell line was propagated using standard cell medium and incubated. After four to five days of incubation, the cells were detached from the dish surface and transferred to aluminum dishes in dilution ratios of 1 : 5 to 1 : 6, where they were incubated for four to five more days. The number of cells in each dish at the end of incubation was about 10^6 on average. The viability of cultures was determined by Trypan Blue exclusion (a measure of the integrity of the cell membrane) to be 91–97%. Trypan Blue is a dye that tends to enter cells if their membranes have sufficiently large pores, indicating lack of cell viability. Cells were treated in FE-DBD for times ranging from 5 to 30 s.

The most important finding of this work was that the melanoma cells were not simply destroyed through a process known as **necrosis**, but that they were induced into **apoptotic cell death**. Apoptosis is a multi-step, multi-pathway cell-death program that is inherent to most cells in the body. In contrast to necrosis, which is a form of traumatic cell death that results from an acute cellular injury, apoptosis confers certain advantages. For example, unlike necrosis, apoptosis results in cell fragmentation called apoptotic bodies that phagocytic cells can engulf and quickly remove before the contents of the cell can spill out onto surrounding cells and cause damage. The removal of apoptotic cell fragments by phagocytes occurs in an orderly manner without eliciting an inflammatory response. Apoptosis can be viewed as a natural part of a cell renewal cycle where cells in the body maintain a certain ratio of cell division to apoptosis. In a growing body, this ratio favors cell proliferation. Once the body stops growing, the ratio of cell division to apoptosis is normally stabilized to keep renewing cells, while maintaining the overall size of the body and its organs. In cancer, the apoptosis to cell division ratio is altered, to favor proliferation of cancer cells.

It was originally believed that cancer cell accumulation is due to an increase in cellular proliferation, but it is now known that it is also significantly due to a decrease in apoptotic cell death. *Many nonsurgical cancer treatments such as chemotherapy and irradiation kill malignant cells primarily by inducing apoptosis. The ability of nonthermal plasma treatment to induce apoptosis can therefore be viewed as a promising step in developing plasma cancer treatment.*

The following events are typically observed in apoptotic cells: the phospholipid phosphatidylserine (PS), which is normally hidden within the plasma membrane, is exposed on the surface; cells develop bubblelike blebs on their surface; cells have their mitochondria breakdown with the release of cytochrome *c*; the chromatin (DNA and protein) in cell nuclei are degraded; cells break into small, membrane-wrapped, fragments; and calls release molecules such as adenosine triphosphate (ATP) and UTP (uridintriphosphate). These nucleotides bind to receptors on wandering phagocytic cells such as macrophages and dendritic cells, and attract them to the dying cells (a "find-me" signal). This signal is bound by other receptors on the phagocytes, which then engulf the cell fragments and secrete cytokines that inhibit inflammation (e.g. IL-10 and TGF-β). Apoptosis may occur through three primary pathways. Two of these pathways, one intrinsic and the other extrinsic, proceed through activation of proteases (enzymes that degrade polypeptides) from a family known as caspases. Intrinsic pathways typically proceed through opening of pores in mitochondrial membranes, release of cytochrome c, and subsequent activation of caspase 9 followed by other caspases. Extrinsic pathways are typically initiated by the binding of molecules TNF-α, Fas ligand, or lymphotoxin to a death receptor followed by activation of caspases beginning with caspase 8. The third caspase-independent pathway proceeds through release of the apoptosis-inducing factor (AIF), a molecule found within mitochondrial membranes.

Several biochemical assays are typically employed to detect and follow apoptosis. Cells in their final stages of apoptosis are often recognized by the TUNEL assay which enzymatically labels DNA strands with exposed 3′-hydroxyl ends exposed in DNA fragmentation. Annexin-V is a calcium-dependent phospholipid-binding protein with high affinity for PS. As mentioned above, PS is a membrane component normally localized to the internal face of the cell membrane. Early in the apoptotic pathways, molecules of PS are translocated to the outer surface of the cell membrane where Annexin-V can readily bind them. Annexin-V is therefore an assay that detects early indicators of apoptosis. Caspase-dependent pathways can be detected through caspase cleavage assays where cleaved caspase is detected from lysed cells.

While the initial plasma-medical studies detected apoptosis in melanoma cells only by the **TUNEL assay** (Fridman et al. 2007a–e), subsequent work by the same group (Sensenig et al. 2011) observed both early apoptosis by **Annexin-V assay** and caspase cleavage (caspase 3) in addition to the detection of late apoptotic cells by the TUNEL assay. This was important not only in confirmation of apoptosis but also in the observation that apoptosis in the DBD-treated melanoma cells occurs via **caspase cleavage**. In addition, it was observed that scavenging of intracellular reactive oxygen species through pre-incubation of the cells with *N*-acetylcysteine (NAC), a common method of increasing levels of intracellular glutathione that scavenges ROS, reduces, or eliminates apoptosis following the plasma treatment. This suggests that the mechanism of apoptosis due to the DBD treatment relies on intracellular generation of ROS.

Later research by the same group went beyond studying apoptosis in cancer cell line. The key to a successful cancer treatment is not only its ability to induce "natural" death of cancer cells but also to **reduce the impact on normal cells** in the body. Chemotherapy and radiation treatments both tend to have a stronger effect on cancer cells than on the surrounding normal cells, and plasma treatment could be viewed as a viable alternative when it comes to more localized treatment if its effects on transformed (cancer) vs normal cells could be selective, see Fridman and Friedman (2013). Initial promising results related to selectivity of the plasma treatment were recently demonstrated using immortalized (in contrast to primary cells taken from the body, immortalized cells typically acquire some mutation to make them capable of dividing indefinitely) but noninvasive and not-tumor-forming mammary epithelial cells MCF10A, and an invasive (tumor-forming) genetic variant of these cells called MCF10ANUET. At doses of 7.8 J cm^{-2}, apoptotic effect on the tumor-forming cells is about twice as strong as the apoptotic effect on the noninvasive cells. Importantly, differences in the tumor-forming and noninvasive cell responses could not be attributed to differences in culturing conditions, cell medium, or other similar environmental conditions since they were identical for the two different cell types. Other groups (Zirnheld et al. 2010) recently considered differential effects on some melanoma and keratinocyte cells. It was shown that percent death of melanoma cells was about six times as large as keratinocyte cell death following treatment with a He plasma jet powered by a 2.5 kV AC source at a frequency of 80–120 kHz. Treatment of colorectal cancer cells by a He dielectric barrier jet also demonstrated (Kim et al. 2010) loss of invasiveness by the tumor cells as well as apoptosis and phosphorylation of β-catenin, which is a transcription factor that controls colorectal tumorigenesis (its phosphorylation causes its rapid degradation in cells).

Important factors influencing **selectivity of plasma treatment of cancerous and relevant "normal" cells** are related to the transport processes across cell membranes and reactions on the cell membranes. These factors have been discussed in Lecture 27, and especially in Section 27.3. One of these factors is related to the **aquaporin.** Cancer cells tend to express more aquaporins on their cytoplasmic membranes, which may cause the H_2O_2 uptake speed in cancer cells to be faster than in normal cells. As a result, plasma treatment kills cancer cells more easily than normal cells. See Keidar et al. (2019) and Keidar (2020), for more details on the subject. Interesting selectivity aspects of plasma treatment of cancerous cells due to biochemical reactions on the cell membranes have been discussed by Bauer and Graves (2016).

32.15 Nonthermal Plasma Treatment of Explanted Tumors in Animal Models

The first comprehensive study to evaluate the **direct effect of nonthermal plasma treatment on malignant tumors** was performed by a group of researchers from Orleans University (Vandamme et al. 2010, 2012). This group employed malignant cells of glioma lineage xenografted (explanted) by subcutaneous injection of the tumor cells suspension (106 cells in 0.1 ml 0.9% NaCl) into the hind legs of Swiss nude female mice about four weeks of age. These cells are known to exhibit resistance to chemotherapy and radiotherapy. FE-DBD was used for the nonthermal plasma treatment. The effect of this treatment on tumor volume and the consequences of treatment on cell proliferation, cell cycle, and apoptosis induction were investigated in cell culture and in the animal. The cells employed in this work were genetically modified by transfection with firefly luciferase gene and, as a result, exhibited bioluminescence that made it possible to observe them even when implanted in the mouse. Monitoring of the growth and activity of the tumor was therefore possible by bioluminescent imaging (BLI).

 The key conclusions regarding apoptosis of the employed cells when treated by the plasma in cell medium were confirmed through Annexin and propidium iodide assays. Moreover, the plasma treatment of the cell cultures confirmed the expected DNA damage as well as the fact that intracellular scavenging of the reactive oxygen species using NAC protected the cells from DNA damage and induction of apoptosis. *The most interesting result, however, was the reduction of the xenografted tumor activity and size, as illustrated in Figure 32.11 (Vandamme et al. 2010, 2012; see review in Fridman and Friedman 2013) The survival of the plasma-treated mice was extended by about 60%. These results are remarkable, particularly given the fact that the tumor cells were several millimeters under the surface of the tissue clearly demonstrating that nonthermal plasma treatment can penetrate at least several millimeters. Vandamme et al. (2012), hypothesized that the possible mechanism of the treatment penetration was like the bystander effect* observed in radiation treatments, where cells directly exposed to the treatment pass signals to unexposed cells. Another possible mechanism could be related to diffusion of active species created through plasma treatment in tissue.

Keidar et al. (2011), also reported surprising efficacy and selectivity against several different types of xenografted cancers. In this study, the researchers employed a helium plasma jet. They compared responses to the plasma treatment of several different cells including lung cancer cells and bronchial epithelial cells as well as murine melanoma (cancer) cells and primary macrophages from mouse bone marrow. The team noted much stronger response to the plasma treatment in increased detachment and reduced survival for the lung cancer compared to the bronchial epithelial cells. For the subcutaneously implanted tumors with bladder cancer cells, the group demonstrated a strong anti-tumor effect of the plasma treatment through mouse skin. In the cases of smaller tumors of c. 5 mm in diameter, the tumor appeared to be completely irradiated and did not grow back after a single treatment; this is a remarkable result. Larger tumors shrunk but did grow back (albeit at a slower rate). Xenografted melanoma tumors also reduced in size during treatment and demonstrated reduced growth after treatment.

32.16 Plasma Control and Suppression of Precancerous Conditions in Dermatology, Clinical FE-DBD Treatment of Actinic Keratosis

Clinical trials involving cancers are contentious, both from an ethical and regulatory perspective: the risk posed by undertreated or untreated malignancies must be addressed and an acceptable solution needs to be developed. Animal

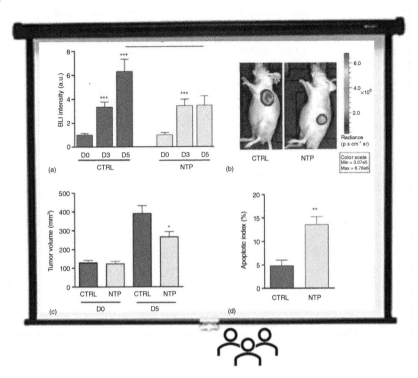

Figure 32.11 FE-DBD nonthermal plasma (NTP) treatment of explanted tumors in animal model. In vivo evaluation of the anti-tumor activity. Plasma treatment was delivered each day for five consecutive days (6 min, 200 Hz). (a) BLI imaging performed during treatment course (D3) and 24 h after the end of treatment protocol (D5) was normalized to total intensity prior to treatment (D0). (b) Representative BLI imaging of CTRL and NTP-treated mice at D5. (c) Tumor volume was determined using a caliper 24 h after the last day of treatment (D5). (d) Apoptosis indexes were determined by immunohistochemical detection of cleaved caspase 3.

models for skin cancers are not always reliable. Fortunately, dermatology has a skin condition that can be used as a "surrogate" target to test the feasibility of using cold plasma to treat at least one type of skin cancer: **squamous cell carcinoma**. The skin condition in question is **actinic keratosis (or solar keratosis)**. It is viewed as a precancerous lesion, on one end of the spectrum of keratinocyte cancers. These red scaly skin growths are commonly found in sun-exposed areas, mostly in older individuals with lighter skin color. There are multiple modalities in use to treat them, yet nonperfect. Developing an effective and well-tolerated treatment for actinic keratosis is itself an important goal. Given its relationship with squamous cell carcinomas of the skin, successfully treating actinic keratosis with cold plasma can be helpful in multiple ways. Success in treating them can justify trial of cold plasma on low-risk cases of squamous cell carcinoma. Actinic keratosis treatment can also be used to optimize dosing and delivery of cold plasma treatment with the expectation that those parameters can be transferrable to treating squamous cell carcinomas at least as a starting point.

 Two different approaches have been reported for cold plasma treatment of actinic keratosis. First, Friedman and his group pioneered the use of a FE-DBD plasma device to perform a so-called **lesion-directed treatment of actinic keratosis**, see Friedman (2020). This means, that individual lesions diagnosed as actinic keratosis were treated. *In this study of 5 patients, 17 actinic keratosis lesions have been treated only one time. One month later 9 lesions resolved fully and 3 improved significantly, which equals a 53% clearance rate and a 70% rate of at least significant improvement or clearance. The treatment was completely painless and there were no reports of after treatment inflammation, scab or blister formation, or any discomfort*, see Friedman et al. (2017).

Later, Wirtz et al. (2018), used a very different strategy: **field-directed plasma treatment of actinic keratosis**. This means, an entire area with multiple actinic keratoses is treated, and the lesion count is used as a measure of success, instead of focusing on individual lesions. This approach has gained ground in dermatology, as it treats very early, visually not detectable actinic keratosis lesions as well. Their study involved seven patients. Target areas were treated twice a week, seven times in total. The authors presented before and after total lesion counts and counts of improved lesions according to an actinic keratosis severity scale. They found that the total lesion count decreased by 23%, and 53% of lesions were given a lower grading marking clinical improvement. Patients tolerated the treatments well. In a step forward, a follow-up, randomized, prospective, rater-blinded study compared cold plasma and a prescription topical medication (diclofenac gel, a well-established and commonly used actinic keratosis treatment), for the field-directed treatment of actinic keratosis. The authors showed that not only was cold plasma statistically significantly more effective than diclofenac, but it also produced less side effects (Koch et al. 2020). All these studies have certainly shown that cold plasma is a legitimate treatment approach for actinic keratosis, and they also helped paving the way to introducing clinical trials examining cold plasma treatment of skin cancer, see Friedman (2020), for more details.

32.17 On a Way to Larger Clinical Studies in Plasma Oncology, Combination Therapies, Clinical Plasma Treatment of Cancer

To realize and fulfill clinical use of nonthermal plasma in the field of oncology, some important preconditions must be fulfilled, above all, regarding its applicability and safety, see Von Woedtke et al. (2022). Beside proving plasma selectivity against cancer cells, which has been discussed above, another important question surrounds a possible **induction of cancer cell growth and proliferation with the final consequence of metastasis.** Such effect should be analyzed if sub-efficient plasma treatment intensity, e.g. in the edge zone of plasma impact, occurs. The first studies give evidence that the latter danger is low, see Bekeschus et al. (2019) and Freund et al. (2020).

Other studies demonstrated that plasma treatment can enhance tissue oxygenation and modulate blood flow. On the one hand, this may support metastasis, which should be further investigated, see von Woedtke et al. (2022). *On the other hand,* **effects of radiotherapy and chemotherapy may be enhanced in combination treatments with plasma.** *Such combination therapies can be effectively applied in the clinical oncology. Several significant positive synergistic effects of plasma with radiotherapy, PEF, or chemotherapeutics have already been demonstrated.* More details on the plasma-including combination

therapy can be found, for example in Brullé et al. (2012), Daeschlein et al. (2013), Masur et al. (2015), Wolff et al. (2019), Pasqual-Melo et al. (2020), Chung et al. (2020), and Zhang et al. (2021a,b). Such combination plasma treatments are under investigation because they offer the chance to reduce effective doses of radiation or chemotherapeutics, respectively, and by this means lower side effects of cancer therapy.

Another effective clinical setup is a **supportive plasma application to treat the operation field of surgical tumor resections to remove possibly remaining tumor cells** *in cases where large-scale surgical tumor removal is impossible. This approach already demonstrated significant success of plasma oncological applications in hospitals.* Thus, Canady et al. (2018), reported the clinical experience using cold plasma jet for Stage IV metastatic colon cancer. The report describes the patient's history, chemotherapy, radiation, surgical treatment, and ex-vivo and intra-operative treatment of the tumor and normal tissue with plasma. They report evaluation of intra-operative temperature, oxygen, and end tidal CO_2 measurements during the surgery and post-surgery analysis of tissue treated by cold plasma. The studies suggest that TRAIl-R1 expression increases in the plasma-treated liver tissues, suggesting that the death receptor molecule may be involved in inducing apoptosis. Three-month postoperative CT scan of the abdomen and pelvis revealed no evidence of recurrent tumor at the resected and cold plasma treatment sites.

In a similar pathway, Metelman and his group from Greifswald, Germany, reported patients with real clinical benefit following application of nonthermal plasma jets, see von Woedtke and Metelmann (2014) and Metelmann et al. (2018a,b, 2022). In most of the clinical cases, the benefits were in terms of palliation, but some of them with not only visible change of the tumor surface but lasting partial tumor remission. These patients with locally advanced squamous cell carcinoma of the oropharynx suffering from open infected ulcerations had been treated with a jet plasma source (kINPen® MED, neoplas tools GmbH, Greifswald, Germany) to inactivate the microbial pathogens causing odor and pain. The palliative therapy program included repeated cycles of three single applications (1 min cm^{-2} from 8 mm) within one week, each followed by an intermittence of one week. Plasma treatment resulted in most of the patients in a significant reduction of odor and of pain medication requirements, an improvement in social function, and a positive emotional effect. In two patients, further observance revealed partial tumor remission for at least nine months beyond palliation. Incisional biopsies at remission demonstrate a moderate amount of apoptotic tumor cells and a desmoplastic reaction of the connective tissue.

Details on these breakthrough clinical studies can be found in von Woedtke and Metelmann (2014), Yoon et al. (2018), Canady et al. (2018), Metelmann et al. (2018a,b, 2022), Keidar et al. (2019), and Keidar (2020).

Clinical tests focused on treatment of the cancerous and precancerous dermal lesions, especially related to studies of Emmert and his group (Bernhardt et al. 2019; Metelmann et al. 2018a,b, 2022) as well as Friedman and his group (Friedman et al. 2020) have been quite successful. These clinical studies have been already discussed above in Sections 32.1 and 32.16. Summarizing, the state of knowledge on nonthermal plasma effects on cancer cells, including immunostimulatory effects (to be discussed below), together with the first successful clinical applications, characterizes cancer treatment as the next challenging but promising field of clinical plasma application (Metelmann et al. 2018a,b, 2022). Further efforts must be made to find out if the nonthermal plasma will be able to mark a "paradigm shift in cancer therapy" as was predicted more than one and a half decades ago, see Keidar et al. (2011), Fridman et al. (2008) and, Fridman and Friedman (2013).

32.18 Plasma in Onco-immunotherapy, Plasma-induced Systemic Tumor-specific Immunity, Immunogenic Cell Death, and Overcoming Immune Suppression

Substantial in vitro and in vivo evidence recently provided that redox-based nonthermal plasma effects causing cell death and discussed above may also have **immunomodulation aspects**, see for more details Bekeschus et al. (2018a,b), Miller et al. (2016), Lin et al. (2017a,b), Lin et al. (2018), Lin et al. (2015), Miller et al. (2014), Lin et al. (2017a,b), Bekeschus et al. (2018a,b), and Smolková et al. (2020). These studies demonstrated that nonthermal plasma affects immune cells, which was mainly investigated so far in conjunction with cancer treatment.

*There are strong hints that nonthermal plasma treatment may lead to **systemic tumor-specific immunity**. Several studies reported that plasma-treated cancer cells succumb to the so-called **immunogenic cell death (ICD)** promoting anticancer immune responses, which was observed before for example in radiotherapy. The ICD is a regulated cell death mechanism aligned with the adaptive immune system. Cells undergoing ICD release so-called **danger-associated molecular patterns (DAMP)** to communicate with immune cells,* which is illustrated in Figure 32.12. The ICD process includes active secretion of such DAMP as ATP or mobilization of calreticulin (CRT) to the outer cell surface. The **ATP, called "find me" signals,** being released from a tumor

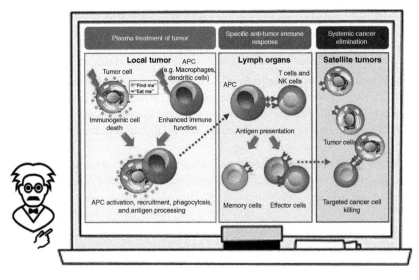

Figure 32.12 Illustration of mechanism of the plasma-stimulated onco-immunotherapy through the immunogenic cell death and triggering the "find me" and "eat me" signals from tumor cells to the antigen-presenting cells (APC, for example, macrophages and dendritic cells).

cell designate to immune cells direction how to find the tumor cells and destroy them. The **CRT, called "eat me" signals**, approach the tumor cell surface to meet the immune cells and "invite them to eat" the tumor cells. In this way, the plasma-induced ICD stimulates immune system suppressed by cancer to selectively destroy the tumor. It can be formulated as plasma overcoming the cancer-related immune suppression.

Thus, the so-called **antigen-presenting immune cells (APC), i.e. macrophages and dendritic cells** are alerted and ready to destroy the tumor. Attracted by ATP and finding surface CRT, the cancer cells are phagocytized with the cell fragments serving as antigens. The antigen-loaded APCs then travel to lymph organs and stimulate the **generation of antigen-specific T-cells and memory cells.** These cells are not only traveling over the entire body making the initially local anti-cancer plasma effect systemic, but also memorize information about the tumor cells for vaccination that is for further protection against the cancer, see Figure 32.12. Detailed in-vitro analysis of the discussed sequence of processes was carried out using the nanosecond-pulsed FE-DBD and can be found in Lin et al. (2017a,b) and Miller et al. (2016). The relevant in vivo animal studies are going to be discussed in the next section. It is important to point out that the plasma treatment dose required to ICD and triggering of the immunotherapeutic sequence of processes is safe (see Lecture 30), not damaging the immune cells, and even stimulating their migration, see Miller et al. (2014). It also can be formulated as plasma overcoming the cancer-related immune suppression. Summarizing, the circulating tumor-suppressing immune cells can effectively target other cancer cells of the same origin, e.g. metastatic cells, that are not plasma treated, see Bekeschus et al. (2018a,b) and Miller et al. (2016).

Such plasma-immunological ICD processes can explain the **abscopal plasma effects**, *i.e. reduction of growth of tumors in animal experiments that were not exposed to plasma treatment but are localized elsewhere in the same animal,* see Mizuno et al. (2017) and Mahdikia et al. (2020). This observed plasma effect opens a highly promising therapeutic approach for cancer treatment in cases, where direct plasma control of a metastatic lesion cannot be realized (Dobrynin et al. 2022; Mohamed et al. 2020). We should mention that this approach may have potential consequences regarding **systemic effects of a local plasma therapy** provided by nonthermal plasma (Von Woedtke et al. 2022).

32.19 FE-DBD Plasma Immunotherapeutic Treatment of Colorectal Tumors in Animal Model, Vaccination in Plasma Cancer Treatment

To validate whether plasma could induce the **ICD in vivo**, CT26 colorectal tumors were established subcutaneously in Balb/c mice and exposed to nanosecond pulsed DBD plasma when they became palpable (Lin et al. 2018). A safe operating FE-DBD (29 kV and 750 Hz) plasma regime was first identified. 10 s of plasma treatment resulted in minimal epidermal damage with no direct changes to the tumor. The 10 s plasma treatment over five days has been performed to induce ICD in the subcutaneous tumors. Three days after the last plasma treatment, tumors were resected, fixed, and sectioned. Immunofluorescence microscopy was performed on tumor sections to identify ICD markers. CRT expression increased in all plasma treatment groups and was maximum (1.6 ± 0.2-fold) following the "plasma multiple spot" regime (different areas of the tumor treated each day). Tumor sections were also stained for the "high mobility group box 1" (HMGB1), another DAMP signal much like ATP, that recruits inflammatory immune cells and mediates signals between APCs. HMGB1 has been observed to translocate from the nucleus to the cytoplasm and extracellular space. Here, an increase in immunofluorescence intensity of HMBG1 was observed in all plasma treatment groups. Increased CRT and HMGB1 were associated with more CD45 (leukocytes, 2.0 ± 0.4-fold) and CD11c+ cells (1.8 ± 0.2-fold) following again the "plasma multiple spots regime." It demonstrates that **emitted DAMPs stimulated the recruitment of immune cells into the tumor environment**. CD45 is expressed by all leukocytes, including T cells, B cells, neutrophils, and NK cells, therefore other immune cell subsets were probably also being also recruited as a downstream consequence of ICD induction.

Summarizing, the "plasma multiple spot treatment" (different areas of the tumor treated each day) enhanced in animal model both emission of DAMPs and recruitment of immune cells in the tumor compared to the other application methods. Therefore, treatment of different spots may be more beneficial compared to repeat treatment of the same area.

To demonstrate that **plasma-induced ICD could stimulate in vivo an adaptive anti-tumor response**, the subcutaneous CT26 tumors expressing the colorectal cancer antigen GUCY2C (CT26-GUCY2C) in Balb/c mice have been treated (Lin et al. 2018). Mice were treated with either plasma alone or with plasma in combination with the Ad5-GUCY2C-S1 vaccine. The Ad5-GUCY2C-S1 vaccine was previously shown to safely induce GUCY2C-specific immune responses and antitumor immunity in mice and has been translated to human clinical trials. Mice were treated with plasma for five consecutive days and one group was vaccinated with the Ad5-GUCY2C-S1 vaccine, one week after the last plasma treatment. An untreated group and a vaccine only group (Ad5-GUCY2C-S1 vaccination) served as negative and vaccine controls. Splenic GUCY2C-specific T-cell responses were analyzed by IFNγ ELISpot assay 28 days after the initial tumor inoculation. *FE-DBD plasma treatment alone had a marginal effect on GUCY2C-specific responses. However, plasma treatment prior to vaccination amplified GUCY2C-specific T-cell responses (167.8 ± 41.5 spots vs 109.7 ± 22.3 spots with vaccine alone). This observation supports the **potential of plasma to increase the immunogenicity of cancer cells** and stimulate canonical pathways required for tumor control.*

 Lin et al. (2018), also demonstrated the quite impressive **vaccination effect with plasma-induced ICD cells providing protection against further tumor challenge in mice**. To demonstrate that the DAMP signals elicited by plasma could enhance immune responses against further cancer, a special vaccination assay has been performed. Balb/c mice were immunized with CT26 cells treated in vitro with plasma at the best ICD-inducing FE-DBD regime (29 kV, 30 Hz, 1 mm gap distance, 10 s). Cells were prepared for inoculation and injected into the left flank as a whole-cell vaccine to allow an immune response to develop. One week after immunization, mice were challenged with live CT26 cancer cells on the opposite flank, and tumor growth was monitored twice a week until day 26 when the study was terminated as a subset of the animals reached IACUC-approved endpoints. CT26 cells treated with media only or Cisplatin (50 μM for 24 h), a non-ICD inducer, were used as controls. *The challenge tumors in the media and Cisplatin groups grew rapidly while tumors in the plasma group developed relatively slowly. The mean tumor volume for the plasma immunized group was significantly smaller compared to that of the media group (414.7 ± 104.3 mm³ vs 847.4 ± 141.5 mm³) or the Cisplatin group (1041.8 ± 208.3 mm³) at day 26. Indeed, 90% of the mice in the plasma-immunized group had tumor volumes smaller than the mean tumor volume of the media group (850 mm³), suggesting that these **mice were partially protected from cancer by the preliminary plasma vaccination**. Moreover, 3 out of the 10 mice in the plasma group did not develop subcutaneous tumors on the challenge site.* These first steps of the plasma vaccination of colorectal tumors make us believe that in future (hopefully nearest future) people will be protected by plasma vaccines against cancer like we are protected by vaccination from so many diseases. It is a good optimistic point to finish this book of lectures, which we started with the billboards from Belgrade, Serbia, claiming "Plasma, That's All You Need!" see, Figure 1.1.

32.20   Problems and Concept Questions

32.20.1   Nonthermal Plasma Dermatology

Analyzing data presented in Section 30.9.12 on depth of penetration of plasma effects in tissue as well as data on direct plasma sterilization of skin presented in Section 31.9, estimate which layers of skin can be treated by plasma dermatology? Discuss differences in direct application of FE-DBD, plasma jets, and plasma-activated water solutions, gels, and creams.

32.20.2   Direct Nonthermal Plasma Applications in Dentistry

Analyzing physics of propagation of the ionization waves, plasma bullets, in the nonthermal plasma jets (see Section 13.10), estimate possibility of the plasma bullets penetration into infected root canals for disinfection and healing. Compare the relevant effects on the plasma bullets with simple afterglow effects of the plasma jets.

32.20.3   Pin-to-hole Microdischarge Plasma Treatment of Ulcerative Colitis

Plasma treatment of ulcerative colitis assumes safe introduction of the plasma device inside of a colon in form of special endoscope. What are the key advantages of application of the PHD (pin-to-hole discharge) in this case with respect to plasma jets? with respect to dielectric barrier discharges (DBD) and coronas?

32.20.4   Plasma Effect on Whole Blood Viscosity (WBV)

Describe typical nonthermal plasma effect on the whole blood viscosity, and potential of its possible applications for the treatment of cardiovascular diseases.

32.20.5   Nonthermal Plasma Control of Low-density-lipoprotein (LDL) Cholesterol in Blood

Explain the effect of selective suppression of the low-density-lipoprotein (LDL) cholesterol in blood using the non-thermal plasma-based systems. Discuss possible approaches to clinical administration of this effect.

32.20.6   Plasma Control of Stem Cells and Tissue Regeneration, a Pathway from Plasma Orthopedics to Plasma Cosmetics

Analyze the data on differentiation of the mesenchymal stem cells (MSC) into bone and cartilage cells related to applications for plasma orthopedics (see Section 32.12) and consider further development of such studies to potential rejuvenation for nonthermal plasma applications in cosmetics.

32.20.7   Selectivity of Nonthermal Plasma Treatment of Cancers

Analyze different potential mechanisms of selective plasma treatment of cancer cells discussed in Sections 27.3 and 32.14 and discuss their specifics regarding different protocols of plasma administration and different types of cancers.

32.20.8   Plasma in Onco-immunotherapy, Plasma-induced Systemic Tumor-specific Immunity

Explain in which way, the plasma-induced immunogenic cell death (ICD) can provide the cancer abatement selectivity and tumor-specific immunity? How does this effect depend on location of the cancer tissue in the body? What are specifics of application of different types of nonthermal plasma (especially, FE-DBD and plasma jets) in stimulation of the ICD?

Afterword and Acknowledgements

An important focus of this book is to inspire researchers, industrial and medical professionals, and especially students, to ignite their interest in plasma science and technology. In this regard in the afterword to this book of lectures, the author thinks about and deeply acknowledges four Professors who brought him to the plasma area and inspire the lifelong momentum to pursue and further develop this exciting field. They were the author's mentors and supervisors from the Moscow Institute of Physics and Technology and Kurchatov Institute of Atomic Energy, who shaped up and stimulate authors education and research from the 1970s to 1990s.

The first one is Professor Leonid I. Rudakov, famous theoretician in plasma physics, world leader in science and applications of the high-current relativistic electron beams, their propagation, and interaction with gases and plasmas. It was Professor Rudakov who started working in January of 1972 with author, then a young student of the IMPT, on propagation and focusing of relativistic electron beams in plasma. It was Professor Rudakov who challenged the young student of MIPT to pass the comprehensive Landau's Theoretical Minimum exams and taught how to find "reasonably" sophisticated solutions of the quite complicated differential equations in plasma physics.

It was Professor Boris B. Kadomtsev, famous theoretician and leader of the Tokamak Program in the Kurchatov Institute of Atomic Energy at that time, head of the Department of Plasma Physics and Chemistry of MIPT, who led the author through the most challenging concepts of theoretical plasma physics, especially those related to the collective phenomena in plasma, plasma magneto-hydro-dynamics, plasma turbulency, plasma waves and instabilities, plasma surface interaction, dusty plasmas, and physics of plasma aerosols. It created a bridge from the fusion plasmas to the area of high-pressure low-temperature plasmas, where the author has been mostly working happily ever after.

The existing field of plasma chemistry, quite young at that time, has been introduced to the author by Professor Vladimir D. Rusanov, famous plasma experimentalist, and one of the founders of the modern plasma diagnostics and plasma chemistry. He inspired author's research in atmospheric pressure plasma-chemical synthesis of NO and ozone, in dissociation of CO_2, decomposition, and conversion of metal compounds, in organic plasma chemistry, in stimulation of processes by vibrational excitation of molecules, in applications in plasma chemistry of microwave discharges, and discharges sustained by relativistic electron beams, and many other areas of plasma chemistry. Author published his first book in 1984 in co-authorship with Professor Rusanov and started under his supervision to teach his first plasma lectures to MIPT students. It's interesting that Professor Rusanov also stimulated in the 1980s the first research and technological steps in plasma biology, including sterilization with DBD plasmas and plasma stimulation of biochemical processes. This research shaped up later first steps of modern plasma medicine.

Plasma Science and Technology: Lectures in Physics, Chemistry, Biology, and Engineering, First Edition. Alexander Fridman.
© 2024 WILEY-VCH GmbH. Published 2024 by WILEY-VCH GmbH.

Last but surely not least author's mentor and supervisor from the Kurchatov Institute of Atomic Energy was Professor Valery A. Legasov, famous inorganic chemist and specialist in physical chemistry, one of the founders of plasma chemistry especially regarding nuclear materials, nuclear/hydrogen energy systems, and nuclear engineering. He introduced author to the practical aspects of plasma-chemical engineering, especially in the directions of large-scale CO_2 dissociation and hydrogen production, separation of isotopes, processing, and refining of metals, hydrogen, and nuclear safety. Professor Valery A. Legasov is also remembered in the world today for his work in containment of the Chernobyl nuclear disaster and for presenting the investigation findings to the International Atomic Energy Agency in Vienna, which was portrayed in the Sky/HBO miniseries Chernobyl in 2019. Author would highly recommend this movie presenting Valery Legasov in a well-balanced way.

The four great scientists presented above were author's mentors and supervisors in Russia, in the Moscow Institute of Physics and Technology and Kurchatov Institute of Atomic Energy. Further author's research in the late 1980s and 1990s in France, in the Orleans University, GREMI, has been mostly inspired by three enthusiastic plasma scientists and friends. Their ideas not only inspired author's research in the 1990s and later but also shaped understanding and way of presentation of the plasma concepts in this book.

The first of them, friend, and inspiring colleague from the Orleans University, is Professor Albin Czernichowski, one of the founders and great enthusiast of the gliding arc discharges, their physics, engineering, and wide range of applications. Professor Albin Czernichowski first inspired authors theoretical works on physics of gliding arcs, and their fast transition from quasi-equilibrium thermal to non-equilibrium non-thermal regimes (FENETRE effect), and then development in the Nyheim Plasma Institute together with Dr. A. Gutsol and A. Rabinovich of the modern non-equilibrium gliding arc discharges stabilized in the reverse vortex "Tornado" flows and their multiple applications.

Professor André Bouchoule, friend, and world's leading scientist in dusty plasmas, founding director of the Plasma Center GREMI in CNRS and Orleans University, introduced author to the key plasma chemical area of microelectronics, to exciting nanoscale physics and chemistry of dusty plasmas, and to plasma nanotechnology. It was Professor André Bouchoule who also stimulated organization of the plasma graduate education in Orleans University, where the author was able to develop special plasma courses. Many elements of these plasma courses were further applied in development of the plasma curriculum in Drexel University, which finally shaped up the structure of this book of lectures.

Professor Jean-Michel Pouvesle is an author's friend, and world's leading scientist in pulsed plasmas, physics of plasma bullets, physics, and engineering of light sources, and recently one of the founders of plasma medicine, who together with Professor Eric Robert led in Orleans University and CNRS the first in-vivo experiments in plasma treatment of cancers. Discussions and early collaborative work with Professor Jean-Michel Pouvesle focused on the nanosecond pulsed DBDs created basis for further author's works with Dr. A. Gutsol, Dr. A. Starikovky, Dr. D. Dobrynin, et al., on uniform regimes of the nanosecond pulsed atmospheric DBDs. These works opened, in turn, a pathway to safe application of FE-DBDs by author together with Dr. A. Gutsol, Dr. G. Friedman, Dr. D. Dobrynin, Dr. G. Fridman, et al., to treat living tissues, and finally to plasma medicine.

Further author's endeavor in plasma science and technology in the late 1990s and the new millennium already in the United States was inspired by Professor Larry Kennedy, friend, now engineering dean emeritus of the University of Illinois at Chicago, and world's leading scientist in combustion and plasma science and engineering. Professor Larry Kennedy together with Professor Bill Rich from Ohio State University significantly supported further development of the author's research in non-equilibrium vibrational kinetics, and further in plasma-assisted and filtration combustion. Professor Larry Kennedy inspired the author's research together with Dr. A. Gutsol, and Professor A. Saveliev on plasma applications for environmental control, especially focused on VOC, NO_x, and SO_2 abatement in large volume exhaust gases. Professor Larry Kennedy co-authored the first book written in English by the author of this lectures in 1908, the third edition of this book has been published in 1921. Professor Larry Kennedy was also inspirational in development of the first in USA comprehensive plasma curriculum, further applied by author in Drexel University under leadership of Prof. S. Guceri (now in the Worcester Polytechnic Institute) and Professor M. Choi (now Chancellor of the University of Missouri), and which is significantly reflected in this book of lectures.

The eight great Professor and Scientist acknowledged above are the major inspirational roots of the author's endeavor in plasma science and technology in Soviet Union, France, and the USA. Surely, it was a big number of other excellent scientists and great minds who inspired, collaborated, and helped to author during his more than a half-century plasma venture. Author sincerely thanks and acknowledges all of them. If they read these lines, they would feel that author addressed these words to them. Also, author deeply appreciates and forever remembers those who during so many years supported plasma science and education at Drexel University. From this perspective the highest gratitude is to John and Chris Nyheim as well as to their family, there is a good reason why the plasma institute of Drexel University is proudly called today – the C. & J. Nyheim Plasma Institute. We are very thankfully to Stanley Kaplan and his family for crucial support of author, plasma research, and our Institute. For supporting the C. & J. Nyheim Plasma Institute today, we deeply acknowledge the Drexel University President John Fry, Senior Vice Provost Aleister Saunders, Deans Sharon Walker, Paul Brandt-Rauf, Associate Dean of Medicine Kenny Simansky, Department Heads Jonathan Spanier, Michael Waring, Steven May, James Tangorra, Richard Hamilton, and many others. We surely remember and acknowledge former Drexel leaders, crucial in establishment and support of the C. & J. Nyheim Plasma Institute, especially President "Taki" Papadakis, Dean Selcuk Guceri (now in the Worcester Polytechnic Institute), Associate Dean Mun Choi (now Chancellor of the University of Missouri), and Senior Vice Provost Ken Blank.

The author greatly appreciates the plasma research support (and therefore indirect support of this book) provided to the C. & J. Nyheim Plasma Institute by the National Science Foundation, U.S. Department of Energy, U.S. Department of Defense (specifically DARPA, DLA, Defense Logistics Agency, TARDEC, Army Research Office, Air Force OSR, SERDP), NASA, NIH, USDA, Center for Produce Safety, W.M. Keck Foundation, Coulter Foundation, as well as our industrial partners, including Chevron, Air Products, Applied Materials, Samsung, Kodak, AA Plasma, GTI, GoJo, Johnson & Johnson, Campbell's, Dole Food Company, and many others.

Several friends and colleagues helped author in the development of this book of lectures, and he sincerely acknowledges their support, fruitful discussions, and recommendations. Between them friends and colleagues from the C. & J. Nyheim Plasma Institute, professors A. Rabinovich, D. Dobrynin, D. Vainchtein, S. Joshi, C. Sales, Drexel University College of Engineering professors Y. Cho, B. Farouk, W. Sun, N. Cernansky, D. Miller, G. Friedman, Y. Gogotsi, professors from Drexel University College of Medicine, College of Art and Sciences, School of Bioengineering, long-time collaborators Profs. & Drs. A. Gutsol, G. Fridman, P. Friedman, V. Vasilets, T. Freeman, M. Keidar, G. Palmese, J. Foster, D. Staack, M. Cooper, T. Farouk, and many others. Last but not least, author is deeply thankful to his family for continuous three-year support and help with this book.

References

Abou-Ghazala, A., Katsuki, S., Schoenbach, K.H., and Moreira, K.R. (2002). *IEEE Transactions on Plasma Science* 30: 1449.

Adamovich, I., Saupe, S., Grassi, M.J. et al. (1993). *Chemical Physics* 174: 219.

Adhikari, B., Adhikari, M., Ghimire, B. et al. (2019). *Scientific Reports* 9: 16080.

Ajo, P., Preis, S., Vornamo, T. et al. (2018). *Journal of Environmental Chemical Engineering* 6: 1569.

Alix, F.R., Neister, S.E., and McLarnon, C.R. (1997). US Patent 5,871,703.

Alves, C. Jr., de Oliveira Vitoriano, J., da Silva, D.L. et al. (2016). *Scientific Reports* 6: 33722.

Ambrazevicius, A. (1983). *Heat Transfer During Quenching of Gases*. Vilnius, Lithuania: Mokslas Publ.

Amnuaysin, N., Korakotchakorn, H., Chittapun, S., and Poolyarat, N. (2018). *Songklanakarin Journal of Science and Technology* 40: 819.

Anikin, N.B., Mintoussov, E.I., Pancheshnyi, S.V. et al. (2003). Nonequilibrium plasma and its application for combustion and hypersonic flow conrol. *41st AIAA Aerospace Sciences Meeting and Exhibit*, Reno, NV (9–12 January). AIAA paper 2003–1053.

Anikin, N.B., Starikovskaia, S.M., and Starikovsky, A.Y. (2004). *Plasma Physics Reports* 30: 1028.

Anikin, N.B., Bozhenkov, S.A., Zatsepin, D.B. et al. (2005). *Encyclopedia of Low-Temperature Plasma*, Chemistry of Low-Temperature Plasma, vol. VIII-I (ed. Y.A. Lebedev, N.A. Plate, and V.E. Fortov), 1–12. Moscow: YANUS-K.

Anpilov, A.M., Barkhudarov, E.M., Bark, Y.B. et al. (2001). *Journal of Physics D: Applied Physics* 34: 993.

Arbatsky, A.E., Vakar, A.K., Vorobiev, A.B. et al. (1988). *Modification of Polymer Films Using Microwave Plasma After-Glow*, vol. 4722/7. Moscow: Kurchatov Institute of Atomic Energy, CNII-Atom-Inorm.

Arbatsky, A.E., Vakar, A.K., Golubev, A.V. et al. (1990). *Soviet Physics, High Energy Chemistry (Khimia Vysokikh Energij)* 24: 256.

Arefi-Khonsari, F., Kurdi, J., Tatoulian, M., and Amouroux, J. (2001). *Surface & Coatings Technology* 142–444: 437.

Arita, M., Nagayoshi, M., Fukuizumi, T. et al. (2005). *Oral Microbiology and Immunology* 20: 206.

Arjunan, K.P., Gutsol, A., Vasilets, V., and Fridman, A. (2007). *18th International Symposium on Plasma Chemistry, ISPC-18*, Kyoto, Japan.

Arjunan, K.P., Friedman, G., Fridman, A., and Clyne, A.M. (2012). *Journal of the Royal Society Interface* 9 (66): 147.

Asinovsky, E.I. and Vasilyak, L.M. (2001). *Encyclopedia of Low-Temperature Plasma*, vol. 2 (ed. V.E. Fortov), 234. Moscow: Nauka (Science).

Asisov, R.I. (1980). *Plasma-Chemical Experiments with Non-Equilibrium Microwave Discharges of Moderate Pressure with Magnetic Field*. Moscow: Kurchatov Institute of Atomic Energy.

Asisov, R.I. et al. (1980). *Soviet Physics, High Energy Chemistry (Khimia Vysokikh Energij)* 14: 366.

Asisov, R.I. et al. (1983). *Soviet Physics, Doklady (Reports of the USSR Academy of Sciences)* 271: 94.

Asisov, R.I. et al. (1986). *Soviet Physics, Doklady (Reports of the USSR Academy of Sciences)* 286: 1143.

Asmussen, J. and Reinhard, D.K. (2002). *Diamond Film Handbook*. Boca Raton, FL: CRC Press.

Assadian, O., Ousey, K.J., Daeschlein, G. et al. (2019). *International Wound Journal* 16: 103.

Atamanov, V.M., Ivanov, A.A., and Nikiforov, V.A. (1979). *Soviet Physics, Journal of Technical Physics* 49: 2311; *Soviet Physics, Plasma Physics (Fizika Plazmy)* **5**: 663.

Attri, P. et al. (2020). *Processes* 8 (8): 1002.

Ayan, H., Fridman, G., Friedman, G. et al. (2007). *18th International Symposium on Plasma Chemistry, ISPC-18*, Kyoto, Japan.

Ayan, H., Fridman, G., Gutsol, A. et al. (2008). *IEEE Transactions on Plasma Science* 36 (2): 504.

Plasma Science and Technology: Lectures in Physics, Chemistry, Biology, and Engineering, First Edition. Alexander Fridman.
© 2024 WILEY-VCH GmbH. Published 2024 by WILEY-VCH GmbH.

Babaritsky, A.I., Diomkin, S.A., Givotov, V.K. et al. (1991). *Non-Equilibrium Approach to Methane Conversion into Acetylene in Microwave Discharge*, vol. 5350/12. Moscow: Kurchatov Institute of Atomic Energy.

Babko-Malyi, S., Battleson, D., Ray, I. et al. (2000). Mercury removal from combustion flue gases by the plasma-enhanced electrostatic precipitator. *Air Quality II: Mercury Trace Elements, and Particular Matter Conference, Proceedings*, McLean, VA.

Babukutty, Y., Prat, R., Endo, K. et al. (1999). *Langmuir* 15: 7055.

Baddur, R. and Timmins, R. (1970). *Application of Plasma in Chemical Processes* (ed. L.S. Polak), 255. Moscow: Mir.

Badie, J.M., Flamant, J., Granier, B. (1997). *13th International Symposium on Plasma Chemistry (ISPC-13)*, Beijing, vol. 4, p. 1748.

Balebanov, A.V., Butylin, B.A., Givitov, V.K. et al. (1985a). *Soviet Physics Doklady* 283: 657.

Balebanov, A.V., Butylin, B.A., Givitov, V.K. et al. (1985b). *Journal of Nuclear Science and Technology, Nuclear-Hydrogen Energy* 3: 46.

Balebanov, A.V., Givitov, V.K., Krasheninnikov, E.G. et al. (1989). *Soviet Physics, High Energy Chemistry (Khimia Vysokikh Energij)* 5: 440.

Baranchicov, E.I., Belenky, G.S., Deminsky, M.A. et al. (1990a). *Plasma-Catalytic SO₂ Oxidation in Air*, vol. IAE-5256/12. Moscow: *Kurchatov Institute of Atomic Energy, Publ.*

Baranchicov, E.I., Denisenko, V.P., Potapkin, B.V. et al. (1990b). *Soviet Physics Doklady* 339: 1081.

Baranchicov, E.I., Belenky, G.S., Deminsky, M.A. et al. (1992). *Radiation Physics and Chemistry* 40: 287.

Barankova, H. and Bardos, L. (2002). *Catalysis Today* 72: 237.

Baronov, G.S., Bronnikov, D.K., Fridman, A. et al. (1989). *Journal of Physics B* 22: 2903.

Basaran, P., Basaran-Akgul, N., and Oksuz, L. (2008). *Food Microbiology* 25: 626.

Basov, N.G., Danilychev, V.A., and Panteleev, V.I. (1977). *Soviet Physics, Doklady (Reports of the USSR Academy of Sciences)* 233: 6.

Basov, N.G., Danilychev, V.A., and Panteleev, V.I. (1978). *Soviet Physics, High Energy Chemistry (Khimia Vysokikh Energij)* 3: 266.

Basting, B. and Marowsky, G. (ed.) (2005). *Excimer Laser Technology*. Berlin: Springer.

Bauer, G. and Graves, D.B. (2016). *Plasma Processes and Polymers* 13: 1157.

Baura, G. (2020). *Medical Device Technologies*. Academic Press.

Baysan, A. and Lynch, E. (2004). *American Journal of Dentistry* 17: 56.

Bazelyan, E.M. and Raizer, Y.P. (2000). *Lightning Physics and Lightning Protection*. Bristol/Philadelphia, PA: IoP Publishing, CRC Press.

Bazelyan, E.M. and Raizer, Y.P. (2017). *Spark Discharge*. New York: Routledge, CRC Press, Taylor & Francis Group.

Becker, K.H., Kogelschatz, U., Schonbach, K.H., and Barker, R.J. (2005). *Non-Equilibrium Air Plasmas at Atmospheric Pressure*, Series in Plasma Physics. Bristol: IOP Publ.

Bekeschus, S., Clemen, R., and Metelmann, H.-R. (2018a). *Clinical Plasma Medicine* 17: 22.

Bekeschus, S., Pouvesle, J.-M., Fridman, A., and Miller, V. (2018b). *Cancer Immunology, "Comprehensive Clinical Plasma Medicine: Cold Physical Plasma for Medical Application"* (ed. H.-R. Metelmann, T. von Woedtke, and K.-D. Weltmann), 409. Berlin: Springer.

Bekeschus, S., Lin, A., Fridman, A. et al. (2018c). *Plasma Chemistry and Plasma Processing* 38: 1.

Bekeschus, S. et al. (2019). *Cancers* 11: 1237.

Bentkover, S.H. (2012). *Facial Plastic Surgery Clinics of North America* 20: 145.

Berg, S., Blom, H.O., Moradi, M. et al. (1989). *Journal of Vacuum Science and Technology* A7: 1225.

Bernhardt, T., Semmler, M.L., Schäfer, M. et al. (2019). *Oxidative Medicine and Cellular Longevity* 1: 10.

Bhandarkar, U., Kortshagen, U., and Girshick, S.L. (2003). *Journal of Physics D: Applied Physics* 36 (12): 1399.

Biberman, L.M., Vorobiev, V.S., and Yakubov, I.T. (1987). *Kinetics of Non-Equilibrium Low-Temperature Plasmas*. New York: Springer-Verlag.

Biederman, H. (2004). *Plasma Polymer Films*. London: Imperial College Press.

Biederman, H. and Osada, Y. (1992). *Plasma Polymerization Process*. Amsterdam: Elsevier.

Bisag, A. et al. (2022). *Plasma Processes and Polymers* 19 (3): 2100133.

Bletzinger, P., Ganguly, B.N., Van Wie, D., and Garscadden, A. (2005). *Journal of Physics D: Applied Physics* 38: 33.

Bochkov, V.D. and Korolev, Y.D. (2000). Pulsed gas-discharge commutation devices. In: *Encyclopedia of Low-Temperature Plasma*, vol. 4 (ed. V.E. Fortov), 203. Moscow: Nauka (Science).

Boeckmann, L. et al. (2020). *Applied Sciences* 10: 6898.

Boenig, H.V. (1988). *Fundamentals of Plasma Chemistry and Technology*. Lancaster, PA: Technomic Publ.

Boerner, H. (2019). *Ball Lightning*. Switzerland: Springer Nature.

Boeuf, J.P. (2003). *Journal of Physics D: Applied Physics* 36: 53.

Boeuf, J.P., Lagmich, Y., Callegari, T., and Pitchford, L. (2007). EHD force in dielectric barrier discharge: parametric study an influence of negative ions, *45th AIAA Aerospace Science Meeting and Exhibit*, Reno, NV, AIAA paper 2007-0183.

Bogaerts, A. (ed.) (2019). *Plasma Catalysis*. Basel, Beijing: MDPI Books.

Bogaerts, A. et al. (2020). *Journal of Physics D: Applied Physics* 53 (44): 443001.

Bogle, M.A., Arndt, K.A., and Dover, J.S. (2007). *Archives of Dermatology* 143: 168.

Bol'shakov, A.A., Cruden, B.A., Mogul, R. et al. (2004). *AIAA Journal* 42: 823.

Bormashenko, E., Shapira, Y., Grynyov, R. et al. (2015). *Journal of Experimental Botany* 66: 4013.

Bouchoule, A. (1993). *Physics World*, vol. 2, 204.

Bouchoule, A. (ed.) (1999). *Dusty Plasmas: Physics, Chemistry and Technological Impacts in Plasma Processing*. New York: Wiley.

Bouchoule, A. and Boufendi, L. (1993). *Plasma Sources Science and Technology* 2: 204.

Boulos, M.I. (1985). *Pure and Applied Chemistry* 57: 1321.

Boulos, M.I. (1992). *Journal of Thermal Spray Technology* 1 (1): 33.

Boulos, M.I., Fauchais, P., and Pfender, E. (1994). *Thermal Plasmas, Fundamentals and Applications*. New York: Plenum Press.

Bourke, P., Ziuzina, D., Boehm, D. et al. (2018). *Trends in Biotechnology* 36: 615.

Bozhenkov, S.A., Starikovskaia, S.M., and Starikovskii, A.Y. (2003). *Combustion and Flame* 133: 133.

Bradley, E.B. (1990). *Molecules and Molecular Lasers for Electrical Engineers*. New York: CRC Press, Taylor & Francis Group.

Brandenburg, R. et al. (2019). *Plasma Processes and Polymers* 16 (1): 1700238.

Brar, J., Jiang, J., Oubarri, A. et al. (2016). *Plasma Medicine* 6 (3–4): 413.

Brehmer, F. et al. (2015). *Journal of the European Academy of Dermatology and Venereology* 29: 148.

Britun, N. and Silva, T. (ed.) (2018). *Plasma Chemistry and Gas Conversion*. London, UK: In-Tech Open.

Bromberg, L., Cohn, D.R., Rabinovich, A., and Heywood, J. (2001). *International Journal of Hydrogen Energy* 26: 1115.

Bronfine, B. (1970). *Application of Plasma in Chemical Processes*, 182. Moscow: Mir (World).

Brovkin, V.G. and Kolesnichenko, Y.F. (1995). *Moscow Physical Society* 5: 23.

Brozek, V., Hrabovsky, M., and Kopecky, V. (1997). *13th International Symposium on Plasma Chemistry (ISPC-13)*, Beijing, vol. 4, p. 1735.

Bruggeman, P. and Leys, C. (2009). *Journal of Physics D: Applied Physics* 42 (5): 053001.

Bruggeman, P.J., Kushner, M.J., Locke, B.R. et al. (2016). *Plasma Sources Science and Technology* 25: 053002.

Brullé, L. et al. (2012). *PLoS One* 7: e52653.

Bugaenko, L.T., Kuzmin, M.G., and Polak, L.S. (1992). *High Energy Chemistry, Ellis Horwood Series in Physical Chemistry*. Ellis Horwood Publ. Ltd.

Busco, G., Robert, E., Chettouh-Hammas, N. et al. (2020). *Free Radical Biology & Medicine* 161: 290.

Bussiahn, R., Brandenburg, R., Gerling, T. et al. (2010). *Applied Physics Letters* 96: 143701.

Campbell, C.A., Snyder, F., Szarko, V. et al. (2006). *Water Treatment Using Plasma Technology"*, SD Report, ed.,. Philadelphia, PA: Drexel University.

Canady, J. et al. (2018). Cold atmospheric plasma (CAP) combined with chemo-radiation and cytoreductive surgery: the first clinical experience for stage IV metastatic colon cancer. In: *Comprehensive Clinical Plasma Medicine: Cold Physical Plasma for Medical Application* (ed. H.-R. Metelmann, T. von Woedtke, and K.-D. Weltmann), 163. Berlin: Springer.

Candler, G.V., Kelley, J.D., Macheret, S.O. et al. (2002). *AIAA Journal* 40: 1803.

Cao, Y., Yang, P., Lu, X. et al. (2011). *Plasma Science and Technology* 13 (1): 93.

Capelli, F. et al. (2021). *Applied Sciences* 11 (9): 4177.

Capitelli, M. (ed.) (1986). *Non-Equilibrium Vibrational Kinetics*, Topics in Current Physics - 39. Berlin: Springer-Verlag.

Capitelli, M. (ed.) (2000). *Plasma Kinetics in Atmospheric Gases*, Springer Series on Atomic, Optical and Plasma Physics, vol. 31. Berlin: Springer-Verlag.

Cha, S. and Park, Y.-S. (2014). *Clinical Plasma Medicine* 2 (1): 4.

Chae, J.O., Desiaterik, Y.N., and Amirov, R.H. (1996). *International Workshop on Plasma Technologies for Pollution Control and Waste Treatment*. Cambridge, MA: Publ., MIT.

Chakravarthy, K., Dobrynin, D., Fridman, G. et al. (2011). *Plasma Medicine* 1 (1): 3.

Chan, J.K. (1996). *Surface Science Reports* 24 (1/2): 1.

Chang, B.S. (2003). *Journal of Electrostatics* 57: 313.

Chen, F.F. and Chang, J.P. (2012). *Lecture Note on Principles of Plasma Processing*. Berlin, New York: Springer-Verlag.

Chen, Z. and Wirz, R.E. (2021). *Cold Atmospheric Plasma (CAP) Technology and Applications*. Morgan & Claypool Publishers.

Chen, W., Huang, J., Du, N. et al. (2012a). *Journal of Applied Physics* 112: 013304.

Chen, M., Zhang, Y., Yao, X. et al. (2012b). *Dental Materials: Publication of the Academy of Dental Materials* 28: 1232.

Chen, M., Zhang, Y., Sky Driver, M. et al. (2013). *Dental Materials: Publication of the Academy of Dental Materials* 29: 871.

Cheng, H., Xu, J., Li, X. et al. (2020a). *Physics of Plasmas* 27: 063514.

Cheng, Z. et al. (2020b). *Physics of Fluids* 32: 111702.

Chernekhovskaia, N.E., Shishlo, V.K., Svistunov, B.D., and Svistunova, A.S. (2004). *4th Army Medical Conference on Intensive Therapy and Profilactic Treatments of Surgical Infections*, Moscow, Russia.

Chernets, N., Zhang, J., Steinbeck, M. et al. (2015). *Tissue Engineering Parts A* 21 (1, 2): 300.

Chesnokova, N.V., Gundorova, R.A., Krvasha, O.I. et al. (2003). Experimental investigation of the influence of gas flow containing nitrogen oxide in treatment of corneal wounds. *News of Russian Academy of Medical Sciences* 5: 40.

Ching, W.K., Colussi, A.J., Sun, H.J. et al. (2001). *Environmental Science & Technology* 35: 4139.

Ching, W.K., Colussi, A.J., and Hoffmann, M.R. (2003). *Environmental Science & Technology* 37: 4901.

Chintala, N., Meyer, R., Hicks, A. et al. (2005). *Journal of Propulsion and Power* 21: 583.

Chintala, N., Bao, A., Lou, G., and Adamovich, I.V. (2006). *Combustion and Flame* 144: 744.

Chiper, A.S., Chen, W., Mejlholm, O. et al. (2011). *Plasma Sources Science and Technology* 20: 025008.

Cho, B.H., Han, G.J., Oh, K.H. et al. (2011). *Journal of Materials Science* 46: 2755.

Choi, J.-H., Nam, S.-H., Song, Y.-S. et al. (2014). *Archives of Dermatological Research* 306: 635.

Chuangsuwanich, A., Assadamongkol, T., and Boonyawan, D. (2016). *The International Journal of Lower Extremity Wounds* 15: 313.

Chung, T.-H. et al. (2020). *Cancers* 12: 219.

Chutsirimongkol, C., Boonyawan, D., Polnikorn, N. et al. (2014). *Plasma Medicine* 4: 79.

Clements, J.S., Sato, M., and Davis, R.H. (1987). *IEEE Transactions on Industry Applications* IA-23 (2): 224.

Conti, S., Fridman, A., and Raoux, S. (1999). *14th International Symposium on Plasma Chemistry (ISPC-14)*, Prague, Czech Republic, vol. 5, p. 2581.

Conti, S., Porshnev, P.I., Fridman, A. et al. (2001). *Experimental Thermal and Fluid Science* 14: 79.

Conway, N.M.J., Ferrari, A.C., Flewitt, A.J. et al. (2000). *Diamond and Related Materials* 9: 765.

Cooper, M., Fridman, G., Staack, D. et al. (2009). *IEEE Transactions on Plasma Science* 37 (6): 866.

Cooper, M., Fridman, G., Fridman, A., and Joshi, S.G. (2010). *Journal of Applied Microbiology* 109: 2039.

Costa Dantas, M.C., do Prado, M., Costa, V.S. et al. (2012). *Journal of Endodontia* 38: 215.

Coulibaly, K., Genet, F., Renou-Gonnord, M.F., and Amouroux, J. (1997). *13th International Symposium on Plasma Chemistry (ISPC-13)*, Beijing, vol. 4, p. 1721.

Critzer, V.J., Kelly-Winterberg, K., South, S.L., and Golden, D.A. (2007). *Journal of Food Protection* 70: 2290.

Cutting, K.F., White, R.J., and Legerstee, R. (2017). *Wound Medicine* 16: 40.

Czernichowski, A. (1994). *Pure and Applied Chemistry* 66 (6): 1301.

D'Agostino, R. (2012). *Plasma Deposition, Treatment, and Etching of Polymers*. Boston, MA, New York: Academic Press.

D'Agostino, R., Favia, P. et al. (2010). *Plasma Processing of Polymers*, NATA Science, Series E, vol. 346. Springer.

D'Amico, K.M. and Smith, A.L.C. (1977). *Journal of Physics D: Applied Physics* 10: 261.

Da Silva, J.S., Amico, S.C., Rodrigues, A.O. et al. (2011). *The International Journal of Oral & Maxillofacial Implants* 26: 237.

Daeschlein, G. et al. (2013). *Experimental Dermatology* 22: 582.

Davalos, R.V., Mir, L.M., and Rubinsky, B. (2005). *Annals of Biomedical Engineering* 33 (2): 223.

Davydov, A.I., Strijakov, A.N., Pekshev, A.V. et al. (2002). *Questions in Obstetrics and Gynaecology (OB/GYN)* 1 (2): 57.

Davydov, A.I., Kuchukhidze, S.T., Shekhter, A.B. et al. (2004). *Problems of Gynecology, Obstetrics and Perinatology* 3 (4): 12.

Dembrovsky, V. (1984). *Plasma Metallurgy: The Principles*, Material Science Monographs. Elsevier Science, Ltd.

Deminsky, M., Givotov, V., Potapkin, B., and Rusanov, V. (2002). *Pure and Applied Chemistry* 74: 413.

Diaz, F. and Seedhouse, E. (2017). *To Mars and Beyond, Fast: How Plasma Propulsion Will Revolutionize Space Exploration.* Springer.

DIN SPEC 91315 (2014). *General Requirements for Medical Plasma Sources.* Berlin: Beuth.

Dobrynin, D. (2011). Physical and chemical mechanisms of direct and controllable plasma interaction with living objects. Ph.D. Dissertation. Drexel University, Philadelphia, PA.

Dobrynin, D., Fridman, G., Friedman, G., and Fridman, A. (2009). *New Journal of Physics* 11 (11): 115020.

Dobrynin, D., Fridman, G., Friedman, G., and Fridman, A. (2010a). *ICOPS-2010, IEEE Xplore*, INSPEC n. 11454243.

Dobrynin, D., Fridman, G., Mukhin, Y.V. et al. (2010b). *IEEE Transactions on Plasma Science* 38 (8): 1878.

Dobrynin, D., Fridman, G., Friedman, G., and Fridman, A. (2012a). *Plasma Medicine* 2 (1–3): 71.

Dobrynin, D., Fridman, A., and Starikovskiy, A.Y. (2012b). *IEEE Transactions on Plasma Science* 40 (9): 2163.

Dobrynin, D., Seepersad, Y., Pekker, M. et al. (2013). *Journal of Physics D: Applied Physics* 46: 105201.

Dobrynin, D., Rakhmanov, R., and Fridman, A. (2019a). *Journal of Physics D: Applied Physics* 52: 39LT01.

Dobrynin, D., Vainchtein, D., Gherardi, M. et al. (2019b). *IEEE Transactions on Plasma Science* 47 (8): 4052.

Dobrynin, D.V., Fridman, A., Lin, A. et al. (2022). Method of vaccination against cancer using plasma treated cancer cells. US Patent 17,567,424.

Doctor, R.D. (2000). *Hydrogen Production in Thermochemical Calcium-Bromine Water Splitting Cycle.* Argonne, IL: Pub. Argonne National Laboratory.

Dragsund, E., Andersen, A.B., and Johannessen, B.O. (2001). *Ballast Water Treatment by Ozonation, 1st Int. Ballast Water Treatment, R&D Symposium*, Global Monograph Series N.5, 21. London: IMO.

Dubinov, A.E., Lazarenko, E.M., and Selemir, V.D. (2000). *IEEE Transactions on Plasma Science* 28: 180.

Durual, S., Pernet, F., Rieder, P. et al. (2011). *Clinical Oral Implants Research* 22: 552.

Duske, K., Koban, I., Kindel, E. et al. (2012). *Journal of Clinical Periodontology* 39: 400.

Easter, G. (2008). *Thermal Spraying – Plasma, Arc and Flame Spray Technologies.* Wexford College Press.

Ebdon, J.R. (1995). *New Methods of Polymer Synthesis.* Glasgow, UK: Blackie.

Efimenko, N.A., Hrupkin, V.I., Marahonich, L.A. et al. (2005). *Journal of Military Medicine (Voenno-Meditsinskii Jurnal)* 5: 51.

Eletsky, A.V. (2000a). *Advances in Physical Sciences (Physics-Uspekhi)* 43 (2): 111.

Eletsky, A.V. (2000b). Production of fullerenes and nano-tubes in low-temperature plasma. In: *Encyclopedia of Low-Temperature Plasma*, vol. 4 (ed. V.E. Fortov). Moscow: Nauka (Science).

Ercan, U.K., Wang, H., Ji, H. et al. (2013). *Plasma Processes and Polymers* 10 (6): 544.

Ercan, U.K., Joshi, S., Ji, H. et al. (2016). *Scientific Reports* 6: 20365.

Esakov, I.I., Grachev, L.P., Khodataev, K.V. et al. (2006). Efficiency of propane-air mixture combustion assisted by deeply undercritical MW discharge in cold high-speed airflow. *44th AIAA Aerospace Science Meeting and Exhibit,* Reno, Nevada, AIAA paper 2006-1212.

Evans, W.H. (2015). *Biochemical Society Transactions* 43: 450.

Evans, D., Rosocha, L.A., Anderson, G.K. et al. (1993). *Journal of Applied Physics* 74: 5378.

Fadhlalmawla, S.A., Mohamed, A.A.H., Almarashi, J.Q.M., and Boutraa, T. (2019). *Plasma Science and Technology* 21: 105503.

Farhat, S., Hinkov, I., and Scott, C.D. (2002). *Journal of Nano-Science and Nano-Technology* 13: 377.

Farouk, T., Farouk, B., Staack, D. et al. (2006). *Plasma Sources Science and Technology* 15: 676.

Fauchais, P. (ed.) (1997). *Progress in Plasma Processing of Materials.* Athens: Begell House Publishers.

Fauchais, P. (2004). *Journal of Physics D: Applied Physics* 37: 86.

Faushais, P., Vardelle, A., and Dusoubs, B. (2001). *Journal of Thermal Spray Technology* 10: 44.

Felten, A., Bittencourt, C., Azioune, A., and Pireau, J.J. (2005). *17th International Symposium on Plasma Chemistry (ISPC-17)*, Toronto, p. 780.

Fernandes, C.P., Vassilakos, N., and Nilner, K. (1992). *Dental Materials: Publication of the Academy of Dental Materials* 8: 354.

Foster, K.W., Moy, R.L., and Fincher, E.F. (2008). *Journal of Cosmetic Dermatology* 7: 169.

Freeman, T., Steinbeck, M., Fridman, G. et al. (2012). Nonthermal DBD plasma enhances skeletal cell differentiation and autopod development. *4th International Conference on Plasma Medicine, ICPM-4,* Orleans, France, p. 17.

Freund, E. et al. (2020). *Frontiers of Physics* 8: 569618.

Fridman, A. (2003). *16th International Symposium on Plasma Chemistry, ISPC-16*, Taormina, Italy, p. 8.

Fridman, A. (2008). *Plasma Chemistry*. New York: Cambridge University Press.

Fridman, G. (2020). *Plasma Medicine* 10 (N3): 1.

Fridman, A. (2023). *15th International Symposium on Advance Plasma Science, ISPlasma 2023*, Gufu, Japan.

Fridman, A. and Friedman, G. (2013). *Plasma Medicine*. Wiley.

Fridman, A. and Kennedy, L.A. (2021). *Plasma Physics and Engineering*, 3e. New York: CRC Press, Taylor & Francis Group.

Fridman, A. and Rudakov, L.I. (1973). *Electrostatic Lenses for Focusing Relativistic Electron Beams*. Moscow: Moscow Institute of Physics and Technology (MIPT).

Fridman, A., Boufendi, L., Bouchoule, A. et al. (1996). *Journal of Applied Physics* 79: 1303.

Fridman, A., Nester, S., Kennedy, L. et al. (1999). *Progress in Energy and Combustion Science* 25: 211.

Fridman, G., Friedman, G., Gutsol, A., and Fridman, A. (2003). *16th International Symposium on Plasma Chemistry, ISPC-16*, Taormina, Italy, p. 703.

Fridman, A., Chirokov, A., and Gutsol, A. (2005a). *Journal of Physics D: Applied Physics* 38 (2): R001.

Fridman, G., Li, M., Lelkes, P.I. et al. (2005b). *IEEE Transactions on Plasma Science* 33: 1061.

Fridman, G., Peddinghaus, L., Fridman, A., et al. (2005c). *17th International Symposium on Plasma Chemistry (ISPC-17)*, Toronto, Canada, p. 1066.

Fridman, A., Gutsol, A., Dolgopolsky, A., and Stessel, E. (2006a). *Energy & Fuels* 20: 1242.

Fridman, G., Peddinghaus, L., Ayan, H. et al. (2006b). *Plasma Chemistry and Plasma Processing* 26: 425.

Fridman, A., Gutsol, A., Cho, Y. (2007a), *Advance in Heat Transfer*, Fridman, A., and Cho, Y. (eds.), vol. 40, Academic Press, Elsevier, Cambridge, MA, pg. 1–24.

Fridman, G., Brooks, A., Balasubramanian, M. et al. (2007b). *Plasma Processing and Polymers* 4: 425.

Fridman, G., Friedman, G., Gutsol, A., and Fridman, A. (2007c). *18th International Symposium on Plasma Chemistry, ISPC-18*, Kyoto, Japan, p. 66.

Fridman, A., Gutsol, A., Gangoli, S. et al. (2007d). *Journal of Propulsion and Power*, S. Macheret (ed.) 22: 1–12.

Fridman, G., Shereshevsky, A., Jost, M.M. et al. (2007e). *Plasma Chemistry and Plasma Processing* 27: 163.

Fridman, G., Friedman, G., Gutsol, A. et al. (2008). *Plasma Processes and Polymers* 5 (6): 503.

Friedlander, S.K. (2000). *Smoke, Dust and Haze: Fundamentals of Aerosol Dynamics*, Topics in Chemical Engineering. Oxford University Press.

Friedman, P.C. (2018). *7th International Conference on Plasma Medicine, ICPM-7*, Philadelphia, PA, USA.

Friedman, P.C. (2020). *International Journal of Dermatology* 59 (4): 15110.

Friedman, J. and Hassan, R. (1984). *The Journal of Prosthetic Dentistry* 52: 504.

Friedman, P.C., Miller, V., Fridman, G. et al. (2017). *Journal of the American Academy of Dermatology* 76 (2): 349.

Friedman, P.C., Miller, V., Fridman, G., and Fridman, A. (2019). *Clinical and Experimental Dermatology* 44: 459.

Friedman, P.C., Fridman, G., and Fridman, A. (2020). *Pediatric Dermatology* 37: 706.

Friedman, P.C., Fridman, G., and Fridman, A. (2022). *Experimental Dermatology* 10: 14695.

Friedrich, J.F. and Meichsner, J. (2022). *Non-Thermal Plasmas for Materials Processing*. Wiley-Scrivener.

Furusato, T. et al. (2012). *IEEE Transactions on Plasma Science* 40 (11): 3105–3115.

Gadri, R.B. (2000). *Surface and Coating Technology* 131: 528.

Gallagher, M.J., Friedman, G., Dolgopolsky, A., et al. (2005). *17th International Symposium on Plasma Chemistry, ISPC-17*, Toronto, Canada, p. 1056.

Gallagher, M., Vaze, N., Gangoli, S. et al. (2007). *IEEE Transactions on Plasma Science* 35: 1501.

Gallagher, M.J. et al. (2010). *Fuel* 89 (6): 1187.

Gan, L., Jiang, J., Duan, J.W. et al. (2021). *Journal of Biophotonics* 14: 202000415.

Gangoli, S., Gallagher, M., Dolgopolsky, A. et al. (2005). *17th International Symposium on Plasma Chemistry, ISPC-17*, Toronto, Canada, p. 1111.

Ganiev, Y.C., Gordeev, V.P., Krasilnikov, A.V. et al. (2000). *Journal of Thermophysics and Heat Transfer* 14: 10.

Garbassi, F., Morra, M., and Occhiello, E. (1994). *Polymer Surfaces, From Physics to Technology*. New York: Wiley.

Garscadden, A. (1994). *Pure and Applied Chemistry* 66: 1319.

Gelker, M., Müller-Goymann, C.C., and Viöl, W. (2018). *Clinical Plasma Medicine* 9: 34.

Ghaffari, A., Neil, D.H., Ardakani, A. et al. (2005). *Nitric Oxide* 12: 129.

Gilman, A.B. (2000). *Interaction of chemically active plasma with surfaces of synthetic materials*. In: *Encyclopedia of Low-Temperature Plasma*, vol. 4 (ed. V.E. Fortov), 393. Moscow: Nauka (Science).

Gilman, A.B. and Potapov, V.K. (1995). *Applied Physics* 3–4: 14.

Giro, G., Tovar, N., Witek, L. et al. (2013). *Journal of Biomedical Materials Research, Part A* 101: 98.

Givotov, V.K., Rusanov, V.D., and Fridman, A. (1984). *Plasma Chemistry*, vol. 11 (ed. B.M. Smirnov). Moscow: Atom-Izdat.

Givotov, V.K., Rusanov, V.D., and Fridman, A. (1985). *Diagnostics of Non-Equilibrium Chemically Active Plasma*. Moscow: Energo-Atom-Izdat.

Givotov, V.K., Potapkin, B.V., and Rusanov, V.D. (2005). Low-temperature plasma chemistry. In: *Encyclopedia of Low-Temperature Plasma*, vol. 8.1 (ed. Y.A. Lebedev, N.A. Plate, and V.E. Fortov), 4. Moscow: Nauka (Science).

Godyak, V.A. (1971). *Soviet Physics, Journal of Technical Physics* 41: 1361.

Gong, W., Zhu, A., Zhou, J. et al. (1997). *13th International Symposium on Plasma Chemistry (ISPC-13)*, Beijing, vol. 4, p. 1578.

Gonzalez, M.J., Sturgill, W.H., Ross, E.V., and Uebelhoer, N.S. (2008). *Lasers in Surgery and Medicine* 40: 124.

González-Mendoza, B. et al. (2019). *Clinical Plasma Medicine* 16: 100094.

Gostev, V. and Dobrynin, D. (2006). Medical micro-plasmatron. *3rd International Workshop on Microplasmas (IWM-2006)*, Greifswald, Germany.

Graves, D.B. (2012). *Journal of Physics D: Applied Physics* 45 (26): 26300.

Griem, H.R. (2005). *Principles of Plasma Spectroscopy*, Cambridge Monographs on Plasma Physics. Cambridge, New York: Cambridge University Press.

Grigorian, A.S., Grudyanov, A.I., Frolova, O.A. et al. (2001). *Application of a new biological factor, exogenous nitric oxide, for the surgical treatment of periodontis. Stomatology (Stomatologia)* 80: 803.

Grishin, S.D. (2000). *Ion and plasma rocket engines*. In: *Encyclopedia of Low-Temperature Plasma*, vol. 4 (ed. V.E. Fortov), 291. Moscow: Nauka (Science).

Grishin, S.D., Leskov, L.V., and Kozlov, N.P. (1975). *Electric-Propulsion Rocket Engines, Machinery*. Moscow: Mashinostroenie.

Grossman, K.R., Cybyk, B., and Van Wie, D. (2003). *AIAA Paper* 2003-0057.

Guo, Q., Wang, Y., Zhang, H. et al. (2017). *Scientific Reports* 7: 16680.

Gutsol, A.F., Givotov, V.K., Potapkin, B.V. et al. (1990). *Soviet Physics, Journal of Technical Physics (JTF)* 60: 62.

Gutsol, A.F., Givotov, V.K., Potapkin, B.V. et al. (1992). *Soviet Physics, High Energy Chemistry (Khimia Vysokikh Energij)* 26: 361.

Gutsol, A., Largo, J., and Hernberg, A. (1999). *14th International Symposium on Plasma Chemistry, ISPC-14*, Prague, vol. 1, p. 227.

Gutsol, A., Tak, G., and Fridman, A. (2005). *17th International Symposium on Plasma Chemistry (ISPC-17)*, Toronto, Canada, p. 1128.

Gutsol, K. et al. (2012). *International Journal of Hydrogen Energy* 37 (2): 1335.

Hammer, T. and Broer, S. (1998). *Plasma Exhaust Aftertreatment*. Warrendale, PA: Society of Automotive Engineers.

Han, G.J., Chung, S.N., Chun, B.H. et al. (2012). *The Journal of Adhesive Dentistry* 14: 461.

Harbec, D., Meunier, J.-L., Guo, L. et al. (2005). *Journal of Physics D: Applied Physics* 37: 2121.

Harkness, J.B.L. and Doctor, R.D. (1993). *Plasma-Chemical Treatment of Hydrogen Sulfide in Natural Gas Processing*. Chicago, IL: Argonne National Laboratory, Gas Research Institute, GRI-93/0118.

Harkness, J.B.L. and Fridman, A. (1999). *The Technical and Economic Feasibility of Using Low Temperature Plasmas to Treat Gaseous Emissions from Pulp Mills and Wood Product Plants*. Research Triangle Park, NC: National Council of the Paper Industry for Air and Stream Improvement (NCASI).

Hartwig, S., Doll, C., Voss, J.O. et al. (2017a). *Journal of Oral and Maxillofacial Surgery* 75: 429.

Hartwig, S., Preissner, S., Voss, J.O. et al. (2017b). *Journal of Cranio-Maxillofacial Surgery* 45: 1724.

Hayakawa, T., Yoshinari, M., and Nemoto, K. (2004). *Biomaterials* 25: 119.

Hayashi, N., Ono, R., and Uchida, S. (2015). *Journal of Photopolymer Science and Technology* 28: 445.

He, X.M., Lee, S.T., Bello, L. et al. (1999). *Journal of Materials Research* 14: 1055.

He, X., Ma, T., Qiu, J. et al. (2004). *Plasma Sources Science and Technology* 13: 446.

He, J., Waring, M., Fridman, A. et al. (2022). *Scientific Reports* 12 (1): 1.

Heberlein, J. (2002). *Pure and Applied Chemistry* 74: 327.

Heesch, E.J.M., Pemen, A.J.M., Huijbrechts, A.H.J. et al. (2000). *IEEE Transactions on Plasma Science* 28: 137.

Heinlin, J. et al. (2013a). *Wound Repair and Regeneration* 21: 800.

Heinlin, J., Isbary, G., Stolz, W. et al. (2013b). *Journal of the European Academy of Dermatology and Venereology* 27: 324.

Held, G. (2016). *Introduction to Light Emitting Diode Technology and Applications*. CRC Press.

Helfritch, D.J., Harmon, G., and Feldman, P. (1996). *Emerging Solutions to VOC and Air Toxics Conference, A&WMA, Proceedings*, Nashville, TN, USA, p. 277.

Henriksen, N.E. and Hansen, F. (2019). *Theories of Molecular Reaction Dynamics: The Microscopic Foundation of Chemical Kinetics*, 2e. Oxford, UK: Oxford University Press.

Herrera, J.E.G. (2021). *Rarefied Gas Flows and Dynamic Plasma Phenomena in Electric Propulsion Systems*. Cuvillier Publisher.

Herrmann, H.W., Henins, I., Park, J., and Selwyn, G.S. (1999). *Physics of Plasmas* 6: 2284.

Hilker, L., von Woedtke, T., Weltmann, K.D., and Wollert, H.-G. (2017). *European Journal of Cardio-Thoracic Surgery* 51: 186.

Hoard, J. and Servati, H. (ed.) (1998). *Plasma Exhaust Aftertreatment*. Warrendale, PA: Society of Automotive Engineers (SAE).

Hocker, H. (1995). *International Textile Bulletin, Veredlung* 41: 18.

Hocker, H. (2002). *Pure and Applied Chemistry* 74: 423.

Hollahan, J.R. and Bell, A.T. (1974). *Techniques and Applications of Plasma Chemistry*. New York: Wiley.

Hollevoet, L., Jardali, F., Gorbanev, Y. et al. (2020). *Angewandte Chemie* 132 (52): 24033–24037.

Honda, Y., Tochikubo, F., and Watanabe, T. (2001). *25th International Conference on Phenomena in Ionized Gases, ICPIG-25*, vol. 4, p. 37, Nagoya, Japan.

Hosseini, S.I., Mohsenimehr, S., Hadian, J. et al. (2018). *Physics of Plasmas* 25: 013525.

Hsiao, M.C., Meritt, B.T., Penetrante, B.M. et al. (1995). *Journal of Applied Physics* 78: 3451.

Hunter, P. and Oyama, S.T. (2000). *Control of Volatile Organic Emissions: Conventional and Emerging Technologies*. New York: Wiley.

Hutchinson, I.H. (2005). *Principles of Plasma Diagnostics*. Cambridge, New York: Cambridge University Press.

Huth, K.C., Saugel, B., Jakob, F.M. et al. (2007). *Journal of Dentistry Research* 86: 451.

Huth, K.C., Quirling, M., Maier, S. et al. (2009). *International Endodontic Journal* 42: 3.

Idlibi, A.N., Al-Marrawi, F., Hannig, M. et al. (2013). *Biofouling* 29: 369.

Ihara, T. et al. (2012). *Journal of Physics D: Applied Physics* 45 (7): 075204.

Inagaki, N. (2014). *Plasma Surface Modification and Plasma Polymerization*. CRC Press.

Isbary, G., Morfill, G., Schmidt, H.U. et al. (2010). *British Journal of Dermatology* 163 (1): 78.

Isbary, G., Heinlin, J., Shimizu, T. et al. (2012). *British Journal of Dermatology* 167 (2): 404.

Isbary, G., Stolz, W., Shimizu, T. et al. (2013). *Clinical Plasma Medicine* 1 (2): 25.

Isbary, G., Shimizu, T., Zimmermann, J.L. et al. (2014). *Clinical Plasma Medicine* 2: 50.

Iskenderova, K. (2003). Cleaning process in high-density plasma chemical vapor deposition reactor. Ph.D. Dissertation. Drexel University, Philadelphia, PA.

Iskenderova, K., Porshnev, P., Gutsol, A. et al. (2001). *15th International Symposium on Plasma Chemistry (ISPC-15)*, Orleans, France, p. 2849.

Ito, M., Ohta, T., and Hori, M. (2012). *Journal of the Korean Physical Society* 60 (6): 937.

Ivanov, A.A., Nikiforov, V.A. (1978), *Plasma Chemistry*, vol. 5, Smirnov, B.M., ed., Atom-Izdat, Moscow, Pg. 197–220.

Ivanov, A.A. and Soboleva, T.K. (1978). *Non-Equilibrium Plasma Chemistry*. Moscow: Atom-Izdat.

Iza, F. and Hopwood, J. (2003). *IEEE Transactions on Plasma Science* 31: 782.

Iza, F., Kim, G.J., Lee, S.M. et al. (2008). *Plasma Processes and Polymers* 5: 322.

Jablonowski, L., Koban, I., Berg, M.H. et al. (2013). *Plasma Processes and Polymers* 10: 499.

Jahn, R.G. (2006). *Physics of Electric Propulsion*. Dover Publications.

Jaisinghani, R. (1999). Bactericidal properties of electrically enhanced HEPA filtration and a bioburden case study. *InterPhex Conference*, New York, NY.

Jasinski, M., Szczucki, P., Dors, M. et al. (2000). *International Symposium High Pressure Low Temperature Plasma Chemistry, HACONE VII*, Griefswald, Germany, p. 496.

Jasinski, M., Mizeraczyk, J., Zakrzewski, Z., and Chang, J.S. (2002a). *Czechoslovak Journal of Physics* 52 (D): 743.

Jasinski, M., Mizeraczyk, J., Zakrzewski, Z. et al. (2002b). *Journal of Physics D: Applied Physics* 35: 1.

Ji, S.H., Choi, K.H., Pengkit, A. et al. (2016). *Archives of Biochemistry and Biophysics* 605: 117.

Jiandong, L., Xianfu, X., Yao, S., and Tianen, T. (1996). *International Workshop on Plasma Technologies for Pollution Control and Waste Treatment*. Cambridge, MA: Publ., MIT.

Jiang, X.L., Tiwari, R., Gitzhofer, F., and Boulos, M.I. (1993). *Journal of Thermal Spray Technology* 2 (3): 265.

Jiang, C., Chen, M.T., Gorur, P.A. et al. (2009a). *Plasma Processes and Polymers* 6 (8): 479.

Jiang, C., Chen, M.T., Gorur, P.A. et al. (2009b). *IEEE Transactions on Plasma Science* 37 (7): 1190.

Johansson, E., Claesson, R., and van Dijken, J.W. (2009). *Journal of Dentistry* 37: 449.

Joshi, R.P., Hu, Q., Schoenbach, K.H., and Beebe, S.J. (2002a). *IEEE Transactions on Plasma Science* 30: 1536.

Joshi, R.P., Qian, J., and Schoenbach, K.H. (2002b). *Journal of Applied Physics* 92: 6245.

Ju, Y. and Sun, W. (2015). *Progress in Energy and Combustion Science* 48: 21.

Jung, J.-M., Yang, Y., Lee, D.H. et al. (2012). *Plasma Chemistry and Plasma Processing* 32: 32.

Kabisov, R.K., Sokolov, V.V., Shekhter, A.B. et al. (2000). Experience in application of exogenous NO therapy for treatment of postoperative wounds and radio-reactions in oncological patients. *Russian Journal of Oncology* 1: 24.

Kabouzi, Y., Nantel-Valiquette, M., Moisan, M. et al. (2005). *17th International Symposium on Plasma Chemistry (ISPC-17)*, Toronto, Canada, p. 1121.

Kado, S., Sekine, Y., Muto, N. et al. (2003). *16th International Symposium on Plasma Chemistry (ISPC-16)*, Orleans, France, p. 569.

Kadomtsev, B.B. and Shafranov, V.D. (2000). *Reviews of Plasma Physics1*, vol. 2. New York: Springer-Verlag.

Kaganovich, I. and Keidar, M. (ed.) (2020). Bulletin of the American Physical Society. *62nd Annual Meeting of the APS Division of Plasma Physics*, vol. 65, number 11.

Kalghatgi, S.U., Cooper, M., Fridman, G. et al. (2007a). Mechanism of blood coagulation by non-thermal atmospheric pressure dielectric barrier discharge plasma. *9th Annual RISC*, Drexel University, Philadelphia, PA.

Kalghatgi, S.U., Cooper, M., Fridman, G. et al. (2007b). *IEEE Transactions on Plasma Science* 35 (5): 1559.

Kalghatgi, S., Friedman, G., Fridman, A., and Clyne, A.M. (2010). *Annals of Biomedical Engineering* 38: 748.

Kalghatgi, S., Kelly, C.M., Cerchar, E. et al. (2011). *PLoS One* 6 (1): 16270.

Kalra, C., Gutsol, A.F., and Fridman, A. (2005). *IEEE Transactions on Plasma Science* 33: 32.

Kanazava, S., Kogoma, M., Moriwaki, T., and Okazaki, S. (1988). *Journal of Physics D: Applied Physics* 21: 838.

Karpenko, E.I. and Messerle, V.E. (2000). Plasma – energetic technologies of utililation of fuels. In: *Encyclopedia of Low-Temperature Plasma*, vol. 4 (ed. V.E. Fortov), 359. Moscow: Nauka (Science).

Katsuki, S., Akiyama, H., Abou-Ghazala, A., and Schoenbach, K.H. (2002a). *IEEE Transactions on Dielectrics and Electrical Insulation* 9: 498.

Katsuki, S., Moreira, K., Dobbs, F. et al. (2002b). *IEEE Journal* 8: 648.

Kawai, H. et al. (2004). *Biomaterials* 25: 1805.

Kee, S.T. et al. (2011). *Clinical Aspects of Electroporation*. Springer.

Keidar, M. (ed.) (2020). *Plasma Cancer Therapy*. Springer.

Keidar, M. and Bellis, I. (2018). *Plasma Engineering: Applications from Aerospace to Bio and Nanothechnology*. Academic Press.

Keidar, M., Walk, R., Shashurin, A. et al. (2011). *British Journal of Cancer* 105: 1295.

Keidar, M., Yan, D., and Sherman, J.H. (2019). *Cold Plasma Cancer Therapy*. JoP E-Books Pub.

Kelly-Wintenberg, K. (2000, 2000). *IEEE Transactions on Plasma Science* 28.

Kelly-Wintenberg, K., Montie, T.C., Brickman, C. et al. (1998). *Journal of Industrial Microbiology and Biotechnology* 20: 69.

Kennedy, L., Fridman, A., Saveliev, A., and Nester, S. (1997). *APS Bulletin* 41 (9): 1828.

Khan, A., Malik, S., Walia, J. et al. (2020). *Journal of Drugs in Dermatology* 19 (12): 1177.

Kharitonov, D.Y., Gogish-Klushin, S.Y., and Novikov, G.I. (1987). *News of Belarus Academy of Sciences, Chemistry* 6: 105.

Kharitonov, D.Y., Gutsevich, E.I., Novikov, G.I., and Fridman, A. (1988). *Mechanism of Pulsed Electrolytic-Spark Oxidation of Aluminum in Concentrated Sulfuric Acid*, vol. 4705/13. Moscow: Kurchatov Institute of Atomic Energy, Atom-Inform.

Kieft, I.E., van der Laan, E.P., and Stoffels, E. (2004). *New Journal of Physics* 6: 149.

Kim, C.-H., Bahn, J.H., Leea, S.-H. et al. (2010). *Journal of Biotechnology* 150: 530.

Kim, B., Yun, H., Jung, S. et al. (2011). *Food Microbiology* 28: 9.

Kim, H.S., Cho, Y.I., Hwang, I.H. et al. (2013). *Separation and Purification Technology* 120: 423.

Kirillov, I.A., Potapkin, B.V., Rusanov, V.D., and Fridman, A. (1983). *Soviet Physics, High Energy Chemistry* 17: 519.

Kirillov, I.A., Potapkin, B.V., Fridman, A. et al. (1984a). *Soviet Physics, High Energy Chemistry* 18: 151.

Kirillov, I.A., Potapkin, B.V., Strelkova, M.I. et al. (1984b). *Soviet Physics, Journal of Applied Mathematics and Technical Physics* 6: 77–80.

Kitazaki, S., Sarinont, T., Koga, K. et al. (2014). *Current Applied Physics* 14: 149.

Kiyan, T. et al. (2008). *IEEE Transactions on Plasma Science* 36 (3): 821–827.

Klebes, M. et al. (2014). *Clinical Plasma Medicine* 2: 22.

Klimov, A.I. and Mishin, G.I. (1990). *Soviet Technical Physics Letters (Pis'ma v JTF)* 16: 960.

Klimov, A.I., Koblov, A.N., Mishin, G.I. et al. (1982). *Soviet Technical Physics Letters (Pis'ma v JTF)* 8: 192.

Klimov, A., Bitiurin, V., Moralev, I. et al. (2006). Non-premixed plasma-assisted combustion of hydrocarbon fuel in high-speed airflow. *44th AIAA Aerospace Science Meeting and Exhibit,* Reno, NV, AIAA paper 2006-617.

Knight, R. and Smith, R. (1998). *Thermal Spray Forming of Materials, ASM International.* 7: 408.

Knight, R., Smith, R., and Apelian, D. (1991). *International Materials Review* 36 (6): 221.

Knight, R., et al. (2005). Thermal spray: past, present and future. *17th International Symposium on Plasma Chemistry (ISPC-17)*, Toronto, Canada, p. 913.

Knizhnik, A.A., Potapkin, B.V., Medvedev, D.D. et al. (1999). *Journal of Technical Physics* 365: 336.

Koban, I., Duske, K., Jablonowski, L. et al. (2011a). *Plasma Processes and Polymers* 8: 975.

Koban, I., Holtfreter, B., Hubner, N.O. et al. (2011b). *Journal of Clinical Periodontology* 38: 956.

Koch, F., Salva, K.A., Wirtz, M. et al. (2020). *Journal of the European Academy of Dermatology and Venereology* 1–3: 16735.

Kogelschatz, U. (2003). *Plasma Chemistry and Plasma Processing* 23 (1): 1.

Kolikov, V. and Rutberg, F. (2016). *Pulsed Electrical Discharges for Medicine and Biology: Techniques, Processes, Applications.* Springer.

Kondratiev, V.N. and Nikitin, E.E. (1980). *Gas-Phase Reactions: Kinetics and Mechanisms.* Berlin, Heidelberg, New York: Springer-Verlag.

Kortshagen, U.R., Sankaran, R.M., Pereira, R.N. et al. (2016). *Chemical Reviews* 116 (18): 11061.

Kouprine, A., Gitzhofer, F., Boulos, M., and Fridman, A. (2003). *Plasma Chemistry and Plasma Processing* 24: 189.

Kramer, P.W., Yeh, Y.-S., and Yasuda, H. (1989). *Journal of Membrane Science* 46: 461.

Krasheninnikov, E.G., Rusanov, V.D., Saniuk, S.V., and Fridman, A. (1986). *Soviet Physics, Journal of Technical Physics* 56: 1104.

Kremer, J., Müller, F., Heininger, A. et al. (2020). *The Journal of Heart and Lung Transplantation* 39: 488.

Kremer, J., Mueller, F., Farag, M. et al. (2021). *The Journal of Heart and Lung Transplantation* 40: 180.

Kristof, J., Aoshima, T., Blajan, M., and Shimizu, K. (2019). *Plasma Science and Technology* 21: 064001.

Krotovskii, G.S., Pekshev, A.V., Zudin, A.M. et al. (2002). *Cardio-Vascular Surgery (Grudnaia i serdechno-sosudistaia hirurgia)* 1: 37.

Kuchukhidze, S.T., Klihdukhov, I.A., Bakhtiarov, K.R., and Pankratov, V.V. (2004). *Problems of Obstetrics and Gynecology (OB/GYN)* 3 (2): 76.

Kudinov, V.V., Pekshev, P.Y., Belashchenko, V.I. et al. (1990). *Coating Spraying by Plasma.* Moscow: Nauka (Science).

Kuo, S. (2019). *Cold Atmospheric Plasmas.* World Scientific.

Kurochkin, Y.V., Polak, L.S., Pustogarov, A.V. et al. (1978). *Soviet Physics, Thermal Physics of High Temperatures, TVT* 16: 1167.

Kushner, M.J. (1988). *Journal of Applied Physics* 63: 2532.

Kushner, M.J. (2004). *Journal of Applied Physics* 95 (3): 846.

Kushner, M.J. (2005). *Journal of Physics D: Applied Physics* 38: 1633.

Kuznetzov, N.M. (1971). *Journal of Technical and Experimental Chemistry (TEC) (Tekh. Eksp. Khim., TEKh)* 7: 22.

Kyzek, S., Holubová, L., Medvecká, V. et al. (2019). *Plasma Chemistry and Plasma Processing* 39: 475.

Labay, C., Hamouda, I., Tampieri, F. et al. (2019). *Scientific Reports* 9 (1): 16160.

Lademann, O., Richter, H., Meinke, M.C. et al. (2011). *Experimental Dermatology* 20: 488.

Landau, L.D. and Lifshitz, E. (2002). *Course of Theoretical Physics, Physical Kinetics*, vol. 10. Oxford, UK: Butterworth-Heinemann.

Landau, L.D. and Lifshitz, E. (2014). *Course of Theoretical Physics, Statistical Physics*, vol. 9. New Delhi, India: Elsevier.

Landau, L.D., Lifshitz, E.M., and Pitaevskii, L.P. (2013). *Course of Theoretical Physics, Electrodynamics of Continuous Media*, vol. 8. New Delhi, India: Elsevier.

Lange, H., Huczko, A., Sioda, M. et al. (2002). *Plasma Chemistry and Plasma Processing* 22: 523.

Laroussi, M. (2005). *Plasma Processes and Polymers* 2: 391.

Laroussi, M. (2009). *IEEE Transactions on Plasma Science* 37: 714.

Laroussi, M. (ed.) (2020). *Cold Plasma Characteristics and Applications in Medicine.* MDPI Pub.

Laroussi, M., Sayler, G., Galscock, B. et al. (1999). *IEEE Transactions on Plasma Science* 27: 34.

Laroussi, M., Alexeff, I., and Kang, W. (2000). *IEEE Transactions on Plasma Science* 28: 184.

Laroussi, M., Richardson, J.P., and Dobbs, F.C. (2002). *Applied Physics Letters* 81: 772.

Laroussi, M., Mendis, D.A., and Rosenberg, M. (2003). *New Journal of Physics* 5: 41.

Laroussi, M., Kong, M.G., Morfill, G., and Stolz, W. (ed.) (2012). *Plasma Medicine*. Cambridge, New York: Cambridge University Press.

Laroussi, M., Bekeschus, S., Keidar, M. et al. (2022). *IEEE Transactions on Radiation and Plasma Medical Sciences* 6 (2): 127.

Lee, H.W., Kim, G.J., Kim, J.M. et al. (2009). *Journal of Endodontics* 35: 587.

Lee, J.K. et al. (2011). *Japanese Journal of Applied Physics* 50: 1–7.

Legasov, V.A. et al. (1978a). *Soviet Physics, Doklady (Reports of the USSR Academy of Sciences)* 238: 66.

Legasov, V.A., Rusanov, V.D., Fridman, A. (1978b), *Plasma Chemistry*, vol. 5, ed. by Smirnov, B.M., Atom-Izdat, Moscow pg. 197–230.

Legasov, V.A., Asisov, R.I., Butylkin, Y.P. et al. (1983). *Nuclear-Hydrogen Energy and Technology -5* (ed. V.A. Legasov), 71. Moscow: Energo-Atom-Izdat.

Legasov, V.A., Belousov, I.G., Givotov, V.K. et al. (1988). *Nuclear – Hydrogen Energy and Technology*, vol. 8 (ed. V.A. Legasov). Moscow: Energo-Atom-Izdat.

Leonov, S., Bityurin, V., Savishenko, N. et al. (2001). The Effect of Plasma-Induced Separation. *AIAA paper* 2001–0640.

Leonov, S., Yarantsev, D.A., Napartovich, A.P., and Kochetov, I.V. (2006). 563.

Lewis, A.J., Joyce, T., Hadaya, M. et al. (2020). *Environmental Science: Water Research & Technology* 6: 1044.

Li, J. and Zhang, G.Q. (2019). *Light Emitting Diodes*. Springer.

Li, L., Jiang, J., Li, J. et al. (2014). *Scientific Reports* 4: 5859.

Li, Y., Wang, T., Meng, Y. et al. (2017). *Plasma Chemistry and Plasma Processing* 37: 1621.

Li, S. et al. (ed.) (2020). *Electroporation Protocols*, 3e. Humana Press.

Lieberman, M.A. and Lichtenberg, A.J. (2005). *Principles of Plasma Discharges and Material Processing*. New York: Wiley.

Likhanskii, A., Shneider, M., Macheret, S., and Miles, R. (2007). Optimization of dielectric barrier discharge plasma actuators driven by repetitive nanosecond pulses. *45th AIAA Aerospace Science Meeting and Exhibit,* Reno, NV, AIAA paper 2007-0183.

Lin, A., Truong, B., Pappas, A. et al. (2015). *Plasma Processes and Polymers* 12 (12): 1392.

Lin, A., Truong, B., Fridman, G. et al. (2017a). *Plasma Medicine* 7 (1): 85.

Lin, A., Truong, B., Patel, S. et al. (2017b). *International Journal of Molecular Sciences* 18 (5): 966.

Lin, A.G., Xiang, B., Merlino, D. et al. (2018). *Onco-Immunology* 7 (9): 1487978.

Lipatov, K.V., Kanorskii, I.D., Shekhter, A.B., and Emelianov, A.Y. (2002). *Annals of Surgery* 1: 58.

Lipner, S.R., Friedman, G., and Scher, R.K. (2017). *Clinical and Experimental Dermatology* 42: 295.

Lisitsyn, I.V., Nomiyama, H., Katsuki, S., and Akiyama, H. (1999). *The Review of Scientific Instruments* 70: 3457.

Liston, E.M. (1993). *Journal of Adhesion Science and Technology* 7: 1091.

Liston, E.M., Martinu, L., Wertheimer, M.R. (1994), *Plasma Surface Modification of Polymers: Relevance to Adhesion*, Strobel, M., Lyons, C., and Mittal, K.L. (eds.), VSP, Netherlands, pg. 199–210.

Liu, C., Dobrynin, D., and Fridman, A. (2014). *Journal of Physics D: Applied Physics* 47 (25): 25/2003.

Liu, C., Chernets, I., Ji, H.F. et al. (2017). *IEEE Transactions on Plasma Science* 45 (4): 683.

Liu, D., Zhang, Y., Xu, M. et al. (2020). *Plasma Processes and Polymers* 17: 1900218.

Liventsov, V.V., Rusanov, V.D., Fridman, A., and Sholin, G.V. (1981). *Soviet Physics, Journal of Technical Physics, Letters* 9: 474.

Liventsov, V.V., Rusanov, V.D., and Fridman, A. (1984). *Soviet Physics Doklady* 275: 1392.

Lloyd, G., Friedman, G., Jafri, S. et al. (2010). *Plasma Processes and Polymers* 7: 194.

Locke, B.R., Sato, M., Sunka, P. et al. (2006). *Industrial and Engineering Chemistry Research* 45: 882.

Long, B.H., Wu, H.H., Long, B.Y. et al. (2005). *Journal of Physics D: Applied Physics* 38: 3491.

Losev, S.A., Sergievska, A.L. et al. (1996). *Soviet Physics Doklady* 346: 192.

Lou, G., Bao, A., Nishihara, M. et al. (2006). Ignition of premixed hydrocarbon-air flows by repetitively pulsed, nano-second pulsed duration introduction plasma. *44th AIAA Aerospace Sciences Meeting and Exhibit* , Reno, NV (January 2006), AIAA paper 2006-1215.

Lou, G., Bao, A., Nishihara, M. et al. (2007). *Proceedings of the Combustion Institute* 31: 31.

Lowry, H., Stepanek, C., Crosswy, L. et al. (1999). Shock Structure of a Spherical Projectile in Weakly Ionized Air. *AIAA paper* 90-0600.

Lu, X., Cao, Y., Yang, P. et al. (2009). *Plasma Science* 37: 668.

Lu, X.P., Reuter, S., Laroussi, M., and Liu, D.W. (2019). *Non-Equilibrium Atmospheric Pressure Plasma Jets*. CRC Press.

Lukes, P. et al. (2014). *Plasma Sources Science and Technology* 23 (1): 015019.

Macheret, S., Ed. (2006), *Journal of Propulsion and Power, Special Issue*, vol.22 pg. 1–5.

Macheret, S. et al. (1978). *Soviet Physics, Journal of Technical Physics Letters (Pis'ma v JTF)* 4: 28.

Macheret, S., Rusanov, V.D., Fridman, A., and Sholin, G.V. (1980a). *Soviet Physics Doklady* 255: 98.

Macheret, S., Rusanov, V.D., Fridman, A., and Sholin, G.V. (1980b). *Soviet Physics, Journal of Technical Physics* 50: 705.

Macheret, S., Rusanov, V.D., and Fridman, A. (1984). *Soviet Physics Doklady* 276: 1420.

Macheret, S., Fridman, A., Adamovich, I. et al. (1994). *Mechanism of non-equilibrium dissociation of diatomic molecules*. AIAA paper 94-1984.

Macheret, S.O., Shneider, M.N., and Miles, R.B. (2000). *2nd Workshop on Magneto-Plasma Aerodynamics in Aerospace Applications*, Moscow.

Macheret, S.O., Shneider, M.N., and Miles, R.B. (2001a). Magnetohydrodynamic Control of Hypersonic Flows and Scramjet Inlets Using Electron Beam Ionization. *AIAA paper 2001-0492*.

Macheret, S.O., Shneider, M.N., and Miles, R.B. (2001b). *AIAA paper 2001-0795*.

Macheret, S.O., Shneider, M.N., and Miles, R.B. (2005). *36th AIAA Plasma dynamics and Lasers Conference*, Toronto, Canada, AIAA-2005-5371.

Mahdikia, H., Saadati, F., Freund, E. et al. (2020). *Oncoimmunology* 10: e1859731.

Malyavko, A. et al. (2020). *Materials Advances* 1 (6): 1494.

Mangolini, L., Thimsen, E., and Kortshagen, U. (2005). *17th International Symposium on Plasma Chemistry (ISPC-17)*, Toronto, Canada, p. 770.

Mankelevich, Y.A. and Suetin, N.V. (2000). *Physical and chemical processes in plasma reactors for deposition of diamond films*. In: *Encyclopedia of Low-Temperature Plasma*, vol. 4 (ed. V.E. Fortov), 404. Moscow: Nauka (Science).

Manos, D.M. and Flamm, D.L. (1989). *Plasma Etching*. New York: Academic Press.

Martinez, E. (ed.) (2020). *Plasma Technology for Biomedical Applications*. Applied Sciences Pub.

Masuda, S. (1993). *Nonthermal Plasma Techniques for Pollution Control: Part B – Electron Beam and Electrical Discharge Processing* (ed. B.M. Penetrante and S.E. Schultheis). Berlin: Springer-Verlag.

Masur, K., von Behr, M., Bekeschus, S. et al. (2015). *Plasma Processes and Polymers* 12: 1377.

Mattachini, F., Sani, E., and Trebbi, G. (1996). *International Workshop on Plasma Technologies for Pollution Control and Waste Treatment*. Cambridge, MA: Publ., MIT.

Matveev, I. (ed.) (2013). *Plasma Assisted Combustion, Gasification, and Pollution Control, vol.1, Methods of Plasma Generation for PAC*. Outskirts Press, Inc.

Matveev, I. (ed.) (2015). *Plasma Assisted Combustion, Gasification, and Pollution Control, vol.2, Combustion and Gasification*. Outskirts Press, Inc.

Matveev, I., Matveeva, S., Gutsol, A., and Fridman, A. (2005). *43rd AIAA Aerospace Sciences Meeting and Exhibit*, Reno, NV (10–13 January 2005), AIAA paper 2005-1191.

Maximov, A.I. (2000). *Encyclopedia of Low-Temperature Plasma*, vol. 4 (ed. V.E. Fortov), 399. Moscow: Nauka (Science).

Maximov, A.I., Gorberg, B.L., and Titov, V.A. (1997). Possibilities and problems of plasma treatment of fabrics and polymer materials. In: *Textile Chemistry – Theory, Technology, Equipment* (ed. A.P. Moryganov), 225–245. New York: NOVA Science Publishers.

McCartney, L. and Lefsrud, M. (2015). *HortTechnology* 25 (3): 313.

McCaughey, M.J. and Kushner, M.J. (1989). *Journal of Applied Physics* 65: 186.

McTaggart, F. (1967). *Plasma Chemistry in Electric Discharges, Topics in Inorganic and General Chemistry*. Elsevier.

Melzer, A. (2019). *Physics of Dusty Plasmas*. Springer.

Meng, Y., Qu, G., Wang, T. et al. (2017). *Plasma Chemistry and Plasma Processing* 37: 1105.

Messerle, V.E. (2004). *High Energy Chemistry (Khimia Vysokikh Energij)* 38: 35.

Messerle, V.E. and Sakipov, Z.B. (1988). *Chemistry Solid Fuels (Khimia Tverdykh Topliv)* 4: 123.

Messerle, V.E., Sakipov, Z.B., Siniarev, G.B., and Trusov, B.G. (1985). *Soviet Physics, High Energy Chemistry (Khimia Vysokikh Energij)* 19: 160.

Mesyats, G.A., Mkheidze, G.P., and Savin, A.A. (2000). *Encyclopedia of Low Temperature Plasma*, vol. 4 (ed. V.E. Fortov), 108. Moscow: Nauka (Science).

Metelmann, H.-R., von Woedtke, T., Bussiahn, R. et al. (2012). *American Journal of Cosmetic Surgery* 29: 52.

Metelmann, H.-R., Seebauer, C., Rutkowski, R. et al. (2018a). *Contributions to Plasma Physics* 58: 415.

Metelmann, H.-R., Von Woedke, T., and Weltmann, K.D. (ed.) (2018b). *Comprehensive Clinical Plasma Medicine*. Springer.

Metelmann, H.-R., Von Woedke, T., Weltmann, K.D., and Emmert, S. (ed.) (2022). *Textbook of Good Clinical Practice in Cold Plasma Therapy*. Springer.

Meyer, S., Gorges, R., and Kreisel, G. (2004). *Thin Solid Films* 450: 276. *Electrochimica Acta* **49**: 3319.

Miletic, M., Mojsilović, S., Okić Dordevic, I. et al. (2013). *Journal of Physics D* 46: 345.

Miller, V., Lin, A., Fridman, G. et al. (2014). *Plasma Processes and Polymers* 11 (12): 1193.

Miller, V., Lin, A., and Fridman, A. (2016). *Plasma Chemistry and Plasma Processing* 36 (1): 259.

Mirpour, S., Fathollah, S., Mansouri, P. et al. (2020). *Scientific Reports* 10: 10440.

Mishin, G.I. (1997). *AIAA Aerospace Sciences Meeting*, AIAA paper 1997-2298.

Mishra, S. (2012). *Thermal Plasma Application in Metallurgy: Plasma Processing of Materials*. LAP LAMBERT Academic Publishing.

Misyn, F.A. and Gostev, V.A. (2000). Cold plasma application in eyelid phlegmon curing. In: *Diagnostics and Treatment of Infectious Diseases*, 12–18. Petrozavodsk, Russia: Petrozavodsk University.

Misyn, F.A., Besedin, E.V., Komkova, O.P., and Gostev, V.A. (2000). Experimental investigation of bactericidal influence of cold plasma and its interaction with cornea. In: *Diagnostics and Treatment of Infectious Diseases*, 19–27. Petrozavodsk, Russia: Petrozavodsk University.

Mizuno, K., Yonetamari, K., Shirakawa, Y. et al. (2017). *Journal of Physics D: Applied Physics* 50: e.12LT01.

Moelleken, M., Jockenhöfer, F., Wiegand, C. et al. (2020). *JDDG: Journal der Deutschen Dermatologischen Gesellschaft* 18: 1094.

Mohamed, H., Esposito, R.A., Kutzler, M.A. et al. (2020). *Plasma Processes and Polymers* 17: e2000051.

Moisan, M., Barbeau, J., Moreau, S. et al. (2001). *International Journal of Pharmaceutics* 226: 1.

Moreau, E. (2007). *Journal of Physics D: Applied Physics* 40 (3): 605.

Moreau, S., Moisan, M., Barbeau, J. et al. (2000). *Journal of Applied Physics* 88: 1166.

Morfill, G.E. and Thomas, H. (1996). *Journal of Vacuum Science and Technology A* 14: 490.

Morgan, R.A. (1985). *Plasma Etching in Semiconductor Fabrication*. New York: Elsevier.

Morozov, A.I. (1978). *Physical Basis of the Space Electric-Jet Engines*. Moscow: AtomIzdat.

Mosse, A.L. and Pechkovsky, V.V. (1973). *Application of Low-Temperature Plasma in Technology of Inorganic Materials*. Minsk, Belarus: Nauka I Technica (Science and Engineering).

Moukhametshina, Z.B., Chekmarev, A.M., Givotov, V.K. et al. (1986). *Soviet Physics, Journal of Technical Physics (JTF)* 56: 757. *Soviet Physics, High Energy Chem. (Khimia Vysokikh Energij)* **20**: 354.

Murphy, A.B. (1997). *Journal and Proceedings. Royal Society of New South Wales* 130: 93.

Murphy, A.B., McAlister, T., Farmer, A.J.D., and Horrigan E.C. (2001). *15th International Symposium on Plasma Chemistry (ISPC-16)*, Orleans, France, p. 619.

Murphy, A.B., Farmer, A.J.D., Horrigan, E.C., and McAllister, T. (2002). *Plasma Chemistry and Plasma Processing* 22: 371.

Mutaf-Yardimci, O., Kennedy, L., Saveliev, A., and Fridman, A. (1998). Plasma exhaust aftertreatment. *SAE*, SP-1395, p. 1.

Mutaf-Yardimci, O., Saveliev, A.V., Fridman, A., and Kennedy, L.A. (1999a). *International Journal of Hydrogen Energy* 23: 1109.

Mutaf-Yardimci, O., Saveliev, A., Fridman, A., and Kennedy, L. (1999b). *Journal of Applied Physics* 87: 1632.

Nagayoshi, M., Fukuizumi, T., Kitamura, C. et al. (2004). *Oral Microbiology and Immunology* 19: 240.

Naidis, G.V. (1997). *Journal of Physics D: Applied Physics* 30: 1214.

Nam, S.H., Lee, H.W., Cho, S.H. et al. (2013). *Journal of Applied Oral Science* 21 (3): 265.

Nau-Hix, C., Multari, N., Singh, R.K. et al. (2021). Field demonstration of a pilot-scale plasma reactor for the rapid removal of poly-and perfluoroalkyl substances in groundwater. *ACS ES&T Water* 1: 680.

Naumova, I.K., Maksimov, A.I., and Khlyustova, A.V. (2011). *Surface Engineering and Applied Electrochemistry* 47: 263.

Neely, W.C., Newhouse, E.I., Clothiaux, E.J., Gross, C.A. (1993), *Non-Thermal Plasma Techniques for Pollution Control: Part B – Electron Beam and Electrical Discharge Processing*, Penetrante, B.M., Schultheis, S.E. (eds.), Springer-Verlag, Berlin, pg. 19–28.

Nelea, V., Luo, L., Demers, C.N. et al. (2005). *Journal of Biomedical Research* 75A: 216.

Nemchinsky, V.A. (1998). *Journal of Physics D: Applied Physics* 31: 3102.

Nemchinsky, V.A. (2002). *IEEE Transactions on Plasma Science* 30 (6): 2113.

Nemchinsky, V.A. (2003). *Journal of Physics D: Applied Physics* 36: 1573.

Nester, S., Demura, A.V., and Fridman, A. (1983). *Disproportioning of the Vibrationally Excited CO Molecules*, vol. 4223/6. Moscow: Kurchatov Institute of Atomic Energy.

Nester, S., Potapkin, B.V., Levitsky, A.A. et al. (1988). *Kinetic and Statistical Modeling of Chemical Reactions in Gas Discharges*. Moscow: Central Research Institute of Informatics in Nuclear Science and Industry, CNII Atom Inform.

Neumann, E., Sowers, A.E., and Jordan, C.A. (ed.) (2001). *Electroporation and Electrofusion in Cell Biology*. Berlin, Heidelberg: Springer-Verlag.

Nikerov, V.A. and Sholin, G.V. (1985). *Kinetics of Degradation Processes*. Moscow: Energo-Atom-Izdat.

Nikitin, E.E. (1974). *Theory of Elementary Atomic and Molecular Processes in Gases, International Series of Monographs on Physics*. UK: Clarendon, Oxford University Press.

Nikitin, E.E. and Umanskii, S.Y. (1984). *Theory of Slow Atomic Collisions*, Springer Series in Chemical Physics, vol. 30. Berlin: Springer-Verlag.

Nojiri, K. (2014). *Dry Etching Technology for Semiconductors*. Springer.

Noriega, E., Shama, G., Laca, A. et al. (2011). *Food Microbiology* 28: 1293.

Novikov, G.I., Bochin, V.P., and Romanovsky, M.K. (1978). *Nuclear – Hydrogen Energy and Technology*, vol. 1 (ed. V.A. Legasov), 231. Moscow: Atom-Izdat.

Nozaki, T. (2002). *55th Gaseous Electronics Conference (GEC-55)*, Minneapolis, MN.

Nozaki, T. and Okazaki, K. (2013). *Catalysis Today* 211: 29.

Nunnally, T., Gutsol, K., Rabinovich, A. et al. (2011). *Journal of Physics D: Applied Physics* 44 (27): 274009.

Nunnally, T., Gutsol, K., Rabinovich, A. et al. (2014). *International Journal of Hydrogen Energy* 39 (6): 12480.

Nzeribe, B.N., Crimi, M., Mededovic-Thagard, S., and Holsen, T.M. (2019). Physico-chemical processes for the treatment of per-and polyfluoroalkyl substances (PFAS). *Critical Reviews in Environmental Science and Technology* 49: 866.

Odeyemi, F. et al. (2012). *IEEE Transactions on Plasma Science* 40 (5): 1362.

Oehr, C. (2005). *17th International Symposium on Plasma Chemistry (ISPC-17)*, Toronto, Canada, p. 9.

Oizumi, M., Suzuki, T., Uchida, M. et al. (1998). *Journal of Medical and Dental Sciences* 45: 135.

Okazaki, K. and Kishida, T. (1999). *14th International Symposium on Plasma Chemistry (ISPC-14)*, Prague, Czech Republic, vol. 5, p. 2283.

Okazaki, S. and Kogoma, M. (1994). *Journal of Physics D: Applied Physics* 27: 1985.

Okazaki, K., Nozaki, T., Uemitsu, Y. et al. (1995). *12th International Symposium on Plasma Chemistry (ISPC-12)*, Minneapolis, MN, vol. 2, p. 581.

Ombrello, T., Qin, X., Ju, Y. et al. (2006a). *44th AIAA Aerospace Sciences Meeting and Exhibit*, Reno, NV (9–12 January 2006), AIAA paper 2006-1214.

Ombrello, T., Qin, X., Ju, Y. et al. (2006b). *AIAA Journal* 44 (1): 142.

Opalinska, T. and Szymanski, A. (1996). *Contributions to Plasma Physics* 36: 63.

Opalska, A., Opalinska, T., and Ochman, P. (2002a). *Acta Agrophysica* 80: 367.

Opalska, A., Opalinska, T., Polaczek, J. et al. (2002b). *International Symposium on High Pressure Low Temperature Plasma Chemistry, HAKONE VIII*, Puhajarve, Estonia, p. 191.

Ostrikov, K. (2005). *Reviews of Modern Physics* 77, 2: 489.

Ostrikov, K., Neyts, E.S., and Meyyappan, M. (2005). *Advances in Physics* 62 (2): 113.

Oviroh, P.O., Akbarzadeh, R., Pan, D. et al. (2019). *Science and Technology of Advanced Materials* 20 (1): 465.

Pakhomov, A.G. et al. (2010). *Advanced Electroporation Techniques in Biology and Medicine*. CRC Press.

Pancheshnyi, S., Lacoste, D.A., Bourdon, A. et al. (2005). *17th International Symposium on Plasma Chemistry (ISPC-17)*, Toronto, Canada, p. 1025.

Park, C.S. (2022). *Advances in Plasma Processes for Polymers*. MDPI AG.

Park, D.-W., Cha, W.-B. (1997), *13th International Symposium on Plasma Chemistry (ISPC-13)*, Beijing, vol. 4, p. 1764.

Park, S.-J., Chen, J., Liu, C., and Eden, J.G. (2001). *Applied Physics Letters* 78: 419.

Park, J.K., Nam, S.H., Kwon, H.C. et al. (2011). *International Endodontic Journal* 44: 170.

Park, Y.S., Chung, S.H., and Shon, W.J. (2013a). *Clinical Oral Implants Research* 24: 586.

Park, D.P., Davis, K., Gilani, S. et al. (2013b). *Current Applied Physics* 13: 19.

Pasqual-Melo, G., Sagwal, S.K., Freund, E. et al. (2020). *International Journal of Molecular Sciences* 21: 1379.

Paton, B.E. et al. (2014). *Plasma Technologies and Equipment in Metallurgy and Casting*. Cambridge International Science Publishing Ltd.

Paur, H.-R. (1999). *Nonthermal Plasma Techniques for Pollution Control* (ed. B.M. Penetrante and S.E. Schultheis), 77. Berlin: Springer-Verlag.

Pawlat, J., Starek, A., Sujak, A. et al. (2018). *Plasma Processes and Polymers* 15: 1700064.

Pawlowski, L. (1995). *The Science and Engineering of Thermal Spray Coatings*. New York: Wiley.

Pei, X., Lu, X., Liu, J. et al. (2012). *Journal of Physics D* 45: 165.

Pei, X., Gidon, D., Yang, D. et al. (2019). *Chemical Engineering Journal* 362: 217–228.

Pekshev, A.V. (2001). Use of a novel biological factor - exogenic nitrogen oxide - in surgical treatment of paradontitis. *Stomatology* 1: 80.

Penetrante, B.M., Hsiao, M.C., Bardsley, J.N. et al. (1996a). *Pure and Applied Chemistry* 68: 1868.

Penetrante, B.M., Hsiao, M.C., Bardsley, J.N. et al. (1996b). *Proceedings of the International Workshop on Plasma Technologies for Pollution Control and Waste Treatment* Beijing (ed. Y.K. Pu and P.P. Woskov), p. 99.

Penetrante, B.M., Brusasco, R.M., Meritt, B.T. et al. (1998). *Plasma Exhaust Aftertreatment*. Warrendale, PA: Society of Automotive Engineers.

Peng, Y., Wang, H., Feng, J., Fang, S., Zhang, M., Wang, F., Chang, Y., Shi, X., Zhao, Q., Liu, J. (2018), Efficacy and safety of argon plasma coagulation for hemorrhagic chronic radiation proctopathy: a systematic review, *Journal: Gastroenterology Research and Practice*, vol. 308, 3087603, PMID: 29681929.

Perni, S., Liu, D.W., Shama, G., and Kong, M.G. (2008a). *Journal of Food Protection* 71: 302.

Perni, S., Shama, G., and Kong, M.J. (2008b). *Journal of Food Protection* 71: 1619.

Philip, N., Saoudi, B., Crevier, M.C. et al. (2002). *IEEE Transactions on Plasma Science* 30: 1429.

Piascik, J.R., Wolter, S.D., and Stoner, B.R. (2011a). *Journal of Biomedical Materials Research - Part B Applied Biomaterials* 98: 114.

Piascik, J.R., Wolter, S.D., and Stoner, B.R. (2011b). *Dental Materials* 27: 99.

Polak, L.S., Ovsiannikov, A.A., Slovetsky, D.I., and Vurzel, F.B. (1975). *Theoretical and Applied Plasma Chemistry*. Moscow: Nauka (Science).

Ponelis, A.A. and Van der Walt, I.J. (2003). *16th International Symposium on Plasma Chemistry (ISPC-16)*, Taormina, Italy, p. 733.

Ponomarev, A.N. (1996). *News Russian Academy of Sciences* 6: 78.

Ponomarev, A.N. (2000). *Encyclopedia of Low-Temperature Plasma*, vol. 4 (ed. V.E. Fortov), 386. Moscow: Nauka (Science).

Ponomarev, A.N. and Vasilets, V.N. (2000). *Encyclopedia of Low-Temperature Plasma*, vol. 3 (ed. V.E. Fortov), 374. Moscow: Nauka (Science).

Potapkin, B.V. et al. (1980). *Soviet Physics, High Energy Chemistry* 14: 547.

Potapkin, B.V., Rusanov, V.D., and Fridman, A. (1983). *Soviet Physics, High Energy Chemistry* 17: 528.

Potapkin, B.V., Rusanov, V.D., and Fridman, A. (1985). *Non-Equilibrium Effects in Thermal Plasma Provided by Selectivity of Transfer Processes*, vol. 4219 (6),. Moscow: Kurchatov Institute of Atomic Energy.

Potapkin, B.V., Deminsky, M., Fridman, A., and Rusanov, V.D. (1993). *Non-Thermal Plasma Techniques for Pollution Control*, NATO ASI Series, vol. G 34, Part A (ed. B.M. Penetrante and S.E. Schultheis). Berlin: Springer-Verlag.

Potapkin, B.V., Deminsky, M., Fridman, A., and Rusanov, V.D. (1995). *Radiation Physics and Chemistry* 45: 1081.

Powel, R.A. (1984). *Dry Etching for Microelectronics*. New York: Elsevier.

Prelas, M.A., Popovici, G., and Bigelow, L.K. (1997). *Handbook on Industrial Diamonds and Diamond Films*. Boca Raton, FL: CRC Press.

Pu, Y.K. and Woskov, P.P. (1996). *International Workshop on Plasma Technologies for Pollution Control and Waste Treatment*, Ed., MIT, Cambridge, MA.

Puac, N., Gherardi, M., and Shiratani, M. (2018). *Plasma Processes and Polymers* 15 (2): 1700174.

Puchkarev, V., Roth, G., and Gunderson, M. (1998). *Plasma Exhaust Aftertreatment*. Warrendale, PA: Society of Automotive Engineers.

Rabinovich, A., Nirenberg, G., Kocagoz, S. et al. (2022). *Plasma Chemistry and Plasma Processing* 42: 35.

Raizer, Y.P. (2011). *Gas Discharge Physics*. Berlin, New York: Springer.

Raizer, Y.P., Shneider, M., and Yatzenko, N. (1995). *Radio-Frequency Capacitive Discharges*. New York: CRC Press.

Ranieri, P. (2018). *Plasma Transfer Processes in Medical and Agricultural Applications*. Philadelphia, PA: Drexel University.

Ranieri, P., Shrivastav, R., Wang, M. et al. (2017). *Plasma Medicine* 7 (3): 283.

Ranieri, P., Mannsberger, A., Liu, C. et al. (2018). *Plasma Medicine* 8 (2): 185.

Ranieri, P., McGovern, G., Tse, H. et al. (2019). Microsecond-pulsed dielectric barrier discharge plasma-treated mist for inactivation of *Escherichia coli In Vitro*. *IEEE Transactions on Plasma Science* 47 (1): 395.

Ranieri, P. et al. (2021). *Plasma Processes and Polymers* 18 (1): 2000162.

Raoux, S., Cheung, D., Fodor, M. et al. (1997). *Plasma Sources Science and Technology* 6: 405.

Raoux, S., Tanaka, T., Bhan, M. et al. (1999). *Journal of Vacuum Science and Technology* B17: 477.

Ratner, B. (1995). *Biosensors & Bioelectronics* 10: 797.

Reshetov, I.V., Kabisov, R.K., Shekhter, A.B. et al. (2000). *Annals of Plastic, Reconstructive and Aesthetic Surgery* 4: 24.

Ritts, A., Lin, J., Li, H. et al. (2010). *European Journal of Oral Sciences* 118: 510.

Robert, E., Saroukh, H., Cachoncinlle, C. et al. (2005). *Pure and Applied Chemistry* 77: 1789.

Robertson, J. (1993a). *Diamond and Related Materials* 2: 984.

Robertson, J. (1993b). *Philosophical Transactions of the Royal Society* A342: 277.

Robertson, J. (1994a). *Pure and Applied Chemistry* 66: 1789.

Robertson, J. (1994b). *Diamond and Related Materials* 3: 361.

Robertson, J. (1997). *Radiation Effects* 142: 63.

Robertson, J. (2002). *Materials Science and Engineering* R37: 129.

Robertson, J. (2004), *Materials Today* Realistic applications of CNTs, vol. 7(10), pg. 46–52.

Rosocha, L.A., Anderson, G.K., Bechtold, L.A. et al. (1993). *Non-Thermal Plasma Techniques for Pollution Control: Part B – Electron Beam and Electrical Discharge Processing* (ed. B.M. Penetrante and S.E. Schultheis). Berlin: Springer-Verlag.

Rotering, H., Al Shakaki, M., Welp, H., and Dell'Aquila, A.M. (2020). *The Annals of Thoracic Surgery* 110: 1302.

Roth, J.R. (1995). *Industrial Plasma Engineering*, vol. 1. Bristol/Philadelphia, PA: Institute of Physics Publishing.

Roth, J.R. (2001). *Industrial Plasma Engineering*, vol. 2. Bristol/Philadelphia, PA: Institute of Physics Publishing.

Roth, J.R. (2003). *Physics of Plasmas* 10 (5): 1166.

Roupassov, D.V., Nikipelov, A.A., Nudnova, M.M., and Starikovskii, A.Y. (2009). *AAIA Journal* 49: 68.

Rubinsky, B. (2009). *Irreversible Electroporation*. Springer.

Rueggeberg, F.A., Caughman, W.F., Curtis, J.W., and Davis, H.C. (1993). *American Journal of Dentistry* 6: 91.

Rupf, S., Lehmann, A., Hannig, M. et al. (2010). *Journal of Medical Microbiology* 59: 206.

Rupf, S., Idlibi, A.N., Marrawi, F.A. et al. (2011). *PLoS One* 6: 25893.

Rusanov, V.D. and Fridman, A. (1978a). *Soviet Physics, Journal of Physical Chemistry (JFCh)* 52: 92.

Rusanov, V.D. and Fridman, A. (1978b). *Soviet Physics, Journal of Technical Physics Letters (Pis'ma v JTF)* 4: 28.

Rusanov, V.D. and Fridman, A. (1984). *Physics of Chemically Active Plasma*. Moscow: Editorial House Nauka (Science).

Rusanov, V.D., Fridman, A., and Sholin, G.V. (1981). *Advances in Physical Sciences (Uspekhi Physicheskikh Nauk)* 134: 185.

Rutberg, F.G., Safronov, A.A., Goryachev, V.L. et al. (1997). *13th International Symposium on Plasma Chemistry (ISPC-13)*, Beijing, China, vol. 4, p. 1727.

Rykalin, N.N. and Sorokin, L.M. (1987). *Metallurgical RF-Plasmatron: Electro- and Gas-Dynamics*. Moscow: Nauka (Science).

Rykalin, N.N., Tsvetkov, Y.V., Petrunichev, V.A., and Glushko, I.K. (1970). *Physics, Engineering and Application of Low-Temperature Plasma*. Alma-Ata, Kazakhstan: Nauka (Science).

Saberi, M., Modarres-Sanavy, S.A.M., Zare, R., and Ghomi, H. (2020). *Journal of Agricultural Science and Technology* 21: 1889.

Saidane, K., Razafinimanana, M., Lange, H. et al. (2004). *Journal of Physics D: Applied Physics* 39: 232.

Sakai, O. and Tachibana, K. (2012). *Plasma Sources Science and Technology* 21 (1): 013001.

Sakai, O., Kishimoto, Y., and Tachibana, K. (2005). *Journal of Physics D: Applied Physics* 38: 431.

Sakakita, H., Ikehara, Y. (2018), "International standardization", in: *Plasma Medical Science*, ed. by S. Toyokuni, et al., Cambridge, MA, USA, pg. 201–225 Academic Press.

Samimi, M., Kim, J.H., Kastner, J. et al. (2007). *Journal of Fluid Mechanics* 578: 305.

Sankaran, R.M. (ed.) (2011). *Plasma Processing of Nanomaterials*. CRC Press.

Sardella, E., Gristina, R., Gilliland, D., Ceccone, G., Senesi, G.S., Rossi, F., d'Agostino, R., Favio, P. (2005), *17th International Symposium on Plasma Chemistry (ISPC-17)*, Toronto, Canada, p. 608.

Sarinont, T., Amano, T., Kitazaki, S. et al. (2014). *Journal of Physics Conference Series* 518: 012017.

Sarinont, T., Amano, T., Koga, K. et al. (2015). Effects of plasma irradiation to seeds of Arabidopsis thaliana and zinnia on their growth. *MRS Online Proceedings Library Archive* 1723: 7–11.

Sarinont, T., Katayama, R., Wada, Y. et al. (2017). *MRS Advances* 2: 995.

Sarron, V., Robert, E., Dozias, A. et al. (2011). *IEEE Transactions on Plasma Science* 39 (11): 2356.

Sato, M., Ohgiyama, T., and Clements, J.S. (1996). *IEEE Transactions on Industry Applications* 32: 106.

Sato, T., Yokoyama, M., and Johkura, K. (2011). *Journal of Physics D: Applied Physics* 44 (37): 372001.

Schnabel, U., Handorf, O., Yarova, K. et al. (2019). *Food* 8: 55.

Schoenbach, K.H., Peterkin, F.E., Alden, R.W., and Beebe, S.J. (1997). *IEEE Transactions on Plasma Science* 25: 284.

Schoenbach, K.H., Joshi, R.P., Stark, R.H. et al. (2000). *IEEE Transactions on Dielectrics and Electrical Insulation* 7: 637.

Schroder, K., Finke, B., Ohl, A. et al. (2010). *Journal of Adhesion Science and Technology* 24: 1191.

Schwabedissen, A., Lacinski, P., Chen, X., and Engelmann, J. (2007). *Contributions to Plasma Physics* 47: 551.

Seeger, W. (2005). *Deutsche Medizinische Wochenschrift* 130 (25,26): 1543.

Selcuk, M., Oksuz, L., and Basaran, P. (2008). *Bioresource Technology* 99: 5104.

Sensenig, R., Kalghatgi, S., Cercha, E. et al. (2011). *Annals of Biomedical Engineering* 39 (2): 674.

Sera, B., Sery, M., Gavril, B., and Gajdova, I. (2017). *Plasma Chemistry and Plasma Processing* 37: 207.

Serov, Y.L. and Yavor, I.P. (1995). *Soviet Physics, Journal of Technical Physics* 40: 248.

Shah, J.R. et al. (2018). *Catalysts* 8: 437.

Shainsky, N., Dobrynin, D., Ercan, U. et al. (2010). *37th IEEE International Conference on Plasma Science, ICOPS*, Norfolk, VA, USA (20–24 June).

Shainsky, N. et al. (2012). *Plasma Processes and Polymers* 9 (6): 555.

Shaji, M., Rabinovich, A., Surace, M. et al. (2023). *Plasma* 6 (1): 45.

Shekhter, A.B., Kabisov, R.K., Pekshev, A.V. et al. (1998). Experimental and clinical validation of plasma dynamic therapy of wounds with nitric oxide. *Bulletin of Experimental Biology and Medicine* 126 (8): 829.

Shekhter, A.B., Serezhenkov, V.A., Rudenko, T.G. et al. (2005). *Nitric Oxide: Biology and Chemistry* 12: 210.

Shibata, Y., Hosaka, M., Kawai, H., and Miyazaki, T. (2002). *The International Journal of Oral & Maxillofacial Implants* 17: 771.

Shibkov, V.M. and Konstantinovskij, R.S. (2005). Kinetical model of ignition of hydrogen-oxygen mixture under conditions of non-equilibrium plasma of the gas discharge. *AIAA Journal*, paper 2005-987.

Shibkov, V.M., Chernikov, V.A., Ershov, A.P. et al. (2001). *AIAA paper*, vol. 2001, p. 3087.

Shibkov, V.M., Aleksandrov, A.F., Chernikov, V.A. et al. (2006). *44th AIAA Aerospace Science Meeting and Exhibit*, Reno, NV, AIAA paper 2006-1216.

Shimizu, N., Ikehara, Y. (2018) "Blood coagulation and regenerative medicine", in: *Plasma Medical Science*, ed. by S. Toyokuni, et al., Academic Press, pg. 225–240.

Shimizu, T. Urayama, T. (2018), "CE marking for medical devices", in: *Plasma Medical Science*, ed. by S. Toyokuni, et al., Academic Press, pg. 240–254.

Shimizu, K., Hayashida, K., and Blajan, M. (2015). *Biointerphases* 10: 029517.

Shimizu, T., Fukui, T., and Sakakita, H. (2022). *Japanese Journal of Applied Physics* 61: 1016.

Shmakin, Y.A. and Marusin, V.V. (1970). *Application of Low-Temperature Plasma in Technology of Inorganic Materials and Powder Metallurgy*. Novosibirsk, Russia: Nauka (Science).

Shneider, M.N., Pekker, M., and Fridman, A. (2012). *IEEE Transactions on Dielectrics and Electrical Insulation* 19: 1579.

Shulutko, A.M., Antropova, N.V., and Kryuger, Y.A. (2004). *Surgery* 12: 43.

Silveira, A.M., Lopes, H.P., Siqueira, J.F. Jr. et al. (2007). *Brazilian Dental Journal* 18: 299.

Simec, M. and Homola, T. (2021). *The European Physical Journal D* 75: 210.

Singh, R., Prasad, P., Mohan, R. et al. (2019). *Journal of Applied Research on Medicinal and Aromatic Plants* 12: 78.

Sivachandiran, L. and Khacef, A. (2017). *RSC Advances* 7: 1822.

Sladek, R.E. and Stoffels, E. (2005). *Journal of Physics D: Applied Physics* 38: 1716.

Sladek, R.E., Stoffels, E., Walraven, R. et al. (2004). *Plasma Science* 32: 1540.

Sladek, R.E., Baede, T.A., and Stoffels, E. (2006). *IEEE Transactions on Plasma Science* 34: 1325.

Sladek, R.E., Filoche, S.K., Sissons, C.H., and Stoffels, E. (2007). *Letters in Applied Microbiology* 45: 318.

Slone, R., Ramavajjala, M., Palekar, V., and Pushkarev, V. (1998). *Plasma Exhaust Aftertreatment*. Warrendale, PA: Society of Automotive Engineers.

Slovetsky, D.I. (1980). *Mechanisms of Chemical Reactions in Non-Equilibrium Plasma*. Moscow: Nauka (Science).

Slovetsky, D.I. (1981). *Plasma Chemistry*, vol. 8 (ed. B.M. Smirnov), 181. Moscow: Energo-Izdat.

Smirnov, B.M. (2001). *Physics of Ionized Gases*. New York: Wiley.

Smolková, B., Frtus, A., Uzhytchak, M. et al. (2020). *International Journal of Molecular Sciences* 21: 6226.

Snitsiriwat, S. et al. (2022). *Journal of Physical Chemistry A* 126 (1): 3.

Snoeckx, R. and Bogaerts, A. (2017). *Chemical Society Reviews* 46: 5805.

Snoeckx, R., Rabinovich, A., Dobrynin, D. et al. (2016). *Plasma Processes and Polymers* 14 (6): 1600115.

Sobacchi, M., Saveliev, A., Fridman, A. et al. (2002). *International Journal of Hydrogen Energy* 27 (6): 635.

Sobacchi, M.G., Saveliev, A.V., Fridman, A. et al. (2003). *Plasma Chemistry and Plasma Processing* 23: 347.

Solonenko, O.P. (1986). *News of Siberian Branch of the USSR Academy of Sciences, Engineering Science* 4 (1): 136.

Solonenko, O.P. (1996). *Thermal Plasma and New Materials Technology*. Cambridge, UK: Cambridge International Science Publishing.

Song, H.P., Kim, B., Choe, J.H. et al. (2009). *Food Microbiology* 26: 432.

Song, J.S., Lee, M.J., Ra, J.E. et al. (2020). *Journal of Physics D: Applied Physics* 53: 314002.

Song, H. et al. (ed.) (2022). *Plasma at the Nanoscale (Micro and Nano Technologies)*. Elsevier.

Staack, D., Farouk, B., Gutsol, A., and Fridman, A. (2005). *Plasma Sources Science and Technology* 14: 700.

Staack, D., Farouk, B., Gutsol, A., and Fridman, A. (2006). *Plasma Sources Science and Technology* 15: 818.

Staack, D., Fridman, A., Gutsol, A. et al. (2008). *Angewandte Chemie* 47 (42): 8020.

Stalder, K.R. and Woloshko, J. (2012). Electrical discharges in conducting liquids: plasma mediated electrosurgical systems. In: *Plasma Medicine* (ed. M. Laroussi et al.), 261–293. Cambridge University Press.

Starikovskaia, S.M. (2006). *Journal of Physics D: Applied Physics* 39 (16): 265.

Starikovskaia, S.M. (2014). *Journal of Physics D: Applied Physics* 47 (35): 353001.

Starikovskaia, S.M., Kukaev, E.N., Kuksin, A.Y. et al. (2004). *Combustion and Flame* 139: 177.

Starikovskaia, S.M., Anikin, N.V., Kosarev, I.N. et al. (2006). *44th AIAA Aerospace Science Meeting and Exhibit*, Reno, NV, AIAA, paper 2006-616.

Starikovskii, A.Y. (2000). *Plasma Physics Reports* 26: 701.

Starikovskii, A.Y. (2003). *Combustion, Explosion, and Shock Waves* 39: 619.

Starikovskii, A.Y., Anikin, N.B., Kosarev, I.N. et al. (2008). *Journal of Propulsion and Power* 24 (6): 1182.

Starikovskii, A.Y., Nikipelov, A.A., Nudnova, M.M., and Roupassov, D.V. (2009). *Plasma Sources Science and Technology* 18 (3): 034015.

Starikovskiy, A.Y. (2011). *IEEE Transactions on Plasma Science* 39: 2602.

Starikovskiy, A. and Alexandrov, N. (2013). *Progress in Energy and Combustion Science* 39 (1): 61.

Starikovskiy, A.Y., Yang, Y., Cho, Y.I., and Fridman, A. (2011). *Plasma Sources Science and Technology* 20: 024003.

Stauss, S., Muneoka, H., and Terashima, K. (2018). *Plasma Sources Science and Technology* 27 (2): 023003.

Steinbeck, M.J., Chernets, N., Zhang, J. et al. (2013). *PLoS One* 8: 82143.

Steinberg, M. (2000). Decarbonization and sequestration for mitigating global warming. *International Symposium "Deep Sea & CO, 2000,"* SRI, Mitaka, Tokyo, p. 4-2-1.

Stix, T.H. (2012). *The Theory of Plasma Waves*. Literary Licensing, LLC.

Stoffels, E. (2003). *Proceedings Gaseous Electronics Conference*, AIP, San Francisco, CA, p. 16.

Stoffels, E. (2006). *Journal of Physics D: Applied Physics* 39: 16.

Stout, P.J. and Kushner, M.J. (1993). *Journal of Vacuum Science and Technology* A11: 2562.

Stratmann, B. et al. (2020). *JAMA Network Open* 3: 2010411.

Stratton, G.R. et al. (2017). *Environmental Science & Technology* 51 (3): 1643.

Stueber, G.J., Clarke, S.A., Bernstein, E.R. et al. (2003). *The Journal of Physical Chemistry* A107: 7775.

Sugano, T. (1985). *Applications of Plasma Processes to VLSI Technology*. New York: Wiley.

Sugiarto, A.T., Ito, S., Ohshima, T. et al. (2003). *Journal of Electrostatics* 58: 135.

Suhr, H. (1973). *Fortschritte der Chemischen Forschung* 36: 39.

Suhr, H. (1974). *Techniques and Applications of Plasma Chemistry* (ed. J.R. Hollahan and A.T. Bell), 57. Wiley-Interscience.

Sun, B., Sato, M., and Clements, J.S. (1997). *Journal of Electrostatics* 39: 189.

Sun, B., Sato, M., and Clements, J.S. (1999). *Journal of Physics D: Applied Physics* 32: 1908.

Surace, M., Fridman, A., Sales, C. et al. (2023). *Plasma* 6 (3): 419–434.

Svelto, O. (2009). *Principles of Lasers*. New York: Springer.

Tajmar, M. (2004). *Advanced Space Propulsion Systems*. Berlin: Springer.

Takaki, K., Takahata, J., Watanabe, S. et al. (2013). *Journal of Physics Conference Series* 418: 012140.

Takashima, K., Adamovich, I., Xiong, Z. et al. (2011). *Physics of Plasmas* 18: 083505.

Tanaka, H. et al. (2011). *Plasma Medicine* 1 (3–4): 265.

Tanaka, H., Mizuno, M., Ishikawa, K. et al. (2014). *IEEE Transactions on Plasma Science* 42 (12): 37.60.

Tanaka, H., Nakamura, K., Mizuno, M. et al. (2016). *Scientific Reports* 6 (1): 1.

Tanaka, H., Bekeschus, S., Yan, D. et al. (2021). *Cancers* 13 (7): 1737.

Tatoulian, M., Bouloussa, O., Moriere, F. et al. (2004). *Langmuir* 20: 10481.

Tong, J., He, R., Zhang, X. et al. (2014). *Plasma Science and Technology* 16: 260.

Toyokuni, S., Ikehara, Y., Kikkawa, F., and Hori, M. (ed.) (2018). *Plasma Medical Science*. Academic Press, Elsevier.

Treanor, C.E., Rich, I.W., and Rehm, R.G. (1968). *The Journal of Chemical Physics* 48: 1798.

Trivedi, M.H., Patel, K., Itokazu, H. et al. (2019). *Plasma Medicine* 9 (1): 23.

Tsvetkov, Y.V. and Panfilov, S.A. (1980). *Low-Temperature Plasma in Reduction Processes*. Moscow: Nauka (Science).

Tumanov, Y.N. (1981). *Electro-Thermal Reactions in Modern Chemical Technology and Metallurgy*. Moscow: Energo-Atom-Izdat.

Tumanov, Y.N. (1989). *Low-Temperature Plasma and High-Frequency Electromagnetic Fields in Production of Nuclear Energy Related Materials*. Moscow: Energo-Atom-Izdat.

Tzeng, C.-C., Kuo, Y.-Y., Huang, T.-F. et al. (1998). *Journal of Hazardous Materials* 58: 207.

Ulrich, C. et al. (2015). *Journal of Wound Care* 24: 196.

Utsumi, F. et al. (2013). *PLoS One* 8 (12): 81576.

Van Rooij, G.J., Akse, H.N., Bongers, W.A., and Van de Sanden, R. (2018). *JoVE (Journal of Visualized Experiments)* 126: e55066.

Vandamme, M., Robert, E., Pesnel, S. et al. (2010). *Plasma Processes and Polymers* 7: 264.

Vandamme, M., Robert, E., Lerondel, S. et al. (2012). *International Journal of Cancer* 130: 2185.

Vandersee, S., Richter, H., Lademann, J. et al. (2014). *Laser Physics Letters* 11: 5701.

Vasilets, V.N. (2005). Modification of physical chemical and biological properties of polymer materials using gas-discharge plasma and vacuum ultraviolet radiation. Doctor of Science Dissertation. Institute of Energy Problems of Chemical Physics, Russian Academy of Sciences, Moscow.

Vasilets, V.N., Gutsol. A.F., Shekhter, A.B., Fridman, A. (2009), *Khimiya Vysokikh Energiy (High Energy Chemistry)*, 43 (3): 229–233.

Vassilakos, N., Fernandes, C.P., and Nilner, K. (1993). *The Journal of Prosthetic Dentistry* 70: 165.

Veprek, S. (1972). *The Journal of Chemical Physics* 57: 952.

Vinogradov, G.K. (1986). *Soviet Physics, High Energy Chemistry (Khimia Vysokikh Energiy)* 20 (3): 195.

Vinogradov, G.K. and Ivanov, Y.A. (1977). *Chemical Reactions in Low-Temperature Plasma* (ed. L.S. Polak), 27. Moscow: Institute of Petrochemical Synthesis of USSR Academy of Sciences.

Virin, L., Dgagaspanian, R., Karachevtsev, G. et al. (1978). *Ion-Molecular Reactions in Gases*. Moscow: Nauka (Sciene).

Volkov, A.G., Hairston, J.S., Patel, D. et al. (2019). *Bioelectrochemistry* 128: 175.

Volodin, N.L., Vurzel, F.B., Polak, L.S., and Enduskin, P.N. (1970a). *Physics, Technology and Applications of Low-Temperature Plasma* (ed. L.S. Polak), 576. Kazakhstan: Alma-Ata.

Volodin, N.L., Vurzel, F.B., Polak, L.S., and Enduskin, P.N. (1970b). *Problems of Petrochemical Synthesis* (ed. L.S. Polak), 11. Ufa, Russia: Nauka (Science).

Volodin, N.L., Vurzel, F.B., Diatlov, V.T. et al. (1971a). *Soviet Physics, High Energy Chemistry (Khimia Vysokikh Energij)* 5: 3.

Volodin, N.L., Vurzel, F.B., Polak, L.S., and Enduskin, P.N. (1971b). *Plasma Chemistry-71* (ed. L.S. Polak), 157. Moscow: Institute of Petrochemical Synthesis of USSR Academy of Sciences.

Von Woedtke, T. and Metelmann, H.-R. (2014). *Clinical Plasma Medicine* 2: 37.

Von Woedtke, T., Laroussi, M., and Gherardi, M. (2022). *Plasma Sources Science and Technology* 31: 054022.

Vurzel, F.B. (1970). *Application of Plasma in Chemical Processes* (ed. L.S. Polak). Moscow: Mir (World).

Vurzel, F.B. and Nazarov, N.F. (2000). Plasma-chemical treatment of powder materials and coating formation. In: *Encyclopedia of Low-Temperature Plasma*, vol. 4 (ed. V.E. Fortov), 349. Moscow, Russia: Nauka (Science).

Walia, K. et al. (2017). *Food Control* 75: 55.

Walschus, U., Hoene, A., Patrzyk, M. et al. (2012). *Journal of Materials Science, Materials in Medicine* 23: 1299.

Walsh, J.L., Iza, F. et al. (2010). *Journal of Physics D: Applied Physics* 43: 075201.

Wang, Y.F., Lee, W.J., Chen, C.Y., and Hsieh, L.T. (1999). *Industrial and Engineering Chemistry Research* 38: 3199.

Watanabe, T. (2003). *16th International Symposium on Plasma Chemistry (ISPC-16)*, Taormina, Italy, p. 734.

Watanabe, Y., Shiratani, M., Kubo, Y. et al. (1988). *Applied Physics Letters* 53: 1263.

Watanabe, T., Taira, T., and Takeuchi, A. (2005). *17th International Symposium on Plasma Chemistry (ISPC-17)*, Toronto, Canada, p. 1153.

Wertheimer, M.R., Dennler, G., and Guimond, S. (2003). *16th International Symposium on Plasma Chemistry (ISPC-16)*, Taormina, Italy, p. 11.

Wirtz, M., Stoffels, I., Dissemond, J. et al. (2018). *Journal of the European Academy of Dermatology and Venereology* 32 (1): 37.

Wolff, C.M., Steuer, A., Stoffels, I. et al. (2019). *Clinical Plasma Medicine* 16: 100096.

Wright, K.C. et al. (2014). *Desalination* 345: 64.

Wu, Z.L., Gao, X., Luo, Z.Y. et al. (2005). *Energy & Fuels* 19: 2279.

Xie, K.-C., Lu, Y.-K., Tian, Y.-J., and Wang, D.-Z. (2002). *Energy Sources* 24: 1093.

Xiong, Z., Cao, Y., Lu, X., and Du, T. (2011). *IEEE Transactions on Plasma Science* 39: 2968.

Yamamoto, T., Ramanathan, K., Lawless, P.A. et al. (1992). *IEEE Transactions on Industry Applications* 28: 528.

Yamamoto, T., Lawless, P.A., Owen, M.K., Ensor, D.S., Boss, C. (1993), *Non-Thermal Plasma Techniques for Pollution Control: Part B – Electron Beam and Electrical Discharge Processing*, Penetrante, B.M., Schultheis, S.E. (eds), Springer-Verlag, Berlin pg. 254–260.

Yamamoto, M., Nishioka, M., and Sadakata, M. (2001). *15th International Symposium on Plasma Chemistry, ISPC-15*, Orleans, France, vol. 2, p. 743.

Yamazaki, H., Ohshima, T., Tsubota, Y. et al. (2011). *Dental Materials Journal* 30: 384.

Yan, K., Yamamoto, T., Kanazawa, S. et al. (2001). *IEEE Transactions on Industry Applications* 37: 1499.

Yan, D., Talbot, A., Nourmohammadi, N. et al. (2015). *Biointerphases* 10 (4): 040801.

Yan, D., Sherman, J.H., and Keidar, M. (2018). *Anti-Cancer Agents in Medicinal Chemistry* 18 (6): 769.

Yang, Y. et al. (2010). *Water Research* 44: 3659.

Yang, B., Chen, J., Yu, Q. et al. (2011). *Journal of Dentistry* 39: 48.

Yang, Y., Cho, Y.I., and Fridman, A. (2012). *Plasma Discharges in Liquid: Water Treatment and Applications*. Boca Raton, FL: CRC Press.

Yasuda, H.K. (2012). *Plasma Polymerization*. Cambridge, MA, USA: Academic Press.

Yavirach, P., Chaijareenont, P., Boonyawan, D. et al. (2009). *Dental Materials Journal* 28 (6): 686.

Yerokhin, A.L., Snizhko, L.O., Gurevina, N.L. et al. (2003). *Journal of Physics D: Applied Physics* 36: 2110.

Yildirim, E.D., Ayan, H., Vasilets, V.N. et al. (2008). *Plasma Processing and Polymers* 5: 58.

Yokayama, T., Kogoma, M., Okazaki, S. et al. (1990). *Journal of Physics D: Applied Physics* 23: 1125.

Yong, H.I., Park, J., Kim, H.-J. et al. (2018). *Plasma Processes and Polymers* 15: 1700050.

Yoon, Y.J., Suh, M.J., Lee, H.Y. et al. (2018). *Free Radical Biology & Medicine* 115: 43.

Yoshida, T. (2005). Thermal plasma spraying. *17th International Symposium on Plasma Chemistry (ISPC-17)*, Toronto, Canada, p. 9.

Yun, H., Kim, B., Jung, S. et al. (2010). *Food Control* 21: 1182.

Zaitsev, V.M. (2003). *Russian Journal of Otorhinolaryngology* 1: 58.

Zhan, J. et al. (2020). *Journal of Hazardous Materials* 387: 121688.

Zhang, B., Liu, B., Renault, T. et al. (2005). *17th International Symposium on Plasma Chemistry, (ISPC-17)*, Toronto, Canada, p. 782.

Zhang, S., Rousseau, A., and Dufour, T. (2017). *RSC Advances* 7: 31244.

Zhang, H. et al. (2021a). *Environmental Science & Technology* 55 (23): 16067.

Zhang, J., Li, B., Xu, S. et al. (2021b). *Plasma Processes and Polymers* 18: e2000226.

Zhou, X., Xiong, Z., Cao, Y. et al. (2010). *IEEE Transactions on Plasma Science* 38: 3370.

Zhukov, M.F. and Solonenko, O.P. (1990). *High Temperature Dusted Jets in the Powder Material Processing*. Novosibirsk: Institute of Thermophysics, Siberian Branch of Russian Academy of Sciences.

Zirnheld, J.L., Zucker, S.N., DiSanto, T.M. et al. (2010). *IEEE Transactions on Plasma Science* 38 (4): 948.

Index